Handbook of Engineering Acoustics

Gerhard Müller • Michael Möser
Editors

Handbook of Engineering Acoustics

Springer

Editors
Prof. Dr. Gerhard Müller
Technische Universität München
Lehrstuhl für Baumechanik
München
Germany

Prof. Dr. Michael Möser
Technische Universität Berlin
Institut für Technische Akustik
Berlin
Germany

ISBN 978-3-642-43550-8 ISBN 978-3-540-69460-1 (eBook)
DOI 10.1007/978-3-540-69460-1
Springer Heidelberg New York Dordrecht London

Printed on acid-free paper

Springer is part of Springer Science+Business Media (www.springer.com)

Preface

The editors of the first English edition are firmly committed to the aims and intentions of their predecessors. They would like to express their gratitude to the authors for their very thorough editing work and compilation of new entries.

Berlin, Germany M. Möser
Munich, Germany G. Müller

From the Preface to the Second Edition of the "Taschenbuch der Technischen Akustik"

The Handbook of Technical Acoustics aims to provide a broad audience with a quick and reliable source of information on engineering questions. It is directed not only toward specialists and students of Technical Acoustics, but rather toward engineers in all domains, e.g., mechanical engineering, process engineering, traffic systems or civil engineering, who deal with the topic of noise reduction.

In light of the increasing significance of environmental protection, a number of articles cover the generation, transmission, insulation, measurement and assessment of air, and structure borne noise; at the same time, questions pertaining to electro-acoustics, roomacoustics and audio-physiology are discussed in detail.

Just as in the first edition, in addition to the basics of technical acoustics and associated domains, great emphasis has been placed on including as much information as possible on material data, empirical values, important measurement results, tested approximation formulas, useful reference values, etc. This will help the reader understand the general context, while at the same time providing data required for the solution of practical problems – without having to engage in time-consuming studies of the literature. The editors extend their sincere thanks, especially to all of the authors who, alongside their professional work, have committed themselves to the tedious task of presenting their respective fields of specialization in a comprehensive, yet compact format.

The editors hope this Handbook of Technical Acoustics will prove to be a useful aid to its readers in their daily work.

Berlin, Germany — M. Heckl
Munich, Germany — H. A. Müller

Contents

Fundamentals

1

M. Möser and G. Müller

1.1 Introduction

Sound consists in mechanical vibrations within the frequency band of human hearing from approximately 16 Hz–16 kHz. Vibrations in air and in liquids are called airborne and liquid-borne sound respectively, vibrations in solids are called structure-borne sound.

For further fundamentals on Technical Acoustics, see, for example, [1–17].

1.2 Level and Measurement Quantities

For airborne sound, the primary quantity is the sound pressure, which is determined using microphones (Chap. 2). More elaborate techniques (for example, the intensity measurement, see Chap. 2) are based on several microphones. For structure-borne sound, the surface displacements ξ, their velocity $v = \partial\xi/\partial t$ or their acceleration $a = \partial v/\partial t$ are determined.

In view of Weber's and Fechner's law, stating that the human perception is proportional to the logarithm of the stimulus, the physical quantities are usually expressed in terms of their levels. The sound pressure level L_p is defined to be

$$L_p = 10\,\lg \frac{p_{\mathrm{rms}}^2}{p_0^2} = 20\,\lg \frac{p_{\mathrm{rms}}}{p_0}, \tag{1.1}$$

where p_{eff} denotes the root mean square value of the signal

$$p_{\mathrm{rms}}^2 = \frac{1}{T}\int_0^T p^2(t)\mathrm{d}t, \tag{1.2}$$

and the standardized reference is $p_0 = 2 \times 10^{-5}\ \mathrm{N/m^2}$. Human beings perceive sound pressures approximately between $p_{\mathrm{rms}} = p_0$ (threshold of audibility) and $p_{\mathrm{rms}} = 200\ \mathrm{N/m^2}$ (threshold of pain). The atmospheric pressure amounts to about $10^5\ \mathrm{N/m^2}$; the sound pressures therefore are comparatively very small. The human perception comprehends approximately $0 < L < 140$ dB. The pseudo unit 'decibel' is used for the outcome of Eq. (1.1). 1 dB corresponds to the threshold of audible difference between two pressures. 10 dB correspond approximately to the doubling of stimulus.

The quantities local velocity v, temporally averaged intensity I and power P (see Sects. 1.4 and 1.5) also are described by their levels. They are defined in such a way, that equal level values result for the case of plane propagating waves with $v = p/\rho_0 c$ and $I = p_{\mathrm{rms}}^2/\rho_0 c$ ($\rho_0 c$ = characteristic impedance of air = 400 kg/s/m^2)

$$L_v = 10\,\lg \frac{v_{\mathrm{rms}}^2}{v_0^2}, \tag{1.3}$$

$$L_I = 10\,\lg \frac{I}{I_0}, \tag{1.4}$$

M. Möser (✉)
Institut für Strömungsmechanik und Technische Akustik, Technische Universität Berlin, Einsteinufer 25, 10587 Berlin, Germany
e-mail: mimoe48@googlemail.com

G. Müller
Lehrstuhl für Baumechanik, Technische Universität München, Arcisstrasse 21, 80333 München, Germany
e-mail: gerhard.mueller@bv.tum.de

G. Müller and M. Möser (eds.), *Handbook of Engineering Acoustics*,
DOI 10.1007/978-3-540-69460-1_1, © Springer-Verlag Berlin Heidelberg 2013

and

$$L_P = 10\lg\frac{P}{P_0}, \quad (1.5)$$

with $v_0 = p_0/\rho_0 c = 5 \cdot 10^{-8}$ m/s, $I_0 = {p_0}^2/\rho_0 c = 10^{-12}$ W/m^2 and $P_0 = I_0 \cdot 1$ m$^2 = 10^{-12}$ W. As mentioned, $L_p = L_v = L_I = L_P$ holds for plane propagating waves (L_P denotes the level of the power through an area of 1 m^2).

Often several incoherent signals ('incoherent' means signals with different frequency components, for example, two cars or two noise sources, etc.) with known partial levels have to be combined to a total level. In such cases, the single mean square values have to be calculated from the inverse of Eq. (1.1)

$$\frac{p^2_{\mathrm{rms}}}{p_0^2} = 10^{L_p/10}. \quad (1.6)$$

The total mean square value equals the sum of partial ones; therefore, the total level equals

$$L_{\mathrm{ges}} = 10\lg\left\{\sum_{i=1}^{N} 10^{L_i/10}\right\}, \quad (1.7)$$

(L_i: partial levels, N: their number). Equation (1.7) says, for example, that a second source with the same level yields a 3 dB increase, that three equal one third octave band levels are 4.8 dB lower than the corresponding octave band level and that a 6 dB signal to noise ratio results in an error of 1 dB.

1.3 Fundamentals of System Theory

Acoustical transducers and vibrating structures may be described by a 'cause-effect-chain', which often is called the 'system'. An excitation $x(t)$ (also named the system input) causes vibrations of a dynamical structure, which can be observed at a certain point using some appropriate quantity $y(t)$ (being the system output). Examples are the input voltage $x(t)$ for a loudspeaker in an arbitrary acoustical environment and the output voltage $y(t)$ of a microphone observing the sound field; or force signals $x(t)$ exciting structures like beams and plates somewhere to result in some response $y(t)$. Literature on system theory can be found for example in [18, 19].

For sufficiently small amplitudes, acoustical transducers behave LINEARLY; this is, for example, true for airborne sound below 130 dB. 'Linear' simply means that the principle of superposition is valid. If the operator L denotes the transformation of the input signal accomplished by the system in question to give the output signal

$$y(t) = L\{x(t)\}, \quad (1.8)$$

then LINEAR transduction is described by the fact

$$L\{a_1x_1(t)+a_2x_2(t)\}=a_1L\{x_1(t)\}+a_2L\{x_2(t)\}, \quad (1.9)$$

(a_1, a_2: constants, x_1, x_2: arbitrary signals).

In addition, systems are called time invariant, if their reaction after a time shift of the input also only results in the same time shift of the output:

$$y\,(t-\tau) = L\{x\,(t-\tau)\}. \quad (1.10)$$

Linear and time-invariant transducers are easily described using their impulse responses $h(t)$ or their transfer functions $H(\omega)$; dropping one of these assumptions causes often a difficult and complex description of the system. The assumptions of linearity and time invariance are of course not self-evident in general, but may be assumed in many cases; appropriate experiments have to be carried out, if needed.

1.3.1 Description with Impulse Response Function

The simple mathematical fact that all signals may be represented by a comb of delta functions (Dirac's comb)

$$x(t) = \int_{-\infty}^{\infty} x(\tau)\delta(t-\tau)\mathrm{d}\tau, \quad (1.11)$$

($\delta(t)$ = Dirac's delta function with $\delta(t{\neq}0) = 0$ and $\int_{-\infty}^{\infty}\delta(t)\mathrm{d}t = 1$) already results in a description of a linear and time invariant system L by its impulse response function. Applying the operator L on Eq. (1.11) yields

$$y(t) = L\left\{\int_{-\infty}^{\infty} x(\tau)\delta(t-\tau)\mathrm{d}\tau\right\}$$
$$= \int_{-\infty}^{\infty} x(\tau)L\{\delta(t-\tau)\}\mathrm{d}\tau$$
$$= \int_{-\infty}^{\infty} x(\tau)h(t-\tau)\mathrm{d}\tau, \tag{1.12}$$

where

$$h(t) = L\{\delta(t)\}, \tag{1.13}$$

denotes the impulse response of the transducer.

The right hand side integral in Eq. (1.12) is called the convolution integral. The total operation defined by Eq. (1.12) on $x(t)$ and $h(t)$ to give the output signal $y(t)$ is termed 'convolution', shortly written as $y(t) = x(t) \cdot h(t)$. The convolution is commutative with respect to x and h, $x(t) \cdot h(t) = h(t) \cdot x(t)$ holds. A direct application of the convolution for room acoustic purposes is described in Chap. 10.

1.3.2 Pure Tones (Time Convention)

Signals consisting of pure tones (with harmonical time dependence of the form $\cos(\omega t)$) are most easily described using their complex amplitudes $\underline{f}$, which is related to the real valued time signal by taking the real part as follows:

$$f(t) = \mathrm{Re}\left\{\underline{f}e^{j\omega t}\right\}. \tag{1.14}$$

In Eq. (1.14), f and $\underline{f}$ may denote arbitrary physical quantities (pressure, velocity, electrical voltage or mechanical stress, for example) that may also depend on spatial variables or others. The complex amplitude is also called 'phasor'. Equation (1.14) permits the inverse mapping of the complex number $\underline{f}$ to the real valued, observable reality $f(t)$. The complex notation benefits from the fact that a single complex symbol contains two informations (namely amplitude and phase). The most important advantage of the use of phasors consists of the much simplified arithmetics. For example, to the take the sum of two sinusoidals with different phases and amplitudes is a time consuming and annoying procedure, compared with the concise sum of two complex numbers.

For linear and time-invariant systems with pure tones as input, the output is given by

$$y(t) = \mathrm{Re}\left\{\underline{x}\int_{-\infty}^{\infty} e^{j\omega\tau}h(t-\tau)\mathrm{d}\tau\right\}$$
$$= \mathrm{Re}\left\{\underline{x}e^{j\omega t}\int_{-\infty}^{\infty} e^{-j\omega u}h(u)\mathrm{d}u\right\}$$
$$= \mathrm{Re}\{\underline{x}\,\underline{H}(\omega)\,e^{j\omega t}\}. \tag{1.15}$$

If the input signal consists of a pure tone, then the output is a pure tone of the same frequency but different amplitude and phase. Therefore, for pure tones the transmission problem is reduced to the complex amplification $\underline{H}(\omega)$. Since the pure shapes of input and output signals are equal, a complete description of the transmission is given by changes in amplitude and phase.

The underlining of complex number to distinguish them from real values henceforth will be omitted, if misunderstanding is impossible.

1.3.3 Description with Frequency Response Function

The transmission problem for arbitrary, non-harmonic input signals $x(t)$ also may be solved from the above principle of the distortion-free transmission of sinusoidals, if the general input signal $x(t)$ is decomposed into a sum of pure tones. To include the most general case, the distance of frequencies that have to be taken into account must be infinitesimally small. The signal therefore must be represented by an integral over frequency

$$f(t) = \frac{1}{2\pi}\int_{-\infty}^{\infty} F(\omega)\,e^{j\omega t}\mathrm{d}\omega = F^{-1}\{F(\omega)\}, \tag{1.16}$$

(the factor $1/2\pi$ arises because the frequency integration $\mathrm{d}f = \mathrm{d}\omega/2\pi$ is intended). Obviously negative frequencies are included in the purely mathematical definition Eq. (1.16). $F(\omega)$ is the amplitude density function with the dimension dim(F) = dim(f)/Hz (dim() = dimension of the argument). The function $f(t)$ is real valued whenever the additional condition $F(-\omega) = F^*(\omega)$ holds (* = conjugate complex).

The spectral density $F(\omega)$ is related through

$$F(\omega) = \int_{-\infty}^{\infty} f(t)\, e^{-j\omega t} \mathrm{d}t, \qquad (1.17)$$

to the signal $f(t)$.

Equation (1.17) defines the operation known as Fourier transform. Equation (1.16) gives the inverse Fourier transform (for details relating to Fourier transforms see, for example, [20, 21]). $F(\omega)$ often also is called the Fourier transform of $f(t)$. The short hand notations with operators are also included in Eqs. (1.16) and (1.17). The Fourier transform can be unambiguously inverted.

Similar to the representation of a given function in form of series of other functions, the Fourier transform does not gain new information. Given information is simply described in a different way. The reason for the extra mathematical effort is, that the 'new representation' $F(\omega)$ allows for a simple treatment of linear and time invariant systems. The transmission is completely described by a complex amplification 'per frequency'. It can be shown that the transform of Eq. (1.12) yields

$$Y(\omega) = H(\omega)X(\omega), \qquad (1.18)$$

where $X(\omega)$ and $Y(\omega)$ denote the transforms of input $x(t)$ and output $y(t)$, respectively.

$H(\omega)$ is the so-called frequency response function and identical with the Fourier transform of the impulse response function

$$H(\omega) = F\{h(t)\}. \qquad (1.19)$$

$H(\omega)$ and $h(t)$ both contain a complete description of the system, they therefore cannot be independent of each other; they must be related as stated in Eq. (1.19).

Equation (1.18) reduces the transmission problem to the multiplication with the complex frequency response function.

The fact that the multiplication 'in frequency' corresponds to the convolution 'in time', is inherent in Eqs. (1.12) and (1.18). The relation

$$F^{-1}\{\mathrm{X}(\omega)\mathrm{H}(\omega)\} = \int_{-\infty}^{\infty} x(\tau)h(t-\tau)\mathrm{d}\tau, \qquad (1.20)$$

is termed the 'convolution theorem'. In principle, time t and frequency ω may be interchanged in Eqs. (1.16) and (1.17). Therefore, the convolution theorem is reciprocal; the multiplication of two time signals corresponds to the convolution of the transforms

$$F\{x(t)\cdot g(t)\} = \frac{1}{2\pi}\int_{-\infty}^{\infty} \mathrm{X}(\upsilon)\cdot G(\omega-\upsilon)\mathrm{d}\upsilon. \qquad (1.21)$$

From Eqs. (1.20) and (1.21) other properties of signals and their transforms can be deduced (energy theorem, Parseval's theorem and correlation functions).

In acoustics, the transform of spatial functions is an important tool for the treatment of radiation problems too (see Sects. 1.6.2, 1.8.3 and 1.9.3).

1.4 Fundamentals of Wave Propagation

The physical state of a gas is described by the density ρ_G, the temperature T_G and by the pressure p_G. Sound consists of (very) small fluctuations of these quantities in space and time; it is therefore possible to write

$$p_G = p_0 + p(x,y,z,t), \qquad (1.22)$$

$$\rho_G = \rho_0 + \rho(x,y,z,t), \qquad (1.23)$$

and

$$T_G = T_0 + T(x,y,z,t), \qquad (1.24)$$

where p_0, ρ_0 and T_0 denote the equilibrium pressure, density and temperature in the medium respectively; p, ρ and T are the sound pressure, sound density and sound temperature, respectively.

The quantities p_G, ρ_G and T_G are related through the gas law (for ideal gases)

$$p_G = R\frac{\rho_G T_G}{M_{\mathrm{mol}}}, \qquad (1.25)$$

(the constant R equals $R = 8{,}314$ Nm/K, M_{mol} is the molar mass of the gas).

The slow heat conduction processes usually can be neglected in acoustics (an exception consists only of narrow tubes at low frequency). Sound phenomena, therefore, are subject to adiabatic compression such that

$$\frac{p_G}{p_0} = \left(\frac{\rho_G}{\rho_0}\right)^{\chi}, \tag{1.26}$$

where $\chi = c_p/c_v$ denotes the ratio of specific heats for constant pressure, c_p, and constant volume, c_v. For a diatomic gas (like air) $\chi = 1.4$ holds.

As mentioned, the sound quantities p, ρ and T in Eqs. (1.22)–(1.24) normally are extraordinarily small compared with the static valuesp_0, ρ_0 und T_0. Products of the acoustical quantities p, ρ and T and higher then the first order in Eqs. (1.25) and (1.26), therefore, play only a role for the highest amplitudes beyond human perception. Under practical conditions, Eqs. (1.25) and (1.26) may be represented by their linearised versions

$$\frac{p}{p_0} = \frac{\rho}{\rho_0} + \frac{T}{T_0}, \tag{1.27}$$

and

$$\rho = p/c^2, \tag{1.28}$$

with

$$c^2 = \chi\frac{p_0}{\rho_0} = \frac{\chi R T_0}{M_{mol}}. \tag{1.29}$$

The most important conclusion from Eqs. (1.27) and (1.28) is, that sound pressure p, sound density ρ and sound temperature T exhibit one and the same signal shape; they differ only by scaling constants. Most often the sound pressure is used to describe the field, since this quantity can be measured easily using microphones.

Changes in density result from spatially varying displacements $\vec{\xi}$ in the elastic continuum consisting of gas or another fluid. Because of mass conservation, the following relation:

$$\rho = -\rho_0 \partial\xi/\partial x, \tag{1.30}$$

holds (one-dimensional field). The increase in density originates simply from closer material packing. On the other hand, due to Newton's law, the displacement is caused by the forces. Applied to a small cube of air Newton's law yields

$$\rho_0 \frac{\partial^2 \xi}{\partial t^2} = \rho_0 \frac{\partial v}{\partial t} = -\frac{\partial p}{\partial x}, \tag{1.31}$$

(v = local velocity). Together with Eq. (1.27), stating that sound pressure p and sound density ρ are proportional to each other, Eqs. (1.30) and (1.31) already describe the wave phenomenon. Upon eliminating ξ in the spatial derivative of Eq. (1.31) with the help of the second time derivative of Eq. (1.30), and finally expressing the sound density by the sound pressure p from Eq. (1.28), one readily finds the one-dimensional wave equation for the sound pressure

$$\frac{\partial^2 p}{\partial t^2} = c^2 \frac{\partial^2 p}{\partial x^2}. \tag{1.32}$$

Obviously, the solutions are waves in the form

$$p = f(t \pm x/c). \tag{1.33}$$

Sound signals in gases propagate without deformation of their shape (non-dispersive waves). The function f in Eq. (1.33) consists of an arbitrary signal shape produced by the source. The minus sign '−' in the argument stands for waves propagating into the positive x-direction; whereas '+' is associated with a wave in the opposite direction, provided the temporal dependence is chosen positive.

Obviously, c denotes the speed with which the wave propagates. c is called the wave or phase speed. Equation (1.2) shows, that the speed of sound c depends on material and temperature only, and is not, for example, influenced by the static air pressure. For example, from

M_{mol} (air) = 0.75 M_{mol} (N_2) + 0.25 M_{mol} (0_2) = 28.8 g one finds $c(\text{air}) = 20.1\sqrt{T/K}$ m/s and therefore c = 343 m/s for T = 293.2 K (corresponding to about 20°C). For hydrogen with M_{mol} (H_2) = 2g the wave speed is c (H_2, 20°C) = 1,310 m/s, and for carbon dioxide with M_{mol} (CO_2) = 44 g c (CO_2, 20°C) = 278.5 m/s result. In general, high density gases have a smaller sound speed than light density ones. Other wave velocities can be found in the literature ([22, 23] for sea water).

A sound field containing only one of the two solutions given in Eq. (1.33) is a PROPAGATING wave (active field). Two propagating waves of equal amplitudes but opposite propagation directions form a STANDING wave (reactive field). For plane propagating waves, Eq. (1.31) shows that sound pressure and local velocity constitute a constant ratio:

$$p(x,t) = \rho_0 c\ v(x,t). \qquad (1.34)$$

The ratio $\rho_0 c$ is called the wave resistance or specific resistance of the medium. The most important conclusion of Eq. (1.34) is that the particle velocity may be determined by simply measuring the sound pressure. This simplification is used for the measurement of sound intensity and sound power in a free field (see Chap. 5). Furthermore, Eq. (1.34) shows that the signal shape f in Eq. (1.33) equals that of the velocity time signal of a source at $x = 0$ for pure propagating waves; it is $p(x,t) = \rho_0 c\ v(t - x/c)$ with $v(t)$ = velocity of a membrane at $x = 0$.

For pure tones with circular frequency ω, Eq. (1.33) becomes

$$p = p_{\pm} \cos(\omega(t \pm x/c)) = p_{\pm} \cos(\omega t \pm kx). \qquad (1.35)$$

The time period T is included in the circular frequency via $\omega = 2\pi/T$, and in the same way the wave number k must include the spatial period λ:

$$k = \frac{\omega}{c} = \frac{2\pi f}{c} = \frac{2\pi}{\lambda}, \qquad (1.36)$$

which is called the wavelength $\lambda = c/f$. For human perception between 16 Hz and 16 kHz the wavelengths of 20 m > λ > 0.02 m are related.

In three dimensions, the above considerations result in

$$\mathrm{div}\vec{v} = \frac{\partial v_x}{\partial x} + \frac{\partial v_y}{\partial y} + \frac{\partial v_z}{\partial v} = -\frac{1}{\rho_0 c^2}\frac{\partial p}{\partial t}, \qquad (1.37)$$

(mass conservation) and in

$$\rho_0 \frac{\partial \vec{v}}{\partial t} = -\mathrm{grad}p$$
$$= -\left(\vec{e}_x \frac{\partial e}{\partial x} + \vec{e}_y \frac{\partial p}{\partial y} + \vec{e}_z \frac{\partial p}{\partial z}\right), \qquad (1.38)$$

(Newton's law). The general operators in Eqs. (1.37) and (1.38) (div and grad) are given in details for Cartesian co-ordinates in addition ($\vec{e}_x, \vec{e}_y, \vec{e}_z$ basic unity vectors). Again, after eliminating the velocity from Eqs. (1.37) and (1.38) the three dimensional wave equation

$$\Delta p = \frac{\partial^2 p}{\partial x^2} + \frac{\partial^2 p}{\partial y^2} + \frac{\partial^2 p}{\partial z^2} = \frac{1}{c^2}\frac{\partial^2 p}{\partial t^2}, \qquad (1.39)$$

results (Δ is the Laplacian, again made explicit for Cartesian co-ordinates).

For pure tones, (1.39) can be formulated using the dimensionless co-ordinates x/λ, y/λ and z/λ only. Therefore, model scale experiments are possible. To do so all dimensions are scaled down while the frequency is scaled up by the same scaling factor. A problem results from the absorption, which depends strongly on frequency, and which is not included in Eq. (1.39) (see also Chap. 10 for scale modelling in room acoustics).

The above equations are valid for a medium at rest. For a moving medium, the convection has to be taken into account. For flow with velocity U_0 in x-direction neither depending on time nor on a spatial co-ordinate, must the time derivatives in Eqs. (1.21)–(1.39) be substituted by the complete derivative

$$\frac{\partial}{\partial t} \to \frac{\mathrm{d}}{\mathrm{d}t} = \frac{\partial}{\partial t} + U_0 \frac{\partial}{\partial x}. \qquad (1.40)$$

For one-dimensional waveguides with flow, the solutions Eq. (1.33) are rewritten as

$$p = f\left(t \mp \frac{x}{c \pm U_0}\right). \qquad (1.41)$$

The speed of sound relative to the origin of the co-ordinate system used increases with the flow speed for waves propagating in the flow direction and decreases accordingly for waves propagating in the opposite direction.

1.5 Sound Energy and Power Transport in Gases

Sound consists physically in oscillatory elastic compression and in oscillatory displacement of in a fluid. Therefore, the medium acts as storage for both potential and kinetic energy as well. Consequently, the total energy density distribution is composed of the two parts

$$w = \frac{1}{2}\frac{p^2}{\rho_0 c^2} + \frac{1}{2}\rho_0|\vec{v}|^2. \qquad (1.42)$$

($|\vec{v}|^2$ denotes the absolute value of vector $\vec{v}$squared). The energy instantaneously stored in a volume V equals

$$E = \iiint w \mathrm{d}V. \qquad (1.43)$$

The energy simply 'travels with the wave'. For example, for a plane propagating wave $w = p^2(x,t)/\rho_0 c^2 = p^2(t \mp x/c)/\rho_0 c^2$ holds (see Eq. (1.42)).

In particular for stationary sources in continuous operation, the energy transport is described using the energy flow for convenience. The net energy flow through an area S equals

$$P = \iint I_n \mathrm{d}S, \qquad (1.44)$$

where I_n denotes the component of the so-called intensity $\vec{I}$ perpendicular to S. The intensity is the product of sound pressure and velocity

$$\vec{I} = p\vec{v}, \qquad (1.45)$$

indicating the power per unit square (the dimension of I is W/m^2). The proper definition of the intensity is given by Eq. (1.44), such that the energy conservation principle delivers the relation

$$\frac{\partial w}{\partial t} = -\mathrm{div}\vec{I} = -\left(\frac{\partial I_x}{\partial x} + \frac{\partial I_y}{\partial y} + \frac{\partial I_z}{\partial z}\right), \qquad (1.46)$$

between energy density and intensity. Eq. (1.45) therefore does not realise a definition. Rather, Eq. (1.45) results from Eq. (1.46) after inserting the fundamental field equations (1.37) and (1.38) in the expression for the energy density Eq. (1.42).

Field quantities as well as energy quantities perform temporal fluctuations. For stationary sources, the temporal mean of the emitted sound power $\overline{P}$ normally is used to describe the source. The following measurement procedure applies for far field conditions (see Sect. 1.6.2) in an anechoic room or in a free field. An (imaginary) surface around the source is divided into N 'small' partitions S_i with the partial intensities $\overline{I}_i$. The sound power $\overline{P}$ then results in

$$\overline{P} = \sum_{i=1}^{N} \overline{I}_i S_i, \qquad (1.47)$$

where the partial intensities

$$\overline{I}_i = \frac{{p_{i,\mathrm{rms}}}^2}{\rho_0 c}, \qquad (1.48)$$

are calculated from the root mean square pressures measured.

The power emitted from a source constitutes an important source quantity, from which the pressure level in a specific acoustical environment (for example in a room with known reverberation time) can be estimated.

1.6 Sound Radiation

1.6.1 Compact Sources

Some sources of technical relevance are characterised by their volume flow Q (='pumped' volume/time). Examples are vehicle exhausts, small explosions, firecrackers, outlet of valves, and small loudspeakers in boxes at low frequency (with excluded 'hydrodynamic short circuiting'). As long as the relevant dimensions of the radiator are small compared to the wavelength, such sources may be represented by an equivalent source consisting of a pulsating sphere (with contracting and expanding surface) of the same volume flow Q. The latter results from the original source using

$$Q = \int_S v \, \mathrm{d}S, \qquad (1.49)$$

where S denotes the surface of the original source. For the pulsating sphere (and for the original source being represented)

$$p = \frac{\rho_0}{4\pi r}\frac{\partial Q(t - r/c)}{\partial t}, \qquad (1.50)$$

holds where r is the distance between the centre of the source and the measurement point. As a consequence of Eq. (1.50), sudden variations of temporal volume flow produce loud noise whereas slow variations are comparatively quiet. The time delay in the argument of Q indicates the time needed for propagation from source to receiver. Note that the sound pressure is proportional to the surface acceleration of the source. For pure tones (or for Fourier transforms), Eq. (1.50) becomes

$$p = \frac{j\omega\rho_0}{4\pi r} Q e^{-jkr}, \qquad (1.51)$$

showing, that the radiation process itself amplifies velocities at high frequencies, a fact being important for the radiation from loudspeakers, for example. Small sources producing net volume flow often are called monopoles.

Another source of practical interest consists of the dipole, which in general produces the sound field of small oscillating rigid bodies. A typical dipole is again a small sphere (radius a) with the radial surface velocity $v_r = v_0 \cos\vartheta$ exciting the sound pressure

$$p(r,\vartheta) = -\frac{\omega^2 2\pi a^3 \rho_0}{4\pi rc} v_0 \cos\vartheta \left(1 + \frac{1}{jkr}\right) e^{-jkr}. \qquad (1.52)$$

A more detailed consideration shows that $2\pi a^3$ represents the sum of the volume V_k replaced by the source (here $V_k = 4\pi a^3/3$) and of the volume V_H oscillating hydrodynamically (here $V_H = 2\pi a^3/3$). Arbitrary, small bodies moving on the whole with velocity v_0 generate the sound pressure

$$p(r,\vartheta) = -\frac{\omega^2 (V_H + V_k)\rho_0}{4\pi rc} v_0 \cos\vartheta \left(1 + \frac{1}{jkr}\right) e^{-jkr}. \qquad (1.53)$$

For a thin disc (radius a, thickness h, $h \ll a$) moving in axial direction, $V_H = 8a^3/3$ holds, for example.

The sequence of 'mathematical' source types – being produced through combinations from monpole combinations – can be continued. Of practical interest is also the quadrupole source (a combination of two dipoles with opposite signs), playing a role for sound generated by flow (see for example [24, 25] and Chap. 20).

1.6.2 Radiation from Plane Structures

The radiation from plates constitutes an important topic. The principal conclusions are adequately transmitted to similar radiation problems of sources with curvilinear surface (see e.g. [26–29]). Moreover, the radiation from plane structures (walls, ceilings) appears often in technical acoustics.

The simplest consideration – the radiation from a radiator consisting of a sinusoidal wave – already reveals the variety of radiation processes by superposition.

For a velocity distribution perpendicular to the radiating plane (see Fig. 1.1) of the form

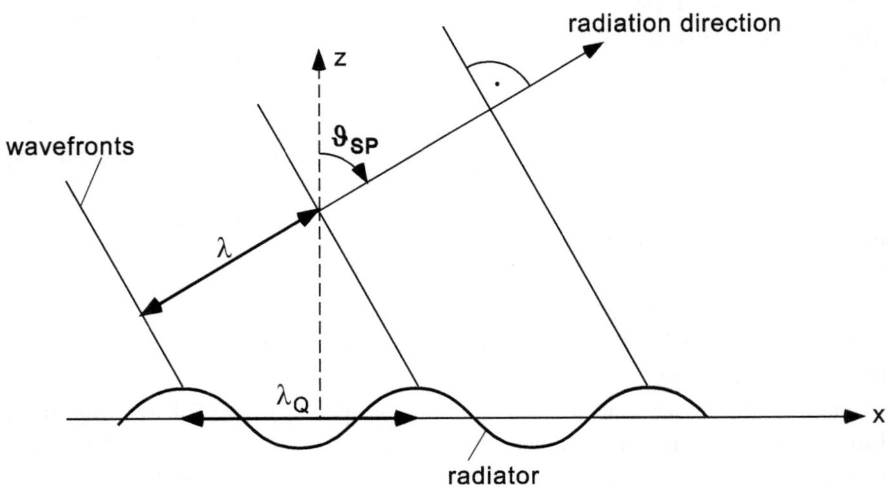

Fig. 1.1 Radiation from planes

$$v = v_0 e^{-jk_Q x}, \quad (1.54)$$

($k_Q = 2\pi/\lambda_Q$ = radiator wave number, λ_Q = radiator wavelength), which, for example, could be produced on a sheet metal (in this case λ_Q denotes the bending wavelength) the sound pressure is

$$p = \frac{k}{k_z} \rho_0 c v_0 e^{-jk_z z} e^{-jk_Q x}, \quad (1.55)$$

where the wave number k in the direction perpendicular to the radiator plane is given by

$$k_z^{\ 2} = k^2 - k_Q^{\ 2} \quad (1.56)$$

due to the wave equation (1.39) ($k = \omega/c = 2\pi\lambda$ wave number in air). The root is extracted so, that physically meaningful result arise.

The following effects are obvious :

1. A radiator with a wavelength λ_Q longer than the wavelength λ of air ($\lambda_Q > \lambda$ and therefore $k_Q < k$, thus leading to a real value for k_z) emits an oblique wave under the angle $\vartheta_{\rm sp}$ with (see Fig. 1.1).

$$\sin \vartheta_{sp} = \frac{\lambda}{\lambda_Q}. \quad (1.57)$$

2. A radiator with a wavelength λ_Q smaller than the wavelength λ of air ($\lambda_Q < \lambda$ and therefore $k_Q > k$, thus leading to an imaginary value for k_z) on the other hand produces a nearfield only, which exists in the neighbourhood of the radiating surface. No temporal mean of power is emitted. The fact, that no noticeable sound field exists, can be interpreted as complete short circuit between neighbouring radiator segments of opposite signs ('peaks and troughs' of the radiator).

These very fundamental ideas are found again in nearly all following remarks.

Indeed, proceeding to the description of arbitrary radiator velocities using (double) Fourier transforms

$$V(k_1, k_2) = \int_{-\infty}^{\infty}\int_{-\infty}^{\infty} v(x, y) e^{-jk_1 x} e^{-jk_2 y} \mathrm{d}x\mathrm{d}y, \quad (1.58)$$

the Fourier transform of the sound pressure becomes

$$P(k_1, k_2) = \frac{k}{k_z} \rho_0 c V(k_1, k_2) e^{-jk_z z}, \quad (1.59)$$

where

$$k_z^2 = k^2 - k_1^2 - k_2^2, \quad (1.60)$$

holds, following the wave equation. The total sound pressure is determined from the inverse transform giving

$$p(x,y,z) = \frac{1}{4\pi^2} \int_{-\infty}^{\infty}\int_{-\infty}^{\infty} P(k_1, k_2) e^{jk_1 x} e^{jk_2 y} \mathrm{d}k_1 \mathrm{d}k_2. \quad (1.61)$$

In the above Eqs. (1.58)–(1.61), the radiator itself is decomposed into wave components, which – depending on their wave number $k_Q^{\ 2} = k_1^{\ 2} + k_2^{\ 2}$ – lead to a pure nearfield or an oblique plane wave radiation under a specific angle, respectively. The descriptions Eqs. (1.60) and (1.61) open the field of acoustical holography [29, 30]. For example, one could measure the sound pressure distribution (amplitude and phase) in one plane and compute the Fourier transform $V(k_1, k_2)$ of the radiator, or even the velocity $v(x, y)$, taking inverse transforms. Moreover, the sound pressure distribution in every other plane can be calculated.

The sound pressure can be calculated directly from the spatial radiator velocity distribution if Eq. (1.59) is inserted into Eq. (1.61) and the convolution theorem Eq. (1.20) is used. That procedure yields

$$p(x,y,z) = \frac{j\omega\rho_0}{2\pi} \int_{-\infty}^{\infty}\int_{-\infty}^{\infty} v(x_Q, y_Q) \frac{e^{-jk\sqrt{(x-x_Q)^2+(y-y_Q)^2+z^2}}}{\sqrt{(x-x_Q)^2+(y-y_Q)^2+z^2}} \mathrm{d}x_Q \mathrm{d}y_Q. \quad (1.62)$$

Equation (1.62) presents the so-called Rayleigh integral, which can be interpreted as decomposition of the radiator into (infinitesimally) small monopole sources and summing their partial sound pressures at the observation point (x,y,z). The root expression depicts the distance between the observation point and the 'actual' source point x_Q, y_Q. Obviously, the Rayleigh integral can be used, if the radiator velocity is known in the whole plane $z = 0$ (or if v can be estimated in some reasonable way). Often, that is not the case. For example, the sound radiation from railway wheels cannot be computed from the Rayleigh integral. The formal reason is that the velocity distribution outside of the wheel itself remains unknown. To artificially set that velocity to zero would mean to omit the short circuiting between front and rear sides, being of significant importance at low frequency, and therefore would produce erroneous results.

Often, only the farfield is of interest. If the largest radiator dimension is set to l (for example, the velocity equals zero outside a circle of diameter l), then the farfield is defined by the three conditions $r/l \gg 1$, $r/\lambda \gg 1$ and $r/l \gg l/\lambda$. The first condition is purely geometrical, the second and third conditions depict the lower and upper frequency limits, for which the farfield conditions hold. If these hold, the Rayleigh integral can be approximated by

$$p(r,\vartheta,\varphi) = \frac{j\omega\rho_0}{2\pi}\frac{e^{-jkR}}{R} V(k_1 = -k\sin\vartheta\cos\varphi, k_2 = -k\sin\vartheta\sin\varphi).$$

As usual, R is the distance between the observation point and the radiator centre, ϑ relates to the z-axis and φ to the x-axis. The directivity pattern in the farfield consists in the visible cut-out ${k_1}^2 + {k_2}^2 < k^2$ of the Fourier-transformed radiator velocity.

Computing the radial velocity component in the far field from Eqs. (1.38) and (1.63), one finds

$$p(R,\vartheta,\varphi) = \rho_0 c v_r(R,\vartheta,\varphi). \tag{1.64}$$

In the farfield, the impedance always equals the impedance of the medium. For a certain angle, the sound field behaves like a plane wave. As mentioned above (see Eqs. (1.47) and (1.48)), this fact is used to determine intensity and power from single pressure measurements only.

1.6.3 The General Radiation Problem

In the general case, the problem consists in calculation of the sound field radiated by a body of arbitrary geometrical shape with a given normal component of the surface velocity.

For some quite special geometries, this problem can be treated analytically. Solutions in form of series are known, for example, for cylindrical bodies (circular or elliptical cross-section) or spheres. The complete description of calculation procedures and results is given in [31–32].

A representative case of two dimensional (independent from the z direction) radiation from a circular cylinder (radius a) will be given. In addition, symmetry $p(\varphi) = p(-\varphi)$ is assumed. Under these conditions, the sound pressure becomes

$$p(r,\varphi) = -j\rho c \sum_{n=0}^{\infty} v_n \frac{H_n^{(2)}(kr)}{H_n^{(2)\prime}(ka)} \cos n\varphi, \tag{1.65}$$

(${H_n}^{(2)}$: Hankel function of the second kind and order n, see e.g. [33]; the prime indicates the derivative with respect to the argument). The coefficients V_n denote the series expansion of the velocity $V(\varphi)$ on the cylinder surface into $\cos(n\varphi)$ functions:

$$V_n = \frac{1}{\pi}\int v(\varphi)\cos n\varphi \mathrm{d}\varphi, n \neq 0, \tag{1.66a}$$

$$v_0 = \frac{1}{2\pi}\int_0^{2\pi} v(\varphi)\mathrm{d}\varphi. \tag{1.66b}$$

The sound power radiated (per unit length of the z axis) equals

$$P = \sum_{n=0}^{\infty} |V_n|^2 \sigma_n, \tag{1.67}$$

with

$$\sigma_n = \frac{2}{ka\pi \, H_n^{(2)'}(ka)}. \tag{1.68}$$

The most important observations are:

- The sound field is composed of modes (another name for the addends in the series expansion Eq. (1.65)). The modal order n indicates the number of radiator wavelength along the cylinder circumference; the modal wavelength λ_n amounts $\lambda_n = 2\pi a/n$.
- The sound power radiated equals the sum of modal powers (the latter equals the power radiated form a

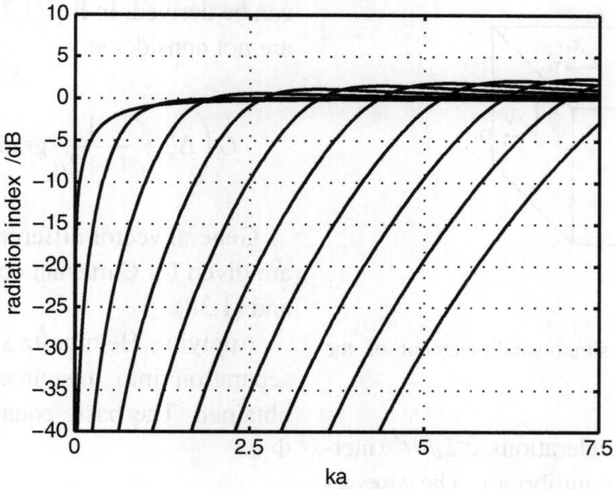

Fig. 1.2 Modal radiation efficiencies for the radiation from a cylinder with radius a in relation to the modal order n, indicating the number of radiator wavelengths along the cylinder circumference

single mode in absence of all other modes). As can be seen from the modal radiation efficiencies in Fig. 1.2, the radiation is large for modes with long wavelengths $\lambda_n > \lambda$ (corresponding to $2\pi a/\lambda > n$) and small for modes with short wavelengths $\lambda_n < \lambda$. The difference to the case of radiation from a plane results from the incomplete mass short circuiting due to the curvilinear surface; therefore, a small radiation still is exhibited in the shortwave region. Similar observations can be made for spherical radiators.

In practice, often the radiation from arbitrarily shaped, complex structures (like motors, machines and vehicles) needs to be calculated. This is done using numerical methods treated in Chap. 3. A fundamental inconvenience inherent in all procedures can already be identified. Small measurement or observation errors in a velocity distribution with short wavelengths only may lead to a much higher radiation than in reality.

1.7 Basic Equations for the Propagation of Sound in Solids

Similar to fluids, the phenomena of sound in solids is caused by very small time- and space dependent deviations $\vec{\zeta}$ from the static values $\vec{\zeta}_0$

$$\vec{\zeta}_G = \vec{\zeta}_0 + \vec{\zeta}(x, y, z, t). \qquad (1.69)$$

The dynamic process can be described via the related changes of temperature, density, stresses, and displacement. Since the stresses and the displacement usually are the values that can be obtained by measurements, the description of structure-borne sound is typically based on them.

Generally, the principle of superposition applies. Due to the mostly small amplitudes in acoustics, the description of the materials might be approached adequately by applying linear material laws.

Considering a Cartesian co-ordinate system, the continuum equations in dynamics result from the equilibrium conditions with respect to the three co-ordinates (see Fig. 1.3):

$$\begin{aligned}
\frac{\partial \sigma_x}{\partial x} + \frac{\partial \tau_{xy}}{\partial y} + \frac{\partial \tau_{xz}}{\partial z} + K_x &= \rho \frac{\partial^2 \zeta_x}{\partial t^2}, \\
\frac{\partial \sigma_y}{\partial y} + \frac{\partial \tau_{yx}}{\partial x} + \frac{\partial \tau_{yz}}{\partial z} + K_y &= \rho \frac{\partial^2 \zeta_y}{\partial t^2}, \qquad (1.70) \\
\frac{\partial \sigma_z}{\partial z} + \frac{\partial \tau_{zx}}{\partial x} + \frac{\partial \tau_{zy}}{\partial y} + K_z &= \rho \frac{\partial^2 \zeta_z}{\partial t^2}.
\end{aligned}$$

The equations describe the equilibrium of a section j of an infinitely small element in the continuum. The shear stresses in direction i τ_{ij}, the normal stresses σ_i, the external volume forces K_i and the product of the

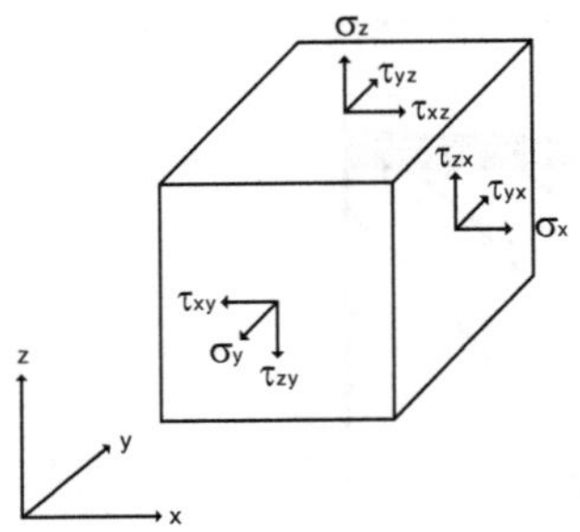

Fig. 1.3 Stresses at an infinitesimal small element of the continuum

material density ρ and the accelerations $\partial^2\zeta_i/\partial t^2$ (inertia forces $\rho\partial^2\zeta_i/\partial t^2$) are in equilibrium. The stresses occurring inside the continuum are related to the strains. Strains are always related to varying displacements. In case of small displacements, which mostly can be assumed in practical applications of structure-borne sound, longitudinal ε_i or shear strains γ_{ij} can be calculated at the basis of the displacements ζ_i:

$$\varepsilon_x = \frac{\partial\zeta_x}{\partial x};\ \varepsilon_y = \frac{\partial\zeta_y}{\partial y};\ \varepsilon_z = \frac{\partial\zeta_z}{\partial z},$$
$$\gamma_{xy} = \frac{\partial\zeta_x}{\partial y} + \frac{\partial\zeta_y}{\partial x};\ \gamma_{xz} = \frac{\partial\zeta_x}{\partial z} + \frac{\partial\zeta_z}{\partial x};\ \gamma_{yz} = \frac{\partial\zeta_y}{\partial z} + \frac{\partial\zeta_z}{\partial y}. \tag{1.71}$$

The stresses can be derived from the strains by taking the material properties into account. For typical applications in structure-borne sound, Hooke's law for the linear elastic isotropic continuum is assumed.

$$\sigma_x = 2G\left[\varepsilon_x + \frac{\mu}{1-2\mu}(\varepsilon_x+\varepsilon_y+\varepsilon_z)\right],$$
$$\sigma_y = 2G\left[\varepsilon_y + \frac{\mu}{1-2\mu}(\varepsilon_x+\varepsilon_y+\varepsilon_z)\right],$$
$$\sigma_z = 2G\left[\varepsilon_z + \frac{\mu}{1-2\mu}(\varepsilon_x+\varepsilon_y+\varepsilon_z)\right],$$
$$\tau_{xy} = \tau_{yx} = G\gamma_{xy};\ \tau_{xz} = \tau_{zx} = G\gamma_{xz};\ \tau_{yz} = \tau_{zy} = G\gamma_{yz}. \tag{1.72}$$

with the shear modulus G and the Poisson's ratio μ.

Describing the relationships given in Eqs. (1.70) – (1.72) by means of tensors, a transformation to other co-ordinate systems (e.g. spherical or cylindrical co-ordinates) can be carried out.

Combining the Eqs. (1.70)–(1.72), the basic equation (Lamé equation) of the elastic isotropic continuum can be derived. In Eq. (1.73), external volume forces K_i are not considered.

$$G\left(\Delta\vec{\zeta} + \frac{1}{1-2\mu}\text{grad div }\vec{\zeta}\right) = \rho\frac{\partial^2\vec{\zeta}}{\partial t^2}. \tag{1.73}$$

General vector differential operators (div and grad) are given for Cartesian co-ordinates in the Eqs. (1.37) and (1.38).

Applying Helmholtz's law to Lame's equation, a separation into a source and a vortex-free part is obtained. The basic equation for the vortex-free part Φ is:

$$G\frac{2-2\mu}{1-2\mu}\cdot\Delta\Phi = \rho\frac{\partial^2\Phi}{\partial t^2}\text{mit }\Phi = \text{div}\vec{\zeta}. \tag{1.74}$$

The basic equation for the source-free part $\vec{\psi}$is:

$$G\Delta\vec{\psi} = \rho\cdot\frac{\partial^2\vec{\psi}}{\partial t^2}\text{mit }\vec{\psi} = \frac{1}{2}\text{rot }\vec{\zeta}. \tag{1.75}$$

The scalar potential Φ describes the dilatation, which is related to a change of the volume. The vector potential $\vec{\psi}$ describes the vector of rotation.

The governing equation for the dilatation Eq. (1.74) corresponds to the three-dimensional wave equation (1.39) for gases and fluids. The corresponding velocity of sound, the wave velocity of the longitudinal waves c_L, is mostly given by the shear modulus G and the Poisson's ratio μ or by the compression modulus E_s:

$$c_L = \sqrt{\frac{G}{\rho}\frac{2-2\mu}{1-2\mu}} = \sqrt{\frac{E_s}{\rho}}, \tag{1.76}$$

with:

$$G = \frac{E_s(1-2\mu)}{2-2\mu}. \tag{1.77}$$

The governing equation for $\vec{\psi}$ Eq. (1.75) leads to the transversal wave velocity c_T:

$$c_T = \sqrt{\frac{G}{\rho}}. \tag{1.78}$$

In pure longitudinal waves (also called compression or dilatation waves), the particles move in the

Table 1.1 Selected sound velocities of gases and fluids (20°C)

Material	c (m/s)	ρ (kg/m^3)
Air	344	1.21
Hydrogen	1,332	0.084
Helium	1,005	0.167
Nitrogen	346	1.17
Oxygen	326	1.34
Carbon dioxide	268	185
Water (distilled)	1,492	1,000

direction of the propagation of the wave as for sound waves in gases and fluids. The energy content and the power transport can be derived in the same way as for sound in gases and fluids (cf. Sect. 1.5). Longitudinal waves are mainly of interest if the structure to investigate is large compared to the wavelength (Table 1.1).

In case of pure transversal waves (also called shear, rotation or distortion waves), the shear deformation causes a change of the shape but no change in the volume. The power transmission per area can be derived as in the case of the longitudinal wave. The sound pressure p in Eq. (1.48) is replaced by the shear stress τ perpendicular to the propagation. Furthermore, the wave velocity in this case is given by c_T.

The transition from solids to gases and fluids can be visualised by the help of Eq. (1.77). In inviscid gases and fluids, the Poisson's ratio μ becomes 0.5. In this case, the shear stiffness G vanishes and shear forces can no longer be transmitted. The wave velocity c_T of transversal waves also becomes 0 Eq. (1.78). The sound field in gases and fluids thus is exclusively determined by longitudinal waves, whereas the sound fields in solids with shear stiffness by longitudinal and transversal (distorsion) waves. A good overview over the phenomena described in the following is given in [34].

1.8 Waves in Solids with Limiting Edges

1.8.1 Waves and Near Fields

In an elastic isotropic homogeneous solid, all vibration patterns can be described as a superposition of longitudinal and transversal waves and – similar to gases – near fields that occur in the vicinity of boundaries in inhomogeneous areas and next to sources (see Fig. 1.1). Near fields typically decay quickly with the distance from the inhomogeneity or source.

Near fields on the surface of solids that decay perpendicular to the direction of propagation are called surface waves. The longitudinal and transversal waves that propagate in all three dimensions of space are called volume waves. Whereas the latter propagate in three dimensions, the surface waves show an exponential decay with the depth. They propagate parallel to the surface and thus in two dimensions. For this reason, the decay caused by the geometry of propagation of the energy of structure-borne sound – which is proportionate to the square of the velocity of vibration – is the following:

1. Point sources at the surface and propagation via volume waves: energy is inversely proportional to the square of the distance from the source
2. Point sources at the surface and propagation via surface waves: energy is inversely proportional to the distance of the source
3. Line sources on the surface and propagation via volume waves: energy is inversely proportional to the distance of the source
4. Line sources on the surface and propagation via surface waves: no decay due to geometry

The partition of the input power in volume waves and near fields depends on the wave number of the excitation. At a point excitation of an unlayered half-space, the energetic partition on the different waves is [35]:

- Longitudinal volume wave 7%
- Transversal volume wave 28%
- Surface waves 67%

In typical structures of technical acoustics, the continuum has free or fixed surfaces and often also layers. The combination of the vortex-free Φ (Eq. 1.74) and the source-free $\vec{\psi}$ (Eq. 1.75) parts and the characteristics of their solution (near-field or volume wave) can be derived from the boundary and coupling conditions (see also Sect. 22).

For practical investigations, simplified equations, e.g. for plates or beams can be taken into account. Those equations are easier to handle than the Lamé equation. In those equations, approaches for the deformation are introduced – typically vertical to the axis or plane of the structure. Those approximations are appropriate for a certain set of geometries and wavelengths.

1.8.2 Rayleigh Wave

The solution of the Lamé equation for a free surface without stresses can be considered like a free vibration without exterior loads. Such a 'homogeneous solution' corresponds for finite structures to the eigen vibrations (resonant vibration). Those vibration patterns can easily be excited. For the elastic isotropic half-space, the homogeneous solution is given by the Rayleigh wave. As this wave is easy to excite, showing typically little geometrical damping, it is in practice especially in soil dynamics of utmost importance.

The Rayleigh wave can be described by the combination of vortex and source-free parts, the composition of which depends on the Poisson's ratio. It is a near field (surface wave) and decays from the free surface to the interior of the half-space exponentially [34, 36]. Depending on the Poisson's ratio, the propagation velocity c_R ranges approx. from 0.87 to 0.96 of the transversal wave velocity c_T .It is approximately given in Eq. (1.79) [34]:

$$c_R \approx c_T \frac{0.874 + 1.12\mu}{1 + \mu}. \quad (1.79)$$

In case of Poisson's ratio of typical materials, the amplitude decays at a distance of one wavelength from the surface to approx. 20–30%.

The particles on the surface move in a vertically oriented elliptical orbit. The displacement in horizontal directions is – depending on the Poisson's ratio– approx. 60–80% of the vertical displacement. Table 1.2 indicates typical longitudinal, transversal and Rayleigh wave velocities for some materials.

1.8.3 Thick Plates, Horizontally Layered Continua

In practice, often structures with parallel delimiting surfaces have to be investigated (e.g. plates, layered soils).

For horizontally layered continua, as well as for thick plates (plate thickness $h > \lambda/6$) the structure-borne sound can be described by the Lamé equation 1.73 and the derived equations (1.74) and (1.75). Here, the number of degrees of freedom cannot be reduced by means of approximations – as is the case for plates or beams (Sect. 1.8.4). For the solution, it is possible to use reference solutions (see Sect. 22), numerical solutions (see Sect. 3) or integral transformations.

By means of a three-dimensional Fourier transform (Cartesian co-ordinates) or, analogously, a two-dimensional Fourier transform combined with a Hankel transform (cylindrical co-ordinates), it is possible to calculate both the source and the vortex-free condition for horizontally layered solids by means of a simple algebraic system of equations [34, 37]. The Fourier transforms or the Hankel transform have to be carried out with respect to spatial and temporal co-ordinates. The solutions of the vortex and source-free displacements are calculated considering the boundary and coupling conditions of the stresses and displacements.

Often it is not necessary to carry out the inverse transform as the spectral results can be interpreted as volume waves or near fields. This is an advantage of this type of solution. In the spectral domain, the spatial co-ordinates x_i are transformed into wave numbers $k_{xi} = 2\pi/\lambda_i$ the time co-ordinate t is transformed into the circular frequency $2\pi f$. In order to make this approach apparent, a two-dimensional case is taken as an example. Based on the wavelength λ_x and the frequency, the 'trace velocity' c_x can be calculated as

$$c_x = \lambda_x \cdot f. \quad (1.80)$$

1. When the trace velocity c_x is larger than the longitudinal wave velocity c_L resp. transversal wave velocity c_T, the corresponding vortex and source-free part of the displacement in the continua describes

Table 1.2 Wave-velocities for approximate calculations with the help of the elastic isotropic continuum (Normal conditions ca. 20°C)

Material	Longitudinal wave velocity	Transversal wavevelocity	Rayleighwave velocity
Aluminium	6,450	3,100	2,900
Led	2,100	730	700
Steel	6,000	3,100	2,900
Concrete	3,400–3,700	1,500–2,000	1,400–1,800

a spatially propagating volume wave. Its direction is inclined an angle ϑ_L respectively ϑ_T with respect to the normal of the surface (see Fig. 1.1) such that

$$\begin{aligned}\sin\vartheta_T &= \frac{c_T}{c_X},\\ \sin\vartheta_L &= \frac{c_L}{c_X}.\end{aligned} \tag{1.81}$$

2. When the trace velocity c_x on the surface is smaller than the corresponding longitudinal wave velocity c_L respectively transversal wave velocity c_T, the respective vortex and source-free part of the displacement in the continua form near fields (surface waves) with the following decay of the field with the distance z from the surface:

$$\begin{aligned}A_L(z) &= A_L(z=0)\cdot e^{-2\pi f\sqrt{\frac{1}{c_L^2}-\frac{1}{c_x^2}}},\\ A_T(z) &= A_T(z=0)\cdot e^{-2\pi f\sqrt{\frac{1}{c_T^2}-\frac{1}{c_x^2}}}.\end{aligned} \tag{1.82}$$

Pulsating loads can always be considered as a superposition of two load functions that are moving in opposite directions along the 'trace'.

Figure 1.4 shows the relationship between the transversal velocity c_T and the phase velocity c_x (trace velocity) for plates of different plate thicknesses h. The thickness is normalized on the wavelength of the transversal wave [38]. The index in the individual curves in the dispersion diagram indicates the number of 'zero crossings' of the vortex and source-free part (L respectively T) over thickness of the plate. For very large plates, the phase velocity approaches the velocity of the Rayleigh wave. For very small heights with respect to the wavelengths, T_0 and L_0 are the only solutions. They can be interpreted as bending waves and longitudinal waves in the plate. Both wave types are described by a deformation pattern that can be derived perpendicular to the plate with the help of simple approximations for the displacements (see Sect. 1.8.4)

Similar to the homogenous half-space, the so-called wave impedances are a good and appropriate tool in order to describe the impedance of the structure. They will be discussed in detail in Sect. 1.9.3.

1.8.4 Thin Plates and Thin Beams

In practice often thin plates and thin beams are chosen as appropriate models. In this case, the displacement is modelled with the approach of 'planar cross-sections'. Furthermore, it is generally assumed that the cross-sections remain perpendicular to the neutral layer of the beam or the plate in the deformed state. Whereas the differential equations for thick plates have to be

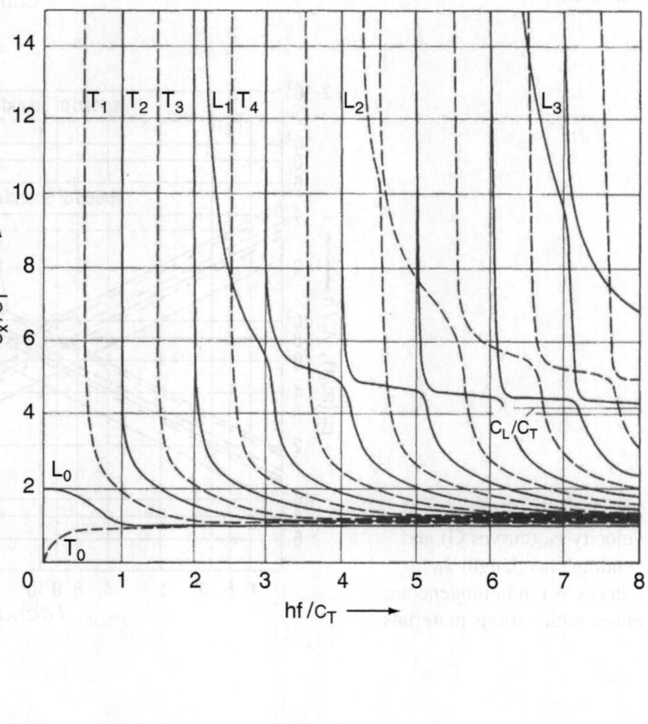

Fig. 1.4 Normalized phase-velocity c_x of free waves in a homogeneous planar plate (thickness h, transversal wave velocity c_T and Poisson's ratio 0.47, [38]

solved by derivations with respect to the three co-ordinates, the co-ordinates perpendicular to the plane of the plate or the beam have not to be considered for thin plates and beams. The propagation of waves perpendicular to the neutral layer can be omitted.

With this approach, simple differential equations of the fourth order can be derived (the plate is located in the x, y-plane):

$$B''\left(\frac{\partial^4}{\partial x^4}+\frac{\partial^4}{\partial x^2\partial y^2}+\frac{\partial^4}{\partial y^4}\right)\zeta+m''\frac{\partial^2\zeta}{\partial t^2}=p$$
$$\text{with}\quad B''=\frac{Eh^3}{12(1-\mu^2)}. \tag{1.83}$$

$\zeta(x,y)$ describes the displacement perpendicular to the plane of the plate, h the thickness of the plate and E the Young's modulus. A beam with a width b can be considered as a special case of the plate having constant displacement in the y direction:

$$B'\frac{\partial^4\zeta}{\partial x^4}+m'\frac{\partial^2\zeta}{\partial t^2}=F'\ \text{with}\,B'=\frac{Ebh^3}{12}. \tag{1.84}$$

In both equations, B'' resp. B' describe the bending stiffness of the plate respectively the beam, p resp. F' the exciting pulsating force per area respectively length, m'' resp. m' describes the mass per area respectively length. Introducing a wave approach in the differential equation, the bending wave speed can be calculated:

$$\text{Beam}\quad c_B=\sqrt[4]{\omega^2B'/m'}, \tag{1.85}$$

$$\text{Plate}\quad c_B=\sqrt[4]{\omega^2B''/m''} \tag{1.86}$$

From the first equation, the bending wavelength can be calculated:

$$\lambda_B=1.35\cdot\sqrt{\frac{h\cdot c_L}{f}}, \tag{1.87}$$

where c_L describes the longitudinal wave speed and f the frequency.

The validity of the approach of plane cross-sections holds in a range given by

$$\lambda_B>6\cdot h\ \text{or}\ f<\sqrt{\frac{E}{\rho}}\cdot\frac{1}{20\,\mathrm{h}}. \tag{1.88}$$

Figure 1.5 shows the calculated bending wave velocities for various materials and plate thicknesses. The limit of validity Eq. (1.88) is introduced in the figure.

In order to expand the validity range given in Eq. (1.88), the shear deformation of the beam as well as the rotational inertia per unit length must be considered. In this case, a planar cross-section is assumed that is no longer perpendicular to the neutral layer of the plate or the beam. In order to describe a beam considering its bending stiffness, the shear stiffness,

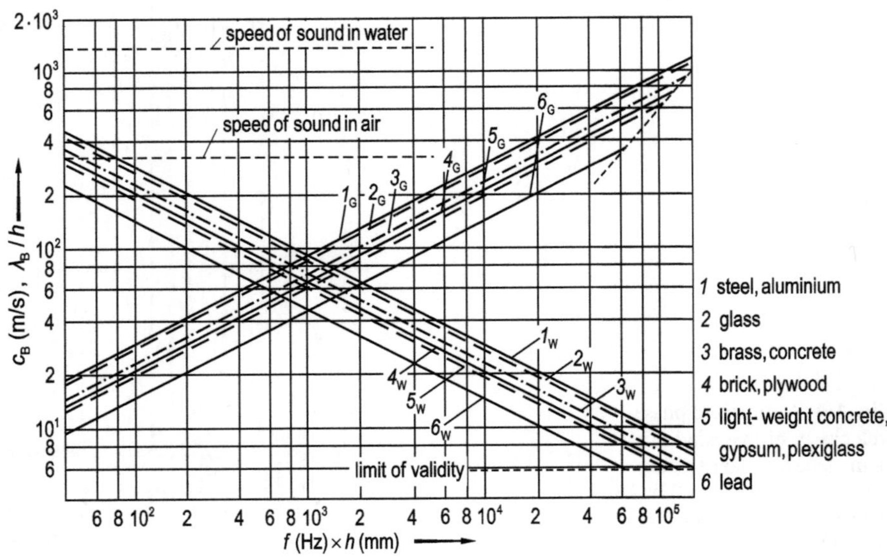

Fig. 1.5 Bending wave velocity c_B (curves G) and bending wavelength λ_B (curves W) in homogeneous plates with various materials

the mass and the rotational inertia, coupled differential equations can be established. The following relationship between the frequency of excitation and wavelength can then be derived [39]:

$$c_{B_K} \approx \frac{c_B}{\sqrt[4]{1 + i^2 \cdot \frac{4\pi^2}{\lambda^2} \cdot \left(1 + \frac{E}{G}\right)}}. \tag{1.89}$$

A closed form of the solution of the wave velocity c_{BK} that is smaller than the bending wave velocity is not possible. Assuming that the wavelength λ can be assessed, Eq. (1.89) gives the possibility to calculate how far the influence of the rotational inertia and the shear stiffness can be neglected.

Due to the frequency dependency of the wave speed, a simple relationship between the shear force F, the bending moment M, the vibration velocity v and the angle velocity ω can only be established for sinusoidal excitations:

$$P = \overline{Fv} + \overline{M\omega} = 2 \cdot c_B S \cdot W'', \tag{1.90}$$

W'' describes the energy density, given by the sum of kinetic and potential energy per volume.

1.8.5 Longitudinal, Torsion, Shear Waves and Waves in Strings

Equations of motions for longitudinal waves in bars have the same structure as the differential equations for torsion and shear waves as well as waves on strings. They are given in Eq. (1.91). The system is described by a 'stiffness value' ES (longitudinal stiffness), T (torsional stiffness), GS (shear stiffness), H (tension in the string) and 'inertia' m' (mass per unit length) resp. Θ (rotational inertia per unit length). The corresponding wave velocities are given in Eq. (1.91).

Longitudinal wave:

$$\mathrm{ES}\frac{\partial^2 \zeta}{\partial x^2} - m'\frac{\partial^2 \zeta}{\partial t^2} = F' \quad c_D = \sqrt{\frac{E}{\rho}},$$

Torsional wave:

$$T\frac{\partial^2 \varphi}{\partial x^2} - \Theta\frac{\partial^2 \varphi}{\partial t^2} = m_\varphi \quad c_\Theta = \sqrt{\frac{T}{\Theta}},$$

Shear wave:

$$\mathrm{GS}\frac{\partial^2 \zeta}{\partial x^2} - m'\frac{\partial^2 \zeta}{\partial t^2} = F' \quad c_T = \sqrt{\frac{G}{\varsigma}},$$

Waves on strings:

$$H\frac{\partial^2 \zeta}{\partial x^2} - m'\frac{\partial^2 \zeta}{\partial t^2} = F' \quad c_S = \sqrt{\frac{H}{m'}} \tag{1.91}$$

Membrane:

$$N\left(\frac{\partial^2 \zeta}{\partial x^2} + \frac{\partial^2 \zeta}{\partial y^2}\right) - m''\frac{\partial^2 \zeta}{\partial t^2} = p \tag{1.92}$$

The transport of power in very large systems is given by:

$$P = \text{inertia} \cdot \text{wavespeed} \frac{\partial\,\text{displacement}}{\partial t}. \tag{1.93}$$

The velocity of propagation c_D in a beam differs from the longitudinal wave velocity c_L. The wave is called 'quasi-longitudinal wave'

$$\frac{c_{D_{\text{Beam}}}}{c_L} = \sqrt{\frac{(1-2\mu)\cdot(1+\mu)}{1-\mu}}. \tag{1.94}$$

In a plate, the corresponding relationship is given by:

$$\frac{c_{D_{\text{Plate}}}}{c_L} = \sqrt{1-\mu^2}. \tag{1.95}$$

The difference to the continuum is due to the lateral contraction. In the case of a pure longitudinal wave, no displacements perpendicular to the direction of propagation occur. In the 'quasi-longitudinal wave', a smaller effective stiffness is due to the tension-free surfaces and the fact that no tensions perpendicular to the propagation of the waves occur.

1.9 Excitation of Structure-Borne Sound

Structure-borne sound is caused by forces or movements. The input impedance describes how 'difficult' it is to excite a structure. The impedance either refers to forces or moments. It describes the relationship between the amplitude of the force $\hat{F}$ respectively the moment $\hat{M}$ acting on the structure to the resulting amplitude of the velocity $\hat{v}$ respectively angle velocity $\hat{\omega}$ at the point of excitation. In practice, typically point impedances are assumed:

$$Z_F = \frac{\hat{F}}{\hat{v}} \quad Z_M = \frac{\hat{M}}{\hat{\omega}}. \tag{1.96}$$

The point impedance is an adequate description if the dimensions of the areas of excitation are smaller than one tenth of the wavelength. The mobility is the inverse of the impedance.

1.9.1 Impedances of Infinite Structures

The impedances of infinite structures are important for the description of input power in systems. Under the assumption that the reflected waves at boundaries are not correlated with the exciting force, the input power calculated by means of those impedances gives a very good approximation also for finite systems – especially in the case of a broad band excitation. The Statistical Energy Analysis SEA (Sect. 3) draws upon these characteristics. An overview of different impedances is given in Sect. 9, and [34]. Refs. [40–42] give additional hints.

1.9.2 Excitation of Limited Systems

At the boundaries of finite systems waves are reflected. Here, wave fields with fixed maxima and minima, called 'standing waves', are formed.

The sound or vibrational pattern of arbitrary systems can be described by superposed eigenfunctions (eigenmodes $\varphi_n(x, y, z)$ multiplied by their amplitudes γ_n)

$$p(x,y,z) = \sum_{i=1}^{\infty} \gamma_n \cdot \varphi_n(x,y,z) \ resp.$$

$$\vec{v}(x,y,z) = \sum_{i=1}^{\infty} \gamma_n \vec{\varphi}_n(x,y,z), \tag{1.96}$$

$$\gamma_n = \frac{A_n^*}{m_n^*(2\pi f)^2} \cdot \frac{1}{\left[1 - \left(\frac{f}{f_n}\right)^2\right] + j\eta \frac{f}{f_n}}. \tag{1.98}$$

Each eigenfunciton has an eigenfrequency f_n. In the absence of exterior loads and neglecting the damping η, the system vibrates in its eigenforms with its corresponding eigenfrequencies.

The modal amplitudes γ_n can be calculated with the help of the equivalent single degree of freedom systems with a modal stiffness (that results from the modal mass m_n^* multiplied by the eigen circular frequency), the eigenfrequency f_n of the undamped structure, the modal excitation A_n^* and the loss factor η. The modal excitation A_n^* depends on the location of the exciting force with respect to the individual modes, often expressed by the deviation of the wave number of excitation from the wave number of the eigenform. For frequencies that are significantly lower than the eigenfrequency, the stiffness (modal stiffness) is relevant for the dynamic response, whereas for frequencies significantly above the eigenfrequency, the mass (modal mass) is governing. For frequencies in the resonant range, the vibration depends on the damping.

Only for simple geometries eigenfrequencies and eigenfunctions can be given in closed form. In case of more complicated geometries, they can be calculated by means of, e.g., a finite-element approach (see Sect. 3).

The eigenfunctions of shells, membranes or plates, etc. have an immediate apparent significance: Their zero crossings can be assigned to the nodes of the so-called Chladni's pattern. For investigations in the higher-frequency range, and especially for excitations with random noise, detailed knowledge about the eigenforms and eigenfrequencies is not necessary. Here, generally, one assumes that the waves reflected from the boundaries are uncorrelated with the excitation and thus do not create any work.

The input power is equal to the power dissipated due to the damping of the eigenforms in resonance. In such cases, it is only of interest how many eigenmodes are vibrating in the vicinity of the resonance and what is their damping (see also Sect. 3). Calculations of the exact values of the eigenfrequencies in numerical calculations are not possible due to the uncertainties of the input values. This is the reason why in case of theoretical predictions of the structure-borne sound with narrow-band excitation – especially in structure with small damping – significant uncertainties of the calculated values have to be expected. In Table 1.3, eigenfunctions and eigenmodes for ideal structures are

Table 1.3 Eigenfunctions and eigenfrequencies of geometrically simple structures

1 .One-dimensional wave propagation (pipes filled with air)

1.1.Pipes with rigid walls (sound pressure)

$\Delta N = l\Delta\omega/(c\pi)$;

With reflecting ends on both sides:

$\varphi_n = \cos\left(\frac{n\pi x}{l}\right); \omega_n = n\pi c/l,$

With absorbing ends on both sides:

$\varphi_n = \sin\left(\frac{n\pi x}{l}\right); \omega_n = n\pi c/l,$

With one absorbing and one reflecting end:

$\varphi_n = \cos\left(\frac{n\pi x}{l} + \pi/2\right); \omega_n = \left(n - \frac{1}{2}\right)\pi c/l.$

1.2 Pipe with soft walls (e.g. elastic water-pipes). In this case a plane waves do not occur

1.3 Vibrations of strings

$\Delta N = l\Delta\omega\sqrt{\frac{m'}{H}}/\pi$;

Clamped on both sides:

$\varphi_n = \sin\left(\frac{n\pi x}{l}\right); \omega_n = n\pi\sqrt{\frac{H}{m\prime}}/l.$

1.4 Longitudinal waves in trusses

$\Delta N = l\Delta\omega\sqrt{\frac{\varrho}{E}}/\pi,$

Free on both sides:

$\varphi_n = \cos\left(\frac{n\pi x}{l}\right); \omega_n = n\pi\sqrt{\frac{E}{\varrho}}/l.$

1.5 Bending waves in beams

$\Delta N = \frac{l\cdot\Delta\omega}{2\pi}\cdot\sqrt[4]{\frac{m'}{\omega^2 B'}}$

Both sides clamped:

$$\varphi_n = \cos h\left(\frac{\beta_n x}{l}\right) - \cos\left(\frac{\beta_n x}{l}\right) - \frac{\cos\beta_n - \cos h\beta_n}{\sin\beta_n - \sin h\beta_n}\left[\sin h\left(\frac{\beta_n x}{l}\right) - \sin\left(\frac{\beta_n x}{l}\right)\right];$$

$$\beta_1 = 1.506\pi; \beta_2 = 2.5\pi; \beta_n = \left(n + \frac{1}{2}\right)\pi; \omega_n = \sqrt{\frac{B\prime}{m\prime}}\left(\frac{\beta_n}{l}\right)^2;$$

Both sides free: like clamped just cos replaced by – cos, sin replaced by – sin

both sides momentum-free supports:

$$\varphi_n = \sin\left(\frac{\beta'_n x}{l}\right);\ \beta'_n = n\pi;\ \omega_n = \sqrt{\frac{B'}{m'}}\left(\frac{\beta'_n}{l}\right)^2,$$

One side clamped one side free:

$$\varphi_n = \cos h\left(\frac{\beta_n x}{l}\right) - \cos\left(\frac{\beta_n x}{l}\right) - \frac{\sinh\beta_n - \sin\beta_n}{\cosh\beta_n + \cos\beta_n}\left[\sin h\left(\frac{\beta_n x}{l}\right) - \sin\left(\frac{\beta_n x}{l}\right)\right];$$

$$\beta_1 = 0.597\pi; \beta_2 = 1.494\pi; \beta_3 = 2.5\pi; \beta_4 = \left(n - \frac{1}{2}\right)\pi; \omega_n = \sqrt{\frac{B\prime}{m\prime}}\left(\frac{\beta_n}{l}\right)^2;$$

1.6 Circular ring

$\varphi_n = \cos n\psi; \omega_0 = \sqrt{E/\varrho}/a; \omega_n = \sqrt{E/\varrho}\cdot\frac{h}{\sqrt{12}a^2}\sqrt{\frac{n^2(n^2-1)^2}{(n^2+1)}}.$

2 Two-dimensional wave equation

2.1 Flat cavity with rigid walls

$\Delta N = \frac{S\omega\Delta\omega}{2\pi c^2},$

Rectangular:

$\varphi_n = \cos\frac{n_1\pi x}{l_1}\cos\frac{n_2\pi y}{l_2}; \omega_n = \pi c\sqrt{\left(\frac{n_1}{l_1}\right)^2 + \left(\frac{n_2}{l_2}\right)^2};$

Circular:

$\varphi_n = \cos n\psi J_n\left(\frac{\omega_n r}{c}\right); \omega_n = \frac{\pi c\gamma_{n,v}}{a};$

$\gamma_{1.0} = 0.586; \gamma_{2.0} = 0.972; \gamma_{0.1} = 1.22; \gamma_{3.0} = 1.34; \gamma_{4.0} = 1.693; \gamma_{1.1} = 1.697.$

(*continued*)

Table 1.3 (continued)

2.2 Membrane clamped on all edges

$\Delta N = S\varrho h\omega\Delta\omega/(2\pi H\prime)$;

Rectangular:

$\varphi_n = \sin\frac{n_1\pi x}{l_1}\sin\frac{n_2\pi y}{l_2}; \omega_n = \pi\sqrt{H'/\varrho h}\sqrt{\left(\frac{n_1}{l_1}\right)^2 + \left(\frac{n_2}{l_2}\right)^2};$

Circular:

$\varphi_n = J_n\left(\pi\gamma\prime_{n,v}\frac{r}{a}\right)\cos n\psi; \omega_n = \pi\sqrt{H'/\varrho h}\gamma\prime_{n,v}/a,$

$\gamma\prime_{0.1} = 0.765; \gamma\prime_{1.1} = 1.22; \gamma\prime_{2.1} = 1.635; \gamma\prime_{0.2} = 1.757; \gamma_{n,v'} \approx \frac{n}{2} + v - \frac{1}{4}.$

2.3 Bending waves on plates

$\Delta N = \sqrt{\varrho h/B}S\frac{\Delta\omega}{4\pi};$

Rectangular, momentum free supports at all edges:

$$\varphi_n = \sin\frac{n_1\pi x}{l_1}\sin\frac{n_2\pi y}{l_2}; \omega_n = \pi^2 c\sqrt{B/\varrho h}\left[\left(\frac{n_1}{l_1}\right)^2 + \left(\frac{n_2}{l_2}\right)^2\right];$$

Circular, all edges clamped

$$\varphi_n = \left[J_n\left(\pi\gamma''_{n,v}\frac{r}{a}\right) - \frac{J_n\left(\pi\gamma''_{n,v}\right)}{J_n\left(j\pi\gamma''_{n,v}\right)}J_n\left(j\pi\gamma''_{n,v}\frac{r}{a}\right)\right]\cos n\psi;$$

$$\omega_n = \pi^2\sqrt{B/\varrho h}\gamma''_{n,v}2/a^2,$$

$$\gamma''_{0.1} = 1.015; \gamma''_{1.1} = 1.468; \gamma''_{2.1} = 1.879; \gamma''_{0.2} = 2.007; \gamma''_{n,v} \approx v + \frac{n}{2}.$$

2.4 Cylinder

$\Delta N = 1.25\sqrt[4]{\frac{\varrho^3}{E^3}}\sqrt{\omega}\cdot a^3\frac{l\cdot\Delta\omega}{h}$ for $\omega < \sqrt{E/\varrho}/a$;

$\Delta N = \sqrt{3\varrho/E}la\,\Delta\omega/h$ for $\omega > \sqrt{E/\varrho}/a$;

Momentum-free support at the edges:

$$\varphi_n = \cos(n_1\psi)\sin(n_2\pi x/l);$$

$$\omega_n^2 = \frac{E\left(1-\mu^2\right)(n_2\pi a/l)^4}{\varrho a^2\left[(n_2\pi a/l)^2 + n_1^2\right]^2} + \frac{Eh^2}{12a^4\varrho}\left\{\left[(n_2\pi a/l)^2 + n_1^2\right]^2 - \frac{n_1^2(4-\mu)-2-\mu}{2(1-\mu)}\right\}.$$

3. Three-dimensional wave equation

3.1 Cavity with air or fluid:

$\Delta N = V\omega^2\frac{\Delta\omega}{2\pi^2 c^3};$

Rectangular with rigid walls:

$$\varphi_n = \cos\frac{n_1\pi x}{l_1}\cos\frac{n_2\pi y}{l_2}\cos\frac{n_3\pi z}{l_3}; \omega_n = \pi c\sqrt{\left(\frac{n_1}{l_1}\right)^2 + \left(\frac{n_2}{l_2}\right)^2 + \left(\frac{n_3}{l_3}\right)^2};$$

Rectangular with soft walls:

$$\varphi_n = \sin\frac{n_1\pi x}{l_1}\sin\frac{n_2\pi y}{l_2}\sin\frac{n_3\pi z}{l_3}; \omega_n = \pi c\sqrt{\left(\frac{n_1}{l_1}\right)^2 + \left(\frac{n_2}{l_2}\right)^2 + \left(\frac{n_3}{l_3}\right)^2};$$

Cylindrical with rigid walls:

$$\varphi_n = J_n\left(\omega_n\frac{r}{c}\right)\cos(n\psi)\cos\frac{n_3\pi z}{l_3}; \omega_n = \pi c\sqrt{\left(\frac{\gamma_{nv}}{a}\right)^2 + \left(\frac{n_3}{l_3}\right)^2};$$

Remarks: l_1, l_2, l_3 length of edges, a radius, ψ angle, H tension of the string, $m\prime$ mass per length, V volume, $B\prime$ bending stiffness of a beam, $H\prime$ tension of the membrane, h thickness, B bending stiffness of a plate, μ Poisson's ratio, ΔN number of modes in the frequency range $\Delta\omega$. In case of one-dimensional propagation n = 1,2,3 otherwise n = 0,1,2,3 given [34, 43].ΔN indicates the total number of the modes up to the frequency ω. In case of two and three-dimensional structures, this value can be used with a good approximation also for arbitrarily shaped structures.

The associated impedances are to be found in Sect. 9 as well as in [34].

1.9.3 Wave Impedances

The wave impedance describes the relationship between the amplitude of an exciting pressure (sinusoidally varying with respect to space and time) to that of the velocity of the excited structure.

$$Z(k,\omega) = \frac{\hat{p}(k,\omega)}{\hat{v}(k,\omega)}. \tag{1.99}$$

$\hat{p}(k,\omega)$ and $\hat{v}(k,\omega)$ describes the pressure resp. velocity, Fourier transformed with respect to space and time.

For the typical equations of motion (1.74, 1.75, 1.83, 1.84, 1.91, 1.92) those values can easily be calculated, since the derivatives with respect to space can be calculated by multiplications with the wave numbers k and the derivatives with respect to time by multiplication with the circular frequency ω.

For the bending waves in beams and plates, it is:

$$Z_p = \frac{B'}{j\omega}(k^4 - \omega^2 m'). \tag{1.100}$$

For longitudinal waves in trusses it is:

$$Z_D = \frac{ES}{j\omega}(k^2 - \omega^2 m'). \tag{1.101}$$

Arbitrary load functions can be described by means of the Fourier transformation (spectrum of wave numbers). The sound radiation of vibrating structures can also be calculated in the spectral domain [44], (see Sect. 22).

1.10 Damping

The transformation of sound energy into heat is termed damping. Damping is caused by transport of heat and viscosity (so-called classic absorption) as well as intramolecular relaxation processes. With very high amplitudes – like in the vicinity of extremely loud sound sources or supersonic boom – non-linear processes can occur that cause a distortion of the wave shape and strong damping. In the case of sound in gases and fluids, the damping plays an important role, especially when the propagation occurs over a large distance or in the case when bodies with very large surfaces are located in the transmission path of sound. The geometrical damping is described in Sect. 6, the absorption in Sect. 8.

In structure-borne sound, the damping is caused by a heat transfer due to molecular or intercrystalline displacements [45, 46], in composite structures due to friction between different parts (see Sect. 9).

1.11 The Principle of Reciprocity

The above-mentioned equations are symmetrical with respect to the spatial variables. This holds under the condition that the structure remains at its position and that the limiting surfaces remain unmoved – apart from small vibrations. In acoustics, this holds generally as long as no flow processes have to be considered. In the case of moving loads that can be investigated by means of moving co-ordinate systems, the spatial variables are no longer symmetrical. In this case, the describing differential equations are no longer self-adjointed.

In some situations occur asymmetric differential equations, e.g., for the moving of cylindrical shells or for the propagation of sound in porous media, which are approximate relations in which the asymmetry occurs due to simplifications [34].

In case of symmetrical processes, the principle of reciprocity can be applied. This principle says: In case a force $\vec{F}$ applied to a position A creates a velocity of $\vec{v}_{1A}$ at a position B, the same force applied to position B creates the same velocity $\vec{v}_{2A} = \vec{v}_{1B}$ at the position A.

In case the point of excitation and the point of observation are exchanged, the relationship between exciting force with respect to the measured velocity remains constant. The direction of the force vector in one case has to be equivalent to the direction of the velocity vector in the other case. The principle of reciprocity can be applied to the couple of force/velocity (in case of equal direction), and to other couples the product of which corresponds to energy or power (pressure volume/flow or moment/rotational velocity).

The reciprocity can be used for the calibration or for the indirect evaluation of sound fields (see Sect. 2).

References

1. Lord Rayleigh (1943) The theory of sound. New York, Dover, S18
2. Morse PM, Ingard U (1968) Theoretical acoustics. McGraw-Hill, New York
3. Meyer E, Neumann EG (Neubearbeitung 1979) Physikalischen und Technische Akustik. Braunschweig, Vieweg
4. Dowling A, Ffowcs-Williams JE (1983) Sound and sources of sound. Horwood, Chichester
5. Kinsler LE, Frey R (1962) Fundamentals of acoustics. Wiley, New York
6. Kurtze G (1964) Physik und Technik der Lärmbekämpfung. Braun, Karlsruhe
7. Pierce AD (1981) Acoustics, an introduction to its physical principles and application. McGraw-Hill, New York
8. Skudrzyk E (1954) Die Grundlagen der Akustik. Springer, Wien
9. Schirmer W (ed) (1989) Lärmbekämpfung. Tribüne, Berlin
10. Fasold W, Kraak W, Schirmer W (1984) Taschenbuch Akustik Teil 1, VEB Technik, 2. Berlin
11. Beranek LL (ed) (1971) Noise and vibration control. McGraw-Hill, New York
12. Landau LD, Lifschitz EM (1966) Lehrbuch der theoretischen Physik Band VI, Hydromechanik. Akademieverlag, Berlin
13. Rschevkin SN (1963) The theory of sound. Pergamon Press, Oxford, London
14. Cremer L, Hubert M (1985) Vorlesungen der Technischen Akustik, 3rd edn. Springer, Berlin
15. White RG, Walker JG (eds) (1982) Noise and vibration. Horwood, New York
16. Beranek LL (1993) Acoustics. Acoustical Society of America, New York
17. Filippi P, Haboult D (1999) Acoustics – basic physics, theory and methods. Academic, London
18. Unbehauen R (1993) Systemtheorie. Oldenbourg-Verlag, München
19. Lüke HD (1979) Signalübertragung. Springer, Berlin
20. Papoulis A (1962) The Fourier-integral and its applications. McGraw-Hill Comp, New York
21. Bracewell R (1978) The Fourier-transform and its applications. McGraw-Hill, New York
22. Beranek LL (1993) Acoustical measurements. Acoustical Society of America, New York
23. Tolstoy I, Clay SS (1966) Ocean acoustics. McGraw-Hill, New York
24. Lighthill MJ (1952) On sound generated aerodynamically. I General Theory. Proc R Soc A 211:564–587
25. Lighthill MJ (1954) On sound generated aerodynamically. II Turbulences as a source of sound. Proc R Soc A 222:1–32
26. Heckl M (1977) Abstrahlung von ebenen Schallquellen. Acoustica 37:155–166
27. Stenzel BO (1958) Leitfaden zur Berechnung von Schallvorgängen. Springer, Berlin
28. Cremer L, Heckl M (1996) Körperschall. Springer, Berlin
29. Möser M (1988) Analyse und Synthese akustischer Spektren. Springer, Berlin
30. Williams EC (1999) Fourier acoustics. Sound radiation and nearfield acoustical holography. Academic, London
31. Morse P, Ingard U (1968) Theoretical acoustics. McGraw-Hill, New York
32. Moon P, Spencer DE (1971) Field theory handbook. Springer, New York
33. Abramowitz M, Stegun IA (1965) Handbook of mathematical functions. Dover, New York
34. Cremer L, Heckl M (1995) Körperschall. Springer, Berlin
35. Miller GF, Pursey H (1954) The field and radiation impedance of mechanical radiators on the surface of a semi-infinite isotropic solid. Proc R Soc London A 223:521–541
36. Studer J, Koller MG (1997) Bodendynamik. Springer, Berlin
37. Wolf JP (1985) Dynamic Soil structure Interaction. Prentice Hall, Englewood Cliffs, NJ
38. Naake HJ, Tamm K (1958) Sound propagation in plates and rods of rubber-elastic materials. Acustica 8:S54–S76
39. Grundmann H, Knittel G (ed) (1983) Einführung in die Baudynamik. Mitteilungen aus dem Inst. Für Bauingenieurwesen, Techn. Universität München
40. Beitz W, Küttner K-H (1986) Dubbel – Taschenbuch für den Maschinenbau. Springer, Berlin
41. Junger M, Feit D (1986) Sound, structures and their interaction, 2nd edn. MIT, Cambridge
42. Heckl M (1969) Köperschallanregung von elastischen Strukturen durch benachbarte Schallquellen. Acustica 21: S149–S161
43. Blevins RD (1979) Formulas for natural frequency and mode shape. Van Nostrand Reinhold Comp, New York
44. Heckl M (1987) Schallabstrahlung in Medien mit Kompressibilität und Schubsteife bei Anregung durch ebene Strahler bzw. Zylinderstrahler. Acustica 64:229–261
45. Mason WP (ed) (1966) Physical acoustics. Vol. II, Part B 8,9 Vol. II, Part A. Academic, New York
46. Junger M, Feit D (1986) Sound, structures and their interaction, 2nd edn. MIT, Cambridge

Acoustic Measurements

2

M. Vorländer

2.1 Introduction

Acoustic measurements are the obvious prerequisite of acoustic investigations, in research as well as in applied acoustics. They are an important tool for the analysis of acoustical problems or for creation of experimental references in theoretical and numerical approaches. The accuracy required can only be reached if certain requirements concerning the instrumentation are met and if the acoustical conditions and the measurement methods are clearly specified. In applied acoustics, acoustic measurements are often difficult to perform and to interpret. Accordingly, one cannot expect that measured results are absolutely reproducible. Typically, the deviations in repeated measurements are of an order of 1 dB, a magnitude that is acceptable in most cases. These uncertainties are caused by changes in the sound field itself or in the measurement instrumentation.

A measurement arrangement can typically be separated into source and receiver components. The receiver component consists of a 'sound level meter' or 'sound analyser,' which displays the total sound level in decibels or any other frequency-dependent data. It can also produce results such as 'spectra' or 'impulse responses'. Furthermore, measurements are based on special measurement environments, which allow several reference sound fields to be created.

In this chapter, the components of acoustic measurement instrumentation are explained, together with the most important measurement quantities and the various techniques of signal processing. Finally, some examples of applications are given.

2.2 Microphones and Loudspeakers

Almost every acoustic measurement arrangement contains microphones that normally convert sound pressures into electrical signals, which in turn can be displayed, stored, and analysed by analogue or digital techniques. Hence, in a wider sense, other electroacoustical or electromechanical transducers can be called microphones, as well, whereas transducers for underwater sound are called hydrophones, and sensors for structural vibrations, accelerometers.

Microphones for airborne sound contain a very thin and flexible diaphragm, which follows the air movement of the local sound field. The vibration of the diaphragm is converted by the electromechanical force interaction into an electrical signal. In the optimal case, this process is linear and frequency independent.

The electroacoustic conversion yields electrical signals that are proportional to one of the specific sound field quantities, i.e., to sound pressure or to sound velocity. This proportionality depends on whether the sound pressure excites only one or both sides of the diaphragm. In the first case, the electromechanical force on the diaphragm is proportional to the sound pressure, while in the second case it is proportional to the sound pressure gradient.

The sensitivity of a pressure microphone is represented by the open-circuit voltage with reference to the sound pressure on the diaphragm. Each microphone placed into the sound field, however, distorts

M. Vorländer (✉)
Institute of Technical Acoustics, RWTH Aachen University, Templergraben 55, 52056 Aachen, Germany
e-mail: mvo@akustik.rwth-aachen.de

G. Müller and M. Möser (eds.), *Handbook of Engineering Acoustics*,
DOI 10.1007/978-3-540-69460-1_2, © Springer-Verlag Berlin Heidelberg 2013

the sound field the more, the larger it is in comparison to the wavelengths of sound. Accordingly, we distinguish between pressure sensitivity and free-field or diffuse-field sensitivity. The latter two are related to the sound pressure in the sound field, without the microphone in place. The frequency responses of pressure and diffuse-field sensitivities differ only slightly, while at high frequencies (>10 kHz) free-field responses are significantly higher by several decibels due to directivity effects.

2.2.1 Condenser Measurement Microphones

Condenser microphones are based on the principle of electrostatic transducers, as will be explained in the following. A condenser microphone is a passive electrostatic transducer, consisting of a mobile diaphragm and a rigid backplate. The relation between the mechanical force and the voltage on the condenser is, at first, non-linear, since two charged plates interact with a quadratic law of force and voltage. This is the reason for applying a constant polarisation voltage U_0 (typically 200 V) over a very large resistance R (>10 GΩ). This voltage creates a constant charge on the condenser. A sound-induced modulation of the distance between diaphragm and backplate results in a change in capacity with constant charge, and thus in a sound-induced AC voltage signal U added to the constant polarisation voltage. For amplitudes that are not too high, the relation between sound pressure and voltage is linear to an excellent approximation. With the measurement microphones used today, this approximation is valid for very high sound pressure levels of up to 140 dB (Figs. 2.1 and 2.2).

The diaphragm typically consists of pure nickel foil of just a few micrometres thick. The dimensions of a standard measurement microphone are: diameter of 1/2-in. (12.7 mm) and distance between diaphragm and backplate about 20 μm. The capacity of this construction is about 20–30 pF. Furthermore, diameters of 1-, 1/4-, or 1/8-in. are in use. Other dimensions, of course, can be found in studio or miniature recording microphones. For all constructions, the microphone is, in a narrow sense, just the capsule containing the diaphragm, the backplate and the volume. Since the impedance of the capsule itself is extremely high (10–100 GΩ), a preamplifier for impedance transformation must be placed near the capsule, so that long cables may be used. Accordingly, the whole arrangement of capsule and preamplifier is called 'microphone'.

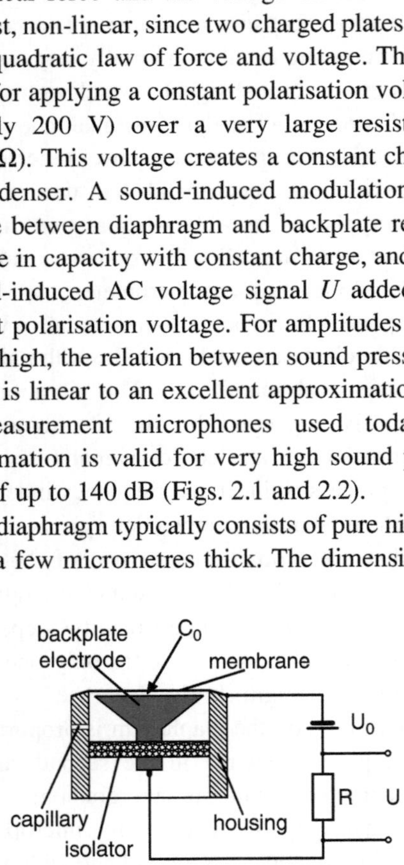

Fig. 2.1 Principle and equivalent circuit of condenser microphones

By using a simple equivalent circuit, the sensitivity of a condenser microphone can be estimated in a first approximation. The significant components on the electrical side are resistance and capacity and, on the mechanical side, the compliance of both the diaphragm connected to the housing and of the air cavity. One important task of microphone design is the optimisation of (a) the compliance and (b) the damping of the mechanical resonance caused by holes in the backplate. The resulting frequency response covers a range from 2 Hz to 22 kHz (1/2-in. microphones), and the sensitivity is constant over this range at approximately

$$\frac{U_{I=0}}{p} = nS\frac{U_0}{d}, \tag{2.1}$$

with the open-circuit receiving voltage $U_{I=0}$ and excitation with the sound pressure p. The variable n is the total compliance (diaphragm stiffness and air cavity), U_0 the polarisation voltage, S the diaphragm surface, and d the distance from the diaphragm to the backplate. The operation range is limited at low frequencies by electrical and mechanical high pass effects, also by capillary tubes for quasi-static pressure equalisation. At high frequencies, the range is limited by the mechanical resonance of diaphragm mass and total stiffness. The sensitivity of measurement microphones is typically between 10 and 50 mV/Pa, often expressed in terms of −40 dB to −26 dB re 1 V/Pa.

The simplest general type of microphone, usually called a point microphone, has a frequency- and directionally independent sensitivity. In these applications, measurement microphones can be used for all frequencies for which the microphone dimensions are small compared with wavelengths.

For 1/2-in. microphones, this simple approach is valid up to approximately 2 kHz. Above this limit, diffraction at the microphone distorts the sound pressure. The total displacement of the diaphragm is then given by the integral over the surface elements excited by the incident sound wave with locally varying sound pressures in amplitude and phase. Accordingly, the sensitivity is dependent on the direction and describable by a directivity function. This behaviour of microphones must be considered at higher frequencies, and the type of sound field must then be specified.

The polarisation U_0 voltage can be ignored if a dielectric material with permanent polarisation (a so called 'electret material') is placed between the condenser electrodes. With electret foils, miniature microphones with dimensions of a few millimetres are available (Table 2.1).

2.2.2 Sound Velocity Measurement

For applied acoustics, sound pressure is the most important quantity. For physically correct and complete investigations of sound fields, however, the additional measurement of sound velocities is mandatory, especially for determining field impedances, for solving coupled vibration–radiation problems and, foremost, for measuring sound intensity (see Sect. 2.2.5).

For measuring sound velocity, gradient microphones or, alternatively, combinations of several pressure microphones can be used, e.g., for determining the pressure gradient vector. Using direct velocity sensors is also possible (Fig. 2.3).

Direct velocity sensors based on hot-wire anemometers [1] have proved suitable in practice. This type of sensor is made of platinum resistors, very thin wires that transfer their thermal energy at the operating temperature to the surrounding air when driven at a temperature of 200–400 °C. If a local air flow (sound velocity) is present, the temperature distribution changes asymmetrically, and two closely mounted wires can be used to detect a local temperature difference and a corresponding voltage difference. The measurement range can be from 100 nm/s to 0.1 m/s.

2.2.3 Vibration Sensors

Vibration sensors are receivers for the measurement of structural vibrations. They should be rigidly connected with the surface under investigation. In principle, they consist of a mechanical resonator of a mass m, a compliance n,, and unavoidable mechanical losses w (Fig. 2.4).

Let x be the vibration amplitude of the measurement object and x' the vibration amplitude of the mass m. Then the law of momentum conservation gives

$$m\frac{d^2x'}{dt^2} + w\,\frac{d}{dt}(x' - x) + \frac{x' - x}{n} = 0, \qquad (2.2)$$

or with the differential amplitude $\xi = x - x'$ and harmonic vibrations:

$$\left(-m\omega^2 + j\omega w + \frac{1}{n}\right)\xi = -m\omega^2 x. \qquad (2.3)$$

This equation describes a basic mechanical resonator of mass, spring and damping with the resonance frequency ω_0. If highly tuned vibration sensors are used, with

$$\omega_0 = \frac{1}{\sqrt{mn}} \gg \omega, \qquad (2.4)$$

the resonator impedance is dominated by the compliance. This simply yields

$$\xi = -mn\omega^2 x = -\left(\frac{\omega}{\omega_0}\right)^2 x, \qquad (2.5)$$

from which follows that the quantity to be observed, ξ, is proportional to the acceleration of the measurement surface, $\omega^2 x$ (Fig. 2.5).

Table 2.1 Technical data for several microphones

Type	Durchmesser mm	Übertragungsfaktor 10^{-3} V/Pa	Frequenzbereich Hz	Dynamikbereich dB(A) re $2 \cdot 10^{-2}$ Pa
Kondensador 1/8□	3,2	1	6,5...140k	55...168
Kondensador 1/4□	6,4	4	4...100k	36...164
Kondensador 1/2□	12,7	12,5	4...100k	36...164
Kondensador 1□	23,8	50	2,6...18k	11...146
Dauerpol. 1/2□	12,7	50	4...16k	15...146
Elektrodyn.	33	2	20...20k	10...150

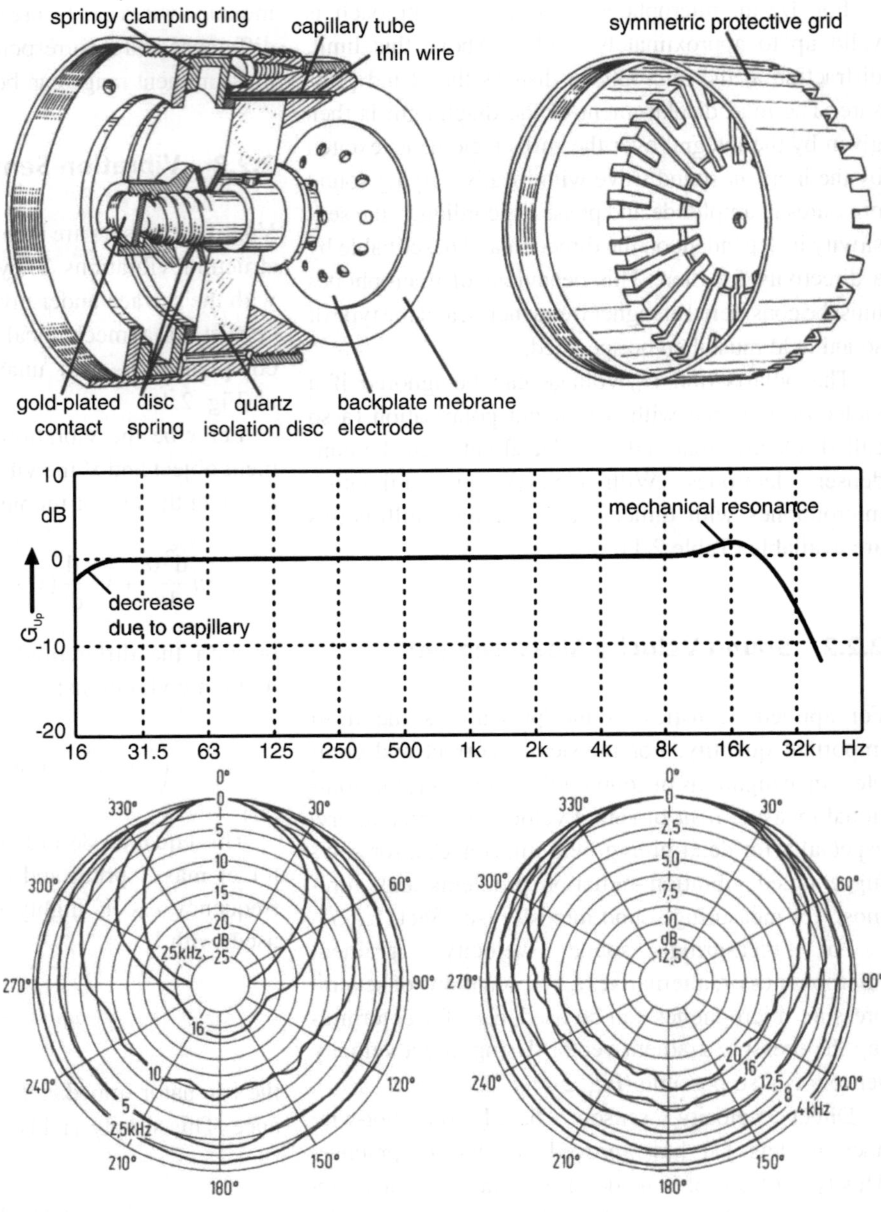

Fig. 2.2 Condenser microphone. (**a**) Construction sketch; (**b**) Frequency response; (**c**) directional characteristics

It is essential that the mass impedance ωm of the acceleration sensor is small compared with the impedance of the measurement object. Therefore, when measuring on small impedances, for instance, those of light and soft construction elements, the limits related to the mass and mounting of the sensor must be taken into account. The maximum permissible sensor mass M can be estimated by

$$M < 0.36\sqrt{10^{\Delta L/10} - 1}\rho c_L h^2/f, \tag{2.6}$$

with c_L, ρ, and h being longitudinal wave speed, density, and thickness of the measurement object (plate), respectively, and ΔL being the maximum permissible error. Depending on the way the sensor is mounted, the connection of sensor and surface can be interpreted as an additional spring. If measurements are to be performed at high frequencies, a very stiff connection is necessary, possibly by using screws. If only low frequencies are of interest, connections with wax or contact pins are sufficient.

In building acoustics, the range of surface displacements and the corresponding accelerations can be extremely large, covering vibrations of thin sheet metal or of massive walls. The sensitivities, masses, mounting and the relevant frequency ranges

Fig. 2.3 Scanning electron microscope image of a velocity sensor (Microflown [1]) according to the hot-wire anemometer. The diameter of the wire (aluminium) is 80 μm

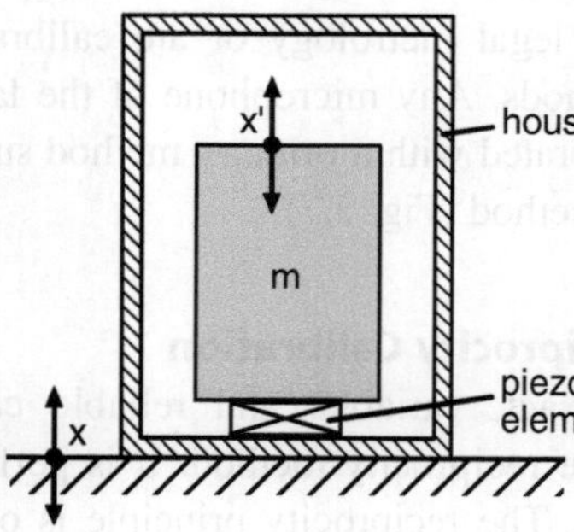

Fig. 2.4 Construction sketch of an accelerometer

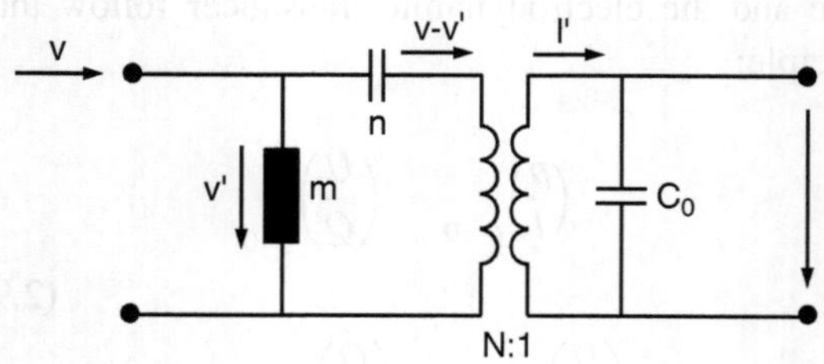

Fig. 2.5 Equivalent circuit of an accelerometer

must be specifically chosen in each case. This is, however, not problematic due to the large variety of commercially available accelerometers.

One very elegant but also quite sophisticated instrumentation is based on optical measurement methods, for instance by using laser-Doppler vibrometers. They are typically based on the principle of a Mach–Zehnder interferometer, which not only makes use of optical interferences caused by phase differences but also of the Doppler frequency shift. Hence, extremely small displacements can be measured accurately. If the laser beam is reflected or scattered by the measurement object, the reflected wave is shifted in phase and frequency. The problem then is to measure the rather small laser light frequency shift by beating a reference beam with the reflected beam and by measuring the intensities with a photo detector. It is thus possible to resolve vibration amplitudes, which are smaller than the light wavelength. This method was successfully applied, for instance, for measurement of microphone diaphragms or eardrums, although the displacement amplitudes are of orders of magnitude of just nanometers.

2.2.4 Microphone Calibration

In applied acoustics, the most frequently used method for calibration of the instrumentation makes use of so called 'pistonphones' or sound calibrators (see Sect. 2.2.4.1 below). Almost everyone who uses a measurement microphone will test its function with a sound calibrator. To achieve this goal, the user must be absolutely sure that the calibrator is working well. The manufacturer of the instrumentation, together with calibration authorities and reference laboratories, are responsible for assuring quality. This chain of measurement and calibration is called 'traceability' of measurement standards, finally connected to a primary standard, including an absolute primary calibration.

The calibration of an electroacoustic transducer can be performed in four different ways, by using: (1) a calculated mechanical or optical effect in the sound field, (2) a theoretically 'known' sound field (see

Sect. 2.2.4.1), (3) a reference microphone (see Sect. 2.2.4.2), and (4) the reciprocity principle (see Sect. 2.2.4.3).

Furthermore, there is a simple procedure for relative calibration and for quality control in production, namely the electrostatic actuator technique. An electrostatic actuator is a plate with an adapter ring, which is mounted directly on the microphone (instead of the protection grid). With an AC voltage on the actuator, the diaphragm is excited by a quasi-electrostatic force. The received voltage at the output terminals of the microphone is approximately proportional to the voltage, which would occur in the pressure sound field in an acoustic coupler. Accordingly, the sensitivity determined with the actuator gives an approximate figure for the pressure sensitivity.

2.2.4.1 Sound Calibrators

A vibrating piston in a pressure cavity (coupler) produces a sound field of known sound pressure (Fig. 2.6). The sound calibrator and the pistonphone are based on this principle. With a displacement amplitude $\hat{\xi}$, the pressure amplitude in the cavity volume is

$$\hat{p} = \frac{\rho_0 c^2}{V_0} S \hat{\xi}. \tag{2.7}$$

We assume that all dimensions are small compared to the wavelength and that the boundaries of the cavity are rigid. V_0 is the cavity volume and S is the piston area. Then the microphone sensitivity is the ratio of the output voltage and the sound pressure amplitudes:

$$M = \frac{\hat{U}}{\hat{p}} = \frac{V_0}{\rho_0 c^2 S} \cdot \frac{\hat{U}}{\hat{\xi}}. \tag{2.8}$$

2.2.4.2 Comparison Methods

Comparison methods are usually very easily applied. We excite a reference microphone and the microphone under test one after the other, and we get the difference between the two sensitivities directly. Knowing the absolute sensitivity of the reference microphone, gives us the sensitivity of the tested microphone. Consequently, the reference microphone or 'standard microphone' must be available. This must have previously been absolutely calibrated with a precision method or, at least, compared with a primary standard. Comparison methods are usually applied by calibration services or similar authorities that are responsible for sound level meter and microphone testing. These measurements are related to (coupler) pressure fields, for instance, for microphones in headphone couplers (artificial ears) for audiometer calibration. In the case of free field or diffuse field calibrations, the devices tested are typically sound level meters for noise immission control and for building acoustics.

The standard measurement microphones are compared with national or internationally agreed upon standards of legal metrology or are calibrated with primary methods. Any microphone of the latter kind must be calibrated with a primary method such as the reciprocity method (Fig. 2.7).

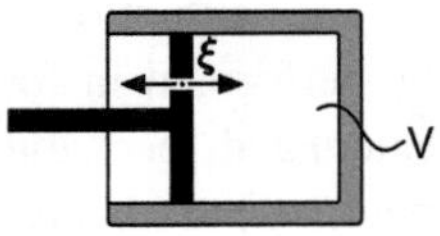

Fig. 2.6 Principle of the pistonphone

2.2.4.3 Reciprocity Calibration

The most exact, variable, and reliable calibration method is the reciprocity method. It is performed in several steps. The reciprocity principle is one of the basic principles of reversible transducers, best explained by electroacoustic four-poles. The electrostatic and the electrodynamic transducer follow that principle:

$$\left(\frac{p}{I}\right)_{Q=0} = \left(\frac{U}{Q}\right)_{I=0}$$
$$\left(\frac{U}{p}\right)_{I=0} = -\left(\frac{Q}{I}\right)_{p=0}, \tag{2.9}$$

where Q = volume velocity or, volume flow in m^3/s. Of specific importance is the second equation, which

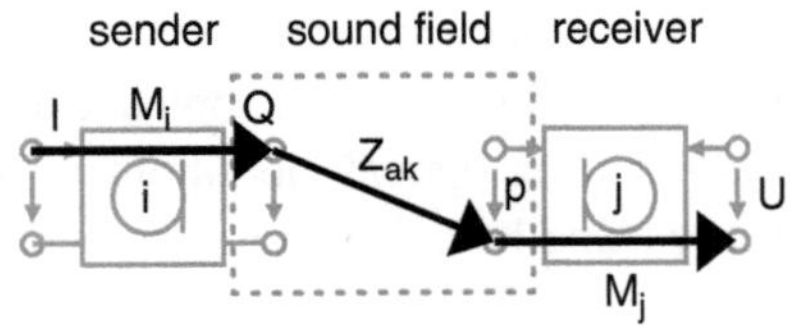

Fig. 2.7 Equivalent circuit with electrical and acoustical two-ports representing an arrangement of two microphones in reciprocity calibration and definition of transfer impedance: $Z_{ij} = M_i \cdot Z_{ak} \cdot M_j$

relates the sensitivity factor of the transduction between the source, Q/I, and the receiver, U/p. Both quantities are generally frequency dependent and complex, for instance $U = \underline{U}(f)$:

$$M = \frac{U_{I=0}}{p} = -\frac{Q_{p=0}}{I}. \tag{2.10}$$

It shall be noted, however, that the sensitivity in the receiving case (sensitivity factor M) is not defined in exactly the same way as the so-called 'sensitivity' sources such as of loudspeakers. The latter is not related to the volume velocity of the microphone diaphragm but to the sound pressure in the far field. Accordingly, it contains the radiation function (Green's function).

A further basic quantity of reciprocity calibration is the electrical transfer impedance Z_{ij} of a transfer four-pole consisting of two microphones (i and j) coupled over an acoustic path. Microphone i is driven as source and microphone j as receiver. U_j is the open-circuit receiving voltage and I_i the source current. By definition, this is

$$Z_{ij} = \frac{U_j}{I_i}. \tag{2.11}$$

The electrical transfer impedance can also be formulated by means of the two sensitivity factors M_i and M_j, which are independent of the direction, and by the acoustic transfer impedance Z_{ac}. This then yields

$$Z_{ij} = \frac{U_j}{I_i} = M_i \cdot Z_{\mathrm{ac}} \cdot M_j. \tag{2.12}$$

With this equation and measurement of the electrical transfer impedance, the product of two sensitivity factors can be determined if the acoustic transfer impedance (Green's function) is known. Equation (2.12) contains two unknowns (M_i and M_j). Two measurements with exchanged source and receiver are redundant due to reciprocity $Z_{ij} = Z_{ji}$, but three measurements on three microphones i, j and k in pairs give the result of the three microphone sensitivity factors M_i, M_j and M_k by calculating

$$M_i = \sqrt{\frac{1}{Z_{\mathrm{ak}}}\frac{Z_{ij}Z_{ik}}{Z_{jk}}}, \tag{2.13}$$

as well as two more equations of the same kind with a cyclic exchange of i, j, and k.

Under free field and diffuse field conditions, the acoustic transfer impedance can be easily calculated. The practical conditions, however, of the signal-to-noise ratio are extremely difficult, due to very low acoustic signal levels. However, the most accurate acoustic measurement is the pressure reciprocity calibration in small acoustic cavities. The cavity is considered small compared with the wavelength. In this simple consideration, the closed cavity with volume V_0 behaves mechanically like a spring:

$$Z_{\mathrm{ac|cavity}} = \frac{\kappa p_0}{j\omega V_0}. \tag{2.14}$$

For precision calibrations, several thermodynamic corrections (atmospheric pressure, heat conduction, etc.) must be added. Furthermore, extensions of Eq. (2.14) concerning the equivalent volumes of the

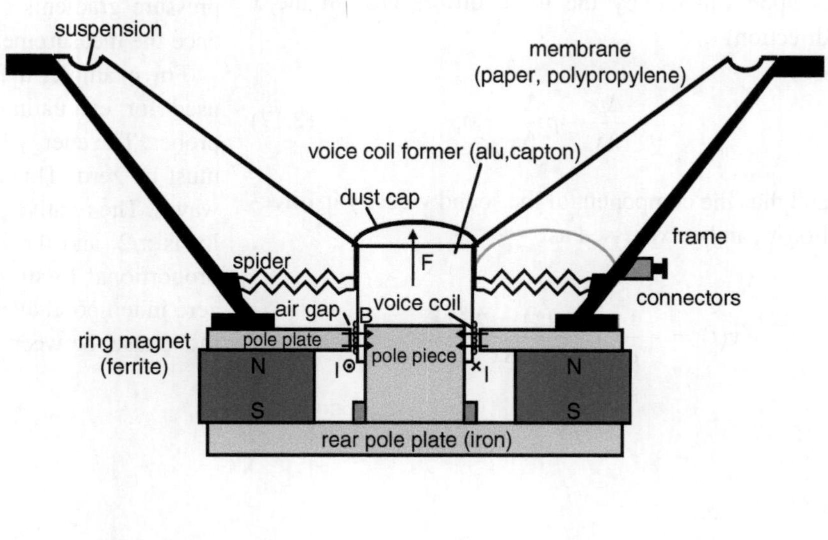

Fig. 2.8 Electrodynamic loudspeaker

microphone diaphragms, higher-order modes and waveguide models in cylindrical couplers must be used.

2.2.5 Intensity Probes

Sound intensity is an important quantity for evaluating and localising sound sources and sound absorbers and for determining physical sound field properties. It can give information about the type of waves and about the acoustic energy flow. Furthermore, it can be used for measuring the sound power of sources, provided the intensity is obtained over a closed surface surrounding the source.

Intensity probes must deliver the sound pressure p and the sound velocity v simultaneously. This is achieved by using a p–u probe, in which a condenser microphone is closely coupled to a velocity sensor according to the ultrasound Doppler principle [2] or to an anemometer (see Sect. 2.2.2). Much more common is the so-called p–p probe consisting of two condenser microphones placed either face to face or side by side. The sound pressure measured by this probe is

$$p(t) = \frac{p_1(t) + p_2(t)}{2}. \qquad (2.15)$$

For determining the sound velocity, the basic force equation

$$\mathrm{grad}p + \rho \frac{\partial v}{\partial t} = 0, \qquad (2.16)$$

is approximated by the finite differences (in the x direction)

$$\frac{\Delta p}{\Delta x} + \rho \frac{\Delta v}{\Delta t} = 0, \qquad (2.17)$$

and thus the component of the sound velocity in direction x can be expressed as

$$v(t) = -\frac{1}{\rho} \int_{-\infty}^{t} \frac{p_2(\tau) - p_1(\tau)}{\Delta x} \mathrm{d}\tau. \qquad (2.18)$$

The component of the active sound intensity in the direction of the pressure gradient is thus:

$$I_r = \overline{p(t)v(t)} = \frac{1}{2\rho\Delta x} \frac{1}{T} \int_0^T [p_1(t) + p_2(t)] \int_0^t [p_1(\tau) - p_2(\tau)] \mathrm{d}\tau \mathrm{d}t, \qquad (2.19)$$

with T = averaging time (see Sect. 2.3.1). By Fourier transformation, the same circumstances can be expressed in the frequency domain by Fourier transformation in terms of the imaginary part ($\Im$) of the cross power spectrum G_{12}:

$$I_r = -\frac{1}{\omega\rho\Delta x} \Im\{G_{12}(f)\}. \qquad (2.20)$$

Determining the sound velocity with velocity sensors or with microphone pairs yields only one component of the sound intensity, which is the nominal direction of the probe. If the direction of sound intensity shall be determined also, spatial probes must be applied, for instance as an arrangement of several microphone pairs, oriented orthogonally or multi-microphone probes on regular polyhedra [3].

However, here it must be considered that more complex probes produce larger errors concerning the acoustic centre and the spatial resolution, since pressure and velocity are not strictly measured at the same point. The construction and calibration of sound intensity probes must be performed very carefully, because the differences in sensitivities and, particularly, in relative phases of the transducers produce apparent pressure gradients or contributions to G_{12} and influence the measurement result.

For example, a fully reactive sound field can be used for evaluating the quality of sound intensity probes. The energy flow and the active sound intensity must be zero. This applies, for example, to standing waves. The relative phase between pressure and velocity is $\pi/2$, and the instantaneous intensity $p(t)\cdot v(t)$ is proportional to $\sin(\omega t)\cdot\cos(\omega t) \propto \sin(2\omega t)$, which is zero in temporal average. This example illustrates that the phase between pressure and velocity must be measured very accurately. The relative phase results in the end from the phase difference between the pressure microphones or between the pressure and the velocity sensor. Therefore, high standards must be set for the transducers and the connected circuits, such as analogue input sections and filters of the analysers. Phase errors are significant in p–p probes, if the pressure gradient is small, i.e., if the microphone distance is too small compared to the wavelength. For quality control of intensity probes and analysers [typically real-time analysers are applied to calculate Eqs. (2.19) or (2.20)], special calibrators and a set of field indicators [4] are available, which give information, for instance, about the smallest measurable sound intensity (residual intensity).

2.2.6 Loudspeakers

The classical measurement loudspeaker principle is based on the dynamic transducer. An electrodynamic loudspeaker consists of a conical membrane, which in its centre is connected to a cylindrical voice coil. The coil is placed in the air gap of a pot magnet of hard magnetic ferrite or of Alnico alloy, in which a radially homogeneous magnetic field is established by a magnetic flux density of some Vs/m^2. The resistance of the coil is typically several Ohms; at higher frequencies the inductance is significant, too. The latter can be reduced by copper rings in the air gap, which at the same time add damping to the mechanical resonance (Fig. 2.8).

Fig. 2.9 Dodecahedron loudspeaker

The movable parts of the loudspeaker (the membrane and voice coil) are supported by springs. This is realised by a spider to keep the voice coil in the nominal position and by a soft mounting of the membrane at its outer perimeter. The membrane is made of a material with high internal damping and low density in order to suppress bending waves. The common material is paper, but today PVC or light metals are also used.

Due to the relatively large mass of the vibrating parts and due to the mechanical system's resonance, the impulse onset of dynamic loudspeakers is typically not acceptable. This limits the possibility of generating short pulses. It can be improved by adequate damping (low impedance of the connected power amplifier). However, with modern methods of digital signal processing and inverse filtering, the system parameters can be optimised and an absolutely linear sound reproduction in a wide frequency range can be obtained. Particularly because loudspeaker equalisation can be realised rather well, loudspeaker optimisation is to be focussed on improvement of the directivity of radiation.

Nonlinear distortions of dynamic loudspeakers result from the nonlinear stiffness of the membrane support, from inhomogeneities of the magnetic field and from the Doppler shift, if signals contain low and high frequencies simultaneously (broadband signals). Since they are relatively small, the dynamic loudspeaker is the most frequently used loudspeaker type.

Improvement of radiation at low frequencies is gained by using loudspeaker boxes; however, with the side effect that the system resonance shifts to higher frequencies. The resonance frequency sets the low frequency limit. Thus it is important to match the loudspeaker system with the box dimensions. The radiation directivity and the axial sound field, estimated from the radiation of a piston mounted in a baffle, can be used as an approximation for the sound field of loudspeaker boxes. Then, however, diffraction at the box edges is an additional factor influencing the far field sound pressure.

2.2.6.1 Special Measurement Loudspeakers

In many acoustic measurements, the goal is to produce plane waves or an approximation of particular types of sound fields. Plane waves can be produced by common loudspeakers quite well, if the measurement area is on the main radiation axis of the far field and on a small area. The result will be better (a) if the loudspeaker membrane area S is small, since the far-field distance

$$r_F \approx \frac{S}{\lambda} = \frac{Sf}{c}, \tag{2.21}$$

in this case is very small and (b) with coaxial multiple loudspeakers, which have a common axis of radiation. Such systems concentrate the radiation on the main axis, but the wavefront in far field is symmetrical and approximately plane and is constant in level (many measurement standards assume plane wave conditions with p and v in phase at a distance of 2 m).

Special directivities are necessary for artificial singers or speakers, for instance, in the measurement of headsets, communication devices or Lavalier microphones. For these purposes, loudspeakers that radiate the sound from a mouth opening and an artificial human diffraction body (head and torso simulator, artificial singer) can be used.

If one does not want any specific directivity pattern but an omnidirectional radiation, particular methods of loudspeaker construction must be followed, as well.

Provided that the loudspeaker dimensions are not very small compared to the wavelength, then an omnidirectional radiation can at least be approximated by loudspeakers in spherical symmetry by means of housings based on regular polyhedra (tetrahedron, cube, dodecahedron, icosahedron with 4, 6, 12, or 20 loudspeaker systems, respectively). The dodecahedron is most commonly used (Fig. 2.9).

The illustration of the pure-tone directivity patterns of dodecahedrons is difficult to interpret, because broadband signals are more likely required in acoustic applications. It is thus reasonable to use three-dimensional illustrations of the directionally radiated sound pressure in terms of frequency-averaged spectra (Fig. 2.10).

2.3 Sound Level Measurement and Rating

Generally, almost every acoustic instrumentation can be divided into a source component and a receiving component. The receiving component is a sound level meter or, more generally, a sound analyser. This component displays either the total sound level in decibels or it performs a frequency analysis and displays a 'spectrum'.

The basic quantity to be considered is the sound pressure level (see Sect. 2.1):

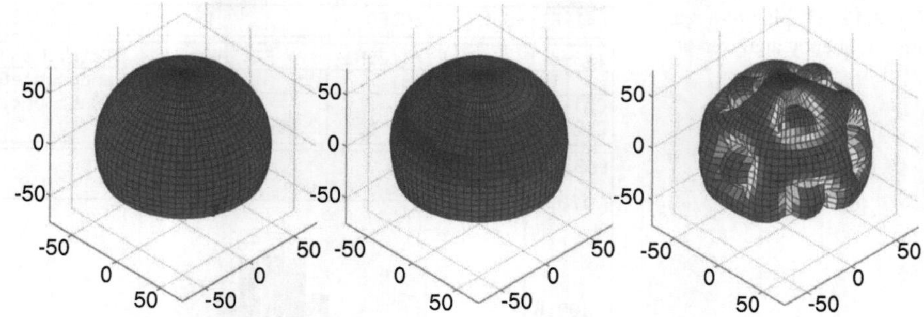

Fig. 2.10 Directivity of a dodecahedron loudspeaker. One-third octave band analysis at centre frequencies of 100 Hz, 1 kHz and 10 kHz (left to right)

$$L = 20\log\left(\frac{\tilde{p}}{p_0}\right); p_0 = 2\times10^{-5}\ \mathrm{N/m^2}, \tag{2.22}$$

where $\tilde{p}$ is the rms sound pressure determined over an averaging time T_m according to

$$\tilde{p} = \sqrt{\frac{1}{T_m}\int_0^{T_m} p^2(t)\mathrm{d}t}. \tag{2.23}$$

from a sound pressure time function $p(t)$. For instance,

$$p(t) = \hat{p}\sin\omega t, \tilde{p} = \frac{\hat{p}}{\sqrt{2}}. \tag{2.24}$$

According to the formula for the effective sound pressure, the term 'rms'= *r*oot [*m*ean (*s*quare)] can be easily remembered.

2.3.1 Averaging Times

It cannot be generally assumed that the sound signal to be measured is periodic (Fig. 2.11). The averaging time is therefore not unique and must be specified beforehand. The duration of the averaging T_m is dependent on whether the signal $p(t)$ is pulse like or stationary. International standards have fixed the application of two main time constants: 125 ms (='FAST') or 1 s (= 'SLOW'). SLOW has the advantage that the sound pressure level is rather stable and is easily readable from the display. Pulsive sound, however, is smoothed significantly. Beside FAST and SLOW, there are other (also unsymmetrical) time averages. Furthermore, it is important to consider a quantity for the evaluation of long-term noise exposition, represented by the energy equivalent sound pressure level L_{eq}. In order to describe the total sound energy exposure, the averaging time can last from several seconds up to several hours. The latter application occurs in noise abatement, in working environments (8 h) and in industrial or urban noise control (day and night periods).

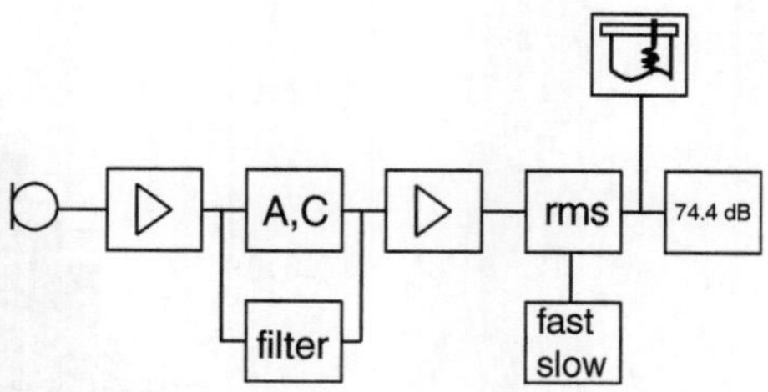

Fig. 2.11 Components of a sound level meter. Left to right: Microphone, preamplifier, A, C, or bandpass filter, amplifier, rms detector (time constants), display or level recorder

2.3.2 Frequency Rating

A second, very important averaging procedure is frequency rating. Here we try to take into account the frequency dependence of human hearing. Historically, the so called 'A-weighting' and indication of 'dB(A)' became the leading quantity. The meaning of dB(A) is that the sound level meter is extended with a standardised bandpass filter which simulates the frequency curve of the equal loudness contour L_N at 20 phon (Table 2.2).

The filters are, however, strongly simplified in order to ensure a feasible realisation with elementary circuits. B and C weighting is also specified to simulate equal loudness contours at higher levels.

The choice of time and frequency rating is specified in each individual case of measurement. An exact indication of the type of sound level in terms of indices is advisable, for instance, L_{AF}, $L_{A,eq}$, or L_{CS}.

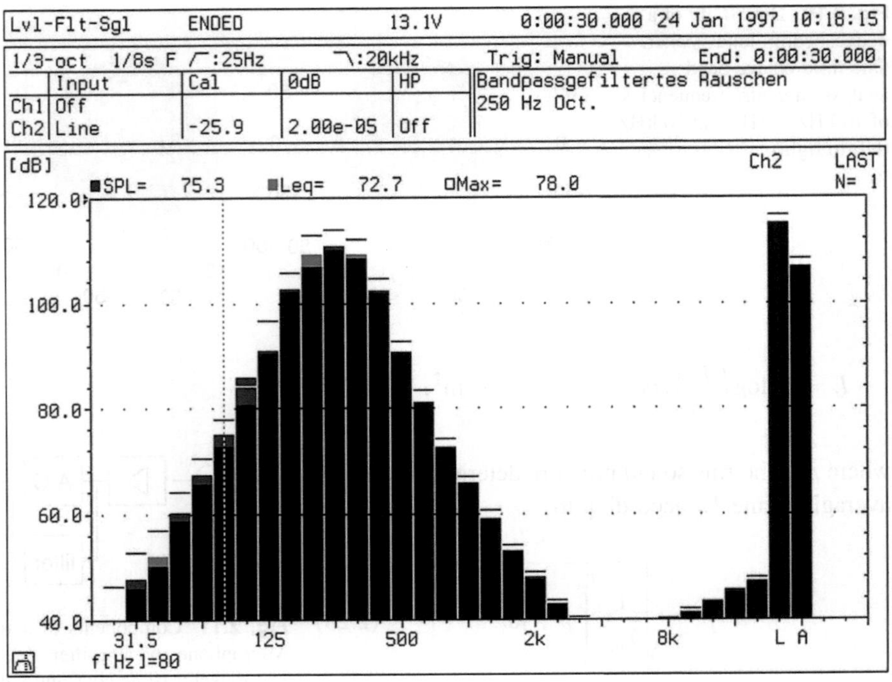

Fig. 2.12 Display of a real-time frequency analyser

2.3.3 Precision

Sound level meters for the measurement of absolute sound levels must fulfil several requirements. Before a qualified sound level meter is purchased, these are checked in type testing or in type approval tests, where acoustic and electric tests are performed (internationally standardised in IEC 60651). The goal of the control system is a correct and reliable result of the sound level, independent of conditions such as temperature or moisture. Another goal is the exact description of the application of the instrument in the sound field (directivity, calibration with pistonphone, etc.). The electrical test includes, in addition to other tests, excitation of the rms detector with several signals to check the correct implementation of the time and frequency ratings. Depending on the quality, the sound level meter is then placed into class 0, 1, 2, or 3 (see Table 2.3).

2.3.4 Bandpass Filters

A modern but more complicated method of sound level measurement is frequency analysis in frequency bands, typically in one-third octave bands or in octave bands. A traditional sound level meter can be extended by a set of bandpass filters to determine the sound pressure level for a specific frequency band. If the sound event is not stationary, however, the bandpass filters must be used at run-time in parallel. This is implemented in a real-time frequency analyser (Fig. 2.12). The midband frequencies of one-third octave bands are defined on a logarithmic frequency scale as follows (here in the example of the base 2 logarithm):

$$\begin{aligned} f_u &= 2^{1/3} \cdot f_l \\ \Delta f &= f_u - f_l = f_l\left(2^{1/3} - 1\right) \\ f_m &= \sqrt{f_l \cdot f_u} \\ f_{m+1} &= 2^{1/3} f_m, \end{aligned} \tag{2.25}$$

with f_l and f_u as lower and upper edge frequency and f_m, f_{m+1} as midband frequencies of the bands m and $m + 1$.

Similarly, for octave bands

$$\begin{aligned} f_u &= 2f_l \\ \Delta f &= f_u - f_l = f_l \\ f_m &= \sqrt{f_l \cdot f_u} = \sqrt{2} f_l \\ f_{m+1} &= f_m \cdot 2. \end{aligned} \tag{2.26}$$

Formerly, one-third octave bands and octave bands were designed using analogue Butterworth filters,

Table 2.2 Table of standard frequencies and A-weighting

Nominal frequency in Hz	Exact frequency (base 10) in Hz	A-weighting
10	10,00	−70.4
12.5	12.59	−63.4
16	15.85	−56.7
20	19.95	−50.5
25	25.12	−44.7
31.5	31.62	−39.4
40	39.81	−34.6
50	50.12	−30.2
63	63.10	−26.2
80	79.43	−22.5
100	100	−19.1
125	125.9	−16.1
160	158.5	−13.4
200	199.5	−10.9
250	251.2	−8.6
315	316.2	−6.6
400	398.1	−4.8
500	501.2	−3.2
630	631.0	−1.9
800	794.3	−0.8
1,000	1,000	0.0
1,250	1,259	+0.6
1,600	1,585	+1.0
2,000	1,995	+1.2
2,500	2,512	+1.3
3,150	3,162	+1.2
4,000	3,981	+1.0
5,000	5,012	+0.5
6,300	6,310	−0.1
8,000	7,943	−1.1
10,000	10,000	−2.5
12,500	12,590	−4.3
16,000	15,850	−6.6
20,000	19,950	−9.3

Table 2.3 Precision classification of sound level meters

Class	Application	Uncertainty limit
0	Laboratory, standard reference	±0.4 dB
1	Laboratory, field measurement	±0.7 dB
2	General field measurement	±1.0 dB
3	Survey measurement	±1.5 dB

whereas today, digital filters have replaced the analogue solutions. Mostly the filters are constructed as IIR filters (IIR = infinite impulse response), also as digital representatives of Butterworth filters. The realisation for real-time applications, i.e. instantaneous filtering without delay, is hardly possible without digital signal processors (DSPs). With this technology, a powerful DSP can refresh many bandpass filters sequentially in real-time with output (in FAST every 125 ms).

An optimum of filters is reached only as a compromise between slope steepness in the frequency domain and temporal onset and offset. For the specification of fractional band filters, international standards are available [5], in which the frequency curves in the filter pass band and in the attenuation band are fixed with exact tolerances and classifications. Specific requirements are established concerning real-time applications with onset and offset times; and group delays of the different band filters must be considered.

In the case of the quick measurement of spectra that might have a random character, frequency curves in rooms or other systems with high modal density, signals of random noise are advantageous. These allow a direct broadband excitation of the system (e.g., a room, a separating wall, or a muffler) and a direct determination of the band-filtered spectrum. The most general random noise is called 'white noise'. It contains all frequencies with the same amplitude. Also useful is 'pink noise', which contains fewer high frequencies with a slope per octave of −3 dB. It is used if predominantly low frequencies are to be excited and if the tweeter of the measurement loudspeaker should not be overloaded. Due to the energy integrative effect of band filters with increasing frequency (one-third octave, octave), pink noise produces a constant (horizontal) result on the display of a real-time frequency analyser, whereas white noise produces a slope of +3 dB/octave.

The sound pressure level measured has random behaviour due to the random excitation. In order to achieve sufficient accuracy, the averaging time T_m of the equivalent sound pressure level L_eq must be chosen adequately. The relation between the standard deviation of the level and the measurement duration is estimated by

$$\sigma_L = \frac{4.34}{\sqrt{B \cdot T_\mathrm{m}}}\,\mathrm{dB}. \tag{2.27}$$

where B = bandwidth of the filter in Hz.

Fig. 2.13 Discrete Fourier transformation of a time series $s(n)$ into a line spectrum $\underline{S}(k)$

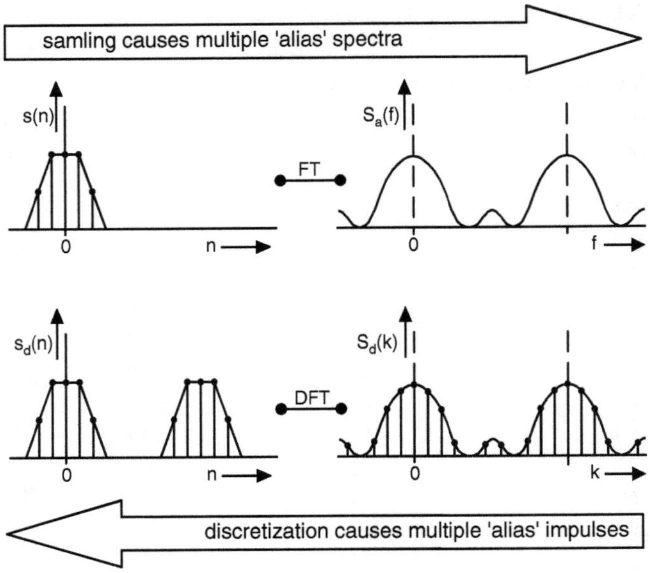

2.4 FFT-Analysis

2.4.1 Sampling of Measurement Signals

In order to obtain signals that can be digitally processed, the voltage signal produced by the microphone must be sampled. This is done by an A/D converter. The fine structure of the process of discretising is dependent on the appropriate time and amplitude resolution. For the hearing range, it is typical to use sampling rates of 44,1 kHz or 48 kHz with a resolution of 16 bits (discretised in steps from −32,768 to +32,767). If the sound event to be measured involves a high dynamic range, A/D converters can be found that allow 20 bit or more resolution, so that a 120 dB dynamic range between peak level and quantisation noise can be covered without an amplification switch. The speed of sampling is dependent on the frequency content of the signal. If the sampling frequency is not sufficient to detect fast variations in the signal, sampling artefacts occur that are found in the frequency domain as overlapping 'alias' spectra (aliasing). To avoid aliasing, low pass filters are used that limit the frequency range to, at most, half of the sampling frequency (Nyquist theorem). With regard to the discrete amplitude time function, we have obtained a satisfyingly accurate image of the analogue signal. All further steps such as filtering, analysis, amplification, and storage can now be performed by signal processing much more effectively and with greater variability (e.g., digital filtering, digital memory devices, Compact Disk, and Digital Audio Tape).

2.4.2 Discrete Fourier Transformation (DFT)

Frequency analysis is a very important tool for measurement techniques on acoustic systems. Provided we have sampled time functions, the question is how this data set can be processed efficiently by Fourier transformation. First, we must take into account that the samples are discrete in time, which means that the result of the Fourier transformation will be periodic (see above). The crucial point is that the spectrum to be calculated can only be discrete, too, since a computer memory can only process and store a finite number of spectral lines. Therefore, we must work with line spectra instead of continuous spectra. Line spectra, however, are related to periodic signals only, which gives a second important prerequisite for digital signal analysis: it must be kept in mind that the numerically determined spectra refer to (apparent) periodic time functions (Fig. 2.13).

The discrete Fourier transformation (DFT) can be calculated as follows:

$$\underline{S}(k) = \sum_{n=0}^{N-1} s(n)\, e^{-j2\pi nk/N}; \quad k = 0, 1, \ldots, N-1. \tag{2.28}$$

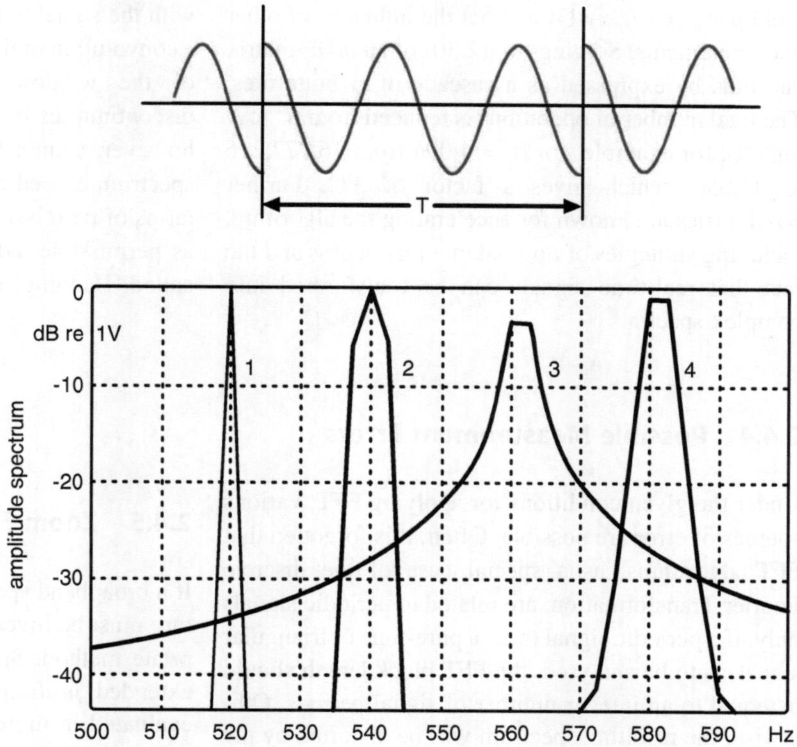

Fig. 2.14 (**a**) Time frame of a pure tone, the period of which is not in an integer number within the time frame; (**b**) section of the envelope of its line spectrum of harmonics with amplitude 1 V and frequency resolution $\Delta f =$ 2.5 Hz, (1) and (2) signal frequency identical with one of the frequency lines, analysed with (1) rectangular window, (2) Hanning window, (3) and (4) signal frequency in the middle between two frequency lines, analysed with (3) rectangular window and (4) Hanning window

Solving Eq. (2.28) for $\underline{S}(k)$, N^2 (complex) multiplications are required.

2.4.3 Fast Fourier Transformation (FFT)

A very powerful variant of the DFT is the Fast Fourier Transformation, FFT. It is not an approximation but a numerically exact solution of Eq. (2.28), which is faster by orders of magnitude. It is, however, only applicable if the number of samples to be transformed is $N = 2^m$ (4, 8, 16, 32, 64, etc.). The reason for the acceleration compared with the DFT is the reduction of the calculations to the necessary minimum. If we express Eq. (2.28) in terms of a matrix operation, with $N = 4$, for example:

$$\begin{pmatrix} S(0) \\ S(1) \\ S(2) \\ S(3) \end{pmatrix} = \begin{pmatrix} W^0 & W^0 & W^0 & W^0 \\ W^0 & W^1 & W^2 & W^3 \\ W^0 & W^2 & W^4 & W^6 \\ W^0 & W^3 & W^6 & W^9 \end{pmatrix} \begin{pmatrix} s(0) \\ s(1) \\ s(2) \\ s(3) \end{pmatrix}. \tag{2.29}$$

we obtain the matrix $\boldsymbol{W}$. It consists of the complex exponential terms of the phasors $2\pi k/N$ to the power of n, and it can be rearranged into a form of high symmetry with quadratic blocks (2×2, 4×4, 8×8, etc.) of zeros. The transformation of the matrix means that the vectors must be rearranged, too. The time sequence $s(n)$ becomes a column vector $x_1(n)$ (which is called bit reversal, see Eq. (2.30)) and the spectrum vector $x_2(k)$ is transformed into the final result $S(k)$.

The algebraic system to be solved only requires the solution of a sparse matrix equation:

$$\begin{pmatrix} S(0) \\ S(2) \\ S(1) \\ S(3) \end{pmatrix} = \begin{pmatrix} x_2(0) \\ x_2(1) \\ x_2(2) \\ x_2(3) \end{pmatrix} = \begin{pmatrix} 1 & W^0 & 0 & 0 \\ 1 & W^2 & 0 & 0 \\ 0 & 0 & 1 & W^1 \\ 0 & 0 & 1 & W^3 \end{pmatrix} \begin{pmatrix} x_1(0) \\ x_1(1) \\ x_1(2) \\ x_1(3) \end{pmatrix}, \tag{2.30}$$

with $x_2(0) = x_1(0) + W^0 x_1(1)$, $x_2(1) = x_1(0) + W^2 x_1(1)$ and so on. The latter operations can be illustrated best in terms of a butterfly algorithm, which means that pairs such as $(x_1(0), x_1(1))$ will yield pairs ($x_2(0)$, $x_2(1)$) without the influence of other vector elements. Solving Eq. (2.30) of an *mxm*-matrix can thus be expressed as a cascade of *m* butterflies. The total number of operations is reduced from N^2 to N ld(N/2), for example, for $N = 4{,}096$ from 16,777,216 to 45,056, which gives a factor of 372. Further possibilities are known for accelerating the algorithm, including strategies of optimal memory access and the fact that real-time signals can be transformed into complex spectra.

2.4.4 Possible Measurement Errors

Under the given conditions for applying FFT, various sources of error are possible. Often, it is forgotten that FFT algorithms, as a special case of the discrete Fourier Transformation, are related to periodic signals only. If a periodic signal (e.g., a pure-tone or triangular signal) is to be analysed, the FFT block length should be equal to an integer number of signal periods. Otherwise, the resulting spectrum will be distorted by the so-called 'leakage effect'. If the signal ends at a 'wrong' sample, the periodic extension will cause a discontinuity. In addition, it is not guaranteed that the fundamental frequency of the signal is represented by exactly one spectrum line (see Fig. 2.15).

If, however, the FFT block length is exactly equal to an integer number of periods, this error will be omitted. Generally, the sampling block of integer number of periods can be mapped to an FFT block length by sampling rate conversion.

An approximate method for error reduction is given by the window technique. A 'window' in this sense is a time function with slope onset and offset multiplied with the signal to be transformed. This corresponds to a convolution of the signal spectrum with the spectrum of the window function. The influence of the discontinuities is thus reduced. The optimal window, however, cannot be specified. The distortions in the spectrum caused by the window can be considered in terms of permissible distortions of spectrum slopes or as permissible side lobes. As a compromise, the so-called 'Hanning' window is often used (Fig. 2.14).

$$w(n) = 2\sin^2\left(\frac{n}{N}\pi\right). \tag{2.31}$$

2.4.5 Zoom FFT

If a broadband spectrum with a particularly fine structure must be investigated, the Zoom FFT is an appropriate method. Specific parts of the spectrum can be extended in frequency resolution and can thus be evaluated in more detail. This method is based on a shift of the most interesting frequency band, f_{low} to f_{high}, symmetrically around zero and an analysis of this 'zoomed' range with full line density. The frequency shift is gained by the multiplication of a real signal with complex phase vectors $\exp(-j\pi(f_{\text{low}} + f_{\text{high}})t)$ and low pass filtering. It is worth mentioning that the sampling rate can be reduced significantly because only the number of frequency lines (bandwidth) between f_{low} and f_{high} must comply with the Nyquist theorem. The general requirement that the line spacing be equal to the temporal block length, however, is still valid. The measurement, therefore, lasts as long as a measurement with the full number of spectral lines.

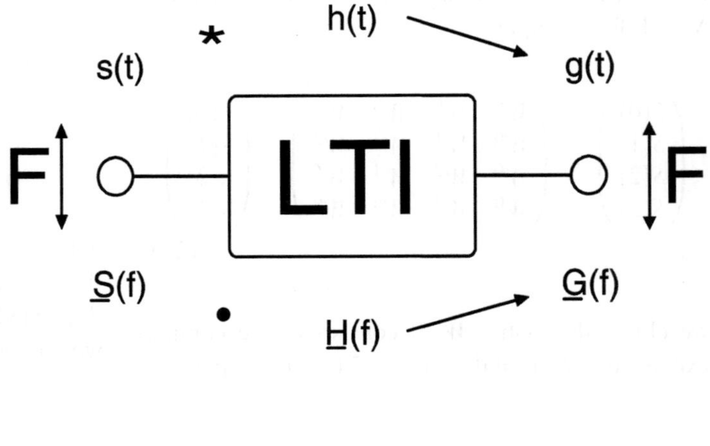

Fig. 2.15 Signal path in linear time-invariant systems (LTI). The input signal $s(t)$ or $\underline{S}(f)$ and the output signal $s'(t)$ or $\underline{S}'(f)$ are related to each other by the response to Dirac pulse (impulse response $h(t)$) or by the response to pure tones (stationary transfer function $\underline{H}(f)$). In the time domain this is realised by convolution (upper path) and in the frequency domain by multiplication (lower path)

2.4.6 Advanced Signal Analysis

The analysis using FFT equipment can recover several signal parameters. Accordingly, it makes the detection of correspondence, similarities or separations of signal components possible. Often, it is not 'just' a Fourier transformation that is performed, but signal theory provides some interesting methods and results can be gained from FFT and some further operations. One example is the so-called 'Cepstrum analysis', which is applied for detecting periodicities in the spectrum, particularly for the analysis of harmonic components in musical acoustics, in machine diagnosis and in speech processing. The Cepstrum is defined as the power spectrum of the (base-10) logarithmic power spectrum:

$$C_s(\tau) = |\mathrm{F}\{\log[|S(f)|^2]\}|^2, \tag{2.32}$$

with

$$S(f) = \mathrm{F}\{s(t)\}; \tag{2.33}$$

the variable, τ, called 'Quefrency' of the Cepstrum. With ***F*** we denote formally a Fourier transformation (by FFT, for instance).

Examples of the application of cross-correlation analysis can be found in the calculation of the 'similarity' or 'coherence' of signals or in the detection of periodicities in signals (auto correlation). The so-called 'correlation integral' for a given measurement duration T

$$k_{xy}(\tau) = \int_{-T/2}^{T/2} x(t)y(t+\tau)\mathrm{d}t \tag{2.34}$$

can be processed in FFT analysers by the product of Fourier transformations in the form

$$K_{xy}(f) = X^*(f) \cdot Y(f), \tag{2.35}$$

with the frequency functions K, X, and Y denoting the Fourier transforms of the time functions k, x, and y), respectively (between $-T/2$ and $T/2$).

Generally with (2.37), the overlap is calculated, that is, the parts of the signals which are similar in dependence on the relative time shift τ.

$K_{xx}(\tau)$ is called the autocorrelation function. $K_{xx}(\tau)$ has its maximum at $\tau = 0$. If for certain time shifts, values comparable to the maximum (in the normalised definition = 1) appear, the signal is considered periodic (see also Sect. 2.5.4). Stochastic signals are internally uncorrelated and show accordingly very small autocorrelations, apart from $\tau = 0$.

From a measurement of two signals over a certain transmission path (for instance, an acoustic transmission line, a vibroacoustic path or an airborne sound path between two points in a room or between two rooms), the complex spectra can be calculated by using an FFT analyser. The characteristics of the transmission path, the complex stationary transfer function $\underline{H}(f)$ (see Sect. 2.5) can be determined from the cross power spectrum $K_{xy}(\tau)$, in spite of the signals being stochastic:

$$\underline{H}(f) = \frac{\underline{Y}(f)}{\underline{X}(f)} = \frac{\underline{Y}(f)}{\underline{X}(f)} \cdot \frac{\underline{X}^*(f)}{\underline{X}^*(f)} = \frac{K_{xy}(f)}{K_{xx}(f)} \tag{2.36}$$

This method of measurement is very effective if one considers stationary random signals from aerodynamic noise sources or any other stochastic process. Measurement tasks in which the excitation signal must be generated are referred to in the description of deterministic periodic signals in Sect. 2.5.

2.5 Measurement of Transfer Functions and Impulse Responses

Today, it is no longer a problem to create special test signals using PC-based equipment, to generate and store signals and send them through a DA converter. This offers more flexibility, better power output control and thus higher measurement precision than methods based on stochastic signals and the cross power spectra. The following relationships are very important for performing measurements in which the generation and modification of source signals are explicitly possible.

The device under test shown in Fig. 2.15 is considered a linear time invariant system (LTI). A measurement of the sound transmission through a muffler mounted in a pipe can serve as a basic LTI example. The muffler is located between two microphone positions. Generally speaking, sound transmission is

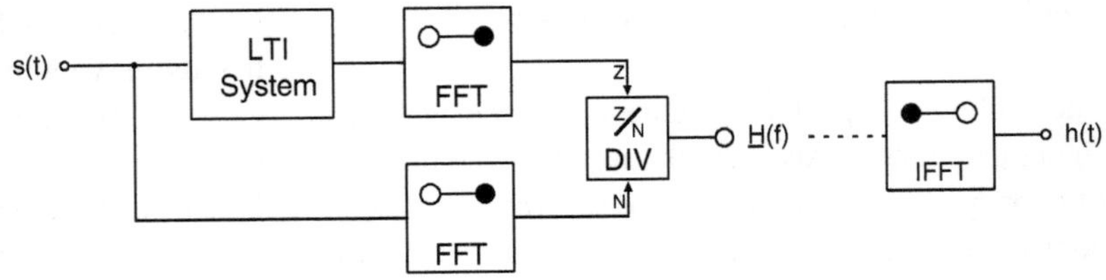

Fig. 2.16 Block diagram of the two-channel-FFT-measurement technique

considered from a source position to a receiver position. The electric, acoustic, electro-acoustic or vibroacoustic path between the two points is given by the LTI system.

The assumptions of LTI are the most important prerequisite of the digital measurement technique described here. They are also a prerequisite for the applicability of Fourier analysis (see Sect. 2.4). The consequences of violating these conditions are summarised in Sect. 2.5.5. The condition 'linearity' means that the system behaviour is invariant in spite of changes in the input power. Time invariance means that the system does not change over time.

In the figure, too, the logical chain of signal theoretic operations both in the time and frequency domain is illustrated. In each step, these domains are coupled unambiguously through Fourier transformation.

The signal path formulated in the time domain reads (convolution)

$$s'(t) = s(t) * h(t) = \int_{-\infty}^{\infty} s(\tau)h(t-\tau)d\tau. \tag{2.37}$$

The signal path formulated in the frequency domain is

$$\underline{S}'(f) = \underline{S}(f) \cdot \underline{H}(f). \tag{2.38}$$

while the elementary equation for determining system properties in the frequency domain is (see Sect. 2.5.1):

$$\underline{H}(f) = \frac{\underline{S}'(f)}{\underline{S}(f)} = \underline{S}'(f) \cdot \frac{1}{\underline{S}(f)}, \tag{2.39}$$

The same is expressed in the time domain by a so-called 'de-convolution'

$$h(t) = s'(t) \; * \; s^{-1}(t), \tag{2.40}$$

with the signal $s^{-1}(t)$ being the signal with the inverse spectrum $1/\underline{S}(f)$. $s^{-1}(t)$ is called 'matched filter' or transversal filter (see Sect. 2.5.2).

Note: For better understanding, in this chapter the signal transformations and steps of signal processing are expressed in continuous form. In digital instrumentation, the signals are, of course, represented in discrete form [compare Eq. (2.28)].

If $\underline{S}(f)$ has a white spectrum, we can also write:

$$s^{-1}(t) = s(-t), \tag{2.41}$$

and Eq. (2.34) can be transformed into

$$h(t) = s'(t) * s(-t) = s'(t) \otimes s(t)$$
$$\equiv \int_{-\infty}^{\infty} s'(\tau)s(t+\tau)\mathrm{d}\tau, \tag{2.42}$$

which means that $h(t)$ can also be expressed by cross-correlation of $s(t)$ and $s'(t)$ (see Sect. 2.5.3).

Obviously, Eqs. (2.39), (2.40) and (2.42) are equivalent in the case of broadband white excitation signals. Differences exist, however, in the phase spectrum of the excitation signal and the corresponding time sequences. This has a somewhat significant influence on the level adjustment of power amplifiers and loudspeakers. It must be faced that a swept sine and a Dirac pulse have the same magnitude spectra. However, their maximum signal amplitudes are extremely different, although their signal energies are equal.

It is important to note that further consequences of digital representation of signals and spectra must be considered. The finite length, T_{rep}, of the deterministic excitation signal and its periodicity is an important factor. The periodicity can well be used for coherent averaging. Periodic signals, however, have a discrete line spectrum with a frequency spacing of

$$\Delta f = \frac{1}{T_{\text{rep}}}. \tag{2.43}$$

Due to the excitation of the system with a periodic signal, the system transfer function is multiplied with a line spectrum. Accordingly, results of the measurement can only be found at these discrete frequency lines. In addition, the results are not averages over frequency intervals (between the lines, for example), but they are the correct measurement results corresponding to results from pure-tone excitation and high-quality narrow-band filters, for instance. Deterministic periodic signal must, therefore, be strictly separated from stochastic or pseudo-stochastic non-periodic noise signals.

To ensure that the system is in a steady state and that all eventual modes are excited properly, the frequency spacing must be sufficiently dense. One can express the same condition by formulating a requirement for the duration of the signal, which must be long enough to let the system load or decay acoustically. In room acoustics, for example, this is achieved when

$$T_{\text{rep}} \geq T_{\text{rev}}. \tag{2.44}$$

T_{rev} denotes the reverberation time. Another example is the measurement of resonant systems of the second order with quality, $Q = 2.2/T$. Equation (2.47) expresses the fact that the half-width of the mode is covered by at least two frequency lines.

The feature of coherent superposition of signals is used in averaging a number of signal periods. The uncorrelated background noise is added incoherently. Hence, the gain in signal-to-noise ratio by N averages is

$$\Delta_{\text{av}} = 10 \lg \mathrm{N}\ \mathrm{dB}. \tag{2.45}$$

More information on the background of digital signal processing for measurements is given in the recent standard ISO 18233 [6].

2.5.1 2-Channel-FFT Technique

Measurement and the corresponding signal processing are performed in the frequency domain. The same holds for the measurement of transfer functions by using a 2-channel-FFT analyser and the cross power spectrum technique. Input and output signals are measured simultaneously, are FFT transformed and are processed by complex spectrum division Eq. (2.39). An important condition is therefore a sufficiently broad bandwidth, i.e. the excitation signal may not contain 'zeros' in the spectrum to avoid problems in the division. Any periodic signal of length 2^m may be used. For reasons of optimised level adjustment, sweeps, chirps, or deterministic noise do indeed have advantages [7] (Fig. 2.16).

After performing the spectrum division, the impulse response can be calculated from inverse Fourier transformation, if necessary:

$$h(t) = \boldsymbol{F}^{-1}\{\underline{H}(\omega)\} = \int_{-\infty}^{\infty} \underline{H}(\omega)\, e^{j\omega t}\, \mathrm{d}\omega. \tag{2.46}$$

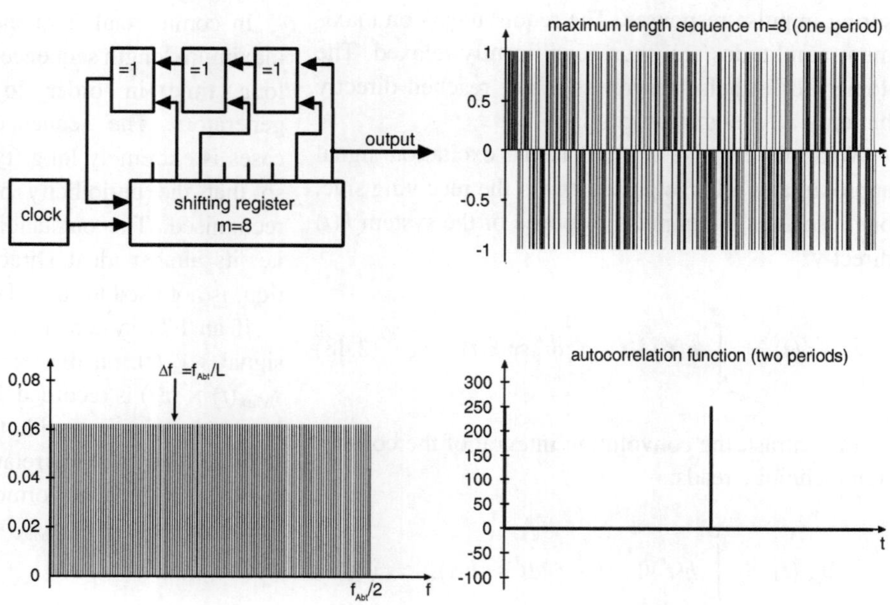

Fig. 2.17 (**a**) Generation of a maximum-length sequence with a shift register (example: sequence of order 8 with length 255); (**b**) one period of the maximum-length sequence; (**c**) two periods of the autocorrelation function

2.5.2 Swept-Sinusoidal Signals

These kinds of excitation signals and the corresponding signal processing are based on the matched filter or transversal filter approach, which includes a direct de-convolution in the time domain Eq. (2.40). The most commonly used signals $s(t)$ are sweeps, chirps, or time-stretched pulses [8].

The matched filter is determined by inverting the signal sequence in time:

$$s^{-1}(t) = s(T_{\text{Rep}} - t). \tag{2.47}$$

A great advantage of this technique is that the problem of signal periods too short in relation to long impulse responses is avoided. A sequence, short compared with the decay time of the system, is sent only once, while the received signal is recorded over a long duration, which theoretically is infinitely long. This corresponds to an excitation of a (short) signal amended with zeros (zero padding).

2.5.3 Correlation Technique

The correlation technique is a special case of impulse measurement techniques. It was originally developed for measuring pulse propagation delays. It can, however, be well used for determining impulse responses and transfer functions. The most important advantage is the possibility of using spread-out signals as measuring impulse responses. The requirements on maximum amplitudes are, thus, significantly relaxed. The domain of impulse responses is not reached directly but only after signal processing.

With direct pulse excitation, the excitation signal approximates a Dirac pulse$\delta(t)$. On the receiving side, one measures the impulse response of the system $h(t)$ directly:

$$s'(t) = \int_{-\infty}^{\infty} h(t')\delta(t - t')\mathrm{d}t' \approx h(t). \tag{2.48}$$

In contrast, the convolution integral of the correlation technique reads:

$$\Phi_{ss'}(t) = \int_{-\infty}^{\infty} h(t')\Phi_{ss}(t - t')\mathrm{d}t' \approx h(t). \tag{2.49}$$

Rather than the signal $s(t)$ itself, it contains the autocorrelation function $\Phi_{ss}(t)$ of the signal and the cross-correlation function $\Phi_{ss'}(t)$ of the received signal $s'(t)$ with the excitation. Now, it is not the signal but its autocorrelation function that must approximate a Dirac pulse. This permits much better conditions for level adjustment of the measurement system and the resulting signal-to-noise ratio. However, this advantage must be paid for by having to process the cross-correlation of the received signal $\Phi_{ss'}(t)$.

2.5.4 Maximum Length Sequences

An important example of correlation signals is maximum length sequences [9]. These are periodic pseudo-random noise signals with an autocorrelation function approximating an almost perfect Dirac pulse. They are generated from a deterministic, exactly reproducible process by using a shift register with feedback. With m denoting the length (order) of the shift register, a variety of $L = 2^m - 1$ states is possible. The periodic sequence of maximum length (maximum-length sequence, m-sequence, MLS) covering all possible states will be achieved under appropriate feedback conditions. For all orders of shift registers, there is at least one feedback rule. In practice, a bipolar excitation signal of '1' corresponds to a positive signal amplitude $+ U_0$ and of '0' to a negative value $-U_0$, accordingly (Fig. 2.17).

In commercial frequency analysers, generators of maximum length sequences have been used for quite a long time in order to replace analogue noise generators. The sequence length chosen in these cases is extremely long (typically $m > 30$, $L > 10^9$), so that the periodicity of the signal is no longer recognised. The outstanding property of the signal, i.e. its almost ideal Dirac-like autocorrelation function, is not used to its full capability.

If an LTI system is excited by a stationary MLS signal $s_{\text{Max}}(t)$, on the receiving side the convolution $s_{\text{Max}}(t) \times h(t)$ is recorded. This signal is to be sampled in synchronisation with the clock, Δt, of the shift register. The cross-correlation with the excitation signal Eq. (2.42) is performed by convolution with the time-inverted signal $s_{\text{Max}}(-t)$:

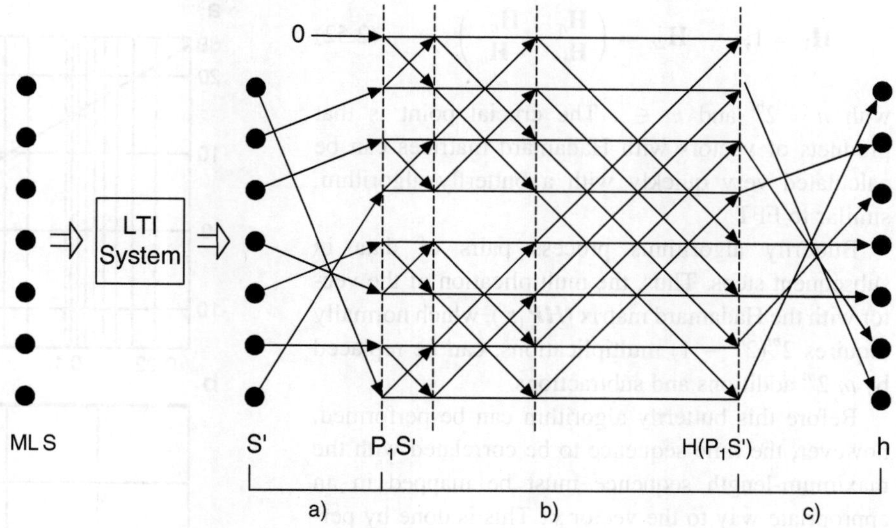

Fig. 2.18 Hadamard transformation of a measured sequence $s'(n)$. (**a**) Permutation of the samples; (**b**) Hadamard Butterfly; (**c**) Back permutation in the final time sequence of the impulse response $h(n)$

$$\begin{aligned} &s_{\mathrm{Max}}(t) * h(t) * s_{\mathrm{Max}}(-t) \\ &= s_{\mathrm{Max}}(t) * s_{\mathrm{Max}}(-t) * h(t) \\ &= \Phi_{\mathrm{Max}}(t) * h(t) \approx \delta(t - iL\Delta t) * h(t), \end{aligned} \quad (2.50)$$

with $i = 0, \pm 1$, and ± 2 denoting the sequence counter of periods $L\Delta t$.

The autocorrelation function of a stationary periodic maximum-length sequence is a series of pulses of height L and a small negative offset of -1. The period is the same as the period of the initial sequence (in the example $L = 255$). Each pulse contains the same energy as the complete maximum-length sequence period. Further enhancement of the signal-to-noise ratio can be achieved by adding N periods. Due to the strict periodicity of MLS, the amplitudes of the signal must be added, while the background noise is uncorrelated to the signal and adds up quadratically (incoherently). Thus, the signal-to-noise ratio increases according to Eq. (2.45).

In what was stated thus far, the maximum-length sequence offers no specific advantages compared with similar deterministic periodic signals with a smooth amplitude broadband spectrum and low crest factor (ratio of peak value to rms value), such as is described in Sects. 2.5.5.1–2.5.5.2 The interesting point with maximum-length sequences in applications of the impulse measurement technique is a fast cross-correlation algorithm in the time domain. For the calculation of the cross -correlation, we can generally apply either the discrete convolution or the FFT convolution. For a block length B, the discrete convolution requires B^2, whereas the FFT convolution requires 'only' B (4 $\log_2 B$ + 1) multiplications of complex numbers.

Since the period $L = 2^m - 1$ of maximum-length sequences is just one sample shorter than the FFT block length, tedious methods for sampling rate conversion must be applied to avoid leakage errors. Much faster is a method for correlation in the time domain, the fast Hadamard transformation (FHT). It is based, like the FFT, on a so-called 'butterfly' algorithm and requires no more than $m\,2^m$ additions and subtractions; typically, one obtains an acceleration by a factor 10 compared with FFT-based algorithms.

2.5.4.1 Hadamard Transformation

The basis of FHT is the representation of the correlation integral Eq. (2.42) in terms of a matrix operation, i.e. by multiplying a vector with a Hadamard matrix:

$$\bar{\mathbf{h}} = \frac{1}{L+1}\mathbf{P}_2\mathbf{H}(\mathbf{P}_1\bar{\mathbf{s}}'). \quad (2.51)$$

The vector s' contains the samples of the measured sequence and the vector $\boldsymbol{h}$ represents the impulse response to be determined. The vectors $\boldsymbol{P}_1$ and $\boldsymbol{P}_2$ represent permutation rules (see below). A Hadamard matrix $\boldsymbol{H}$ is closely related to maximum-length sequences. If we arrange maximum-length sequences in rows one below the other and shifted by one column, we create a matrix that can be transformed into a Hadamard matrix by permutating the columns. Hadamard matrices show interesting features. For instance, they contain a special internal basic structure, which is repeated in several scales (self-similarity). The generation rule for Hadamard matrices of a 'Sylvester type' is recursive:

$$\mathbf{H}_1 = 1, \qquad \mathbf{H}_{2n} = \begin{pmatrix} \mathbf{H}_n & \mathbf{H}_n \\ \mathbf{H}_n & -\mathbf{H}_n \end{pmatrix}, \tag{2.52}$$

with $n = 2^m$ and $m \in$. The crucial point is that products of vectors with Hadamard matrices can be calculated very quickly with a butterfly algorithm, similar to FFT.

Butterfly algorithms process pairs of data in subsequent steps. Thus, the multiplication of the vector with the Hadamard matrix ($\boldsymbol{HP}_1\boldsymbol{s}'$), which normally requires $2^m(2^m - 1)$ multiplications, can be replaced by $m\ 2^m$ additions and subtractions.

Before this butterfly algorithm can be performed, however, the time sequence to be correlated with the maximum-length sequence must be mapped in an appropriate way to the vector $\boldsymbol{s}'$. This is done by permutation of the samples in the data structure. The permutation rules ($\boldsymbol{P}_1$ and $\boldsymbol{P}_2$, see above) are derived from the maximum-length sequence. The complete measurement procedure can be summarised as shown in Fig. 2.18.

An example of applying the correlation technique is the measurement of room impulse responses (reverberation time measurement). The property of impulse compression by the Hadamard transformation and the option to average a number of periods permits measurements at low signal level (for instance, with inaudible signals) and in the presence of an audience during a performance. With increasing averages and increasing signal-to-noise ratios, the time interval useful for the evaluation of room acoustical criteria is becoming larger and larger.

Another advantage of the MLS Hadamard transformation is that it is a very simple method for colouration of the excitation signal. This can be useful for the frequency-dependent optimisation of the level adjustment (for instance, for minimisation of the crest factor) or for implicit equalisation of components of the measurement arrangement. Because of the fact that convolution and cross-correlation are similar except for a time reversal of the signal, a convolution of a maximum-length sequence $m(t)$ with a filter impulse response $f(t)$ can be expressed in terms of a correlation:

$$m'(-t) = m(-t) * f(-t) = m(t) \otimes f(-t) = \mathrm{FHT}[f(-t)], \tag{2.53}$$

(convolution) (correlation) (FHT)

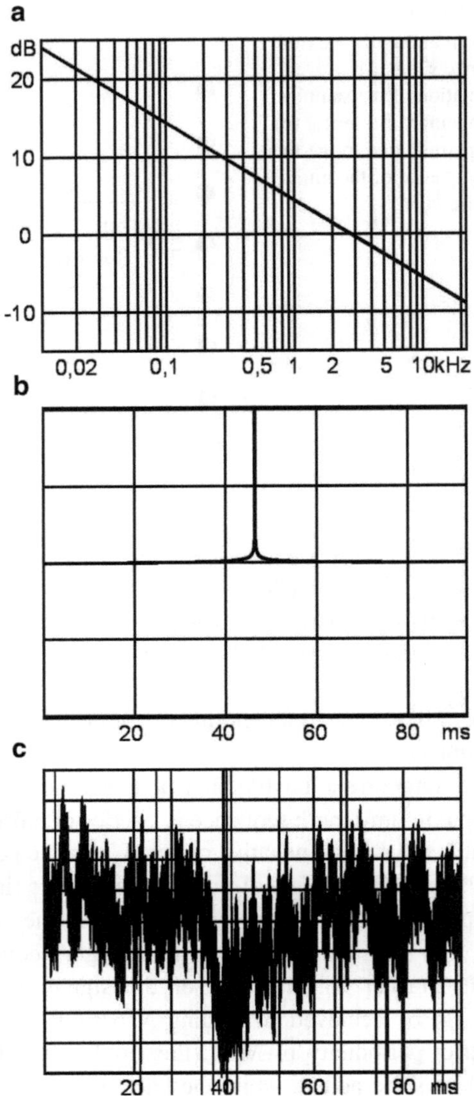

Fig. 2.19 (**a**) 'Desired' spectrum of a filter $F(f)$; (**b**) corresponding impulse response of the filter; (**c**) corresponding maximum-length sequence with colouration by the chosen filter. Example: 'Pink filter' (linear low-pass function of first order, −3 dB/Octave)

The time-reversed filtered maximum-length sequence $m'(-t)$ is nothing but the Hadamard transformation of the time-reversed filter impulse response $f(-t)$. In order to obtain the filtered maximum-length sequence, the desired filter impulse response must be measured or synthesised, time inverted, then Hadamard-transformed and time inverted again. A Hadamard transformation of this equalised sequence then gives not the ordinary auto-

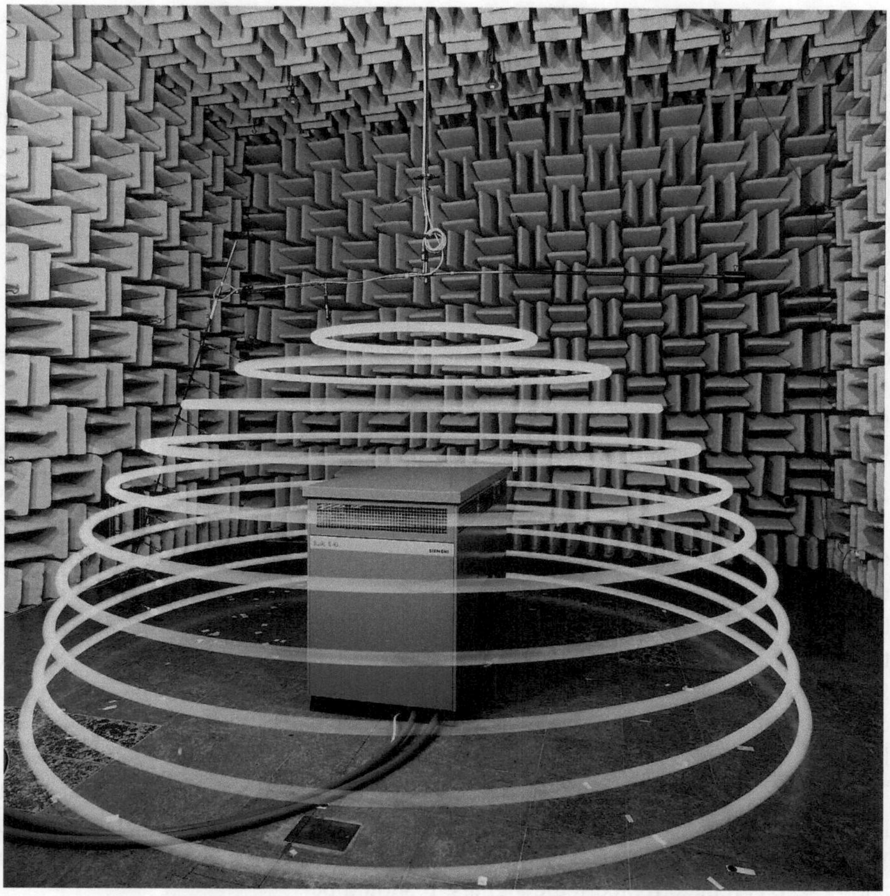

Fig. 2.20 Hemi-anechoic chamber (PTB Braunschweig). Example of the surface scanning method for measurement of sound power of a heat pump. The microphone paths of the scanning surface are illustrated by using moving lamps and long-time exposure

correlation function (series of Dirac pulses), but provides directly the previously chosen filter impulse response *f*(*t*) (Fig. 2.19).

2.5.5 Sources of Errors in Digital Measurement Techniques

The validity of LTI systems is the most important prerequisite for the usage of cross-spectrum, FFT, or correlation measurement techniques as alternatives for the application of impulsive signals. If non-linearities or time variations cannot be neglected, measurement errors will occur. This holds for the system under test as well as for the measurement instrumentation.

2.5.5.1 Background Noise

Background noise is normally not correlated with the excitation sequence. Thus, impulses as well as pure-tone or broadband stochastic extraneous signals are spread out by the cross-correlation over the complete measurement duration and are noticeable in the measured impulse response only with their mean power. The parts of the impulse response being dominated by background noise can easily be deleted or filled with 'zeros' (window technique, see Sect. 2.4.4). In order to avoid time aliasing, the early part of the impulse response and the necessary decay are to be considered.

2.5.5.2 Non-Linearities

Weak non-linearities can often be tolerated. They appear in the measured impulse response as an apparent noise floor and can hardly be separated from background noise. Accordingly, they are treated like background noise by using the window technique. Non-linearities can be detected by observing the effective signal-to-noise ratio improvement by coherent averaging and by checking if this gain in dynamic range is asymptotically limited. The effect of non-linearities can be reduced by reducing the signal amplitude and the dynamic range can then be further increased by averaging. Experience has shown that with all methods listed above, signal-to-noise ratios of more than 70 dB can be easily reached. In the case of higher requirements, however, problems must be faced. A further increase of the signal-to-noise ratio is limited by non-linearities in the components of the instrumentation, depending on a clever choice of the excitation signal and on optimal adaptation, on the system to be measured, on power amplifiers, loudspeakers or Sample&Hold devices and A/D converters. Currently, signals with a low crest factor are to be found. Maximum-length sequences are, at first, superior since they have a crest factor of 0 dB. Sweeps and chirps, for example, have a crest factor of 3 dB. Under extreme conditions (signal-to-noise ratio <70 dB), however, maximum-length sequences are not the best choice, due to low-pass components in the digital measurement chain and overshoot effects at the sequence flanks of up to 8 dB. Therefore, their theoretic crest factor cannot be reached in practice. Sweeps or similar signals still keep their theoretical crest factor of 3 dB in electronic realisation and, thus, leave the maximum-length sequences behind.

The FFT and MLS technique are both related to periodic signal processing, whereas the straightforward sweep de-convolution, according to Eqs. (2.40) and (2.47), is related to aperiodic signals. It should be noted that the signal processing related to swept-sinusoidal signals could be performed by the FFT technique as well (see Sect. 2.5.1); thus, periodic or aperiodic algorithms are applicable apparently in free choice. However, the choice of aperiodic signal processing has several practical advantages [7], although related to heavier computational load. Direct aperiodic processing requires de-convolution in the time domain (according to Eq. (2.40) with a sequence of length N, N^2 multiplications must be processed). Using a sweep with increasing frequency, the response to harmonic components will appear before the main excitation at that (harmonic) frequency. Accordingly, the harmonic distortion products in the excitation will appear at negative time in the impulse response and may easily be removed by using a time window.

Another important aspect of differences in aperiodic or periodic de-convolution concerns the noise floor in the impulse responses. Periodic processing by FFT or MLS results in an impulse response with a noise floor that is approximately constant, up to the time where the first distortion products appear. Aperiodic direct de-convolution, however, produces a decaying noise tail that is increasingly low-pass filtered towards its end. This results from the fact that the last part of the impulse response stems from steady-state noise convolved with the excitation sweep in reverse order [see Eq. (2.41)]. One should therefore not interpret the decreasing noise floor as a reverberant tail of the impulse response.

2.5.5.3 Time Variances

Time variances do not result in apparent background noise but they do change the signal shape slightly and can, therefore, hardly be detected. Two kinds of time variances are to be discussed: (a) fast variations, which are noticeable within one measurement period, and (b) slow effects, which come into play only in cases of longer averaging. In both cases, a phase distortion in sequences and between sequences is the reason for measurement errors. The coherent averaging and FFT or cross-correlation is affected.

At least some rules or guidelines can be used to avoid the effects of time variances. The result is quite correct if, for instance, it is ensured that maximum temperature drift (degrees Celsius) in a room during the measurement is not larger than

$$\Delta\vartheta < \frac{300}{fT}, \tag{2.54}$$

during a decay measurement with a low level and long averaging time (reverberation time T, one-third octave band or octave midband frequency f). Similar rules of thumb are available concerning the influence of wind.

2.6 Measurement Facilities

Connected with various tasks of acoustical measurement and testing, several measurement arrangements, test mock-ups and test rooms are standardised. To obtain acoustic characteristics, acoustic measurements are performed in research regarding biological and medical developments as well as numerous test methods of material testing. The acoustic test arrangements or rooms are implemented to create well-defined acoustic environments.

2.6.1 Anechoic Chamber

Spherical waves should propagate as freely as possible, undisturbed by reflections and diffraction (see Sect. 2.1). Measurement rooms fulfilling these conditions are called 'anechoic chambers' or 'free-field rooms' (Fig. 2.20).

Their walls must absorb sound by 99.9% in order to provide a level reduction of reflections by 30 dB [10]. This requirement is met by mounting wedge-like porous material on the walls and ceiling, either with or without airspace behind. Depending on the dimensions of the wedges, the desired absorption coefficient reached is above 50%. The floor can be treated in the same way (in this case, one enters the room on a net) or the floor stays acoustically 'hard' (hemi free-field room).

Besides good free-field conditions, anechoic chambers should have good insulation against background noise. Accordingly, more sophisticated solutions include a vibration-isolated foundation. In this case the room stands on springs, and the low resonance frequency (depending on room mass and spring stiffness (typically <10 Hz)) hinders vibrations of the surrounding building from outside to propagate into the room.

The qualification test of anechoic chambers is performed by the measurement of free field wave propagation, i.e., of the $1/r$ law, by using various pure-tone signals. Eventually, deviations from the $1/r$ curve that occur indicate insufficient absorption and, thus, room modes.

Applications for anechoic chambers are all 'free-field' measurements on loudspeakers, microphones, hearing aids, noise sources, etc. The frequency response can be determined as well as radiation characteristics and the emitted sound power.

2.6.2 Reverberation Room

Diffuse-field conditions can be found in closed rooms, according to the explanations in section **XXX ((Room acoustics))**. The conditions are, however, not perfectly

Fig. 2.21 Reverberation room (PTB Braunschweig) with treatment for increased sound diffusivity. **(a)** Structures on the walls; **(b)** Rotating diffuser in the room space

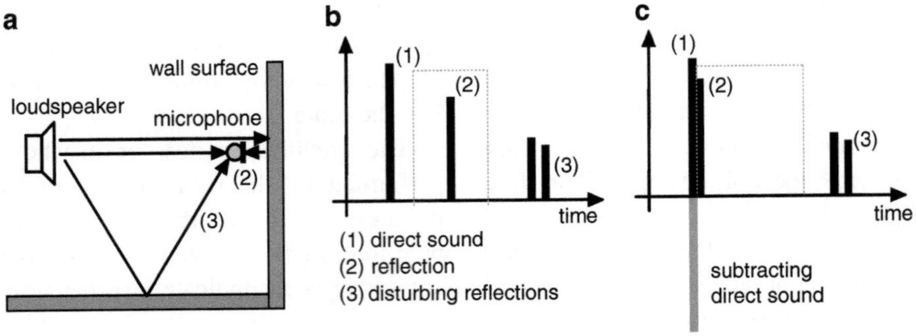

Fig. 2.22 In situ measurement method for wall impedances and reflection factors. (**a**) Measurement set-up with reflection paths; (**b**) Impulse response of this arrangement; (**c**) Impulse response of the arrangement with the microphone placed close to the surface with time-windowing and direct sound subtraction. The time function of the remaining reflection (2) and Fourier transformation yields the reflection factor $\underline{R}(f)$

fulfilled due to the statistical superposition of wall reflections or, if the description in the frequency domain is preferred, of room modes.

Prerequisites for diffuse sound fields, however, can be achieved in good approximation if the room boundaries absorb very little energy and if the absorption is uniformly distributed on the walls. The sound field is in this case nearly homogeneous and isotropic. As a further action for 'mixing' the sound, non-rectangular rooms and diffusers are used (Fig. 2.21).

The *reverberation room* is in fact the opposite of the anechoic chamber. Its walls should preferably reflect the sound totally. This can be obtained up to a remaining partial absorption of 1–2%, by suitable properties of the air medium and the absorption in a certain layer at the boundaries, even in the case of heavy and smoothly painted walls. At high frequencies, the air absorption is dominant anyway.

For improvement of sound diffusion, the walls of the reverberation room are fitted with sound scattering structures, and some (fixed or rotating) diffusers are placed in the room.

The reverberation times are, therefore, frequency dependent and in a range from above 10 s at low frequencies to about 1 s at high frequencies. Applications for reverberation rooms are measurements of absorption coefficients of materials, of sound power and of diffuse-field sensitivities of transducers.

For the qualification of a reverberation room (quality approval), various theoretical and experimental tests can be performed. At first, the lower frequency limit, f_{gr}, can be calculated (V = room volume in m^3, T = reverberation time in s).

$$f_{\text{gr}} \approx 2,000\sqrt{\frac{T}{V}} \text{ Hz.} \tag{2.55}$$

Below the critical frequency, the conditions of a diffuse sound field are not valid. The modal density is too low and strong spatial variations of energy density are present. One important way to improve the sound field is to add absorption material in order to achieve modal damping and a corresponding larger modal overlap.

In spite of apparently ideal diffuse field conditions, measurements in reverberation rooms must be performed at several source and microphone positions and then averaged, since the sound field is not perfectly homogeneous. Qualification tests of reverberation rooms, therefore, are not based on room geometry but on a selection of source and microphone positions, which yield results with the smallest possible variations around the average value. It must be taken into account that measurements in reverberation rooms are related to broadband incoherent signals analysed in frequency bands and based upon averages $< L >$ of local diffuse-field sound pressure levels L_i by energetic averaging:

$$\langle L \rangle = 10 \log \frac{1}{N} \sum_{i=1}^{N} 10^{L_i/10} \text{dB.} \tag{2.56}$$

Furthermore, in the case of absorption measurements, the sound field in the reverberation room is disturbed by the sample (see section **XXX**, room acoustics). The requirements of uniform absorption at the room boundaries are no longer fulfilled.

The identification of appropriate positions and averaging procedures is then of particular importance.

2.7 Sample Applications

2.7.1 Absorption and Impedance

In principle, every method capable of separating incident from reflected waves can be used for measuring the impedances or reflection factors of materials. Accordingly, impulse measurement methods are generally qualified, at least under certain conditions (plane-wave approximation, not-grazing incidence, smooth surfaces and homogeneous impedance distribution on the surface, sufficient time delay between direct sound (1), reflection (2) and disturbing reflections (3), etc.) (Fig. 2.22).

One method requires subtracting the direct sound (part c) which is previously determined under free-field conditions, a time window (broken line) and a Fourier transformation of the remaining reflection component (2), which results directly in the frequency response of the reflection factor.

For obtaining correct results, however, the above-mentioned conditions must be fulfilled. If one or more of these conditions are violated or not even approximately fulfilled, the method should not be applied. Nevertheless, it is the basis for a standardised test method for sound absorption and sound insulation of noise barriers at traffic lines and of road surfaces. In order to account for the simplifications involved in the measurement standard, the results are not denoted as a classical reflection factor, *R*, or impedance, *Z*, but by the term 'reflection loss'.

2.7.1.1 Classical Method (Kundt Tube)

The magnitude of the sound pressure along the tube axis is

$$|p(x)| = \hat{p}\sqrt{1 + |R|^2 + 2|R|\cos(2kx + \gamma)}. \quad (2.57)$$

By investigating the standing wave one can determine

$$|p|_{max} = \hat{p}(1 + |R|) \quad \text{and} \quad |p|_{min} = \hat{p}(1 - |R|), \quad (2.58)$$

and thus, $|R|$:

$$|R| = \frac{|p|_{max} - |p|_{min}}{|p|_{max} + |p|_{min}}. \quad (2.59)$$

If $d_{min} = |x_{min}|$ denotes the distance of the first minimum to the wall, the phase

$$\gamma = \pi\left(\frac{d_{min}}{\lambda/4} - 1\right), \quad (2.60)$$

can be determined and, according to

$$Z = \rho_0 c \frac{1 + R}{1 - R}, \quad (2.61)$$

the complex sample impedance, *Z*. The minima and maxima of the sound pressure of the standing wave are called 'nodes' and 'antinodes'. Positions with maximum sound pressure have minimum sound velocity and vice versa.

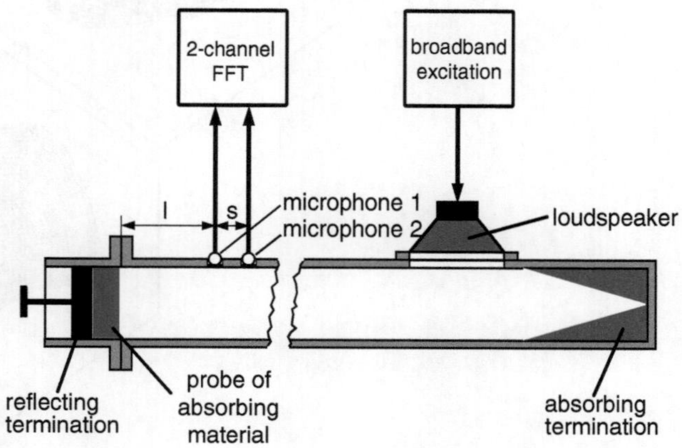

Fig. 2.23 Impedance tube of the 'two-microphone method' for measurement of impedances and reflection factors of material samples and other termination impedances

In many cases, the intensity loss of reflections is of interest. This is characterised by the absorption coefficient

$$\alpha = \frac{\text{not reflected Intensity}}{\text{incident Intensity}} = 1 - |R|^2. \qquad (2.62)$$

2.7.1.2 Two-Microphone Method

With the so-called two-microphone method or 'transfer function method', impedances and reflection factors can be determined in the broadband approach and much faster (Fig. 2.23). The instrumentation, however, is more complex and the signal evaluation requires a frequency analysis. In this method, it is assumed that the travelling incident wave and the reflected wave can be separated into the spectra $S_i(f)$, $S_r(f)$ and the reflection factor and that these spectra can be mapped to the transfer function between two points, 1 and 2. For the measured broadband spectra, $S_1(f)$ and $S_2(f)$ yield

$$S_1 = e^{jkl} S_i + e^{-jkl} S_r, \qquad (2.63)$$

$$S_2 = e^{jkl} e^{jks} S_i + e^{-jkl} e^{-jks} S_r, \qquad (2.64)$$

where l is the distance between the sample and the nearer microphone position and s is the distance between the microphone positions. By solving these equations, the signal amplitudes and the reflection factor can be calculated:

$$S_i = \mathrm{e}^{-jkl} \frac{S_2 - \mathrm{e}^{-jks} S_1}{\mathrm{e}^{jks} - \mathrm{e}^{-jks}}, \qquad (2.65)$$

$$S_r = \mathrm{e}^{jkl} \frac{\mathrm{e}^{jks} S_1 - S_2}{\mathrm{e}^{jks} - \mathrm{e}^{-jks}}, \qquad (2.66)$$

$$R = \frac{S_r}{S_i} = \mathrm{e}^{j2kl} \frac{\mathrm{e}^{jks} S_1 - S_2}{S_2 - \mathrm{e}^{-jks} S_1}. \qquad (2.67)$$

If S_1 is cancelled in the last equation, a representation independent of the absolute spectra S_1 and S_2,

$$R = \mathrm{e}^{j2kl} \frac{\mathrm{e}^{jks} - H_{12}}{H_{12} - \mathrm{e}^{-jks}}, \qquad (2.68)$$

can be discussed, which contains only the complex transfer function $H_{12} = S_2/S_1$.

Requirements on the measurement tube and the instrumentation are specified in detail. The measurement uncertainty is quite small (a few percent), provided the sample is mounted carefully.

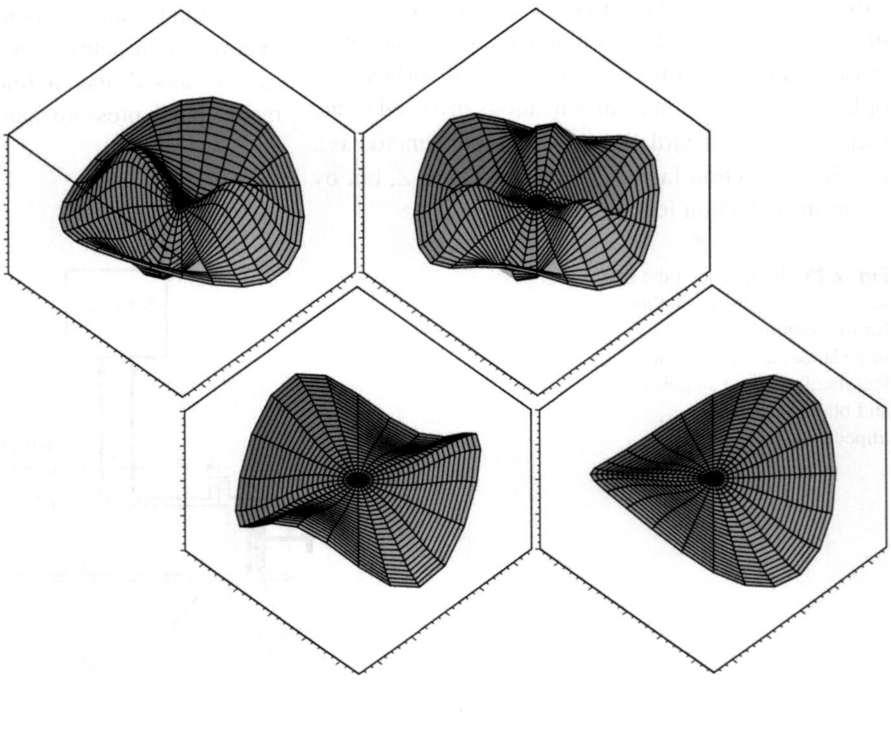

Fig. 2.24 Illustration (animation) of wave forms of a circular membrane. Measurement of vibration velocity and displacement on a grid (mesh) and subsequent processing by modal analysis (four eigenfrequencies)

2.7.2 Modal Analysis

Modal analysis is a measurement and evaluation method that provides extended and detailed information. Also here, transfer functions are measured. They are defined between an excitation point and numerous mesh-like located points on structures or in rooms. Such a point mesh must be defined according to the geometry of the object under test, and it must be sufficiently dense according to the highest measurement frequency. As the upper limit of the distance between two mesh points, $\lambda/6$ is typically used. The transfer functions can be determined on measurements of sound pressure, acceleration or displacement. After the measurement, a computerised evaluation of the measurement data is performed in order to separate the system into specific characteristics in space and time, such as eigenfrequencies (modes) and damping of modes. These modes are mapped to a resonator model based on a set of parameters. Finding the optimal sufficient model and its parameters with best fit to the measured data is the basic task in the method of modal analysis (Fig. 2.24).

Finally, a correlation of the measured peaks with a spatial wave distribution on the mesh opens up the possibility of illustrating each mode at each eigenfrequency in graphical plots and in wave animation. With this method, vibration systems can be analysed and systematically improved.

2.7.3 Reciprocal Measurement of Sound Radiation

Measurement of transfer functions [Green's functions $G(v_n|p_1)$] of coupled vibroacoustic problems is essential in noise control. This kind of transfer function is defined as the ratio of the sound pressure measured at a certain field point and the excitation velocity of a vibrating body. However, in practical noise control, the objects of interest are not mathematically simple elements, so that calculation methods cannot be used at all, except numerical methods such as FEM, BEM, etc. For discussion of some more details of the theoretical foundations, it should be emphasised that two measurement methods can yield the same results if the reciprocity between sound radiation and sound reception is exploited [11] (see also Sect. 2.2.4.2) (Fig. 2.25).

$$\frac{p_1}{v_n \mathrm{d}S} = j\omega\rho_0 G(v_n|p_1) = \frac{p_2}{Q}. \tag{2.69}$$

A surface element dS of a radiating body is considered as point source on an otherwise rigid body. The transfer function between the normal component of the velocity on dS and the sound pressure at the field point is identical with the transfer function between the volume flow of a (volume) source placed at the field point and the sound pressure near the rigidly fixed surface element. Provided the problem was solved for a specific surface element, the complete solution consists in a superposition of all surface contributions. Needless to say, the transfer functions are complex and all modern digital measurement methods with FFT, sweeps or correlation techniques can be applied.

The fact that the measurement can be performed in two directions opens enormous advantages, particularly if sound velocities of vibration surface elements cannot be detected with optical sensors or accelerometers but rather easily by piezoceramic pressure sensors, for instance, in measurements of sound radiation from components or machines under extreme conditions of temperature, moisture, or with highly

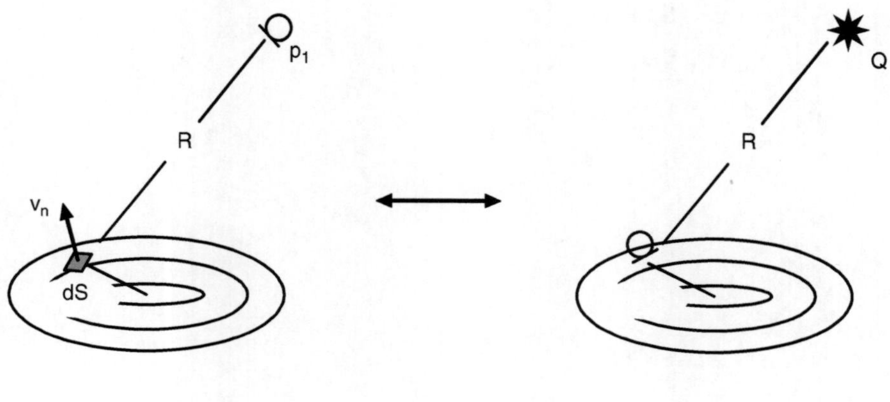

Fig. 2.25 Reciprocity principle: Equivalence of sound radiation (left) and sound reception (right). The transfer function for the distance R is called Green's function; it can be interpreted as acoustic transfer impedance

reactive gases. Another example is the sound radiation of tyres. Measurement by scanning the sound pressure near the tyre is much easier than measuring the normal velocity of the tyre surface.

References

1. De Bree H-E (2003) An overview of microflown technologies. Acta Acustica United Acustica 89:163
2. Bjor O-H (1982) Schnellemikrofon für Intensitätsmessungen. FASE/DAGA 82 – Fortschritte der Akustik, Göttingen, 629
3. Kuttruff H, Schmitz A (1994) Measurement of sound intensity by means of multi-microphone probes. Acustica 80:388
4. IEC 61043 (1993) Electroacoustics – Instruments for the measurement of sound intensity – Measurements with pairs of pressure sensing microphones
5. IEC 61260 (1995) Electroacoustics – Octave-band and fractional-octave-band filters
6. ISO 18233 (to be published) Acoustics – Application of new measurement methods in building acoustics
7. Müller S, Massarani P (2001) Transfer-function measurement with sweeps. J Audio Eng Soc 49:443
8. Aoshima N (1980) Computer-generated pulse signal applied for sound measurement. J Acoust Soc Am 69:179
9. Rife D, Vanderkooy J (1989) Transfer-function measurement with maximum-length sequences. J Audio Eng Soc 37:419
10. Mechel FP (1998) Schallabsorber, Band III. S. Hirzel Verlag, Stuttgart, Kap, 4
11. Fahy FJ (1995) The vibro-acoustic reciprocity principle and applications to noise control. Acustica 81:544

Numerical Acoustics

3

Björn A.T. Petersson

3.1 Introduction

Numerical acoustics – horrible concept! It is not a question of number crunching, it is an issue of constructively translating the knowledge in the branch of classical physics – acoustics – into quantitative result.

Classical acoustics – the days of Galiliei Galileo and further – dealt with quite idealised systems but despite this, managed to furnish remarkable findings through the analytical study of differential and integral equations. With the, seemingly, ever increasing speed and storage capacity of modern computers, it is almost self-evident and unavoidable that complex and fully realistic systems are addressed, also in an acoustic respect. From one aspect, this is quite natural since noise – the unwanted sound – is always there. From another, more constructive aspect it is less natural but despite this more important what can we do about or with sound.

The increased realism and perhaps detail of treated systems, however, should not give us false confidence in the results. The old truth of 'garbage in–garbage out' is even more prominent in the computerised age of numerical analysis of vibro-acoustics.

Accordingly, it is important to select the calculation tool or methodology carefully and not just fling oneself at the nearest or dearest or the 'in house' most commonly used scheme to churn out some results. The tool or methodology must be chosen such that it accommodates the problem without numerical strain.

B.A.T. Petersson (✉)
Institut für Strömungsmechanik und Technische Akustik, Technische Universität Berlin, Einsteinufer 25, 10587 Berlin, Germany
e-mail: b.a.t.petersson@tu-berlin.de

The tool or methodology must be chosen such that it aims at the quantities sought. If, for instance, the spatial average velocity of a multi-resonant room is sought, a finite element Odyssé is perhaps not the most suitable thing to embark on. Similarly, if, at a suspected fundamental resonance frequency, the stress at a junction of two beam members of a truss is required, an asymptotic or statistically based computational scheme is more than probable of missing the target. In the selection of the tool, therefore, the time spent on pre-preprocessing the problem is usually very well invested time in order to, at least, clarify what is the main concern – a quantitative result or a qualitative.

This chapter aims at an overview of some of the available numerical tools. 'Some of', refers to the fact that it is literally impossible, in a chapter, to comprise all the flora of numerical schemes that have been proposed and demonstrated viable in the literature; nor is it desirable. The chapter must therefore rather be received as a first point of reply. Within the flora of numerical tools four basic principles form the categorisation:

- Discretisation of differential equations
- Discretisation of boundary value problems and integral equations
- Statistical formulations, and
- Asymptotic formulations

Within each category, there is, of course, a myriad of versions and hybrides, but it is far beyond the scope to list, let alone, describe them. Instead it is herein attempted to give the fundamentals of each category and then to exemplify each with a successful application.

G. Müller and M. Möser (eds.), *Handbook of Engineering Acoustics*,
DOI 10.1007/978-3-540-69460-1_3, © Springer-Verlag Berlin Heidelberg 2013

3.2 Discretisation of Differential Equations

For linear acoustic and vibro-acoustics problem, i.e. for situations without extreme amplitudes or frequencies, the principle of superposition is valid. For most of the acoustic or vibro-acoustic problem encountered in engineering practice, accordingly, linear differential or integral equations are applicable. The step from a continuous functional description to a discretisation is not that big. By such a transcription, the result is a large number of linear algebraic equations. Thereby, the power of linear algebra can be released.

Although the original target for the finite element method was the static strength of elastic structures, it possesses all the features required to solve sound and vibration problems. The basis for the FE method (although formulation is more adequate) is the variation principle for energies, often encountered in the literature as Hamilton's principle.

In a first case, the analysis concerns a fluid volume, V, with rigid boundaries such that no energy is transported across. Following Hamilton's principle, the function to minimise reads

$$L = U - T = \frac{1}{2}\iiint\limits_V \rho c^2 (\text{div}\ \xi)^2 \text{d}V - \frac{1}{2}\iiint\limits_V \rho \left(\frac{\partial \xi}{\partial t}\right)^2 \text{d}V, \tag{3.1}$$

where U is the potential energy and T the kinetic. In Eq. (3.1), moreover, ρ is the density of the fluid, c, the wavespeed and ξ, the displacement vector. Equation (3.1) can be rewritten in terms of the pressure leading to,

$$L = \frac{1}{2}\iiint\limits_V \left(\frac{p^2}{\rho c^2} - \frac{(\text{grad}\ p)^2}{\rho \omega^2}\right) \text{d}V. \tag{3.2}$$

It should be noted that in Eqs. (3.1) and (3.2), there are no restrictions on c and ρ to be constant but they can vary spatially. For a single frequency, the volume, V, can be subdivided into element volumes, V_n, such that the function to minimise is turned into,

$$L = \sum_{n=1}^{N} \int_{V_n} \left[\frac{p_n^2}{2\rho c^2} - \frac{(\text{grad}\, p_n)^2}{2\rho\omega^2}\right] \text{d}V_n, \tag{3.3}$$

In this form, p_n designates the pressure in the element volume, V_n. Many recipes on how to divide the volume have been suggested, rectangular, tetrahedrical, wedge shaped and curved. In most cases the choice is dependent on the domain of the problem but personal taste also plays a role. The standard works on FEM by Zienkiewicz [Zienkiewicz, 1977] and Bathe and Wilson [Bathe & Wilson, 1976] as well as [Bathe, 1982] and [Huebner, 1975] constitute excellent sources of reference for the selection of element shapes. Also the handbooks to commercially available FE codes usually furnish this guidance. Once the element selection is made and the original volume is subdivided, the discretised space contains a great number of coupling points at which the pressure, in the present case, must be made continuous. This is done by the values at the *nodes*, see Fig. 3.1. With the *nodal* values denoted p_V, a relation between the continuous pressure function, $p_n(x,y,z)$, within the element, n, and

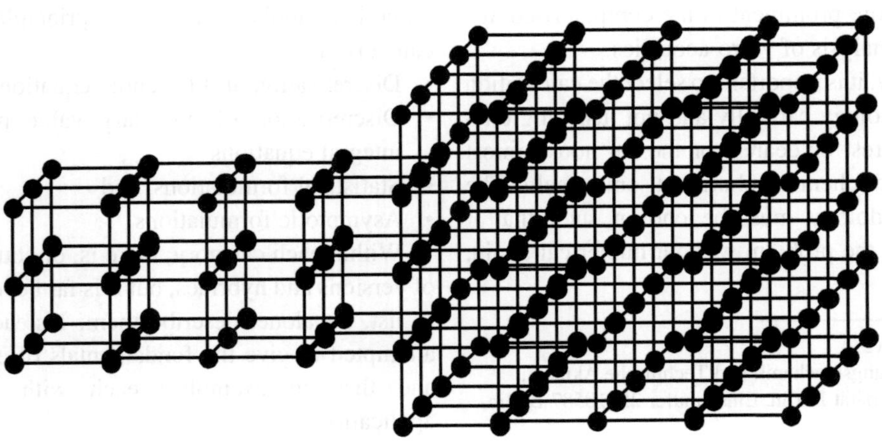

Fig. 3.1 FE mesh for an intersection of two ducts of differing cross-sections

those nodal values can be designed. This design is termed the *shape function* which is an interpolating relationship, say $a_{n\nu}(x,y,z)$, chosen in such a way that the pressure in an element can be described by,

$$p_n(x,y,z) = \sum_\nu a_{n\nu}(x,y,z)P_\nu. \tag{3.4}$$

Although the summation in Eq. (3.4) is extended over all nodal points ν, there are only a few terms that are non-zero since $a_{n\nu}(x,y,z)$ is zero outside the element n. The non-zero ones refer to the values inside or at the boundary of an element. It is thus clear that the shape function must be chosen such that the nodal values equal the physical pressures p at the points (x_ν,y_ν,z_ν).

Used in the standard codes are a variety of shape functions, ranging from the simplest, linear to the more elaborate sinusoidal. The shape function is also related to the element size in that a low-order element such as the linear can only approximate a small spatial region of the physical problem whilst a high-order one can encompass a larger region. Typically, six linear elements are required per wavelength and it must be assured that its size is significantly smaller than any other dimension of the problem such as, for instance, the source region. It is furthermore to be noted that elements must be chosen such that the conditions at discontinuities are correctly described. As a prominent example, one may take the flexural wave field at a discontinuity where in order to fulfil equilibrium of forces, the shape function has to be of at least third order ($F \propto \partial^3\xi/\partial x^3$). Whenever an 'exact' solution exists for a spatial domain, this domain is set off as a *substructure* with an element covering the entire domain. This is, for example, the case in a duct in the plane wave region and the FE procedure incorporates these extended substructures.

With Eq. (3.4) introduced in the discretised version of the function to minimise,

$$L = \frac{1}{2\rho}\sum_n\left\{\frac{1}{c^2}\int_{V_n}\left[\sum_\nu P_\nu a_{n\nu}(x,y,z)\right]^2 dV_n - \frac{1}{\omega^2}\int_{V_n}\left[\sum_\nu P_\nu \operatorname{grad} a_{n\nu}(x,y,z)\right]^2 dV_n\right\} \tag{3.5}$$

For transparency in the presentation of the method, the selection of element type and the inherent choice of shape functions is suppressed in the following.

With the function – *the Lagrangian* – expressed in terms of the finite number of nodal values, P_ν, the next step in the procedure is to minimise Eq. (3.5) which is the same as finding, $\partial L/\partial P_\nu = 0$ for each nodal point , ν. In this way, the following linear set of equations Lagrange's equations is established,

$$\sum_n\sum_{\nu'} P_{\nu'}\left\{\frac{1}{c^2}\int_{V_n} a_{n\nu}a_{n\nu'}dV_n - \frac{1}{\omega^2}\int_{V_n}\operatorname{grad} a_{n\nu}\operatorname{grad} a_{n\nu'}dV_n\right\} = 0,$$

where the nodal sound pressures, P_ν, form the unknowns. With a matrix formalism, the matrix of coefficients is sparse implying that even for thousands of unknowns, the solution is rather straightforward.

The above procedure can easily be extended to include also sound sources and sinks supplying and extracting energy to and from the field, respectively. The Lagrangian in Eq. (3.1) is then only augmented by an external work function,

$$W = -\int_S p\xi dS.$$

This added function simply describes the energy flowing in or out of the fluid domain through an area S over which the displacement ξ is known.

For most problems encountered in engineering practice, there are boundary conditions associated, i. e. physical, prescribed constraints at the boundary of the domain studied. For the fluid domain considered so far, either the values P_ν at some or all nodal points are prescribed or, which is equally common, differences of adjacent nodal values establishing derivatives such as, for example, the velocity. The specific boundary condition called the Sommerfeldt's radiation condition, which is suitable for large fluid domains, is inconveniently handled with standard finite elements. For such cases, the so-called infinite element has been proposed [Lysmer and Kuhlemeyer] cf. [Bettes].

To shed some light on the use of FE in conjunction with vibration, the axial vibration of a rod, see Fig. 3.2, can be considered. Hamilton's principle requires that,

$$\int_{t_1}^{t_2}\{\delta(U-T)+\delta W\}dt = 0, \tag{3.6}$$

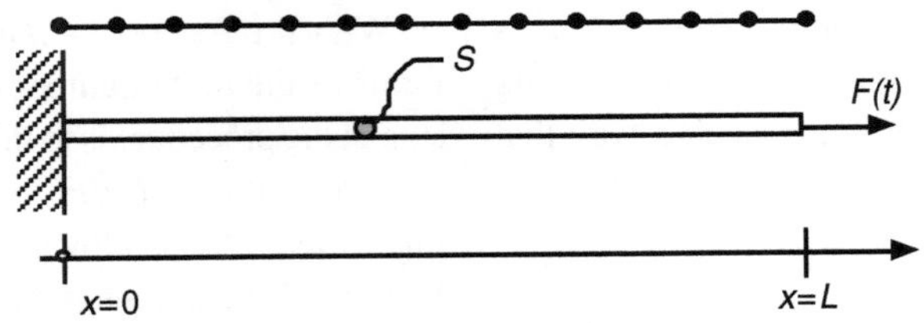

Fig. 3.2 Axially vibrating rod and its FE-mesh

where, again, U and T are the potential and kinetic energies, respectively, and W is the external work, e.g. supplied by the force. The symbol δ stands for the first variation which for most cases encountered in engineering practice leads to the above-mentioned minimisation of the integrand. With u, in this case, denoting the cross-sectional displacement, positive in the positive x-direction, the kinetic energy reads,

$$T = \frac{1}{2}\int_0^L \rho S\left(\frac{\partial u}{\mathrm{d}t}\right)^2 \mathrm{d}x,$$

and the potential energy can be written as,

$$U = \frac{1}{2}\int_0^L ES\left(\frac{\partial u}{\mathrm{d}x}\right)^2 \mathrm{d}x,$$

where E is the Young's modulus of the material. The virtual work supplied by the force, δW, is obtained as,

$$\delta W = F\delta u(x = L).$$

Upon subdividing the rod in a number of small elements – *the finite elements* – there will be two degrees of freedom for each one. Therefore, let the displacement of any point along the element be given by,

$$u(x) = u(0)w_1(x) + u(l)w_2(x). \tag{3.7}$$

Here the element length is denoted l and w_r ($r \in [1,2]$) are the shape functions. Upon assuming that the displacement over the element can be described by a linear relation,

$$w(x) = a_0 + a_1 x,$$

it is seen that the shape functions must fulfil $w_1(0)=w_2(l)=1$ and $w_1(l)=w_2(0)=0$. These two conditions readily leads to the shape functions

$$w_1(x) = 1 - x/l, w_2(x) = x/l,$$

respectively. This results in unity weighting of each degree of freedom at its nodal point. Thus, having the displacement along an arbitrary element, the kinetic energy of an element is given by,

$$T_e = \frac{1}{2}\rho S\int_0^l \left(\frac{\mathrm{d}u(x)}{\mathrm{d}t}\right)^2 \mathrm{d}x$$
$$= \frac{1}{2}\rho S\int_0^l \left(\frac{\mathrm{d}u(0)}{\mathrm{d}t}w_1(x) + \frac{\mathrm{d}u(l)}{\mathrm{d}t}w_2(x)\right)^2 \mathrm{d}x.$$

Similarly, the element's potential energy is obtained as,

$$U_e = \frac{1}{2}ES\int_0^l u'(x)^2 \mathrm{d}x$$
$$= \frac{1}{2}ES\int_0^l (u(0)w_1{}'(x) + u(l)w_2{}'(x))^2 \mathrm{d}x.$$

Although there are no derivatives involved than degrees of freedom in the example with an axially vibrating rod, it must be observed that the choice of shape function should accommodate continuity at each node in derivatives up to one order less than the highest order involved in the energy expressions.

With expressions for the energies of an element, the system energy or, directly, the Lagrangian is straightforwardly obtained by summation as shown above in Eq. (3.5). Upon substituting the element energies, expressed through a set of independent degrees of freedom say, q_i, and its time derivatives, $\dot{q}_i$, i.e., the nodal values into Eq. (3.6) the Lagrange's equations are obtained,

$$\frac{\mathrm{d}}{\mathrm{d}t}\left(\frac{\partial T}{\partial \dot{q}_i}\right) + \frac{\partial D}{\partial \dot{q}_i} + \frac{\partial U}{\partial q_i} = f_i. \tag{3.8}$$

Therein, D is a dissipative function which can account for any energy loss associated with internal non-conservative forces and f_i represents the generalised forces defined as the work done for a unit displacement q_i. For the q_ith degree of freedom, the instantaneous rate of energy loss is,

$$D_i = D\dot{q}_i^2.$$

To prepare for the computations, the equations involved in the FE analysis are usually matrix formulated. For the rod, the set of degrees of freedom for an element can be written as a column vector,

$$\{q\}_e^T = \{u(0) \quad u(l)\},$$

which means that Eq. (3.7) becomes,

$$u(x) = \{W\}^T\{q\}_e,$$

where $\{W\}^T = \{w_1\ w_2\}$ is the column vector for the nodal weighting factors. Thus, the element energies can now be rewritten as,

$$T_e = \frac{1}{2}\{\dot{q}\}_e^T[M]_e\{\dot{q}\}_e,$$

whereby the 'mass' matrix for the element is given by,

$$[M]_e = \rho S \int_0^l \{W\}^T\{W\}\mathrm{d}x,$$

and

$$U_e = \frac{1}{2}\{q\}_e^T[K]_e\{q\}_e,$$

in which the 'stiffness' matrix is defined by,

$$[K]_e = ES \int_0^l \frac{\partial^2}{\partial x^2}\{W\}^T \frac{\partial^2}{\partial x^2}\{W\}\mathrm{d}x.$$

With the often simple expressions for the shape functions, the integrations for the mass and stiffness matrices are easily evaluated. It should be noted that the terms mass and stiffness in conjunction with the matrices, are only partially correct since all elements will necessarily have physically appropriate dimensions.

Regarding boundary conditions, only those with derivatives of up to one order less than the highest order involved in the energy expressions must be explicitly satisfied. This is simply done by removing the weighting factors that do not fulfil them.

In the example with the rod, the boundary condition at the origin is automatically fulfilled. Also the work done by applied forces or force distributions must be taken into account. This is done by integrating any force distribution multiplied by the displacement over the elements, i.e.

$$f_e = \int_0^l F(x)w(x)\mathrm{d}x = \int_0^l F(x)\{W\}_e^T\{q\}_e\mathrm{d}x.$$

Any dissipation may be introduced in the same way but this also involves a selection of model for the dissipative mechanism, e.g. [Lazan] and [Nashif et al.].

Once all element matrices are developed, the system matrices are obtained by means of summation as described for the acoustic case above. In the matrix formulation, this means that a topology matrix is introduced, linking the degrees of freedom of the element to those of the system globally. Since each element for the rod only has two degrees of freedom, the topology matrix will be of the form,

$$[\psi]_e = \begin{bmatrix} \cdots & 0\,1 & 0\,0 & \cdots \\ \cdots & 0\,0 & 1\,0 & \cdots \end{bmatrix},$$

such that the element displacements are given by $\{q\}_e = [\psi]_e\{q_i\}$.

With this introduction, the kinetic energy of the system is obtained from

$$T = \frac{1}{2}\{\dot{q}_i\}^T[M]\{\dot{q}_i\},$$

where now, the 'mass' matrix is modified to include the topology

$$[M] = \sum_{e=1}^{N}[\psi]_e^T[M]_e[\psi]_e.$$

Similarly, the potential energy is expressed as,

$$U = \frac{1}{2}\{q_i\}^T[K]\{q_i\},$$

with the global 'stiffness' matrix given by

$$[K] = \sum_{e=1}^{N}[\psi]_e^T[K]_e[\psi]_e.$$

For both the mass and the stiffness matrix, the summation, of course, ranges over all the N elements.

If it is assumed that the dissipative mechanism can be modelled by a linear viscous model, application of the Lagarnge's equation results in $2N$ coupled equations of motion,

$$[M]\{\ddot{q}_i\} + [C]\{\dot{q}_i\} + [K]\{q_i\} = \{f_i\},$$

for which there are numerous computational routines [Bishop et al., Gourlay and Watson, Jennings].

3.3 Integral Equations

For linear radiation and scattering problems, the Kirchoff–Helmholtz equation establishes a general description of a sound field containing sources in a closed or unbounded space. Restricting the discussion to harmonic analyses, the equation for the sound pressure reads,

$$p(\mathbf{r}) = \frac{1}{4\pi}\int_V Q\frac{e^{jk|\mathbf{r}-\mathbf{r}_Q|}}{|\mathbf{r}-\mathbf{r}_Q|}\mathrm{d}V - \frac{j\omega\rho}{4\pi} \times \int_S \mathbf{v}_S\mathbf{n}\frac{e^{jk|\mathbf{r}-\mathbf{r}_{S_v}|}}{|\mathbf{r}-\mathbf{r}_{S_v}|}\mathrm{d}S + \frac{1}{4\pi} \times \int_S p_S\frac{\partial}{\partial n}\frac{e^{jk|\mathbf{r}-\mathbf{r}_{S_v}|}}{|\mathbf{r}-\mathbf{r}_{S_v}|}\mathrm{d}S, \qquad (3.9)$$

for a point $\mathbf{r}$ in the space. In this expression, Q represents the source strength of an enclosed source at position $\mathbf{r}_Q$ which must have the feature that they are acoustically transparent. $\mathbf{v}_S$ is the velocity at the surface S with normal vector $\mathbf{n}$ and p_S the pressure at that surface.

Usually, when the radiation from a surface is considered, its velocity field is known or computed separately. Unknown and to be determined are the pressures at the receiver position and at the surface. For a scattering problem, similarly, the source strength Q or the amplitude of the incoming wave as well as the boundary conditions at the scatterer are considered known and the pressures are the quantities sought. The boundary conditions at the scatterer are commonly given by the impedance of its surface, which must be locally reacting, i.e. the dynamics of a point (a small area) on the surface is not influenced by those of the adjacent ones. For non-locally reacting boundaries, the complexity of the problem grows markedly.

Irrespective of the type of problem, the numerical procedure for the solution of the integral equation implied involves a discretisation of the surface in elements, as done in an FE analysis. In contrast, however, the fluid space remains a single, continuous subsystem. Both interior and exterior problems are treated in the same way with a single closed surface although the conditions at the outer part of the surface are different, see Fig. 3.3.

Also when scattering is considered, the investigated domain is enclosed by a surface but in this case the surface must be branched out to encompass also the scatterer, see Fig. 3.4. From a discretisation of the bounding surface, the associated integrals in Eq. (3.9) are rewritten as,

$$p(\mathbf{r}) = \frac{1}{4\pi}\int_V Q\frac{e^{jk|\mathbf{r}-\mathbf{r}_Q|}}{|\mathbf{r}-\mathbf{r}_Q|}\mathrm{d}V - \frac{j\omega\rho}{4\pi} \times \sum_v \mathbf{v}_{S_v}\mathbf{n}\int_{S_v}\frac{e^{jk|\mathbf{r}-\mathbf{r}_{S_v}|}}{|\mathbf{r}-\mathbf{r}_{S_v}|}\mathrm{d}S_v + \frac{1}{4\pi} \times \sum_v p_{S_v}\int_{S_v}\frac{\partial}{\partial n}\frac{e^{jk|\mathbf{r}-\mathbf{r}_{S_v}|}}{|\mathbf{r}-\mathbf{r}_{S_v}|}\mathrm{d}S_v,$$

which means that the surface integrals are replaced by sums of integrals over surface elements – the boundary elements [Brebbia] – which are so small that the surface velocity as well as pressure can be approximated as constant and hence removed from the integration.

Upon computing the integrals, this is straightforward for observation points inside the continuous domain but when points on the surface are considered a singularity arises as $\mathbf{r} \to \mathbf{r}_{S_v}$. In such a case the principal value of the integrals must be taken which means that an arbitrarily small surface around this point is omitted. With the element integrals calculated and denoted $I^v_{\rho v}(|r_\rho - r_v|)$ and $I^p_{\rho v}(|r_\rho - r_v|)$, respectively, the linear set of equations to solve is,

$$p(\mathbf{r}_\rho) = \frac{1}{4\pi}\int_V Q\frac{e^{jk|\mathbf{r}_\rho-\mathbf{r}_Q|}}{|\mathbf{r}_\rho-\mathbf{r}_Q|}\mathrm{d}V - \frac{j\omega\rho}{4\pi} \times \sum_v v^{\perp}_{S_v} I^v_{\rho v}(|\mathbf{r}_\rho - \mathbf{r}_v|) + \frac{1}{4\pi} \times \sum_v p_{S_v} I^p_{\rho v}(|\mathbf{r}_\rho - \mathbf{r}_v|).$$

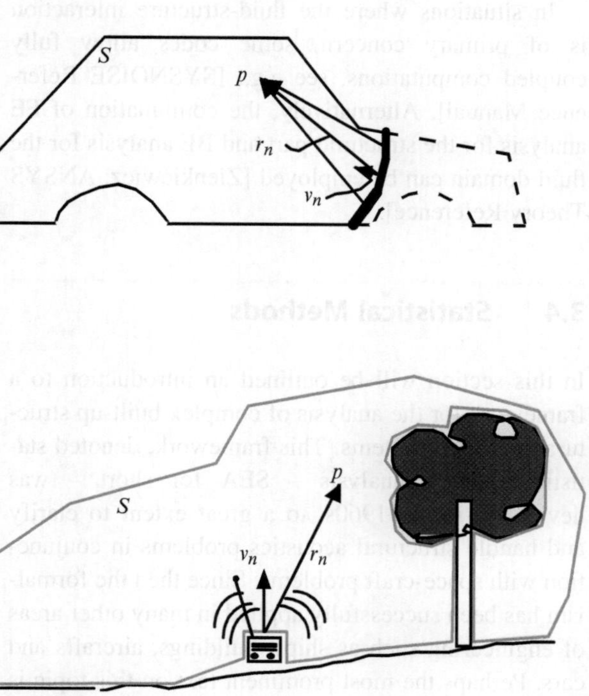

Fig. 3.3 (a) Interior sound radiation (b) exterior.sound radiation

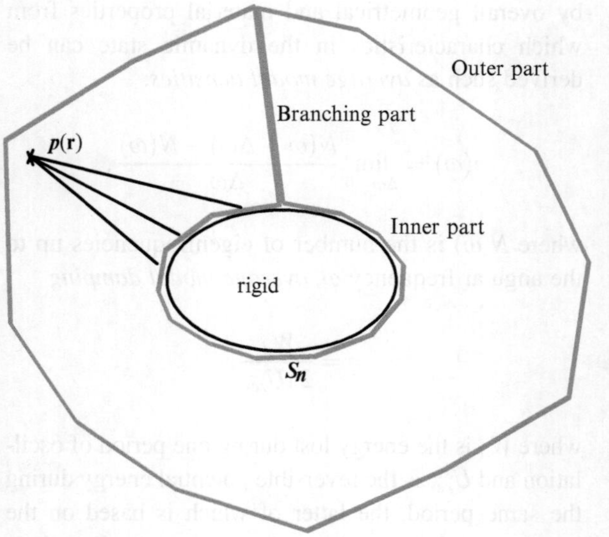

Fig. 3.4 Scattering

This expression can now be used twice, once to solve for the unknown pressures at the surface elements and then to compute the pressure at the observation point. In the first step, ρ equals v and the resulting pressures are subsequently substituted back into the sums in the second step.

For cases where the radiators or scatterers have dimensions bigger than the wavelength, it is necessary to modify the procedure. This is so since another singularity arises as the frequency approaches an eigenfrequency of the radiator or scatterer volume. Two primary modified procedures have been proposed and are employed in commercially available codes. The first [Schenk] takes into account a few additional points at which the pressure vanishes such as at points outside the computational domain but interior to the radiator or scatterer. Thereby an over determined system of equations is obtained which then is solved in the least square sense [Press et al.]. As long as the extra calculation points are not situated on nodal lines of the 'interior modes', the procedure furnishes unique results. Alternatively, the second approach [Burton and Miller] can be employed, which combines Eq. (3.9) and its normal derivative, giving the normal pressure gradient. For the latter is noted [Filippi], however, that the differentiated form gives rise to strong singularities associated with the Green's functions.

For most codes of the boundary element method (BEM), see, e.g. [SYSNOISE Reference Manual], the surface velocity distribution over an element need not be constant but can be prescribed through higher order functions as is the case in an FE analysis. Furthermore and in contrast to the FE analysis, the matrix established is small but is far from sparsely occupied.

In situations where the fluid-structure interaction is of primary concern, some codes allow fully coupled computations, see e.g., [SYSNOISE Reference Manual]. Alternatively, the combination of FE analysis for the structural part and BE analysis for the fluid domain can be employed [Zienkiewicz, ANSYS Theory Reference].

3.4 Statistical Methods

In this section will be outlined an introduction to a framework for the analysis of complex built-up structural and fluid systems. This framework, denoted statistical energy analysis – SEA for short – was developed in the 1960s, to a great extent to clarify and handle structural acoustics problems in conjunction with space-craft problems. Since then the formalism has been successfully applied in many other areas of engineering such as ships, buildings, aircrafts and cars. Perhaps the most prominent text on this topic is that by Lyon [Lyon 1975] which gives a valuable description of the framework by explaining each word in its name. The explanations are cited here:

> "Statistical emphasises that the systems being studied are presumed to be drawn from statistical populations having known distributions of their dynamical parameters. Energy denotes the primary variable of interest. Other dynamic variables such as displacement, pressure, etc are found from energy of vibrations. The term Analysis is used to emphasise that SEA is a framework of study rather than a particular technique."

Hence, the statistical philosophy introduced in SEA must be seen as a compilation of methods or ways to study structural acoustic questions where basic knowledge must not be set aside for a routine scheme.

Apart from the textbook mentioned above or it's later edition [Lyon and DeJong], a few other references can be given which either exemplify the use of SEA in engineering practice [Hsu et al.] or bring out the limitations for the framework [Fahy]. A rather comprehensive and advanced discussion of SEA is given in [Plunt] where the objects under study are ships.

The natural area of operation for SEA is the analysis of the high frequency behaviour of complex structures. Usually the analysis involves coupling between two or more simply identifiable but complex structural and/or fluid systems. In this context, high frequencies refer to the frequency region where the system under consideration possesses a large number of eigenfrequencies in a limited frequency band.

It is of course justified to ask why introduce a statistical point of view for the analysis of the dynamics of systems that are normally considered to be rather deterministic. The answer lies in the many difficulties relating to the application of numerical or semi-analytical methods for complex systems having a large number of eigenfrequencies in the frequency range of interest. Among those we may list:

- Uncertainties in the input data (e.g. boundary conditions)
- Unknown excitation (e.g. excitation points are not well defined or specified)
- Variations in the fabrication (e.g. two cars of the same type will not be identical)

Since the designer often wants to make predictions early at the design stage when details of the structural configuration are missing or incomplete, a simple 'macroscopic' analysis is motivated. One of the advantages with SEA is namely the representation of the structural and fluid systems. These are represented by overall geometrical and material properties from which characteristics in the dynamic state can be derived such as *average modal densities*:

$$n(\omega) = \lim_{\Delta\omega \to 0} \frac{N(\omega + \Delta\omega) - N(\omega)}{\Delta\omega},$$

where $N(\omega)$ is the number of eigenfrequencies up to the angular frequency ω, *average modal damping*

$$\eta = \frac{W_d}{2\pi U_{rev}},$$

where W_d is the energy lost during one period of oscillation and U_{rev} is the reversible potential energy during the same period, the latter of which is based on the *average modal energies* and *average coupling data.* The coupling between structural and/or fluid members of the systems – the subsystems – leads to energy flows between them in order to maintain an energy balance in the presence of dissipative losses. In turn, the dynamic field variables, of interest to the engineer, are obtained as spatial and temporal averages which are directly related to the total energy of the subsystem.

The most obvious disadvantage with SEA is that the energy quantities obtained for the different subsystems are statistical estimates of the true energy and accordingly involve some uncertainty. Usually the uncertainty will be less pronounced at high frequencies where the number of resonant eigenfrequencies included is high for all subsystems. This means that there will be a low frequency limit in practice but no principle limitation exists as long as the subsystems have resonant eigenfrequencies.

The subsystems in SEA are presumed to be finite, linear elastic structures or fluid cavities which can be described by their uncoupled eigenfrequencies, eigenfunctions (mode shapes) and losses. It is assumed that each eigenfunction can be modelled by a simple oscillator – a mass-spring system – and that the interaction between two multi-modal subsystems can be represented by the coupling between two sets of oscillators, see Fig. 3.5.

No attempt will be made in this section to fully compile all the theoretical considerations underlying SEA, but the most necessary fundamentals required for the understanding of the procedure will be given.

In SEA, the complex system under study may be subdivided into a number of subsystems. Some can be directly excited by sources whereas others are indirectly excited through different couplings to the former ones (junctions, discontinuities, transfer from one medium to another). Accordingly, a basic question in SEA is how to subdivide the system into subsystems. This division can be intuitively made at the boundaries of the different structural and fluid members. However, it is very important to bear in mind that for complex systems, there will be different wave types present which realise different groups of modes. This implies that in addition to the previously mentioned structural or fluid subsystems a subdivision must be undertaken also with respect to these groups of modes. Hence, of a physical structural element two or more subsystems may have to be formed where one involves only flexural motion, another only different in-plane longitudinal waves and a third only in plane transverse waves.

The interaction between two simple oscillators which are assumed to be linear and coupled by linear, non-dissipative elements(no energy lost at the connecting boundaries) constitute the fundamental component of an SEA model, from which some important theorems in SEA can be deduced. To summarise it is appropriate to start from the basic equation,

$$\Pi'_{21} = B(E_1 - E_2). \tag{3.10}$$

Consider the coupled system in Fig. 3.5 where, if the energies in the two subsystems are E_1 and E_2, the net energy flow between them is found from Eq. (3.10) above. In this equation, B is a function of the coupling strength. This coupling is solely governed by the properties of the subsystems and the eventual coupling elements.

From this equation it can be stated:

- The energy flow is proportional to the actual energies of the two subsystems
- The coupling function or proportionality is positive definite and symmetric in the system parameters; therefore the system is reciprocal and the energy flows from the subsystem with the higher energy to the one with the lower
- If only one subsystem is directly excited, the highest possible energy for the indirectly driven susbsystem is that of the first

As can be noted, the analogy with thermodynamics is not far away. Thus the picture with an energy flow between a hot body and a cold one in contact may be helpful although it should not be taken too literally.

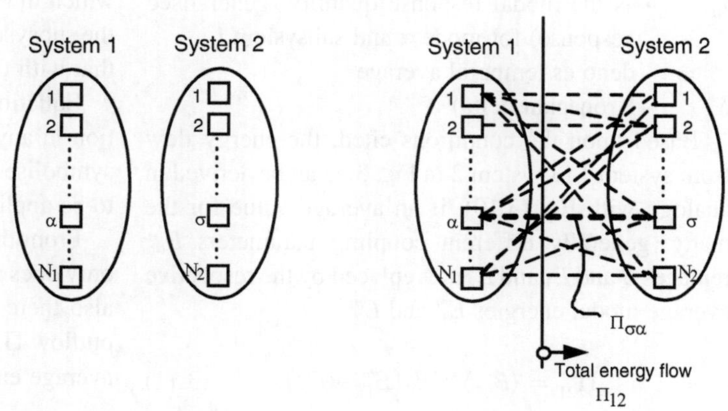

Fig. 3.5 Illustration of interaction and coupling between multi-modal systems. After [Lyon]

Consider again the model in Fig. 3.5. The basic assumptions when SEA is used in conjunction with such a coupling problem can be cited from [Lyon]:

"1. Each mode is assumed to have a natural frequency $\omega_{i\alpha}$ that is uniformly probable over the frequency interval $\Delta\omega$. This means that each subsystem is a member of a population of systems that are generally physically similar, but different enough to have randomly distributed parameters. The assumption is based on the fact that nominally identical structures or acoustical spaces will have uncertainties in modal parameter, particularly at higher frequencies.

2. We assume that every mode in a subsystem is equally energetic and that its amplitudes $Y_{i\alpha}(t) = \int \rho_i y_i \psi_{i\alpha} dx_i / M_i$ are incoherent, that is,

$$\langle Y_{i\alpha} Y_{i\beta} \rangle_t = \delta_{\alpha\beta} \langle Y_{i\alpha}{}^2 \rangle.$$

This assumption requires that we select mode groups for which this should be approximately correct, at least, and is an important guide to proper SEA modelling. It also implies that the excitation functions are drawn from random populations of functions that have certain similarities (such as equal frequency and wave number spectra) but are individually incoherent."

(In the expressions cited the following notation is used:

ρ_i – is the density distribution of subsystem i

y_i – is the physical response quantity for subsystem i

$\psi_{i\alpha}$ – is the eigenfunction associated with the αth eigenfrequency of subsystem i

x_i – is the local spatial co-ordinate for subsystem i

M_i – is the total mass of subsystem i

$Y_{i\alpha}$ – is the modal response quantity (generalised response) for mode α and subsystem i

$< >_t$ – denotes temporal average

δ – Kroneckar delta)

Based upon the conditions cited, the energy flow from system 1 to system 2 in Fig. 3.5 can be derived in analogy with Eq. (3.10), if an average value for the many, generally different coupling parameters $B_{\sigma\alpha}$ replaces B and E_1 and E_2 are replaced by the respective average modal energies E_1^m and E_2^m

$$\Pi'_{21} = \langle B_{\sigma\alpha} \rangle N_1 N_2 \left(E_1^m - E_2^m \right). \tag{3.11}$$

In the above equation, $< B_{\sigma\alpha} >$ denotes the average modal coupling and N_1, N_2 the number of modes (eigenfunctions) in the two subsystems, respectively.

The average modal energies are defined by

$$E_i^m = \frac{E_i^{tot}}{n(\omega)\Delta\omega}, \tag{3.12}$$

where $n(\omega)$ is the average modal density introduced above and E_i^{tot} is the total vibrational energy of subsystem i.

Since the energy flow between the subsystems, from one specific subsystem seen, can be interpreted as a loss of energy, the formalism can be developed by defining a coupling loss factor as

$$\eta_{21} = \langle B_{\sigma\alpha} \rangle N_2 / \omega, \tag{3.13a}$$

which says that the loss from system 1 is proportional to the number of modes in system 2, i.e. how many energy reservoirs there are available in system 2. Since the complete system is reciprocal, we can similarly define

$$\eta_{12} = \langle B_{\sigma\alpha} \rangle N_1 / \omega, \tag{3.13b}$$

and from the two relations in Eq. (13) the reciprocity relation

$$\eta_{12} N_2 = \eta_{21} N_1, \tag{3.14}$$

can be deduced.

Introducing Eq. (3.13a) in Eq. (3.11) yields

$$\Pi'_{21} = \omega \eta_{21} N_1 \left(E_1^m - E_2^m \right), \tag{3.15}$$

which in turn shows that the energy flow will be from the subsystem with the higher average modal energy to that with the lower.

Equation (3.15) constitutes the fundamental relation in any SEA calculation. The diagram in Fig. 3.6 symbolises the general situation in which Eq. (3.15) is to be applied.

From the diagram in Fig. 3.6, one may see that not only does a flow between the two subsystems exist but also there may be both independent inflow $\Pi_{i,in}$ and outflow $\Pi_{i,diss}$, to and from each subsystem. Hence the average energy quantities sought for the prediction of

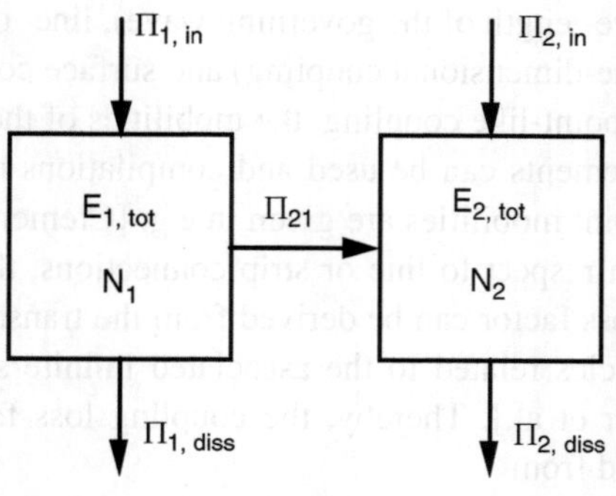

Fig. 3.6 Illustration of energy flows to, from and in between two coupled subsystems

the average response of a subsystem must be solved from a system of linear equations.

A number of questions arise concerning the subdivision of a physical system into different, coupled subsystems. A basic feature of a subsystem must be that its response is determined by resonant modes. It is consequently not meaningful nor appropriate to apply an SEA approach to a system (or subsystem) below its fundamental eigenfrequency. However, coupling via non-resonant elements may be admitted.

The sources of input power and the groups of eigenfunctions to which they are coupled must be identified. Moreover, the different groups of eigenfunctions with similar properties must be selected and represented by separate subsystems. As mentioned previously, these considerations usually mean that one physical part of the system must be represented by several SEA subsystems.

The junctions between the subsystems must be assigned correct coupling loss factors. Very often the junctions are chosen to coincide with the geometrical boundaries of the physical parts but this is not always necessary or approriate. A comprehensive discussion of the considerations involved in SEA modelling can be found in [Plunt] where also some valuable hints on how to practically proceed from a physical object to an SEA model are outlined.

The average modal density introduced previously will usually be estimated with sufficient accuracy from closed form expressions for elementary fluid and structural elements such as beams, plates, cylinders, cavities etc. Compilations of such expressions are found in many text books, e.g. [Lyon] and [Cremer et al.]. It must be emphasised, however, that the closed form expressions normally are equivalent to the asymptote found at high frequencies. Therefore, the actual modal density, in the low and mid-frequency regions, may be different from the one predicted. At low frequencies, where the number of modes in the frequency interval (band) of interest is small, the estimates can be improved by measurements on the system if a sample of it exists and methods for such investigations are sketched in [Lyon and DeJong].

If, in Eq. (3.14), the average modal density is introduced, the reciprocity relation becomes

$$\eta_{21} n_1 = \eta_{12} n_2,$$

and the basic energy flow can be rewritten as

$$\Pi'_{21} = \omega \eta_{21} \Delta\omega \left(n_1 E_1^m - n_2 \frac{\eta_{12}}{\eta_{21}} E_2^m \right).$$

In addition to the energy flow due to the coupling between subsystems also the energy flow out from the system under consideration – *the dissipation* – must be taken into account. This energy loss can be due to e.g. internal losses (conversion into heat) or to radiation or coupling to connected systems not included in the SEA model. The dissipation is represented by an internal loss factor, η_{ii}, for each subsystem i.

Unfortunately, reliable prediction methods for the dissipation are not available today and the internal loss factors, therefore, either must be measured or rely upon experience. Theories exist for different kinds of mechanisms leading to losses but so far they are only applicable under rather specific and restrictive conditions. Moreover, experience may give the right order of magnitude in an overall sense but frequency trends are hard to foresee. This means that measurements rather often is what remains and a few experimental techniques are described in [C&H] and [Plunt 1991] [Cremer et al.].

The description of *the coupling* is as shown above formalised by using coupling loss factors. Often these loss factors can be estimated theoretically whereby the duality of junction properties for infinite subsystems and statistical ensembles of finite subsystems is utilised [Lyon].

In principle three different types of junctions can be distinguished:

- Fluid cavity to fluid cavity (e.g. coupling through a partition wall)
- Structural element to fluid cavity (e.g. panel to cavity)
- Structural element to structural element (e.g. wall to floor).

For the first-type of junction, using the transmission efficiency τ, the coupling loss factor can be derived to be

$$\eta_{21} = \frac{\tau c_1 S}{4\omega V_1}, \tag{3.16}$$

where c_1 is the longitudinal wave speed of the fluid, S is the area of the partition and V_1 the volume of the sending space.

The coupling loss factor for the second type of junction is related to the real part of the impedance Z which the structure sees looking into the fluid space. This leads to the ratio

$$\eta_{fs} = \frac{\mathrm{Re}[Z]}{\omega M_s}, \tag{3.17}$$

which hence implies a comparison of the input fluid impedance and the mass impedance of the structure.

For the specific case of a surface coupling, Eq. (3.17) can be explicitly written as

$$\eta_{fs} = \frac{2\rho_f c_f \sigma}{\omega m}, \tag{3.18}$$

where the radiation efficiency, σ, of the structure has been introduced. ρ_f, c_f denotes the density and wave speed of the fluid and m is the structural mass per unit area.

Theoretical expressions for the input impedance (often designated radiation impedance) and the radiation efficiency have been derived for a variety of structures in contact with air or water and are readily found in the acoustic literature [Cremer et al., Junger and Feit, Heckl].

The coupling loss factor for the third type of coupling can be derived in a number of ways. The most appropriate manner depends on the specific junction but often use is made of the infinite systems, associated with the actual structural elements. The most frequent classes of structural coupling are point-like (contact areas of dimensions smaller than the wave-length of the governing wave), line- or strip-like (one-dimensional coupling) and surface coupling.

For point-like coupling, the mobilities of the structural elements can be used and compilations of ordinary point mobilities are given in e.g. [Cremer et al.].

With respect to line or strip connections, the coupling loss factor can be derived from the transmission efficiencies related to the associated infinite systems [Cremer et al.]. Thereby, the coupling loss factor is obtained from

$$\eta_{21} = \frac{c_{g1} L}{2\omega S_1} \tau_{21}, \tag{3.19}$$

where τ_{21} denotes the transmission efficiency from element 1 to element 2, c_{g1} is the group velocity for waves in element 1 (the group velocity introduced here is the velocity with which the power is transferred which generally is different from the phase speed or ordinary wave speed, for dispersive waves), S_1 the area of subsystem 1 and L the length of the line interface. Moreover, the reciprocity relation obtained for normal incidence, $\tau_{12} = \tau_{21}$, is not valid in the case of 'random' incidence but is replaced by

$$k_1 \tau_{21} = k_2 \tau_{12},$$

where k_1 and k_2 are the wave numbers for the two structural elements respectively. The coupling of, for instance, plates over surfaces generally is more cumbersome to handle. Theoretical studies of waves in layered media describe the features of this class of coupling but it is beyond the scope here to further expand on the pertinent results. For most of the commercially available codes the different junctions are catalogued and the associated coupling loss factors compiled.

The origin of wave propagation or, in the SEA philosophy, energy flow is of course input from one or more external sources. From an SEA point of view one may therefore consider the input power as prescribed although all questions relating to transmission are far from answered.

The basic energy balance Eq. (3.15) is valid for two coupled, multi-modal systems when the underlying assumptions are fulfilled. The normal situation however, is that a lot more than two subsystems are involved. The block diagram in Fig. 3.7 illustrates a general situation. As is seen from the figure, the general case includes many external source inputs as well

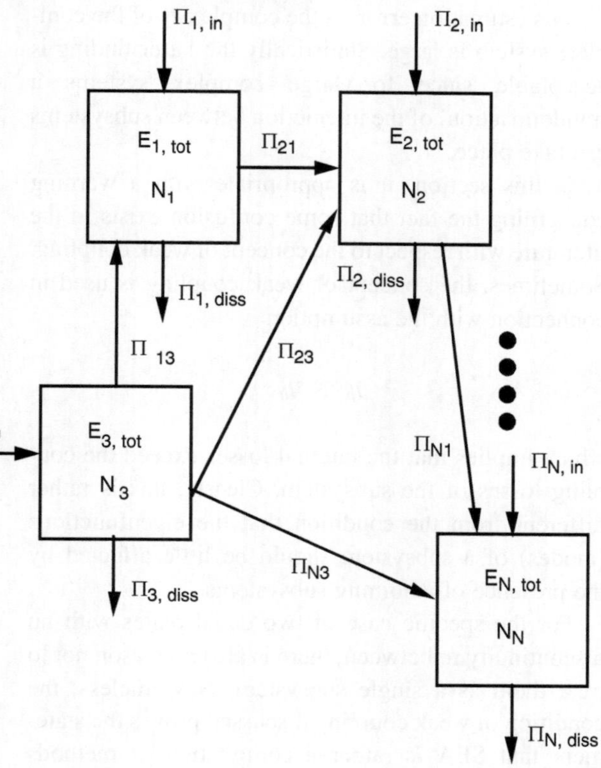

Fig. 3.7 Block diagram of an SEA model with several subsystems. After [Lyon]

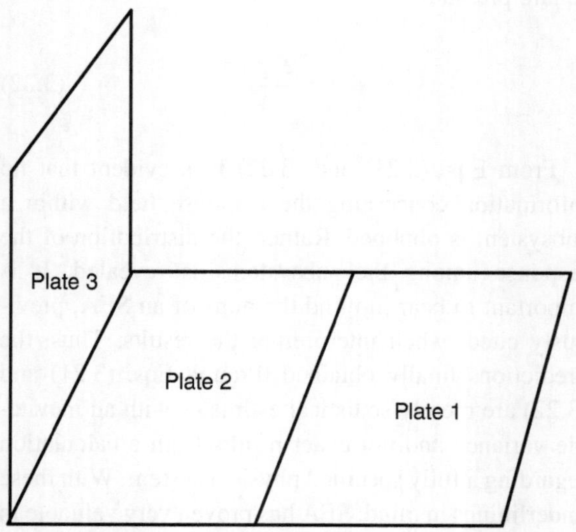

Fig. 3.8 A three plate-element system. After [Plunt]

as dissipation from all subsystems. Intuitively, it seems to be a straightforward task to generalise the two-subsystem case to N subsystems and obtain a system of N linear equations and indeed there are no formal difficulties. There is, however, a physical implication that must be taken into account before doing so. This implication is rather clearly demonstrated by considering the system in Fig. 3.8. In the two-subsystem case previously discussed, e.g. the two horizontally connected plates in Fig. 3.8 with the vertical plate excluded, there is no ambiguity as to what is meant by the coupling loss factor. With respect to the three-plate system the question arises: Should the rest of the system be connected or disconnected when one specific junction is considered? Obviously, it should make a difference when we are seeking the coupling loss factor at the L-junction if the horizontal plate consists of one or two elements (compare Eq. (3.19) with S_1 being either the area of one- or two-plate elements). We may, therefore, conclude that in order to proceed straightforwardly from the two-subsystem case to that with N subsystems, the coupling losses at all junctions (connections between subsystems) must be small. This is equivalent to the prescription of weak coupling between the different subsystems. Consequently, the existence of adjoining subsystems must not markedly affect the properties of the eigenfunctions of a subsystem (in a statistical sense).

At a first sight, the above-mentioned condition may seem rather drastic but comparisons made between SEA predictions and experimental results show that the condition of weak coupling does not impose serious estimation errors if the complexity of the complete system is large. Statistically the latter finding is acceptable since for large complex systems a 'randomisation' of the interaction between subsystems can take place.

In this section, it is appropriate with a warning concerning the fact that some confusion exists in the literature with respect to the concept of weak coupling. Sometimes, the concept of weak coupling is used in connection with the assumption

$$\eta_{ji} \ll \eta_{ii},$$

which implies that the internal losses exceed the coupling losses of the subsystem. Clearly, this is rather different from the condition that the eigenfunctions (modes) of a subsystem should be little affected by the presence of adjoining subsystems.

For the specific case of two equal plates with no discontinuity in between, there is also no reason not to treat them as a single subsystem. Nevertheless, the condition of weak coupling discussed proves the statement that SEA is rather a compilation of methods where basic knowledge in structural acoustics is required for judging the applicability of the different assumptions made in an analysis.

On the basis of the weak coupling condition being sufficiently well fulfilled, the generalisation of the energy balance equation becomes

$$\begin{Bmatrix} \Pi_1^{in} \\ \Pi_2^{in} \\ \vdots \\ \Pi_N^{in} \end{Bmatrix} = \omega \begin{bmatrix} n_1\eta_1^{\text{tot}} & -n_2\eta_{12} & \cdots & -n_N\eta_{1N} \\ -n_1\eta_{21} & n_2\eta_2^{\text{tot}} & & \vdots \\ \vdots & & \ddots & \vdots \\ -n_1\eta_{N1} & \cdots & \cdots & -n_N\eta_{NN}^{\text{tot}} \end{bmatrix} \times \begin{Bmatrix} E_1^m \\ E_2^m \\ \vdots \\ E_N^m \end{Bmatrix}, \tag{3.20}$$

where η_i^{tot} includes all the types of losses from a subsystem, i.e.

$$n_i^{\text{tot}} = \eta_{ii} + \sum_j^N \eta_{ji},$$

is the outflow from the ith subsystem.

The loss factor matrix is symmetric and positive definite $(\{x\}^t[A]\{x\} > 0;\quad \text{every}\quad \{x\})$ which means that the solution of Eq. (3.20) can be established by means of standard computer routines. If, in addition, some systematics is introduced in the numbering of the subsystems the loss factor matrix can be made banded (the elements are clustered around the main diagonal) which greatly facilitate the numerical handling.

From the solution of Eq. (3.20), a number of average modal energies are obtained. By multiplying with the actual number of modes in the frequency band of interest the total energy of the subsystem within that band is estimated.

The spatial average, mean square response of a structural element is proportional to the kinetic energy. Therefore, with the notations of this chapter, the response estimates sought can be written as

$$\langle \tilde{v}_i^2 \rangle = E_i / M_i \tag{3.21}$$

where M_i is the total mass of the ith subsystem. Similarly for a fluid space or subsystem the response estimates are given by the spatial average, mean square pressure as

$$\langle \tilde{p}_i^2 \rangle = \frac{E_i \rho_i c_i^2}{V_i}. \tag{3.22}$$

From Eqs. (3.21) and (3.22) it is evident that no information concerning the response field within a subsystem is obtained. Rather, the distribution of the response among the subsystems is revealed. It is important to bear in mind the aims of an SEA, previously cited, when interpreting the results. Thus, the predictions finally obtained through Eqs. (3.21) and (3.22) are merely statistical estimates with an inevitable variance and not exact results from a calculation regarding a fully specified physical system. With these underlinings in mind, SEA has proven very valuable in most branches of noise and vibration control.

3.5 Asymptotic Methods

Instead of considering a system as drawn from a population of similar systems and applying a statistical reasoning, the asymptotic methods such as asymptotic modal analysis (AMA) [Dowell] and the mean value method (MVM) [Skudrzyk] approaches the acoustic or dynamic behaviour of a given system from the multi-resonant range. Hereby is addressed the Helmholtz number range in which determinstic methods such as FEM struggle.

In asymptotic modal analysis forms the modal decomposition of the dynamic behaviour of a subsystem the basis for the analysis. The external excitation, moreover, is assumed to be multiple random point forces or point sound sources, be it a directly or an indirectly driven subsystem, since, for a large number of modes, the excitation applied to the subsystem tends to be spatially incoherent. A second ingredient in the AMA is the approximation of the modal sums by integrals. In principle, therefore, the AMA delivers the same spatially averaged response as an SEA So far developed, AMA is not complete as method due to the lack of methodology to couple the acoustics or dynamics of different subsystems and thus cannot provide any insight into the behaviour of a complete system.

Also in the MVM, the basis is the modal decomposition of the dynamic behaviour and the summation implied is replaced by an integral. From this integral, the mean value of the response is computed which is equal to the response of the infinite system that corresponds to the finite subsystem considered. The infinite system response is demonstrated equal to the geometric mean of adjacent resonance maxima and anti-resonance minima, hence not only just a high frequency asymptote but also the value the response at low and intermediate frequencies tends to when the dissipation is large. The MVM further involves the upper and lower envelopes to the response thus establishing bandwidth for the response fluctuations, see also [Langley]. Like AMA, the MVM is not complete as it inherently devices no means to couple subsystems but gives only the asymptotic behaviour of a subsystem. Attempts have, however, been made to extend it [Girard and Defosse] where the potential is highlighted but no general methodology is presented.

4 The Effects of Sound on Humans

C. Maschke and U. Widmann

4.1 Physiological Aspects

4.1.1 The Ear

Anatomically, the ear is divided into the outer ear, middle ear and inner ear (Fig. 4.1). The outer ear comprises the auricle and the ear canal. A thin membrane called the eardrum separates the outer ear from the middle ear.

The middle ear is an air-filled chamber containing three tiny bones called the ossicles (hammer, anvil and stirrup), which conduct sound. The air pressure in the middle ear must be adapted to changes in the outside air pressure. This is the purpose of the Eustachian tubes, which connect the middle ear to the nasal cavity. Swallowing and yawning are two ways in which the pressure is equalised. The handle of the hammer is attached to the eardrum and conducts sound vibrations of the eardrum to the anvil and the stirrup, which is attached to the oval window of the inner ear. The ossicles are attached to two muscles that tighten the eardrum and the stirrup. They can reduce the conduction of sound by means of the acoustic reflex.

Movements of the oval window are transferred to the lymph fluid in the inner ear (cochlea). The inner ear is a coil of two and a half turns, divided into three chambers called scalae (Fig. 4.2). The scala vestibuli, the scala tympani and the scala media are separated by the basilar membrane and Reissner's membrane, respectively. The scala media is filled with endolymph. The two other scalae, which meet at the top of the cochlea (helicotrema), contain perilymph. The perilymph mostly contains sodium ions, while the endolymph contains potassium ions. These different concentrations of ions acts as a battery for the bioelectrical processes that take place in the sensory cells.

Located on the basilar membrane is the organ of Corti, in which sensory cells called hair cells are embedded. They consist of the cell body and sensory hairs called stereocilia. When movements of the lymph fluid deflect the basilar membrane, the stereocilia move sideways, sending chemical messengers known as neurotransmitters into the synaptic cleft. If the concentration of neurotransmitters is sufficient, bioelectrical impulses called spikes are fired in the adjacent nerve cells. The number of spikes fired per unit of time in the auditory pathway determines the perceived loudness (e.g. [1]).

Depending on the frequency of the sounds and on the damping properties of membrane, the wave movements in the inner ear peak at different points on the basilar membrane. Travelling waves triggered by high frequencies peak near the oval window. Lower frequencies peak near the end of the coil. This transformation of frequency into location enables us to detect differences in pitch [2].

However, the physical properties of the basilar membrane alone do not explain the auditory system's high resolution of frequencies. Current research indicates that active processes take place in the cochlea. Experiments by Brownell [3] showed that

C. Maschke
Landesamt für Umwelt, Gesundheit und Verbraucherschutz, Seeburger Chaussee 2, 14476 Potsdam, Germany
e-mail: christian.maschke@lugv.brandenburg.de

U. Widmann (✉)
AUDI AG, 85045 Ingolstadt, Germany
e-mail: ulrich.widman@audi.de

G. Müller and M. Möser (eds.), *Handbook of Engineering Acoustics*,
DOI 10.1007/978-3-540-69460-1_4, © Springer-Verlag Berlin Heidelberg 2013

Table 4.1 Stimuli and sensations

Dominant stimuli	Cognitive parameters
Sound pressure level (dB)	Loudness (sone)
	Loudness level (phon)
Frequency (Hz)	Critical band rate (Bark)
	Ratio pitch (mel)
Degree of modulation (%)	Roughness (asper)
Modulation frequency (Hz)	
Frequency (Hz)	Sharpness (acum)
Degree of modulation (%)	Fluctuation strength (vacil)
Modulation frequency (Hz)	
Spectral components (Pa)	Pitch strength
	Tonality (tu)
Impulse duration (s)	Subjective duration of impetus (IU)
Sound pressure level (dB) Frequency (Hz)	Density (dasy)

The *left column* shows dominant physical parameters (stimuli), in the *right column* the psycho-acoustic hearing sensation are listed, (proposed) units in *brackets*

the outer hair cells are able to contract in the kilohertz range and thus amplify the deflection of the basilar membrane in a small area.

4.1.2 The Auditory Pathway

The auditory pathway comprises nerve fibres that transmit neural impulses from the inner ear to the auditory cortex (afferent auditory pathway) and nerve fibres that transmit impulses from higher processing levels to lower ones and back to the inner ear (efferent auditory pathway).

The afferent auditory pathway is especially important for the extraaural affect of sound. Nerve fibres branch from the afferent auditory pathway to various processing levels, thus providing direct transitions to other functional systems. This is the direct way of sound activation.

The afferent pathway is shown in very simplified form along with its processing levels and the transitions to other functional systems in Fig. 4.3.

The first stage of processing after the inner ear (1) is the cochlear nucleus (2). The auditory pathway divides here and leads to various areas of the brain.

One cord leads to the lateral superior olive (3). The main cord leads to the olive on the opposite side to the stimulated ear (contralateral side). A third cord leaves

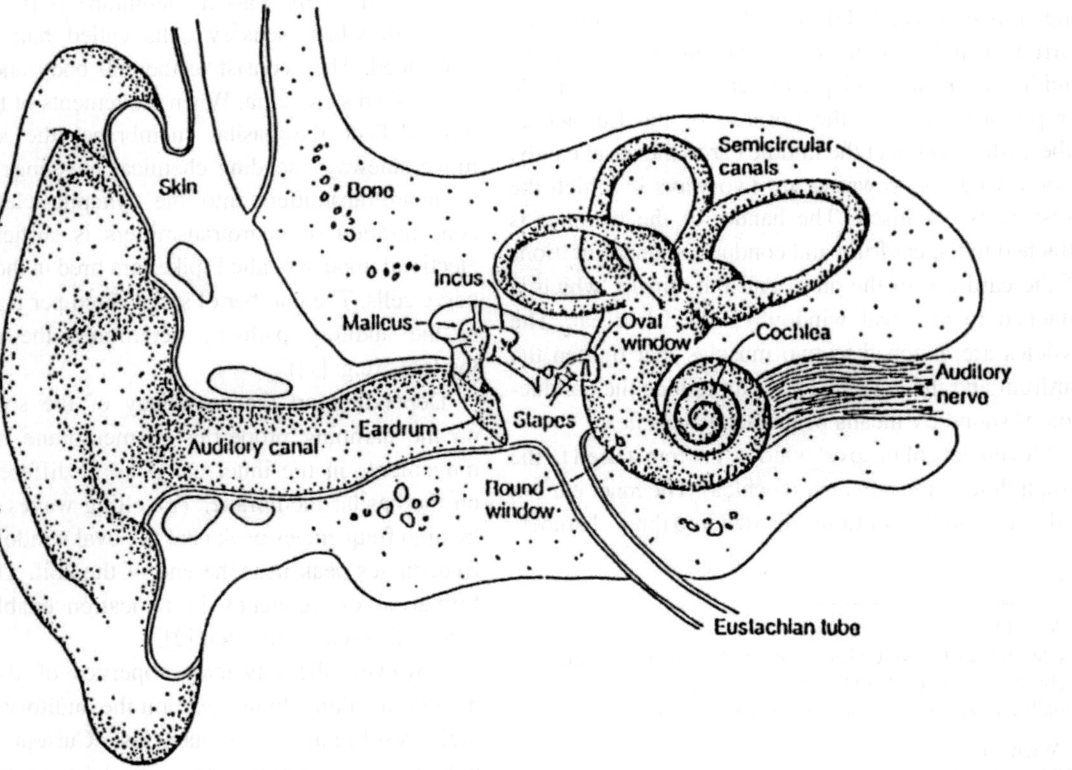

Fig. 4.1 Anatomy of the outer, middle and inner ear (after: [4])

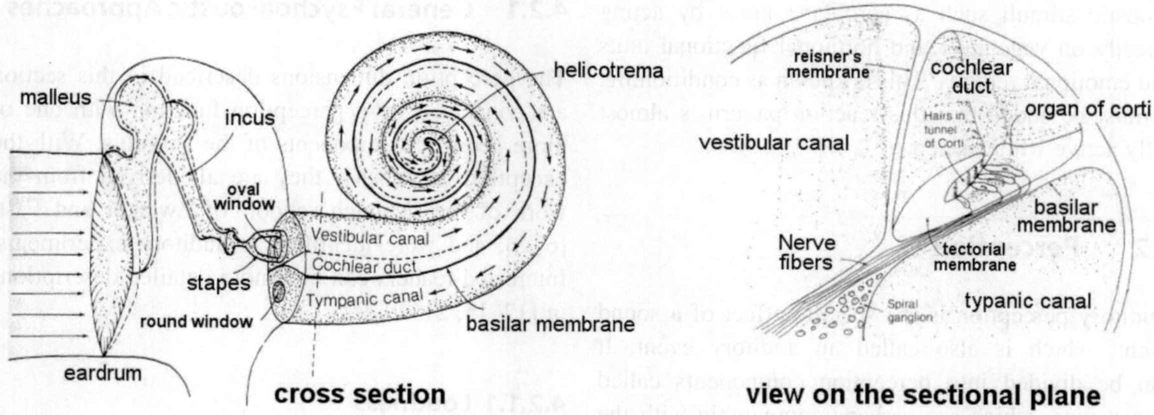

Fig. 4.2 Schematic drawing of the middle and inner ear (after [4])

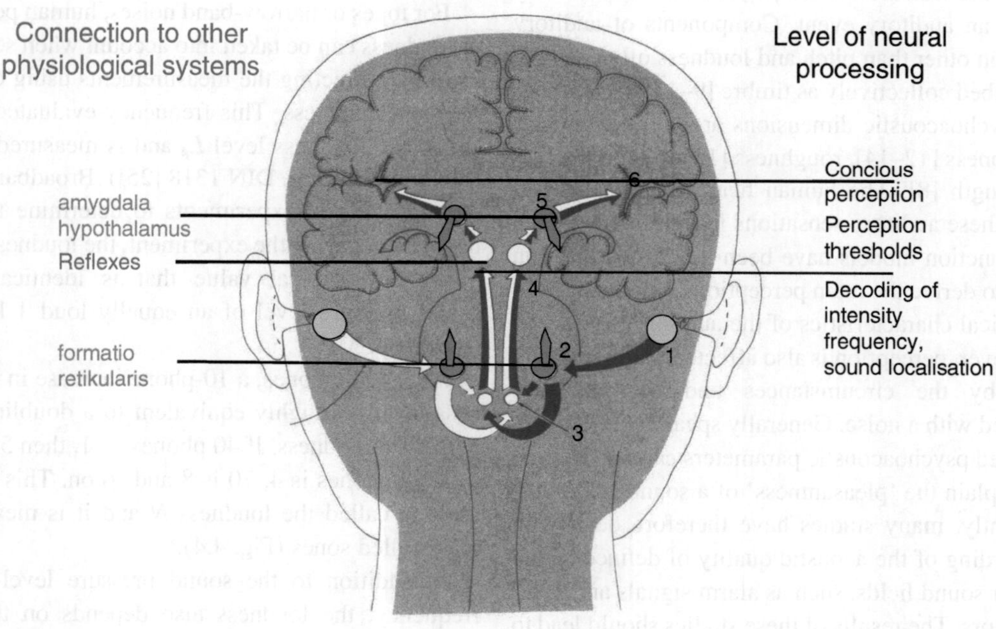

Fig. 4.3 Afferent pathway (after [5]) and stages of processing

the auditory pathway and ends in the reticular formation. This is a formation of cells that stretches from the spinal cord to the midbrain. The reticular formation controls the activation state and the sleep/waking cycle.

The auditory pathway continues via the lateral lemniscus to the inferior colliculus (4). This is where the resolution of frequency and intensity takes place. The sound is localised and reflexes may be triggered. The auditory cortex (6) is the final stage of the afferent pathway. It is reached via the medial geniculate body (5) and is responsible for conscious perception of sound events.

In the medial geniculate body, there are direct branches from the auditory pathway to the amygdala and the hypothalamus. The amygdala is characterised by its unusual learning capacity as regards aversive aural stimulation (fear conditioning). If the stimulus is frequently repeated, the amygdala can alter so that the entire organism reacts sensitively to aversive noises [7]. The final stage then consists of a very fast and coarse processing pattern, which reacts to known acoustic stimuli such as aeroplane noise by acting directly on vegetative and hormonal functional units and emotional regions. This is known as conditioning. It must be added that this reaction pattern is almost fully active while asleep.

4.2 Perception

Auditory perception is the specific effect of a sound event, which is also called an auditory event. It can be divided into perception components called dimensions, which are perhaps comparable with the four taste sensations: bitter, sweet, salty and sour.

The dominant perception components of pitch and loudness are well researched [8], but these do not fully describe an auditory event. Components of auditory perception other than pitch and loudness often used to be described collectively as timbre [9–11]. Nowadays, other psychoacoustic dimensions are defined, including sharpness [12–14], roughness [15, 16] and fluctuation strength [8]. The human hearing apparatus can process these auditory sensations individually of each other. Function models have been developed that can be used to derive the main perception component from the physical characteristics of the auditory event.

However, perception is also affected to a significant degree by the circumstances and the emotions associated with a noise. Generally speaking, the aforementioned psychoacoustic parameters cannot satisfactorily explain the 'pleasantness' of a sound [17].

Recently, many studies have therefore dealt with the recording of the acoustic quality of defined sound events or sound fields, such as alarm signals and vehicle interiors. The results of these studies should lead to optimisation of sound immissions in terms of acceptance and effect. The study of product sound quality has become particularly established in the automobile industry [18–20].

Its primary aim is for unbiased customers to associate sound with important criteria such as the reliability and value of the product. Improving spoken communication (for example, inside the vehicle) is another important development aim of sound design. Speech quality essentially depends on the level and spectrum of both speech and noise disturbance. However, articulation, hearing capacity, eye contact between the speaker and the listener and echoes inside the space also play a part.

4.2.1 General Psychoacoustic Approaches

The perception dimensions described in this section are connected by a perception function with one or more acoustic components of the stimulus. With the exception of density, they are all derived from the work of the 'Munich school' of Zwicker and Fastl [6, 8], and were identified in auditory experiments. Interested readers can find more detailed descriptions in [12–15, 21–24, 92].

4.2.1.1 Loudness

The perception of loudness depends on the sound level, the frequency and the bandwidth of the sound event, as well as on masking effects.

For tones or narrow-band noises, human perception of loudness can be taken into account when setting the level by correcting the measurements using curves of the same loudness. This frequency-evaluated level is called the loudness level L_s and is measured in units called phones (e.g. DIN 1318 [25]). Broadband noises require auditory experiments to determine the loudness level. During the experiment, the loudness level is given a numerical value that is identical to the sound pressure level of an equally loud 1 kHz tone (see Chap. 5).

Above 40 phones, a 10-phone increase in the loudness level is roughly equivalent to a doubling of the perceived loudness. If 40 phones = 1, then 50 phones is 2, 60 phones is 4, 70 is 8 and so on. This loudness scale is called the loudness N and it is measured in units called sones (Fig. 4.4).

In addition to the sound pressure level and the frequency, the loudness also depends on the bandwidth of the signal. Thus, an increase in the bandwidth leads to an increase in loudness, if the frequency range of the sound event exceeds the critical band. The critical bands (Δf_G)can be approached relatively well above 500 Hz using one-third octave bands.

The loudness of a tone or a noise can be reduced by a second event of the same loudness (attenuation) or only the louder sound event is perceived (masking). In order to investigate factors determining masking, the masking threshold is measured. The masking threshold states the sound pressure level that the test sound (normally a sine tone) must have in order to be perceived as well as the disturbance sound, in other words, so that it can just be heard (Fig. 4.5).

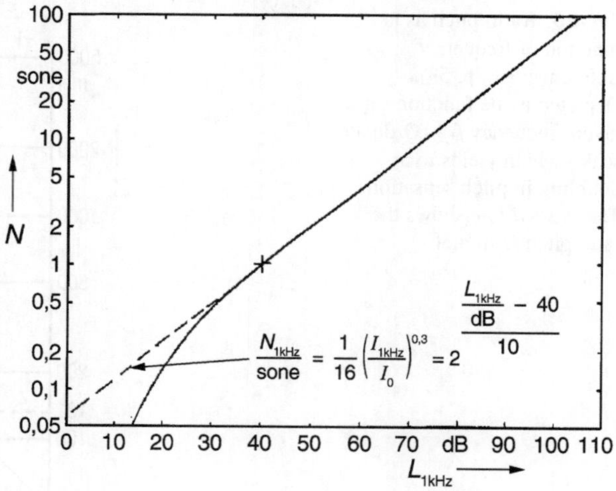

Fig. 4.4 Loudness function of a 1 kHz tone (*solid line*). Above 40 dB an increase of 10 dB corresponds to a doubling in loudness. Below 40 dB smaller/tiny differences in sound pressure level correspond to a loudness doubling (after [6], p. 81)

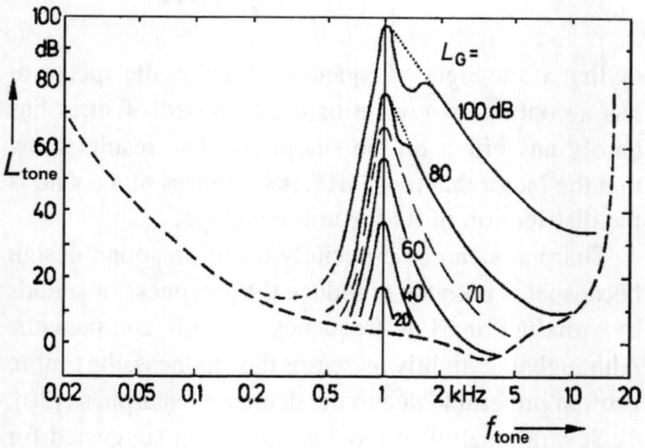

Fig.4.5 Level of test tones just masked by critical band-wide noise with centre frequency of 1 kHz with level L_G. Below maximum, the masked threshold climbs very steeply and than towards higher frequencies the level-dependant non linear spread of the upper masking slope is depicted. (reference: [6], p. 41)

In order to determine the loudness of a sound event from its physical characteristics, critical bands, masking and attenuation must be taken into account. One technique that takes comprehensive account of the way the human auditory system works for stationary noises is Zwicker's loudness calculation method [81, 82, 95]. The German standard DIN 45631 [26] for PCs contains a current version of the loudness calculation [97]. Widmann [27] provides C source code for further use.

The loudness model must be expanded for non-stationary sounds [96] because temporal masking effects, particularly post-masking [28] must be taken into account.

Current loudness models for non-stationary sounds differ most significantly in the way post-masking is simulated ([29]; Fastl and Schmid 2001) [83, 84]. Standardisation in this area is not yet complete.

4.2.1.2 Pitch

Besides the level, pitch perception of sinusoidal tones depends most on the frequency. Using experiments with pure tones, the following perception function can be determined (Fig. 4.6). For a 125 Hz tone, the ratio pitch H is defined as 125 mel.

Up to around 1 kHz, a doubling of the frequency is perceived as a doubling of the pitch. Above that, greater increases are necessary for a perceived doubling of the pitch.

Complex sounds with multiple spectral pitches could be perceived by the auditory system with an additional pitch. This interesting phenomenon is called virtual pitch [6].

4.2.1.3 Pitch Strength

Pitch strength, referred to by some authors as tonality [24], is the perception of the pitch of a sound. A sinusoidal tone is perceived with high tonality. Other signals, such as sounds or high-pass noise, are perceived less tonally, or hardly at all. It has not yet been possible to formulate a generally applicable function model for this perception dimension. Only for simple sounds are there regulations for determining the tonality of noises (e.g., the German standard DIN 45681[30]). This is why no unit has yet been assigned for tonality.

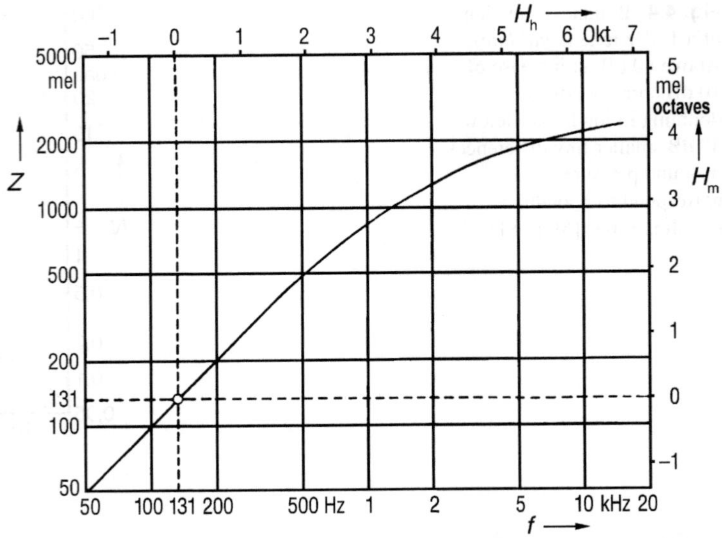

Fig.4.6 Ratio pitch as a function of frequency (reference: [6], p. 58), Depicted is the function for a given frequency $f_{1/2}$ (Ordinate scale) which yields to a doubling in pitch sensation. The *dashed line* shows the ratio pitch *H* in mel

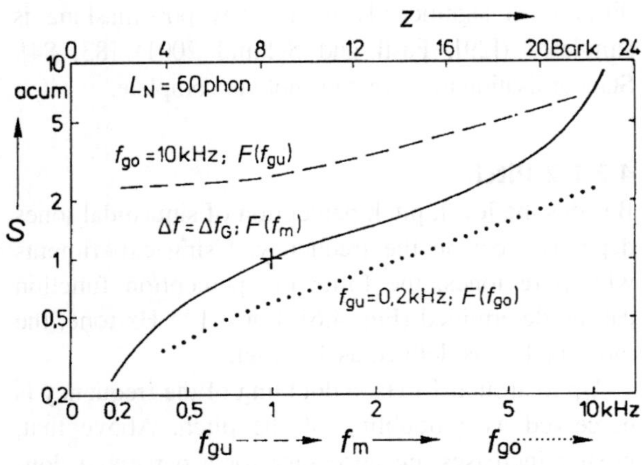

Fig. 4.7 Sharpness of narrow-band noise (*solid*), low pass filtered noise (*dotted*) and high pass fildered noise (*dashed*) as a function of centre frequency (f_m), the upper frequency bandary (f_{go}) and the lower frequency bandary (f_{gu}) (reference: [6], p. 84)

Generally speaking, the strength of complex sounds can only be derived from auditory experiments.

4.2.1.4 Sharpness

The sharpness of a sound event depends on its frequency composition (Fig. 4.7). Fundamentally, the sharpness of a signal is the higher the more high frequencies it contains. Narrow-band noise ($\Delta f \leq \Delta f_G$) with a mean frequency of 1 kHz and a sound pressure level of 60 dB is defined as having a sharpness of 1 acum.

The sharpness of broader-band sounds depends on the band limit at the low frequency and, more severely on that at the higher frequency. Whether the spectrum has a continuous course or is composed of lines has hardly any effect on the sharpness. The results show that the factor that most affects sharpness of a sound is the distribution of its spectral envelope.

Sharpness can be especially useful in sound design because it is possible to reduce the sharpness of sounds by mixing in low-frequency sound components. Although this slightly increases the loudness, the timbre is often preferable due to the decreased sharpness [20].

Several analytical models have been suggested for sharpness ([13, 14]; Aures 1984; [8]) and are now available in current simulations.

4.2.1.5 Tonality

For measuring tonality according to Aures [15, 21], two components are extracted from the amplitude spectrum of a signal. One of them contains all the tonal, i.e., narrow-band components, and the other, the noise components. The difference between the overall level of the two spectra in dB is used as a measure for tonality, after the pitch and masking phenomena are corrected. Tonality is believed to be connected to the sensory pleasantness of a sound that Aures [15], decomposes into four qualities: roughness, sharpness, tonality and loudness.

4.2.1.6 Roughness

Roughness is a perception dimension that occurs most often with frequency-modulated and amplitude-modulated sounds.

For a 1 kHz tone with a level of 60 dB, amplitude modulated with a modulation frequency of 70 Hz and a degree of modulation of 1, a roughness of 1 asper is defined.

The perceived roughness of modulated tones is highly dependent on the carrier frequency, the modulation frequency and the degree of modulation (see Fig. 4.8). It is less dependent on the sound pressure level. The sound pressure level has to be increased by circa 40 dB in order to double the roughness. Frequency modulation leads to higher perceived roughness than amplitude modulation.

4.2.1.7 Fluctuation Strength

Sound signals with a temporally fluctuating envelope, such as amplitude- or frequency-modulated sounds, with a maximum modulation frequency of 20 Hz, are perceived without roughness, but with fluctuation instead. A 1 kHz tone with a sound pressure level of 60 dB, amplitude-modulated with a degree of modulation of 1 and a modulation frequency of 4 Hz is therefore assigned a fluctuation strength of 1 vacil [8]. A modulation frequency of 4 Hz results in the maximum fluctuation strength, both for amplitude modulation and frequency modulation (see Fig. 4.9).

Perception of fluctuation strength is especially important in terms of the unpleasantness of sounds. It is especially strong in alarm signals, which for their purpose must be loud, sharp and tonal (Figs. 4.10 and 4.11).

Due to the temporal masking effects, there are masking period patterns for modulated sounds. The modulation depth of the masking period pattern plays a crucial part in the explanation of auditory sensations such as fluctuation strength and roughness (see Sect. 4.2.1.6) [16, 32–34].

Masking period patterns are particularly important in practice, because the non-linear diversification of

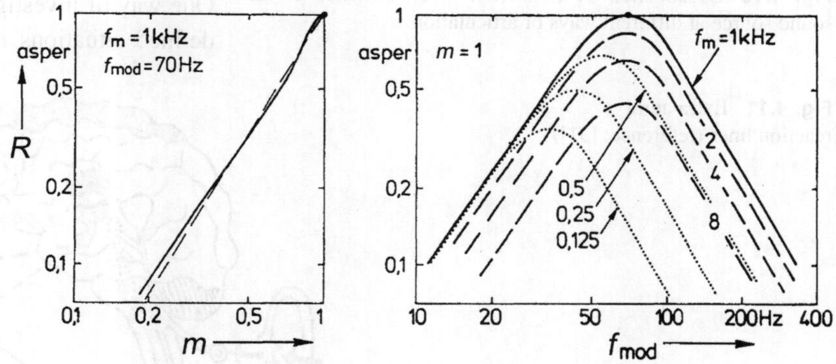

Fig. 4.8 Roughness R (reference: [6], Sp 107). Roughness of an amplitude modulated tone as a function of degree of modulation m and the modulation f_{mod}. The centre frequency f_m varied

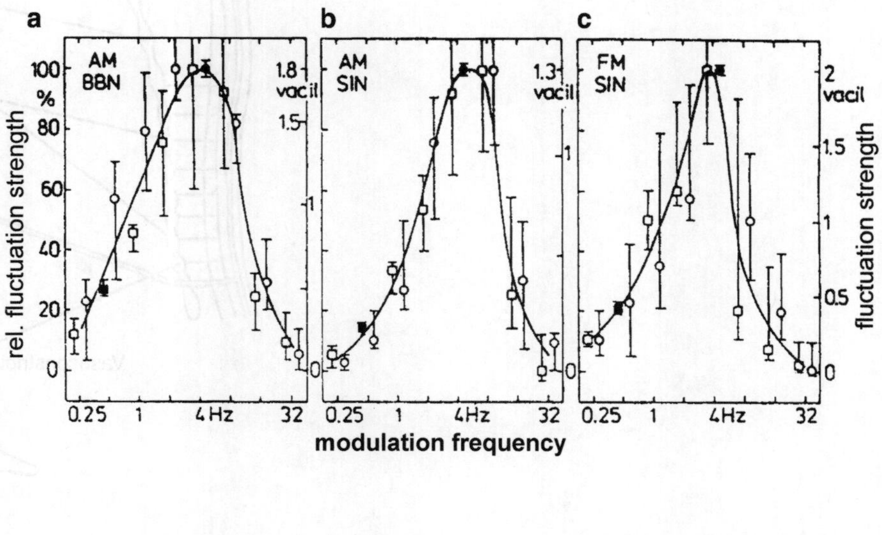

Fig. 4.9 Fluctuation strenght of an amplitude-modulated broad-band noise (**a**), an amplitude modulated tone (**b**) and a frequency modulated tone (**c**) as a function of modulation frequency. The centre frequency f_m is varied (reference: [8], p. 223)

the upper masking edge produces periodic changes in low-frequency sound components, especially at medium and high frequencies. Thus, low-frequency sounds (for example, speech) can modulate periodically as sound disturbance, thus impairing spoken communication. Other examples include 'booming' noises that can occur at certain speeds and opening angles of car windows and sunroofs.

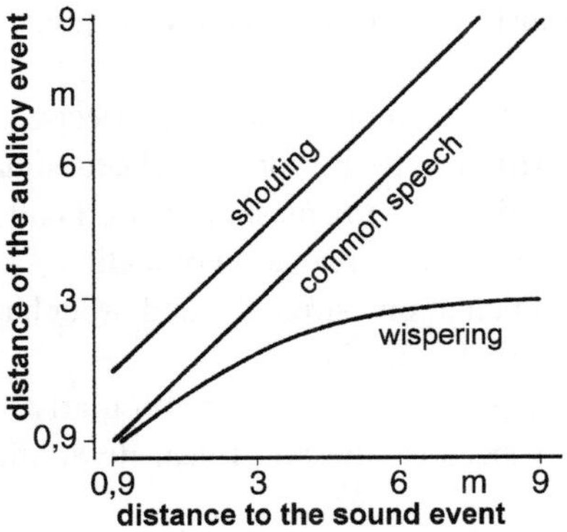

Fig. 4.10 Localization as a function of the distance of the sound source at different ways of articulation

4.2.1.8 Other Psychoacoustic Perception Dimensions

A study by Heldmann [22] defines the psychoacoustic perception dimension of impetus. The suggested unit is IU (impetus unit). One impetus unit is assigned to a 1-kHz tone pulse with an impulse duration of 20 ms and an impulse level of 73 dB; the 20 ms time-window (envelop) is delimited by Gaussian flanks of 3.5 ms width.

Density was defined as a psychoacoustic perception dimension by Guirao and Stevens [35]. They identified a positive connection between density and both the frequency and the sound pressure level. The suggested unit is the 'dasy', the Greek word for density. According to this study, a 1 kHz tone with a sound pressure level of 40 dB has a density of 1 dasy. There is not yet a recognised calculation method for this perception factor.

4.2.2 Specific Psychological Approaches

One way of investigating dimensions of perception in defined situations is to use psychologically oriented

Fig. 4.11 Extra aural reaction lines (reference: [31])

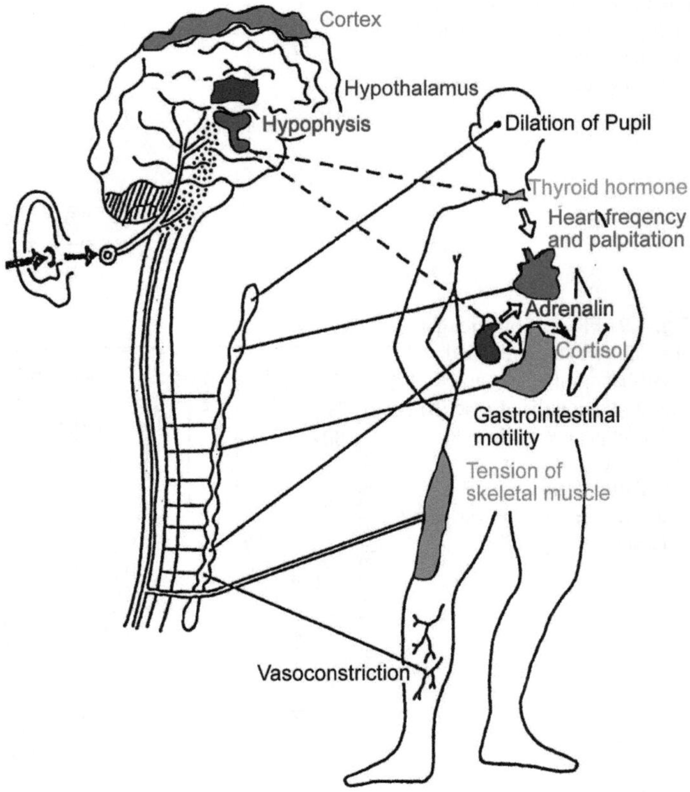

approaches. Besides research into fundamentals, sound quality of commercial product is in the foreground, as shown in recent works by Kuwano et al. [36, 37] and Namba et al. [38]. The investigation method used is often a 'semantic differential' (Hashimoto 1996; [37, 39]). The results then take the form of noise attributes such as luxury and power.

4.2.2.1 Semantic Differential

In this type of experiment, subjects must rate noises on a scale between two opposite adjectives (such as sharp and dull). Seven-stage scales are mostly used. The evaluation of the experiments, performed with at least ten pairs of adjectives, employs dimension analyses (factor- or cluster analysis); the components (factors) are identified and the nontrivial variations in the data set are explained.

A fundamental problem of this method is connecting the identified factors with the stimulus parameters of the noises.

The first experiments on noise perception using the semantic differential were performed in 1958 by Solomon [40]. In recent years, many studies have recorded the acoustic quality of individual sound events and sound fields, for example, warning signals or car interiors [91].

4.2.3 Localisation

The auditory system is able to identify the place of origin of an auditory event. This is called localisation. Generally speaking, the direction information is derived from the differences in the level and frequency of sound pressure distribution in the left and right ears. There are three main types of localisation:

- Localisation on the horizontal plane
- Localisation on the medial plane
- Localisation and diffusion of the auditory event; auditory distance perception.

Familiarity with the sound event is very important for auditory distance perception. With speech, the perceived distance to the auditory event corresponds well to the distance of the sound source. With unusual types of speech (such as whispering) there are already significant discrepancies (see Fig. 4.10). A more detailed description of localisation can be found in [41].

If two sound events merge into a single auditory event, the perceived location of the auditory event is determined by the sound that reaches the ear first. Cremer calls this effect the law of the first wave front [42]. Without this property of the auditory system, acoustic orientation in the interior would be almost impossible.

4.3 Health Hazards Associated with Noise

High sound pressure levels can damage the hearing and cause dangerous neurovegetative reaction patterns. However, noise is not just a physical stimulus, but also an individual experience. Insufficient handling can also lead to inadequate reaction patterns and ultimately to regulatory disorders. Regulatory disorders can be considered as adverse effects constituting a transitional stage from health to illness.

Including individual experience takes into consideration the fact that the human organism is a biopsychosocial unit [43]. Consequently, health impairments are not only physically demonstrable injuries but also functional disorders of psychological and biological processes that cannot be separated from each other. Impairments of psychobiological regulation often manifest themselves as somatoform disorders [44]. These include the reflection of disorders in mental and psychological processes (such as stress, chronic noise effects, suppressed emotions, permanent anger, frequent rage, social and temporal conflicts, helplessness, despair and inability to combat problems), and physical complaints (such as headaches, backache, exhaustion, digestive disorders, cardiovascular disorders, asthma, skin disorders and impotence).

4.3.1 Aural Disorders

Deafness, communication problems and tinnitus are the most striking hearing disorders caused by noise.

If the human ear is subjected to sound of sufficiently high intensity, increased metabolic processes take place that are difficult to compensate. If excessive sound intensity or duration causes a physiologically excessive metabolic process, the hair cells show signs of fatigue, which can cause temporary or permanent damage. Consequently, following subjection to sound and depending on its intensity and duration, the sound receptors become less sensitive, which manifests itself in a temporary increase in the auditory threshold. The difference between the auditory threshold before and after exposure is called the TTS (temporary threshold shift).

A reduction in hearing ability with age is known as age-related deafness, and the medical term for this is presbycusis. Age-related deafness is a gradual process, which generally begins in Western industrial societies at the age of about 30 years.

4.3.1.1 Communication Problems

Speech perception is a very sensitive indicator of the disturbing effect of noise. The extent to which communication sound is masked by disruptive sound is measured using special methods such as the articulation index or the speech interference index (SII). The disturbance not only depends on the level difference and the frequency spectra of the two sounds but also on the clarity of articulation, the information content of the text and the listener's prior understanding of it, the possibility of visual contact and the acoustic conditions of the surrounding space [45]. Children and grown-ups with deficient hearing are much more impaired in their speech perception than people with normal hearing [90]. Broadband ambient noise hardly affects speech perception at all if the noise disturbance level is at least 10 dB(A) below the speech level [90].

4.3.1.2 Hearing Impairment (Noise-induced Deafness)

Constant or frequent exposure to high-intensity sound can lead to an irreversible shift in the auditory threshold (deafness, noise-induced permanent threshold shift: NIPTS). The hearing impairment is measured as the difference between the auditory threshold of the damaged ear and the normal auditory threshold (see [46, 47]). If the loss of hearing exceeds a specified value, it is classified as a hearing impairment or noise-induced deafness. The German [48] defines deafness as an audio-measurable loss of hearing in the inner ear if the reduction in hearing exceeds 40 dB at 3,000 Hz.

Noise-induced deafness is still (one of the most frequent recognised occupational illnesses. Not only must noise at work be taken into account but also that at leisure times, visiting discos or listening to personal stereos. Considering the actual sound levels in discos and exposure times young adults experience, it is to be expected that after 10 years, around 10% of young people today will suffer a loss of hearing of at least 10 dB(A) [49]. Since 40-year-old men can already expect an age-related loss of hearing, also of 10 dB, 10% of 40-year-olds can expect to suffer a hearing loss of 20 dB or more, which significantly impairs communication.

Besides gradual, noise-induced loss of hearing, short-term excess noise at extreme sound intensities can also cause deafness. Toy guns, blank guns and fireworks deserve particular mention in this context. The peak level is sometimes well above the damage threshold for single events, $L_{peak} = 140$ dB.

Hearing loss must be considered as a serious social handicap. Difficulty in understanding speech first becomes apparent in loud environments (self-service restaurants, parties, loud events), and later in situations such as church services, theatre performances and public meetings. People with impaired hearing can partially compensate by lip reading and thus remain unaware of their shortcoming.

However, loss of hearing is not only caused by excessive exposure to noise. Illnesses, ototoxic drugs, hereditary factors and inflammation of the middle ear during childhood can also cause deafness.

4.3.1.3 Tinnitus

Many people suffer from tinnitus, a ringing in the ears that can be extremely unpleasant. According to two studies by the American National Center for Health Statistics, tinnitus is suffered by 32% of the population, including 2% with serious cases. Similar data has been collected in the United Kingdom [50]. Tinnitus can often occur in a previously healthy ear as a result of excessive acoustic exposure, in connection with sudden deafness, and is generally a sign for at least a temporary loss of hearing. The causes of tinnitus are not fully known, and explanations range from vascular deficiency and damage to the hair cells in the inner ear to disorders of the superior auditory pathway [51]. Tinnitus occurs more frequently in times of stress.

Drugs to promote the circulation are usually prescribed, but these normally only provide temporary relief or none at all [50]. In recent years, there have been increasing attempts to counter tinnitus using psychophysiological methods. Attempts have also been made using temporary maskers (for example, noise transmitted through earpieces) to desensitise the perception process.

4.3.2 Extraaural Impairment

Parallel to the specific effects of sound on perception, it can also have a non-specific effect on the human body. These processes initially take place in order to adapt the organism to altered situations. Consequently, vegetative reactions in the peripheral circulatory system can be measured, such as decreased galvanic skin resistance, skin temperature and finger pulse amplitude, or changes in heartbeat frequency [52–54] or change concentrations of stress hormones such as adrenalin and cortisol in bodily fluids (see Fig. 4.11).

The pathogenetic concept, which associates noise effects with health risks, is derived from known stress models. Medically speaking, there is a distinction between eustress and distress. Eustress is the type of stress that enhances performance and health, while distress is a type of stress with pathological manifestations. Eustress is generally temporary. Long-term stress or frequent short-term psychobiological distress can cause functional disorders. Consequently, it cannot be said that there is a specific extraaural noise-induced illness. However, noise acts as a stress factor and promotes illnesses where stress is a contributory factor, e.g., cardiovascular disorders and also mental disorders (neuroses).

4.3.2.1 Sleep and the Consequences of Sleeping Disturbances

Sleep is not a state of general muscular, sensory, vegetative and mental relaxation, but instead has a very complex dynamic. The characteristic features of human sleep are periods, dynamics, altered motor and sensory functions, as well as an altered state of consciousness. Using recordings from electroencephalograms (EEG), electromyograms (EMG) and electrooculograms (EOG), sleep can be divided into four non-REM phases as well as REM sleep (named after the rapid eye movements that occur during this phase). The length of the various sleep phases during a night's sleep largely depend on age. REM accounts for up to 60% of newborn babies' sleep, but falls to around 20% in adults.

The period from the onset of sleep until REM sleep is called REM latency and the intervals between REM phases are known as sleep cycles.

Figure 4.12 shows the sleep cycles of a young, healthy subject, as well as the concentrations of plasma cortisol and growth hormones during the night.

The most important periods of regeneration are stages III and IV of non-REM sleep (delta sleep), as well as REM sleep. Delta sleep (deep sleep) is for physical regeneration, while REM sleep (dream sleep) is for mental and emotional regeneration, as well as for the continuing process of individual learning, by transferring data from the short term to the long-term memory.

As well as the biological component of sleep, the psychological aspect must also be taken into account. Any disturbance to night-time sleep is experienced as something unpleasant, a violation of our intimacy. Waking up during night-time sleep is an unpleasant experience and evokes negative emotional conditions.

Effects of Noise on Sleep

Sleep disorders are very serious complaints. Disorders caused by noise can be divided into primary and secondary reactions, according to the time they occur.

Primary reactions include brief changes on the EEG (arousal), flattening of the current depth of sleep (phase changes) or even waking up, fragmented sleep, changes in the distributions of sleep phases, increased latency periods (especially sleep onset latency), shortening of the overall duration of sleep, increased duration of periods of high muscular tension (movements), but also vegetative reactions such as changes in breathing rate, hormone secretion and peripheral circulation.

Secondary reactions are reversible impairments of general well-being after waking. These include impairment of the physical and mental condition, experience of sleep, well-being, performance and concentration.

Disturbance of sleep caused by noise is associated with changes to physiological factors. The sensitivity of indicator systems decreases in the following order: EEG, vegetative-hormonal system (heart rate, peripheral circulation, hormone secretion) and motor system behaviour. Frequent or continuous sound stimuli during sleep activate the nervous system, manifesting

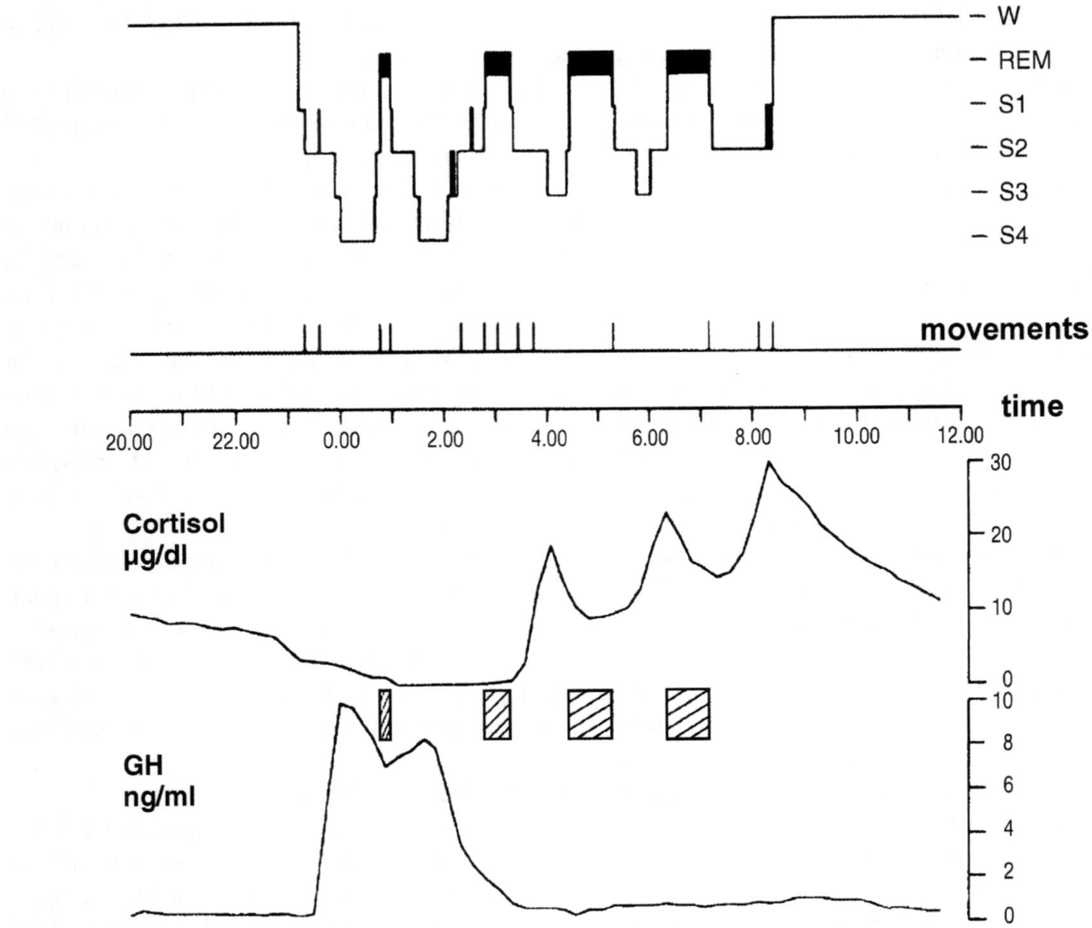

Fig. 4.12 Typical sleep cycles of a young, healthy adult and the nightly behaviour of the plasma-cortisol concentration and the growth hormones (reference: [55])

itself in the EEG during intermittent noise as fragmented sleep (disturbed sleep cycles), or as superficial sleep in the case of more or less constant noise. Both types of noise shorten deep sleep (stages III and IV) and sometimes REM phases, and disrupt the periods of sleep. As well as stimulation processes of the central nervous system, changes to the secretion of hormones are a characteristic of sleep disorders. Stress-oriented studies into traffic noise [56–59] indicate that traffic noise affects nocturnal cortisol secretion. At the same time, the quality of sleep and the test subjects' well-being on waking deteriorated.

Sound-induced activation of the nervous system can even lead to waking up. Apart from the physical characteristics of noise disturbance – especially discontinuity – its content is also important in terms of its effect on the sleeper. The auditory system's alarm function can wake the sleeper even in the case of slight noises if they contain information that is unfamiliar or indicates danger (see Sect. 4.3.3). Conversely, accustomisation to chronically familiar noises can go so far that the sleeper wakes when they do not occur (for example, when a train normally scheduled to pass by is cancelled). With less unusual noises, the number of sleepers who wake up only rises significantly above a level of 40 dB(A). Children are especially good at sleeping through noises with sound levels above 90 dB (A). The waking effect not only depends on the sound level but also on the deviation from the background noise level.

Frequent disturbance of physiologically programmed functions must always be regarded as unhealthy. This also applies to awakenings. Noise-induced waking periods must be classified as abnormal and a health risk in the long term. However, a general disruption to physiological functions can also be seen below the awakening threshold. It therefore makes little sense to derive a healthy level for the protection of sleep solely from a mean experimental awakening threshold.

According to the recommendations of the World Health Organization [5], an equivalent continuous nightly sound level of $L_{eq,indoor} = 30$ dB(A) and a maximum level of $L_{max,indoor} = 45$ dB(A) should not be exceeded, if sleeping disorders are to be prevented. Comparable recommendations have also been issued by the interdisciplinary committee on the effects of noise at the German Federal Environment Ministry [88]. The committee believes that an equivalent continuous nightly sound level of 30 dB(A) at the sleeper's ear and a maximum level below 40 dB(A) are largely sufficient to prevent sleeping disorders. These guidelines not only protect the average person from noise-induced sleeping disorders but also guarantee a nocturnal ambient quality that is also suitable for people who are more sensitive to noise.

Performance impairments are the most serious and frequently cited effects of noise. Any mental performance or physical task requiring special concentration can be impaired at an average sound level as low as 45 dB(A) [60]. This impairment is worsened by any kind of peculiarity of the noise stimulus, such as intermittent, unpredictable noise, irregular sound level fluctuations, high-frequency components or particular tone and information content (for example, speech).

4.3.2.2 Impaired Concentration and Performance

In many situations of exposure, noise-induced deterioration in performance is compensated by increased effort, such as additional concentration, so that the performance may even increase temporarily. However, many studies have shown that noise can have an affect beyond the period of exposure, which may manifest itself in the form of increased fatigue or reduced ability to concentrate and to work [89].

4.3.3 Annoyance

Noise is not merely the action of a physical stimulus, but also an experience. The experience of noise and the associated alteration of functions can become permanently ingrained as annoyance in the human memory. Annoyance therefore means the expression of negative emotions in connection with particular stimuli from a person's external and internal environment. Annoyance manifests itself in emotions such as discomfort, fear, insecurity, anger, uncertainty, a feeling of restricted freedom, irritability or defencelessness.

The evaluation of annoyance includes not only noise-related variables (mediators) but also variables called moderators, which refer to the individual or group exposed to noise. As a conscious perception process, annoyance also manifests itself in changes to the vegetative and hormonal regulation process [61–63]. Intense, continuous annoyance must be classified as a health risk.

Functional relationships between annoyance and noise exposure have been studied many times around the world, and have been summarised, for example, in a meta-analysis by Miedema [64]. He investigated traffic noise (air, road and rail traffic) and stationary sources (industry, marshalling yards and shooting ranges). The relationship between exposure and effect can be seen in Fig. 4.13.

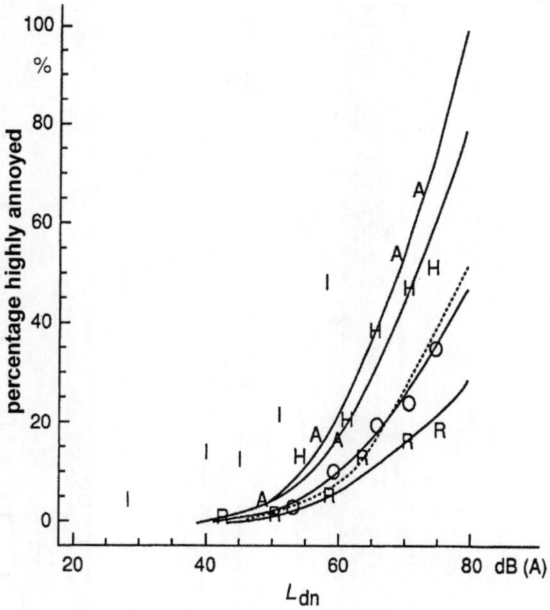

Fig. 4.13 Percentage of strongly annoyed people as a function of L_{dn}: Dots belonging to the same study are visualised by a solid line (*A* aircraft, *H* highway, *O* other road traffic, *R* railway, *I* impulse sources). The synthesis curve of Schulz [65] is depicted as a *dotted line* (reference: [64])

Immission values derived from these exposure–effect relationships are fundamentally different from environmental medical or hygienic thresholds and limits for non-sensory environmental factors. While pollutant hygiene experts use the NOAEL (no observable adverse effect level), ADI (acceptable daily intake), MIC (maximum immission concentration) and similar exposure-related factors to attempt to establish 'zero risks' and individual biological acceptance, noise prevention is limited to the statistical evaluation of annoyance assessments. The aim of protection is to sensibly minimise the number of people exposed and the intensity of annoyance [66].

Generally speaking, a 10–15% range of very sensitive people is seen as a nominal threshold for noise-induced annoyance, since the number of very sensitive people in the population is also between 10 and 15% [67] (Fig. 4.14).

The German Federal Immissions Law defines annoyance as a 'damaging environmental influence' if it is 'substantial'. In an attempt to develop environmental physical criteria for the seriousness of annoyance, residents affected by traffic noise were asked what they considered to be 'substantial', with an annoyance situation being classified as serious if the percentage of residents classing it as such was more than 25%. The proposal made by Hörmann [69] is in a similar region: if more than 25% are strong annoyed, then immediate action is necessary, between 10 and 25%, long-term action is needed, and at 10% or less, no immission protection measures need be initiated.

4.3.4 Cardiovascular Illnesses

Noise affects the neuroendocrine system either directly via processes of the central nervous system or indirectly through subjective experience (disturbance, annoyance). This in turn affects metabolic processes and the regulation of essential bodily functions. These include blood pressure, cardiac activity, blood lipids (cholesterol, triglycerides, free fatty acids),

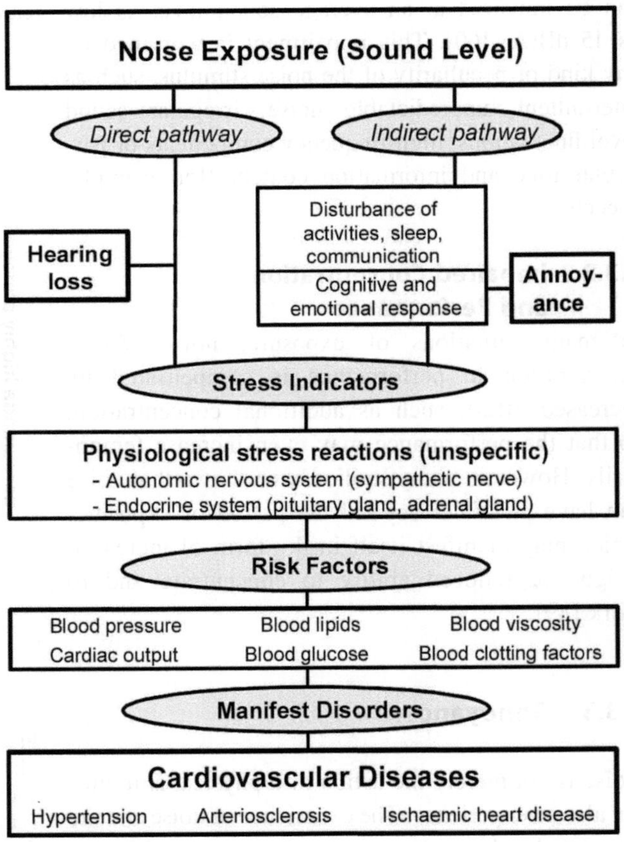

Fig. 4.14 Effect mechanism of the noise induced cardiovascular risk (reference: [68])

blood glucose levels and haemostatic factors (such as fibrinogen) that affect the blood flow (plasma viscosity) [70]. Since these are classical (endogenic) risk factors for cardiovascular disease, noise is regarded as an (exogenic) risk factor for the development of high blood pressure and heart disease including arteriosclerosis and myocardial infarction [48].

According to older studies, it is to be feared that residents of areas with daytime traffic noise immission levels of more than 65 dB(A) have a higher risk of heart disease, namely in the region of 20–30% [71]. Newer epidemiological studies reveal that the nightly noise exposure is related to health impairments more narrowly than the noise exposure on the day ([57]; Maschke et al. 2003, garup 2008).

4.4 Non-acoustic Factors (Moderators)

Because noise is a psychophysical stimulus, the effects of noise cannot be explained by the intensity of the noise, i.e., the sound level, alone. Population studies show, for example, that no more than a third of the variation between annoyance assessments in field conditions can be explained by the equivalent continuous sound level [72]. An increase in the prediction of individual reactions to exposure using psychoacoustically motivated noise indicators was proved in complex acoustic situations, for example, after noise-reduction measures were taken or in Alpine regions [73].

The individual differences can be further clarified using the concept of non-acoustic factors. Noise researchers now believe that situational, personal and social factors influence the effect of noise pollution without themselves being significantly affected by it.

One of the most important situational moderator variables is the time of day when noise pollution occurs. Results of international field experiments show that people who live near noise sources want quiet mainly at night, in the late evening and at weekends (especially Sundays), and that their annoyance at a noise source increases when it is not only active during the day but also in the evening and at night. Thus, Fields [74] draws the conclusion from a major American study that a 24-h day can be roughly divided into four periods of different sensitivity: night (midnight to 5 a.m.), day (9 a.m–4 p.m.) and two transitional periods. He justifies this division primarily with people's different intended activities at various times of day. Hecht et al. [75] reached similar results. They processed the daily courses of various body functions published during the last 35 years, and came to the conclusion that the human organism reacts most sensitively to noise at night and during the transitional period between day and night.

One of the most important personal moderator variables is the individual's sensitivity to noise. McKennell [76] was able to demonstrate that people who describe themselves as sensitive to noise react much more strongly to aeroplane noise than those who describe themselves as less sensitive.

Demographic variable such as age, sex, education and house ownership, on the other hand, generally have no systematic influence on the effects of noise. The only exceptions are slight effects reported when the noise pollution changes [77].

Even if personal and social factors can rarely be clearly separated from each other, Guski [78, 79] suggests devoting special attention to factors that are of a social nature and therefore affect whole groups of people. These factors play a central part in the development of reactions to noise exposure. The main ones are

- General assessment of a noise source
- Trust in those responsible for noise protection
- History of exposure to the noise
- Expectations of residents

Great importance should be placed on these factors when taking action to reduce noise or when building or expanding noise-intensive projects such as airports. This must be done positively. Distrust of the responsible authorities and institutes among residents affected by noise, for example, also generates distrust of planned or implemented noise-protection measures. Regardless of the physical success of the protective measures, annoyance can still remain at a high level. However, increased acceptance of a noise source cannot replace sound engineering or control measures.

References

1. Zwislocki JI (1969) Temporal summation of loudness: An analysis. J Acoust Soc Am 46:431
2. v. Békésy G (1960) Experiments in hearing. McGraw-Hill, New York
3. Brownell WE (1986) Outer hair cell motility and cochlear frequency selectivity. In: Moore BCJ, Patterson RD (eds) Auditory frequency selectivity. Plenum, New York

4. Lindsay PH, Norman DA (1977) Human information processing, 2nd edn. Academic, New York
5. WHO 2000 (2000) Noise and health. Regional Office for Europe, World Health Organization (WHO)
6. Zwicker E (1982) Psychoakustik. Springer, Berlin
7. Spreng M (2000) Central nervous system activation by noise. Noise Health 2(7):49–57
8. Zwicker E, Fastl H (1990) Psychoacoustics. Springer, Berlin
9. Benedini K (1979) Ein Funktionsschema zur Beschreibung von Klangfarbenunterschieden. Biol Cybernet 34:111–117
10. Letowski T (1992) Timbre, tone color and sound quality: Concepts and Definitions. Arch Acoust 17–30
11. Schauten JF (1968) The perception of timbre. Reports of the 6th ICA Tokyo, 35–40
12. Aures W (1985) Ein Berechnungsverfahren der Rauhigkeit. Acustica 58:268–281
13. v. Bismarck G (1974) Sharpness as an attribute of the timbre of steady sounds. Acustica 30(5):159–172
14. v. Bismarck G (1974) Timbre of steady sounds: A factorial investigation of its verbal attributes. Acustica 30(5): 146–159
15. Aures W (1985) Berechnungsverfahren für den sensorischen Wohlklang beliebiger Schallsignale. Acustica 59:130–141
16. Fastl H (1977) Roughness and temporal masking patterns of sinusoidally amplitude-modulated broadband noise. In: Evans EF, Wilson JP (eds) Psychophysics and physiology of hearing. Academic, London, pp 403–414
17. Johannsen K (1997) Zusammenhangsanalyse zwischen physikalischen Merkmalen und Hauptkomponenten der Beurteilungsattribute von Umweltgeräuschen. Dissertation, TU-Berlin
18. Blauert J (1986) Cognitive and aesthetic aspects of noise engineering. Proc. Inter-Noise'86, vol I, 5–14
19. Blauert J, Jekosch U (1997) Sound quality evaluation – a multi-layered problem. Acustica-Acta Acustica 87
20. Widmann U (1998) Aurally adequate evaluation of sounds. In: Proc. Euro-Noise'98, vol I, 29–46
21. Aures W (1985) Der sensorische Wohlklang als Funktion psychoakustischer Empfindungsgrößen. Acustica 58:282–290
22. Heldmann K (1994) Wahrnehmung, gehörgerechte Analyse und Merkmalsextraktion technischer Schalle. Dissertation, TU-München
23. Terhardt E (1998) Akustische Kommunikation – Grundlagen mit Hörbeispielen. Springer, Berlin
24. Terhardt E, Stoll G (1981) Skalierung des Wohlklangs (der sensorischen Konsonanz) von 17 Umweltschallen und Untersuchung der beteiligten Hörparameter. Acustica 48: 247–253
25. DIN_1318 (09–1970) Lautstärkepegel; Begriffe, Messverfahren. Beuth, Berlin
26. DIN 45631 Berechnung der Lautheit und des Lautstärkepegels aus dem Geräuschspektrum, Verfahren nach E. Zwicker. Beuth, Berlin
27. Widmann U (1994) Krach gemessen – Gehörbezogene Geräuschbewertung. Heise Verlag, Hannover
28. Zwicker E (1984) Dependence of post-masking on masker duration and its temporal effects in loudness. J Acoust Soc Am 75:219–223
29. Widmann U, Lippold R, Fastl H (1998) Ein Computerprogramm zur Simulation der Nachverde-ckung für Anwendungen in akustischen Meßsystemen. In: Fortschritte der Akustik, DAGA'98, Zürich
30. DIN 45681 (01–1992) Bestimmung der Tonhaltigkeit von Geräuschen und Ermittlung eines Tonzuschlages für die Beurteilung von Geräuschimmissionen. Beuth, Berlin
31. Jansen G (1992) Auswirkungen von Lärm auf _den Menschen. VGB Kraftwerkstechnik 72(1):60–64
32. Fastl H (1982) Fluctuation strength and temporal masking patterns of amplitude modulated broadband noise. Hear Res 8:59–69
33. Fastl H (1982) Beschreibung dynamischer Hörempfindungen anhand von Mithörschwellen-Mustern. Hochschulverlag, Freiburg
34. Widmann U, Fastl H (1998) Calculating roughness using time-varying specific loudness spectra. In: Proc. Noise Con'98, Ypsilanti (USA)
35. Guski R (1996) Psychological methods for evaluating sound quality and assessing acoustic information. Acta Acustica 83:765–774
36. Kuwano S, Namba S et al (1997) Evaluation of the impression of danger signals – comparison between Japanese and German subjects. In: Results of the 7th Oldenburg Symposium on Psychological Acoustics, 115–128
37. Kuwano S, Namba S, Hato T (1994) Psychologische Bewertung von Lärm in Personenwagen; Analyse nach Nationalität, Alter und Geschlecht. Z Lärmbekämpfung 41:78–83
38. Namba S, Kuwano S, Koyasu M (1993) The measurement of temporal stream of hearing by continuous judgements – Ii the case of the evaluation of helicopter noise. J Acoust Soc Jpn (E) 14(5):341–352
39. Takao H, Hashimoto T (1994) Die subjektive Bewertung der Innengeräusche im fahrenden Auto – Auswahl der Adjektivpaare zur Klangbewertung mit dem semantischen Differential. Z Lärmbekämpfung 41:72–77
40. Solomon L (1958) Semantic approach to the perception of complex sounds. J Acoust Soc Am 30:421–425
41. Blauert J (1996) Spatial hearing. The psychophysics of human sound localisation. MIT, Cambridge, MA
42. Cremer L (1948) Die wissenschaftlichen Grundlagen der Raumakustik. Bd. 1. Hirzel-Verlag, Stuttgart
43. v. Uexküll T (1990) Über die Notwendigkeit einer Reform des Medizinstudiums. Berliner Ärzte 27/7, 11–17
44. Hecht K, Balzer H-W, Rosenkranz J (1998) Neue Regulationsdiagnostik zum objektiven Nachweis psychosomatischer Prämorbidität und Morbidität (sogenannte Modekrankheiten). Thüringisches Ärzteblatt 8
45. Lazarus H, Lazarus-Mainka G, Schubeius M (1985) Sprachliche Kommunikation unter Lärm. Kiehl Verlag, Ludwigshafen
46. DIN 45620 (1981) Audiometer – Begriffe, Anforderungen, Prüfung. Beuth, Berlin
47. DIN 45630–2 (09–1967) Grundlagen der Schallmessung; Normalkurven gleicher Lautstärkepegel. Beuth, Berlin
48. VDI 2058 Blatt 2 (1988) Beurteilung von Lärm hinsichtlich Gehörgefährdung. Verein Deutscher Ingenieure, Düsseldorf
49. Zenner HP, Struwe V et al (1999) Gehörschäden durch Freizeitlärm. HNO 47:236–248
50. De Camp U (1989) Beeinflussung des Tinnitus durch psychologische Methoden. Dissertation, TU-Berlin

51. Feldmann H (1971) Homolateral and contralateral masking of tinitus by noisebands and by pur tones. Audiology 10:138–144
52. Jansen G, Klosterkötter W (1980) Lärm und Lärmwirkungen. Ein Beitrag zur Klärung von Begriffen. Bundesminister des Innern (Hrsg.), Bonn
53. Neus H, Schirmer G et al (1980) Zur Reaktion der Fingerpulsamplitude auf Belärmung. Int Arch Occup Environ Health 47:9–19
54. Rebentisch E, Lange-Asschenfeld H, Ising H (1984) Gesundheitsgefahren durch Lärm. Kenntnisstand der Wirkungen von Arbeitslärm, Umweltlärm und lauter Musik. BGA-Schriften 1/94. MMV Medizin Verlag München, München
55. Born J, Fehm HL (2000) The neuroendocrine recovery function of sleep. Noise Health 7:25–37
56. Braun C (1998) Nächtlicher Straßenverkehrslärm und Stresshormonausscheidung beim Menschen. Dissertation, Humboldt-Universität, Berlin
57. Ising H, Ising M (2001) Stressreaktionen von Kindern durch LKW-Lärm. Umweltmedizinischer Informationsdienst 1
58. Maschke C, Breinl S et al (1992) Der Einfluß von Nachtfluglärm auf den Schlaf und die Katecholaminausscheidung. Bundesgesundheitsblatt 35:119–122
59. Maschke C, Arndt D et al (Hrsg.) (1995) Nachtfluglärmwirkungen auf Anwohner. Schriftenreihe des Vereins für Wasser-, Boden- und Lufthygiene 96, 1–140
60. Sust C (1987) Geräusche mittlerer Intensität – Bestandsaufnahme ihrer Auswirkungen. Schriftenreihe der Bundesanstalt für Arbeitsschutz, Dortmund, Fb 497. Wirtschaftsverlag NRW, Bremerhaven
61. Cannon WB (1928) Neural organisation of emotional expression. In: Mirchinson C (eds) Feeling and emotions. Worcester (USA)
62. Felker B, Hubbard JR (1998) Influence of mental stress on the endocrine system. In: Hubbard JR, Workman EA (eds) Handbook of stress medicine – An organ system approach. CRC, London, pp 69–86
63. Simonov PW (1975) Widerspiegelungstheorie und Psychophysiologie der Emotionen. Verlag Volk und Gesundheit, Berlin
64. Miedema HME (1993) Response functions for environmental noise in residential areas. Nederlands Instituut voor Praeventieve Gezondheidszorg TNO, NIPG-Publikatienummer 92.021, Leiden (Niederlande)
65. Schulz TJ (1978) Synthesis of social surveys on noise annoyance. J Acoust Soc Am 64:377–405
66. Schuschke G, Maschke C (2001) Lärm als Umweltfaktor. In: Dott, Merk et al. (Hrsg.) Lehrbuch der Umweltmedizin. Wissenschaftliche Verlagsgesellschaft mbH, Stuttgart
67. Griefahn B (1985) Schlafverhalten und Geräusche. Enke Verlag, Stuttgart
68. Babisch (2000) Gesundheitliche Wirkungen von Umweltlärm. Lärmbekämpfung 47(3):95–102
69. Hörmann H (1974) Der Begriff der Anpassung in der Lärmforschung. Psychol Beiträge 16:152–167
70. Friedman M, Rosenman RH (1974) Type A behavior and your heart. Alfred A Knopf Inc., New York
71. Ising H, Babisch W, Kruppa B (1998) Ergebnisse epidemiologischer Forschung im Bereich Lärm. In: Gesundheitsrisiken durch Lärm. Tagungsband zum Symposium, Bonn 1998, 35–49. Bundesministerium für Umwelt, Naturschutz und Reaktorsicherheit (Hrsg.), Bonn
72. Guski R (1987) Lärm: Wirkungen unerwünschter Geräusche. Hans Huber, Stuttgart
73. Lercher P, Widmann U (1998) Der Beitrag verschiedener Akustik-Indikatoren für eine erweiterte Belästigungsanalyse in einer komplexen akustischen Situation nach Lärmschutzmaßnahmen. In: Fortschritte der Akustik, DAGA'98, Zürich
74. Fields JM (1985) The timing of noise-sensitive activities in residential areas. Nasa Contractor Report 177937. Hampton, VA (USA)
75. Hecht K, Maschke C et al (1999) Lärmmedizinisches Gutachten DA-Erweiterung Hamburg. Institut für Stressforschung (ISF), Berlin
76. McKennell AC (1963) Aircraft noise annoyance around Heathrow airport. Her Majesty's Stationary Office, London
77. Hatfield J, Job RFS et al (1998) Demographic variables may have a greater modifying effect on reaction to noise when noise exposure changes. Noise-Effects'98, 7th Int. Congress on Noise as a Public Health Problem, 2, 527–530
78. Guski R (1998) Psychological determinants of train noise annoyance. Euro-Noise 98(1):573–576
79. Guski R (1998) Fluglärmwirkungen – auch eine Sache des Vertrauens. Fortschritte der Akustik, DAGA'98, 28–29
80. Barkhausen H (1927) Ein neuer Schallpegelmesser für die Praxis. VDI-Z 71:1471–1474
81. Fastl H (1991) Beurteilung und Messung der wahrgenommenen äquivalenten Dauerlautheit. Z Lärmbekämpfung 38:98–103
82. Fastl H (1997) Gehörbezogene Geräuschbeurteilung. Fortschritte der Akustik, DAGA'97, Kiel, 57–64
83. Fastl H, Schmid W (1997) On the accuracy of loudness analysis systems. Proc. Inter-Noise'97, II:981–986
84. Fastl H, Schmid W (1998) Vergleich von Lautheits-Zeitmustern verschiedener Lautheits-Analysesysteme. Fortschritte der Akustik, DAGA'98
85. Fastl H, Widmann U (1993) Kalibriersignale für Meßsysteme zur Nachbildung von Lautheit, Schärfe, Schwankungsstärke und Rauhigkeit. Fortschritte der Akustik, DAGA'93
86. Fastl H, Widmann U et al. (1991) Zur Lärmminderung durch Geschwindigkeitsbeschränkungen. Fortschritte der Akustik, DAGA'91, DPG-GmbH, Bad Honnef, 751–754
87. Hashimoto T (1994) Die japanische Forschung zur Bewertung von Innengeräuschen im Pkw. Z Lärmbekämpfung 41:69–71
88. Interdisziplinärer Arbeitskreis für Lärmwirkungsfragen beim Umweltbundesamt (1982) Beeinträchtigung des Schlafes durch Lärm. Z Lärmbekämpfung 29:13–16
89. Interdisziplinärer Arbeitskreis für Lärmwirkungsfragen beim Umweltbundesamt (1983) Wirkungen von Lärm auf die Arbeiteffektivität. Z Lärmbekämpfung 30:1–3
90. Interdisziplinärer Arbeitskreis für Lärmwirkungsfragen beim Umweltbundesamt (1985) Beeinträchtigung der Kommunikation durch Lärm. Z Lärmbekämpfung 32:95–99
91. Kuwano S, Namba S (1978) On the loudness of road traffic noise of longer duration (20 min) in relation to instantanous judgement. J Acoust Soc Am 64:127–128

92. Schick A (1990) Schallbewertung. Springer, Berlin
93. WHO Ottawa (1986) Charta zur Gesundheitsförderung. In: Abelin T, Brzezinski ZJ (eds) Measurement in health promotion and protection. Kopenhagen (WHO Regional Publications) European Series No 22, 653–658
94. Widmann U (1992) Ein Modell der psychoakustischen Lästigkeit von Schallen und seine Anwendung in der Praxis der Lärmbeurteilung. Dissertation, TU München
95. Zwicker E (1960) Ein Verfahren zur Berechnung der Lautstärke. Acustica 10:304–308
96. Zwicker E (1978) Procedure for calculating loudness of temporally variable sounds. J Acoust Soc Am 62:675–682, Erratum: J Acoust Soc Am 63:283 (1978)
97. Zwicker E, Fastl H et al (1991) Program for calculating loudness according to DIN 45631 (ISO 532B). J Acoust Jpn (E) 12:39–42

Noise Emission Assessment

5

G. Hübner and E. Schorer

5.1 Basic Description and Standardisation

The acoustical efficiency of machines varies in the range of 10^{-9} to 10^{-5}. This means even high power machines generate sound powers of a few Watts only. Due to the high sensitivity of the human ear however, such low sound powers create close to the machine loudnesses higher than 100 phon (64 sone). Consequently, the assessment of machinery noise emission requires relations to these subjective properties.

5.1.1 Basic Physical Quantities

5.1.1.1 Introduction

The sound emission of machines and equipment is described by the sound power totally radiated into the surrounding environment. This sound power is weighted by human perception quantities and expressed on a level scale. The main emission quantity for machines[1] is characterized by the A-weighted sound power level.

The sound power of machines practically, independent of the acoustical properties of the relevant environmental properties, qualifies the sound power as an intrinsic quantity of the machine. Earlier descriptions of machine noise emission using sound pressure levels are functions of the measurement distance and depend in addition on the acoustical environmental quality.

For certain problems, the sound power information is necessary but not sufficient, these require additional data such as:

1. The distribution of the sound power as a function of frequency (octave, third-octave or FFT band spectra).
2. The time history for noises with heavy time fluctuations, e.g., by giving relevant data of the relevant maxima and minima.
3. The spatial variation of the sound power, e.g., by giving its directivity, especially if this variation is significant.

These additional data supplement the sound power that is based only on temporal and spatial averages.

Sound power data, completed by such information, enable the solution of the following problems:

(a) Comparison of the noise emission of machines of the same and of different types and sizes.
(b) The noise emission comparison within given prescribed limit data.
(c) The (approximate) calculation of the machine's sound pressure levels relevant for a given distance and environment.
(d) The sound power of several machines or of different parts of one machine by adding up various noncorrelated partial powers.
(e) The determination of the noises caused by one or several machines and transmitted through walls, ceilings, windows or hoods.

[1] In the following, the term "machine" is used for both machinery and equipment.

G. Hübner
Im Heidewinkel 18e, Berlin 13619, Germany

E. Schorer (✉)
Müller-BBM GmbH, Robert-Koch-Strasse 11, 82152 Planegg, Germany
e-mail: Edwin.Schorer@MuellerBBM.de

G. Müller and M. Möser (eds.), *Handbook of Engineering Acoustics*,
DOI 10.1007/978-3-540-69460-1_5, © Springer-Verlag Berlin Heidelberg 2013

(f) The development of noise-controlled machines.

For some groups of machines, it is usual to declare the sound emission by sound pressure quantities, too, e.g., by the so-called emission sound pressure level for moving sound sources (vehicles) or for very large ("noncompact") industrial plants. Such a noise declaration requires well-defined environmental conditions, measurement distances, and positions as well as operating and mounting conditions. In general, such emission data do not allow to calculate precise sound pressure values in other distances and positions and cannot be added with a reasonable uncertainty. Sound pressure measurements on a single position or as a mean value yielded from measurements on a line around the machine contours correspond to a sectional part of the relevant sound power only.

5.1.1.2 Sound Power, Sound Power Level, A-Weighted Sound Power, Sound Power in Frequency Bands

The sound power P of a machine is given by the sound energy per second radiated into the environment under prescribed operating and mounting conditions.

The sound power level L_W is defined by

$$L_W = 10 \lg \frac{P}{P_0} \mathrm{dB}, \tag{5.1}$$

with $P_0 = 10^{-12}$ W.

In general, the sound power of a machine covers all audible frequency components between 16 Hz and 16 kHz. Approximating the characteristics of the human ear, the sound power frequency components are weighted according to the A-weighting curve (see Sect. 2.3.2) yielding the so-called A-weighted sound power P_A, respectively, the A-weighted sound power level

$$L_{WA} = 10 \lg \frac{P_A}{P_0} \mathrm{dB}. \tag{5.2}$$

The (unweighted) sound power can be expressed in frequency bands using relevant filter systems such as octave, third octave, or narrow band devices. For these sound power spectra, the reference value $P_0 = 10^{-12}$ W is used too.

5.1.1.2.1 Sound Power Description for Sound Sources Under Free-Field Conditions

The sound power P of a machine operating under free-field conditions is determined by sound field quantities effective on a surface S enveloping the machine (see Fig. 5.1) following

$$P = \oiint_S \overline{pv}^t \cdot \mathrm{d}S = \oiint_S I_n \mathrm{d}S, \tag{5.3}$$

where $p(t)$ and the vector $v(t)$ are the instantaneous values of sound pressure and sound velocity at any enveloping surface S position. dS is the vectorial surface element of S and the scalar product $v \cdot dS$ indicates the projection of v to the dS-direction, which is the v-component v_n perpendicular to the surface S being of interest only (see Fig. 5.2). The bar over the product $\overline{pv}^t$ indicates the time averaging where this product prescribes the sound intensity vector

$$I = \overline{pv}^t, \tag{5.4}$$

and I_n is its component perpendicular to S, respectively, I_n is the I-component in the dS-direction. So I_n describes the sound energy passing per second the surface element dS per area unit.

If the surface S can be situated in an adequately large distance (far field) from the source (see Sect. 5.1.2.2.2), the sound intensity component I_n can be approximated by

$$I_n \approx \frac{1}{pc}\tilde{p}^2, \tag{5.5}$$

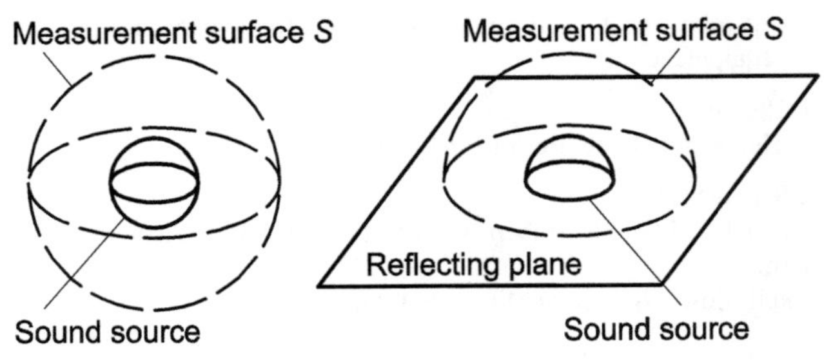

Fig. 5.1 Examples for measurement surfaces using the enveloping surface method. **a** Sound source with a spherical acoustic emission; **b** Sound source on a reflecting plane

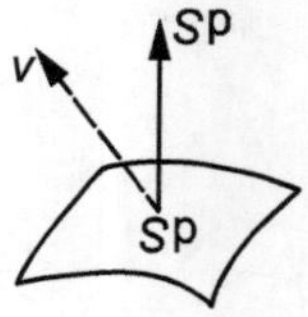

Fig. 5.2 Sound power flow through a surface element. Sound velocity v and vectorial surface element dS

and consequently we have an approximated sound power

$$P \approx \oiint_S \frac{1}{\rho c} \tilde{p}^2 \cdot \mathrm{d}S. \qquad (5.6)$$

The relevant level quantities follow as

$$L_W \approx \overline{L_p} + L_S + K_0, \qquad (5.7)$$

where $\overline{L_p}$ is the level of the square of the rms sound pressure averaged over S:

$$\overline{L_p} = 10 \lg \frac{\overline{\tilde{p}^2}^S}{p_0^2} \mathrm{dB}, \qquad (5.8)$$

called measurement surface sound pressure level with $p_0{}^2 = \left(2 \times 10^{-5}\right)^2 \mathrm{Pa}^2$,
and L_S is the measurement surface index

$$L_S = 10 \lg \frac{S}{S_0} \mathrm{dB}, \qquad (5.9)$$

with $S_0 = 1\ \mathrm{m}^2$, and finally, K_0 expresses a certain meteorological correction

$$K_0 = -10 \lg \frac{\rho c}{(\rho c)_0} \mathrm{dB}, \qquad (5.10)$$

using the reference value $(\rho c)_0 = 400\ \mathrm{Ns/m}^3$.

The correction K_0 takes into consideration the deviation of the characteristic acoustical impedance ρc under different meteorological conditions relevant for the actual measurement situation in relation to $400\ \mathrm{Ns/m}^3$.

For air with temperature Θ in Celsius and static pressure B in kPa, we have

$$K_0 = -10 \lg \left[\sqrt{\frac{313.51}{273 + \Theta}} \cdot \frac{B}{101.325} \right] \mathrm{dB}. \qquad (5.11)$$

For temperatures Θ and static pressures B existing in usual climates, the Eq. (5.11) yielded $|K_0| < 1$ dB excluding greater geographic heights H such as $H > 1{,}000$ m (see Sect. 5.1.2.2.2 too).

The sound power according to Eq. (5.7) is related to the meteorological condition relevant for the measurement and varies with measurement distance and shape of the measurement surface.

In the far field, the level of the A-weighted sound power L_{WA} can be expressed analogously to Eq. (5.7) by

$$L_{WA} \approx \bar{L}_{pA} + L_S + K_0. \qquad (5.12)$$

The exact value of the sound power L_W is determined by measuring the mean value of the intensity components I_{n} taken over S

$$\bar{L}_I = 10 \lg \frac{\bar{I}_{\mathrm{n}}^S}{I_0} \mathrm{dB}, \qquad (5.13)$$

with $I_0 = 10^{-12}\ \mathrm{W/m}^2$ and followed by

$$L_W = \bar{L}_I + L_S, \qquad (5.14)$$

where this L_W is the true sound power generated under the actual meteorological condition at the time of the measurement. This quantity is independent of the measurement distance – near field or far field – and of the shape of the enveloping measurement surface (for more details, see Sect. 5.1.2.2.6).

5.1.1.2.2 Sound Power Description for Sound Sources Under Approximate Free Field Conditions in the Presence of Parasitic Noises

Frequently the measurement of the sound power of a machine under practical environmental conditions must be carried out under nonideal conditions such as given by a free field disturbed by reflections and in the presence of background noises. In these cases, the previous equations must be modified or larger measurement uncertainties must be tolerated Eq. (5.7).

Under such practical conditions, the best results can be expected from the sound intensity method according to Eq. (5.14) requiring a minimum of modification.

In such nonideal conditions, the sound intensity $I_{\mathrm{n},\Sigma}$ at any position on the measurement surface is the total

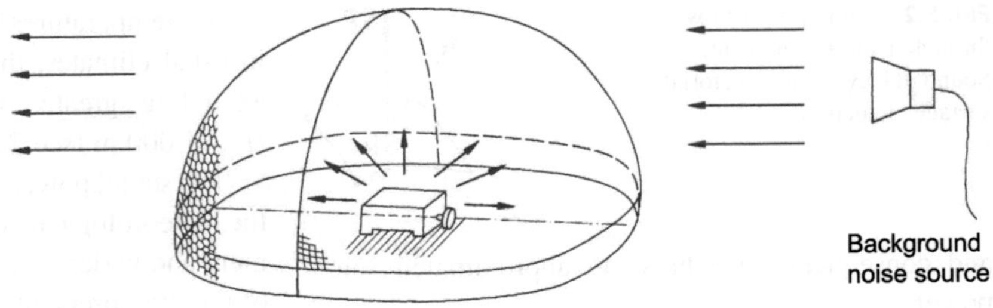

Fig. 5.3 Typical sound field situation under practical noise measurement conditions

sum of the intensities $I_{n,Q}, I_{n,par}, I_{n,int}$ designating the perpendicular intensity components caused by the source under test, by parasite noises (reflections and background noises) and by interaction effects:

$$I_{n,\Sigma} = I_{n,Q} + I_{n,par} + I_{n,int}. \tag{5.15}$$

The interaction component $I_{n,int}$ can be neglected for parasitic noises being not correlated with the sound of the source under test. Under the condition of stationary noises and of negligible sound absorption within S, we have

$$\oiint_S I_{n,\Sigma} \cdot dS = \oiint_S I_{n,Q} \cdot dS = P, \tag{5.16}$$

because

$$\oiint_S I_{n,par} \cdot dS = 0. \tag{5.17}$$

Equation (5.17) follows from the energy law for acoustical fields, it describes the balance of the parasitic sound, which incidences into the measuring surface on one side and leaves the measurement surface in the opposite direction (see Fig. 5.3).

Regarding the approximate sound power determination using sound pressure squared measurements according to Eqs. (5.6) and (5.7) for nonideal measurement conditions Eq. (5.15) is replaced by

$$\tilde{p}_\Sigma^2 = \tilde{p}_Q^2 + \tilde{p}_{par}^2 + \tilde{p}_{int}^2, \tag{5.18}$$

where the resulting rms the sound pressure square $\tilde{p}_\Sigma^2$ is the sum of the relevant sound pressure $\tilde{p}_Q^2$ radiated from the source under test and the sound pressures caused by the parasitic noise and interaction effects.

Assuming a neglected interaction component $\tilde{p}_{int}^2$, too, it must be stated that the enveloping integral over $\tilde{p}_{par}^2$ does not vanish due to the always positive sign of $\tilde{p}_{par}^2$ at each position on S

$$\oiint_S \frac{1}{\rho c} \tilde{p}_{par}^2 \cdot dS \neq 0, \tag{5.19}$$

leading to the basic difference to Eq. (5.16) so that

$$\oiint_S \frac{1}{\rho c} \tilde{p}_\Sigma^2 \cdot dS \neq P. \tag{5.20}$$

The quantity determined by $\oiint_S \frac{1}{\rho c} \tilde{p}_\Sigma^2 \cdot dS$ is contaminated by the parasitic noises and requires relevant corrections. This is a general disadvantage of the sound pressure method in comparison with the intensity method. The latter method compensates the parasitic noises automatically both for the parasitic noises $I_{n,back}$ caused by background noises from other disturbing sound sources and by diffuse components $I_{n,dif}$ from room reflections, which are compensated at each single measurement position on S, too.

$$I_{n,dif} = \frac{1}{T} \int_0^T (p \cdot v_n)_{dif} \cdot dt = 0. \tag{5.21}$$

Consequently, the sound power of machines installed under practical conditions for stationary noises can be determined by intensity measurement directly without any corrections in general.

5.1.1.2.3 Sound Power Description for Sound Sources Under Reverberant Conditions

The sound power level L_W of a sound source mounted in a reverberant room qualified according to ISO 3741[2] is determined by

[2] The [1–32] lists the international ISO and IEC Standards relevant for Sect. 5.1 together with its full titles.

$$L_W = L_m + \left[10 \lg \frac{A}{1 \text{ m}^2} - 6 + K_0 + K_{01} + K_{02}\right] \text{dB}, \tag{5.22}$$

in which L_m is the level of the spatial and temporal average of the squared sound pressure $p(t)^2$ measured in the qualified parts of the reverberant room and A is the equivalent absorption area (see ISO 3741, Chaps. 9 and 11). The value of K_0 is determined by Eq. (5.11) too. The so-called "Waterhouse-Term" is given by

$$K_{01} = 10 \lg\left(1 + \frac{S_R \cdot c}{8Vf_m}\right) \text{ and } K_{02} = -4.34\frac{A}{S_R}, \tag{5.23}$$

where S_R is the area of walls, ceiling and floor, and V the volume of the reverberant room, f_m is the center frequency of the relevant band width and c the sound velocity in the medium (air) during the measurement.

The Waterhouse-Term takes into consideration that the diffuse fields near the source and walls are not ideally established. These spatial sections are to be excluded from the measurement positions. For most frequencies, the positive value of K_{01} covers generally a range up to 2 dB.

Absorption area A and correction K_{01} are frequency dependent. Therefore, the equations are related to one specific band mid-frequency and the overall sound power and the A-weighted are to be calculated as the source of all band sound powers unweighted or A-weighted.

Equation (5.22) describes the sound power level under meteorological conditions relevant for the measurement.

5.1.1.2.4 Sound Power Description for a Sound Source Radiating Its Structure Borne Noise into the Surrounding Air

If a machine radiates the airborne sound power totally or partially by structure-borne vibrations of its outer surface, this relevant portion can be determined by

$$L_W = \bar{L}_v + \left[10 \lg \frac{S_1}{m^2} + 10 \lg \sigma + 10 \lg \frac{\rho c}{(\rho c)_0}\right] \text{dB}, \tag{5.24}$$

where $\bar{L}_v$ is the level of the averaged perpendicular outer surface vibration velocity squared component v_n, related to $v_0{}^2 = \left(5 \cdot 10^{-8}\right)^2 \text{ m}^2/\text{s}^2$

$$\bar{L}_v = 10 \lg \frac{\overline{\tilde{v}_n^2}^{S_1}}{v_0^2} \text{dB}, \tag{5.25}$$

and S_1 is the area of the outer surface of the machine, and finally σ describes the sound radiation factor of the machine.

For common machine types, the value of σ is more or less unknown, but in most cases $\sigma \leq 1$ can be assumed. Therefore, the relationship

$$L_W \leq \bar{L}_v + \left[10 \lg \frac{S_1}{m^2} + 10 \lg \frac{\rho c}{400 \frac{Ns}{m^3}}\right] \text{dB}, \tag{5.26}$$

sets an upper limit for the sound power caused by structure-borne vibrations. This may be a very helpful tool for a distinction between noises caused by the vibration of the machine's outer surface together with a measured total sound power.

5.1.1.3 Measurement Surface Sound Pressure Level and Measurement Surface Index

As described in Sect. 5.1.1.2.1, the measurement surface sound pressure level

$$\overline{L_p}^S = 10 \lg \frac{\overline{\tilde{p}^2}^S}{p_0^2} \text{dB}, \tag{5.8}$$

is calculated from the rms sound pressure squared $\tilde{p}^2$ taken and averaged over the measurement surface S and measured under ideal free-field conditions.

Under nonideal conditions, the measured rms sound pressure squared $\left(\tilde{p}^2\right)'$, respectively, its level

$$\overline{L_p'}^S = 10 \lg \frac{\overline{\left(\tilde{p}^2\right)'}^S}{p_0^2} \text{dB}, \tag{5.27}$$

must be corrected accordingly

$$\overline{L_p}^S = \overline{L_p'}^S - K_1 - K_2, \tag{5.28}$$

where the corrections K_1 and K_2 require additional measurements as described in detail in ISO 3744 and ISO 3746 [1–33]. In general, the values of K_1 and K_2 are positive, so that $\overline{L_p}^S \leq \overline{L_p'}^S$.

The determination of K_1 and K_2 includes additional uncertainties. Therefore, in the frame of the total uncertainties being prescribed by the relevant measurement standard, the amounts of the corrections are limited, e.g., for a class 2 sound power measurement according to ISO 3744, it is not allowed to use values K_1 and K_2 exceeding 1.3 dB, respectively, 4 dB. By this rule, the field of application of this standard is limited, too. In measurements in 1 m distance carried out under usual environmental conditions K_2-values vary in the range of 2–5 dB. Consequently, such measurements can often be qualified as class 3 only indicating greater uncertainties (ISO 3746).

The measurement of the surface sound pressure level $\bar{L}_p$ together with its measurement surface area index L_S prescribes the sound power of a machine under meteorological condition with $K_0 = 0$ with the same approximation as L_W for the A-weighted sound power level by Eq. (5.12).

$$L_{WA} \approx \bar{L}_{pA} + L_S. \tag{5.29}$$

For sound pressure determined sound powers a parallelepiped, hemisphere or sphere usually is prescribed by machinery-specific standards and measurement distances of 1 m are preferred.

5.1.1.4 Additional Data Describing Noises Varying in Time and Space

In cases where

(a) The noise fluctuates in time within a larger span
(b) The noise quantities show strong variations over the measurement surface especially under free-field conditions

The description of the noise emission of a machine by using the sound power should be complemented.

For fluctuations in time a distinction must be between short time variation such as for punch machines, for shooting noises, or for specific household appliances and long-term variations existing, e.g. in practical situations of traffic noises or for several situations at the work places regarded over one work day.

In the first case of the short time variations (for $\Delta t < 1$ s), the measured data depends significantly on the speed of the measurement equipment indication where its response is described as "slow", "fast" or by "impulse".

The difference of measurements carried out for both "slow" and "impulse" settings gives an information of the noise "impulsiveness" (see Sect. 2.3.1).

For long-term varying noises, it is recommended to subdivide the total duration of interest into smaller sections typical for working cycles and to determine for each of these sections the relevant mean values separately, where the "fast" indication is preferred (see Fig. 5.4c). An assessment of the noise of the total duration can be calculated by averaging the working cycles data.

If under free-field conditions, the difference on the measurement surface between maximum and minimum sound pressure level is greater than 10 dB, it is usual to call this sound radiation directional. But under practical machine installations, this difference

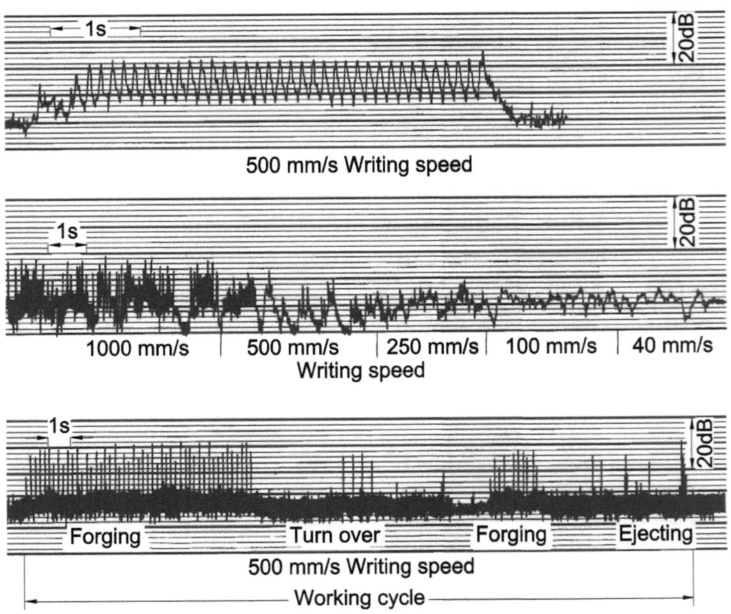

Fig. 5.4 Noise emission varying with time. **a** Impulsive noise of a punch machine; **b** Same impulsive noise measured with different writing speeds of the level recorder; **c** Noise emissions during a working cycle

in typical rooms is smoothed by the effects of room reflections if sound pressure measurements are used. Sound intensity measurements can help in these cases if the measurement surface is free from noise coming directly from sound sources outside the measurement surface of the machine under test.

Situations of sound sources with high directivity are usual, e.g., for large machine sets consisting of several single machines and for in-line loudspeakers at high frequencies.

5.1.1.5 Emission Sound Pressure Level

If work places are situated in a fixed position near a machine, e.g., as an operating place, a so-called emission sound pressure level is defined by the A-weighted-sound pressure level L_{pA} taken under free field conditions and a well defined operating condition (ISO11200 series). According to this definition the actual immission sound pressure level at the working location under usual environmental conditions deviates significantly from this free-field emission sound pressure level in general where unfortunately the actual immission value is higher than the free-field one. In so far the emission sound pressure level is an artificial quantity. Furthermore, the machinery specific relevant standards prescribe the emission sound pressure level under certain operating and mounting conditions, which may deviate from these conditions under true working situations. In several cases, these standardized conditions are chosen primarily under the consideration of a good reproducibility of the measurement result but seldom for the maximum sound radiation.

Consequently, sound pressure measurements taken under practical environmental conditions must be corrected to approximate the free-field emission sound pressure level. The international standardization takes this into consideration and offers several methods to determine such corrections which are described by five parts of ISO 11200 (ISO 11201–11205, [1–32])

ISO 11200 gives general advice for the problem and for handling the other parts.

ISO 11201 describes a precision method assuming exact free-field conditions. The result yielded by this method is the reference method for all other parts and needs no correction.

ISO 11202 describes a "survey"-method, that is a method determining a result with a low effort but with higher uncertainty.

ISO 11203 gives advice how to determine L_{pA} as a mean value around the machine calculated from the given relevant A-weighted sound power level being related to the adequate working distance. This information is useful in cases where the operator usually moves around or if a machinery specific working position cannot be defined. The measurement uncertainty in this case is assumed the same as indicated by sound power uncertainty class having used.

ISO 11204[3] demands the greatest effort compared with all other methods but provides a class 2 measurement uncertainty. The great effort of this method is caused by several additional measurements and calculations.

ISO 11205 is the newest part of the 11200 series using the intensity measurement technique. Thereby, the A-weighted total intensity vector I is determined by the measurement of its 3 A-weighted orthogonal intensity components. This method gives a very good approximation for the emission sound pressure level by $L_{pA} \approx 10\,\lg\frac{|I_A|}{I_0}$ dB. This was shown by an article issued for EuroNoise, München 1998 [33] and by a paper presented at Internoise 99, Fort Lauderdale, 1999 [34]. The great advantage of this method is the low effort for the measurements under practical environmental conditions not requiring any correction.

5.1.1.6 Further Emission Quantities

Before the standardized sound power measurement procedures were developed and established the so-called "radius referred sound pressure level", L_{Rm} was used describing the sound emission of machines. Under correct measurement conditions, these quantities are approximate sound power levels L_W as given by Eq. (5.7) but being related to reference values different from 10^{-12} W. For a hemispherical sound radiation, the L_{Rm} is given by

$$L_{Rm} = L_W - 8\ \mathrm{dB} - 20\,\lg\frac{R}{m}\mathrm{dB}, \qquad (5.30)$$

[3] See especially [33, 34].

where R is the reference radius of an equivalent hemispherical measurement surface. For the distances $R = 1$ m, 3 m or 10 m the L_{Rm} are systematically smaller values than L_W.

At present, the noise emission of moving sources, especially of railbound vehicles and motor vehicles is still determined by a sound pressure level measured at a certain distance of 25 m or at 7.5 m during the passing. But for construction vehicles and moving construction machines, the sound power level characterization is already used.

The tonality of a machinery noise often is regarded as an additional characteristic for a good noise-controlled machine. This quality can be described by measuring the relevant third-octave band spectrum and estimating the elevation between the level of the frequency band containing the tone to the two adjacent band levels. Such an elevation of 5 dB and more at frequencies higher than 300 Hz indicates the presence of an audible tone. Some machinery specific measurements procedures require an additional charge to the sound power level if a tonal noise is identified.

The directivity of the noise radiation of a machine can be described by the difference ΔL calculated from the maximum and minimum value of L_p, respectively, L_{pA} detected over the total measurement surface under free field conditions.

Finally, a directivity D_i for each individual i-th measurement position is defined by the difference of the sound pressure level measured under free-field conditions at the i-th position L_{pi} related to the averaged sound power level for the same measurement surface $\bar{L}_p$, preferably for such surfaces with a hemispherical shape

$$D_i = L_{pi} - \bar{L}_p. \tag{5.31}$$

5.1.2 Measurement Procedures[4] for Emission Quantities

5.1.2.1 Introduction

In general, the main goal of measuring the sound emission of a machine is the determination of its sound power levels. Yet, some additional characteristics will be obtained by these measurements automatically. Four different groups of procedures to determine the sound power using airborne sound quantities are standardized:

(a) The enveloping surface sound pressure method for measuring the rms sound pressure values under free-field or approximate free-field conditions (see Sect. 5.1.2.2.2 and ISO 3744, ISO 3745, ISO 3746 [1–32]).
(b) The enveloping surface sound intensity method for measuring the time averaged sound intensity component perpendicular to the total measurement surface (see Sect. 5.1.2.2.5) This procedure can be applied in (nearly) any environment ("in situ") i.e., the whole range from free field to reverberant conditions (ISO 9614 part 1, 2 and 3 [1–32]).
(c) The reverberant field procedure to be carried out in a qualified reverberant room (see Sect. 5.1.2.2.3, ISO 3741 [1–32]).
(d) The comparison procedure requires the use of a calibrated reference sound source having a known sound power level. This procedure can be applied under any environmental conditions ("in situ") but requires low background noise levels and some specific spatial positions and source dimension must be regarded (ISO 3747 [1–32]).

The standardized sound power measurement procedures distinguish two groups of requirements: frame documents and machinery-specific test codes.

(a) Frame documents ISO 3740-, ISO 9614-, and ISO 7849-series [1–32] covering general acoustical requirements for all kinds of stationary machines and include in detail
    1. Definitions and terminology
    2. Qualification criteria for the measurement environment
    3. Microphone and source array
    4. Quality of measurement equipment
    5. Calculation procedure for sound power determination
    6. Information to be recorded

[4] A summary of issues giving explanations, fundamentals, relevant investigations, and data for the measurement methods of noise emission quantities being presented by the relevant international standards both for sound pressure and for sound intensity methods is listed under references [35–73].

(b) Machinery-specific documents including
1. Definition of the machinery family under test
2. Mounting and operating conditions prescribed for the sound emission test
3. Selection of the most appropriate frame documents procedure
4. Selection of a prescribed shape and distance of the measurement surface for the sound pressure procedure
5. Additional quantities, if relevant (e.g., directivity, tonality, impulsiveness, extreme variations in time…)
6. Local definition of the work place if emission sound pressure level is to be determined

5.1.2.2 Frame Measurement Procedures

5.1.2.2.5 General

ISO 3740 [1–32] gives a general overview and advice for handling the procedures of the ISO 3740 series. Table 5.1 gives an overview for the different fields of application for the ISO 3740 series frame documents. The frame measurement documents distinguish three classes of uncertainty: Precision (class 1), engineering (class 2), and survey (class 3). At the beginning of all these documents, the upper limits for the procedures uncertainty in function of the band mid-frequencies and for the A-weighted sound power levels are given in a table. These uncertainties are expressed by the standard deviations of reproducibility and related to the acoustical frame procedure only. Uncertainties caused by the machinery-specific variations of mounting and operating must be taken into consideration additionally.

The precision method having the smallest measurement uncertainties requires a significantly higher number of measurement positions, higher environmental quality, and stability of the sound radiation and for the measurement effort compared, with the class 2 and 3 procedures having greater measurement uncertainties. The higher these requests certain acoustical "boundary conditions" exclude the application of the one or the other procedure (see Table 5.2).

5.1.2.2.6 Sound Pressure Enveloping Surface Measurement Under Approximated Free Field Conditions

The sound power level L_W, A-weighted or in frequency bands, using rms sound pressure squared measurement on an surface enveloping the machine, is determined under approximated free-field conditions by (see Eq. (5.7) and ISO 3744, ISO 3745, ISO 3746 [1–31])

$$L_W = \bar{L}_p + L_S + K_0, \tag{5.32}$$

where $\bar{L}_p$ is the measurement surface sound pressure level determined under exact free-field conditions. Regarding the sound pressure levels L'_{pi} being measured under the nonideal environmental conditions, a correction according to Eq. (5.28) must be determined by additional measurements

$$\overline{L_p} = \overline{L'_p} - K_1 - K_2. \tag{5.33}$$

The value of the corrections K_1 and K_2 decides on the adequacy of the sound field in respect to the

Table 5.1 ISO Sound pressure standard documents for noise emission measurements

Environment	Entirely anechoic laboratory room			ISO 3745
	Free-field above reflecting plane (semi-anechoic)	ISO 3746	ISO 3744	ISO 3745
	Large room or open air	ISO 3746	ISO 3744	
	Environment difficult to record	ISO 3746		
	Special reverberation room		ISO 3743	
	Laboratory reverberation room			ISO 3741 ISO 3742
		Control class(survey) accuracy class 3	Operation class (engineering)accuracy class 2	Precision class(precision/ laboratory)accuracy class 1

Table 5.2 Connection of sound pressure measuring methods with measured variables and requirements on the environment

	Enveloping surface method	Reverberation room method	Measurement with reference sound source
Determinable quantities	*Measuring surface sound pressure level, measuring surface measure, sound power level* as A-weighted (1 value indication) and as spectrum of the band levels of different bandwidths, standard emission value at the workplace	*Sound power level* as spectrum of band levels of different bandwidths; A-weighted total levels by calculation from band levels	*Sound power level* as for reverberation room method, *measuring surface sound pressure level and measuring surface measure* in anechoic room as for enveloping surface method
Nondeterminable quantities		Measuring surface sound pressure level, measuring surface measure, standard (local distribution of sound radiation) maximum and minimum values for very variable noises, emission value at the workplace	In reverberant rooms: like reverberation room method, otherwise: like enveloping surface method, emission value at the workplace
Permissible types of noise	All types of noise, also impulsive and tonal	Preferably stationary noise, for tonal noise the measurement is more difficult	In reverberant rooms: like reverberation room method, otherwise: like enveloping surface method
Non-permissible types of noise		Impulsive noises, for tonal noises the measurement is less difficult	In reverberant rooms: like reverberation room method, otherwise: like enveloping surface method
Requirements on the environment	In the area of the measuring surface; "free" sound field. Special criteria have to be fulfilled. Realized by: 1. All-round anechoic rooms 2. Anechoic rooms with reflecting floor or 3. Very large room for installation of machinery with low extraneous noise levels	The reverberation field must be within the measurement volume. Special requirements have to be fulfilled. Realised by: standard "reverberation room" (usually room volume > 200 m, stipulated range for the sound absorption coefficients depending on the frequency etc.)	No restrictions, apart from extraneous noise levels which are not too high compared with the sound pressure levels of the reference sound source at the measuring position
Restrictions on the dimension of the sound source	In the open: no, in anechoic rooms: volume of the source < 1% volume of the room	Volume of the sound source < 0.5–1% volume of the reverberation room	For sources with the dimension >2 m, the measurement is more difficult owing to more reference sound source positions

requirements of a sufficient good free-field approximation and a sufficient background noise difference.

The measurement surface sound pressure is the spatial average of the measured rms sound pressure squared for all measurement positions of the enveloping measurement surface being corrected according to K_1 and K_2. These microphone positions cover this surface uniformly and completely but can be positioned on one part of the total surface only if a relevant sound field symmetry exists.

For a class 2 measurement, the number N of microphone locations is sufficient if the difference between the maximum and the minimum of the sound pressure levels L'_{pi} in dB measured on S is smaller than N. Otherwise, the number N must be increased.

In general, the measurement surface follows approximately the machine's outer contour in a fixed distance. Practical reasons require simple measurement shapes such as parallelepiped, hemisphere, or sphere. A measurement distance of 1 m is preferred usually (see Fig. 5.5).

If each microphone position is correlated with the same portion of $S_i = S/N$ of the measurement surface S, the $\overline{L'_{pi}}$ is calculated as

$$\overline{L'_p} = 10\,\mathrm{lg}\left[\frac{1}{N}\sum_{i=1}^{N} 10^{0.1 \cdot L'_{pi}}\right]\ \mathrm{dB}. \tag{5.34}$$

For a nonuniform distribution of the microphone array, we have

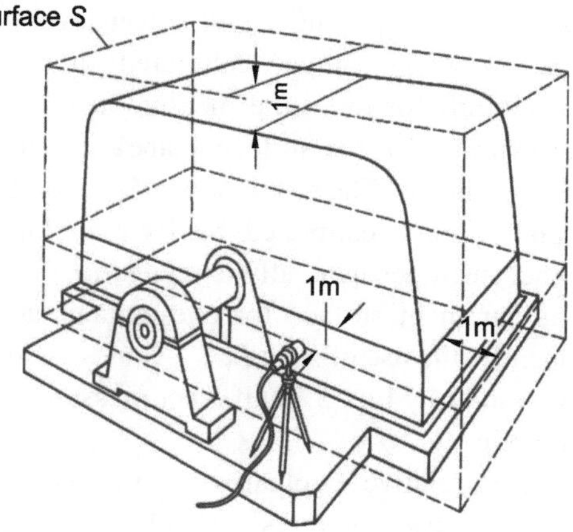

Fig. 5.5 Measurement surface for the noise measurement on a machine

$$\overline{L_p'} = 10\lg\left[\frac{1}{S}\sum_{i=1}^{N} S_i \cdot 10^{0.1 \cdot L_{pi}'}\right] \text{dB}. \quad (5.35)$$

If the difference max/min of the L_{pi}' along S is smaller than 5 dB, Eq. (5.34) can be replaced by the simple arithmetic mean of the level quantities L_{pi}'.

In general, the time averaging is taken by the measurement equipment automatically. But integration time must be regarded adequately in relation to the time history of the noise. The measurement equipment has to fulfill the requirements of IEC 60651, IEC 61672, IEC 60942, IEC 61094, IEC 60804, and IEC 60942 [1–32] for calibration.

The measurement surface area index L_S is determined according to Eq. (5.9) for the specifically chosen measurement shape.

The background noise correction K_1 is calculated by

$$K_1 = -10\ \lg\left(1 - 10^{-0.1\Delta L}\right)\text{dB}, \quad (5.36)$$

where $\Delta L = \overline{L_p'} - \overline{L_p''}$ is the difference between the levels of the averaged rms sound pressure square determined for a measurement of the machinery noise together with background noise $\overline{L_p'}$ and for the background noise alone $\overline{L''_p}$ when switching off the machine under test. For a class 2 measurement, values of $K_1 \leq 1.3$ dB are allowed.

The *environment correction* K_2 can be determined according to three different procedures (see ISO 3744, [1–32])

(a) By an absolute comparison test using a calibrated reference sound source

(b) By a relative comparison test with a small test sound source

(c) By the determination of the equivalent sound absorption area A of the relevant room. A can be determined by reverberation time measurement or estimating by tabulated absorption data. For a class 2 measurement, K_2 is limited to 4 dB.

The procedure (c) can be applied in closed rooms only and (a) and (b) in closed rooms and in the open landscape. The amount of K_2 gives a criterion for the adequacy of the sound field situation on the chosen measurement surface in respect to the requirements of an approximate free field.

Determining K_2 for the same room according to the three procedures will yield values that differ in general. Furthermore, the procedure (c) yields same K_2 if in the same room the sound source is located in the middle of the room or close to a wall or to a large reflecting object, which in general is not correct.

The characteristics of the reference sound source used for the absolute comparison tests are standardized by ISO 6926–2 [1–32]. This source should be situated (see ISO 3747 [1–32]) at the same location and measured on the same measurement surface S as the machine under test. If the machine cannot be moved, the reference source should be placed as close as possible to (on or besides) the machine. In order to calculate the amount of the correction K_2 according to procedure (a) the uncorrected sound pressure level, $\overline{L_p'}$ of the reference sound source must be determined for the actual measurement surface S, when switching off the machine under test and for sufficient low background noise level.

Then K_2 is given by

$$K_2 = \overline{L_p'} + 10\lg\frac{S}{\text{m}^2}\text{dB} - L_{W,\text{cal}}, \quad (5.37)$$

whereby $L_{W,cal}$ is the sound power level of the reference sound source calibrated under free-field conditions for the same measurement surface (same distance and same surface shape) carried out in frequency bands. The sound power $L_{W,cal}$ of the reference sound source calibrated by its manufacturer or by other institutes generally is related to a hemispherical measurement surface for one certain radius. Consequently, the use of such $L_{W,cal}$-values yields erroneous K_2-values by Eq. (5.37) if other measurement surfaces are relevant.

The relative comparison test, also called "two surfaces test", requires a small sound source with a stable sound radiation to be located in the same way as described before. During this test the machine under test is switched off and removed. At first, the uncorrected $\overline{L'_p}$ is determined for the artificial source on the same measurement surface S as is the actual machine. Then another $\overline{L'_{p1}}$ is determined for the test source for a different measurement surface S_1 being geometrically similar to S but having a different measurement distance. The difference

$$\Delta L = \overline{L'_{p1}} - \overline{L'_p}, \tag{5.38}$$

together with the ratio S/S_1 gives information about the K_2 value. For a semi-reverberant sound field (see Fig. 5.6) the actual K_2 on S can be calculated as

$$K_2 = 10\,\mathrm{lg}\frac{S_1/S - 1}{1 - M}\,\mathrm{dB} \text{ with } M = 10^{0.1 \cdot \Delta L}. \tag{5.39}$$

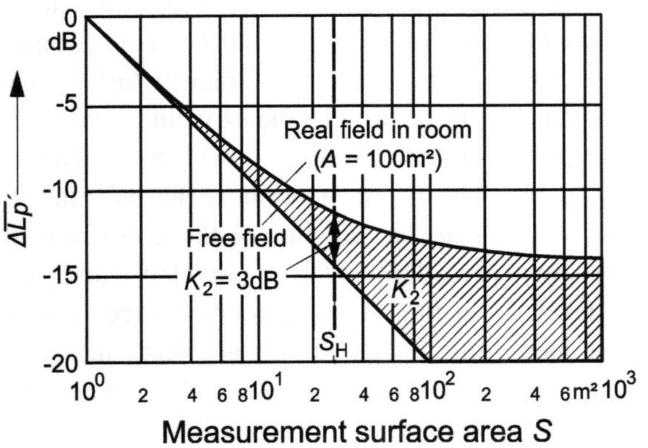

Fig. 5.6 Environmental correction K_2 of the semi-reverberant sound field with the absorption area $A = 100$ m², $\Delta\overline{L'_p} = f(S)$: Measurement surface sound pressure level versus measurement surface area S, S_H : Reverberant area

More details to this method are given under [35–53].

The reverberant time method yields a mean K_2-value related to the entire measurement room. Therefore, the K_2 of this method cannot distinguish between different positions in the relevant room and consequently allows only a rough K_2-determination. For a semi-reverberant sound field

$$K_2 = 10 \text{ lg}\left[1 + \frac{4S}{A}\right] \text{dB}, \tag{5.40}$$

whereby A is determined by measuring the reverberation time T in seconds and with $A/\mathrm{m}^2 = 0.16 \cdot \frac{V/\mathrm{m}^3}{T/\mathrm{s}}$ (see Chap. 11) and V the room volume expressed in m (see Fig. 5.7). Usually, the absorption area A and, consequently, the environment correction K_2 is frequency dependent, which requires its determination at all band mid-frequencies in the entire range of the audible frequencies in general (Table 5.3).

Any true qualification of the sound field under in situ conditions is related to that location in the room and the measurement distance only where the measurement surface being used is situated. Smaller K_2-values result from measurement surfaces closer to the machine.

Uncertainty of Sound Power Determinated In Situ According to Sound Pressure Enveloping Surface Method

The sound power determined according to the sound pressure squared method based on Eq. (5.32) deviates from the true sound power of the sound source

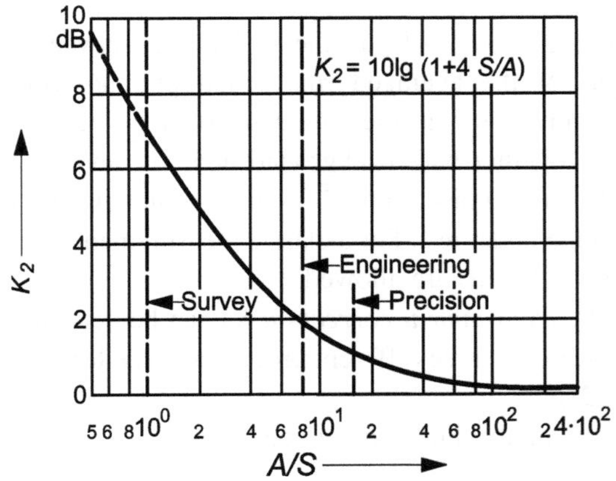

Fig. 5.7 Environmental correction K_2 of a semi-reverberant sound field

Table 5.3 Qualification according to ISO

Class	Qualification		
Precision class (Accuracy class 1) Precision/Laboratory	For the free field ("anechoic")	$K_2 \leqslant 1$ dB	$K_1 \leqslant 1.3$ dB
	For the free field above the reflecting surface ("semi-anechoic")	$K_2 \leqslant 1.5$ dB	$K_1 \leqslant 1.3$ dB
Operation class (Accuracy class 2) Engineering	Approximated free field above the reflecting surface	$K_{2A} \leqslant 2$ dB	$K_{1A} \leqslant 1.3$ dB
Control class (Accuracy class 3) Survey		$K_{2A} \leqslant 7$ dB	$K_{1A} \leqslant 3$ dB

systematically as well as randomly caused by several influences. These deviations depend on the different measurement parameters, e.g. on the measurement distance, the shape of the measurement surface, the number and the array of measurement positions, the level of the background noise, the environmental influences and on the radiation pattern of the sound field emitted by the machine.

An error analysis by which the total measurement uncertainty Δ_{tot} is split up in to five partial errors gives the opportunity to relate the causes for errors to measurement and source parameters (see [35–53], especially JASA 54, No. 4).

$$\Delta_{tot} = \Delta_0 \cdot \Delta_1 \cdot \Delta_2 \cdot \Delta_3 \cdot \Delta_4, \tag{5.41}$$

here Δ_0 is caused by variation of the radiated sound power of the machine to be tested. Reasons for this may be a change in the machine's noise generation caused, e.g., by changes of the machine temperature or by uncontrolled variations of meteorological influences. Obviously, this is no true measurement uncertainty but is included in the finally determined sound power result.

Δ_1 the near field error, is defined as the difference between the results of the two sound power procedures described by Eqs. (5.7) and (5.14). Following a decision of ISO/TC43 Acoustics, the sound power determined by the intensity measurement Eq. (5.14) is regarded as the true sound power. The sound power determined according to the sound pressure squared method Eq. (5.7) deviates due to two effects: the so-called angle error and the impedance error (see Fig. 5.8). The reasons are, on the one hand, the use of approximating the intensity as $|I| \sim \frac{1}{\rho c}\tilde{p}^2$ and, on the other hand, taking $|I|$ instead of the component I_n perpendicular to the measurement surface (angle error). The near field error can be reduced by choosing a large measurement distance and by taking a more convenient shape of the measurement surface.

Δ_2 the finity error caused by the limited number N of measurement positions, related to the infinite number of positions theoretically required by the integral of the Eqs. (5.3) and (5.6).

Δ_3 the environmental error caused by the uncertainties in determining the corrections K_1 and K_2.

Δ_4 the real measurement error caused by instrumentation quality, handling the instrumentation, and handling the measurement procedure.

These partial errors were investigated in detail [35–53] and yield the following knowledges:

(a) The sound power determined by sound pressure squared measurements is greater than the true sound power determined by sound intensity measurements.
(b) This difference remains below 5 dB, for frequencies higher than 100 Hz and measurement distances larger than 0.25 m. Under these conditions, a majority of machines show such L_{WA}-differences smaller than 2 dB.
(c) If the shape of the measurement surface follows the contour of the outer machine, the error Δ_1 can be minimized.
(d) Smaller and medium sized machines often have directivities similar to spherical sources between zero and first order.
(e) For higher frequencies ($f > 4$ kHz), the sound power uncertainty is more influenced by the microphone diameter than by most of the other parameters. Microphones with diameters $\leq$ ½" should be preferred.
(f) If the ratio $A/S > 10$ [see Eq. (5.40)], the error caused by an uncertain K_2 is < 1 dB.

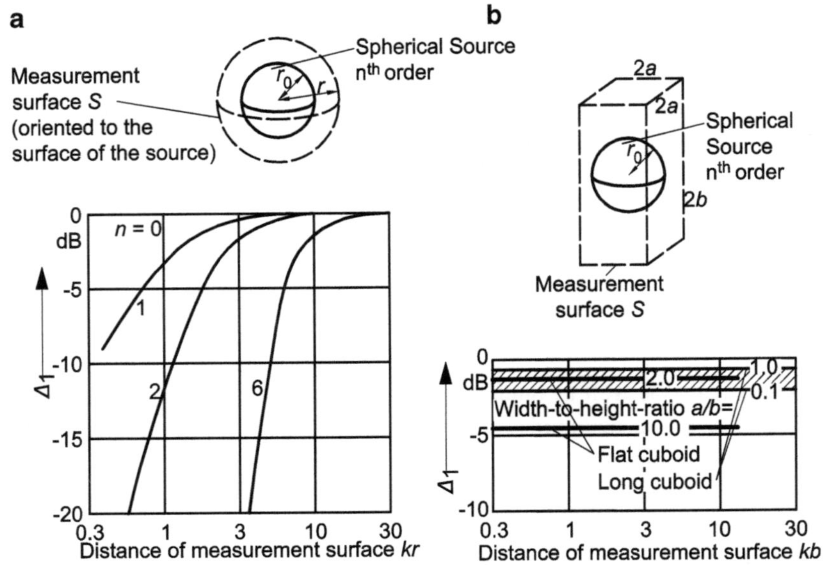

Fig. 5.8 Near field error Δ_1. **a** Impedance error for measurement surfaces oriented to the source surface; **b** Angle error for measurement surfaces not oriented to the source surface

5.1.2.2.7 Measurements in Reverberation Rooms

The sound power of a machine can be determined in a reverberant room, which is qualified according to ISO 3741 [1–31] *for a class 1 measurement* following Eq. (5.22). Expressing A by $A/\mathrm{m}^2 = 0.16 \cdot \frac{V/\mathrm{m}^3}{T/\mathrm{s}}$ and introducing Eq. (5.23), we have

$$L_W = L_{\mathrm{pm}} + \left[-10\,\lg\frac{T}{s} + 10\,\lg\frac{V}{m^3} + 10\lg\left(1 + \frac{S_R \cdot c}{8Vf_m}\right) - 4.34\frac{A}{S_R} - 10\,\lg\frac{B}{101.325\,\mathrm{kPa}}\right]\mathrm{dB}. \quad (5.42)$$

As explained in Sect. 5.1.1.2.3, the L_{pm} is the level of the sound pressure squared averaged over time and space. The spatial averaging can be carried out in the measurement room either by a discrete array of the microphones or along several (circular or spiral) paths by an automatically moved microphone. The number of microphone positions, respectively, the length and diameter of the circular paths depends on

(a) The quality of the reverberant room;
(b) The spectral distribution of the machine noise. Noises with tonal components require greater effort, e.g., by using higher numbers of microphone positions and by additional source locations;
(c) The measurement uncertainties required

The number of microphone and source positions is determined by an initial measurement by which for six different microphone positions the standard deviation

$$s = \sqrt{\frac{1}{5}\sum_{i=1}^{6}\left(L_{\mathrm{pi}} - L_{\mathrm{pm}}\right)^2}, \quad (5.43)$$

must be determined. As an example, a tonal component yields $s = 5.57$ dB for an ideal diffuse field [35–53]. Noises with tonal components create in ISO 3741-qualified reverberant rooms standard deviations s up to 15 dB. Depending on the value of s, three different noise characters are distinguished

Range I: $s \leq 1.5$ dB: broad band noise

Range II: 1.5 dB $< s \leq 3$ dB: noise with some narrow band components

Range III: $s > 3$ dB: noise with significant tonal components

This assessment includes both the quality of the actual reverberant room and the noise characteristics. This resulting quality can be improved by using the so-called moving vanes by which the spatial maximum and minimum of L_{pi} are smoothed by a sound field mixture [35–53]. By this procedure, the s-values, respectively, its range number can be reduced.

For Range II the number of required microphone positions N depending on the band mid frequencies f_m are between $N = 6$–15 (see Table 5 of ISO 3741) and the number of source positions N_S *for class 1 measurements* are given by

$$N_S \geq \frac{K^0}{2}\left\{0.79\left(\frac{T/s}{A/m^2}\right)\left(\frac{1{,}000}{f_m/\mathrm{Hz}}\right)^2 + \frac{1}{N}\right\}, \quad (5.44)$$

where the constant K^0 is prescribed by ISO 3741 between 5 and 25 [1].

For situation of Range III, the corresponding positions of microphones and source must be significantly increased compared with those for the ranges I and II.

The length l of the path of a continuously moved microphone is given by

$$l = N \cdot \frac{\lambda_m}{2}, \tag{5.45}$$

where λ_m is the wavelength of the band mid-frequency f_m. All these requirements for N and N_S are given to guarantee sound power levels of class 1 (precision) quality.

Reverberant Measurement According Class 2

These measurements following ISO 3743 [1] are carried out in a specifically designed "reverberant room" having a constant reverberation time between 500 Hz and 8 kHz within a given tolerance

$$T = K_T \times T_N, \tag{5.46}$$

where

$$K_T = 1 + \frac{257}{f_m \cdot V^{\frac{1}{3}}} \text{ and } 0.5\,\text{s} \leq T_N \leq 1.0\,\text{s}, \tag{5.47}$$

This procedure gives the advantage to determine $L_{pA,i}$, respectively, L_{WA} directly, means without measuring band levels as necessary for the class 1.

The class 2 reverberant room is qualified by one single test where the sound power of a test source is measured twice, one time in this room and second in a class 1 reverberant room. The difference between both results must not exceed 3–5 dB depending on the frequency band. The numbers N and N_S are determined from measured standard deviations similar to the class 1 procedure.

For a longer period in the past, this method was preferred for determining the sound power levels of household appliances.

Sound Power Determination in Reverberant Rooms: Uncertainties and Some General Rules

The relevant ISO standards (ISO 3741, ISO 3743) prescribing the determination of the sound power of sound sources in reverberant rooms give uncertainty values expressed by upper limits of standard deviations of reproducibility as function of the frequencies. For the class 1 procedure (precision), these standard deviations are between 0.5 and 3.0 dB and for the engineering procedure between 1.5 and 3.0 dB. The ISO standard gives several directions and rules which, are summarized below:

(a) Only stationary noises are allowed. Hence, impulsive noises are excluded at present, but newer developments may open the door for such noise classes, too;
(b) The volume of the sources under test should not exceed 1% of the room's volume V;
(c) The minimum of the room volume V depends on the lowest frequency $f_{\min}$ to be measured. For $f_{\min} = 100$ Hz a $V_{\min}$ must exceed 200 m^3;
(d) Certain edge ratios of the room dimension are required. The mean absorption coefficient should be in the range of 6–16%;
(e) The background noise level in any relevant frequency band should be at least 6 dB, if possible 12 dB below the levels of the noise from the source under test;
(f) The measurement equipment should fulfil the requirements of relevant IEC standards [22–32].

5.1.2.2.8 Sound Power Measurement With the Help of a Calibrated Reference Sound Source (Absolute Comparison Test)

For this procedure, a qualified measurement environment is not necessary. It can be carried out under "in situ" condition.

The characteristics of the reference sound source are prescribed by ISO 6926–2 [10]. The main requirements for usable reference sound sources are:

(a) A broad band noise radiation;
(b) Directivity $D_i \leq 3$ dB;
(c) A time-constant noise radiation. Variation in time should be limited by ± 0.5 dB;
(d) Its size should not be greater than 0.5 m in each direction;
(e) Its sound power level should be as high as possible;
(f) Its sound power should be calibrated in frequency bands by a recognized institute.

At first, the uncorrected surface sound pressure level $\overline{L'_p}$ of the machine under test is determined on the chosen measurement surface S. After that, the machine is removed from its location and the reference sound source be installed at the former place of the machine. Then the uncorrected surface sound pressure level $\overline{L'_{p,\mathrm{RSS}}}$ of the reference sound source on the same measurement surface S as used for the machine is determined. When determining $\overline{L'_p}$ and $\overline{L'_{p,\mathrm{RSS}}}$, the background noise influences should be negligible or corrected if possible. Then

$$L_W = L_{W,RSS} + \overline{L'_p} - \overline{L'_{p,\mathrm{RSS}}}, \tag{5.48}$$

where $L_{W,\mathrm{RSS}}$ is the sound power from the reference sound source calibrated for the same measurement surface as used for this comparison test. According to the different spectra of the machine under test and the reference sound source, the measurements according to Eq. (5.48) must be carried out for each frequency band of interest, generally in octave or one-third octave bands. A warning must be expressed with respect to the $L_{W,\mathrm{RSS}}$-value being given by the reference sound source manufacturer, which is generally related to a hemispherical calibration (see Sect. 5.1.2.2.2: K_2-determination).

If the machine cannot be removed from its location, the measurement of the reference sound source should be carried out for an "equivalent location" near the machine or the reference sound source is installed next to the machine ("juxtaposition method"). Such locations change the calibrated sound power of the reference sound source and consequently yield a slightly falsified L_W-result. For larger machines of a length exceeding 2 m, the measurements are to be carried out for several reference sound source positions besides and above the machine and a relevant mean value should be determined.

5.1.2.2.9 Sound Intensity Measurements[5]

The sound power of a machine using the sound intensity technique is determined according to ISO 9614 part 1, 2, or 3 [13–15]. The procedure is very similar to that for the pressure squared procedure according to ISO 3745, ISO 3744, ISO3746 but instead of determining the field quantity $\frac{1}{\rho c}\tilde{p}^2$ the sound intensity component $I_n = \overline{p.v_n}^t$ perpendicular to the enveloping measurement surface S has to be determined on S. Then

$$L_W = \overline{L_{I_n}} + L_S, \tag{5.49}$$

where $\overline{L_{I_n}}$ is the level of the spatial mean of the I_n-component taken over the measurement surface S and L_S is the surface area index as defined by Eq. (5.9).

The intensity method to determine the sound power of a machine has several great advantages compared with the sound pressure squared method. These main aspects are:

(a) The intensity method according to Eq. (5.49) yields a more precise value of the machine's sound power by eliminating the near field error automatically.
(b) The influence of parasitic noise caused by both undesirable environmental sound reflections and background noise from other sound sources is eliminated automatically, too. Relevant corrections are not necessary.
(c) The intensity measurements do not require specific measurement room qualifications and, consequently, can be carried out "in situ" where machines are installed usually. However, the ideal laboratory situations as given by semi-anechoic and reverberant rooms are covered by this technique, too (see Fig. 5.9).

The handling of the sound intensity procedure and the requirements for the sound intensity measurement equipment are prescribed by ISO 9614, parts 1, 2 and 3 [13–15]. Following these prescriptions, a certain upper limit for the sound power's uncertainty is guaranteed. There ISO standards give rules, e.g., for the number of measurement positions N for a discrete probe array, for the measurement distance d, for the required dynamic of the instrumentation set, and for the moving speed of the relevant scanning procedures.

The large field of application of the intensity procedure covering very different acoustical situations requires, in turn, very different measurement parameters. In order to guarantee certain upper limits for the yielded sound power level, it is necessary to adjust these measurement parameters to the acoustical situations according to the specific measurement. According to a proposal of G. Hübner [56–69], this adjustment, which later was established as a part of ISO 9614, should be carried out in two steps as follows (see Fig. 5.10):

[5] For more information see [1–32] and especially [56–69].

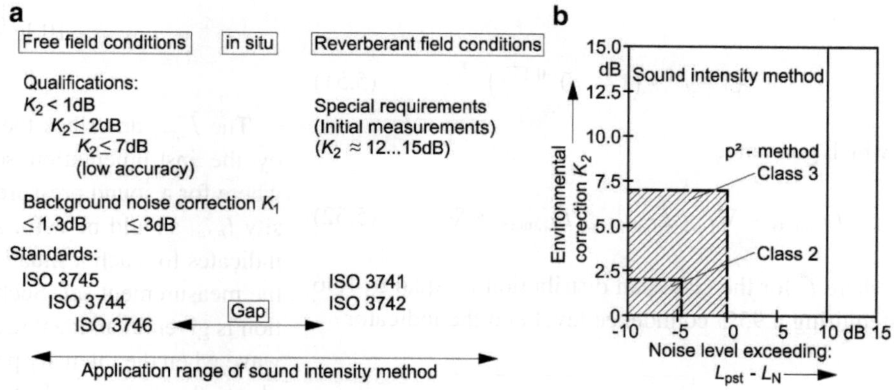

Fig. 5.9 Application range of measurement methods for sound power determination. **a** Comparisons of p^2- and I_n – method; **b** Application range of p^2- and I_n – enveloping surface methods

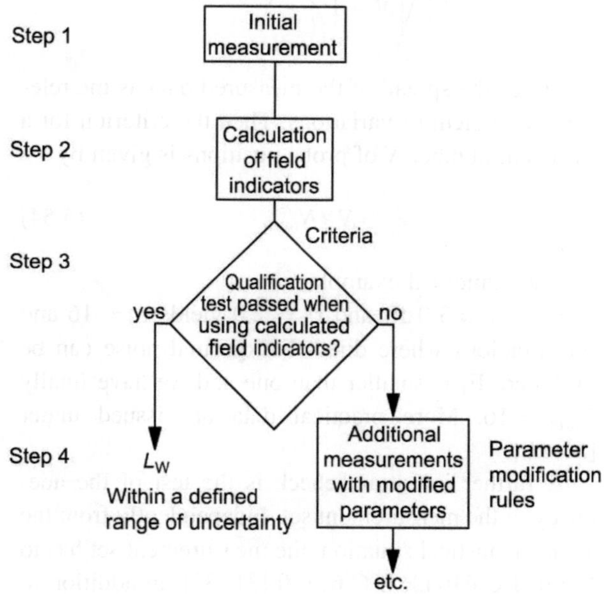

Fig. 5.10 Flow chart of a sound power measurement using the sound intensity method

(a) By an initial measurement on the chosen measurement surface, the relevant acoustical situation must be determined by certain data, called sound field indicators, which describe the relevant sound field situation objectively.

(b) Then these field indicators are introduced in so-called "criteria" for testing the adequacy of the chosen measurement parameters, such as the number N of probe positions, dynamics of the measurement equipment in relation to the actual sound field "dynamics" etc. If one or several of these criteria are not fulfilled, the field indicators give directions as to which of the measurement parameters should be changed and in which direction this is to be done. With the new measurement parameters, the initial measurement has to be repeated until the criteria are met. After having carried out these parameter adjustments, the final set of measurement parameters is determined and the sound power can be presented within a well-defined range of uncertainty.

The background of this parameter adjustment was published in several papers [56–69]. Details for its handling are prescribed by ISO 9614. Some information should be given as follows:

The number N of probe positions is determined by statistical laws describing the spread of randomly distributed measured values, by its confidence limits ∇ in relation to its mean value $\overline{L_{I_n}}$. Assuming statistical independence of these measured data, the following formula is used

$$N_{\min} = C \cdot F_4^2, \tag{5.50}$$

where

$$C = t^{*2} \cdot \left(1 - 10^{-0.1\nabla_1}\right)^{-2} \quad (5.51)$$

which guarantees

$$\bar{L}_{I_n,\text{mess}} - \nabla_1 \leq \bar{L}_{I_n,\text{true}} \leq \bar{L}_{I_n,\text{mess}} + \nabla_2, \quad (5.52)$$

where t^* for the Gaussian distribution is equal to 1.96 requiring a 95% confidence level and the indicator

$$F_4 = \frac{1}{|\bar{I}_n|}\sqrt{\frac{1}{N-1}\sum_{i=1}^{N}\left(I_{n,i} - \bar{I}_n\right)^2} \quad (5.53)$$

describes the spread of the measured data as the relevant coefficient of variations. Then the criterion for a sufficient number N of probe positions is given by

$$N \geqslant N_{\min}. \quad (5.54)$$

One numerical example:

For $\nabla_1 = 3.0\,\text{dB}$ and $t^* = 2.0$ yields $C = 16$ and for situations where direct background noise can be excluded, F_4 is smaller than one and we have finally $N_{\min} = 16$. More practical data are issued under [56–69].

A further important check is the test of the adequacy of the measurement set. Independently from the actual acoustical situation, the measurement set has to fulfill IEC 61043, IEC 62370 [31, 32]. In addition to this general requirement, a further indicator F_3 is introduced and must be determined by the initial measurement, too.

$$F_3 = \bar{L}_p - \bar{L}_{I_n} - 10\,\lg\frac{\rho c}{400\,\frac{\text{NS}}{\text{m}^3}}\,\text{dB}. \quad (5.55)$$

This indicator is given by the level difference of the surface average of the $\tilde{p}^2$ and I_n measured on the actual measurement surface S and "corrected" by K_0 (for more details of this criterion, see under [56–69]). Then the criterion of the instrumentation set adequacy is given by

$$L_D \geqslant F_3 + K, \quad (5.56)$$

where for class 2 measurements $K = 10$ dB and the so-called instrumentation dynamic is expressed by

$$L_D = -10\,\lg\frac{|I_{n,\text{res}}|}{\frac{1}{\rho c}\tilde{p}^2_{\text{res}}} \cdot \text{sgn}I_{n,\text{res}}. \quad (5.57)$$

The $I_{n,\text{res}}$ describes the residual intensity detected by the instrumentation set in the specific situation where for a sound pressure squared $\tilde{p}^2_{\text{res}} \neq 0$ the intensity $I_{n,\text{res}}$ should be zero. A measured value $I_{n,\text{res}} \neq 0$ indicates for such a situation a certain insufficiency of the measurement set. Such a specific acoustical situation is given in an ideal reverberant field or for a plane wave when the intensity probe is orientated perpendicular to the direction of its propagation. In practice, the frequency-dependent L_D is given by the manufacturer of the measurement set and can be determined by a specific calibrator.

5.1.2.2.10 Noise Declaration and Comparison of Noise Emission Data[6]

Noise emission data are required for comparing noise emitted by different products, for assessing noise emission against noise limits, for planning workshop noise levels, as well as for checking noise reduction achievements. Declared noise emission data are based on sound power or emission sound pressure levels L, as determined from measurements. As long as these values are yielded by different measurement procedures (previous chapters) and under different operating and meteorological conditions, their comparison and verification makes several requirements. Such requirements, dealing primarily with the handling of the procedures' uncertainty by means of the international standard ISO 4871 "Acoustics-Declaration and verification of noise emission data of machinery and equipment" completed by the implementation of a meteorologically standardized noise power level. (see e.g. ISO 3745 and ISO 9614.)

The noise emission declaration is given by the manufacturer or the supplier in technical documents or other literature concerning the products. According to ISO 4871, these declared data L_d have the form either of

- A single number or of
- A dual number noise emission value.

[6] For more information see [1–32] and especially [70–73]

Both alternative values inform about the uncertainty K^* of the measurement uncertainty expressed in decibels and stated as

$$L_d = L + K^*, \tag{5.58}$$

where L_d may be given as a single number of the sum of both terms or given separately by two numbers.

The measured value L can be obtained from one individual machine or as the mean value $\overline{L}$ representing the "noise quality" of a batch of a machine family manufactured according to the same technical specification.

The K^* should take into account the uncertainty with respect to the accuracy of the relevant measurement method considering reproducibility. For batches K^* includes the standard deviation of production as well. If no prior knowledge is available, the uncertainty K^*, in decibel, may be determined as follows

$$K^* \geqslant 1.5 \cdot \sigma_M, \tag{5.59}$$

with $\sigma_M = 2.5\,\text{dB}$ for an engineering grade measurement procedure, which is based on 5% risk of rejection. More details for determining K^* are given by ISO 7574 "Acoustics-Statistical[7] methods for determining and verifying stated noise emission values of machinery and equipment" part 1, 2, 4.

The declared noise emission value L_d for a single machine is verified either by $L_1 \leq L_d$ or by $L_1 \leq (L + K^*)$, where L_1 is the noise emission value measured for the verification one time on the relevant machine, initiated by the user or his representative.

The declared noise emission value L_d for a batch of machines, having measured L_1 on one machine of the relevant family only, the whole batch is verified either by $(L_d - L_1) \geqslant 3\,\text{dB}$ or by $[(L_d + K^*) - L_1] \geqslant 3\,\text{dB}$.

For more details, references are given to ISO 4871 and especially to ISO 7574.

The sound power levels defined in Sects. 5.1.1 and 5.1.2 are related to the actual meteorological conditions during the measurement (static pressure B_1 and temperature Θ_1). If this sound power is designated P_1, another result P_2 can be expected if the same machine is measured once more under different meteorological conditions B_2 and Θ_2. Such a deviation required a *meteorological normalized sound power* level $L_{W,N}$. ISO TC43 defines normalized meteorological conditions as $\Theta_N = 23°\text{C}$ corresponding to $T_1 = 273\ \text{K} + 23\ \text{K} = 296\ \text{K}$ and $B_N = 1.0132 \times 10^5\,\text{Pa}$. These values correspond to $\rho c = 411\,\text{Ns/m}^3$.

Consequently, a formula is applied to convert the values $L_{W,1}$ or $L_{W,2}$ to the normalised one $L_{W,N}$. The different noise generation mechanisms – airborne sound caused by a vibrating machine's outer surface and/or noise caused aerodynamically – are governed by different physical laws. As a compromise applicable for a certain limited (Θ, B range, the following formula was defined by ISO TC 43

$$\Delta L_W = -10\,\lg\frac{B}{B_0}\,\text{dB} + 15\,\lg\frac{T}{T_1}\,\text{dB}, \tag{5.60}$$

and finally

$$L_{W,N} = L_W + \Delta L_W. \tag{5.61}$$

For comparison of the sound power levels of machines as well as for acoustical planning, the use of normalized sound power levels related to a normalized meteorological condition is essential.

5.1.2.3 Machinery Specific Emission Test Codes

The basis of modern noise emission measurement standards are one or several of the frame documents of this series ISO 3740, ISO 7849, ISO 9614, or ISO 11200. For application on a specific family of machines, these frame documents require certain essential supplements which are represented by noise test codes covering the following aspects:

(a) A precise definition of the relevant family of machinery;
(b) The operation and mounting condition which should be used for the emission measurement;
(c) The shape of the measurement surface and the measurement distance to be used in general;
(d) The selection of the basic measurement procedures (i.e. frame documents applicable for this kind of machinery);
(e) If relevant, an indication dealing with tonality and impulsiveness.

[7] The following tests based on statistical laws exclude bias occurring due to the use of measurement procedures different from declaration and verification.

A small selection of such machinery specific noise test codes being available and issued by ISO and IEC is given as examples:
Rotating electrical machines, e.g. by ISO 1680
Transformers and reactors
Combustion engines
Compressors
Construction site machines
Electrical switch gears
Heating equipments
Office machines
Turbines
Cooling equipments
Lawn mowers
Household appliances
Pneumatic tools
Machine tools, driver pneumatically and electrically
Pumps

5.2 Measurement of Noise Emissions In Situ

Acoustic measurements of larger machines and plants are usually carried out locally at the installation or in machine test facilities but hardly in special acoustic test rooms. Such measurements, however, are often affected by environmental conditions that are not ideal from the acoustical point of view, because undesired sound reflections and high background noise levels at the measuring position are difficult to control, and so are variable meteorological conditions. Still, a number of reasons can make these so-called "in-situ" measurements seem the desirable or only possible methods.

For the acoustic acceptance test of a new or modified machine, it is preferable to carry out the measurements at the final installation, if the mounting and connection of the machine to surrounding equipment at the installation location cannot be adequately simulated at the manufacturer, if, for example, the connection of the machine to the foundation in the test lab of the manufacturer is less solid than in the final one. Besides that, the manufacturer often cannot operate the machines with the required load. Finally, the interaction of several machines connected to each other in the final operation, can result in a completely new type of noise emission different from the sum of the individual emissions.

A common task in the planning of noise control measures for industrial plants is the acoustic analysis of a larger plant, where the sound power levels of the individual aggregates have to be determined during full plant operation. In this case, the in-situ situation is given by the task, and the setting of special operating conditions, which would simplify the measurements (e.g., shutdown of strong disturbing noise sources or continuous operation of the unit to be measured, which normally runs intermittently or process-controlled), is not possible.

Before an acoustic measurement of an aggregate, which is part of a larger system, can be carried out, the operating conditions and the operating duration of the unit must be determined. In addition to the pertinent information from the plant operator, an understanding of the technical processes within the plant will be required, at least to such an extent that the task of the aggregate in that process is definitely known. Only with this knowledge, it is possible, for example, to predict the sound emission of a machine at load conditions different from the measured ones.

If owing to a high background noise level, a restricted measuring accuracy must be expected, multiple measuring methods can be used for a sound source. For example, in addition to an enveloping surface measurement of a centrifugal fan, an airborne noise measurement at the cooling air outlet of the fan motor and a structure-borne sound measurement of the console and the spiral casing of the fan can be carried out in order to determine the sound power levels of these sound sources separately and thus check the plausibility of the total measuring result.

For sound sources with an unsteady operating noise or intermittent operation, Sect. 5.1.1.4 should be observed. In case that very short-term sound events – especially those which are not easily reproducible – are to be evaluated, it is recommended to record the measuring signals by means of a (preferably digital) storage medium for a later analysis of the data in the laboratory.

5.2.1 Practical Aspects in the Application of Existing Measuring Methods

5.2.1.1 Airborne Sound Measurements with Omnidirectional Microphones

Today, the standard measuring microphone for Class 1 sound level meters is a ½-inch condenser microphone with free-field equalization and a sensitivity of 50 mV/Pa (see Chap. 2). It is a pressure-responsive receiver with a spherical response pattern. Besides the classical condenser microphones which require a polarization voltage, pre-polarized microphone cartridges are common which, utilizing electret field generation, do not need an external polarization voltage. One-inch microphones are used for the measurement of very low sound levels <30 dB, where an extremely low inherent noise of the microphone is essential. For measurements of extremely high frequencies (ultrasound up to 200 kHz) or at extremely high sound levels (up to 180 dB), 1/4 or 1/8 inch microphones have to be used.

When they are exposed to sound from the front in the free sound field, microphones for sound level measurements according to IEC 60651 [22] must have an even frequency response. This is achieved by the so-called free-field equalization of the microphone, which results in a frequency-proportional damping of the microphone membrane by means of an appropriate perforation in the counter electrode of the microphone (see Chap. 2, Fig. 2.2a). As for the mechanic membrane resonance, this attenuation also compensates for the level increase caused by the disturbance of the sound field ("pressure boost action") by the microphone membrane itself at wave lengths in the range of the microphone diameter. For sound exposure from the side or the rear, such a level increase does not occur. In this case, the free-field compensation of the microphone results in a decrease of the frequency response toward higher frequencies, i.e., at very high frequencies the microphone response characteristic is not any longer isotropical, but peaks toward the front (see Chap. 2, Fig. 2.2c). For a ½-inch microphone, the free-field equalization starts at frequencies above 2 kHz, the maximum attenuation for sound coming from the back depends on the microphone make, but is normally less than 10 dB at 20 kHz. In the manufacturer's test certificate of each microphone cartridge, the free-field equalization is stated as the difference between free-field frequency response and calibration grid frequency response. For measurements of sound sources with strong high pitched spectral components, the directional sound reception of the free-field microphone must be taken into account in order to avoid measuring errors owing to directivity. For omnidirectional sound measurements, e.g. in the reverberation room, diffuse-field corrected microphone cartridges are available.

Due to its principle of operation, the transverse sensitivity of a condenser microphone is very low to extraneous environmental impacts, such as temperature differences, vibrations, and above all magnetic fields. Thus, sound measurements with a condenser microphone can be carried out without any problem, for example, in very strong magnetic fields such as in Magnetic Resonance Tomographs.

In principle, the high-impedance input of the microphone preamplifier is a very sensitive receptor of electric stray fields, but normally this input is sufficiently shielded by the microphone cartridge directly screwed onto it. However, for sound measurements in very strong electric alternating field, e.g., near electric arc furnaces or inside or close to large electric machines, it is recommended to measure besides the sound also the electrostatically caused background noise level by replacing the microphone cartridge on the preamplifier by a dummy capacitor (capacity and casing equivalent to the microphone) at the measuring location and thus check the electric noise immunity of the entire measuring chain.

Attenuators consisting of a capacitive voltage divider, which are screwed between microphone cartridge and preamplifier, are impermeable to the DC-polarization voltage and thus only work in combination with a pre-polarized microphones. Besides, it has to be taken into account that such attenuators only increase the level range of the sound level meter, but not that of the microphone, i.e., the maximum permissible sound level of the microphone cartridge definitely determines the maximum measurable sound pressure level.

In order to reduce turbulence noise caused by air flow at the microphone, measurements in the open or near air flow, e.g., at an intake or outlet, must always be carried out with a wind screen. The usual ball-shaped wind screen, which consists of a suitable porous plastic foam, has a diameter of 100 mm and causes negligible additional damping, see Fig. 5.12.

Weatherproof (outdoor) microphones for use in long-term measuring stations differ from the usual ½-inch measuring microphones by a rain cap, which drains off water from above to the side of the cap, and a reinforced wind screen with protection spikes against birds. Moreover, they are usually equipped with a remote calibration device, an electric heating for the cartridge and the pre-amplifier and a line driver circuit for the transmission of the measuring signals over the required distance. Weatherproof microphones are either equalized for sound incidence from above (application for aircraft noise) or for lateral sound incidence (general environmental measurements).

For measurements of sound emission of wind turbines (WTGS) according to IEC 61400–11 [Chaet. 20, 157], the installation of the microphone at ground level on a sound-reflecting plane of specified size (interfacial microphone) with a special hemispherical wind screen is required. A second wind guard, e.g., consisting of a hemispherical wire frame with a plastic foam layer is needed in order to obtain a sufficient signal-to-noise ratio at high wind speeds and low frequencies. When using such a windscreen arrangement, the frequency response of the system must be recorded.

5.2.1.2 Airborne Sound Measurements with Directional Microphones

The preferred direction for sound reception of directional microphones is from front. Thus, an improvement of the signal-to-background noise ratio is possible in the free sound field, if the sound to be measured and the disturbing sound comes from different directions. Directional microphones for acoustic measurements with large measuring distances must have directivities as high as possible. Tubular directional microphones such as the MKH 815 or MKH 816 made by Sennheiser Electronic GmbH & Co. KG, Germany, proofed to be very effective. They consist of a condenser pressure-gradient receiver in an RF-circuit fitted to a directional tube, which works according to the interference principle [74]. These microphones, designed for broadcasting and TV-sound recordings from greater distances, have a lobar response pattern, see Fig. 5.11. Power supply is made through the microphone cable by means of phantom power supply.

As shown in Fig. 5.11, the directivity depends on the frequency. In the 8 kHz octave, the frontal aperture angle for a level decreases to -3 dB is $\pm 15°$ and rises continually toward lower frequencies to up to $\pm 60°$ at the 125 Hz octave centre frequency. The attenuation for sound coming from the rear ($180 \pm 30°$) is 25 dB in the 8 kHz octave, 20 dB between the 4 kHz and 500 Hz octave, and then drops to only 5 dB in the 63 Hz octave.

The transfer factor of a tubular directional microphone depends on frequency, with decreasing sensitivity toward lower frequencies. Thus, for evaluating sound level measurements with a directional microphone, the results must be corrected with regard to the frequency response. The correction factor is determined in 1/3-octave bands or in octave bands by simultaneous measurement of a test noise with a directional microphone and a calibrated microphone with omnidirectional characteristic in the free sound field. The frequency response correction function of the directional microphones is the difference of the results

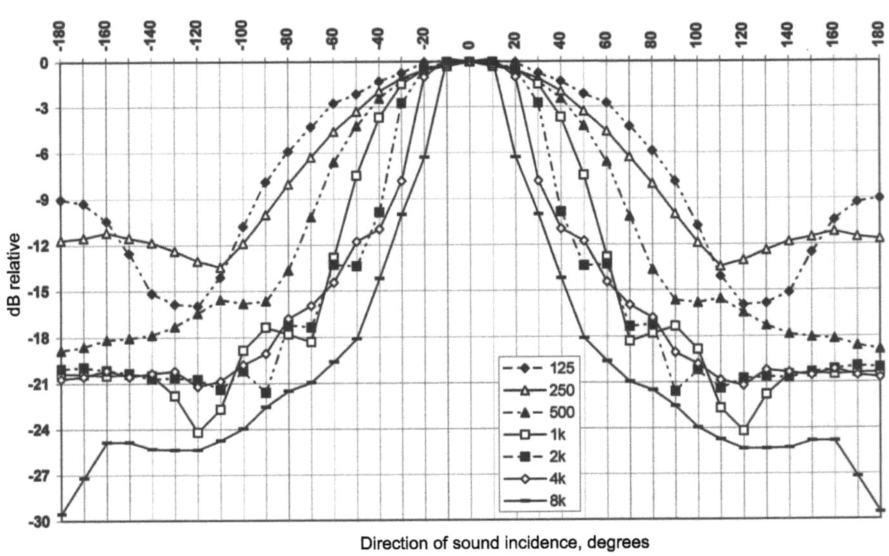

Fig. 5.11 Directional response pattern of the directional microphone MKH 816, manufactured by Fa. Sennheiser electronic, measured with octave-band noise. 0° corresponds to frontal sound incidence, the directivity is axially symmetric (lobe)

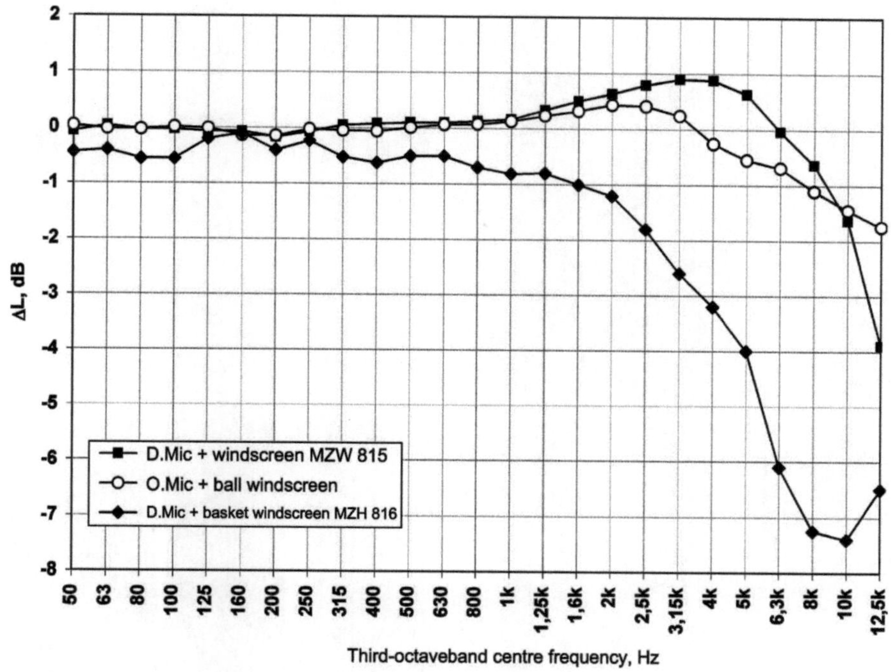

Fig. 5.12 Deviation of the sound pressure level received by a ½-inch condenser microphone with omnidirectional response pattern (O. Mic) and a directional microphone MKH 816 (D. Mic) caused by wind screens, measured with $^1/_3$-octave band noise

obtained with omnidirectional and with directional microphone, respectively. It must be individually determined for each directional microphone.

In the open, tubular directional microphones are very sensitive to wind and, thus, they can be used there only with a wind screen. A wind screening basket with or without a cover made of long-haired polyester fleece may be very effective for news-reporting work in the open. However, due to its bulkiness and its rather high additional attenuation at high frequencies – see Fig. 5.12 – it is considered impracticable for sound level measurements. A plastic foam wind screen with little additional attenuation, which is pulled over the microphone, is more suitable. For the determination of the frequency correction function of the directional microphone, the wind screen must also be taken into account.

By means of directional microphones, a better source separation can be achieved, e.g., for sound emission measurements of individual aggregates in the open or for sound measurements at points of interest within residential areas. For a plausibility check of the results, omnidirectional measurements should be carried out in parallel. In practice, especially for the measuring methods described in Sects. 5.2.3.1 and 5.2.3.2 for determining the sound emission of extensive sound sources, the use of a directional microphone is required in order to avoid the noise from disturbing sources sometimes in the immediate vicinity.

However, for continuous measurements in the open directional microphones are not suitable as they are not weather proof and temperature resistant.

For the localization of individual sound sources, e.g., of a motor vehicle (rolling noise, driving noise, exhaust noise, flow noise) a distinctly higher resolution is required than that provided by a tubular directional microphone. In this case, microphone arrays can be applied: The sound field is scanned by several microphones, which are arranged in regular specific distances and their output signals are added up. The sound reaching the microphone array frontally is reinforced as it has the same phase angle, the sound coming from the side is reduced because of its different phase angle or even extinguished. This results in a main directional lobe for sound incidence from the 0° direction and in less distinct side lobes for other sound incidence angles [75]. The directivity index 10 lg $\Gamma^2(\varphi)$ of microphone array (N individual microphones) in distance d depending on the angle φ between the incidence of the sound and of the array normal is defined as follows:

$$10\lg\Gamma^2(\varphi) = 10\lg\left(\frac{\sin(N\,\pi\,d\sin(\varphi)/\lambda)}{N\sin(\pi\,d\sin(\varphi)/\lambda)}\right)^2. \quad (5.62)$$

A directivity only exists for wavelengths $\lambda > 2d$. Depending on the geometric arrangement of the

Fig. 5.13 Vertical linear microphone array applied in the pass-by noise measurement of rail vehicles

Fig. 5.14 Cross-shaped microphone array

microphones in a row or in a cross or spiral (see Figs. 5.13 and 5.14), a one-dimensional (directional plane) or a two-dimensional focus of the transfer function results. The shape of the side lobes can be influenced by the geometric arrangement of the microphones and by the use of weighting functions in the signal processing (Dolph-Chebyschev, [75]).

If the time delay between the microphones for a sound incidence direction of $\varphi \neq 0$ is compensated in the evaluation, the array behaves like rotated physically by the angle φ. Thus, the direction of maximum sensitivity can be changed in the calculation, i.e. the focus can be changed without moving the microphone arrangement. A stationary sound source arrangement can now be scanned by changing the main reception direction step by step. For the measurement of moving sound sources, e.g., a passing vehicle, this method can be used to "follow" the sound source in order to obtain, for example, a longer averaging time [76].

5.2.1.3 Sound Pressure Measurements with Pressure Transducers

Sound pressure level measurements in pipelines and vessels, which are under high pressure or contain a medium different from air and may, therefore, not be opened, cannot be measured with standard microphones. Such measurements can be carried out with piezoelectric pressure transducers up to static pressures of 200 bar and temperatures of 200°C (or even higher when adapters for liquid-cooling are used), if it is possible to install the necessary threaded hole in the wall of the pipe before the measurements. A disadvantage is the comparatively low sensitivity of such pressure sensors. Normally only sound pressure levels of > 100 dB can be measured.

Owing to the flush position of the pressure sensor in the pipe wall, a correction of the measuring result is required, as for the determination of the sound power level in a pipe the average sound pressure level for the pipe cross-sectional area has to be determined. The

Table 5.4 Correction factors in dB for the determination of the average sound pressure level in pipes filled with air ($t = 20°C$) from measurements with flush mounted pressure transducers depending on octave centre frequency and nominal width of the pipe

Nominal width	63 Hz	125 Hz	250 Hz	500 Hz	1,000 Hz	2,000 Hz	4,000 Hz	8,000 Hz
DN 100	0	0	0	0	0	−1.4	−4.2	−6.4
DN 200	0	0	0	0	−1.4	−4.2	−6.4	−8.6
DN 300	0	0	0	0	−3.1	−5.5	−7.7	−9.8
DN 400	0	0	0	−1.4	−4.2	−6.4	−8.6	−10.7
DN 500	0	0	0	−2.4	−4.9	−7.1	−9.3	−11.4
DN 600	0	0	0	−3.1	−5.5	−7.7	−9.8	−11.9
DN 800	0	0	0	−4.2	−6.4	−8.6	−10.7	−12.8
DN 1000	0	0	0	−4.9	−7.1	−9.3	−11.4	−13.5

sound power level L_{Wi} in the pipe can be calculated according to Sect. 21.2.2. In case of a sound pressure measurement at a flush wall position, the correction factor K_d can also be estimated as follows:

$$K_d = 0 \qquad \text{for } k_o a < 1.5$$
$$K_d = -7\lg\left(\frac{2k_o a}{\sqrt{3}} - \frac{1}{k_o a}\right) \qquad \text{for } k_o a \geqslant 1.5. \quad (5.63)$$

$k_o = 2\pi f/c$ is the wave number of the medium in the pipe, a is the radius of the pipe. Correction factors for pipes filled with air at ambient temperature calculated in this way are shown in Table 5.4.

For lower frequencies and for pipe diameters small compared to the wavelength ($k_o a < 1.5$), the measurement results with the sensors installed in the wall widely correspond to measurements in the reverberation room if the reflections at the pipe end are considered. For $k_o a \geq 1.5$ and taking into account, the recommended correction factor between the measurements with flush-mounted sensors in the pipe wall and the measurements in the reverberation room, for a pipe with a nominal width of 400 mm (sound exposure of the channel from the front side) a difference in the sound power levels of ± 3 dB at max. results.

5.2.1.4 Structure-Borne Sound Measurements

5.2.1.4.11 Acceleration Measurements

The acceleration sensor used most frequently and available in a huge number of special models consists of a piezoelectric transducer installed between a seismic mass and the sensor basis, see Fig. 2.4. Owing to the piezoelectric effect, in case of vibrations the alternating force acting upon the transducer between the seismic mass leads to a charge transfer. This charge transfer can be picked off at the output terminals. Therefore, piezoelectric acceleration sensors are charge sources. Before further signal processing, their output signal must be conditioned by a charge amplifier [77].

The charge amplifiers required for each acceleration sensor can often lead to a complex and confusing measuring set-up, especially in multichannel measurements. Therefore, apart from the conventional charge-type accelerometers today acceleration sensors with on-board charge amplifiers are commonly used, which are supplied with a constant current via the signal line. Such sensors can be directly connected to measuring instruments with a corresponding supply.

A seismic acceleration sensor consists of a spring-mass system tuned to the upper end of the transfer frequency range. Depending on the size of the sensor, the natural resonance f_{res} of such a sensor is normally far above 10 kHz. The useful frequency range ends at approx. 30% of f_{res}, provided that an amplitude deviation of the linear frequency response of 10% is permissible. The attachment of the sensor to the measuring object is another spring-mass system which shifts the eigenfrequency of the whole arrangement toward lower frequencies (see Fig. 5.15). Therefore, the attachment method must be selected according to the measuring task.

For measurements of low-frequency vibrations up to approx. 1 kHz at measuring positions with very intense accelerations at high frequencies, e.g., the wheel set bearings of a rail vehicle, the acceleration sensor can be installed on a mechanical low-pass filter. This low-pass filter is a screw base with an elastic isolating layer, which is installed between the structure and the sensor and which reduces disturbing high-frequency vibrations in order to avoid overload of the sensor and the following electronic systems [77].

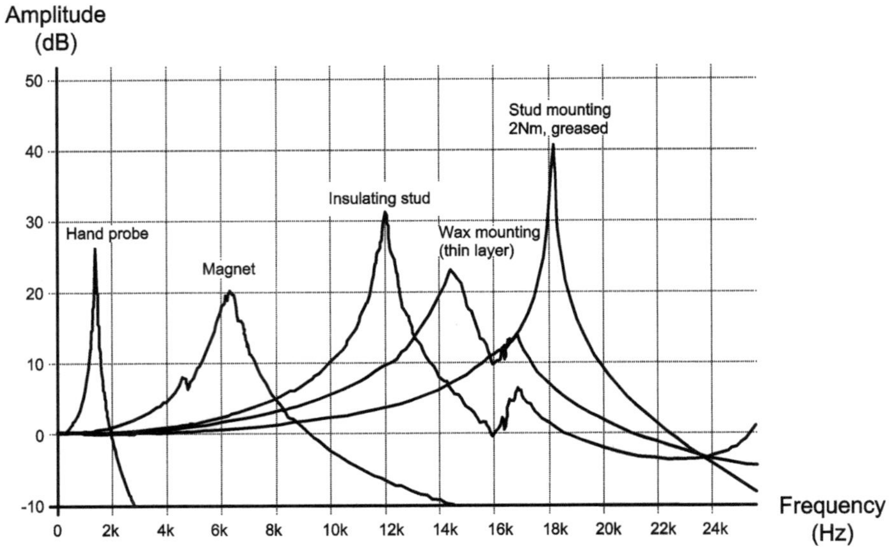

Fig. 5.15 Mounted resonance frequency of the accelerometer B&K 4370 obtained using several mounting techniques (acc. to Brüel & Kjaer information)

Especially for measurements at light or soft structures, the retroactive effect from the sensor mass to the vibrating structure has to be taken into account. The measuring error ΔL owing to the attenuation of the structure by the sensor mass can be estimated by rearranging Eq. (5.6) as follows:

$$\Delta L = 10\ \lg\left(1 + \frac{(2\pi f \cdot M)^2}{(2.3 \cdot c_L \cdot \rho \cdot h^2)^2}\right)\ \mathrm{dB}, \qquad (5.64)$$

with f frequency, Hz

M sensor mass in kg

c_L longitudinal wave velocity in the plate material, m/s

ρ density of the plate, kg/m

h thickness of the plate, m

Not for each measuring task a sensor with a sufficiently small mass is available. In this case, a frequency-depending correction of the measuring results by up to 6 dB is admissible according to the above equation.

Owing to the required firm connection, the acceleration sensor quickly takes on the temperature of the structure to be investigated. Therefore, for vibration measurements at very cold or hot parts, the permissible operating temperature range of the sensor has to be observed. Because of temperature dependence of the parameters, the specified lowest temperature for common acceleration sensors is usually specified to about −70°C. The upper limit temperature for common charge types is determined by the piezoelectric transducer material and usually is at approx. 250°C. For higher temperatures, an irreversible change in the sensitivity of the piezoelectric elements will occur. For measurements at very hot structures, water- or air-cooled mounting adaptors are required [77]. Special high-temperature sensors are available for temperatures up to 400°C.

For acceleration sensors with on-board charge amplifiers, the upper limit temperature is not determined by the piezoelectric sensor element, but by the electronic semiconductor components, and is at about 125°C.

The cross sensitivity of piezoelectric acceleration sensors to disturbances such as magnetic fields, radioactivity, humidity, or high sound levels is specified in the data sheet of the sensor and is normally small enough so that under common technical measuring conditions the measured results are reliable. However, before carrying out measurements under extreme measuring conditions (e.g., in the presence of ionizing radiation or very strong magnetic fields), the expected interference should be estimated by means of the specified cross sensitivity, the lowest measurable acceleration and the remaining dynamic range should be estimated beforehand.

5.2.1.4.12 Noncontact Vibration Measuring Methods

By means of laser vibrometers, a noncontact and thus completely nonreactive vibration measurement can be carried out, see also Sect. 2.2.3. Measurements, e.g., at hot components or components under high voltage can be carried out safely and simply with a laser vibrometer. Unlike accelerometers, laser vibrometers do not measure an absolute quantity of motion, but the velocity difference of the laser light source and the measuring object. That means that it cannot be taken from the measuring result, if the vibrometer or the measuring object has moved. In practice, this is not a problem as long as the vibrometer head can be mounted on a static basis (foundation, laboratory floor). However, for example for measurements on vehicles special attention is required. If necessary, the motion of the vibrometer in the measuring direction can be cancelled out by means of an additional accelerometer measurement on the mounting surface of the vibrometer and superposition of the output signals of vibrometer and accelerometer in proper phase relation.

5.2.2 In-Situ Measurement of the Noise Emission of Individual Sound Sources

5.2.2.1 Machines

As explained above and shown in Fig. 5.9, for the determination of the sound power of machines under common environmental conditions the enveloping surface method according to ISO 9614, which works with *sound intensity measurements* (see Sect. 5.1.2.2.5), can be used and in most cases is also sufficiently accurate.

In contrast, for the *enveloping surface sound pressure method*, the determination of the correction factors K_1 und K_2 under in-situ conditions is required (see Sect. 5.1.2.2.2). Considering the numerical values of these correction factors, it will be decided if the determination of the sound power by means of sound pressure measurements is actually possible regarding the acoustic conditions at the measuring position and the acceptable measurement uncertainties (see Fig. 5.9). Many rooms for the installation of machinery have an acoustic "retroactive effect of the room," which leads to K_2 values (environmental correction) between 2 and 5 dB(A) at a measuring distance of 1 m, so that in view of the K_2-criterion alone the class 2 measurement (engineering grade) must be cancelled; results obtained then correspond to the rather inexact class 3. If the background noise level is relatively high compared with the noise of the machine to be measured (criterion K_1, see Sect. 5.1.2.2.2), the sound pressure measurement is not allowed even according to accuracy class 3.

If K_1 and K_2 values are obtained which are only a little above the stated limits, they can be reduced – possibly sufficiently – by:

(a) Reducing the measuring surface S, i.e., by choosing smaller (but $d \geq 0.25$ m) measuring distances d
(b) Extending the sound absorptive surface of the measuring room
(c) Screening/(partial) enclosure of the background noise sources, if they cannot be completely switched off

The correction of the measuring results owing to reflections and extraneous noise is a principle in the determination of the sound power level. However, it also enables a clear separation of the machine noise emission the manufacturer is responsible for, and influences those are responsible for, who design the acoustic environment of the machine and the acoustic properties of the rooms where the machines are installed.

5.2.2.2 Measurements at Openings, Chimneys, Pipes, and Ducts

In addition to enveloping surface measurements, noise measurements with an omnidirectional microphone at openings without or with only very low air flow velocity can be carried out immediately at the opening surface, if required by the background noise level situation. In this case, sound field correction values according to Fig. 5.16 must be applied.

When measuring the sound emission of flue gas outlets, the measuring microphone must be held out of the gas flow in order to avoid the generation of disturbing sound due to turbulence at the microphone. For exhaust-gas pipes of vehicles, the so-called "Q point" is often specified as standard measuring position (ISO 3095) [1], see Fig. 5.17. The microphone is held at a distance of 0.86 m above the opening and at a distance of 0.5 m from the outside of the pipe, which corresponds to a distance of 1 m between the edge of the opening and the microphone, see Fig. 5.17.

In order to calculate the sound power level of the exhaust gas opening from this measurement, normally the hemisphere surface $2\pi r^2$ is required. The radius r can be calculated depending on the diameter D_R of the exhaust-gas pipe as follows:

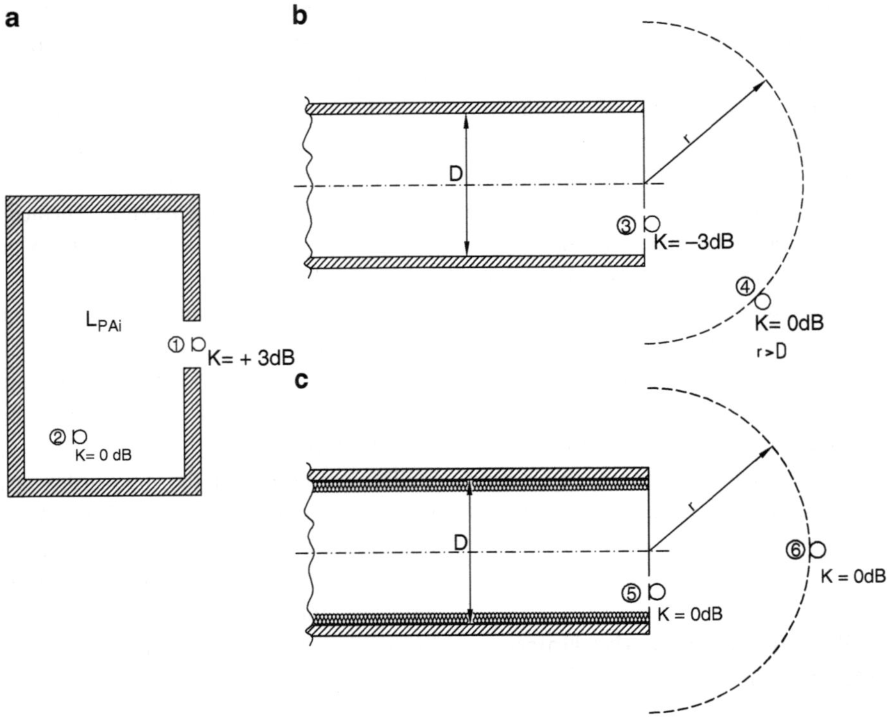

Fig. 5.16 Sound field corrections for measurements with an omnidirectional microphone. **a** Determination of the diffuse sound pressure level in a room by measurement inside the room (2) or in the semi-diffuse sound field (1) of an orifice in the room's boundary surface. **b** Determination of the sound power level of a pipe vent radiating sound into the open by measurement of the sound pressure level in the free field (4) on an enveloping surface ($S_4 = 2\pi r^2, r \geq D$) or in the semi-diffuse sound field (3) of the orifice ($S_3 = D^2\pi/4$). **c** Determination of the sound power level radiated by an acoustically lined duct vent by measurement of the sound pressure level in the free field (6) on an enveloping surface ($S_6 = 2\pi r^2, r \geq D$) or in the free sound field (5) of the orifice ($S_5 = D^2\pi/4$).

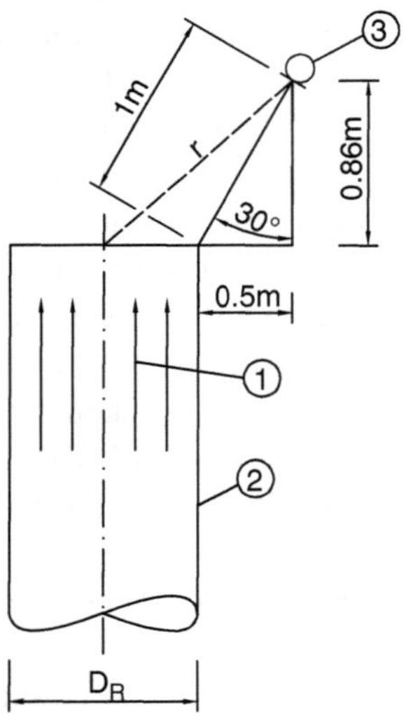

Fig. 5.17 'Q point' microphone position (3) for the determination of the sound emission of a flue gas duct (2) outside the gas flow (1). The radius r of the enveloping surface depends on the pipe diameter D_R and can be calculated according to Eq. (5.65)

$$r = \sqrt{\left(\frac{D_R/\text{m}}{2} + 0.5\right)^2} + 0.74\,\text{m}. \qquad (5.65)$$

Due to their restricted accessibility, the sound emission of chimney openings can rarely be determined by a direct measurement. It can be calculated from the sound power level determined at the chimney base according to the method described in VDI 3733, Appendix D [78] – e.g., by measurement with a hot-gas probe according to Fig. 5.18.

In general, the sound power level L_{Wi} in the interior of a channel or a pipeline can be calculated from the average sound pressure level $\bar{L}$ of the channel cross section, the cross-sectional area S in m^2, the characteristic sound impedance of the flow medium and, if necessary, a correction factor K_d according to (20.38). The correction factor K_d for higher modes in the pipe depends on the respective measuring method, see also Sect. 5.2.1.3.

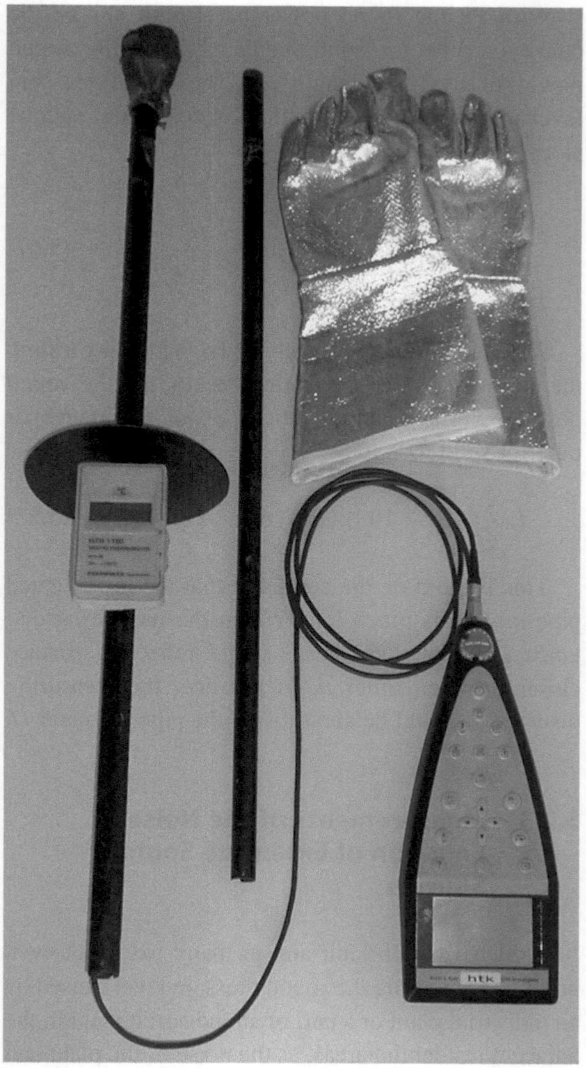

Fig. 5.18 This heat-protected microphone probe (with hand heat shield, on-board thermometer and extension rod) is equipped with a standard ½-inch condenser microphone and is suited for short-term measurements inside flue gas ducts or similar piping with gas temperatures up to 600°C

In order to determine the sound power level in a pipe, the sound pressure level within the pipe can be measured with a standard microphone if the pipe is filled with air and nearly pressureless, if there is no flow or only with a very low velocity (approx. <5 m/s) and if the temperature in the pipe permits such a measurement. For the measurement, an opening (blind flange, sensor stud or similar) of at least the microphone diameter must be available in the pipe wall, so that the microphone can be placed in the interior. During the measurements, any remaining opening has to be covered with cloth or the like to avoid flow noise caused by air flowing in or out.

In channels with flow turbulent pressure, fluctuation occurs at the microphone (pseudo sound), which can falsify the measuring results considerably. In order to reduce these turbulent pressure fluctuations, the microphone can be equipped with a turbulence screen (Friedrich probe) [79]. This screen consists of a metal pipe with a length of 540 mm and an interior diameter of 13 mm which has at one side a slot 400 mm long and 1 mm wide, covered with a flow resistance consisting of a four-layer metal fabric ($Z = 1c$–$3c$). The entire probe including the microphone is placed in the channel. For in-situ measurements, an opening with a diameter of approx. 100 mm in the pipe wall and a fixture for adjusting the probe in the pipe from the outside is required. As a rule, permanent measuring installations in test stands have installed probe fixtures. At the front (in the direction of the sound source), the Friedrich probe ends in a nose cone, the far end holds a ½ inch microphone with preamplifier. This microphone screen recommended in ISO 5136 [80] can be applied in measuring channels with a diameter range of $0.15 \text{ m} \leq D_R \leq 2.0 \text{ m}$ for flow velocities up to 30 m/s. The sound pressure levels L_{pi} received with the microphone screen have to be offset against the combined frequency correction:

$$C = C_1 + C_2 + C_3 + C_4, \tag{5.66}$$

with:

C_1 free field correction of the microphone in dB;

C_2 frequency response correction of the microphone screen in dB;

C_3 flow velocity frequency response correction in dB;

C_4 modal frequency response correction in dB.

The transfer constant C_2 of the microphone screen can be determined experimentally according to the method for directional microphones explained in ISO 5136, see Sect. 5.2.1.2. For the determination of the remaining correction factors, we would like to refer to the standard [80].

The main application of the Friedrich probe are noise measurements at fans. However, the application is restricted to flow velocities ≤ 30 m/s and pipe diameters ≥ 150 mm. These restrictions can be overcome by using the wall slot probe according to a proposal of M. Hubert [81]. In this case, the microphone including the tube attachment is outside of the pipe. In this way, the flow area is not blocked and, so, the subsequent disturbance of the flow and sound field is avoided. When using the Hubert probe, frequency response corrections must be made just as for the Friedrich probe. However, a decisive disadvantage of the Hubert probe is the required slit of 400 × 1 mm in the pipe wall, which practically restricts the use of the Hubert probe to stationary test stands.

With both the Friedrich probe and the Hubert probe, standardized measurements can be carried out only in channels with a circular cross-section, made of sheet steel of a thickness of at least 4 mm or of other material of corresponding weight per unit area and stiffness, so that the inner walls are sufficiently sound reflecting. The channel extensions to either side of the probe location must be at least four channel diameters or a length of at least one half the wavelength long (center frequency of the lowest frequency band of interest), and furthermore, the channel must be terminated anechoically: see [80].

Short-time measurements in nearly pressureless hot-gas channels, like the flue gas channels and heat-recovery boilers of gas turbine plants, can be carried out using heat-protected microphone probes according to Fig. 5.18. For such measurements, an orifice or blind flange of at least 50 mm diameter must be available at the measuring position [82].

The sound emission of pipelines and channels can be determined from the inner sound power taking into account the transmission loss of the pipe wall and the attenuation along the pipe, see Sect. 20.2 and VDI 3733, Appendix B [78]. The sound emission of pneumatic conveyer pipelines for solid media is a special case (see Sect. 20.11) because of the collisions of particles with the pipe wall. Such a pipeline is in fact a line sound source with many incoherent individual sound sources. Figure 5.19 shows the situation for a sound measurement of such a pipe with an omnidirectional microphone.

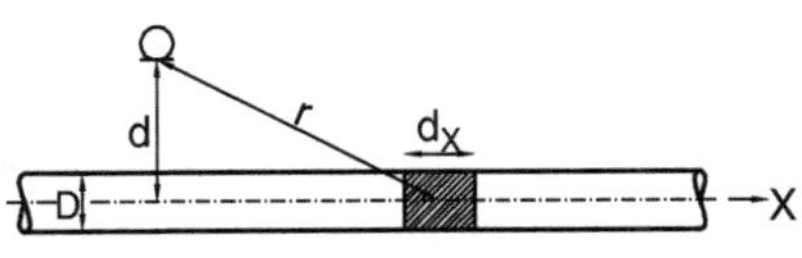

Fig. 5.19 Sound measurement at a conveyor pipeline for bulk material with an omnidirectional microphone

With P', the sound power per length unit of the conveyor pipe for solid media, $P'dx$ is the sound power of the pipe length dx. The sound pressure level measured with an omnidirectional microphone at an infinite pipe is:

$$p^2 = \int_{-\infty}^{+\infty} \frac{P'dx}{4\pi r^2} = \frac{P'}{4\pi} \int_{-\infty}^{+\infty} \frac{dx}{d^2 + x^2}. \tag{5.67}$$

By evaluating the integral and taking the logarithm, the following simple equation results for the sound power level per length unit, $L_{W'}$, of the conveyor pipe for solid media:

$$L_{W'} = L_p + 10\,\lg(2\pi d/\mathrm{m}) - 2\,\mathrm{dB/m}. \tag{5.68}$$

This is valid on the condition that the investigated pipe section is much longer than the measuring distance and that there is no sound reflecting surface closer than ten times d. In practice, the measuring distance d should be about twice the pipe diameter D.

5.2.3 Measurements of the Noise Emission of Extensive Sound Sources

It is often very difficult and in many cases not even possible to measure the sound pressure level caused by an industrial plant or a part of an industrial plant in the adjacent residential areas, as the noise of the plant and the extraneous noise cannot be separated. The extraneous noise can be caused by traffic, by other industrial plants or by other parts of the same industrial plant. These difficulties can be avoided by initially determining the sound emission of the plant and then calculating the sound pressure level at a point of interest – the sound immission – by means of a sound propagation calculation (see Chap. 7).

In theory, the sound power of an industrial plant can be determined in the same way as for individual machines by using the enveloping surface method. An assumed enveloping surface is put around the plant and the sound pressure level is measured at many points that are evenly distributed on this assumed surface. However, this would require much time and energy, especially regarding elevated measuring positions above the plant.

Therefore, measuring methods were developed which enable the determination of the immission-relevant sound power level of extensive industrial plants with reduced effort. Two of these methods – the "window" method and the "perimeter path" method – are based on the consideration that for the sound immission in the environment of the plant only the sound power emitted approximately horizontally or in a small angle of elevation is important. Therefore, measuring positions at the edge of the plant are sufficient and measurements above the plant are not necessary. The sound propagation situation between an industrial plant and an immission position for downwind conditions is shown schematically in Fig. 5.20.

Assuming an average wind speed gradient of $7\,\frac{\text{m/s}}{\text{m}}$ independent from height, the sound rays are circular paths with a radius of curvature of $a = 5{,}000$ m. The elevation angle β of the sound ray emitted from the industrial plant can be calculated from the average height of the plant, the height of the immission position above the plant and the distance between plant and immission position. It will normally be 2–10 degrees.

Figure 5.20 shows typical microphone positions for different measuring methods. For the 'fence measurement,' the measuring position is mostly near the ground, the distance to the plant is not exactly defined (Fig. 5.20, measuring points 'F'). For the 'window' method, usually the measuring positions are at greater distances to the plant and at greater heights than those in the 'perimeter path' method (Fig. 5.20, measuring points 'W' and 'P').

The 'fence measurement' is not a defined method with fixed rules, but describes the occasionally used enveloping surface method for the determination of the sound power level of individual machines to large plants in a very simplified form: At a few or only one easily accessible measuring position (i.e., which is near the ground) on an assumed enveloping surface around the plant the sound pressure level is measured with a hand-held sound level meter. The measuring height is usually approx. 1.2 m. The sound power level of the plant can be calculated as the sum of the measured sound pressure level and the surface index of the assumed enveloping surface. The term 'fence measurement' results from the occasional desire to be able to measure at the fence of industrial plants and to conclude from those measuring results on the sound immision of parts of the plant or even the whole plant. The factory fence is a measuring position of special interest, as here measurements can be carried out in a public area at any time and even without knowledge and/or consent of the operating company.

As the layout of the fence in relation to the plant is not generally defined, there may be very different measuring positions and various errors may occur, therefore. Regarding measuring positions near a plant, the sound emission of upper plant components, which is essential for sound immision, usually cannot be recorded as the measuring position near the ground is in a 'sound shadow', screened by the plant itself. An average sound power determined by such a measurement would possibly be considerably lower than the actual sound power. On the other hand, also errors with a positive sign can occur for measurements near the plant if essential individual sound sources are in the vicinity of the measuring position.

But also fence measuring positions near the ground which are at a larger distance to the plant are not suitable for evaluating the sound immission as there are normally remarkable sources of error. On the one hand, owing to the small measuring height only the sound propagation near the ground is recorded. The attenuation of this sound with distance is larger than that at a greater height. On the other hand, the share of background noise by road traffic or other plants can already be very high at these measuring positions. Therefore, 'fence' measuring positions can be used

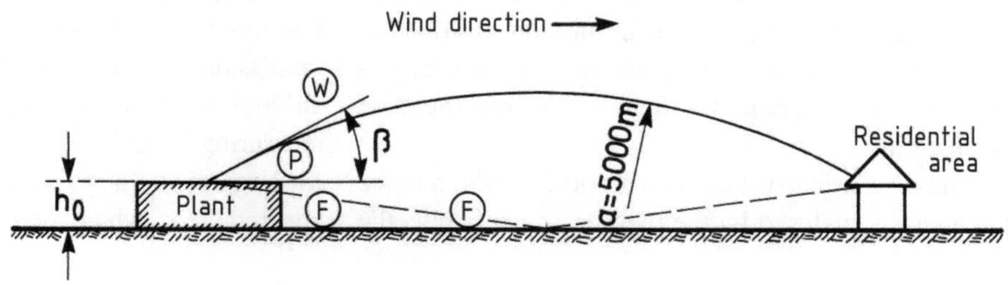

Fig. 5.20 Sound emission of an industrial plant under down-wind conditions; microphone positions for a 'fence' measurement (F), for the 'perimeter path' method (P) and for the 'window' method (W)

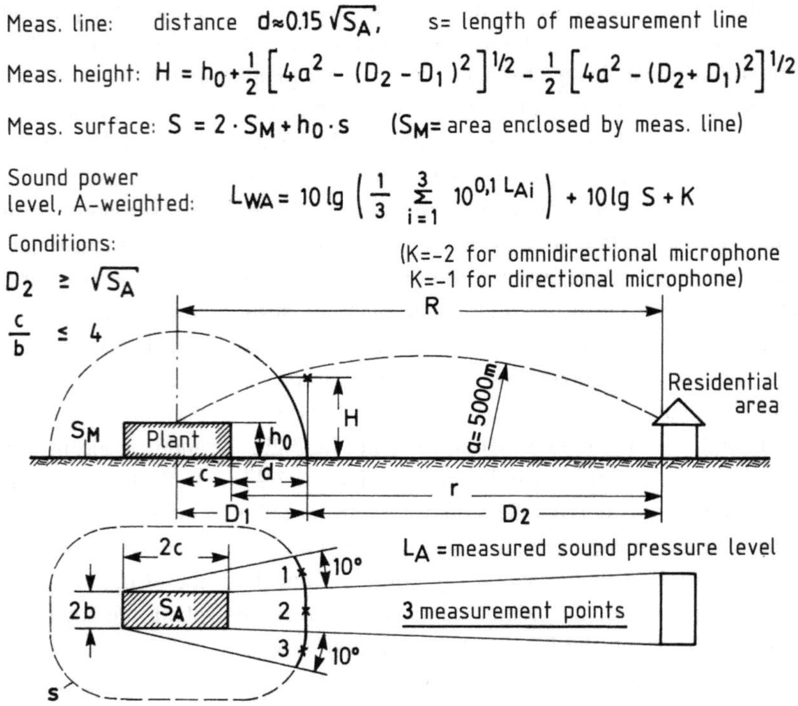

Fig. 5.21 'Window' measurement method

only to get a general idea and not for determining the immission-relevant sound power level of a plant [83].

5.2.3.1 Window Method

Figure 5.21 gives an overview of the 'window method', which goes back to a proposal by Meurers [84]. The principle of this method is to measure only sound that leaves the plant in direction to the immission position through an assumed 'window' in the enveloping surface around the plant. The sound pressure level is measured at three positions in the window and the results are averaged energetically. The measuring distance from the edge of the plant should be about 15% the average extension of the plant.

Apart from the areal expansion and the average building height h_0 of the plant, the measuring height H also depends on the distance D_2 between measuring position and immission position of interest. This means that for testing several immission positions at different distances from the plant, more than one measurement at different microphone heights must be carried out.

The measuring surface area S corresponds to twice the area S_M enclosed by the measuring line s plus the circumferential surface $h_0 \cdot s$. The immission-relevant sound power level is the sum of the energetic average value of the three measured sound pressure levels, the measuring surface area index and a correction factor K for correction of the geometric near field error ($K = -2$ dB for measurements with an omnidirectional microphone) or the near field error and the directivity of the microphone ($K = -2$ dB $+ 1$ dB $= -1$ dB for measurements with a directional microphone). A correction for the (small) measuring error owing to absorption by air is not intended. The window method can only be applied if the distance to the immission position is large compared to the size of the plant and if the ratio of length to width of the plant footprint is not very large ($c/b \leq 4$).

The evaluation of the measurements can be carried out comparatively easy and does not require a computer.

An important disadvantage of the window method is that the measuring position depends on the respective immission position, i.e., for each immission position an individual measurement with the corresponding measuring height and distance is required. Moreover, for large plants large measuring distances and heights are required, which sometimes are not accessible.

Therefore, the window method is hardly ever used in practice, but the more universal perimeter path method is preferred.

5.2.3.2 Perimeter Path Method

Figure 5.22 shows the conditions for the perimeter path method according to Stüber and Fritz [85]. The measuring path around the plant is chosen in a way that the average distance $\bar{R}$ between the plant borderline and the measuring path is approx. 5% of the average extension of the plant, but at least 5 m. Then measuring positions are chosen on the measuring path in the distance d, with

$$d \leq 2\bar{R}. \tag{5.69}$$

The measuring height H_1 should be chosen according to the following rule:

$$H_1 = h_0 + 0.025\sqrt{S_m}, \tag{5.70}$$

whereby h_0 is the characteristic average building height of the plant and S_m is the area enclosed by the measuring path.

The measuring height is independent of the distance to the immission position. For extremely high plants, a second microphone height H_2 can be chosen in order to increase the measuring accuracy. If the plant has noise sources which are located very high above the characteristic height of the plant so that they cannot be recorded by the measurement along the measuring path, the sound power levels of such sound sources should be recorded separately by use of a suitable method and the results should then be added to those of the perimeter path method.

The measurement of the sound pressure level is normally carried out by means of a directional microphone positioned horizontally toward the plant which can be moved from measuring position to measuring position along the measuring path by means of a mobile tripod, see Fig. 5.23. Measurement and evaluation are carried out in octave bandwidths.

The immission-relevant octave sound power level of the plant is determined according to

$$L_W = \bar{L}_p + \Delta L_S + \Delta L_F + \Delta L_M + \Delta L_\alpha, \tag{5.71}$$

with $\bar{L}_p$ energetic average of all sound pressure levels measured along the measuring path, in dB

ΔL_S measuring surface area index, in dB

ΔL_F near field correction, in dB

ΔL_M correction for the frequency response of the microphone, in dB

ΔL_α air absorption factor, in dB

The individual correction factors can be calculated as follows:

$$\Delta L_S = 10\lg\left(\frac{2S_m + h_0 l}{S_0}\right)\ \text{dB}, \tag{5.72}$$

in this equation, l is the length of the measuring path in m and S_0 is the reference area 1 m^2

$$\Delta L_F = \lg\frac{\bar{R}}{4\sqrt{S_p}}\ \text{dB}, \tag{5.73}$$

with the plant area S_p.

ΔL_M: The microphone correction is 0 dB for omnidirectional microphones. For directional microphones, a frequency-dependent correction factor according to Sect. 5.2.1.2 has to be applied.

$$\Delta L_\alpha = 0,5\alpha\sqrt{S_m}\ \text{dB}, \tag{5.74}$$

typical values for the air absorption factor α are given in Sect. 7.3.

The A-weighted total sound power level of the plant L_{WA} can be calculated by summation of the octave bands according to the following equation:

$$L_{WA} = 10\lg\left(\sum 10^{0.1(L_{Wj}+C_j)}\right)\ \text{dB}, \tag{5.75}$$

whereby C_j is the correction factor for A-weighting of the jth octave band.

An efficient evaluation of measurements according to the perimeter path method can be made with computer support. By separately evaluating the sound pressure levels measured at the measuring lines 1–8 (see Fig. 5.22), apart from the immission-relevant sound power level also an eventual horizontal directional effect of the plant can be detected in the form of a directivity factor for each of the sectors 1–4.

The perimeter path method is a measuring method that can be used universally for plants with horizontal extensions from 10 m to approx. 300 m. Regarding both the footprint of the plant and the direction and distance of the immission positions of interest, there

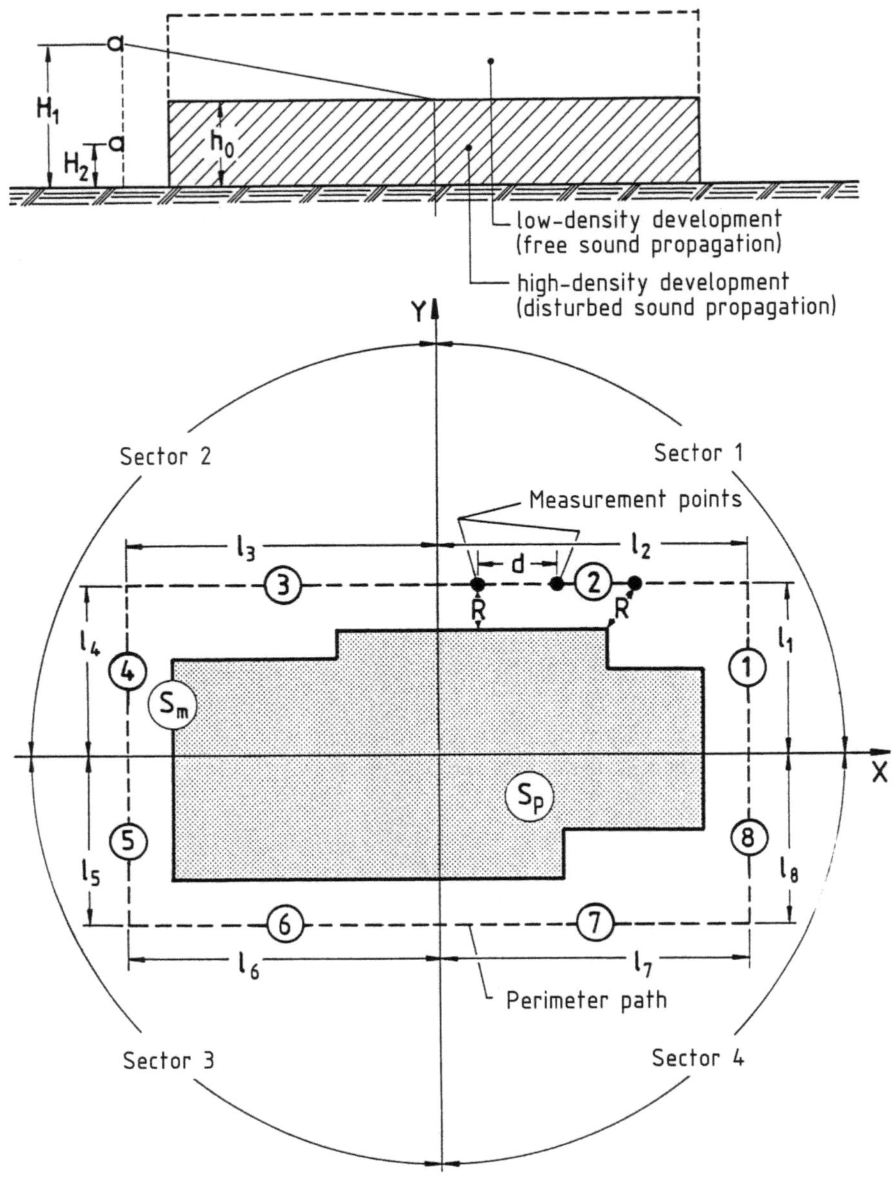

Fig. 5.22 'Perimeter path' measurement method

are no restrictions regarding the use of the method. An investigation of the achievable accuracy in [85] shows that the down-wind average levels in the vicinity of industrial plants with many individual sound sources can be calculated with an accuracy of ± 2 dB(A) from the sound emission values determined with the perimeter path method. Just as for the window method, there is a larger measurement uncertainty for plants with only a few sound sources and a significant horizontal directivity. For the perimeter path method, the noise of the plant has to be constant over time.

The perimeter path method is a standard method described in ISO 8297 [86] of Grade 2 accuracy.

5.2.3.3 Approximation Method

For very quiet open plants like refineries or petrochemical plants, measurements outside of the plant to determine the immission-relevant sound power level (e.g., by means of the perimeter path method) may not always be possible as extraneous noise from adjacent plants would govern the sound level measured. In this case, the immission-relevant

Fig. 5.23 Application of the 'perimeter path' method for the determination of the immission-relevant sound power level of a petrochemical plant

A-weighted sound power level of the open-air plant can be estimated approximately from the average A-weighted sound pressure level within the plant.

$$L_{WA} = \bar{L}_{pA} - 9\ \text{dB} + 10\,\text{lg}\left(\frac{S_p + 2h_0U}{S_0}\right)\ \text{dB}, \quad (5.76)$$

with $\bar{L}_{pA}$ average A-weighted sound pressure level in the plant: Starting at the border of the plant, the A-weighted sound pressure levels are measured at positions which are distributed evenly over the plant area S_p in a height of approx. 1.5 m above the ground and in a scanning pattern of 5–10 m. The distance to the individual sound sources determining the level must constantly be larger than 2 m. In principle, no measurements are carried out in elevated locations and in closed or partly closed buildings. $\bar{L}_{pA}$ is the energetic average of all measuring values.

S_p plant area in m^2; the plant area is defined as built-up area of a production plant or an auxiliary plant including development areas (roads, intermediate storage areas, and pipe lines) of the plant, but without the boundary areas (see Fig. 5.22).

S_O reference area 1 m^2

h_O characteristic height of the plant in m, see Sect. 5.2.3.2.

U circumference of the plant in m, i. e., length of the borderline around the plant area S_p.

The sound power level of the individual sound sources determining the level, which are situated above the densely built-up area, have to be determined separately and added to the sound power level of the plant determined from the average A-weighted sound pressure level.

According to an investigation of 21 different plants [87], the immission-relevant sound power level of typical production and auxiliary plants of refineries and petrochemical plants can be determined with the method described above with an accuracy of ± 3 dB (A) (95% confidence interval).

References

1. ISO 3095: Acoustics – Measurement of noise emitted by railbound vehicles
2. ISO 3740: Acoustics – Determination of sound power levels of noise sources – Guidelines for the use of basic standards and for the preparation of noise test codes
3. ISO 3741: Acoustics – Determination of sound power levels of noise sources – precision methods for sources in reverberation rooms
4. ISO 3743: Acoustics – Determination of sound power levels of noise sources – Engineering methods for special reverberation test rooms
5. ISO 3744: Acoustics – Determination of sound power levels of noise sources – Engineering methods for free-field conditions over a reflecting plane
6. ISO 3745: Acoustics – Determination of sound power levels of noise sources – Precision methods for anechoic and semi anechoic rooms
7. ISO 3746: Acoustics – Determination of sound power levels of noise sources – Survey method
8. ISO 3747: Acoustics – Determination of sound power levels of noise sources – Survey Method using a reference sound source
9. ISO 4871: Acoustics – Declaration and verification of noise emission values of machinery and equipment
10. ISO 6926: Acoustics – Requirements for the performance and calibration of reference sound sources used for the determination of sound power levels
11. ISO 7574: Acoustics – Statistical methods for determining and verifying stated noise emission values of machinery and equipment – Part 1–4
12. ISO/TR 7849: Acoustics – Estimation of airborne noise emitted by machinery using vibration measurement
13. ISO 9614–1: Acoustics – Determination of sound power levels of noise sources using sound intensity – Part 1: Measurement at discrete points
14. ISO 9614–2: Acoustics – Determination of sound power levels of noise sources using sound intensity – Part 2: Measurement by scanning
15. ISO 9614–3: Acoustics – Determination of sound power levels of noise sources using sound intensity – Part 3: Precision method for measurement by scanning
16. ISO 11200: Acoustics – Noise emitted by machinery and equipment – Guidelines for the use of basic standards for the determination of emission sound pressure levels at a work station and at other specified positions
17. ISO 11201: Acoustics – Noise emitted by machinery and equipment – Measurement of emission sound pressure levels at a work station and at other specified positions – Engineering method in an essentially free field over a reflecting plane
18. ISO 11202: Acoustics – Noise emitted by machinery and equipment – Measurement of emission sound pressure levels at a work station and at other specified positions – Survey method in situ
19. ISO 11203: Acoustics – Noise emitted by machinery and equipment – Determination of emission sound pressure levels at a work station and at other specified positions from the sound power level
20. ISO 11204: Acoustics – Noise emitted by machinery and equipment – Measurement of emission sound pressure levels at a work station and at other specified positions – Method requiring environmental corrections
21. ISO 11205: Acoustics – Noise emitted by machinery and equipment – Engineering method for the determination of emission sound pressure levels in situ at the work station and at other specified positions using sound intensity
22. IEC 60651: Sound level meters
23. IEC 60804: Integrating-averaging sound level meters
24. IEC 61672–1: 2002 Electroacoustics – Sound level meters – Part 1: Specifications
25. IEC 61672–2: 2002 Electroacoustics – Sound level meters – Part 2: Pattern evaluation tests
26. IEC 61672–3: 2002 Electroacoustics – Sound level meters – Part 3: Periodic tests
27. IEC 60942: 2003 Electroacoustics – Sound calibrators
28. IEC 61094–1: 2000 Measurement microphones
29. IEC 61183: 1994 Electroacoustics – Random-incidence and diffuse-field calibration of sound level meters
30. IEC 60942: Electroacoustics – sound calibrators
31. IEC 61043: 1993 Electroacoustics – Instruments for the measurements of sound intensity – Measurements with pairs of pressure sensing microphones
32. IEC/TS 62370: 2004 Electroacoustics – Instruments for the measurement of sound intensity – Electromagnetic and electrostatic compatibility requirements and test procedures
33. Hübner G, Gerlach A (1998) Determination of emission sound pressure levels using three-component sound intensity measurements. EuroNoise 98, München, Proceedings:807–812
34. Hübner G, Gerlach A (1999) Further investigations on the accuracy of the emission sound pressure level determined by a three-component intensity measurement procedure. Inter-Noise 99, Fort Lauderdale, Proceedings:1517–1522
35. Hübner G (1973) Analysis of errors in the measurement of machine noise. J Acoust Soc Am 54(4):967–977
36. Hübner G et al (1976) Investigations for the establishment of International Standards for the measurement of noise emitted by machines – Never aspects of uncertainties in determination of the sound power. Inter-Noise 76, Washington, Proceedings:405–410
37. Hübner G (1977) Qualification procedures for free field conditions for sound power determination of sound sources and methods for the determination of the appropriate environmental correction. J Acoust Soc Am 61(2):456–464
38. Hübner G (1986) Experiences with the ISO 3740's Series with Sound Power Measurement Codes. Inter-Noise 86, Boston, Proceedings:1365–1370
39. Hübner G (1992) How to prepare Noise Emission Test Codes for a specific Machinery Family. Inter-Noise 92, Toronto, Proceedings:1249–1254
40. Hübner G, Wu J (1993) Round Robin Test Determining the Sound Power by Sound Pressure. Measurements according

ISO 3744 – First Results. Inter-Noise 93, Leuven, Proceedings:323–328
41. Hübner G (1994) National sound robin test determining the sound power by sound pressure measurements – further results. Inter-Noise 94, Yokohama, Proceedings:1769–1774
42. Hübner G (1997) Final results of a national round robin test determining the sound power of machines/equipment. Inter-Noise 97, Budapest, Proceedings:1317–1322
43. Hübner G, Wittstock V (1998) Schalleistungsbestimmung nach dem Hallfeldverfahren bei statischen Drücken zwischen 1 und 100 kPa. Fortschritte der Akustik (DAGA, 98), Zürich, 178–179
44. Schroeder MR (1954) Eigenfrequenzstatistik und Anregungsstatistik in Räumen. Acoustica 4:456–468
45. Schroeder MR, Kuttruff KH (1962) On frequency response curves in rooms. Comparison of experimental, theoretical and Monte Carlo results for the average frequency spacing between maxima. J Acoust Soc Am 34:76–80
46. Schroeder MR (1969) Effects of frequency and space averaging in the transmission response of multimode media. J Acoust Soc Am 46:277–283
47. Ebbing CE (1968) Experimental evaluation of moving sound diffusors for reverberation rooms. Paper presented at the 76th meeting of the Acoustical Society of America
48. Tichy J (1979) The effect of rotating vanes on the sound field in reverberation chambers, Paper presented at th 80th Meeting of the Acoustical Society of America, Houston
49. Tichy J, Baade PK (1974) The effect of rotating diffusors and sampling techniques on sound pressure averaging in reverberation rooms. J Acoust Soc Am 56:137–143
50. Lang WW (1971) Determination of sound power emitted by small noise sources in reverberant rooms. Proc. 7th ICA, Budapest, paper 19 N 1
51. Maling GC (1971) Guidelines for determination of average sound power radiated by discrete frequency sources in a reverberant room. Proc. 7th ICA, Budapest, paper 19 N 2
52. Pedersen OJ, Jensen JO (1971) Measurement of sound power levels in a small room with special sound absorption properties. Proc. 7th ICA, Budapest, paper 20 A 5
53. Ebbing CE, Maling GC (1973) Reverberation room qualification for determination of sound power of sources of discrete frequency sound. J Acoust Soc Am 54:935–949
54. Hübner G (2002) Remarks on the "Guide to the Expression of Uncertainty in Measurement" (GUM). Published by ISO and other international organisations-floating error analysis for an acoustic example. Inter-Noise 02, Dearborn, Proceedings
55. Hübner G (2003) A contribution for reduction of the measurement effort in determining the sound power of machines. EuroNoise, Naples, Proceedings
56. Hübner G (1982) Analysis of errors and field of application for sound intensity measurement for the determination of sound power of sound sources under "in situ" measurement conditions. Inter-Noise 82, San Francisco, Proceedings:691–694
57. Hübner G et al (1983) Determination of sound power sources under in situ conditions using intensity method – Field of application, suppression of parasitic noise, reflecting effects. Inter-Noise 83, Edinburgh, Proceedings:1043–1046
58. Hübner G (1983) Determination of sound power radiated by structural components of machinery using vibration measurements – A direct finite element method. 11. ICA Paris, Lyon, Revue d´acoustique 10:163–166
59. Hübner G (1984) Development of requirements for an intensity measurement code determining sound power level of machines under (worst) in situ conditions. Inter-Noise 84, Honolulu, Proceedings:1093–1098
60. Hübner G (1985) Recent Developments of Sound Power Determination for Machines by Using Sound Intensity-Measurements – A Survey of Procedure and Accuracy Aspects. Inter-Noise 85, München, Proceedings:57–68
61. Hübner G, Meynerts E (1985) Application of Sound Intensity Method for the Sound Power Determination of Machines in the Presence of Time Fluctuating Extraneous Noise -"Reflection Effect". Inter-Noise 85, München, Proceedings:319–322
62. Hübner G (1987) Sound Intensity Measurement Method – Errors in Determining the Sound Power Levels of Machines and its Correlation with Sound Field Indicators. Inter-Noise 87, Peking, Proceedings:1227–1230
63. Hübner G (1989) Sound power determination of machines using sound intensity measurements – Reduction of number of measurement positions in cases of "Hot areas". Inter-Noise 88, Avignon, Proceedings:1113–1116
64. Hübner G (1989) The use of sound field indicators for the measurement of the sound intensity determined sound power. Inter-Noise 89, Newport Beach, Proceedings:1015–1020
65. Hübner G, Gao-Sollinger Y (1993) Investigations of the Limits of Sound Intensity Scanning Method Determining Sound Powers. Inter-Noise 93, Leuven, Proceedings:351–356
66. Hübner G (1994) Handling of Criteria to Limit the Measurement Uncertainty of the Sound Intensity Scanning Method Determining Sound Powers. Inter-Noise 94, Yokohama, Proceedings:1805–1808
67. Hübner G (1996) Indicators for a method obtaining sound power by using a scanning intensity probe. 4th International Congress on Sound and Vibration, St. Petersburg, Proceedeings:1881–1888
68. Hübner G, Wittstock V (1999) Sound power determination using sound intensity scanning procedure – An investigation to check the adequacy of sound field indicators and criteria limiting the measurement uncertainty. Conference Proceedings on CD-ROM, 137th Meeting of the Acoustical Society of America and the 2nd Convention of the European Acoustics Association: Forum Acusticum integrating the 25th German Acoustics DAGA Conference, Berlin
69. Hübner G, Wittstock V (1999) Further investigations to check the adequacy of sound field indicators to be used for sound intensity technique determining the sound power level. Inter-Noise 99, Fort Lauderdale, Proceedings:1523–1528
70. Hübner G, Wittstock V (1998) Investigation of the sound power generation of cylinders moving in gases under different static pressures – first results. EuroNoise 98, München, Proceedings:868–870
71. Hübner G, Wittstock V (1998) Determination of sound power levels in a reverberant room varying the static

pressure between 1 and 100 kPa. Inter-Noise 98, Christchurch, New Zealand, Proceedings
72. Hübner G (1999) Sound power related to normalised meteorological conditions. Inter-Noise 99, Fort Lauderdale, Proceedings:1529–1534
73. Hübner G (2000) Accuracy considerations on the metrological correction for a normalised sound Power level. Inter-Noise 00, Nizza, Proceedings
74. Tamm K, Kurtze G (1954) Ein neuartiges Mikrofon großer Richtungsselektivität. Acustica 4:469–470
75. Möser M (1988) Analyse und Synthese akustischer Spektren, Springer, Berlin
76. Barsikow B, King WF III (1988) On removing the doppler frequency shift from array measurements of railwaynoise. J Sound Vibr 120(1):190–196
77. Serridge M, Licht TR (1987) Piezoelectric accelerometer and vibration preamplifier handbook. Brüel & Kjaer
78. VDI 3733, Noise at pipes (bilingual German/English)
79. Friedrich J (1967) Ein quasi schall-unempfindliches Mikrofon für Geräuschmessungen in turbulenten Luftströmungen. Rundfunk- und fernsehtechnisches Zentralamt, Technische Mitteilungen Nr.11
80. ISO 5136: Acoustics – Determination of sound power radiated into a duct by fans and other air-moving devices – In-duct method
81. Bommes L (1979) Minderung des Drehklanglärms bei einem Radialventilator kleiner Schnelläufigkeit., Forschungsbericht Nr. 2895 des Landes NRW, Westdeutscher Verlag
82. Böhm A, Danner J, Gilg J und Goldemund K (1996) Stand der Lärmbekämpfungstechnik bei Gasturbinen, Forschungsbericht 10503102/11, VGB PowerTech Essen
83. Schorer E (1991) Vom Schalleistungspegel einer Anlage zur Geräuschimmission in der Wohnnachbarschaft – Vor- und Nachteile der Rundum-Methode, des Zaun- oder des Fensterverfahrens. VDI-Berichte Nr 900:125–142
84. Meurers H (1975) Vorschläge zur Prognose mit Nachweismöglichkeiten über die Einhaltung vorgegebener Immissions-Anforderungen durch Kontrollen im Emissionsbereich (Kurzfassung). Chemie-Ing.-Technik, 47. Jahrg. Nr 15: 641
85. Stüber B, Fritz K (1986) Ermittlung der Schalleistung großflächiger Industrieanlagen (Rundum-Messverfahren). UBA-Forschungsbericht 10502607
86. ISO 8297: Acoustics – Determination of sound power levels of multisource industrial plants for evaluation of sound pressure levels in the environment; Grade 2 of accuracy
87. Stüber B, Lang F (1983) Bestimmung des immissionswirksamen A-Schalleistungspegels einer Freianlage durch Schallmessungen innerhalb der Anlage. DGMK Projekt 308, DGMK Hamburg

Sound Propagation Outdoors

6

L. Schreiber and T. Beckenbauer

6.1 Preliminary Remarks

The sound pressure produced by a sound source outdoors at a location in the vicinity (point of sound immission, receiver point) depends on the properties of the sound source (sound power spectrum, directivity), the geometry of the sound field (relative position of receiver point and sound source to each other, to the ground and to obstacles in the sound field), on ground effects, and on meteorological conditions.

For simple cases, the sound propagation has been investigated repeatedly by means of numerical methods as discussed in Chap. 3 (see for example [1]). On account of the long computation times, these methods are rarely used for sound immission prognosis. Moreover, in practice, frequently a multitude of sound sources have to be considered and the boundary conditions can be described only approximately, for instance, concerning industrial plants and traffic facilities. Furthermore, wind speed, wind direction, temperature, air humidity, atmospheric air pressure, turbulence, and natural cover are subject to continuous temporal and local fluctuations. Therefore, statements on sound immissions that can be expected from a sound source can only be based on statistics. The variations will also grow with the distance between sound source and receiver. In such cases, relatively simple calculation methods, which result from a combination of measuring experience and idealizing modeling, prove of value.

In the following, all effects influencing sound propagation will be described at first. Subsequently, it will be shown in which way the interaction of these effects is to be taken into account in calculating the sound immission. Unless otherwise noted, the following simplifying assumptions are applied:

(a) *Linearity*: The nonlinear members of the wave Eq. (1.32) can be disregarded. This assumption is always met except for the close proximity of extremely strong sound sources.

(b) *Nearly plane sound waves*: Sound pressure p and sound velocity v of a wave are connected to each other according to the equation

$$p(t) = \rho c v(t), \tag{6.1}$$

where ρc is the characteristic impedance of air. As v is the velocity component in propagation direction, this requirement is always fulfilled except for the close proximity of a sound source (see Sect. 1.4).

(c) *Incoherence*: Unless otherwise stated, it is assumed that for the superposition of multiple sound waves i with their effective sound pressures p_i the resulting effective sound pressure is

$$p_\mathrm{res} = \sqrt{\sum p_\mathrm{i}^2}, \tag{6.2}$$

and, consequently, the resulting sound pressure level is

$$L_\mathrm{res} = 10\,\mathrm{lg} \sum 10^{0.1 L_\mathrm{i}}\ \mathrm{dB}. \tag{6.3}$$

Concerning broadband noise or noise averaged over time, this assumption is met in most cases.

L. Schreiber • T. Beckenbauer (✉)
Müller-BBM GmbH, Robert-Koch-Strasse 11, 82152 Planegg, Germany
e-mail: Thomas.Beckenbauer@MuellerBBM.de

G. Müller and M. Möser (eds.), *Handbook of Engineering Acoustics*,
DOI 10.1007/978-3-540-69460-1_6,

(d) *Permanent noise*: The description is based on power-oriented considerations and time means of the sound pressure square. Therefore, the conclusions are true only for stationary conditions and not for single short sound events (impulses). For sound sources with time-variant sound power levels and for unsteadily distributed impulses (e.g., noise from shooting ranges) they are true on an time average.

(e) *Point sound sources:* The following discussions are confined to point sound sources at first. All sound sources can be regarded as such if dimensions of the sound source are small with respect to the distance between the sound source and the receiver point. Spacious sound sources are regarded as being built up of point sound sources.

6.2 Lossless Sound Propagation

Increasing the distance to the sound source yields an increase of the surface S the sound power P is distributed on in outdoor conditions. Assuming an unobstructed sound propagation vertically with S sound intensity

$$I = \frac{dP}{dS}, \tag{6.4}$$

decreases continuously as does the sound pressure in a lossless medium depending on the distance to the source.

6.2.1 Unlimited Sound Field

In a lossless, homogenous medium, a sound point source in the origin produces at a distance d the sound pressure square

$$p^2(d,\varphi,\vartheta) = \frac{\rho c}{d^2} \cdot \frac{\delta P(\varphi,\vartheta)}{\delta \Omega}, \tag{6.5}$$

whereby $\delta P\ (\varphi,\vartheta)\ /\ \delta\Omega$ is the sound power radiated per solid angle unit in the direction φ, ϑ. The total sound power radiated by the sound source is

$$P = \int_0^{4\pi} \frac{\delta P}{\delta \Omega} d\Omega. \tag{6.6}$$

Concerning an omnidirectional sound source, the sound pressure square is then

$$p^2(d) = \frac{P\rho c}{4\pi d^2}, \tag{6.7}$$

and the sound pressure level

$$L_p(d) = L_P - \left(20\lg\frac{d}{d_0} + 11\right)dB, \tag{6.8}$$

with $d_0 = 1$ m and L_P sound power level of the sound source.

The difference

$$A_{\text{div}} = \left(20\lg\frac{d}{d_0} + 11\right)dB, \tag{6.9}$$

between sound power level and sound pressure level is given in Fig. 6.1. In order to distinguish from additional attenuations (level reductions) caused by losses during sound propagation, this is called “attenuation of sound due to geometrical divergence” (ISO 9613–2 [2]), “level reduction due to geometry,” or “level reduction by divergence.”

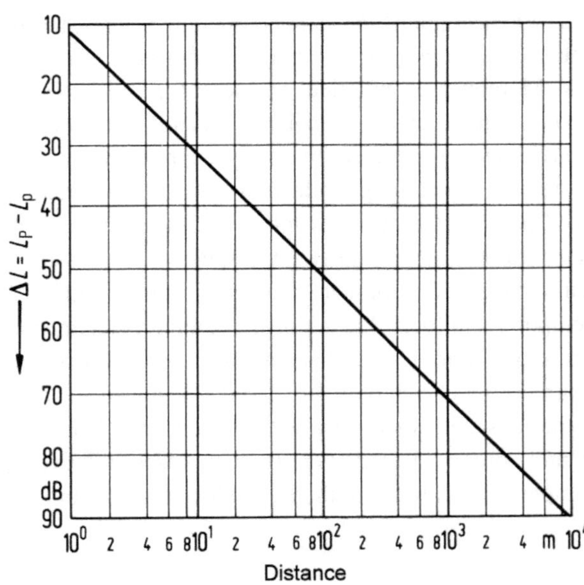

Fig. 6.1 Difference between sound power level L_P and sound pressure level L_p of an omnidirectional sound source as a function of distance for lossless sound propagation. Sound radiation only in the upper half space (sound source on the ground) reduces the values by about 3 dB

In the case of a directional sound source, the total sound power in Eqs. (6.7) and (6.8) has to be replaced by the sound power

$$P_{\varphi,\vartheta} = 4\pi \frac{\delta P(\varphi, \vartheta)}{\delta \Omega}, \quad (6.10)$$

that is, the power of an omnidirectional sound source which in the direction (φ, ϑ) produces the same sound pressure as the directional sound source.

In the following, only omnidirectional sources are discussed at first.

6.2.2 Sound Source Over Ground, Reflection

When a sound source S with sound power P is located over plane ground showing the sound absorption coefficient α_{gr} and the reflection coefficient $\rho_{gr} = 1 - \alpha_{gr}$ both direct sound and sound reflected at the ground reach the receiver point. The reflected sound thereby appears to come from a mirror sound source S' with the sound power $\rho_{gr}P$ (Fig. 6.2); the sound pressure rises correspondingly.

When the vertical distance of the sound source from the ground is small compared to the horizontal distance, the sound source is treated as emitting from the ground. Original and mirror sound sources are nearly equidistant from the receiver point. It is thereby assumed that $\rho_{gr} \approx 1$. Sound power is then radiated only in the upper half space and the sound pressure level is 3 dB higher than in Eq. (6.8). At low frequencies, however, original sound source and mirror sound source can no longer be taken as incoherent (see Sect. 6.4.1).

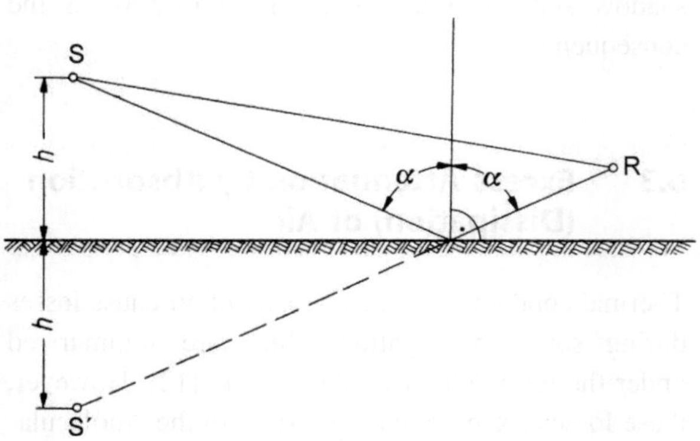

Fig. 6.2 Reflection of a sound source at the ground or at a different (plane) surface

The increase of sound level by reflection has also to be taken into account if there are other surfaces (walls, buildings, and large objects) in the proximity of sound source or receiver. Specular reflection can always be applied in calculations if the reflecting surfaces and their radii of curvature are large compared to the wavelength.

In the example shown in Fig. 6.3, there is another house wall behind the receiver. Original sound source S and the first mirror sound source S^1 are reflected along the wall. The sound pressure square at the receiver point is twice as high as in the example in Fig. 6.2, thus yielding a further increase of the sound pressure level by (approximately) 3 dB.

The increase of the sound level by reflection is of particular importance if sound is reflected into a sound shadow (see Sect. 6.2.3), i.e., into an area which is shielded from sound coming directly from the sound source.

6.2.3 Screening by Barriers

Behind a barrier impervious to sound (wall, row of houses, and hill), which is large with respect to wavelength, a sound shadow is formed due to the fact that the impinging sound wave is restrained from propagating and, as far as it is not absorbed, it will be reflected in other directions. However, the shadow forming is not total because a part of the sound wave will be diffracted at the edges of the obstacle (Fig. 6.4).

The sound pressure in the shadow zone can be calculated by means of the diffraction theory (see, for example, [3, 4]). A barrier impervious to sound with a long straight diffraction edge normal to the sound propagation direction produces a level reduction of the point source, the barrier attenuation of

$$D_z = \left(20 \lg \frac{\sqrt{2\pi N}}{\tan(h)\sqrt{2\pi N}} + 5 \right) \text{dB}, \quad (6.11)$$

at the receiver point compared to a freely propagating sound wave. Here, N is the Fresnel number:

$$N = \pm \frac{2}{\lambda}(a + b - d), \quad (6.12)$$

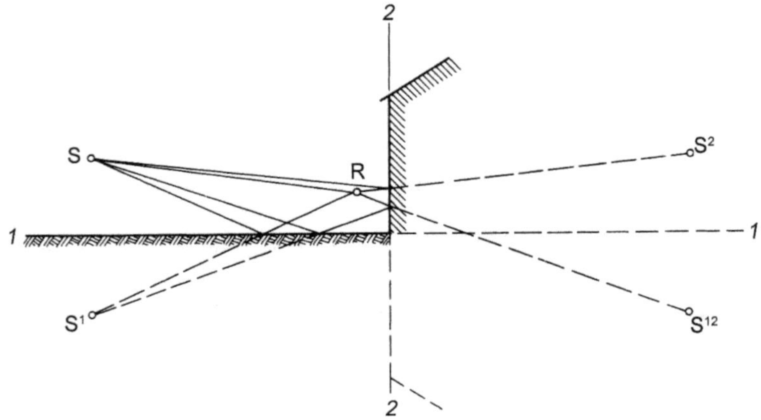

Fig. 6.3 The original sound source as well as the mirrored sound source are reflected at the wall of the house behind the receiver. The sound pressure level at the receiver increases another 3 dB in comparison to the situation shown in Fig. 6.2

Fig. 6.4 Behind a barrier impervious to sound, a sound shadow is formed only by parts of the sound diffracted at the edge of the barrier

where λ is the wavelength, $a + b$ the shortest path between sound source and receiver over the top edge of the barrier, d the distance sound source–receiver point, and h the distance between the diffraction edge and the line SR (see Fig. 6.5).

Equation (6.11) applies for $N \geq -0.2$.

Unless the line sound source–receiver is subtended (i.e., if the sound source can be seen from the receiver's point of view) the negative algebraic sign has to be applied in Eq. (6.12).

For small Fresnel numbers, the values measured by Maekawa [5] and given in Fig. 6.5 deviate slightly from those resulting from Eq. (6.12).

In practice, severe deviations from values calculated by means of Eq. (6.11) or determined from Fig. 6.5 for small shadow angles (ϑ in Fig. 6.5) can be frequently found. These are causes of meteorological and ground effects.

The barrier attenuation in the sound shadow close to the diffraction edge can be slightly increased through sound-absorbing top pieces [6–8]. Field measurements along sound barriers with such pieces atop show that noise reductions cannot be expected of more than 1 dB close to the border of the shadow and at distances greater than 50 m [9, 10].

6.2.4 Diffuse Scattering

Small barriers do not form pronounced shadows but scatter the impinging sound wave diffusely, i.e., reflect it at a great number of angles.

If there is a great deal of barriers between sound source and receiver – for example, natural cover (see Sect. 6.4.2) or machines and pipes of a petrochemical plant – this scattering produces an additional level reduction (excess attenuation) [11]. Even in air diffuse scattering can occur due to inhomogeneities which can cause an additional level reduction in free-field conditions. However, if sound energy reaches the shadow zone by scattering, a level increase is the consequence.

6.3 Excess Attenuation by Absorption (Dissipation) of Air

Thermal conductivity and viscosity of air cause losses during sound propagation which are summarized under the term "classical absorption" [12]. However, these losses are marginal compared to the "molecular absorption" which is due to relaxation processes of the air molecules.

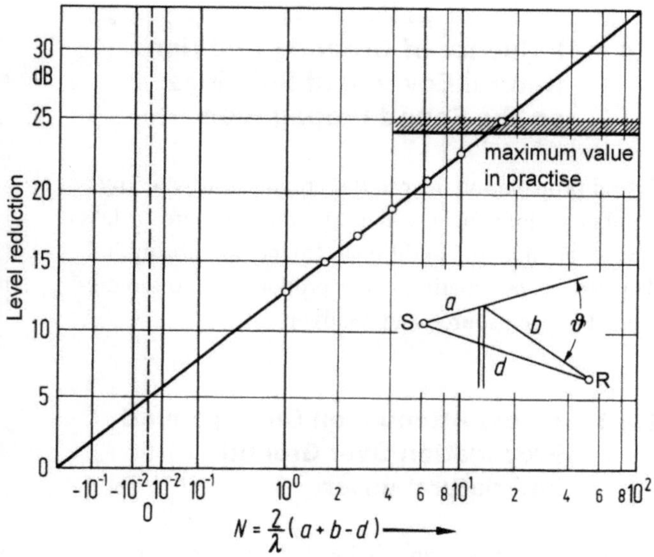

Fig. 6.5 Level reduction compared to free-field conditions as a function of the Fresnel number N for a large barrier vertical with the line sound source and receiver (see Maekawa [5])

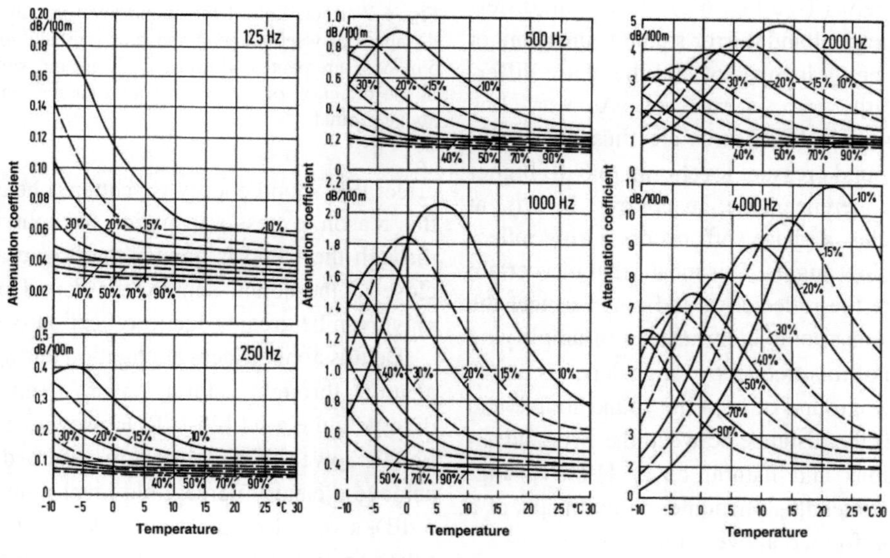

Fig. 6.6 Attenuation coefficient of air as a function of temperature for six different frequencies, parameter is the relative humidity [13]

The level reduction by dissipation is proportional to the path covered. It depends on relative humidity and to a great extent on frequency (see Fig. 6.6). At frequencies below 100 Hz, it is virtually negligible. Analytical formulas for the calculation of the attenuation coefficient (excess attenuation per pathlength unit) can be found in [14].

For calculations of the sound propagation of noise of airborne aircraft, the aircraft industry takes a little higher values [15] which are derived from field measurements.

The influence of fog, rain, and snowfall on sound propagation is unimportant. Only for extremely dense (artificial) fog, laboratory experiments showed a significant excess attenuation [16], whereas field measurements did not yield a statistically significant excess attenuation [17, 18]. When the auditory range of sound sources during foggy weather or a snow cover sometimes seems to be wider than normal, this might be due to the lower background noise level.

6.4 Influence of Ground Conditions, Natural Cover, and Buildings on the Sound Propagation

Sound propagation near to the ground is affected by a number of influences which all cause a more or less severe attenuation additional to the air absorption. This excess attenuation is not proportional to the distance between source and receiver.

6.4.1 Excess Attenuation During Sound Propagation Over Ground and Natural Cover

In case the sound source is located above ground, the sound reflected from ground interferes with the direct sound at the receiver (see Fig. 6.2). If the path difference is small, original and mirror sound source cannot be regarded as being incoherent and the phase difference between the two is relevant. At very low frequencies, the sound pressures are thus adding in-phase and the sound pressure level is up to 6 dB higher than in the incoherence case. In contrast to this, at higher frequencies, a phase shift occurs during reflection at very shallow angles. This means that in a certain frequency range the reflected sound could cancel the direct sound almost completely instead of amplifying it. The position of this frequency range depends on (1) the height above ground of both the sound source and the receiver, (2) the distance between the two, and (3) ground conditions and natural cover [19–24]. Figure 6.7 demonstrates this phenomenon by means of a measurement performed above sandy ground.

At larger distances from the sound source, the ray reflected from ground is attenuated by absorption and scattering to such an extent that a constant excess attenuation of 3 dB can be added to the calculated sound pressure level due to a spherical wave propagation.

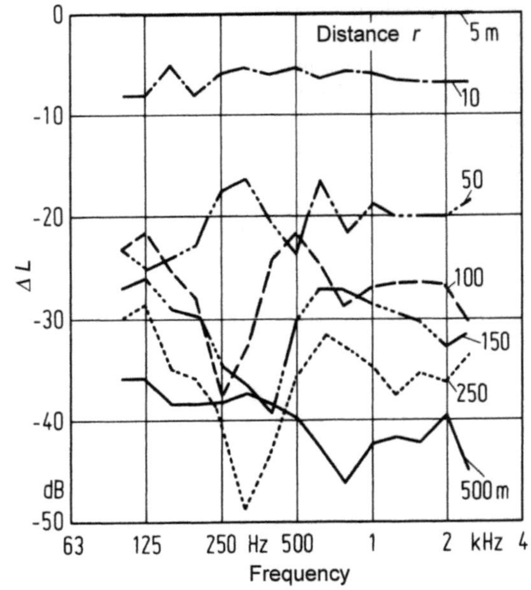

Fig. 6.7 Reduction of the sound level measured for different distances between receiver and source related to the distance of 5 m. Measurement conditions: sandy ground, sound propagation in the direction of the wind, sound source, and receiver 5 m above ground

6.4.2 Excess Attenuation During Sound Propagation Through Natural Cover

Sound is scattered multiply during propagation through natural cover (wood and scrub). Parts of the scattered sound energy get lost by absorption on the ground or in the leaves of trees which is related to the effect that sound energy is scattered heavenward. For this reason, excess attenuation for sound propagation through the wood is greater than open air. However, close to the sound source, the sound pressure level may even be greater due to reflection (see Fig. 6.8).

Details about excess attenuation during sound propagation through natural cover given in literature diverge quite a bit [25–30]. However, excess attenuation is lower than commonly assumed. In order to achieve a clearly perceptible level reduction (at least 5 dB), a very deep planting (at least 50 m) with dense foliation is required. During the leafless season, the level reduction by deciduous woodland is marginal. Because scattering plays an important role for the excess attenuation in woods, it is definitely not quite correct to base calculations on a damping which is proportional to the length of the path. However, most of the authors still specify an excess attenuation in dB/m. According to Hoover ([28], quoted in [31]), it is acceptable to calculate with

$$\alpha_{\text{wood}} = 0.01\left(\frac{f}{1\,\text{Hz}}\right)^{1/3}\ \text{dB/m}, \qquad (6.13)$$

as an average for various woods.

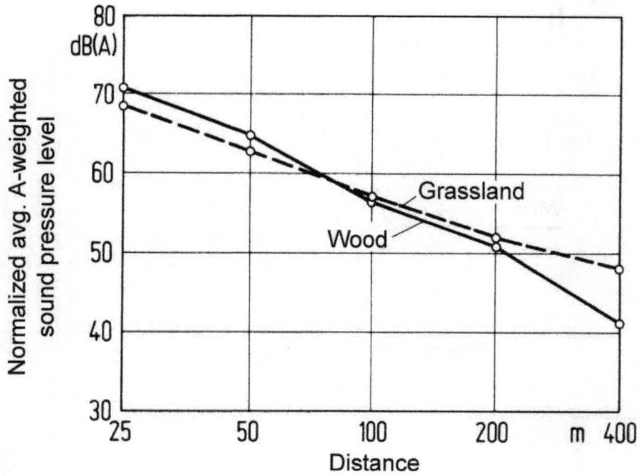

Fig. 6.8 Energy equivalent mean level normalized to the density of vehicles as a function of the distance to highways (averaged over a large number of measurements) measured in the plane in open terrain and in wood, microphone height 4 m

6.4.3 Excess Attenuation by Buildings

In terrain covered with single buildings, the A-weighted sound pressure level of traffic or industrial noise decreases as an average by 10–15 dB in addition to the level decrease for free sound propagation. The dependency on distance is related to the building density.

6.5 Influence of Inhomogeneities of Air

At large distances from the sound source measured sound levels can considerably deviate downward or even upward from those which could be expected theoretically. Then the level reduction due to geometry, the dissipation in air, and the ground absorption is underestimated or overrated. This is attributed to the fact that air is not a homogeneous motionless medium.

6.5.1 Wind-Speed Gradient

The strongest influence on sound propagation outdoors can be expected from the wind direction. The common experience points out that sound propagates further on downwind condition than it does on headwind condition.

This is due to the fact that the motion of air at the ground is slowed down by friction and obstacles and that the wind speed, therefore, increases with the height above ground. As the sound propagation velocity relative to the ground is composed of the sound speed in the motionless air and the air speed, sound waves are propagating faster with increasing height on downwind condition and slower on headwind condition. Therefore, the sound rays are diffracted toward the ground on downwind condition and are diffracted off the ground on headwind condition. This is shown in an exaggerated manner in Fig. 6.9. Due to the diffraction toward the ground, excess attenuation by natural cover and buildings as well as screening by barriers can be compensated on the whole or in part.

Contrary to this, on downwind condition, a shadowing zone is formed (Fig. 6.9) which is not reached by direct sound at all. In the case of constant wind-speed gradient $c' = dc/dz$ (z is the height above ground), the sound rays form catenary curves [32] which can be approximated by arcs of a circle with the radius

$$R = \frac{c_0}{c' \cos \varphi}. \tag{6.14}$$

where φ is the angle between the direction sound source–receiver point and the direction where the wind is blowing from. For example, a wind-speed gradient of 0.1 m/s per m height difference results in a radius of curvature of about 3.4 km. The distance x_s of the shadow border for a receiver in the height z_R from a sound source at the height z_S is approximately

$$x_s = \sqrt{\frac{2c_0}{c' \cos \varphi}} (\sqrt{z_R} + \sqrt{z_S}). \tag{6.15}$$

For a receiver at 5 m height, the shadow zone of a sound source located also at 5 m height above ground begins at a distance of approximately 370 m from the sound source, given that the wind-speed gradient is 0.1 m/s per m.

Generally, the shadow zone is not well defined. There is a transient zone where the excess attenuation decreases more or less rapidly with distance depending on the angle of the wind direction. In the shadow zone, the excess attenuation can be up to 30 dB.

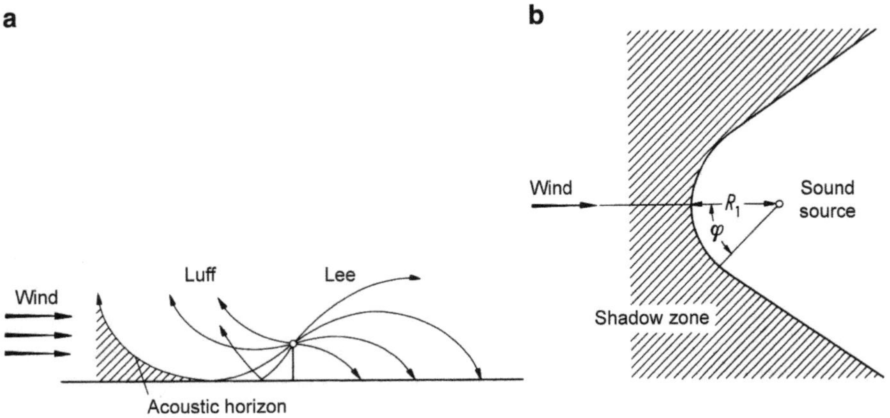

Fig. 6.9 Refraction of the sound waves and forming of a sound shadow by increase of wind speed with height (wind-speed gradient)

Generally, the wind-speed gradient depends on height. In Fig. 6.10, some characteristic runs of the curve are shown.

6.5.2 Temperature Gradient

Because the sound speed is proportional to the square root of the absolute temperature, sound propagation is also affected by thermal layering in the atmosphere. During clear nights when cooling of the ground by thermal radiation temperature inversion is frequently observed, temperature, and, thus also, sound speed, increase with height and the sound waves are refracted toward the ground, which is comparable with the sound propagation on downwind condition. During the day, when the ground is heated up by solarization, temperature above ground decreases with increasing height. The sound rays are refracted upward and (at still air) a circular shadow border is formed around the sound source. Independent from the propagation direction, no sound rays reach the ground if the sound source is beyond a certain distance. Figure 6.11 shows two diurnal cycles of the temperature gradient at ground level.

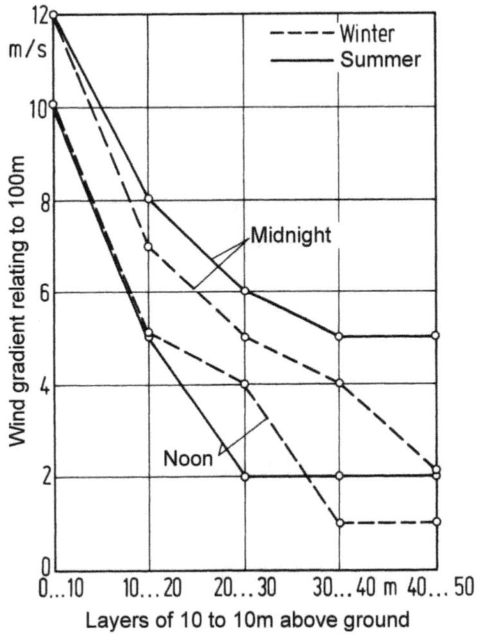

Fig. 6.10 Mean decrease of the wind-speed gradient as a function of height [18]

6.5.3 Turbulence

During gusty wind, which means turbulent atmosphere, there also occurs an additional excess attenuation in wind direction which is unsteady and causes fluctuations of the sound level which increase with distance.

6.6 Sound Immission Calculation

6.6.1 Preliminary Remarks

For all types of facilities (industry, business, traffic, sports and recreation, and shooting ranges), precalculations of the sound immissions expected from these sound sources (sound immission prediction) and of the noise protection measures required

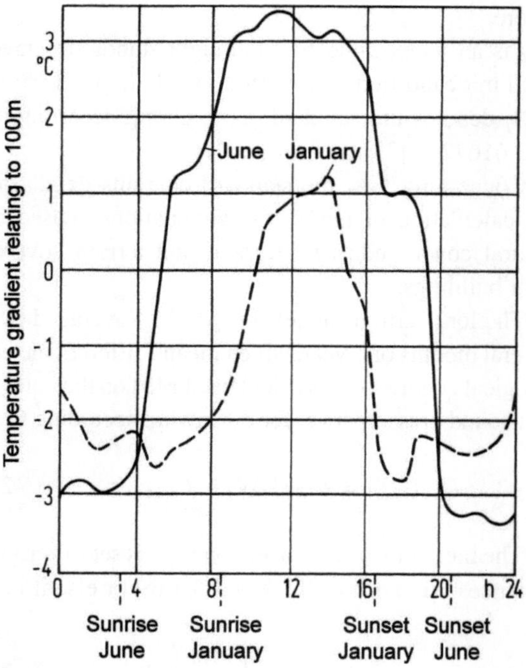

Fig. 6.11 Temperature gradient during the course of a day (averaged) for a layer of 7–17 m above ground for June and January [18]

are a demand in the framework of planning permission processes. In many cases, measurements of the sound immission have to be replaced by calculations on the whole or in part, because the noise from the facility cannot be measured at the point of sound immission due to excessive background noise. In order to pursue a consistent approach for such calculations, adequate methods have been defined in many administrative regulations, standards, and directives. Normally, the equivalent continuous noise level (mean level) L_{eq} has to be calculated at first. Based on this quantity, the noise rating level L_r is calculated by applying various noise character penalties or reliefs (see Chap. 6). In this chapter, only the calculation of the mean level according to ISO 9613–2 [2] is discussed without going into further details. Its application is stipulated in most of the regulations and directives.

6.6.2 Calculation According to ISO 9613–2

For the calculations, sound sources whose dimensions are small with respect to the distance d from the receiver point are considered point sources. Bigger sound sources are divided into smaller part sound sources. One has to take into account that the distance d has always to be referred to the center of the (part) sound source.

Calculations are carried out in octave bands.

The equivalent continuous downwind octave-band sound pressure level L_{fT}(DW) which occurs at a receiver point on downwind condition is calculated for each sound source and its mirror sound source according to

$$L_{fT}(\mathrm{DW}) = L_W + D_c - A, \qquad (6.16)$$

where L_W is the octave-band sound power level of the sound source, $D_c = D_I + D_\Omega$ the directivity correction, and A the octave-band attenuation during the sound propagation from the sound source to the receiver point.

The directivity index D_I specifies the difference in dB by which the sound pressure level of the sound source during free sound propagation in the direction of the receiver point is higher than that of an omnidirectional sound source with same sound power. Examples for directional sound sources are railway trains and openings of funnels. D_I is a function of the radiation direction and the frequency. The directivity index of an omnidirectional sound source is 0 dB for all directions. However, the majority of sound sources radiates almost omnidirectional and D_I is set to 0 dB.

The index $D_\Omega = 10\lg(4\pi/\Omega)$ dB is taken into account if a sound source which is supposed to be omnidirectional but radiates due to its location only in the solid angle Ω instead of 4π. For example, if it is located at ground level, then $\Omega = 2\pi$ and $D_\Omega = 3$ dB (as a rule). If it is located in addition in front of a wall, then $D_\Omega = 6$ dB.

The attenuation term A is composed of several attenuations:

$$A = A_{\mathrm{div}} + A_{\mathrm{atm}} + A_{\mathrm{gr}} + A_{\mathrm{bar}} + A_{\mathrm{misc}}. \qquad (6.17)$$

Several attenuations are discussed as follows:

- $A_{\mathrm{div}} = \left(20\ \lg\frac{d}{1m} + 11\right)dB$, geometrical divergence.
- $A_{\mathrm{atm}} = \alpha d/1{,}000$, atmospheric absorption, with α absorption coefficient of the air in dB/km.
- $A_{\mathrm{gr}} = 4.8dB - \frac{2h_m}{d}\left(17 + \frac{300\,m}{d}\right)dB \geq 0\,dB$, ground effect, with h_m mean height of the sound propagation path above ground. The value thus calculated

holds only if the A-weighted sound level must be calculated and if the sound propagates over mainly sound-absorbing ground and the sound is not a tone. Otherwise, A_{gr} has to be calculated separately for each octave band according to a more complex method, which is also defined in the standard. Based on this method, A_{gr} already implies the sound reflected at the ground and D_Ω is set to zero.

– $A_{bar} = D_z - A_{gr} \geq 0\ dB$, screening; herein, the barrier attenuation D_z for diffraction over a single barrier edge is

$$D_z = 10\lg\left(3 + \frac{C_2}{\lambda} z K_{met}\right) dB < 20\, dB, \quad (6.18)$$

where $C_2 = 20$ or (only if the ground reflections are taken into account separately by means of mirror sound sources) $C_2 = 40$ and the pathlength difference

$$z = \sqrt{(d_{ss} + d_{sr})^2 + a^2} - d, \quad (6.19)$$

where d_{ss} is the distance from the source to the (first) diffraction edge, d_{sr} the distance from the (second) diffraction edge to the receiver point, a the component distance between source and receiver point running parallel to the barrier edge and

$$K_{met} = \exp\left(-\frac{1}{2{,}000}\sqrt{\frac{d_{ss} d_{sr} d}{2z}}\right) \text{for } z > 0, \quad (6.20)$$

a correction factor for meteorological effects.

– A_{misc}, additional types of attenuation.

The total equivalent continuous A-weighted downwind sound pressure level is obtained by summing the octave-band sound pressure levels according to Eq. (6.16) for each octave band:

$$L_{AT}(DW) = 10\lg \times \left\{\sum_{j=1}^{8} 10^{\left(L_{fT}(j) + A_f(j)\right)/10dB}\right\} dB, \quad (6.21)$$

where

j is an index indicating the eight standard octave-band midband frequencies from 63 Hz to 8 kHz.

A_f denotes the standard A-weighting according to IEC 61672–1 [33].

Concerning this, the standard contains details for the calculation of the excess attenuations caused by natural cover, industrial terrain, and terrain covered with buildings.

The long-term mean level L_{AT}(LT) – averaged over several months or 1 year, which means different meteorological conditions – is calculated based on the equivalent sound pressure level for downwind according to

$$L_{AT}(LT) = L_{AT}(DW) - C_{met}. \quad (6.22)$$

The meteorological correction C_{met} is set to zero for distances up to $d_p = 10\ [(h_r + h_s)/1\ m]$ or else it is

$$C_{met} = C_0\left(1 - 10\frac{h_r + h_s}{d_p}\right) dB. \quad (6.23)$$

where h_r, h_s are the heights of receiver point and source, d_p the horizontal distance between source and receiver, and C_0 a constant whose value is fixed by the local authorities or can be derived from the local propagation classes statistics [34]. In most cases, one can calculate with 2 dB for daytime and 3 dB for nighttime.

6.6.3 Limitations

The estimated accuracy of the values calculated for L_{AT}(LT) for broadband noise is ±1 dB concerning situations without reflections and screening and mean rayheights $h_m = (h_r + h_s)/2$ between 5 and 30 m; otherwise, it is ±3 dB. The specification of an upper distance limit for the validity is omitted. For distances beyond 1 km, only spare experiences are available.

The standard does not apply to air traffic noise and pressure waves through blastings. For the latter, results of measurements of the sound propagation in wooded terrain and over water across distances of up to 10 km are available [35]. According to this investigation, the sound level decreases by approximately 15 dB with every doubling of the distance in the range from 1 to 10 km. By contrast, the decrease is not more than about 5 dB across water surfaces. Over expanses of water, multiple reflections at the surface during wind and temperature inversion cause higher sound levels than calculated for all types of noise.

References

1. Attenborough K et al (1995) Benchmark cases for outdoor sound propagation models. J Acoustic Soc Amer 97/1:173–191
2. ISO 9613–2 (1996) Acoustics: Attenuation of sound during propagation outdoors – Part 2: General method of calculation
3. Redfearn SW (1940) Some acoustical source-observer problems. Philos Mag 30:223–236
4. Kurze UJ, Anderson GS (1971) Sound attenuation by barriers. Appl Acoust 4:56–74
5. Maekawa Z (1965) Noise reduction by screens. Kongressbericht 5. Int. Kongress für Akustik. Lüttich (Belgien) F 13
6. Möser M, Volz R (1999) Improvement of sound barriers using head pieces with finite impedance. J Acoust Soc Amer 106/6:3049–3060
7. Ohkubo T, Fujiwara K (1999) Efficiency of a noise barrier with an acoustically soft cylindrical edge for practical use. J Acoust Soc Amer 105/6:3326–3335
8. Volz R, Möser M (2000) Aufsätze für Schallschirme – Messungen an einer Lärmschutzwand und Aufsätze für Schallschirme – verschieden abgestimmte Resonatoren. Fortschritte der Akustik DAGA 2000:444–447
9. Ullrich S (1998) Vorschläge und Versuche zur Steigerung der Minderungswirkung einfacher Lärmschutzwände – eine Literaturauswertung. Straße + Autobahn 7/98:347–354
10. Beckenbauer T et al (2001) Absorbierende Schallschutzwand-Aufsätze. Forschritte der Akustik DAGA
11. Jovicic S (1971) Untersuchungen zur Vorausberechnung des Schallpegels in Betriebsgebäuden. Durchgeführt im Auftrag des Ministers für Arbeit, Gesundheit und Soziales des Landes Nordrhein-Westfalen. Müller-BBM GmbH, Bericht Nr. 2151, München
12. Cremer L (1950) Die wissenschaftlichen Grundlagen der Raumakustik. Bd.III: Wellentheoretische Raumakustik. Hirzel, Leipzig
13. Harris CM (1966) Absorption of air versus humidity and temperature. J Acoust Soc Amer 40:148–159
14. ISO 9613–1 (1993) Acoustics: Attenuation of sound during propagation outdoors – Part 1: Calculation of the absorption of sound by the atmosphere
15. Aerospace recommended practice ARP 866 A: Standard values for atmospheric absorption as a function of temperature and humidity for use in evaluating aircraft flyover noise. Society of automotive engineers Inc, Warrendale PA, 1964-08-31, revised 1975-15-03
16. Knudsen VO (1946) The propagation of sound in the atmosphere – Attenuation and fluctuation. J Acoust Soc Amer 18:90–96
17. Wiener FM (1961) Sound propagation over ocean waters in fog. J Acoust Soc Amer 33:1200–1205
18. Sieg H (1940) Über die Schallausbreitung im Freien und ihre Abhängigkeit von den Wetterbedingungen. ENT 17:193–208
19. Parkin PH, Scholes WE (1965) The horizontal propagation of sound from a jet engine close to ground, at Radlett. J Sound Vib 1:1–13
20. Parkin PH, Scholes WE (1965) The horizontal propagation of sound from a jet engine close to ground, at Hatfield. J Sound Vib 2:353–374
21. Ingard U (1969) On sound transmission anomalies in the atmosphere. J Acoust Soc Amer 45:1038–1039
22. Ingard U (1953) A review of the influence of meteorological conditions on sound propagation. J Acoust Soc Amer 25:405–411
23. Attenborough K (1988) Review of ground effects on outdoor sound propagation from continuous broad band sources. Appl Acoust 24:289–319
24. Attenborough K (1999) A comparison of engineering models for prediction of ground effects. 237th Meeting of the Acoustical Society of America and the 2nd Convention of European Acoustics Associations, Berlin, March 14–19, 1999, Collected papers
25. Eyring CF (1946) Jungle acoustics. J Acoust Soc Amer 18:257–270
26. Embleton TFW (1963) Sound propagation in homogeneous deciduous evergreen woods. J Acoust Soc Amer 35:1119–1125
27. Dneproskaya IA, Jofe V, Levitas FI (1963) On the attenuation of sound as it propagates through the air. Sov Phys Acoust 8:235–239
28. Hoover RM (1961) Tree zones as barriers for the control of noise due to aircraft operations. Bolt, Beranek and Newman Inc. Rep. 844, Cambridge (MA)
29. Aylor D (1972) Noise reduction by vegetation and ground. J Acoust Soc Amer 51:197–205
30. Wiener FM, Keast DN (1959) Experimental study of the propagation of sound over ground. J Acoust Soc Amer 724–733
31. Beranek LL (ed) (1971) Noise and vibration control (Chapter 7: Sound propagation outdoors). McGraw-Hill, New York
32. Tedrick RX (1962) Determination of zones subject to meteorological focussing. 4. Int. Kongress für Akustik (012), Kopenhagen
33. IEC 61672–1 (2002) Electroacoustics – Sound level meters – Part 1: Specifications
34. Manier G (1975) Ausbreitungsklassen und Temperaturgradienten. Meteorologische Rundschau 28:6–11
35. Schomer PD (2001) A statistical description of blast sound propagation. Noise Control Eng J 49(2):79–87

Sound Insulation in Buildings

7

K. Gösele and E. Schröder

Sound insulation between the different rooms inside a building or to the outside is a very complex problem. First, the airborne sound insulation of ceilings, walls, doors and windows is important. Second, a sufficient structure-borne sound insulation, also called impact sound insulation, for the ceilings, has to be provided especially. Finally, the service equipment should be sufficiently quiet.

These three problems will be discussed in the following.

7.1 Airborne Sound Insulation

7.1.1 Characteristics

The airborne sound insulation of a building element is its ability to prevent that airborne sound power incident upon the building element that is transmitted to the adjacent rooms. The airborne sound insulation between two rooms usually depends on the design of the separating building elements (partition, ceiling, etc.) between them. Their insulating performance is described by the sound reduction index R, also called transmission loss TL, defined as

$$\mathrm{TL} = R = -10\lg\frac{W_2}{W_1}\,\mathrm{dB}, \tag{7.1}$$

where W_1 is the sound power incident upon the building element and W_2 is that transmitted to the adjacent room. If sound transmission does not only take place via the separating building element, but also partly via flanking elements – as is usually the case for buildings – the apparent sound reduction index R' is applied [44, 61]. The apostrophe means that sound transmission does not only take place via the separating partition itself but also via flanking elements and other flanking paths:

$$R' = -10\lg\frac{W_{\mathrm{tot}}}{W_1}\,\mathrm{dB}, \tag{7.2}$$

W_{tot} is the total sound power transmitted to the adjacent room, which consists of the transmitted sound power both of the separating component W_2 and of the flanking components and other possible transmission paths (e.g., air ducts).

If this definition is applied to the sound transmission between two rooms via a separating building element and diffuse sound fields are assumed, the following equation is valid:

$$R' = L_1 - L_2 + 10\lg S_s/A\ \mathrm{dB}. \tag{7.3}$$

In this equation, L_1 is the sound pressure level in the source room, L_2 is the sound pressure level in the receiving room, S_s is the surface of the separating building element and A is the equivalent sound absorption area of the receiving room.

For a rough estimate of R' for a furnished small living room, the logarithmic term in Eq. (7.3) may be neglected.

K. Gösele • E. Schröder (✉)
Müller-BBM GmbH, Robert-Koch-Strasse 11, 82152 Planegg, Germany
e-mail: Elmar.Schroeder@MuellerBBM.de

G. Müller and M. Möser (eds.), *Handbook of Engineering Acoustics*,
DOI 10.1007/978-3-540-69460-1_7, © Springer-Verlag Berlin Heidelberg 2013

In some cases, it is not possible or not practical to apply definition Eq. (7.3). For example, it is not possible if the sound incident and sound radiating surface do not share a common surface of the separating building element, as it is the case, e.g., when the rooms are staggered. In this case, the so-called normalised level difference D_n is used:

$$D_n = L_1 - L_2 + 10\lg A_0/A \text{ dB}. \quad (7.4)$$

D_n is a sound level difference, which is free of the randomness of the interior decoration of a room, as it refers to a certain equivalent sound absorption area A_0 of the receiving room.

The reference equivalent sound absorption area A_0 usually has a size of 10 m [44]. This value approximately corresponds to the equivalent sound absorption area of a furnished small living room.

However, in individual cases, e.g., in a large hall, it may be reasonable to use another value A_0, which is adapted to the stated, desired sound absorption in the room.

The normalised level difference is also stated for measurements of small components at the test stand (surface <1 m^2), e.g., outdoor air intakes, transfer air devices, roller shutter housings, etc., and is then called element normalised level difference $D_{n,e}$ [49]. Furthermore, it is used to describe sound transmission from one room to another via flanking elements [58–60], and is then called normalised flanking level difference $D_{n,f}$.

The sound reduction index and the normalised level difference are independent from the measuring direction, i.e., independent from which room is chosen as source or receiving room.

Another quantity to characterise sound insulation between rooms based on the reverberation time T is the standardised level difference D_{nT} according to

$$D_{nT} = L_1 - L_2 + 10\lg T/T_0 \text{ dB}. \quad (7.5)$$

Here, for living rooms, the measured reverberation time T refers to a reference reverberation time in the receiving room of $T_0 = 0.5$ s, which is equivalent to $A_0 = 0.32$ V. It is assumed that furnished living rooms have a reverberation time of 0.5 s, nearly independent from volume and frequency. If the reverberation times of source and receiving room are different from each other, the standardised level difference depends on the direction of sound transmission.

The sound insulation depends on the frequency. For a simple characterisation, the following single number quantities are used and the values are calculated over the frequency range from 100 Hz up to 3,150 Hz:

- *Weighted sound reduction index* R_w or R'_w (comparison with an curve of reference values) [64]
- *Average sound reduction index R or* R_m (linear averaging over the above-mentioned frequency range, if R is plotted over a logarithmic frequency scale)

Of these parameters, only the weighted sound reduction index R_w or R'_w are still in use. In order to determine the single number values, a reference curve B is moved in steps of 1 dB compared with the measured values, as long as the sum of the negative deviations compared with the shifted reference curve B_v is as large as possible in the individual one-third-octave bands, but not larger than 32 dB (see Fig. 7.1). The R-value of this shifted reference curve B_v at 500 Hz is the weighted sound reduction index R_w; see also [1, 54].

This method, which may seem like a random method at first sight, takes into consideration the varying sensitivity of the human ear to different frequencies according to the A-weighting and the noise spectrum of typical noise in living rooms in the frequency range of 100 Hz up to 3,150 Hz [2].

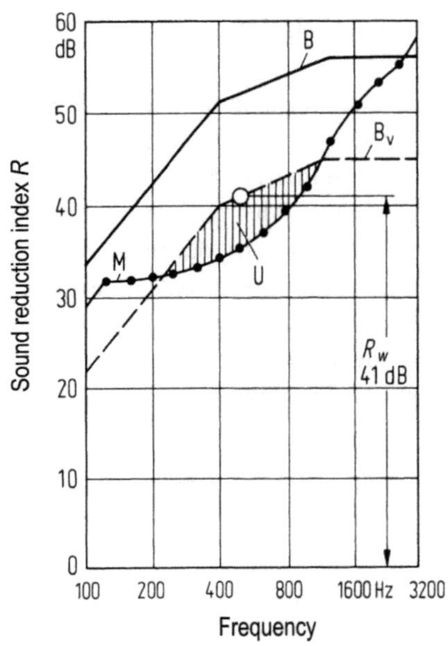

Fig. 7.1 Definition of the sound reduction index R_w or R'_w. B reference curve, B_v shifted reference curve, M measured values and U unfavourable deviations of M compared with B_v

Recently, two other noise spectra that represent certain types were used with the aim to obtain a single number value for judging the insulation efficiency. This enables an assessment of sound insulation, e.g., of one building component for different types of noise (e.g., traffic noise and music). The characterisation of sound insulation regarding these types of noise can be made with so-called spectrum adaptation terms, which have to be added to the weighted sound reduction index and can be stated for example as follows: $R'_w(C; C_{tr}) = 42\ (0; -5)$ dB.

Also, an extended frequency range up to 50 Hz and/or up to 5,000 Hz can be considered, which is stated in the form of an index to the spectrum adaptation term (e.g., $C_{100\text{–}5{,}000}$) (Table 7.1).

7.1.2 Measurements

The measurements can be carried out in the test stand or in existing buildings. In accordance with the Eqs. (7.3)–(7.5), airborne sound is generated in one of the two rooms and the sound pressure levels L_1 and L_2 are measured, usually in one-third-octave bands. The reverberation times T (in s) are determined for the same frequency bands and the equivalent sound absorption areas $A = 0.16\ V\ T^{-1}$ are determined from the reverberation times assuming a diffuse sound field (V is the volume of the receiving room in m). The source sound can be a broadband noise (sine sweep, pink or white noise) or a noise with one-third-octave or octave bandwidth. In buildings, the measurements are carried out in the frequency range of 100 Hz up to 3,150 Hz; in test stands in the frequency range of 100 Hz up to 5,000 Hz, by extending the frequency range from 50 to 5,000 Hz, additional information can be obtained.

More detailed information is given in the standards ISO 140 [41–53]. A survey measurement method, which can be applied in rooms up to 150 m^3, is described in [57].

7.1.3 Behaviour of Single Partitions

7.1.3.1 Basic Behaviour

For open porous walls, e.g., walls made of unrendered pumice concrete, sound transmission can occur via the narrow air ducts inside the wall. In this case, the sound insulation is usually very small and is determined by the flow resistance of the wall material. By means of an air-tight layer, e.g., plaster, applied to the wall, this transmission path can be suppressed, furthermore it suffices to seal only to one side of the wall. The air-tight wall will still be excited to bending vibrations, which in turn send sound radiation into the adjacent room. Nevertheless, sealing both sides of the wall will better the sound insulation.

The isolation behaviour of homogeneous plates becomes particularly clear when the product of the sound reduction index R and wall thickness is plotted versus the frequency of the exciting radiation. Here, the three areas A, B and C can be distinguished (see Fig. 7.2):

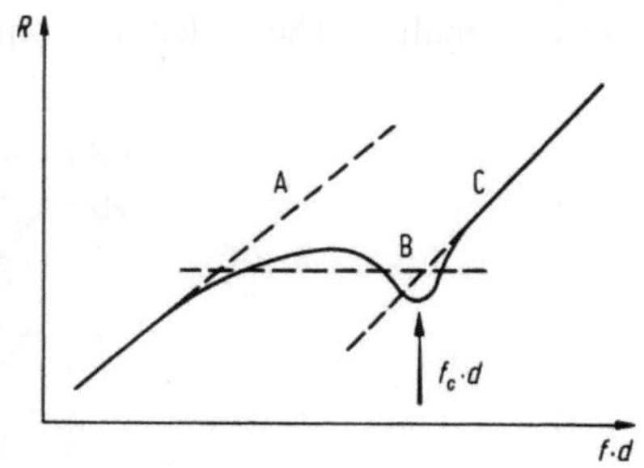

Fig. 7.2 Different areas A, B and C in the course of the sound reduction index R depending on the product of frequency and thickness of the plate

Table 7.1 Spectrum adaptation term for different noise sources [56]

Noise source	Type of noise sources	Spectrum adaptation term
A-weighted pink noise	Living activities (talking, music and radio) Children playingRailway traffic at medium and high speedHighway road traffic > 80 km h^{-1}Jet aircraft at a short distanceFactories emitting mainly medium and high frequency noise	C
Urban road traffic noise spectrum	Urban road trafficRail traffic with low speedsAircraft, propeller drivenJet aircraft at a large distanceDisco musicFactories emitting mainly low and medium frequency noise	C_{tr}

Area A For low frequencies, the factor which is most important for R is the mass per unit area m'' of the plate. According to Heckl [3], the transmission loss R can be calculated as follows:

$$R = \left(20\ \lg\frac{\pi f\ m''}{\rho c} - 3\right)\text{dB}, \tag{7.6}$$

(ρ density of air and c sound velocity in air)

Area B In a second area, a resonance-like dip of sound insulation occurs with a minimum near the critical frequency f_c. According to Cremer [4], this critical frequency can be calculated as follows:

$$f_c = \frac{c^2}{2\pi}\sqrt{\frac{m''}{B'}}, \tag{7.7}$$

$$f_c \approx \frac{6.4 \times 10^4}{h}\sqrt{\frac{\rho_P}{E}}\text{Hz}. \tag{7.8}$$

In this equation, B' is the bending stiffness (per unit of width) of the plate (in N m), h is the thickness of the plate (in m), ρ_P is the density of the plate material (in kg m^{-3}) and E is the Young's modulus (in N m^{-2}).

Above all, the depth of the minimum depends on the ratio l/λ_c (l length of the plate and λ_c wavelength of airborne sound at the critical frequency) and on the material damping of the plate. For thick plates, l/λ_c is comparatively small. The distinct minimum degenerates and becomes nearly flat; see Fig. 7.5 and [5, 6].

Area C The third area runs above the limiting frequency in a straight line, with a slope of 25 dB per frequency decade. The sound reduction index can be calculated according to Heckl [3], in accordance with practical results [7], as:

$$R = \left(20\lg\frac{\pi f\ m''}{\rho c} + 10\lg\frac{2\eta}{\pi}\frac{f}{f_c}\right)\text{dB}. \tag{7.9}$$

In this equation, sound insulation also depends on the total loss factor η of the plate; the higher η, the higher is the transmission loss. The total loss factor consists of the energy loss at the boundaries, due to material damping and due to acoustical radiation.

For the sound insulation behaviour, the location of the critical frequency is important. Some values for the critical frequency of different plates and walls are stated in Fig. 7.3. If possible, they will be shifted to the upper frequency limit or, if this is not possible, to the lower frequency limit. An essential parameter for the critical frequency is the thickness of the component. However, for the same thickness the critical frequency can also be influenced by the choice of material, depending on whether the ratio ρ_P/E is large (rubber and lead sheet) or small (glass and steel sheet).

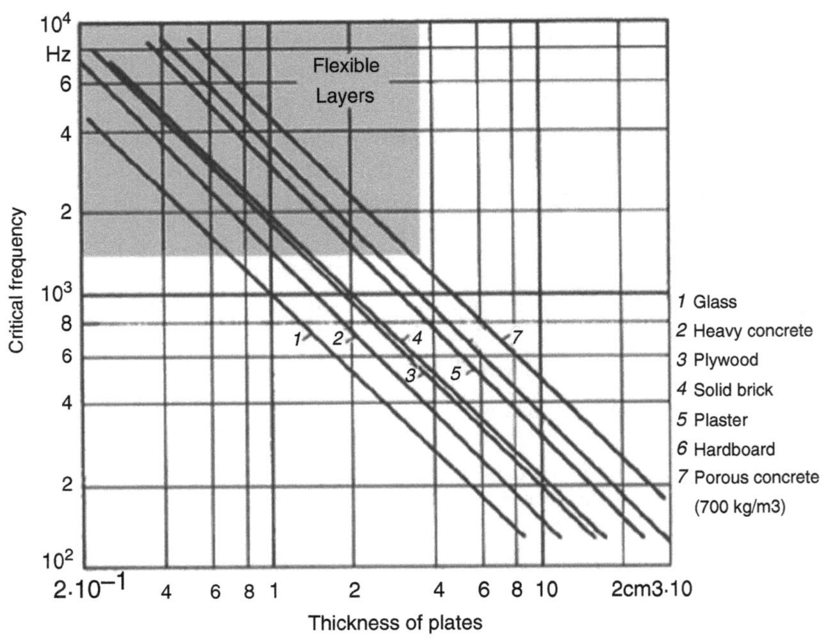

Fig. 7.3 Critical frequency for plates made of different materials, depending on the thickness of the plate

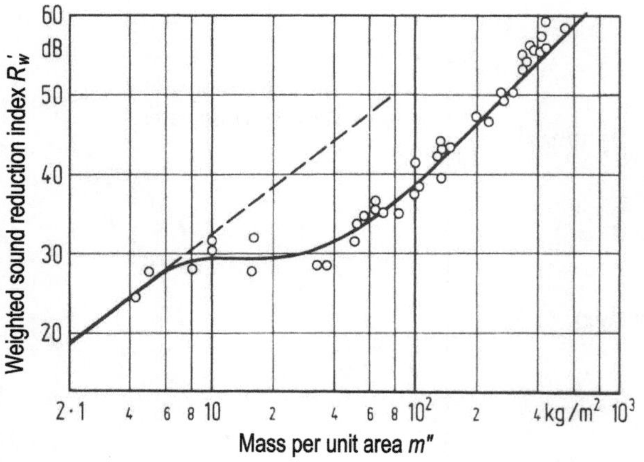

Fig. 7.4 Dependency of the apparent sound reduction index R'_w of single-layer monolithic partitions from the mass per unit area m'' of common building materials with standard flanking paths

7.1.3.2 Behaviour of Existing Partitions

The airborne sound insulation index R'_w of single monolithic partitions is shown in Fig. 7.4 as a function of its mass per unit area m''. Despite the influence of the Young's modulus, for common building materials R_w only depends on the mass per unit area, as the ratio ρ_P/E in Eq. (7.8) is very similar for most building materials.

Only for extremely flexible materials (rubber and lead sheet) or rigid materials (wood), this rule is not valid [5]. Owing to the dominating influence of the mass per unit area on R_w, this is called "mass law" for single-layer plates. This mass law was stated first by Berger [8].

The frequency dependency of R' for nearly homogeneous partitions with a mass per unit area of more than 30 kg m^{-2} is shown in Fig. 7.5. At low frequencies, R' hardly depends on the frequency [5, 6], but will then increase above the critical frequency by 25 dB per frequency decade.

There are several deviations from this law.

Larger cavities in walls and ceilings can lead to resonances of the created shell elements and to a deterioration of sound insulation. For more details, see [5]. Therefore, in such a case the mass law is only valid in a restricted way.

Smaller cavities: If the openings are arranged in an acoustically unfavourable way (staggered openings), this leads to a strong reduction of the Young's modulus of the wall perpendicular to its surface and thus to disturbing thickness resonances of the wall with a decrease in the transmission loss by approximately 10 dB and more [9, 10]. For higher frequencies, the

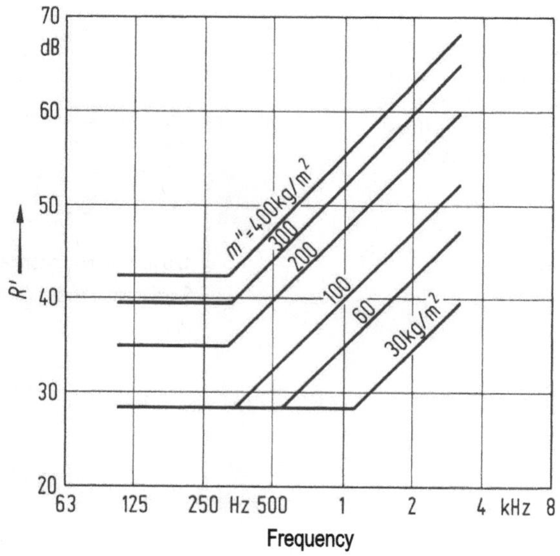

Fig. 7.5 Idealised frequency course of the sound transmission index R' of homogeneous single-layer walls

individual limiting elements of the cavities can show resonances, which according to Heckl [11] can lead to a deterioration of the sound insulation.

Multilayered walls: Above all those with coverings made of materials of medium stiffness (rigid foam plate and wood–wool fibreboards), can behave more unfavourably than expected with regard to their mass per unit area. An example is shown in Fig. 7.6. This effect, which is surprising at first, can be explained by resonances of the outer layers (masses) and the intermediate layer (spring), together with the wall itself (mass).

Actually, these are multi-layer partitions according to Sect. 7.1.4 but are dimensioned in an unfavourable way.

7.1.4 Behaviour of Double-Layer Partitions

7.1.4.1 Basic Behaviour

From the acoustic point of view, double-layer partitions are made of two solid panels separated by a soft resilient intermediate layer, usually an air layer, if possible damped with a sound-absorbing material. The behaviour of such a construction can be explained approximately by the oscillating system shown in Fig. 7.7, where two masses are connected by a spring.

The basic course of the sound reduction index R depending on the frequency is shown in Fig. 7.7 in comparison with the sound insulation by a single

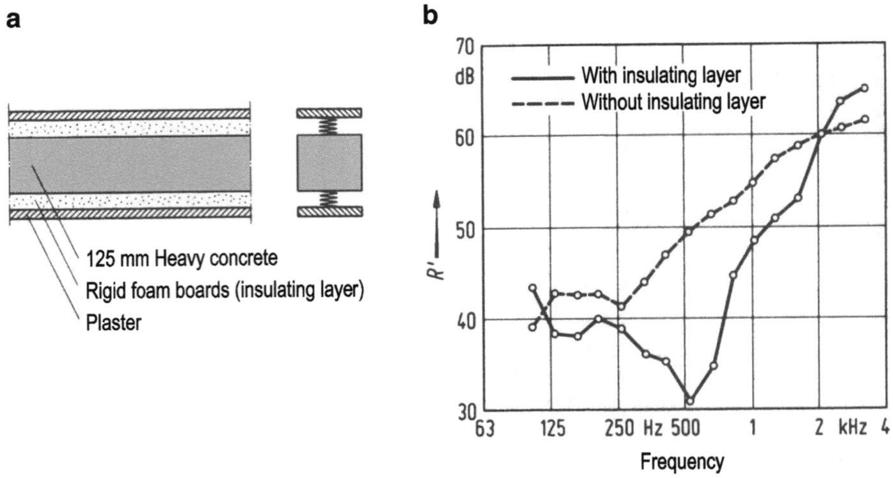

Fig. 7.6 Deterioration of sound insulation of single-shell walls by plaster on insulating layers of unfavourable stiffness (resonance effect)

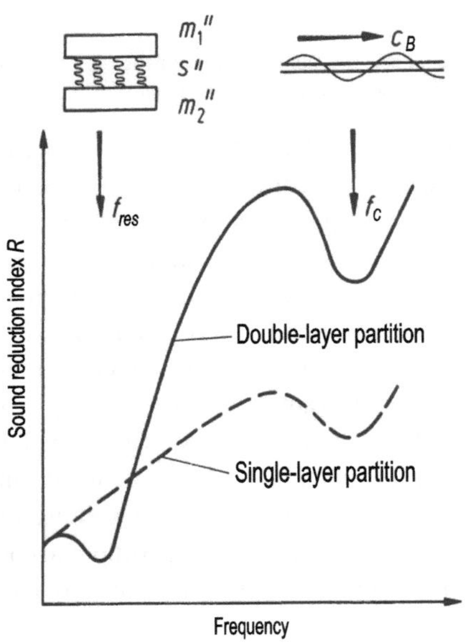

Fig. 7.7 Principle course of the sound insulation R of a single-layer and a double-layer partition made of thin flexible plates depending on the frequency f

layer. At low frequencies, the double-layer construction with intermediate layer brings no improvement in the sound insulation, whereas at the resonance frequency f_{res}, a considerable deterioration of sound insulation occurs. According to the mass–spring–mass model shown in Fig. 7.7, f_{res}[1] results as follows:

– If the shells are connected with each other via an intermediate layer across the entire surface,

$$f_{res} \approx 190\sqrt{\frac{s''}{m''}}, \tag{7.10}$$

– For air as intermediate layer (filled with sound-absorbing material, see Sect. 7.1.4.2),

$$f_{res} \approx 190\sqrt{\frac{0.11}{d \times m''}}\text{Hz} \approx \frac{65}{\sqrt{d \times m''}}\text{Hz}. \tag{7.11}$$

In these equations, $m'' = m_1'' \cdot m_2''/(m''_1 + m''_2)$, m_1'' and m_2'' are the masses per unit area of the two layers (in kg m^{-2}), d is the thickness of the air layer (in m) and s'' is the dynamic stiffness of the intermediate layer (in MN m^{-3}).

Above f_{res}, the sound insulation strongly increases with the frequency. The increase ΔR compared with the sound reduction index of a single layer can be described as follows:

$$\Delta R = 40\lg(f/f_{res})\text{dB};\ f > f_{res}. \tag{7.12}$$

This improvement – calculated according to the idealised model shown in Fig. 7.7 – is only valid for a restricted frequency range. For high frequencies, the critical frequency f_c of the layers becomes very distinct and leads to a relative sound insulation minimum. Besides, for high frequencies ($f > c/(4d)$) the stiffness

[1] The resonance frequency f_{res} results from a modified mass-spring-mass model for the calculation of sound insulation of double-layer constructions [12]. The relations given in (21.10) and (21.11) are valid for thin, flexible layers.

of the intermediate air layer increases, which limits the increase in sound insulation.

These influences can be taken into account by an enhanced model, which accounts for the actual conditions [12]. In this case, the following equation for the sound reduction index R of a double-layer partition (without structure-borne connections and gap filled with sound-absorbing material) results:

$$R = R_1 + R_2 + 20\lg\frac{4\pi f\cdot d}{c}\ \text{dB}\quad \text{for}\quad f > f_{\text{res}}\ \text{and}\ f < \frac{c}{4d}, \tag{7.13}$$

$$R = R_1 + R_2 + 6\text{dB}\quad \text{for}\ f > \frac{c}{4d}, \tag{7.14}$$

R_1 and R_2 transmission loss of the first or second layer,

c sound velocity in air (in m s^{-1}), and

d distance of the layers (in m).

There is usually a satisfactory correspondence between calculation and measurement for the investigated double-layer partitions; see Fig. 7.8 for a double-layer partition with flexible layers and Fig. 7.9 for a double-layer partition with rigid layers. In the latter case, there are deviations above 500 Hz, which are caused by (a few) sound bridges.

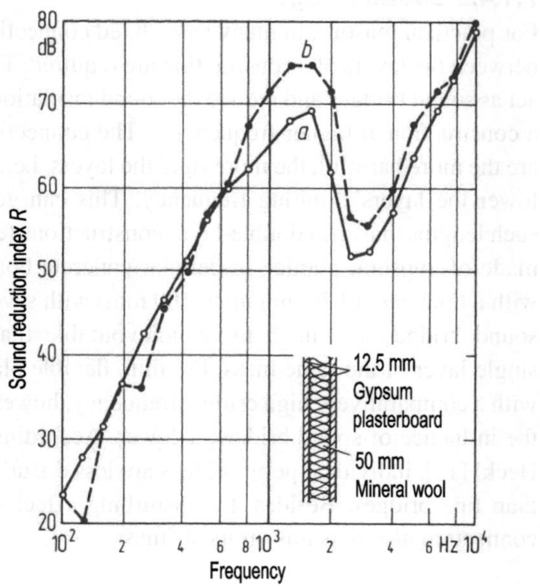

Fig. 7.8 Comparison of measurement a and calculation b for a double-layer wall made of thin flexible plates at the test stand (without flanking paths)

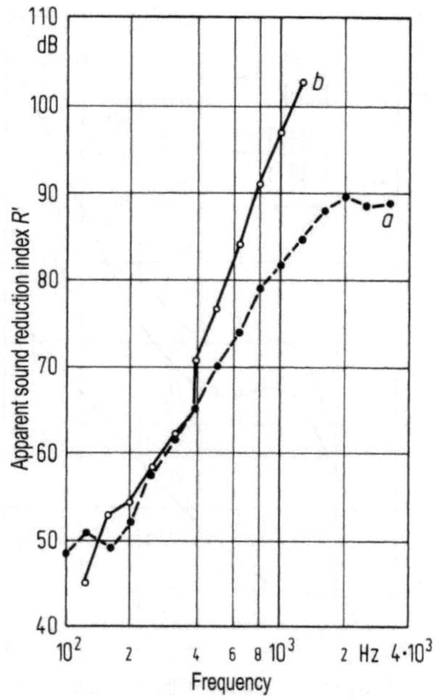

Fig. 7.9 Comparison of measurement a and calculation b for a double-layer partition wall of a house with solid shells on site. Wall shells: 175 mm cellular concrete, distance of the shells: 40 mm (with mineral fibre boards)

Sometimes, deviations may also occur at low frequencies. The reason for such deviations may be the correspondence of the plate resonances of both wall layers. Figure 7.10 gives a rough overview of the sound insulation of double-layer partitions with sound-absorbing material in the cavity and without structure-borne connections.

7.1.4.2 Design of the Intermediate Layer

In order to calculate the resonance frequency f_{res}, not only the mass per unit area of the layers but also the dynamic stiffness s'' of the intermediate layer must be known.

When the layers are connected by an intermediate layer, obviously the dynamic stiffness of the intermediate layer must be used for the calculation. Values of common materials are given in Table 7.6. Besides, it is important to know whether the intermediate layer is glued to the outer layers on both sides or only on one side (influence of contact stiffness). Multiple intermediate layers without adhesive reduce the dynamic stiffness because of the additional contact stiffness between the layers.

The most important intermediate layer for double-layer constructions is air without a fixed connection of the layers via the intermediate layer. Air has the lowest possible stiffness s'', if the material in the cavity has a

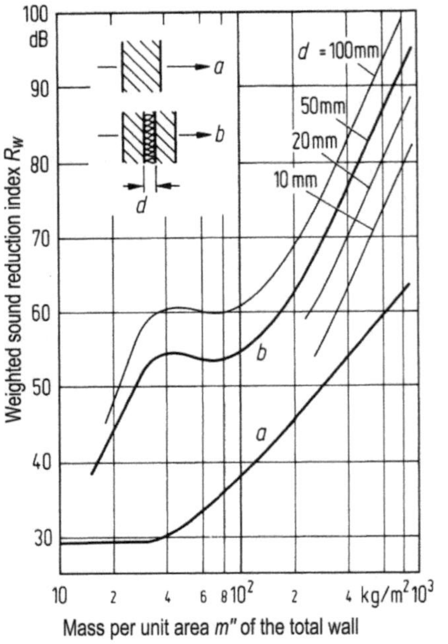

Fig. 7.10 Overview of the achievable airborne sound insulation index R_w of double-layer walls (with separate shells) depending on the mass per unit area m'' of the whole wall and the distance of the shells d

Table 7.6 Dynamic stiffness of insulating layers (values averaged and rounded)

Insulating material	Thickness (mm)	Dynamic stiffness s'' (MN m^{-3})
Mineral fibre boards	10	20
Mineral fibre boards	20	10
Coconut fibre mats	12	40
Granulated cork mats	7	150
Polystyrene rigid foam boards, depending on the manufacturer	15	10–200
Wood wool lightweight building boards	25	200
Cork boards	10	500
Sand filling	25	300

flow resistance of at least $r = 5$ kPa s m^{-2}, which is usually realised in the form of a flexible porous absorber, e.g., mineral fibre or open-cell foam. In this case, dynamic stiffness is

$$s'' = \frac{0.11}{d}\ \mathrm{MN \cdot m^{-3}}, \tag{7.15}$$

with d (in m) being the thickness of the air layer. Without a flow resistance in the air cavity, an

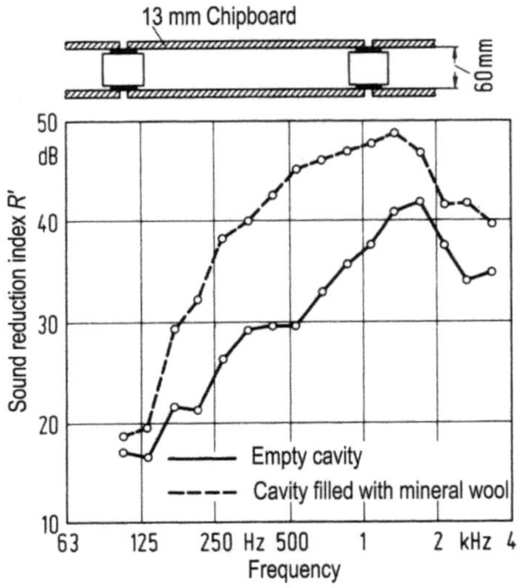

Fig. 7.11 Example for improving the sound reduction index R' of a double-layer wall by inserting a "flow resistance" in the cavity

essentially higher stiffness of the air layer leads to cavity resonances with little damping; see Fig. 7.11. According to Meyer [13], this imperfection of an undamped air layer can be partly alleviated by a so-called edge damping (= installation of a porous material at the edge of the cavity), but cannot be completely removed [14].

7.1.4.3 Sound Bridges

For practical reasons, in many cases fixed connections between the layers of a construction are required. They act as sound bridges and reduce the sound insulation of a construction at higher frequencies. The connections are the more harmful, the more rigid the layers, i.e., the lower the layers' limiting frequency. This can go to such lengths that such double-layer constructions (e.g., made of gypsum, pumice or cellular concrete boards with a thickness of 50 mm up to 100 mm) with several sound bridges are much more unfavourable than a single layer of the same mass. For thin, flexible plates with a comparatively high critical frequency, however, the influence of sound bridges is lower. According to Heckl [15], individual point bridges are less disturbing than line bridges. Besides, the disturbing effect of a connection also depends on its stiffness.

7.1.4.4 Importance of Flexible Layers

The bending stiffness of the layers is very important regarding the sound insulation of double-layer constructions. This can be explained by the different radiation behaviour of layers below and above their critical frequency, if free bending waves are excited. Such excitations are caused by airborne sound, but, above all, by sound bridges or fixed connections with the side walls. For this reason, a critical frequency f_c of the layers which is as high as possible is aimed at ("flexible" layers). Conversely, a sufficiently high mass per unit area of the layers is required, so that for a restricted distance of the layers the resonance frequency f_{res} according to Sect. 7.1.4.1 is not too high. Both conditions can be fulfilled when the layers are weighed down at the inside with a suitable material (Fig. 7.12), so that an increase of the mass per unit area and *no* significant increase of the bending stiffness result. Tests have shown that it is very favourable if the material also has a high material damping. Practical use: Gluing of individual pieces of plate made of brick, concrete, sheet steel, or sheet lead or heavy layers made of plastic onto the layers.

The use of two thin plates is more favourable than the use of one thick plate, as the two plates have a higher critical frequency for the same mass and, in case of materials with low internal damping, also higher total damping.

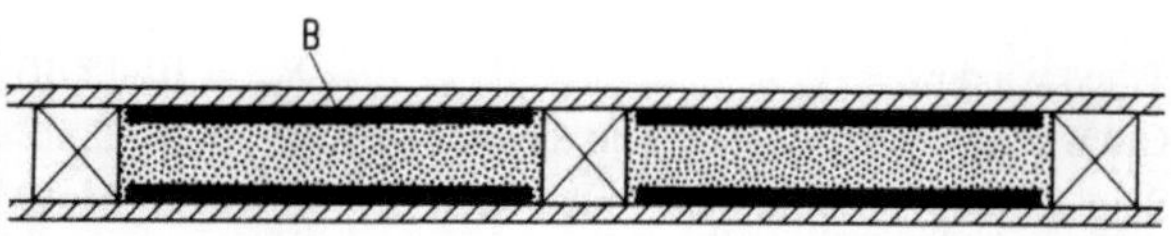

Fig. 7.12 Improving the sound insulation of double-layer wall panels by weighing them down with B

7.1.4.5 Double Walls in Practice

Double walls made of rigid layers, which are self-supporting, e.g., made of gypsum or cellular concrete plates with a thickness of 60 mm up to 100 mm, have an airborne sound reduction index between 47 and 50 dB. In the middle of the frequency range of interest, they have a resonance dip in the sound insulation curve, which is caused by the critical frequency. By an increase of structure-borne sound insulation of the layers, the sound insulation can be improved, e.g., by connecting the layers to the flanking partitions with bitumen strips or by weighing down the wall layers at the inside with loose sand [16].

With flexible layers, usually chipboards or gypsum boards, the values stated in Table 7.2 can be achieved [17]. In case of connecting the layers with stands, an increase of sound insulation by means of higher structure-borne sound damping and/or insulation of the stands is possible. This can be achieved, e.g., by special supports or by damping the stands with insulating boards [18] (Table 7.2).

7.1.4.6 Acoustical Linings

It is often the task to improve airborne sound insulation of existing massive walls. This can be achieved by so-called acoustical linings, i.e., light, flexible layers (e.g., chipboard, gypsum plasterboard or plaster), which are fixed to the wall by means of a supporting frame with a total thickness of at least 50 mm. The possible improvement becomes lesser with an increasing mass per unit area of the massive wall. Typical values for the weighted sound reduction improvement

Table 7.2 Airborne sound insulation of gypsum plasterboard walls with flexible layers

Wall construction shells	Connection between the layers	Distance of the shells (mm)	Total wall thickness (mm)	m'' per layer (kg m^{-2})	Weighted sound reduction index R_w[a] (dB)
Chipboards, gypsum plaster boards	Via wooden studs	60–80	80–100	approx. 8.5	37–39
Gypsum plaster boards, 1 layer per shell	Via common metal studs CW 50 × 0.6	50	75	8.5	39
Gypsum plaster boards, 1 layer per shell	Via common metal studs CW 100 × 0.6	100	125	8.5	41
Gypsum plaster boards, 2 layers per shell	Via common metal studs CW 50 × 0.6	50	100	17	46
Gypsum plaster boards, 2 layers per shell	No connection (staggered metal studs CW 50 × 0.6)	105	155	17	58

[a] R_w without considering flanking paths, e.g., flanking partitions.

Table 7.3 Example for acoustical linings and typical weighted sound reduction improvement indices ΔR_w

Lining made of two layers with approx. $m'' = 7$ kg m^{-2} each	Weighted sound reduction improvement index ΔR_w	
Supporting construction with a depth of at least 50 mm and damped cavity	Light solid walls approx. $m'' = 120$ kg m^{-2}	Heavy solid walls approx. $m'' = 400$ kg m^{-2}
Detached wooden or metal studs in front of the wall (no contact to wall)	18 dB	12 dB
Sandwich panels made of a gypsum plaster board and a resilient insulating layer (e.g., mineral fibre boards), which are glued to the wall	16 dB	11 dB
Minimum values for detached linings with a resonance frequency of $f_{res} \leq 80$ Hz according to [63]	16 dB	7 dB

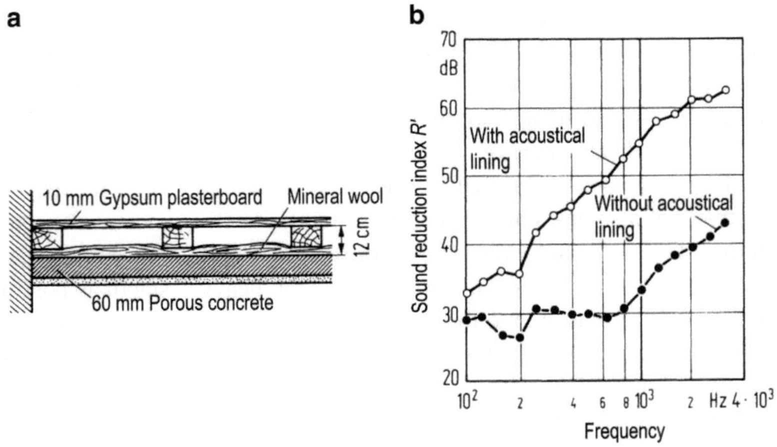

Fig. 7.13 Example for improving the airborne sound insulation of a lightweight wall by means of an acoustical lining

index ΔR_w with different substructures are given in Table 7.3.

The improvement of the sound insulation for substructures with a connection to the wall is usually lower than for free-standing substructures or composite boards, which are only fixed punctually to the wall. Examples are shown in Fig. 7.13. Measured values for different coverings on direct and flanking partitions are stated in [19, 61]. Laboratory measurements of acoustical linings can be executed according to [52].

During construction, the achievable improvement by means of acoustical linings is often limited to distinctly lower values than shown in Table 7.3, owing to the sound transmission via flanking building elements (see Sect. 7.1.5).

7.1.4.7 Windows

In connection with single and double walls, windows also have to be discussed. For them, the described basic laws are valid as for other separating elements. However, leakages are a great problem, especially for highly sound-insulating windows (see Sect. 7.1.6). Typical weighted airborne sound reduction indices of functioning windows are:

Tilted window	$R_w = 10$–12 dB
Old window without sealings in the rabbets	$R_w = 20$–28 dB
Single window with normal double glazing	$R_w = 32$–34 dB
Single window with heavy double glazing	$R_w = 38$–40 dB
Single window with highly sound-insulating laminated sheet glass	$R_w = 44$–47 dB
Linked casement with double frames	$R_w = 37$–40 dB
Highly sound-insulating linked casement with double frame	$R_w = 45$ dB
Box-type window, depending on glazing and frame	$R_w = 45$–60 dB

For highly sound-insulating windows ($R_w > 45$ dB), it is possible that more sound is transmitted via the frame than via the panes. In this case, a separation (decoupling) of frame and panes with regard to structure-borne sound is required. Calculated values for sound insulation of windows are stated in [39].

For double façades, the following improvement of sound insulation by means of the outside façade can be achieved, depending on the size of the supply and waste air openings and the absorbing lining of the space in between:

For closed windows of the inner façade	$\Delta R_w = 3\text{–}11$ dB
For tilted windows of the inner façade	$\Delta R_w = 6\text{–}14$ dB

7.1.5 Flanking Sound Transmission in Solid Buildings

With suitable double-layer partitions or ceilings, very high sound insulation can be achieved in the laboratory (e.g., $R_w = 65$ dB up to 75 dB). On site, for terraced houses with double-layer partitions between the individual houses such values can be achieved with a gap running over the whole depth of the house, so that the houses are completely decoupled. In other buildings with continuous flanking walls and ceilings, such high transmission losses cannot be achieved, as apart from the transmission via a separating partition between both rooms, there is also a transmission via flanking partitions (see Fig. 7.14). Bending vibrations are excited in these partitions in the source room. These vibrations are transmitted to the flanking partitions in the adjacent room on the right or the left, at the top or the bottom, usually with a low damping between about 3 and 20 dB. This transmission limits the transmission loss achievable between adjacent rooms in common solid buildings to values of about $R'_w = 50$ dB up to 57 dB, in worse cases also to lower values. More details are stated in [20].

In order to describe sound transmission via the individual transmission paths, according to Fig. 7.14, for each transmission path ij (e.g., Ff) a flanking sound reduction index R_{ij} (e.g., R_{Ff}) can be defined:

$$R_{ij} = -10\lg\left(\frac{W_{ij}}{W_1}\right)\,\mathrm{dB}. \tag{7.16}$$

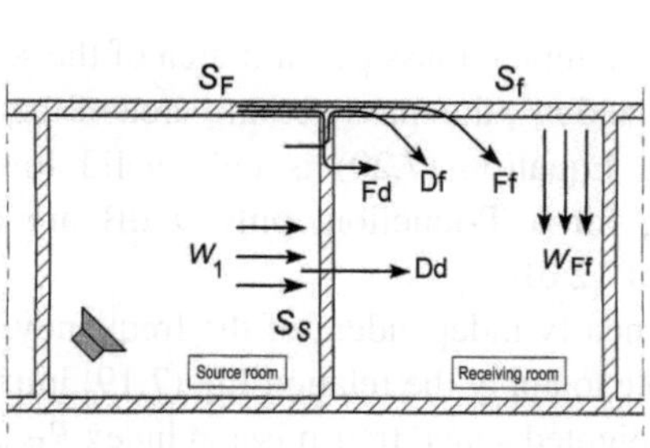

Fig. 7.14 Airborne sound transmission paths according to [61]

This flanking sound reduction index R_{Ff} is e.g., the ratio of the sound power incident W_1 exiting the separating partition with the reference area S_s and the sound power W_{Ff} radiated by the flanking partition f in the receiving room, owing to the sound exiting the flanking partition F in the source room. The flanking sound reduction index is defined as the ratio of the radiated sound power W_{ij} and the sound power W_1 exiting the reference surface S_s of the separating partition, in order to be able to state the contribution of the individual transmission paths to the total transmission.

The total sound power W_{tot} radiated to the receiving room results from the sum of the radiated sound power of the n flanking partitions (paths Df and Ff) and the separating partition (paths Dd and Fd). With the definition of the apparent sound reduction index R' Eq. (7.3), the following equation results:

$$R' = -10\lg\left(10^{-\frac{R_{\mathrm{Dd}}}{10}} + \sum_{F=1}^{n} 10^{-\frac{R_{\mathrm{Fd}}}{10}} + \sum_{f=1}^{n} 10^{-\frac{R_{\mathrm{Df}}}{10}} + \sum_{F=f=1}^{n} 10^{-\frac{R_{\mathrm{Ff}}}{10}}\right)\mathrm{dB}, \tag{7.17}$$

with R_{Dd} being the transmission loss for direct sound transmission.

Thus, for two adjacent rooms with usually $n = 4$ flanking walls/ceilings, 13 transmission paths result from this equation.

The described method to calculate R' corresponds to the calculation model for structure-borne sound transmission according to ISO 12354–1 [61]. Here, a detailed model is presented for a frequency-dependent calculation according to Eq. (7.17), as is a simplified model, in which the single-number quantities are used in Eq. (7.17), i.e., the weighted sound reduction index and weighted flanking sound reduction index.

7.1.5.1 Determination of the Flanking Sound Reduction Index

According to [61], the flanking sound reduction index for the transmission path ij can, in general, be determined as follows for any junction and different walls in the source and receiving room:

$$R_{ij} = \frac{R_i}{2} + \Delta R_i + \frac{R_j}{2} + \Delta R_j + \bar{D}_{v,ij} + 10\lg\left(\frac{S_s}{\sqrt{S_i S_j}}\right)\mathrm{dB}, \tag{7.18}$$

with R_i sound reduction index of the excited element i in the source room

R_j – sound reduction index of the radiating element j in the receiving room

ΔR_i, ΔR_j – sound reduction improvement index by means of acoustical linings for the element i or j (see Sect. 7.1.4.6)

$\bar{D}_{v,ij}$ direction averaged velocity level difference at the junction between the building components i and j

S_s – surface of the separating element

S_i, S_j – surface of the element i or j

R_i and R_j are the sound reduction indices for direct sound transmission. The direction averaged velocity level difference $\bar{D}_{v,ij}$ describes the decrease of the bending vibrations when crossing the junction between element i and j.

The acoustic data should mainly result from test stand measurements, but calculations are also possible. The values R_i, R_j and $\bar{D}_{v,ij}$ have to be converted into values on site (in situ). The different installation conditions at the test stand and on site, which are mainly described by the loss factor, lead to different sound reduction indices R. The adjustment is made according to the loss factors measured (or assumed) at the test stand and expected on site. The loss factors are determined from measurements of structure-borne sound reverberation times. For solid building elements, this method has proved successful above the critical frequency [7]. The value $\bar{D}_{v,ij}$ is determined from the vibration reduction index K_{ij}, taking into consideration the junction length l_{ij} between the elements i and j (common edge length) and their damping characteristics (equivalent absorption lengths) in situ.

The kind of junction has a great influence on the flanking sound insulation: is it a double-T junction or a T-junction, is the connection rigid or a soft intermediate layer; it also depends on the mass ratio between the elements. The vibration reduction indices for several building elements are stated in [61].

This approach has also been applied to determine flanking sound insulation for building elements with critical frequencies in or above the frequency range of interest, e.g., for wood panelling etc. Because of the different behaviour of these elements regarding the bending radiation of either forced (in case of direct sound transmission) or of free (in case of flanking sound transmission) bending waves, there are significant deviations between measurement and calculation. Further problems occur because of the uncertainty of the kind of the actual junction (rigid junction, without connection, or anything in between). These problems were often solved by transferring the different behaviour of the partitions in the case of flanking transmission compared to direct transmission to the vibration reduction index. Unfortunately, these approaches assume laboratory measurements and the vibration reduction index loses physical significance.

7.1.5.2 Simplified Determination of Flanking Sound Insulation

The above-described detailed calculation to determine the sound reduction index from the direct and the flanking sound reduction index can be simplified taking into consideration the actual condition in solid construction, as shown in the following.

In solid construction, an approximate calculation of the flanking sound transmission index R_{Ff} (also R_{L}) can be carried out according to Eq. (7.18). In this calculation, it is assumed that the flanking partitions in the source and the receiving room are identical and the surface of the separating partition is equal to the surface of each flanking partition [21]:

$$R_{\mathrm{Ff}} \approx R_{\mathrm{f}} + \bar{D}_{v,\mathrm{Ff}}, \qquad (7.19)$$

R_{f} is the sound transmission index of the flanking partition for direct sound transmission (an approximate value can be taken from Fig. 7.15); $\bar{D}_{v,\mathrm{Ff}}$ can be calculated according to [21] for a rigid connection between flanking components and separating components as follows:

$$\bar{D}_{v,\mathrm{Ff}} = \left(20\lg\frac{m''_{\mathrm{t}}}{m''_{\mathrm{f}}} + 12\right)\mathrm{dB}, \qquad (7.20)$$

with m''_t being the mass per unit area of the separating partition and m''_{f} the mass per unit area of the flanking partition. Equation (7.20) is only valid for a cross junction; for a T-junction, only 9 dB are assessed instead of 12 dB.

D_v is nearly independent of the frequency. Thus, it is possible to apply the relation Eq. (7.19) immediately on the weighted sound transmission index $R_{\mathrm{Ff},w}$ or $R_{\mathrm{f},w}$, which simplifies the calculation considerably.

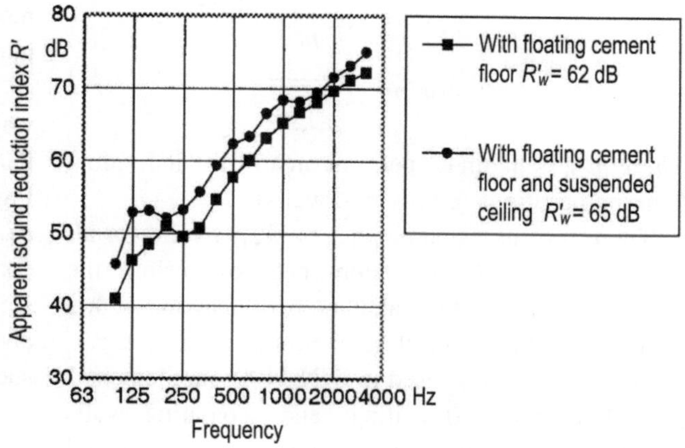

Fig. 7.15 Influence of a flexible suspended ceiling on the sound reduction index of a concrete slab with floating floor in the ceiling test stand with flanking paths

The flanking sound transmission indices R_{Fd}, R_{Df} and R_{Ff} have nearly the same value. The expected improvement of the sound insulation of a wall with one acoustical lining, which can be achieved with a second acoustical lining on the other side of the wall, can be derived by the calculation methods above. As with a second acoustical lining, R_{Df} is cancelled, and which has nearly the same value as R_{Ff}, the transmission loss R'_w is only improved by 3 dB at the most. The same is valid for ceilings with a suspended ceiling and a floating floor.

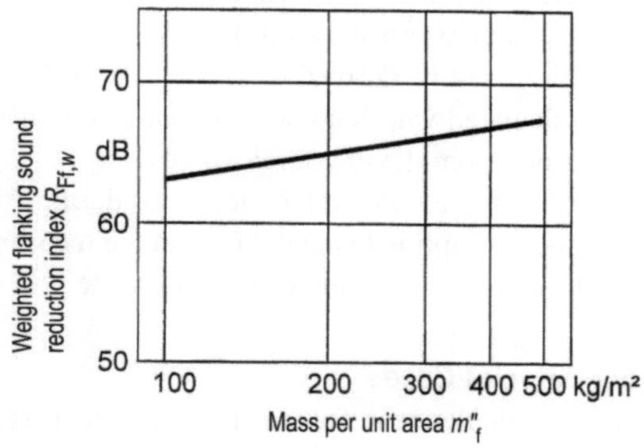

Fig. 7.16 Little dependence of the weighted flanking sound reduction index $R_{\mathrm{Ff},w}$ from the mass per unit area m''_{f} of the flanking wall in solid buildings

7.1.5.3 Influence of the Mass Per Unit Area of Flanking Partitions

The flanking sound insulation in solid constructions with rigid junctions depends on two quantities:

- On the mass per unit area of the flanking walls and
- On the separation of ceiling or wall

In general, there is the opinion that for a high flanking sound insulation the mass of the flanking partition, especially, should be as high as possible. However, calculation shows that this is not the case. Figure 7.16 renders this dependence according to the calculations in Eqs. (7.20) and (7.21). In the range between $m''_f = 100$ kg m^{-2} and 400 kg m^{-2}, which is the range of interest in civil engineering, the dependence increases by only 3 dB. This surprisingly low influence is due to the fact, that although for light flanking walls the vibrations of the walls are very large, for the same separating ceiling the velocity level difference at the junction D_v is larger than for heavier flanking walls. Both opposing influences lead to the fact that the flanking sound insulation only depends on m''_{f} in a small way. In contrast, the mass per unit area m''_{t} of the separating ceiling has much more influence. For the above variation of 1–4, a difference of $R_{\mathrm{Ff},w}$ of 12 dB would result. It follows for solid constructions with rigid junctions of the partitions: The flanking partitions should not be heavy, but the separating partition, i.e., the separating ceilings and walls of a flat, respectively.

7.1.5.4 Simplified Determination of R'_w

The results of the previous section can be roughly summarised as follows: Owing to the *existence* of the flanking walls of a ceiling with floating floor, the airborne sound insulation deteriorates by approximately 15–20 dB. However, the amount of this deterioration is influenced by m''_{f} only in a small way. On account of this small influence of the mass per unit area m''_{f}, the calculation of R'_w can be simplified. A linear averaging of the values m''_{f} of flanking partitions to a value $m''_{\mathrm{f,\ average}}$ can be made:

$$m''_{\text{f,average}} = \frac{\sum_{i=1}^{n} m''_{\text{f}i}}{n} \quad (7.21)$$

where m''_{fi} is the mass per unit area of the individual flanking partitions (e.g., $n = 4$ walls).

Then, for this value $m''_{\text{f, average}}$, $R_{\text{Ff},w}$, valid for all four walls, will be determined by using the simplifications for the calculation of flanking sound insulation indices according to Sect. 7.1.5.2. The calculation method was used in DIN 4109, supplement 1 [38]. For about 50 ceilings and separating walls investigated during construction, these calculated values were compared with measured values, and the resulting deviations are shown in Fig. 7.17.

Two things can be derived:

(a) The calculated values do not show any systematic deviations from the measured values.
(b) The deviations are tolerable, considering the extremely simplified calculation and the measurement inaccuracies on the construction site.

7.1.5.5 Special Cases

The above calculation refers to the case when walls and ceilings are solidly built and connected rigidly. This was largely the case for constructions in the past. In recent years, however, other solutions have been applied – above all for outer walls. One possibility is that solid separating walls between two flats are only loosely connected to the outer walls with a so-called blunt joint, and thus, there is no rigid connection. This leads to an essentially reduced flanking insulation in the horizontal direction. Another novelty is that the flanking walls are fitted with a lining, which leads to a resonance in the middle of the frequency range of interest; see Fig. 7.18.

In the past decade, resonances inside certain outer walls made of perforated brick caused a reduced flanking insulation. This is shown in an example in Fig. 7.19, where the flanking sound reduction index R_{Ff} of a certain vertically perforated brick wall is compared to the calculated values for a solid wall of similar mass. In general, it can be said that, for the above-mentioned reasons, the flanking transmission of such outer walls determines the achievable sound

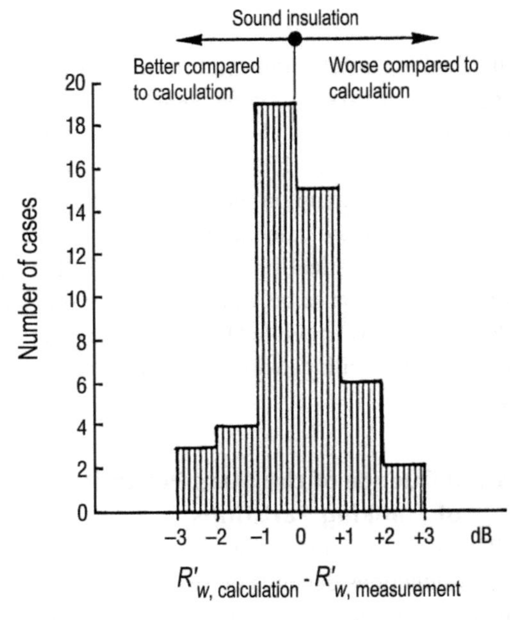

Fig. 7.17 Frequency distribution of the deviation of the airborne sound insulation index R'_w calculated according to DIN 4109, amendment 1 [38] and measured on site (measured values by Lang, Eisenberg, Schulze and Gösele)

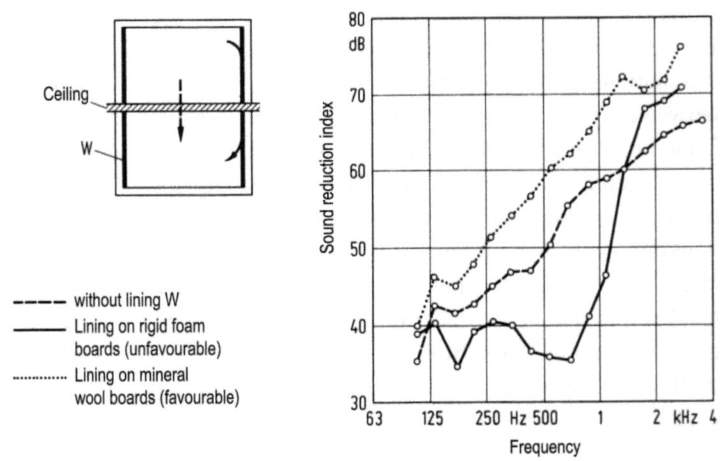

Fig. 7.18 Influence of a suitable and unsuitable wall covering on the flanking sound insulation (sound insulation between two rooms, one of top of the other, with concrete ceiling and floating floor)

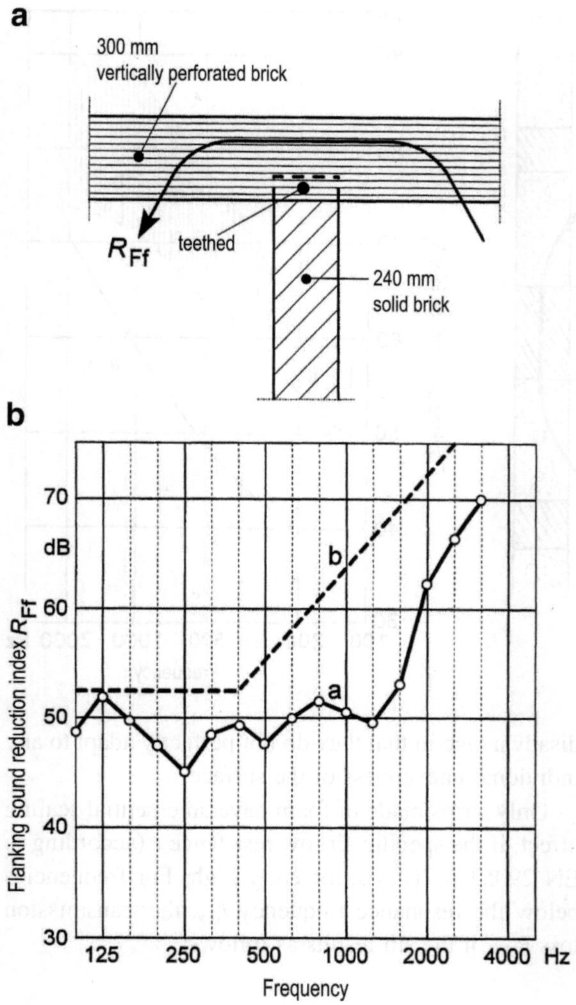

Fig. 7.19 Reduced flanking sound transmission index R_{Ff} of an acoustically unfavourable outer brick wall according to measurements of GSA Limburg [22]. (**a**) 300 mm vertically perforated bricks (**b**) calculation for a solid homogeneous wall (without any cavities) of the same mass

insulation R'_w to a large extent. Also in these special cases, a very simple additional calculation can be applied.

7.1.5.6 Potential Improvements

(a) Increase of the mass per unit area of the flanking walls.
The above comments show that a certain increase of the mass per unit area of the walls (apart from the separating wall to the adjacent flat) does not lead to an essential improvement in flanking insulation. If an improvement is intended by changing the mass, this must be done for the separating walls and ceilings. Here, an increase of the thickness of the concrete ceiling from 150 to 300 mm, raises from $R'_w = 55$–61 dB.

(b) Acoustical lining of the flanking walls.
Another possibility is to add an acoustical lining to the flanking walls. An example is shown in Fig. 7.18. However, note the limit, as the lining becomes ineffective near its resonance frequency (approximately 100 Hz) and even counterproductive at the resonance frequency. Above all, the main argument against such linings is the high efforts and the reduction of living space.

(c) Structure-borne sound insulation.
Another method is a structure-borne sound insulation by additional measures inside the walls. The effect of such measures is shown in Fig. 7.20 for very old timber framing walls where, for a mass per unit area of about 150–200 kg m^{-2}, by means of friction between individual parts of the filling an improvement of R_L of about 15–25 dB was achieved.

7.1.6 Leakages

For certain building elements, such as doors and windows, and also for constructions where e.g., the walls are built from individually removable plates, the airborne sound insulation is decisively influenced by sound transmission via joints and other leakages. There are only a few quantitative documents on that subject.

7.1.6.1 Simple Slits

In the following, the sound reduction index of a slit is the sound reduction index R_{ST} of an (assumed) high sound insulating partition with a size of 1 m and with a slit with a length of 1 m. Sufficiently below the resonance frequency f_{res} of the slit, according to Gomperts [23] the following equation is valid:

$$R_{ST} = [20 \lg l/b + 10 \lg f/f_0 - 3] \text{dB}, \qquad (7.22)$$

Terms: see Fig. 7.21; $f_0 = 100$ Hz.

Thus, the sound insulation is better, the deeper the slit (higher inertia). Conversely, each slit has very disturbing resonances as soon as the length (including the so-called end correction) is equal to $\lambda/2, \lambda, \ldots$; see

Fig. 7.20 Longitudinal transmission loss R_L of half-timbered walls. (**a**) filled with bricks, approx. 70 years old (**b**) with different filling materials, more than 100 years old (range of dispersion for five different buildings)

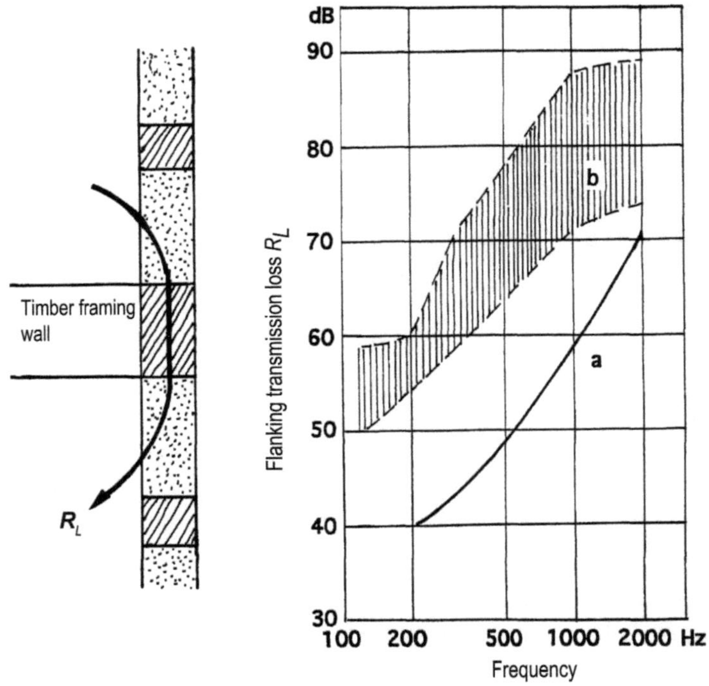

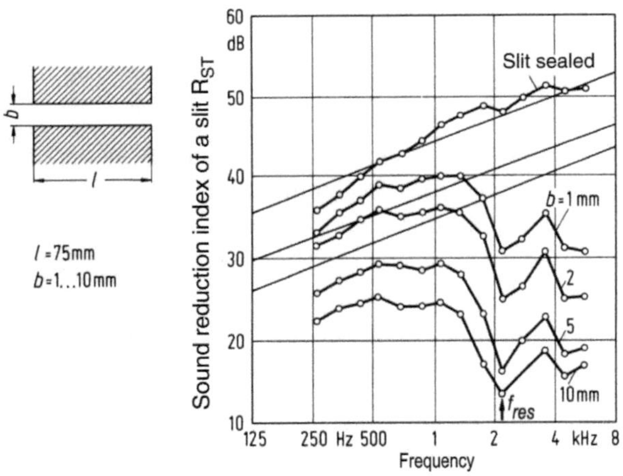

Fig. 7.21 Slit sound insulation of different width b. Length of the slit 1 m

Fig. 7.21 for $f > 0.5\,f_{\text{res}}$. This reduced sound insulation occurring at higher frequencies is the more distinct and, in the frequency range of interest, the deeper the joint [24]. Resonances often reduce sound insulation of untight windows and doors.

7.1.6.2 Sealing Materials

For sealing slits, either so-called lip-seals or strips made of foam are used. For low contact forces, the lip-seals are more favourable. However, they have the disadvantage in that they do not perfectly adapt to any individual unevenness of the surface.

Only strips made of foam have an essential sealing effect, if the specific airflow resistance r (according to EN 29053 [67]) is sufficiently high. For frequencies below the resonance frequency f_{res}, the transmission loss R_{ST} of the slit results as follows:

$$R_{\text{ST}} = \left[20\lg\frac{lr}{br_0} + 10\lg\frac{f}{f_0} - 1\right]\text{dB}. \tag{7.23}$$

Reference value $r_0 = 10$ Pa s m^{-1}.

Common open-cell foams have only very small flow resistances (below 100 Pa s m^{-1}), as long as they are not strongly compressed. As soon as they are compressed, their insulation increases strongly (Fig. 7.22).

7.1.6.3 Acoustic Filters

A high sound insulation for existing open slits can be achieved by coupling cavities to the slit [24, 25]; see Fig. 7.23. Thus, an acoustic filter results. The resonance frequency f_{res1} of that filter can be calculated as follows:

$$f_{\text{res1}} = 1.1 \times 10^4\sqrt{\frac{b}{lS}}\text{Hz}. \tag{7.24}$$

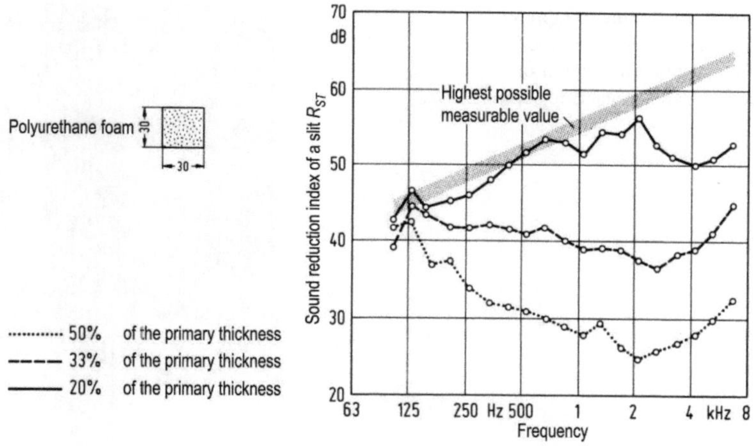

Fig. 7.22 Slit sound insulation of a slit sealed with an open-cell foam, depending on the compression of the foam

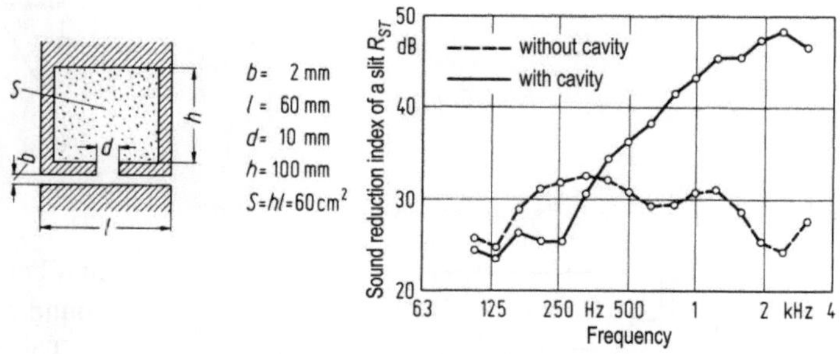

Fig. 7.23 Influence of a cavity connected to the slit on the slit sound insulation R_{ST} of a slit with a width of 2 mm

Above this resonance, the improvement of the insulation compared with a simple slit begins. This basic principle can also be advantageously applied in other forms, e.g., in the form of multiple-stage filters [25]. Such arrangements have the advantage of a higher insulation, above all concerning high frequencies, which are particularly disturbing with regard to slits.

7.2 Impact Sound Insulation

7.2.1 Characterisation and Measurement of Ceilings

In residential buildings, hotels, hospitals, etc., often disturbing impact sound is produced by, e.g., walking on ceilings, moving chairs or operating kitchen appliances. This can be explained by the fact that massive ceilings (without any special measures) can be excited to strong bending vibrations even by low forces. Impact sound insulation, the ceiling to be tested is excited with a usually electrically operated standard tapping machine (see Fig. 7.24). The sound level L in the room below this ceiling is measured for each one-third octave band. It is converted to a room with an equivalent sound absorption area of $A_0 = 10\ \text{m}^2$ as follows:

$$L'_n = L + 10\lg A/A_0 \text{dB}, \qquad (7.25)$$

with A being the equivalent sound absorption area of the receiving room and L'_n being the normalised impact sound pressure level in buildings, according to [47] (see Fig. 7.25). For measurements with an additional transmission of the impact sound via the flanking partitions, as in most buildings, L_n is provided with an apostrophe L'_n. In the test stand, the impact sound pressure level is measured without sound transmission via the flanking partitions, according to [46], and is called L_n.

Another quantity to characterise the impact sound insulation of ceilings on site is the standardised impact sound pressure level L'_{nT} [47]:

$$L'_{nT} = L - 10\lg \frac{T}{T_0} \text{dB}. \qquad (7.26)$$

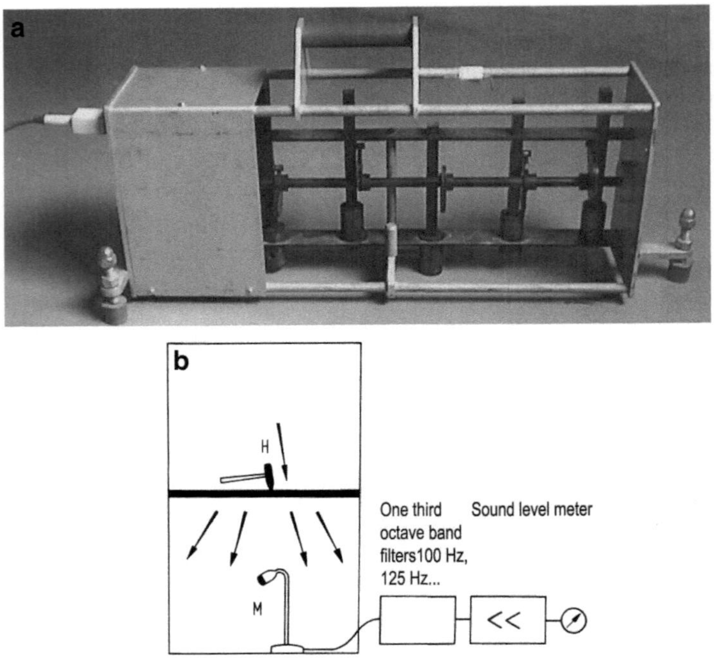

Fig. 7.24 Standard tapping machine to determine the impact sound insulation of ceilings and measuring set-up

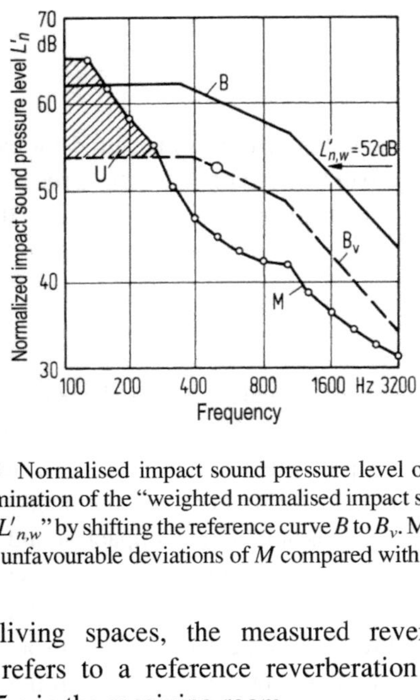

Fig. 7.25 Normalised impact sound pressure level of a ceiling and determination of the "weighted normalised impact sound pressure level $L'_{n,w}$" by shifting the reference curve B to B_v. M measured values; U unfavourable deviations of M compared with B_v

For living spaces, the measured reverberation time T refers to a reference reverberation time of $T_0 = 0.5$ s in the receiving room.

The impact sound insulation depends on the frequency. In order to gain a single number quantity, the reference curve B (Fig. 7.25) is used, which is shifted in vertical direction as long as the sum of the exceeding U is as high as possible compared with the shifted reference curve but not larger than 32 dB. The standardised impact sound pressure level of the shifted reference curve at 500 Hz is then used as a single number quantity and called the weighted impact sound pressure level $L'_{n,w}$.

The excitation of a ceiling with a standard tapping machine, whose steel hammers create impulsive strokes, corresponds to the noise of hard objects of hammers of the same mass ($m = 500$ g) falling onto the ceiling, similar a knocking noise. The noise of walking on the ceiling is usually of lower frequency, as the cushioning of the shoes has a similar effect as a (slightly) elastic surface layer. However, such a layer reduces the high frequencies; see Sect. 7.2.4. In order to judge the measured levels created by a standard tapping machine on a ceiling with respect to the excitation by a footfall, the spectrum adaptation value C_I was introduced [55]. The sum of C_I and the weighted normalised impact sound pressure level is then proportional to the unweighted linear impact sound level.[2] It can be stated in addition to the weighted impact sound pressure level, e.g., $L'_{n,w}\ (C_I) = 53(-3)$ dB. For effective floor coverings on heavy solid ceilings, C_I is nil, for ceilings with low-frequent peaks such as wooden beam ceilings C_I is a positive value (up to

[2] Investigations have shown that for footfall, the unweighted linear impact sound level rather corresponds to the A-weighted overall level than to the impact sound pressure level. $L'_{n,w} + C_I$ can thus be regarded as an index for the impact sound insulation concerning footfall.

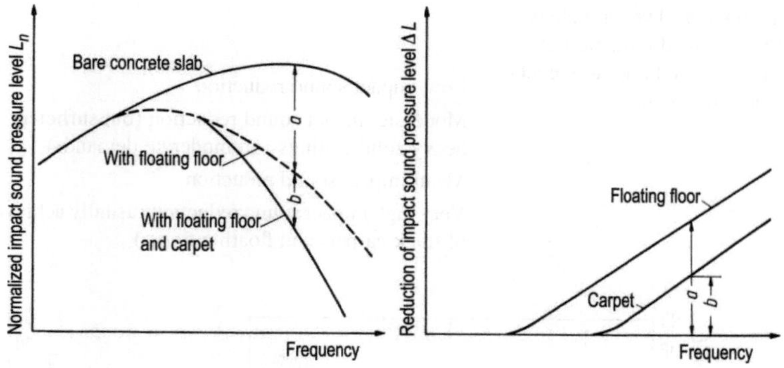

Fig. 7.26 Definition of reduction of impact sound pressure level ΔL of floor coverings

2 dB), whereas for concrete slabs C_I is a negative value (down to −14 dB).

Examples for the resulting values are given in the following:

	Weighted impact sound pressure level $L'_{n,w}$ [dB]
Common concrete slab (without covering)	73–83
Maximum permissible value for dwellings in apartment houses, according to German standard [38]	53
Increased noise control for dwellings in apartment houses, according to German guideline [40]	46
Very high noise control for dwellings in apartment houses, according to German guideline [40]	39

For characterising ceilings without floor covering ("bare concrete slab"), another index is introduced. With this index, the potential improvement of the impact sound insulation of a ceiling by means of a floor covering is taken into account. Two ceilings, which, as bare concrete slabs, have the same standardised impact sound pressure level, can behave in a different way after applying the same floor covering. This is considered with the equivalent weighted impact sound pressure level $L_{n,\text{eq},0,w}$, according to [55].

7.2.2 Characterisation of Floor Coverings

The impact noise insulation of solid ceilings is mainly improved by suitable floor coverings, e.g., so-called floating floors, raised floors and hollow floors as well as flexible coverings. The insulation effect is characterised by measuring the normalised impact sound pressure level of a ceiling *without* ($L_{n,0}$) and *with* (L_n) the floor covering to be tested versus the frequency. The difference

$$\Delta L = L_{n,0} - L_n, \tag{7.27}$$

is called improvement of impact sound insulation or reduction of impact sound pressure level of a floor covering; see Fig. 7.26. From the reduction of impact sound pressure level depending on the frequency in the range of 100 Hz up to 3,150 Hz, a single number quantity is calculated [55], which is called weighted impact sound improvement index ΔL_w. It describes the reduction of the weighted impact sound pressure level of a reference floor (stated in [55]), if the floor covering is applied. The range for common floor coverings is as follows (Table 7.4):

From the equivalent weighted impact sound pressure level of bare concrete slab – see Sects. 7.2.1 and 7.2.3 as well as Fig. 7.28 – and the weighted impact sound improvement index ΔL_w of a floor covering, the weighted impact sound pressure level $L'_{n,w}$ of a finished ceiling can be calculated with sufficient accuracy by subtraction:

$$L'_{n,w} = L_{n,\text{eq},0,w} - \Delta L_w. \tag{7.28}$$

More details on this subject are stated in [26].

Also for impact sound insulation, a spectrum adaptation value $C_{I,\Delta}$ can be stated. A single-number quantity ΔL_{lin} for the unweighted linear impact sound pressure level can be calculated as follows: $\Delta L_{\text{lin}} = \Delta L_w + C_{I,\Delta}$ [55].

Table 7.4 Typical values of the weighted reduction of impact sound pressure level of floor coverings

	Weighted reduction of impact sound pressure level ΔL_w
Low impact sound reduction	0–10 dB
Moderate impact sound reduction (but sufficient for especially heavy solid ceilings and moderate demands)	10–18 dB
Mean impact sound reduction	19–23 dB
Very high impact sound reduction (usually achievable by means of thick carpets and floating floors)	24–35 dB

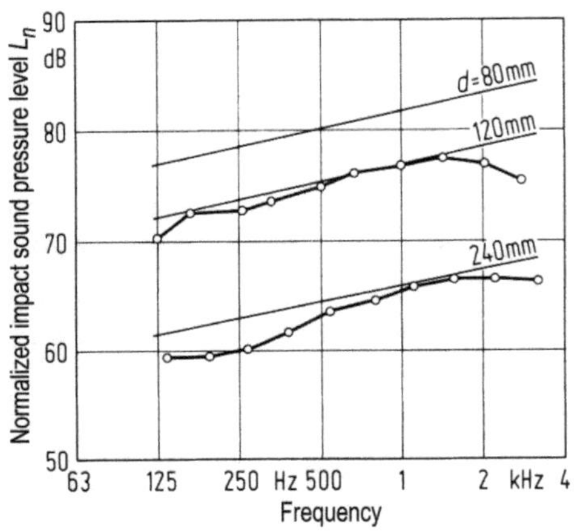

Fig. 7.27 Normalised impact sound pressure level of reinforced bare concrete slabs depending on their thickness

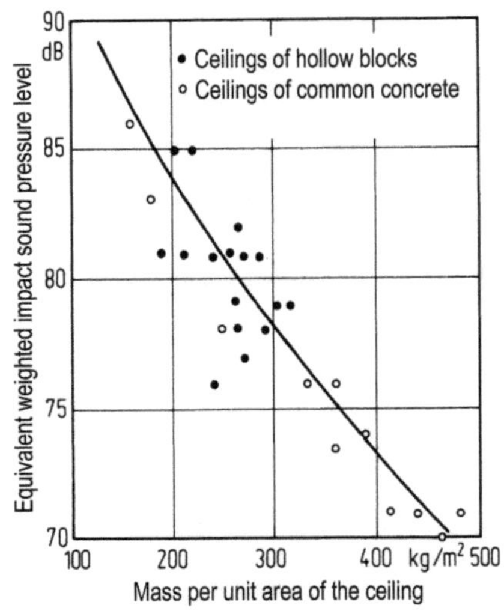

Fig. 7.28 Equivalent weighted impact sound pressure level of solid single-layer ceilings depending on their mass per unit area

7.2.3 Behaviour of Ceilings Without Floor Covering

The impact sound behaviour of single-layer homogeneous floor plates was calculated by Cremer [27]. Accordingly, the normalised impact sound pressure level L_n increases with the frequency by 5 dB per frequency decade. A doubling of the floor thickness leads to a reduction of L_n by about 10 dB. Figure 7.27 shows the corresponding values.

Both homogeneous and not homogeneous single-layer solid floors approximately follow a "mass law" including impact noise control, which is shown in Fig. 7.28, as equivalent weighted impact sound pressure level versus the mass per unit area of the floor.

However, from this behaviour, it cannot be concluded that the mass per unit area of the floor is the only determining quantity regarding its impact noise insulation. As in case of the airborne sound insulation, bending stiffness is important in the same way. As long as the same material is used for floors, i.e., common concrete, the mass, which is easy to determine, can be used as characterising quantity. However, there are two cases where such a simplification is no longer valid: for coffered ceilings with distinct ribs and for lightweight floors made of a much lighter material than common concrete. An example for coffered ceilings is shown in Fig. 7.29, where an impact sound transmission reduced by about 10 dB compared to a homogeneous ceiling plate of the same mass results for a coffered ceiling, owing to the increased bending stiffness of the ribs for the decisively low and middle frequencies. The other deviation from the mass law occurs for cellular concrete ceilings and also for thick wooden plates, which owing their smaller density and thus higher thickness compared with a ceiling of the same mass made of common concrete, have a smaller normalised impact sound level in agreement with the theory of Cremer [27].

The impact sound insulation of floors can also be improved by a suspended ceiling. However, this improvement is limited since, besides direct sound transmission, there will also be a transmission of ceiling vibrations to the flanking walls, if these are solid; see Fig. 7.30. Thus, the normalised impact sound pressure level is, at best, reduced by a suspended ceiling to the values stated in Fig. 7.30. Experience shows that such suspended ceilings are only suitable for light solid floors, provided the walls are solid.

If light walls made of flexible layers are used in buildings with frame constructions, the flanking transmission is highly reduced and the restricted effect, as shown in Fig. 7.30, does not occur. Here, suspended ceilings designed in the right way (airtight, with mineral wool or similar material in the cavity) lead to considerable improvements in impact sound insulation, so that they can replace a floating floor on the upper side of the floor to a great extent.

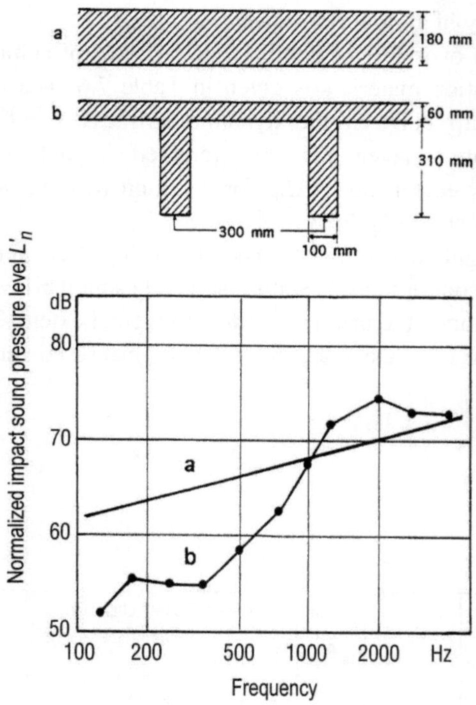

Fig. 7.29 Example for the more favourable impact sound behaviour of a concrete slab braced with ribs in the form of a coffered ceiling (**b**) compared with a concrete slab of the same mass, but without ribs (**a**)

7.2.4 Behaviour of Flexible Coverings

Flexible coverings are impact sound insulating owing to the elastic layer, with the spring being arranged between the mass of the impacting hammer (or the feet) and the floor. The lower the effective dynamic stiffness of the covering, the better is its insulation; see Cremer [27]. Especially effective are carpets. Common flexible coverings such as linoleum, coverings made of plastic or parquet have only a slightly insulating effect. However, this effect can be improved essentially by a padding of a thickness of only a few millimetres (made of cork, plastic foam, felt or similar material). Weighted impact sound insulation values for different coverings are given in Table 7.5.

7.2.5 Behaviour of Floating Floors

Solid slabs of any type, which are directly applied to the bare floor, do not lead to any considerable improvement in impact sound insulation. Only when the solid slab is applied to a resilient insulating layer is a large insulating effect achieved.

According to Cremer [28], the insulating effect starts above a resonance frequency f_{res} of the floating floor, which can be calculated as follows:

$$f_{res} = 160\sqrt{\frac{s''}{m''_e}}; \qquad (7.29)$$

with s'' being the dynamic stiffness of the insulating layer [66] (in MN m^{-3}) and m_e'' being the mass per unit area of the floating floor (in kg m^{-2}).

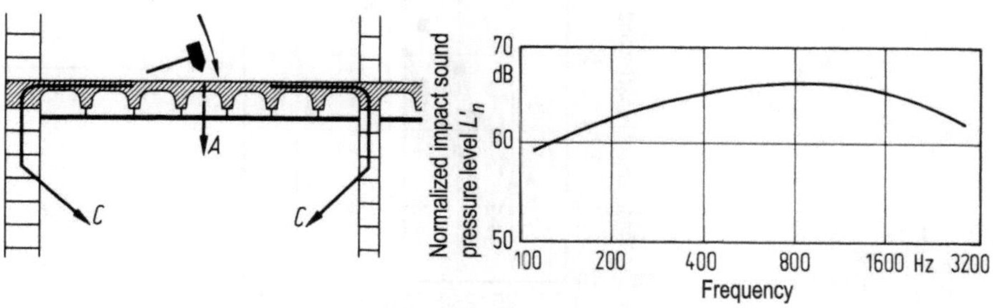

Fig. 7.30 Normalised impact sound pressure level of a solid ceiling to be achieved at best by a suspended ceiling in solid buildings

Thus, according to [28], the following impact sound reduction ΔL results:

$$\Delta L = 40\lg\left(f/f_{\mathrm{res}}\right)\ \mathrm{dB}. \qquad (7.30)$$

Table 7.5 Weighted reduction of impact sound pressure level of common floor coverings

Floor construction	ΔL_w in dB
Flexible coverings	
Linoleum, PVC coating without padding	3–7
Linoleum on cork layer with a thickness of 2 mm	15
Laminate on soft wooden fibre boards with a thickness of 7 mm	18
PVC coating with felt with a thickness of 3 mm	15–19
Needle felt coating	18–22
Carpet, thick design	25–35
Floating cement floors	
On corrugated cardboard	18
On rigid foam boards, stiff	approx. 18
On rigid foam boards, soft	approx. 25
On mineral fibre boards	27–33
Hollow floors	
Without floor covering	10–15
Without floor covering, with impact sound decoupling of the pedestals	20–31
With carpet	21–28
Raised floors	
Without floor covering	12–18
With carpet	25–29

Equation (7.30) was derived by Cremer for an infinitely extended floating floor, where only the transmission near the tapping machine is considered. This assumption is fulfilled for floating floors with a large loss factor (Fig. 7.31). For floating floors with a small loss factor, above the critical frequency, the transmission takes place via bending waves of the complete floating slab, so that Eq. (7.30) is not valid any longer (Fig. 7.32).

The dynamic stiffness s'' of the insulating layers is the sum of the structural stiffness and air stiffness according to Eq. (7.15). Here, it is assumed that the insulating layer consists of a material with a sufficiently high flow resistance.

An overview of the dynamic stiffness of common insulation materials is given in Table 7.6. The relationship between the dynamic stiffness s'' of the insulating layer and the weighted impact sound improvement index ΔL_w for common floating floors is shown in Fig. 7.33.

Rigid connections between a floating floor and the bare floor or the side walls – so-called sound bridges – deteriorate the insulation effect to a great extent. This can be shown, e.g., by the following measured values:

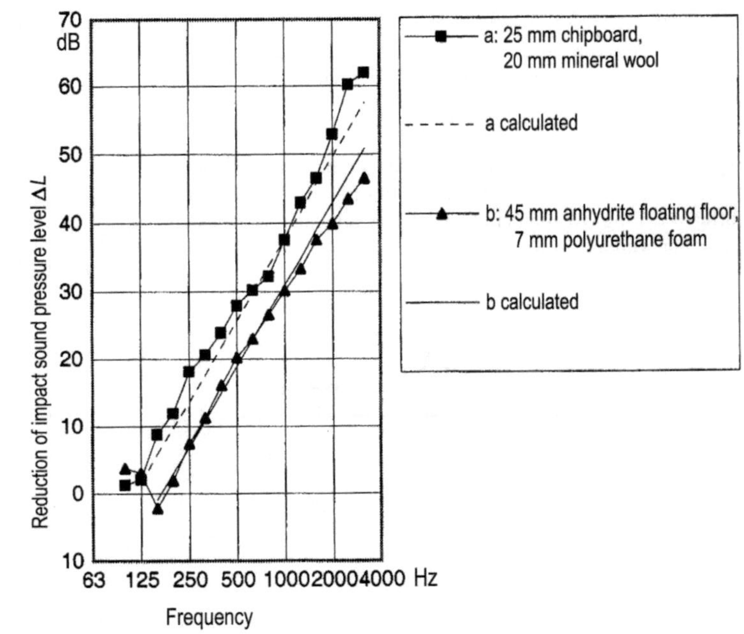

Fig. 7.31 Reduction of impact sound pressure level ΔL by a dry floating floor and an anhydrite floating floor on a thin impact sound insulation

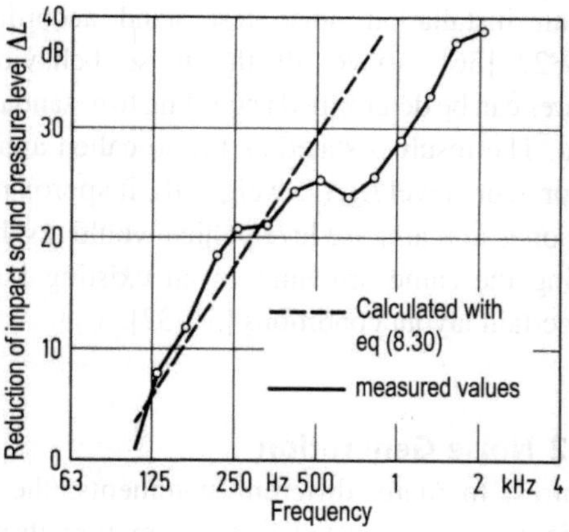

Fig. 7.32 Reduction of impact sound pressure level ΔL by floating floors (35 mm cement floor on 15 mm mineral fibre boards)

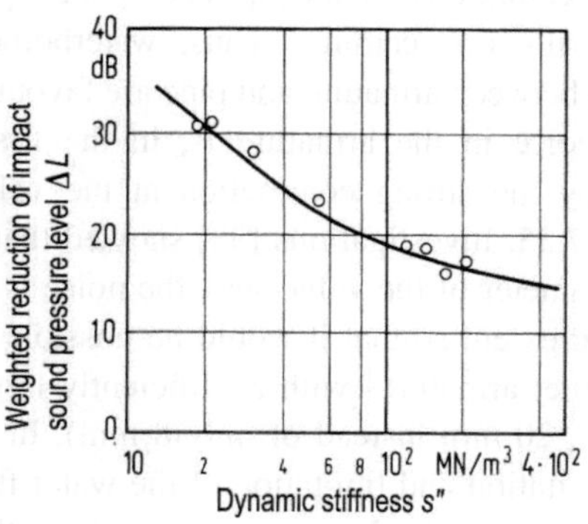

Fig. 7.33 Connection between the weighted reduction of impact sound pressure level ΔL_w of a floating floor and the dynamic stiffness of the impact sound insulation

	Weighted normalised impact sound pressure level
Without sound bridge	52 dB
One sound bridge	63 dB
Ten sound bridges	70 dB
Ceiling without floating floor	80 dB

The effect of sound bridges can be calculated according to Cremer [29]. The measurements agree very well with the calculation [30]. Sound bridges between floating floor and bare floor have a stronger effect than those between floating floor and walls, especially at high frequencies. This defect can be remedied by using a flexible floor covering or, in case of a solid floor covering, by putting a thin insulating layer (e.g., felt board) under the floor covering.

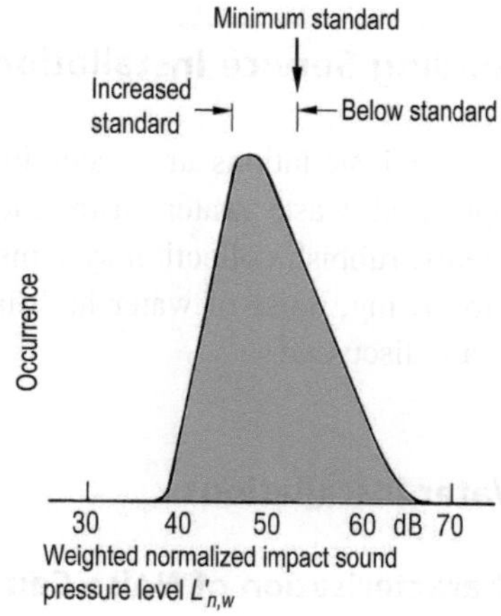

Fig. 7.34 Approximate frequency distribution of the weighted normalised impact sound pressure level $L'_{n,w}$ for new living space ceilings in Germany, usually with carpet. Requirement according to the German standard DIN 4109 [38]

7.2.6 Behaviour of Common Solid Floors with Floating Floors

The frequency distribution of the weighted normalised impact sound pressure level in Fig. 7.34 shows the impact sound behaviour of solid concrete floors in apartment houses in Germany, which are usually provided with a floating floor. As "normal" for a new building and corresponding to the average, an weighted normalised impact sound pressure level of about $L'_{n,w} = 48$ dB can be assumed.

7.2.7 Calculation of Impact Sound Insulation

According to supplement 1 of DIN 4109 [38], the impact sound insulation between rooms can be calculated with sufficient accuracy. Deviations on site can normally be explained by shortcomings in the execution of the work, especially by sound bridges.

In a similar way as described in Sect. 7.1.5 for airborne sound insulation, within the scope of the European standardisation a method to calculate the impact sound insulation between rooms was developed with an universally valid approach, DIN EN ISO 12354–2 [62]. Here, the direct impact sound transmission via the excited floor and the impact sound transmission via the individual flanking partitions are considered.

7.3 Building Service Installations

Building service installations are water installations (water supply and waste water installations), lifts, heating systems, rubbish collection systems, garages, etc. In the following, noise of water installations and lift systems are discussed.

7.3.1 Water Installations

7.3.1.1 Characterisation of Noise Caused by Armatures

One must distinguish between measurements on-site and measurements of armatures, waterborne sound silencers and other isolators in the laboratory.

Measurements on-site are carried out according to ISO 10052 [57] or EN 16032 [65]. The pressure level L when operating an armature in the nearest room outside the respective apartment is measured. Depending on local demands, the sound pressure level can be frequency A- or C-weighted, time fast- or slow-weighted, mean level or maximum level taken. In the same way as the sound pressure level, the measured value can be normalised to an equivalent sound absorption area of $A_0 = 10\ \text{m}^2$ or a reverberation time of $T_0 = 0.5$ s.

$$L_n = L + 10\lg\frac{A}{A_0}\text{dB}, \quad (7.31)$$

$$L_{nT} = L + 10\lg\frac{T}{T_0}\text{dB}. \quad (7.32)$$

In an installation noise test stand according to ISO 3822 [56], above all the noise behaviour of armatures can be determined according to a standardised method. The result is stated as the so-called appliance sound pressure level L_{ap} (A-weighted). It approximately corresponds to that noise level which would result from operating the same armature in an existing building under certain layout conditions [31, 32].

7.3.1.2 Noise Generation

In contrast to many different statements, the outlet noise is always generated in the armature itself and not in the pipe. This is even valid when the pipe causes an unfavourable flow.

The noise generated in the armature is transmitted by the water column within the pipe and by the pipe proper to the walls and ceilings. Thus, waterborne sound silencers between armature and pipe are favourable.

The noise in the armature is, in the first place, caused by the strong contraction at the valve seat; see Fig. 7.35. Investigations [33] showed that, with a larger diameter of the valve seat, the noise is reduced to a great extent so that it would be possible to build nearly quiet armatures with a sufficiently large valve seat (e.g., 20 mm instead of only 8 mm). In order to avoid cavitation and limitation of the water flow, it is necessary to install a flow resistance at the outlet of the armature, which must be a low-noise design. Effective are simple perforated metal sheets or small tubes [33].

The noise of the armature depends on the maximum possible flow; see Fig. 7.36. The noise increases by 12 dB(A) when doubling the flow.

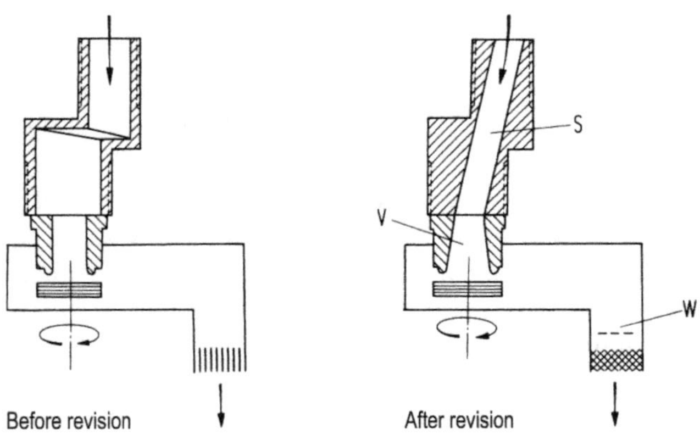

Fig. 7.35 Measures at an outlet armature, which lead to a noise reduction of approx. 20 dB(A) for the same flow rate. V larger valve seat diameter, W larger and low-noise flow resistance at the outlet, *S* so-called S-connection at the pipe designed in a favourable way with regard to flow

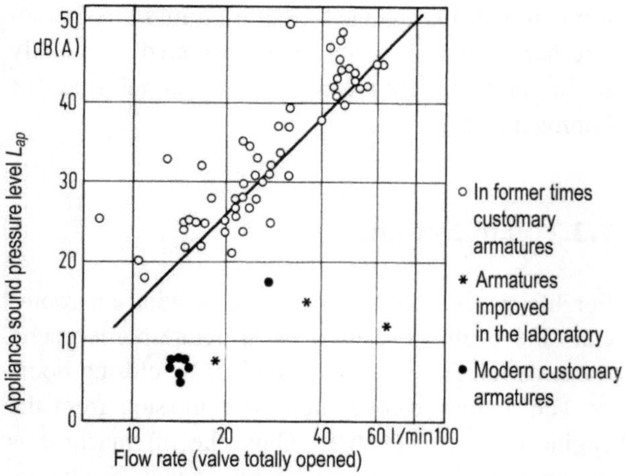

Fig. 7.36 Dependency of the armature noise from the maximum water flow rate through the armature

For conventional armatures today, the noise problem seems to be solved. The behaviour of modern mixers, however, is less favourable. Above all, owing to the quicker opening and closing processes, which are possible with these mixers, noise peaks occur. Apart from that, nowadays, mainly complaints are made regarding splashing and waste water noise, which is especially disturbing and has nothing to do with armatures.

7.3.1.3 Pipe Insulation

Another possibility to reduce the propagation of armature noise is to fix the pipes with a structure-borne sound insulation. For this purpose, there are different types of clamp insulations, which result in sound level reductions between 5 and 15 dB when tested in the laboratory. However, in practice they had no or only little effect. According to [34], these failures are caused by other fix connections between pipe and wall and above all between armature and wall, which have not been considered. If connections between armature and wall are avoided, a desired improvement of 10–15 dB can be achieved [34].

7.3.1.4 Structure-Borne Sound Transmission in the Building

Previously, almost exclusively noise reduction by means of the respective armature was investigated. The question of noise transmission within the building structure was only considered to a minor degree. This can be explained by the fact that, it had been assumed originally that a transmission would take place via the pipes and not via the building structure to the next floor. However, thorough measurements in buildings [35] showed that this is not the case in present buildings, probably owing to a sound insulating effect caused by various pipe junctions.

Independent of these findings, up to then only one constructional regulation regarding installation noise had been made in Germany: according to DIN 4109 [38] walls to which pipes and armatures are fixed should have a mass per unit area of at least 220 kg m^{-2}. This measure is very effective for the horizontally adjacent rooms; there it becomes more quiet. However, detailed investigations [36] showed that for the rooms below, it is hardly important whether the walls are heavier or lighter. As for airborne sound transmission along the walls, this has to do with the fact that also here, the junction insulation between wall and ceiling lessens when the wall is heavier. Astonishingly, the structure-borne sound excitation of a wall on a massive ceiling results in about the same airborne sound level in the room diagonally below, like an excitation of the massive ceiling (without covering) itself. Thus, it is understandable why water pipe noise is so disturbing. It is as if the ceiling would be directly excited via the apartment below.

A reduced transmission in the building structure can be achieved if the installation wall itself is designed in a comparatively light way but heavier near the point of excitation [37]. Figure 7.37 shows an example (model test) of how much structure-borne sound transmission to the ceiling is reduced if the point of excitation is designed heavier than the remaining wall. This was achieved by using heavier stones at the point of excitation than for those of the remaining wall. The improvement was achieved by essentially increasing the input resistance of the wall without reducing the butt joint insulation towards the ceiling.

With this effect of a heavier point of excitation, it can also be explained that the installation noise on the floor below is reduced by about 10 dB(A) if the installation wall is designed as a light post wall with gypsum boards and not as the normal brickwork wall with a weight of about 220 kg m^{-2}; see Fig. 7.38. The difference can be explained by the fact that for a post wall, the light layer is weighed down comparatively more at the point of excitation by the wash basin or the toilet bowl than for a brickwork wall.

In summary, it can be said that, astonishingly, structure-borne sound excitation and transmission for

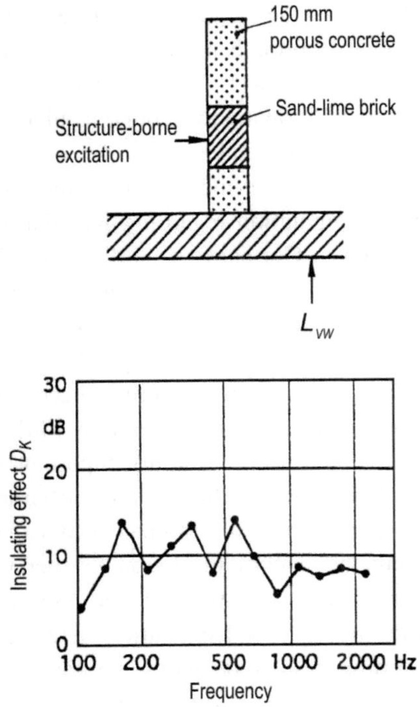

Fig. 7.37 Insulating effect D_K of an inhomogeneously constructed installation wall (heavy sand-lime bricks at the point of excitation and light porous concrete bricks for the remaining wall); model test, values converted to large-scale construction

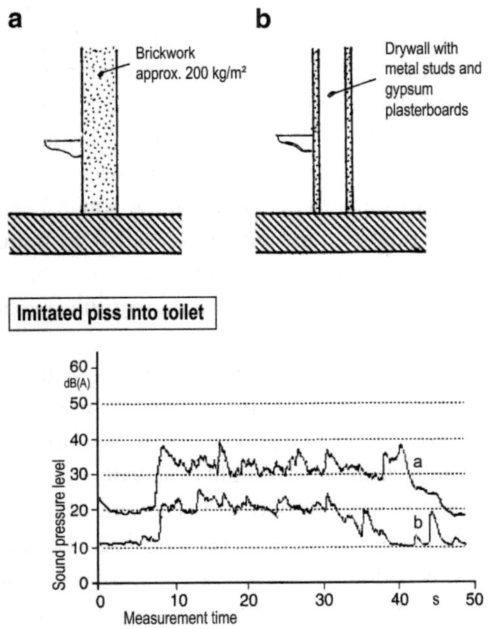

Fig. 7.38 Installation noise in the room diagonally below the bathroom for imitated pissing water. (**a**) wall-mounted toilet at massive wall (**b**) wall-mounted toilet at dry wall with light gypsum boards

installation walls and the further transmission of structure-borne sound have not been considered sufficiently so far and that there are considerable possibilities for improvement.

7.3.2 Lift Systems

For lift systems, disturbing noise in adjacent rooms can, above all, be caused by structure-borne sound transmission from the lift machine, by closing doors as well as by airborne sound transmission from the engine room or lift shaft. Thus, the lift machine is mounted with a structure-borne sound insulation. Doors can be designed in a low-noise way, e.g., by using soft-start. In order to reduce airborne sound transmission, constructional measures are required. In solid buildings, according to the respective conditions, primarily the walls of the well and the walls of the engine room and if necessary the flanking walls and ceilings must be designed with a high mass per unit area. Information on noise control can be found in the VDI regulations for lift systems with machine room [68] and without machine room [69].

References

1. Cremer L (1961) Der Sinn der Sollkurven. In: Schallschutz von Bauteilen. Berlin: Ernst u. Sohn, 1
2. Gösele K (1965) Zur Bewertung der Schalldämmung von Bauteilen. Acustica 15:264–270
3. Heckl M (1960) Die Schalldämmung von homogenen Einfachwänden endlicher Größe. Acustica 10:98–108
4. Cremer L (1942) Theorie der Schalldämmung dünner Wände bei schrägem Einfall. Akust Z:81–104
5. Gösele K (1968) Zur Luftschalldämmung von einschaligen Wänden und Decken. Acustica 20:334
6. Watters G (1959) Transmission-loss of some Masonry Walls. J Acoust 31:898
7. Meier A (2000) Die Bedeutung des Verlustfaktors bei der Bestimmung der Schalldämmung im Prüfstand. Diss. RWTH Aachen
8. Berger R (1911) Über die Schalldurchlässigkeit. Diss. TH München
9. Gösele K (1990) Verringerung der Luftschalldämmung von Wänden durch Dickenresonanzen. Bauphysik 12:187
10. Lang J (1985) Wirtschaftliche Erfüllung des normgerechten Schallschutzes im Wohnungsbau. Forschungsbericht 5160/WS Fachverband der Stein- und Keramischen Industrie Österreichs, Wien
11. Heckl M, Lewit M (1990) Luftschalldämmung von Vielschichtplatten mit zahlreichen Schallbrücken. Fortschritte der Akustik DAGA:199–202

12. Gösele K (1980) Berechnung der Luftschalldämmung von doppelschaligen Bauteilen. Acustica 45:208
13. Meyer E (1935) Die Mehrfachwand als akustisch-mechanische Drosselkette. ENT 12:393
14. Gösele K (1977) Einfluß der Hohlraumdämpfung auf die Schalldämmung von doppelschaligen Wänden. Acustica 38:159
15. Heckl M (1959) Untersuchungen über die Luftschalldämmung von Doppelwänden mit Schallbrücken. Congress-Report III of the IIIrd ICA-Congress, 1010
16. Gösele K, Gösele U (1969) Schalldämmende Doppelwände aus biegesteifen Schalen. FBW-Blätter 1967. Folge 1. Betonstein-Zeitung 35, Nr. 5, 29617
17. Scholl W, Brandstetter D (2000) Neue Schalldämmwerte bei Gipskartonplatten-Metallständerwänden. Bauphysik 22:101
18. Gösele K, Kurz R (2001) Zur Schalldämmung von GK-Ständerwänden, 1. Wirkung der Hohlraumdämpfung. Fortschritte der Akustik DAGA
19. IBP-Forschungsbericht B-BA 2/2003 (2003) Verbesserung der Längsdämmung durch Vorsatzschalen. Frauenhofer IRB Verlag
20. Gösele K (1968) Untersuchungen zur Schall-Längsleitung. In: Heft 56 der Schriftenreihe, Berichte aus der Bauforschung 23
21. Gösele K (1984) Berechnung der Schalldämmung in Massivbauten unter Berücksichtigung der Schall-Längsleitung. Bauphysik 6(79–84):121–126
22. GSA Limburg (1993) Schall-Längsdämmung porosierter Außenmauerwerke in Abhängigkeit von der Stoßstellen-Ausbildung. Forschungsbericht B I 5 – 80188–13 für das Bundesamt für Raumordnung und Städtebau
23. Gomperts MC (1964) The Sound Insulation of Circular and Slit-Shaped Apertures. Acustica 14:1–16
24. Gösele K (1969) Schalldämmung von Türen. In: Schriftenreihe "Berichte aus der Bauforschung, Heft 63, 1
25. Olson HF (1941) Tone Guard. J Acoust Soc Amer 12:374–377
26. Gösele K (1964) Die Beurteilung des Schallschutzes von Rohdecken. Ges Ing 85:261
27. Cremer H, Cremer L (1948) Theorie der Entstehung des Trittschalls. Frequenz 1:61
28. Cremer L (1952) Näherungsweise Berechnung der von einem schwimmenden Estrich zu erwartenden Verbesserung. Fortschr und Forsch im Bauwesen 2:123
29. Cremer L (1954) Berechnung der Wirkung von Schallbrücken. Acustica 4:273
30. Gösele K (1964) Schallbrücken bei schwimmenden Estrichen und anderen schwimmend verlegten Belägen. Schriftenreihe Berichte aus der Bauforschung. Heft 35:23
31. Gösele K, Voigtsberger CA (1967) Zur Messung des Geräuschverhaltens von Armaturen im Laboratorium. Heizung – Lüftung – Haustechnik 18:230
32. Schneider P (1967) Eine Bezugsarmatur und deren Anwendung zur Messung und Bewertung von Installationsgeräuschen. Sanitär- und Heizungstechnik 32:7
33. Gösele K, Voigtsberger CA (1970) Grundlagen zur Geräuschminderung bei Wasserauslaufarmaturen. Ges Ing 91:108–117
34. Gösele K, Voigtsberger CA (1975) Verminderung von Installationsgeräuschen durch körperschall-isolierte Rohrleitungen. HLN
35. Gösele K, Voigtsberger CA (1980) Der Einfluß der Bauart und der Grundrißgestaltung auf das entstehende Installationsgeräusch in Bauten. Ges Ing 101:79
36. Gösele K, Engel V (1995) Körperschalldämmung von Sanitärräumen. Bauforschung in der Praxis, Band 11
37. Gösele K (1998) Schalldämmende Installationswände. Neue Wege zur Verringerung der Installationsgeräusche. Fraunhofer IRB-Verlag

Standards and Regulations

38. DIN 4109 (1989) Schallschutz im Hochbau: Anforderungen und Nachweise. Beiblatt 1: Ausführungsbeispiele und Rechenverfahren
39. DIN 4109 (2003) Beiblatt 1/A1: Ausführungsbeispiele und Rechenverfahren, Änderung A1
40. VDI 4100 (2007) Noise control in dwellings – criteria for planning and assessment
41. ISO 140–1 (1997) Acoustics – Measurement of sound insulation in buildings and of building elements – Part 1: Requirements for laboratory test facilities with suppressed flanking transmission. Amendment 1 (2004) Specific requirements on the frame of the test opening for lightweight twin leaf partitions
42. ISO 140–2 (1991) Acoustics – Measurement of sound insulation in buildings and of building elements – Part 2: determination, verification and application of precision data
43. ISO 140–3 (1995) Acoustics – Measurement of sound insulation in buildings and of building elements – Part 3: Laboratory measurements of airborne sound insulation of building elements. Amendment 1 (2004) Installation guidelines for lightweight twin leaf partitions
44. ISO 140–4 (1998) Acoustics – Measurement of sound insulation in buildings and of building elements – Part 4: Field measurements of airborne sound insulation between rooms
45. ISO 140–5 (1998) Acoustics – Measurement of sound insulation in buildings and of building elements – Part 5: Field measurements of airborne sound insulation of façade elements and façades
46. ISO 140–6 (1998) Acoustics – Measurement of sound insulation in buildings and of building elements – Part 6: Laboratory measurements of impact sound insulation of floors
47. ISO 140–7 (1998) Acoustics – Measurement of sound insulation in buildings and of building elements – Part 7: Field measurements of impact sound insulation of floors
48. ISO 140–8 (1997) Acoustics – Measurement of sound insulation in buildings and of building elements – Part 8: Laboratory measurements of the reduction of transmitted impact noise by floor coverings on a heavyweight standard floor
49. ISO 140–10 (1991) Acoustics – Measurement of sound insulation in buildings and of building elements – Part 10: laboratory measurement of airborne sound insulation of small building elements
50. ISO 140–11 (2005) Acoustics – Measurement of sound insulation in buildings and of building elements – Part 11: Laboratory measurements of the reduction of transmitted

impact sound by floor coverings on lightweight reference floors
51. ISO 140–14 (2004) Acoustics – Measurement of sound insulation in buildings and of building elements – Part 14: Guidelines for special situations in the field
52. ISO 140–16 (2006) Acoustics – Measurement of sound insulation in buildings and of building elements – Part 16: Laboratory measurement of the sound reduction index improvement by additional lining
53. ISO 140–18 (2006) Acoustics – Measurement of sound insulation in buildings and of building elements – Part 18: Laboratory measurement of sound generated by rainfall on building elements
54. ISO 717–1 (1996) Acoustics – Rating of sound insulation in buildings and of building elements – Part 1 – Airborne sound insulation. Amendment 1 (2006) Rounding rules related to single number ratings and single number quantities
55. ISO 717–2 (1996) Acoustics – Rating of sound insulation in buildings and of building elements – Part 2 – Impact sound insulation. Amendment 1 (2006) Rounding rules related to single number ratings and single number quantities
56. ISO 3822–1 (1999) Acoustics – Laboratory tests on noise emission from appliances and equipment used in water supply installations – Part 1: Method of measurement Amendment 1 (2006) Measurement uncertainty
57. ISO 10052 (2004) Acoustics – Field measurements of airborne and impact sound insulation and of service equipment sound – Survey method
58. ISO 10848–1 (2006) Acoustics – Laboratory measurement of the flanking transmission of airborne and impact sound between adjoining rooms – Part 1: Frame document
59. ISO 10848–2 (2006) Acoustics – Laboratory measurement of the flanking transmission of airborne and impact sound between adjoining rooms – Part 2: Application to light elements when the junction has a small influence Technical Corrigendum 1 (2007)
60. ISO 10848–3 (2003) Acoustics – Laboratory measurement of the flanking transmission of airborne and impact sound between adjoining rooms – Part 3: Application to light elements when the junction has a substantial influence
61. EN 12354–1 (2000) Building acoustics – Estimation of acoustic performance of buildings from the performance of products – Part 1: Airborne sound insulation between rooms
62. EN 12354–2 (2000) Building acoustics – Estimation of acoustic performance of buildings from the performance of elements – Part 2: Impact sound insulation between rooms
63. EN 12354–3 (2000) Building acoustics – Estimation of acoustic performance of buildings from the performance of elements – Part 3: Airborne sound insulation against outdoor sound
64. EN 12354–4 (2001) Building acoustics – Estimation of acoustic performance of buildings from the performance of products – Part 4: Transmission of indoor sound to the outside
65. ISO 16032 (2004) Acoustics – Measurement of sound pressure level from service equipment in buildings – Engineering method
66. EN 29052–1 (1991) Acoustics – Determination of dynamic stiffness; Part 1: materials used under floating floors in dwellings
67. EN 29053 (1993) Acoustics – Materials for acoustical applications; determination of airflow resistance
68. VDI 2566 (2001) Supplement 1: Acoustical design for lifts with a machine room
69. VDI 2566 (2004) Supplement 2: Acoustical design for lift systems without machine room

Sound Absorbers

8

H.V. Fuchs and M. Möser

8.1 Introduction

Sound absorption indicates the transformation of sound energy into heat. It is, for instance, employed to design the acoustics in rooms. The noise emitted by machinery and plants shall be reduced before arriving at a workplace; auditoria such as lecture rooms or concert halls require a certain reverberation time. Such design goals are realised by installing absorbing components at the walls with well-defined absorption characteristics, which are adjusted for corresponding demands. Sound absorbers also play an important role in acoustic capsules, ducts and screens to avoid sound immission from noise intensive environments into the neighbourhood.

This chapter not only intends to explain the well-known passive and reactive sound-absorbing materials and components and their basic principles but also unfolds a variety of innovative possibilities available today for noise control and the design of room acoustics. This includes, amongst others, the description of closed, slit and micro-perforated absorbing plates of various materials.

Conventional silencers and sound absorbers made of fibrous or porous materials are appropriate for the reduction of high-frequency components of various noise sources. Yet, in noise control and room acoustics the actual problem often is to fight low frequencies, which are usually difficult to reduce due to the required large volumes of passive absorbers. The following sections therefore refer to this problem and offer novel solutions for it.

According to the aim of this handbook, the focus is on practical applications rather than on theoretical completeness. The concept, design and application of the described sound absorbers are exemplified using realised projects as a reference. A slightly more detailed description of advanced sound-absorbing technologies for various practical demands can be found in a series of publications [1], which arose from drafts of this contribution.

8.2 Sound Absorption for Noise Control and Room Acoustics

Airborne sound can be transmitted to the listener's ear in many different ways. An incident sound wave with sound power P_i, sound pressure p_i, sound velocity v_i and frequency f hitting an obstacle, which is large compared to the wavelength λ can be partially reflected (P_r) or scattered and diffracted, transmitted (P_t) or conducted as structure-borne sound (P_s), but can also be absorbed (P_a). The incident sound power P_i thus splits into four components as shown in Fig. 8.1

$$P_i = P_r + P_t + P_s + P_a. \qquad (8.1)$$

If, for instance, the obstacle is a wall (or a ceiling), where the mass per unit area m''_W is large compared to the oscillating mass per unit area m''_A of the air

H.V. Fuchs (✉)
Research Centre of Applied System Safety and Industrial Medicine, Terrassenstraße 21 C, D-14129 Berlin, Germany
e-mail: hvfuchs@hotmail.com

M. Möser
Institut für Strömungsmechanik und Technische Akustik, Technische Universität Berlin, Einsteinufer 25, 10587 Berlin, Germany
e-mail: mimoe48@googlemail.com

G. Müller and M. Möser (eds.), *Handbook of Engineering Acoustics*,
DOI 10.1007/978-3-540-69460-1_8, © Springer-Verlag Berlin Heidelberg 2013

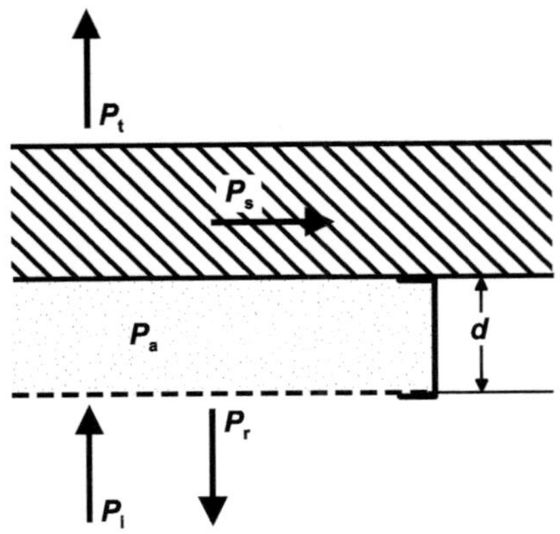

Fig. 8.1 Paths of the power P_i in a sound wave incident on an absorbing obstacle

$$m''_W \gg m''_A = \frac{1}{\omega}\frac{p_i}{v_i} = \frac{1}{\omega} Z_0 = \frac{\varrho_0 \lambda}{2\pi}, \qquad (8.2)$$

using the specific impedance of air

$$Z_0 = \varrho_0 c_0 = 408\ \mathrm{Pa\,s\,m^{-1}}, \qquad (8.3)$$

(for 20°C and 10^5 Pa), the angular frequency $\omega = 2\pi f$, the density $\varrho_0 = 1.2\,\mathrm{kg\,m^{-3}}$ and the sound speed $c_0 = 344\ \mathrm{m\ s^{-1}}$ of air, only a small portion of the sound power is transmitted or conducted. The main part will be reflected to the source or into the room, if not otherwise an absorbing material or component is mounted, which absorbs a major portion of P_i, i.e. P_i is dissipated (transformed into heat).

If such an absorber is to be characterised with respect to its effectiveness P_t and P_s can be added to P_a on the source side:

$$\alpha = \frac{P_\mathrm{a} + P_\mathrm{t} + P_\mathrm{s}}{P_\mathrm{i}} = \frac{P_\mathrm{i} - P_\mathrm{r}}{P_\mathrm{i}} = 1 - \rho. \qquad (8.4)$$

The value of the absorption coefficient α and the sound power reflection coefficient ρ ranges between 0 and 1. The effect of reflection can also be expressed as the ratio of amplitudes of the reflected and the incident sound wave, which results in the, generally complex, sound pressure reflection factor r

$$\rho = \frac{P_r}{P_i} = \frac{p_r^2}{p_i^2} = |r|^2 = 1 - \alpha. \qquad (8.5)$$

According to ([2], Chaps. 2–4), r can also be deduced from the complex wall impedance W, which

Table 8.1 Level difference ΔL of a plane standing wave in front of a more or less absorbing obstacle ([3], p. 459), corresponding absorption coefficient α and magnitude of the sound pressure reflection coefficient $|r$

α	ΔL (dB)	$\lvert r\rvert$
0.99	2	0.11
0.60	13	0.63
0.20	25	0.89
0.01	50	0.99

completely describes the acoustic properties of a structure or a component. For normal sound incidence using the sound pressure p_W and the sound velocity v_W in front of the wall (W' real part, W'' imaginary part of W), the wall impedance is given by

$$W = \frac{p_W}{v_W} = W' + jW''; \qquad (8.6)$$

$$r = \frac{W - \varrho_0 c_0}{W + \varrho_0 c_0}; \qquad (8.7)$$

$$\alpha = 1 - |r|^2 = \frac{4W'\varrho_0 c_0}{(W' + \varrho_0 c_0)^2 + W''^2}. \qquad (8.8)$$

Equation (8.8) is called the ‘matching law’: the absorption coefficient reaches its maximum, when the imaginary part of the wall impedance is zero. But the maximum possible value of unity is reached only if the real part of the wall impedance is equal to $\varrho_0 c_0$. For ‘mismatched’ absorbers with partial reflection ($\rho < 1$), the sound field in front of the reflector is composed of a forward progressive and a reflected wave in x-direction

$$\begin{aligned} p &= p_0 e^{-jkx} + r p_0 e^{jkx} \\ &= (1 - r) p_0 e e^{-jkx} + r p_0 \left(e^{-jkx} + e^{jkx}\right), \end{aligned} \qquad (8.9)$$

where $k = \frac{\omega}{c_0}$ is the wave number. For matching with $r = 0$, only the forward progressing wave occurs with a spatially constant level distribution. For total reflection $r = 1$ the sound field consists of a standing wave with a strongly fluctuating level in x-direction. Generally, the level difference $L = L_{\max} - L_{\min}$ forms a measure of the absorption coefficient (see [3] and Table 8.1). The extrema for α can be realised by a smoothly plastered or tiled masonry ($\alpha \approx 0.01$) or by a special wall lining of an anechoic room ($\alpha \approx 0.99$). The absorbing surface areas S_i of most of the sound-absorbing materials and components used in buildings having absorption coefficients α_i between 0.2 and 0.6 up to 0.8 (values for α_i can be found in tables e.g. in [4–6]) are added to result in the equivalent absorption area A_s of all bare room surfaces (also known as room absorption). Apart from these furniture, installation objects and acoustic modules suspended from the ceiling or standing on the floor as single elements (A_j) as well as occupants contribute to the equivalent absorption area A_I like additionally installed objects:

$$A_S = \sum_i \alpha_i S_i; \quad A_I = \sum_j A_j. \tag{8.10}$$

The total equivalent absorption area A (m^2) of the room is then given by

$$A = A_S + A_I + 4Vm, \tag{8.11}$$

where $V\,(m^3)$ is the room volume and m the attenuation of the sound wave between two reflections along its propagation path (see Table 8.2).

At least seven different areas of application can be defined, where sound absorption is of major importance:

1. In front of lightly absorbing surfaces ($\alpha < 0.2$), the sound field is strongly space dependent according to Eq. (8.9) and Table 8.1. The localisation of sound sources is extremely difficult and the clarity of music and the speech intelligibility are poor. Apart from changing the architectural design (e.g. giving planes with windows or walls additional slant) and adding protruding or suspended reflectors, the only improvement in these cases is the cancellation of the detrimental reflections by absorption.
2. If, in contrast to that the, reverberation time T(s)

$$T = 0.163\frac{V}{A}, \tag{8.12}$$

Table 8.2 Damping constant m for sound propagation in rooms (for 20 and 50% relative humidity), absorption coefficient α_a under free-field conditions (for 10 and 70%) and thickness of the acoustic boundary layer δ at 20 for air

F (Hz)	<250	500	1 k	2 k	4 k	8 k
m (10^{-3} m^{-1})	<0.08	0.25	0.75	2.5	7.5	25
α_a (dB) km^{-1}	<1	2	4	8	20	50
δ (μm)	>95	67	47	34	24	17

in a theatre or a church with large volume V(m^3) is too high due to the equivalent absorption area A(m^2) according to Eq. (8.11) being too small, the resulting speech intelligibility is poor. The designer of room acoustics has to look for areas S_i suitable for this purpose because the absorption due to installed objects and the audience is mostly given. The attenuation of the sound wave between two reflections (m) along its propagation path decreases towards lower frequencies (Table 8.2). Thus, large as well as small rooms demand absorbers attenuating primarily low rather than mid and high frequencies, which are usually absorbed by various installation objects or the audience anyway.
3. In arenas, conference rooms or offices, restaurants, class rooms, auditoria, etc. where many people raise their voices simultaneously, the conversation can be problematic, if A is not sufficiently large. This fact can be deduced from the reverberation radius r_R(m) [7]

$$r_R = 0.14\sqrt{A\frac{\nu P_1}{P_{tot}}} = 0.057\sqrt{\frac{\nu P_1 V}{nPT}}, \tag{8.13}$$

which represents the distance from the source, where the sound pressure level of the direct sound field is equal to that of the diffuse field, which results from the sound power P_{tot} of a number of n sources emitting P and multiple reflections. The conditions for a single speaker with a sound power P_1 to be intelligible can be improved if the speaker is located not in the middle of the room (directivity index $\nu = 1$) but in front of a large reflecting wall ($\nu = 2$) or close to an edge ($\nu = 4$) or even in a corner of the room ($\nu = 8$). Similar improvements can be made using loudspeakers with a large directivity index ν, which are directed to certain important areas of the room [7].

According to Eq. (8.13), it seems that with an increasing number of speakers P_{tot} the introduced absorbing area A also increases proportionally. Experience teaches, however, that understanding the interlocutor is more difficult the more persons congregate and talk because the participants only bring absorption for frequencies above 250 Hz. When the low frequencies remain undamped and the reverberation time for them increases, 'booming' noise fills the room, which sort of 'masks' the higher frequencies, which are so important for intelligibility [8]. This leads to the fact that all speakers tend to raise their voices even more, which in turn worsens communication. To tackle this problem (especially in small rooms), special low frequency absorbers for frequencies down to 63 Hz need to be installed, which was proven in several redevelopment projects–often to the surprise of the users [9].

4. In small and medium sized rehearsal or lecture rooms, a similar communication problem occurs for conductors or teachers. Missing absorption at low frequencies can cause a poor clarity of bass instruments as well as human voices [10, 11], which are important to be heard clearly by an ensemble for a good ensemble play. Mutual hearing between the different instruments in orchestra pits and rehearsal rooms can be so poor that musicians (need to) articulate their instruments louder than it would be necessary for a good resulting overall performance. Using appropriate sound absorbers can improve the working conditions at these high demand workplaces significantly as successful improvements like in [12] show.
5. For sound sources with a given constant sound power level L_W, the averaged sound pressure level $\bar{L}$ in the room can be reduced using sound-absorbing fittings or linings:

$$\bar{L} = L_\mathrm{W} - 10 lgA + 6\,\mathrm{dB}\ . \tag{8.14}$$

It is vital in that case that the frequency characteristics of the absorption (A) is matched, as well as possible, to the excitation spectrum of the corresponding sound source(s) (L_W). It should be noted that room acoustic treatment has no effect within the reverberation radius according to Eq. (8.13). Yet, most of the investments are spent on treatments, where according to

$$\Delta\bar{L} = -10\ \lg\frac{A_2}{A_1}, \tag{8.15}$$

a doubling of A yields only a room level reduction of 3 dB, while workplaces located close to loud machinery can hardly have a benefit from these room acoustic measures.

6. The sound pressure level L_i transmitted into the interior of a building through an exterior partition wall is given by

$$L_\mathrm{i} = L_\mathrm{e} - R + 10 lgS - 10 lgA, \tag{8.16}$$

where L_e is the external sound pressure level, R is the sound reduction index, S is the surface area of the partition and A is the equivalent absorption area of the receiving room. Large partition areas S with a low sound reduction index R (e.g. windows and glass facades) result in a higher interior sound pressure level L_i. A large weighted sound reduction index of multiple-leaf partitions is often twined with a low sound reduction below 100 Hz. Thus, the low frequency portion of traffic noise, noise from discotheques or industrial noise typically remains as the actual disturbance, when windows are closed. Lightweight structures with a low bending stiffness, which are used now and then in machine and building construction, lose about 6 dB per octave of their sound reduction towards lower frequencies according to the mass law [13]

$$R = 20\lg m''_W + 20\lg f - 45\,\mathrm{dB}, \tag{8.17}$$

which is otherwise determined by their mass per unit area $m''_W[\mathrm{kg\,m^{-2}}]$.

7. The valid regulations, standards and measurement procedures concerning the emission, transmission and immission of sound in buildings generally do not respect the frequency range below 100 Hz. For the design of silencers for ventilation systems, it is, on the other hand, obvious to match their maximum damping to the given noise spectrum $L_{W,e}$ of the plant or machinery. Often, the damping at higher frequencies is overdone. The low frequency portion of the immitted sound pressure level L_i remains high also during propagation towards large distances s(m) according to

$$L_i = L_{W,e} + 10\lg v - 20\lg s - \sum_i D_i - 11\,\mathrm{dB}, \tag{8.18}$$

using the directivity index v as in case 3, because all damping elements along the propagation path s and additional damping due to possibly present obstacles D_i have substantially larger values for higher than for lower frequencies. The viscothermal damping in air, for example, for sound propagation under free-field conditions

$$D_a = \alpha_a s, \tag{8.19}$$

according to Table 8.2 is more than 10 dB km^{-1} above 2.5 kHz, whereas it can be neglected below 250 Hz. The absorbing material used in silencers, which needs to be mounted in ducts or exhaust stacks, therefore must have a large absorption coefficient α especially at low frequencies, in order to achieve a high insertion loss D_e far below the 'beaming frequency' [14] according to Piening's equation (see also Sect. 8.10.3)

$$D_e = 1.5\alpha \frac{U}{S} l[\text{dB}], \tag{8.20}$$

with a given length l(m), the circumferential length of the sound-absorbing lining U(m) S(m^2) of the silencer configuration.

8.3 Passive Absorbers

The absorber species with the largest market volume, which is also used in the widest range of applications, follows the basic principle to ideally represent a matched resistance W for incident sound waves according to Eq. (8.8) with a value close to $W = \varrho_0 c_0$. The absorbing material consisting of fibres or open-pore foams is characterised by the absorber impedance [13]

$$W_\alpha = \varrho_0 c_0 \frac{\sqrt{\chi}}{\sigma} \sqrt{1 - \mathrm{j}\frac{\sigma \Xi}{\omega \varrho_0 \chi}}, \tag{8.21}$$

and the complex wave number

$$k_\alpha = \frac{\omega}{c_0} \sqrt{\chi} \sqrt{1 - \mathrm{j}\frac{\sigma \Xi}{\omega \varrho_0 \chi}}. \tag{8.22}$$

The parameters in Eqs. (8.21) and (8.22) are

- The porosity σ, which is the ratio of the acoustically effective air volume V_{air}, enclosed inside the absorber and the total volume of the absorber V_{tot}

$$\sigma = \frac{V_{air}}{V_{tot}} < 1\,, \tag{8.23}$$

- The structure coefficient χ, which is the ratio of the compressed air volume V_c and the accelerated air volume V_{acc}

$$\chi = \frac{V_c}{V_{acc}} \geq 1, \tag{8.24}$$

- The flow resistance per unit length Ξ (also called the flow resistivity), which is the ratio of the pressure difference Δp for a low-speed steady flow of velocity v through the absorbing layer of thickness Δx

$$\Xi = \frac{\Delta p}{v \Delta x}\,. \tag{8.25}$$

For very large layer thicknesses, the wall impedance W is equal to the absorber impedance W_α of the porous medium. For small flow resistivities or high frequencies Eqs. (8.21) and (8.8) can then be simplified as

$$\Xi \ll \omega \varrho_0 \rightarrow W = \varrho_0 c_0 \frac{\sqrt{\chi}}{\sigma}; \qquad \alpha = \frac{4}{2 + \frac{\sigma}{\sqrt{\chi}} + \frac{\sqrt{\chi}}{\sigma}}. \tag{8.26}$$

For fibrous materials usually used for acoustic purposes where σ and χ are close to unity W approximates $\varrho_0 c_0$ and α approximates unity ('matching'). In this limiting case, a plane sound wave would decay exponentially along the propagation direction inside the material. The sound pressure level decreases proportionally to the propagation direction x [15]

$$L(x) = L(0) - \frac{4.35\sigma}{\sqrt{\chi}} \frac{\Xi x}{\varrho_0 c_0}. \tag{8.27}$$

For large layer thicknesses only, the matching between air and absorber would be important.

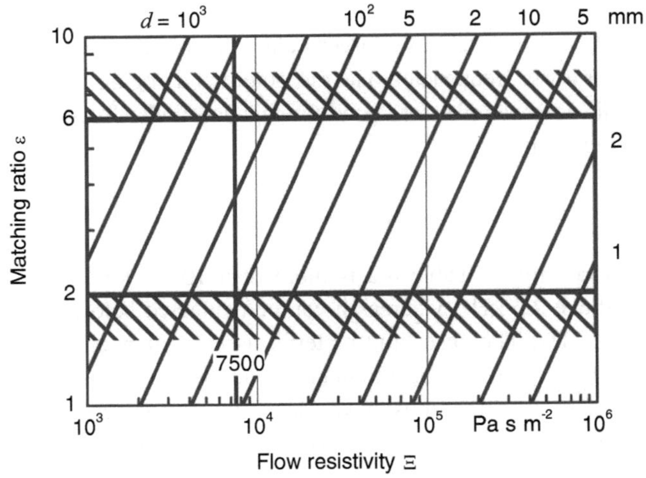

Fig. 8.2 Matching ratio ε as a function of flow resistivity Ξ for varying layer thickness d [13]

It would be sufficient to make Ξ as small as possible according to Eq. (8.26), for example, far below 750 Pa s m^{-2} for 100 Hz. In the practice of noise control and room acoustics mainly material with flow resistivities of far more than Ξ >7,500 Pa s m^{-2} are used.

For normal finite layer thicknesses, the sound should, on the one hand, easily enter the absorber, thus Ξ should not be too large. On the other hand, Ξ should be large enough to provide enough resistance in order to damp the waves propagating inside the absorber. The characteristic performance parameters for an acoustic component is the flow resistance (Ξd or this value normalised to $\varrho_0 c_0$). In general, the range between

$$800 < \Xi d < 2,400 \text{ Pa s m}^{-1} \quad \text{or}$$
$$2 < \varepsilon = \frac{\Xi d}{\varrho_0 c_0} \frac{\sigma}{\sqrt{\chi}} < 6. \tag{8.28}$$

has turned out to be optimal. The 'matching ratio' ε is depicted in Fig. 8.2 as a function of Ξ using the layer thickness d as parameter with $\sigma \approx \chi \approx 1$ [14]. The slightly schematic and normalised representation in Fig. 8.3 shows that for a layer thickness $d \ll \lambda$ large absorption coefficients cannot be realised. For $d \geq \lambda/8$

$$\alpha \geq 80\% \quad \text{for} \quad d \geq \frac{42.5}{f} 10^3 (\text{mm}), \tag{8.29}$$

can be expected, but only for $d \geq \lambda/4$ absorption coefficients of $\alpha > 0.9$ are obtained.

This very simple dimensioning rule for nearly all homogeneous porous/fibrous materials used as sound absorbers or silencers suggests an almost universal usefulness. It should be kept in mind, however, that the optimum Ξ according to Fig. 8.2 for maximum absorption at 100 Hz lies in the range between 1,600 and 4,800 Pasm^{-2}, which is again below the flow resistivity of usual absorption materials. Such fluffy material would not be employed even when carefully protected or wrapped for room acoustic purposes. Apart from this restriction for practical realisations, the solid or dashed curves in Fig. 8.3 may be used as reference for passive absorbers for diffuse or normal sound incidence for comparison even if one deals with other materials and structures of a given thickness d.

8.3.1 Fibrous Materials

The absorbers dealt with so far are preferably and mainly produced of artificial mineral fibres. They are called passive because they cannot be excited by sound waves to perform oscillations, despite their usual small density $\varrho_A \leqslant 60\,\text{kg}\,\text{m}^{-3}$. Their structure – fragile and sensitive as it may be with respect to mechanical stress – is usually heavy enough to be considered as rigid when excited by arbitrary airborne sound fields in the audible range. It can be summarised that especially fibrous materials with a layer thickness of 50–100 mm represent nearly perfect sound absorbers for the frequency range above approximately 500 or 250 Hz. The required absorption in that frequency range can be estimated fairly easily according to Eqs. (8.10)–(8.15). To efficiently absorb also in the kHz range, a thick woven carpet or a cloth

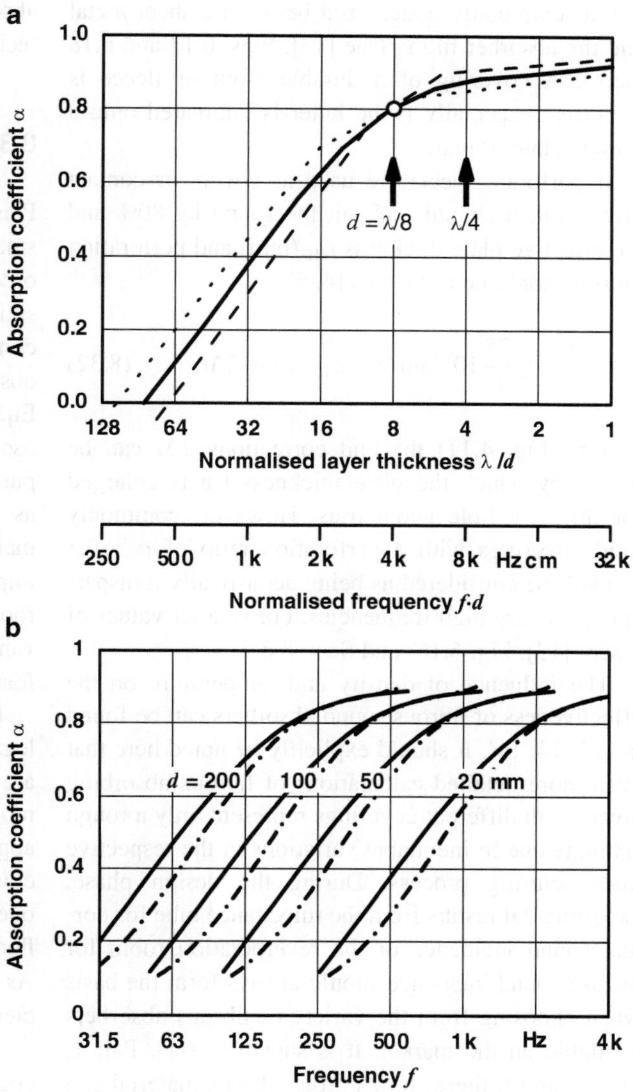

Fig. 8.3 Design characteristics of porous/fibrous absorbers with optimum matching ratio $\varepsilon = 2$ (*dashed line*) to 6 (*dotted line*) for diffuse (*continuous line*) and normal (*dashed dotted line*) sound incidence

wall cover of 5–10 mm thickness is sufficient with a flow resistivity of ideally more than 10^5 Pasm^{-2}.

To protect all fibrous absorbers against abrasion, a tight fibre fleece is needed, which has also to be matched to the optimum flow resistance according to Eq. (8.28). Flow resistivities of different commonly used cloth can e.g. be found in ([5], Table 4.2). Additional protection using a foil in front of the absorber should not disturb the intrusion of the sound wave into the absorber and therefore needs to have a mass per unit area m''_F not too large compared to the oscillating mass of the air according to Eq. (8.2), and it should not be too thick (t thickness) either:

$$m''_{\mathrm{F}} = \varrho_{\mathrm{F}} t \ll m''_{\mathrm{A}} = \frac{\varrho_0 c_0}{2\pi f}. \qquad (8.30)$$

If the sound power transmission coefficient of the foil $\tau_F = P_t/P_i$ should have at least 80%, the mass per unit area of the foil needs to stay small [4, 5]

$$m''_F \leqslant \frac{90}{f} \left(\mathrm{kg\,m^{-2}}\right). \qquad (8.31)$$

For $f > 250$ Hz this results in $m''_F < 360$ gm^{-2} but for $f > 2{,}500$ Hz this requires $m''_F < 36$ gm^{-2}. This estimation is only valid if the foil is suspended freely and not, as commonly done, fixed between a sheet metal and the absorber filling (see [13], Figs. 6.17 and 6.18 there). A covering of a durable cloth or fleece is preferred, especially if the latter is laminated onto a fibrous plate or mat.

If perforate sheets are used as covers or contact protection, it should transmit the sound by 80% and the effective plate thickness t_{eff} (mm) and perforation ratio σ should be chosen as [4, 5]

$$\frac{t_{\text{eff}}}{\sigma} \leq \frac{75}{f} 10^3 (\text{mm}); \quad t_{\text{eff}} = t + 2\Delta t. \tag{8.32}$$

In ([5], Fig. 4.11) the end corrections $2\Delta t$ can be found, by which the plate thickness t acts enlarged for different hole geometries. However, commonly used coverings with a perforation ratio of $\sigma > 0.3$ can still be considered as being acoustically transparent up to very high frequencies. For smaller values of σ see ([13], Fig. 6.16) and Sect. 8.6.2.

The influence of density and temperature on the effectiveness of fibrous sound absorbers can be found in [13, 14, 16]. It should explicitly be noted here that even more detailed calculations of fibrous absorbing layers with different coverings represent only a rough estimate due to inevitable variations in the respective manufacturing process. During the design phase, experimental results from the impedance tube for normal sound incidence or the reverberation room for diffuse sound incidence should always form the basis when choosing from the variety of fibrous absorbers available on the market. It is shown in ([1], Part 1, Figs. 5 and 6 there) how porous/fibrous material can absorb sound, when their flow resistance is not perfectly matched to that of air.

8.3.2 Open-Pore Foams

Plastic foams, whose fine skeleton structure builds small pores in the sub-millimetre range, act in a first-order approximation according to Eqs. (8.21)–(8.29) similar to fibrous absorbers as shown in Fig. 8.3. For certain soft foams, an oscillation of the material can be observed at lower frequencies, where according to Eq. (8.2) a substantial air mass is in motion. This oscillation is also advantageous for sound reduction purposes. High flexibility, easy handling, and shaping as well as durable fixing techniques with other materials using permanent gluing make foams an important sound absorber in noise control and in room acoustics; they can be located e.g. in the turning vanes of large air ducts as aerodynamically shaped foam parts [17].

For certain applications in room acoustics, where legal requirements for flame retardance allow it, foams are easier to handle, are more flexible to install and more attractive than fibres. A durable cloth with appropriate flow resistance is often appropriate as cover. When covered with a perforate sheet metal, one can even step on foam linings installed on the floor of noise capsules or anechoic chambers [18]. As an example, Fig. 8.4 shows the absorption coefficient of a foam made of a special gypsum [19].

The trend to use organic material (e.g. sea grass, cocos fibres, wood chips) or material from animals

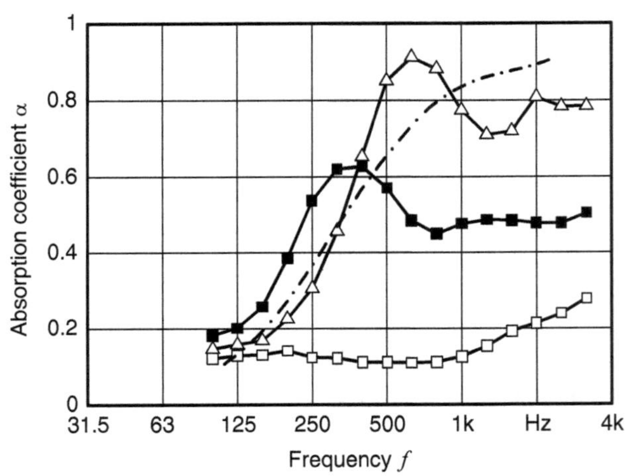

Fig. 8.4 Absorption coefficient α at normal sound incidence of a 50 mm plaster foam [20] with a density of 120 kg m^{-3} *open square* raw material and *closed square*, *open triangle* treated with different needles. For comparison: *dashed dotted line* porous/fibrous absorber of same thickness according to Fig. 8.3

(e.g. sheep wool) as an environmentally attractive alternative for artificial fibres has ceased after a short boom. It can nevertheless be stated that all porous or fibrous materials with optimum flow resistance according to Eq. (8.28) can be used as damping material. A filling of mineral fibres in a splitter silencer, which is sensitive to dirt can first be covered with a fleece or cloth and can then be mounted with a thinner (and more expensive) layer of stainless steel wool with a perforate sheet metal in front. A splitter enwrapped in that way can be cleaned from dust and dirt of the fluid with, for example, a water jet or compressed air. It is by no means necessary to realise porosities in the μm-range ([21], Tables 19–7), if not together with the sound absorption the heat insulation should be optimised because the thickness of the acoustic (viscous) boundary layer δ (μm) at a plane obstacle given by

$$\delta = \sqrt{\frac{\eta}{\varrho 0 \omega}} = \frac{1,500}{\sqrt{f}} \qquad (8.33)$$

with the dynamic viscosity in air ($\eta = 0.018$ kg m^{-1} s^{-1} at 20 °C) for medium and low frequencies f in Hz is also only in the sub-millimeter range (see Table 8.2)

8.3.3 Porous Construction Materials

There are some unintended damping effects in buildings which are caused by edges, niches and air gaps even if these were intended to serve a different purpose like increasing the diffusivity of a sound field. Ventilation ducts and other installations can have a considerable influence on the design of room acoustics. On the other hand, there are some components at walls and ceilings, which can deliberately serve sound-absorbing purposes as well as static ones. These are, for instance, components made of lightweight or porous concrete or clay or specially shaped hollow building blocks. If their porosity is not sealed by a closed plaster or some other cover an absorption coefficient α close to unity can be expected for $d \approx \lambda/4$ even with a suboptimal flow resistance Ξd in such inhomogeneous porous media. As shown in ([1], Part 1, Fig. 12), the damping decreases for accumulated petrified lava with $\chi \approx 4$, a sound speed of $\underline{c} \approx 170$ ms^{-1} and a thickness $d = 120$ mm at approximately 800 Hz corresponding to $d \approx \lambda/2$, whereas the second maximum occurs not before $d \approx 3\lambda/4$. If, however, the material is replaced by an open cellular porous, fine structured glass foam [20], an absorption characteristics similar to a layer of mineral wool can be achieved after some optimisation, see Fig. 8.5.

8.4 Reactive Absorbers in Closed Spaces

Due to their dominating presence on the market, passive absorbers occupy the major parts of all cited standard textbooks on sound absorbers and silencers, because their basic principle, calculation and application can be described straightforwardly. The present section, however, describes reactive absorbers, which not only absorb incident sound waves according to Fig. 8.1 and Eq. (8.4) but also interact with the sound field impinging on them. This interaction can best be observed when eigenresonances of small rooms are damped where at least one dimension of the room is smaller than about 5 m (Fig. 8.5).

In the frequency range between 200 and 50 Hz, occasionally even down to 31 Hz, standing waves (so-called 'models') dominate the sound field similar to the one-dimensional case (see paragraph after Eq. (8.9) and Table 8.1). Figure 8.6 shows the transfer function between diagonal corners of a 5 × 4 × 3 m rectangular room with nearly rigid walls [21]. Hardly more than ten dominating resonances can be observed at ([2], Chap. 8)

$$f_{n_x,n_y,n_z} = \frac{c_0}{2}\sqrt{\left(\frac{n_x}{l_x}\right)^2 + \left(\frac{n_y}{l_y}\right)^2 + \left(\frac{n_z}{l_z}\right)^2}; \qquad n_x, n_y, n_z = 0, 1, 2, \ldots. \qquad (8.34)$$

The spatial distribution of the sound pressure level in a plane 1.3 m above the floor for the 1,1,0-mode at 55 Hz shoes maximum differences of $\Delta L > 30$ dB between the centre of the room and the four edges. If the minimum absorption at the walls is taken into account using a damping constant δ_n for each single mode n and calculated from the reverberation time $T_n(s)$ (for 60 dB) according to ([2],Chap. 9)

$$\delta_n = \frac{6.91}{T_n}, \qquad (8.35)$$

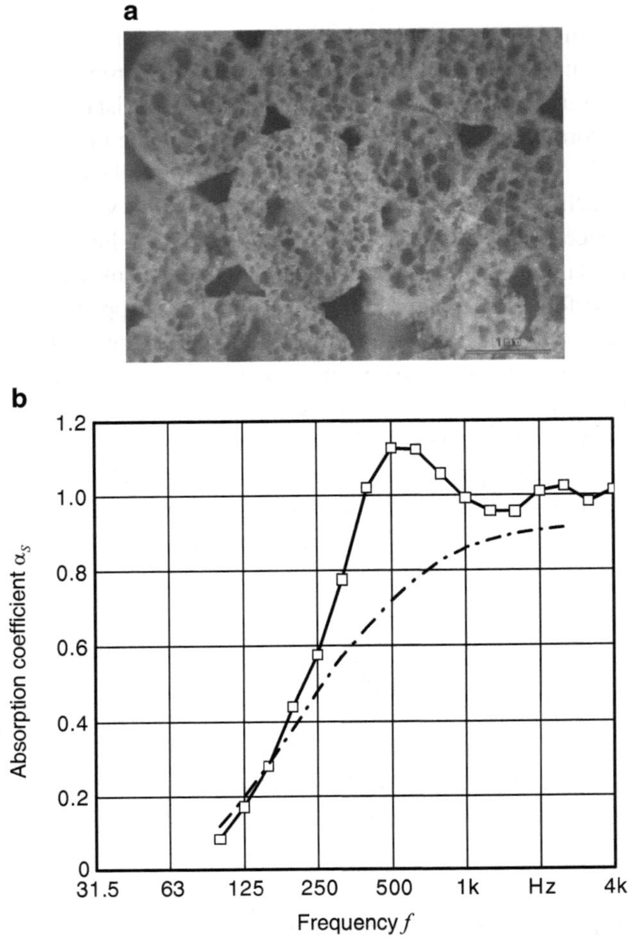

Fig. 8.5 (**a**) Microscopic section and (**b**) absorption coefficient α_s *open square* measured in the reverberation room of a sintered open-pore glass foam [20] with a density of 260 kg m^{-3} and a thickness of 50 mm. For comparison, *dashed dotted line* porous/fibrous absorber of same thickness for diffuse sound incidence according to Fig. 8.3

the sound field in this reference room can be calculated in good agreement with numerous measurements at the lowest frequencies. Every rigid room, like asymmetrically shaped sound capsules for noisy machinery, sound studios for recording and mixing of audio productions and reverberation rooms for the determination of the absorption coefficient of components and the sound power of sound sources, even 'anechoic' chambers show a similar behaviour at the lowest frequencies [22]: all sound sources actuated in the room are selectively amplified at the room modes, which creates booming noise. The sound and the radiation characteristics of the source are influenced by the room. Acoustic measurements are only possible using special treatment, which is described in more detail in [23].

For a rectangular room with dimensions $l_x \geq l_y \geq l_z$ (or a cubic room), the lowest resonance frequency is given by

$$f_1 = \frac{c_0}{2l_x} \quad \text{or} \quad f_1 = \frac{c_0}{2\sqrt[3]{V}} \tag{8.36}$$

Below this limiting frequency, the room acts as a compressible air volume. Above f_1, the modal sound field dominates. Even with sinusoidal excitation, it is difficult to excite the room between two resonances according to Eq. (8.34). Above a certain limiting frequency f_s the interval between two resonances becomes very small, thus making single modes difficult to distinguish. Hence, the modal density increases so that, for example, more than 20 modes are contained in a third-octave bandwidth. The sound field can therefore be regarded as being uniform (or 'diffuse') enough so that the standards of statistical room and building acoustic measurements can be applied. The increase of the total number of resonances N between 0 and f is given by [24, 25]

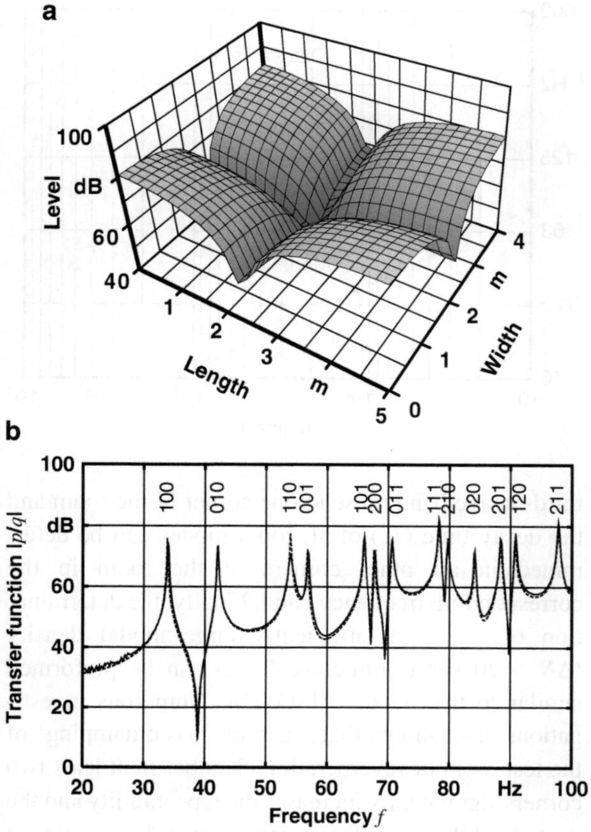

Fig. 8.6 a Calculated sound pressure level distribution of the $(1, 1, 0)$-mode ($f = 55$ Hz) in a plane 1.3 m above the floor and b calculated (*dashed line*) and measured (*continuous line*) transfer function between diagonal corners of a lightly damped rectangular room ($V = 5$ m $\times$ 4 m $\times$ 3 m $= 60$ m^3) [21]

$$N = \frac{4\pi}{3c_0^3} f^3 V + \frac{\pi}{4c_0^2} f^2 S + \frac{1}{8c_0} fL, \tag{8.37}$$

where $V = l_x l_y l_z$ (m^3) is the volume of the room, $S = 2\,(l_x l_y + l_x l_z + l_y l_z)$ (m^2) the total surface area and $L = 4\,(l_x + l_y + l_z)$ (m) the total length of the edges of a rectangular room. To evaluate the diffuseness of a room, the numbers of modes ΔN in a certain frequency band Δf around a centre frequency f_c (Hz), the so-called modal density, is of interest which can be estimated by

$$\Delta N = \left(\tfrac{4\pi f_c^2}{c_0^3} V + \tfrac{\pi}{2c_0^2} f_c S + \tfrac{L}{8c_0}\right) \Delta f. \tag{8.38}$$

Usually, a constant relative bandwidth $\Delta f/f_c$ is used in measurements. For commonly used bandwidths, the modal density can be calculated as

Table 8.3 Constants for the calculation of the modal density in a room according to Eq. (8.39) for commonly used relative bandwidths

$\Delta f/f_c$	C_3	C_2	C_1
$1/\sqrt{2}$ (Octave)	8.89	1.11	0.087
$1/3\sqrt{2}$ (Third-octave)	2.96	0.37	0.029
$1/12\sqrt{2}$ (Twelfth-octave)	0.74	0.09	0.007

$$\Delta N = C_3 \left(\frac{f_c}{c_0}\right)^3 V + C_2 \left(\frac{f_c}{c_0}\right)^2 S + C_1 \frac{f_c}{c_0} L, \tag{8.39}$$

using the constants given in Table 8.3. To a first-order approximation, Eqs. (8.38) and (8.39) are also valid for non-rectangular rooms, though not for flat-shaped rooms.

The upper limiting frequency f_s beyond which the sound field can be regarded as diffuse is given slightly different in [25, 26],

$$f_s = \frac{3c_0}{\sqrt[3]{V}} \quad \text{and} \quad f_s = \frac{2c_0}{\sqrt[3]{V}}. \tag{8.40}$$

This uncertainty, also mentioned implicitly in ([7], Chap. 2.5, p. 261), is indicated by the grey area in the simplified presentation of Fig. 8.7. For rectangular rooms which are exceptionally used in building acoustics laboratory measurements even the limiting frequency according to [25] can be regarded as being optimistic. Even measurement engineers with high expertise do not trust their experiments in a 300 m^3 reverberation chamber below 200 Hz. Suppressing these in many respects unwanted room modes by using, for example, passive edge absorbers known from sound engineering (so-called bass traps) would require a lot of volume. More compact absorbers, which withdraw sound energy from the modal sound field as resonance systems do not only dissipate energy but also have a reactive effect on the sound field. Their contribution to the sound field in the room can be described by a second modal sound field opposite to the original source field [21, 27]. For a calculation of the total sound field, the exact position of these concentrated absorbers and their (complex) wall admittance W (for normal sound incidence) must be known.

The exact calculation of the space-dependent sound field is impossible and furthermore not necessary for absorbers extending over a surface area (see Sect. 8.5.3), which shall be excited by differently

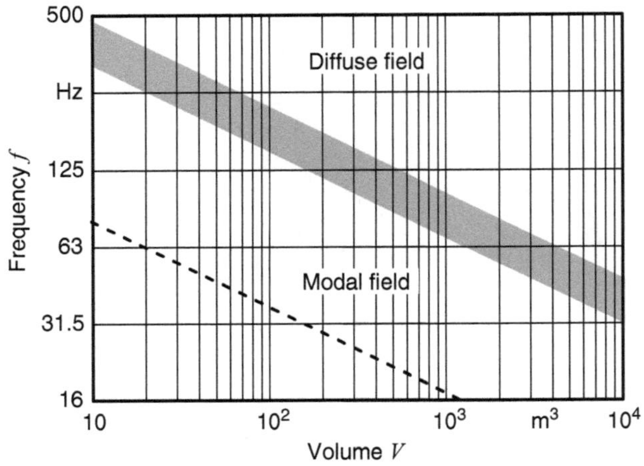

Fig. 8.7 Frequency range for a mainly modal or diffuse sound field of a cubic room versus its volume. Transition area (*grey*) according to Eq. (8.40), *dashed line* first eigenresonance of the room according to Eq. (8.36)

structured pressure fields of as many room modes as possible simultaneously. These kinds of absorbers are also difficult to test with plane waves in the impedance tube. It should be noted that all resonance-type absorbers do affect not only the sound pressure level but also the structure of the sound field close to the absorbers and thus can also interact with each other when mounted adjacently.

During the development of special low-frequency absorbers and for performance comparisons of different constructions, a measurement procedure has been established in rooms with a very small modal density ($\Delta N < 5$ per third-octave band). The reverberation time introduced earlier for the modal damping factor in Eq. (8.35) is measured with sinusoidal excitation first without ($T_{n,0}$) and then with ($T_{n,w}$) the test samples at certain positions in the corners or edges of a room similar to that of a reverberation chamber according to ([2], Chap. 11, p.258) at certain measurement locations carefully chosen respecting the spatial distribution of the modes ([1], Part 2, Fig. 3). Similar to the reverberation chamber technique described in [26], an 'effective' absorption coefficient

$$\alpha_e = 0.163 \frac{V}{S\alpha} \left(\frac{1}{T_{n,w}} - \frac{1}{T_{n,0}} \right) \tag{8.41}$$

can thus be derived according to [22] using the room volume V (m^3) and the surface area of the absorber S (m^2). The validation limits of this kind of technique as described in ([1], Part 2, Sect. 5) should, however, be kept in mind.

For slightly higher modal densities ($5 < \Delta N < 20$ per third-octave band), the room can be excited with third-octave band noise in one corner of the room and the decay time (T_n) of all room modes can be determined in all other corners of the room in the corresponding frequency band. Finally, the determination of α_s for a sufficiently large modal density ($\Delta N > 20$ per third-octave band) can be performed similar to the standard ISO 354. Numerous investigations have shown that a certain 'basic damping' of the test room or reverberation chamber in at least two corners significantly increases the repeatability and the reproducibility in different rooms up to frequencies of about 200 Hz [23]. It should again be noted that in the lower frequency range, which is so important for noise control and room acoustics, a frequency-dependent $\alpha(f)$, however measured, represents a characterisation of absorbers that can only be used with adequate experience. Useful product benchmarks for test rooms and the arrangement of the test samples within them at the lower end of the frequency range can only be set up with even stricter requirements than those of the standards initially developed for higher frequencies.

8.5 Panel Absorbers

Section 8.3.1 already dealt with protecting foil covers as additional airtight layers in the context of the commonly used coverings against abrasion of fibrous absorbers. Their mass per unit area should not exceed a certain value (see Eq. (8.31)), so that the sound penetration into the porous material is affected as little as possible. It is described in Sect. 8.6.2 how broadband absorbers for mid-frequencies can be created by

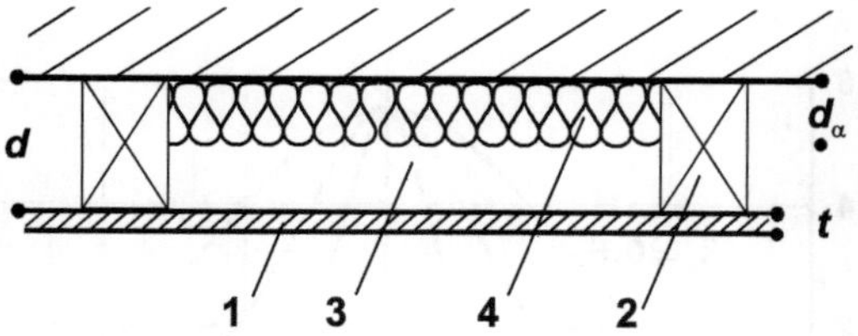

Fig. 8.8 Classical plate resonator consisting of *1*, a closed layer of mass per unit area m'', *2*, a rigid frame, *3*, a closed air volume of thickness d and *4*, a loosely inserted damping layer of thickness d_α

using partial coverage forming, for instance, a slit cover in front of a densely packed porous or fibrous material. This section treats reactive absorbers with rigid layers, whose mass per unit area m'' is large compared to the oscillating air mass of the incident sound field according to Eq. (8.2). Such a large mass can only react with the sound field if it is excitable as part of a resonating system. The simplest case is a plate mounted on a frame a distance d in front of a rigid wall (Fig. 8.8). A thin layer of thickness d consisting of a fibrous or porous damping material should be located within the air gap, which is compressed by the plate displacement. Ideally, the flow resistance Ξd of the layer should be equal to values according to Eq. (8.28) [24]. This layer should be mounted in such a way that it does not touch the plate so that the plate vibration is not affected by it but damped indirectly.

8.5.1 Foil Absorbers

If the heavy layer *1* in Fig. 8.8 has no stiffness by itself, the incident sound wave sketched in Fig. 8.1 impinges on the wave impedance given by Eq. (8.6)

$$W = r + W_{\mathrm{m}} + W_{S}; W_{\mathrm{m}} = j\omega m'' = j\omega\varrho_t t, \qquad (8.42)$$

using a somewhat difficult to quantify friction coefficient r (Pasm^{-1}) [13] approximately given by $r = \Xi d_\alpha/3$, the mass per unit area m'' ($\mathrm{kg\ m}^{-2}$) and the plate thickness t (mm). For an air cushion with a thickness d small compared to the quarter wavelength $\lambda/4$, the impedance can be simplified to yield

$$W_S = -j\varrho_0 c_0 cot\frac{\omega d}{c_0} \approx -j\frac{\varrho_0 c_0^2}{\omega d} = -j\frac{s''}{\omega}, \qquad (8.43)$$

using the stiffness per unit area $s''[\mathrm{Pam}^{-1}]$. The strongest reaction of this resonator occurs if the imaginary part of W vanishes, which is the case at the resonance frequency f_{R} (Hz) with d (mm) for

$$f_R = \frac{1}{2\pi}\sqrt{\frac{s''}{m''}} \approx \frac{c_0}{2\pi}\sqrt{\frac{\varrho 0}{m''d}} \approx \frac{1,900}{\sqrt{m''d}}. \qquad (8.44)$$

Using Eq. (8.44) and normalising to $\varrho_0 c_0$, W can be written as

$$\frac{W}{\varrho_0 c_0} = \frac{r}{\varrho_0 c_0} + j\frac{\sqrt{m''s''}}{\varrho_0 c_0}\left(\frac{f}{f_{\mathrm{R}}} - \frac{f_{\mathrm{R}}}{f}\right)$$
$$= r' + \mathrm{j}Z'_{\mathrm{R}}F. \qquad (8.45)$$

The normalised specific resonator resistance

$$Z'_{\mathrm{R}} = \frac{Z_{\mathrm{R}}}{\varrho_0 c_0} = \frac{\sqrt{m''s''}}{\varrho_0 c_0} = \sqrt{\frac{m''}{\varrho_0 d}}, \qquad (8.46)$$

is a function of the mass and the spring stiffness of the resonator and multiplied by the frequency mistuning F, it determines the absorption coefficient α for normal sound incidence according to Eq. (8.8)

$$\alpha = \frac{4r'}{(r'+1)^2 + (Z'_{\mathrm{R}}F)^2} = \frac{\alpha_{\max}}{1 + \left(\frac{Z'_{\mathrm{R}}}{r'+1}F\right)^2};$$
$$F = \frac{f}{f_{\mathrm{R}}} - \frac{f_{\mathrm{R}}}{f}. \qquad (8.47)$$

Equation (8.47) shows three things:

1. The maximum possible absorption coefficient $\alpha_{\mathrm{R}} = 1$ can only be realised with optimal damping ($r' = 1$ or $r = \varrho_0 c_0$) at the resonance frequency ($F = 0$ or $f = f_{\mathrm{R}}$).
2. The smaller the friction resistance r' the stronger α decays on both sides of f_{R} with increasing $|F|$ independently of the value of absorption at resonance α_{R}.
3. Whilst the influence of r' on the bandwidth can only be changed by a factor of 5 ($r' \approx 0.2$ compared to $r' \approx 1$) the specific resistance Z_{R} appearing in the product with F represents a parameter for the achievable broadband characteristics of such reactive absorbers, which can vary by several orders of

Fig. 8.9 Absorption coefficient α of a simple mass-spring system versus normalised frequency f/f_R for varying normalised specific resistance Z_R. (**a**) light damping, ($r' = 0.2$); (**b**) optimum damping ($r' = 1$); (**c**) heavy damping ($r' = 5$)

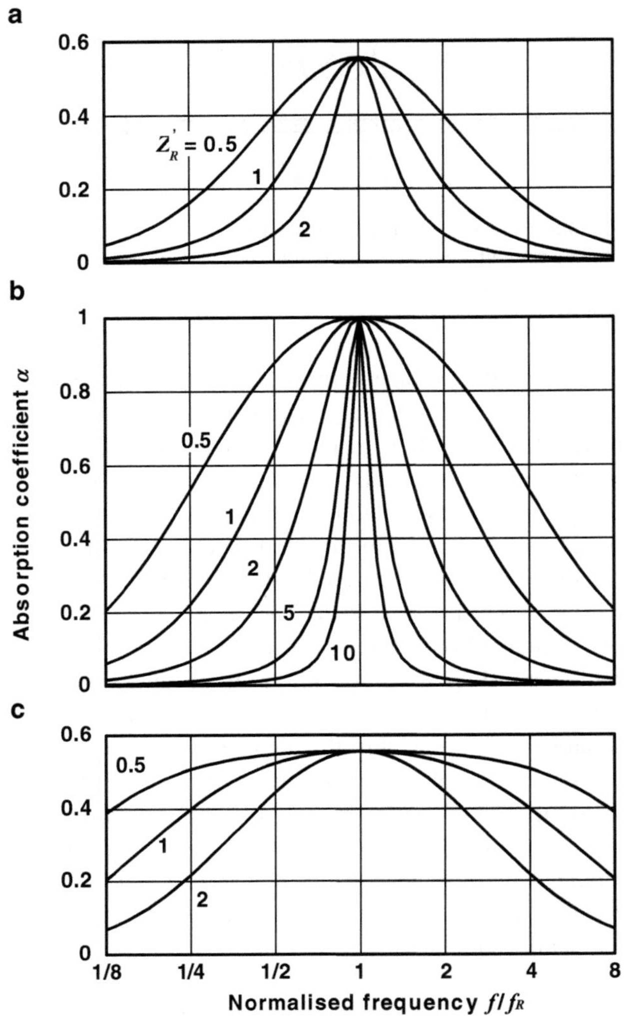

magnitude. This fact is shown in Fig. 8.9 using a frequency scale normalised to f_R.

The optimum design of a mass-spring system is achieved by choosing the right amount for the specific resistance Z_R. The most important design rule is to choose both m'' and s'' as small as possible, independent of f_R. It is thereby confirmed that low frequency absorbers cannot be realised by large masses only. Not only from an acoustic but also from a constructional point of view should the wall distance be not too large and not too small compared to the incident wavelength λ [5]

$$\frac{3{,}400}{f} = \frac{\lambda}{100} < d < \frac{\lambda}{12} = \frac{28}{f}10^3(\mathrm{mm}). \qquad (8.48)$$

Very thick wall claddings are generally undesirable. It should be aimed to use small values for m'' and obtain small values for Z_R. But this is in contrast to the main design rule Eq. (8.44), by which conventional low-frequency absorbers are usually narrow–banded or reach α-values below 0.5. See ([4], Table7.1) for conventional plywood, chipboard or plaster plates with or without filling in the air gap.

The design of resonators consisting of plastic foils or metal membranes for mid-frequencies is slightly more convenient when several layers are used. If the resonances of three layers of foil are spaced far apart by choosing their mass per unit area and separation correspondingly, then measurement and calculation according to [28] show separate α-maxima in Fig. 8.10. If the resonances are closer together, the absorption of the corresponding multi-layered arrangement is slightly more spread than that of the single

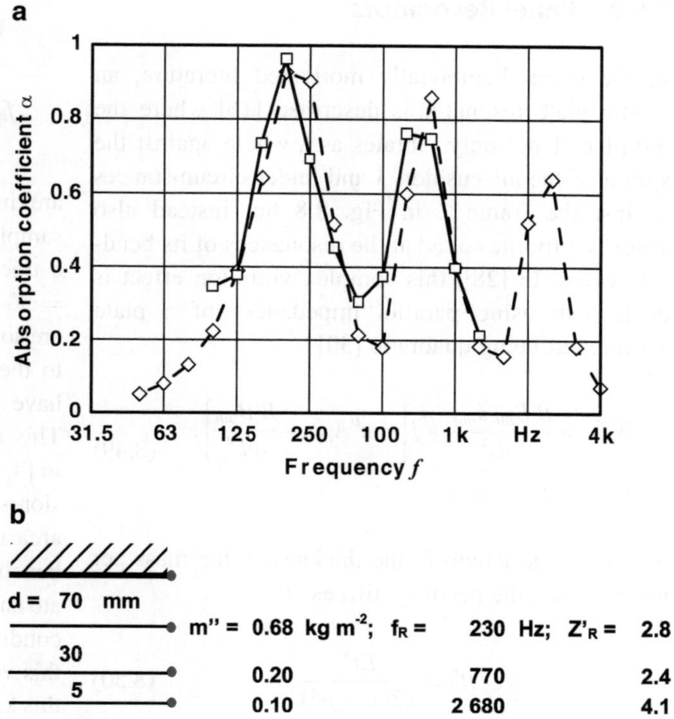

Fig. 8.10 Absorption coefficient α of a threefold arrangement of foils in front of a rigid wall. *Open triangle* calculation; *open square* measurement in an impedance tube

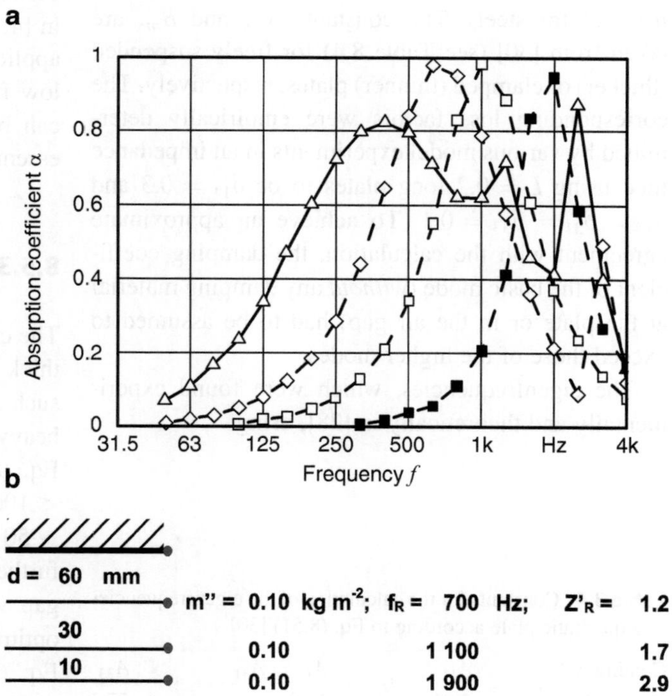

Fig. 8.11 Calculated absorption coefficient α of foils in front of a rigid wall at normal sound incidence [28]; *dashed line* single arrangement, *continuous line* threefold arrangement with air gaps *d* as indicated

layers. An absorber is shown in Fig. 8.11, which absorbs more than 60% between 200 and just above 2,000 Hz.

A foil absorber with a moulded cup structure, described in [29], reaches approximately the absorption represented in Fig. 8.11. In the meantime, this absorber was replaced by the micro-perforated foils described in Sect. 8.9.2.

8.5.2 Panel Resonators

In the more theoretically motivated literature, an elastic plate resonator is described [16] where the top plate 1 not only vibrates as a whole against the spring of the air cushion 3 and under circumstances against the frame 2 in Fig. 8.8 but instead also absorbs airborne sound at the resonances of its bending waves. In [28], this complex vibration effect is dealt with using parallel impedances of a plate assumed as being quadratic [30]

$$W_{mn} = \frac{B' B_{mn} \delta_{mn}}{\omega L^4} + j\left[\omega m'' A_{mn} - \frac{B' B_{mn}}{\omega L^4}\right]; \qquad m, n = 1, 3, 5, \ldots. \tag{8.49}$$

with the edge length L, the thickness t, the mass per unit area m'', the bending stiffness B'

$$B' = \frac{Et^3}{12(1-\mu^2)}, \tag{8.50}$$

the elastic modulus E and the Poisson ratio μ (e.g. $\mu = 0.3$ for steel). The constants A_{mn} and B_{mn} are taken from [30] (see Table 8.4) for freely suspended (thicker) or clamped (thinner) plates, respectively. The corresponding loss factors were empirically determined by various model experiments in an impedance tube using $L = 0.2$ long plates to be $\delta_{11} = 0.3$ and $\delta_{13} = \delta_{31} = \delta_{33} = 0.1$. To achieve an approximate agreement with the calculation, the damping coefficient of the basic mode (*without* any damping material at the plate or in the air gap) had to be assumed to exceed those of the higher modes.

The eigenfrequencies, which were found experimentally and theoretically in [28], using

Table 8.4 Constants for the calculation of the eigenfrequencies of a quadratic plate according to Eq. (8.51) [30]

Condition	A_{11}	$A_{13} = A_{31}$	A_{33}
Clamped	2.02	10.8	57.1
Free	1.52	13.7	123
B_{11}	$B_{13} = B_{31}$		B_{33}
2,640	1.9×10^5		2.8×10^6
592	1.3×10^5		3.9×10^6

$$W = W_{mn} - j\frac{\varrho_0 c_0^2}{\omega d}; \qquad f_{mn} = \frac{c_0}{2\pi}\sqrt{\frac{\varrho_0}{m'' d}\left(\frac{1}{A_{mn}} + \frac{B' d}{\varrho_0 c_0^2 L^4}\frac{B_{mn}}{A_{mn}}\right)}, \tag{8.51}$$

are in close agreement for small ($L = 0.2$ m) test samples even for multiple layers of aluminium up to a thickness of $t = 0.8$ mm. For air gaps between $d = 30$–50 mm, the resonances, being far above 125 Hz, are so well separated and narrow banded that this leads to the conclusion such plate resonators would never have gained major importance in applied acoustics. This is confirmed by the practical experience stated in [4, 5], according to which the smallest plate dimension should not be chosen smaller than 0.5 m surface area not below 0.4 m^2. Only then one may hope to benefit from the mass-spring resonance with appropriate damping in the air gap, provided that the 'clamped' condition at the edges of each panel would at all allow this very restricted vibration. Even then the design of this kind of resonance absorber is considered as uncertain due to a variety of influences of the fixation between *1* and *2* in Fig. 8.8, and it is recommended in [4, 5] to rely on experimental data in any concrete application. A plate resonator which better reaches low frequencies and is more broadband in character can be designed if the structure is altered in some essential details.

8.5.3 Compound Panel Absorbers

The compound panel absorber consists of a 0.5–3 mm thick steel plate, which is suspended at the edges in such a way that it can be excited to vibrate. For such heavy plates ($5 < m'' < 25$ kg m^{-2}) according to Eq. (8.44) or (8.51), if thicknesses of only $50 < d < 100$ mm are used for the frequency range $100 > f > 50$ Hz or even lower, it becomes evident without further notice that a loose damping material in the air gap without contact to the plate cannot produce an optimum damping of all plate resonances, see Eq. (8.47) and Fig. 8.9. Starting with the densely stuffed foil absorber in ([2], Fig. 61), it can be assumed that a compound structure of a front panel (with very low internal damping as in steel) in an intimate connection to an elastic material with high internal damping, what does not affect the vibrations, is advantageous. This is best achieved when the plate 'floats' on an elastomeric layer covering the entire surface area.

If, for instance, the latter according to Fig. 8.12 consists of a soft foam plate *2* (as described in Sect. 8.3.2), which has roughly the dimensions of the front plate *1* or even larger (as described in the application in Sect. 8.10.2) both layers can be excited by the sound field surrounding this compound panel in front of a back wall *5* (or a sufficiently heavy second layer *1'* to form a baffle) to perform a multitude of always strongly damped vibrations.

Such a universally applicable acoustic module realises the mass-spring system according to Sect. 8.5.1. For most applications, the demand for additional honeycomb or other sub-structures or frames can be dropped, since the highly damped plate *2* replaces the air gap. The resonance frequency of this compound system,

$$f_d = \frac{c_d}{2\pi}\sqrt{\frac{\varrho_d}{\varrho_t d}} = f_R\sqrt{\frac{E_d}{E_0}}, \tag{8.52}$$

is only slightly shifted compared to f_R in Eq. (8.44) if the speed of the strain wave c_d in the damping plate is decreased in the same order of magnitude compared to c_0 as $\sqrt{\varrho_d}$ is increased compared to $\sqrt{\varrho_0}$ or in other words, if the elastic modulus of the damping layer differs only slightly from $E_0 = 0.14\ 10^6$ Pa (for air at 20°C) e.g. $0.1\ 10^6\ E < 0.8\ 10^6$ Pa for a soft foam. Compared to the arrangement shown in Fig. 8.8, the compound plate can vibrate more freely in all of its own modes if the damping layer 2 takes part in the vibration and, at the same time, acts like an efficient damping layer attached to the vibrating structure.

The mathematical description of free plate vibrations of finite extent ('Chladni's-Figures') is not at all trivial (see e.g. [31–33]). Since a model for a broadband resonator is dealt with here, the resonance frequencies of a clamped rectangular plate may suffice as approximation:

$$\begin{aligned} f_{m_x,m_y} &= \frac{\pi}{2}\sqrt{\frac{B'}{m''}}\left[\left(\frac{m_x}{L_x}\right)^2 + \left(\frac{m_y}{L_y}\right)^2\right] \\ &= 0.45 c_t t\left[\left(\frac{m_x}{L_x}\right)^2 + \left(\frac{m_y}{L_y}\right)^2\right]; \\ &m_x, m_y = 1, 2, 3, \ldots. \end{aligned} \tag{8.53}$$

For a $t = 2$ mm thick and $L_x = 1.5$ m × $L_y = 1$ m large steel plate with a strain wave propagation speed of $c_t \approx 5.100\ \mathrm{m_s}^{-1}$, the lowest eigenfrequency is approximately $f_{1,1} = 6.6$ Hz, which is far below the resonance of the mass-spring according to Eq. (8.43) of $f_R = 48$ Hz for $d = 100$ mm. In contrast to room resonances Eq. (8.39), the number of eigenfrequencies in a frequency band Δf according to [34]

$$\Delta N = 1.75\frac{S_\alpha}{c_t t}\Delta f; \quad S_{\alpha =} L_x L_y, \tag{8.54}$$

does not increase with the centre frequency. Nevertheless, nine eigenfrequencies result for the example given above in the 50 Hz-octave- band and even 18 eigenfrequencies in the 100 Hz octave- band. In any case, these are enough to supply a plate resonance for any of the room modes according to Sect. 8.4.

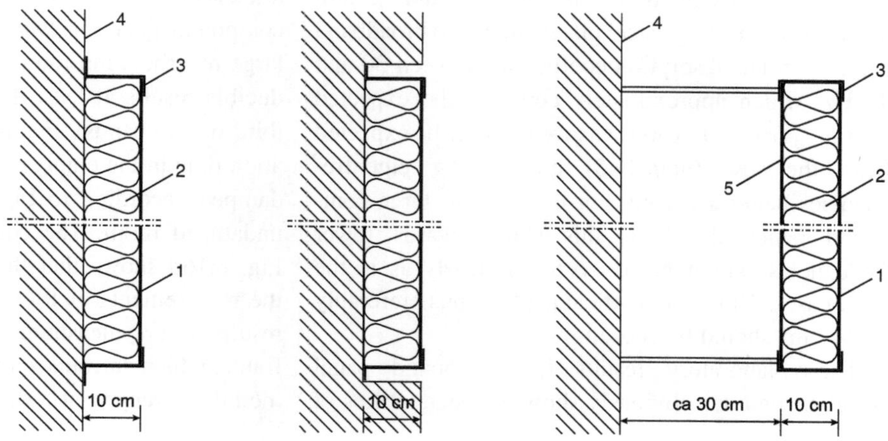

Fig. 8.12 Compound panel absorber (schematic): *1* freely vibrating plate (e.g. 0.5–3 mm steel), *2* porous damping layer, *3* mounting bracket, *4* solid structure, *5* back plate (e.g. 2–3 mm steel)

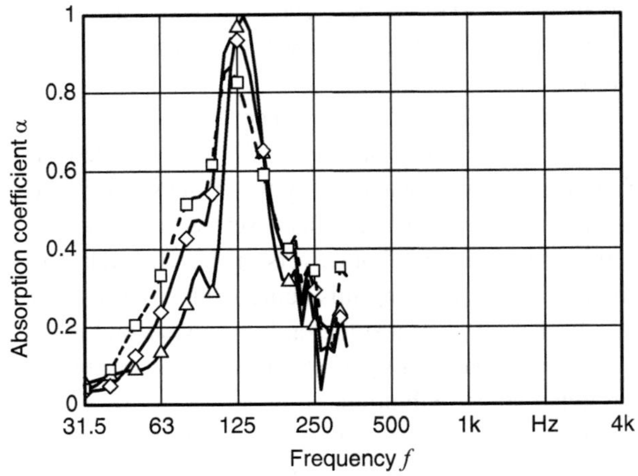

Fig. 8.13 Absorption coefficient α, measured in the impedance tube, of a 1.70×0.65 m large, 0.2 mm thick stainless steel plate as a solid 'lid' on a rigid 'tub' of depth $d = 100$ and partly filled with mineral fibres (50 kg m^{-3}; $\Xi = 2.18 \times 10^4$ Pasm^{-2}) of varying thickness according to Fig. 8.8; *open square* $d_\alpha \approx 88$; *open diamond* $d_\alpha \approx 50$, *open triangle* $d_\alpha \approx 0$ mm

Figure 8.13 shows the absorption coefficient of a conventional plate resonator according to Fig. 8.8 consisting of a stainless steel plate ($t = 0.2$ mm) in front of an air gap ($d = 100$ mm). The resonance frequency $f_R \approx 150$ Hz according to Eq. (8.44) is only slightly shifted as expected but the absorption coefficient increases significantly towards lower frequencies when an optimum flow resistance Ξ $d = 1{,}090$ or 1,740 Pa sm^{-1} is installed in the air gap.

Absorption coefficient measurements of compound panel absorbers, which were cut to the cross-sectional area of the impedance tube (1.7×0.65 m leaving an air gap at the edges of 5–20 mm) are presented in ([1], Part 2, Figs. 10–12). The excitation in the impedance tube using plane waves is similar to that of Fig. 8.14, where six compound panel absorbers, each of 1.5×1 m size, were placed parallel in pairs at opposite boundaries of a rectangular room, if the effective absorption coefficient according to Eq. (8.41) is determined at the axial modes, which are normal to the absorbers surface areas. The results for 1 and 2.5 mm thick and 1.5 m^2 large steel plates show two broadband maxima in the absorption coefficient between 30 and 90 Hz, which approximately confirms the expected shift of f_d from 80 to 50 Hz as well as may be expected from the few (namely:5) contributing eigenfrequencies being involved. For the 0.75 m^2 large compound panel absorbers, this shift towards lower frequencies cannot be observed as clearly as in the other cases. Thus, for practical applications preferably $S_\alpha > 1$ m^2 should be chosen.

These laboratory results for compound panel absorber modules obviously show the peculiarities of low-frequency absorbers in small rooms mentioned in Sect. 8.4, which have to be dealt with in practice. Nevertheless, they allow us to quantify the influence of geometric and material parameters on the absorption coefficient of such absorbers and allow to compare different products. For the transfer of new absorber technologies into practise, it is vital to have reverberation room data at hand as well. As usual the test samples must be on a surface of about 12 m^2 as illustrated in Fig. 8.15. The problems discussed in Sect. 8.4 require a test room qualified for frequencies below 100 Hz.

Because the commonly used inclination of opposite boundaries and of additionally installed diffusers do not contribute to the homogenisation of the sound field at low frequencies, only a suitable damping of the room modes as demonstrated in [23] achieves the desired effect.

For the measurement of the absorption coefficient, three compound panel absorber modules at a time with dimensions $1.5 \times 1.0 \times 0.1$ m with 1 or 25 mm thick compound plates in two lower corners of a 392 m^3 large reverberation room were proven to yield reproducible results at least down to 63 Hz (measured in third-octave bands). Figure 8.16a shows the reverberation time in the empty reverberation room when it is damped according to Fig. 8.15 as compared to the undamped room. The equivalent absorption area in Fig. 8.16b shows that the damped room still fulfils the requirements of ISO 354. Figure 8.17 shows the results of a α_s-measurement using a 0.1 m high rigid frame, which surrounds six compound panel absorber modules arranged at a distance of 0.2 m each with a

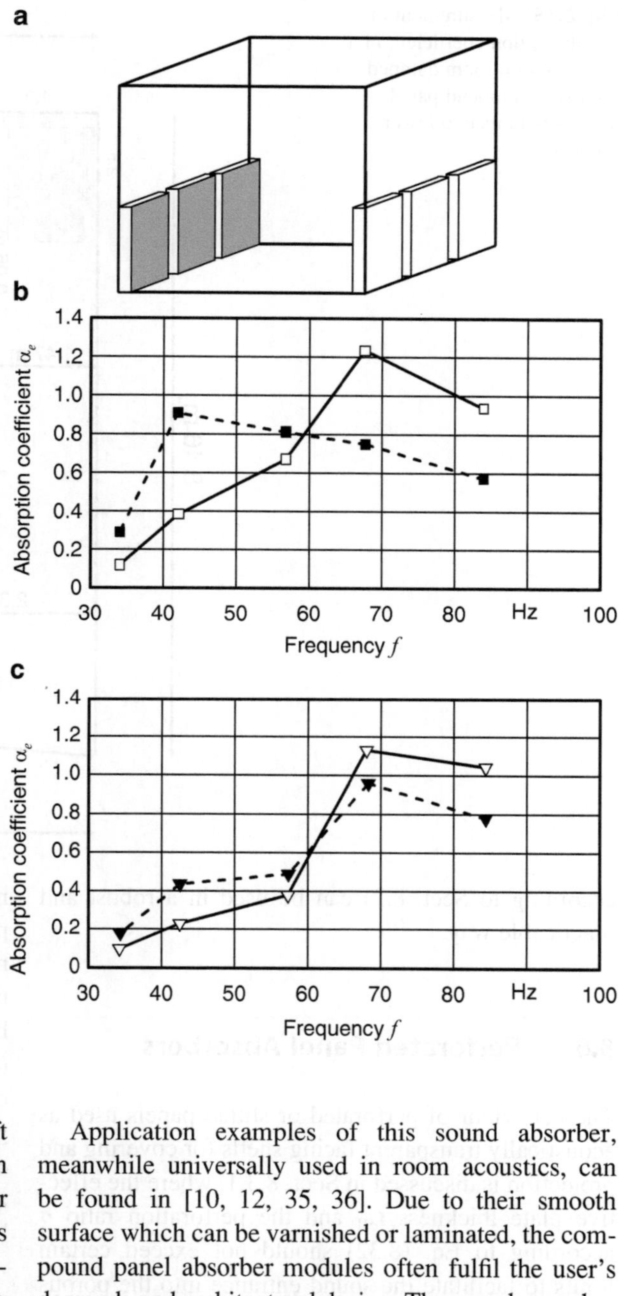

Fig. 8.14 (a) Arrangement of six compound panel absorbers and b,c corresponding effective absorption coefficient α_e measured at the lowest axial modes [22] with $d = 100$ mm; *open square* $L_x = 1.5$ m; $L_y = 1.0$ m; $t = 1.0$ mm; *down triangle* $L_x = 1.0$ m; $L_y = 0.75$ m; $t = 1.0$ mm; *filled square* $L_x = 1.5$ m; $t = 2.5$ mm; *filled down triangle* $L_x = 1.0$ m; $L_y = 0.75$ m; $t = 2.5$ mm

1 mm steel plate in compound with a 0.1 m thick soft foam plate as sketched in Fig. 8.15. The absorption coefficient in Fig. 8.17 related to the grey absorber sample surface area marked in Fig. 8.15 (9 m^2 in this case) shows a broadband efficiency around the resonance frequency $63 < f_d < 125$ Hz and a 'tail' falling off slowly towards the kHz-range, which can be explained by the edges in this arrangement, which are open to approximately 60%. For thicker steel plates, the maximum is shifted to frequency ranges where the reverberation room is no longer usable for α-measurements below 63 Hz even when damped as described.

Application examples of this sound absorber, meanwhile universally used in room acoustics, can be found in [10, 12, 35, 36]. Due to their smooth surface which can be varnished or laminated, the compound panel absorber modules often fulfil the user's demands and architectural design. They can be used as pinboards, whiteboards, projection or mirror surfaces which also justifies their minor space requirement [37]. Due to their relatively small thickness, compound panel absorber modules can also be 'hidden' behind acoustically transparent linings, ceilings or floors [38]. Thus, low-frequency absorbers for the manifold problems in room and building acoustics

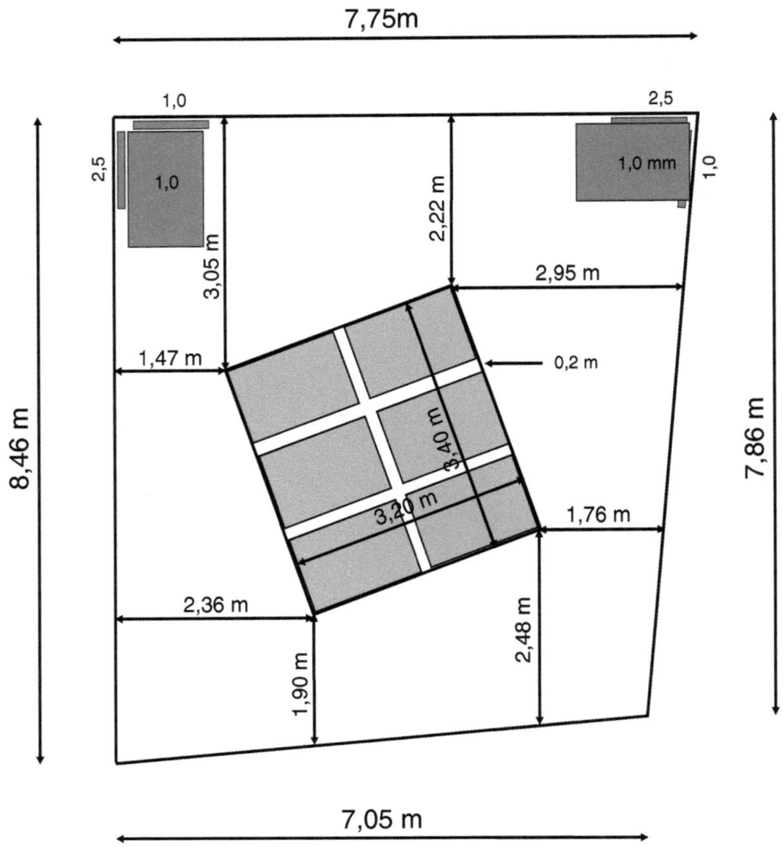

Fig. 8.15 Measurement of the absorption coefficient in a reverberation room damped with six compound panel absorbers in its two lower corners

according to Sect. 8.2) can be used in a robust and practicable way.

8.6 Perforated Panel Absorbers

The behaviour of perforated or slitted panels used as acoustically transparent facing shells for covering and protection is discussed in Sect. 8.3.1, where the effective plate thickness t_{eff} and the perforation ratio σ according to Eq. (8.32) should not exceed certain limits to facilitate the sound entrance into the porous material as the actual sound absorber.

Here, only reactive absorbers are of interest, where the mass in the holes or slits of differently perforated plates or membranes is not small compared to the mass of the air according to Eq. (8.2) but may still be excited by an incident sound wave. Such a mass, which can also be enlarged by the oscillating air near the holes, can only interact with the sound field if it is excited as part of a resonance system, similar to the plate resonator. The simplest form is given by a suitably perforated plate mounted at a distance d in front of a rigid wall (Fig. 8.18) and supported by a frame structure, which acoustically encloses the air cushion behind the plate. In contrast to the plate resonator (Fig. 8.8), the damping of this 'air-in-air'-oscillator can be obtained not only by a loose filling of the air gap using damping material but also, and even more effectively, by spanning an optimum flow resistance according to Eq. (8.28) closely in front of or behind the holes using, for instance, fibre fleece or cloth.

8.6.1 Perforate Surface Absorbers

These absorbers can again be described using Eqs. (8.44)–(8.47), where r' stands for the dimensionless flow resistance normalised by $\varrho_0 c_0$ ($r' = \Xi d/\varrho_0 c_0$ for a given flow resistivity Ξ) and m'' is understood as the acoustic mass m''_H (S_{H} total surface area of the holes) transformed to the absorbers surface area S_{A}

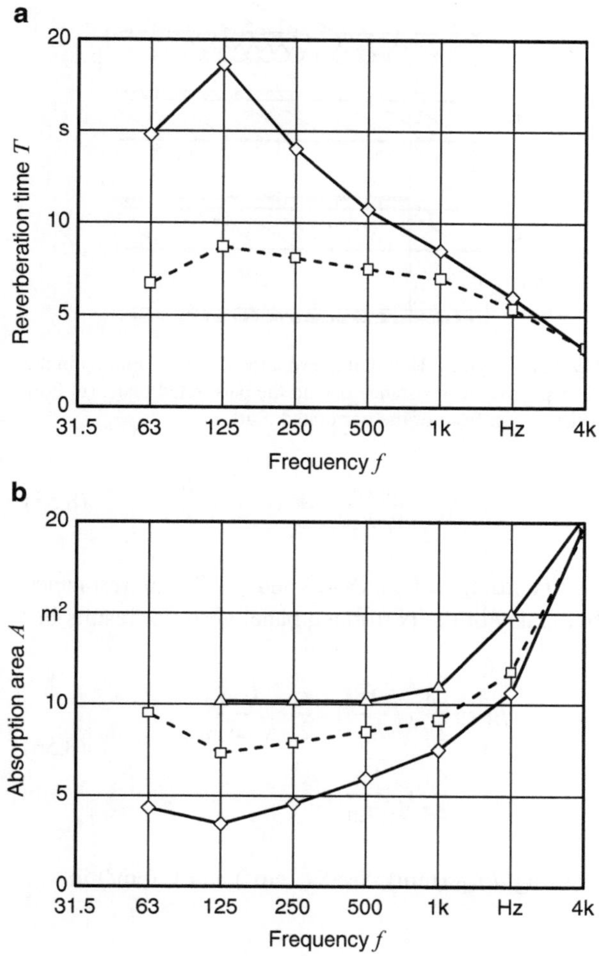

Fig. 8.16 (**a**) Reverberation time T and (**b**) equivalent absorption area A in a reverberation room ($V = 392\ \mathrm{m}^3$) *open diamond* without and *open square* with damping in its two lower corners according to Fig. 8.15. For comparison: *open triangle* maximum allowed absorption area according to ISO 354

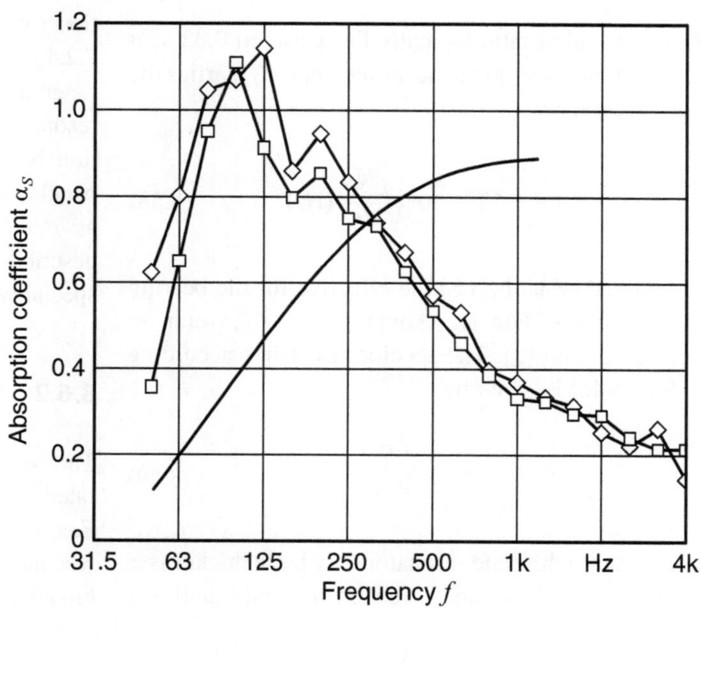

Fig. 8.17 Absorption coefficient α_s of compound panel absorbers ($1.5 \times 1 \times 0.1$ m, 1 mm steel), measured in the reverberation room according to Fig. 8.15 and normalised to $S_\alpha = 9\ \mathrm{m}^2$ absorption area. For comparison: porous/fibrous absorber of same thickness according to Fig. 8.3 *continuous line*; *open square* melamine resin foam, *open diamond* polyester fibre fleece

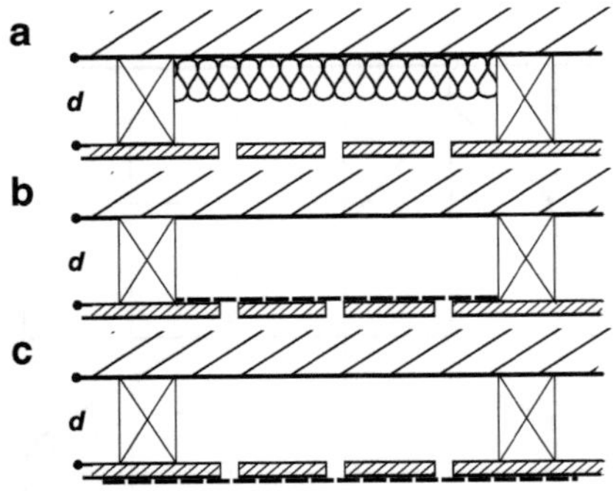

Fig. 8.18 Typical Helmholtz resonator with (**a**) damping in the air gap, (**b**) flow resistance behind the perforated plate, (**c**) flow resistance in front of the perforated plate

$$m''_{\mathrm{H}} = \frac{\varrho_0 t_{\mathrm{eff}}}{\sigma} \quad \text{with} \quad \sigma = \frac{S_{\mathrm{H}}}{S_{\mathrm{A}}}. \tag{8.55}$$

According to Eqs. (8.42) and (8.43), the resonance frequency of the perforated panel absorber results in

$$f_{\mathrm{H}} = \frac{c_0}{2\pi}\sqrt{\frac{\sigma}{d t_{\mathrm{eff}}}} = \frac{c_0}{2\pi}\sqrt{\frac{S_{\mathrm{H}}}{d S_{\mathrm{A}} t_{\mathrm{eff}}}} = \frac{c_0}{2\pi}\sqrt{\frac{S_{\mathrm{H}}}{V t_{\mathrm{eff}}}}, \tag{8.56}$$

or using $d;t_{\mathrm{eff}}$ (mm), S_{H}; S_{A} (cm^2) and V (cm^3) in

$$f_{\mathrm{H}} = 54 \times 10^3 \sqrt{\frac{\sigma}{d t_{\mathrm{eff}}}} (\mathrm{Hz}). \tag{8.57}$$

The perforation ratio typically lies between $0.02 < \sigma < 0.2$. If only a single concentrated hole S_{H} carries the excited air mass

$$f_{\mathrm{H}} = 17 \times 10^3 \sqrt{\frac{S_{\mathrm{H}}}{V t_{\mathrm{eff}}}} (\mathrm{Hz}), \tag{8.58}$$

is obtained (V is the volume effective for the bearing of the mass). For an estimation of t_{eff}, refer to Sect. 8.3.1 and [5]. The specific resistance according to Eq. (8.46) is given by

$$Z'_{\mathrm{H}} = \sqrt{\frac{t_{\mathrm{eff}}}{d\sigma}}. \tag{8.59}$$

Similar to the plate resonator only large thicknesses d lead to low resonance frequencies and small Z_{H}, whereas very small holes and thick plates result in narrow-band low-frequency absorbers even for optimum damping $r' = 1$. It should therefore be attempted for this hollow chamber resonator to couple it to other vibration mechanisms, which broaden its absorption characteristics (the considerations in ([4], p. 141) also tend towards that direction). In [28], a variety of hole resonators with and without partitions subdividing their hollow spaces are examined in good agreement between theory and experiment when plate and cavity resonances are correctly modelled.

It becomes apparent that for a thickness of 50 mm the bandwidth of absorption always stays small even for the mid-frequencies, as long as the resonances are far apart. If they get closer together, only one of the mechanisms dominates ([28], Fig. 4–7). But if the Helmholtz-resonance and the first plate resonance (f_{H} according to Eq. (8.57) and f_{11},f_{13} according to Eq. (8.51)) are optimally designed, spaced apart about one octave away from each other, they do not interact ([28], Fig. 8). A sufficient damping is needed to 'merge' the single maxima to a sufficiently broad-band absorption spectrum.

An overview over the commonly used hole geometries is given in ([4], Fig. 41, p. 296) for relatively thick and therefore not excitable wood or plaster plates, where the perforated surface area can vary between 2 and 30%, the oscillating air mass in the holes between 30 and 330 gm^{-2} and the resonance frequency according to Eq. (8.57) between 420 and 1,460 Hz. Absorption characteristics measured in the reverberation room can be found in ([4], Table 7.2). Example 7.2.4 in [4] shows the difficulties to cover the frequency range below 250 Hz with that type of Helmholtz resonators. Even with a thickness of 240 mm the absorption below 200 Hz decrease significantly. Perforated panel absorbers are commonly used in room acoustics as mid-frequency absorbers. The following section describes a design and optimisation procedure for a special type of board-band slit absorbers.

8.6.2 Slitted Panel Absorbers

The design of conventional Helmholtz or perforated panel absorbers is usually performed using Eqs. (8.55)–(8.58) resulting in relatively narrow band resonance absorbers to be verified by experiment. If a broadband space saving mid-frequency absorber is to be optimised, a closer look at the mechanisms and components of the cavity resonator described in Sect. 8.6.1 is valuable. For its optimisation, an intimate interaction between the oscillating air in the slits with the porous or fibrous flow resistance behind it is advantageous similar to that which takes place at the edges of the compound panel absorber. Furthermore, the distribution of the slits within the absorber surface area S_A becomes relevant not only with respect to the end correction as part of t_{eff} according to Eq. (8.32). Finally, the eigenfrequencies in the cavity formed between the slitted panel and the rigid wall play an important role in the extended resonance frequency range.

If slit width b and slit distance a are both individually chosen as design parameters rather than only a perforation ratio $\sigma = b/a$ (as is the case in [2, 5, 6, 13]), new possibilities for optimisations result. For the description of the functional model of slit absorbers the geometric (forms of the slits) and material parameters (absorbing layer) can be identified in Fig. 8.19. The air mass per unit area in the slits m_s'' including the corresponding end correction [5] (but only at the front side in this case) is given similar to that of a Helmholtz resonator described in Sect. 8.6.1

$$m''_S = t_S \varrho_0 \quad \text{with} \quad t_S = t + \Delta t \,. \tag{8.60}$$

The impedance of the absorbing layer of thickness d_α with respect to the free slit surface area is given by [13]

$$W = \sigma W_A \coth \Gamma_A d_\alpha . \tag{8.61}$$

The impedance W_A and the propagation constant Γ_A of the absorbing layer can be calculated using [16] as

$$W_A = \varrho_0 c_0 \sqrt{(E + 0.86) - j\frac{0.11}{E}}\,; \Gamma_A = \frac{2\pi f}{c_0}\sqrt{(E - 1.24) + j\frac{0.22}{E}} \tag{8.62}$$

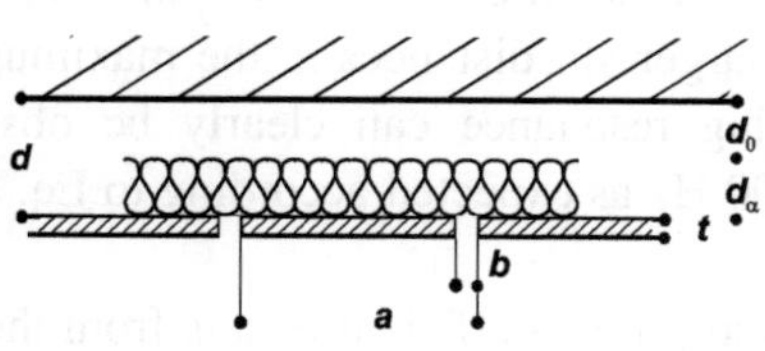

Fig. 8.19 Structure of a slit absorber with parallel slits

and

$$E = \frac{\varrho_0 f}{\Xi} \tag{8.63}$$

for $\Xi > 7{,}500$ Pa s m^{-2} with sufficient precision. For open-cellular melamine resin foam with skeleton oscillations, it is reasonable to include the mass per unit volume ϱ_α as an additional mass

$$E = \frac{\varrho_0 f}{\Xi} - j\frac{\varrho_0}{2\pi \varrho_\alpha} . \tag{8.64}$$

Assuming that the sound field in the absorber propagates in the same way as it would propagate behind a diffracting grid, the wall impedance of the slit absorber is derived in [39] including the air mass in the slits and the end correction as

$$W_S = \frac{1}{\sigma}\left(j\omega m''_S + \sigma W_A \coth \Gamma_A d_\alpha \times + \frac{a^2}{b\pi^3} W_A \Gamma_A \left(\sin \pi \frac{b}{a}\right)^{\frac{3}{2}} \right) . \tag{8.65}$$

On the one hand, a resonance system is formed by the coupling between the spring-like wall impedance of the absorbing layer and the air mass in the slits. On the other hand, the effective spring stiffness and damping of the absorbing layer are increased due to the third part of the sum in Eq. (8.65). This is the reason for the lower resonance frequency and higher bandwidth of slit absorbers compared to damped or undamped Helmholtz resonators of the same thickness.

Figure 8.20 shows the absorption coefficient of an absorber with different slit geometries but about the same perforation ratio $\sigma \approx 0.02$ calculated according to Eqs. (8.8) and (8.65) and measured in the impedance tube. In Fig. 8.21, a slit absorber with $\sigma \approx 0.02$ with one single centred slit, on the one hand, and a varying damping layer behind a single circumferential slit is compared to two conventional Helmholtz resonators with optimum damping and to the porous absorber alone.

With respect to their practical application, these novel slit absorbers excel by a high and broadband absorption mainly in the mid-frequency range. They allow space-saving constructions, special requirements

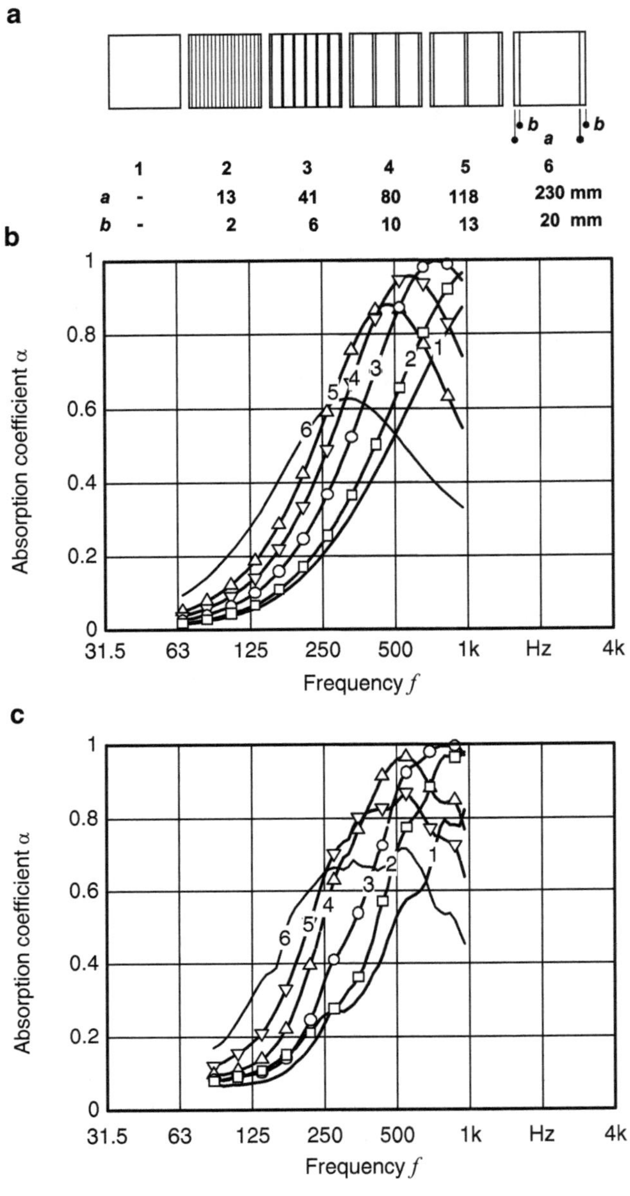

Fig. 8.20 Absorption coefficient α for slitted covers (1 mm steel) with different slit widths b and slit distances a, but about the same perforation ratio in front of 50 mm open cellular melamine resin foam ($\rho_\alpha \approx 10\ \mathrm{kg\ m^{-3}}$; $\Xi \approx 10\ \mathrm{kPa\ s\ m^{-2}}$) **a** shape of different covers (*front view*); **b** Calculation; **c** Measurement (in an impedance tube 250 mm × 250 mm)

for the shape or the fixation of the slitted covers are not necessary. This results in a variety of new surface structures. The shift of the absorption maximum of the passive absorber according to Fig. 8.21 by nearly two octaves towards lower frequencies by only a partial coverage comes up to meet the actual demands as described in Sect. 8.2.

So far, the cover between the slits was assumed to be rigid, so that no bending modes occur and no oscillations of the cover may compress the absorbing layer. Even more possibilities result from the combination with other resonance principles, for example with foils with low bending stiffness in front of an absorbing layer according to Sect. 8.5.1 or with plates with finite bending stiffness according to Sect. 8.5.3. Figure 8.22, for instance, shows the absorption coefficient for diffuse sound incidence for a slit absorber tiled with steel plates of different size. For larger slit distances a, the maximum of the mass spring resonance can clearly be observed at around 100 Hz as expected according to Eq. (8.44) or (8.52).

The absorption coefficient results from the impedance of a simple mass spring system

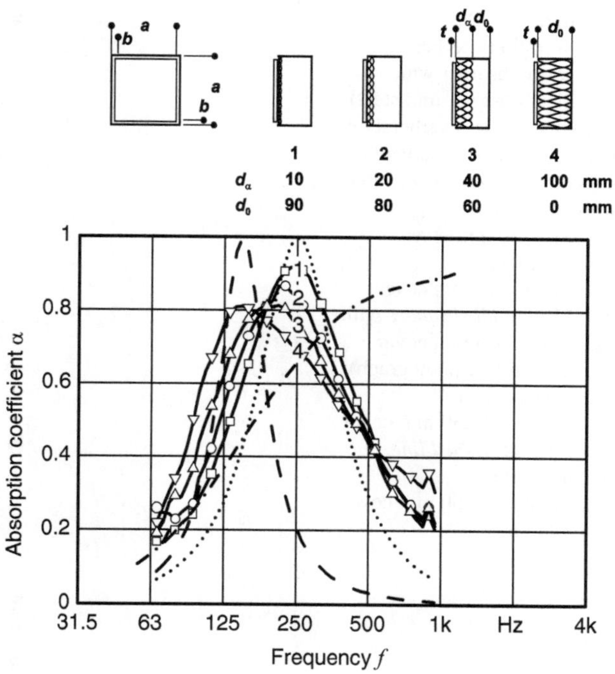

Fig. 8.21 Absorption coefficient α of slitted absorbers with constant perforation ratio $\sigma \approx 2\%$ consisting of a 199 × 199 × 1 mm steel plate with a 1 mm circumferential slit in front of a melamine resin foam ($\rho_\alpha \approx 10$ kg^{-3}; Ξ 10 k Pa s m^{-2}) measured in the impedance tube (200 mm × 200 mm) *dashed dotted line* without cover according to Sect. 8.3.1; *dotted line* with a 4 mm slit in the centre, calculated according to Sect. 8.6.1 for optimum damping ($r' = 1$); *dashed line* same with a 32 mm hole in the centre

$$W_P = \frac{1}{1-\sigma}(j\omega m''_P + W_A \coth \Gamma_A d_A), \qquad (8.66)$$

with the mass per unit area m''_P of the oscillating plate. Here, the sound propagation behind the plate in the homogeneous absorbing layer with relatively high flow resistivity $\Xi \approx 10$ k Pa s m^{-2} and relatively large slit distances $a = 1{,}250$ mm is neglected. The resulting impedance due to the parallel setup of W_P with W_s according to Eq. (8.65) is given by [28]

$$W_{res} = \frac{W_P W_S}{W_P + W_S}, \qquad (8.67)$$

which again demonstrates an impressive shift towards lower frequencies (see Fig. 8.22) of the acoustic efficiency of a porous or fibrous layer (see Sect. 8.3).

8.6.3 Membrane Absorbers

In certain applications, the use of not only fibrous but also porous damping material such as soft foams is forbidden due to health regulations, for hygienic reasons, fire regulations or durability. For heating, ventilation and air-conditioning as, for example, in hospitals, senior residences or production sites requiring supreme clean air conditions and for air-processing plants with, for example, soils or aggressive fluids in the flow ducts or exhaust stacks silencers completely made of aluminium or steel have been approved, which are hermetically sealed from the flow. The remarkable stiffness and resistivity of these membrane absorber modules make good use of a lightweight honeycomb structure, upon which two relatively thin ($0.05 < t < 1$mm) plates are a panned (Fig. 8.23) one- or two-sided.

The strong subdivision of the otherwise empty cavity by a honeycomb structure enables the local reaction of the membrane, so that for diffuse or oblique sound incidence (e.g. when used in a splitter silencer), the propagation of sound in the cavity parallel to the membrane surface is suppressed. If the distance e of the partitions normal to the propagation direction is given by

$$e \leqslant \frac{\lambda}{8} = \frac{42.5}{f} 10^3 (\text{mm}), \qquad (8.68)$$

with f (Hz), even this fibreless absorber is always reacting locally [2] with a wall impedance W according to Eq. (8.6). The membrane absorber requires only a fraction of the thickness d of a passive absorber for maximum absorption at the lower

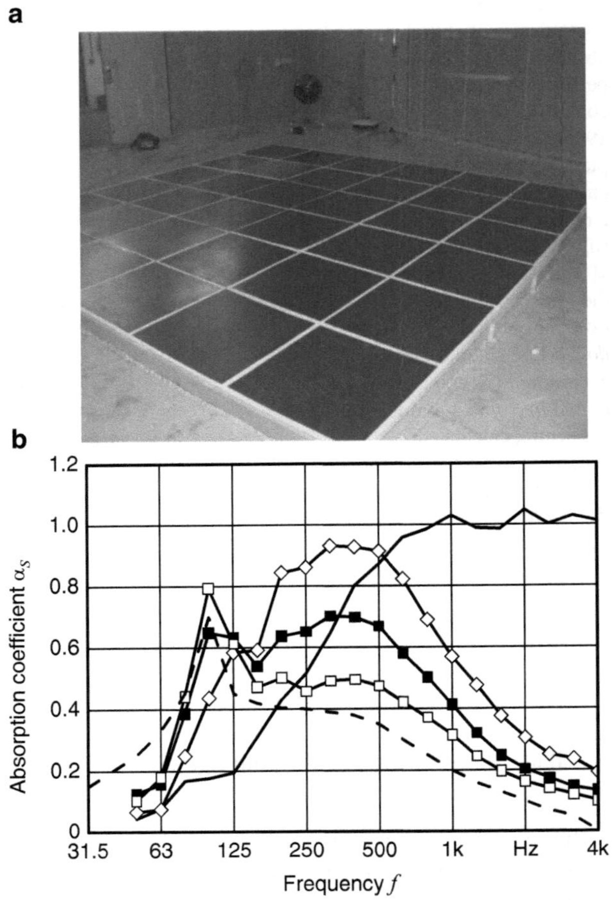

Fig. 8.22 (**a**) Design and (**b**) absorption coefficient α_s of a slit absorber with a resilient cover (1 mm steel) measured in the reverberation room damped as described in Sect. 8.5.3; *continuous line* without cover $d = d_\alpha = 50$ mm *open diamonds* with 312 mm × 312 mm cover $b = 15$ mm; *filled square* with 625 mm × 625 mm cover $b = 28$ mm (see photograph); *open square* with 1,250 mm × 1,250 mm cover $b = 50$ mm; *dashed line* calculated for 1,250 mm × 1,250 mm cover $b = 50$ mm

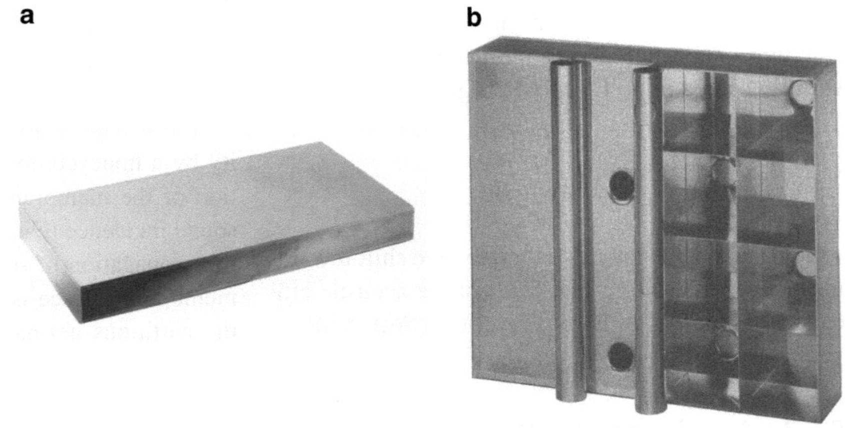

Fig. 8.23 (**a**) Model of a two-sided membrane absorber (**b**) with partially unfolded perforated and cover membranes

frequencies but still requires larger chamber thicknesses d to obtain a sufficiently broadband performance. Thus, an approximately constant ratio e/d between 1 and 2 also satisfies the requirements for the static stiffness of the component. In practice, cubically shaped chambers of, for example, $L_xL_yd = d^3 = V \approx 1{,}000\ \text{cm}^3$ for maximum absorption at 250 Hz have succeeded.

For each cavity, the perforated membrane on the inside, which is fixed softly on the honeycomb, provides a hole or a slit to create a kind of Helmholtz resonator. The size of the hole and the chamber is approximately tuned using Eq. (8.57) to mark a lower limiting frequency in faint analogy to Eq. (8.29) for passive absorbers. For circular holes which are not smaller than 5 mm and for a membrane thickness t usually very small compared to the hole diameter d_H, the end correction $2\,t \approx 0.85 d_H$ according to Sect. 8.3.1 and [5] becomes particularly important. For $V = 1{,}000$ cm^3; $d_H = 10$ mm, $S_H = 0.78$ cm^2; $t = 0.2$ mm, $t_{eff} = 8.7$ mm using Eq. (8.57) $f_H \approx 160$ Hz is obtained and according to Eq. (8.59) $Z'_H \approx 3.3$. These parameters according to Fig. 8.9 hint upon a relatively broadband absorber if the damping inherent in this resonator is not too small.

For the unperforated aluminium membrane, the first plate resonance according to Eq. (8.44) would be approximately given by $f_R = 258$ Hz. In reality, however, the compression of the air cushion of the Helmholtz resonator is slightly increased by the compliance of the membrane and the compression of the same cushion activated by the plate resonator is slightly increased by the 'leakage' through the hole. This problem is dealt with in [40] experimentally and theoretically as well. Figure 8.24 shows in good agreement with a detailed calculation (including the boundary conditions at the perforated membrane) that two main maxima dominate the absorption characteristics of the membrane absorber (still without a cover membrane): f_H at approximately 125 Hz and f_{11} at approximately 270 Hz. A side maximum can be observed at $f_{13} \approx 650$ Hz.

A cylindrical silencer, which is composed of a polygon of membrane absorbers, shows similar characteristics in the insertion loss measured according to ISO 7235 in Fig. 8.25. If the holes of the perforated membranes are closed, a plate resonator with a significantly reduced damping remains. If a cover membrane is positioned adjacent in front of the perforated membrane without touching it, the absorption maximum is slightly shifted towards lower frequencies. Obviously, the additional mass couples to this complex oscillator. Higher order modes of the perforated membrane, however, usually disappear in this 'sealed' configuration. If the cover membrane is suspended on soft stripes, a significant improvement of the absorption characteristics at higher frequencies can be observed as shown in [40]. In [41], photographs of 'Chladny'-Figures of a f_{15}-Mode are shown which prove that the cover membrane can perform vibrations similar to that of the compound panel absorber (Sect. 8.5.3).

It is characteristic for the membrane absorber that it works even if the cover membrane is mounted close to the holes and thus strongly deforms the oscillating air mass. But since the membrane absorber also works with a larger gap and even without cover membrane in the narrow gap between perforated membrane and cover membrane with a correspondingly increased friction coefficient at the wall, as can be observed in the damping mechanism of bending waves in double-leaf partitions [42] forced oscillations do not seem to play an important role.

It is known that part of the damping of conventional Helmholtz resonators can be caused by sharp edges at the holes. This effect can play an even stronger role for the extremely thin membranes herein because the fluid particles at the whole edge

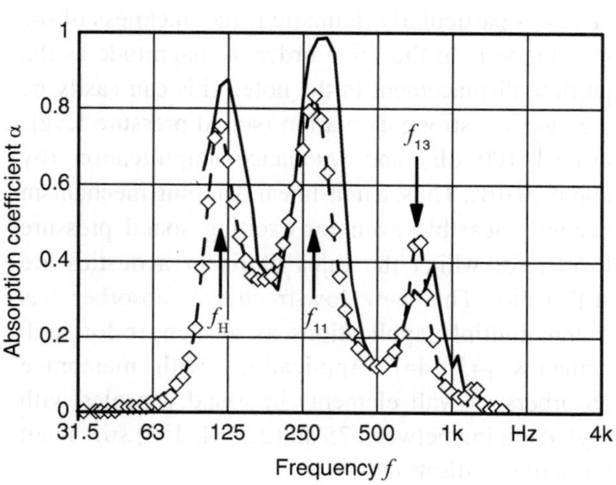

Fig. 8.24 Absorption coefficient α of a Helmholtz resonator (without cover membrane) for normal sound incidence; *continuous line* experiment; *open diamond* calculation

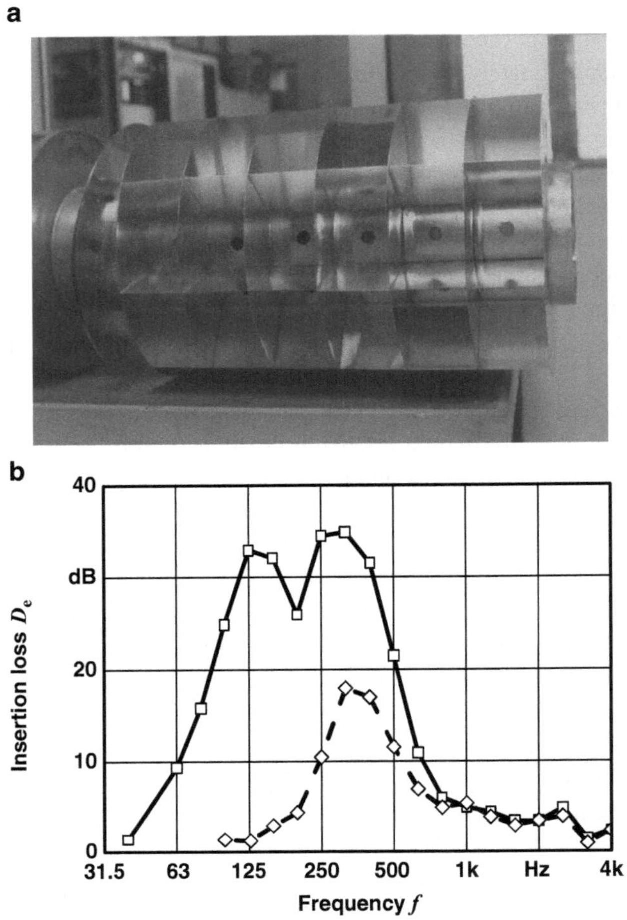

Fig. 8.25 (**a**) View (without cover) and (**b**) insertion loss D_e (without cover membrane) of a duct silencer composed of membrane absorbers; *open square* holes open; *open diamond* holes closed

must pass through a 180° deflection instead of only one or two 90° deflections as for thicker plates. The unsteady flow in this area of discontinuities may separate from the boundaries even for relatively low sound velocities so that free shear layers with large friction losses may develop. The cut-off edge effect becomes particularly dramatic if the thickness of the membrane is in the same order of magnitude as the particle displacement in the hole. This can easily be the case for strong excitation (sound pressure levels around 100 dB) and resonance amplification (by about 20 dB). Thus, a non-linear damping mechanism becomes feasible even for exciting sound pressure levels, for which the rules of linear acoustics are still valid. This new low-frequency absorber has found multiple applications as a silencer for high demands [43, 44]. Applications with membrane absorbers as wall elements in sound capsules with high damping between 25 and 125 Hz [45] still await attractive realisations.

8.7 Interference Silencers

The previous sections dealt with low-frequency broadband absorbers, which were primarily designed for room acoustic purposes. In contrast to that, silencers and sound capsules often have to be tuned to different, sometimes even narrow-band noise spectra depending on the noise source and have to withstand extreme environmental conditions such as mechanical, chemical or thermal attacks. Here, for instance, cavity resonators of different constructions with walls made of highest quality stainless steel have succeeded (also membrane absorbers according to Sect. 8.6.3). Their performance in ducts (even without any damping material) is based upon different interference mechanisms, which may also result in a reflection of sound energy back to the noise source. Usually, a number of interference silencers must be combined if their efficiency is narrow band in character.

8.7.1 Quarter-Wavelength Resonators

The basic principle of a purely reflecting silencer can be described considering a simple change in cross section of a duct according to Fig. 8.26a [15]. If the dimensions of both cross-sectional areas S_1 and S_2 are small compared to the wavelength and if P_a and P_s in Eq. (8.4) are set to zero, using Eqs. (8.4)–(8.6) with

$$W = \varrho_0 c_0 m\,; \quad r = \frac{m-1}{m+1}; \quad m = \frac{S_1}{S_2}, \qquad (8.69)$$

and the wave resistance $\rho_0 c_0$ of the medium, the reflection coefficient ρ and the sound reduction index R result in

$$\varrho = 1 - \frac{P_t}{P_i}; \quad \frac{P_i}{P_t} = \frac{1}{1-\varrho},$$
$$R = 10\,lg\frac{P_i}{P_t} = 10\,lg\frac{1}{1-r^2} = 10\,\lg\frac{(m+1)^2}{4m}. \qquad (8.70)$$

Thus, low frequencies as, for instance, in an air outlet of a large wall surface or a ceiling ($S_2 \gg S_1$) may be strongly reflected:

$$R = 10\,lgm - 6\,\text{dB} \quad \text{for} \quad m \gg 1. \qquad (8.71)$$

This, however, is only valid for plane wave propagating in the parts where the cross section is constant. If the receiving room interacts with the duct due to its eigenresonances, the sound transmission shows corresponding troughs and (between two resonances) peaks as shown experimentally and theoretically in [46].

If separated by a distance l an abrupt expansion of a duct is followed by a likewise abrupt contraction as sketched in Fig. 8.26b, the reflection is repeated at this point but this time with an opposite sign which results in [15]

$$R = 10\lg\left[1 + \left(\left(\frac{m^2-1}{2m}\right)\sin 2\pi\frac{l}{\lambda}\right)^2\right], \qquad (8.72)$$

with maxima of the sound reduction index

$$R_{\max} \approx 10lgm - 6\,\text{dB} \quad \text{for} \quad m \gg 1 \qquad (8.73)$$

at the frequencies

$$f_n = \frac{c_0}{4l}(2n-1); \quad n = 1, 2, 3, \ldots. \qquad (8.74)$$

Such an expansion chamber resonator was investigated in [47] as a water-borne sound silencer with $m = 20$ (Fig. 8.27). 'Expansion chambers' find but rarely applications in ventilating and exhaust ducts.

'Side-branches' according to Fig. 8.26c with a cross section comparable to that of the main duct have more often been approved in practice as so-called side-branch resonators. Progressive and reflected plane wave in the side branch are superimposed to that in the main duct at frequencies according to Eq. (8.74) in such a way that the transmitted wave P_t is considerably attenuated. Similar to the Helmholtz resonator (Sect. 8.6.1), the oscillating air mass at the duct connection causes a certain end correction of the effective length l depending on the duct radius r

$$\Delta l \approx 0.6r \quad \text{or} \quad 0.85r, \qquad (8.75)$$

whether the duct terminates in free space or is mounted in a large duct wall. In order to make these cavity resonators applicable for a broad frequency band, chambers of different lengths can be combined and their walls can be lined with damping material.

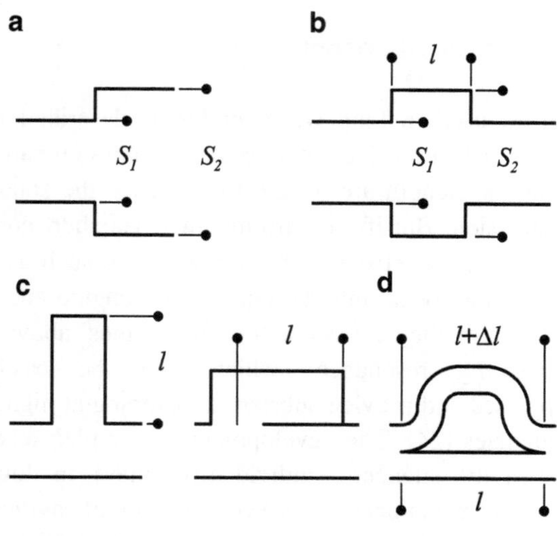

Fig. 8.26 Basic principles of reactive interference silencers. (**a**) Simple change in cross section; (**b**) Expansion chamber; (**c**) Side branch resonators; (**d**) By-pass duct

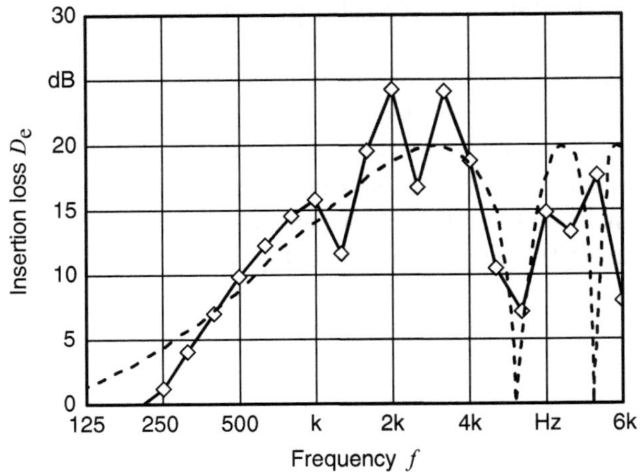

Fig. 8.27 Insertion loss D_e of a rigid expansion chamber in a water pipe [47] with $m = 20$, $l = 25$ mm; *open diamond* measurement, *dashed line* calculation according to Eq. (8.72)

8.7.2 Half-Wavelength Resonators

The interference principle described in Sect. 8.7.1 can also be realised using a 'by-pass' according to Fig. 8.26d, which issues the incident sound power P_i into two cross sections of the same area and superimposes a wave with opposite sign to the progressing wave behind the by-pass at frequencies

$$f_n = \frac{c_0}{2l}(2n-1); \quad n = 1, 2, 3, \ldots. \tag{8.76}$$

This one-dimensional destructive interference principle is rarely realised due to the large necessary mechanical effort.

8.7.3 Pipe Silencers

Cavities used in long wave guides as described in Sects. 8.7.1 and 8.7.2 which remain small compared to the wavelength are unable to influence the sound transmission. But if they are mounted via short connection pieces between a pulsating source such as a piston pump or an internal combustion engine and a duct system, they can act as 'buffer volumes' above a mass-spring resonance, which may be easily suppressed and provide substantial damping at higher frequencies [24]. The development of complex reactive cavity silencers, individually tuned to loud engines or machinery act as a combination of cavities, side-branches and perforated duct elements with often a plurality of flow bends as, for example, sketched in Fig. 8.28 interacting with the source and a coupled

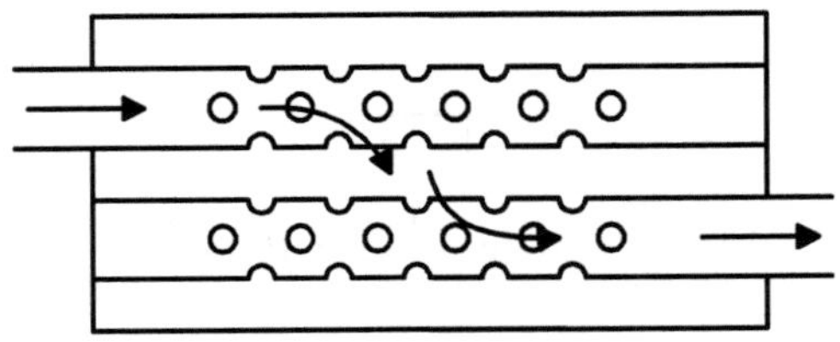

Fig. 8.28 Silencer in the exhaust line of an internal combustion engine (schematic)

duct system, has become a special research field in noise control. Linear and non-linear theories as well as numerical methods can deal with the variety of geometric parameters, effects of temperature and flow on the damping [48, 49].

These innovative silencers for the use in exhaust stacks can do without any porous or fibrous material. They revolutionized a specific industrial market segment [50], usually completely made of stainless steel allowing easy cleaning on demand. The cylindrical reactive silencers are produced with diameters of up to 1 m, aiming at lower frequencies. They consist of chambers co-centric to the main duct, which are connected by a perforated metal cylinder (Fig. 8.29). The input impedance of a single annular chamber is given by [49]

$$W_R = \frac{\varrho_0 \omega^2}{n_x \pi c_0} + j \times \left(\frac{\omega \varrho_0 t_{\mathrm{eff}}}{n_x S_h} - \frac{\varrho_0 c_0}{S_c \left(tan\frac{\omega}{c_0} L_a + tan\frac{\omega}{c_0} L_b \right)} \right), \tag{8.77}$$

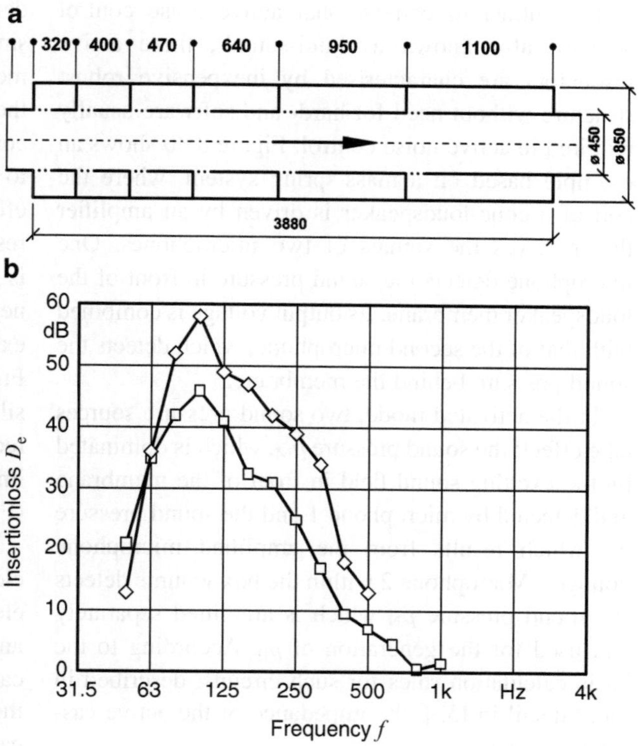

Fig. 8.29 (a) Longitudinal section and (b) insertion loss D_e of a cleanable reactive silencer with six annular chambers. *Open square* measured in a test stand; *open diamond* calculation

where n_x denotes the number of holes in the perforated cylinder, L_a and L_b are the chamber lengths, $S_c = \pi r_a^2 - \pi r_i^2$ is the cross-sectional area of the chamber, t the thickness of the perforated cylinder, r is the radius of the holes, $S_h = \pi r^2$ the area of the holes and $t_{eff} = t + 1.7r$ is the effective length of the oscillating fluid mass in the holes. The first term in Eq. (8.77) represents the friction of the air in the holes, the second the mass of the oscillating fluid mass in the holes and the third represents the compliance of the air volume enclosed in the chamber. For elongated chambers, the silencer mainly acts as two quarter-wave tubes with the partial lengths L_a and L_b according to Fig. 8.29.

The silencer shown in Fig. 8.29a was actually inserted in the chimney of a power plant with coal dust burning. The burnt exhaust gases which still contain residual dust even after several filters are dissipated into the surrounds over a 40 m high chimney with a diameter of 450 mm at a temperature of 180°C and 10 ms^{-1} flow velocity. To meet the requirements, additional damping of up to 30 dB in the octave bands between 63 and 250 Hz was necessary. The carrier of the plant can clean the perforated parts and annular chambers with a simple high pressure steam cleaner annually as part of the regular maintenance program.

8.8 Active Resonators

The oscillators described in Sects. 8.5–8.7 consisting of concentrated elements (mass, spring, friction) and modal components (vibrating plates, waveguides, cavities) can be further enhanced and improved in their acoustic performance by integrating electromechanic actuators. Some of these innovative active – or better activated – sound absorbers are a welcome alternative especially of the passive and reactive components at low frequencies due to their compact structure and high efficiency. Equipped with low cost electro-acoustic standard elements, which only use a DC-source of, for example, 24 V and, depending on the noise level to be reduced, consume no more than 8 W of electrical power, active resonators gained practical importance in noise control as series products, for example, integrated in air-conditioning devices [51] and ventilation systems [52].

In contrast to conventional active noise control devices, also known as 'anti-sound', these active resonators are characterised by inexpensive robust structure without need for hard- and software usually required in active noise control. Figure 8.30 shows an example based on a mass-spring system, where the coil of a cone-loudspeaker is driven by an amplifier that receives the signals of two microphones. One microphone detects the sound pressure in front of the loudspeaker membrane. Its output voltage is combined with that of the second microphone, which detects the sound pressure behind the membrane.

In the activated mode, two sound pressure sources take effect: the sound pressure p_M, which is dominated by the exciting sound field in front of the membrane and detected by microphone 1 and the sound pressure p_0, which results from the amplified microphone voltages. Microphone 2 within the box volume detects the sound pressure p_N which is amplified separately and used for the generation of p_0. According to the basic calculation rules for such circuits, described in more detail in [53], the impedance of the active cassette results in

$$W = \frac{1}{1+V_0}\left(r + j\omega M + \frac{1-V_N}{j\omega N}\right), \tag{8.78}$$

where r denotes the friction coefficient, M the mass and N the compliance of the passive system. New possibilities for tuning the impedance to the corresponding noise spectrum using separate amplifications V_0 and V_N arise from Eq. (8.78). By controlling V_0, the specific resistance of the resonator defined in Sect. 8.5.1 can be actively controlled. A smaller impedance in the range around the given mechanical resonance band has a positive effect on the absorption. This is shown by the straight line compared to the dashed line in Fig. 8.31. In contrast to that, using the amplification V_N actively controls the effective compliance of the box volume and thus the resonance frequency of the mass-spring system. This is also shown in Fig. 8.31 for a positive (□) and a negative (◇) value of V_N in good agreement with experiments in the impedance tube ([1], Part 4, Fig. 9.8a). If the insertion loss of a 25 × 25 × 16 cm silencer cassette, mounted at the side of a 25 × 25 cm rectangular duct, is determined [52] the practical importance of such a variability becomes apparent (Fig. 8.32).

Of course, the amplification is limited by the electro-acoustic stability of the control circuit. Duct elements coupled to the active resonator can also have an influence on its efficiency. Compared to silencer cassettes with only one feedback path ($V_N = 0$) [51], the absorption maximum of that depicted in Fig. 8.30 can be easily shifted by one octave without exchanging mechanical components. An automated tuning in compliance with the corresponding operation condition of the sound source using signals of simple control elements or e.g. tacho signals can be envisaged.

An active side branch resonator according to Sect. 8.7.1 used as a silencer for heating systems is described in [54]. Here, an active silencer cassette equipped with only one sensor terminates a quarter-wave length tube. First, the impedance with a rigid termination may be formulated for the undamped case

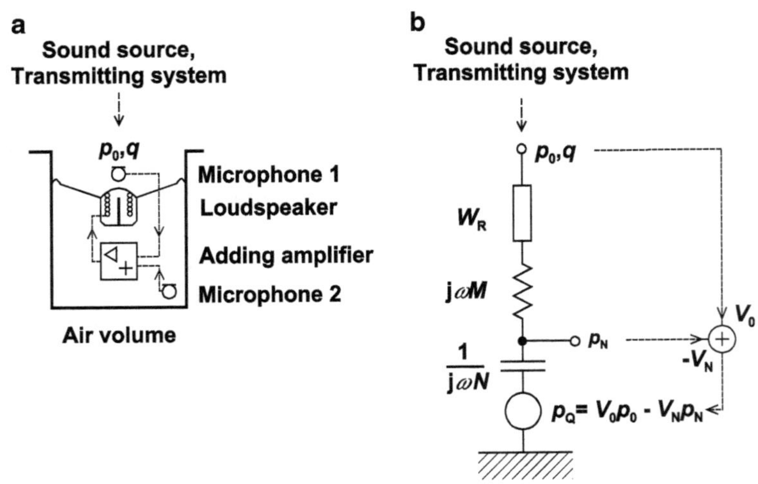

Fig. 8.30 (a) Structure, components and (**b**) Electrical equivalent circuit of an activated and electronically tunable acoustic mass-spring system

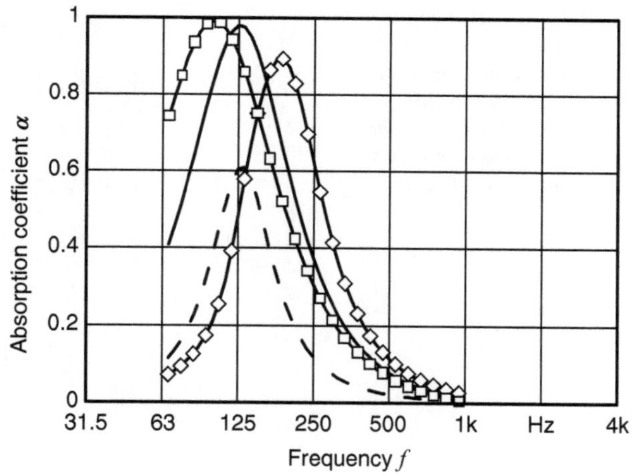

Fig. 8.31 Absorption coefficient α of the mass-spring system shown in Fig. 8.30, calculated for normal sound incidence using Eqs. (8.6) and (8.78); *dashed line* without amplification ($V_0 = V_N = 0$, passive); *open diamond* with amplification (V_0, V_N negative); *solid line* with amplification (V_0); *open square* with amplification (V_0, V_N positive)

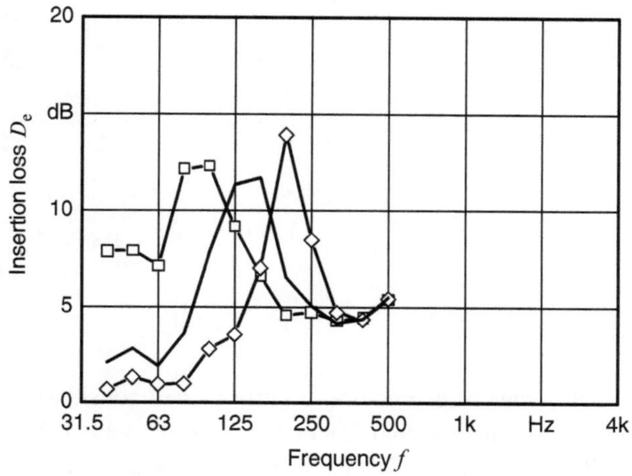

Fig. 8.32 Insertion loss D_e measured in a 25 × 25 cm rectangular duct of an active mass-spring resonator according to Fig. 8.30 with amplification – V_0 only; *open diamond* V_0, V_N negative; *open square* V_0, V_N positive

according to Eq. (8.43). The imaginary part of the impedance vanishes at the resonance frequencies derived from Eq. (8.74) so that at least the requirement for high absorption or damping is fulfilled. For the more general case of an arbitrary termination impedance W_L of the cavity, the impedance at the duct connection is given by [55]

$$W_0 = \frac{W_L cos\frac{\omega L}{c_0} + j\varrho_0 c_0 sin\frac{\omega L}{c_0}}{j\frac{W_L}{\varrho_0 c_0} sin\frac{\omega L}{c_0} + cos\frac{\omega L}{c_0}}. \quad (8.79)$$

The transition from Eqs. (8.79)–(8.43) is easily comprehensible for very high impedances (rigid termination). To describe the active side branch resonators, the impedance of the active mass-spring system Eq. (8.78) has to be inserted into Eq. (8.79) for W_L. The result of a prototype consisting of two side branches connected to a main duct is shown in Fig. 8.33. By activating the control ($V_0 > 0$), the basic frequency of the passive side branch resonator is shifted from 250 to 63 Hz. Thus, activation saves substantially in length, which would be necessary for a passive side branch resonator. Generally, active components once again qualify for the development of highly integrated compact low-frequency absorbers. Their practical application is not necessarily limited to mass-spring systems and quarter-wavelength tubes.

8.9 Micro-Perforated Absorbers

A survey of conventional materials for sound absorbers and structures thereof was given in the previous sections. Not all but many of them employ fibrous or porous materials which passively react to an incident sound field (Sect. 3). On the other hand,

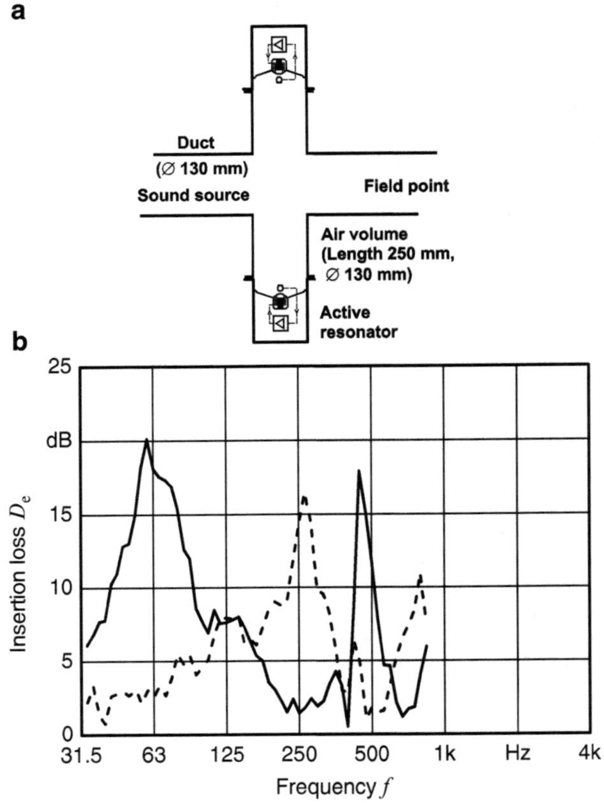

Fig. 8.33 **a** Sketch and **b** insertion loss D_e, measured in a 13 cm exhaust tube of an active quarter-wave side branch resonator pair; *dashed line* without and *continuous line* with amplification V_0

several types of resonators recently gained considerable importance which interact differently with the exciting sound field (Sects. 8.4–8.8). It does not matter if the latter are excited to vibrations in plates, foils or membranes (Sects. 8.5 and 8.8) or in air volumes of differently shaped cavities (Sects. 8.6 and 8.7): Their efficiency can be raised or optimised by inserting or adding a smaller or larger amount of passive damping material. Kurtze 35 years ago [56] was able to prove that ceiling or wall panels made of passive layers as thick as possible according to Sect. 8.3 and Fig. 8.34a behind a perforated sheet with a minimum of 15% perforation ratio could be easily and economically be replaced by similarly perforated metal cassettes, wood or plaster plates with a significantly thinner fleece or cloth lining in front or behind (see Fig. 8.34b).

For a fleece cover with different flow resistances, it is shown in Fig. 8.35 how a broad absorption maximum develops at $d \approx \lambda/4$, a relative minimum is passed at $d \approx \lambda/2$ but at $d \approx \lambda/8$ again $\alpha \approx 80\%$ can be reached for an optimum flow resistance similar to the homogeneous passive absorber according to Fig. 8.3 and Eq. (8.28). Thus, if the distance d is

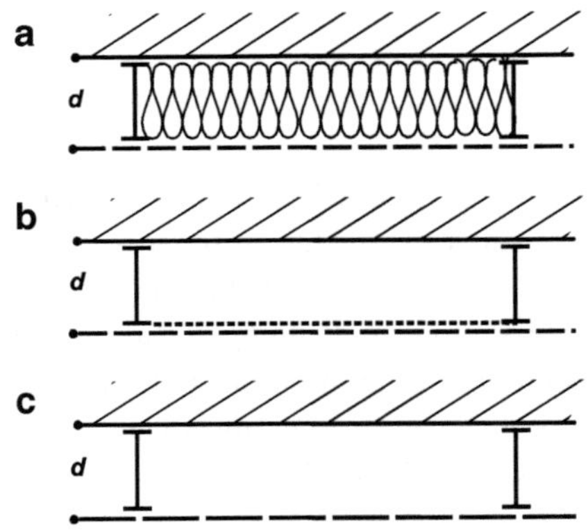

Fig. 8.34 Acoustic panels for walls and ceilings. (**a**) Perforated plates with more than 15% perforation; cavity at least partially filled with porous/fibrous damping material; (**b**) Perforated plates as in a, but covered with a thin porous/fibrous flow resistance; (**c**) Micro-perforated plates or foils with approximately 1% perforation without any damping in front of or behind it

large enough sound absorbers can be realised, especially for room-acoustic purposes, which are broadband even at lower frequencies. Yet, the well-known perforation or slit design of conventional optical and

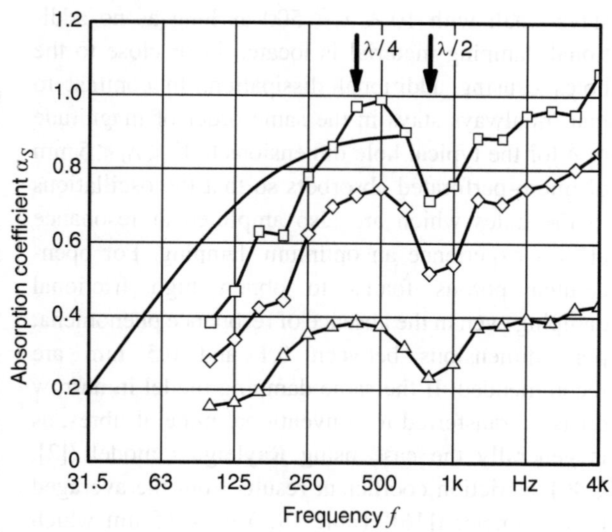

Fig. 8.35 Absorption coefficient α_S measured in a reverberation room of a ceiling with perforated sheet metal cassettes according to Fig. 8.34a and b with $d = 200$ mm, *continuous line* porous/fibrous absorber of same thickness d according to Fig. 8.3; *open square* 7 mm fleece (1 kg m^{-2}); *open diamond* 5 mm fleece (0.5 kg m^{-2}); *open triangle* 0.6 mm fleeced

protective covers still remains visible as typical for conventional porous/fibrous absorbers if the fleece is pointing towards the room, because a certain air flow passage and thus soiling of the perforated cover unavoidably occurs after some time, even when an 'acoustic plaster' is additionally applied on the front-fleece.

With respect to ergonomics and hygienic, Maa [57] brought the development of completely innovative acoustic components on the way with his idea of micro-perforated plates, which do no longer rely on any porous or fibrous damping materials for dissipating acoustic energy (see Fig. 8.34c). Their acoustic efficiency can be adjusted nearly independently of the chosen plate material, which for the first time allows the production of transparent or translucent sound absorbers, made for example of acrylic glass, polycarbonate or even of plane glass [58].

In all variants, meanwhile approved in countless applications, the air oscillates many adjacent holes shown in Fig. 8.36a (a, b) or slits as a mass together with the air in the air gap (d) in front of a usually rigid wall as spring in a faint analogy to the mass-spring system of Helmholtz resonators described in Sect. 8.6. Compared to conventional perforated panel absorbers (Sect. 8.6.1) and slit absorbers (Sect. 8.6.2) [39], however, only a relatively small perforation ratio σ is chosen (preferably in the order of 1%) for micro-perforated absorbers. Even more importantly, the smallest dimension of the holes or slits ($2r_0$) is always made so small that it reaches the same order of

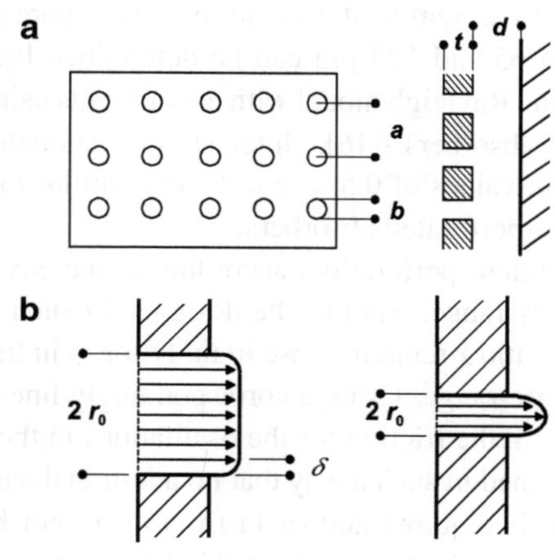

Fig. 8.36 Basic principle of micro-perforated absorbers: **a** Top view and sectional drawing (*schematic*); **b** velocity distribution of oscillating air in large (*left*) and small (*right*) holes or slits

magnitude as the acoustic boundary layer in air (see Fig. 8.36b and Eq. (8.33)).

For all porous absorbers, where the oscillating air is damped by friction, the ratio of the pore dimension normal to the oscillation direction and the size of the boundary layer δ play an important role. For cylindrical holes of radius r_0 (mm), the dimensionless ratio

$$x = \frac{r_0}{\delta} = 0.65 r_0 \sqrt{f}, \tag{8.80}$$

with f (Hz) enables a qualitative statement on how effectively the friction at the walls can damp the oscillations in the holes. For conventional perforated panel absorbers with $2 < r_0 < 25$ mm, the friction stays small with $10 < x < 500$ as long as no additional damping material is located in or close to the holes causing additional dissipation. In contrast to that, r_0 always stays in the same order of magnitude as δ for the typical hole dimensions $0.05 < r_0 < 5$ mm of micro-perforated absorbers so that the oscillations in the holes which are also amplified at resonance always experience an optimum damping. For open-cellular porous foams to obtain high frictional damping even in the absence of resonance phenomena, pore dimensions between 0.1 and 0.5 mm are recommended. If the same damping model in narrow ducts is transferred to conventional mineral fibres, as is generally the case using Rayleigh's model ([2], § 40), a friction coefficient results from the averaged fibre diameter ([16], Table 19.7) of 4–15 μm which considerably deviates from the optimum value of $x \approx 1$. As a matter of fact, an effective pore radius between 65 and 125 μm can be determined by comparing the Rayleigh model with experiments using real fibrous absorbers [16]. Interestingly enough, this results in values of $0.5 < x < 5$ very similar to those of micro-perforated absorbers.

The micro-perforation, according to the envisaged frequency range, can thus be designed in such a way that the ratio x remains close to unity for r_0 in the sub-millimetre-range. Using a correspondingly fine perforation (r_0), the friction for the oscillations in the holes can be tuned in such a way that no additional damping material is required neither in front of, in nor behind the holes or in the cavity behind them to achieve optimum damping in a broad range of frequencies. Using this *inherent* damping and a principle which is completely defined by geometric parameters, micro-perforated absorbers can be calculated precisely by their design parameters and can be tuned to almost any given sound spectrum.

For heat-conducting plates made of metal or glass additional losses due to heat-dissipation in a thermal boundary layer of the same order of magnitude as the acoustic boundary layer ([7], § 30) can be identified. Using the same geometric parameters, micro-perforated absorbers made, for example, of glass should show a larger inherent absorption than those made of acrylic glass. In rough agreement with other authors (e.g. [2]), Maa introduced additional losses to the boundary layer coefficient x which add to the dynamic viscosity η using the value 0.024 g^{-1} s^{-1} so that

$$x = 0.42 r_0 \sqrt{f}, \tag{8.81}$$

replaces Eq. (8.80) for micro-perforated components with substantial heat-conduction.

8.9.1 Micro-Perforated Plates

The theoretical treatment of micro-perforated absorbers and its antecedents in which Velizhanina has played a prominent role was thoroughly expressed in [59]. The wall impedance of a micro-perforated plate with cylindrical holes in the approximation of Maa [57] as shown in Fig. 8.36 according to Eq. (8.6) and normalised to the specific impedance of air is given by

$$\frac{W}{\rho_0 c_0} = r' + j\left(\omega m' - \cot\frac{\omega d}{c_0}\right). \tag{8.82}$$

Here, the term cot $\omega d/c_0$ in Eq. (8.82) describes the fact that the hollow space between the perforated plate and the wall represents strictly speaking a cavity resonator for higher frequencies, in contrast to the simple mass-spring system, which was dealt with in Sects. 8.5.1 and 8.6.1 as a model for the resonance absorber with concentrated elements ($d \ll \lambda$). This would enable a maximum of the oscillation damped with r' for $d = \lambda/4$ provided that the mass per unit area of the oscillating air in the holes m' normalised by $\varrho_0 c_0$ is not too large. On the other hand, cot $\omega d/c_0$ tends to infinity for $d = \lambda/2$ so that at the corresponding frequency and higher harmonics no oscillation and thus in the sense of this model also no absorption occurs. For small angular frequencies ω, the approximation

$$cot\frac{\omega d}{c_0} \approx \frac{c_0}{\omega d}, \tag{8.83}$$

can be used and the frequency of maximum absorption tends towards lower frequencies compared to the rough approximation given by Eq. (8.56). The main difference to a conventional Helmholtz resonator is obviously given by the frequency dependent form of r' and m' in Eq. (8.82) (due to the boundary layer thickness x)

$$m' = \frac{t}{c_0\sigma}K_m;$$
$$K_m = 1 + \frac{1}{\sqrt{9+0.5x^2}} + 1.7\frac{r_0}{t}; \tag{8.84}$$

$$r' = \frac{8\eta}{\varrho_0 c_0}\frac{t}{\sigma r_0^2}K_r \approx 0.34(0.78)10^{-3}\frac{t}{\sigma r_0^2}K_r;$$
$$K_r = \sqrt{1+0.031x^2} + 0.35x\frac{r_0}{t}, \tag{8.85}$$

where the corresponding last term of the sum in the multipliers K_m and K_r which are characteristic for micro-perforated absorbers can be easily identified as an end correction which – similar as for conventional Helmholtz resonators – increases the oscillating mass by the ratio r_0/t but loses importance for small holes (r_0 [mm]) and thick plates (t [mm])

Using the approximations Eqs. (8.83) and (8.84), the micro-perforated absorbers can be characterised analogous to the simple oscillator according to Sect. 8.6.1 by their main resonance frequency

$$f_{\text{MPA}} = 54 \times 10^3\sqrt{\frac{\sigma}{dtK_m}}[\text{Hz}], \tag{8.86}$$

and their normalised impedance

$$Z'_{\text{MPA}} = \sqrt{\frac{tK_m}{d\sigma}}. \tag{8.87}$$

All dimensions are given in millimetre and the correction factor K_{m} is estimated according to Eq. (8.84) which depends on x using Eq. (8.80) or (8.81), respectively, and in turn on the design frequency. Using the ratio $(r' + 1)/Z_{\text{MPA}}$, an expression for the relative bandwidth of the absorber can be derived according to the model described in Sect. 8.5.1. A design tool easy to use has thus become available to tune micro-perforated absorbers for any given demand.

In various publications [59–65], the possibility to absorb sound in nearly invisible holes using micro-perforated surfaces of different materials was made accessible for practical applications. The influence of the geometric parameters a, b, d, t in Fig. 8.36 on the absorption is theoretically and experimentally described in ([1], Part 5). If the holes are chosen too narrow, the modified Helmholtz resonator according to Eqs. (8.84)–(8.87) is 'overdamped'. If they are too large additional damping material, e.g. a fleece, is needed to provide the friction losses required for optimum absorption.

Moving from normal to oblique or diffuse sound incidence, the absorption maximum for $\theta > 0$ is shifted towards slightly higher frequencies and is somewhat reduced according to [57]

$$\alpha = \frac{4r'\cos\theta}{(r'\cos\theta+1)^2 + \left(\omega m'\cos\theta - \cot\frac{\omega d\cos\theta}{c_0}\right)^2}. \tag{8.88}$$

The relative bandwidth in the diffuse field is but slightly increased because not only r' is smaller but also Z'_{MPA}. While theory and experiment are in good agreement at normal sound incidence (<5% difference in $\alpha(0)$), it can often be observed that the theoretical values are slightly exceeded especially for higher frequencies in reverberation room measurements. If the approximation according to Maa is used, one is thus usually on the safe side. This can be quite different if micro-perforated absorbers are used at grazing sound incidence, for example, as resonators in splitter silencers and if the cavity behind the perforated plate is not sub-divided into partitions to make the lining locally reacting. Using a convex curvature towards the room, as depicted in ([59], Fig. 11) the bandwidth of maximum efficiency can be increased, so that it is possible to absorb more than 50% of the incident sound energy over a band of more than two octaves.

If the bandwidth of a single layer of a micro-perforated absorber is not sufficient, two or three layers of micro-perforated plates can be combined preferably with increasing distance in such a way that the higher frequencies are absorbed in the upper layers and the lower frequencies in the lower plates. A practical example [28] with a bandwidth of four octaves is shown in Fig. 8.37.

As was already discussed in Sect. 8.5 on plate resonators, fairly broadband sound absorbers can also be realised by arranging differently tuned modules (e.g. wooden panels with varying wall distance) side by side. If, for instance, the wall distance d of a suspended ceiling made of micro-perforated sheet metal elements is varied [59], some broadband effect can be achieved. Figure 8.38 shows the influence of the wall distance d on the absorption coefficient using

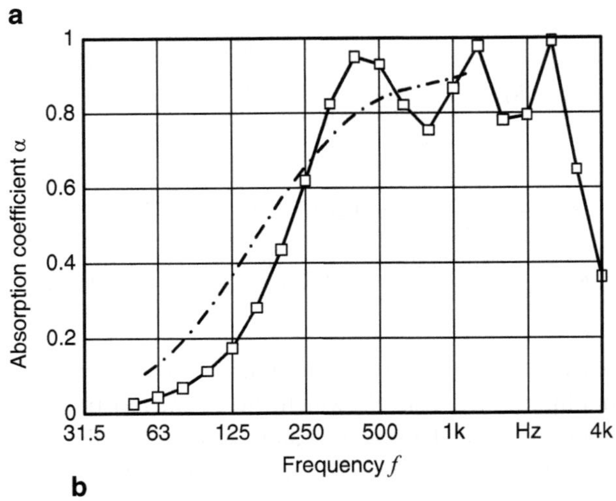

Fig. 8.37 Absorption coefficient α of three layers of micro-perforated aluminium plates calculated for normal sound incidence [28]. For comparison: *dashed dotted line* porous/fibrous absorber of same thickness according to Fig. 8.3

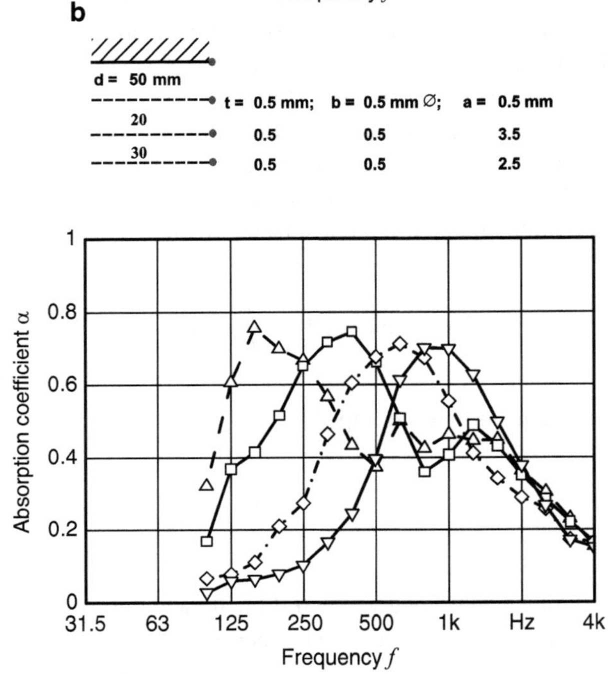

Fig. 8.38 Absorption coefficient α_s of a micro-perforated suspended ceiling with $t = 0.5$ mm, $b = 0.5$ mm, $a = 5$ mm, measured in the reverberation room according to ISO 354 for different wall distances; ∇ $d = 50$ mm, *open diamond* 100 mm, *open square* 200 mm, *open triangle* 400 mm

sheet metal of thickness $t = 0.5$ mm and hole diameters of 0.5 mm with a perforation ratio of less than 1%.

It was noted above that the effective air mass in the holes is inversely proportional to the perforation ratio σ, see Eq. (8.84). By using only small values for σ and large values t to tune the absorption to low frequencies, r' can be increased at the same time according to Eq. (8.85). Yet, the bandwidth will be limited, because Z_{MPA}' also increases according to Eq. (8.87). Finally, an important limitation of this strategy should be mentioned: if m' in the holes is in the same order of magnitude as the normalised mass per unit area $m''/\varrho_0 c_0$ or even larger than that, the designed absorber cannot be excited. That is, if

$$\frac{m'}{m''/\varrho_0 c_0} = \frac{\varrho_0 K_m}{\varrho \sigma} < 1, \tag{8.89}$$

with the density ϱ of a micro-perforated structure with perforation ratio σ becomes a lot larger than unity, it acts like a plate resonator as described in Sect. 8.5.2.

8.9.2 Micro-Perforated Foils

The development of a whole family of quite different types of micro-perforated acoustic elements started in 1993, after initial drilling tests with prototypes of aluminium, by cutting holes and slits into acrylic glass plates using a suitable programmed single laser beam. Hereafter, drilling plastics with a multi-spindle machine was productive. As suspended ceiling elements sheet metals thinner than 1 mm were micro-perforated using punching tools shortly after that. For the application of thin sheet metal and plastic parts for sound attenuation in vehicles, a special slitting technique followed by a roller process made it to series production in the meantime.

Plastic foils can best be micro-perforated by hot needles on a roller; for optimum design, a multitude of holes with $d \approx t$ is needed which amounts to about 250,000/m^2. Since cavity resonances in such a structure with a perforation ratio of 1% can easily be excited, due to the cot-term in Eq. (8.82), the main maximum of the micro-perforated absorber (Fig. 8.39) can be moved further towards lower frequencies for increasing wall distance *d*. Moreover, a relatively large 'ridge' at mid frequencies and a long 'tail' at higher frequencies may be obtained. The resulting large bandwidth of this special Helmholtz resonator where the undamped cavity is made excitable for incident sound waves by a micro-perforated structure resembles that of slit absorbers in Sect. 8.6.2 if its cavity is densely filled with damping material. Due to the condition in Eq. (8.89), such low frequencies can, however, only be reached by such thin foils if they are kept in form by a supporting grid and vibrations of the foil are thus suppressed. If, for example, the frequency range between 250 and 4,000 Hz is targeted which is the dominating band in 'fun-pools' or 'bottling plants', the micro-perforated absorber variant with two layers as shown in Fig. 8.40 offers suitable absorption as the result of a room-acoustical treatment under the roof area of a swimming pool clearly shows, achieving a reduction to half the reverberation time [65].

Transparent micro-perforated absorber foils recommended themselves as linings or venetian blinds especially in front of glass windows or facades [66] if higher costs for facing shells made of acrylic glass [38, 58, 59] should forbid. Translucent, printed or coloured micro-perforated absorber foils are suitable as a low frequency sound-permeable cover for compound panel absorber modules according to Sect. 8.5.3 [11]. Various possibilities to improve noise control and acoustic comfort are offered by spanned ceilings made of PVC foils with a micro-perforation.

8.9.3 Micro-Perforated Structures

All micro-perforated components discussed in Sects. 8.9.1 and 8.9.2 are always air-in-air resonators whose thickness determines according to Eqs. (8.83) and (8.86) not only their resonance frequencies but also according to Eqs. (8.47) and (8.87) the bandwidth of their maximum performance. Acoustically activated foils also stand the test as mid-to-high frequency absorbers in factory halls, atria and auditoria even if they do not form enclosed cavities with a rigid

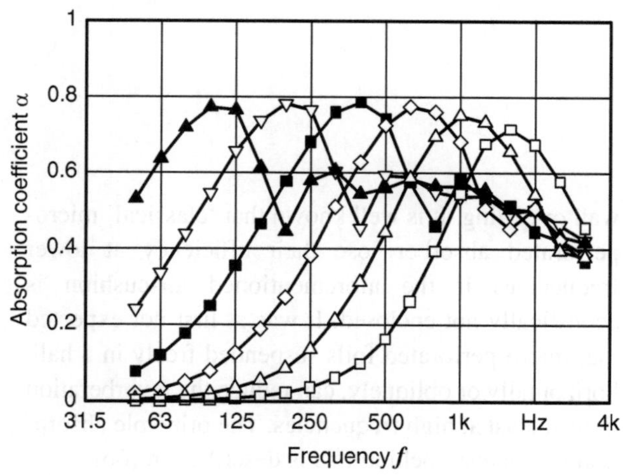

Fig. 8.39 Absorption coefficient α (at normal sound incidence without heat conduction), calculated according to Sect. 8.1 for hole diameter $b = 0.2$ mm, hole distance $a = 2$ mm, foil thickness $t = 0.2$ mm, at distances □ $d = 25$ mm, *open triangle* 50 mm, *open diamond* 100 mm, *filled square* 200 mm, *down open triangle* 400 mm, *closed filled triangle* 800 mm

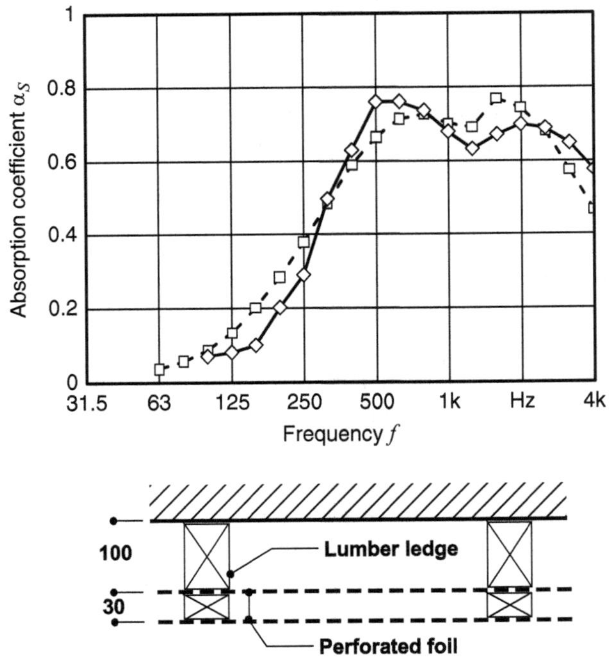

Fig. 8.40 Absorption coefficient α_s of a two-layered micro-perforated foil absorber with a wall distance of $d = 100$ and 130 mm; *open diamond* measurement, *open square* calculation for diffuse sound incidence

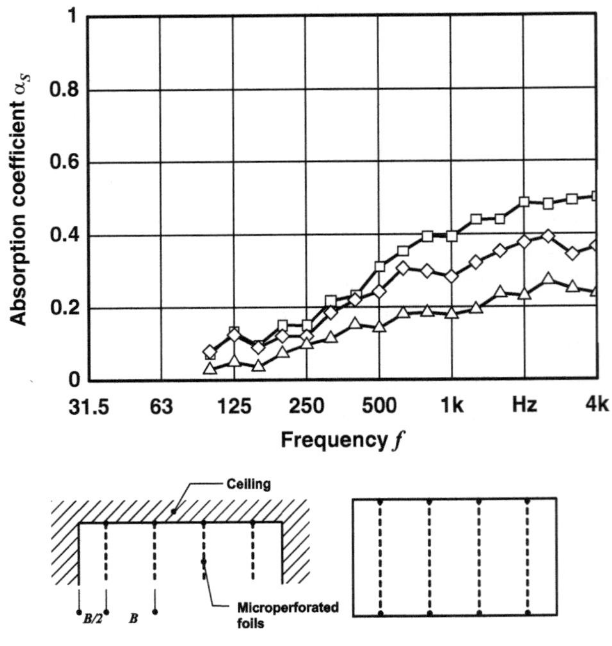

Fig. 8.41 Absorption coefficient α_s for diffuse sound incidence on freely or parallel suspended micro-perforated foils in front of a rigid surface measured in a frame at *open square* $B = 170$ mm, *open diamond* 330 mm, *open triangle* 525 mm

wall or ceiling. It is well known that 'classical' micro-perforated absorber lose their efficiency at lower frequencies if the aforementioned air-cushion is acoustically not enclosed. It was at first not expected that micro-perforated foils suspended freely in a hall, horizontally or obliquely, can reduce the reverberation time at least at high frequencies. The principle absorption mechanism behind this is described in [66].

If micro-perforated planar structures are suspended vertically from a ceiling or as a suspended ceiling simply in parallel, they show a substantial sound absorption at high frequencies in a diffuse field, which is again strongly influenced by all geometric parameters of such a construction (Fig. 8.41). The absorption is in any case higher

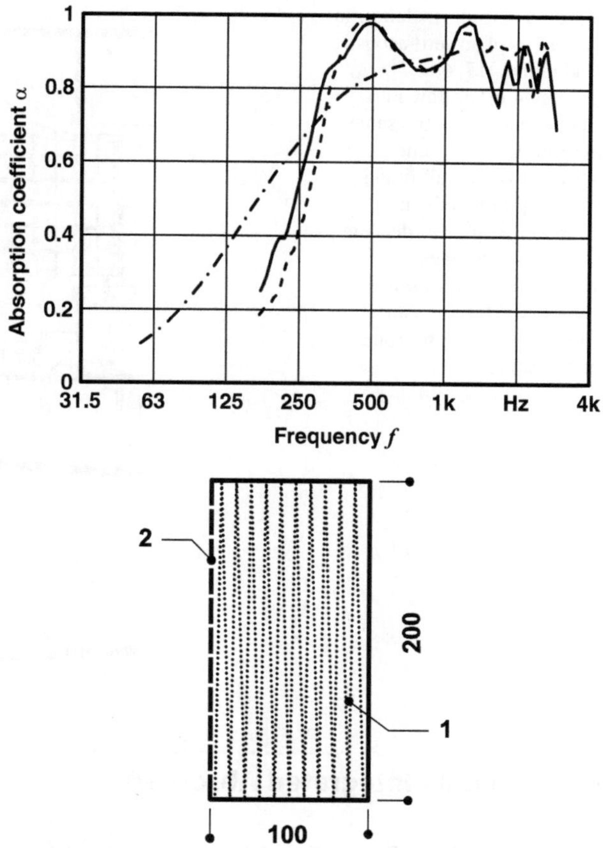

Fig. 8.42 Absorption coefficient α measured at normal sound-incidence of a silencer-package consisting of about 20 layers of micro-perforated foils *1* with perforated sheet metal cover *2*; *continuous line* folded regularly as sketched and *dashed line* folded irregularly. For comparison: *dashed dotted line* porous/fibrous absorber of same thickness according to Fig. 8.3

than for the same construction made of fleece or cloth.

A substantial additional absorption in a room may also be obtained if ventilation ducts are micro-perforated [67]. If air leaks through the holes due to a certain pressure in the duct and thus supporting the ventilation of the room, the absorption is even slightly increased. Additionally, by this the mid- and low-frequency noise emitted by the fan and carried through the duct by the flow can be reduced by a few dB/m. For the external as well as for the internal damping mechanisms, the resting or moving air in the duct acts as a concentrically compressible air cushion, where in one case the externally incident waves and in the other case the approximately plane internally progressing waves excite this substantially modified Helmholtz resonator to oscillate and thus to absorb sound.

Finally, micro-perforated sheet metals or foils cannot only be used to build ducts or volumes to absorb as much sound in their envelopes as possible. It is also advantageous to fill silencers, which conventionally use porous/fibrous material according to Sect. 8.3, with micro-perforated surface structures. Instead of very closely arranged adjacent pores and fibres in the μm-range extremely thin micro-perforated foils which are folded, rolled or 'stuffed' into the cavity may span a sufficiently large number of a more or less uniform distribution of holes or ducts between the partitions of different size.

Silencers, which are completely micro-perforated as schematically sketched in Fig. 8.42, can offer certain advantages: the large internal and external micro-perforated structures (e.g. made of stainless steel) can catch hard strokes or endure continuous vibrations better than relatively brittle mineral fibres or hard foams according to Sect. 8.3. Their nearly closed and smooth surface is less sensitive to soiling and repel liquids. Figure 8.42 shows the absorption coefficient of two prototypes with 100 mm thickness measured in an impedance tube.

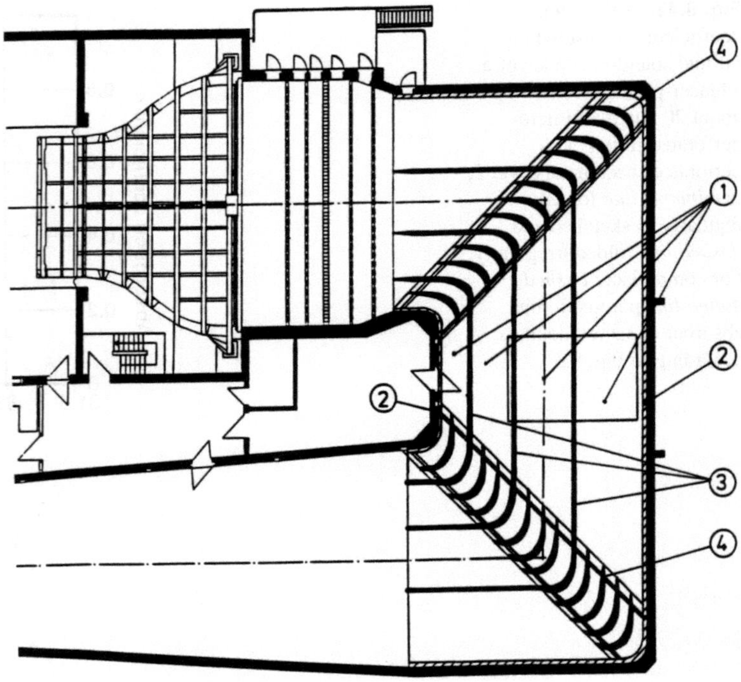

Fig. 8.43 Bending silencer in the corners of an aero-acoustic wind tunnel [17, 68]: *1* Separation of air flow into unequal ducts with the same insertion loss and same pressure loss, *2* wall lining consisting of membrane absorbers, *3* splitter silencers with a 'back-to-back' arrangement of membrane absorbers, *4* turning vanes with profiled and absorbing foam coating

8.10 Highly Integrated Absorbers

In the previous sections, an overview over the different mechanisms and constructions of conventional but also innovative market-ready sound absorbers was given, and the various different physical damping mechanisms served as a logical guideline for the more fundamental explanations. The practice of noise control and even more in room acoustic design is to solve the task of realising a certain weighted sound pressure level as, for example, in Eq. (8.18) or to exactly adjust or safely keep a desired reverberation characteristic as in Eq. (8.12). It is thus important to achieve a given target under the given installation, operating or usage conditions with an effective and valuable choice or even better an optimal combination of suitable absorbers. The expectations can be met by the experienced designer or consultant the better the larger the variety of absorbers he or she has to choose from. Under the given opulence of offerings even for innovative absorber types, however, he or she is not able to rely only on the specifications of the manufacturer. The proposed measurement procedure for low frequencies in a damped reverberation room [23] is not commonly accepted so far. But urgent problems cannot wait until corresponding new standards are passed. New technologies will, no doubt, be employed even without updated standards if they only stand the test in practice.

To eliminate, for instance, the noise of a 3 MW fan of an aero-acoustic wind tunnel at the test room, it is not very wise to press the air through conventional narrow silencer-packages stuffed with fibres. Mid and high frequencies can be damped instead using coated foam profiles [17], which are integrated into the turning vanes to also respect fluid dynamics (see Sect. 8.3.2). The low frequencies can also be absorbed without substantial pressure loss using membrane absorbers as described in Sect. 8.6.3, which according to Fig. 8.43 optimally guide the flow around the corner by their smooth metallic surface along the walls and partitions of the two 180° bends. Since the completion of this pilot project, bent silencers of this or a similar construction are a standard recommendation for silencer design in order to use a hefty 'bending bonus' of up to 15 dB compared to Eq. (8.20) [69].

If, for example shown in [38], a factory hall classified as a historical monument with a lot of glass and stucco shall be converted into an open-plan office or multi-purpose hall, suitable sound absorbers can be mounted invisibly (Fig. 8.44) in the studs of the construction under the floor, behind large projection screens but also on the preserved crane rails and

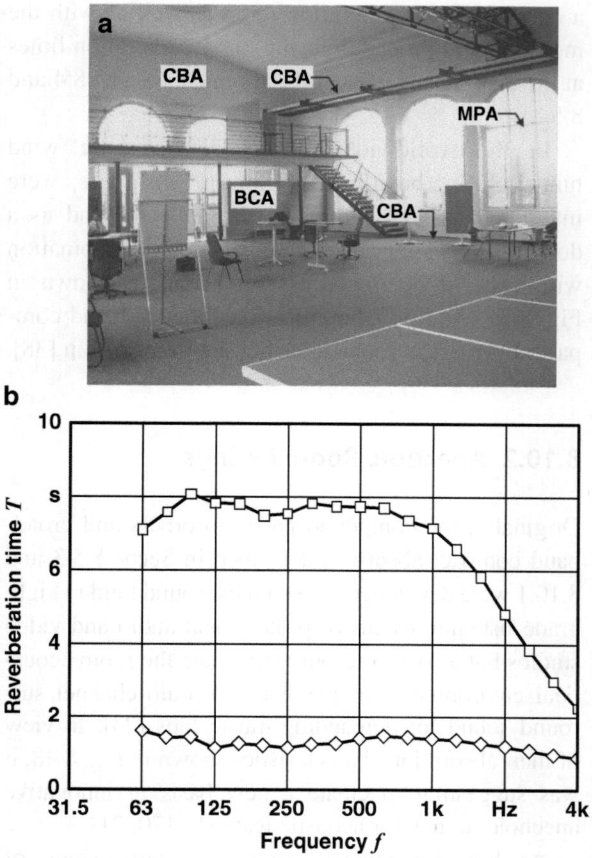

Fig. 8.44 (a) Acoustic design (with invisibly installed absorber-modules) and b reverberation time T in a factory hall classified as a historical monument converted into a training centre [38]; *open square* original condition, *open diamond* final condition

even in front of large windows. All that is needed is to say goodbye to the conventional 20–100 mm thick fibre mat usually covered by a more or less attractive perforated cover as acoustic lining. If it is explained to architects, building owners and users that the room absorption can be invisibly integrated into non-bearing components and outfits or that absorbers may themselves become robust construction elements [38], such planning in due time would perhaps remove the existing aversions against acoustic treatment in environments hard as steel and clear as glass.

But only when the complete audible range from 30 Hz up to a few kHz- range can be completely absorbed by one and the same thin bounding surface (see Sect. 8.10.1), some of the doubters might be convinced of the importance of acoustic measures at high-demand workplaces and in communication rooms. Using plane compact absorbers, where low frequency absorbers according to Sect. 8.5.3 and high-frequency absorbers according to Sect. 8.3.2 are arranged consecutively one behind the other, even anechoic chambers for acoustic precision measurements can be designed with minimum consumption of valuable space (see Sect. 8.10.2).

Since 1996, a successful innovative noise reduction concept is on the market [50]. It replaces the common solution, that is, silencers installed in separate housings *in front of* loud exhaust stacks, by a novel kind of lining integrated *into* the exhaust stack employing stainless steel panels to absorb the dominating low-frequency content of the noise (see Sect. 8.12).

8.10.1 Broadband Compact Absorbers

Figure 8.18 already showed a low-frequency absorber designed for a broad frequency range, where free vibrations of a ‘floating’ steel plate are combined with the damping mechanism of a soft foam plate, which remains accessible for sound at its edges. For this compound panel absorber, the elastic compound between the oscillating plate and the damping layer plays an important role. If a second porous layer is mounted similarly in front of the steel plate as shown in Fig. 8.45, the latter unfolds the absorption at higher frequencies as described in Sect. 8.3.2. With a thickness of 150 mm this optimizes the absorption above 250 Hz as it can be expected using Eq. (8.28). As exemplified in Fig. 8.46, the absorption is strongly increased in reverberation room measurements even over the whole frequency range of interest. Obviously, the damping potential of this combined reactive–passive broadband compact absorber reaches an optimum in this configuration where the softly embedded plate can freely vibrate.

Here, the absolute values of the measured values are not so important. At higher frequencies, values for α larger than 1 can be explained as is commonly known by the conventional arrangement of diffusers in the reverberation room, diffraction at the edges of the test specimen but also with the usage of Sabine’s equation given by

$$\alpha_s = 0.163\frac{V}{S_\alpha}\left(\frac{1}{T_m}-\frac{1}{T_o}\right), \qquad (8.90)$$

with the room volume V (m^3) the surface of the test specimen S_α (m^2) and the reverberation times T_m (s) with and T_o (s) without the test specimen. In contrast to absorption coefficients measured in the impedance tube at normal sound incidence, α values up to 1.2 are nothing to be worried about ([7], § 25).

The results in Fig. 8.46 show that by using a broadband compact absorber which is only 250 mm thick one can cover in practice the whole audio spectrum of interest. The influence of thicker vibrating plates (up to 2.5 mm) on absorption cannot be quantified in such a conditioned reverberation room as well as with the measurement procedure using the reverberation times at the eigenresonances of the room (see Sects. 8.4 and 8.5.3).

In the worldwide quietest aero-acoustic wind tunnel [68], broadband compact absorbers were integrated as wall linings at the collector and as a double-sided silencer in the vanes (in combination with caps of open porous soft foam) as shown in Fig. 8.47. Additional applications of broadband compact absorbers in room acoustics are described in [38].

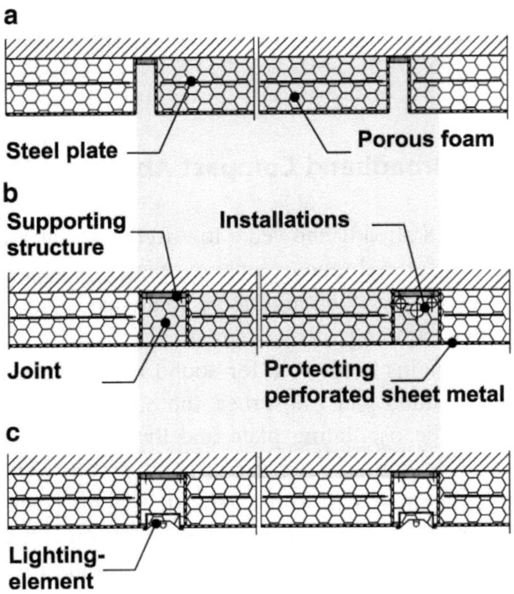

Fig. 8.45 Broadband compact absorber with 1–2.5 mm thick steel plate embedded between 10 and 15 cm thick open porous soft foam plates with **a** open joints **b** fittings covering wires, ducts, etc., **c** integrated lighting elements

8.10.2 Anechoic Room Linings

Originally, the compound panel absorbers and broadband compact absorbers described in Sects. 8.5.3 and 8.10.1 were developed to design the sound field in high-grade listening rooms of professional audio and video studios but also as a resource to create the room acoustical environment for playback of multi-channel surround sound for demanding music fans [23]. In view of their absorption characteristics shown in Fig. 8.48, it was suggesting to create a new basis of innovative anechoic linings for acoustic test cells [70, 71].

At low frequencies where the dimensions of anechoic rooms are comparable to the wavelength a non-uniformly distributed sound field as described in Sect. 8.4 exists. The innovative broadband compact absorber lining takes this into account by distributing the backing resonator plates also *not* uniformly as recommended by the relevant standard ISO 3745 but placing the absorbers with the thickest plates preferably at the edges and corners of the room in contrast to

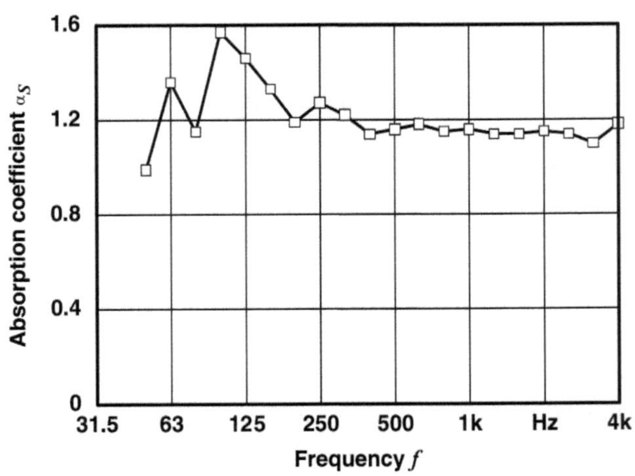

Fig. 8.46 Absorption coefficient α_S of six broadband compact absorber-modules with 1 mm thick steel plate measured in the reverberation room damped as shown in Fig. 8.15, normalised by $S_\alpha \approx 11$ m^2

conventional linings with mineral fibre wedges. At these positions, the room modes can be damped most efficiently. In certain areas, e.g., in front of doors, the heavy vibrating plates can be omitted. Between the broadband compact absorber modules with a preferred surface area of larger than 1 m^2, which are always mounted at a certain distance, ducts and wiring or other installations for the purpose of the room usage can be comfortably integrated. Even lighting elements can be flush-mounted in the lining at the ceiling (see Fig. 8.45b, c).

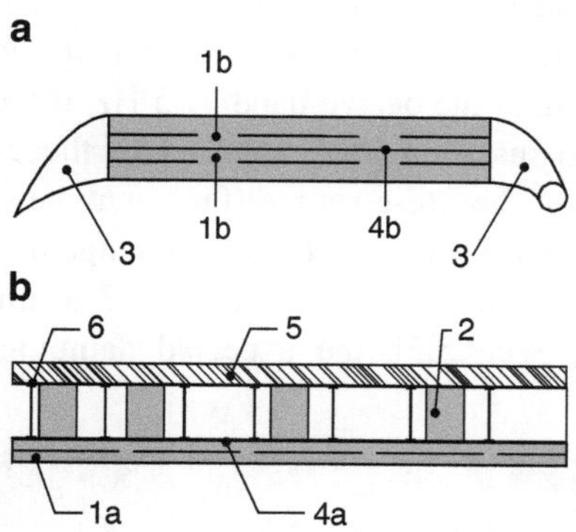

Fig. 8.47 Integration of broadband compact absorber-modules into (**a**) the splitter silencer and (**b**) the wall linings of an aero-acoustic wind tunnel of a car manufacturer [68] *1a* 300 mm thick broadband compact absorber modules *1b* 250 mm thick broadband compact absorber-modules, *2* foam bulkhead for dampening of the cavity *3* foam profiles to optimise the guidance of the flow and the insertion loss *4a* 3 mm thick steel plate *4b* 2 mm thick steel plate *5* concrete wall or ceiling

The broadband compact absorber lining in its simplest form leads to an utterly low reaction of the room on the direct field of a broadband radiating sound source. Its absorption coefficient at normal sound incidence does not necessarily reach 99% as postulated by the cited standard. Qualification tests on postulated absorbers with a size in the metre range, which act at low frequencies due to vibrating plates are difficult to perform. It can also be doubted whether this requirement is sensible at all [72]. It is obviously derived from the situation in a plane wave field (as in the impedance tube), where the standing wave field is a result of forward and backward progressing waves. According to Eq. (8.8) and Table 8.1, a ripple of only ±1 dB results if the absorption coefficient is 0.99 In reality, however, spherical waves are more likely to propagate especially in anechoic rooms of which the amplitude is not constant, but changes continuously along the propagation path with approximately −20 lg r [73].

Yet, in some special cases, where investigations of narrow-band radiating sources are performed, it is sensible to combine the broadband compact absorber lining with a structured porous absorber [74]. If the

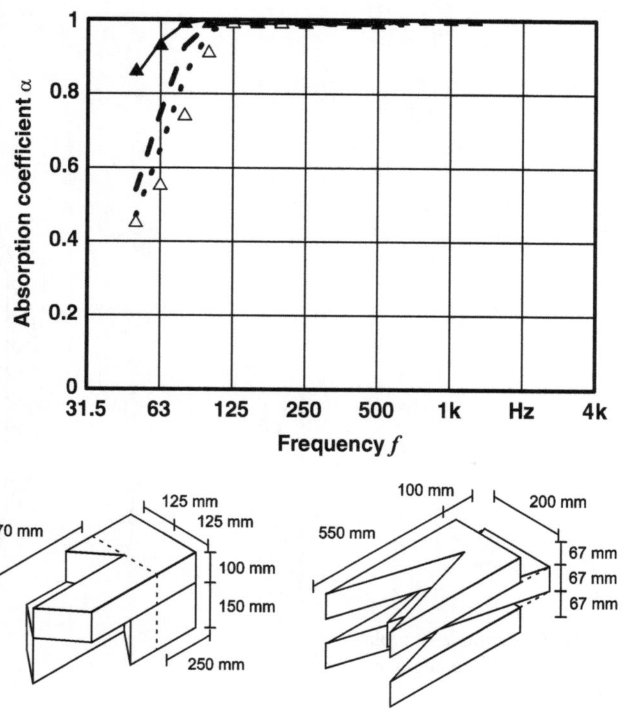

Fig. 8.48 Absorption coefficient α of differently structured porous or fibrous sound absorbers measured in the impedance tube with 200 × 200 mm cross-sectional area: Asymmetrically structured absorber (*left*): *dotted line* 520 mm, *dashed line* 650 mm, 780 mm, Wedges (*right*): *open triangle* 680 mm, *filled triangle* 1075 mm [70]

entrance of the sound into the porous layer is facilitated by a special shape as sketched in Fig. 8.48 the absorption coefficient of conventional absorbers can even be outranged or the same absorption can be obtained at low and high frequencies with significantly thinner lining. This is achieved by the fact that these innovative asymmetrically structured absorbers not only absorb passively according to Sect. 8.3 but interact positively with the sound field due to their specific material parameters.

8.10.3 Angular Stack Silencers

Pipelines and exhaust stacks are ideal wave guides as long as they are new, i.e. when they are not soiled. Frequently exhaust stacks are used, which contain a thin ‘inlay’ made of high-grade material (stainless steel), which is held by an external tube bearing the static and dynamic loads. If these internal components guiding the flow and the sound in the exhaust stack are not shaped cylindrically but formed as a polygon, the sound can excite the plane internal surface elements to perform vibrations. The polygonally designed angular stack silencers (Fig. 8.49) can be tuned by choosing their geometric and material parameters in such a way that starting in the octave band 31.5 Hz, if required, a broadband insertion loss, matched to the excitation spectrum of the noise source (fan), may be obtained. Using the α values in the frequency range between 30 and 300 Hz measured in a 0.65 × 1.7 m impedance tube [70, Sect. 5.3], the expected damping can be

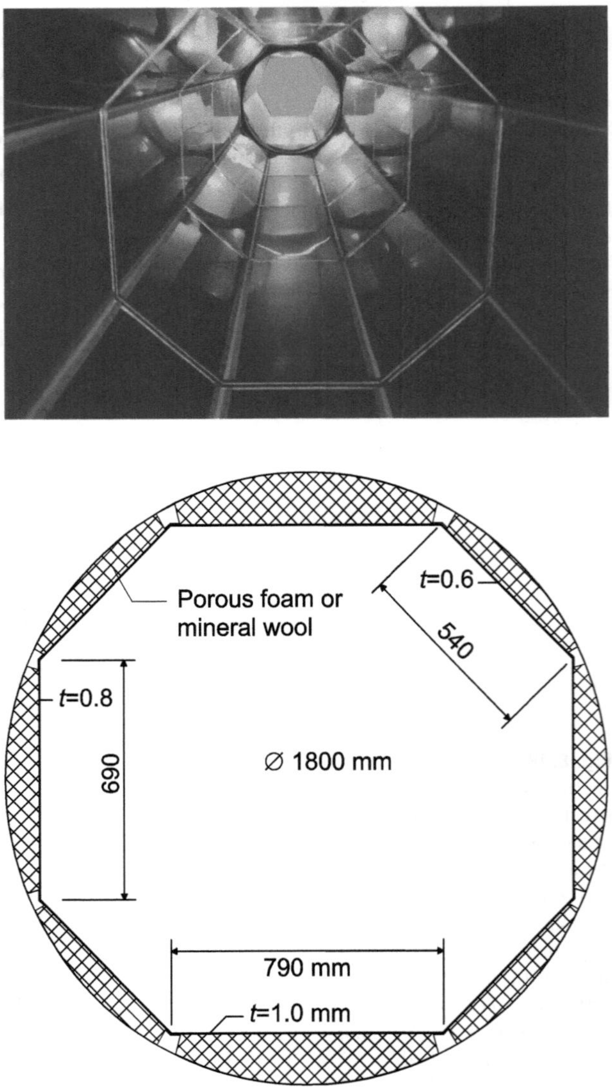

Fig. 8.49 Example of a plate arrangement in the angular stack silencer [75]

roughly estimated using Piening's equation (8.20). Meanwhile, the layout of the angular stack silencer can quickly and effectively be done with a computer program. The absorption coefficient of the plate resonances is calculated by using the impedances of their eigenvibrations Eq. (8.49):

$$W_P = \frac{1}{\sum\limits_m \sum\limits_n \frac{1}{W_{mn}}}; \quad m, n = 1, 3, 5, \ldots \qquad (8.91)$$

The transfer impedance W_T takes into account the effect of the air volume according to Eq. (8.43) and the impedance of the termination behind the air volume [29]. Wave resistance, thickness of the absorber and propagation constant in the absorber are respected as in Eq. (8.61). Using $W = W_P + W_T$, the absorption coefficient α is then calculated for normal sound incidence as given in Eq. (8.8).

The sketch in Fig. 8.49 shows an arrangement of an angular stack silencer tuned to 50–200 Hz consisting of two plates with 0.8 and 1 mm thickness and four plates with 0.6 mm thickness. With the aid of Piening's equation adapted for the angular stack silencer of length l, the insertion loss D can be estimated by

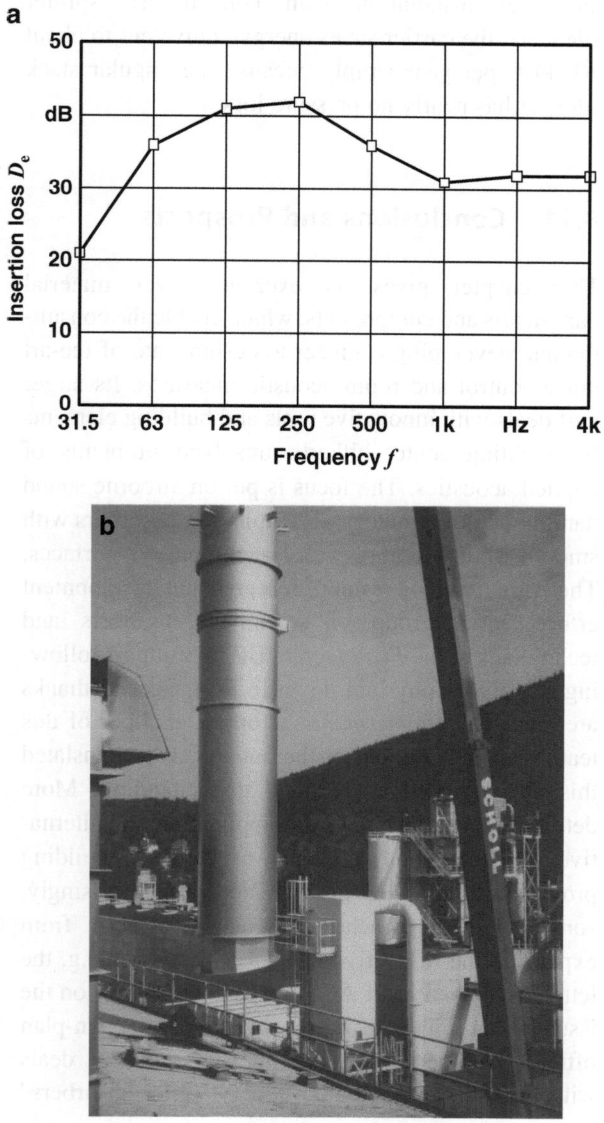

Fig. 8.50 **a** Insertion loss D_e and **b** assembly of the first (5 m long) exhaust stack with angular stack silencer [76]

$$D = \frac{1.5l}{S} \sum_i n_i \alpha_i U_i; \quad i = 1, 2, 3, \ldots \quad (8.92)$$

Figure 8.50 shows the measured insertion loss of an exhaust stack with such an integrated silencer [76]. In the photograph, taken before the final installation, the polygonal fitting of the exhaust stack can easily be seen. For this relatively high damping requirement especially at 63 and 125 Hz, the complete length of the exhaust stack (31 m) was made good use of. Angular stack silencers are installed along 20 m with an inner diameter of 1.6 m and an outer diameter of 1.8 m and additional porous absorbers are installed along 11 m of the stack length. The silencer is cleaned two or three times a year with a water jet. Compared to an older installation with conventional splitter silencers, the carrier saves energy equivalent to about 30000 € per year simply because the angular stack silencer has nearly no pressure loss.

8.11 Conclusions and Prospects

This chapter gives an overview over material parameters and components, which enable the consulting and developing engineer to design state-of-the-art noise control and room acoustic measures. Its larger part deals with innovative tools and building elements for tackling acute, hitherto unsolved problems of applied acoustics. The focus is put on airborne sound damping at low frequencies employing absorbers with smooth and non-abrasive, largely closed surfaces. These are mainly a result of research and development efforts in a group of scientists, engineers and technicians at the *Fraunhofer* IBP in Stuttgart following the philosophy laid down in [77]. Special thanks are due to *S. Zimmermann*, a former member of this team and PhD student of the authors, who translated this chapter with a profound understanding. More detailed reports on manifold applications of alternative absorbers and silencers in concrete building projects are compiled in [78]. Not very surprisingly, some novel approaches provoke resistance from experts in the respective field of activity, see e.g. the letters to the editor of ACTA ACUSTICA [79] on the issue of an adequate noise abatement in open-plan offices. Another somewhat puzzling concept deals with more recently created novel "corner absorbers" [80] for improving the acoustic comfort in spaces for intense communication, e.g. restaurants, education and conference rooms – yet another area where the relevance of the low-frequency problem [37] has been overlooked so far. The collaboration with numerous industrial licensees and development partners is deeply appreciated, who enabled the introduction of ideas and prototypes into the market in accordance with the strategies outlined in [81].

References

1. Fuchs HV et al. Schallabsorber und Schalldämpfer. Innovatorium für Maßahmen zur Lärmbekämpfung und Raumakustik; Parts 1–6. *Bauphysik*, 24(2):102–113, 2002. 24(4):218–277, 2002. 24(5):286–295, 2002.24(6):361–367, 2002. 25(2):80–88, 2003. 25(5):261–270, 2003
2. Cremer L and Müller HA (1982) Principals and Applications of Room Acoustics, Volume 2. Applied Science, London, New York
3. Tennhardt HP (1984) In: Fasold et al. (eds) Taschenbuch der Akustik, chapter Messung von Nachhallzeit, Schallabsorptionsgrad und von Materialkennwerten poröser Absorber. Verlag Technik, Berlin
4. Fasold W, Sonntag E, Winkler H (1987) Bau- und Raumakustik. Verlag für Bauwesen, Berlin
5. Fasold W, Veres E (2003) Schallschutz und Raumakustik in der Praxis. Verlag für Bauwesen, Berlin
6. Kuttruff H (2000) Room acoustics. E & F N Spon, London
7. Cremer L and Müller HA (1982) Principals and Applications of Room Acoustics, Volume 1. Applied Science, London, New York
8. Zwicker E, Fastl H (1990) Psychoacoustics. Springer, Berlin
9. Fuchs HV et al (2001) Creating low-noise environments in communication rooms. Appl Acoust 62(12):1375–1396
10. Drotleff H et al (2004) Acoustic improvements of the working conditions for musicians in orchestra pits. In: Proceedings CFA/DAGA, Strasbourg, 525–526
11. Drotleff H et al (2004) New room acoustic design concept for rehearsal rooms. In *Proceedings CFA/DAGA, Strasbourg*, 483–484
12. Zha X, Fuchs HV, Drotle H (2002) Improving the acoustic working conditions for musicians in small spaces. Appl Acoust 63(2):203–221
13. Lotze E (2006) In: Schirmer W (ed) Technischer Lärmschutz, chapter Luftschallabsorption. VDIVerlag, Düsseldorf
14. Frommhold W (1996) Technischer Lärmschutz, chapter Absorptionsschalld ämpfer. VDI-Verlag, Düsseldorf
15. Möser M (2004) Engineering Acoustics. Springer, Berlin
16. Mechel FP (1994) In: Heckl M, Müller HA (eds) Taschenbuch der Technischen Akustik, chapter Schallabsorption. Springer, Berlin
17. Künstner R, Potthoff J, Essers U (1995) The aero-acoustic wind tunnel of Stuttgart University. In: SAE Conf., Detroit, Paper 950 625

18. Pfeiffer G et al (1997) Modern testing techniques in BMW powertrain development – three new special test rigs. ATZ Worldwide 99(7/8):12–15
19. König N (1993) Schaumgips oder Gipsschaum? Eigenschaften und Einsatzmöglichkeiten eines neuen Baustoffes. Bauphysik 15(2):33–36
20. Gödeke H, Fuchs HV (1998) REAPOR – Sintered open-pore glass foam as a high-strength sound absorber. Glastech Ber Glass Sci Technol 71(9):282–284
21. Zimmermann S (2003) Control Performance of Active Absorbers in Enclosed, Harmonic Sound Fields. PhD thesis, Technische Universität Berlin (D83), GCA, Herdecke
22. Zha X et al (1999) Measurement of an effective absorption coefficient below 100 Hz. Acoust Bull 1/2:5–10
23. Fuchs HV, Zha X, Pommerer M (2000) Qualifying freefield and reverberation rooms for frequencies below 100 Hz. Appl Acoust 59(4):303–322
24. Bies DA, Hansen CH (1996) Engineering Noise Control. E&FN Spon, London
25. Morse PM, Ingard KU (1968) Theoretical acoustics. McGraw-Hill, New York
26. DIN 52 212. Bestimmung des Absorptionsgrades im Hallraum, 1961
27. Cummings A (1992) The effects of a resonator array on the soundeld in a cavity. J Sound Vibr 154(1):25–44
28. Zhou X, Heinz R, Fuchs HV (1998) Zur Berechnung geschichteter Platten- und Lochplatten-Resonatoren. Bauphysik 20(3):87–95
29. Kiesewetter N (1980) Schallabsorption durch Platten-Resonanzen. Gesundheitsingenieur 101(1):57–62
30. Ford RD, McCormick MA (1969) Panel sound absorbers. J Sound Vibr 10(3):411–423
31. Chladni EEF (1787) *Entdeckungen über die Theorie des Klanges*. Leipzig
32. Hurlebaus S, Gaul L, Wang JTS (2001) An exact series solution for calculating the eigenfrequencies of orthotropic plates with completely free boundary. J Sound Vibr 244(5):747–759
33. Lord Rayleigh (1945) The Theory of Sound. Dover, New York, 2 edn of 1896 reprinted with corrections
34. Schirmer W (1996) Technischer Lärmschutz, chapter Schwingungen und Schallabstrahlung von festen Körpern. VDI-Verlag, Düsseldorf
35. Leistner M and Fuchs HV (2004) Supplementary acoustic measures in the conference centre of the federal ministry of economy and labour. In: Proceedings CFA/DAGA, Strasbourg, 487–488
36. Zha X et al (2004) Room acoustics for 4 uses – Großes Haus Staatstheater Mainz. In: Proceedings CFA/DAGA, Strasbourg, 523–524
37. Fuchs HV, Zha X, Drotleff H (2005) Relevance and treatment of the low frequency domain for noise control and acoustic comfort in rooms. ACTA ACUSTICA 91 (4):920–928
38. Drotleff H, Zhou X (2001) Attractive room acoustic design for multipurpose halls. ACTA ACUSTICA 87(4):500–504
39. Leistner P, Fuchs HV (2001) Schlitzförmige Schallabsorber. Bauphysik 23(6):333–337
40. Frommhold W, Fuchs HV, Sheng S (1994) Acoustic performance of membrane absorbers. J Sound Vibr 170(5): 621–636
41. Hunecke J and Zhou X (1992) *VDI Berichte 938*, chapter Resonanz- und Dämpfungsmechanismen in Membran-Absorbern, 187–196. VDIVerlag, Düsseldorf
42. Trochidis A (1982) Körperschalldämpfung mittels Gas- oder Flüssigkeitsschichten. ACTA ACUSTICA 51(4):201–212
43. Ackermann U, Fuchs HV (1989) Noise reduction in an exhaust stack of a papermill. Noise Control Eng J 33(2): 57–60
44. Fuchs HV (1993) Sound absorbers for heavily fouling exhaust gas systems. Cement Lime Gypsum ZKG 46B(7): E185–E191
45. Vér IL (1992) In: Beranek LL, Ver IL (eds) Noise and Vibration Control Engineering, chapter Enclosures and wrappings. Wiley, New York
46. Teige K, Brandstätt P, Frommhold W (1996) Zur akustischen Anregung kleiner Räume durch Luftauslässe. Zeitschrift für Lärmbekämpfung 43(3):74–83
47. Fuchs HV (1993) Generation and control of noise in water supply installations, Part 3, Rating and abating procedures. Appl Acoust 39(3):165–190
48. Galaitsis AG, Vér IL (1992) In: Beranek LL, Ver IL (eds) Noise and Vibration Control Engineering, chapter Passive silencers and lined ducts. Wiley, New York
49. Munjal M (1987) Acoustics of Ducts and Mufflers. Wiley, New York
50. Fuchs HV, Eckoldt D, Hemsing J (1999) Alternative sound absorbers for industrial use: Acousticians on the quest for alternative attenuators. VGB Power Tech 79(3):76–78
51. Leistner P, Meneghin G, Sklenak B (2000) Aktive Schalldämpfer für Raumklimageräte. Heizung Lüftung/Klima Haustechnik 51(7):42–45
52. Leistner P, Castor F (2000) Aktive Schalldämpfer für *Absauganlagen*. Luft- und Kältetechnik 36(8):366–368
53. Lenk A (1977) Elektromechanische Systeme. Systeme mit konzentrierten Parametern, volume 1. Verlag Technik, Berlin
54. Bay K-H, Krämer MM, and Brandstätt P (2004) Compact silencer for heating systems. In: Proceedings CFA/DAGA, Strasbourg
55. Krüger J and Leistner P (1998) Wirksamkeit und Stabilität eines neuartigen aktiven Schalldämpfers. ACTA ACUSTICA, 84(4):658–667
56. Kurtze G (1977) Wirtschaftliche Gestaltung von Schallschluckdecken. VDIZeitschrift 119(24):1193–1197
57. Maa D-Y (1975) Theory and design of microperforated panel sound absorbing constructions. Scientia Sinica 18(1):55–71 (In Chinese)
58. Fuchs HV, Zha X (1997) Acrylic-glass sound absorbers in the plenum of the Deutscher Bundestag. Appl Acoust 51(2):211–217
59. Fuchs HV, Zha X (1995) Einsatz mikro-perforierter Platten als Schallabsorber mit inhärenter Dämpfung. ACTA ACUSTICA 81(2):107–116
60. Fuchs HV (1997) Metal ceilings for optimum sound absorption. Environ Eng, 10(7)
61. Fuchs HV (2000) Helmholtz resonators revisited. ACTA ACUSTICA 86(3):581–583
62. Fuchs HV et al (2001) Raum-Akustik mit System. Glas-Verarbeitung 8(3):59–64
63. Maa D-Y (1987) Microperforated panel wideband absorbers. Noise Control Eng J 29:77–84

64. Wack R and Fuchs HV (2004) On the use of micro-perforated sails in assembly rooms. In: Proceedings CFA/DAGA, Strasbourg, pages 485–486
65. Zha X et al (1998) Microperforated absorbers for noise control in enclosures. Proceedings Euro-Noise, pp 699–704
66. Fuchs HV, Zha X (2006) Micro-perforated structures as sound absorbers – a review and outlook. ACTA ACUSTICA 92(1):139–146
67. Leistner P, Hettler S (2004) Sound absorption of microperforated duct systems. HVAC Res 10:265–274
68. Brandstätt P, Fuchs HV, Roller M (2002) Novel silencers and absorbers for wind tunnels and acoustic test cells. Noise Control Eng J 50(2):41–49
69. Eckoldt D (1995) Neuartiger Umlenk-Schalldämpfer auf dem Dach. Luft- und Kältetechnik 31(4):188–189
70. Fuchs HV (2010) Schallabsorber und Schalldämpfer. Springer, Berlin
71. Fuchs HV, Zha X, Babuke G (2004) Broadband compact absorbers for anechoic linings. In: Proceedings CFA/DAGA, Strasbourg, pages 955, 956
72. Eckoldt D, Fuchs HV, Frommhold W (1994) Alternative Schallabsorber für reexionsarme Messräume. Zeitschrift für Lärmbekämpfung 41(6):162–170
73. Zha X, Fuchs HV, Späh M (1998) Ein neues Konzept für akustische Freifeldräume. Rundfunktechnische Mitteilungen 42(3):81–91
74. Dreyer W et al (2003) The new Volkswagen Acoustics Centre in Wolfsburg. ATZ Worldwide 105(3):11–15
75. Fuchs HV (2001) Alternative fibreless absorbers – New tools and materials for noise control and acoustic comfort. ACTA ACUSTICA 87(3):414–422
76. Eckoldt D, Fuchs HV (1999) Erfahrungen mit in den Schornstein integrierten Schalldämpfern. Zeitschrift für Lärmbekämpfung 46(6):214
77. Fuchs HV (2002) Innovative sound absorption products – New tools and materials for noise control and acoustic comfort. In: Pandalalai (ed) Recent Research Developments – Sound & Vibration, part 1:203–239. Transworld Research Network, Kerala
78. Fuchs HV (2012) Applied Acoustics: Concepts, Absorbers, and Silencers for Acoustical Comfort and Noise Control. Springer, Berlin
79. Fuchs HV (2006) Letter to the editor on "Simple model for the acoustical design of open-plan offices". ACTA ACUSTICA 92(1):181
80. Fuchs HV, Lamprecht J, Zha X (2011) Zur Steigerung der Wirkung passiver Absorber: Schall in Raumkanten schlucken! Gesundheits-Ingenieur 132(5):240–250
81. Fuchs HV (2001) From advanced acoustic research to novel silencing procedures and innovative sound treatments. ACTA ACUSTICA 87(3):407–413

Structure-Borne Sound, Insulation and Damping

9

G. Muller

9.1 Introduction

In noise and vibration control, the most important tasks are the minimization of the excitation and of the propagation of sound in structures. Furthermore, the transfer from structure-borne sound into airborne sound – the vibration-induced radiation of sound – should be kept as low as possible. For this purpose, various possibilities are available:

(a) Reduction of forces, momentums, pressure, movements, etc. that excite the structure; reduction of high frequent structure-borne sound by avoiding quick changes in flows or loads; avoiding of self-excitation mechanisms, e.g., stick-slip processes; reduction of slackness; elimination of loose parts (that can cause rattling), etc.
(b) Insulation of structure-borne sound, which means reflection of structure-borne sound at certain locations by means of elastic layers (springs), changes of the transmitting medium, changes of the dimensions in the transfer path, blocking masses or other discontinuities.
(c) Enlargement of the distance, making use of the fact that with increasing distance the energy is distributed over a larger area and thus the energy density decreases.
(d) Damping of structure-borne sound that means transfer of vibration energy into heat, e.g., by the use of appropriate damping materials or by friction at contact surfaces.
(e) Reduction of the radiation, e.g., by reducing the radiating surfaces or the radiation efficiency.
(f) Active control.

Of course, for optimal results, in practical applications the various methods have to be combined appropriately.

As in bounded structures structure-borne sound results from a complicated combination of different wave types (see Sect. 1.8), it is difficult to handle the issues related with insulation and damping of structure-borne sound in a general way. Therefore, just some important special cases are considered in the following.

In case of 'compact' bodies, e.g., compact machines, and in case of low frequencies the structure's dimensions are often considerably smaller than the wavelengths; this means that the individual structures can be either considered as rigid masses or as massless springs.

Beamlike or surface structures also permit simplifications. As in these cases, at least one dimension typically is considerably smaller than the wavelengths, it can often be assumed that the response is determined by bending waves. Other wave types just have to be considered in exceptional cases.

Just in the case of very high frequencies – especially in the ultrasonic range – it is necessary to consider all different wave types; this case will not be

†In Memoriam G. Müller

G. Muller (✉)
Lehrstuhl für Baumechanik, Technische Universität München, Arcisstrasse 21, 80333 München, Germany
e-mail: gerhard.mueller@bv.tum.de

G. Müller and M. Möser (eds.), *Handbook of Engineering Acoustics*,
DOI 10.1007/978-3-540-69460-1_9, © Springer-Verlag Berlin Heidelberg 2013

considered in the following (we refer here to the summarising literature [1–4] and to Chap. 21).

9.2 Insulation at Low Frequencies (Elastic Mounts)

In practical applications, the elastic mount is one of the most important elements for the insulation of structure-borne sound. It consists of elastic elements (rubber- or metal springs, soft cork, foams, etc.) that decouple either the system of excitation (e.g. machine) or the system, which has to be insulated from structure-borne sound (e.g., building area with high requirements), see Fig. 9.1. By means of an elastic mount, the force of excitation will be reduced for frequencies 1.4 times higher than the resonant frequency f_0. However, elastic mounts are related with amplifications near to the resonance frequencies. Due to this frequency dependent behaviour, the time function of the force below the mount caused by a knocking sound is altered. It is less abrupt and 'smoother' with a lower tuning frequency. However, the total impulse transmitted via the elastic mount remains unchanged.

The first task when designing an elastic mount is the choice of an appropriate elasticity, determining the tuning frequency f_0 sufficiently below the frequency range of interest. The second task is the choice of an appropriate material for maximising the insulation and avoiding unwanted frequency dependent minima of the insertion loss [5–8].

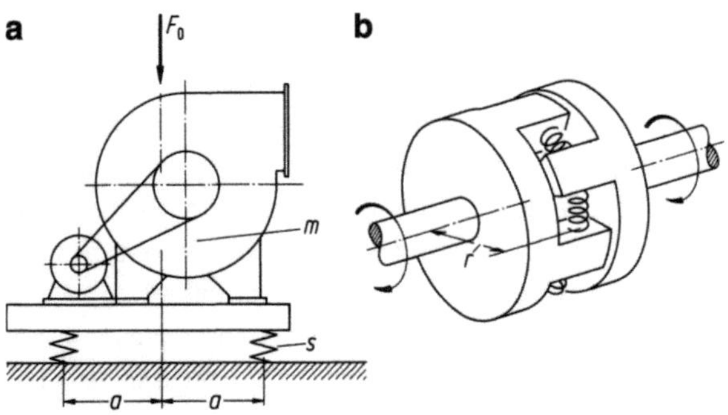

Fig. 9.1 Examples for elastic supports (**a**) vertical movement; (**b**) Rotational movement

9.2.1 Tuning Frequency

The stiffness s of the chosen springs and the effective mass m of the elastically mounted system are the decisive parameters for the insulation effect of an elastic mount. For a simple system with one degree of freedom, the tuning frequency is given by (see Figs. 9.1 and 9.7):

$$f_0 = \frac{1}{2\pi}\sqrt{s/m}\quad \text{Hz}. \qquad (9.1a)$$

For rotating movements (Fig. 9.1b), the tuning frequency is given by:

$$f_0 = \frac{1}{2\pi}\sqrt{s_r/\Theta}\quad \text{Hz}, \qquad (9.1b)$$

(s – sum of the stiffness of the elastic elements in N/m, m – effective mass in kg, s_r – rotatory stiffness Nm/rad, Θ – moment of inertia in kgm^2).

In case of plate-like structures with a mass/area m'' in kg/m^2 mounted on a continuous elastic support with a stiffness/area s'' in N/m^3, the tuning frequency results in:

$$f_0 = \frac{1}{2\pi}\sqrt{s''/m''}\quad \text{Hz}. \qquad (9.2)$$

Here, the bending stiffness of the plate-like structure is neglected.

In case the isolated machine is positioned directly on the springs and the total mass of the machine vibrates (dynamic mass = maschine's mass), the following simple relationship between the tuning frequency and the static deformation of the spring elements ξ (see Fig. 9.2) can serve for a first estimate:

$$f_0 = 5/\sqrt{\xi/\text{cm}}\quad \text{Hz}. \qquad (9.3)$$

Here, the stiffness is derived from the static deformation. However, it has to be considered that the dynamic stiffness of springs often is significantly higher than their static stiffness, so that Eq. (9.3) often returns too low values.

Insulation of structure-borne sound by simple elastic mounts often is not sufficient. The insulation for very noisy machines in the vicinity of apartments or working places can be improved by placing them on double elastic mounts. The systems thus obtained have two tuning frequencies, which can be calculated according to

$$f_{\mathrm{I,II}}^2 = \frac{1}{2}\left[\left(f_1^2 + f_2^2 + f_3^2\right) \pm \sqrt{\left(f_1^2 + f_2^2 + f_3^2\right)^2 - 4f_1^2 f_2^2}\right]. \quad (9.4)$$

Figure 9.3 indicates the respective parameters.

9.2.2 Elastic Mounts in Practice

9.2.2.1 Stiffness

In the linear case – which can often be assumed – stresses and strains are proportional. The spring stiffness s is then given by a load independent constant (F – load, ξ – deformation):

$$s = F/\xi. \quad (9.5a)$$

In the special case of homogenous spring elements of cross-sectional surface S and thickness d consisting of a single material with the Young's modulus E, s is given as

$$s = ES/d. \quad (9.5b)$$

Here, the Young's modulus is assumed independently of the load. With this assumption Eq. (9.3) respectively Fig. 9.2 is obtained.

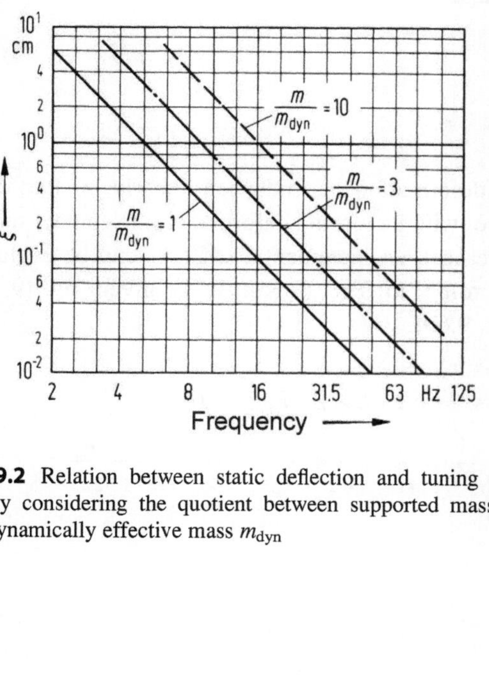

Fig. 9.2 Relation between static deflection and tuning frequency considering the quotient between supported mass m and dynamically effective mass m_{dyn}

Often, this requirement is not fulfilled, e.g., in case of tuning frequencies below the frequency of excitation, which necessitates highly loaded spring elements. Then the spring becomes stiffer with increasing loads, so that non-linear spring characteristics are obtained. In those cases, the tuning frequency is determined as the tangent of the relevant spring characteristics the corresponding stiffness [see Eq. (9.1a)] and given by:

$$s = \frac{\Delta F}{\Delta \xi}. \quad (9.5c)$$

Generally, the stiffness given in Eq. (9.5c) is larger than the stiffness resulting from the total load and the total displacement. Therefore, often the tuning frequency decreases with increasing mass down to a certain value and possibly increases again for higher stresses. The minimal achievable tuning frequency is given in Fig. 9.4 and Table 9.2 for some cases (for rubber sheets bonded between steel plates this value is approximately around 5 Hz, for compressive loads and 3 Hz for shear loads). Table 9.1 and Figs. 9.4 and 9.5 give examples for spring stiffness' and spring characteristics. E.g. in Table 9.1[1] the spring stiffness per cm length is given for various rubber stiffness' (in Shore A). In case an element with B = 50 mm, H = 45 mm, L = 120 mm and $Sh\ A$ = 55 is loaded with 2,000 N (pressure load), the resulting stress amounts 2,000/(BL) = 33 N/cm^2. The corresponding spring stiffness according to Eq. (9.1a) is given by 1,160 (L/cm) $\approx 1.4 \times 10^6$ N/m.

The characteristics of Ω – springs, which are often applied at long machines or line-like structures, are given in Fig. 9.4. Compared to punctual spring elements, they have the advantage of providing some load distribution. Here, the tuning frequency results from the spring stiffness per length s' in N/m^2 and the relevant mass per length m' in kg/m:

$$f_0 = \frac{1}{2\pi}\sqrt{\frac{s'}{m'}}\ \mathrm{Hz}. \quad (9.6a)$$

Figure 9.5 shows the spring characteristics for two-dimensional planar mounts that are applied below foundations. Table 9.2 contains further data about

[1] Loads above the values given in the table should not be chosen for rubber sheets bonded between steel plates.

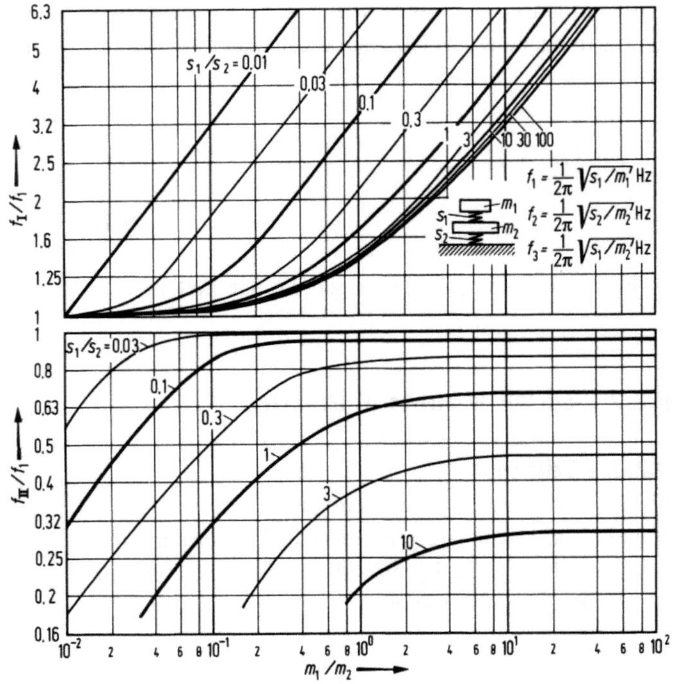

Fig. 9.3 Diagram for the evaluation of tuning frequencies for double elastic mounts

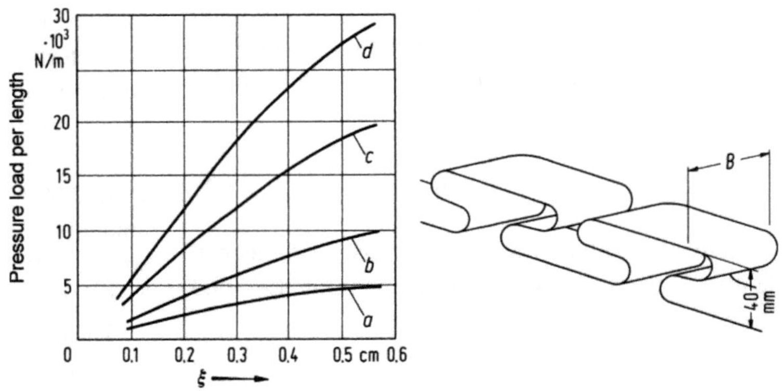

Fig. 9.4 Characteristics of Ω-springs

Curve	*a*	*b*	*c*	*d*
Width *B* (cm)	3	3	10	10
Load (kN/m)	5	10	20	30
Smallest tuning frequency (Hz)	7	7	7	7
Stiffness in the linear range (N/cm²)	100	200	400	600

stiffness. It indicates the Young's modulus. Often the stiffness per surface area (Young's modulus per thickness corresponding to Eq. (9.5b) is given. Table 9.2 does not contain values of the Young's modulus for solid rubber, which means for plates without wholes, ribs or slubs, where the flexibility is determined by the lateral deformation of the material. The dependency of the stiffness on the form factor accounts for this effect (Fig. 9.6a).

In case of a rigid contact of two non-planar bodies the contact stiffness is relevant. Though in this case, the deformation depends non-linearly on the acting force (with increasing deformation the contact surface increases), an incremental stiffness can be evaluated according to Eq. (9.5c). It is given by [9, 10] (Fig. 9.6b).

$$s = \left(6R * E*^2\right)^{1/3} F_{\text{con}}^{1/3} = 2aE*,$$
$$\frac{1}{R*} = \frac{1}{R_1} + \frac{1}{R_2}; \frac{1}{E*} = \frac{1-\mu_1^2}{E_1} + \frac{1-\mu_2^2}{E_2}. \quad (9.6b)$$

Table 9.1 Stiffness and tuning frequency for rubber-metal parts under loading

Dimensions (mm)			Shore-hardness (Sh A)	Compressional stress (N/cm²)				
B	*H*	*h*		10	20	30	40	50
50	45	25	40	640 (18)	680 (13)	700 (11)	720 (9)	720 (8)
50	45	25	55	1,150 (24)	1,160 (17)	1,160 (14)	1,160 (12)	1,200 (11)
50	45	25	65	1,640 (29)	1,660 (21)	1,660 (17)	1,800 (15)	1,900 (14)
50	70	50	40	240 (11)	240 (7.5)	240 (6)	240 (5.5)	240 (5)
50	70	50	55	380 (14)	400 (10)	400 (8)	400 (7)	420 (6.5)
50	70	50	65	600 (17.5)	600 (12)	600 (10)	600 (8.5)	640 (8)
100	60	30	40	2,000 (22.5)	2,000 (16)	2,000 (13)	2,000 (11)	2,000 (10)
100	60	30	55	2,600 (30)	2,700 (21.5)	2,700 (18)	2,700 (15)	2,800 (14)
100	60	30	65	5,200 (36)	5,400 (26)	5,400 (21.5)	5,500 (19.5)	5,800 (17)
Dimensions (mm)			Shore-hardness Sh A	Shear stress (N/cm²)				
B	*H*	*h*		5	10	20		
50	45	25	40	80 (9)	–	–		
50	45	25	55	150 (12.5)	70 (6)	60 (4)		
50	45	25	65	220 (15)	150 (8.5)	120 (5.5)		
50	70	50	40	30 (5.5)	24 (3.5)	–		
50	70	50	55	55 (7.5)	50 (5)	35 (3)		
50	70	50	65	90 (9.5)	80 6.5)	70 (4)		
100	60	30	40	200 (10)	170 (6.5)	140 (4)		
100	60	30	55	350 (13)	320 (9)	300 (6)		
100	60	30	65	520 (16)	520 (11.5)	520 (8)		

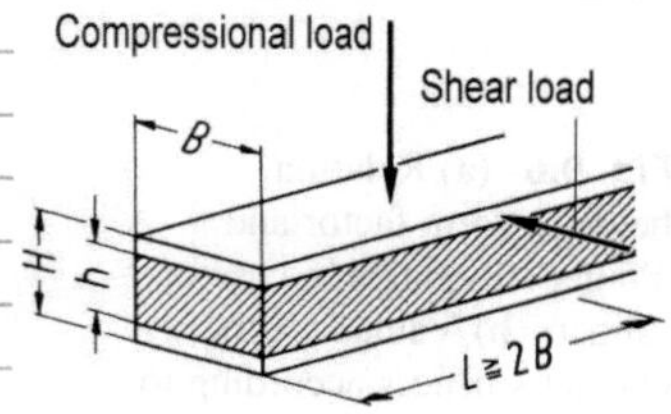

Remark: The values without brackets give the stiffness in N/cm² per cm length. The values in brackets give the tuning frequencies in Hz, in case the compressional, resp. shear load is just caused by the weight of the isolated component

Table 9.2 Young's modulus, maximal permissible load and lowest tuning frequency of some materials (thickness between 3 and 6 cm)

	Young's modulus (N/mm²)	Maximal load (N/mm²)	f_0 (Hz)
Mineral fibre plates	0.15–0.4	0.01	20
Coconut fibre plates	0.25–0.5	0.01	20
Plates made out of balls in foam plastic	0.3–0.6		
Closed porous foams	0.45–0.7		
Foamed polystyrene	0.3–3.0	0.01	
Expanded polystyrene	5.0–24.0	0.05–0.25	10–25
PUR Elastomer[a] G	0.18–0.36	0.01	17
PUR Elastomer R	0.35–0.75	0.025	12
PUR Elastomer L	0.35–1.1	0.05	12
PUR Elastomer M	1.0–2.0	0.10	12
PUR Elastomer P	2.2–3.6	0.20	12
PUR Elastomer V	4.5–6.5	0.40	12
Rubber fibre plates. 6010 BA	0.35–1.7	0.05	22
Rubber fibre plates. 6010 SH	1.0–2.5	0.1	18
Soft fibre plates	10.0	0.01	20
Mineral bound woodwool plates	6.0–17.0		
Cork plates, soft	10.0	0.05	20
Cork plates, medium	15.0	0.25	20
Cark plates, stiff	30.0	1.0	20
Elastomer barings[b] (without reinforcement)	50.0	2.4–5.0	< 8

[a]Sylomer
[b]Form factor 2 (see Fig. 9.6a)

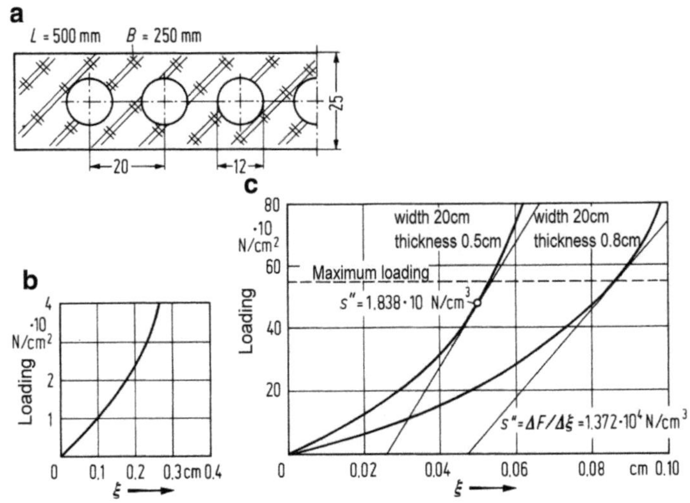

Fig. 9.5 Stress–strain relationship for plane supports. (**a**, **b**) Perforated rubber plate; (**c**) Rubber plate without holes

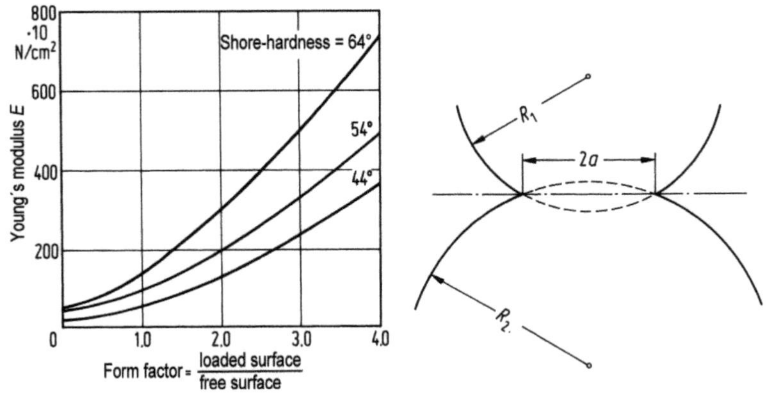

Fig. 9.6 (**a**) Relation between form factor and Young's modulus in case of rubber; (**b**) Values for the contact stiffness according to Eq. (9.6a)

(R_1, R_2 – radii of curvature of both bodies in the contact area; one of the two radii can also be infinite, E_1; E_2, μ_1, μ_2 – Young's moduli and Poisson values, a – radius of the contact area, F_{con} – contact force).

In the design of elastic mounts, a careful assessment of the load applied to the individual elements is crucial. That means that equally designed elements should be positioned in equal distances from the axes of gravity of the machine. If this is not possible, elements with different stiffnesses can be applied. Furthermore, all connections like, e.g., tubes and shafts have to be decoupled by means of elastic connecters. Especially in the case of low tuning frequencies, this is often complicated: E.g., a tuning frequency of 5 Hz is related with a static displacement of 1 cm. An oscillating force that amounts to 10% of the machinery weight leads to a movement with an amplitude of approximately 1 mm. In case the tuning frequency is in the vicinity of the frequency of excitation, the oscillation might become considerably larger (resonance case). Elastic connectors of tubes and shafts have to withstand such amplitudes typically relevant for fatigue, which can be even amplified due to leverage effects.

9.2.2.2 Dynamic Mass

When applying the model of a single degree of freedom system [Eq. (9.1a)], it is assumed that the elastically mounted machine consists of a rigid mass in which each point moves with the same vertical amplitude or the same angle of rotation. Compact machines like engines or machine tools fulfil this requirement in the often most relevant low frequency range. In this case, the mass of the machine is identical to the dynamic mass. However, there are numerous constructions where large static forces are transmitted via elastic mounts, but just relatively small masses determine the tuning frequency. The mounts can even strut the constructions. For example, elevators are acting with their total weight including the cabin and the counterweight on the spring elements. The tuning frequency of the elastic mount, however, is determined by the mass, which is directly located above the springs. The same holds in case of elastic mounts of rails (Fig. 9.8). Also, here the spring elements below the rails are deformed by the total mass of the train. Due to the soft springs between the bogie and the wagon, the tuning frequency is determined by the mass of the wheel sets and a piece of the track. In case of very long or large equipment (elastically mounted beams, plates, crane runways, printing machines or buildings), the dynamically relevant mass can be considerably smaller than the total mass. This is caused by the fact that distant parts are not rigidly coupled (this holds for distances of above half a wavelength of the corresponding waves). Here, typically bending waves are considered. In engineering practice, Eq. (9.1a) still can be applied by introducing a dynamically active mass $m_{\rm dyn}$ in which just the part of the mass is considered that shows the vibration of the upper surface of the elastic mount. It is difficult to calculate this value, which, in addition, is frequency dependent. It has to be considered that the lower limit of the tuning frequency resulting from the static load (Fig. 9.4, Table 9.2) is then increased by $\sqrt{m/m_{\rm dyn}}$.

In case the impedance of the elastically mounted structure above the spring is known, the dynamic mass can also be estimated by:

$$m_{\rm dyn} \approx |Z_m|/\omega. \tag{9.6c}$$

This formula just holds for mass impedances yet not for spring impedances.

9.2.2.3 Special Mounts

Elastic mounts on systems that carry out rotating movements, especially elastic shaft couplings, can be treated similar to simple mass spring systems [Eq. (9.1b)]. However, in practice high static loads occur (e.g., the transmitted moment) and as frequently not sufficient data describing the dynamic behaviour of the torsion springs are available, the evaluation of s_r and Θ often is difficult. If the total vibration results from a superposition of different types of movement, the evaluation of one tuning frequency is not possible. In such cases, the frequencies have to be calculated by the methodof Finite Elements or with the help of MDOF (multi degree of freedom systems).

Especially in the vehicle industry hydraulic, electrodynamic or pneumatic active mounts are used. Here actively forces are induced which are optimised such that the amplitudes of movement or the transmitted oscillating forces are minimised (see Chap. 12).

9.2.3 Insulation Related with Elastic Mounts

The simplest description of the effect of an elastic mount is given by the level difference ΔL_K. This value returns the difference of the structure borne sound´s third- or octave-band spectra above and below the elastic mounts. The most interesting description of the obtained insulation is given by the insertion loss ΔL_I. It is defined by the reduction of the original level of structure-borne sound due to the insertion of the elastic mounts. The ratio of the forces that act above and below the elastic mount that means F_0/F_1 respectively 20 lg (F_0/F_1) is a further value. All three values have their specific justification. They must be never confounded as they can differ considerably (Fig. 9.8).

For the calculation of the individual values, the assumption of sinusoidally varying time functions (harmonic excitation) is recommended. That means, a spectral analysis of all values has to be carried out. Furthermore, point impedances [11] which relate the force of excitation F to the resulting velocity v at the point of excitation must be introduced [see Eq. (9.21a)][2]:

$$Z = F/v. \tag{9.7}$$

In case of a pure sinusoidal (harmonic) excitation, F and v are complex magnitudes and mostly frequency dependent. For some cases, Z is given in Table 9.3 [11–13].

[2] Analogously for rotating movements the moment impedance can be defined as the relationship between the moment of excitation and the corresponding angular velocity.

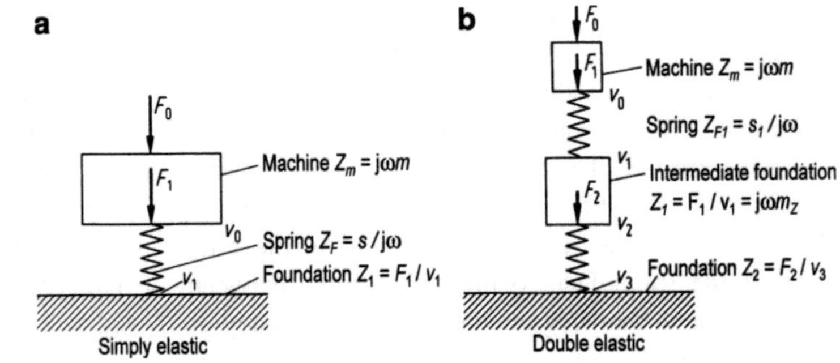

Fig. 9.7 Definitions regarding Eqs. (9.8a)–(9.10)

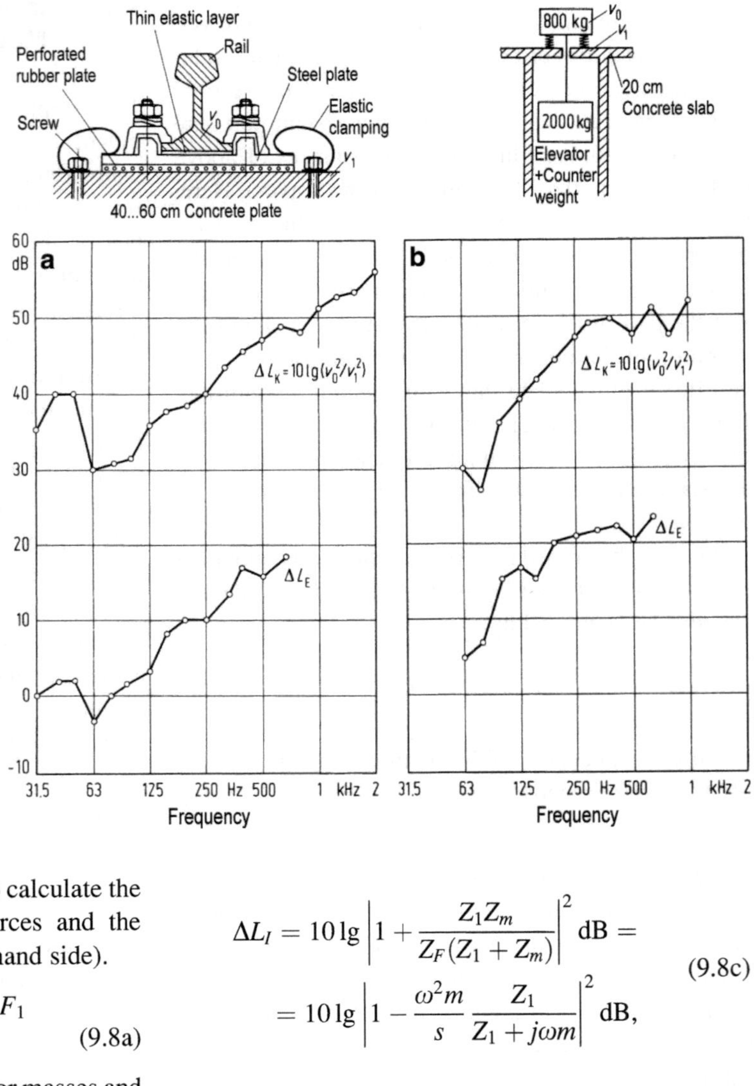

Fig. 9.8 Measurement result at elastics supports. (**a**) Level difference ΔL_K and insertion loss ΔL_E for a elastically mounted subway rail; excitation by a passing train; (**b**) Level difference ΔL_K insertion loss ΔL_E for an elastically mounted elevator machine; excitation by normal use

The concept of impedances permits to calculate the relationship between the oscillating forces and the velocities of vibration (see Fig. 9.7 left hand side).

$$F_0 - F_1 = Z_m v_0;\ (v_0 - v_1) Z_F = F_1;\ F_1 = Z_1 v_1. \tag{9.8a}$$

Introducing the impedance formulas for masses and springs returns:

$$\Delta L_K = 10\lg v_0^2/v_1^2\,\text{dB} = 10\lg|1 + Z_1/Z_F|^2\,\text{dB} = 10\lg|1 + j\omega Z_1/s|^2\,\text{dB}, \tag{9.8b}$$

$$\Delta L_I = 10\lg\left|1 + \frac{Z_1 Z_m}{Z_F(Z_1 + Z_m)}\right|^2 \text{dB} = 10\lg\left|1 - \frac{\omega^2 m}{s}\frac{Z_1}{Z_1 + j\omega m}\right|^2 \text{dB}, \tag{9.8c}$$

$$10\lg\frac{F_0^2}{F_1^2}\text{dB} = 10\ \lg\left|1 + \frac{Z_m}{Z_F} + \frac{Z_m}{Z_1}\right|^2 \text{dB}, = 10\ \lg\left|1 - \frac{\omega^2 m}{s} + \frac{j\omega m}{Z_1}\right|^2 \text{dB}. \tag{9.8d}$$

The equations show that there can be enormous quantitative differences between the various values. This is especially the case (see Fig. 9.8) if $|Z_1| \gg |Z_m|$, corresponding to a lightweight machine on a heavy foundation. In this case, the elastic mount provides a certain insertion loss. However, due to the decoupling of the device, the machine's vibrations (above the springs) will increase significantly so that we can obtain $\Delta L_K \gg \Delta L_I$.

The insertion loss of already installed elastic mounts can be assessed by applying a force above the mount and measuring the level difference ΔL_K. Then a further measurement is carried out by exciting at the foundation in the 'wrong' direction. The lower value ΔL_K obtained returns a rough approximation for the insertion loss (the lower value is taken separately in each frequency band).

The equations and the figures show that the spring stiffness should comply both conditions $s/\omega \ll |Z_m|$ and $s/\omega \ll |Z_1|$; that means the spring element must be much softer than the supported mass and it should be also softer than the supporting structure. Furthermore, the elastic element should have very little mass. It can be shown that above the first spring resonance that is around

$$f_K = \pi f_0 = \sqrt{m/m_F} \quad \text{Hz}, \tag{9.9}$$

significant reductions of the insulation occur (m_F – total mass of the applied springs; f_0 is given by Eq. (9.1a)). Spring resonances can occur in case of steel springs already around 100 Hz. The impact of spring resonances can be reduced by introducing damping on the spring. Rubber springs or additional damping by applying of anti-drone coating onto steel spring are more advantageous than viscous dampers installed parallel to the springs.

Relationships to those for elastic mounts can be established for double elastic mounts. Equation (9.8a) then has to be replaced by

$$\begin{aligned} F_0 - F_1 = Z_m v_0; (v_0 - v_1) Z_{F1} = F_1; \\ F_1 = Z_1 v_1; v_2 = A v_1; \\ (v_2 - v_3) Z_{F2} = F_2; F_2 = Z_2 v_3. \end{aligned} \tag{9.10}$$

The respective definitions are given in Fig. 9.7b. A corresponds to the change of amplitude in the intermediate foundation. Typically $A \approx 1$.

The level differences and the insertion loss derived from Eq. (9.10) show the effect of both occurring resonances (see Eq. (9.4) and Fig. 9.3). At high frequencies, double elastic mounts show a high insulation that typically is limited by bypasses (tubes, cables, etc.) or by the transmission of airborne sound.

9.3 Insulation of Structure-Borne Sound

9.3.1 Decrease with Distance

Similar to airborne sound, structure-borne sound is also reduced with increasing distance between emitter and receiver. Apart from the damping caused by the transfer of vibration energy into heat, increasing distance is related with the fact that the energy is distributed on an increasing surface, so that the 'energy density' becomes smaller. This is the reason why one-dimensional structures (long beams, tubes, etc.) almost show no decay with increasing distance if no energy is deviated via mounting points, etc. Typically elastically mounted tubes transmit structure-borne sound almost without attenuation over long distances.

Apart from a relatively small nearfield where the relations differ, in large two-dimensional structures (large plates or – grids) the decay of structure-borne sound is proportional to the circumference of an envelope line. This leads to a decay of 3 dB when doubling the distance.

Three-dimensional structures (large bodies or large buildings) show a decrease which is inversely proportional to the envelope surface, which means 6 dB when doubling the distance (see Sect. 1.8.1).

Those rules show that it is wrong to indicate the distance-related decay by a value given in dB/m, which would correspond to an exponential decay. E. g., in case of a well-compressed soil under a dynamic force (induced by a vibrator or a machine with an unbalanced rotating mass, etc.) even at low frequencies (below 200 Hz) a level decrease of some dB/m can occur, whereas in the same soil and in the same frequency range the decrease in the vicinity of a large source (e.g., underground train) on a large distance from the source can be significantly lower than 1 dB/m. Similar observations can be made in buildings, where in the vicinity of the point of excitation the structure-borne sound is reduced around 4–6 dB per floor, whereas at a larger distance (above the third to the fourth floor), a reduction of just 2–3 dB per floor is observed.

9.3.2 Material Changes in the Cross-section and Deviations

When waves hit on a discontinuity in the material or in the dimensions of the wave guide, the energy is partly reflected so that the transmitted energy is reduced. This process has been investigated extensively especially for longitudinal or bending waves in beams and plates [Eq. (9.11a)] in order to characterise their transmission. The corresponding descriptive values are defined as follows:

Transmission factor:

$$T = v_2/v_{1+}. \tag{9.11a}$$

Transmission efficiency:

$$\tau = P_2/P_1 = \left|\frac{m'_2 c_2 v_2^2}{m'_1 c_1 v_{1+}^2}\right| = |T^2| \frac{m'_2 c_2}{m'_1 c_1}. \tag{9.11b}$$

Transmission loss:

$$R = 10 \lg(1/\tau) \text{ dB}. \tag{9.11c}$$

The level difference at both sides of the discontinuity is given by:

$$\Delta L = 10 \lg \frac{v_1^2}{v_2^2} \text{dB} \approx R + \left[10 \lg \frac{\rho_2 S_2 c_2}{\rho_1 S_1 c_1} + 10 \lg(2-\tau)\right] \text{dB}. \tag{9.11d}$$

(P_1, v_{1+} – power respectively velocity, impinging on the discontinuity, P_2, v_2 – corresponding values radiated from the discontinuity, m'_1, c_1, respectively m'_2, c_2 – mass per length and group velocity in front, respectively behind the discontinuity).

In these definitions, it is assumed that the propagation is not further influenced or hindered. This means behind the discontinuity, no reflections of the waves occur. The transmission into a finite weakly damped beam or plate will be treated in Sect. 4.4. Formulas for the transmission efficiency τ are given in Table 9.4 and shown in Fig. 9.9 for the most important cases.

If beams or plates are interrupted by an elastic layer or by an additional mass,[3] two positions of discontinuities lie close together; their effect will be noted just above a lower frequency limit $f < f_u$. Approximate formulas for f_u and the transmission efficiency τ are given in Table 9.4. The formulas do not indicate all effects, e.g., frequencies in which $\tau = 0$ are not taken into account. Furthermore, the transitions from bending waves to longitudinal waves and vice versa, which occur in unsymmetrical or skew arrangements, are not considered. Those phenomena are described in the literature [11].

9.4 Damping of Structure-Borne Sound

Sound energy that is transformed into heat is not audible any longer. Thus, the concerted installation of damping is a very effective and largely applied method for the attenuation of structure-borne sound. This is done by applying appropriate materials or constructions that transfer the sound energy – optimally near to the sources – into heat.

Additional damping can be provided by applying materials with sufficient interior damping, by friction at contact surfaces or by adding damping materials like high polymers or sand to weakly damped constructions (e.g. metal constructions). As far as linear processes are considered, the damping of structure-borne sound can be characterised by the loss factor

$$\eta = \frac{W_v}{2\pi W_r}, \tag{9.12}$$

(W_v – energy loss per cycle of vibration, W_r – restituted energy per cycle of vibration)

Thus, values $\eta > 1$ are possible. There are materials whose loss factors are relatively independent of frequency, temperature and the type of waves (not generally valid).

[3] In case additional masses are applied, a rigid connection has to be guaranteed. Glued or welded connections are recommended whereas screwed connections are mostly not sufficiently stiff at high frequencies.

Table 9.3 Formulas for impedances

Rigid mass	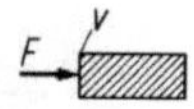	$Z = j\omega m$
Massless spring	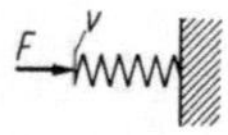	$Z = s/j\omega$
Infinitely long beam, Longitudinal wave		$Z = S\sqrt{E\rho}$
Short beam, longitudinal wave rigid calmping	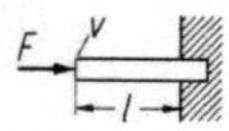	$Z = -jS\sqrt{E\rho}\ \cot\left(\sqrt{\rho/E}\,\omega/l\right)$
Short beam, longitudinal wave without support	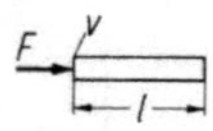	$Z = jS\sqrt{E\rho}\ \tan\left(\sqrt{\rho/E}\,\omega/l\right)$
Infinitely long beam, bending wave	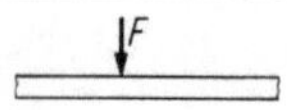	$Z = 2\rho S c_B(1+j)$
Infinitely long beam, bending wave	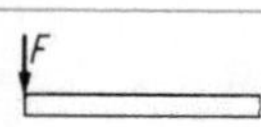	$Z = \frac{1}{2}\rho S c_B(1+j)$
Short beam, bending wave	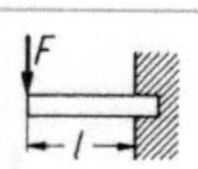	$Z = \frac{BK_B^3}{j\omega}\frac{1+\cos hk_B l\cos k_B l}{\sin k_B l\cos hk_B l-\sin hk_B l\cos k_B l}$
Infinitely large plate, bending wave	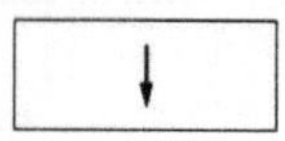	$Z = 8\sqrt{D\rho h} = 4h^2\sqrt{E\rho/3}$
Finite plate	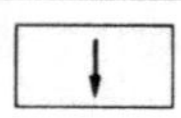	$Z = \frac{\rho h S_p}{4j\omega}\left\{\sum_n \frac{\varphi_n^2(x_0;y_0)}{\omega_n^2(1+j\eta_n)-\omega^2}\right\}^{-1}$
Infinitely large plate, 'inplane' movement	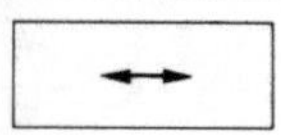	$\frac{1}{Z} = \frac{\omega(3-\mu)}{16hG} + j\frac{\omega}{8\pi hG}\left[(1-\mu)\ln\frac{2c_L}{\omega\alpha} + 2\ln\frac{2c_S}{\omega\alpha}\right]$
Elastic halfspace	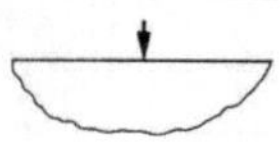	$\frac{1}{Z} \approx 0.19(1-\mu)\frac{\omega^2}{G\sqrt{G/\rho}} + j\frac{1-\mu}{\pi}\frac{\omega}{G\alpha}$
Approximate formula	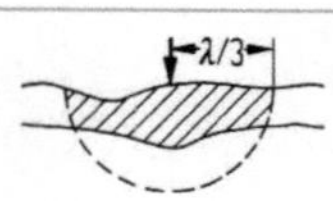	$\lvert Z\rvert \approx \omega m_\lambda$

Remark: The dimension of the impedance is Ns/m = kg/s. S cross-section of the beam, S_P surface area of the plate, E Young's modulus (in case of damped material, this value is complex, see Eq. (9.13), B bending stiffness of the beam, D bending stiffness of the plate, h plate thickness, $k_B = \frac{\omega}{c_B} = (\omega^2\rho S/B)$, G shear modulus, μ Poisson value, c_L longitudinal wave velocity, c_S shear wave velocity, a radius of the exited surface $((a \ll \lambda))$, η loss factor, φ $(x0, y0)$ value of the eigenfunction at the position of the excitation, ω_n eigenfrequency see Table 1.3, m_λ mass of an area, which is located at a distance of less than $\lambda/3$ from the point of excitation. Here λ is the shortest wavelength (typically bending wave). It is assumed, that the local elasticity, which is represent by the imaginary part of the formula can be neglected for an elastic half space

Table 9.4 Transmission factors for longitudinal and bending waves

a. Longitudinal waves change in the cross-section	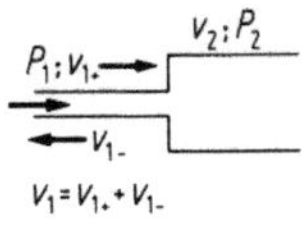	$\tau = 4[\sigma^{1/2} + \sigma^{-1/2}]^{-2}$
Material change	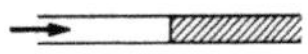	$\tau = 4\left[\left(\frac{E_1\varrho_1}{E_2\varrho_2}\right)^{1/2} + \left(\frac{E_1\varrho_1}{E_2\varrho_2}\right)^{-1/2}\right]^{-2}$
Elastic layer	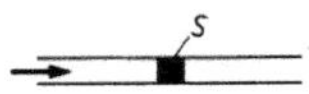	$\tau = [1 + (f/f_u)^2]^{-1}; \text{mit} f_u \approx \frac{1}{\pi}\frac{s}{S_1\sqrt{E_1\varrho_1}}$
Blocking mass	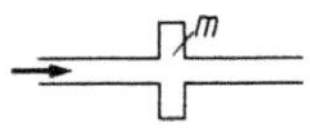	$\tau = [1 + (f/f_u)^2]^{-1}; \text{with} f_u = \frac{1}{\pi}\frac{S_1\sqrt{E_1\rho_1}}{m}$
b Bending waves change in the cross-section	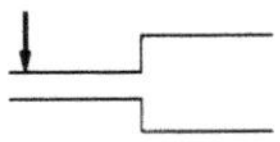	$\tau = \left[\frac{\sigma^{-5/4}+\sigma^{-3/4}+\sigma^{3/4}+\sigma^{5/4}}{\frac{1}{2}\sigma^{-2}+\sigma^{-1/2}+1+\sigma^{1/2}+\frac{1}{2}\sigma^{2}}\right]^2$
Material change	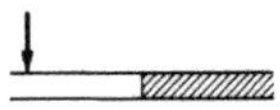	$\tau = \left[\frac{2\sqrt{\chi\psi}(1+\chi)(1+\psi)}{\chi(1+\psi)^2+2\psi(1+\chi^2)}\right]^2$
Corner	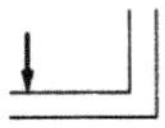	$\tau \approx 2[\sigma^{-5/4} + \sigma^{5/4}]^{-2}$
Intersection	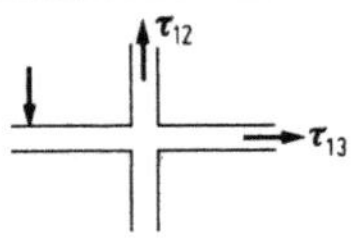	$\tau_{12} \approx \frac{1}{2}[\sigma^{-5/4} + \sigma^{5/4}]^{-2}$ $\tau_{13} \approx \frac{1}{2}[1 + 2\sigma^{5/2} + \sigma^{5}]^{-1}$
Branches	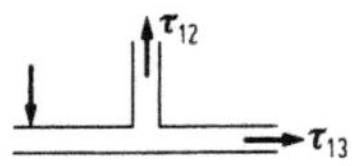	$\tau_{12} \approx [\sqrt{2}\sigma^{-5/4} + \sigma^{5/4}/\sqrt{2}]^{-2}$ $\tau_{13} \approx [2 + 2\sigma^{5/2} + \sigma^{5}]^{-1}$
Elastic layer	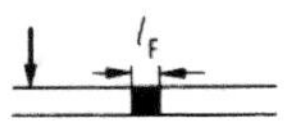	$\tau \approx [1 + (f/f_u)^3]^{-1}$ $\text{mit} f_u = \left(\frac{G_F^2}{1.8\pi^2\varrho_1\sqrt{E_1\varrho_1}h_1/l_F^2}\right)^{1/3}$ $= \left(\frac{G_F^2}{2\pi^3\varrho_1\sqrt{E_1\varrho_1}K_1 l_F^2}\right)^{1/3}$
Blocking mass	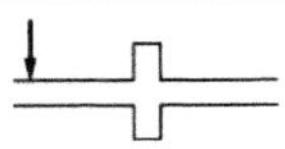	$\tau = 1 \text{ for } f < 0.5 f_s$ $\tau = [1 + f/f_u]^{-1} \text{ for } f > 2f_s$ $f_s = \frac{1}{2\pi}\frac{K_1}{K^2}\sqrt{\frac{E_1}{\varrho_1}}; f_u \approx \frac{2\varrho_1 S_1^2 K_1\sqrt{E_1\varrho_1}}{\pi m^2}$

Remark: In case of plates $\sigma = h_2/h_1$; beams $\sigma = S_2/S_1$; E_1, E_2, ρ_1, ρ_2, h_1, h_2, S_1, S_2, K_1, K_2, c_1, c_2 Young's modulus, density, thickness, cross-sectional area, radius of inertia, group velocity at both sides of the coupling points (in case of plates $K_1 = h_1/\sqrt{12}$), s stiffness of the intermediate layer, m mass of the blocking mass, G_F shear modulus of the intermediate layer, l_F length of the intermediate layer, K radius of inertia of the blocking mass

$$K = \sqrt[4]{\frac{\rho_2 E_1 K_1^2}{\rho_1 E_2 K_2^2}}; \quad \psi\frac{K_2 S_2}{K_1 S_1} = \sqrt{\frac{E_2\rho_2}{E_1\rho_1}}$$

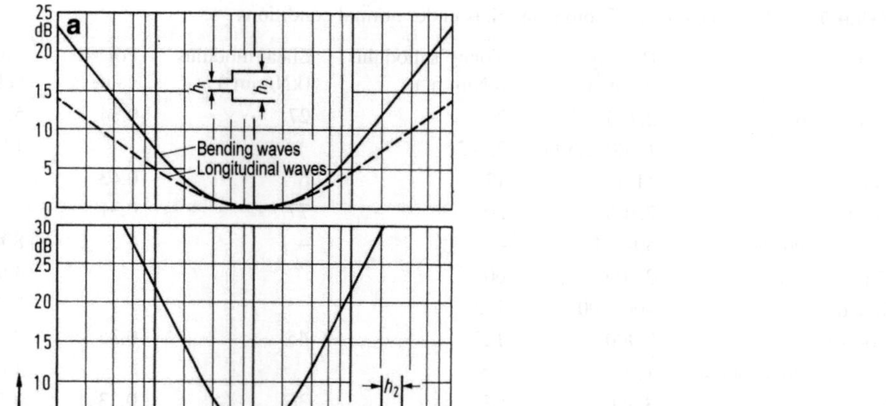

Fig. 9.9 (**a**) Isolation of structure-borne sound at long beams made out of a homogeneous material; above: longitudinal and bending waves; rest: just bending waves; (**b**) Insulation of structure-borne sound caused by elastic layers and blocking masses for long beams f_u, see Table 9.4

Table 9.5 Relation between loss factor and other damping values

Half power band width of single resonances	$\Delta f = \eta f$
Reverberation time	$T = 2.2/(\eta f)$
Log decrement	$\Lambda = \eta\pi$
Phase angle between strain and stress	$\varphi = \arctan\,\eta$
Q-factor	$Q = 1/\eta$
Level attenuation in case of longitudinal waves on very long beams	$D = 27.2\eta/\lambda_L(dB/m)$
Level attenuation in case of bending waves on very long beams	$D = 13.6\eta/\lambda_B(\mathrm{dB/m})$
Mean square velocity of a finite system (plate, cylinder, etc.), in case of a broadband excitation covering several resonances	$v_{\mathrm{eff}}^2 = \frac{P}{\omega m\eta}$

Remark: λ_L longitudinal wavelength in m; λ_B bending wavelength in m; P inserted structure borne sound power; m total mass of the system

In theoretical considerations – assuming harmonic processes with a time dependency $e^{j\omega t}$ – the loss factor can be interpreted as the imaginary part of the corresponding Young's modulus resp. shear-modulus. Thus, the damping is already considered in the corresponding equations by the stiffness parameters:

$$\underline{E} = E(1 + j\eta); \underline{G} = G(1 + j\eta),$$
$$\underline{B} = B(1 + j\eta). \tag{9.13}$$

At the basis of these calculations, the relationship between the loss factor and the respective measured values as shown in Table 9.5 can be derived.

Table 9.6 Mechanical data of some materials under normal conditions

Material	Density (kg/m^3)	Young's modulus (kN/mm^2)	Shear modulus (kN/mm^2)	Poisson value	Longitudinal wave velocity (m/s)	Loss factor
Aluminum	2,700	72	27	0.34	5,200	$>10^{-4}$
Asphalt	1,800–2,300	7.7–21			1,900–3,200	0.05–0.3
Lead	11,300	17	6	0.43	1,250	10^{-3}–10^{-2}
Steel	7,800	200	77	0.31	5,100	$\approx 10^{-4}$
Fibre structures	50–150	–	–	–	80–300	≈0.1
Glass	2,500	60			4,900	$\approx 10^{-3}$
Wood	400–800	1–5			2,000–3,000	$\approx 10^{-2}$
Copper	8,900	125	45	0.35	3,700	$\approx 2 \times 10^{-3}$
Light weight concrete	1,300	3.8			1,800	$\approx 10^{-2}$
Brass	8,500	95	36	0.33	3,200	$<10^{-3}$
Nickel	8,900	205	77	0.3	4,800	$<10^{-3}$
Plexiglas	1,150	5.6			2,000	$\approx 2 \times 10^{-2}$
Sand	1,300–1,800	–	–	–	100–300	0.05–0.2
Heavy concrete	2,300	26			3,500	$4\text{–}8 \times 10^{-3}$
Bricks	1,900–1,100	≈26			2,500–3,000	$\approx 10^{-2}$
Zinc	7,100	13	5	0.33	1,350	$<10^{-3}$
Tin	7,300	4.4	1.6	0.39	780	$\approx 10^{-3}$

Remark: $1 kN/mm^2 = 10^9 N/m^2$

9.4.1 Loss Factor of Various Materials and Constructions

In Table 9.6, loss factors and other important mechanical data of various materials are listed. Apart from asphalt, the listed materials show a more or less constant loss factor for typical parameter variations (20 Hz to 10 kHz; −30°C to +100°C). One has to consider that the corresponding measurements are carried out at homogeneous specimens and under laboratory conditions, which means without energy dissipation to the exterior. Furthermore, the effects of microstructural changes are not considered. For the detection of those effects in material research, the loss factor is as an important measurement value. In practice, all constructions consist of various parts that are affixed somewhere, so that in any case friction on contact surfaces and related dissipation of energy occurs. Due to these inevitable losses, loss factors can hardly be below 10^{-3}. Bells or gongs are an exception from this.

In practical applications, the following approximate values for loss factors may be applied. The indicated values do not consider specially designed additional measures for damping (anti-drone coating, filling with sand, etc.):

(a) Concrete or masonry buildings $\eta \approx 10^{-2}$, frequency independent;

(b) Metal constructions consisting of few thick parts (e.g., exterior shell of a ship):

$$\eta \approx 3 \times 10^{-3} \quad \text{for } f < 500\,\text{Hz},$$

$$\eta \approx 10^{-3} \quad \text{for } f > 1{,}000\,\text{Hz};$$

(c) Metal constructions consisting of many thick and some thin parts (e.g. engine, car)

$$\eta \approx 10^{-2}.$$

(d) Metal constructions consisting of many thin parts (small complicated machines):

$$\eta \approx 5 \times 10^{-2} \quad \text{for } f < 500\ \text{Hz},$$

$$\eta \approx 10^{-2} \quad \text{for } f > 1{,}000\ \text{Hz}.$$

Apart from metals and building materials, high polymer synthetics are important. The loss factors of these materials are typically strongly dependent on frequency and on temperature. An example is given in Fig. 9.10 [14]. At low temperatures and at high frequencies, the curve show a 'frozen condition' with loss factor below 10^{-1}. For damping of structure-borne sound, the 'transition zone' is more interesting,

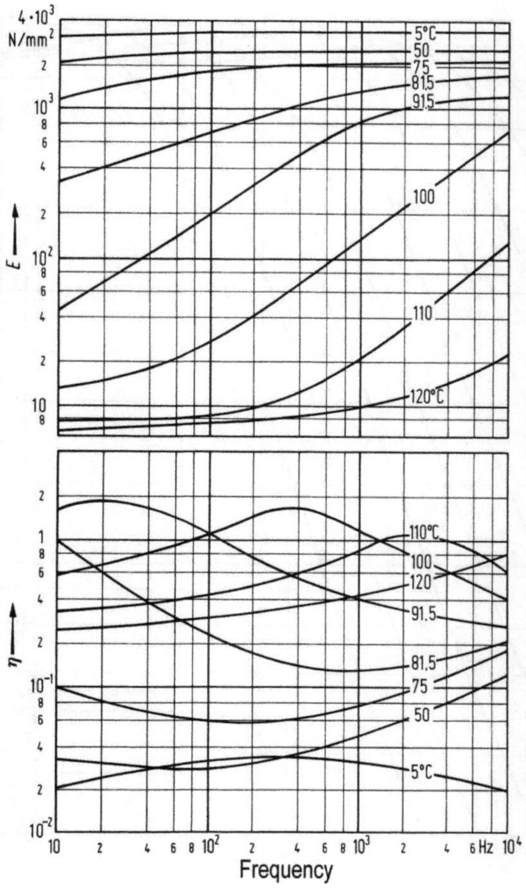

Fig. 9.10 Young's modulus and loss factor of polyvinyl chloride (freezing temperature approximately 25°C)

in which the Young's modulus changes quickly and the loss factor is relatively high. For practical applications in vibration control, the subsequent liquefaction is not of interest.

It would be optimal to provide a high loss factor over a large temperature range. Unfortunately, e.g., for applications [15] at composite materials one can observe that the higher the maximum loss factor, the smaller the temperature ranges. A similar relationship is observed with special alloying that can also lead to high loss factors $\eta \approx 10^{-2}$ for metals.

9.4.2 Combinations of Materials with Large and Small Loss Factors

Materials with high permissible stresses (metals) typically show low damping, whereas materials with high damping (e.g., high polymer in the transition zone) often have low permissible stresses. This is the reason why often both types of materials are combined. In practical applications, metal plates, tubes, beams, etc. are provided with highly damping layers made out of synthetic material. This can be done either before the materials are further processed (e.g., compound plates) or after they are installed (e.g., anti-drone coating). By designing such materials, it is necessary to consider that a maximum part of the sound energy is introduced into the damping material. For example, it makes little sense to apply the damping material at a location where the amplitudes of the induced strains are small or to apply a material with a high loss factor, which, due to its very low Young's modulus, will not contain a significant part of the vibration energy. For complicated structures (bent or ripped plates, etc.), it is not always possible to calculate the losses resulting from a combination of the carrying structure and the damping material. However, for homogenous plates and beams the relationships can be given [16–20]. For propagating bending waves and layers attached on one side, the loss factor can be taken from Fig. 9.11 (valid for $f < \sqrt{\rho_1/E_1}E_2/(4\rho_2 d_1)$). One can see that the product $E_2\eta_2$ should be very large and that the efficiency increases approximately with the square of the thickness ratio d_2/d_1. Thus stiff, thick surfaces with a high loss factor should be applied. Typical data for a well-designed, anti-drone coating applied to steel plates[4] are $E_2/E_1 \approx 3 \times 10^{-3}$; $E_2\eta_2 \approx 10^3\,\mathrm{N/mm^2}$; $d_2 = 2d_1$; $\eta \approx 0.08$. In compound plates (sandwich-plates) (Fig. 9.12), the evaluation of the loss factor is slightly more complicated. Assuming the values of Fig. 9.12, the relationship is given by

$$\eta = \eta_2 \frac{YX}{1 + (2+Y)X + (1+Y)(1+\eta_2^2)X^2}, \tag{9.14}$$

with

$$\frac{1}{Y} = \frac{E_1 d_1^3 + E_3 d_3^3}{12 d_{13}^2}\left(\frac{1}{E_1 d_1} + \frac{1}{E_3 d_3}\right);$$

$$X = \frac{G_2}{k^2 d_2}\left(\frac{1}{E_1 d_1} + \frac{1}{E_3 d_3}\right),$$

$$k^2 = \omega\sqrt{\frac{12(\rho_1 d_1 + \rho_3 d_3)}{(E_1 d_1^3 + E_3 d_3^3)(1 + XY/(1+X))}}. \tag{9.15}$$

[4] The loss factor related with the (relatively unimportant) longitudinal waves in metal or compound plates with anti-drone coating are around 0.01–0.1 of the loss factor related with bending waves.

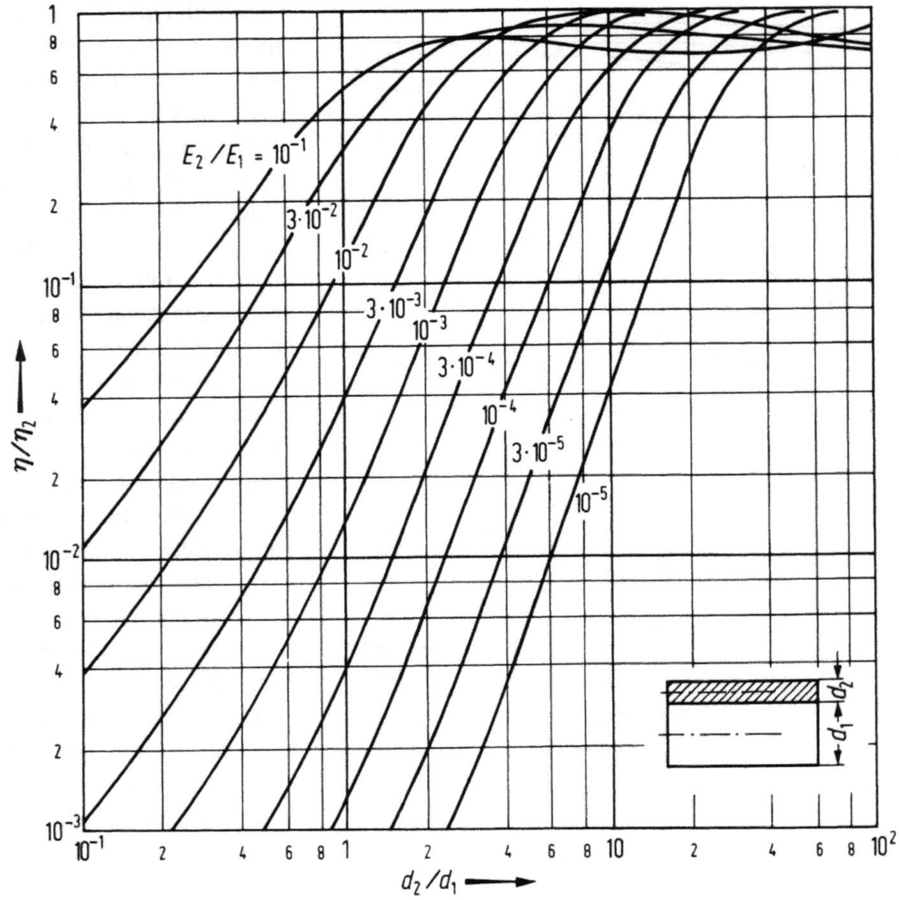

Fig. 9.11 Loss factor η of a plate with an anti-drone coating. E_1, d_1 Young's modulus and thickness of the plate; E_2, η_2, d_2 Young's modulus (real part), loss factor and thickness of the anti-drone coating

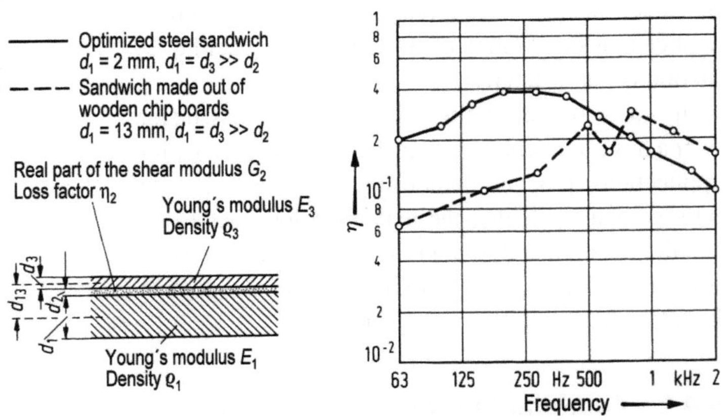

Fig. 9.12 Composition of a sandwich plate and measurement results

The loss factors described by these formulas typically show a frequency dependency with a relatively large frequency range for the maximum.

As in Eq. (9.15), X depends on k^2 and k^2 on X generally iterations procedures are applied with a starting value $X = 0$ in 2–5 consecutive iterations.

It should be mentioned that in case of $d_1 > d_3$ the loss factor at low frequencies [Eq. (9.15)] can be slightly increased by cutting the counter plate in distances of approximately 20 cm (details see [21, 22]).

9.4.3 Damping at Contact Surfaces

Damping at contact surfaces (joints, etc.) is related with the following mechanisms:

(a) Gas pumping
When the small distance d between two parallel surfaces is changed periodically, a pumping action on the gas between occurs. The gas velocity is of the order of magnitude $v\lambda/d$ (v – relative velocity of the contact areas, λ – wavelength of the vibration, d – distance of the contact areas). In this tangential flow, viscous losses occur, causing damping of structure-borne sound. Gas pumping occurs mainly in case of thin plates (e.g., double plates). For the loss factor, the following relationship [23, 24] holds:

$$\eta \approx \frac{\omega\rho}{m''}\sqrt{2\nu/\omega}(\lambda/2\pi d)^2, \tag{9.16}$$

(m'' – mass per unit area of the thicker plate; ρ, ν – density and kinematic viscosity of the material between the plates, $\sqrt{2\nu/\omega}$ results in the acoustic boundary layer)

(b) Friction, microslip
When two contact surfaces touch each other and move in tangential direction, friction occurs. In case of small amplitudes of movements, which are typical for structure-borne sound at higher frequencies, the so called 'microslip' [25] happens. In this case, the interior of the large number of small contact zones that form the coupling is deformed elastically. Friction occurs at the delimiters of the zones. This is the reason why the contact zones between structures that vibrate tangentially with respect to each behave similar to linear shear springs with an imaginary part given by the loss factor η. $\underline{s} = s\,(1 + j\eta)$ [26], (see Fig. 9.13).
In case of large amplitudes, which sometimes occur at low frequencies, the structures move as a total with respect to each other and thus create non-linear coulomb friction [27, 28].

(c) Friction Lubricating film
A film of oil, fat or other lubricants leads to relations similar to compound plates (see Sect. 4.2). Instead of the complex shear modulus, the approach $\underline{G_2} = j\omega\nu_{2\rho 2}$ has to be taken (ν_2, ρ_2 – kinematic viscosity and density of material in the interface).
This leads to:

$$\eta = \frac{YX_\nu}{1 + (1 + Y)X_\nu^2}, \tag{9.17}$$

with

$$X_\nu = \frac{\omega\nu_2\rho_2}{k^2 d_2}\left(\frac{1}{E_1 d_1} + \frac{1}{E_3 d_3}\right),$$

(d) Filling Material, etc
When the vibrating structure is delimited by a filling material (e.g., sand), fibres or another medium with high interior losses, structure-borne sound is 'radiated' into this medium and transformed into heat. The achievable loss factor amounts approximately:

$$\eta \approx \frac{\rho_s d_s \mathrm{Im}\{A\}}{m'' + \rho_s d_s \mathrm{Re}\{A\}}, \quad A = \frac{\underline{c}_s}{\omega d_s}\tan\frac{\omega d_s}{\underline{c}_s}. \tag{9.18}$$

(m'' – mass per area of the vibrating plate, ρ_s, d_s – density and thickness of the filling material, $\underline{c}_s = c_s(1 + j\eta_s)$ complex wave velocity in the filling material).
Equation (9.18) is not only valid for fillings but also for other thick damping layers, under the condition that the calculated loss factor according to Eq. (9.18) is larger than the loss factor derived from Fig. 9.11 [29].

Fig. 9.13 Shear stiffness per area and loss factor of the shear stiffness for dry contact as a functions of the pressure

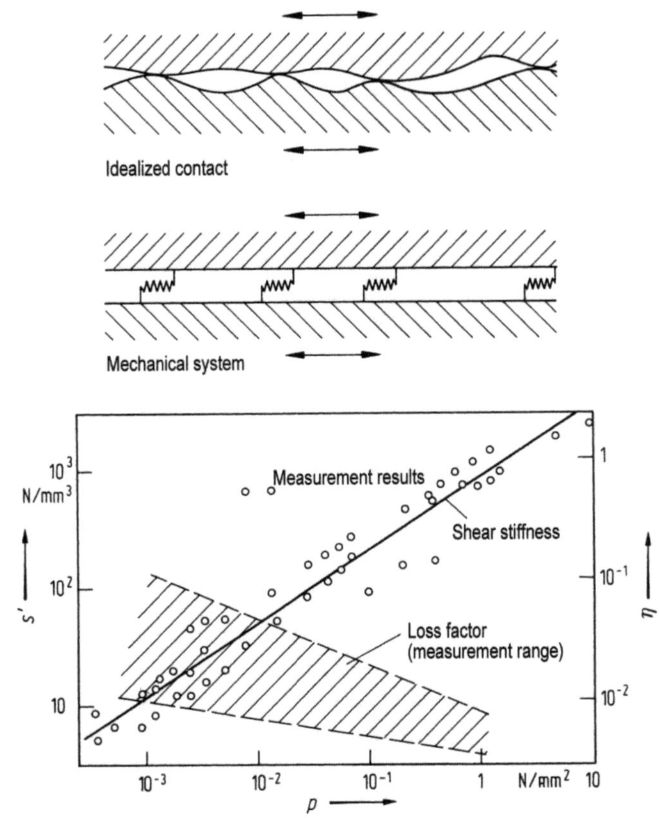

9.4.4 Combination of Insulation and Damping

9.4.4.1 Attenuation of Structure-Borne Sound by Additional Damping

Often the question arises whether an improvement of the loss factor can also be obtained by additional measures for damping. Here various cases shall be distinguished:

(a) In the case, a body is excited by pure tones with frequencies of excitation that coincide with the system's resonance frequencies an increase of the loss factor from η_v to η_n will cause a reduction of the level of structure-borne sound of:

$$\Delta L \approx 20\,\lg(\eta_n/\eta_v)\mathrm{dB}, \qquad (9.19a)$$

(b) In practical applications, the pure resonance excitation as described under (a) is very seldom. Very often the excitation consists of a broadband noise with various frequencies. If the resonant frequency is not in the frequency range of excitation (e.g., in case of small thick plates and low frequencies) additional damping will not lead to a reduction, as the level of structure-borne sound then mainly depends on the mass and on the stiffness, but not on the damping. However, if there is one or more resonant frequencies in the frequency range of excitation – which is the typical case –the levels at the resonances will be reduced. As the resonance curves become 'wider', the attenuation of structure-borne sound will result in:

$$\Delta L \approx 10\,\lg(\eta_n/\eta_v)\ \mathrm{dB}. \qquad (9.19b)$$

In case of a steady state excitation (random noise), the reduction of the level will be perceived as a reduction of loudness. In case of an impulse excitation (single knocks), it will be rather perceived as a reduction of the duration of the noise. In practical applications, the achievable level reduction by additionally applied damping amounts approximately 5–10 dB. Due to altered characteristics of the radiation (see Sect. 5), it might happen that the attenuation of the structure-borne sound is smaller than the attenuation of the airborne sound radiated.

(c) In case plates (respectively beams, tubes, etc.) are large enough that the number of wavelengths fitting on the structure multiplied with the loss factor η_n results in a value larger than 0.2, the observations described above are not always valid. In this case, resonances do not dominate the response. The structure behaves as if it was infinitely large. This means that in case of a localized excitation, there will be an additional level decrease with increasing distance compared to the undamped plate, which can be taken from Table 9.5.

9.4.4.2 Attenuation by the Combination of Insulation and Damping, Statistical Energy Analysis

One of the most effective methods for the attenuation of structure-borne sound consists of a combination of insulation and damping. This means that at a certain distance from the source, measures for insulation (elastic layers, deviations, etc.) are applied. The energy is partly reflected there. Additional damping will be provided in the area, where multiple reflections create the highest energy density. The idea is to concentrate the energy related with structure-borne sound on a limited zone, in which – very effectively – a transition into heat can be obtained.

As in practical applications, bending waves are very important; they shall be discussed in the following. Figure 9.14 shows the power balance for two finite beams under a stationary load [13, 30–32]:

$$P + P_{21} = P_{v1} + P_{12}, \quad 0 + P_{12} = P_{v2} + P_{21}. \qquad (9.20a)$$

(P – externally induced power,[5] P_{v1} respectively P_{v2} – power transferred into heat in the two beams, P_1 respectively P_2 – power arriving at the contact points, P_{12} – P_{21} – power transferred into the adjacent structure).

[5] Under the given assumption and if the excitation is linked with a broadband punctual load $\tilde{F}$(effective value the input power can be calculated by: $P = \tilde{F}^2 \mathrm{Re}\{1/Z\}$), (see Table 9.3) is valid.

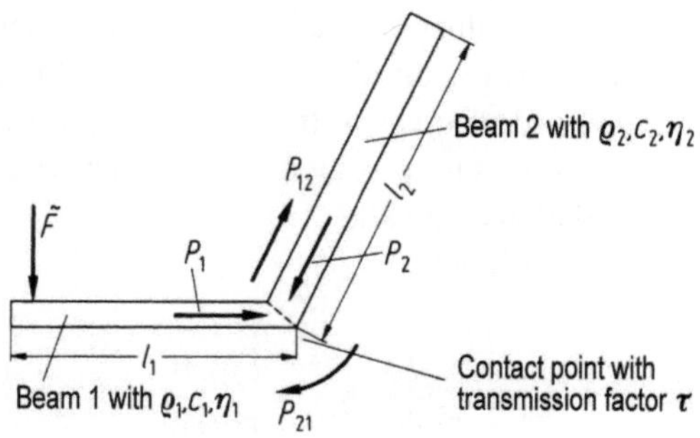

Fig. 9.14 Definitions concerning Eqs. (9.20a) and (9.21a)

In this approach, 'statistical conditions' are assumed; that means frequency bands in which at least four resonances of the beams occur are considered. Furthermore, the level of vibration on the beams must not vary too much lengthwise, and the beams are not exactly equally long. Under those conditions, Eqs. (9.20b) and (9.21a) return good approximations:

$$\begin{aligned} P_{12} &= \tau P_1; P_{21} = \tau P_2; \\ P_{v1} &= \rho_1 S_1 l_1 \omega \eta_1 \tilde{v}_1^2; P_{v2} = \rho_2 S_2 l_2 \omega \eta_2 \tilde{v}_2^2; \\ P_1 &= \rho_1 S_1 c_1 \tilde{v}_1^2; P_2 = \rho_2 S_2 c_2 \tilde{v}_2^2. \end{aligned} \qquad (9.20b)$$

Therefore:

$$\frac{\tilde{v}_2^2}{\tilde{v}_1^2} = \frac{\rho_1 c_1 S_1}{\rho_2 c_2 S_2} \frac{1}{1 + \omega l_2 \eta_2 / \tau c_2} = \frac{m_1 \Delta N_2}{m_2 \Delta N_1} \frac{1}{1 + \eta_2/\eta_{21}}, \qquad (9.21a)$$

$$\tilde{v}_2^2 = \frac{P}{\rho_2 S_2 l_2 \omega \left[\eta_1 \frac{l_1 c_2}{l_2 c_1} + \eta_2 (1 + \omega l_1 \eta_1 / \tau c_1)\right]} = \frac{P}{m_2 \omega \left[\eta_1 \frac{\Delta N_1}{\Delta N_2} + \eta_2 (1 + \eta_1/\eta_{12})\right]}. \qquad (9.21b)$$

The first forms of Eqs. (9.21a) and (9.21b) result directly from Eqs. (9.20a) and (9.20b). The second form describes the generalization, which is obtained under the application of the principle of reciprocity or the Statistical Energy Analysis (SEA) [13, 30–32]. It is valid for arbitrary combinations of one-, two- or three-dimensional systems (beams, plates, volumes). Here m_1 respectively m_2 describe the total mass of the respective systems and ΔN_1 and ΔN_2 the number of eigenmodes in the frequency band of excitation (see

Table 1.3). The values η_{21} respectively η_{12} describe the power transmission from system two to one, respectively, and vice versa. In case the transmission efficiency and the group velocities c_g are known, the following relations are valid [13]

in two dimensions:

$$\eta_{21} = \frac{c_{g2} L \tau}{\pi \omega S_2}; \eta_{12} = \frac{c_{g1} L \tau}{\pi \omega S_1},$$

in three dimensions:

$$\eta_{21} = \frac{c_{g2} S \tau}{4 \omega V_2}; \eta_{12} = \frac{c_{g1} S \tau}{4 \omega V_1}, \quad (9.22)$$

(S – area separating the volumes V_1 and V_2 L – length of the separating line between the areas S_1 and S_2).

One of the most important conclusions from Eqs. (9.21a) and (9.21b) is based on the observation that a measure for insulation is just effective if a sufficient damping is available. Equation (9.21a) shows that the level difference of structure-borne sound depends very strongly on η_2/η_{21} respectively η_2/τ, that means on the relation between the power loss due to damping to the transmitted power. In the case of $\eta_{21} > \eta_2$, that means if $\tau > \eta_2 \lambda_2 \omega / c_2$ insulation will not play a role.

Similarly Eq. (9.21b) shows that not only the transmission efficiency but also the relation η_1/τ is decisive. Small levels of vibrations will just occur, if the insulation is effective ($\tau \ll 1$) and the damping is high. As in problems of structure-borne sound, often the insulation is not very high (changes of cross-sections, etc.) it might be possible that $\eta_{21} > \eta_2$. In the case of strong coupling and small damping, Eq. (9.21a) results in a simpler relationship

$$\frac{\tilde{v}_2^2}{\tilde{v}_1^2} \approx \frac{m_1 \Delta N_2}{m_2 \Delta N_1}. \quad (9.23)$$

For practical applications, one can see that strongly coupled light weighted plates might thus have higher levels of vibrations than an adjacent heavy plate. For practical applications, Eq. (9.23) gives an upper limit for the value, $\tilde{v}_2^2$ which can be very helpful for quick estimations.

9.5 Radiation of Structure-Borne Sound

Structure-borne sound would be rather unimportant for noise and vibration control if it were not radiated. Problems of acoustic fatigue just occur at extremely high levels of vibration. The radiation of simple bodies is already described in Sect. 1.6. There, also the radiation efficiency σ is introduced [Eq. (1.67)]. Considering airborne sound under normal conditions [11] the sound power level L_p (re 10^{-12} W) and the average velocity level are related by:

$$10 \lg \sigma \text{ dB} = L_p - L_v - 10 \lg\left(\frac{S}{m^2}\right) \text{dB}. \quad (9.24)$$

The radiation efficiency of compact bodies (engines, gearboxes and pumps with thick boxes, but not machines with light-weighted cladding) with average dimensions l is about $\sigma \approx 1$in the frequency range $k_O l > 3$ or $l > \lambda_O/2$ and $\sigma < (k_O l)^2/8$ at lower frequencies (λ_O – wavelength of the surrounding medium, typically air, $k_0 = 2\pi/\lambda_0$).

In plate-like structures, the relationships are more complicated. Here, the radiation depends strongly on λ_B/λ_O (λ_B – bending wavelength). It, furthermore, depends on the type of excitation, on its supports (e. g., clamping at the delimiters) and on the existence of discontinuities. The wavelengths of the bending wave and the sound wave in air have different frequency dependencies. For the so-called coincidence frequency f_g, the bending wavelength and the wavelength in air are equal ($\lambda_B = \lambda_O$). The related effects are treated in Sect. 7.1.3 (Fig. 7.3). For radiation in water, f_g is about a factor 18 higher than in air.

At frequencies smaller than the coincidence frequency, one can take the simplified, yet correct, picture of a hydrodynamic shortcut. In the vicinity of the plates the air is not compressed, but shifted between the wave troughs and peaks; there the sound decreases rapidly with the distance from the radiator. As no compression in the air occurs, no sound power is radiated into the far field.

Sound is just radiated from areas at the border at which 'no neighbour' provides the hydrodynamic shortcut.

Some practically interesting conclusions are:

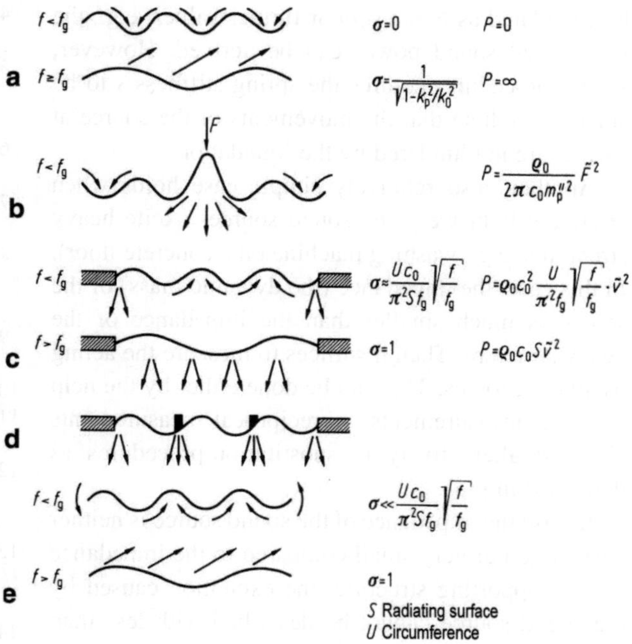

Fig. 9.15 Radiation of plates. (**a**) Infinitely large plate; (**b**) Punctually excited large plate with damping; (**c**) Clamped plate; (**d**) Plate with stiffness, like (**c**) here U is the circumference of all the individual surfaces between the stiffeners and the delimeter; (**e**) Free plate. $k_p = 2\pi/\lambda_B$ bending wave number, m_p'' mass of the plate per area

(a) With a given average velocity, additional stiffeners create higher radiation. This can be derived from the fact that the 'active' delimiter U results from the total delimiter of the single areas, which means, from the delimiter of the plate plus twice the length of the stiffeners.
(b) Generally a plate clamped at its border radiates more sound than a freely vibrating plate, where near to the delimiter an additional shortcut between the front and the back is possible. An extreme shortcut exists between front and back at perforated plates, which radiate very little sound.
(c) In case of finite plates, excited punctually at frequencies $f < f_g$ sound power is radiated from the delimiters and from the point of excitation (see Fig. 9.15). The sound power that is radiated from the delimiters is proportional to the mean square velocity of the plates. It thus decreases with increasing damping. The sound power level that is created at the point of excitation is proportional to the square of the force of excitation and almost independent from damping. With increasing damping, just the radiation from the vicinity of the point of excitation remains. This is the reason why additional damping often results in a considerable reduction of the structure-borne sound but just a relatively small reduction of the radiated airborne sound.

As the radiation efficiency is nearly constant for $f > f_g$, the sound insulation of thick walls can be assessed by means of vibration measurements as well.

In the following, L_{p1} describes the sound pressure level in the source room, L_{v2} the velocity level at the receiving room. For $f > f_g$, the sound insulation can be approximated by:

$$R = L_{p1} - L_{v2} - 6\,\text{dB}. \tag{9.25}$$

9.6 Characterization of Sources

In the scope of the Statistical Energy Analyses (SEA), (see Chap. 3) all sources are specified by their power. The sound power of sources of airborne sound is relatively independent from the surroundings (see Chap. 5). Therefore, their description can be carried out effectively just by emitted the sound power. As in case of structure-borne sound, the power transmission is related to different media and various wave types and as the power input depends on impedances, here the description becomes much more complicated.

A frequent assessment of sources of structure-borne sound is based on measuring the velocity of vibration, respectively, the rotational velocity at the feet of the source placed on very soft springs. Then the force can be calculated as $F = vs/j\omega$ or further values, e.g., the transmitted sound power can be derived. However, such a procedure requires the spring stiffness s to be small enough so that the movements of the source at its feet are not hindered by the foundation.

Another, also relatively simple case holds when relatively light weighted sound sources excite heavy structures (e.g., washing machine on a concrete floor). In this case, the impedance (the dynamic mass) of the source is much smaller than the impedance of the heavy structure. Then it suffices to measure the acting oscillating forces. This can be done either by the help of direct measurements, or reciprocal measurements [33] and, alternatively, by substitution procedures, as described in [34].

In case the impedance of the sound source is neither very large nor very small compared to the impedance of the supporting structure, the excitation caused by the sound source cannot be described with less than two values. One of these values is, e.g., the vibration velocity at the support, characterizing the excitation. The other value is the impedance characterizing the (passive) reaction of the supporting structure under the exterior excitation. Possible combinations of two such values can be found in [35], where source descriptors and coupling functions are proposed.

A further proposal consists of describing the structure-borne sound power similar to the typical procedures for reverberant chambers (see Chap. 2). In this case, a large plate is the equivalent to the reverberant system. The power is given by Eqs. (9.22) and (9.20b) with $P_{21} = P_{12} = 0$. This procedure is relatively simple, with the slight disadvantage that the results can be applied only to structures whose input impedance equals approximately that of the 'reverberant plate'.

Further investigations are necessary in order to obtain a simple, robust and reliable procedure for the characterization of the structure-borne sound emission that can be broadly accepted. Further hints are given in Chap. 22.

References

1. Achenbach JD (1973) Wave propagation in elastic solids. North-Holland, Amsterdam
2. Beltzer AI (1988) Acoustics of solids. Springer, Berlin
3. Auld BA (1973) Acoustics fields and waves in solids. Vol I a. II. Wiley, New York
4. Mason PM (ed) (1966) Physical acoustics, vol I–III. Academic, New York
5. Rausch E (1959) Maschinenfundamente und andere dynamisch beanspruchte Baukonstruktionen. VDI-Verlag, Düsseldorf, Ergänzungsband, 1968
6. Harris CM, Crede DE (eds) (1961) Shock and vibration handbook Vol 2, sect. 30-35. Mc-Graw-Hill, New York
7. Hartzt H (1937) Schwingungstechnische Gestaltungen von Maschinengründungen. Werner Genest, Berlin
8. (1979) Handbook of noise and vibration control. Sect. 3b, 3c, 4th edn. Trade and Technical Press, Morden, Surrey (England), 586–652
9. Hertz H (1882) Über die Berührung fester, elastischer Körper. J reine angew Math 92:156–171
10. Johnson KL (1985) Contact mechanics. Cambridge 93,220
11. Cremer L, Heckl M, Petersson BAT (2005) Structure borne sound. Springer, Berlin
12. Ljunggren S (1984) Generation of waves in an elastic plate by a torsional moment and horizontal force. J Sound Vib 93:161–187
13. Wöhle W (1984) Statistische Energieanalyse der Schalltransmission. Abschn. 1.10. In: Fasold/Kraak/Schirmer (Hrsg.) Taschenbuch Akustik. VEB Verlag Technik Berlin
14. Becker GW, Oberst H (1965) Über das dynamische Verhalten linearer, vernetzter und gefüllter Kunststoffe. Kolloid Z 148:6–16
15. Linhardt F, Oberst H (1961) Über die Temperaturabhängigkeit schwingungsdämpfender Kunststoffe. Acustica 11:255–264
16. Oberst H (1952) Über die Dämpfung der Biegeschwingungen dünner Bleche durch festhaftende Beläge. Acustica 2:181–194
17. Kerwin EM (1961) Damping of flexural waves in plates by spaced damping treatments having spaces of finite stiffness. In: Proc. 3rd ICA Congress. Elsevier, 412–415
18. Ross D, Ungar EE, Kerwin EM (1959) Damping of plate flexural vibrations by means of viscoelastic lamina. In: Ruzicke JE (ed) Structural damping. Amer Soc Mech Engineers
19. Kurtze G (1959) Bending wave propagation in multilayer plates. J Acoust Soc Am 31:1181–1201
20. Tartakovskii BD, Rybak SA (1962) On vibration of layered plates with losses. 4th ICA Congress, Kopenhagen, Paper P43
21. Parfitt GG (1962) The effect of cuts in damping tapes. 4th ICA Congress, Kopenhagen, Paper P 21
22. Zeinetdinova RU, Naumkina NI, Tartakovskii BD (1978) Effectiveness of a vibration-absorbing coating with a cut constraining layer. Sov Phys Acoust 24:347–348
23. Maidanik G (1960) Energy dissipation associated with gas pumping at structural joints. J Acoust Soc Am 40:1064–1072
24. Möser M (1980) Körperschalldämpfung durch Reibung in der zwischen zwei Platten befindlichen Luftschicht. Acustica 46:210–217
25. Mindlin RD, Dersiewicz M (1953) Elastic spheres in contact under oblique forces. Trans ASME Ser E J Appl Mech 20:327–335
26. Schober U (1990) Untersuchung der Körperschalldämpfung durch Fügestellen in Motoren. DAGA 90. DPG, Bad Honnef, 349–352

27. Schierling R (1984) Reibungsdämpfung in Konstruktionsfugen betrachtet an verschraubten Balkensystemen. Diss. TH Darmstadt
28. Gaul L (1981) Zur Dämmung und Dämpfung von Biegewellen an Fügestellen. Ing Arch 51:101–110
29. Albrecht A, Möser M (1989) Die dämpfende Wirkung dicker Entdröhnschichten auf Platten. Z Lärmbekämpfung 36:73–79
30. Lyon RH, Maidanik G (1962) Power flow between linearly coupled oscillators. J Acoust Soc Am 34:623–639
31. Scharton TO, Lyon RH (1968) Power flows and energy sharing in random vibration. J Acoust Soc Am 43:1332–1343
32. Cremer L, Heckl M, Ungar ED (1973) Structure-borne sound. Kap. V, 8. Springer, Berlin
33. Heckl M (1985) Anwendung des Satzes von der wechselseitigen Energie. Acustica 58:111–117
34. ten Wolde T, Gadefelt GR (1987) Development of standard measurement methods for structure-borne sound emission. Noise Control Eng J 28:5–14
35. Mondot JM, Petersson B (1987) Characterization of structure-borne sound sources: The source descriptor and the coupling function. J Sound Vib 114:507–518
36. Trochidis A (1982) Körperschalldämpfung mittels Gas- oder Flüssigkeitsschichten. Acustica 51:201–212
37. Petuelli G (1983) Theoretische und experimentelle Bestimmung der Steifigkeits- und Dämpfungseigenschaften normalbelasteter Fügestellen. Diss. RWTH Aachen

10 Room Acoustics

Heinrich Kuttruff and Eckard Mommertz

The traditional task of room acoustics is to create or formulate conditions which ensure the best possible propagation of sound in a room from a sound source to a listener. Thus, objects of room acoustics are in particular assembly halls of all kinds, such as auditoria and lecture halls, conference rooms, theaters, concert halls or churches. Already at this point, it has to be pointed out that these conditions essentially depend on the question if speech or music should be transmitted; in the first case, the criterion for transmission quality is good speech intelligibility, in the other case, however, the success of room-acoustical efforts depends on other factors that cannot be quantified that easily, not least it also depends on the hearing habits of the listeners. In any case, absolutely "good acoustics" of a room do not exist.

In recent times, more emphasis is given to another task of room acoustics, that has been neglected up to now: the assessment of noise propagation in workrooms (factory buildings, offices etc.). This development is based on the realization that the noise level at a working place depends only partly on the properties of the noise source (its sound power, spectral composition, and directivity), but also partly on the acoustical properties of the respective room. The same is valid for the effective measures of noise control in workrooms.

H. Kuttruff (✉)
Institute of Technical Acoustics, RWTH Aachen University, Templergraben 55, 52056 Aachen, Germany
e-mail: kutruff@akustik.rwth-aachen.de

E. Mommertz
Müller-BBM GmbH, Robert-Koch-Strasse 11, 82152 Planegg, Germany
e-mail: Eckard.Mommertz@MuellerBBM.de

10.1 Basic Facts of Sound Propagation in Rooms

10.1.1 Preliminary Remarks

Apart from simple cases, a complete, exact, and detailed presentation of sound propagation in closed rooms is not possible for several reasons; on the one hand, rooms have usually complicated shapes and varied wall designs, so that even their mathematic–physical description is difficult and often hopelessly complicated. On the other hand, the sound field in a very simple room already consists of numerous components, which must be calculated separately.

Moreover with a physically complete description of the sound field, nothing decisive would be achieved. Statements concerning the acoustic properties or "acoustics" of a room are only possible if we also know how sound fields of such a complicated spatial and temporal structure are perceived by our sense of hearing and transferred into subjective impressions. Although our knowledge today in this field is still very incomplete, we already know that the listener can certainly not "hear" all details of sound transmission, but that a special hearing impression results from a combination of objective facts.

Therefore, it is not only a provisional solution but it also thoroughly corresponds to the functioning of our sense of hearing, when for a quantitative assessment of the acoustic properties of a room, a complete characterization of the sound field is neglected in favor of a more global approach, where special combinations or mean values of sound field data are considered. According to that, an important task of room acoustics

G. Müller and M. Möser (eds.), *Handbook of Engineering Acoustics*,
DOI 10.1007/978-3-540-69460-1_10, © Springer-Verlag Berlin Heidelberg 2013

is to define objective sound field parameters, which correspond as clearly as possible to hearing impressions that can be classified and differentiated from each other.

Nevertheless, certain knowledge of the basic physical facts of sound propagation in rooms is necessary for a profound understanding of problems in room acoustics. This knowledge will be provided in the following sections.

10.1.2 Wave Theory of Room Acoustics

The exact description of sound fields in a room is based upon the concept of normal modes, which can be imagined as three-dimensional standing waves. Each of them is associated with a so-called eigenfrequency or resonance frequency. Mathematically, each normal mode corresponds to an eigenfunction or characteristic function, which is a solution of the acoustical wave equation satisfying proper boundary conditions.

The steady state response of a room to a sinusoidal sound excitation is composed of numerous normal modes. However, the contribution of a particular normal mode is only significant as long as the driving frequency is close to the eigenfrequency associated with it. Therefore, each contribution has the character of a room resonance. The width of these resonance curves is closely related to the reverberation time of the room (see Sect. 10.1.4).

Unfortunately, only for simple room shapes the eigenfunctions and eigenfrequencies can be calculated and represented in a closed formula. Another difficulty is the large number of eigenfunctions needed. Consider a frequency interval from 0 to f Hertz, then the number of eigenfrequencies in this interval is

$$N_E \approx \frac{4\pi}{3}\left(\frac{f}{c}\right)^3 V, \tag{10.1}$$

whereby V is room volume in m^3. From this formula, it follows that the number of eigenfrequencies per Hertz is

$$\frac{\mathrm{d}N_E}{\mathrm{d}f} = 4\pi\frac{f^2}{c^3}V. \tag{10.2}$$

At very low frequencies, the density of eigenfrequencies is so low that the associated room resonances are well separated. At higher frequencies, however, the eigenfrequencies are so closely arranged along the frequency axis that the resonance curves show strong overlap. Both frequency ranges are separated by the so-called Schroeder frequency [1]

$$f_s = 2{,}000\sqrt{\frac{T}{V}}, \tag{10.3}$$

with T denoting the reverberation time, defined in Sect. 10.1.4. By combining Eqs. (10.1) and (10.3), one obtains the number of eigenfrequencies below the Schroeder frequency f_s, i.e. the number of well-separated resonances

$$N_E(f < f_s) = 850\sqrt{\frac{T^3}{V}}. \tag{10.4}$$

(In Eqs. (10.3) and (10.4), T must be expressed in seconds, V in m^3, and f in Hz).

The contents of these equations may be illustrated by an example: consider a room with a volume of 1,000 m^3 and a reverberation time of 1 s. The total number N_E of eigenfrequencies in the range from 0 to 5,000 Hz is more than 13 million; their density at 1,000 Hz is 305 per Hertz. The Schroeder frequency of that room is $f_s = 63$ Hz with 27 eigenfrequencies lying below this limit.

Obviously, the typical situation in room acoustics is that of strongly overlapping room resonances ($f > f_s$). In this case, the sound pressure amplitude $\tilde{p}$ varies in a quasi-stochastical manner when the driving frequency or the receiver position is changed. They follow a Rayleigh distribution, i.e. the probability of encountering a particular pressure amplitude $\tilde{p}$ with the uncertainty $d\tilde{p}$ is

$$W(\tilde{p})d\tilde{p} = \frac{1}{\sigma_p^2}\exp(-\tilde{p}^2 / 2\sigma_p^2)\tilde{p}d\tilde{p}, \tag{10.5}$$

σ_p = standard deviation of the sound pressure amplitude from the mean value. This probability density is shown in Fig. 10.1.

Figure 10.2 shows a section of a so-called *frequency curve* of a room, i.e. the steady state sound pressure level as a function of driving frequency.

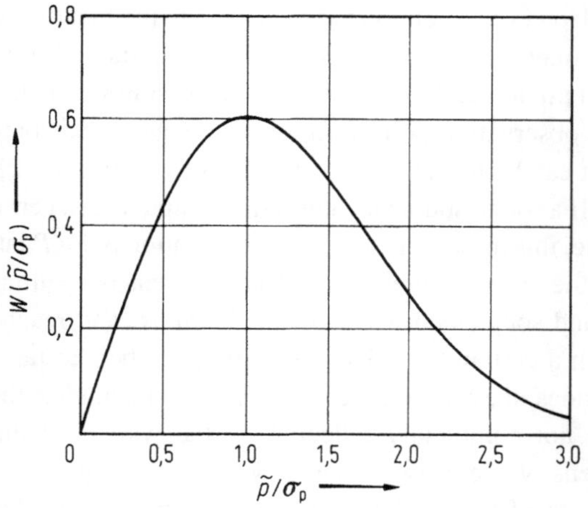

Fig. 10.1 Distribution of sound pressure amplitudes in a room related to σ_P (spatial or with respect to different frequencies)

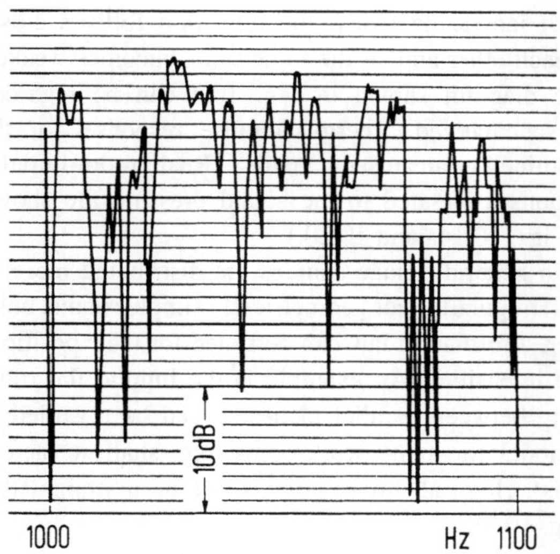

Fig. 10.2 Part of a "frequency curve" between 1,000 and 1,100 Hz, measured in a small auditorium

These irregular variations are due to the varying interaction of numerous eigenmodes that increase or diminish according to their mutual, more or less accidental phase relations. The average range of variation of such a frequency curve is approximately 10 dB, the average separation of adjacent maxima is [1]:

$$\langle \Delta f \rangle \approx 4/T. \quad (10.6)$$

If at time $t = 0$ a sound source is switched off, the decaying sound field is the sum of eigenmodes at the eigenfrequencies $f_n = \omega_n/2\pi$ and with the damping constants δ_n:

$$p(t) = \sum_n B_n \exp[j(\omega_n t - \varphi_n) - \delta_n t], t > 0. \quad (10.7)$$

The coefficients B_n and the phase angles φ_n depend on the kind and position of the sound source, on the observation point, and the spectrum of the exciting sound signal. This decay process, called *reverberation*, is of central importance in room acoustics.

Often the damping constants can be replaced without large errors by their mean value δ, the energy density of sound w then decays according to a simple exponential rule:

$$w(t) = w_0 e^{-2\delta t}, t > 0. \quad (10.8)$$

Since the validity of Eqs. (10.5)–(10.8) requires that the observation point is sufficiently separated from the sound source, the contribution of direct sound to the total sound pressure can be neglected (see Sect. 1.4).

10.1.3 Geometrical Room Acoustics

A less formal description of spatial sound fields focuses on the sound ray instead of the sound wave. The sound ray is regarded as an infinitely small part of a spherical wave. Accordingly, the sound intensity along a sound ray changes inversely proportional to the square of its distance from the origin. Just like in the field of geometrical optics, in acoustics the idea of a ray only makes sense if the dimensions of the room and its walls are large in comparison with the wavelength. As there are no inhomogeneous media in room acoustics, the propagation of sound rays is straight. There is no noticeable refraction of sound rays; diffraction is completely neglected.

The sound types dealt with in room acoustics (speech, music, noise) nearly always have a very broad and often also a fast-changing spectrum. If at one point two or more sound rays, which normally have had different ways, overlap with each other, the different sound components can be regarded as incoherent. This means that all phase differences can be disregarded and the energy density at the respective point is the sum of energy densities of the individual components (energy addition). Therefore, geometric room acoustics is restricted to energy propagation in a room.

10.1.3.1 Sound Reflection at Plane Surfaces, Image Sources

If a sound ray falls on a smooth, reflecting surface, the direction of the reflected ray can be calculated or constructed according to the law of specular reflection. This construction is especially easy if the reflecting surface is plane (Fig. 10.3). In this case, after its reflection the sound ray seems to come from a virtual sound source Q'. Concerning the reflecting surface, this virtual sound source is situated in a mirror image position to the original sound source Q. The normally incomplete reflection is considered by reducing the intensity of the reflected ray by a fraction $\rho = 1 - \alpha$ in comparison with the intensity of the incident ray. In this equation, α is the absorption coefficient of the wall. If the absorption coefficient can be assumed to be independent from the angle of incidence, the sound power of the image sound source will be reduced accordingly.

If there are several reflecting walls, i.e. especially in a closed room, image sources must be constructed for each one. In addition, each of the image sources of first order must be reflected by other walls. Thus, image sources of second order and, if the process is continued, of higher order are created, which are situated more and more distant from the room center. Let N denote the total number of walls, then the total number of image sources up to and including the kth order is

$$n_k \approx N \frac{(N-1)^k}{N-2}. \tag{10.9}$$

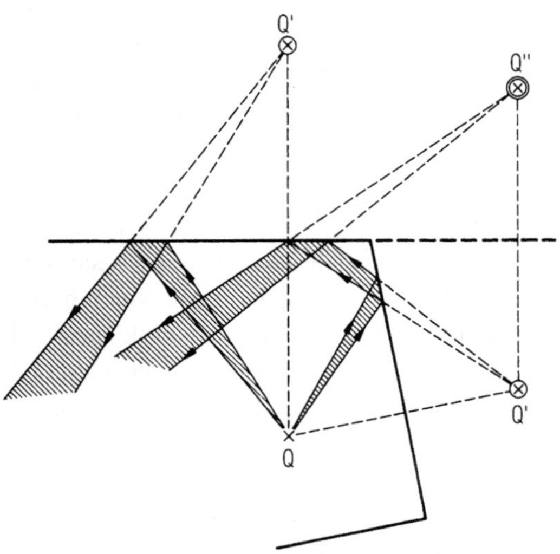

Fig. 10.3 First and second order image sources

However, attention has to be paid to the fact that not all image sources are "visible" from a certain observation point, i.e. that there is no real transmission path to the observation point from every image sound source that can be formally constructed (see e.g. Borish [2]).

If a sound source creates a short impulse at a certain time, this impulse reaches the observation point P later as the *direct sound*. The impulses created by image sound sources correspond to *reflections* and reach P with a certain time delay according to their distance; besides, they are weaker than the direct sound as they are not completely reflected by the walls and their intensity decreases inversely proportional to the square of their delays. Figure 10.4 shows the reflections as vertical lines depending on their delay with respect to the direct sound. Their lengths correspond to the respective sound pressure or intensity level. With increasing delay, the reflection follows closer to each other and, at the same time, becomes weaker. The diagram shown in Fig. 10.4 can be regarded as schematized "energy impulse response" of the examined transmission path. (Real room-impulse responses related to the sound pressure are shown in Figs. 10.14a and 10.20.)

In general, image sources are distributed more or less irregularly. For geometrically simple rooms, however, a formation law can be stated for their positions and intensities. An example is the infinite flat room lying between parallel planes. It can be regarded as a model for many workrooms (e.g. shallow factory halls). Here, all image sources are on a straight line (Fig. 10.5).

Assuming that both the original sound source Q radiating in all directions with the sound power P and the receiver E are in the middle between ceiling

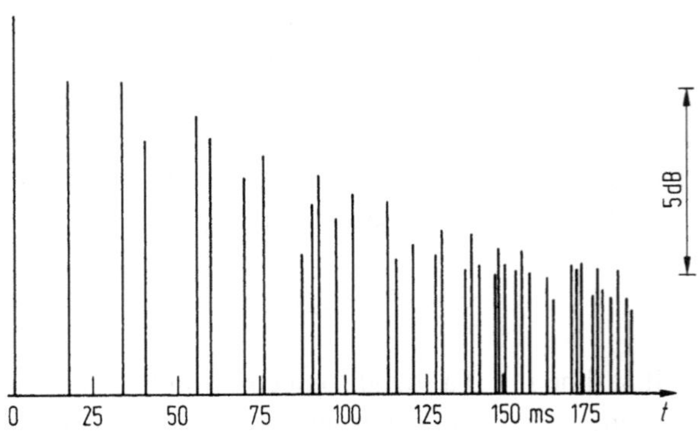

Fig. 10.4 Temporal distribution of reflections in a room with plane boundary surfaces

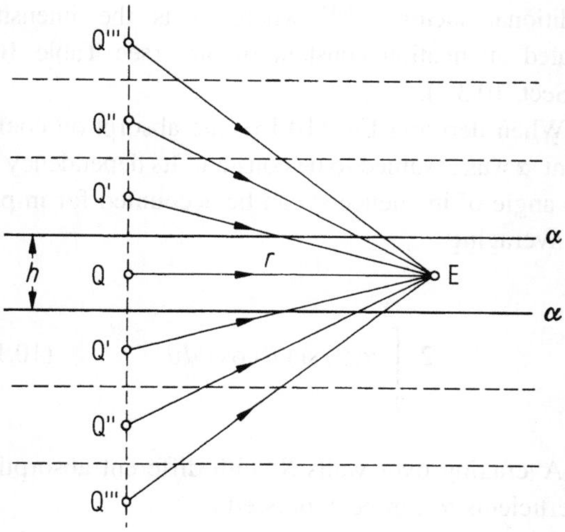

Fig. 10.5 Image sources in a flat room with infinite horizontal dimensions

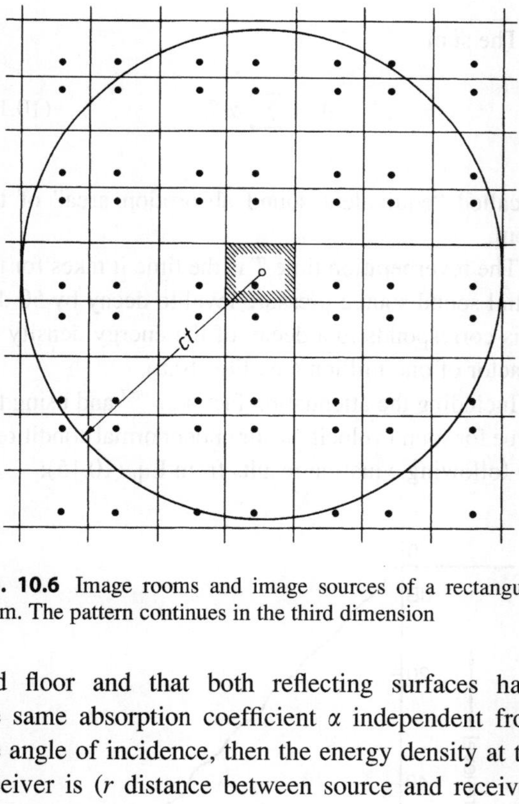

Fig. 10.6 Image rooms and image sources of a rectangular room. The pattern continues in the third dimension

and floor and that both reflecting surfaces have the same absorption coefficient α independent from the angle of incidence, then the energy density at the receiver is (r distance between source and receiver, h height of the room)

$$w = \frac{P}{4\pi c}\left[\frac{1}{r^2} + 2\sum_{n=1}^{\infty}\frac{(1-\alpha)^n}{r^2 + n^2h^2}\right]. \qquad (10.10)$$

In a rectangular room, the image sources create a regular three-dimensional pattern (Fig. 10.6). Like in the flat room, there are no "invisible" or "virtual" image sources.

10.1.3.2 Sound Propagation in Rooms with Diffusely Reflecting Walls

Often, walls are not plain, but have surface irregularities with dimensions that are in the same order as the wavelength. Thus, they do not reflect the incident sound in a certain direction, but rather scatter it in all directions. This is called *diffuse reflection*. In the limiting case of completely diffuse reflection, which can be regarded as the opposite of a specular reflection at a plane wall, the intensity of the scattered sound is proportional to the cosine of the angle of reflection (*Lambert's law*). This idealized case, like that of a plane wall, is often a good approach to the actual reflection conditions.

It is obvious that the concept of sound rays fails in this case. However, this does not exclude the application of geometric acoustical methods. For example, it can easily be accepted that the sound energy, which hits a wall element dS every second, consists of the direct contribution of the sound source and of the radiation of the other wall elements dS' (Fig. 10.7). If we relate this energy to the unit area, the following equation is valid [3]

$$B(\boldsymbol{r},t) = \iint (\boldsymbol{r}\prime) B\left(\boldsymbol{r}\prime, t - \frac{R}{c}\right) K(\boldsymbol{r},\boldsymbol{r}\prime)\mathrm{d}S' + B_d(R,t), \qquad (10.11)$$

thereby R is the distance between the wall elements dS and dS' at $\boldsymbol{r}$ and $\boldsymbol{r}'$; as before, ρ is the reflection coefficient. If Lambert's law is valid, the kernel $K(\boldsymbol{r},\boldsymbol{r}')$ of this integral equation for the "irradiation strength" B is:

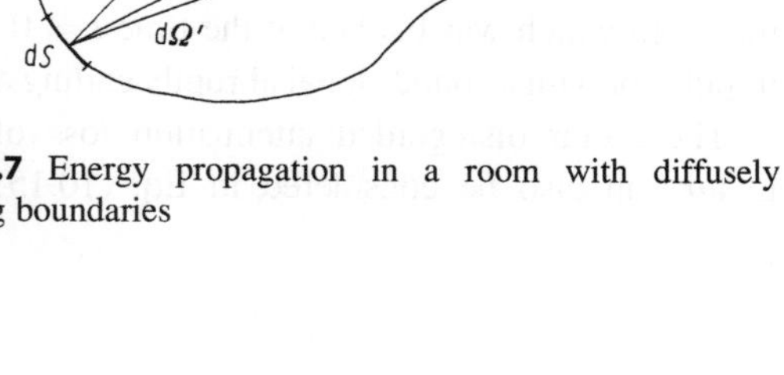

Fig. 10.7 Energy propagation in a room with diffusely reflecting boundaries

$$K(\mathbf{r},\mathbf{r}') = \frac{1}{R^2}\cos\vartheta\cos\vartheta', \tag{10.12}$$

ϑ and ϑ' are the angles formed by the ray going from $\mathbf{r}$ to $\mathbf{r}'$ and the corresponding wall normal [4]. Equation (22.10) is often referred to as "radiosity integral." If the wall reflections are partly diffuse and partly specular, the method described here can be complemented with the image source model according to Sect. 10.1.3.1 [5].

10.1.4 Reverberation Time and Steady State Energy Density in Rooms with Diffuse Sound Fields

The description of sound propagation is very simple when a *diffuse sound field* in the room can be assumed; on statistical average, at each point in the room the same amount of energy per second is coming from all directions. Then, the sound intensity vanishes everywhere. Also, the energy density w is locally constant, and the energy incident on each wall element per second and per unit area is

$$B = \frac{c}{4}w. \tag{10.13}$$

With this relation, the average number $\bar{n}$ of reflections per second can be derived as

$$\bar{n} = \frac{cS}{4V}, \tag{10.14}$$

S being the total wall surface.

As the energy transported along the ray diminishes at each reflection by the fraction $\rho = 1 - \alpha$, the total energy in the room after the period t decreases to

$$E(t) = E_0(1-\alpha)^{\bar{n}t} = E_0\exp\left[\frac{cSt}{4V}\ln(1-\alpha)\right], t>0. \tag{10.15}$$

This equation thus describes the time dependence of the reverberation, i.e. the decay of the sound energy in the room which was excited at the time $t = 0$ with an impulse or with a sound signal abruptly ending at $t = 0$.

The so-far disregarded attenuation loss of sound in air can also be considered in Eq. (10.15) by an additional factor e^{-mct} where m is the intensity-related attenuation constant of air (see Table 10.3 in Sect. 10.3.2).

When deriving Eq. (10.15), the absorption coefficient α was assumed to be constant. Its dependency on the angle of incidence ϑ can be accounted for in part by averaging

$$\alpha' = 2\int_0^{\pi/2}\alpha(\vartheta)\sin\vartheta\cos\vartheta d\vartheta. \tag{10.16}$$

Averaging over walls S_i with different absorption coefficients α_i can be expressed as

$$\bar{\alpha} = \frac{1}{S}\sum_i S_i\alpha_i. \tag{10.17}$$

The sum

$$A = \sum_i \alpha_i S_i, \tag{10.18}$$

is called "equivalent sound absorption area" of the room.

The reverberation time T is the time it takes for the initial sound source pressure level to decay by 50 dB. This corresponds to a decay of the energy density by a factor of one million (see Fig. 10.8).

Including the attenuation factor e^{-mct} and using the value for sound velocity in air under normal conditions, the following equation results from Eq. (10.15):

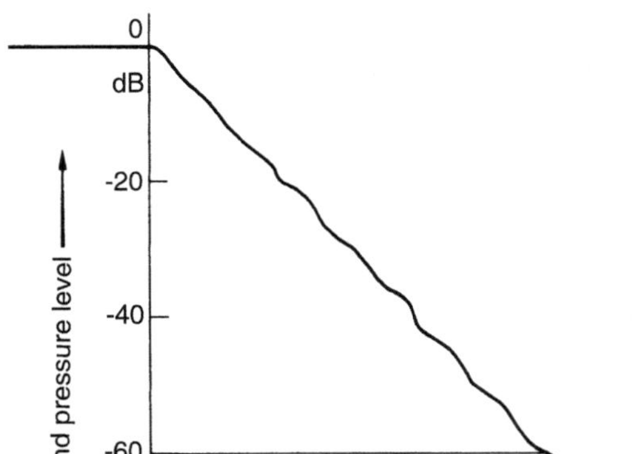

Fig. 10.8 Definition of the reverberation time T

$$T = 0.163\frac{V}{4mV - S \cdot \ln(1-\bar{\alpha})}\left(\frac{\mathrm{s}}{\mathrm{m}}\right). \quad (10.19)$$

This relation, together with Eq. (10.17), is called Eyring's formula. If the average absorption coefficient $\bar{\alpha}$ is small compared to unity, it turns the previous expression into the simpler Sabine´s formula

$$T = 0.163\frac{V}{S\bar{\alpha} + 4mV}\left(\frac{\mathrm{s}}{\mathrm{m}}\right) = 0.163\frac{V}{A + 4mV}\left(\frac{\mathrm{s}}{\mathrm{m}}\right). \quad (10.20)$$

This is the formula usually employed in practical calculations of reverberation. For small rooms, the term 4 mV may be neglected.

If a steady state sound source emits the constant power P, this leads to an equilibrium which is marked by the equality of the power emitted and the power absorbed by the walls. The latter portion can be expressed in Eq. (10.13) by the stationary energy density w_s for which the following equation holds:

$$w_s = \frac{4P}{\bar{\alpha}cS} \quad \text{for } \bar{\alpha} \ll 1. \quad (10.21)$$

With Z_0 being the characteristic impedance of air, the corresponding time-averaged square of the sound pressure can be calculated as:

$$\langle p^2 \rangle = \frac{4Z_0 P}{S\bar{\alpha}}. \quad (10.22)$$

Thus, energy density and sound-pressure level are locally constant, apart from certain deviations which are not predicted by the present theory. Partly these deviations are due to interferences between the normal modes described in Sect. 10.1.2. More information on this effect can be found in [7]. Furthermore, standing waves may occur next to a wall on account of the phase relation between incident and reflected waves imposed by the boundary conditions [8]. The energy density w_s according to Eq. (10.21) is just one part of the total energy density in a room, namely that created by wall reflections, i.e. by the reverberant field. Another part is associated with the direct sound field of the source:

$$w_\mathrm{d} = \frac{P}{4\pi c r^2}. \quad (10.23)$$

In a certain distance, the *"reverberation distance"*

$$r_\mathrm{h} = 0.057\sqrt{V/T}, \quad (10.24)$$

(r_h in m, V in m^3) both energy densities have the same value. For directed sound radiation, the reverberation distance in the main radiation direction is increased by a factor $\sqrt{\gamma}$ with γ denoting the *directivity factor* of the sound source, i.e. the ratio of the maximum intensity $I_{\max}$ to the average intensity in a certain distance r:

$$\gamma = 4\pi r^2 \frac{I_{\max}}{P}. \quad (10.25)$$

By applying the reverberation distance, the total energy density w can be described by

$$w = w_\mathrm{d}\left(1 + \frac{r^2}{r_\mathrm{d}^2}\right). \quad (10.26)$$

10.2 The Subjective Effect of Three-Dimensional Sound Fields

As described in Sect. 10.1.1, if only its sound field is known, it is not possible to draw conclusions concerning the room-acoustical qualities. For this, additional knowledge is required, like how the reflections individually or together influence the listener's perception of sound signals.

Such knowledge can be gained by systematic psycho-acoustic tests with synthetic sound fields [9, 10]. A setup to create such sound fields is shown schematically in Fig. 10.9. One or several frontal loudspeakers create the direct sound; reflections from any directions can be created by additional loudspeakers, arranged accordingly with adjustable delays and attenuators. Reverberation can be added by passing the original signal through a reverberator, nowadays realized by digital methods, and finally emitted incoherently from several loudspeakers.

Another possibility of determining the subjective effect of complex sound fields is based on the comparison of hearing impressions from various real or simulated halls (see Sect. 10.3.3). The application of factor analysis permits a separation into individual perceptive categories and, within certain limits, also

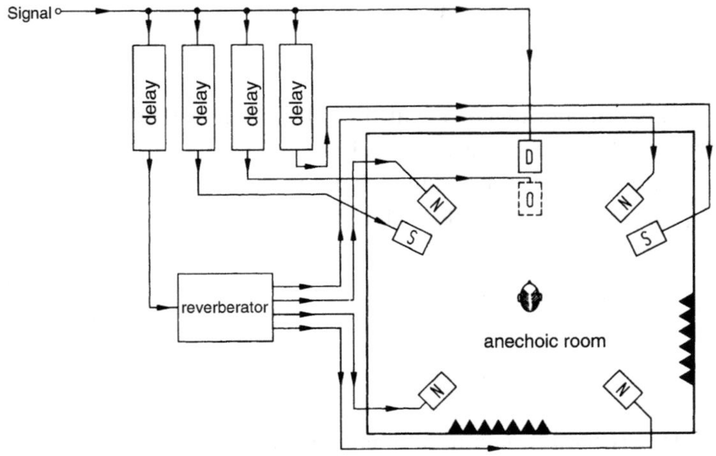

Fig. 10.9 Schematic set-up to generate synthetic sound fields in an anechoic room. *D* direct sound, *O* ceiling reflection, *S* side wall reflection, *N* reverberation

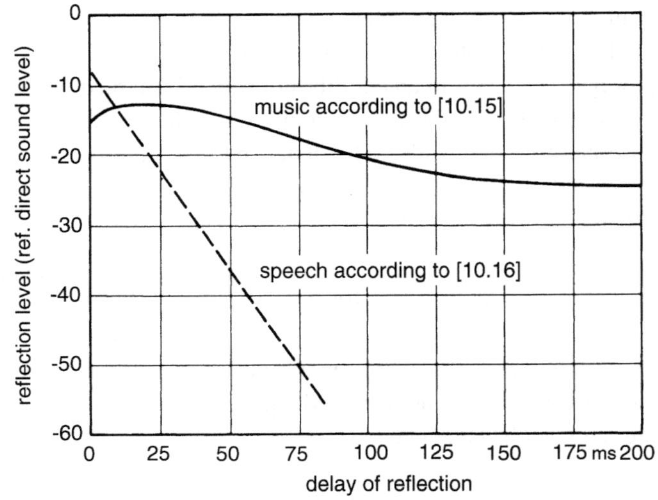

Fig. 10.10 Threshold of absolute perceptibility of a single reflection as a function of the delay for equal direction of incidence

their assignment to objective sound field properties in the respective rooms [11, 12].

10.2.1 Individual Reflections

A reflection must have a certain minimum intensity if it is to be perceived beside the direct sound. This value, which is called *threshold of absolute perceptibility*, depends on the delay between direct sound and reflection, on the direction from which it reaches the listener and on the type of the sound signal. For the same direction of direct sound and reflection, Fig. 10.10 shows the result of corresponding investigations. According to that, our sense of hearing is more sensitive to reflections regarding speech than to reflections regarding music. With an increasing delay, the threshold decreases. Further results can be found in the corresponding literature [13–16]. It should only be mentioned that reflections from lateral directions are perceived more easily than frontal reflections or reflections coming from above. The discrimination threshold for reflections, i.e. the minimum level difference which leads to a change in the hearing impression is at best ± 1.5 dB. For very high or very low reflection levels, it is higher [10].

A weak and slightly delayed reflection arriving from the same direction as the direct sound can only be sensed as an increase of loudness – if at all. On the other hand, if it hits the listener from a lateral direction, it leads to an apparent spatial extension of the sound source. This "spatial impression" is stronger the closer the angle between direct sound and reflection is to 90°. In all these cases, the *law of the first wave front* can be applied [17]. According to this law, the sound source is assigned to the direction of the direct sound.

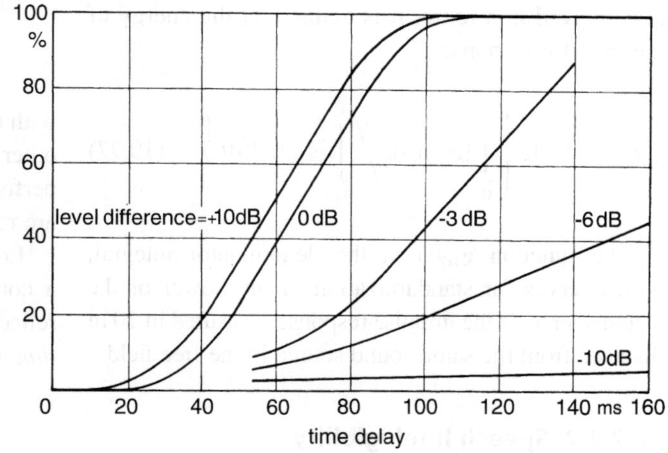

Fig. 10.11 Percentage of listeners disturbed by a single reflection as a function of time delay. The test signal is speech at a rate of 5.3 syllables per second. Parameter is the level difference between reflection and direct sound in dB

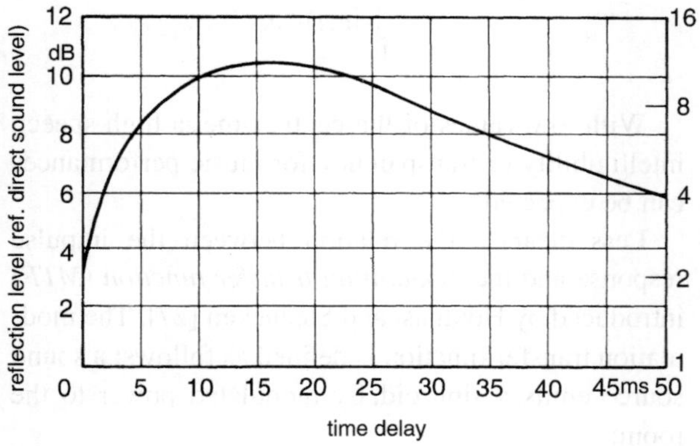

Fig. 10.12 Relative level of a reflection which just leads to an equal loudness perception or disturbs localization of the direct sound (speech)

Sufficiently intense and strongly delayed reflections are perceived as an *echo,* i.e. as a normally disturbing repetition of the direct sound signal. For speech, Fig. 10.11 shows the percentage of listeners feeling disturbed by the reflection depending on the delay [18]. Unexpectedly, this percentage increases only slightly when the reflection level related to the direct sound increases from 0 dB to +10 dB. This fact, known as *Haas effect* is also important for the design of sound reinforcement systems (see Chap. 13). It is shown in a different way in Fig. 10.12 with the relative reflection level as ordinate. If the level of the reflection exceeds a critical value, the majority of the listeners feel disturbed by the echo [19]. If the directions of direct sound and reflection are not equal, the disturbance can also be that the sound is not heard from the direction of the original sound source. Generally, for sound signals created by music, our sense of hearing is less sensitive to echoes than for speech [20].

The superposition of the direct sound and a slightly delayed reflection can also be perceived as a coloration. Especially unpleasant are periodic sequences of reflections, which for larger delays are perceived as flutter echoes. Concerning the criteria for such disturbances, see also the corresponding literature [21, 22].

10.2.2 Sequences of Reflections

The facts described in the above section are suitable for the assessment of room impulse responses dominated by a single reflection apart from direct sound. For more complicated impulse responses consisting of many reflections (see Fig. 10.4), the subjective effect of each reflection depends on its embedment, so that a separate assessment is normally not possible. Instead, more general measures are applied which are derived from the impulse response of the room. By means of these measures, statements, e.g. about *speech intelligibility* or the *transparency* of music can be made. Considering the incident directions of sound, also statements about the subjective *spaciousness (apparent sound width)* of the respective sound field are possible. Furthermore, criteria for the occurrence of disturbing echoes can be derived from a measured or calculated impulse response.

10.2.2.1 Strength Factor

A first criterion, the strength factor G, serves as an indication of how loud a sound source or performance is perceived in a room. It is defined by the energy of the impulse response

$$G = 10 \cdot \lg\left\{ \int_0^\infty [g(t)]^2 \mathrm{d}t \Big/ \int_0^\infty [g_{10}(t)]^2 \mathrm{d}t \right\}. \quad (10.27)$$

The function g_{10}(t) in the denominator integral, which serves for standardization of the power of the sound source, is the impulse response measured in 10 m distance from the same sound source in the free field.

10.2.2.2 Speech Intelligibility and Transparency

The criteria described in the following are mostly based on the often confirmed experience that moderately delayed reflections mainly support direct sound. Thus, they improve sound transmission and can be regarded as "useful." Although there is no clear difference between useful and "bad" reflections, the cut is often marked by a sharp line in the time delay. According to this, the useful and the rather disturbing part of the energy contained in an impulse response is:

$$E_N = \int_0^{t_0} [g(t)]^2 \mathrm{d}t, \quad (10.28)$$

and

$$E_S = \int_{t_0}^\infty [g(t)]^2 \mathrm{d}t. \quad (10.29)$$

In these equations, g(t) is the impulse response of the room beginning at $t = 0$ with the direct sound, i.e. the sound pressure at the observation point after excitation of the room with an impulse of very short duration. On this basis, Thiele [23] introduced the *definition*

$$D_{50} = \frac{E_N}{E_N + E_S}, \quad (10.30)$$

with $t_0 = 50$ ms in order to describe speech intelligibility. If it exceeds the value of 0.5, a syllable intelligibility of more than 90% can be expected. The *clarity index* according to Reichardt, Abdel Alim and Schmidt [24]

$$C = 10 \lg\left(\frac{E_N}{E_S}\right) \mathrm{dB} \quad (10.31)$$

with $t_0 = 80$ ms (in literature, often called C_{80}), on the other hand, characterizes the *transparency* of music performances; values in the range of -3 dB up to 0 dB are regarded as favorable for symphonic music.

Lochner and Burger [25] were the first to use a continuous transition from useful and detrimental reflections. All the more, this applies to the *centre time* introduced by Kürer [26]:

$$t_s = \frac{\int_0^\infty [g(t)]^2 t \mathrm{d}t}{\int_0^\infty [g(t)]^2 \mathrm{d}t}. \quad (10.32)$$

With low values of the centre time, a high speech intelligibility or transparency for music performances can be expected.

Less clear is the relation between the impulse response and the *modulation transfer function* (*MTF*) introduced by Houtgast and Steeneken [27]. The modulation transfer function is defined as follows: a sound source emits a sinusoidally modulated power to the room:

$$P(t) = P_0(1 + \cos \Omega t). \quad (10.33)$$

The reverberation of the room as well as possibly existing disturbing noise lead to a smoothing of the intensity variations at the point of observation by a factor m (and also to a delay by the time t_0):

$$I(t) = I_0\{1 + m \cos[\Omega(t - t_0)]\} \quad (10.34)$$

This factor, which depends on the modulation frequency Ω, but also on the type of the sound signal, is the modulation transfer function. For white noise as exciting signal, it can directly be calculated from the impulse response [28]:

$$m(\Omega) = \left| \int_0^\infty [g(t)]^2 e^{j\Omega t} \mathrm{d}t \Big/ \int_0^\infty [g(t)]^2 \mathrm{d}t \right|. \quad (10.35)$$

By means of a comparatively complicated evaluation, a single number criterion is determined from the modulation transfer function, the *speech transmission index* (STI), which correlates very well with speech intelligibility. Also, the *rapid speech transmission index* (*RASTI*) is important in practice, which is determined in an abbreviated evaluation procedure and is described in more detail in Sect. 10.5.

10.2.2.3 Spatial Impression

As already mentioned in Sect. 10.2.1, a reflection that is slightly delayed contributes the more to the subjective impression of spaciousness, the more it arrives from lateral directions. This is especially important for concert halls. According to Barron [29], who carried out the first systematic investigations regarding this subject, this does not only apply to an individual reflection, but to all reflections with delays between 5 and 80 ms. Accordingly, the *lateral efficiency* which is defined by the following equation

$$\mathrm{LFC} = \frac{\int_{5\,\mathrm{ms}}^{80\,\mathrm{ms}} [g(t)]^2 \cos\theta \mathrm{d}t}{\int_{0}^{80\,\mathrm{ms}} [g(t)]^2 \mathrm{d}t}, \qquad (10.36)$$

can be used to describe the impression of spaciousness connected with the respective sound field. In this equation, the angle of incidence of a certain part of energy is described by the angle θ between the incidence direction and the axis connecting both ears (see Fig. 10.13). For an alternative definition of the lateral efficiency, which is called LF, the numerator in Eq. (10.36) contains the square of the cosine function. In contrast to LFC, LF can be measured directly using a microphone with gradient characteristics.

Most important for the spatial impression is the fact that reflections coming from the sides create different sound signals at both ears. The difference can be characterized by a short-term cross correlation function:

$$\varphi_{rl} = \frac{\int_0^{t_0} g_r(t) g_l(t) dt}{\left[\int_0^{t_0} [g_r(t)]^2 dt \int_0^{t_0} [g_l(t)]^2 dt\right]^{1/2}}, \qquad (10.37)$$

with $t_0 = 100$ ms. Here, $g_r(t)$ and $g_l(t)$ are the impulse responses occurring at both ears of the listener. The maximum of this function in the interval $|\tau| < 1$ ms is called *Interaural Cross Correlation, IACC* [30]. The smaller this maximum, the higher is the perceived spaciousness of the corresponding sound field. Apart from this quantity, different versions of the IACC are used occasionally.

More recently, it has been suspected that the impression of spaciousness consists at least of two components, that is the "apparent source width" (*ASW*) that is described by the above-mentioned measures and the "listener envelopment." [31] According to Bradley and Soulodre [32], the latter is characterized best by the lateral energy contained in the later parts of the impulse response:

$$\mathrm{LG}_{80}^{\infty} = 10\log_{10}\left[\frac{\int_{80\,\mathrm{ms}}^{\infty} [g(t)\cos\theta]^2 \mathrm{d}t}{\int_0^{\infty s} [g_{10}(t)]^2 \mathrm{d}t}\right]. \qquad (10.38)$$

In this equation again, g_{10} is the impulse response measured in the free field in 10 m distance.

10.2.2.4 Echoes

Disturbing echoes can be detected very well by the echo criterion of Dietsch und Kraak [33]. It can be determined from the energy impulse response as follows – at first, the build-up function of the centre time

$$t_s(\tau) = \frac{\int_0^{\tau} |g(t)|^n t \mathrm{d}t}{\int_0^{\tau} |g(t)|^n \mathrm{d}t}. \qquad (10.39)$$

is calculated with the exponent n (see Table 10.1). (For $n = 2$ and $\tau = \infty$ this formula provides the centre time defined in Eq. (10.32).) With this function, the

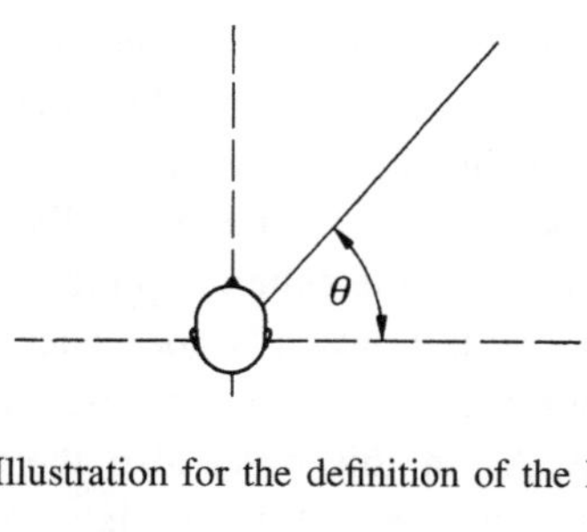

Fig. 10.13 Illustration for the definition of the lateral energy fraction

Table 10.1 Characteristic data for the echo criterion according to Dietsch and Kraak [32]

Type of signal	Exponent n	Time difference Δt (ms)	$(\Delta t_s/\Delta\tau)_{critical}$	Bandwidth of test signal (Hz)
Speech	2/3	9	1.0	700...1,400
Music	1	14	1.8	700...2,800

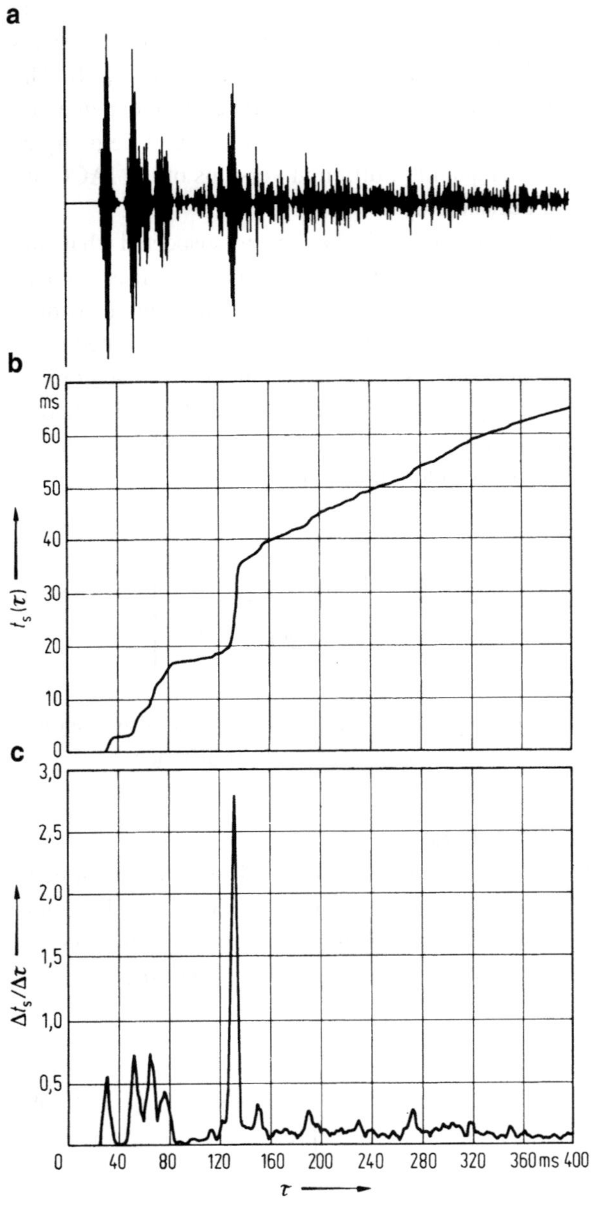

Fig. 10.14 Echo criteria of Dietsch and Kraak. **a** First 400 ms of measured room impulse response; **b** corresponding build-up function for $n = 2$; **c** difference quotient with $\Delta t = 5$ ms

difference quotient $\Delta t_s/\Delta\tau$ is calculated. In Fig. 10.14, the necessary steps are shown with an example. If the maximum value of the difference quotient exceeds a certain value, which also depends on whether music or speech is concerned, a perceptible echo must be expected. Table 10.1 shows the critical maximum values of the difference quotient for speech and music as well as the recommended values of the exponent n and the time interval $\Delta\tau$, both resulting from extensive listening tests. Moreover, experience shows that it is sufficient to limit the bandwidth of the test signal to one or two octaves.

10.2.3 Reverberation

The most characteristic acoustic feature of a room probably is its reverberation, i.e. the gradual sound decay after the sudden stop of a sound source or after an impulse-like sound. As both the whole sound power and the spectral content of the signal change continuously during speech and music performances, the hearing impression of the listener depends on the kind and duration of the reverberation of the room. Sabine [34] was the first who clearly recognized these facts. Moreover, he not only described the reverberation quantitatively but also investigated its dependence on characteristics of the room (size, absorption), and thus is rightly considered to be the pioneer of modern room acoustics.

Whether the reverberation of a room, for which the relations mentioned in Sect. 10.1.4 are valid, has a positive or negative effect, mainly depends on its duration, on the frequency dependence of the reverberation time, and the kind of use. Moreover, the size of a room plays a certain role as one instinctively expects a longer reverberation in larger rooms than in smaller rooms and considerable deviations from these expectations lead to unusual or perhaps even unnatural sensations.

The optimum reverberation time for a certain kind of performance and room size can only be determined empirically, i.e. by means of specific listening tests or opinions on existing halls. Early results of such listening tests for speech were already published in 1925 (see [35]). However, the general opinion on optimum reverberation times nowadays goes back to surveys carried out among visitors, musicians, music critics, and the like. As in this case, also the differences in judgment and personal taste play an important role, it is no wonder that only ranges of favorable reverberation times can be determined, but no distinct optimum values. Moreover, the subjective ability to distinguish different reverberation times is not more than 5% [36], so that it does not make sense to determine the reverberation time of large rooms more exactly than to 0.05–0.1 s. The ranges of reverberation time, which are favorable for different performances and room types, are stated in Sect. 10.4.

The description of the sound decay by the reverberation time of course only makes sense if the decay process is approximately exponential, see Eq. (10.15). According to Atal, Schroeder and Sessler [37], in all other cases the early decay determines how the reverberation of speech or music is perceived by the listener. This insight led to the introduction of the *Early Decay Time, EDT* by Jordan [38]: The Early Decay Time is six times the duration of the sound level decrease by 10 dB after having switched off the sound source.

10.3 Room-Acoustical Planning Methods

The aim of each room-acoustical design is to get a picture of the acoustical characteristics of the planned room (sound distribution, reverberation time, etc.). Moreover, possible acoustical shortcomings should be recognized at an early planning stage and appropriate measures be developed to avoid them. For this, different planning instruments can be used which are described in the following.

10.3.1 Graphical Construction of Sound Rays

It is possible to obtain a general idea of the sound supply and also of the potential echo risk by simple geometric-acoustic considerations, especially by the graphic construction of sound paths in a room. The construction of sound paths is especially easy if the room is limited by plane, specularly reflecting surfaces. With the help of image sources (see Sect. 10.1.3.1), it is possible to determine all first-order sound reflections from those walls which are perpendicular to the plane of the drawing. However, in case of multiple reflections, this method becomes rather involved, and in case of curved surfaces it fails completely. Here, for each single sound ray the normal at the point of incidence must be determined.

If r_i is the total length of a ray leading from the sound source to the receiver and the length of the shortest connecting line between the two points, then the delay of the respective reflection compared with the direct sound is given by

$$t_i = (r_i - r_0)/c, \qquad (10.40)$$

whereas the level difference is at least

$$\Delta L = 20\lg(r_i/r_0). \qquad (10.41)$$

The latter equation, which only considers the level decrease due to spherical wave propagation, is only valid for reflections at surfaces that are plane and smooth in terms of the wave length. If the room shapes are more complicated, the graphic illustration of sound paths may become intricate, as in this case methods of descriptive geometry must be applied. Here, the sound transmission should be investigated by means of room models (Sect. 10.3.4) or by mathematical simulation (Sect. 10.3.3). This also applies if consideration of the first reflections only turns out to be insufficient and a more complete picture of the sound distribution would be desirable.

10.3.2 Calculation of the Reverberation Time

Corresponding to its importance, it is recommended to calculate the reverberation time for different frequency ranges already at an early design stage, and to adapt this calculation from time to time corresponding to the planning progress. For this, in general, Sabine's equation (10.20) might suffice, although it slightly overestimates the reverberation times.

In most cases, the calculations are made for octave bands with mid-frequencies from 125 to 4,000 Hz. Geometrical input data like room volume and the surface areas of the room boundaries are taken from architectural drawings. Sound absorption coefficients of wall materials are found in published tables (e.g. [39–41]) or possibly in test certificates of manufacturers of sound-absorbing materials. Table 10.2 lists data of some typical materials. Discrete objects are taken into account by

Table 10.2 Sound absorption coefficients of materials (exemplary planning values)

Material	Frequency (Hz)					
	125	250	500	1,000	2,000	4,000
Concrete, plaster, stone	0.02	0.02	0.03	0.04	0.05	0.05
Planks, parquet on battens	0.10	0.08	0.06	0.05	0,05	0.05
Linoleum on felt layer	0.02	0.05	0.10	0.15	0.07	0.05
Carpet, approx. 5 mm thick	0.03	0.04	0.06	0.20	0.30	0.40
Window, door	0.12	0.08	0.06	0.05	0.05	0.05
Gypsum board wall (50 mm studs)	0.25	0.11	0.07	0.06	0.06	0.06
8 mm plywood, wall distance 60 mm, filled with 30 mm mineral fibre	0.50	0.15	0.07	0.05	0.05	0.05
9.5 mm perforated gypsum board, open area approx. 15%, wall distance 60 mm, filled with 30 mm mineral fibre	0.40	0.95	0.90	0.70	0.65	0.65
0.5 mm perforated metal sheets, open area approx. 15%, wall distance 60 mm, filled with 30 mm mineral fibre	0.45	0.7	0.75	0.85	0.8	0.6
20 mm tiles of mineral fibre, laminated with painted fleece, ceiling distance 300 mm	0.50	0.70	0.74	0.90	0.93	0.85
Audience	0.50	0.70	0.85	0.95	0.95	0.90

Table 10.3 Intensity attenuation constant m (in $10^{-3}\ \mathrm{m}^{-1}$) of air at 20°C and normal atmospheric pressure (according to Bass et al. [6])

Relative humidity (%)	Frequency (Hz)					
	500	1,000	2,000	4,000	6,000	8,000
40	0.60	1.07	2.58	8.40	17.71	30.00
50	0.63	1.08	2.28	6.84	14.26	24.29
60	0.64	1.11	2.14	5.91	12.08	20.52
70	0.64	1.15	2.08	5.32	10.62	17.91

their absorption area, which is added to the total absorption area of the room (see Eq. (10.18)). Air absorption must only be regarded for larger rooms and higher frequencies. Values of the attenuation constant m depending on the air humidity are listed in Table 10.3.

An important element of uncertainty is the sound absorption of single persons or audience areas, as it depends on the clothes, audience density, and arrangement of seats as well as on the inclination of the audience surface area. The absorption coefficients stated in the last line of Table 10.2 are recommended by Cremer and Müller [42] as average planning values. For concert halls or opera houses, it may be recommendable to measure the absorption of the seating in the reverberation room with and without persons, e.g. according to the method stated by Kath and Kuhl [43]. Alternatively, the absorption coefficients determined by Beranek and Hidaka (see below) by means of reverberation time measurements in various concert halls and opera houses can be used. These absorption coefficients are listed in Table 10.5 and can also be used for other halls. The sound absorption of the surfaces facing the corridors is taken into account by adding a 0.5-m wide strip at the perimeter of each seating block to the projected surface.

The sound absorption by single persons is taken into consideration by their absorption area. Some typical values are listed in Table 10.4. Often, at an early planning stage the design concept is known, whereas the wall material to be used is not yet defined. In such a case, it is recommended first of all to estimate the reverberation times to be expected on the basis of experience values. For this, a method described by Beranek and Hidaka [44] is suitable, which can at least be used for concert halls and comparable rooms. This method only distinguishes between two kinds of surfaces: the surface occupied by the audience S_p, considering the influence of the boundary as described above, and the sum of all other surfaces S_r ("residual area"). The corresponding absorption coefficients α_p and α_r are listed in Table 10.5. With these values, the reverberation time can be calculated on the basis of equation (10.20) (with $m = 0$) according to:

Table 10.4 Absorption areas of single persons

	Absorption area (in m^2)					
	125	250	500	1,000	2,000	4,000 Hz
Standing single person[a]	0.15	0.25	0.60	0.95	1.15	1.15
Sitting single person[a]	0.15	0.25	0.55	0.80	0.90	0.90
Musician with instrument[b]	0.60	0.95	1.05	1.10	1.10	1.10

[a]According to [41]
[b]According to [43]

Table 10.5 Absorption coefficients of unoccupied and occupied seating areas and residual absorption coefficients, determined from reverberation measurement in concert halls (according to Beranek and Hidaka [44])

Type of seats		Frequency					
		125 Hz	250 Hz	500 Hz	1,000 Hz	2,000 Hz	4,000 Hz
Heavily upholstered	Unoccupied	0.70	0.76	0.81	0.84	0.84	0.81
	Occupied	0.72	0.80	0.86	0.89	0.90	0.90
Medium upholstered	Unoccupied	0.54	0.62	0.68	0.70	0.68	0.66
	Occupied	0.62	0.72	0.80	0.83	0.84	0.85
Lightly upholstered	Unoccupied	0.36	0.47	0.57	0.62	0.62	0.60
	Occupied	0.51	0.64	0.75	0.80	0.82	0.83
Residual absorption of halls lined with wood (thickness < 3 cm) or other light materials		0.16	0.13	0.10	0.09	0.08	0.08
Residual absorption of halls lined with heavy material, e.g. with concrete, plaster more than 2.5 cm thick		0.12	0.10	0.08	0.08	0.08	0.08

Table 10.6 Coupled rooms

	Listener is in the Source room	Coupled room
Source room is more reverberant	Due to the larger absorption in the coupled room, the coupling area can be regarded as absorbent	The reverberation is heard from the coupling area. The reverberation time mainly depends on the acoustical properties of the source room
Coupled room is more reverberant	When the sound emission abruptly stops, reverberation might be heard from the coupling area	No particularities

$$T = \left(0.163\frac{s}{m}\right)\frac{V}{S_p\alpha_p + S_r\alpha_r}. \tag{10.42}$$

Special conditions may occur when the room cannot be considered as one space but is divided into one or more volumes, which are coupled to the main room (stage towers in theaters, areas underneath deep balconies, churches with several naves etc.) [42]. The most important cases are shown in Table 10.6 assuming that a sound source is situated only in one of the coupled spaces.

10.3.3 Computer Simulation of Sound Propagation

Nowadays, computer simulation of sound propagation in rooms is a very flexible prediction tool. It can be used to determine room-acoustical criteria and to "auralize" sound propagation in non-existing rooms, i. e. to create audible impressions based on objective room data (see Sect. 10.3.5). Meanwhile, such room-acoustical simulation programs are commercially available. However, it has to be taken into account that realistic predictions require appropriate algorithms as well as sufficient experience of the user. This was shown, for example, by two international round robin tests of room-acoustical simulation programs [45, 46].

In the following, some basics of room-acoustical simulations are described with the focus on geometric-acoustic methods. The numerical solution of the wave equation, e.g. by the help of the boundary element method, is not considered because the application of this method is limited to small rooms and/or low frequencies owing to the required computation time.

10.3.3.1 Room Modelling

The basis of a room-acoustical simulation is a digitized room model. Three-dimensional CAD-plans elaborated by architects are normally less suitable, as such plans contain too much details. Thus, architectural plans usually serve only as a basis for an acoustical room model. In such a room model, the individual surfaces are normally described by plane surfaces limited by polygons. However, it is also possible to take curved surfaces into account [47]. Moreover, the position and orientation as well as acoustical characteristics (frequency-dependent directivity) of sound source and receivers are described analytically or by means of measured data. Figure 10.15 shows an acoustical computer model for illustration.

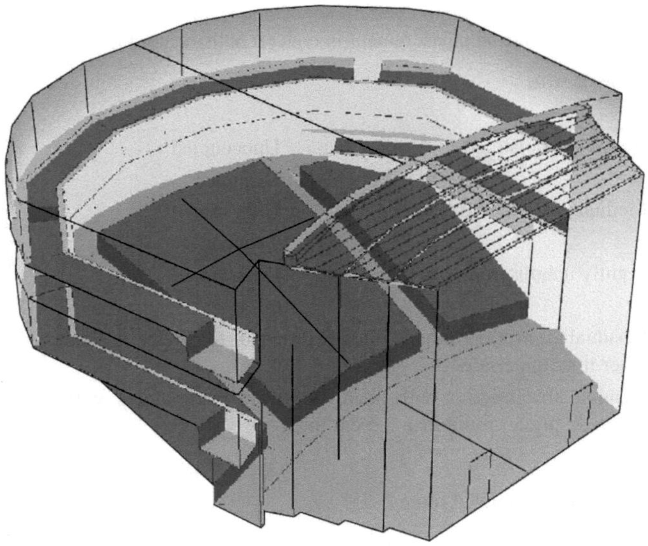

Fig. 10.15 Computer model of a small concert hall ($V = 2{,}200\ \text{m}^3$)

The next step is to assign acoustical characteristics to the surfaces of the room. Within the scope of geometrical considerations, in the first place the frequency-dependent absorption coefficient is assigned, which can be taken from literature or manufacturer's data. Moreover, it has to be considered that only "smooth" surfaces reflect sound specularly. However, if the surfaces show irregularities as, for example, recesses, projections, and the like that are neither very large nor very small in comparison with the wave lengths of interest, which is mostly the case, the scattering properties of the surfaces must be included at least approximately. Quantitatively, this can be made by indicating a scattering coefficient, i.e. the ratio of not specularly to totally reflected energy. Up to now, no data tables or simple calculation rules for the scattering coefficient are available. However, there are methods that allow to measure the frequency-dependent-scattering coefficient [48].

10.3.3.2 Ray Tracing

If the simulation only serves to predict room-acoustical energy criteria like reverberation time, strength factor, or other energy parameters, the ray tracing method can be used [49–52].

Here, a sound source is imagined to send out numerous sound particles, that propagate along straight lines. If necessary, the directivity of the sound source can be taken into consideration. Now, each particle is traced while propagating through the room. If it hits a wall, it is reflected either specularly or diffusely. To decide this, a random number between 0 and 1 is generated. If this number is smaller than the scattering coefficient, the particle is scattered. The new direction can then be determined by another pair of random numbers, the distribution of which corresponds to the scattering characteristics of the wall, e.g. to Lambert's law. The absorption coefficient α can be accounted for in two ways: either by reducing the energy of the particle by a factor $1 - \alpha$, or by interpreting α as "absorption probability", i.e. by generating another random number that decides whether the particle will proceed or end its trip.

The results are collected by means of previously defined detector areas or volumes. Whenever a particle hits such a detector, its arrival time, its energy and, if necessary, also the direction is registered. After tracing all particles, the results are classified according to their arrival time. Thus, for each detector, a chronological energy distribution is obtained that represents an approach toward the energy-impulse response (Fig. 10.16).

From these results, it is possible to calculate certain characteristic parameters of the room like definition, strength factor, clarity index, centre time as well as reverberation times (e.g. *EDT*, T_{20}) described in Sect. 10.2.2. If the direction where the sound particles come from is also considered, the lateral energy defined in Eq. (10.36) or the *IACC* defined in Eq. (10.37) can be determined.

Because of the frequency-dependent scattering properties of the surfaces, it is necessary to carry out different calculations for the various frequency ranges.

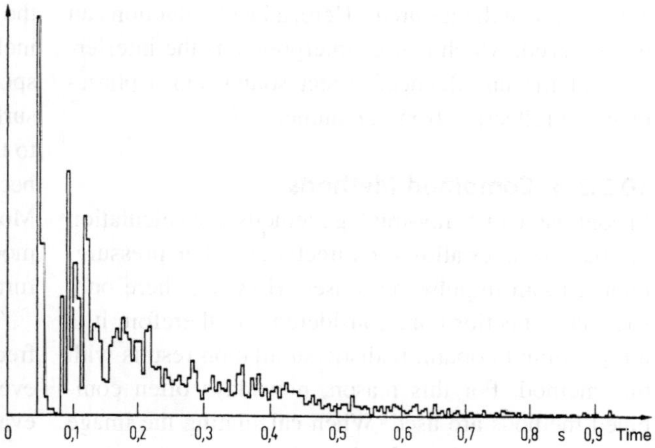

Fig. 10.16 Energy impulse response, obtained by ray tracing (time interval 5 ms) [52]

A prerequisite for realistic results is an adequate consideration of the frequency-dependent scattering coefficients of the walls. If this is the case, by means of the sound particle method it is also possible to determine irregular and locally different decays, e.g. with bended decay curves. Moreover, this method is also suitable for the determination of sound propagation in factory buildings by additionally modeling the scattering at fittings and machines [53].

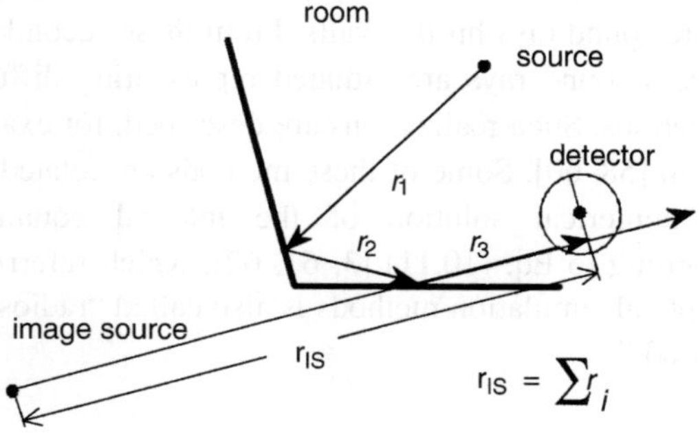

Fig. 10.17 Predicting valid mirror-image sources by means of a ray tracing process

10.3.3.3 Determination of Image Sources

As described above, the sound particle method only gives a probability distribution of the energy-time curve for an impulse-like excitation. Thus, the results are no pressure-related impulse responses as they are needed for auralization (see Sect. 10.3.5). In this respect, a simulation of the sound propagation on the basis of the image source model mentioned in Sect. 10.1.3.1 would be more suitable. However, this model requires unacceptably long computation times, except if only rectangular rooms are considered [54]. In all other cases, a time-consuming determination of the few "visible" image sources is necessary (see Sect. 10.1.3.1). This can be avoided by determining only the visible image sources by a shortened ray-tracing procedure, which precedes the actual sound field calculation [55, 56]. For this purpose, sound particles are also traced in the room, but only geometrical reflections are considered. If a sound particle hits a detector, a potential image source is located on the axis of the incoming sound ray and at a distance corresponding to the path length of the particle (see Fig. 10.17). As normally more than one particle hits the same wall sequence, it also has to be checked if the image source had already been detected before.

Finally, the impulse response for a given source–receiver combination results from superposition of the contributions of all valid image sources by taking into account their delay, the energy decrease caused by distance and reflection losses, and the directivity of the source and/or receiver.

In principle, it is also possible to consider angle-dependent complex reflection factors which, however, are usually not known. For this reason, the magnitude of the reflection factor neglecting its phase way be used, derived from the absorption coefficient for random sound incidence. This is justified since the exact phase shifts of the reflections are usually not relevant for the perception and in addition the angle dependence is small. However, there are also exceptions where the angle dependence as well as the phase shifts play an important role. This is the case when direct sound and early lateral reflections propagate at grazing incidence above large audience areas. Here, a level reduction can be observed, which can be interpreted as the interference of the uninfluenced direct sound and a phase-reversed reflection from the audience [57].

10.3.3.4 Combined Methods

In contrast to pure ray-tracing methods, the calculation of image sources allows to directly calculate pressure-related room impulse responses. However, here only specular reflections are considered and therefore it is not possible to obtain realistic simulation results with this method. For this reason, nowadays often combined methods are used. When calculating the image sources, for each reflection the scattered energy is extracted and secondary sources are generated where sound rays hit the walls. From these secondary sources sound rays are radiated representing diffuse reflections. Such realizations are described, for example, in [58–60]. Some of these methods are related to the numerical solution of the integral equation according to Eq. (10.11) [3, 61, 62], which referring to optical simulation methods is also called "radiosity method."

10.3.4 Physical Model Tests

In contrast to computer simulations, physical models have already been used for a long time for the prediction of sound propagation. Already in the nineteenth century, investigations of two-dimensional room sections were carried out using water waves and later also optically by means of Toeplers "Schlieren" method (see [42]).

Optical models make use of the fact that the propagation laws for sound waves and light waves are similar to some extent. In the simplest case, a mirror foil is glued on sound-reflecting surfaces, whereas sound-absorbent surfaces can be painted matt black. The stationary energy distribution, e.g. in the audience area, can be photographed. Single sound transmission paths can be made visible by means of laser rays.

Recent developments in acoustical measurement techniques and digital signal processing make it possible to carry out accurate room-acoustical model tests with sound waves. In acoustic scale modeling, the sound wavelength has to be reduced in accordance with the model scale, e.g. 1:20. (The sound frequencies in the model are increased accordingly.) In addition, the sound-absorption coefficients of the surfaces including the audience in the model have to correspond at the model frequencies to those of the real surfaces at the original frequencies. The same applies to the air attenuation, which is especially problematic because of its complicated frequency dependence. Moreover, the bandwidth and directivity of the model sources must correspond to the original. For further details see, e.g. [63, 64].

The application is simplified when the model frequencies are limited to roughly 30 kHz. Thus, even at a scale of 1:20 the 1 kHz octave band can be evaluated. The surface modeling is especially simple for concert halls since their surfaces are usually sound reflecting (except for the sound absorbent audience areas).

The advantage of model measurements in comparison with computer simulations is that sound diffraction is considered. Thus, measured impulse responses can also be investigated very well concerning early reflections or sound concentrations. But also for investigations of specific effects, like sound fields in coupled rooms, model measurements are well suited. However, for more exact predictions of the reverberation time model measurements are less accurate because of reasons described above.

10.3.5 Auralization

The auditory impression to be expected in a room can also be demonstrated directly by binaural auralization. For this purpose, binaural room impulse responses which contain the directivity characteristics of the human ear, are convolved with a "dry" sound signal and are presented, e.g. via headphones.

Binaural room impulse responses can be measured e.g. in acoustic scale models by means of a miniature dummy head [65]. They can be simulated as well with computer models, by convolving the direct sound and each reflection with the directional-dependent outer ear impulse responses (e.g. [66, 67]). The outer ear impulse responses have to be measured in advance for discrete incidence directions using dummy heads or individual persons.

The convolution of a "dry" speech or music signal can be realized in real time by powerful signal processors. The results can be presented by using specially equalized headphones. Alternatively, loudspeakers in combination with crosstalk cancelation can be used in an anechoic environment [68].

An overview of different aspects of auralization is given by Kleiner et al. [69] and Vorländer [70].

10.4 Principles of Room-Acoustical Planning

The desired room-acoustical conditions of a room essentially depend on its intended use. For example, in concert halls for classical music usually a longer reverberation as well as a good acoustical spaciousness are important, whereas in lecture halls, speech theaters, etc. a good speech intelligibility is required.

Consequently, the first step toward a successful room-acoustical planning is to determine the intended uses in close cooperation with the building owner, the users, and the architects. On such occasions, it often becomes obvious that a room must be suitable for different kinds of music as well as for speech. Here, it is important either to make a compromise corresponding to the main uses or to adapt the acoustical characteristics to the respective kind of performance by means of variable mechanical or electroacoustical measures (see Sect. 10.4.5).

In the following, some general acoustical demands are stated. Furthermore, specific room-acoustical requirements for some room groups and room uses are described and measures are stated to meet these requirements. In Sect. 10.4.8 work spaces and other rooms are considered, which are not used for acoustical performances, but where noise reduction is most important.

10.4.1 General Remarks

10.4.1.1 Background Noise Level

A condition for good speech communication, unspoiled pleasure of listening to music, or undisturbed sound recordings, is a sufficiently low background noise level. This level mostly consists of external noise, noise from adjacent rooms or of H/V devices. Noise control measures for these sources are usually considered within the building acoustical planning and the planning of the H/V devices (see Chap. 8). However, from the room acoustical point of view, it is important to determine the maximum permissible background noise levels corresponding to the use of the room. In concert halls and opera houses, the background noise level should normally not exceed 25 dB(A). In conference rooms, class rooms, or lecture halls a value of 35 dB(A) should be aimed at. Additionally, in case of noise-sensitive rooms, like recording studios, concert halls, or theaters, it may be necessary to make also demands on the spectral composition of the disturbing noise.

10.4.1.2 Arrangement of the Audience

The strength of the direct sound first of all depends on the distance of the listener's seat to the sound source. From this, consequences for the design of the seating layout can be derived. The acoustical demands relating to this are often consistent with visual aspects. If, for example, in a theater, the mimicry must be perceivable, the distance toward the front edge of the stage should not exceed 24 m; expressive movements of actors still can be seen from a distances of up to 32 m (opera).

Moreover, in large halls it has to be considered that sound, which propagates quasihorizontally over the heads of the listeners, experiences a remarkable frequency-dependent additional attenuation [71–73]. This effect can be reduced considerably by inclining the seating area. It is especially favorable to increase this inclination constantly or by sectors, so that the angle of incidence toward the audience only varies little. This can be achieved if the inclination follows a logarithmic spiral [42].

10.4.1.3 Room Volume and Shape

The necessary room volume decisively depends on the desired reverberation time and on the planned seating capacity. For this reason, orientating values for the required volume per seat can be stated for each individual use of the room. The desired volume per seat is shown for some room functions in Table 10.7. With regard to the fact that the reverberation time should not depend very much on the occupancy – especially in concert hall, theaters, and multipurpose halls – the sound absorption of the seating should vary only little with or without persons.

Considering the room shape, there are proven geometries for different uses but there is no optimal room shape. At least, surfaces for useful reflections

Table 10.7 Orientating values for the volume per seat for different uses

Room type and use	Volume per seat m³/seat
Lecture halls, lyric theaters, congress halls	4...6
Multipurpose halls for speech and music	4...7
Opera houses	6...8
Concert halls for chamber music	6...10
Concert halls for symphonic music	8...12
Churches	10...15
Orchestral rehearsal rooms	30...50

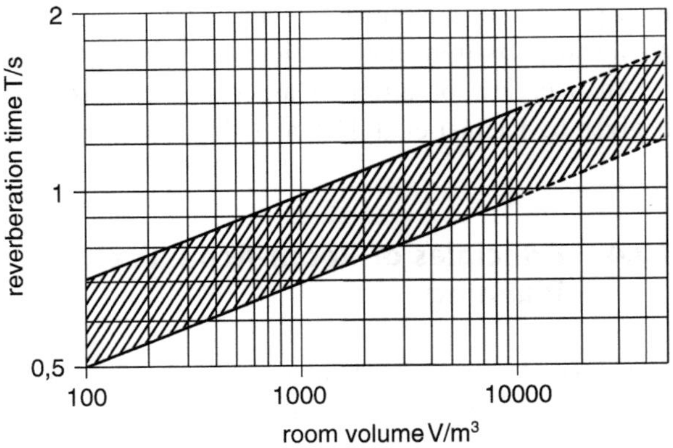

Fig. 10.18 Range of desirable reverberation times in rooms for speech

are important and surfaces which might create detrimental reflections shall be avoided. In any case echoes, flutter echoes, and sound foci in the source and receiver area should be avoided. This means parallel walls, cylindrical ground plans, etc. are generally disadvantageous. By means of secondary measures (sound diffusion, re-direction, absorption), these disturbing effects can at least be reduced. However, it should be considered that the efficiency of such measures is frequency dependent.

10.4.2 Rooms for Speech

The acoustic demands made on a room for speech (class room, lecture hall or auditorium, congress hall, theater, etc.) are basically a good speech intelligibility. This requires a sufficient loudness of the speaker, an unimpeded direct sound propagation, and high reflection energy within the first 50 ms (path difference $\leq$17 m) after the arrival of direct sound as well as a comparatively short reverberation time.

These conditions are partly contradictory as there can be no early sound reflections if there is no reverberation. Moreover, for a speaker a strongly damped room often seems to be unnatural. Orientating values for optimal reverberation times for speech performances in rooms of various volumes and mid-frequencies are shown in Fig. 10.18. In rooms where mainly electroacoustic sound systems or audiovisual media are used, reverberation times at the lower end of the shown tolerance range should be aimed at. Toward lower frequencies, the reverberation times should be slightly shorter, since otherwise the higher components of the speech spectrum, which are important for the speech intelligibility, could be masked.

In order to achieve the desired reverberation times, room volumes of approximately 5 m^3 per person are recommended. In case the volume per seat is much larger, more extensive sound-absorbing measures are required thus leading to a reduction in loudness. Accordingly, this should be avoided in spacious halls for speech performances, e.g. in theaters.

For such rooms, a sufficient strength of direct sound is especially important, which can be achieved by an inclination of the seating rows.

Strong first reflections that support efficiently direct sound are generated by reflecting walls or ceiling areas, especially near a sound source, as, for example, the rear and side walls of the podium or the whole proscenium area in theaters. In special cases also the use of reflectors above musicians and actors can be effective. Moreover, a favorable design of the ceiling is very important. The distribution of sound-absorbing areas necessary for the adjustment of the reverberation time should be such as not to cause detrimental reflections. So, preferably they are planned in the rear part of the room as well as in the lateral ceiling area near the walls.

For lecture rooms in schools and universities as well as for seminar rooms, but also for conference rooms, etc. the remarks given above are valid. According to recent research results, in class rooms, especially in elementary schools, nursery schools, or group rooms in kindergartens, even shorter reverberation times of less than 0.5 s should be aimed at together with a sufficiently low background noise level (see e.g. [74]). These rooms often have volumes of less than 300 m^3 and a length of less than 10 m.

Because of the small distance between speaker and listener, measures to direct the sound propagation are often of minor importance. However, a concentrated arrangement of absorbing surfaces – for example, a completely absorbing ceiling – in connection with parallel, smooth, and reflecting walls should be avoided, as in this case disturbing periodic reflection sequences have to be expected.

10.4.3 Concert Halls

In concert halls, it is of minor importance that the audience is able to follow the temporal structure of the presented music in detail, but, on the contrary, a certain temporal melting of successive tones and sounds is absolutely necessary. The same is valid for the spatial perception of sound; of course, an orchestra has a considerable spatial extension, and the various instrument groups are situated at different positions on the podium, but the individual instruments shall not be clearly localized by the audience. So, the acoustical properties of a concert hall as well as the arrangement of the podium must result in a certain temporal and spatial mixing of sounds.

First of all, this includes a sufficiently long reverberation time, which is in a range of approximately 1.7–2.1 s in recognized good halls for symphonic music (see also Table 10.8). Some authors say that the increase of the reverberation time toward lower frequencies is responsible for a "warm" sound of music. The mentioned reverberation times can be reached with a volume per seat of approximately 10 m^3/person, provided that there are no absorbing surfaces apart from the seating.

In order to obtain the above-mentioned mixing of sounds, not only a sufficiently long reverberation time is necessary but also the design of the surfaces is important. This means that the lateral walls surrounding the orchestra, as well as the ceiling above the podium must not direct the whole sound energy toward the audience, but have to reflect part of it toward the podium area. This is favorable for the mutual contact between the musicians [82]. However, based on numerous empirical facts, it results that the ceiling height above the podium should not exceed 10 m [83].

In general, directing the sound by reflecting surfaces is less important in concert halls than it is in rooms as mentioned in Sect. 10.4.2. In order to achieve a well-balanced sound propagation, slightly diffuse reflecting ceilings are favorable, like coffered ceilings, which can be seen in many traditional concert halls. Sculptural decorations, statuettes, columns, etc. are often assumed to have a positive effect.

Moreover, the subjective spaciousness, dealt with in Sect. 10.2.2.3, which is created by early reflections from lateral directions, is of great importance for the acoustical quality of a concert hall [84, 85]. For the creation of these reflections, naturally the lateral walls of the hall play an essential role. It can be assumed that the shoebox shape of many traditional concert halls with their small widths (18–26 m) has influenced the idea of many concertgoers and musicians how a concert hall must sound.

This does not mean that concert halls with a totally different ground plan are acoustically worse as, for example, shown by the Berliner Philharmonie, the

Table 10.8 Reverberation time of some concert halls with audience

Concert hall	Volume (m^3)	Number of seats	Year of inauguration (reopening)	Reverberation time (s) 125 Hz	500 Hz	2,000 Hz	Source
Musikvereinssaal, Wien	14,600	2,000	1870	2.1	1.9	1.5	[75]
Liederhalle Stuttgart	16,000	2,000	1956		1.6	1.7	[76]
Chiang Kai Shek Memorial, Taipeh	16,700	2,077	1987	1.95	2.0	1.9	[77]
Symphony Hall, Boston	18,800	2,630	1900	1.95	1.85	1.65	[78]
Concertgebauw, Amsterdam	19,000	2,200	1887	2.2	2.05	1.8	[78]
Concert Hall Athens	19,000	2,000	1992	1.9	1.9	1.7	[79]
New Gewandhaus Leipzig	21,000	1,900	1884(1981)	1.95	2	1.9	[80]
Philharmonie, Berlin	24,500	2,230	1963	2.4	1.95	1.9	[81]
Carnegie Hall, New York	24,250	2,800	1891	2.3	1.8	1.6	[78]
Concert Hall "De Doelen", Rotterdam	27,000	2,220	1979	2.3	2.1	2.2	[42]

Table 10.9 Reverberation time of some opera houses with audience

Opera house	Volume (m^3)	Number of seats	Inauguration (reopening)	Reverberation time (s) 125 Hz	500 Hz	2,000 Hz	Source
Staatsoper unter den Linden, Berlin	7,000	1,490	1956	1.2	1.0	1.0	[86]
La Scala, Mailand	10,000	2,290/400	1778 (1946)	1.5	1.2	0.9	[87]
Festspielhaus Bayreuth	11,000	1,800	1876	1.9	1.5	1.3	[75]
Deutsche Oper, Berlin	11,000	1,900	1962	1.7	1.5	1.2	[88]
Nationaltheater, Taipeh	11,200	1,522	1987	1.5	1.4	1.3	[77]
Staatsoper, Wien	11,600	1,658/560	1869 (1955)	1.7	1.5	1.2	[75]
Semperoper, Dresden	12,500	1,290	1878 (1985)	1.9	1.7	1.5	[89]
Neues Festspielhaus, Salzburg	14,000		1960	1.7	1.5	1.5	[75]
Festspielhaus Baden-Baden	19,600	2,300	1998	2.2	1.8	1.7	[90]
Metropolitan Opera, New York	30,500	3,800	1966	2.2	1.7	1.7	[91]

Gewandhaus in Leipzig or the concert hall in Athens. In these halls, reflecting surfaces were created between the terraced audience areas, which direct early sound reflections from lateral directions toward the audience.

The requirements made on halls for chamber music are basically similar to those made on halls for symphonic music. However, due to smaller ensembles and thus a reduced sound source power, the room volume should not exceed 8,000 m^3. The reverberation times to be aimed at are shorter and range between approximately 1.4–1.7 s.

Contrary to the above-mentioned halls, orchestra rehearsal halls normally are considerably smaller. In order to avoid an excessive sound pressure level, rehearsal halls should have a volume of at least approximately 30 m^3/musician and a short reverberation time. Moreover, shorter reverberation times are favorable for the critical listening at rehearsals. As in this case, the demands are very different and also depend on the skill of the musicians; variable acoustical measures are recommended for rehearsal halls in addition to fixed absorbing, reflecting, and sound-diffusing surfaces.

10.4.4 Opera Houses

Ideally, the acoustic quality of an opera house should guarantee excellent speech intelligibility and, at the same time, should ensure the full and well-balanced sound of music known in good concert halls. Unfortunately, it is not possible to combine these two requirements completely. In practice, this means that the reverberation time complies either with the one or the other requirement.

An overview of the reverberation times of some opera houses is given in Table 10.9. This table shows that older opera houses have a reverberation time around 1 s or less at mid-frequencies, i.e. there is relatively little reverberation. The reason for this probably was the aim to get as many listeners as possible in a hall of given size. Furthermore, the good speech intelligibility was favorable for the formerly existing styles as well as perhaps also for the need of entertainment by the audience. However, more modern and newer opera houses show a clear tendency to longer reverberation times. Perhaps nowadays, the operagoer attaches greater importance to a full sound of music than to the intelligibility of the texts, with which the expert is familiar, anyhow.

Indeed, in opera houses, more attention has to be paid to the unimpeded propagation of direct sound than in a concert hall, particularly as an acoustically favorable arrangement of the seating also ensures a good sight toward the stage. However, here the provision with lateral reflections is less important since the orchestra is situated in a lowered orchestra pit, so that only little sound is reflected at the lateral walls of the auditorium. Significant reflecting surfaces are the lateral walls and the ceiling of the proscenium, which should be almost horizontal with respect to the longitudinal axis of the auditorium. The ceiling of the auditorium is of special importance, particularly for the sound supply of the balconies, whereas the likewise important stage area in most cases cannot be considered in the acoustic planning, since it depends very much on the individual stage scenery.

Finally, it should be mentioned that in principle similar requirements apply to theaters for musicals; however, the use of electroacoustic devices and effects prevails. Therefore, lower reverberation times are required than for modern opera houses.

10.4.5 Multipurpose Halls and Rooms with Variable Reverberation

Often a hall shall be used for different kinds of performances. In this case, a compromise has to be found as far as the reverberation time is concerned, i.e. a compromise between the relatively large values, which would be optimal for musical performances, and the short reverberation times required for a good speech intelligibility. Such a compromise could be that the reverberation time at mid-frequencies is at the lower end of the range desired for classical music. Additionally, the creation of strong early reflections, low frequency absorbers (e.g. wooden claddings), and in particular a sound-enhancement system adjusted to the room-acoustical situation will contribute to the desired speech intelligibility. Moreover, in these rooms a seating providing a sound absorption that varies only little with and without persons is favorable. The compromise regarding the requirements of speech transmission consists of a higher transparency of sound as well as in shorter reverberation times at low frequencies than for a pure concert hall.

Such compromises, which are not very satisfying in the end, can be avoided by measures allowing to modify the acoustics and thus to adapt it to the respective use of the room. These measures include turnable, hinged wall, or ceiling elements absorption devices which can be hidden or rolled up or reverberantion chambers which can be coupled to the main hall. However, all these measures are only useful if a correct operation can be guaranteed. This is one of the reasons why such measures are mainly applied in studios or music rehearsal rooms. However, the change of the reverberation, which can be achieved in large halls with audience, is usually limited due to the absorption by the audience. A modern alternative to changing the reverberation by variable absorption surfaces is to adjust acoustics by appropriate acoustic enhancement systems. In this case, the "natural" reverberation of the hall is chosen so as to be adequate for speech performances. By means of an acoustic

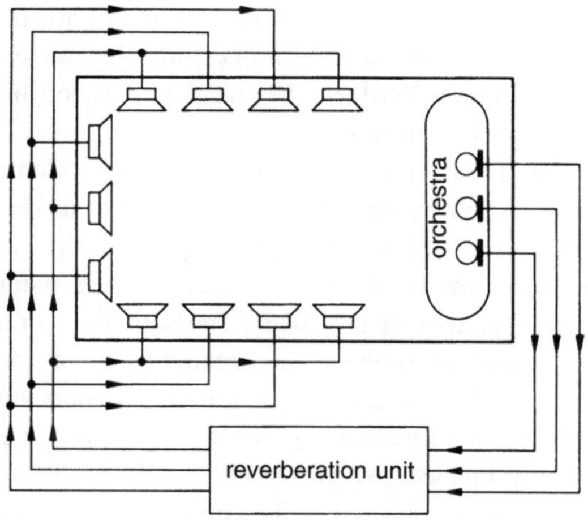

Fig. 10.19 Electroacoustic reverberation enhancement

enhancement system it can be changed, so that a sufficient or even optimal value for musical performances is reached.

The easiest way might be to record the sound signals with microphones, close to the sound sources, and to enhance the reverberation acoustically or electronically (see Fig. 10.19). In the first case, the sound signal is radiated by a loudspeaker in a separate room which has the desired longer reverberation time and is recorded once again in this room [92]. Much easier and more flexible are electroacoustic room-enhancement systems with a combination of feedback delays, which have several outputs delivering incoherent signals. Loudspeakers in the hall then emit these signals. By means of adequate delays, it has to be ensured that the loudspeaker signal does not reach any place before the direct sound.

Other systems for an electroacoustic increase of the reverberation time benefit from the usually undesired acoustic feedback. However, in order to suppress coloration, a large number of electroacoustic transmission paths is necessary that either work independently from each other or that are acoustically coupled. The Assisted Resonance System, for example [93], which was originally planned to improve the acoustical conditions in the Royal Festival Hall in London, works with up to 200 separate channels consisting of microphones, loudspeakers, and acoustic narrow bandwidth resonators. In contrast, in case of the multichannel system by Franssen [94] (MCR: Multi Channel Reverberation), each channel covers the whole frequency range. Therefore, a careful equalization is required. In each case, the amplification is critical. If it is too low, there will be no considerable increase of reverberation, on the other hand if it is too high there is a risk of instability.

A totally different system is the Acoustic Control System (ACS) by Berkhout [95, 96]. Here, artificial image sources of low and high order of a desired and previously simulated hall are added to the natural reflection pattern of the room. The signals recorded by numerous microphones are treated by a processor in a way that, if they are emitted by adequately distributed loudspeakers in the existing hall, wave fronts are created, which approximately correspond to those in the desired hall. Together with special loudspeaker arrangements, the system also can be used to improve the acoustical conditions on the podium. Moreover, an increase of the reverberation time can be reached by a controlled feedback. Coloration should be avoided by making the transmission between microphone and loudspeaker time variant. The latter also applies to LARES [97] and VIVACE [113]. This system only uses few microphones, but a high number of loudspeakers and, for example, can also be used in small rehearsal rooms.

In summary, it can be said that systems for actively influencing room acoustics have already been used successfully and their acceptance has significantly been increased in recent years. For a more detailed overview see, for example, Kleiner and Svensson [98].

10.4.6 Churches

The acoustical demands made on churches are contradictory. On the one hand, for the sake of good intelligibility of the spoken word, a short reverberation time is required here, whereas for organ music in large churches reverberation times of 3–4 s would be desirable. In many older churches, the task of the acoustician to create a sufficient speech intelligibility even in case of few visitors is complicated due to the fact that these churches are subject to preservation, which excludes the application of usual acoustical measures. In modern churches, sometimes a certain regulation of reverberation can be achieved by using perforated bricks on walls backed with an absorbent felt or by absorbing wooden claddings. However, these measures must not be excessive, since the visitors unconsciously expect a certain reverberation of a church. Especially helpful are reflecting surfaces near the sound sources as well as a carefully planned sound reinforcement system. More detailed general rules cannot be given here, as the room-acoustical measures depend too much on the kind of service and on architectonic facts (e.g. see also [99, 100]).

10.4.7 Rooms with Mainly Electroacoustic Use

Nearly every event or assembly hall is equipped with a sound reinforcement system either for improving speech transmission or for producing electroacoustic effects. If the sound reinforcement system in the first place has a supporting function, the electroacoustic planning has to be adjusted especially to the room-acoustical situation. However, nowadays in many rooms electroacoustic presentations or the reproduction of speech or music is important. For example, this concerns large assembly halls, like congress or plenary halls, arenas for rock concerts or sport events, cinemas, studios for electronic music, control rooms, and even the living room of the hi-fi-enthusiast. Here, it is the task of room acoustics to favor a pure sound quality for loudspeaker transmission, which especially requires adequately low reverberation times. Moreover, disturbing reflections should be avoided by coordinating the electroacoustical and room-acoustical design. Because of the different uses and room sizes, the individual demands considerably differ from one another. Especially low reverberation times are aimed at in modern cinemas, which, for example, should be between 0.4 and 0.6 s to reach the so-called THX-standard for rooms of about 2,000 m^3. Besides a sound-absorbing seat upholstery and carpeting, this requires extensive sound-absorbing measures on walls and ceiling.

Moreover, special demands on the room-acoustical situation are made in the control rooms of recording studios or other listening rooms, e.g. for product development and sound design. Although here, the desired room-acoustical conditions also depend on the exact use, room size, and shape in general it can be said that room volumes of less than 40 m^3 should be avoided, and that at a size of up to approximately 300 m^3, reverberation times of 0.3–0.4 s in the whole frequency range should be aimed at. The rooms should be symmetric with regard to the listening position and should have proportions that, at low frequencies, avoid a coincidence of eigenfrequencies. For improving the reflection pattern, apart from adequate broadband absorbing measures also sound scattering claddings are especially important. Reflections from smooth, reflecting surfaces that can hardly be avoided (e.g. control room window) should not be directed directly to the listener's ears (e.g. [101]).

10.4.8 Workspaces and Other Rooms

This category includes, for example, production halls, plants, or offices. Even if the acoustical demands made on such rooms can be very different in detail, room-acoustical measures mostly aim at restraining sound transmission and thus reducing noise levels caused by machines, office machines, and/or persons. At the same time, normally sufficient speech intelligibility over short distances should be guaranteed.

There are often no diffuse field conditions in the considered rooms, the ratio of two room dimensions is larger than three (long or flat room), the absorption is not distributed equally or the room is heavily damped. For this reason, the prediction of the reverberation time and its interpretation with respect to the attainable noise reduction is often critical. In larger rooms with a floor area of more than approximately 200 m^2 and without diffuse sound field conditions a more detailed assessment can be derived from the so-called sound propagation curve. This curve shows the sound pressure level as a function of the distance from a noise source. It can be measured or predicted by means of approximate calculations or computer simulations (e.g. [53, 102]). The sound propagation curve is also used to calculate adequate parameters, as, for example, the average level decrease per doubling distance DL_2 (see e.g. sound absorbing[103, 104]).

Constructional measures are sound absorbing ceiling systems. Sometimes also absorbing measures in wall areas are required. Moreover, fittings that shadow the direct sound transmission or act as diffusers are favorable. In connection with absorbing ceiling systems, sound barriers are helpful which should be situated as near as possible to the noisy area or to the area to be protected. However, their efficiency considerably depends on the mounting conditions. Absorbing shields are used in connection with sound absorbing ceilings in open plan offices, call centers, or customer areas of banks, where privacy toward neighboring working places is demanded. In this connection, the positive masking effect of background noise (e.g. caused by ventilation, office equipment, music, or noise emitted by loudspeakers) should also be pointed out. Thus, the noise level of these sources should not be too low or unacceptably high (see e.g.[105]).

Moreover, other rooms should also be considered that are not directly working rooms but where absorbing measures make sense or are even necessary in order to avoid extreme reverberation. These rooms are, for example, halls of railway stations, terminals in airports, entrance halls, corridors, staircases, or canteens. In these rooms, noise is often mainly caused by the persons themselves. Here, it has to be considered that in a reverberant and therefore noisier environment the vocal effort is raised reflex-like (Lombard effect). Therefore, with an adequate room absorption, for example, by sound absorbing ceiling systems, also a considerable reduction of sound emission can be expected.

In this connection, also gymnasiums and swimming halls have to be mentioned. In order to reduce noise emission and to ensure sufficient speech intelligibility, the reverberation times should be below approximately 1.8 s. In these rooms, mostly large areas of the ceiling are absorptive. Flutter echoes can be avoided by covering parts of the walls with sound-absorbing claddings. In gymnasiums with additional uses, also the acoustical requirements for e.g. lectures or music performances have to be considered.

10.5 Room-Acoustical Measurements

Room-acoustical measurements are necessary for the objective verification and documentation of acoustic conditions in completed rooms or halls. Before rebuilding or reconstructing, it is suitable to document the acoustical conditions by means of measurements. In this way, an exact planning basis can be obtained, which makes the dimensioning of room acoustical measures easier and later allows an objective comparison.

The most important room acoustical parameter is the reverberation time. Often the determination of the reverberation time in the unoccupied hall is sufficient, the absorption by the audience can be considered by calculations. Since for common room shapes the reverberation time does not noticeably depend on the measuring position, only a few measuring points are sufficient. Their distance to the sound source should be at least twice the reverberation distance according to Eq. (10.24). Moreover, an omnidirectional radiation of the sound source is of minor importance.

In contrast, the, "*Early Decay Time*" (EDT) (see Sect. 10.2.3) or, e.g. the parameters for speech intelligibility, clarity, and spaciousness (see Sect. 10.2.2), depend on the temporal and partly also on the directional distribution of early sound reflections, i.e. they change from place to place. In order to avoid differences in the results caused by the orientation of the sound source, omnidirectional sound sources should be used.

During all room-acoustical measurements, the room is excited by a test signal and its "response" is registered by the help of a measuring microphone. As these measurements are relative measurements, the use of standardized sources and calibrated microphones is not necessary. The latter normally have an omnidirectional response, except if the parameters to be determined also demand directional information. A real-time analyzer or a PC nowadays usually carries out the generation and evaluation of the signals. After adequate pre-amplification, the signals registered by the measuring microphone are transmitted to the digital computer for further processing via an analog– digital converter.

For measurements of the reverberation time, random noise may be used as test signal which first excites the room stationary and then, at a certain moment, is abruptly switched off. For orientating measurements, the excitation of the room can also be made by a pistol shot. The frequency dependency of the reverberation can be determined by filtering the received signal at octave or one-third octave bands between 125 und 4,000 Hz. In some cases, also already filtered test signals are used. When exciting the room by random noise, at least three independent measurements for each measuring position have to be carried out because of the stochastic signal character. The results have to be averaged energetically. According to Schroeder [106], it is alternatively possible to determine the exact mean value or the expected value $\langle E(t)\rangle$ by integrating the energy impulse response as follows:

$$\langle E(t)\rangle = \int_t^{\infty} [g(t')]^2 \mathrm{d}t'. \qquad (10.43)$$

So, by this procedure an exact determination of the reverberation time is possible, for which the level range between −5 and −35 dB (or −25 dB) is used, with reference to the stationary level. The corresponding results are described by T_{30} (or T_{20}). This method is most suitable if the EDT according to Sect. 10.2.3 is of interest. Of course, the upper limit ∞ in Eq. (10.43) has to be replaced by a maximum integration time, which must be chosen in a way that on the one hand the decay curve is not cut too early and, on the other hand, not too much noise is integrated. Concerning further details, for example, also regarding measurements at very short reverberation times, see ISO 3382 [107].

All other parameters described in Sect. 10.2.2 are determined by the room impulse response. According to the definition, the impulse response results from excitation of the room with a short impulse with duration considerably shorter than the reciprocal value of the highest frequency of interest. However, for such a direct measurement, the signal-to-noise ratio is often insufficient. This difficulty can be avoided by using special excitation signals (maximum-length sequences [108, 109], or so-called sine sweeps [110, 111]). Because of their autocorrelation characteristics, these signals can be eliminated, so that the results represent the impulse responses (see Fig. 10.20). With these methods, even measurements in occupied halls can be carried out. If the interaural coherence (i.e. the IACC) according to Eq. (10.37) should be determined from the measurement, the receiver has to be a dummy head. The same is valid if the impulse responses shall be used as basis of an auralization. For the determination of the lateral efficiency (see Sect. 10.2.2.3), a gradient microphone or a microphone pair [112] has to be used, similar as for intensity measurements (see Sect. 2.2.5).

Finally, it should be mentioned that the Speech Transmission Index (see Sect. 10.2.2.2) could also be determined by direct measurement of the complete modulation transfer function (MTF). Houtgast und Steeneken [27] have also developed a shortened measurement method, where the values of the MTF in the octave bands with mid frequencies 500 Hz and

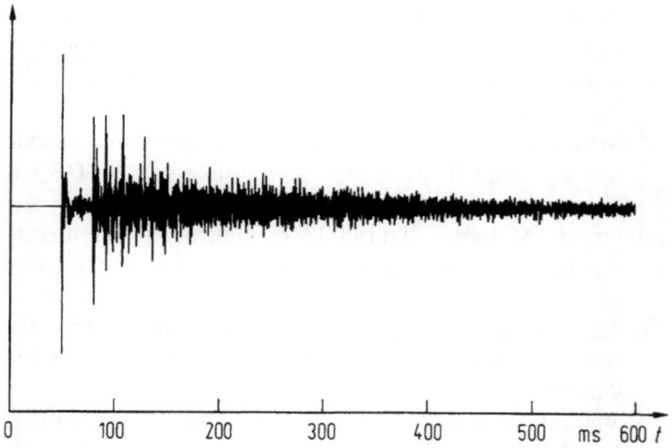

Fig. 10.20 Measured room impulse response of the transmission between two points in a room

2,000 Hz for four respectively five modulation frequencies between 0.7 and 11.2 Hz are measured automatically. From the results, the Rapid Speech Transmission Index (RASTI) is determined according to a special arithmetic rule, which characterizes speech intelligibility influenced by the acoustics of the room as well as by disturbing noise.

References

1. Schroeder MR, Kuttruff H (1962) On frequency response curves in rooms. Comparison of experimental, theoretical and Monte Carlo results for the average frequency spacing between maxima. J Acoust Soc Amer 34:76–80
2. Borish S (1985) Extension of the image model to arbitrary polyhedra. J Acoust Soc Amer 75:1827–1836
3. Kuttruff H (1971) Simulated reverberation curves in rectangular rooms with diffuse sound fields. Acustica 25:333–342, (1976) Reverberation and effective absorption in rooms with diffuse wall reflections. Acustica 35:141–153 (in German
4. Carroll MM, Miles RN (1978) Steady-state sound in an enclosure with diffusely reflecting boundary. J Acoust Soc Amer 64:1425–1436
5. Korany N, Blauert J, Abdel Alim O (2001) Acoustic simulation of rooms with boundaries of partially specular reflectivity. Appl Acoust 62:875–887
6. Bass HE, Sutherland LC, Zuckerwar AJ, Blackstock DT, Hester DM (1995) Atmospheric absorption of sound: Further developments. J Acoust Soc Am 97:680–683
7. Andres HG (1965/66) Über ein Gesetz die räumlichen Zufallsschwankung von Rauschpegeln in Räumen und seine Anwendung auf Schalleistungsmessungen. Acustica 16:279–294
8. Waterhouse RV (1955) Interference patterns in reverberant sound fields. J Acoust Soc Amer 27:247–258
9. Meyer E, Burgtorf W, Damaske P (1965) Eine Apparatur zur elektroakustischen Nachbildung von Schallfeldern. Acustica 15:339–344
10. Reichardt W, Schmidt W (1967) The detectability of changes in sound field parameters for music. Acustica 18:274–282 (in German)
11. Schroeder MR, Gottlob D, Siebrasse KF (1974) Comparative study of European concert halls: Correlation of subjective preference with geometry and acoustic parameters. J Acoust Soc Amer 56:1195–1201
12. Gottlob D, Siebrasse F, Schroeder MR (1975) Fortschr. d. Akustik – DAGA '75, Physik-Verlag, Weinheim
13. Burgtorf W (1961) Untersuchungen zur Wahrnehmbarkeit verzögerter Schallsignale. Acustica 11:97–111
14. Seraphim H-P (1961) Über die Wahrnehmbarkeit mehrerer Rückwürfe von Sprachschall. Acustica 11:80–91
15. Burgtorf W, Oehlschlägel HK (1964) Untersuchungen über die richtungsabhängige Wahrnehmbarkeit verzögerter Schallsignale. Acustica 14:254–266
16. Schubert P (1969) Die Wahrnehmbarkeit von Rückwürfen bei Musik. Z Hochfrequenztechn u Elektroakust 78: 230–245
17. Cremer L (1948) Die wissenschaftlichen Grundlagen der Raumakustik, Band 1: Geometrische Raumakustik, 1.Aufl., S Hirzel, Stuttgart
18. Haas H (1951) Über den Einfluß eines Einfachechos auf die Hörsamkeit von Sprache. Acustica 1:49–58
19. Meyer E, Schodder G R (1952) Über den Einfluß von Schallrückwürfen auf Richtungslokalisation und Lautstärke von Sprache. Nachr. Akad. Wissensch. Göttingen, Math.-Phys. Kl. No. 6: 31–42
20. Muncey RW, Nickson AFB, Dubout P (1953) The acceptability of speech and music with a single artificial echo. Acustica 3:168–173
21. Atal BS, Schroeder MR, Kuttruff H (1962) Perception of coloration in filtered Gaussian noise. Short time spectral analysis of the ear. Proc. 4th Intern. Congr. Acoustics, Copenhagen, Paper H 31
22. Bilsen FA (1967/68) Thresholds of perception pitch. Conclusions concerning coloration in room acoustics and correlation in the hearing organ. Acustica 19:27–31
23. Thiele R (1953) Richtungsverteilung und Zeitfolge der Schallrückwürfe in Räumen. Acustica 3:291–302
24. Reichardt W, Abdel Alim O, Schmidt W (1974) Abhängigkeit der Grenzen zwischen brauchbarer und unbrauchbarer Durchsichtigkeit von der Art des Musikmotives, der Nachhallzeit und der Nachhalleinsatzzeit. Appl Acoust 7:243–264
25. Lochner JPA, Burger JF (1961) Optimum reverberation time for speech rooms based on hearing characteristics. Acustica 11:195–200
26. Kürer R (1969) Zur Gewinnung von Einzahlkriterien bei Impulsmessungen in der Raumakustik. Acustica 21: 370–372
27. Houtgast T, Steeneken HJM (1973) The modulation transfer function in room acoustics as a predictor of speech intelligibility. Acustica 28:66–73; (1984) A multi-language Evaluation of the RASTI-method for estimation speech intelligibility in auditoria
28. Schroeder MR (1981) Modulation transfer functions: definition and measurement. Acustica 49:179–182
29. Barron M (1974) The effects of early reflections on subjective acoustical quality in concert halls. PhD-Thesis, University of Southampton

30. Damaske P, Ando Y (1972) Interaural crosscorrelation for multichannel loudspeaker reproduction. Acustica 27: 232–238
31. Morimoto M, Maekawa Z (1989) Auditory spaciousness and envelopment. 13th Intern. Congr. Acoustics, Belgrad, 215–218
32. Bradley JS, Soulodre GA (1995) Objective measures of listener envelopment. J Acoust Soc Amer 98:2590–2597
33. Dietsch L, Kraak W (1986) An objective criterion for the measurement of echo disturbances during presentation of music and speech. Acustica 60:205–216 (in German)
34. Sabine WC (1922) Collected Papers on acoustics. Harvard University Press, Cambridge
35. Knudsen VO (1929) The hearing of speech in auditoriums. J Acoust Soc Amer 1:56–82
36. Seraphim H-P (1958) Untersuchungen über die Unterschiedsschwelle exponentiellen Abklingens von Rauschbandimpulsen. Acustica 8:280–284
37. Atal BS, Schroeder MR, Sessler GM (1965) Subjective reverberation time and its relation to sound decay. Proc. 5th Intern. Congr. Acoustics, Liege, Paper G 32
38. Jordan VL (1970) Acoustical criteria for auditoriums and their relation to model techniques. J Acoust Soc Amer 47: 408–412
39. Normenausschuß D (1968) Schallabsorptionstabelle. Beuth, Berlin
40. Harris CM (1994) Noise control in buildings. Mc Graw Hill, New York
41. Fasold W, Veres E (2003) Schallschutz und Raumakustik in der Praxis: Planungsbeispiele und konstruktive Lösungen, 2nd edn. Beuth, Berlin
42. Cremer L, Müller HA, Schultz TJ (1982) Principles and applications of room acoustics. Applied Science Publishers, London
43. Kath U, Kuhl W (1965) Messungen zur Schallabsorption von Polsterstühlen mit und ohne Personen. Acustica 15: 127–131
44. Beranek L, Hidaka T (1998) Sound absorption in concert halls by seats, occupied and unoccupied, and the hall's interior surfaces. J Acoust Soc Am 104(6): 3169–3177
45. Vorländer M (1995) International Round Robin on Room Acoustics Computer Simulations, Proc 15th Intern. Congr. Acoustics, Trondheim, 689–692
46. Bork I (2000) A comparison of room simulation software – the 2nd round robin on room acoustical computer simulation. Acustica/acta acustica: 943–956
47. Mommertz E, Müller K (1995) Berücksichtigung gekrümmter Wandflächen im raumakustischen Schallteilchenverfahren. Fortschr. d. Akustik – DAGA '95, DPG-GmbH, Bad Honnef
48. Vorländer M, Mommertz E (2000) Definition and measurement of random-incidence scattering coefficient. Appl Acoust 60:187–199
49. Schroeder M R, Atal B S, Bird C (1962) Digital computers in room acoustics. Proc 4 Intern. Congr. Acoustics, Copenhagen, paper M 21
50. Krokstad A, Strøm S, Sørsdal S (1968) Calculating the acoustical room response by the use of a ray tracing technique. J Sound Vibr 8:118–124, (1983) Fifteen years experience with computerized ray tracing. Appl Acoust 16: 291–312
51. Stephenson U (1985) An acoustic computer simulation technique for calculating parameters relevant to subjective acoustical impression in concert halls. Acustica 59:1–20 (in German)
52. Vorländer M (1988) A ray-tracing program for the calculation of sound fields in rooms. Acustica 65:138–148 (in German)
53. Ondet AM, Barbry JL (1988) Modelling of sound propagation in fitted workshops using ray tracing. J Acoust Soc Am 87:787–796
54. Allen SP, Berkley DA (1979) Image method for efficiently simulating small-room acoustics. J Acoust Soc Amer 65: 943–950
55. Vian SP, Van Maercke D (1986) Calculation of the room impulse response using a ray-tracing method. Proc. 12th Intern. Congr. Acoustics, Vancouver, 74–78
56. Vorländer M (1989) Simulation of the transient and steady-state sound propagation in rooms using a new combined ray-tracing/image-source algorithm. J Acoust Soc Amer 86:172–178
57. Mommertz E (1995) Untersuchung akustischer Wandeigenschaften und Modellierung der Schallrückwürfe in der binauralen Raumsimulation. Dissertation RWTH Aachen, Shaker Verlag, Aachen
58. Heinz R (1993) Binaural spatial simulation using a combined method – separate simulation of the geometrical and the diffuse components. Acustica 79:207–220 (in German)
59. Naylor GM (1993) Odeon– another hybrid room acoustical model. Appl Acoust 38:131–143
60. Dalenbäck B-I L (1996) Room acoustic prediction based on a unified treatment of diffuse and specular reflection. J Soc Acoust Am 100:899–909
61. Kuttruff, H (2009) Room Acoustics, 5th edn. Taylor & Francis, New York
62. Lewers T (1993) A combined beam tracing and radient exchange computer model of room acoustics. Appl Acoust 38:161–178
63. Brebeck P, Bücklein R, Krauth E, Spandöck F (1967) Akustisch ähnliche Modelle als Hilfsmittel für die Raumakustik. Acustica 18:213–226
64. Tennhardt H-P (1984) A method of measurement with a model, which is used in research on balancing auditions in musical performances, as exemplified in the acoustics of the large hall in the Neue Gewandhaus Leipzig. Acustica 56:126–135 (in German)
65. Xiang N, Blauert J (1993) Binaural scale modelling for auralization and prediction of acoustics in auditoria. Appl Acoust 38
66. Martin J, Vian SP (1989) Binaural sound simulation of concert halls by a beam tracing method. Proc. 13th Int. Congr. Acoustics, Belgrad, 253–256
67. Kuttruff H, Vorländer M, Classen T (1990) On the auditory assessment of the acoustics of simulated halls. Acustica 70:230–231 (in German)
68. Köring J, Schmitz A (1993) Simplifying cancellation of Cross-talk for playback of head-related recordings in a two-speaker system. Acustica 79:221–232
69. Kleiner M, Dalenbäck B-I, Svensson P (1993) Auralization – an overview. J Audio Eng Soc 41:861–875
70. Vorländer M (2007) Auralization: Fundamentals of Acoustics, Modelling, Simulation, Algorithms and Acoustic Virtual Reality, Springer, Berlin Heidelberg

71. Meyer E, Kuttruff H, Schulte F (1965) Versuche zur Schallausbreitung über Publikum. Acustica 15:175–182
72. Schultz TJ, Watters BG (1964) Propagation of sound across audience seating. J Acoust Soc Amer 36:885–902
73. Mommertz E (1993) Some measurements of the propagation of acoustic waves skimming over the audience and seats. Acustica 79:42–52 (in German)
74. Bradley JS, Reich RD, Norcross SG (1999) On the combined effects of signal-to-noise ratio and room acoustics on speech intelligibility. J Acoust Soc Am 106:1820–1828
75. Bruckmeyer F (1962) Handbuch der Schalltechnik im Hochbau. Deuticke, Wien
76. Cremer L, Keidel L, Müller HA (1956) Die akustischen Eigenschaften des großen und des mittleren Saales der neuen Liederhalle in Stuttgart. Acustica 6:466–474
77. Kuttruff H (1989) Acoustical design of the Chiang Kai Shek Cultural Centre in Taipei. Appl Acoust 27:27–46
78. Beranek LL (1996) Concert and opera halls: how they sound. Acoustical Society of America, Woodbury
79. Opitz U (1996): The Athens Concert hall: A multipurpose hall for Concert and Opera events or a new solution ? Acoustical Symposium, Turin
80. Fasold W (1982) Akustische Maßnahmen im Neuen Gewandhaus Leipzig. Bauforschg., Baupraxis H. 117
81. Cremer L (1964) Die raum- und bauakustischen Maßnahmen beim Wiederaufbau der Berliner Philharmonie. Schalltechn 24:1–11
82. Meyer J (2009) Acoustics and the Performance of Music: Manual for Acousticians, Audio Engineers, Musicians, Architects and Musical Instrument Makers, 5th edn. Springer, New York
83. Marshall AH, Gottlob D, Alrutz H (1978) Acoustical conditions preferred for ensemble. J Acoust Soc Amer 64:1437–1442
84. Jordan VL (1969) Room acoustics and architectural acoustics development in recent years. Appl Acoust 2:59–81, (1975) Auditoria acoustics: Development in recent years. Appl Acoust 8: 217–235
85. Barron M (2009) Auditorium acoustics and architectural design, Spon, London
86. Reichardt W (1961) Die Akustik des Zuschauerraums der Staatsoper Berlin, Unter den Linden. Z Hochfrequenztech u Elektroakust 70:119
87. Furrer W, Lauber A (1972) Raum- und Bauakustik. Lärmabwehr, Birkhäuser, Basel, Stuttgart
88. Cremer L, Nutsch J, Zemke HJ (1962) Die akustischen Maßnahmen beim Wiederaufbau der deutschen Oper Berlin. Acustica 12:428–432
89. Kraak W (1990) Personal communication
90. Müller Kh (1998) Raumakustische Gestaltung des Festspielhauses Baden-Baden. Bericht zur 20. Tonmeistertagung, Karlsruhe, 79–88
91. Jordan VL (1980) Acoustical design of concert halls and theatres. Applied Science Publishers, London
92. Meyer E, Kuttruff H (1964) Zur Raumakustik einer großen Festhalle. Acustica 14:138–147
93. Parkin PH, Morgan K (1970) "Assisted Resonance" in the Royal Festival Hall London: 1965–1969. J Acoust Soc Amer 48:1025–1035
94. Franssen NV (1968) Sur l'amplification des champs acoustiques. Acustica 20:315–323
95. Berkhout AJ (1988) A holographic approach to acoustic control. J Audio Eng Soc 36:977–995
96. Berkhout AJ, de Vries D, Vogel PJ (1993) Acoustic control by wavefield synthesis. J Acoust Soc 93:2764–2778
97. Griesinger D (1996) Beyond MLS – Occupied hall measurement with FFT techniques, 101st AES convention, preprint 4403
98. Kleiner M, Svensson P (1995) Review of active systems in room acoustics and electroacoustics. Proc. Active 95, Newport Beach: 39–54
99. Lottermoser W (1983) Orgeln, Kirchen und Akustik, Band 2. Bochinsky, Frankfurt am Main
100. Meyer J (2000) Zur Raumakustik in Johann Sebastian Bachs Kirchen. Bericht zur 21. Tonmeistertagung, Hannover, 1064–1077
101. Everest F A, Pohlmann K C (2008) Master handbook of acoustics, 5th edn. McGraw Hill, New York
102. Kuttruff H (1985) Stationary sound propagation in flat enclosures. Acustica 57:62–70 (in German)
103. Kurze UJ (1997) Sound propagation in work spaces. In: Encyclopedia of acoustics Vol. III, Malcom J. Crocker (ed), Wiley, New York 1881–1187
104. ISO 11690–3 (1999) Acoustics – Recommended practice for the design of low-noise workplaces containing machinery – Part 3: Sound propagation and noise prediction in workrooms
105. Bradley JS (2003) The acoustical design of conventional open plan offices. Canad Acoust 31:23–31
106. Schroeder MR (1965) New method of measuring reverberation time. J Acoust Soc Amer 37:409–412
107. ISO 3382 (1997) Measurement of the reverberation time of rooms with reference to other acoustical parameters
108. Alrutz H, Schroeder M R (1983) A fast hadamard transform method for the evaluation of measurements using pseudorandom test signals. Proc. 11th Intern. Congr. Acoustics, Paris, Vol. 6, 235–238
109. Borish J, Angell JB (1983) An efficient algorithm for measuring the impulse response using pseudorandom noise. J Audio Eng Soc 31:478–487
110. Griesinger D (1995) Design and performance of multichannel time variant reverberation enhancement systems. Proc. Active 95, Newport Beach: 1203–1212
111. Müller S, Massarani P (2001) Transfer function measurement with sweeps. J Audio Eng Soc 49:443–471
112. Kleiner M (1989) A new way of measuring the lateral energy fraction. Appl Acoust 27:321–327
113. Engel G (2011) Acoustic Enhancement Systems for classical concerts, 8th International conference on Auditorium acoustics, Dublin, Proc. of the Institute of Acoustics

Silencers

11

U. Kurze and E. Riedel

11.1 Scope

11.1.1 Areas of Application

Large size silencers are attached to the intake and exhaust of large industrial plants, e.g. forced ventilation systems for mining industry, intake of cooling towers (Fig. 11.1) or flue gas stacks of power plants to protect the neighbourhood from plant noise. Large silencers are also required for ventilation openings of rooms with high internal sound pressure levels, e.g. industrial production halls or subway ventilation ducts. In order to achieve a broad-band sound attenuation, it is often necessary to have silencer elements with dimensions comparable to the quarter-wavelength of the sound to be attenuated. Therefore, silencers for sound of about 1 m wavelength must be partitioned into many elements.

Smaller silencers consisting of few elements are applied in ducts and on their openings, on individual machinery and on enclosures for machines that are equipped with sound radiating openings for cooling, for air intake and exhaust, or for material flow. Silencers are widely used in heating, ventilation and air conditioning (HVAC) systems to reduce ventilation noise and cross-talk between rooms. Finally, silencers are necessary to suppress blow-down noise from valves and pneumatic tools or machines.

11.1.2 Construction Types

Wide openings and ducts in buildings, plants and enclosures are partitioned by silencer elements.[1] The lateral partitioning in one direction by parallel baffles yields relatively narrow slits. The partitioning in two orthogonal directions results in parallel ducts, which are acoustically similar to the slits. Other ways of partitioning large cross sections [1] are hardly ever used.

Smaller and narrow openings of machines attached to free space or to tubes (Fig. 11.2) and ducts between rooms can be equipped with silencers formed by absorptive or reactive duct linings (Fig. 11.3). The acoustic efficiency of such linings can be achieved by passive elements of suitable shape and material or by active elements involving electro-mechanical transducers. Distinction is made between porous and impervious linings with respect to gas flow. Silencers are most effective for high frequency sound when positioned at elbows or other duct elements diverting the flow direction (Figs. 11.4 and 11.5).

A further type of silencer is the blow-off or throttle silencer that affects the flow through an opening by a considerable flow resistance. Throttle silencers range from large size constructions for gas and steam valves (Fig. 11.6) to small screw-on elements for pneumatic systems (Fig. 11.7).

E. Riedel (✉)
BBM Akustik Technologie, Robert-Koch-Strasse 11, 82152 Planegg, Germany
e-mail: info.planegg@bbm-akustik.de

[1] Constructional elements in such ducts or openings mainly employed for guiding or filtering the gas flow are also designed to act as silencers.

G. Müller and M. Möser (eds.), *Handbook of Engineering Acoustics*,
DOI 10.1007/978-3-540-69460-1_11,

Fig. 11.1 Cooling tower of power plant equipped with parallel baffle silencers (Source: BBM-Gerb Akustik)

Fig. 11.2 In-line silencer with splitter

11.1.3 Requirements and Features

The design of silencers mainly depends on the frequency spectrum of the sound source and on the operating conditions. Table 11.1 contains examples for parallel baffle silencers.

Basically, absorptive ducted silencers can be composed of large absorbing chambers. However, the space required and the costs are rather high. An example for the more appropriate application is the absorbent wall lining of a subway ventilation duct where attenuation is needed for rolling noise with high-frequency and tonal components [2]. Special requirements are best met by tailored solutions. Examples are given in Table 11.2.

Stability of silencers is important in view of environmental conditions and vibration excitation by the sound source. It is needed for supporting components

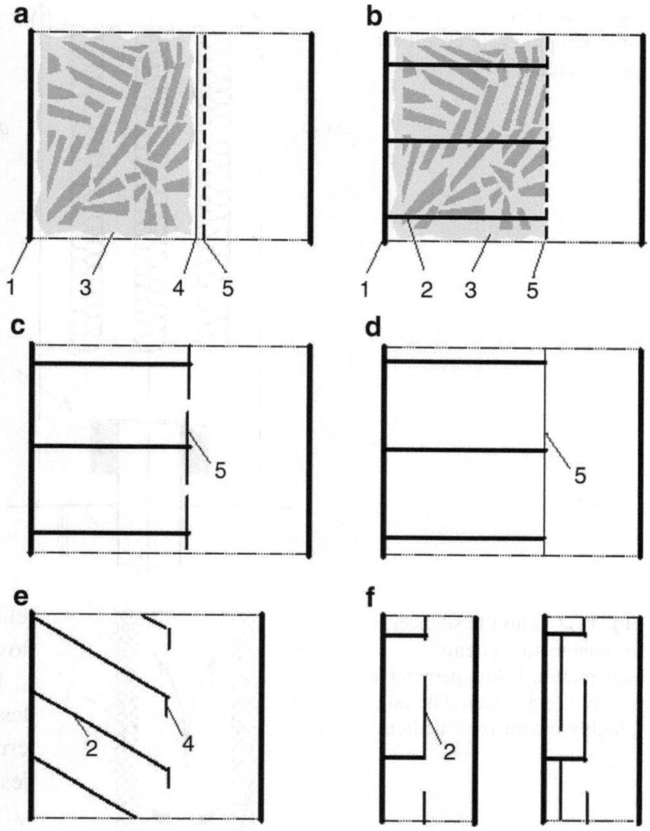

Fig. 11.3 Wall linings of absorptive and reflective silencers (schematic); 1 Wall of silencer or plane of symmetry, 2 Bulkheads, 3 Porous or fibrous absorbent material (e.g. PU foam, mineral glass or metal wool), 4 Cover consisting of fleece, foil, mesh wire, perforated plate with high porosity, depending on operational requirements, 5 top layer. **a** homogeneous absorber; **b** partitioned lining; **c** Helmholtz resonator formed by perforated or slotted plate of low porosity; **d** Helmholtz resonator formed by foil cover, **e** quarter-wavelength resonator as inclined side branch, **f** quarter-wavelength resonator as folded side branch (two cases with single and double folding)

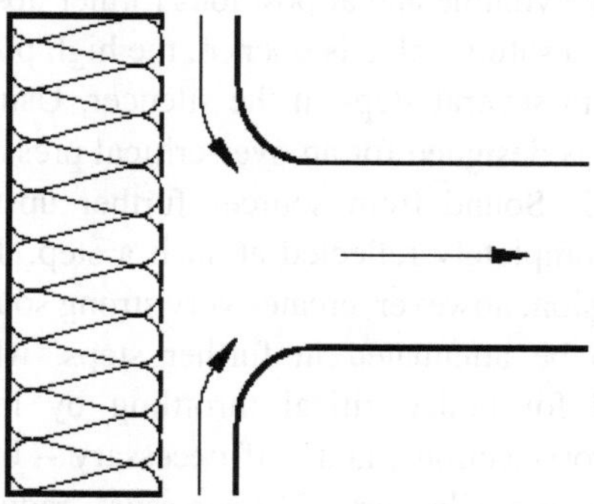

Fig. 11.4 Intake duct with absorbent baffle

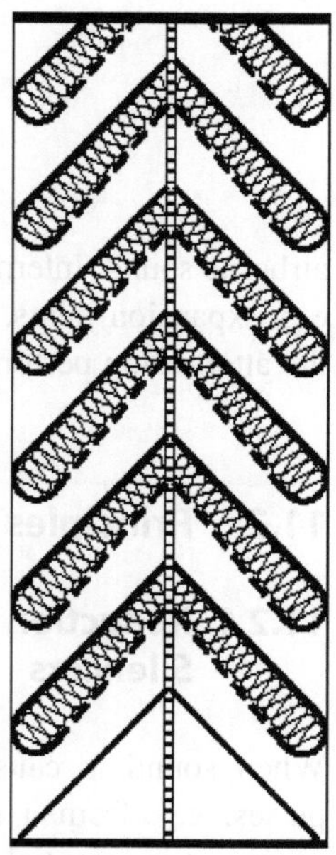

Fig. 11.5 Louver with absorbent angles

and absorptive material with respect to its structure and position (see Sect. 11.3.2.1).

Requirements for constructions without fibres and with low risk of contamination can be met by resonator and reflective silencers. Welded tubular silencers are suitable as special sturdy constructions attached to compressors and valves.

An important feature for the design of silencers is flanking transmission of structure-borne sound. As shown schematically in Fig. 11.8, in particular the outer shell of a silencers must be taken into account as transmission path of sound waves and radiator of

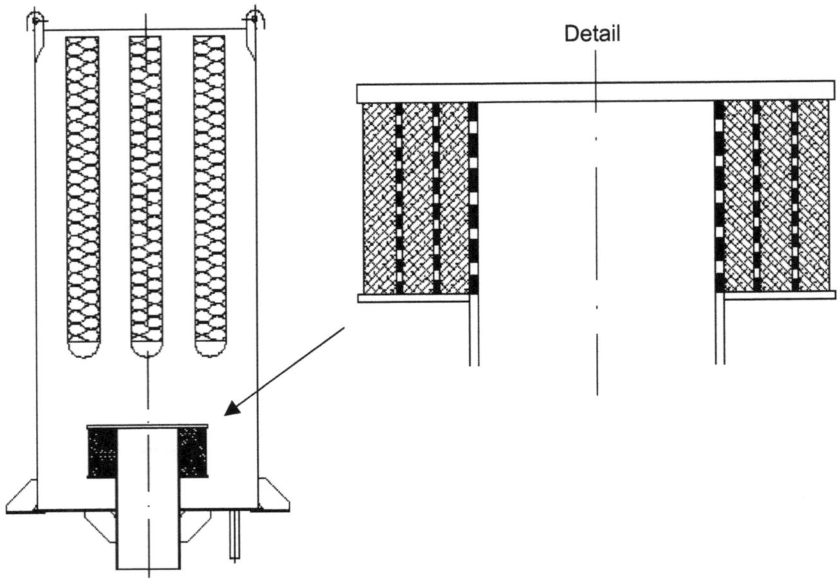

Fig. 11.6 Blow-off silencer for steam pipe with absorbent top (schematic)

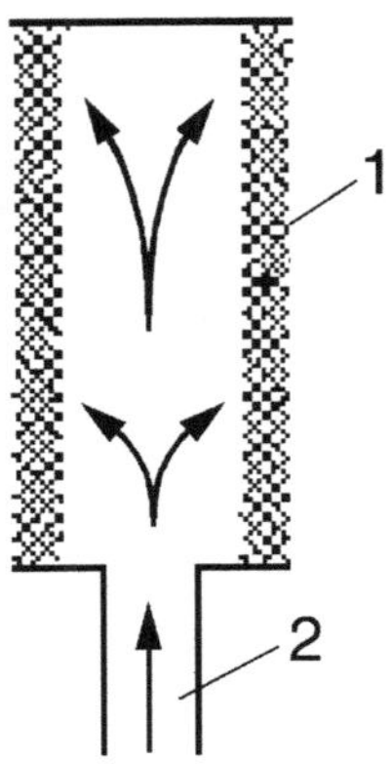

Fig. 11.7 Throttle silencer for pneumatic systems (schematic), 1 flow-permeable material (e.g. sintered metal), 2 highly pressurized medium

airborne sound. Internally, un-damped ducts and slots, e.g. expansion joints, and structure-borne sound limit the attenuation performance of a silencer.

11.2 Principles of Performance

11.2.1 Reduction of Pulsations by Throttle Silencers

When sound is caused by single or periodic flow pulses, e.g. from a safety valve or a reciprocating engine, it is essential to attenuate the pulsation of the flow by means of an equalization, which avoids the creation of sound as much as possible. This can be achieved by throttle silencers, which consist of a volume storing potential energy and a flow resistance at relatively large openings providing a small average flow velocity and, hence, little sound.

In the range of low frequencies, the volume can be described in terms of its stiffness s and the opening in terms of its flow resistance r. A silencer must be designed as a low-pass filter with low cut-off frequency $s/(2\pi r)$ to allow for flow passage at low non-linearities. Attention must be paid to the creation of sound at the inlet to the volume and at positions further upstream.

When a safety valve is opened, the high pressure is reduced in several steps in the silencer. Usually, the first step is designed for an over-critical pressure ratio $p_1/p_2 > 2$. Sound from sources further upstream is almost completely reflected at such a step. The flow transmission, however, creates very strong sound. This needs to be attenuated in further steps, which are designed for under-critical throttling by means of larger cross-sections, and – if necessary – by means of absorptive silencers. The pressure reduction in gases goes along with a temperature reduction and involves the danger of icing.

Exhaust pipes from reciprocating engines and other periodic flow sources do not require the expansion of very high pressure gases, but the oscillating pressure can still be high enough to create shock fronts caused by the fact that the velocity of sound increases with increasing instantaneous pressure. Along its propagation path in pipes, sound energy is transferred by this non-linearity from low-frequency pulsations to higher frequency sound. Any element in the pipe with a flow resistance, e.g. a catalytic converter, performs as a low-pass filter.

Table 11.1 Examples for parallel baffle or splitter silencers

Example	Spectrum	Operation conditions	Construction
Mining ventilation	Fan noise with tonal components	Humid and warm air favours the growth of moss, splitters must be removable at regular intervals for cleaning	Highly robust
Cooling tower	Low-frequency fan noise, higher frequency water splashing noise	Exposed to the atmosphere (wind, precipitation, UV radiation)	Robust
HVAC plant	Tonal components of fan noise, flow noise from in-line components	Surface non-abrasive, fire hazard, hygienic requirements	Light
Exhaust and flue gas ducts	Low-frequency combustion noise, fan noise, flow noise from in-line components	High temperatures, aggressive flue gas, contamination, possibilities for cleaning or replacement	Robust
Absorbent section of silencer for safety valve	High-frequency flow noise	Exposed to the atmosphere (wind, precipitation, UV radiation)	Highly robust
Exhaust air from production plants (paper manufacturing, tires, foils, fibres)	Fan noise, flow noise, noise from plant components	Contamination, frequent cleaning or replacement	Light
Quiet wind tunnel (vehicle tests with circulating air)	Fan noise, flow noise from bends	Redirection of airflow	Robust

Table 11.2 Examples for absorptive silencers without baffles

Example	Spectrum	Operation conditions	Construction
Plate with absorbent cover in front of opening	Broad-band and tonal components of fan noise	Exposed to the atmosphere (wind, precipitation, UV radiation)	Robust
Absorbent pipe lining at the intake of a compressor	Broad-band and tonal components of flow noise	Contamination	Exchangeable absorbent material

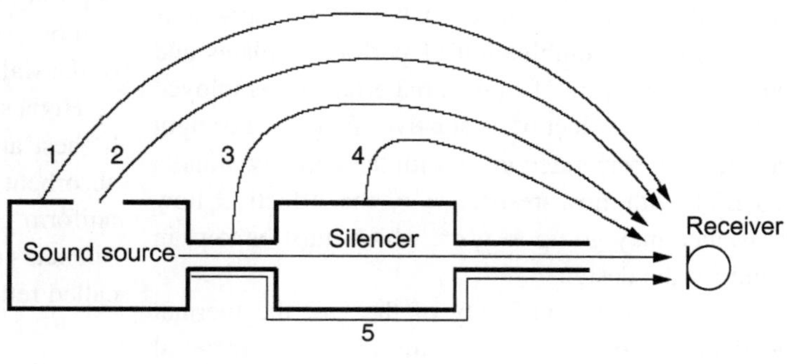

Fig. 11.8 Flanking transmission of sound via 1 radiation from the housing of the source, 2 radiation from openings at the source, 3 radiation from duct walls in front of the silencer, 4 radiation from the shell of the silencer, 5 structure-borne sound

Plant equipment not permitting a substantial back pressure may not be furnished with a throttle silencer, which is performing by the combination of a volume and a flow resistance. The application of absorptive or reflective silencers is necessary.

11.2.2 Absorption in Porous or Fibrous Structures

The viscosity of the gas in a structure with fine pores results in a flow resistance that is inversely proportional to the diameter of the pores. A similar description holds for the flow resistance of fine fibres. The flow resistance is very similarly effective for constant and oscillating flow. It causes the conversion from kinetic energy into heat. Pressure oscillations going along with flow oscillations consequently result in heating of the gas. If the heat is not carried away, as observed in the absorbent lining of test rooms for very loud equipment, the interior of the lining may turn to coal.

Example: A bare hand is not sensitive to a sound field with a sound pressure level of 150 dB. But mineral wool in the hand becomes hot after a short time.

The market requires mineral wool with fine fibres primarily for thermal insulation. With a flow resistivity of about 10 kPa s/m^2, a layer of 0.1 m thickness of such mineral wool is often matched to acoustical requirements for high absorption. For optimum performance of industrial silencers at low frequencies, however, thicker layers are needed. Fine fibres are not suitable for such layers because the flow resistance is too high and the mechanical stability too low. Improvements can be obtained by the use of thicker basalt wool. But the iron content results in oxidation and, consequently, in reduced stability and limited lifetime of basalt wool as compared to glass fibres.

Stable open-porous structures are produced as sinter metal for filter applications. The dimensions are restricted to a few millimetres in thickness and a few square-centimetres in area. In addition, material prices are high and restrict the application to special cases.

Special constructions for sound absorbers of high mechanical stability consist of perforated plates frit together with fine stainless steel fibres or of porous aluminium plates, which are available with a well-controlled total flow resistance of about 1 kPa s/m. Woven structures made from natural fibres, metal or plastic material are often not fine enough to provide a sufficient flow resistance or not strong enough to be used without a carrier material. Constructions employing the combination of perforated plates and one or more layers of fibrous materials are employed outside the flow duct of dissipative silencers. For tight contact of a perforated layer with the porosity σ and a material with flow resistance r, the effective flow resistance may go up to r/σ^2, which must be kept in mind for the design.

Aside from the viscosity of the gas, the thermal conductivity of the gas and the absorbent material provides a small contribution to the attenuation of sound [3]. Only under adiabatic conditions, as in the free sound field, or under isothermal conditions, as to be assumed in very fine metallic structures, the compression and expansion in a sound field is free of losses. Under intermediate conditions, relaxational processes between the pressure and temperature field cause small losses. For practical applications, such effects cannot efficiently be used in silencer design, e.g. to compensate for low friction losses in the range of low frequencies where particle displacement is small even at some distance from a rigid wall.

The lateral connection of absorbent material to flow ducts results in a frequency characteristic of attenuation

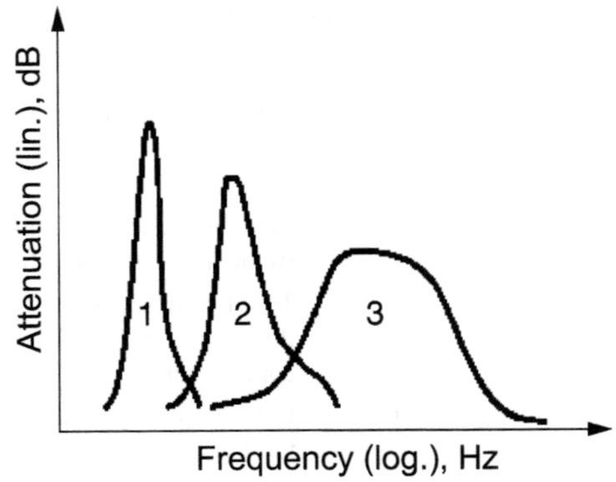

Fig. 11.9 Typical frequency characteristic of the attenuation by lined ducts for constant volume of the lining consisting of 1 Helmholtz resonators, 2 quarter-wavelength resonators, 3 homogeneous absorbent material

that covers a relatively broad frequency band (Fig. 11.9). Similar to the average gas flow, low frequency oscillations of the gas are subject to small or moderate losses during transmission through the duct. But also high frequency sound waves can propagate along the duct axis without much attenuation by the duct walls. Sound can propagate like a beam above a frequency where about two wavelengths fit between opposing walls. At medium frequencies, a high dissipation of sound can be obtained by appropriate matching of the wall lining to the sound field in the duct.

Highest attenuation by absorption goes along with highest attenuation by reflection at both ends of the absorbent duct section (and cannot be accomplished uniform passive linings over a wide frequency band). Silencers designed for such performance are generally called reflective (or reactive) silencers.

11.2.3 Absorption by Non-linearities

The flow resistance of a perforated plate or a wire mesh contains one portion that increases with increasing flow resistance and another portion that belongs to a very low flow velocity and is called the acoustic flow resistance to be determined according to DIN 52313. The latter is attributed to the viscosity of air acting on the structure, the former to the hydraulic pressure loss coefficient that causes the conversion of kinetic energy into turbulence and finally via viscous losses into heat.

There is a strong coupling between static and dynamic flow resistance, which may be treated as a weak non-linearity. It can be used to absorb sound at vortices created by the flow past or through rough structures formed by perforated plates or woven structures. Such structures are employed at the boundary of flow ducts. Together with cavities behind the structures, they provide for flow guidance and sound absorption. The flow guidance reduces the pressure loss and thereby the excitation of broad-band flow noise. In addition, tonal components created by the flow at the neck of resonators can be suppressed by a flow resistance in the excitation plane.

A strong non-linearity is associated with the effect of dissipation at a shock front. This occurs in pipes carrying sound of very large amplitudes when the steepening of the wave front is stronger than the attenuation. The magnitude of the dissipation in a shock front is determined by the conservation of mass, energy and momentum at both sides of the shock front exclusively. Consequently, the effect provides no particular means for the design of silencers.

11.2.4 Reflection

In order to reflect sound in a duct back to the source, the wave impedance in the direction of the duct must be highly different from the wave impedance in the source direction. This can happen under two conditions: either a substantial variation in cross section of the duct results in a strong variation of sound particle velocity, or a pronounced pressure minimum in the duct occurs in most of the duct cross section, passively caused by a side branch resonator or actively created by a sound source with opposite phase.

Basically, contractions of the duct cross section can be just as effective as expansions. Contractions have the advantage of not requiring much space, but the disadvantage of a limited attenuation band width and of higher regenerated flow noise. Therefore, the application of contractions is limited to cases of sources tolerating high back pressure, of high sound pressure levels making flow noise negligible and of limited space inside pipes. Typical applications are automotive mufflers.

At the expense of a large volume, all the shortcomings can be avoided with pipe expansions. The expansion is ineffective at a length that equals integer multiples of half a wavelength in the direction of the pipe axis, which transform the termination impedance to the entrance of the pipe expansion (Fig. 11.10). In between, where the expansion length equals an odd integer of a quarter wavelength of sound, the reciprocal of the termination impedance is transformed to its entrance, which makes the expansion pipe most effective. Expansion pipes are sometimes laterally filled with sound absorbing material in order to prevent the collapse of attenuation.

For a given sound particle velocity, the sound pressure in front of a resonator adopts a minimum at resonance. The spatial extent is limited, for grazing sound incidence along the duct axis not as pronounced as for sound incidence perpendicular to the entrance, but in any case limited to about half a wavelength. In addition, there is a spectral limitation. The larger the effective mass at the resonance frequency, the smaller is the frequency bandwidth. The minimum mass belongs to the quarter-wavelength resonator, which therefore requires the largest volume. Such resonators symmetrically inserted in the walls of ducts not more

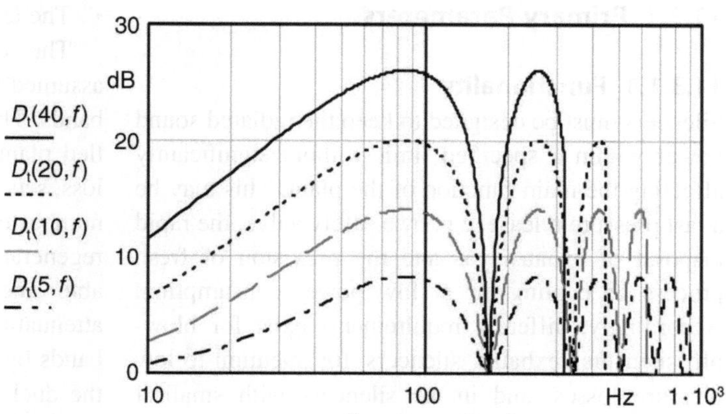

Fig. 11.10 Transmission loss D_t, in dB, vs. frequency f, in Hz, for a 1 m long expansion silencer in a pipe of 0.1 m diameter (without flow, sound velocity 340 m/s), plotted in the frequency range valid for one-dimensional transmission theory; parameter is the area ratio S_2/S_1 of the expansion

than a wavelength wide provide a high reflection loss over the frequency range of about half an octave. Other resonators designed with constrictions in a top layer by means of perforated plates or slots, with foil covers and with combinations thereof are generally effective in narrower frequency bands (Fig. 11.9).

The damping of resonators by fibrous material in the volume may increase the absorption of sound but not the height or bandwidth of reflection. Damping can also be applied close to the oscillating mass of the top layer by friction between layered constructions. Such measures may be suitable to avoid flutter noise of vibrating covers.

11.2.5 Regeneration of Sound

The effectiveness of silencers can be limited by regenerated sound. It occurs particularly at constrictions which cause pressure losses of the gas flow (see Sect. 11.5.2) and at resonators which are excited by the gas flow. At very high sound pressures, there may be low frequency sound energy shifted to higher harmonics. High particle velocities in constrictions between wider ducts are particularly responsible for such effects.

In pipes carrying high pressure sound, it is most important to attenuate the low frequency components by expansions or side branches in order to inhibit the creation of shock waves. Side branches must not cause periodic vortex shedding, which may excite resonators.

11.3 Design Parameters and Principles

11.3.1 Primary Parameters

11.3.1.1 Functionality

Silencers must be designed to keep the radiated sound power within a specified limit without significantly affecting the main function of the plant. This may be a fast pressure release, e.g. by a safety valve, the rapid disposal of exhaust gas and the provision of fresh process or cooling air at low power consumption. Accordingly, different requirements exist for blow-off silencers, exhaust silencers for medium-to-low pressure losses, and intake silencers with smallest pressure losses. Also different acoustical requirements result from the permissible radiated sound power. In blow-off silencers, the original sound source is replaced by quieter sound sources. Exhaust silencers are put on equipment with more or less specified sound emission, while intake silencers for fresh or cooling air sometimes cause pressure losses, which must be compensated by forced draft systems with fans as relevant additional sound sources.

11.3.1.2 Blow-Off Silencers

Parameters for the design of blow-off silencers are

- The pressure in the pipe upstream (p_v) and downstream (p_1) of a valve,
- The mass flow rate, and
- The permissible A-weighted sound power level.

Situations with high pressure and also with large mass flow often require several expansion stages where the first stage is designed for an over-critical pressure ratio $p_1/p_2 > 2$ that determines the sound creation (see Sect. 14.5.2.1). Subsequent stages belong to expanded cross sections, which provide for a gradual pressure decay and sound absorption by eddies in the turbulent flow. If necessary, an absorptive silencer is attached downstream and the high-frequency content of sound is reduced by an absorbent elbow. The pressure release requires the silencer to have a good mechanical stability. But also the design of the absorptive silencer is determined by mechanical stability criteria for the maximum permissible Mach number of the flow.

11.3.1.3 Exhaust Silencers

Parameters for the design of exhaust silencers are

- The required insertion loss in octave or one-third-octave bands,
- The mass flow rate,
- The permissible pressure loss, and
- The temperature.

The sound power level of the un-muffled plant is assumed to be specified in octave or one-third-octave bands. The permissible sound power level of the muffled plant, which results from the required insertion loss, sets a limit to the (temperature dependent) Mach number of the flow in the silencer. It determines the regenerated sound and, thereby, the maximum achievable attenuation. The performance of a silencer attenuating the original sound in critical frequency bands by means of great length and special inserts in the duct by more than the required insertion loss is finally limited by the regenerated noise. Such a design is sometimes applied with tubular reactive silencers for small flow rates and uncritical pressure losses.

Generally, the consideration of requirements for a pressure drop permissible or as small as possible, the opposite design is chosen and the regenerated noise limited to a level 10 dB below the permissible sound power level in the critical frequency bands. The limitation of the pressure loss requires larger free duct cross sections. Without additional measures, this will result in a smaller attenuation. An extension of the silencer length is effective at low frequencies only where the duct width is small compared to the wavelength of sound. If requirements for insertion loss have to be met at higher frequencies, a sub-division is needed into parallel partial ducts together with a lateral extension of the cross section that allows for linings of the partial ducts with absorbers or resonators.

At a given frequency, the speed of sound and its wavelength increases with increasing temperature. Consequently, the ratio of geometrical dimensions and wavelength of sound decreases with increasing temperature. The attenuation of low frequencies becomes increasingly difficult and at high frequencies somewhat easier. Rather important is the consideration of temperature for the tuning of resonators and the selection of absorber material. Bonded mineral wool loses stability at temperatures above 250°C when the binder burns out. Basalt wool with a higher content of iron oxide becomes brittle and decays into dust at temperatures above 500°C. Only special basalt fibre is stable up to 750°C. Higher temperature ranges require fibrous materials, which may cause cancer and are generally not used in absorbent silencers.

Exhaust silencers need consideration of the thermal expansion of supporting elements. Major gaps permitting the sound to propagate parallel to the absorbent duct result in a limited attenuation performance and must be avoided by suitable constructions. The attenuation limit may be estimated from ten times the logarithm of the attenuated and un-attenuated cross sectional areas.

11.3.1.4 Intake Silencers

Parameters determining the design of silencers at the intake of power sensitive equipment are

- The insertion loss required in octave or one-third-octave bands and
- The permissible pressure drop

Regenerated noise may be used for checking special applications. Attenuation and pressure loss simultaneously vary with the length and cross section of silencers. This, however, applies to a fraction of the performance only. The other part is determined by acoustical and aero-dynamical discontinuities internally and at the ends of the silencer. The design may benefit from the fact that often the attenuation is primarily related to cross section and length, while pressure loss mainly occurs at discontinuities.

11.3.1.5 Parallel Baffle Silencers

When the duct cross section is sufficiently large, the pressure loss is the quantity relevant for determining its relative blockage by a centre body or by several parallel baffles (or splitters). This is in a first step done under sole consideration of the pressure losses at the ends of the duct inserts. They can be reduced by means of caps with semi-circular profile at the upstream end and with trapezoidal shape at the downstream end. In the second step, the distance between baffles or their thickness is selected from acoustical and economical points of view. In the range of low frequencies where the absorption coefficient is about proportional to the baffle thickness and, consequently, the silencer attenuation is almost independent of baffle thickness, thick baffles are preferred for economical reasons. At high frequencies, the transmission via beaming of sound between baffles is inhibited by thin baffles. The lateral displacement of subsequent rows of baffles and the use of baffles of different thickness results in similar pressure losses so that the choice of an appropriate measure to inhibit beaming is mainly determined from the view point of economy. Software programs for optimisation of silencer design finally allow to take all aspects into account regarding the effects of discontinuities and flanking transmission on insertion loss, pressure loss, and regenerated sound.

When the duct cross section is restricted to an extent that high frequency beaming already occurs in a relevant frequency range, the insertion of a silencer with splitters requires a lateral expansion of the duct together with transition elements for suitable flow conditions. To avoid flow separation and non-uniform pressure distribution at splitters, the flaring angle at the inlet should not exceed 15° for an expansion in one dimension and 10° for two dimensions. The downstream end is less critical.

When the mass flow is relatively small, it is possible to maintain a circular cross section inside the silencer. The outside wall can be designed as an absorber with sufficient volume to perform at low frequencies. Splitters in the tube can be made as rectangular baffles or a circular centre body. For larger mass flow, rectangular ducts with rectangular baffles are used exclusively. With respect to low pressure loss, baffles of half the regular thickness may be applied at the walls. For economical reasons, they are seldom used.

In order to cover wide frequency ranges with uniform baffles, the baffles are designed non-symmetrical and variable in the axial direction. Half the length, e.g., is uncovered and covered by sheet metal or made from shorter and longer side branches, respectively. The baffles should be mounted symmetrically for maximum attenuation in particular frequency bands. Asymmetrical configurations provide more broad-band attenuation for particular applications.

11.3.2 Further Operational Requirements

11.3.2.1 Mechanical Stability

Accounting for operational conditions, the casing, flanges and supports for silencers shall be designed for a minimum lifetime of 5 years. Relatively thin sheet metal constructions are sufficient for an HVAC plant. When the components are exposed to flow velocities of more than 20 m/s, mechanical stability is increasingly important. Special measures can be taken for protection against effects of weather, acids in exhaust gases, and differences in voltage potentials of different materials. They include the selection of particular material, e.g. aluminium, or the application of protecting covers, e.g. rubber.

When requirements are low – as in HVAC plants – the stability of fibrous absorbers can be enhanced by fleece covering the surfaces. At high impact, e.g. behind safety valves in steam pipes at flow velocities from 100 to 200 m/s and temperatures up to 600°C, or when the surface is mechanically damaged, large amounts of particles are carried away through erosion. Occasionally, absorption elements are entirely depleted. This can be avoided by protecting perforated plates in front of fine mesh wire or glass fibre felt.

11.3.2.2 Abrasion Resistance of Absorbers

Foam material is more resistant to abrasion than fibrous material but causes a fire hazard when made of plastic. Special materials, e.g. foams based on melamin resin or sinter metal, are better suitable but more expensive and restricted to special applications. The abrasion of fibrous material and its spreading can be reduced by covering foils or fleeces.

Foils are used for air-tight sealing. They have to be very light not to impair the transparency of higher frequency sound. A low weight of not more than 50 g/m^2 is generally achieved by plastic foils only. Their stability and temperature limit must be observed. Occasionally, the durability of foils exposed to ultra-violet light must be considered for intake silencers. When a foil shall be protected from mechanical damage by a perforated plate, full attachment or sticking to the perforated plate must be avoided. Otherwise, its transparency for sound is substantially reduced. It should always be kept in mind that light plastic foils inside and outside of sealed elements may rupture due to static and dynamic pressure differences. In critical cases, this consideration yields to the exclusion of fibrous absorbers.

11.3.2.3 Fire Hazard

Constructional specifications for HVAC equipment and particularly for technical plants with flow carrying oil aerosols or organic substances such as flour or milk powder include fire resistance of ("non-combustible") silencer materials. This shall avoid that fire is started or transmitted. Resonator-type silencers without absorbent material meet such requirements, but products made from mineral fibres and melamin resin as well. Furthermore, precaution is required by means of appropriate design and configuration of silencers against the deposit of burning or explosive substances.

11.3.2.4 Hygienic Requirements

High hygienic requirements must be observed for hospitals and food processing plants. Absorbers with open pores and open resonators are not permissible where viable particles (bacteria) can settle and grow, especially the air temperature and humidity is elevated. Unless absorbers with closed surface are used, air-tight foils are necessary covers. The surfaces shall be cleanable and for that purpose sturdy and without un-accessible slots. Silencer elements should be exchangeable.

11.3.2.5 Air Pollution

Fibrous absorbers may cause some small air pollution. This is generally irrelevant for human health in HVAC equipment commonly equipped with material of a high index for avoiding cancer of 40 or more; however, it may be important for clean rooms. In such applications, protection by foils is sufficient as a rule.

11.3.2.6 Deposits and Cleaning

Particles carried by the gas flow may settle on the silencer and reduce its efficiency. Open porous or rough surfaces of absorptive silencers are mostly endangered by sticking particles. Perforated plates and mesh wire may be rapidly clogged by accumulating fly-ashes, humid particles of wood pulp and the like. The settling of particles on foils reduces the transmissibility of sound. In critical cases, baffles with wide uncovered side branch resonators have been successfully applied. They allow for the deposit of substantial amounts of particles without decay of acoustical performance. Alternatively, special foils may be used to which such particles poorly stick and, therefore, do not settle in thick layers and can be removed easily.

If a need is seen for regular cleaning or exchange of silencers or baffles, this must be considered with the design. Pressurized air, steam jets, brushes and solvents or decontamination fluids are used for cleaning. It can be done either in situ or with the baffles extracted.

11.3.2.7 Start-Up and Shut-Down of Plants

Pressure, temperature and humidity can vary during starting-up and closing-down of plants and stress silencers. Air-tight foils enveloping absorbers and metal construction have to go through expansions and constrictions. Condensation of fluids, e.g. after transition of the dew point, must not be collected in the silencer but drained by appropriate design.

11.3.3 Guidelines for Economical Constructions

In addition to meeting the requirements for functionality, the methodology of design lists requirements for a construction aiming at ease of manufacturing and maintenance, resistance to corrosion, suitable for assembly, high safety and low energy consumption. In the past, low production costs were particularly important, but presently, there is an increasing awareness for operational costs, which must be included in the overall calculation. For example, the reduction of pressure losses for the silencers in a power plant may cause a small increase in efficiency only, but can result in a cost reduction way beyond the stock price for the silencers.

The space available for silencers is often of decisive importance. It determines the pressure losses or the measures to reduce them. The attenuation needed at low frequencies is essential for the volume required, while the ratio of length and diameter is determined by the available space. Depending on the type of silencer, the costs increase with increasing size. Increasing operational requirements generally result in increasing weight and price of a silencer.

11.4 Experience Data

11.4.1 Splitter Silencer

11.4.1.1 Absorbent Splitters

Laboratory data on the insertion loss of baffles in normal commercial usage are shown in Fig. 11.11. The thickness of 200 mm is frequently chosen to obtain high attenuation at medium frequencies around 500 Hz. Thicker splitters are needed to achieve a broad-band maximum attenuation at lower frequencies. The relevant parameter is the ratio of splitter thickness to wavelength of sound, of which the dependence on temperature must be borne in mind for exhaust lines.

The airway width between splitters depends on the permissible pressure loss. The smaller the width the further is the attenuation maximum extended toward high frequencies. For a width of 100 mm and air at 20°C, the range of sound beam formation lies above 2,000 Hz, where the wavelength is less than twice the width between splitters and the absorber has little influence on the propagation of the fundamental mode.

The insertion loss has been measured for splitters homogeneously filled with mineral wool providing a flow resistivity of about 12 kPa s/m^2. Absorbers made from melamin resin rather than fibrous material are often inhomogeneous and therefore exhibit a slightly different frequency characteristic of attenuation.

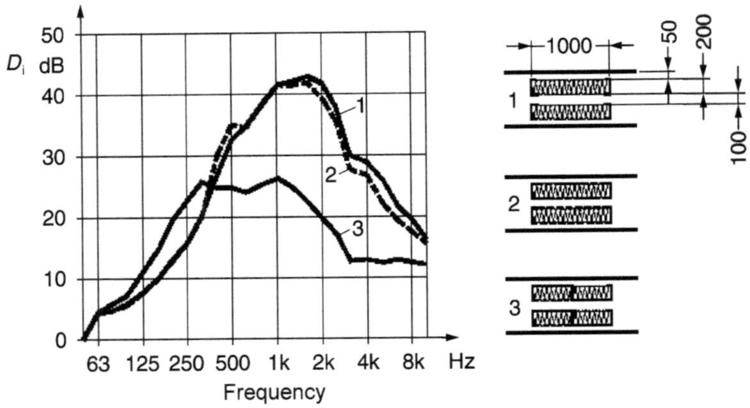

Fig. 11.11 Insertion loss of commercially available absorbent splitters according to laboratory testing; 1 without cover, 2 with perforated plate cover, 3 with alternating cover of thin sheet metal

Differences also appear with PU foam due to vibrations of the skeleton material.

By comparison of the measured data in Fig. 11.11, it can be seen that a cover of perforated plate applied for protection against higher flow velocities has little influence on the attenuation. Similarly ineffective is a cover of fleece, cloth, perforated or stretched sheet metal as long as its flow resistance is less than 200 Pa s/m. Covers of plastic foil or sheet metal, however, require particular attention. The latter are specifically used to enhance the attenuation – as shown in Fig. 11.11 in the frequency range around 250 Hz – alas, at the expense of a reduced attenuation at higher frequencies. The partial covering of absorbent splitters is normal commercial usage.

Rectangular splitters are equipped with profiles at both ends. Semicircular profiles are proved particularly for the upstream side. At the downstream side, similar or conical profiles with a flaring angle of more than 15° are less efficient.

Absorbent rectangular splitters receive a basic structural strength from an enveloping sheet metal folded at its rim. Perforated covers or stretched sheet metal significantly enhances the stability. Galvanized steel is commonly used, aluminium and stainless steel are used for special applications. Rubber covers are employed for better protection against chemically aggressive liquids. Long baffles are stabilized by bulkheads. The acoustical effect of such partitions is limited to the range of low frequencies where the distance of bulkheads is less than half a wavelength. When the distance is smaller than the thickness of the splitter, the attenuation is reduced in this frequency range.

The installation of splitters is done on supporting rails with seals and extension joints as specified by the manufacturer, thus ensuring the consistency with laboratory performance. The symmetrical installation of splitters where absorber sections and sheet metal protected sections face each other, respectively, results in acoustically particularly narrow ducts providing a high attenuation in particular frequency bands. The non-symmetrical installation shown in Fig. 11.11 results in less attenuation per unit length, but is effective over longer sections of the silencer. This allows for meeting different requirements on the frequency characteristic of the insertion loss.

Absorbent splitters installed in HVAC plants and in intake ducts requiring low pressure loss are insensitive in attenuation to airflow, but not in exhaust and transmission pipes carrying flow at a velocity of more then 20 m/s.

11.4.1.2 Resonator Splitters

Laboratory data on the insertion loss of baffles with side branch quarter-wavelength resonators tuned to 630 Hz are shown in Fig. 11.12. A narrow range of high attenuation shows up at this frequency, while only smaller discontinuity attenuation is effective at lower and higher frequencies. As shown in Fig. 11.12, the width between the splitters is smaller than the splitter thickness. The width of the side branches is about equal to their depth. A bulkhead is necessary to separate the two sides of the splitter. When tuning of resonators is needed for lower frequencies, the branches of quarter-wavelength resonators are getting

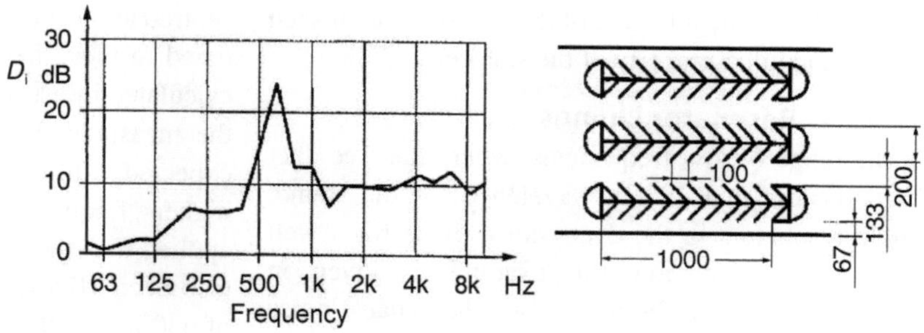

Fig. 11.12 Insertion loss of a splitter silencer with quarter-wavelength resonators

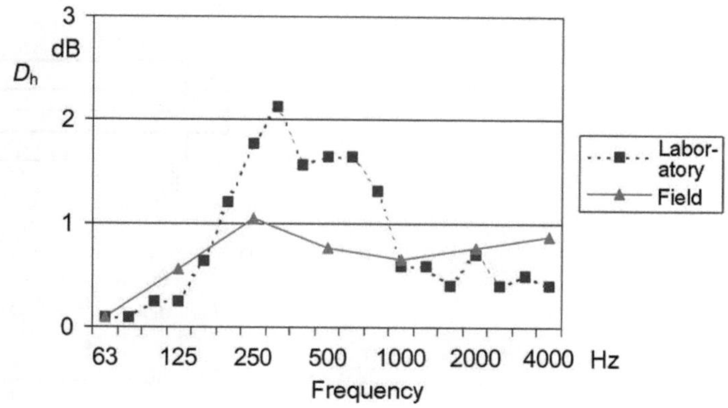

Fig. 11.13 Attenuation per half duct width D_h for a reactive splitter silencer containing two types of differently tuned quarter-wavelength resonators, results from laboratory and field measurements

too long. The quarter-wavelength resonators have to be replaced by Helmholtz resonators at the expense of attenuation bandwidth. Helmholtz resonators with open coupling areas to the air passage may be detuned by flow and particle deposits at the neck. Even more sensitive to deposits are resonators covered with thin foils. By use of special materials (Teflon), the deposit of particles can be reduced or easily removed.

Quarter-wavelength side branch resonators of the type shown in Fig. 11.12 (top view), when installed in a horizontal duct, may be positioned with branches inclined in the direction of flow for smallest pressure drop. Then particles in the flue gas will drop to the bottom of the chambers without blocking them and render them un-effective, even when deposited in a larger amount. In vertical ducts, splitters need to be positioned in a way that deposits of particles can slide down on the inclined branches. In this case, small guide vanes are necessary to reduce the pressure loss. Covers of perforated plates or stretched metal are rapidly blocked by deposits and cannot be used.

The subdivision of splitters by bulkheads must be carried out with care and results in considerable weight. Accordingly, very stable supporting structures are needed. The application in flue gas ducts requires large expansion joints, which may impair the efficiency of splitters by flanking transmission of sound. Neglecting such viewpoints may result in differences between laboratory and field measurements as exhibited in Fig. 11.13.

11.4.2 Lining of Ducts

11.4.2.1 Absorbent Linings

The lining of ducts with sound absorbing material is state-of-the-art in noise control. Insertion loss and pressure loss data are available for ducts of circular, quadratic and rectangular cross section, obtained from software programs and laboratory tests that are suitable for practical application. Vibrations of duct walls must be taken into consideration, which limit the performance of linings in circular ducts due to flanking transmission of structure-borne sound, but results in some low-frequency excess attenuation of rectangular ducts. In all cases, flanking transmission of airborne sound needs attention.

In straight ducts, the attenuation performance rapidly decays beyond the cut-on frequency of beam formation, where the free width of the duct is about 1.5 times the wavelength of sound. In ducts with elbows, absorbent material in the vicinity of the elbows efficiently attenuates sound at high frequencies.

Pressure losses are a feature of the construction of such elbows and not a property of the silencer.

11.4.2.2 Resonator Linings

In the range of low frequencies, where the free duct width is less than half a wavelength of the sound, single resonators in the duct wall provide for a well predictable attenuation. An example is given in Fig. 11.14. Without the resonators, the exhaust noise shows tonal components in the one-third octave bands centred at 80 and 160 Hz. After insertion of resonators tuned to these frequencies, the peaks disappear. The calculated insertion loss is sufficiently consistent with the measured data at the design frequencies, but is expectedly different in the range of low levels and at higher frequencies. The one-dimensional theory applied for the prediction of silencer performance (see Sect. 10.5.1.3) is limited to the frequency range of 630 Hz for the silencer of Fig. 11.14.

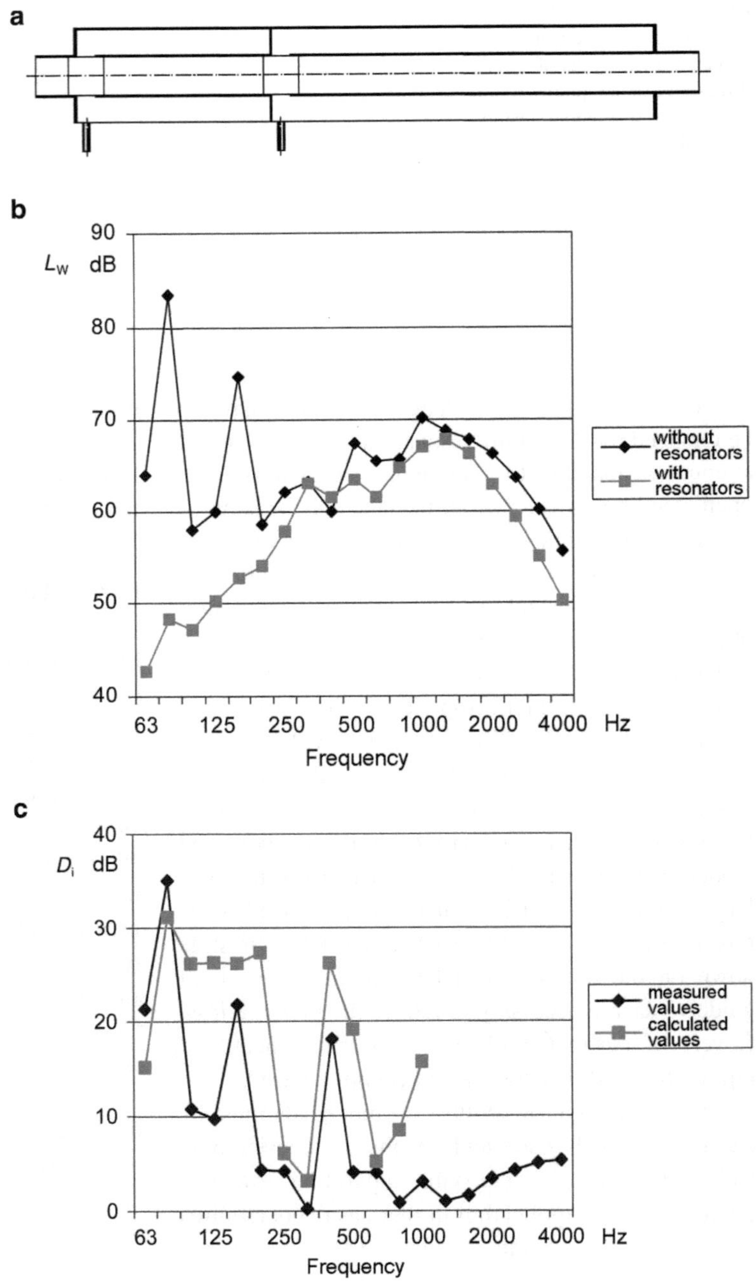

Fig 11.14 Pipe (free diameter 132.5 mm) with side-branch resonators; (**a**) schematic, (**b**) level of transmitted sound power with and without resonators, (**c**) insertion loss

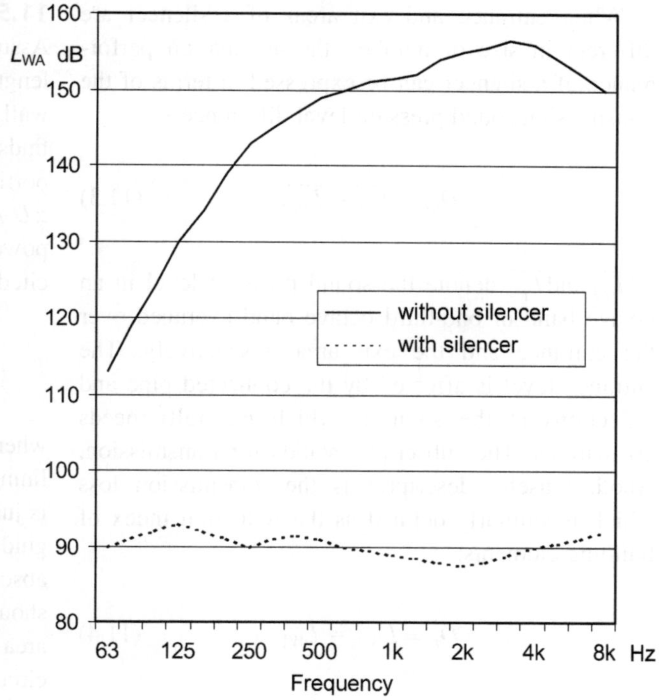

Fig.11.15 A-weighted octave-band sound power level during blow-off of steam (mass flow rate 50 kg/s, temperature 400°C, pressure before valve 18 Mpa, pressure before silencer 1 Mpa); blow-off silencer with expansion and absorbent section as shown in Fig. 11.6

11.4.3 Blow-Off silencer

The blow-off of large amounts of steam from a pipe downstream of a valve results in very loud high-frequency sound, as shown by the example given in Fig. 11.15 with an A-weighted sound power level of 165 dB. A typical commercially available blow-off silencer, consisting of a pressure release section formed by perforated layers or woven structures and an absorbent section according to Fig. 11.6, reduces the radiated sound approximately to a pink noise with an A-weighted sound power level of 104 dB. The over-critical expansion at the first perforated plate replaces the original sound source at the valve. The high level reduction is obtained by perforated layers and woven structures in the under-critical expansion section and the subsequent absorptive silencer.

11.5 Calculation Procedures

11.5.1 Sound Attenuation

11.5.1.1 Definition of Attenuation Measures

When a silencer is inserted in a pipe or attached to an opening, the reduction in sound pressure level at a specified immission point is called the insertion sound pressure level difference:

$$D_{ip} = L_{p\mathrm{II}} - L_{p\mathrm{I}}. \tag{11.1}$$

$L_{p\mathrm{I}}$ and $L_{p\mathrm{II}}$ denote the sound pressure level in an octave band or one-third octave band with and without the silencer installed [4]. The subscript i stands for insertion and the subscript p for sound pressure. In ISO 11820 an additional subscript S stands for the area at the intake or exhaust side of the silencer, respectively. If these areas are different, an insertion loss is not defined strictly. When no immission point is specified, the preferred quantity for description of silencer performance is the insertion loss

$$D_i = L_{W\mathrm{II}} - L_{W\mathrm{I}}. \tag{11.2}$$

$L_{W\mathrm{I}}$ and $L_{W\mathrm{II}}$ denote the sound power level in an octave band or one-third octave band with and without the silencer installed. The sound power levels may be determined from measurements of the sound pressure or the sound intensity on an enveloping surface or by reverberation room measurements. Again, entrance and exit areas of the silencer must be equal in order to apply Eq. (11.2). The information about directivity from the open end is lost, as opposed to results from measurements of the insertion sound pressure level difference.

When entrance and exit areas of a silencer are different in size or number, the attenuation performance of a silencer can be expressed in terms of the transmission sound pressure level difference:

$$D_{tp} = \overline{L_{p1}} - \overline{L_{p2}}, \quad (11.3)$$

$\overline{L_{p1}}$ and $\overline{L_{p2}}$ denote the sound pressure level in an octave band or one-third octave band averaged over the entrance and the exit area, respectively. The entrance level is affected by the connected pipe and reflections at the silencer, which generally needs estimations. The subscript t stands for transmission. Another useful descriptor is the transmission loss which is similarly defined as the reduction index of building elements:

$$D_t = L_{W+} - L_{WI}. \quad (11.4)$$

Again, L_{WI} denotes the level of the transmitted sound power in an octave band or one-third octave band with the silencer installed; L_{W+} denotes the corresponding level of the incident sound power. A reflection at the silencers changes the incident sound power only in the case of an additional reflection in the entrance pipe.

In general, the insertion loss is to be determined preferably from measurements and the transmission loss preferably from calculations. In the case of negligible reflections in the entrance and exit pipes and equal cross sectional areas, the insertion loss and the transmission loss are about equal.

When the reaction of the exit on the entrance of a silencer is small – which is generally the case for absorptive silencers – the insertion loss of a homogeneous silencer can be split up into a discontinuity attenuation at both ends, D_s, and a propagation loss, D_a, describing the decrease in sound pressure level per unit length:

$$D_i = D_s + D_a l, \quad (11.5)$$

where l denotes the length of the silencer. D_s depends upon the field distribution in the intake pipe, which is determined by a number of propagating modes and by a mismatch of the lowest order modes in the intake pipe and in the silencer.

11.5.1.2 Simple Estimates

Assuming that the power flow in a silencer element of length dz is split up between the equivalent absorbing wall area $\alpha U \mathrm{d}z$ and the downstream area of size S, one finds the decay of sound power per unit element proportional to the propagating sound power via the factor $\alpha U/S$. This yields an exponential decay of sound power described by Piening's propagation loss ([5], cited in [6]):

$$D_a \propto \frac{U}{S}\alpha, \quad (11.6)$$

where U denotes the perimeter of the absorbent wall lining and α is the absorption coefficient. The relation is just a simple approximation but provides for a good guidance to the design of absorptive silencers: The absorbing perimeter and the absorption coefficient should be as large as possible, while the cross-sectional area S should be as small as possible. For pipes with circular cross section and full absorbent lining the ratio $U/S = 2/r$, where r is the radius of the pipe. For rectangular ducts of cross section $B \times H$ lined at the longer sides B, the ratio approaches the maximum value $U/S = 2/H$ in the limit $H \ll B$.

The consideration of a complex propagation constant Γ for the description of the pressure distribution $p(z) \propto e^{-\Gamma z}$ along the duct includes the propagation loss $D_a = \mathrm{Re}(\Gamma) \cdot 8.7\,\mathrm{dB}$ via its real part $\mathrm{Re}(\Gamma)$. From a simplifying balance of sound volume flow for a duct element, one finds [7]:

$$\Gamma = jk\sqrt{1 + \frac{1}{jk}\frac{U}{S}\frac{\rho c}{Z}}, \quad (11.7)$$

where jk denotes the propagation coefficient for a plane, un-attenuated wave of wavenumber k, Z is the complex wall impedance of a wall locally reacting along the perimeter U and ρc is the characteristic impedance of plane waves in air. The approximation of Eq. (11.7) is primarily valid for large magnitudes of Z. The series expansion of the square root shows that a large attenuation can be achieved by large values of the real part of the wall admittance $1/Z$, which can be realized by resonators.

However, there is a second possibility to obtain a real part of Γ. If the wall admittance reacts like a mass, the square of the imaginary unit in the denominator of

Eq. (11.7) yields a negative term. As long as the magnitude of this term is larger than 1 – and this is limited to the range of low frequencies and large wall admittances – the square root of a negative value yields a complex number. In this range, the co-vibrating mass of the wall more than compensates the stiffness of the air in the duct and inhibits the storage of potential energy, which is necessary for wave propagation. This results in a positive real part of Γ. In a duct that is small compared to the wavelength of sound, a limp wall yields a propagation loss (even without sound radiation from this wall to the environment). The effect shows up at low frequencies in sheet metal ducts. It can be exploited in narrow frequency bands above the resonance frequency of duct linings formed by resonators without any absorbent material. Within the approximation, however, the effect is overestimated at the resonance frequency.

Far below the resonance frequency and generally in the range of low frequencies, the normalized admittance of the duct wall is $\rho c/Z \approx jkd$, where d is the thickness of a locally reacting wall. The radical in Eq. (11.7) is determined by the ratio of total cross sectional diameter of the lined duct and the diameter of the free cross section. The square root of this ratio determines the propagation speed of axial waves that is slower than in rigid ducts, resulting in a discontinuity attenuation

$$D_s = 20\ \lg\left(\frac{1}{2}\left|\sqrt{-j\Gamma/k} + \sqrt{jk/\Gamma}\right|\right) \mathrm{dB}, \qquad (11.8)$$

which is about half as much as can be obtained from a change in cross section from area S_1 to area S_2, e.g. at the ends of parallel baffles [8]:

$$D_s = 20\ \lg\left(\frac{1}{2}\left[\sqrt{\frac{S_1}{S_2}} + \sqrt{\frac{S_2}{S_1}}\right]\right) \mathrm{dB}. \qquad (11.9)$$

For an area ratio $S_2/S_1 = 2$, Eq. (11.9) yields 0.5 dB only and can be neglected in practical application. It requires multiple reflections and large changes in cross section to provide relevant attenuation in straight narrow ducts. However, discontinuity attenuation becomes relevant in bent and wide ducts with cross-sectional dimensions of more than half a wavelength.

As to the question of suitable absorbent material, the evaluation of Eq. (11.7) shows that the propagation loss slightly increases with increasing flow resistance at frequencies way below the first resonance in the lining, while it is inversely proportional to the flow resistance in the vicinity of the resonance. In the intermediate frequency range, the flow resistance plays a minor role. This frequency range often contains the fundamental frequency of fan blade passage, the tone of which needs most attenuation. For splitters of thickness $2d$ forming air passages of width H, the propagation loss can be approximately calculated from

$$D_a \approx 2.2\,k\,\frac{2d}{H}\ \mathrm{dB}. \qquad (11.10)$$

The performance of splitter silencers is thus mainly determined by the (temperature dependent) wave number k, i.e. the number of wavelength along the silencer, and by the ratio of splitter thickness and width of air passages. The selection of material is mainly determined by requirements on mechanical stability, temperature, hygiene, cost and others. For HVAC application, light-weight fibrous materials with 30 kg/m^3 and low flow resistivity are sufficient, while heavier fibrous absorbers with more than 80 kg/m^3 are used in the exhaust duct of gas turbine installations. The maximum attenuation obtainable from such heavy absorbers in the vicinity of the quarter-wavelength resonance at $kd \approx \pi/2$ is no longer 7 dB per length H, as calculated from Eq. (11.10), but about 4 dB. The propagation loss in the vicinity of the resonance is similarly reduced.

The simple approximations are based on a close interaction of the sound field in the duct with the absorbent or vibrating duct wall. This assumption is no longer valid for higher frequencies where 1.5 or more wavelengths of sound fit between opposed walls. Sound can then propagate like a beam along the duct axis and suffers very little attenuation. At frequencies beyond $kH \approx 9$, the fundamental mode decays approximately with $D_a \propto 1/f^2$. However, with increasing frequency there is an increasing number of higher order modes, which suffer stronger attenuation. Field experience with densely filled splitters can be approximated by the relation

$$D_aH \approx \left(4 - 4.4\,\lg\left(\frac{kH}{9}\right)\right)\text{dB}, \quad (11.11)$$

with a minimum attenuation $D_a l \approx 10\ \lg(\pi l/2H)$dB due to ray geometry. To increase the silencer performance beyond this limit at high frequencies, it is necessary to avoid or intercept the free beaming. The former is done by splitters reducing the free duct width, the latter by setting splitters at an angle or off line. Simple acoustical calculations are available for narrow passages between straight splitters only. But this covers the field of main practical interest since angles and non-aligned splitters generally result in a disadvantageous pressure loss.

The effect of airflow on the propagation loss can be well estimated from the travelling time of a sound particle in the silencer. In the downstream direction, the travelling time is reduced by a factor $(1 - M)$, where M is the Mach number; in the upstream direction, the Mach number is negative and the travelling time increases correspondingly. The influence of airflow on the frequency characteristic of attenuation can be taken into account by shifting the frequency scale by a factor $(1 \pm M)$. The downstream attenuation thus increases at high frequencies and decreases at low frequencies. Further corrections due to diffraction of sound in the flow profile [9], which partially compensate the effect of travel time, and the increase in flow resistance of the top layer of absorbent material due to grazing airflow are mostly neglected. In total, airflow with Mach numbers $|M| < 0.05$ is acoustically negligible and $|M| > 0.15$ is not permissible from the viewpoints of mechanical stability and regenerated noise.

11.5.1.3 Transmission Line Theory

Sound propagation in narrow ducts is well described by approximations for one-dimensional transmission lines. They account for waves propagating forward and backward between inlet and exit. In absorptive silencers of sufficient length, the reaction of the exit on the inlet is negligible due to sufficient propagation loss, but not in ducts with rigid walls.

The transmission line theory originally developed with matrix calculus for electrical lines [10] has been applied for many years on silencers [11] and is now well established in the special literature [12, 13]. This includes various publications on special components for automotive applications, e.g. perforated tubes representing non-compact elements with extended coupling of sound fields. Such elements are not discussed in the following.

The application of transmission line theory on sound propagation in pipes or ducts is based on two assumptions:

- The sum of the volume velocities q emanating from a node is zero.
- The pressure distribution in a cross section and at a discontinuity of a pipe is sufficiently described by a single value p which continuously varies along the longitudinal coordinate z.

A node can be any reference point for the calculation of sound in a duct with partitioned or otherwise locally reacting walls, at a discontinuity or at a branch of a pipe system.

In transmission line theory any cylindrical pipe or duct with rigid wall is acoustically described by its cross-sectional area S and its length l exclusively. The pipe may be straight or bent. Even sharp corners are allowed as long as the cross section does not change much.

Waves travelling in the pipe forwards and backwards with the wavenumber k are described by the transfer matrix T:

$$T = \begin{pmatrix} \cos kl & j\frac{\rho c}{S}\sin kl \\ j\frac{S}{\rho c}\sin kl & \cos kl \end{pmatrix}. \quad (11.12)$$

The transfer matrix connects sound pressures p and volume velocities q at the inlet side 1 and the exit side 2:

$$\begin{pmatrix} p_1 \\ q_1 \end{pmatrix} = T\begin{pmatrix} p_2 \\ q_2 \end{pmatrix}. \quad (11.13)$$

The volume velocities propagating away from the source have a positive sign, those coming back are negative. The determinant of the T-matrix is $|T| = 1$. For consideration of superimposed flow k is replaced by

$$k_c = \frac{k}{1 - M^2}, \quad (11.14)$$

and the elements of the T-matrix are:

$$\begin{pmatrix} T_{11} & T_{12} \\ T_{21} & T_{22} \end{pmatrix} = e^{-jMk_c l}\begin{pmatrix} \cos k_c l & j\frac{\rho c}{S}\sin k_c l \\ j\frac{S}{\rho c}\sin k_c l & \cos k_c l \end{pmatrix}. \quad (11.15)$$

Pipe elements in series are described by the product of their T-matrices. A pipe line with anechoic termination, entrance area S_1 and exit area S_2 (see e.g. Fig. 11.10) has the transmission loss:

$$D_t = \left[20\lg\left|T_{11} + \frac{T_{12}S_2}{\rho c} + \frac{T_{21}\rho c}{S_1} + T_{22}\frac{S_2}{S_1}\right| - 6 + 10\lg\frac{S_1}{S_2}\right] \text{dB}. \tag{11.16}$$

For a pipe with absorbent lining, the wavenumber k has to be replaced by Γ/j, where the complex propagation constant Γ can be calculated approximately from Eq. (11.7) for a locally reacting wall.

A single side branch of a pipe extending over a short distance in z – direction and rigidly closed at its end[2] may be considered as a pipe section in series with the pipeline described by the transfer matrix

$$T = \begin{pmatrix} 1 & 0 \\ \frac{S^{(A)}}{Z} & 1 \end{pmatrix}, \tag{11.17}$$

where $S^{(A)}/Z = T_{21}^{(A)}/T_{11}^{(A)}$ is the admittance of the side branch multiplied by its entrance area $S^{(A)}$ and $T_{21}^{(A)}$ and $T_{11}^{(A)}$, are elements of the T-matrix of the side branch. Providing reflection of sound essentially without hindrance of the gas flow, side branches are silencer elements of particular importance. For high performance in selected frequency bands, side branches are designed as tuned resonators. The interaction of pipe sections with different resonators, however, may be difficult and requires numerical evaluation.

In principle, reflection of sound can be obtained from constrictions as well. But they are hardly ever applicable in pipes due to intolerable pressure loss. The T-matrix of a constriction with cross-sectional area S follows directly from Eq. (11.12). For an orifice with $kl \ll 1$, the effectiveness of the air mass in the centre shows up in the term $T_{12} \approx j\rho ckl/S = j\omega\rho l/S$. Airflow through the orifice causes a dynamic flow resistance $\rho cM\zeta/2$.[3] The effective thickness of the orifice is determined by end corrections Δl, which account for non-propagating modes by an additional mass.

An end correction must always be added to the length l of the narrower of two connected pipes. Its value depends on the shape and the size of the pipes [14, 15]. For pipes with circular, quadratic or similar cross section and without flow, the end correction is approximately:

$$\Delta l = \frac{\pi}{2}d(1 - 1.47\varepsilon^{0.5} + 0.47\,\varepsilon^{1.5})$$

where

$$\varepsilon = \left(\frac{d}{b}\right)^2, \tag{11.18}$$

d (or U/π) denotes the diameter of the narrower pipe, b (or U/π) the diameter of the wider pipe or the distance between holes in a plane perforated layer. Airflow reduces the end correction. At the entrance of a side branch to a pipe carrying airflow, the end correction is neglected in a first-order approximation.

In order to account for the increase of side branch cross-sectional area, a special end correction may be applied. For side branches with rotational symmetry from a circular pipe, the first resonance between the pipe radius r_i and the outer radius $r_a < 5\,r_i$ can be calculated from the quarter-wavelength resonance in a cylindrical tube of length $r_a - r_i + \Delta l$, where

$$\Delta l = \frac{1}{2}\frac{(r_a - r_i)^2}{r_a + r_i}, \tag{11.19}$$

instead of evaluating the relevant Bessel function.

The transmission line theory is suitable to determine the performance of in-line silencers, particularly those without absorbent material. It is less suitable for the calculation of splitter silencers with open or covered resonators.

11.5.1.4 More Precise Calculations for Absorptive Silencers

Goals and Limitations

The reasoning for more precise calculations of silencers includes

- The experience that rough estimates occasionally deviate from measured results, thus requiring retrofit of hardware or additional safety margins in the design

[2] A branch that is not small compared to the wavelength can be treated as an expansion of the pipe as long as the total diameter is less than half a wavelength.

[3] Tangential flow past the perforated plate in front of a side branch results in a fraction of the dynamic flow resistance which adds to the damping. This effect is more important than the variation of propagation velocity of sound waves.

- The assumption that simple geometries of silencers allow for analytic solutions of the wave equation, and
- The confidence that even complicated shapes and materials are calculable by advanced numerical methods

Correspondingly, scientific investigations on this subject are still going on. However, it should be kept in mind that

- There is a considerable uncertainty about the baseline data on flow and sound in view of spatial and spectral distribution and as a function of loading the source by the silencer
- The calculation effort is often out of proportion with the practical results
- Large requirements on the information about material data, which may not be available or vary with time due to contamination, erosion, settling etc., and
- The uncertainties in modelling flow sub-layers, flow profiles, inhomogenities, displacements, extension joints, flanking transmission of structure-borne sound and many others

Circular Silencers

Results calculated for rectangular ducts and splitters can be found in several publications [2, 16, 17]. In the following, a detailed calculus for circular silencers with lined walls and centre body is described exclusively as an example for a construction frequently applied and suitable for reliable modelling (see Fig. 11.16). Consistent with prevailing usage, a fibrous absorber is assumed. For plastic foam material, it would be necessary to take structural vibrations into account [18]. Furthermore, the theory presented is limited to cylindrical ducts and does not include variations in cross sectional area [19, 29, 30].

Without airflow, the linear sound field equations account for the inertia and the friction of air by the relation

$$j\omega\rho'\vec{v} = -\text{grad}\ p, \tag{11.20}$$

and for the compressibility by the relation

$$\frac{j\omega}{K}p = -\text{div}\ \vec{v}. \tag{11.21}$$

where

p is the sound pressure, in Pa

$\vec{v}$ is the vector of the particle velocity, in m/s

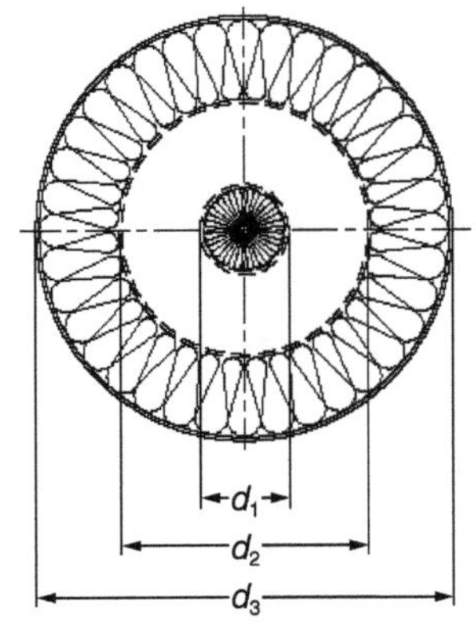

Fig. 11.16 Circular silencer with axial symmetry

ω is the circular frequency, in 1/s

ρ' is the complex density with the real part for the inertia and the imaginary part for the friction, kg/m^3

K is the complex modulus for compression with the real part for isentropical compression and the imaginary part for relaxation processes, in Pa.

For non-isotropic material Eq. (11.20) yields

$$j\omega\rho'_w v_w = -\frac{\partial p}{\partial w}, \tag{11.22}$$

$$j\omega\rho'_z v_z = -\frac{\partial p}{\partial z}, \tag{11.23}$$

whereby w denotes the radial and z the axial cylindrical coordinate. In this coordinate system, Eq. (11.21) yields

$$\frac{j\omega}{K}p = -\frac{1}{w}\frac{\partial(wv_w)}{\partial w} - \frac{\partial v_z}{\partial z}. \tag{11.24}$$

From Eqs. (11.22)–(11.24) follows the differential equation for p:

$$\frac{j\omega}{K}p = \frac{1}{j\omega\rho'_w}\left(\frac{\partial^2 p}{\partial w^2} + \frac{1}{w}\frac{\partial p}{\partial w}\right) + \frac{1}{j\omega\rho'_z}\frac{\partial^2 p}{\partial z^2}. \tag{11.25}$$

In axial direction, the fundamental mode of the pressure distribution decays according to

$$p(z) \propto e^{-\Gamma z}, \tag{11.26}$$

where Γ is the propagation constant to be calculated. It is determined by the lateral distribution, which according to Eqs. (11.25) and (11.26) is described by the differential equation

$$\frac{\partial^2 p}{\partial w^2} + \frac{1}{w}\frac{\partial p}{\partial w} + \left(\frac{\omega^2 \rho'_w}{K} + \frac{\rho'_w}{\rho'_z}\Gamma^2\right)p = 0. \quad (11.27)$$

Solutions in terms of the Bessel functions $J_0(\gamma w)$ and $Y_0(\gamma w)$ involve the radial propagation constant

$$\gamma = \sqrt{\frac{\omega^2 \rho'_w}{K} + \frac{\rho'_w}{\rho'_z}\Gamma^2}. \quad (11.28)$$

The general expression for the radial pressure distribution is

$$p(w) = p_0 J_0(\gamma\omega) + q_0 Y_0(\gamma w), \quad (11.29)$$

and for the radial velocity distribution according to Eq. (11.22):

$$\begin{aligned} v_w &= -\frac{1}{j\omega\rho'_w}\frac{\partial p}{\partial w} \\ &= \frac{\gamma}{j\omega\rho'_w}(p_0 J_1(\gamma w) + q_0 Y_1(\gamma w)). \end{aligned} \quad (11.30)$$

The constants p_0 and q_0 follow from boundary conditions. For the configuration shown in Fig. 11.16, the velocity at the centre is

$$v_{wi}(0) = 0. \quad (11.31)$$

Since the Bessel function Y_1 has a pole at the centre, it follows that $q_{i0} = 0$. Pressure and velocity at the surface $w = w_i = d_1/2$ of the centre body are continuous, thus

$$\begin{aligned} p_i(w_i) &= p_{i0} J_0(\gamma_i w_i) \\ &= p_{a0} J_0(\gamma_a w_i) + q_{a0} Y_0(\gamma_a w_i) \\ &= p_a(w_i), \end{aligned} \quad (11.32)$$

and

$$\begin{aligned} v_i(w_i) &= \frac{\gamma_i}{j\omega\rho'_{wi}} p_{i0} J_1(\gamma_i w_i) \\ &= \frac{\gamma_a}{j\omega\rho_0}(p_{a0} J_1(\gamma_a w_i) + q_{a0} Y_1(\gamma_a w_i)) \\ &= v_a(w_i). \end{aligned} \quad (11.33)$$

Similarly for the inner side of the outside duct lining at $w = w_a = d_2/2$:

$$\begin{aligned} p_a(w_a) &= p_{a0} J_0(\gamma_a w_a) + q_{a0} Y_0(\gamma_a w_a) \\ &= p_{e0} J_0(\gamma_e w_a) + q_{e0} Y_0(\gamma_e w_a) \\ &= p_e(w_i), \end{aligned} \quad (11.34)$$

and

$$\begin{aligned} v_a(w_a) &= \frac{\gamma_a}{j\omega\rho_0}(p_{a0} J_1(\gamma_a w_a) + q_{a0} Y_1(\gamma_a w_a)) \\ &= \frac{\gamma_e}{j\omega\rho_0}(p_{e0} J_1(\gamma_e w_a) + q_{e0} Y_1(\gamma_e w_a)) \\ &= v_e(w_a). \end{aligned} \quad (11.35)$$

At the outer end of the lining, $w = w_e = d_3/2$ the radial velocity is zero:

$$v_e(w_e) = \frac{\gamma_e}{j\omega\rho'_{we}}(p_{e0} J_1(\gamma_e w_e) + q_{e0} Y_1(\gamma_e w_e)) = 0. \quad (11.36)$$

This determines the ratio

$$\frac{q_{e0}}{p_{e0}} = -\frac{J_1(\gamma_e w_e)}{Y_1(\gamma_e w_e)}. \quad (11.37)$$

Admittances G_i and G_a can be defined as boundary conditions for the free air passage:

$$G_i = -\frac{v_i(w_i)}{p_i(w_i)} = -\frac{\gamma_i}{j\omega\rho'_{wi}}\frac{J_1(\gamma_i w_i)}{J_0(\gamma_i w_i)}, \quad (11.38)$$

$$\begin{aligned} G_a &= \frac{v_e(w_a)}{p_e(w_a)} = \frac{\gamma_e}{j\omega\rho'_{wa}}\frac{J_1(\gamma_e w_a) + \frac{q_{e0}}{p_{e0}} Y_1(\gamma_e w_a)}{J_0(\gamma_e w_a) + \frac{q_{e0}}{p_{e0}} Y_0(\gamma_e w_a)}, \\ &= \frac{\gamma_e}{j\omega\rho'_{wa}}\frac{J_1(\gamma_e w_a) Y_1(\gamma_e w_e) - J_1(\gamma_e w_e) Y_1(\gamma_e w_a)}{J_0(\gamma_e w_a) Y_1(\gamma_e w_e) - J_1(\gamma_e w_e) Y_0(\gamma_e w_a)}. \end{aligned} \quad (11.39)$$

The reciprocal of the admittances may be extended by impedances of thin top layer, e.g. perforated plates, fleece or foils [20]. The relations remaining from Eqs. (11.32) to (11.35) can be combined with the admittances of Eqs. (11.38) and (11.39) to yield the characteristic equation for γ_a in terms of the ratio q_{a0}/p_{a0}:

$$\frac{q_{a0}}{p_{a0}} = -\frac{G_i J_0(\gamma_a w_i) + \frac{\gamma_a}{j\omega\rho_0} J_1(\gamma_a w_i)}{G_i Y_0(\gamma_a w_i) + \frac{\gamma_a}{j\omega\rho_0} Y_1(\gamma_a w_i)}$$
$$= -\frac{G_a J_0(\gamma_a w_a) - \frac{\gamma_a}{j\omega\rho_0} J_1(\gamma_a w_a)}{G_a Y_0(\gamma_a w_a) - \frac{\gamma_a}{j\omega\rho_0} Y_1(\gamma_a w_a)}. \quad (11.40)$$

This is a transcendental equation for γ_a or Γ with numerous solutions for various modes. The solution with smallest real part of Γ is of particular interest but is not available in closed form.

For the numerical evaluation of Eq. (11.40), various procedures can be applied. Common solvers start from suitable initial values of the propagation constant $\Gamma = jk(\tau - j\sigma)$ for the fundamental mode in the ranges $\tau \approx 1$ and $0 < \sigma < 1$ and then apply a Newton/Raphson procedure. However, problems arise when two solutions of similar attenuation coefficient but different phase coefficient are close together and form two intersecting lines of the frequency characteristic. If the intersection is not identified, attenuations calculated for higher frequencies in the range of half wavelengths that exceed the distance between the lining and the centre body are not the ones for the fundamental mode.

A special case is the silencer with locally reacting lining. Such a lining resulting from the partitioning by bulkheads at distances of less than a quarter-wavelength has an effective axial density ρ'_z that is large compared to the radial density ρ'_w. According to Eq. (11.28), the propagation constants γ_i and γ_a become independent of Γ and, thus, of γ_a. The admittances G_i and G_a no longer depend on Γ, which reduces the problems with iterative solutions.

A further special case is the lined pipe without centre body. For $w_1 = 0$, the left-hand side of eq. (11.40) is zero and the transcendental equation for γ_a is reduced to:

$$G_a = \frac{\gamma_a}{j\omega\rho_0} \frac{J_1(\gamma_a w_a)}{J_0(\gamma_a w_a)}. \quad (11.41)$$

The ratio of Bessel function can be determined either from an B/A-algorithm following from the theory of continuants [21] or from continued fractions. Both methods yield the same approximation for a specified number of modes. The first two modes are well approximated by the solution

$$\Gamma = jk \times \sqrt{1 - \left(\frac{2}{kw_a}\right)^2 \frac{24 + 9j\beta \mp \sqrt{576 + 144j\beta - 57\beta^2}}{12 + j\beta}}, \quad (11.42)$$

where $\beta = G_a\rho_0\omega w_a$ is the normalized wall admittance and $k = \omega/c$ is the wavenumber. In most cases, the negative sign in front of the inner square root is valid for the fundamental mode. In some cases, the positive sign yields the smaller real part of Γ.

When flow with the Mach number M shall be taken into account, the sound pressure field in the air passage is – instead of Eq. (11.27) – described by Eq. (11.33):

$$\frac{1}{k^2}\frac{\partial^2 p}{\partial w^2} + \frac{\partial p}{k\partial w}\left(\frac{1}{kw} - \frac{2j\Gamma/k}{1 - Mj\Gamma/k}\frac{\partial M}{k\partial w}\right) + p\left[(1 - Mj\Gamma/k)^2 + (\Gamma/k)^2\right] = 0. \quad (11.43)$$

The flow profile can be approximated by

$$M(w) = 1.08\bar{M}\left(\frac{2}{w_a - w_i}\right)^{\frac{2}{7}} (w_a - w)^{\frac{1}{7}}(w - w_i)^{\frac{1}{7}}, \quad (11.44)$$

where $\bar{M}$ is the mean value of the mach number averaged over the free cross section. A finite difference approximation of Eq. (11.43) is [31]:

$$\left[\frac{4}{k(w_a - w_i)}\right]^2 (p_0 - 2p_1 + p_2) + \frac{2}{k(w_a - w_i)}(p_2 - p_0) \times \left(\frac{1}{k\left[w_i + \frac{1}{8}(w_a - w_i)\right]} - \frac{2k_z/k}{1 - M_1 k_z/k}\frac{(\mathrm{d}M/\mathrm{d}w)_1}{k(w_a - w_i)}\right) + p_1\left[(1 - M_1 k_z/k)^2 - (k_z/k)^2\right] \approx 0, \quad (11.45)$$

where the subscripts 0, 1, and 2 refer to 3 of 6 points in the duct cross section shown in Fig. 11.17. Similar approximations hold for the points 2–4. The pressure p_0 outside the free air passage is related to the pressure p_1 by

$$\frac{4}{k(w_a - w_i)}(p_1 - p_0) \approx -j\rho_0 c v_{wi}(r_i)(1 - M_i k_z/k), \quad (11.46)$$

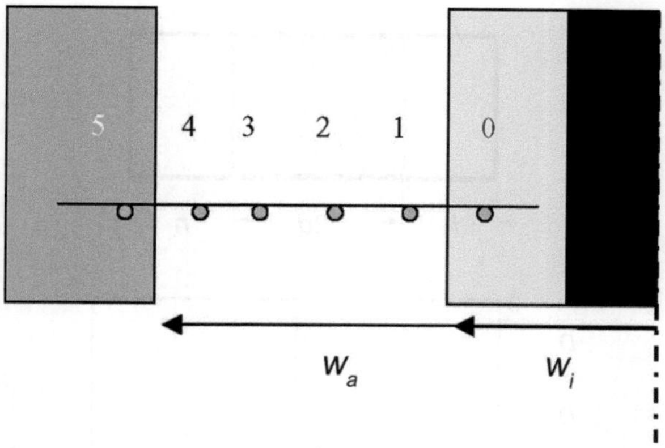

Fig. 11.17 Duct cross section with points for finite difference scheme

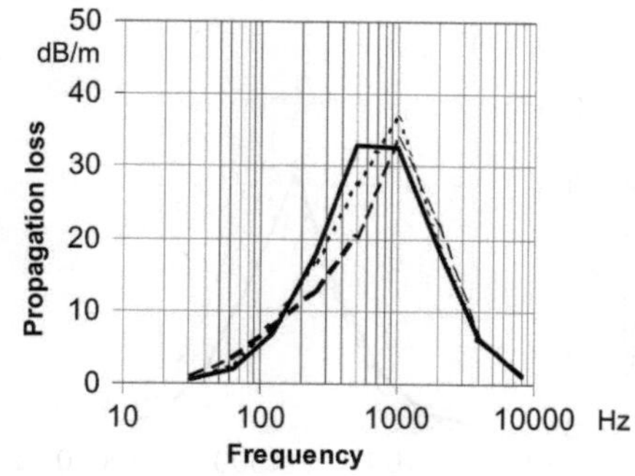

Fig. 11.18 Calculated propagation loss for circular silencer with centre body; $d_1 = 0.1$ m, $d_2 = 0.3$ m and $d_3 = 0.5$ m; isotropic material covered with foil 50 g/m and flow resistance 200 Pa s/m; absorbent material with flow resistivity *(continuous line)* 10 kPa s/m, *(dotted lines)* 20 kPa s/m, *(dashed lines)* 40 kPa s/m

and by the boundary condition

$$\frac{p_1 - p_0}{p_1 + p_0} \approx \frac{jk(w_a - w_i)\rho_0 c G_i}{8}. \qquad (11.47)$$

Similar relations hold for the pressure p_5 and the boundary condition Y_a at the outside lining.

The reason for taking six reference points in the positions shown in Fig. 11.17 is as follows:

- Non-symmetrical boundary conditions (centre body and outside lining) are covered.
- The flow profile can be taken into account. Uncertainties about the appropriate boundary condition of continuous particle velocity or particle displacement are avoided by taking the average flow velocity $M(w_i) = M(w_a) = 0$.
- Four points in the free duct cross section result in a quartic equation for the propagation constant, which is the limit for closed solutions.
- The potential for calculating four different modes increases the precision of the approximation for the lowest order mode.

Figures 11.18–11.21 show results calculated by the finite difference approximations. Figure 11.18 refers to a typical silencer with bulkheads spaced at 0.5 m and filled with absorbent material of different flow resistivity. The material is covered with a thin foil and a small additional flow resistance is assumed for the surface. In the octave bands from 250 to 500 Hz, the propagation loss decreases with increasing resistivity of the absorbent material.

The dependence is inverted at 1,000 Hz.[4] In the frequency range above 2,000 Hz, the calculated attenuation is almost independent of the material flow resistivity.

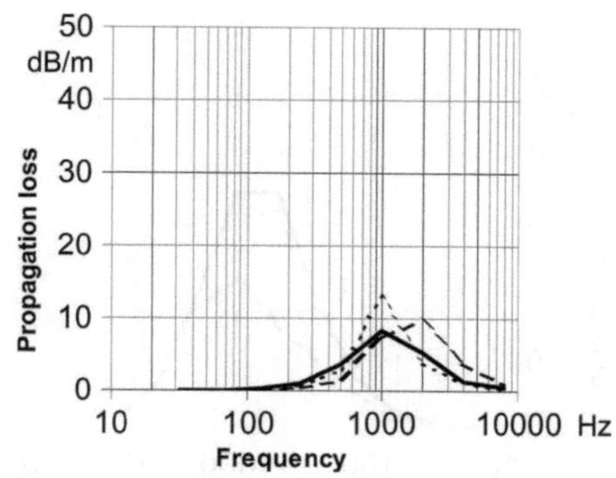

Fig. 11.19 Calculated propagation loss for circular silencer without outside lining; $d_1 = 0.1$ m, $d_2 = 0.3$ m and $d_3 = 0.5$ m; absorbent material with flow resistivity 10 kPa s/m; *(continuous line)* isotropic material covered with foil 50 g/m and flow resistance 200 Pa s/m, *(dotted lines)* locally reacting material covered with foil 50 g/m and flow resistance 200 Pa s/m, *(dashed lines)* locally reacting material without cover

Figure 11.19 shows results calculated for the case of a pipe without lining but with a centre body filled with material of low flow resistivity. There is just a small maximum of the propagation loss in the octave band of 1,000 Hz.

Results shown in Fig. 11.20 for the case of a lined pipe without centre body demonstrate the importance

[4] The results are calculated for the fundamental mode in a sound field with rotational symmetry. Anti-symmetric sound fields may be less attenuated by lower flow resistivity of the material.

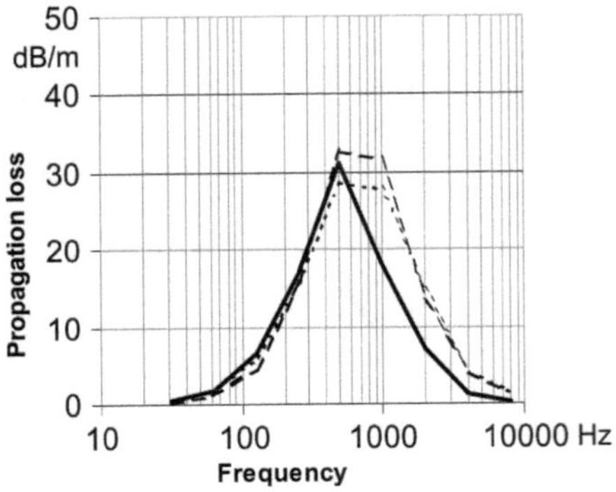

Fig. 11.20 Calculated propagation loss for circular silencer without centre body; $d_1 = 0.1$ m, $d_2 = 0.3$ m and $d_3 = 0.5$ m; isotropic absorbent material with flow resistivity 10 kPa s/m; (*continuous line*) covered with foil 50 g/m and flow resistance 200 Pa s/m, (*dotted lines*) covered with flow resistance 200 Pa s/m only, (*dashed lines*) without cover

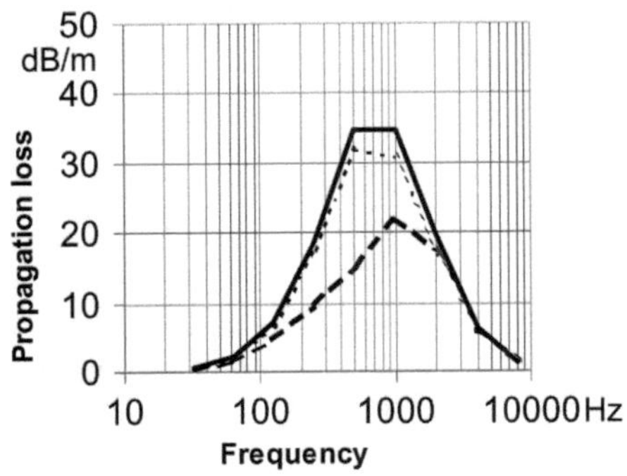

Fig. 11.21 Calculated propagation loss for circular silencer with centre body; $d_1 = 0.1$ m, $d_2 = 0.3$ m and $d_3 = 0.5$ m; isotropic absorbent material with flow resistivity 10 kPa s/m, covered with foil 50 g/m and flow resistance 200 Pa s/m, (*continuous line*) 20°C, flow velocity 27 m/s (*dotted line*) 20°C, flow velocity −27 m/s, (*dashed line*) 500°C, without flow

of the wall lining of narrow pipes, which provides significantly more attenuation than the centre body alone. The covering foil reduces the attenuation in the octave bands at and above 1,000 Hz.

Figure 11.21 exhibits the small influence of flow with a mean Mach number 0.1 and −0.1, but the substantial influence of an elevated temperature of 500°C.

Splitters with Isotropic Absorbent Material

Theory shows for the range of medium frequencies that materials with low flow resistivity should be chosen to obtain high attenuation. However, this is valid for splitters in a symmetrical sound field only. Measurements on splitters without centre panel and with low flow resistance gave relatively low

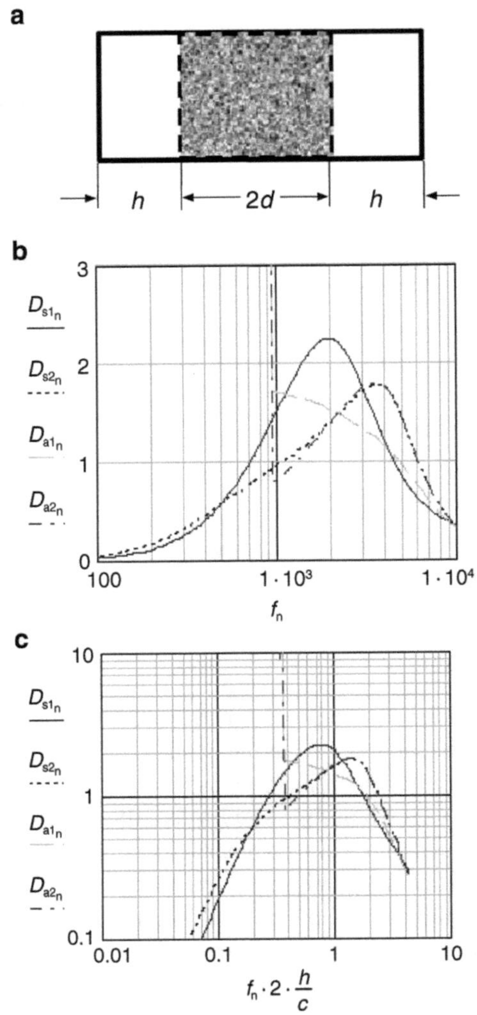

Fig. 11.22 Calculated propagation loss per free duct width $h = 0.1$ m for an absorbent splitter of thickness $2d = 0.08$ m in the centre of a rigid duct of width $2(d + h)$ for symmetrical and anti-symmetrical sound fields in the free air passages at 400°C without flow; the splitter contains isotropic material with a flow resistivity of 10 kPa s/m (*continuous line* for symmetrical excitation; *dashed line* for anti-symmetrical excitation) and of 50 kPa s/m (*dotted line* for symmetrical excitation; *dashed dotted line* for anti-symmetrical excitation) covered by a material with flow resistance of 200 P s/m

attenuation – depending on the exciting sound field [22]. Often the measured values are somewhat higher than calculated for the symmetrical sound incidence in the frequency below the range of maximum attenuation, but the maximum itself is substantially lower.

A calculated example is given in Fig. 11.22. The upper part shows the propagation loss multiplied by a length h, which is the width of the air passage on either side of an absorbent baffle in a rigid duct. The lower part gives the same information on a logarithmic ordinate over the Helmholtz number formed by the frequency, the free duct width h and the speed of sound c. The symmetrical fundamental mode exists in the entire frequency range. The propagation of an anti-symmetrical mode in a duct of width $2(h + d)$ is restricted to the frequency range above $f_c = 0.5c/[2 \times (h + d)]$ and accordingly indicated in Fig. 11.22. The lower part of the figure allows for extrapolations with a power law. However, this is limited to the range of low attenuation where typically higher order modes are more important. As already described for circular silencers, splitters with low flow resistance in symmetrical sound fields yield low attenuation at low frequencies, high attenuation at medium frequencies and slightly lower attenuation at high frequencies as compared to splitters with higher flow resistance. With increasing flow resistance, the attenuation maximum is shifted to higher frequencies. It decreases in magnitude with increasing flow resistance of the top layer of the material.[5] For the case of anti-symmetrical excitation, Fig. 11.22 shows smaller attenuation above the cut-on frequency f_c than for symmetrical excitation. The effect is pronounced for absorbent material with low flow resistance.

In practical cases, little is known about the distribution of sound on various directions or on modes in wide ducts. The simplifying assumption of an equal distribution on all propagating modes is questionable. One may rather assume for aerodynamic sources that the highest propagating mode is relatively strong excited. The uncertainty about modal distribution limits the quality of attenuation prediction in splitter silencers with low density absorbing material. In practice, higher density materials are preferred.

11.5.2 Pressure Reduction

11.5.2.1 Pressure Release by Perforated Plates

Over-Critical Pressure Release

The flow rate Q of gas in a pipe of cross-sectional area S – or many partial areas of a perforated pipe – is determined by the density ρ and the flow velocity v (p. 405, [23]):

$$Q = \rho v S. \tag{11.48}$$

The maximum velocity of stationary flow is the sound velocity, which can be somewhat higher in the pipe than the velocity of sound c_0 in the pressurized vessel (p. 405, [23]):

$$v_1 = c_0 \sqrt{\frac{2}{\gamma + 1}}, \tag{11.49}$$

where γ denotes the ratio of specific heats,

$$c = \sqrt{\gamma R T} \tag{11.50}$$

is the sound velocity at absolute temperature T, R is specific gas constant,[6] and the subscript 0 refers to the vessel in front of the pipe. The cross-sectional area S_1 required for the mass flow rate results from the general equation of state for ideal gases,

$$\rho = \frac{p}{RT}, \tag{11.51}$$

from Poisson's equation for adiabatic processes,

$$\frac{p_1}{p_0} = \left(\frac{\rho_1}{\rho_0}\right)^{\gamma}, \tag{11.52}$$

from the pressure p_0 and the temperature T_0 inside the vessel, and from the pressure p_1 at the exit of the pipe:

$$S_1 \geq \frac{Q}{p_0} \sqrt{RT} \left(\frac{p_1}{p_0}\right)^{-\frac{1}{\gamma}} \left(\frac{\gamma + 1}{2\gamma}\right)^{\frac{1}{2}}. \tag{11.53}$$

The pressure in the released gas can drop from the value p_0 to a minimum critical value

$$p_1 = p_* = p_0 \left(\frac{2}{\gamma + 1}\right)^{\frac{\gamma}{\gamma - 1}}, \tag{11.54}$$

[5] Numerous calculated data for absorbent splitters without cover material are given in [16].

[6] For steam $R = 461$ Nm/kg K; for air $R = 287$ Nm/kg K.

i.e., for $\gamma = 1.4$ at 53% of p_0 at the most. This determines the needed area:

$$S_1 \geq \frac{Q}{p_0}\sqrt{\frac{RT_0}{\gamma}}\left(\frac{\gamma+1}{2}\right)^{\frac{\gamma+1}{2(\gamma-1)}}. \quad (11.55)$$

For $\gamma = 1.4$, this result is consistent with the common calculation based on the constant $k_v = 5.17$. For smaller values of γ, the result from Eq. (11.52) is up to 20% smaller, for larger values up to 10% higher.

For individual and non-interacting holes, the flow produces extensive noise with a sound power level of approximately

$$L_W = \left[93 + 10\ \lg\frac{Qc^2}{1W}\right]\text{dB}. \quad (11.56)$$

Under-Critical Pressure Release

A further under-critical pressure release over a perforated pipe can be described by the pressure loss coefficient ζ_n:

$$\Delta p_n = p_n - p_{n-1} = \zeta_n\frac{Q_n}{2}v_n^2. \quad (11.57)$$

The pressure loss can be expressed in terms of the Mach number v_n/c of the released gas:

$$\Delta p_n = \zeta_n\frac{\gamma}{2}\frac{v_n^2}{c^2}p_n. \quad (11.58)$$

The pressure loss coefficient ζ_n of a perforated plate results from the geometry. For regular holes with sharp edges, which determine a porosity σ, $\zeta_n = \zeta/\sigma^2$, where ([24], p. 321)

$$\zeta = \left(\sqrt{\frac{1-\sigma}{2}} + 1 - \sigma\right)^2. \quad (11.59)$$

Values of 2.9 are achievable, i.e. a substantial single-stage pressure release can be obtained by high Mach numbers in the opening exclusively, which means additional noise creation.

In order to keep the sound generation small, the gas flow must be kept within small Mach numbers $v_n/(c\sigma)$. The limitation to a Mach number 0.2 yields a sound level reduction by about 40 dB compared to the over-critical situation. However, according to Eq. (11.59), a Mach number 0.2 allows for a pressure reduction of not more than 8% in one stage.

11.5.2.2 Pressure Release via Porous Materials

In fine pores or mesh wires, the sound generation is no longer determined by the flow velocity in individual holes. Equalization of pressure differences results in a sound power, which increases with about the fourth power of the flow velocity.

The pressure loss coefficient depends on the Reynold's number

$$\text{Re} = \frac{v_a\delta}{v}, \quad (11.60)$$

where $v_a = v_1/\sigma$ is the oncoming velocity of the gas flow, δ is the fibre diameter and v is the kinematic viscosity. For Re ≥ 400, e.g. for round wires of 0.28 mm diameter exposed to steam of 1 MPa and 275°C ($v = 5\ \text{mm}^2/\text{s}$) at $v_a = 20\text{m/s}$ (Re $= 1,120$), the pressure loss coefficient of mesh wire is

$$\zeta_a = 1.3(1-\sigma) + \left(\frac{1}{\sigma} - 1\right)^2. \quad (11.61)$$

Low porosities result in high pressure loss coefficients. The porosity of single layers of mesh wire can be calculated from the wire gauge δ and the number of wires per unit length z':

$$\sigma = 1 - z'\delta. \quad (11.62)$$

For multi-layer mesh wire, the individual pressure loss coefficients add up to the total value ζ_{ges}.

11.5.2.3 Pressure Loss

The reduction of flow energy in a silencer is described in terms of the total pressure loss coefficient

$$\zeta = \frac{\Delta p_t}{\frac{\rho}{2}v_1^2}, \quad (11.63)$$

where

Δp_t – is the total pressure loss, in Pa,

ρ – is the gas density, in kg/m^3, and

v_1 – is the average face velocity in front of the silencer with cross sectional area S.

With reference to the space-averaged maximum velocity v_{max} in the silencer with minimum cross sectional area S_{min} the total pressure loss is calculated from

$$\Delta p_t = \zeta \frac{\rho}{2} v_{max}^2 \left(\frac{S_{min}}{S_1} \right). \tag{11.64}$$

The total pressure loss results from discontinuities at both ends and inside the silencer as well as from friction losses at the surfaces "wetted" by the flow. For HVAC equipment, the friction losses are calculated – similar to Eq. (11.6) – from

$$\zeta_f = 0.0125 \frac{U_f l}{S_f} \left(\frac{S_f}{S_1} \right)^2, \tag{11.65}$$

where

0.0125 – is the friction factor for commercially rough steel pipes, which also applies to absorbent surfaces of ducts,[7]

U_f – is the total "wetted" surface (not only the absorbent)

S_f – is the corresponding surface and

l_f – is the corresponding length of the silencer

Circular silencers without centre body provide a particularly low pressure loss. Silencers with a circular centre body or with a rectangular splitter inside are about equivalent if the ratio U_f/S_f is the same. A very thin splitter already results in a pressure loss equal to that of a circular centre body filling 40% of the cross-sectional area. Acoustically, a circular centre body is better than a splitter due to the higher potential for absorption. However, the mechanical design of splitters has advantages.

For parallel baffle silencers with baffles of thickness $2d$ spaced at a distance $2h$, the pressure loss coefficient is approximately (not accounting for rigid duct walls) calculated from

$$\zeta_f = 0.0125 \frac{l}{h} \left(1 + \frac{d}{h} \right)^2. \tag{11.66}$$

Baffles partitioned by bulkheads without smooth cover, e.g. quarter-wavelength resonators, provide for pressure losses depending on the flow direction and the shape of the resonator covering, which need to be determined experimentally. In the direction favourable for flow, a factor 0.013 is possible to achieve.

Losses at entrance discontinuities depend on the cross distribution and the swirl of the gas flow and on the geometry of the silencer. For axial flow without swirl past parallel baffles is according to [1]:

$$\zeta_s = \left(\frac{d}{h} \right)^2 \left[0.5 \zeta_1 \left(\frac{h}{d} + 1 \right) + \zeta_2 \right], \tag{11.67}$$

where

ζ_1 – is the shape factor on the inlet side; for rectangular baffles[8] $\zeta_1 = 1$, for semicircular profile $\zeta_1 = 0.1$.

ζ_2 – is the shape factor for the outlet side; for rectangular baffles, $\zeta_1 = 1$, for semicircular profile $\zeta_1 \approx 0.7$, for narrow wedges smaller.

Losses at discontinuities of baffles in series are small, even with gaps in between as large as the air passage between the baffles. However, when a second set of baffles is not aligned, additional losses occur which exceed those from a single set.

Total pressure loss coefficients for various elements combined with silencers, e.g. spark protectors or conical transition ducts, have been collected by Idel'Chik [24].

11.5.3 Flow Noise

According to Stüber [27], flow noise increases with increasing pressure drop at single-stage throttling devices in pipes or ducts.[9] In the range of small pressure differences, which is most relevant for silencers,

[7] According to measurements by Kurze/Donner [25], the value is 0.02 for perforated plate. According to VDI 2081 [25], the average value is 0.016 for HVAC equipment, but the exponent in Eq. (11.63) is about 2.9.

[8] Based on former experience reported in VDI 2081 (1983), Kurze/Donner [26] propose the value $\zeta_1 = 1.2$. Recent experimental data for the range $0.3 < h/(d+h) < 0.6$ for 100–300 mm thick baffles for HVAC equipment reported in VDI 2081 (2001) are in the range from $\zeta_1 = 1$–1.2.

[9] Flow noise is occasionally reported in the literature for smooth pipes as a function of flow velocity and cross-sectional area. However, there are no measured data or physical evidence for the boundary layer as a substantial noise source.

the sound power is proportional to the 2.5th power of the pressure loss coefficient and to the 6th power of the flow velocity.[10] The frequency spectrum of the sound radiated into free space from an opening with diameter D is broadband with a maximum at the Strouhal number $fD/v \approx 0.2$. For industrial applications with low Mach number $M < 0.2$ and $D > 0.1$ m, the maximum frequency is generally at low frequencies $f < 140$ Hz. At higher frequencies, the spectrum rolls off at about $1/f$.

A silencer is acting as a throttling device mainly by its end. For the downstream end of rectangular baffles, laboratory measurements of octave-band sound power levels can be approximated by

$$L_{W,\mathrm{oct}} = B + 10\ \lg\left(\frac{pcS}{W_0}M^6\right)\mathrm{dB} + C_1 + C_2, \tag{11.68}$$

where

B – is a constant (the dependence on the pressure loss coefficient is not yet sufficiently established) for rectangular baffles, $B = 58$ dB

p – static pressure, in Pa

c – speed of sound, in m/s

S – cross-sectional area between baffles, in m^2

M – Mach number of flow in the air passage

W_0 – reference sound power, 1 pW

C_1 – correction term accounting for the reaction of the duct walls at low frequencies, corresponding to the reflection loss at the open end of the duct, in dB

C_2 – correction term for high frequencies, in dB, typically[11]

$$C_2 = -10\lg\left[1 + \left(\frac{f\delta}{v}\right)^2\right]\mathrm{dB\ and}\ \delta \approx 0.02\,\mathrm{m}. \tag{11.69}$$

A-weighting of the spectrum renders the correction term C_1 unimportant for industrial silencers. The correction term C_2 is rather sensitive on minor shape variations of the baffle ends. From Eqs. (11.65) and (11.66), the A-weighted overall level of the sound power radiated from an area of 1 m^2 is calculated as[12]

$$L_{WA} = \left(-23 + 67\ \lg\frac{v}{v_0}\right)\mathrm{dB}, \tag{11.70}$$

where v is the velocity of sound in the air passage between baffles and $v_0 = 1$ m/s. At higher temperatures, the sound power level decreases at a rate of $-25\ \lg(T/T_0)$dB, where T is the absolute temperature, in K, and $T_0 = 293$ K is the ambient temperature.[13]

11.6 Measurement Procedures

11.6.1 Standards

Laboratory measurements to determine the insertion loss of silencers of all kinds (except from automotive silencers) and of other devices in flow ducts are described in ISO 7235 [28]. This standard defines insertion loss measurements with and without flow, the determination of flow noise, and the pressure loss. A special measurement procedure suitable for small in-line silencers without flow is described in ISO 11691 [21] and also included in the recent edition of ISO 7235.

For in-situ measurements on silencers, ISO 11820 [4] specifies acoustic and superficial flow measurements. Important for acoustic measurements is the appropriate relation of the practical application to 1 of the 20 cases schematically shown in Table 11.3. Sixteen cases deal with transmission loss measurements for silencers upstream and downstream in a duct, in a reverberant room, in other rooms without diffuse sound field, and in free space. Four cases refer to insertion loss measurements in such sound fields with and without the silencer. The standard specifies the data evaluation

[10] Consistent with the flow noise from air outlets described in VDI 2081 [25].

[11] The frequency characteristic of flow noise in straight air conduits is described in VDI 2081 by the asymptotic value $-26\ \lg f$ instead of $-20\ \lg f$. Stüber [27] proposes roughly $-12\lg f$.

[12] In VDI 2081 [25], the noise in straight air conduits is described by $L_{WA} = (-25 + 70\ \lg v + 10\lg S)$dB. In the range from 1 to 22 m/s, the difference to Eq. (11.67) is less than 2 dB. Thus, it is likely that similar sources of sound are described.

[13] Silencers are commonly designed to provide a level of flow noise way below the level of total plant noise. Therefore, just a few field data are available. Eqs. (11.65)–(11.67) refer to laboratory data.

Table 11.3 General scheme for transmission and insertion loss measurements of silencers. See Fig. 11.1 [4]

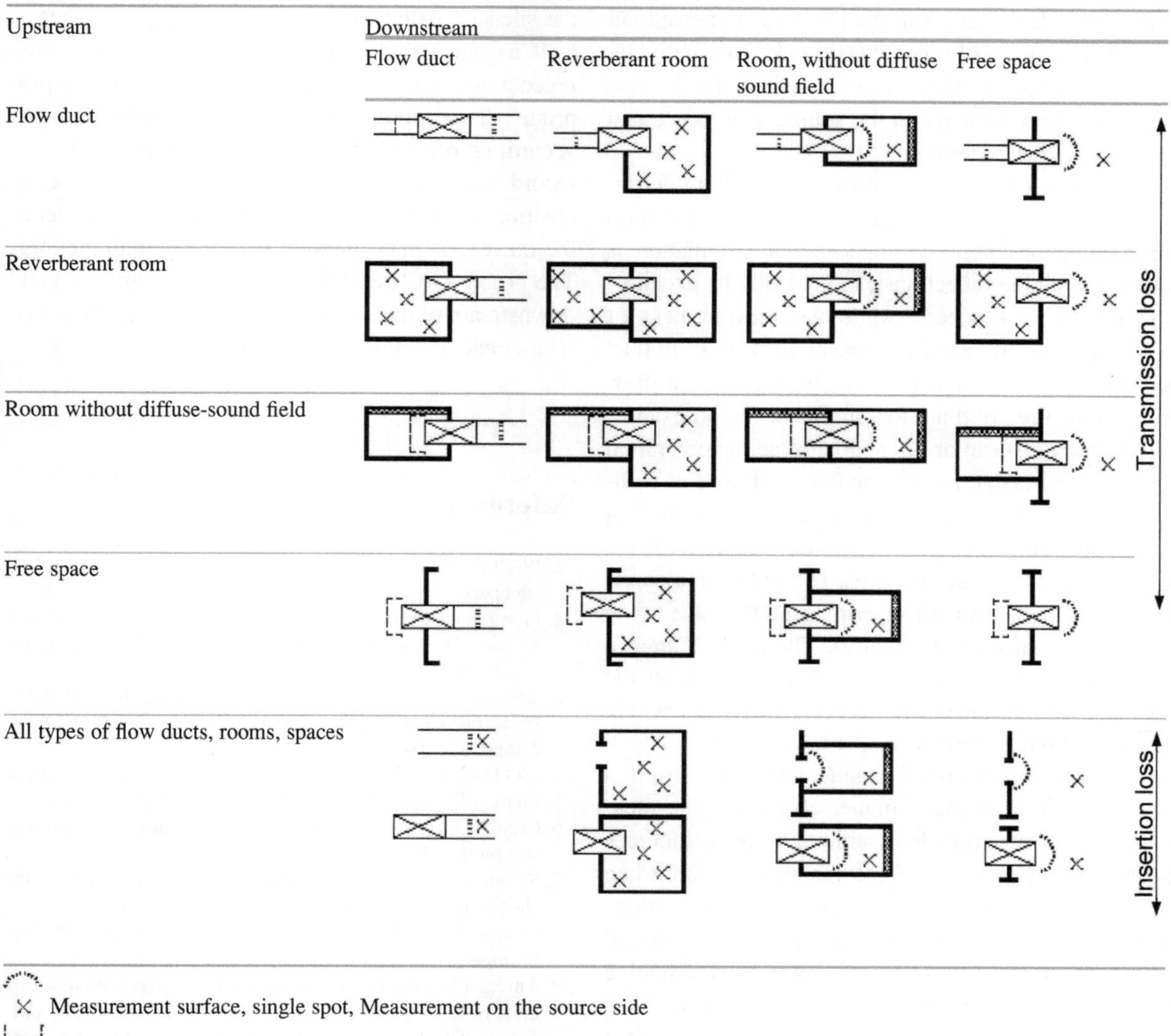

Note: the sound source is located on the left side of the silencer. The direction of flow is optional

from measured sound pressure levels, reverberation times and geometrical properties. In addition, correction terms are given, which should be estimated or agreed upon for the particular situation, within a limit of 3 dB as a rule.

11.6.2 Laboratory Measurements

Measurements in specialized laboratories primarily aim at determining the minimum transmission loss of a silencer. Measurements of the pressure loss and of regenerated noise are restricted to special cases of lacking experience. By suitable configuration of the source and receiver side, transmission loss data can be determined from insertion loss measurements. The minimum attenuation in a narrow duct is related to the excitation of plane sound waves.

At the source side, the connection of the loudspeaker to a pipe of 0.4 m diameter is successfully applied to maintain plane waves up to frequencies of 500 Hz. A transition duct is applied to match the pipe to the inlet of a silencer with different size. To avoid multiple reflections between the source and the silencer under test, some attenuation is needed in the connecting duct without too much damping the source signal. This can be simply provided by a lateral connection of the loudspeaker to the flow duct, which already needs a silencer in the line to a connected fan for measurements of regenerated noise. Alternatively, a short silencer similar in cross section to the one under test may be used on the source side to act as a filter for higher order modes.

At the receiver side, either an anechoic termination or a configuration suitable for suppressing multiple reflections between silencer and termination is needed. Multiple reflections can generally be excluded for absorptive silencers, while the measurement of reactive silencers requires special measures. In-duct measurements are practically limited to no-flow conditions. Preferred are measurements on an exterior enveloping surface or in a connected reverberation room. Under stable environmental conditions at sufficiently low exterior noise, measurements with the test object and with an empty substitution duct provide the insertion loss without problems. Difficulties may arise from measurements with superimposed flow due to the noise regenerated in the silencer. The level difference of 10 dB between signal and noise required by the standard may result in the need for very high source levels for high performance silencers.

The determination of regenerated noise from a silencer requires a much higher effort than the measurement of insertion loss without flow. It includes substantial attenuation of the fan noise, appropriate guidance of the flow at the receiver side, corrections for the reflection at the duct exit and for the potential influence of the reverberation room. Very effective silencers must be inserted between the fan and the silencer under test. The exit duct cross-sectional area must be substantially larger than the free cross sectional area in the silencer.

Laboratory test conditions for total pressure loss specified in ISO 7235 [28] are aiming at optimum reproducibility – just as those for attenuation measurements. This means the determination of minimum values. While such a goal includes a safety margin for noise consideration, it is just the opposite for pressure loss.

11.6.3 Field Measurements

The application of silencers may result in reduced field performance due to various effects, which must be avoided by careful planning, design and construction. This includes flanking transmission via the shell of the silencer, extension joints, supporting structures as well as unequal flow distribution causing increased regenerated noise and mechanical loading. Appropriate inspections and acoustical measurements according to ISO 11820 [4] may be supplemented by recordings of the sound pressure level at a number of positions along the propagation paths in the silencer as the most important measure to identify malfunctions. The probing of lateral flow distribution upstream and downstream of the silencer by means of a Prandtl tube with connected flow velocity meter is informative.

References

1. ISO 14163 (1999) Acoustics: Guidelines for noise control by silencers
2. Frommhold W (1996) Absorptionsschalldämpfer. In Schirmer W (ed) Technischer Schallschutz. VDI-Verlag Düsseldorf, chapter 9
3. Brennan MJ, To WM (2001) Acoustic properties of rigid-frame porous materials – An engineering perspective. Appl Acoust 62:793–811
4. ISO 11820 (1997) Acoustics: measurements on silencers in situ
5. Piening W (1937) VDI-Z. 81, 776 (cited in [6])
6. Cremer L (1940) Nachhallzeit und Dämpfungsmaß bei streifendem Einfall. Akustische 5(2):57–76
7. Sivian LJ (1937) Sound propagation in ducts lined with absorbing materials. J Acoust Soc Am 9:135–140
8. Cremer L (1975) Vorlesungen über technische Akustik. Springer, Berlin, 147
9. Aurégan Y, Starobinski R, Pagneux V (2001) Influence of grazing flow and dissipation effects on the acoustic boundary conditions at a lined wall. J Acoust Soc Am 109 (1):59–64
10. Klein W (1957) Matrizen. In: Rint C (ed) Handbuch der Hochfrequenz- und Elektrotechnik Band III. Verlag für Radio-Foto-Kinotechnik, Berlin, 134 et seqq
11. Kurze U (1969) Schallausbreitung im Kanal mit periodischer Wandstruktur. Acustica 21:74–85
12. Munjal M (1987) Acoustics of ducts and mufflers. Wiley, New York
13. Galaitsis AG, Ver IL (1992) Passive silencers and lined ducts. In: Beranek LL, Ver IL (eds) Noise and vibration control engineering. Wiley, New York, chapter 10
14. Mechel FP (1994) Schallabsorption. In: Heckl M, Müller HA (eds) Taschenbuch der Technischen Akustik, 2. Aufl. Springer, Berlin, chapter 19
15. Aurégan Y, Debray A, Starobinski R (2001) Low frequency sound propagation in a coaxial cylindrical duct: application to sudden area expansions and to dissipative silencers. J Sound Vib 243(3):461–473
16. Wöhle W (1984) Schallausbreitung in absorbierend ausgekleideten Kanälen. In: Fasold W, Kraak W, Schirmer

W (eds) Taschenbuch Akustik Part 1. VEB Verlag Technik, Berlin, chapter 1.8
17. Mechel FP (1994) Schalldämpfer. In: Heckl M, Müller HA (eds) Taschenbuch der Technischen Akustik, 2. Aufl. Springer, Berlin, chapter 20
18. Kang YJ, Jung IH (2001) Sound propagation in circular ducts lined with noise control foams. J Sound Vib 239 (2):255–273
19. Aly ME, Badawy MTS (2001) Sound propagation in lined annular-variable area ducts using bulk-reacting liners. Appl Acoust 62:769–778
20. Munjal ML, Thawani PT (1997) Effect of protective layer on the performance of absorptive ducts. Noise Control Eng J 45(1):14–18
21. ISO 11691 (1997) Acoustics: Measurement of insertion loss of ducted silencers without flow – laboratory survey method
22. Cummings A, Normaz N (1993) Acoustic attenuation in dissipative splitter silencers containing mean fluid flow. J Sound Vib 168(2):209–227
23. Landau LD, Lifschitz EM (1966) Lehrbuch der theoretischen Physik, vol VI, Hydrodynamik. Akademie-Verlag, Berlin, Chapter IX
24. Idel'Chik IE (1994) Handbook of hydraulic resistance, 3rd edn. CRC, London
25. VDI 2081 (2001) Page 1: Geräuscherzeugung und Lärmminderung in raumlufttechnischen Anlagen. Beuth, Berlin
26. Kurze UJ, Donner U (1989) Schalldämpfer für staubhaltige Luft. Schriftenreihe der Bundesanstalt für Arbeitsschutz – Forschung – Fb 574
27. Stüber B, Mühle C, Fritz KR (1994) Strömungsgeräusche. In: Heckl M, Müller HA (eds) Taschenbuch der Technischen Akustik, 2. Aufl. Springer, Berlin, chapter 9
28. ISO 7235 (2002) Acoustics: Measurement procedures for ducted silencers – Insertion loss, flow noise and total pressure loss
29. Selamet A, Radavich PM (1997) The effect of length on the acoustic attenuation performance of concentric expansion chambers: an analytical, computational and experimental investigation. J Sound Vib 201:407–426
30. Kirby R (2001) Simplified techniques for predicting the transmission loss of a circular dissipative silencer. J Sound Vib 243(3):403–426
31. Kurze U (2004) Calculation of silencer with circular cross section and centre body. Fortschritte der Akustk DAGA 2004

Active Control of Sound and Vibrations 12

Joachim Scheuren

12.1 Introduction

"Noise Reduction by Anti-Sound." For more than four decades, this picturesque and catchy motto claims more than the physical existence, it claims the promising and useful application of some sound field whose characteristics allow at least partially the destruction of another annoying – and thus noisy – sound field. This claim and the envisaged methodical expansion of acoustical engineering were and are highly welcome. This is because the control of the increasing noise impact on our life strongly depends on any new possibility which may contribute to reduce the noise in our environment.

Of course it was not possible to keep and realize all what had been announced and promised – sometimes in a rather careless way – by the apologists of active control. Nevertheless, previous successes as well as the justified hopes were able to uphold continuous interest in applying and further developing the method.

The above-mentioned possibility of controlling wave fields is based on the simple physical principle of interference which describes how parts of the two wave fields may cancel if they are out of phase. It was common practice to characterize such intentional, electronically controlled interference by the attribute "active."

Active control measures thus are targeted to modify given mechanical field quantities (the so-called primary field) by the superposition of additionally generated, coherent secondary fields which are generated by electrically driven actuators, the so called anti- or secondary sources.

This approach is not at all restricted to airborne sound fields. It is equally valid for arbitrary media, i.e., for fluids and gases as well as for structure-borne and vibrational fields in elastic solids. Therefore, besides direct compensation of air-borne sound, active measures may also suppress the radiation of air-borne sound by controlling structural vibrations.

The successful compensation of fields opens further, completely new possibilities. If we succeed in negatively reproducing (and thus cancelling) given sound and vibration fields, then we should equally succeed in generating and thus realizing new arbitrary field distributions. Then, instead of approximating a vanishing (zero-) field, the more general task is to approximate an arbitrary nonzero field distribution. Active sound reduction thus turns to active sound generation and active noise control to active sound design (ASD).

The high fascination of active control of sound and vibration fields is not only based on the conceptual elegance of the approach which, in suitable applications like sound fields in ducts or small volumes, was able to lead to attractive results. It equally impresses by the novel combination of classical acoustics with the innovative and rapidly developing domains of digital electronics and integrated electromechanical transducers.

However, any successful application of the technology in practice depends on its robustness and

J. Scheuren (✉)
Müller-BBM GmbH, Robert-Koch-Strasse 11, 82152 Planegg, Germany
e-mail: Joachim.Scheuren@MuellerBBM.de

G. Müller and M. Möser (eds.), *Handbook of Engineering Acoustics*,
DOI 10.1007/978-3-540-69460-1_12, © Springer-Verlag Berlin Heidelberg 2013

economic efficiency rather than on its elegance. And this, in many cases, still interferes with the required electronic and electromechanic efforts. Together with an extremely confusing patent situation which, in the past, was caused by systematic protection of all imaginable solutions this has lead to limited applications in a few special cases only. Also, only a small number of research results appeared to have a realistic chance of practical implementation. It can be assumed, however, that technological progress together with expiring patents will result in a slow but continuous growth of active methods in practice.

In the following, an overview of the present state of the art will be given for active sound and vibration control technology. Besides the most important and promising applications, the basic mechanisms as well as some advice on how to design and apply active methods shall be given. Being subject to rapid changes in development, technical details will be outlined only briefly. For more detailed treatments and discussions, the extensive literature, in particular introductory books [1–6] and surveys [7–16] as well as the proceedings of regular conferences [17–24] and the compilation of literature and patent surveys in [25, 26] should be consulted.

12.2 Some Historical Comments on Technical Development

The first written formulation of the idea of active noise abatement as "noise cancelation by (electromechanically) controlled interference" [27] can be found in the patents of P. Lueg from 1933 to 1934 [28–30]. Among others, they contain those two approaches which – some decades later – have been in the center of first practical efforts: the control of arbitrary sound fields in the vicinity of an appropriately placed secondary source and the restriction to one-dimensional wave propagation.

First results for such a control were published by Olson in 1953. He investigated arrangements which in principle can be reduced to the basic control loop given in Fig. 12.1 [31, 32].

Driving the loudspeaker with an inversely amplified microphone signal is aimed to compensate any sound pressure fluctuations at the location of the microphone. Due to differing propagation directions in the primary and secondary sound fields, the region

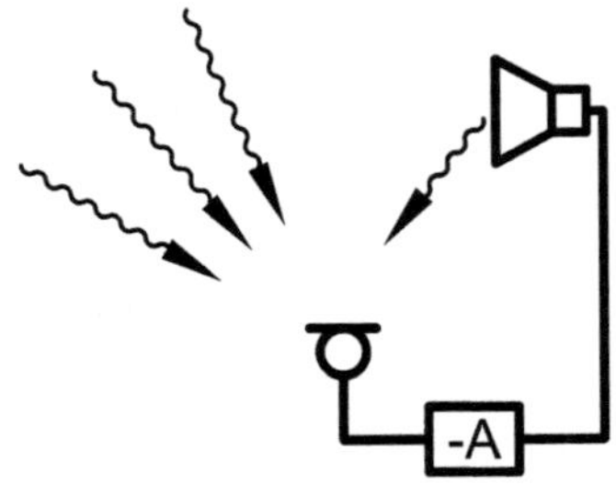

Fig. 12.1 Sound field control at the location of the microphone

of reduced sound levels is limited to some neighborhood around the microphones.

As their extension is proportional to the wavelength of the sound field, the spatial effectiveness of the anti-source is increased at low frequencies. This is an important feature of all active measures which is further emphasized by the fact that any signal processing requirements are met easier for lower sampling rates. Also, the difficulties with passive methods at low frequencies enhance the attractivity of active approaches.

As a representative result, Olson was able to reduce the sound level by some 10–20 dB within 1–3 octaves at distances up to half a meter and at mid-frequencies of some 60 Hz.

The limitation in sound level reduction is caused by instabilities occurring at high amplifications. Nevertheless, in later years this simple approach was maintained with some success for various applications. Leventhall et al. [33] have achieved sound insulations of some 20 dB by controlling the sound pressure in one-dimensional waveguides.

Although Olson was able to point out many potential applications of the method, a long way was to be gone to practical applications. This was equally caused by restrictions in fast signal processing as well as by a lack of thorough understanding of the physical mechanisms involved. The resulting underestimation of related difficulties, which can be illustrated by the confidence in simple successes for electrical transformers [34], has finally led to a temporary decline of activities.

It took another 10–15 years to stimulate new interest for the possibilities of active sound control. The temporary restriction to simple or one-dimensional fields together with the highly improved possibilities of electronic signal processing was able to promise encouraging successes.

Thus, the basic investigations of Swinbanks [35] or Jessel and his group [36] were soon followed by respectable experimental results. Typical achievements were sound level reductions of up to 50 dB for single frequency sounds [37, 38] and 10–15 dB for broadband sounds [38, 39].

A comparable but independent development could be observed in the domain of technical vibration control. Early attempts to reduce the excitation of structures by mechanically driven out-of-phase sources can be found in [7]. Since Olson had pointed to vibrational applications of his concept, Rockwell and Lawther succeeded in actively damping bending wave modes of beams by up to 30 dB [40]. Comparable, mainly theoretical investigations were carried out by Tartakowskij and his co-workers [41].

Again, it took another 10 years before the prospect of any realizability of practical measures caused growing interest. As can be seen from the literature given and the references quoted there, this interest was equally concentrated on mechanical engineering [42], space technology [43], and civil engineering [44].

Despite the close relationship between the tasks of active vibration control and active sound control, the development of both domains has hardly referred to each other. It thus took long from 1967, where the aim to reduce sounds by active control of radiating structures had been formulated [45], to concrete efforts in applying this principle [18, 46].

It then was in the eighties of the last century that the interest in active sound and vibration control and its possibilities significantly grew. Consequently, many large-scale but basic research and application-oriented development projects systematically investigated the principal realizability of the approach to nearly all, even remote fields of technical applications. However, the number of proven practical feasibilities did not correspond to the comparably low number of practical implementations.

This was mainly due to doubts in the long-term robustness of the components involved together with considerable costs to provide appropriate electronic and electro-mechanic equipment. However, any related disillusion should not obscure the fact that in fitting applications active control of mechanical and acoustical wave fields may have essential advantages and therefore can be classified as a most useful technological approach. This shall be demonstrated in the remainder of this chapter.

12.3 Structure of the General Problem

The problem of active noise and vibration control may be generally described by the structural diagram of Fig. 12.2. The source Q may represent one or even more simultaneous sound sources, thus generating the primary field quantities put together in the vector y_P. The basic mechanism of active control can be found in the fact that in some interactions points these primary field quantities y_P are superimposed by secondary field quantities y_S.

While the resulting field quantities y propagate within the mechanical system under consideration, they are subjected to multiple changes and modifications. These can be represented by the matrix of block C with output quantities z which represent the target of any active control measure. These target quantities z shall approximate and follow some given, desired quantities z_S as accurately as possible.

Depending on the application and related effort, the specification of z and z_S may realize such different design objectives as:

- Cancelation of a field or intensity component ($z = 0$) in one or several points
- Optimal approximation of one or several noise spectra at defined frequencies in given points or domains
- Minimization of the mean sound or vibration energy in a given domain
- Minimization of radiated sound power

Also, other characteristics of the sounds to be considered like:

- Adjusting tonal spectral components to given amplitude values.
- Realizing given values of some psychoacoustic noise parameters (e.g., loudness or roughness).

can be obtained by appropriate selection of the target quantities z_S and the desired quantities z.

From the variety of possible definitions of target quantities it can be seen that they are not always easily accessible for direct measurements. Therefore it can be necessary to estimate the effective target quantities from measurements of other additional field quantities which are not elements of z. The complete set of measured quantities as evaluated by the signal processing unit G are put together in the feedback signal vector z_R.

The third arrow starting from C describes possible feedback effects of the generated field to the source

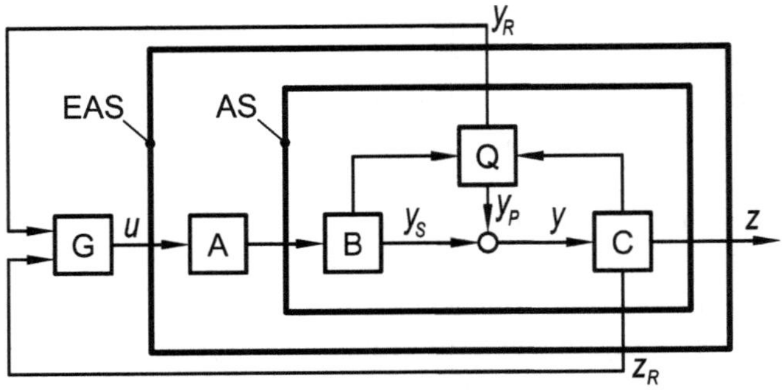

Fig. 12.2 Structural diagram of the general problem

mechanisms in Q. This allows the description of such sound and vibration generating mechanisms which are caused by unstable feedback effects between different mechanisms involved [11, 47].

In particular, the closed feedback loop from Q via C back to Q describes the structure and the mechanism of mechanical self-excitation which is the origin of many vibration and wave fields. By intervening in this mechanism, e.g., by adding an additional feedback loop to compensate the original one, the generation of such instabilities can be prevented (see Sect. 12.4.6).

The remaining blocks represented by A are all actuators that are needed to excite the secondary field quantities y_S and by B all mechanical transfer paths to the interaction points while the signal processing unit G provides all signals u to drive the electromechanical transducers in A.

The input signals of G which provide all information to evaluate optimal control signals u may be distinguished with respect to their content of information. In analogy to Fig. 12.2, y_R contains all signals with some (a priori) information on the sources and on the primary field quantities in y_P in the interaction points. For this reason they are often referred to as reference signals. On the other hand, z_R represents such signals which allow for estimates of the target quantities.

Apart from the paths described so far, the path from B to Q in Fig. 12.2 also takes into account the feedback effect of the secondary field to the sources. This will allow besides possible changes for any distortion of the primary field, information given in y_R caused by any secondary field action via Q. Without appropriate countermeasures, this undesired feedback may cause erroneous secondary fields and – even worse – severe instabilities as illustrated in the context of Fig. 12.1.

Typical measures against such instabilities can be selective measurements to suppress any influence of the secondary field like, e.g., directive measurements or measurements of structural quantities representing the primary field without being influenced by the secondary field. Also, instabilities may be avoided by feedback compensation as described in the next section.

The general model of Fig. 12.2 may represent and describe very different arrangements. This can be demonstrated by an example. If Q represents a combustion engine, z could equally represent the sound power radiated from some exhaust system or radiated directly from the engine. Also, the structure-borne sound power introduced into some foundation may be a useful definition of the target z. To drive the secondary transducers, y_R may contain an rpm signal as well as the sound pressure in the exhaust pipe just before (upstream of) the secondary source. Depending on the target, z_R may represent quantities like sound pressure in an exhaust pipe after (downstream of) the secondary sources, sound pressure in preselected points of the car interior or exterior and field quantities in a given base structure.

Apart from any devices to measure y_R, z and z_R, the blocks B, C and Q solely describe acoustical (or mechanical) systems which may be combined into one single acoustical system AS or into a mechanical system MS. By integrating the electromechanical transducer contained in A, systems may be extended to an electroacoustical system EAS or an electromechanical system EMS.

This formalism clearly demonstrates how the classical task of any noise and vibration reduction, or of selected control of an acoustical target quantity in AS or MS by appropriate (passive) acoustical/mechanical measures, may be transferred into the task of active control of an electro-acoustical/mechanical system EAS or EMS, just by applying electro-acoustical/mechanical transducers.

The structure of Fig. 12.2 and the functionalities and media described by the systems A, B, C and Q identify the problem at hand as a general control problem with distributed parameters [4, 48–52]. This structure includes some simple standard structures of common problems in signal processing and control theory as special cases.

In the terminology of the latter y_P represents the disturbances that must be compensated as well as possible by the anti-sources contained in A such that the target quantities contained in z are minimized. If y_R contains sufficient information on the temporal behavior of y_P, the required control signal u can be derived solely from y_R. This finally leads to a standard feed-forward control problem.

In contrast to this, the exclusive evaluation of the target quantity $z = y$ to be controlled leads to a standard feed-back problem with disturbing noise where the target signal $y = 0$ may be prescribed as desired signal.

In any application, the system design should always be aimed to compensate the measurable (by y_R) and thus predictable influence of disturbances by pure feed-forward controls. Only then does it make sense to act against the remaining components of the control signal y by a feed-back controller [53].

The definition of the controller's structure and its appropriate parameters should be based on the many methods of analog and digital control theory with one or more control signals as well as on design rules for signal processing systems as described in [4] or in the relevant system and control theory and signal processing literature.

12.4 Principal Considerations on Working Mechanisms of Active Systems

Any arbitrary change of given wave fields requires their previous compensation. This is because any change or replacement of field distributions cannot be effective unless the original field can be compensated or reduced at least. For this reason, apart from direct control of the sound-generating mechanism, all active control approaches are based on the negative (anti-phase) reproduction of mechanical sound and vibration fields. Here, field reproduction generally means the approximation of a given primary field quantity $f_p(x,t)$ by a secondary field quantity $f_s(x,t)$ for an infinite number of points t in time and $x = [x_1, x_2, x_3]^T$ in space.

However, as the generation of $f_s(x,t)$ is limited by a finite number of control elements and controllable degrees of freedom, the demand for equality of f_p and f_s cannot be fulfilled for all values of x and t but only for a finite subset.

Any field distribution may be described by rather different parameters. Therefore, the definition of this subset is not necessarily restricted to the space/time domain, but can be based on any parametric description like modal decomposition (modal transformation) or superposition of plane waves (Fourier transformation). The reproduction of waves allows the control of their propagation while the reproduction of modal amplitudes directly aims at the active reduction of the related vibration amplitudes.

To get the relevant field information from as few parameters as possible directly corresponds to the important requirement to enable the related field control by as few actuators as possible. Thus, the decision whether to use a modal or a wave-based approach may be directly deduced from the concrete type of problem. As long as only a few modes of a structure contribute to the disturbing noise component, the modal approach limiting all active measures to the relevant modes may be recommended. By contrast, for one dimensional wave fields e.g., it is advisable to characterize the field by forward and backward propagating waves.

As both approaches describe or – for a finite number of parameters – approximate the same field, they mutually imply each other. Therefore, any immediate control of propagating waves may suppress the modes and thus the resonances of the structure, if propagation loops closed by mechanical feedback are cut off by absorption of incoming waves [54]. For this reason, the concept of propagating waves has been adopted by vibration theory which traditionally is dominated by modal concepts. Here, first applications were focused around large space structures [43, 55].

Apart from concentrating on modes and plane waves, other concepts have been used to theoretically describe the possibilities of active field control. Among these, the direct control of power flow [56] and the control of actively realized impedances [57] shall be mentioned here. Impedance control takes advantage from the fact that its effect only depends on the local impedance of the medium under consideration.

Therefore, any modeling of wave propagation becomes unnecessary. But this advantage may be compensated by the difficulty of finding local criteria which sufficiently describe the global field behavior. An example of how high-damping values may be based on the minimization of wall impedances by using a skillfully chosen arrangement can be found in [58].

A more direct link to the global behavior is found for the targeted control of power flows. Because all related power flow strongly depends on any secondary measures, respectively, it seems to be advantageous to control power flows indirectly only by controlling other target quantities like the amplitudes of propagating waves [59]. In the following, with respect to the most important concepts of field modeling and superposition, a systematic guidance to active system design will be given.

12.4.1 Concepts for Active System Design

As active sound and vibration control is based on the superposition of appropriate secondary fields, the essential task in designing active systems is to define favorable, i.e., most effective secondary sources together with their interaction points. This is because the kind, the number, and the location of secondary sources crucially determine the controllability of the system under consideration (EAS or EMS), i.e., the quality of approximation of the target quantities z to their given desired values.

It is good practice therefore to base the controllability of target quantities on physical considerations, e.g., by systematical analysis of the primary field on its propagation path from the source to the target points, and by investigating the effects of additional secondary sources.

This approach fully corresponds to the proven concept of passive noise control which is given by investigating the propagation path of the sound to be controlled from the source to the receiving point in order to find the optimal location for effective noise control measures. Thus, the following sections of this chapter will exemplarily explore, i.e., by how many sources at which locations and to what effect, sound and vibrations may be controlled

- at source, where the field is introduced
- on the propagation path of the field or, finally
- at a given target volume.

Also, defining the measurement chain that acquires the target quantities to be controlled as well as the available input signals z_R and y_R may crucially influence the success of an active measure. This is because this success is directly related to the quality of estimating the desired behavior or, in the terminology of control theory, to the observability of the system (EAS or EMS) under consideration. With respect to the delay time required by the signal processing unit G, it is important to register appropriate and, above all, early signals y_R and z_R with sufficient lead time.

From these principles, a design strategy for active systems may be derived:

1. Target formulation for an active measure in terms of a finite number of parameters. As an example, these parameters may be given by the complex amplitudes of certain waves or modes or just by the complex amplitudes of the sound pressure at given points.
2. Specification of such a source arrangement (A) that is best capable of achieving the target defined under 1 with acceptable effort.
3. Specification of a measurement arrangement (within C) that again is most suited for achieving the target defined under 1 with acceptable effort.
4. Specification of the structure and the parameters for the signal-processing unit G to optimally implement the control possibilities given in 1.3.

The proven principle to try to fight any sound and vibration as close as possible by the source stays most valuable for active systems too. Besides preventing the introduction of energy by reflecting or absorbing it at the excitation point, in some cases the generation of the field itself can be prevented efficiently.

In particular, if the acoustical or vibrational field is generated by self-excited limit cycles caused by unstable interaction, simple measures may force stability and thus suppress the generating mechanism.

In the following, the most important physical concepts of active sound control are shortly illustrated. Because of some limitations resulting from idealized assumptions, these concepts do not always refer to arrangements realizable in practice. However, they do offer some physical insight which helps to estimate necessary source efforts or limitations in frequency. Also, the most important question as to the occurring flows of energies and/or power can be answered in this context.

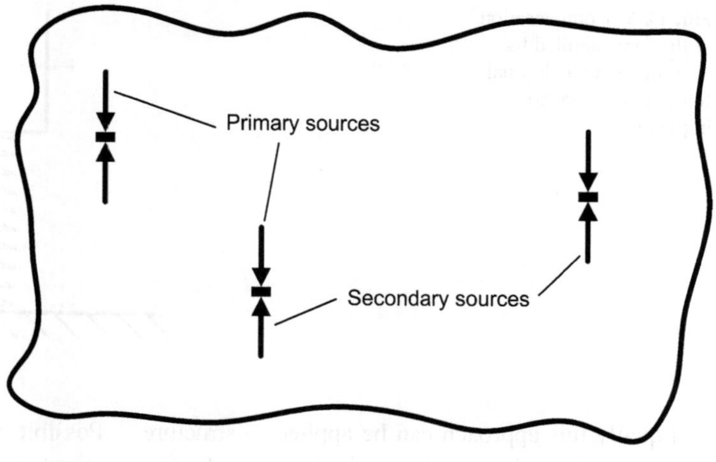

Fig. 12.3 Principle sketch of source reproduction

As said before, active approximation of arbitrary fields requires first, by principle, the compensation of the existing primary field. Therefore, the following considerations focus on such field compensation. This can further be justified by stating that any approximation of arbitrary fields formally can be treated like the approximation of a vanishing zero field.

12.4.2 Source Reproduction

The control of multidimensional wave propagation as well as the compensation of vibrational fields with high modal density often require a high number of sensors and actuators hardly to be realized in practice (see Sects. 12.4.3 and 12.4.4). Therefore, it often is preferrable to try to counteract the field by a copy of the original sources.

The sketch given in Fig. 12.3 illustrates this symbolically by using primary (acting downwards, ↓) and secondary (out of phase by 180° and therefore acting upwards, ↑) source arrows which e.g., may represent loudspeakers in a room or forces acting on a plate.

The smaller the distance between two identical sources in antiphase is, the more the source-generated fields are cancelled; the sources mutually hinder themselves in radiating sound. Besides simple, easily reproducible sources with low directivity, effective field reductions require therefore distances between primary and secondary source of less than half a wavelength [1, 60]. And this applies not only to free fields but also equally to closed spaces with high modal densities [1].

To the extent of successful source reproduction (e.g., in a certain frequency interval), the related field reproduction can be global, i.e., successful within the complete volume under consideration. As can be seen from Fig. 12.3, such global compensation can be achieved with locally arranged sensors because the reproduction of sources in amplitude and phase can be controlled by local measurements.

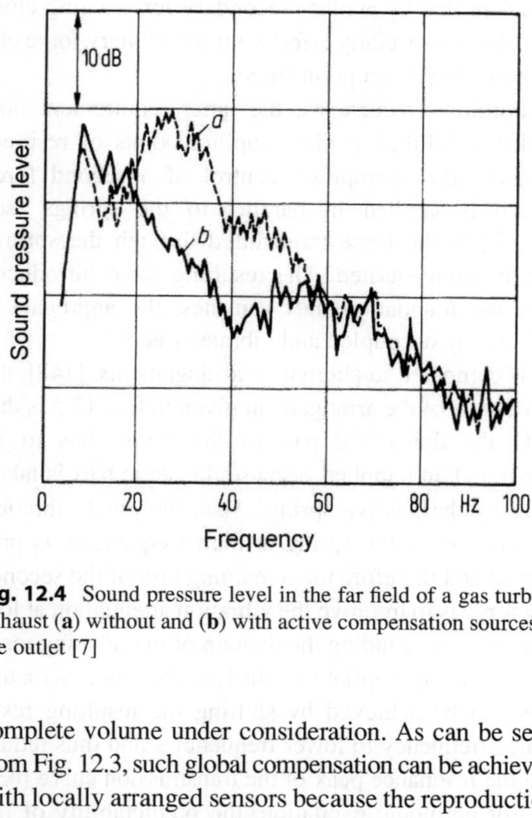

Fig. 12.4 Sound pressure level in the far field of a gas turbine exhaust (**a**) without and (**b**) with active compensation sources at the outlet [7]

As in the first reliable out of lab realization at all, such suppression of sound radiation has been successfully applied in practice to the exhaust outlet of a gas turbine [61, 62]. By applying 72 bass loudspeakers on a circle around the outlet it was possible to reduce in the far field the sound pressure level between 20 and 50 Hz by more than 10 dB (see Fig. 12.4).

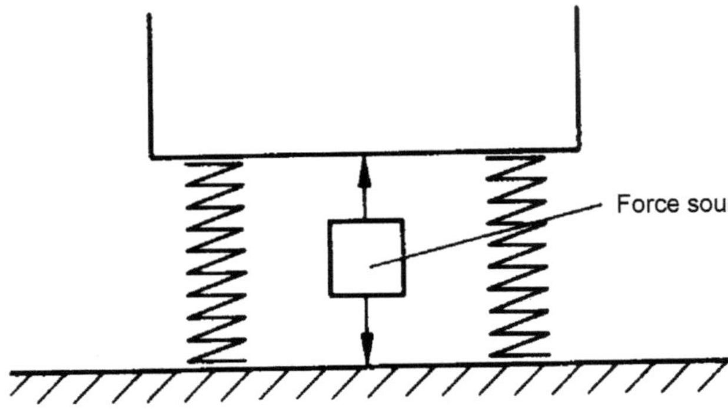

Fig. 12.5 Compensation of the force applied by a spring by an additional force source applied in parallel

Equally this approach can be applied to structure-borne sound problems if the original primary force is compensated by another secondary force acting close by. This is especially effective if the primary force can be considered as a point force.

For low frequencies, the latter requirement normally is fulfilled at the coupling points of resilient mounts. By appropriate control of a second force which is applied in parallel to the springs (see Fig. 12.5) the force transmitted through the springs can be compensated. The resulting force introduced into the foundation thus vanishes, the aggregate is effectively decoupled and vibrates freely.

If compared to alternative arrangements [148], the advantage of the arrangement given in Fig. 12.5 is that only the dynamical part of the forces has to be generated and applied because the static part is taken over by the passive spring. Also, the good vibration attenuation of the spring at high frequencies is preserved and therefore the remaining task of the secondary force is to improve the vibration attenuation at low frequencies including the domain of mass/spring resonance. By appropriate control of the force actuator, this can be achieved by shifting the resulting resonance frequency to lower frequencies and thus reducing the resonance peak of the transmission curve [63].

For harmonic excitation, the predictability of the signal wave form allows attenuation values of vibration which in many cases of passive mounts cannot be achieved. An impressive example will be given in Sect. 12.6.

Thus, active vibration isolation systems, often called active mounts, offer the well-founded perspective of becoming an indispensable engineering tool for vibration control in special cases of high demand. Possible applications go from mounts for complete vibration suppression of highly sensitive processes [63, 64], where active mounts are frequently used today already, to improved mounts for engines and other vibration sources. In addition the operating staff of vibration intensive machinery thus can be protected. Further hints to two active engine mount systems realized in standard cars are found in [65].

General information on the possibilities and on the design of active mounts are given in [2, 3, 6] and in a recent VDI guideline [66].

As concrete example, we refer to the application of secondary forces to the car body of an ICE passenger train car in close proximity to the bogie's secondary spring. By this active measure, the low frequency noise components excited by wheel harmonics under special track conditions could be essentially reduced in the passenger compartment around 90 Hz. Comparing the spatial sound pressure distributions of Fig. 12.6 (measured at the rolling test site of Deutsche Bahn AG at a speed of 200 km/h) shows that the related reductions were up to 20 dB for some seats. The average reduction for a group of six seats was up to 12 dB [67, 68].

The use of sliced piezo elements placed in the interior volume of the secondary springs resulted from a systematic feasibility study investigating various active concepts for the bogie and the passenger compartment with respect to their applicability and effectiveness. It turned out that immediate control of the sound field in the passenger compartment by loudspeakers would also be possible. However, in summary, the global compensation by compensating forces applied to the excitation points was shown to be advantageous.

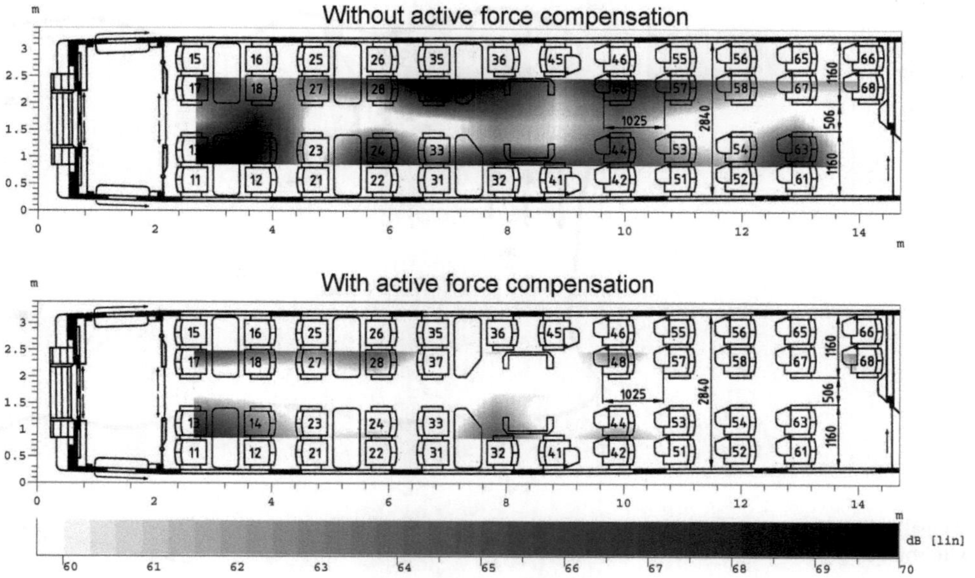

Fig. 12.6 Spatial sound pressure distribution in the passenger compartment of an ICE passenger train car without (*upper*) and with (*lower*) active force compensation at the bogie's secondary springs

12.4.3 Active Control of Wave Propagation

Apart from stability problems, the propagation delay times within the secondary sound field and the electronic controller may limit the performance of the simple standard feedback controller mentioned in Sect. 12.2. It therefore appears useful to take advantage of the finite velocity of wave propagation or the related delay time of the primary wave field to be controlled, respectively. This allows a temporal prediction if the field quantities can be measured before arrival at the point of superposition.

It is possible then to use the delay time of the propagating wave to determine the optimal waveform for an appropriately placed interfering field source. As can be seen from the principal sketch of Fig. 12.7, the propagation of the wave can be hindered essentially or even suppressed completely.

The possibility of determining field quantities from their values on a preceding incoming wavefront follows from Huygens' principle. This principle in the formulation of Kirchhoff has been the starting point for basic considerations first published by Malyuzhinets [69] and Jessel [36].

According to this principle, any field within a volume *V* generated by exterior sources outside *V* can be equally generated by substitute sources *Q** distributed on the closed surface *S* which limits the volume *V*. Figures. 12.8 and 12.9 show this for the case of *V* being free of any interior sources.

Apart from the change in sign, this is exactly what active control aims to do: to define source distributions -*Q** which completely counteracts and thus eliminates a given field within a volume *V* according to Fig. 12.8.

However, this requirement is not unique, i.e., it may be fulfilled by various field distributions. This allows additional requirements like defining that, according to Fig. 12.9, the sources *Q** should provide a field that reproduces the given field within *V* but vanishes outside *V*.

For airborne sound fields, such sources are given by monopoles and dipoles continuously distributed over *S*; they fully compensate the original sound field inside *V* without any feedback effect to the exterior volume, i.e., without changing the field outside *V* (see Fig. 12.10).

Any arrangement driven this way describes an active absorber with antisources -*Q** just receiving but leaving unchanged the complete incoming wave field. Also, the power transmitted from the primary field to the secondary sources is extracted by these secondary sources and, if necessary, reintroduced at some other location. Any original power absorption within *V* will now be taken over by the substitute sources -*Q**.

Other than in the arrangement of Fig. 12.10 which not only absorbs incoming power but also emits power

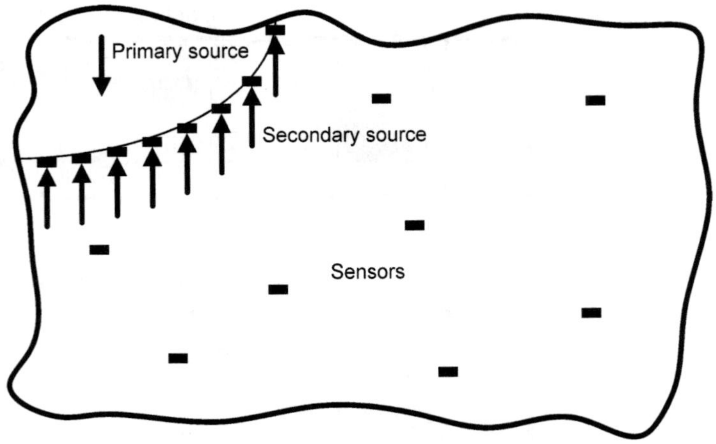

Fig. 12.7 Principal illustration of active control of wave propagation

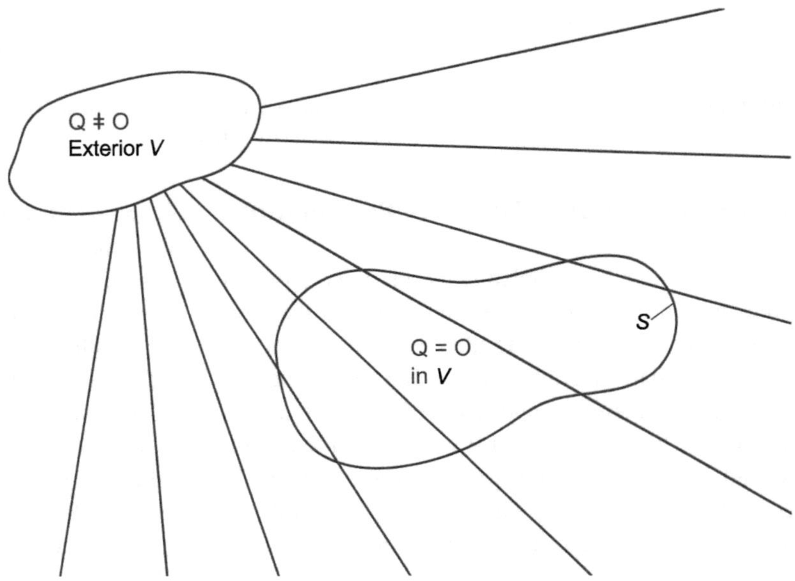

Fig. 12.8 Original field distribution with sources *Q*

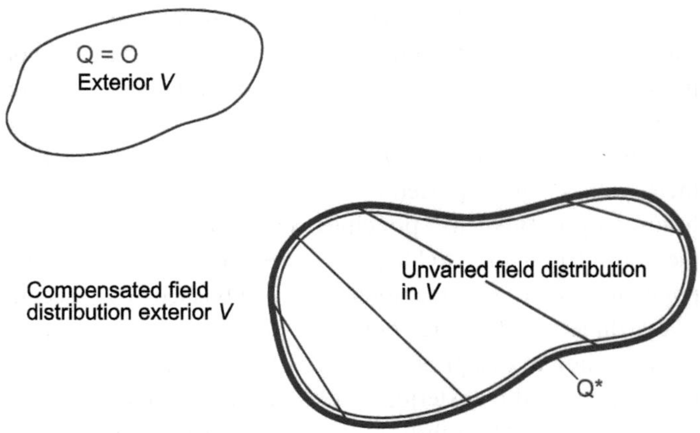

Fig. 12.9 Reproduction of field distribution by substitute sources *Q**

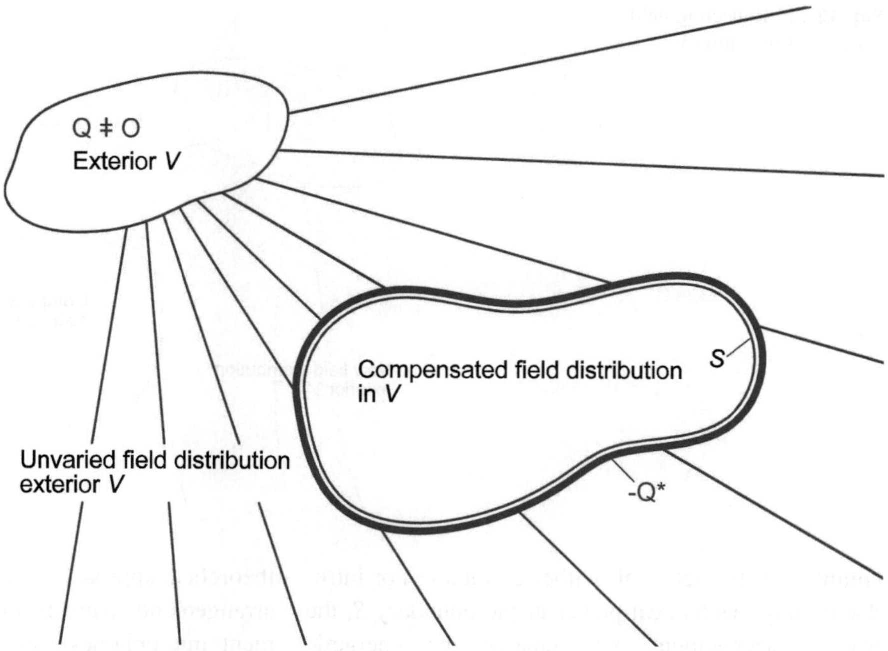

Fig. 12.10 Compensation of the field in V without feedback to the outer volume outside V

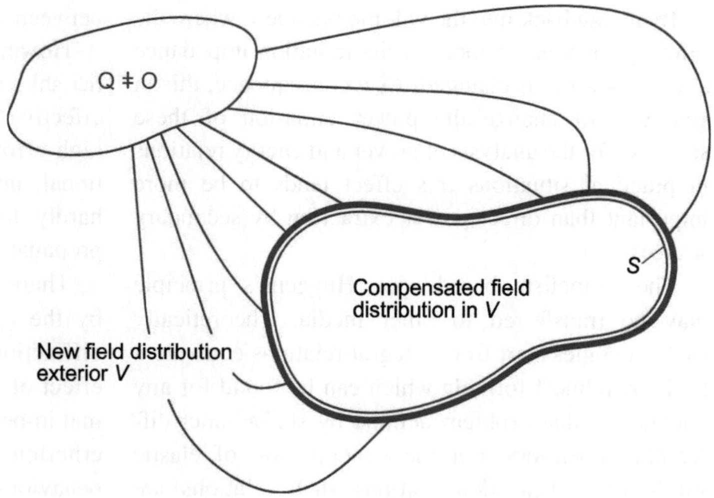

Fig. 12.11 Purely absorbing field compensation in V

in parts of the surface S in order to leave the outer field unchanged, another source distribution could be characterized by sources $-Q^*$ on S which only absorb power. An exemplary field caused by such an active absorber is given in Fig. 12.11.

Here, the energy relations are a bit more complicated and strongly determined by the concrete case. The field generated by the antisources acts back on the primary sources and thus changes their radiative environment. Some simple and illustrative examples of such energetic interaction can be found in [8, 54, 70, 71]

The reproduction of the field within V can also be achieved with simpler source arrangements but then without additional restrictions on the field outside V. This compensation no longer is free of feedback, the field outside V changes and the arrangement thus incorporates an active reflector scattering back the incoming wave field (see Fig. 12.12).

These considerations show that secondary source distributions on the boundary surface S may be configured such that they can absorb (by extracting incoming power) or reflect (by redistributing incoming power) primary wave fields. Even if this reflection is

Fig. 12.12 Reflecting field compensation within V

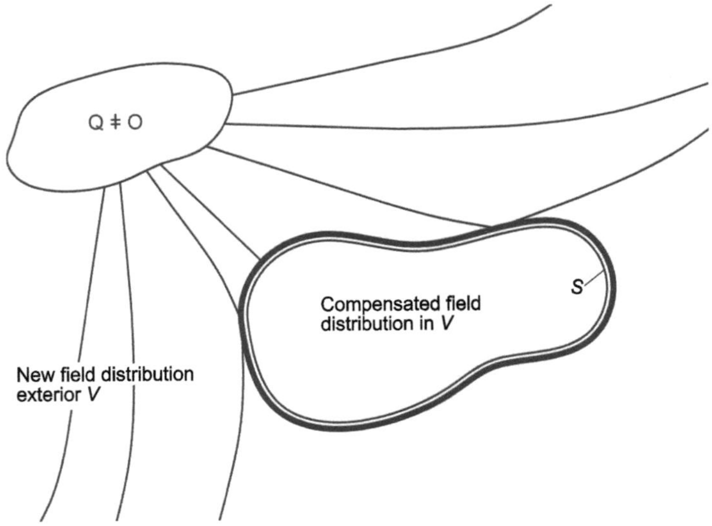

complete in the sense of neither extracting nor introducing any mechanical power at the boundary S, the overall arrangement may change the energetic relations.

By acting back into the volume outside V where the primary sources are located, the radiation impedance at these sources is changed. As a consequence, this in general will change the power emission of these sources. For the analysis of power and energy relations in practical situations this effect tends to be more important than direct power extraction by secondary sources.

The formalism based upon Huygens's principle may be transferred to other media. Theoretically such analogies start from integral relations equivalent to the Kirchhoff formula which can be found for any boundary value problem defined by self-adjunct differential operators. For the general case of elastic solids with shear deformations such relations are found in [72]. For the case of flexural waves in thin plates which are important with reference to acoustic radiation, an analogous equation has been derived in [73].

Although such considerations essentially contributed to the theoretical analysis of active systems, they rarely served as an immediate starting point for practical realizations. This may mainly be due to the difficulty to predict the effects of discretization and of the spatial extension of the secondary sources.

Nevertheless, basic experiments in shielding the multidimensional propagation of sound fields with loudspeakers have been based on this idealized theoretical approach so far. As an example, laboratory arrangements using 16 loudspeakers and 24 measurement microphones were able to obtain in parts of the shielded domain level reductions of up to 30 dB between 200 and 500 Hz [74].

However, like with other experiments, e.g., acoustic shielding of transformers, it became clear that effective level reductions in considerable areas require high efforts. Therefore, although proven to be functional, any practical applications of active barriers are hardly to be expected for three-dimensional wave propagation.

There is, however, one important exception given by the approach described in [75] to decrease the diffraction angle and thus to increase the shielding effect of sound barriers by actively realizing an optimal impedance at the edge of the barrier. As the local criterion "impedance" sufficiently describes the global behavior of the barrier with respect to diffraction it is not necessary to model the spatial propagation of waves in three dimensional space.

Figure 12.13 shows a representative measurement point beyond the barrier where the sound pressure level may be reduced at frequencies between 200 and 600 Hz by some 4–8 dB if an acoustical soft edge of the barrier is realized by active means. This gives an additional improvement by some 2–5 dB in comparison to a barrier with a sound absorbing cylindrical edge.

The restrictions put forth against controlling multidimensional wave propagation do not apply to the one dimensional special case. In particular, if, below some

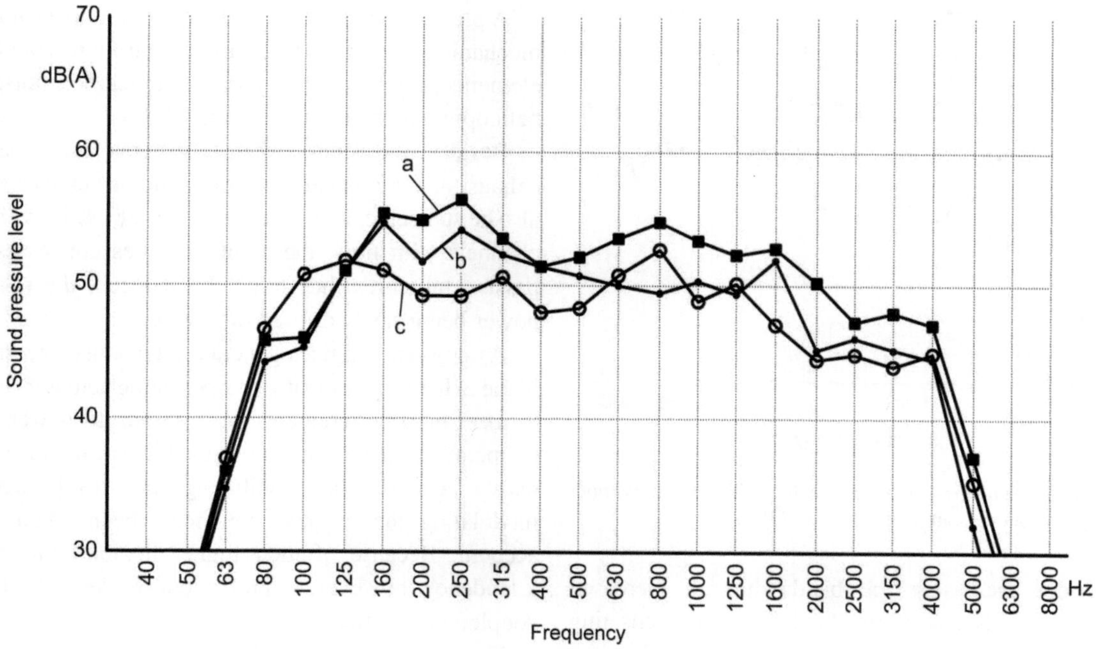

Fig. 12.13 Comparison of sound pressure levels behind a sound barrier (**a**) without and (**b**) with passive absorbing edge and (**c**) with active realisation of a soft edge [75]

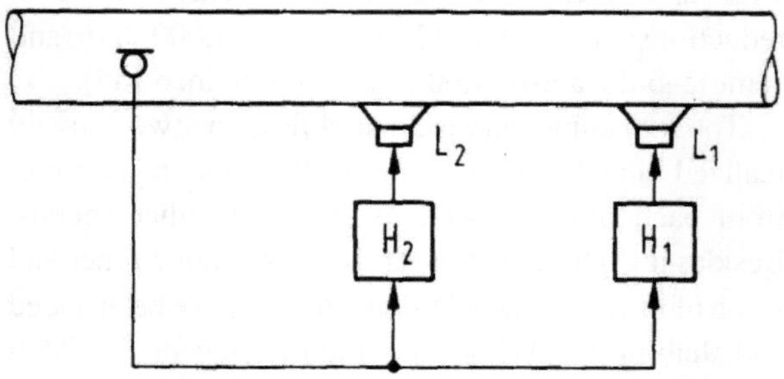

Fig. 12.14 Active sound field control in channels

cut off frequencies, only a limited number of wave types or modes is propagative, this propagation may be influenced effectively with lesser efforts.

The conditions in channels are most favorable if the related frequencies do not exceed the lowest cut off frequency. Then, only plane waves propagate and it doesn't matter where the generating volume flow is introduced to the channel. Instead of monopoles and dipoles distributed over the cross-section of the channel it is sufficient to apply independently controlled loudspeakers to the channel wall as shown in Fig. 12.14.

Approximating a dipole by two closely placed monopoles, results in a limited bandwidth. Swinbanks has shown how this bandwidth may be increased by additional loudspeakers.

Figure 12.14 includes the special case of one loudspeaker only ($H_2 = 0$) which initially was preferred for experimental implementations. The loudspeaker L_1 is driven such that the incoming wave field is reflected and the channel is kept free behind L_1. In analogy to the considerations at the beginning of this section, this arrangement implementing an active reflector may be extended to an active absorber by using and driving L_2 such that the wave reflected by L_1 will be absorbed by L_2 [35, 60].

After the realizability of these approaches had been proven experimentally [38] many attempts have been made to improve the results with respect to level reduction and to bandwidth and to consecutive transfers in practical applications. This may be demonstrated here by the work of La Fontaine/ Shepherd [76], who came up with broadband (30–650 Hz) level reductions of 20 dB. Roure [77] used an adaptive filter to obtain the results of Fig. 12.15 for simultaneous flow.

At low frequencies the limitation of active sound attenuation is caused by turbulent pressure fluctuations at the input microphone, while at high frequencies the effectiveness is reduced by the higher modes.

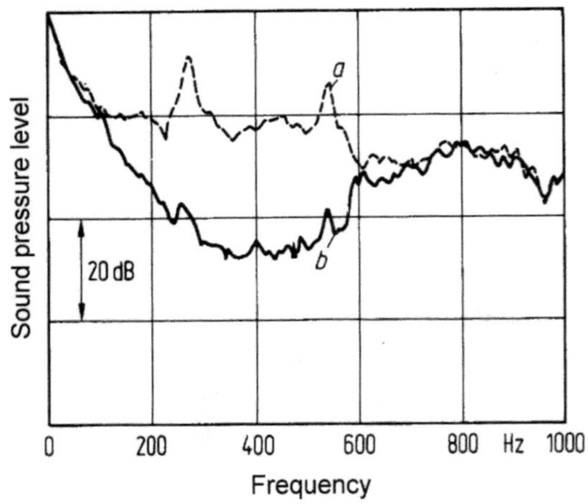

Fig. 12.15 Sound pressure level in a duct with flow (**a**) without and (**b**) with active sound attenuation [77]

Besides increasing the bandwidth, an increased number of loudspeakers may be used for compensating additional modes. In the past, this has not only been investigated theoretically but has also been practically realized in industrial plants[11].

Typical applications for active measures with one-dimensional wave propagation are air conditioning and exhaust systems. As an example, it is referred again to the gas turbine of Sect. 12.4.2 (Fig. 12.4). In this application, the suppression of external radiation means, with reference to the duct channel, a total reflection at the exhaust outlet. Further examples for active systems in ventilation systems are found in [78–80]. Also, direct suppression of ventilation noise by loudspeakers close to the source has been proven first in [81].

The control of wave propagation is not restricted to airborne sound waves. Figure 12.16 illustrates this for bending waves in beams by giving an example of where incoming waves may even be absorbed by appropriately driven secondary sources [60, 70]. The reduction of the reflection coefficient by 32 dB (in the mean) could be obtained by an electrodynamical shaker at the free end of a beam and by suitable reproduction of the passively reflected primary field. Thus, 99.95% of the incoming power could be absorbed.

This also illustrates the possibility (mentioned at the beginning of Sect. 12.4) of active structural damping resulting in multimodal resonance suppression. This approach has some advantages if compared to direct control of modal amplitudes, and may thus be helpful for some structural dynamic applications.

A practical application showing how suppression of mechanical vibration transmission through beamlike elements may be used to reduce the gearbox noise in helicopter cabins is described in [82].

Besides industrial exhaust systems the smaller exhaust and air-intake systems of motor vehicles may also be subject to active measures [83–88]. Like with all channels with flow, the additional pressure losses of passive measures thus might be reduced, the related power being saved or used otherwise.

As powerful engines generate high sound pressures in the exhaust system, any compensating actuators have to meet high requirements with respect to power and temperature. Therefore, besides loudspeakers, alternative actuators like oscillating flaps [87] directly modulating the exhaust flow have been taken into account. Because of their higher flexibility there is a tendency, however, to prefer loudspeakers skillfully coupled via airstream-cooled pipes.

Figure 12.17 shows that a series-proven active silencer (curve 2) has a better noise-reducing performance than a passive silencer (curve 1). Compared with the system without silencer (curve 5) level reductions of some 6–12 dB below 4,000 rpm and some 2–6 dB above 4,000 rpm are obtained [85].

The remaining curves 3 and 4 describe two actively realized sound variants clearly differing in character from each other as well as from the other sounds. Besides a slight reduction of the third engine order and some of its multiples, additional orders have been added and thus raised. While the amplification of the 2.5th and 3.5th order generated a low frequency oscillation resulting in an eight-cylinder-like sound (curve 3), the sportive character of the six-cylinder engine could be further emphasized by raising the 4.5th and 7.5th order (curve 4, [85]).

Besides – in a wide sense – free realizability of sounds the active system was able to save nearly 50% of volume and weight if compared to a standard muffler. However, although the system showed promising mechanical and acoustical robustness, the related cost prevented it from being introduced on a large scale until now.

Another application of wave propagation control could be the suppression of sound propagation in fluid-filled pipes [89], especially because the higher impedance of fluids enables a more effective coupling of actuators.

Finally, for completeness sake, a totally different application area should be mentioned. The acoustic

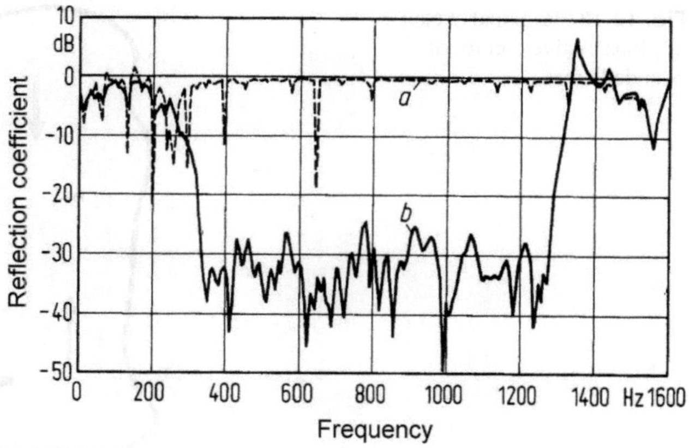

Fig. 12.16 Reflection coefficient (**a**) without and (**b**) with active absorber at the free end of a beam [60, 70]

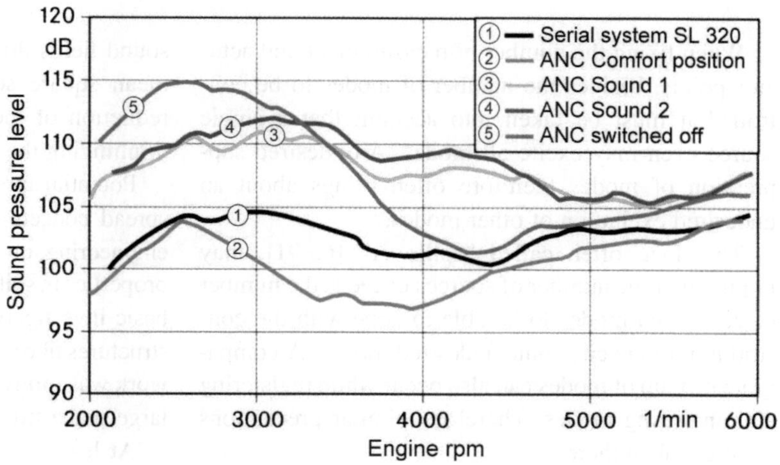

Fig. 12.17 Total exhaust noise level of a passenger car: Comparison of various sound variants of an active muffler with a passive standard muffler [85]

properties of rooms strongly depend on the reflective properties of the surrounding walls; considerations of influencing these by active elements are found in [90].

12.4.4 Active Control of Enclosed Domains

The concepts described so far dealt with directly controlling the excitation and the propagation of waves independent of the space defining the targeted area or volume. These approaches fail where the sources cannot be accessed, or where the sound field is characterized by many multidirectional waves. Such fields typically can be found in bounded areas with reflecting boundaries, thus being capable of resonances. In most such cases, it is easier to restrict active-control measures to a subspace or to base it on a modal decomposition.

Figure 12.18 illustrates this for the case that the effect of several sources distributed over a volume can neither be controlled at the excitation points nor be shielded actively. The only possibility remaining then is to control by applying secondary sources appropriately distributed over the accessible volume.

Another choice to be made then is whether field control is intended for the entire space (global field control) or for a selected subspace only (local field control). It is immaterial then whether this space is given as a closed volume (e.g., an aircraft cabin) or as structural element (e.g., beam or plate).

If the whole space is considered, modal description of the field has the advantage of providing independent parameters for field description and – control: the complex modal amplitudes.

Any targeted suppression and modification of single modal amplitudes assumes that they can be registered as well as excited. The simple example that a mode cannot be registered nor modified in its nodal points or lines illustrates that this requirement must be fulfilled by appropriate selection of all sensing and actuating points first.

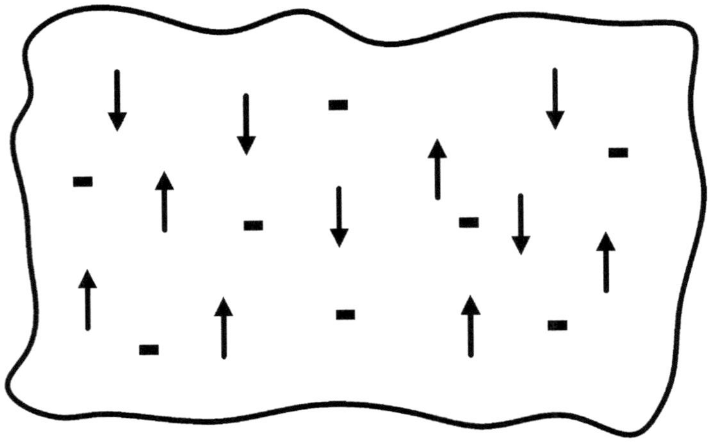

Fig. 12.18 Principal sketch of global active control of bounded areas

When fixing the number of measurement and actuator points, besides the number of modes to be controlled it must be taken into account that a single source even may excite all modes. Any desired suppression of modes therefore often brings about an undesired excitation of other modes.

This fact, often called "spillover" [6, 91], may require that the number of sources exceeds the number of controlled modes to be able to cope with the constraint not to excite some undesired modes. A comparable mix-up of modes can also occur while registering and analyzing modes. Therefore, similar precautions must be taken there.

If undesired modes are to be included into the considerations, the total number of modes in general exceeds the number of degrees of freedom given by the transducers. The thus over determined system of equations as well as the corresponding control law have a unique solution provided the mean square error of the modal amplitudes is minimized.

As a special case, this includes the possibility of controlling uncoupled modes vibrating with clearly separate frequencies by a single force only if this force is driven by the required line spectrum [60].

The modal approach fails as a conceptional basis for active measures if the number of relevant modes or the modal density do not allow a separate treatment. Then, like in the open space global field, compensation only can be obtained by reproducing or shielding the sources. Otherwise, the effect is limited to small zones around one or several measurement points.

This immediately leads to the pragmatic approach of directly minimizing the mean energy in all measurement points without any modal separation. For sound fields this means a reduction of the registrable mean square sound pressure while for structures a reduction of the radiated power can be obtained by minimizing the mean square velocity.

Potential applications of these approaches are widely spread concentrating, within the domain of vibration engineering, on large structures to ensure their dynamic properties in spite of extreme weight reductions. Besides basic investigations [6, 44, 52] how to stiffen selected structures like aircraft wings [92] by active means, much work was motivated by the perspective of being used in large space structures [6, 43, 55, 93, 94].

At least as long as all structural elements may be modeled as homogeneous continua the concept of wave propagation has some advantages over the modal approach like local limitation of registration and excitation or a smaller number of degrees of freedom [54, 95, 96].

For airborne sound fields in closed volumes it can be stated from the above considerations that their control is so much the easier the smaller the dimensions of considered volumes in comparison to the wavelength are. As higher modes for small dimensions only build up at higher frequencies, active measures can be limited to some fewer modes which depend on the highest frequency to be considered. If harmonic waveforms of the field quantities then allow a good prediction of their future behavior, all conditions for effective active field control are met [97].

An early, most useful application for active measures was given by active earphones because they enclose an extremely small volume which, below some kHz, is characterized by a basic mode with spatially constant sound pressure only. Within the last 30 years this has lead to various standard products being offered today by various established manufacturers of headsets. These solutions differ between pure ear protectors and headsets with improved low-frequency broadband sound insulation (typically below 1 kHz) against undesired exterior noises. By such active measures, the reception of any desired signal (music or speech) can be essentially improved.

As arbitrarily incoming broadband sound signals are generally nonpredictable, these ear-related systems require feedback control where the signal delay time determines the upper frequency limit. By integrating a loudspeaker, all active ear protectors are converted into ear phones. Also, some error microphone must be provided.

Figure 12.19 shows that the sound insulation of a headset for pilots can be improved by adding an active noise suppression system by up to 35 dB below some 600 Hz.

For larger volumes, things are more complicated because the larger the volume the lower the frequency where spatial dependence of the sound field has to be taken into account. Then, any active field compensation necessarily requires the reproduction of this spatial dependence in magnitude and phase. This generally requires several loudspeakers and their number must be the larger the higher the upper frequency limit has been defined.

Attempts to control the interior noise of cars by active means have been made since 1980 and led 10 years later to first successful demonstrations (see e.g., [1, 99]). It subsequently turned out that for passenger cars an equipment of 4–6 loudspeakers is appropriate and manageable and that this enables global reductions and/or changes of the sound field below some 300 Hz which can be registered at all seats and which are widely independent of head positions and head movements.

For engine-related sounds having a harmonic spectrum of multiples of half engine orders, the predictable sinusoidal waveform of the noise components allows good compensation and, in consequence, good sound modification. Figure 12.20 illustrates typical input signals and components.

The first and most important step in designing such a system is to find out whether the loudspeakers and power amplifiers of the standard audio system are capable of providing the required sound field (which is the case for most of today's systems) or if additional secondary sound sources are needed. These loudspeakers are driven by signals which an adaptive signal processing unit determines from a given rpm-signal and error signals (difference between actual and desired signal). The respective algorithm realized in the signal processing unit aims to minimize the error signal or to approximate the target signal, respectively. In most cases, the present sound pressure value (actual

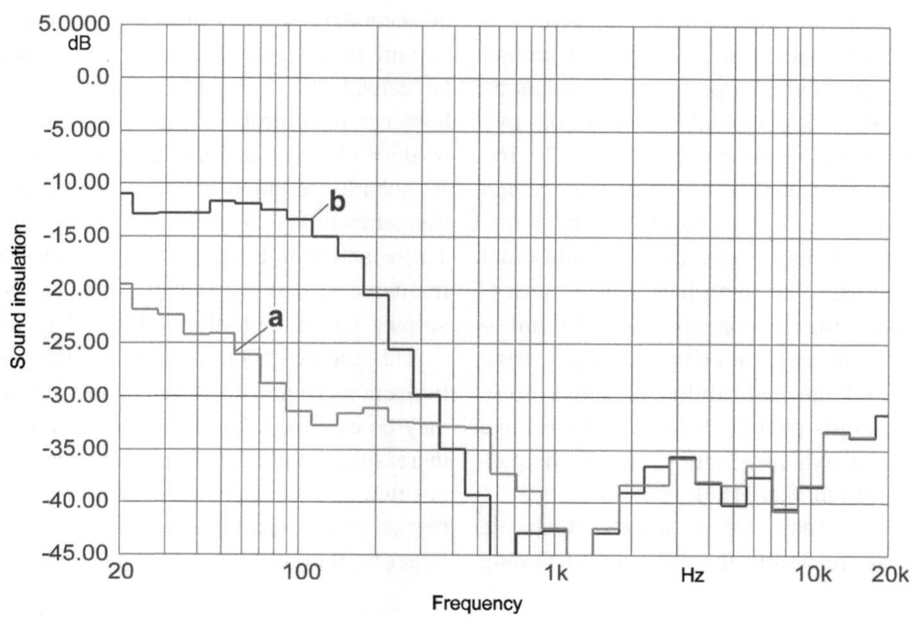

Fig. 12.19 Sound insulation of a pilot's headset (**a**) with and (**b**) without active noise compensation [98]

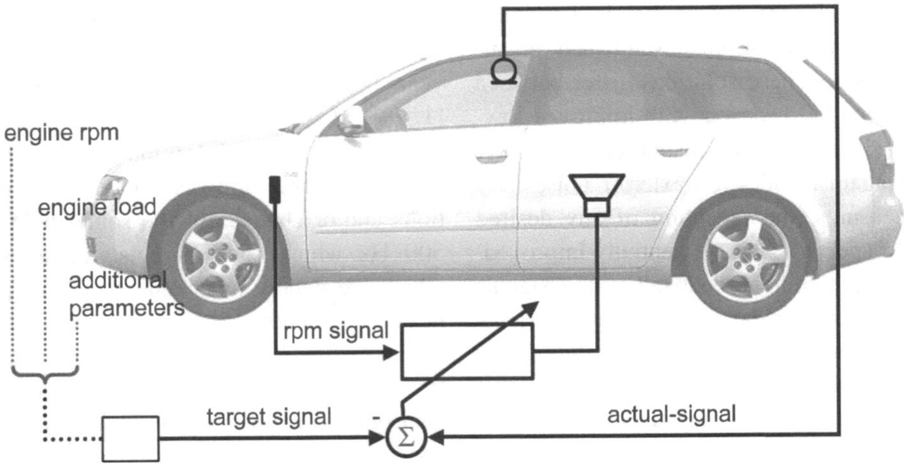

Fig. 12.20 Principal arrangement of a system for active control of car interior sound

signal) is registered via six microphones appropriately distributed over the car's roof area.

In case of a pure noise suppression system, the target signal is zero. By minimizing the mean square error, the signal processing algorithm then equally minimizes the signal itself. By defining differing target signals, it is also possible to realize differing interior sounds in the car [16, 100]. However, to be realistic, such target signals have to be evaluated from important operational parameters like engine rpm or engine load.

An example for the noise reduction which can be achieved by such an arrangement is given in Fig. 12.21. It can be seen that the dominant second engine order of a four cylinder passenger car could be reduced essentially to a comparable level at the two front seats. The reductions in level were up to 20 dB.

By increasing the number of loudspeakers and microphones involved, this approach can be transferred to larger volumes, especially if sinusoidal noise components are to be controlled. This is particularly true for the noise components caused by turbo-propeller engines in aircraft cabins. Starting from proven evidence of the realizability of active noise cancelation in aircrafts [101–103], such systems have been developed for mass production and can be purchased on the world market [104].

Other than for tonal engine sounds stochastic driving sounds are related to the difficulty of finding coherent input signals which provide enough delay time to evaluate suitable secondary signals. Therefore, any broadband level reductions obtained in practical experiments were limited so far to some 5 dB.

Again, higher level reductions may be obtained by limitation of the frequency range to be considered. For low frequency rolling noise with a dominance around some 40 Hz, level reductions of some 10 dB could be obtained with a ready for production system using skillfully a combination of feed-forward and feedback control [65, 105].

As stated above, for global field control the number of secondary sources increases with the size of the volume to be controlled and the highest frequency to be considered. For many problems it is sufficient, however, to control the respective sound and vibration field locally only, in a smaller part of the total volume. As such a limitation in space – for a constant upper frequency limit – needs less secondary sources or – for a constant number of sources – allows higher upper frequencies, it is worthwhile to limit the considered volume according to the sketch of Fig. 12.22.

This can be concretized by an example of how the frequency range for active car interior sound control may be extended. Of course, this can be achieved by increasing the number of loudspeakers and sensors at positions which have to be checked carefully with respect to the specific situation. However, restricting larger head movements of the passengers and allowing

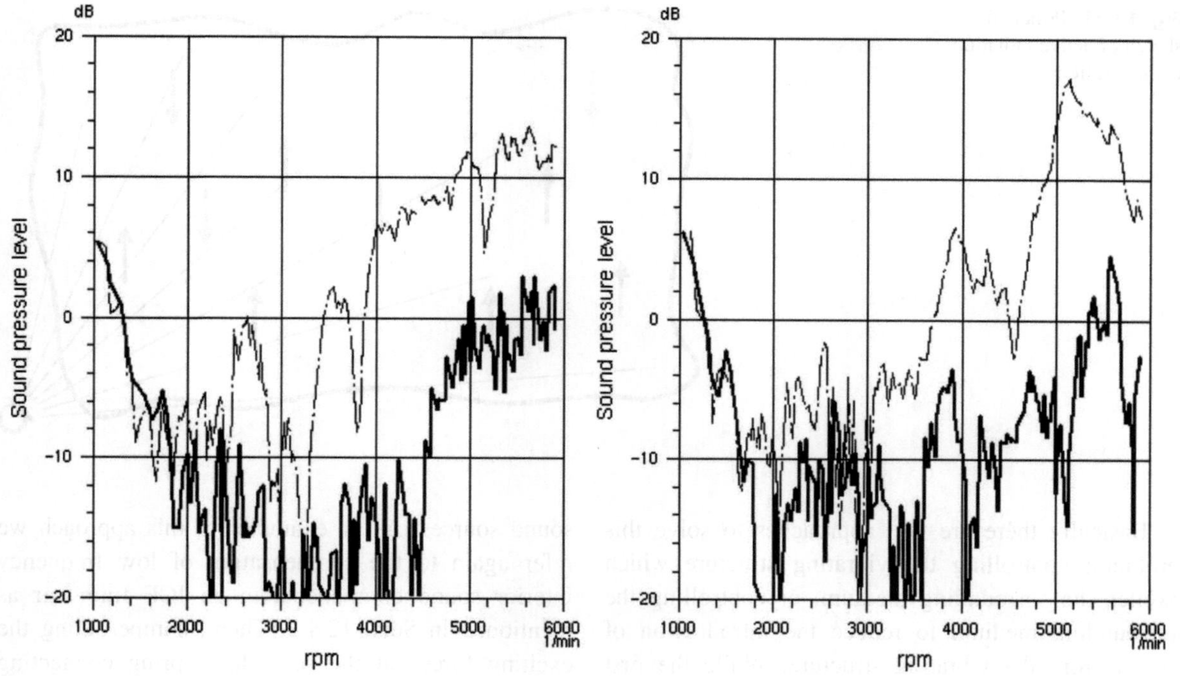

Fig. 12.21 rpm dependence of the sound pressure caused by the second engine order at the driver's (*left*) and co-driver's (*right*) seat without (*upper curve*) and with (*lower curve*) active noise cancelation (ANC)

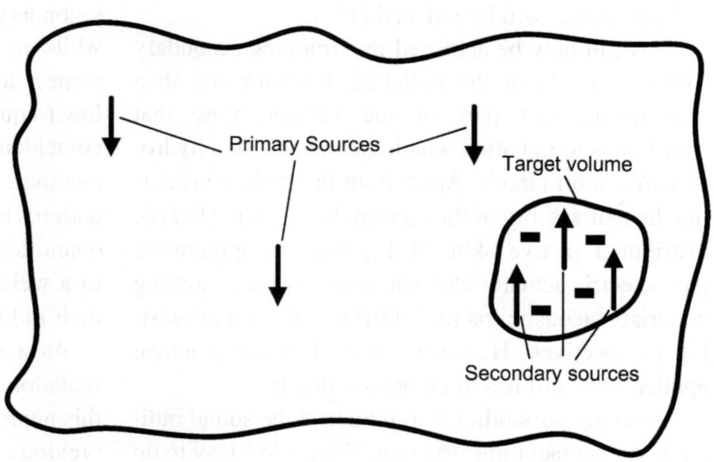

Fig. 12.22 Principal illustration of local control for partial volumes

small movements only, it is sufficient to keep the number of sources constant and to control small volumes only around the heads.

For experimental sound tests in testing vehicles, this is absolutely sufficient and it thus became customary to expand the frequency range for car interior sound control in such special (test) situations by close-to-the-ear positioning of error microphones from some 300 Hz to some 600 Hz.

12.4.5 Active Compensation of Sound Radiation

Many technical noise control problems deal with sound transition through and sound radiation from vibrating structures. In such cases, it is obvious to try to control the radiation itself instead of controlling the radiated sound field. The principal sketch of Fig. 12.23 illustrates this [2, 3].

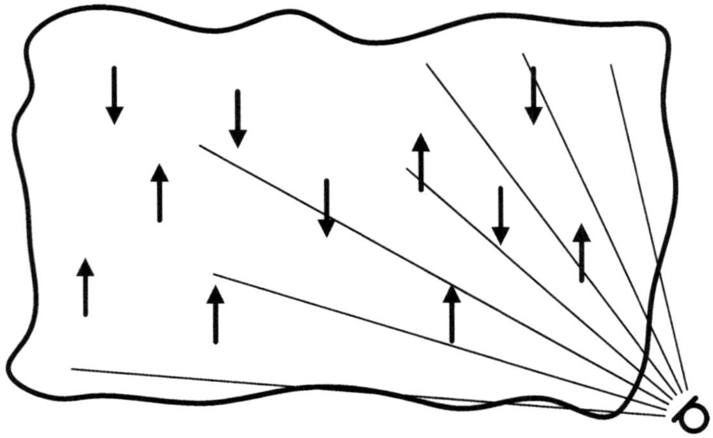

Fig. 12.23 Principal sketch of active radiation compensation

Basically there are two approaches to solve this problem; controlling the vibrating structure which excites the surrounding medium, or controlling the surrounding medium to reduce the introduction of power from the vibrating structure. While the first approach acts on the vibration pattern of the structure, the second approach aims at changing the radiation impedance seen by the structure in a way that the total radiated power will be reduced [106].

This can only be achieved by arranging secondary sources directly at the radiating structure and then reproducing that part of the volume flow that contributes to radiation which then results in a hydrodynamic short circuit. Apart from basically considering the realizability of the concept by discrete [107] or distributed (active skin, [108]) sources, integrating piezoelectric actuator elements into passive absorbing materials (acoustic foams, [106]) might open interesting perspectives. However, their effective practical applicability still has to be proven finally.

The other possibility is to influence the sound radiation via the radiating structure. The easiest way to do this is to reduce the vibrations of the radiating structure in total, e.g., by minimizing the mean square velocity which is proportional to the kinetic energy of the structure. With this definition of solely structure-related target quantities the complete task has been restricted to the structure and therefore must be analyzed systematically in structural terms according to Sects. 12.4.2–12.4.4.

This certainly makes sense if the structural vibrations can be controlled globally by a few sources only, e.g., by reproducing all effective structure-borne sound sources. As an example for this approach we refer again to the compensation of low frequency interior sound components in an ICE train car as mentioned in Sect. 12.4.2. There, compensating the exciting forces at the secondary spring connecting points resulted in significant airborne sound reductions.

In particular, if such simplifying limitations of the secondary source effort are not possible it is worthwhile to differentiate the structural vibrations with respect to their radiation efficiencies. Especially for low frequencies, below the critical frequency defining coincidence with the wavelength of the surrounding medium, the radiation efficiency of various vibration patterns may be very different. Any definition of target quantities directly related to radiation inevitably leads to a weighting of the vibration patterns according to their radiative coupling to the surrounding medium.

As a result, only the modes which contribute to radiation are controlled. It is therefore evident that this approach needs fewer secondary sources than the previous one in order to minimize the mean square velocity of the radiating structure. This applies the more, the less modes contribute to radiation.

Because this active measure fully concentrates on the structural acoustic coupling with the surrounding medium, it usually is called "Active Structural Acoustic Control" or, in short, ASAC.

The difficulty with this approach can be seen in the necessity to register all radiation relevant parameters by appropriate sensors. In some cases, this can be achieved by microphones placed appropriately in the radiated sound field, but often this is not feasible. It is therefore desirable to derive the radiation-relevant parameters, e.g., the complex amplitude of an efficiently radiating mode, from structural measurements only.

As each vibration mode corresponds to a characteristic wave number, this derivation can be reduced to evaluating the wave number spectrum of the radiating vibrational distribution by measurement where the mean value ($k = 0$) of this distribution specifies the total volume flow.

A detailed explanation of these relations may be found in [2] and [3], a recent overview in [106] as well as in the literature given there.

As a representative example for this section we again refer to active control of aircraft cabin interior noise caused by propeller-driven pressure waves acting on the fuselage. Instead of using loudspeakers distributed all over the aircraft cabin, this sound field also can be reduced by force actuators acting on the fuselage.

Figure 12.24 gives the result of such a measure fitted by standard into an aircraft. Here, 42 electromagnetic force actuators (active tuned vibration absorbers, ATVA) tuned to multiples of the blade passage frequency (bpf) act such on the structure that the sound pressure, which is registered by 80 microphones, is minimized at the bpf and its first harmonics [109].

As can be seen from the comparison of Table 12.1 which was taken from some other aircraft, the radiation suppressing system acting on the structure (ASAC) comes up with better results at all four frequencies than a comparable system based on loudspeakers (ANC, [104, 106]).

Table 12.1 Comparison of level reductions (in dB) in a propeller driven aircraft [104]

	1 bpf	2 bpf	3 bpf	4 bpf
ASAC	10.5	7.6	4.4	3.0
ANC	8.0	6.6	3.6	0.4

12.4.6 Stabilization of Self-Excited Systems

Many sound and vibration fields emerge from an unstable interaction of different physical processes mutually affecting themselves. Increase of one field quantity results in the increase of another second field quantity which then causes further increase of the first again until nonlinear restrictions enable a stationary limit cycle in the end.

Typical examples are flutter vibrations caused by flow [110], compressor surge [111] or vibrations resulting from mutual interaction between combustion and air/fuel supply in combustion chambers [112–114].

In the third section it was pointed out that this mutual excitation can be described by two coupled systems (Q and B in Fig. 12.2) in a feedback loop. This arrangement immediately suggests a concept for stabilization: the compensation of unstable feedback by an additional system G (Fig. 12.25).

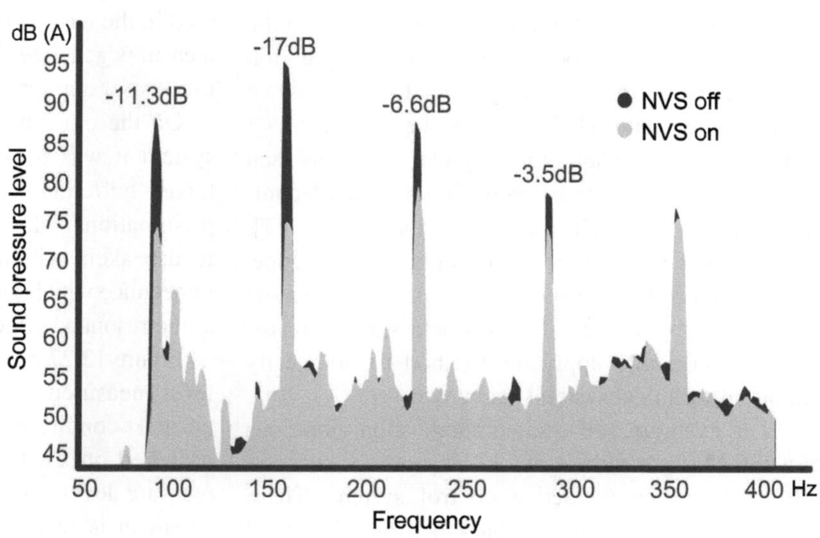

Fig. 12.24 Sound spectra in a turbo-prop driven aircraft without (NVS out) and with (NVS in) active suppression of sound radiation [109]

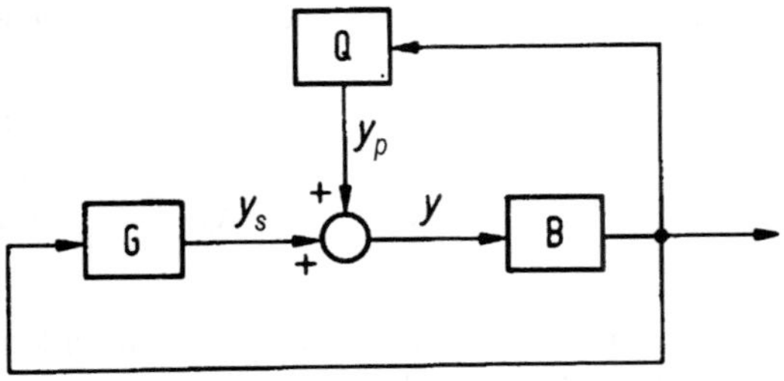

Fig. 12.25 Structural diagram illustrating the compensation of unstable feedback

Unlike the field reproductions considered so far, G does not have to provide a negative copy of the primary field quantity y_P. Because y_P itself depends on y and thus also on y_S, it is sufficient to ensure stability of the total resulting system. Therefore, the transfer characteristics of G can be followed from requirements to the position of zeroes of the characteristic equation

$$B \cdot (Q + G) = 1. \qquad (12.1)$$

In practice this means that amplitude and phase of the secondary field quantity just have to stay within wide stability intervals instead of fulfilling specific values [11, 115]. Also, early interaction with the sound-generating mechanism will avoid large amplitudes. Therefore, like powerful signal processing units, powerful actuators may turn out to be unnecessary. The secondary field quantities remain small but may have huge effects [116].

This simple concept of stabilization may directly be transferred to physical reality. At first this can be followed from the good agreement of computation and experiment for simple systems like the blown Helmholtz resonator [115] or the Rijke-tube [116].

Besides these rather simple systems, the approach described also seems to be promising for other technical problems, especially in flow acoustics [11, 117]. However, to cope with the needs of industrial practice, the requirements to sensors, controllers and actuators are considerably increased. Nevertheless, it was possible to successfully apply the method to practically important technical systems and processes.

As an example, self-excited combustion vibrations of a 170 MW gas turbine at 433 Hz could be successfully reduced by an active control system. These vibrations are caused by feedback between the sound field in the combustion chamber, the supply of fuel and air to the burner and the burning reaction itself. Sound pressure fluctuations cause pulsations in fuel supply which in turn force the flame to fluctuations in the heat release rate. Fluctuating heat release leads to additional pressure fluctuations which then further stimulate the sound field.

Applying the stabilizing method means to monitor the pressure fluctuations in the combustion chamber and to apply them to a controller which then drives a special valve in the fuel supply. This valve modulates the mass flow of fuel such that the pulsations resulting from the cycle of self-excitation are compensated [118].

Figure 12.26 shows the pressure amplitudes measured in the combustion chamber of a gas turbine with and without such active control. It can be seen that the vibrations at 433 Hz in the combustion chamber could be reduced by more than 80%, from some 210 mbar (177 dB) to some 30 mbar (160 dB).

Another example is active control of low-frequency pressure and velocity fluctuations which often are a real problem in open jet wind tunnels. This so-called wind tunnel buffeting is excited by vortex sheds from the nozzle convecting downstream. Reaching the collector, they generate a pressure disturbance which then may excite an acoustic mode of the duct. At resonance, a standing wave occurs inside the duct which then in turn triggers further vortex generation.

As wind tunnel buffeting may seriously distort acoustic and aerodynamic measurements, different measures to diminish this problem have been considered in the past. Unfortunately, many of these passive measures generate disturbing noises and therefore are useless for applications in aeroacoustic wind tunnels.

On the other hand, by applying an active control system it was possible to significantly reduce wind tunnel buffeting without generating additional noise perturbations. The system uses appropriately driven loudspeakers within the tunnel causing a change in impedance which then influences the feedback process at the resonance frequencies of the channel.

Figure 12.27 gives the spectrum of a sound pressure level measured in the test section of a tunnel under normal conditions and with the control system switched on. At the resonance frequency, the sound pressure level interfering with the operational measurement is seen to be reduced by more than 20 dB

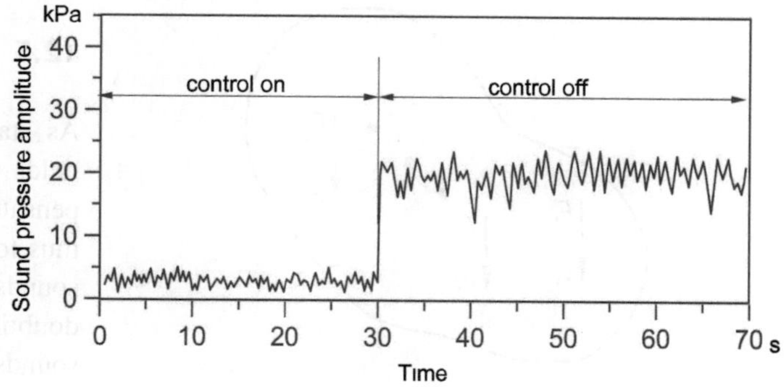

Fig. 12.26 Sound pressure amplitudes in the combustion chamber of a gas turbine (170 MW, some 17 bar statistical combustion chamber pressure) with (0–30 s) und without (30–70 s) stabilizing control [118]

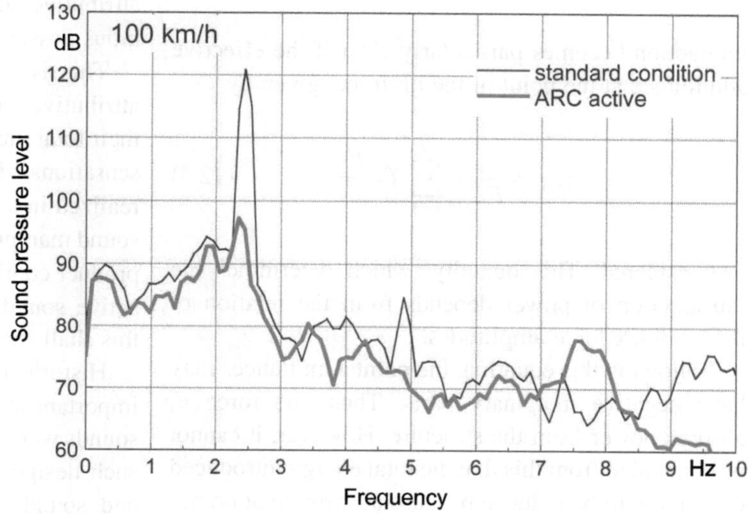

Fig. 12.27 Sound pressure level spectrum in the test section of a wind tunnel at a flow velocity of 100 km/h under normal standard conditions and with active control ARC [119]

when the active system is applied (Active Resonance Control, ARC, [119].

12.4.7 Energy and Power Considerations

In general only little can be said about energy and power relations for active methods of field control. Unlike successful control of sound-generating mechanisms, where the control of field quantities correlates with control of power radiation, the energy requirements related to interference of reproduced fields are not evident. This is because linear superposition of coherent field quantities does not cause additive superposition of the related quadratic power quantities.

This may be shortly illustrated in the following for the example sketched in Fig. 12.28, where N point forces F_i, $1 \leq i \leq N$, act on an arbitrary structure.

The total power introduced to the structure by the forces is given by

$$P = \sum_{i=1}^{N} \frac{|F_i|^2}{2} \left[\mathrm{Re}\{\Upsilon_{ii}\} + 2 \sum_{k=1}^{i-1} \mathrm{Re}\left\{\frac{F_k}{F_i}\right\} \mathrm{Re}\{\Upsilon_{ik}\} \right], \tag{12.2}$$

whereby v_i is the velocity at the point of action of the i-th force F_i and

$$\Upsilon_{ik} = \frac{v_i}{F_k}, \tag{12.3}$$

generally describes the transfer admittance from the excitation point k to the receiving point i. The point admittances are included in this definition as special cases $i = k$.

It is obvious from the second term of Eq. (12.2) that the power introduced does not only depend on the amplitudes $|F_i|$ of the forces and the admittances Υ_{ik} of the structure but also on the differences in amplitude and phase between the force amplitudes. This

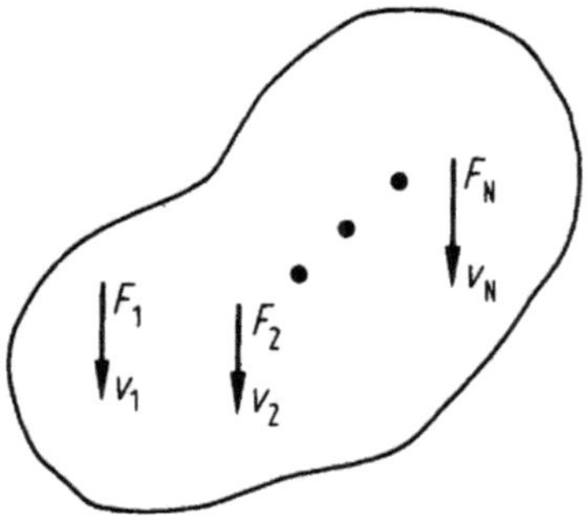

Fig. 12.28 Application of several point forces to a structure

interaction becomes particularly clear if the effective admittance at the point of the ith force, given by

$$\Upsilon_{i,\text{eff}} = \frac{v_i}{F_i} = \sum_{k=1}^{N} \Upsilon_{ik} \frac{F_k}{F_i}, \tag{12.4}$$

is considered. This quantity, which determines the introduction of power depends from the relation of the complex force amplitudes.

Following this equation, the point admittances may have negative imaginary parts. Then, the force F_i absorbs power from the structure. However, it cannot be concluded from this that the total energy introduced or stored will be reduced because absorption at point i may be coupled with increased power injection at other points.

These considerations illustrate that the effects of secondary sources on energy relations can be explained by two mechanisms: by control of the radiation impedance seen by the primary source and by injection or absorption of power.

For this reason, any assessment of an active measure based on local power balances must consider the totality of all sources involved or – similar to the active absorber of Sect. 12.4.3 – be sure that the given arrangement of secondary sources is free of any feedback. Any exclusive consideration of the power absorbed by anti-sources may generally not be sufficient. Illustrative examples confirming this generally may be found in [8, 54, 60, 71, 120].

Active absorption, i.e., power extraction by electroacoustic and electromechanic actuators, is not only a theoretical possibility but has also been proven by practical measurements. For an appropriately driven loudspeaker, such proving evidence was described in [121].

12.5 Active Sound Design

As stated in the introduction, the compensation of fields also gives the possibility to replace the compensated sounds by other sound impressions and thus to change freely – within wide limits – existing sounds. Even if – apart from the related cost – doubting the acceptance of "artificial" electroacoustic sounds may argue against bringing it on a product level: as a tool, useful to preliminary realize and test attributive sounds, active sound control offers enormous possibilities.

This is because any reliable assessment of such attributive sounds requires their reception within their total multimodal context, where other perceptive sensations add to the related sound. As this cannot be realized in a studio, the importance of electroacoustic sound manipulations which change given sounds at the product continuously grows. And this is exactly what active sound design (ASD) can do. In the following this shall shortly be illustrated for car interior sounds.

Historically, the sound of a car is one of the most important attributes for the user. Consequently, such sounds were designed more and more consciously. But such design needs a clear specification of the noises and sounds to be targeted. For success, known or somehow given sounds may be manipulated, new sounds may be defined, both then may be analyzed and assessed psychoacoustically in hearing experiments.

However, because hearing sensation and sound impressions are strongly influenced by the driving experience, such new car sounds can finally not be assessed in the lab. Therefore any final assessment must be based on comparable in-situ tests in driving cars. The availability of related prototypes with various, thoroughly tuned sound variations generally demands very high efforts.

Here, active design and realization of rpm-related car interior sounds provides a development tool which varies the sounds of engine-related components without modifying the components themselves. Thus enabling subjectively justified tests and specifications, differing sounds can be assessed and further developed in the context of a full driving experience without building prototypes [122].

As an example, Fig. 12.29 shows for the car, previously mentioned in Fig. 12.21, how the sound

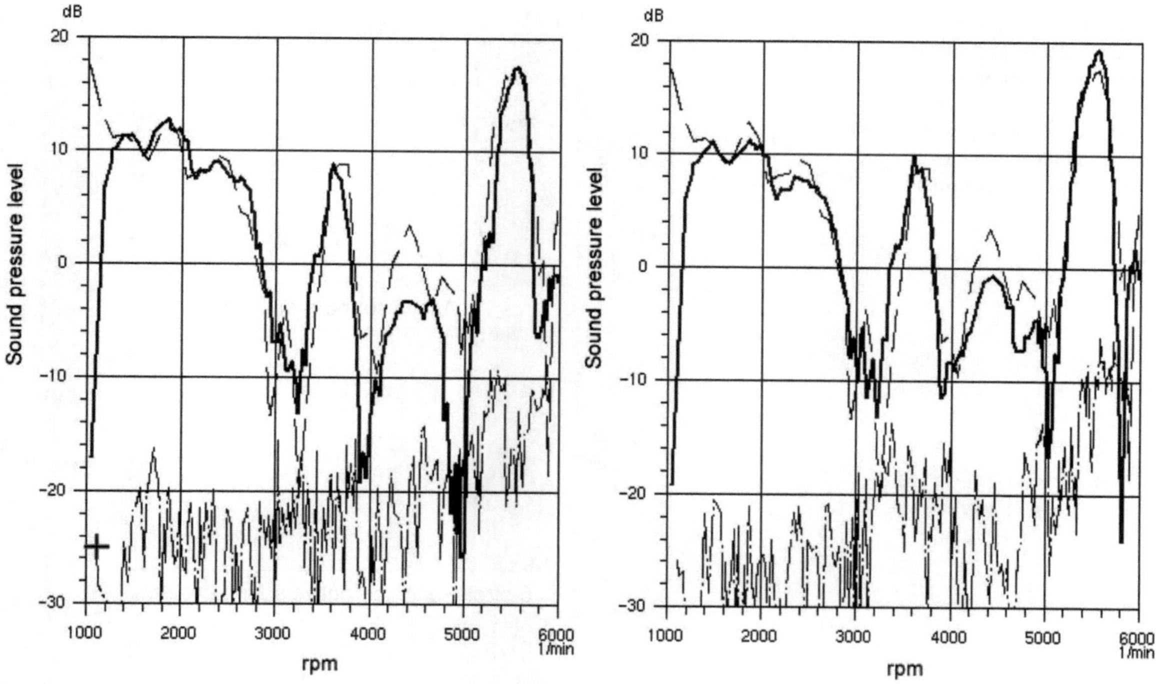

Fig. 12.29 rpm dependence of the sound pressure related to the third engine order at the driver's (left) and co-driver's (right) seat without (*lower dash-dotted line*) and with (*upper continuous line*) active sound design (ASD). The dashed upper line describes the target to be met by ASD

character of this four cylinder engine can be modified [100]. It can be seen that the third engine order which typically can be neglected for four cylinder engines (dash-dotted line) well approximates a given rpm curve (dashed line) by using ASD. Spectra dominated by the third engine order are typical for six cylinder engines. Thus, by comparing Fig. 12.29 with Fig. 12.21 where the suppression of the second order is demonstrated, one can conclude that the sound impression of a six cylinder engine can be generated by a four cylinder engine properly controlled.

This can also be demonstrated for the order analysis of another car equipped with ASD, Fig. 12.30. Here it can be seen again how the reduction of the second engine order in the middle figure enhances the third, sixth, and ninth engine order which are typical for six cylinder engines (see the lower figure).

Exterior noise of cars also can be modified by active mans. For exhaust mufflers this had been shown already in Fig. 12.17. Further realizability of comparable results for engine air intake systems is demonstrated in [86].

Besides evidence by measurements the authenticity of actively realized engine sounds in cars is proven by many subjective assessments. Even if sound variations in practice work with very fine gradations only; together with the authenticity, the examples given clearly demonstrate the possibilities of active sound design when specifying and developing product sounds.

12.6 Aspects of Signal Processing

Realizing effective control of mechanical vibration and wave fields requires not only appropriate selection and arrangement of the actuators but also their proper driving.

Unlike directly influencing the sound-generating mechanism as described in Sect. 12.4.6, all signals provided by the signal processing unit G must exactly meet their desired values in amplitude and phase to obtain considerable level reductions by interference. This can be illustrated easily by considering the superposition of two harmonic vibrations with small deviations α in amplitude and δ in phase. The level reduction with respect to one of these vibrations alone is given by

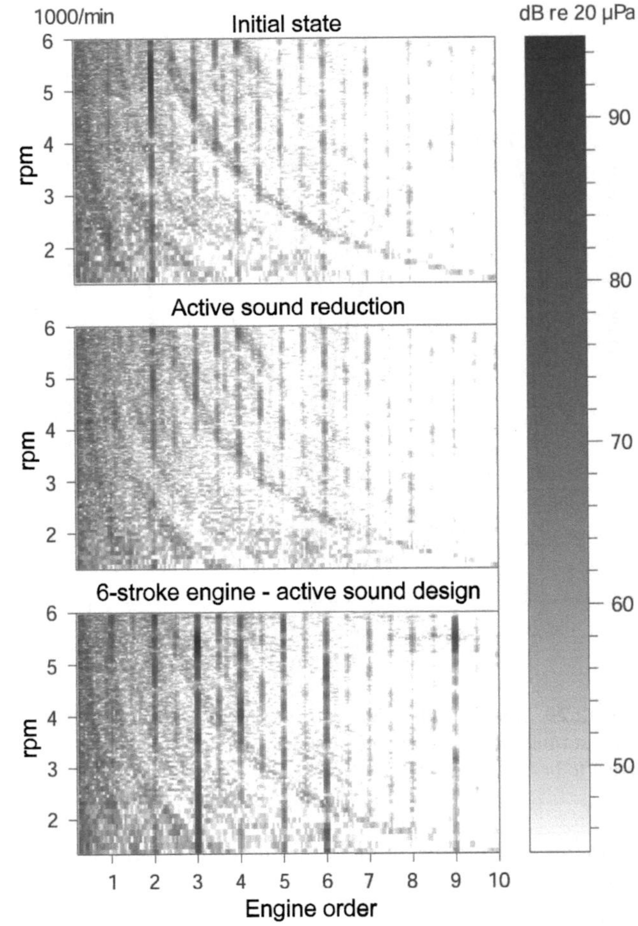

Fig. 12.30 rpm dependence of engine order related sound pressures (order analysis) at the driver's seat of a car at the initial state (*above*), with active sound reduction (ANC, *middle*) and with active sound design (ASD, *below*)

$$\Delta L = -10\ \lg\left[\alpha^2 + 4(1+\alpha)\sin^2\frac{\delta}{2}\right]\text{dB}. \qquad (12.5)$$

From this equation, it can be seen that an error in amplitude of 10% ($\alpha = 0.1$) and in phase of $\delta = 10°$ reduces the level reduction ΔL to 13.6 dB only. For $\delta = 20°$ and constant $\alpha = 0.1$ only 8.5 dB remain.

Together with the fact that large errors in phase even may increase the resulting level this shows that realizing good approximations of the primary field quantities is an important issue. The precision of this approximation is limited by what information on the temporal behavior of the primary fields can be taken from the input signals y_R and z_R.

To quantitatively describe this information the stochastic frequency domain relation between these input signals and the primary field quantities y_P can be taken into account. This relation is given by the coherence function γ^2 and allows to specify an upper limit for the level reduction which, independent of the signal treatment, cannot be exceeded.

If S_g describes the power spectral density of the field resulting from the superposition of primary and secondary field and S_p the power spectral density of the primary field alone, the minimally obtainable lower bound for the ratio of S_g and S_p is given by [4, 123]

$$\frac{S_g}{S_p} = \left(1 - \gamma^2\right). \qquad (12.6)$$

For a coherence of 90% ($\gamma^2 = 0.9$), the maximal level reduction amounts to 10 dB, for 99% coherence to 20 dB therefore, considerable reductions of primary fields require a strong correlation between the primary and the compensating fields. Hence, after a careful specification of the source arrangements, is the selection of such signals a second very important issue in the designing of active measures or in assessing their performance.

Together with the transducers, the original mechanical part of the system and the signal processing components (like low pass filters, measurement, and power amplifiers), the source and measurement arrangements define a global system M whose input and output signals are connected to the signal processing unit G according to Fig. 12.31.

The remaining task in specifying such G is to assure that all input information available is exploited to drive the selected actuators optimally with respect to the target quantities.

This problem has to be solved in two steps. First, the structure of G must be specified in terms of a sufficient number of suitably linked parameters. This structure describes the relations between input and output quantities in general terms, e.g., by a specific form of equations or a structural diagram. Only then any concrete realization of the optimal transfer behavior between input and output quantities can get started by specifying all related parameter values.

Any definition of the structure of G should take advantage from given knowledge about the real structure of M. For example, this may essentially facilitate the approximation of any relation between measured quantities and the primary field quantities or the consideration of internal feedback. Bad structural adjustment between M and G reduces the quality of signal adjustment ultimately obtainable by optimal parameter selection. An illustrative example for this difficulty is found in the difficulty to approximate recursive (IIR) structures by non-recursive (FIR) filters and vice versa [124–126].

An important application of structural adjustment of G to M is given by the compensation of closed (via M) feedback loops which cause actions of the outputs u on the inputs x of G. Such compensation can be achieved by additional feedback loops realized within G [13].

With this measure, the stability problems caused by feedback and mentioned in Sect. 12.2 can be reduced. Therefore, the compensation of feedback (feedback cancelation) is widely used for tests and for applications of active methods [4, 9, 60, 78]. Also, it is shown in [60] how feedforward and feedback loops can be identified in the frequency domain from different input/output measurements while, in [78] a possibility of adaptive identification of both loops is presented.

There are many ways of defining the structure and the parameters when designing the signal processing unit. Unfortunately, it is not possible to deal thoroughly with them here, especially because there is a general lack of systematic procedures. Instead, only some important terms and principles shall be mentioned. Further considerations of system design [4, 48, 53, 124–126] and of concrete applications [1–5, 25] are found in the literature.

The global properties of M, like the transfer behavior, observability and/or controllability [48, 50, 53], not only depend on the total mechanical and acoustical arrangement but also on the number and position of the actuators. If, in addition, the coherence requirements Eq. (12.6) together with the fact, that only causal relations between x and u can be realized, are taken into account, the importance of selecting appropriate locations for measurements and actuations is obvious.

Optimal transfer properties of G may be obtained by minimizing certain field quantities put together in the target quantities z. Here, quadratic criteria are to be preferred because they combine good theoretical insight with easy implementation. Above all, optimal linear filter theory [127] can be applied to support and assess the design of G.

Then, continuous monitoring of the remaining error signal allows an equally continuous adaptation of the transfer behavior aiming at further error reduction.

This adaptive approach, schematically given in Fig. 12.32, ends up with optimal transfer characteristics and the related transfer elements therefore are called adaptive filters. Their ability to automatically

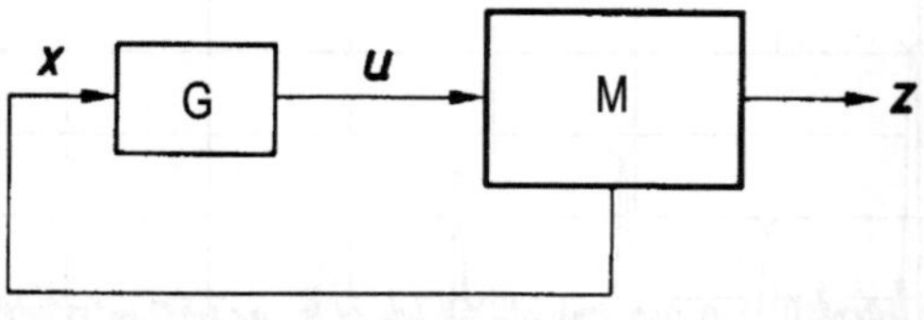

Fig. 12.31 Principal sketch of the total system and its respective separation into an electromechanical system M and a signal processing unit G

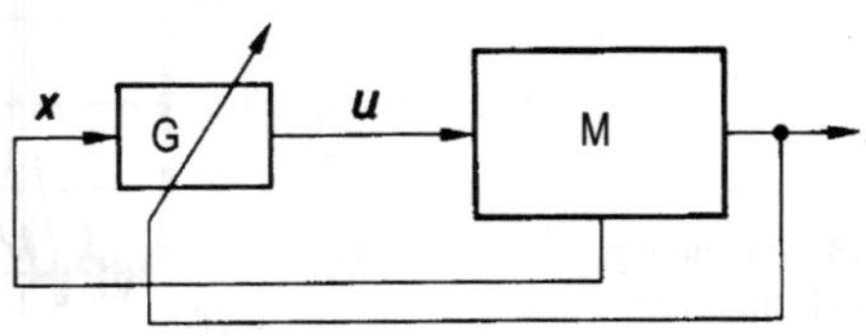

Fig. 12.32 Principal arrangement with an adaptive signal processing unit

compensate any fluctuations of important system parameters like temperature, flow velocity, or rpm has given them a dominant role in applying active methods.

The theory of adaptive signal processing is highly advanced and may be used for designing active systems. Comprehensive treatments are found in [128–131], typical examples for concrete realizations of adaptive algorithms in [4, 5, 78]. Early difficulties related to adaptive implementations of IIR-filters could be overcome by well-converging algorithms with real time capabilities [78, 132].

Of course, the particular choice and efficiency of such computational rules strongly depend on the statistical signal properties. Many sound and vibration sources, e.g., all rotational machinery, generate periodic time histories whose future behavior can be well predicted from their past history. Based on given rpm information this allows the development of efficient, well-converging algorithms which, following a computational proposal first formulated in [128], can be implemented in compact form.

Many of the intuitively derived algorithms for a single input and one output quantity as well as their multidimensional extensions can be considered as generalizations of the LMS-algorithm with filtered input signal (filtered-x-LMS-algorithm). This enables global convergence properties with even many input and output quantities [4, 5, 129].

The predictability of revolutionary signals enables adaptive algorithms to come up with high-level reductions. This can be demonstrated by a laboratory experiment of active vibration isolation. As shown in Fig. 12.5, four electrodynamic vibration sources (shakers) had been arranged in parallel to four passive springs at the four coupling points between an electric motor and its foundation. These shakers were driven such that the forces introduced into the foundation at the coupling points and measured as error signals were minimized.

Figure 12.33 shows the result without and with application of secondary forces [133]. As can be seen, adaptive control is able to fully compensate the force caused by some unbalance to the level of broadband noise. The related force level reduction at the basic frequency is 81.6 dB. Also, all further harmonics of the line spectrum generated by the electric motor are reduced accordingly while the amplitude at some 100 Hz, uncorrelated to the line spectrum, remains unchanged.

Despite all advantages offered by adaptive control schemes, the benefits of using or combining simple basic signal processing elements to more complex control structures have retained continuous interest in feedback control loops. Examples how this approach may be used to suppress structural vibration and radiation can be found in [134–136].

12.7 Electromechanical Transducers as Actuators

Before the 1980s, any limitation in realizing active measures was mainly caused by limitations of electronic signal processing. After the immense development of digital signal processors in the last 20–30 years this does hardly apply today. Instead, present limitations are rather caused by restrictions of electro mechanic transducers.

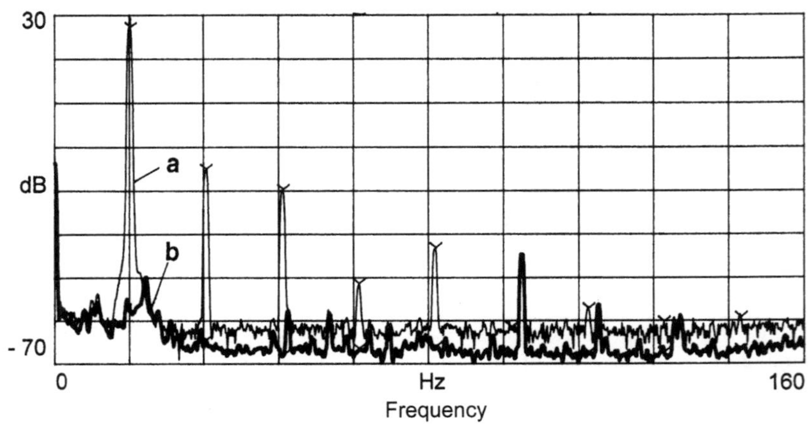

Fig. 12.33 Frequency spectrum of the force introduced into the foundation (**a**) without and (**b**) with active isolation

Such transducers typically are characterized by low efficiencies. This is particularly true for low frequencies as the frequency transfer characteristics of closed loudspeaker boxes fall off with 6 dB/octave below the mechanical resonance frequency. The resulting intention to tune the system to lower frequencies directly leads to larger volumes because otherwise the resonance frequency will be determined by the stiffness of the air enclosed.

If the application provides limited space for transducers only, this tends to be the decisive restriction. For this reason, providing powerful transducers is an essential condition for increasing applications of active measures.

Also, the actuating devices used to directly control structure-borne sound may refer to proven principles. Thus, any application of forces in practice often is realized by electrodynamic or electromagnetic vibration exciters.

Besides that, other transducer principles, not being considered for airborne sound excitation because of insufficient deflections, may be very efficient for structural applications. Both the piezoelectric as well as the magnetostrictive effect allow for applying very strong forces [137–139]. The small deflections related to these effects, however, limit their application to cases where the force application points are subject only to small travel distances.

In special cases this limitation may be overcome, Fig. 12.6 illustrates this for the case of active compensation of force introduction at an ICE bogie. This force being supplied by sliced piezo elements and being arranged together with their seismic masses within the secondary springs has been freely applied to the car body. Despite this combination of high force potential and compact arrangement, problems with linearity finally led to the insight that preference would have been given to electrodynamic actuators in case of practical applications.

For surface structures, active control is not tied to the application of point forces. In particular, with rods and plates it is possible to integrate suitable actuators into or onto these. This application of distributed actuators, e.g., by coating with piezoelectric foils, is not limited to flexural and longitudinal excitation [2, 3, 6, 106]. By using corresponding shaping, it can further be achieved that registration and excitation are confined to specific modes.

The integration of electrically shapable elements into structures gives rise to multiple perspectives and hopes as put into words like "intelligent" or "adaptive" structures. Despite some evidence of realizability mainly based on controlling single modes and waves so far, the practical potential of surface distributed or structure integrated actuators cannot be predicted reliably yet. If such approaches would mature to an applicable technology in the future, they could serve as a promising basis for active control of sound radiation and transmission from or through structures.

12.8 Further Applications

Without claiming completeness, this overview shall give some completing examples for the applicability of active methods being disregarded before.

It was in the early days of active sound control that people tried to take advantage of the predictability of sound fields with single discrete frequencies to effectively cancel them. An ideal technical realization to test the new method was seen in those days in electric transformers with their discrete harmonic frequency spectrum [34].

However, the dimensions of transformers cause rather complicated, moreover load-dependent spatial radiation patterns which therefore require spatially variable secondary fields. Experimental successes therefore were confined to distinct volumes [140, 141], particularly spatial angles [142] or simplified reference radiators [143].

For this reason, any global compensation by secondary sources of the sound field radiated from transformers has been judged unsatisfactory on account of complexity and the obtainable level reductions. Nevertheless, [106] reports on a system which successfully compensates the basic frequency by a loudspeaker and higher harmonics by piezoceramic actuators fixed to the casing. Also, active measures at transformers may complement passive measures, e.g., by supplying indispensable openings of shielding elements with active sound attenuators.

For open windows of buildings, comparable solutions could not be found. Although quite a few corresponding investigations had been made in the first years of active sound control research, they finally failed because of the difficulty of reproducing three dimensional incoming wave fields with acceptable efforts.

Things are more handsome if the sound insulation between two rooms shall be improved by acting on the separating structure. Corresponding investigations for (closed) windows are found in [144, 145].

In vibration control, all approaches requiring electronic control are classified as active methods anyway: magnetic bearings and lightweight robot devices. Active magnetic bearings use appropriately controlled magnetic forces for position control and, because of reduced friction losses, are mainly applied in rotating systems. Their advantages as well as further applications and problems are described in [146] and in the references given there.

Progress with robot devices had to cope with increased moving speed of more and more lightweight arms. Therefore, their control could not only be restricted to their positioning. Instead, the dynamics of the structure to be moved had to be taken into account or even to be compensated by considering them in the signals controlling the robot movements.

In the end of this section reference shall be made to applications where, instead of the sound and vibration field quantities themselves, the dynamic properties of the elements involved (e.g., the stiffness of a spring) are controlled. Typical examples of this approach, which often is referred to as semi active, are shock absorbers and actively supported dynamic vibration absorbers [6].

12.9 Summary and Perspectives

This chapter attempts to present the state of the art of active noise and vibration control methods. Special emphasis has been given to essential physical mechanisms because any use of highest development standards in algorithms and hardware even is limited by the fundamental limits of physics. Furthermore, thorough knowledge of any physical possibilities is the best platform for systematic exploration of the most promising concepts towards satisfactory solutions.

Besides emphasizing these basics, the many successful application exemplify the surpassing of provisional laboratory experiments and tests.

Like any other technical method, active control of sound and vibrations offers a general approach capable of solving many, yet not all, problems of noise and vibration. This may have been overseen by some early, mostly enthusiastic promises stirring unrealizable hopes.

However, the results as well as partial results so far justify further confidence. If further progress is made with materials and concepts to generate sound and vibrations, with electronic units to process signals and with clear and robust control concepts for multiple inputs and outputs, it should be possible to improve development and handling and to reduce effort and cost of active systems to an acceptable level.

Against this background, some confidence and optimism may be justified that the fascinating approach of active sound and vibration control may further flourish in the future by a growing number of fitting applications.

References

1. Nelson PA (1992) Active control of sound. Academic, London
2. Fuller CR, Elliott SJ, Nelson PA (1996) Active control of vibration. Academic, London
3. Hansen CH, Snyder SD (1997) Active control of noise and vibration. E&FN Sponn, London
4. Elliott SJ (2001) Signal processing for active control. Academic, London
5. Kuo SM, Morgan DR (1996) Active noise control systems – algorithms and DSP implementations. Wiley, New York
6. Preumont A (2002) Vibration control of active structures – an introduction. Kluwer Academic Publishers, Dordrecht
7. Swinbanks MA (1984) The active control of sound and vibration and some applications in industry. Proc Instn Mech Eng 198A(13):281–288
8. Ffowcs Williams JE (1984) Anti-sound (review lecture). In: proceedings of the Royal Society London, vol A395, 63–88
9. Swinbanks MA (1985) Active noise and vibration control. Fortschritte der Akustik – DAGA Stuttgart, DPG-GmbH, Bad Honnef, pp 87–101
10. Guicking D (1987) Zur Entwicklung der aktiven Schwingungs- und Lärmbekämpfung. Fortschritte der Akustik - DAGA, Aachen, DPG-GmbH, Bad Honnef. 501–504
11. Ffowcs Williams JE (1988) Active control of "noisy systems". Proc Int Noise Avignon 1:5–20
12. Guicking D (1989) Aktiver Lärmschutz – Erfolge. Probleme und Perspektiven. Fortschritte der Akustik – DAGA Duisburg, DPG-GmbH, Bad Honnef, pp 23–36
13. Scheuren J (1990) Möglichkeiten der aktiven Lärmminderung und Schwingungsabwehr – eine Übersicht. Chap. 2 in "Aktive Beeinflussung der Ausbreitung von Biegewellen", Dissertation, TU Berlin

14. Scheuren J (1995) Aktive Lärm- und Schwingungsminderung. Overview Lecture, Fortschritte der Akustik – DAGA Saarbrücken, Deutsche Gesellschaft für Akustik, Oldenburg, 105–120
15. Scheuren J (1998) Active Control of Structural Vibrations: General Considerations and New Results. Overview Lecture, Proceedings Inter-Noise, Christchurch, New Zealand
16. Schirmacher R (2002) Von der aktiven Geräuschminderung zum Active Sound Design. Overview Lecture, Fortschritte der Akustik – DAGA Bochum, Deutsche Gesellschaft für Akustik, Oldenburg, 8–15
17. (1991) Active Control of Sound and Vibration. In: Proceedings of the International Symposium of the Acoustical Society of Japan, Tokio
18. Rogers CA, Fuller CR (ed) (1991) First Conference on Recent Advances in Active Control of Sound and Vibration, Blacksburg, VA, USA, Conference Proceedings, Technomic Publishing Company, Inc., 1991, Lancaster, Pennsylvania, USA
19. Burdisso RA (ed) (1993) Second Conference on Recent Advances in Active Control of Sound and Vibration, Blacksburg, VA, USA,. Conference Proceedings, Technomic Publishing Company, Inc., 1993, Lancaster, Pennsylvania, USA
20. Sommerfeldt S, Hamada H (ed) (1995) Active 95. In: proceedings of the International Symposium on Active Control of Sound and Vibration, New Port Beach, California, USA
21. Elliott S, Horvath G (ed) (1997) Active 97. In: Proceedings of the International Symposium on Active Control of Sound and Vibration, Budapest, Ungarn
22. International Workshop on Active Noise and Vibration Control (1999) im Rahmen des Forum Acusticum 1999 in Berlin. Kurzfassungen in Acustica/Acta Acustica, Vol 85, Supplement 1, Januar 1999 Vorträge auf CD: Collected Papers from the Joint Meeting "Berlin 99"
23. Douglas S (ed) (1999) Active 99. In: Proceedings of the International Symposium on Active Control of Sound and Vibration, Fort Lauderdale, Florida, USA
24. Gardonio P, Rafaely B (ed) (2002) Active 2002. In: Proceedings of the International Symposium on Active Control of Sound and Vibration, Institute of Sound and Vibration Research, University of Southampton, UK
25. Guicking D (2003) Guicking's Online Reference Bibliography on Active Noise and Vibration Control. GORBI, Version 1.1, 4th edition, Januar, CD ROM (Email: mail@guicking.de)
26. Guicking D (2001) Guicking's Online Patent Information on Active Noise and Vibration Control. GOPI, Version 1.1, Februar, CD ROM (Email: mail@guicking.de)
27. Bschorr OF (1971) Lärmauslöschung durch gesteuerte Interferenz. Proc.7.Int.Congr.Acoust., Budapest, 381–384
28. Lueg P (1933) Verfahren zur Dämpfung von Schallschwingungen. Deutsches Reichspatent Nr. 655508
29. Lueg P (1934) Process of silencing sound oscillations. U.S. Patent Nr. 2043416
30. Guicking D (1976/77) Paul Lueg – der Erfinder der aktiven Lärmbekämpfung. Acustica 36:287–293
31. Olson HF, May EG (1953) Electronic sound absorber. J Acoust Soc Am 25(6):1130–1136
32. Olson HF (1956) Electronic Control of Noise,Vibration, and reverberation. J Acoust Soc Am 28(5):966–972
33. Leventhall HG et al (1983) The tight-coupled monopole active attenuator in a duct. Noise Control Eng J 19(1): 16–20
34. Conover WB (1956) Fighting noise with noise. Noise Control 92:78–92
35. Swinbanks MA (1973) The active control of sound propagation in long ducts. J Sound Vibr 27(3):411–436
36. Jessel MJM, Mangiante GA (1972) Active sound absorbers in an air duct. J Sound Vibr 23(3):383–390
37. Canevet G, Mangiante G (1974) Absorption acoustique et anti-bruit à une dimension. Acustica 30:40–48
38. Poole JHB, Leventhall HG (1976) An experimental study of Swinbanks' method of active attenuation of sound in ducts. J Sound Vibr 49(2):257–266
39. Poole JHB, Leventhall HG (1978) Active attenuation of noise in ducts. J Sound Vibr 57(2):308–309
40. Rockwell TH, Lawther JM (1964) Theoretical and experimental results on active vibration dampers. J Acoust Soc Am 36(8):1507–1515
41. Knyazev AS, Tartakovskii BD (1965) Application of electromechanical feedback for the damping of flexural vibrations in rods. Sov Phys Acoust 11(2):150–154
42. Porter B, Bradshaw A (1974) Synthesis of active dynamic controllers for vibratory systems. J Mech Eng Sci 16(2): 95–100
43. Balas MJ (1982) Trends in large space structure control theory: fondest hopes, wildest dreams. IEEE Trans Automatic Control 27(3):522–535
44. Leipholz H (ed) (1979) Structural control. Proc.Int.IUTAM Symposium, Waterloo, Ontario
45. Knyazev AS, Tartakovskii BD (1967) Abatement of radiation from flexurally vibrating plates by means of active local vibration dampers. Soviet Phys Acoust 13(1): 115–117
46. Fuller CR (1990) Active control of sound transmission/ radiation from elastic plates by vibration inputs. J Sound Vibr 136(1):1–15
47. Ffowcs Williams JE (1987) Active control of unsteady flow. Proc. Inter-Noise, Peking, 7–12
48. Föllinger O (1994) Regelungstechnik. 8. Aufl., Hüthig, Heidelberg
49. Franke D (1987) Systeme mit örtlich verteilten Parametern. Springer, Berlin
50. Aström KJ, Wittenmark BJ (1990) Computer controlled systems, theory and design, 2nd edn. Prentice-Hall, Englewood Cliffs
51. Robinson (1971) A survey of optimal control of distributed parameter systems. Automatica 7:371–388
52. Kappel F, Kunisch K, Schappacher W (Ed.) (1983) Control theory for distributed parameter systems and applications. Lecture notes in control and information sciences, vol 54, Springer, Berlin
53. Isermann R (1987/1988) Digitale Regelsyssteme I und II. 2.Auflage. Springer, Berlin
54. Scheuren J (1990) Aktive Beeinflussung der Wellenausbreitung I: Theoretische Überlegungen zur aktiven Beeinflussung der Ausbreitung von Luft- und Körperschall. Acustica 71:243–256

55. Atluri SN, Amos AK (eds) (1988) Large space structures: dynamics and control. Springer, Berlin
56. Redman-White W, Nelson PA, Curtis ARD (1987) Experiments on the active control of flexural wave power flow. J Sound Vibr 112(1):187–191
57. Guicking D, Melcher J, Wimmel R (1989) Active impedance control in mechanical structures. Acustica 69:39–52
58. Krüger J, Leistner P (1998) Wirksamkeit und Stabilität eines neuartigen Schalldämpfers. Acustica 84:658–667
59. Scheuren J (1990) Active attenuation of bending waves in beams. Proceedings of the Institute of Acoustics, vol 12, Part 1, 623–629
60. Scheuren J (1990) Aktive Beeinflussung der Ausbreitung von Biegewellen. Dissertation, TU Berlin
61. Ffowcs Williams JE (1981) The silent noise of a gas turbine. Spectrum, British Science News, Nr.175/1, 9–11
62. Swinbanks MA (1982) The active control of low frequency sound in a gas turbine compressor installation. Proc. Inter-Noise, San Francisco, 423–426
63. Schubert DW (1991) Characteristics of an active vibration isolation system using absolute velocity feedback and force actuation Proceedings Active 91, 448–463
64. Heiland P (1994) Aktive Schwingungsisolierung. Konstruktion, Band 46. Heft 10:353–358
65. Sano H, Yamashita T, Nakamura M (2002) Recent applications of active noise and vibration control to automobiles. Proceedings Active 2002, 29–42
66. VDI-Richtlinie "Aktive Schwingungsisolierung". In Vorbereitung
67. Schirmacher R et al. (1997) Active Noise and Vibration Control for a High Speed Railcar: A Case Study. Proceedings of Active 97 in Budapest, Ungarn, 557–564, Documenta Acustica of the European Acoustics Association, Lüttich, Belgien
68. Schirmacher R, Hölzl G, Scheuren J (1998) Untersuchung zur aktiven Minderung tieffrequenter Vibrationen am ICE1. Fortschritte der Akustik – DAGA, Deutsche Gesellschaft für Akustik, Oldenburg, pp 210–211
69. Zavadskaya MP, Popov AV, Egelski BL (1975) An approximate solution of the problem of active suppression of sound fields by the Malyuzhinets method. Soviet Phys Acoust 21(6):541–544
70. Scheuren J (1990) Aktive Beeinflussung der Wellenausbreitung II: Realisierungsmöglichkeiten einer aktiven Beeinflussung der Ausbreitung von Biegewellen. Acustica 72:33–46
71. Scheuren J (1990) Aktive Reflexion von Biegewellen. Fortschr. der Akustik – DAGA, Wien, DPG-GmbH, Bad Honnef, 345–348
72. Achenbach JD (1973) Wave propagation in elastic solids. North Holland, Amsterdam
73. Scheuren J (1984) Application of the method of integral equations to the vibration of plates. Proc.2.Int.Conf.on Recent Advances in Struct. Dynamics, Southampton, 171–177
74. Piraux J, Mazzanti S (19859 Commande automatique d'un dispositif d'atténuation acoustique active dans l'espace. IUTAM-Symposium Aero- and Hydroacoustics, Comte-Bellot u. Ffowcs-Williams (ed), Lyon/Frankreich
75. Ohnishi K.et al. (1999) Development of the Noise Barrier using Active Controlled Acoustical Soft Edge. Proceedings Active 99, 595–606
76. LaFontaine RF, Shepherd IC (1983) An experimental study of a broadband active attenuator for cancellation of random noise in ducts. J Sound Vibr 91(3):351–362
77. Roure A (1985) Self-adaptive broadband active sound control system. J Sound Vibr 101(3):429–441
78. Eriksson LJ, Allie MC, Greiner RA (1987) The selection and application of an IIR adaptive filter for use in active sound attenuation. IEEE/ASSP 35(4):433–437
79. Eriksson LJ, Allie MC (1988) A practical system for active attenuation in ducts. Sound Vibr 2:30–34
80. Pelton H, Wise S, Sims W (1994) Active HVAC Noise control systems provide acoustical comfort. Sound Vibr 28:14–18
81. Koopmann GH, Fox DJ, Neise W (1988) Active source cancellation of the blade tone fundamental and harmonics in centrifugal fans. J Sound Vibr 126(2):209–220
82. Babesel M, Maier R, Hoffmann F (2001) Reduction of interior noise in helicoptersby using active gearbox struts – results of flight tests. Proceedings of 27th European Rotorcraft Forum, Moskau
83. Zintel G, Lehringer F (1992) Aktive Pegelminderung des Auspuffgeräusches von Kraftfahrzeugen. Fortschritte der Akustik – DAGA, Deutsche Gesellschaft für Akustik, Oldenburg, 921–924
84. Pricken F (2000) Active Noise Cancellation in Future Air Intake Systems. SAE Paper Nr. 2000-01-0026
85. Heil B et al (2001) Variable Gestaltung des Abgasmündungsgeräuschs am Beispiel eines V6-Motors. Motortechnische Zeitschrift (MTZ), Jahrg. 62. Heft 10:786–794
86. Schirmacher R (2002) Active design of automotive engine sound. AES 112th convention, München
87. Fohr F et al (2002) Active exhaust line for truck diesel engine. Proceedings Inter-Noise, Dearborn, USA, Paper N366
88. Applications of active noise and vibration control in vehicles (September 2002) Noise&Vibration Worldwide, 11–15
89. Fuller C, Brevart B (1995) Active control of coupled wave propagation and associated power in fluid filled elastic long pipes. Proceedings Active 95, 3–14
90. Guicking D, Karcher K, Rollwage M (1985) Coherent active methods for applications in room acoustics. J Acoust Soc Am 78(4):1426–1434
91. Meirovitch L (1988) Control of distributed structures. Large Space Struct:195–212
92. Meirovitch L, Silverberg LM (1984) Active vibration suppression of a cantilever wing. J Sound Vibr 97(3):489–498
93. Balas MJ (1978) Active control of flexible systems. J Optimization Theory Appl 25(3):415–436
94. Meirovitch L, Bennighof JK (1986) Modal control of travelling waves in flexible structures. J Sound Vibr 111(1): 131–144
95. Curtis A et al (1987) Active suppression of acoustic resonance. J Acoust Soc Am 81(3):624–631
96. v.Flotow AH (1988) The acoustic limit of control of structural dynamics. Large Space Struct:213–237

97. Nelson PA et al (1987) The active minimization of harmonic enclosed sound fields, I: Theory, II: A computer simulation, III: Experimental verification. J Sound Vibr 117(1):1–58
98. Niehoff W (2003) Fa. Sennheiser: Persönliche Mitteilung über das im Piloten-Headset HMEC 300 eingebaute System zur Geräuschunterdrückung
99. McDonald AM, Elliott SJ, Stokes MA (2002) Active noise and vibration control within the automobile. Fortschritte der Akustik – DAGA Bochum, Deutsche Gesellschaft für Akustik, Oldenburg, pp 147–156
100. Scheuren J, Widmann U, Winkler J (1999) Active Noise Control and Sound Quality Design in Motor Vehicles. Proceedings of the 1999 SAE Noise and Vibration Conference, Traverse City, Michigan, USA, SAE Paper 1999-01-1846, 1473–1479
101. Bullmore AJ, Nelson PA, Elliott SJ (1990) Theoretical studies of the active control of propeller-induced cabin noise. J Sound Vibr 140(2):191–217
102. Elliott SJ et al (1990) In-flight experiments on the active control of propeller-induced cabin noise. J Sound Vibr 140(2):219–238
103. Dorling CM et al (1999) A demonstration of active noise reduction in an aircraft cabin. J Sound Vibr 128(2): 358–360
104. Ross C, Purver M (1997) Active cabin noise control. Active 97, Proceedings of the International Symposium on Active Control of Sound and Vibration, New Port Beach, California, USA, XXXIX-XLVI
105. Sano H et al (2001) Active control system for low-frequency road noise combined with an audio system. IEEE Trans Speech Audio Process 9(7):755–763
106. Fuller C (2002) Active Control of sound radiation from structures: progress and future directions. Active 2002, Proceedings of the International Symposium on Active Control of Sound and Vibration, Fort Lauderdale, Florida, USA, 3–27
107. Deffayet C, Nelson P (1988) Active control of low frequency harmonic sound radiated by a finite panel. J Acoust Soc Am 84:2192–2199
108. Johnson B, Fuller C (2000) Broadband control of plate radiation using a piezoelectric, double amplifier active skin and structural acoustic sensing. J Acoust Soc Am 107:876–884
109. Ross C (2003) Fa. Ultra Electronics: Persönliche Mitteilung über das in einer Bombardier Q 400 eingebaute System zur Geräusch- und Schwingungsunterdrückung
110. Huang XY (1987) Active control of aerofoil flutter. AIAA J 25:1126–1132
111. Ffowcs Williams JE, Huang XY (1989) Active stabilization of compressor surge. J Fluid Mech 204:245–262
112. Candel SM, Poinsot TJ (1988) Interactions between acoustics and combustion. Proc. Institute of Acoustics., vol 10, 103–153
113. Dowling AP et al (1988) Active control of combustion instabilities. Proceedings of the Institute of Acoustics, vol 10, 873–875
114. Poinsot T et al. (1987) Suppression of combustion instabilities by active control. AIAA-Paper 87–1876
115. Möser M (1989) Aktive Kontrolle einfacher, selbsterregter Resonatoren. Acustica 69:175–184
116. Heckl M (1988) Active Control of the noise from a Rijke tube. J Sound Vibr 124:117–133
117. Ffowcs Williams JE (1986) The aerodynamic potential of anti-sound. AIAA'86, 15th Congress of the International Council of the aeronautical sciences, ICAS-86-0.1, London
118. Seume J et al (1998) Application of active combustion instability control to a heavy duty gas turbine. Trans ASME J Eng Gas Turbines Power 120(4):721–726
119. Wicken G, v.Heesen W, Wallmann S (2000) Wind tunnel pulsations and their active suppression. SAE Paper Nr. 2000-01-0869
120. Ross CF, Yorke AV (1986) Energy flow in active control systems. Colloque Euromech 213 Méthodes actives de controle du bruit et des vibrations, Marseille
121. Ford RD (1983) Where does the power go ? Proc. I.C.A., Paris, Vol.8, 277–280
122. Scheuren J, Schirmacher R, Hobelsberger J (2002) Active design of automotive engine sound. Proceedings Inter-Noise, Dearborn, USA, Paper N629
123. Bendat JS, Piersol AG (1993) Engineering applications of correlation and spectral analysis, 2nd edn. Wiley, New York
124. Oppenheim AV, Schafer RW (1975) Digital signal processing. Prentice Hall, Englewood Cliffs
125. Stearns SD (1999) Digitale Verarbeitung analoger Signale, 7th edn. Oldenbourg Verlag, München
126. Oppenheim AV, Schafer RW (1999) Zeitdiskrete Signalverarbeitung, 3rd edn. Oldenbourg Verlag, München
127. Papoulis A (2002) Probability, random variables and stochastic processes, 4th edn. McGraw-Hill, New York
128. Widrow B et al (1975) Adaptive noise cancelling: principles and applications. Proc IEEE 63(12):1692–1716
129. Widrow B, Stearns SD (1985) Adaptive signal processing. Prentice-Hall, Englewood Cliffs
130. Haykin S (2002) Adaptive filter theory, 4th edn. Prentice-Hall, Englewood Cliffs
131. Aström KJ, Wittenmark BJ (1994) Adaptive control, 2nd edn. Addison-Wesley, Reading (Mass)
132. Schirmacher R (1995) Schnelle Algorithmen für adaptive IIR-Filter und ihre Anwendung in der aktiven Schallfeldbeeinflussung. Dissertation, Göttingen
133. Scheuren J (1995) Experiments with Active Vibration Isolation. Proceedings of Active 95 in Newport Beach, California, 79–88, Noise Control Foundation, Poughkeepsie, NY, USA
134. Petitjean B, Legrain L (1996) Feedback controllers for active vibration suppression. J Struct Control 3(1–2): 111–127
135. Elliott SJ, Gardonio P, Sors T, Brennan M (2002) Active vibroacoustic control with multiple local feedback loops. J Acoust Soc Am 111(2):908–915
136. Maury C, Gardonio P, Elliott SJ (2001) Active control of the flow-induced noise transmitted through a panel. AIAA J 39(10):1860–1867
137. Janocha H (1992) Aktoren. Springer, Berlin
138. Janocha H (ed) (1999) Adaptronics and smart structures: basics, materials, designs and applications. Springer, Berlin
139. Srinivasan A, McFarland M (2000) Smart structures: analysis and design. Cambridge University Press, Cambridge

140. Ross CF (1978) Experiments on the active control of transformer noise. J Sound Vibr 61(4):473–480
141. Berge T, Pettersen O, Soersdal S (1988) Active cancellation of transformer noise: field measurements. Appl Acoust 23:309–320
142. Hesselmann N (1978) Investigation of noise reduction on a 100 kVA transformer tank by means of active methods. Appl Acoust 11:27–34
143. Angevine OL (1981) Active acoustic attenuation of electric transformer noise. Proc. Inter-Noise, Amsterdam, 303–306
144. Jakob A, Möser M (2003) Active control of double-glazed windows I and II. Appl Acoust 64:163–196
145. Jakob A, Möser M (2003) A modal model for actively controlled double-glazed windows. Acta Acustica 89: 479–493
146. Gasch R, Nordmann N, Pfützner H (2002) Rotordynamik, 2nd edn. Springer, Berlin
147. Smith RA, Chaplin GB (1983) The implications of synchronized cancellation for vibration. Proc. Inter-Noise, Edinburgh, 403–406

Sound Reinforcement

13

Harald Frisch

13.1 Introduction

Sound reinforcement includes a large number of different systems, which differ strongly with regard to their technology and their appearance. However, apart from the different applications and the connected technical design of the systems, they have a common target: transforming electrical signals processed by a chain of electronic devices into acoustic signals by means of loudspeakers and reproducing them in a quality demanded by the respective purpose. The quality may range from 'nearly telephone' for alarm systems in open air or in acoustically difficult rooms to 'nearly original' for modern hi-fi systems in living rooms. Between these two extremes, nearly every grade of quality can be realised with sound reinforcement. As everywhere, also for sound reinforcement, there is a trend to achieve perfection, i.e., away from 'telephone' towards 'original'. This trend is also enforced by a increasingly demanding audience. Being accustomed to hi-fi systems at home, the familiar sound quality is desired also in large event rooms.

13.2 Sound Reinforcement Systems for Speech and Music

Systems for sound reinforcement of speech and music are applied in open air and in rooms of any size. They are applied in all those places, where a large number of listeners have to be supplied with acoustic information over larger distances. The reason to distinguish between sound-reinforcement systems for speech and music is a technical and, thus, also a financial one: for speech transmission of good quality, a restricted frequency range of approximately 125 Hz up to 4,000 Hz is already sufficient. Loudspeakers, which can reproduce this limited frequency range linearly, are much cheaper than those working in music-transmission systems in the range from 20 Hz up to 16 kHz or even higher than 20 kHz.

Apart from the frequency response of these systems, the maximum sound pressures needed for speech and music differ strongly. For music transmission of good quality, approximately ten times higher sound pressure peaks are achieved than for speech transmission. The sound amplifying power required for that (and thus also the power-handling capacity of the loudspeakers) is higher by the factor 100, which also leads to a steep increase of the costs.

For the technical reasons stated above, sound-reinforcement systems good for speech can transmit music only in a very restricted way. Thus, they should only be used when it is clear from the beginning that music transmission in a good or very good quality is not required. Nowadays, this is usually only valid for churches, where owing to the often-difficult room-acoustic conditions (long reverberation time, danger of echoes by very late reflections), a comparatively high technical effort is also required for speech-amplifying systems.

In most other cases, the event spectrum is designed in a way that the sound-reinforcement system must transmit speech and music in the same quality. This, increasingly applies also to formerly 'pure' speech

H. Frisch (✉)
Müller-BBM GmbH, Robert-Koch-Straße 11, 82152 Planegg, Germany
e-mail: Harald.Frisch@MuellerBBM.de

G. Müller and M. Möser (eds.), *Handbook of Engineering Acoustics*,
DOI 10.1007/978-3-540-69460-1_13, © Springer-Verlag Berlin Heidelberg 2013

amplifiers or to announcement systems in sports facilities. The massive use of spoken advertisements combined with music in sports events of any type speeds up the trend of installing sound-reinforcement systems capable of high-quality music transmission. Therefore, in the further discussion of sound-reinforcement system, no distinctions are made between speech and music transmission.

13.2.1 Sound Reinforcement Systems in Rooms

For planning a sound-reinforcement system, the acoustic data of the rooms must be considered. Besides architectural and constructional factors, they also determine the suitable locations of the loudspeakers. While architecture and construction mainly influence the type of the loudspeakers (size, weight, etc.), the acoustics of a room (reverberation time, acoustic quality of the limiting surfaces of the room) determine the type of the sound-reinforcement system. The sound-reinforcement system can be described as the interaction of several loudspeakers in a room. Here, the loudspeakers can be placed centrally (at one point) or decentrally (at many points) in the room.

13.2.1.1 Central Sound Reinforcement

Apart from very simple systems for small lecture halls, where a single loudspeaker with suitable directivity may be sufficient, a central sound-reinforcement system, as well as a decentral sound-reinforcement system, consists of several individual loudspeakers. In contrast to a decentral sound-reinforcement system, the loudspeakers of a central system are integrated into one unit (loudspeaker cluster) and thus supply the total area actually from one point in the room. The majority of central sound-reinforcement systems is applied where the form of the room (distance between loudspeakers and audience), the room acoustics (reverberation time, which is not too long) and the architecture allow for such a system (Fig. 13.1).

As a variation of the central sound reinforcement, a central sound reinforcement with additional supporting loudspeakers is widely spread. It is used in rooms where the distance between loudspeakers and audience is too large or the path is obstructed for sound propagation (Fig. 13.2). In this case, on the one hand, it is ensured by the supporting loudspeakers that nearly the same sound pressure level is achieved at all seats, and, on the other hand, the direct sound of the transmitted signal is increased at the more distant seats in the room. In order to prevent an acoustic localisation of the supporting loudspeakers, they must be delayed compared to the main loudspeaker by means of time-delay devices in such a way that the sound of the supporting loudspeaker reaches the listener shortly after the sound of the main loudspeaker (Haas effect) [1].

For a simple assessment of the required time delay, the following equation is valid:

$$\Delta T = (\mathrm{SE} - \mathrm{LE}) \cdot 3 + x, \qquad (13.1)$$

where

SE is the distance of the sound source to be located from the receiver (listener) in m;

LE is the distance of the loudspeaker to be delayed from the receiver in m;

ΔT is the transit-time delay to be adjusted in ms; and

$x = 10 - 20$ ms

13.2.1.2 Decentral Sound Reinforcement

Decentral sound-reinforcement systems play only a minor role in speech and music transmissions. The main field of application are sound reinforcement systems for speech in churches, small or medium lecture halls and multipurpose rooms.

A reason for using such systems is their good suitability for rooms with very long reverberation times (churches) and their visual inconspicuousness owing to the application of a large number of small loudspeakers of comparatively low performance.

Possible locations for installing the loudspeakers are side walls and supporting pillars as well as the ceiling. A basic principle when choosing the right location should be to bring the loudspeakers as near as possible to the audience. Thus, for small high rooms (churches), the side walls will be selected, and for lower, wider rooms (gymnasiums, canteens, etc.) the ceiling. If the visual–acoustic assignment shall be preserved with a decentral sound-reinforcement system (the listener hears the sound coming from the direction of the sound source), comparatively high efforts are required. Each individual loudspeaker of the system must be operated corresponding to its distance from the sound source with an individual time delay. By applying time-delay devices in decentral sound-reinforcement

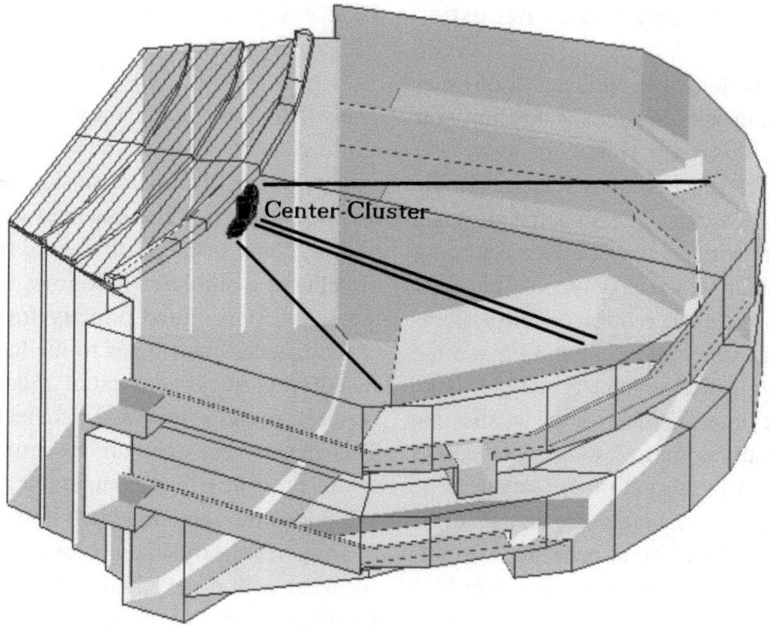

Fig. 13.1 Principle of a central sound reinforcement system

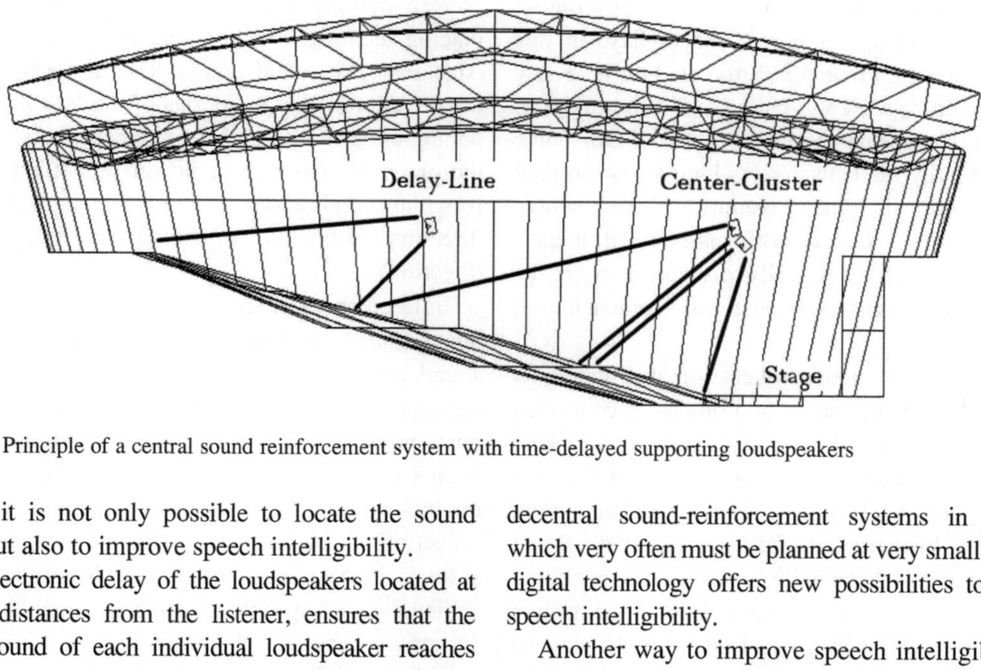

Fig. 13.2 Principle of a central sound reinforcement system with time-delayed supporting loudspeakers

systems, it is not only possible to locate the sound source, but also to improve speech intelligibility.

The electronic delay of the loudspeakers located at different distances from the listener, ensures that the emitted sound of each individual loudspeaker reaches the listener nearly simultaneously. Thus, an echo deteriorating speech intelligibility by sound from more distant loudspeakers 'arriving too late' is avoided. As, at present, comparatively cheap, but high-quality digital time-delay devices can be found in the market, there is nothing to stop the application of such supportive devices also for systems of the low price range. Above all, for decentral sound-reinforcement systems in churches, which very often must be planned at very small expenses, digital technology offers new possibilities to improve speech intelligibility.

Another way to improve speech intelligibility and naturalness of sound transmission also in problematic churches is the use of central instead of decentral systems. By using a few centrally positioned high-quality horn loudspeaker systems instead of a decentral arrangement of many comparatively simple (and often poor sounding) sound columns, usually remarkable improvements in quality can be achieved.

13.2.1.3 Sound Reinforcement with Acoustic Localisation

In the two previous sections, the application of time-delay devices to obtain an acoustic localisation was discussed. However, this only works for immobile sound sources, e.g., a speaker at a lectern. If the sound source moves, however, like an actor on stage, a sound reinforcement with an acoustic localisation of the sound source for all seats is only possible with a high technical effort. Sound-reinforcement systems according to the delta stereophony method make the localisation of moving sound sources possible [2–8].

As stated in the previous sections, for the localisation of an electro-acoustically amplified sound source, it is absolutely required that the sound of the source to be located arrives at the listener before the sound of the loudspeaker. However, if the sound pressure level difference between source and loudspeaker is larger than approx. 8 up to 10 dB, i.e., the loudness of the sound coming from the delayed loudspeaker is twice the loudness of the sound arriving directly at the listener from the (unamplified) source, the acoustic localisation breaks down despite the proper time delay. In this case, the source itself (speaker, actor, etc.) must already be amplified on stage by a so-called source loudspeaker. This source loudspeaker takes the place of the source and is located by the listeners. As long as the original source remains in direct vicinity of the source loudspeaker, the visual–acoustic assignment is preserved. If the source gradually disappears from the source loudspeaker assigned to it, the assignment breaks down.

If several source loudspeakers are distributed on stage, it is possible by means of a processor-controlled adjustment of level and delay time of these loudspeakers to maintain the localisation of a moving sound source [8]. The required level and delay-time conditions among the different loudspeakers are very complex. Small disturbances in this 'balance', e.g., owing to an operating error by the staff, wrong positioning of the source loudspeakers and the like, can cause that the desired visual–acoustic assignment is only achieved partly, only at certain seats in the hall or not at all. Sound-reinforcement systems designed for acoustic localisation of movable sources make very high demands on the technical abilities and the acoustic understanding of the operating staff and are therefore rarely used (e.g., 'Seebühne Bregenz' at Lake Constance and 'Seebühne Mörbisch' at the Neusiedlersee, both in Austria).

13.2.2 Open-Air Sound Reinforcement Systems

The field of application for open-air sound-reinforcement systems ranges from announcement systems on platforms to music-transmission systems for rock concerts. The essential difference compared with systems installed in rooms results from a lack of reflecting surfaces. It follows from this that only the sound transmitted directly from the loudspeaker to the listener contributes to the loudness impression. Besides, in the open, often much larger distances between sound source and listeners must be covered. This means that open-air systems must emit higher sound pressures than comparable systems in a room to achieve the same loudness impression.

In general, this is achieved by applying loudspeakers with a very high efficiency (10% and more for frequencies >500 Hz) and a very strong sound focussing (e.g., horn loudspeakers with 40° horizontal and 20° vertical sound emission) as well as by a correspondingly high number of loudspeakers. By placing identical loudspeakers on top of each other (stacking), the sound impression can be increased in the direction of the main axis of the 'stacks' by 6 dB per doubling of the number of loudspeakers. However, the price that has to be paid for this effect is a very irregular sound radiation apart from the main radiation direction (dropouts depending on frequency and direction).

In large open-air spaces, the regular level decrease due to increasing distance from the loudspeaker (6 dB sound pressure level decrease for a doubling of the distance) is augmented by weather and by interference phenomena owing to surface conditions. Further details on this subject are stated in Chap. 17. Owing to the missing diffuse sound (caused by sound reflections at ceilings and walls in a room), central systems in open air usually have less difficulties to achieve a good speech intelligibility than comparable systems in rooms. The main criterion for a good speech intelligibility in the open is the level-ratio of direct sound to background noise. The direct sound arriving at the listener should be by 15 up to 25 dB above the A-weighted background noise level [9].

This demand can be met with modern central sound-reinforcement systems also for large distances with the corresponding technical effort involved. If, for example, for a background noise level of 60 dB(A), a sound pressure level of 60 + 25 = 85 dB should be created at a distance of 180 m from the central system, a loudspeaker must be used, which can create a sound pressure level at a distance of 1 m of

$$85\ \mathrm{dB} + 20\lg\frac{180\ \mathrm{m}}{1\ \mathrm{m}} = 130\ \mathrm{dB}, \qquad (13.2)$$

(influences by surface and weather conditions are neglected here). However, this means that at a distance of 2 m, there are still 130 − 6 = 124 dB, at a distance of 4 m 118 dB, etc. These are values, which are partly far beyond the limits, representing a risk for the ear [9].

For systems of this type used for the most distant part of the open-air space, attention must be paid so that the audience cannot get too near to these loudspeakers. Thus, these loudspeakers are usually placed very high above the audience or the stage. Together with a distinct directivity and an exact adjustment of these loudspeakers towards the areas to be supplied with sound, it is possible to create sufficiently high sound pressure levels also at a larger distance from the central loudspeaker arrangement, without annoying or even putting at risk the audience near the loudspeakers with a too high level. Apart from that, it is possible to achieve a relatively even sound-level distribution over the total area by carefully choosing the types of loudspeakers and their directivity (Fig. 13.3).

The mentioned sound pressure level of 85 dB at a distance of 180 m is, of course, much too low for open-air concerts in the pop and rock business. In these cases, a central sound reinforcement with time-delayed supporting loudspeakers is more suitable.

13.3 Systems to Simulate Room-Acoustic Properties

The purpose of these kinds of systems is to optimise the room-acoustic conditions of a hall for different types of events. Here, the applied loudspeakers may not be located on no account. The audience and musicians may not notice at no time or even have the feeling that loudspeakers are involved. For this reason, well-working simulation systems are nearly 'inaudible'. They have only a supporting, but not amplifying function. Thus, they get the musicians and the audience to believe that the room acoustics are optimised for the present application.

13.3.1 Extension of the Reverberation Time

A method to extend the reverberation time for low frequencies ('assisted resonance system') proposed by Parkin and Morgan was installed first in the Royal Festival Hall, London, in the mid-sixties [10]. Another step in the development of systems extending the reverberation time electro-acoustically was taken by the MCR system of the company Philips (MCR Multi-Channel Reverberation) in the late seventies of the past century [11–14]. The electronic principle of this system is acoustic feedback. Here, by means of a large number of microphones and loudspeakers installed in a room and by the connected filters and amplifiers, a limited extension of the reverberation time is achieved. A problem here is an increasing colouration of

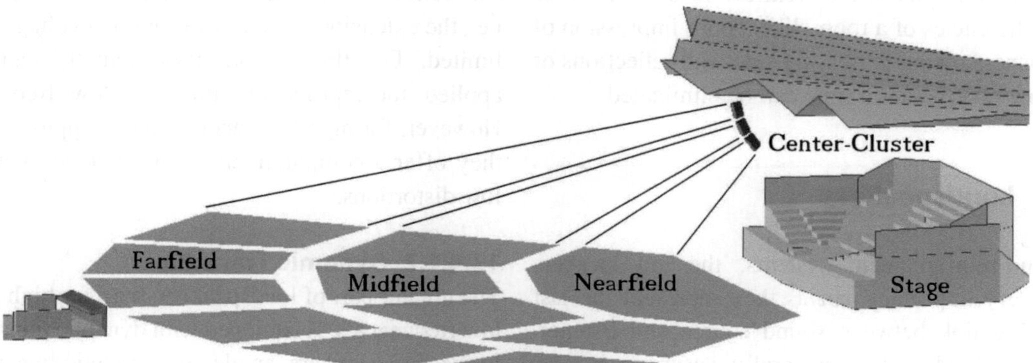

Fig. 13.3 Central sound reinforcement system open-air

the sound characteristics with an extension of the reverberation time [15, 16].

13.3.2 Manipulation of Room-Acoustics

However, in many multipurpose halls it is not enough to adjust the reverberation time to the desired value in order to realise a range of events from theatre up to symphony concerts. When assessing the acoustic quality of a room, the existence of early reflections as well as their arrangement in time and location plays a large part. They influence room-acoustic criteria such as clarity, distinctness and spaciousness, while the course of the reverberation time essentially causes acoustic impressions like brilliance, warmth and envelopment. Systems changing the room impression were already developed and applied in the fifties of the last century. Owing to the still imperfect technology at that time (magnetic sound recording on continuous tape and playback by means of several play heads arranged one after the other to simulated discrete reflections), the systems known as 'Ambiphonie' were not successful for a long time [17].

The modern room-acoustic manipulation systems are more successful. Because of the incomparable progress of electronics and electro-acoustics during more than 50 years, at present very sophisticated manipulations of room-acoustics are possible. Systems, which can both change the reverberation time in very wide limits and simulate room reflections, are, for example, the Acoustic Control System (ACS) [18, 19] and LARES [20–23]. By means of these systems, e.g., an orchestra shell on stage can be simulated by applying electronic means without any larger interventions in the architecture of a room, or other deficiencies of a room, like a poor impression of spaciousness owing to missing side wall reflections or a too low reverberation time can be eliminated.

13.4 Loudspeakers

In sound-reinforcement systems, the loudspeakers play a very important part as they represent the last connecting link between sound source and listener. The term 'loudspeaker' generally describes a very large number of different transducers transforming electric energy into acoustic energy. However, the term is also used for a system consisting of two or more transducers in a common case or box. In order to clearly separate these two terms, the term 'driver' is used for an individual transducer and the term 'loudspeaker' for a system of transducers. Thus, a loudspeaker consists of one or several drivers, which are summarised to an independent unit in a box, or which are adjusted for a special application with a horn.

13.4.1 Types of Electromechanical Transducers

There are many different mechanisms to convert electric energy into acoustic energy. Apart from dielectric transducers (condensers) and the occasional use of venturi-modulated air flows and electro-magnetically modulated gas-plasma systems in hi-fi loudspeakers of the top price range, more down-to-earth types of transducers prevail in sound-reinforcement technology. Such types are the piezoelectric (crystal) and the dynamic (electro-magnetic) transducer. The latter is the most common one.

13.4.1.1 Piezoelectric Transducer

For the piezoelectric transducer, which is only occasionally used in sound-reinforcement systems, the piezoelectric effect discovered in the nineteenth century is applied. Pierre and Jaques Currier discovered that a mechanic deformation of special types of crystals leads to a charge separation, which can be measured at the surface of the crystal. On the other hand, a piezoelectric crystal can be deformed by applying an electric voltage. This movement is transferred to a membrane, which sets the air vibrating. Piezoelectric transducers only have a very low 'lift', i.e., the extension forced by the applied voltage is very limited. For this reason, they actually cannot be applied for reproducing mid and low frequencies. However, for high frequencies (above approx. 5 kHz), they offer a comparatively high efficiency with very low distortions.

13.4.1.2 Dynamic Transducer

The greater part of loudspeakers, both for high and for low frequencies, is equipped with dynamic transducers. In these transducers, an electro-magnetic linear motor is used for operating the membranes. For this purpose, a wire wound to a coil (voice coil) is surrounded by a magnetic field created by a permanent magnet. If direct current flows through the coil, a second electro-magnetic field is created around the coil. The polarity of this field depends on the flow direction of the current through the coil. The electro-magnetic field interacts with the magnetic field of the permanent magnet – a force takes effect on the current-carrying coil (Lorentz force).

If the permanent magnet is assumed to be immobile and the coil freely moving, the coil is moved by the Lorenz force. If a modulated alternating current is applied, the coil starts to move forward and backwards depending on the changing polarity of the current.

Figure 13.4 shows a section of the dynamic driver. The voice coil (b) is placed freely movable in an air gap of the magnet (a). It is placed on a cylindrical base made of specially treated paper or plastic and is connected via this base with a membrane (c). The usually cone-shaped membrane couples the movements of the voice coil with the surrounding air.

13.4.2 Directivity of a Cone Driver

The directivity of a cone driver depends on the emitted frequency. The dependence is determined by the relation between cone diameter and wavelength λ of the emitted frequency. Figure 13.5 shows some polar diagrams of a typical cone driver for different ratios of diameter and wavelength λ [24]. For low frequencies, the wavelength is large compared with the cone diameter and the driver emits spherically (Fig. 13.5 on the left side, at the top). With an increasing frequency, the wavelength is reduced and the cone driver emits the sound increasingly directionally.

13.4.3 Low-Frequency Loudspeakers

In order to effectively reproduce low frequencies, a large air volume must be moved. This can be achieved by a large deflection of the membrane or by a large membrane surface. As piezoelectric (and also dielectric) transducers only allow very small deflections, in sound-reinforcement systems, only dynamic drivers are used as low-frequency drivers. Low-frequency drivers are usually installed in a casing. Thus, the acoustic short circuit of a freely suspended low-frequency driver occurring at low frequencies is prevented. Thus, a so-called 'directly emitting' loudspeaker is created. A directly emitting system is a loudspeaker where the membrane of the driver is directly coupled with the surrounding air. In contrast to this stands the second system often used for sound-reinforcement system: the low-frequency horn loudspeaker. Low frequency horns have a higher efficiency than 'direct emitters' and also have a certain directivity for low frequencies.

In front of the actual driver, a horn is installed. The cross-sectional area of this horn increases exponentially with its length (exponential horn) [25]. The lowest frequency to be emitted by a horn depends on the membrane diameter of the driver, the cross-sectional area of the horn mouth and the length of the horn. For low-frequency horns, however, very impractical lengths of several metres result. Therefore, in sound-reinforcement systems so-called 'folded' horns are used for the low-frequency range. Figure 13.6 shows the sketch of a folded horn as sectional and frontal view. As indicated by the name, for this type of loudspeaker, the horn is folded in itself and thus the physical extension of the loudspeaker is drastically reduced.

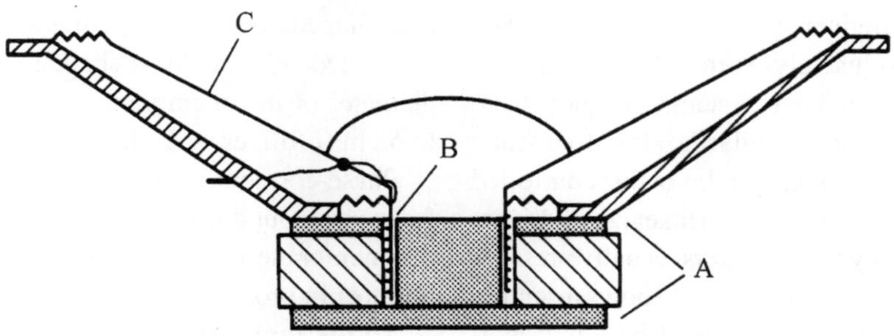

Fig. 13.4 Section of a dynamic transducer

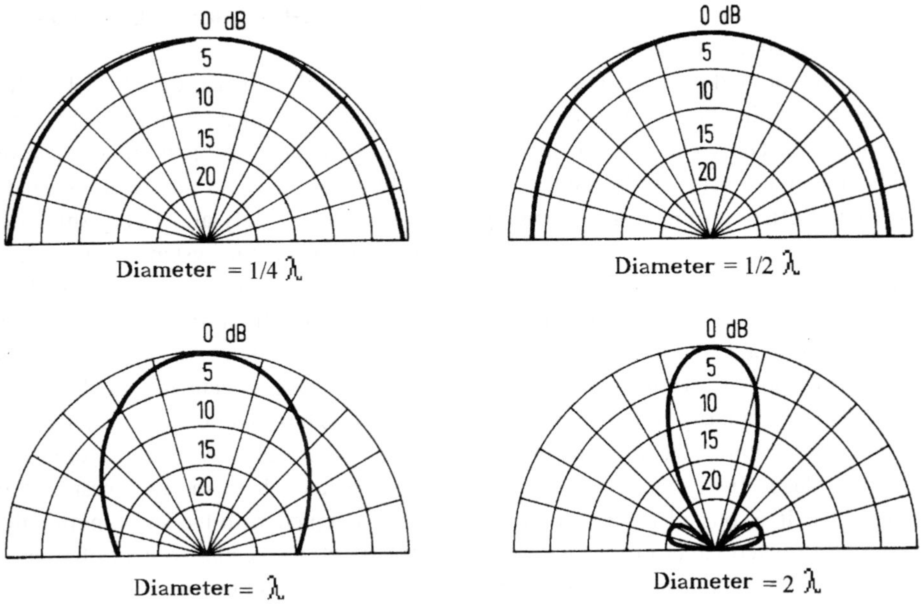

Fig. 13.5 Directivity of a conic transducer mounted in an 'infinite' wall

Fig. 13.6 Folded low-frequency-horn (section and front view)

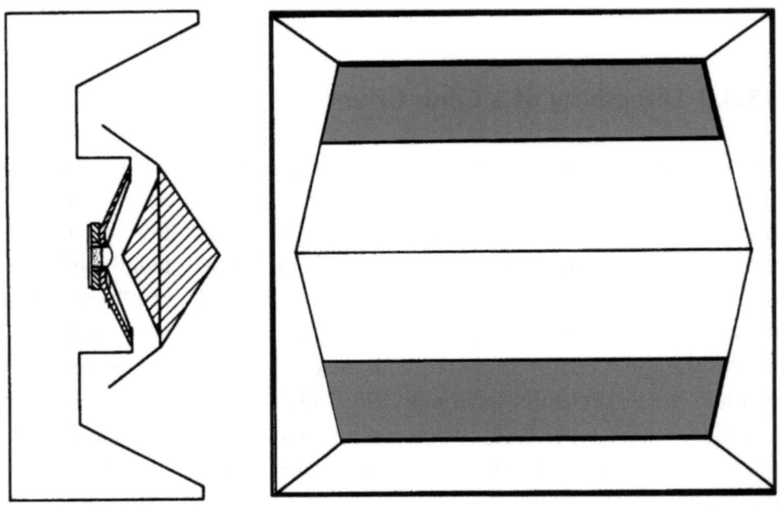

13.4.4 Mid-High-Frequency Loudspeakers

For the mid-high-frequency range (from approx. 800 Hz), nearly exclusively horn loudspeakers are used in sound-reinforcement systems, as they have a much larger efficiency similar to low-frequency horns and a directivity which can be better controlled than for 'directly emitting' loudspeakers. As drivers for mid-high-frequency loudspeakers, usually pressure chamber drivers are used. Figure 13.7 shows a section through a common pressure chamber driver. In principle, this driver works like the transducer described in Sect. 13.4.1.2 according to the principle of an electromagnetic linear motor. The voice coil (a) dips into the ring-shaped air gap of a permanent magnet (b) and is connected with a dome-shaped membrane (c). The diameter of this membrane is usually between 2.5 up to 5 cm (1 'driver or 2' driver).

Phase-correction wedges (d) make sure that high-frequency sound waves coming from different parts of the membrane are superimposed in the throat of the horn (e) in proper phase.

To the throat, a funnel, which is called 'horn', is coupled. Thus, the radiation impedance affecting the

Fig. 13.7 Section of a pressure chamber driver

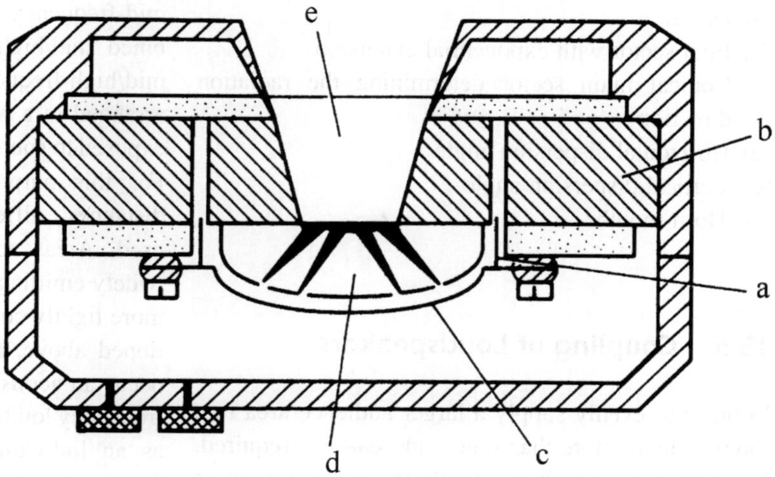

Fig. 13.8 Constant-directivity-horn

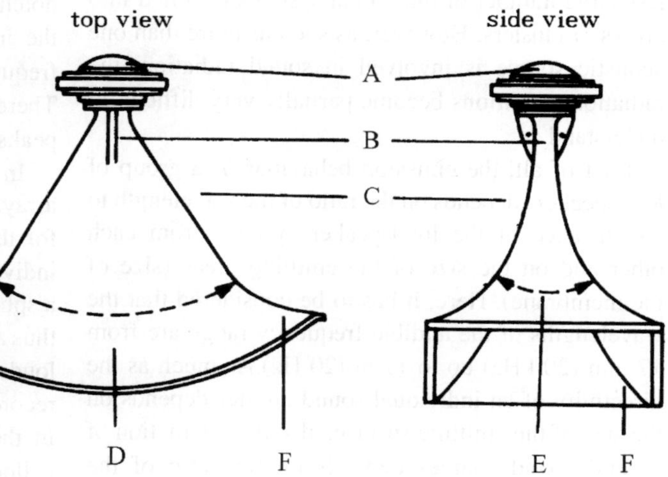

membrane of the driver, i.e., the resistance against which the membrane must work, is increased by the area ratio S_M/S_{TH}. Here, S_M is the area of horn-mouth and S_{TH} the area of the horn-throat [9, 25]. An increase of the membrane load increases the emitted acoustic active power and thus the efficiency. Apart from the high efficiency, the directed sound emission is the most important feature of horn loudspeakers.

In order to determine the dispersion angle, two points on the left and right or above and below the middle axis of the horn loudspeaker (main radiation direction) are determined, where the sound pressure is by 6 dB lower than along the main radiation direction. The angle between the so-called −6 dB points is called dispersion angle. According to the size and form of the horn, these angles vary between 40° and 120° horizontally or 20° and 40° vertically. These different dispersion angles are achieved by different degrees of opening (funnel constant) of the horn funnels in the vertical and horizontal direction.

'Normal' exponential horns, i.e., funnels with a cross-sectional surface increasing exponentially with the distance to the flared throat, have the unpleasant feature that their dispersion angle strongly decreases with an increasing frequency. This so-called 'high-frequency beaming' can be reduced when the shape of the horn opening is varied via the length of the funnel, i.e., that the horn consists of several parts with different funnel constants.

'Constant directivity' horns consist of several of such parts both horizontally and vertically and have a constant dispersion angle nearly over the total frequency range. For this reason, in modern sound-reinforcement systems, nearly exclusively horns of this type are used. The construction of a typical 'constant directivity' horn is shown in Fig. 13.8 with

(a) Driver
(b) First sector with exponential extension
(c) Conical main sector determining the radiation direction
(d) Horizontal dispersion angle
(e) Vertical dispersion angle
(f) Horn-mouth

13.5 Coupling of Loudspeakers

In order to evenly supply a larger audience area in a room, usually more than one loudspeaker is required. In such a case, which is actually the norm, a more or less large number of individual systems is united into arrays or clusters. However, as soon as more than one acoustic source is involved in sound radiation, the radiation conditions become partially very difficult to understand.

First of all, the emission behaviour of a group of loudspeakers depends on the ratio of the wavelength to the distance of the loudspeaker systems from each other and on the size of the emitting areas (size of the membrane). Here, it has to be considered that the wavelengths of the audible frequency range are from 17 mm (20 kHz) up to 17 m (20 Hz).As much as the directivity of an individual sound emitter depends on the size of the emitting surface, also the interaction of several sound sources depends on the ratio of the emitted wavelength to the distance of the individual sources from each other.

If this distance is distinctly smaller than half a wavelength, an acoustic coupling of individual sources results. Here, all individual systems will then work together as one sound source. This acoustic coupling can especially easily be created for low-frequency loudspeakers, as in this frequency range the distance of the low-frequency drivers in the array is always smaller than half a wavelength. For higher frequencies, the distance of the individual mid-high-frequency loudspeakers in the array is always larger than half the wavelength of the frequencies to be emitted. An acoustic coupling is not possible here. The emission behaviour is then marked by strongly frequency-dependent interference patterns.

An array, that must also emit mid and high frequencies apart from low frequencies, always consists of a more or less large number of low-frequency, mid-frequency and high-frequency systems (or combined mid/high-frequency systems). The individual mid/high-frequency systems of the array should be arranged in a way that each part of the room is only exposed to sound coming from one system, if possible. For this purpose, for audience areas of different distances, different types of loudspeakers can be used. Areas nearer to the array are supplied with widely emitting loudspeakers. For more distant areas, more tightly concentrating systems are used. As mentioned above, for low frequencies there will always occur an acoustic coupling. That means that all low-frequency loudspeakers of the array will work together as an individual sound source. For mid and high frequencies, however, partially disturbing peaks and notches will occur owing to interferences, as in reality the individual areas covered by the mid- and high-frequency loudspeaker cannot be clearly separated. There will always be overlapping areas where these peaks and notches will occur.

In order to calculate the emission behaviour of an array, simulation programs are applied. The basis for these programs are so-called balloon data of the individual systems. The balloon data are recorded on a spherical raster grid around the loudspeaker and are thus a three-dimensional image of the directivity of the loudspeaker (Fig. 13.9). The measuring data must be recorded in the so-called far field of the loudspeaker. In the far field of the loudspeaker, the form of the balloon does not change any longer with increasing distance. In the near field, however, the directivity depends on the distance to the acoustic centre of the source. The limit between near and far field depends on the wavelength and the size of the emitting area. The far field starts at

$$r > \frac{L^2}{2\lambda}, \tag{13.3}$$

where

L is the largest dimension of the emitting area (in m) and

λ is the wavelength of the emitted frequency (in m)

Owing to the physical circumstances regarding size of the loudspeaker and wavelength of the frequency to be emitted, common loudspeaker systems actually always emit in the far field. However, if an array is created from many individual systems, the far field can start possibly very distant from the centre of the array.

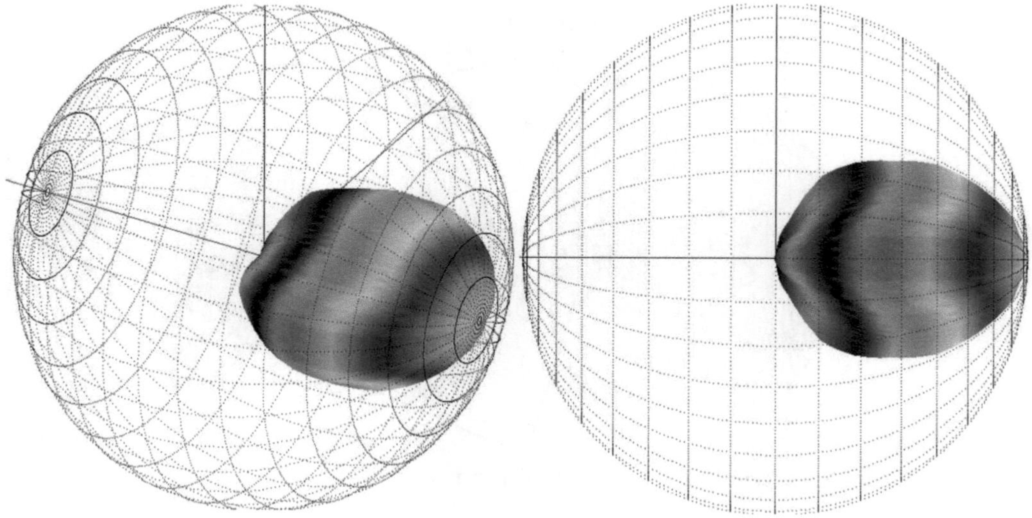

Fig. 13.9 Directivity balloon of a loudspeaker (*left*: 3-D; *right*: vertical section)

13.5.1 Special Arrays

13.5.1.1 Loudspeaker Columns

Loudspeaker columns, which have been known for decades, are the simplest form of a discrete line array. Here, all individual sources are arranged on a line and adjusted in parallel. The horizontal emission behaviour of such a line is determined by the directivity of the individual system. Vertically, it is determined by the strongly frequency-dependent interference of the systems arranged in line. Therefore, such simple loudspeaker lines often have an extremely frequency-dependent vertical directivity with unpleasant side effects like directivity and frequency-dependent dropouts and peaks (side lobes) [26–28] (Fig. 13.10). The use of such simple loudspeaker columns is actually not in keeping with the times. However, the increasing influence of digital and processor technology in sound reinforcement systems leads to a renaissance of the old loudspeaker column with a facelift.

If each loudspeaker within a column is driven with individual filtering, delay and level, a better vertical directivity with essentially less undesired side effects (side lobes) can be achieved. Also, an inclination of the main lobe and even a division into several main lobes with different inclinations in one loudspeaker column is possible with modern digital technology.

13.5.1.2 Line Source

A line source is a continuous, infinitely long sound source. The acoustic wave emitted by such an ideal line source is cylindrical and maintains this mode of propagation independent of frequency and distance.

As real loudspeakers are not infinitely long, but have a limited length, the sound propagation of a real line source must be divided into a near field and a far field. Like for a 'normal' spherically emitting loudspeaker, the border between near field and far field can be calculated according to (13.3). Within the near field, the sound spreads in the form of a cylindrical wave, beyond the near field, the wave becomes spherical. In order to obtain a line array with cylindrical sound propagation from stacked individual systems, the individual transducers must emit a coherent sound wave (connected in phase). This can be achieved very easily at low frequencies by lining up common low-frequency loudspeakers. For mid and high frequencies, however, special waveguides must ensure that all shares of sound emitted by the membrane of the transducer emerge from this waveguide, which is usually in the form of a gap, in an even wave front and at the same time. If it can be achieved that this even wave is created over the total height of the loudspeaker casing, a line array, which operates as a real line source, can be created by stacking several of these loudspeakers one on top of the other [29, 30] (Fig. 13.11).

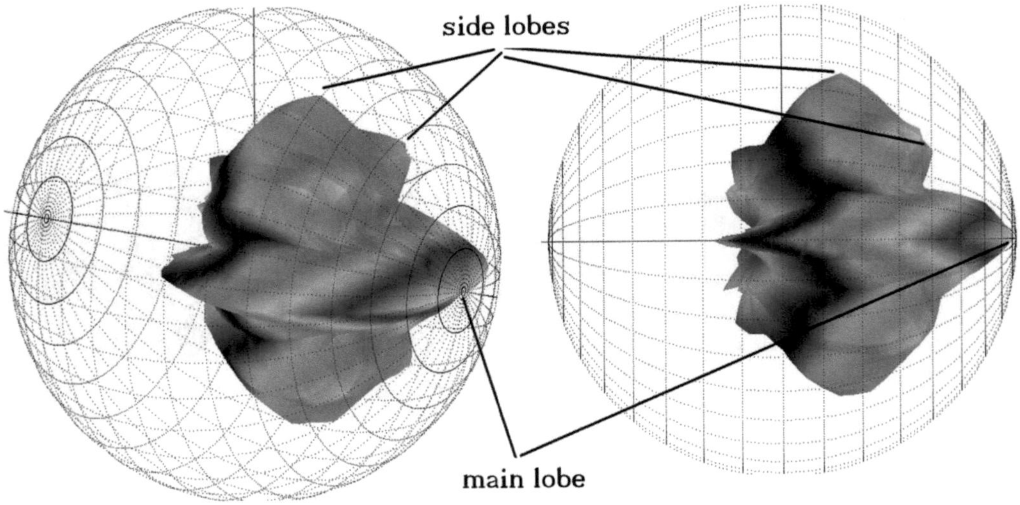

Fig. 13.10 Directivity of a discrete line array with side lobes (*left*: 3-D; *right*: vertical section)

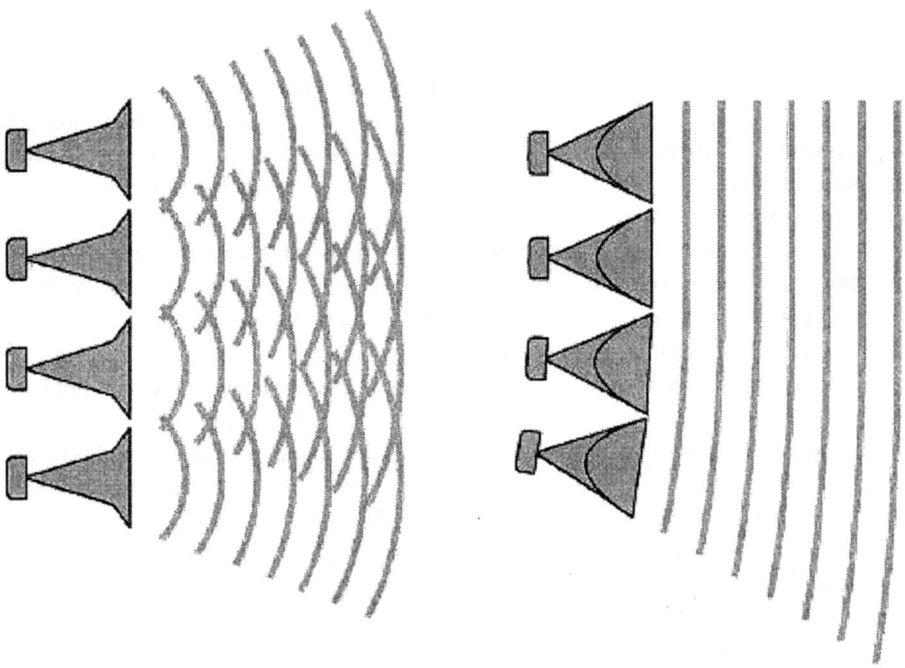

Fig. 13.11 *Left*: discrete line array with spheric radiating transducers and acoustic interference; *Right*: Line Array with Waveguide radiating a coherent soundwave without acoustic interference

According to the length of the array, the near field is extended in a different way. Within the near field, a coherent wave front spreads as a section of a cylindrical wave. This wave has a very precise directivity without lateral side lobes, like for common discrete line arrays. Another advantage is that the surface of a cylindrical wave increases only depending on the distance r to the source and not by the factor r^2 like for a spherical wave (Fig. 13.12). From this results the very advantageous feature of a line source that the sound pressure level decreases with a doubling of the distance only by 3 dB and not by 6 dB (like for conventional, spherically emitting systems). As already mentioned, the cylindrical wave front of a line source turns into a spherical wave front in the far field. According to (13.3), this occurs, e.g., for

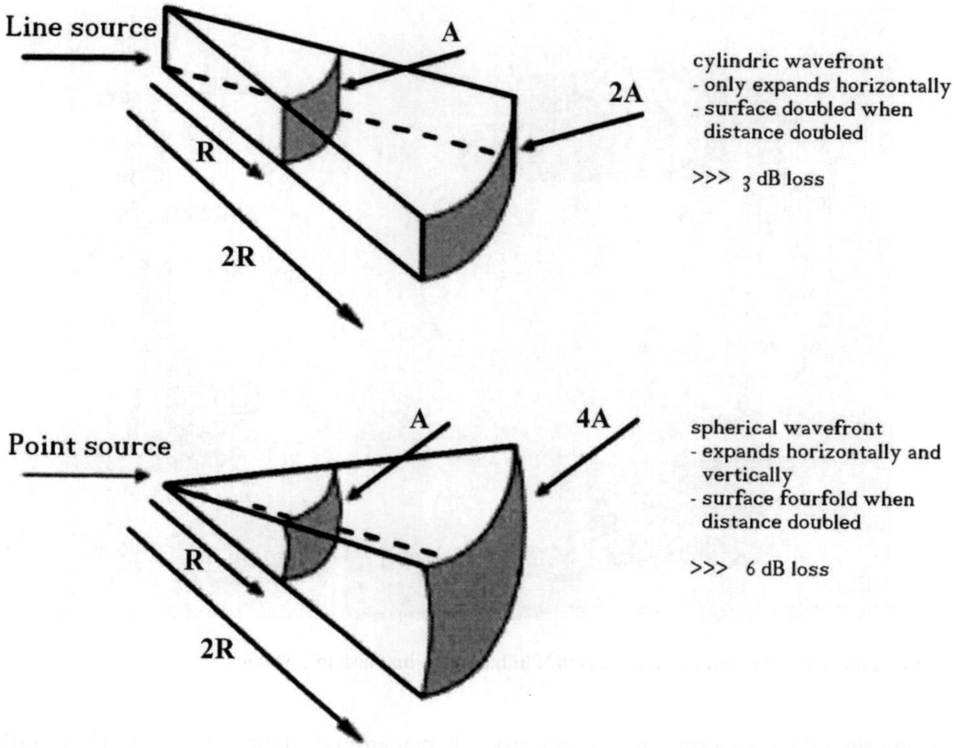

Fig. 13.12 Comparison of the surface enlargement of a cylindric wave and a spheric wave when doubling the distance

a line emitter with a length of 2.7 m at $r > 21.5$ m for 1 kHz and at $r > 2.15$ m for 100 Hz.

In the far field, the cylindrical wave expands to a spherical wave. The beamwidth at −6 dB BW_{-6dB} can be assessed according to (13.4) as follows:

$$BW_{-6\,dB} \approx 2\sin^{-1}\left(\frac{1.9\lambda}{L\pi}\right). \qquad (13.4)$$

With our example of a line source with a length of 2.7 m, the vertical beam width in the far field would be approx. 8.8° for 1 kHz. Due to the described advantages, such as an exactly scalable directivity, strongly reduced interference effects and lack of undesired side lobes as well as a level loss of only 3 dB for a doubling of the distance, modern line array systems working as line sound sources can be recommended for solving problems in demanding sound-reinforcement systems. Often, such systems can be used without applying additional supporting loudspeakers (delay lines). Further details on the subject of loudspeakers and their application in sound-reinforcement systems can be found in the extensive technical literature on this subject [9, 17, 24, 29–33] (Fig. 13.13).

13.6 Microphones

The beginning of the electro-acoustic transmission chain is the microphone. The task of a microphone can be explained quickly and easily. It should transform tiny fluctuations of air (sound) into an alternating voltage proportional to these fluctuations. According to their application and type, microphones have different acoustic properties. Apart from the transfer constant (frequency response), one of the most important microphone features for the use in sound-reinforcement systems is the directivity. According to the reception principle (pressure, pressure gradient or interference receiver), both features are influenced differently.

13.6.1 Pressure Receiver

For pressure receivers, the air-pressure fluctuations on the microphone membrane are used to create an alternating voltage. Microphones designed according

Fig. 13.13 Test of a dVDOSC System (L-acoustics) in 'Philharmonie im Gasteig'; Munich

to this reception principle have a spherical directivity (polar pattern). For high frequencies, when the wavelength is in the order of the membrane size of the microphone, a pressure boost effect can occur on the membrane of a pressure receiver. The sound pressure in this frequency range cannot be relieved unimpeded. It builds up in front of the membrane. This frequency-dependent pressure increase for vertical sound incidence leads to a different frequency response compared to the response for lateral sound incidence. A resulting, noticeable emphasis of high frequencies must be compensated by a corresponding equalisation. Pressure receivers are applied when large areas of a room should be recorded with a minimum of coloration. They are not very suitable for the use in sound-reinforcement systems as owing to their omnidirectionality, they tend to feedbacks.

13.6.2 Pressure-Gradient Receivers

Microphones whose membranes are operated by the velocity affecting it, are called pressure-gradient receivers. For these types of microphones, the membrane is accessible to the sound wave from both sides. Unlike the pressure receiver described above, it has a more or less distinctive directivity (cardioid, supercardioid, hypercardioid, figure of eight). These different polar patterns are achieved by influencing the ratio of pressure received in front of and behind the membrane (Fig. 13.14). Owing to these possible different directivities, pressure-gradient receivers are the most common types of microphones applied in sound-reinforcement systems. With a suitable choice of directivity it is possible to reduce greatly the tendency to feedback. Pressure gradient receivers are very sensitive to air flows. This shows in an especially unpleasant way in the so-called proximity effect, which leads to a strong increase of low-frequency range when the distance between sound source and microphone is very small. Often, this effect is used, e.g., for giving a voice more 'substance'. However, it always has a negative effect on speech intelligibility. Figure 13.15 shows the proximity effect by means of a microphone-frequency response depending on the distance from the sound source. In order to counteract this effect, these microphones are often equipped with switchable filters to suppress low-frequency accentuation.

13.6.3 Interference Receivers

Interference receivers are used to achieve an even stronger directivity than that of a hypercardioid

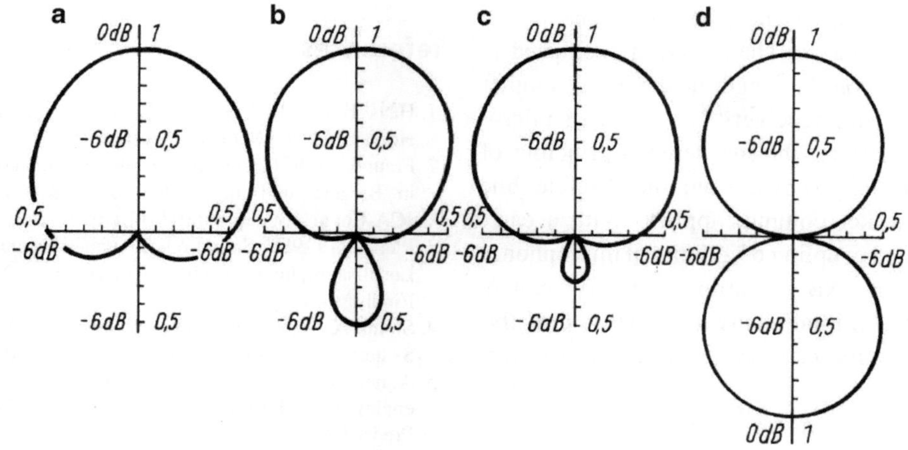

Fig. 13.14 Polar pattern of microphones: (**a**) cardioid; (**b**) hypercardioid, (**c**) supercardioid; (**d**) figure of eight

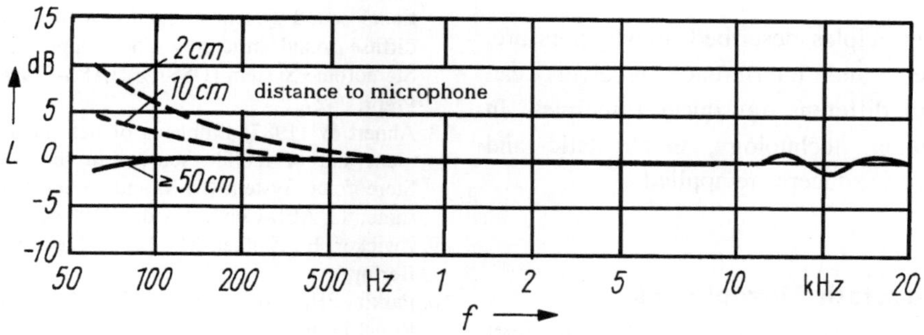

Fig. 13.15 Proximity effect at various distances

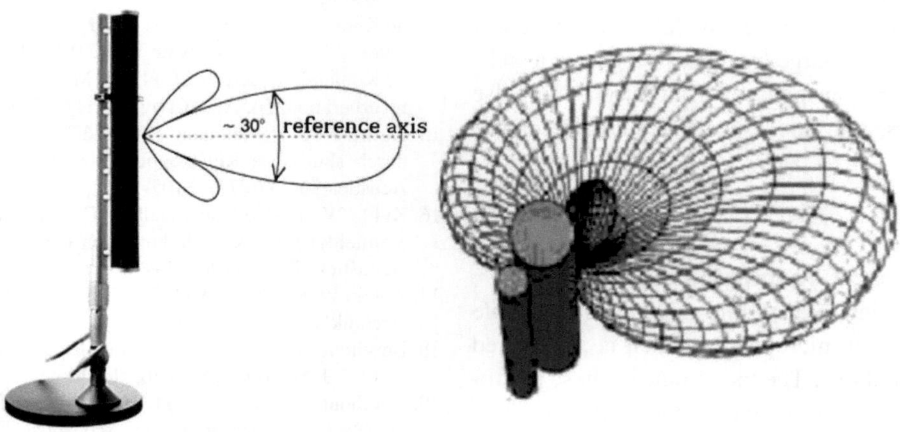

Fig. 13.16 Cardioid-Level-Microphone KEM970 (Microtech Gefell)

microphone. The most common form of these types of microphones is the shotgun microphone which consists of a slotted tube with a pressure-gradient capsule at one end. The air-flow resistance of the slots in the tube increases in the direction of the capsule. With this arrangement, sound waves hitting the tube laterally will be attenuated by interference or partly completely obliterated. Sound hitting the tube directly at the front reaches the membrane nearly without any attenuation. The directivity achieved in this way depends on the length of the tube. Another type of interference receivers are so-called microphone columns. Here, the interference principle of stacked transducers known from the discrete line arrays (loudspeaker column) applies. With a carefully coordinated coupling of individual microphones along a common axis, an arrangement is created, which has a useful rotationally asymmetric directivity. The directivity can be described as disciod. (Fig. 13.16).

13.6.4 Different Types of Transducers

The reception principles described above (pressure, pressure gradient and interference receiver) can be realised with different transducer principles. In sound-reinforcement technology, electrostatic and electro-dynamic transducers are applied.

13.6.4.1 Electrostatic Microphones

This type of transducer is represented by condenser microphones. Here, a change in the sound pressure causes a change of the distance between the movable electrodes. This induced change in the capacity of the condenser results in a voltage modulation. Condenser microphones have a large frequency range and a well-balanced frequency response with a very good impulse response. Thus, they are especially suitable for use in studios.

13.6.4.2 Electro-Dynamic Microphones

A main representative of this transducer principle is the moving-coil microphone, which is also called dynamic microphone. The membrane of these microphones is firmly connected to the coil, which is located elastically centred in a magnetic field. When the membrane moves, a voltage proportional to that movement is induced in the coil. Owing to its large robustness and its low sensitivity to overmodulation, dynamic microphones are very successfully used for soloists.

References

1. Haas H (1951) Über den Einfluß des Einfachechos auf Hörsamkeit von Sprache, Acustica:49–58
2. Freund A (1971) Akustikverbesserungsanlage der Seebühne in Bregenz und im Großen Kreml-Saal in Moskau. 7. JCA-Congr., Budapest (1971) Beitr. 19A5
3. Plenge G (1983) Sound reinforcement System with correct Localisation image in a big congress centre. 73. AES-Conv., Eindhoven Prepr. 1980
4. Steinke G (1985) New developments with Delta Stereofony System. 77. AES-Conv., Hamburg Prepr. 2187
5. Ahnert W (1985) Simulation of complex soundfields in enclosures and open-air-theatres. 77. AES-Conv., Hamburg Prepr. 2186
6. Steinke G, Fels P, Hoeg W, Sellin, U (1990) Das Delta-Stereofonie-System im neuen Kultur- und Tagungszentrum STADEUM. Audio Professional 5/6-(1990). S. 58–64
7. Steinke G, Ahnert W, Fels P, Hoeg W (1987) True directional sound system orientated to the original sound and diffuse sound structures - new applications of the Delta Stereofonie System (DSS). 82. AES-Conv., London (1987) Prepr. 2427
8. Ahnert W (1987) Problems of near-field sound reinforcement and of mobile sources in the operation of Delta Stereofonie System (DSS) and computer processing of the same. 82. AES-Conv., London (1987), Prepr. 2426
9. Zwicker E, Zollner M (1984) Elektroakustik. Springer, Berlin
10. Parkin PH, Morgan K (1965) Assisted resonance in the Royal Festival Hall, London. J Sound Vib 2:S.74–S.85
11. Variable Akustik im Zuschauerraum – MCR. Philips Offprint
12. Franssen NV (1968) Sur l'amplification des champs acoustique. Acustica 20:315
13. de Koning SH, Franssen NV (1969) Amplification of sound fields. 37. AES-Conv., New York (1969) Prepr. 697
14. de Koning SH, Keijser LC (1977) Multiple amplification of reverberation. 56. AES-Conv., Paris (1977) Prepr. 1208
15. Kuttruff H, Hesselmann H (1976/77) Zur Klangfärbung durch akustische Rückkopplung bei Lautsprecheranlagen. Acustica 36(1976/77) S.105–112
16. Kuhl W (1968) Notwendige Eigenfrequenzdichte zur Vermeidung der Klangfärbung von Nachhall. 6. Int. Conf. Acoustics, Tokyo (1968) E-2-8
17. Ahnert W, Reichardt W (1981) Grundlagen der Beschallungstechnik. S. Hirzel, Stuttgart
18. Berkhout AJ (1988) A holographic approach to acoustic control. J Audio Eng Soc, 36, No. 12
19. Berkhout AJ, de Vries D, Hemingway JR, Griffioen A (1988) Experience with the Acoustical Control System "ACS". Sound Reinforcement Conference, Nashville, 5.-8. Mai 1988
20. Griesinger D (1991) Improving room acoustics through time-variant synthetic reverberation. Lexicon Inc, New York
21. Griesinger D (1995) How loud is my reverberation? Lexicon Inc, New York

22. Barbar S (1995) Further developments in the design, implementation and performance of time-variant acoustic enhancement systems. Lexicon Inc, New York
23. Griesinger D (1996) Spaciousness and envelopment in musical acoustics. Lexicon Inc, New York
24. Eargle J (1986) JBL sound system design reference manual. JBL Professional, Northridge, CA
25. Cremer L (1971) Vorlesungen über Technische Akustik. Springer, Berlin
26. Wolfe I, Malter L (1930) Directional radiation of sound. JASA Vol.2, No.2 (1930) S. 201
27. Olson HF (1957) Acoustical engineering. D. Van Nostrand Company, New Jersey
28. Beranek L (1954) Acoustics. McGraw Hill, New York
29. Heil C, Urban M (1992) Soundfields radiated by multiple sound sources arrays. 92. AES-Conv., Wien (1992), Prepr. 3269
30. Bauman P, Urban M, Heil C (2001) Wavefront Sculpture Technology. 111. AES-Conv., New York (2001) Prepr. 5488
31. Davis D, Davis C (1975) Sound System engineering. Howard W. Sams & Co, New York
32. Davis G, Jones R (1987) Yamaha sound reinforcement handbook. Hal Leonard Publishing, Milwaukee
33. Goertz A (2001) Theoretische Grundlagen zur Anwendung von Line-Arrays in der modernen Beschallungstechnik. DAGA 2001, Hamburg

Sources of Sound

14

Ulrich Kurze

14.1 Generation of Sound

14.1.1 Broad-Band Sound

14.1.1.1 Sound Generation by Flow

More or less broadband sound is generated in air or other gases (general in fluids) by (see Fig. 14.1)

- Fluctuating volume flow
- Oscillating forces due to interaction of the fluid with solid boundaries
- Oscillating forces due to interaction of the fluid with moving bodies, and
- Viscous stress in the fluid flow

According to Lighthill's theory, one may distinguish sound sources in a free flow in terms of monopoles, dipoles, and quadrupoles [1]. The volume flow itself is irrelevant for monopole sound generation. It is the variation in time of the volume determined by the fluctuating mass flow, which generates sound. For a free jet, the spectral distribution of sound shows a maximum at the Strouhal number $fD/c \approx 0.1$, where f is the frequency, D is a characteristic length (e.g., the nozzle diameter), and c is the speed of sound in the jet [2]. At low frequencies, the spectral density increases with f^2, and at high frequencies, it decays as $1/f^2$.

The interaction of gases and rigid bodies is most relevant for the generation of airborne sound, while the interaction of liquids and solid bodies can be often neglected due to other more effective sound sources. A small fraction of the power involved in the fluid flow or the motion of solid bodies results in a considerable sound power (see Fig. 14.2). In the speed range U of fluids or solid bodies, that is small compared to the speed of sound c, i.e., for a Mach number $M = U/c \ll 1$, the sound power increases with the fourth to eighth power of the Mach number. The lowest exponent applies to small-scale equalization processes, e.g., in porous material. The highest exponent is related to free-stream conditions. Sound power generated by oscillating forces increases with an exponent 5–6, depending on the type of boundary or solid body. Velocities above the speed of sound result in shock waves, which underlie special mechanisms.

The unsteady inflow of fine air bubbles in a faucet is an example for the efficient generation of acoustic monopoles that cause clearly audible sound at low Mach numbers of the fluid flow. Coarse structures, which are characterized by a pressure loss coefficient, result in a speed dependence of sound generation closely related to that of dipoles which are less efficient than monopoles at low Mach numbers, but increase in efficiency at a higher rate with increasing flow velocity. Sound generation in free-stream jets needs consideration mainly for rockets and low-bypass jet engines. In industrial applications, it is suppressed by silencers.

Impulsive sound is generated in fluids when (see Fig. 14.3)

- A solid body moves at supersonic speed, e.g., an aircraft, a bullet or the end of a whip
- The sudden release of a fluid through a valve in a pipe causes a high sound pressure (waterhammer) and together with the steepening of wave fronts the creation of N-waves in the fluid; or when

U. Kurze (✉)
Müller-BBM GmbH, Robert-Koch-Strasse 11, 82152 Planegg, Germany
e-mail: ukurze@yahoo.de

G. Müller and M. Möser (eds.), *Handbook of Engineering Acoustics*,
DOI 10.1007/978-3-540-69460-1_14, © Springer-Verlag Berlin Heidelberg 2013

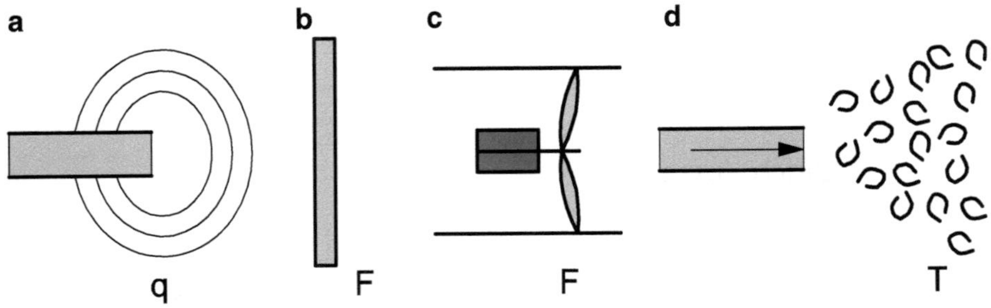

Fig. 14.1 Sound generation by (**a**) fluctuating volume flow emanating from the open end of a pipe, characterized by the volume velocity q; (**b**) and (**c**) interaction of the fluid flow with solid bodies, characterized by oscillating forces F; (**d**) free jet flow, characterized by Lighthill's stress tensor T

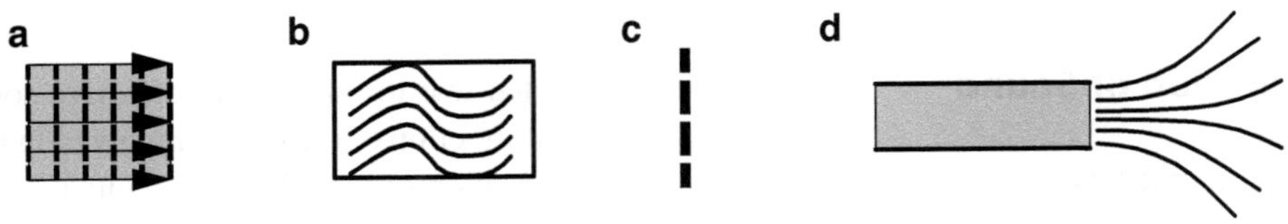

Fig. 14.2 Sound power W due to interaction of flow and solid bodies increases with Mach number M of the flow (**a**) in fine porous material as $W \propto M^4$, (**b**) and (**c**) in coarse materials and structures as $W \propto M^{5\cdots6}$, (**d**) in free-stream jets at low Mach numbers as $W \propto M^{7\cdots8}$

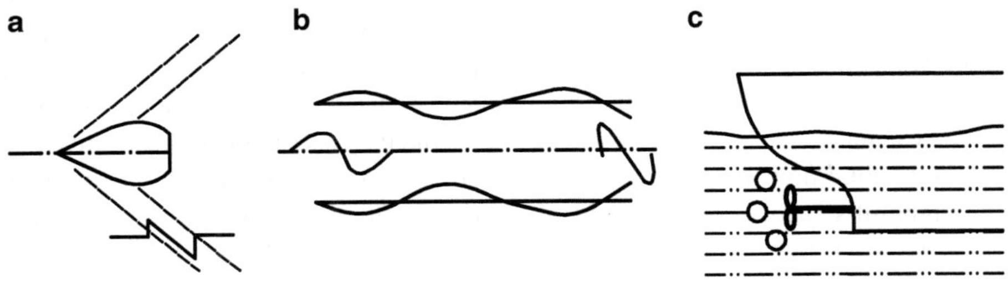

Fig. 14.3 Strong sources of sound due to interaction of fluid-borne sound and adjacent structures; (**a**) supersonic flying object; (**b**) waterhammer due to steepening of wave fronts in elastic pipes; (**c**) cavitating propeller

- Local underpressure in a fluid results in the expansion of gas bubbles, which subsequently collapse during overpressure (cavitation)

High peak pressures related to such impulsive sounds cause strong stress in adjacent construction materials. Water pipes and marine propellers may suffer severe damage.

As sketched in Fig. 14.3a, the spectral maximum of a supersonic bang depends on the effective length of the flying object. A bullet decelerates along the flight path. Consequently, the shock front is bent toward the flight axis. Aircrafts may fly along a path which causes focussing of the shock front and creation of a particularly strong sonic bang.

An impacted water pipe carries a fast longitudinal wave which first arrives at an observer with a comparatively small amplitude [3, 4]. It is followed by a wave of larger amplitude that is caused by the fluid pressure expanding the pipe wall, propagated after conversion into longitudinal and bending waves according to Poisson contraction, and is again radiated into the fluid. The propagation velocity of this wave is lower than the speed of sound in unbounded water due to the flexibility of the pipe walls.

In general, cavitation is the rupture of fluids and subsequent effects [5]. It occurs with energy deposit and pressure reduction below ambient pressure. Acoustic cavitation goes together with strong structural vibrations. The local pressure in the fluid shortly drops below the vapour pressure of the fluid, thus forming bubbles. The subsequent decay of the bubbles near a rigid boundary results in a jet pointing toward the surface, which creates a substantial stress in the surface material. This may cause damage or may be used for cleaning.

Weaker sound pulses emanate from a fluid surface hit by droplets of the fluid. The sound intensity depends on the mass and the velocity of the droplets, i.e., on the momentum, and on the number of droplets and the angle of impact. What is admired as a waterfall in nature is considered as a noise source in the vicinity of a cooling tower.

14.1.1.2 Structure-Borne Sound

The dynamic interaction of solid bodies results in structure-borne sound. This includes (see Fig. 14.4)

- Displacement-induced vibrations, such as those caused by the rolling of rail vehicle wheels on rails due to more or less corrugated running surfaces, and
- Impact-induced vibrations, such as those caused by the slamming of car doors or in bends of pipes carrying granular material.

Corrugations of the running surface of rail vehicle wheels at wavelengths in the critical range from 1 to 10 mm are primarily generated by the interaction with cast-iron block brakes at amplitudes of the order of 1 μm. Running surfaces of wheels braked by modern plastic blocks or of disk-braked cars are substantially smoother. Corrugations on rail running surfaces sometimes exhibit an almost periodical pattern in the same range of wavelengths. In a first-order approximation, it may be assumed that the effective combined roughness in a frequency band results from the sum of the squared roughnesses of wheels and rails. However, rail grinding may reduce the rolling noise even for relatively rough wheels thus indicating that the sound generation may depend on some correlation between wheel and rail roughness.

While the railroad example is based on displacement-induced vibrations, statistical energy analysis (SEA)

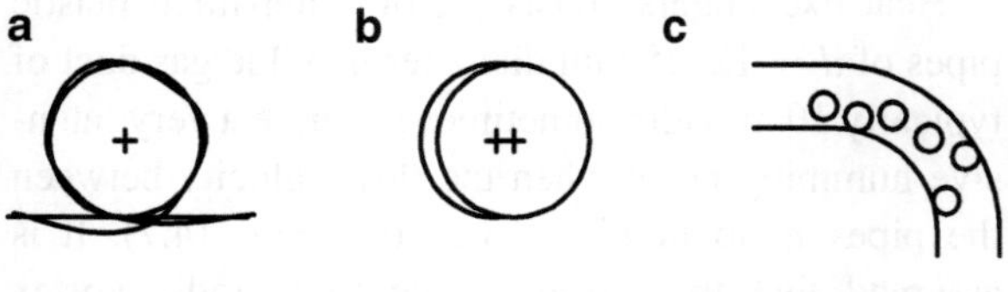

Fig. 14.4 Structure-borne sound caused by (**a**) displacement induced vibrations of railway wheels; (**b**) force induced vibrations of an excentric shaft; (**c**) impact induced vibrations of a pipe bend due to the flow of granular material

starts from force-induced broadband dynamic processes due to its original application on rocket engines. However, this application deals with the excitation of a structure by pressure fluctuations in a fluid. All common structural excitations by interaction with solid bodies refer to sinusoidal or narrow-band processes.

A momentum $m\,\Delta v = F\,\Delta t$ can be described by the rate of change Δv of the velocity of the participating masses m or by the duration Δt of the interaction between the bodies at force F. Thus, the description of a vibration source in terms of momenta involves aspects of both kinematical and force-induced vibrations.

14.1.1.3 Chemical Processes

In addition to mechanical causes for the creation of sound, there are other sources of energy. They include in particular,

- Fast chemical reactions like explosions that generate a pressure pulse and a volume flow, e.g., the muzzle report from a weapon, and
- Slower chemical reactions in combination with gas flow as in combustion processes.

Explosions are described in terms of the maximum pressure generated during the release of a specified energy. For a point source, this value of an outgoing shock wave decays with the third power of the reciprocal distance from it [6]. The subsequent relaxation process occurs with a time function comparable to that of the highly damped resonance of a bursting balloon which is controlled by the mass and radiation impedance of the surrounding air and the stiffness of the contained gas. Accordingly, the spectral distribution has a maximum at the resonance frequency and decays toward both lower and higher frequencies (see Fig. 14.5).

The muzzle report of small arms exhibits a pronounced directivity. In the frequency band around 1,000 Hz, which often determines the A-weighted overall sound pressure level, the level in the forward direction of the bullet is typically 6 dB higher and in the backward direction 6 dB lower than that in the perpendicular direction. Since the muzzle diameter is small compared to the wavelength of sound, the directivity must be controlled by the gas flow. Reflections from overhead barriers and other safety devices on a shooting range modify the directivity of an entire plant.

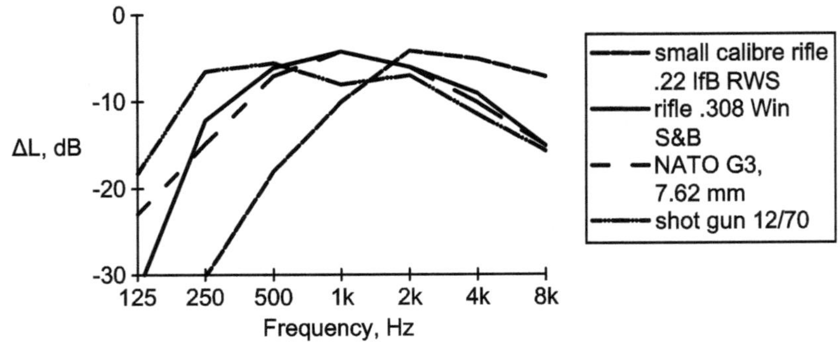

Fig. 14.5 Octave-band sound pressure level (SPL) of the muzzle report from various rifles re overall SPL

The sound generation of burners is also about proportional to the energy supplied, i.e., to the heating power, and depends on the swirl of the gas flow [7]. The maximum of the broadband spectral distribution depends on the diameter of the nozzle and of mixing zones. However, resonances in the feeding gas pipes and in the burner chamber may substantially affect the burning process and, consequently, the sound generation.

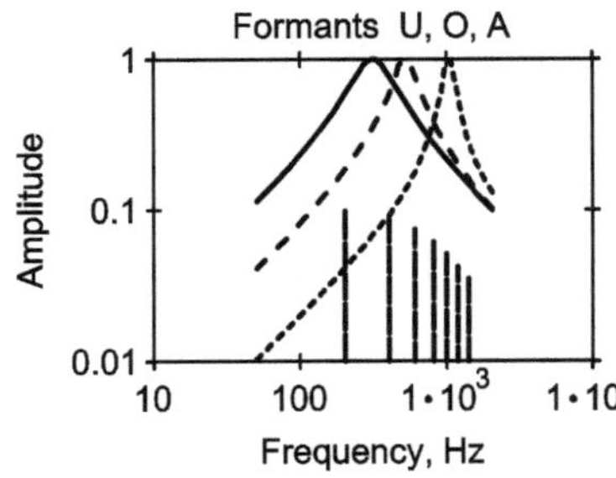

Fig. 14.6 Formation of vowels from harmonics (*perpendicular lines*) with resonators (*schematical*)

14.1.1.4 Electrical Discharge

Single discharge, such as a flash in the atmosphere or a spark starting electronic welding, or continuous discharge during arc welding causes the heating of air or other surrounding gases and, consequently, a substantial rate of change in volume. The strong fluctuation of discharge paths is known from flashes. In order to obtain a high reproducibility of spark sources employed for acoustical model tests, the arc path is pre-ionized. The variation of sound generated during arc welding can be used for quality control of the welding process [8].

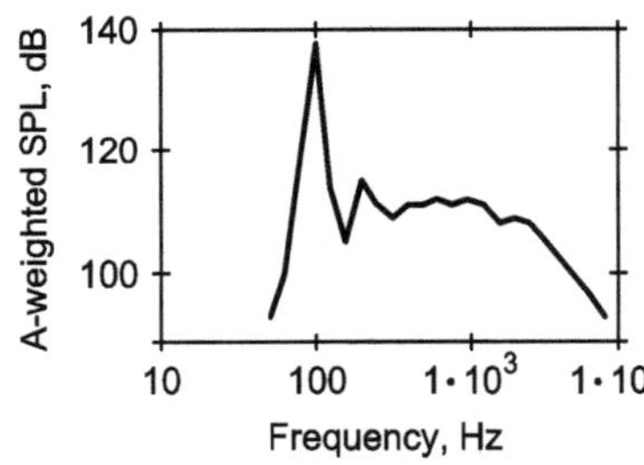

Fig. 14.7 Example for the humming noise from a heat exchanger in a large thermal power plant [9]

14.1.2 Narrow-Band Sound

14.1.2.1 Flow-Generated Sound

Pure tones or harmonic sounds are generated when

- The volume flow of a fluid is controlled to become periodical, as in the human vocal tract or in a siren
- Forces are effective periodically, e.g., between impeller and stator of a fan, or
- There is a feedback loop from the sound field to the source, particularly in the presence of a resonator, e.g., an organ pipe which controls the airflow past the labium or room resonances of a heat exchanger which control the vortex shedding from pipe bundles.

Mass and tension of the vocal cords determine the periodicity of breathing air flow through the glottis. The mouth including tongue and lips form the volume and the opening of a resonator, which determines the formants of vowels, i.e., the relative strength of higher harmonics (see Fig. 14.6).

Heat exchangers consisting of water-filled plastic pipes of $d = 12$–25 mm diameter in a flue gas duct of typically 10 m width sometimes generate a very intensive humming noise when the flow velocity between the pipes is about $U = 7$ m/s (see Fig. 14.7). It is assumed that this noise is due to periodic vortex shedding at a Strouhal's number $fd/U \approx 0.2$, which yields the frequency $f = 100$ Hz, as shown in Fig. 14.7. At this frequency, about six half wavelengths fit with the duct width. That lateral resonances indeed are involved in the sound generation is supported by the observation that it is possible to suppress the humming noise by bulkheads. Unfortunately, in spite of many research activities, there is no definitive knowledge about the occurrence and inhibition of the humming noise and expensive retrofitting of a plant is sometimes needed [9].

14.1.2.2 Structure-Borne Sound

In structural dynamics force-induced vibrations are mostly considered as the source of structure-borne sound caused by

- The imbalance of rotating masses
- Alternating frictional forces between moving bodies, or by
- Variation in effective stiffness, e.g. during teeth meshing in gear boxes

In addition, periodic impact excitation may play a role, as for a circular blade saw.

In all these cases, pure tones or narrow-band noise are created as structure-borne sound and subsequently radiated from connected components as airborne sound. Again, the intensity of sound generation depends on the participation of resonators.

For fixed revolutions of rotating masses, resonances of the system can be taken into consideration and must be strictly avoided by design. When variable revolution is required, ranges of resonances must be passed quickly. If this cannot be done, e.g., with a ship engine during high sea state, the revolution – of the propeller – must be limited.

The classic model for the generation of vibration by friction is shown in Fig. 14.8. A driven belt carries a mass, which is pulled by the belt due to frictional forces and restrained by a spring. If the frictional force is proportional to the belt velocity, there is no vibration but a displacement of the mass and the spring only. However, a dry Coulomb friction causes the mass to move alternating in stick and slip mode. The forced motion of the stick mode follows a free oscillation in the slip mode. As shown in Fig. 14.8, the resulting time function is very similar to that of the pure sine wave, which is in this example effective for 60% of the time. Therefore, a bowed string vibrates at about the same frequency as the plucked string. Even more detailed considerations of the frictional characteristic involved in the transition from high friction during the stick mode to low friction during the slip mode allow for this conclusion [10].

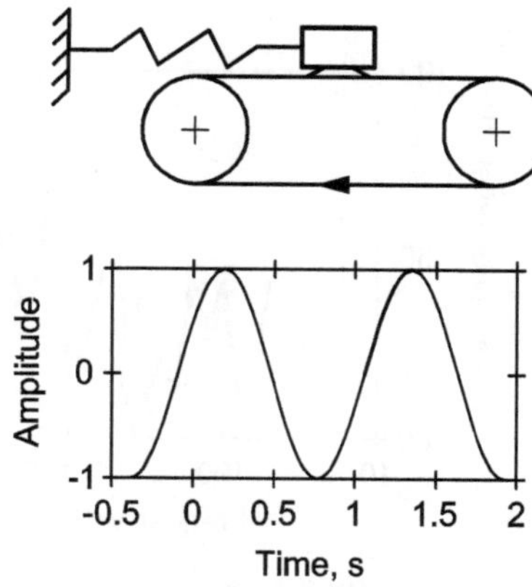

Fig. 14.8 Model for vibration excitation by dry friction and example for the time function of vibrations

There are numerous technical examples for vibration excitation due to dry friction, among others, break squeal (high stiffness) and clatter of windshield wipers (low stiffness).

14.1.2.3 Alternating Electromagnetic Fields

Electric power supply causes the humming of transformers due to magnetic strain, the crackling noise near power lines due to Corona effect and the mechanical noise of the maglev train resulting from the vehicle passage over the grooves of the long stator.

The fundamental frequency of the transformer hum is twice the line frequency. Under strong magnetization of the transformer kernel, several harmonics appear which are relevant for the A-weighted overall SPL of the radiated sound. Frequency controlled electric motors of modern trains generate striking complex tones.

The maglev train system of type Transrapid operates on a track equipped with long stators that are partitioned periodically at 0.258 m. Therefore, the driving speed in km/h is about equal to the operating frequency in Hertz. A tone at the operating frequency is audible along the track when it is energized and clearly measurable on the levitation frame (see Fig. 14.9).

Due to three phases of the stator current and the corresponding partitioning of the grooves the magnetic field oscillates at three times the operating frequency. This can be seen clearly from two harmonics at the levitation frame and – modified by the track impedance – also at the track.

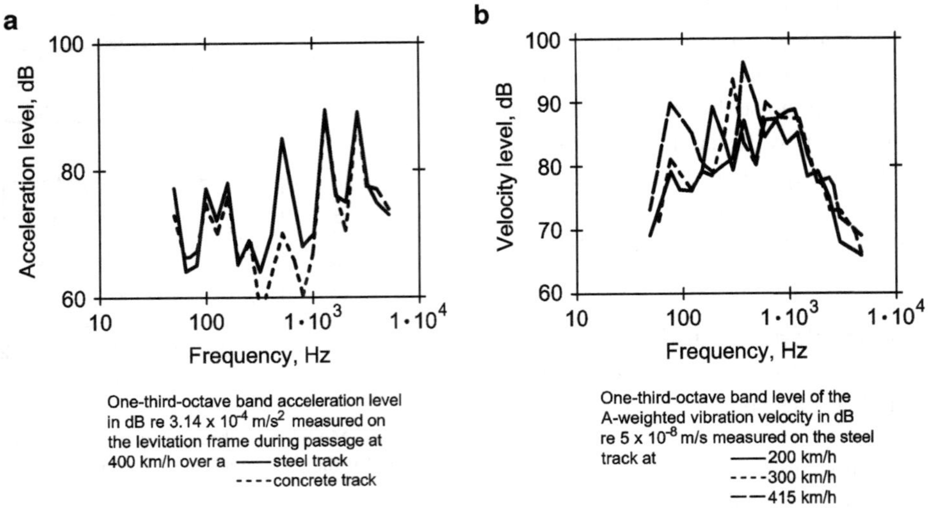

Fig. 14.9 Structure-borne sound measured on the train (*left side*) and on the track (*right side*) of the maglev system TR 07

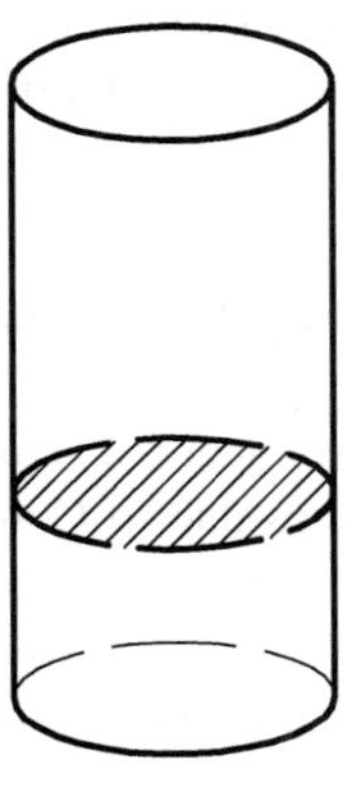

Fig. 14.10 Rijke tube with heated grid in lower half

14.1.2.4 Heat Source

In principle, a stationary heat source can act as a sound source, as this is shown in the laboratory by means of a heated grid positioned in an open tube (see Fig. 14.10). The air column in the tube oscillates in its fundamental mode if the tube axis is positioned vertically and the grid is located in the lower half of the tube – but not in the upper half. The phenomenon has been explained by Lord Rayleigh by the requirement that energy must be supplied to the sound field during the phase of overpressure in order to excite the oscillation. During the phase of overpressure, the convection of the heated air acts in the same direction as the displacement of sound particles when the grid is located in the lower half of the tube.

14.2 Source Term and Inner Impedance/Admittance

14.2.1 Model for Linear Acoustics

Many sound sources can be described in a simplified way either by a dynamic source force and an inner impedance or by a dynamic source velocity and an inner admittance. Both descriptions are equivalent but each one allows for improved lucidity in the limiting cases of

- A vanishing inner impedance (infinite admittance or mobility), when the dynamic source force equals the effective force at the source surface or
- A vanishing inner admittance or mobility (infinite impedance), when the dynamic source velocity equals the effective velocity at the source surface.

In general, the impedances and admittances are complex and are represented by the inertia of masses, frictional losses, and the stiffness of springs. When forces and velocities are represented by vectors, impedances and admittances become matrices. Forces may be generalized to include moments and velocities to include angular velocities.

Descriptions of structural dynamics are often limited to the forces and velocities (or accelerations) measurable at the source surface. The consideration of impedances is taken as borrowed from electrical engineering. However, accounting for impedances is an essential foundation of linear acoustics where all dynamic quantities are assumed as being small compared to the static quantities. The consideration of impedances allows for the distinction between passive components, which act independent of the sound or vibration amplitude, and active sources, which represent the effect of weak nonlinearities of the system at particular static conditions. Magnitude and phase or real and imaginary part of impedances may be determined as a function of frequency from measurements under external excitation or different dynamic loading of the source. The strength of an active source is typically proportional to the static load, the flow velocity, or a certain power of such quantities.

The applicability of the impedance concept is unlimited for compact components which are not larger than one-third of the wavelength of sound. For larger components, the stiffness of the excitation area must be taken into account. It depends on the size of the area. Thus, an impedance hammer yields the stiffness of an extended structure, e.g., a wall, which mainly depends on the shape and the size of the hammer head. Only the relatively small real part of the impedance, that determines the transmitted sound power, is independent of the head area at low frequencies.

In electrical engineering, the term impedance is defined by the ratio of voltage U and current I. According to the electro-mechanical analogy type II b (after Reichardt [11]), the particle velocity v in a sound field corresponds to the voltage U and the dynamic force F corresponds to the current I. Therefore, the electrical impedance is replaced by the reciprocal of the mechanical impedance, which is the admittance. The advantage of this analogy is the maintained topology of elements in series and in parallel when electrical and mechanical descriptions are exchanged. The schemes shown in Fig. 14.11 for velocity and force sources follow this analogy.

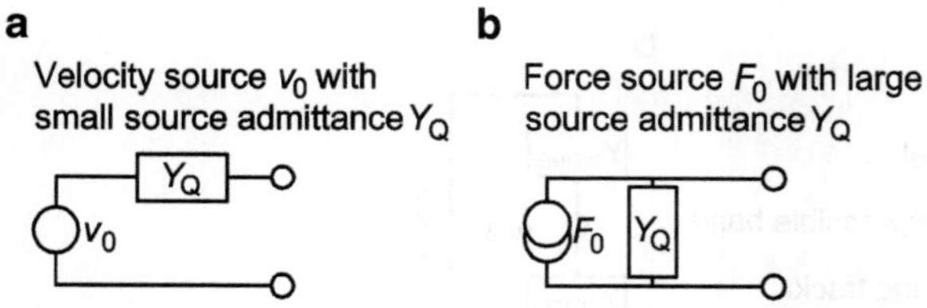

Fig. 14.11 Schemes of sources of different mobility in electro-mechanical analogy for unchanged topology

When a velocity source is switched off, the circle in the scheme is replaced by a connecting line. What remains is a small admittance. When a force source is switched off, the double circle vanishes and a small impedance remains.

Equivalences of electrical transformers may be used in schemes of mechanical elements to relate forces with moments and velocities with angular velocities. Extensions of the one-dimensional model to several dimensions are possible but no longer intuitive.

14.2.2 Matching of Power Output

Maximum power output of a sound source occurs for matching of the internal and external impedances (or admittances). This is the case when the impedances (or admittances) are equal in magnitude and opposite in phase (see Fig. 14.12).

The generation of airborne sound by dynamic forces is related to sources of rather small inner impedance. Consequently, the excitation of vibrations in rigid bodies of high impedance is very weak compared to the excitation of sound in the surrounding air. Note that the inertia of the mass of air co-vibrating with the source is part of the source impedance.

The structure-borne sound generated by wheels rolling over a rough surface is related to a very high source impedance and is efficiently transmitted in the floor by heavy wheels. The delivery service in a shop, for example, may be very annoying for residents in the same building.

When source and load form a mass-spring system with equal magnitudes of the imaginary parts at resonance and small losses, the conditions for power matching are almost met. About one-half of the available power is delivered to the environment and may cause a substantial noise and vibration impact. The other half may cause damage of the source – although beyond the range of linear acoustics.

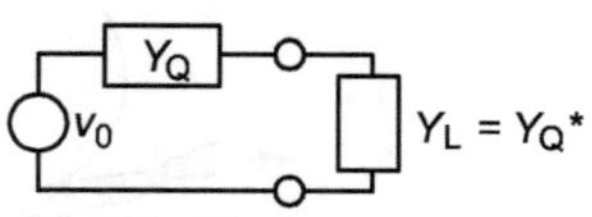

Fig. 14.12 Matching of the load admittance Y_L by the complex conjugate of the inner admittance Y_Q for maximum power output

14.2.3 Airborne Sound in Ducts

In free space, a small monopole sound source is loaded by the mass admittance of a spherical field and the comparatively small plane wave radiation admittance. In a duct, the sound field is composed of many modes, which provide a stiffness controlled load impedance below the cut-on frequency of wave propagation. Just below the cut-on frequency, the opposite phases of load and source impedance provide for a comparatively strong sound radiation, which depends on the position of the source close to or away from a node of the particular mode (see Fig. 9.24). The excitation of tube or duct walls, which have a high impedance, becomes important only when the airborne sound is substantially reduced by silencers.

14.2.4 Rolling Sound

14.2.4.1 Wheel/Rail

The roughness or corrugation of the running surfaces of wheel and rail are modelled by the varying thickness of an incompressible band, which is pulled through the contact area at the speed of rolling. It is shown together with the admittance model in Fig. 14.13.

The figure shows the topological consistency of the admittance model with the mechanical model. In addition, it contains an external link to a fixed reference point, which, however, is irrelevant for the dynamical behaviour. Essential is the information contained that

- The same force (current in the electrical model) acts on wheel and rail
- The deflections or vibration velocities (voltage in the electrical model) splits proportional to the admittances and
- A system exists capable of resonances involving the mass of the wheel and the stiffness of the rail on its bedding

A very stiff contact spring between wheel and rail has a very small admittance at low and medium frequencies and is then negligible in series with the much larger admittances of wheel and track. More important for the refinement of the model are other details and components of wheels and the track.

All of these admittances are small compared to the plane wave admittance of airborne sound in the surrounding air and therefore poorly matched for energy transfer. In addition, the radiation efficiency of components with dimensions small compared to the wavelength of sound in air is small as long as the admittance of the loading air mass is small at low frequencies. Significant sound radiation is restricted to resonances of vibrating bodies and to frequency ranges where the spatial distribution of vibrations provides for an acoustic loading of the vibrating surface with a substantial real part of the radiation admittance. For railroad wheels of 0.9 m diameter, this occurs at frequencies in the range of 2 kHz, where vibration modes of wheel web contain areas of opposite phase that are separated by distances of more than a quarter wavelength of sound in air.

14.2.4.2 Tire/Road

The sound generation from rubber wheels rolling on road surfaces is substantially more complex than that of steel wheels on steel tracks. At least three different mechanisms need consideration:

- The roughness of the running surfaces, i.e., the profile of tires and the texture of road surfaces, which causes the vibration of the tire and, subsequently, the radiation of airborne sound mostly out of the wedges formed by the rim of the wheel and the road surface
- The varying size of the effective contact area due to the profile of the rolling tire, which can be seen as a parametric source of structure-borne sound
- The so-called air pumping due to the displacement of air in the gaps of the profile, which directly generates airborne sound

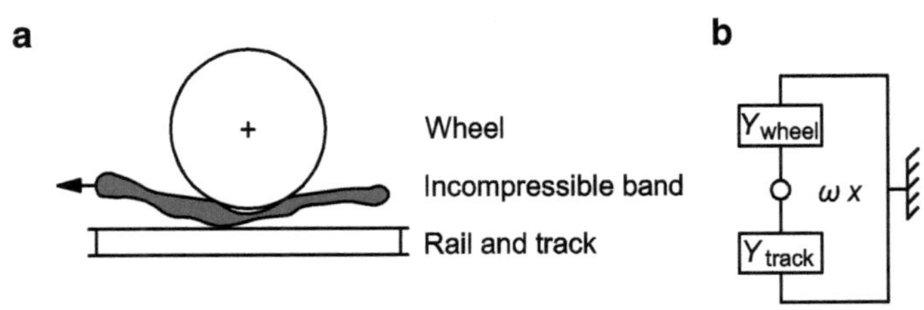

Fig. 14.13 Model for wheel-rail-vibrations due to the roughness of running surfaces; (**a**) mechanical model with incompressible band of varying thickness x, (**b**) admittance model with vibration velocity source ωx

In addition, there is a smacking sound when adhesion between smooth tire and road surface ceases. Differences between sound radiation from the front and rear end of the contact zone are sometimes attributed to the snapping of profiles.

An impedance model is of particular interest for the airborne sound. A velocity source may be employed to model the vibration excitation of the tire due to the roughness of tire and road surface. The radiation of airborne sound is affected by the geometrical properties of the tire parallel to the airborne sound admittance of the road surface, which has to be taken into account at least for open porous road surfaces. Cavity resonances inside the tire profile are discussed in connection with air pumping, but the effects are not clarified.

The linearity of the impedance model allows for an extension of its application from the frequency domain to the time domain in accordance with the general property of linear systems that the Fourier transform of a transfer function (defined in the frequency domain) is the impulse response (defined in the time domain) of the system. Impulse responses have been measured at extension joints of road bridges. Results are shown in Fig. 14.14 by two spectra determined from the pass-by sound pressure level L_{AFmax} of a car running at 80 km/h on a centerline at 3 m distance from microphones next to a multiple-element extension joint and at a position 25 m away. In the relevant frequency band around 800 Hz, the spectra exhibit a similar distribution but a difference in level by more than 10 dB. Thus, the extension joints excite the tire similar to the regular road surface – although much stronger – in the frequency range relevant for the A-weighted overall SPL. Instead of many small

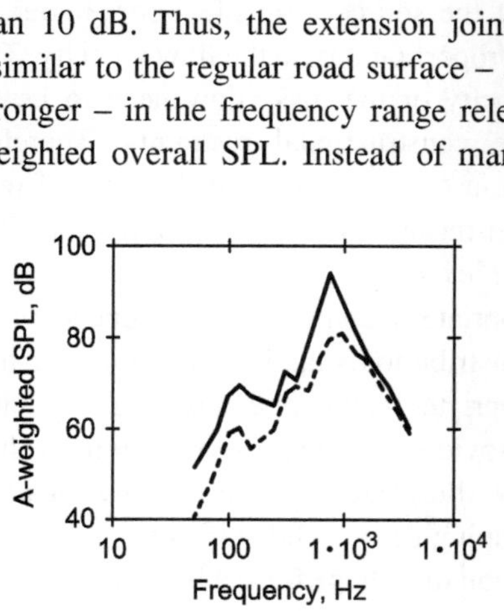

Fig. 14.14 A-weighted one-third-octave-band spectra of the sound pressure level L_{AFmax} in 3 m distance from the centre of the lane during pass-by of a car at 80 km/h; *straight lines* above the extension joints, *dashed lines* 25 m further down the road

excitations due to the roughness of a regular road surface, an extension joint yields one strong impulse. Effects of air pumping are not to be expected at extension joints and may explain the similarity of levels in the frequency range above 1,250 Hz.

The maximum of radiated sound in the frequency bands centred at 800 and 1,000 Hz is likely to be related to the facts that tire vibrations of opposite phase in the front and rear end of the contact zone do not cancel each other and that the tire width is sufficient to provide for a high radiation admittance of the wedge-shaped spaces between tire and road surface. The evaluation of numerous results from measurements covering various running surfaces, types of tires, and velocities aims at the development of a model accounting for all relevant sources of rolling noise in terms of source strength, impedance, and radiation efficiency [4].

14.2.5 Gears

Essential sources of gear noise are

- Teeth errors (surface roughness – similar to rolling – as well as imperfect involute surfaces and spacing caused by the manufacturing process) and
- Variation of the effective mesh stiffness due to alternate meshing of n and $n + 1$ teeth.

The scheme of Fig. 14.15 may be used to describe the generation and transmission of structure-borne sound for a pair of parallel-axis gears. The tooth meshing causes forces, which produce translatory motion of the gears, the shafting and the bearings according to their translatory admittance Y_t, and moments, which produce the rotatory motion of the gears, the shafting and the bearings according to their rotatory admittance Y_r. In the model, the coupling of translatory and rotatory motions is taken into account by ideal transformers.

As opposed to Fig. 14.13, where the contact stiffness was neglected, the mean mesh stiffness $s_=$ in Fig. 14.15 is of central importance. It determines resonances of the system composed of pinion and gear. As a rule, a pair of gears is operated far below the resonance frequency. Driving points above resonance are chosen for extreme gear ratios only. Operation at resonance under load must be avoided.

Corresponding to the vibration source of rolling in Figs.14.13 and 14.15 shows the teeth errors x, which

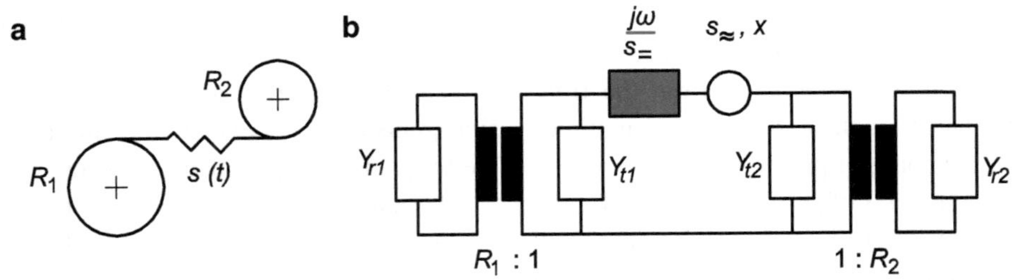

Fig. 14.15 Models for the vibration excitation of two meshing parallel-axis gears with base cylinder radii R_1 and R_2, (**a**) mechanical model including the time-dependent stiffness $s(t)$; (**b**) admittance model with translatory admittances Y_t and rotatory admittances Y_r of gears, shafting, and bearings, admittance of the mean mesh stiffness $s_=$ including the source of vibrations due to variations in mesh stiffness $s_{\approx}$ and teeth errors x

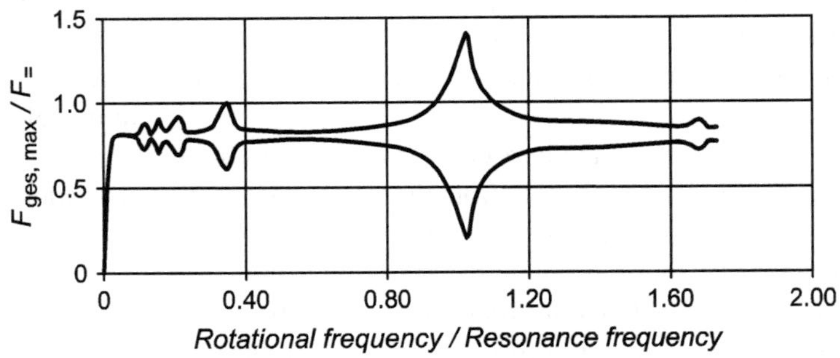

Fig. 14.16 Relative load magnification during run-up of single-stage spur gears, calculated after Troeder et al. [12] for small variation of meshing stiffness $s_{max}/s_{min} = 1.07$, overlap 1.5 and loss factor $D = 0.03$

refers to the spectral distribution resulting from the wavenumber distribution of errors along the perimeter and the speed of rotation. Furthermore, a dynamic displacement by parametric excitation due to a Fourier component of the dynamic stiffness component $s_{\approx}$ is taken into account, which results from the product of relative dynamic stiffness $s_{\approx}/s_=$ and the mean tooth displacement. Within the validity of linear acoustics, it follows from the first terms of a Taylor series for force, stiffness, and displacement that the dynamic stiffness component is proportional to the mean transmitted force $F_=$ and inversely proportional to the square of the mean mesh stiffness.

For standard involute teeth of spur gears, alternately one and two teeth are meshing. The variation of relative dynamic stiffness is substantial but can be reduced by special high teeth or by helical gears when this provides for almost constant integer overlap of teeth. However, this is not fully achievable since the overlap depends on the load. When harmonics of the relative dynamic stiffness coincide with the system resonance, a load magnification and corresponding increase in sound generation can be observed at frequencies far below the main resonance (see Fig. 14.16).

14.3 Measures for Sound Reduction

14.3.1 Reduction of the Source Terms

14.3.1.1 Flow-Generated Sound

Strategies for constructional and operational measures aiming at the source term of airborne sound follow from its proportionality to the flow velocity. The mean flow velocity or the mass flow need to be taken as invariable constructional restraints. Therefore, the strategy must aim at reducing the ratio of maximum value over mean value of the flow velocity. This goal can be achieved by

- Appropriate shaping of intake ducts for uniform flow distribution, e.g., by suitable intake nozzles
- Appropriate shaping or embedding of exhaust ducts that provide for a larger mixing zone with secondary air, thus reducing free jet noise or noise from jets impinging on solid surfaces
- Selection of a shape for fan blades that provides for a uniform radial load distribution
- Selection of the shape and distance of guide vanes for fans that provide for minimum temporal and spatial variation of the flow

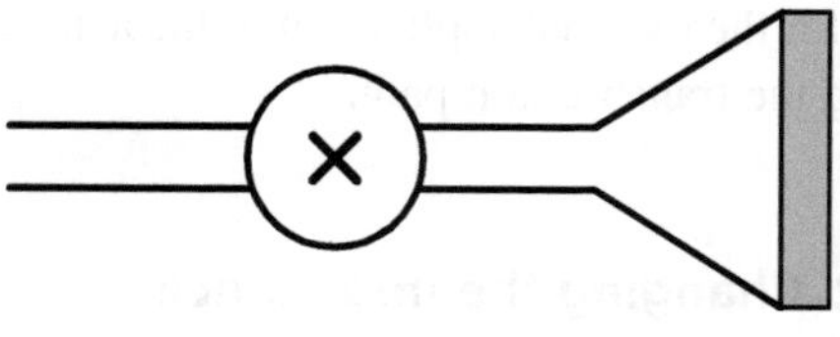

Fig. 14.17 Sinter metal plate in the steam exit plane of a glass pot sealing plant to avoid valve noise (schematical) [14]

- Avoiding or appropriate shaping of inserts in flow ducts
- Avoiding sharp bends and constrictions in flow ducts, but also by
- Controlling of fan speed according to the actually required cooling power and not for the potential extreme case.

Substantial noise reduction has been observed for jet engines with high by-pass ratio, a windfall profit of power enhancement, that can be used in other areas as well, e.g., quiet compressed air pistols [13, 14]. Screech noise emanating from neighbouring valves in branches of a steam pipe could be avoided by increased distances between the valves. Noise from a plant for sealing glass pots with caps was mainly generated by a pressure reduction valve in a narrow steam pipe (see Fig. 14.17). After insertion of a sinter metal plate of suitable flow resistance in the large exit area, the pressure loss at the valve was substantially reduced and a reduction of flow noise by more than 25 dB in A-weighted SPL was achieved.

In order to avoid Karman vortices and aeolean tones, the periodic vortex shedding can be inhibited by spirals or other irregularities mounted on rods, pipes, chimneys, and similar constructional elements exposed to flow [13].

Pressure reduction in pipes for gases or steam does not always require single-stage control valves but can be achieved by multi-stage perforated plates or by mesh wire for spatially expanded pressure release. The noise reduction obtained in practical cases is remarkable. Similar means can be applied in pipes carrying fluids to prevent cavitation [13].

14.3.1.2 Structure-Borne Sound

The following measures are taken to reduce the source strength of structure-borne sound:

- Smoothing of surfaces (except from undulations needed by functionality or irrelevant for acoustical reasons in the range of low wave numbers), providing temporal dilatation, and spatial equalization of transitions, e.g., grinding of rail surfaces
- Smoothing of the time function of forces acting between moving parts (except from variations needed by functionality or irrelevant for acoustical reasons in the range of low frequencies), providing temporal dilatation and reduction of oscillating forces, e.g., by oblique cutting, helical gears, rhombic plates on cross bars for elastic joints of road constructions; but also by lubrication, which reduces the time dependence of frictional processes or shifts them entirely into the viscous range, and
- Reduction of impacts (clatter, rattle) by means of narrow tolerances for construction elements, low drop heights of work pieces, and damping of unnecessary motion. Impacts can be fully avoided by applying a bending machine instead of a hammer, pressing nuts instead of hammering them, or employing a shredder with slowly rotating rollers instead of a cutting device.

14.3.1.3 Waves Propagating in Closed Loops

According to Cremer [15], an essential criterion for the generation of waves or oscillations is the correct phase connection in a closed loop. Accordingly, measures in the propagation path along the loop, which cause a phase shift, may not only attenuate but completely eliminate the source. Whether such a measure is successful or just shifts the source to another frequency band depends on the total gain in the feedback loop. As a rule, a sufficient attenuation effective over a broad frequency band is more reliable than detuning of the loop.

An example for the suppression of low-frequency sound has been studied in a closed-loop wind tunnel. Even for weak excitation of sound by the fan, there is a range of low frequencies where sound waves superimpose at the same phase after each loop through the tunnel. This range of amplification is limited to exclusive propagation of the fundamental mode, which requires low frequencies with half wavelength larger than the tunnel diameter. Mass and stiffness of the air determine two degrees of freedom required for oscillation or wave propagation. When a small mass is effective parallel to the stiffness of the air at the tunnel wall, then the mass is moved rather than the air is compressed at frequencies below resonance of mass and stiffness. Under such conditions, the stiffness of air no longer serves as the second reservoir of alternating energy (or the second degree of freedom) required for longitudinal wave propagation in the tunnel and sound decays exponentially with distance from the source. This effect can be used by opening a number of holes in the tunnel wall. It employs the mass of the air or of light membranes in the holes, but does not depend on sound transmission to the outside. A similar effect of low frequency sound attenuation is known from rectangular ventilation ducts made of light weight sheet metal. The mass character of the wall impedance inhibits wave propagation, while solid duct walls do not.

Another similar effect has been observed in a special water-filled pipe [16]. Regular water pipes have a circular cross-section and are therefore very stiff. Sound is well transmitted. In a water-filled pipe of rectangular cross-section, however, low-frequency sound can suffer a substantial attenuation. Due to the large stiffness of water and the long wavelength of fluid-borne sound, the effect can be observed also for not very light-weight walls over a large frequency range.

Below the frequency range in which the sound propagation can be controlled by massive duct walls, elastic elements – so-called bellows – can be employed. They perform as reflective silencers for fluid-borne sound and, at the same time, reduce the propagation of structure-borne sound along the duct walls.

14.3.1.4 Active Measures

The active reduction of the source strength by means of controlled electro-mechanical actors is done close to the source via injection of velocity or force of opposite phase. For airborne sound from a velocity source, a loudspeaker can be used for the injection in order to convert a monopole source into a dipole source with considerably reduced radiation efficiency. In practice, large volume flows of sound, e.g., from a reciprocating engine or a compressor, cannot be compensated by means of loudspeakers but require duplicate engines that run at opposite phase.

As a rule, the treatment of a force source requires to open the transmission path and to insert an element for the injection of the opposing force. This element may be a piezo crystal or – for larger deflections at lower static loads – an electro-dynamical transducer. The provision of an auxiliary mass for the reaction force allows for the external application of the actor without opening the transmission path.

14.3.2 Changing the Impedance

As shown by the examples, the point of intersection between source and load impedance is not always clear. It may be positioned very close to the source before wave propagation starts, or at the connection of other construction elements so that elements providing for the transmission loss of structure-borne sound or for the attenuation of airborne sound are part of the source. In accordance with the general statement that sound reduction close to the source is most effective, the term source impedance covers the wider range in the following.

The reduction of airborne sound from a low-impedance source is primarily achieved with massive walls or stiff enclosures of high impedance. As a rule, it is limited not only by openings required for heat transfer, material flow, or manual access, but also by leakages. If necessary for improved noise control, the openings must be equipped with silencers and seals.

The reduction of fluid- or structure-borne sound is achieved by resilient elements, i.e., by low-impedance elements. They provide for a mismatch with adjacent elements of high impedance. Resilient elements include bellows, elastic hangers of pipes as well as single and multiple (containing massive intermediate elements) elastic supports for machinery. A speciality is the resilient internal lining of pump casings, originally designed for protection from corrosion, but also acoustically very efficient due to reduction of the effective source impedance.

The elastic support of machinery is always going along with a resonance of the mass of the machine and the dynamic stiffness of the elastic support as well as some effects of connected construction elements. In general, vibration excitation in the frequency range of such resonances cannot be excluded. To avoid unacceptable amplification at the resonances, steel springs are specially damped while rubber material usually provides sufficient natural damping.

The position of a machine on a cantilever beam, a thin floor, or a scaffold is often affected by a considerable load admittance. In these cases, the insertion of a resilient support is not very effective unless the foundation impedance is increased by stiffeners, additional supports, or additional massive elements. All of these measures are usually applicable in plants. For vehicles, however, which require a low-weight construction, stiffeners can be applied exclusively. Since such elements are not reacting as compact elements in the entire frequency range of interest, noise control in vehicles is generally more difficult.

14.3.3 Isolation, Detuning and Damping of Resonators

Occasionally, equipment or parts of it radiate sound without being necessarily connected to the sound source, e.g., the instrument control board of a compressor station [13] or the noise barrier on a railway bridge. In such cases, noise control can be achieved by spatial separation or by transmission loss of structure-borne sound.

The generation of flow-generated pure tones often depends on particular conditions for the geometry of the resonator, on the direction of flow, and on the flow velocity. This is well known from whistling on a key. Whistling sounds are also generated by machine tools when holes are close to rotating cutting devices. Covers on such holes or a reduction in size may solve the problem.

As a rule, safer and effective measures for a broad frequency range result from the damping of resonators. For damping of airborne sound, absorbers are applied as thin layers in the neck or at the mouth of resonators and as thicker layers in the volume of resonators or at the walls of larger spaces. For damping of structure-borne sound, it is occasionally sufficient to use the viscosity of air between two metal sheets attached to each other or separated by a small distance. More efficient are layers of viscous materials [17].

Circular saw blades represent resonators of high quality, which are excited to screech noise already by airflow when idling, but even stronger by mechanical impacts when cutting. The source strength increases with increasing cutting speed and is particularly annoying for wood and aluminium material. Results from investigations on the positioning of the saw blade in a device self-adjusting to a very small clearance – similar to the disk brakes on vehicles – have been successfully demonstrated. However, only damped compound saw blades represent the state of- the art [13].

References

1. Dowling AP, Williams JE (1983) Sound and sources of sound. Ellis Horwood Ltd., Australia
2. Powell A (1997) Aerodynamic and jet noise. In: Crocker MJ (ed.) Encyclopedia of Acoustics, Chapter 28, Wiley, New York
3. Tijsseling AS (1996) Fluid-structure interaction in liquid-filled pipe systems: a review. J Fluids Struct 10:109–146
4. Beckenbauer Th (2003) Reifen-Fahrbahngeräusche: Minderungspotenziale der Straßenoberfläche. Plenarvortrag DAGA, Aachen
5. Lauterborn W (1997) Cavitation. In: Crocker MJ (ed) Encyclopedia of acoustics, Chapter 25. Wiley, New York
6. Raspet R (1997) Shock waves, blast waves, and sonic booms. In: Crocker MJ (ed) Encyclopedia of Acoustics, Chapter 31. Wiley, New York
7. Cabelli A et al (1987) Control of noise from an industrial gas fired burner. Noise Control Eng J 29(2):38–44
8. Prezelj J, Cudina M (2003) Noise as a signal for on-line estimation and monitoring of welding process. Acta Acustica United Acustica 89:280–286
9. Gilg J (1998) Akustische Resonanzen in Wärmetauschern. VGB-Workshop, Essen
10. Woodhouse J (2003) Bowed string simulation using a thermal friction model. Acta Acustica United Acustica 89:355–368
11. Reichardt W (1960) Grundlagen der Elektroakustik. 3. Aufl. Geest & Portig, Leipzig
12. Troeder Ch, Peeken H, Diekhans G (1978) Schwingungsverhalten von Zahnradgetrieben. VDI-Berichte Nr 320: 273–288
13. Ingemansson S (1994–1998) Noise Control – Principles and Practice. Noise News International Vol. 2 (1994) No. 2, 107–115, No. 3, 185–193, No. 4, 243–249, Vol. 3 (1995) No. 1, 45–51, No. 2, 119–127, No. 3, 177–183, Vol. 4 (1996) No. 1, 39–45, Vol. 6 (1998) No. 3, 157–165, No. 4, 217–223
14. Kurze UJ et al (1987) Lärmminderung am Arbeitsplatz III – Beispielsammlung. Schriftenreihe der Bundesanstalt für Arbeitsschutz – Forschungsanwendung – Fa 14, Dortmund
15. Cremer L (1975) Vorlesungen über Technische Akustik. 2. Aufl. Springer, Berlin
16. Kurze UJ, Hofmann P (1997) Grenzen der Leitungstheorie für wassergefüllte Rohrleitungen. DAGA 97:445–446
17. Schommer A (1982) Lärmarm Konstruieren VI – Körperschalldämpfung durch Kunststoffschichten an Strukturen aus Stahl. Bundesanstalt für Arbeitsschutz und Unfallforschung, Forschungsbericht Nr. 312, Dortmund

Road Traffic Noise

15

Thomas Beckenbauer

15.1 Importance of Road Traffic as Noise Source

Road traffic is the most interfering noise source in developed countries. According to a publication of the European Union (EU) at the end of the twentieth century [1], about 40% of the population in 15 EU member states is exposed to road traffic noise at mean levels exceeding 55 dB(A). Nearly 80 million people, 20% of the population, are exposed to levels exceeding 65 dB(A) during daytime and more than 30% of the population is exposed to levels exceeding 55 dB(A) during night time. Such high noise levels cause health risks and social disorders (aggressiveness, protest, and helplessness), interference of communication and disturbance of sleep; the long- and short-term consequences cause adverse cardiovascular effects, detrimental hormonal responses (stress hormones), and possible disturbance of the human metabolism (nutrition) and the immune system. Even performance at work and school could be impaired.

The Green Paper on Future Noise Policy in the EU, published by the European Commission in 1996 [1], quoted that the social costs of transport noise range between 0.2 and 2% of the total GDP of the EU. A recent study [2] showed that the social costs of traffic noise amount to approximately 40 billion Euro per year in the EU. This corresponds to a loss of 0.4% of the total GDP. Ninety percent of these costs are caused by road traffic comprising passenger cars and trucks. Thus, road traffic noise has multiple effects on both human beings and the society.

Heavy duty vehicles (trucks) radiate non-A-weighted sound power, which is equivalent to the sound power of about ten passenger cars at low speeds and four passenger cars at high speeds.

Moreover, the noise impact of road traffic covers a large area if no noise control measures are taken. Provided that sound propagation is not obstructed, heavily trafficked roads with a volume of more than 100,000 vehicles per 24 h cause equivalent sound pressure levels during night time in a more than 4 km-wide noise corridor along the roadside, which exceed 40 dB(A). This value is true for the prevention of traffic noise in residential areas in Germany at night and is considered to be an optimum for the reception-related target of an ambitious future noise policy in Europe [3].

In future, the traffic volumes for the different transport modes are expected to increase significantly. This undoubtedly means an increase in the number of noise sources and in noise emission. Particularly, the traffic situation in Central Europe suffered a dramatic and permanent change caused by the fall of the frontiers with the countries in Eastern and Southeast Europe in the early 1990s. This region became the most important turn-plate for goods and through traffic in Europe. Since 1997, the goods road traffic volume increased from 236 billion tkm (freight in tons times distance in km) to 330 billion tkm in 2004 [4]. For the year 2015, a growth of more than 40 billions tkm up to 374 billion tkm is predicted. This corresponds to an increase of 58% in the period from 1997 until 2015. The passenger road transport volume (number of persons times

T. Beckenbauer (✉)
Müller-BBM GmbH, Robert-Koch-Straße 11, 82152 Planegg, Germany
e-mail: Thomas.Beckenbauer@MuellerBBM.de

G. Müller and M. Möser (eds.), *Handbook of Engineering Acoustics*,
DOI 10.1007/978-3-540-69460-1_15, © Springer-Verlag Berlin Heidelberg 2013

distance) will increase as well in this period, certainly on a lower level of about half of the goods road traffic volume.

15.2 Noise Emission of Single Vehicles

15.2.1 Terms and Definitions

The technical system "road traffic" is divided into three subsystems which are directly contributing to the noise emission of road traffic:

- Vehicle
- Tire
- Road

It is the interaction of the tire rolling on a road surface which produces noise that can be heard in the distance. Therefore, it is more reasonable to deal with the intrinsic sound sources of motorized vehicles that are:

- Engine
- Powertrain
- Air intake and exhaust gas system
- Tire/road contact
- Incident airflow of the body and of the parts attached to the vehicle

In order to describe the noise emission of a vehicle, the above-mentioned sound sources can be combined in three principal components:

- Propulsion noise
- Tire/road noise (rolling noise)
- Airflow noise

where propulsion is meant to cover engine, powertrain, air intake, and exhaust gas system. Compared to the propulsion and the tire/road noise, the aerodynamic noise which is excited by the incident airflow around the vehicle body can be disregarded because it does not dominate the total noise level of passenger cars below 130 kmh and that of commercial vehicles below 100 kmh.

Four main categories of motorized vehicles are distinguished in this chapter. The main aspects the categorization is based on are gross vehicle weight and number of wheels, which cover indirect factors of the noise emission of single vehicles (see Table 15.1) as well. From an acoustical point of view, the driving condition of a vehicle has also to be taken into account. There are important differences in noise production, depending on whether a vehicle is driven at constant or variable speed, whether it runs on flat or on up- and downhill road. Differences in sound emission of vehicles driven at conditions other than the constant speed condition are due to a higher engine load at a particular ignition rate, lower gear ratios, and bigger forces on mechanical components. Concerning measurements of driving vehicles under well-controlled conditions, two situations are often distinguished:

- Pass by
- Coast by

The pass-by situation is characterized by the fact that the engine is on during the passing of the vehicle, whereas coast by instead indicates the engine-off situation. According to the measuring configuration used for statistical pass-by evaluations of single vehicles which are moving in a steady traffic flow and the configuration used for acoustic type approval measurements, the terms pass-by level and coast-by level are often considered to be specified for a distance of 7.5 m to the center of the lane and for a height of

Table 15.1 Vehicle categories

Category	Name	Description	Category according to the ECE/EU type approval procedure
1	Light vehicles	Passenger cars, delivery vans up to 3.5 tons, SUVs, MPVs including trailers and Campmobiles	M1 and N1
2	Medium heavy vehicles	Medium heavy vehicles, delivery vans of more than 3.5 tons, buses, touring cars, etc. with two axles and twin tire mounting on rear axle	M2, M3 and N2, and N3
3	Heavy vehicles	Heavy duty vehicles, touring cars, buses, with three or more axles	M2 and N2 with trailer, M3 and N3
4	Powered two-wheelers	Mopeds, tricycles, or quads	
		(a) with 50 cc	L1, L2, and L6
		(b) with more than 50 cc	L3, L4, L5, and L7

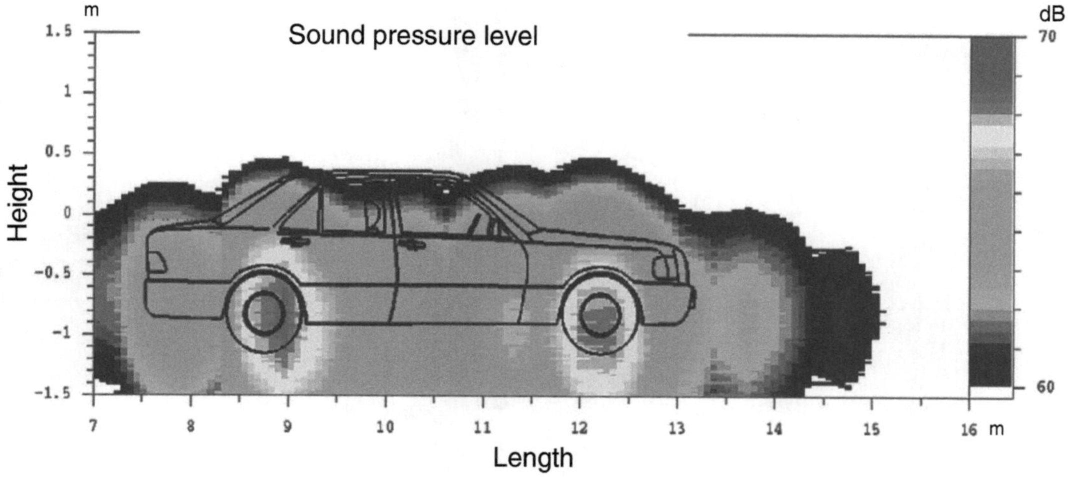

Fig. 15.1 Spatial distribution of the sound pressure level caused by a rolling vehicle in the frequency band between 280 and 4,500 Hz

1.2 m above the lane. Unless otherwise stated, the terms "pass-by level" and "coast-by level" used in this chapter fulfill this specification.

15.2.2 Vehicle Noise

In real traffic situations, the propulsion noise covers a range of about 35 dB [5]. A portion of 15 dB depends on peculiarities of vehicles which identify the range from the most silent passenger car to the noisiest truck. The 20 dB level depends on the engine speed and load in normal driving condition. The tire/road noise covers a range of about 35 dB for the speed varying between 10 and 150 km/h. Another ca. 15 dB of noise level differences are related to the tire/road combination, thus giving a total dynamic range for the tire/road noise of 50 dB. Comparing absolute noise levels, it turns out that tire/road interaction is an important noise source for almost all vehicles at speeds exceeding 50 km/h. By means of acoustic imaging techniques, it can be shown that the noise caused by the rolling tire exceeds the noise from other noise sources by at least 10 dB [6]. Figure 15.1 shows an example of such an acoustic image taken by means of the microphone array technique. This is true for passenger cars at high speeds as well as at low speeds in urban areas.

The sound power level L_W is not only a universal acoustical descriptor for the strength of a noise source but also allows the straightforward comparison of the emission potential of different sound sources. In Fig. 15.2, the A-weighted sound power levels of tire/road noise, propulsion noise, and the total noise are plotted versus the speed of the vehicle. On an average, the curves shown in Fig. 15.2 are valid for vehicles driving at constant speed on a dry and flat road surface made of stone mastic asphalt 0/11 or dense asphalt concrete 0/11 which is not older than 2 years and in good condition. The values are related to an air temperature of 20°C and are due to the average European vehicle fleet in different categories. This means that the light vehicle category 1 is characterized by using nonstudded tires, an average tire width of 187 mm, a portion of 19% diesel cars, and a portion of 10.5% delivery vans. The medium heavy vehicle category 2 comprises trucks with two axles, whereas the heavy vehicle category 3 contains trucks with four axles. The powered two-wheeler category 4 contains 35% motorcycles with illegally replaced exhaust silencer systems.

The noise produced by motorized vehicles depends on driving speed, as is clear from Fig. 15.2. However, the increase of the sound power level with speed is different for propulsion and for rolling noise. Moreover, the gradient is different for different speed ranges (except for the propulsion noise of light vehicles). Starting at high gradients and low speeds, the increase of the sound power level flattens out with increasing speed.

The noise emission of light motor vehicles is clearly dominated by the rolling noise, even at low speeds. In contrast, the total noise emission of duty vehicles is dominated by propulsion noise at low

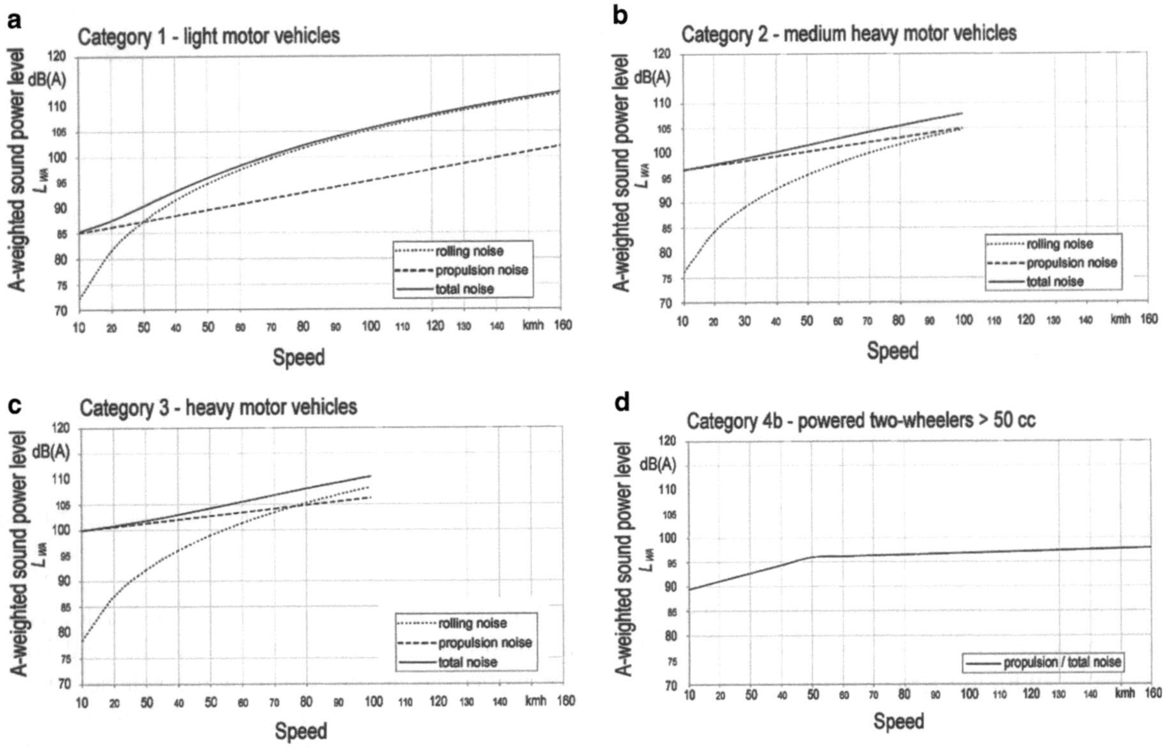

Fig. 15.2 Total A-weighted sound power level L_{tot} and sound power levels of propulsion noise and tire/road noise depending on speed of the vehicle; (**a**) light vehicles (passenger cars), (**b**) medium heavy duty vehicles, (**c**) heavy duty vehicles, and (**d**) powered two-wheelers with more than 50 cc cylinder capacity and normal driving condition. All data derived from [7]

speeds. At higher speeds, propulsion noise is slightly exceeded by rolling noise in category 3. For heavy vehicles, there is a breakeven point between the sound power levels of the two noise components at 100 km/h for medium heavy vehicles and 75 km/h for heavy vehicles. The higher speed value for medium heavy vehicles is due to the rolling noise, which is much lower for this vehicle category (two axles) than for the other (four axles), whereas the propulsion noise levels differ by not more than 5 dB. Powered two-wheelers do not produce a significant portion of tire/road noise; in other words, the total noise is mainly propulsion noise. The total A-weighted sound power levels given in Fig. 15.2 represent average values for the European vehicle fleet in different vehicle categories. In practice, the pass-by levels of single vehicles vary more or less from these average values. Figure 15.3 shows this variation in terms of relative numbers of vehicles that vary from the average value by a distinct level difference. For light vehicles (left), it is clear that 90% show level differences of not more than ±2 dB. Heavy vehicles yield a slightly bigger variation.

In addition to technical differences between vehicles, tires, and road surfaces, there are meteorological and operational factors affecting the noise emission of vehicles:

- Precipitation yielding wet driving conditions
- Slope of the road yielding an engine load which is different from that on a flat road
- Operational conditions of the vehicle such as accelerated motion

According to the findings in [8], the effect of wet road conditions on vehicle noise is significant. If there is a closed water film on the road surface, the vehicle noise levels change substantially at frequencies above 2 kHz. Figure 15.4 shows the spectral level differences of statistical pass-by levels for light vehicles (vehicle category 1) between the wet and the dry condition of a stone mastic asphalt depending on speed. Level differences are substantially bigger for low speeds than for high speeds. Accelerating and decelerating vehicles as well as vehicles driving along ascending or descending roads produce more propulsion noise than driving in steady condition and on a flat road,

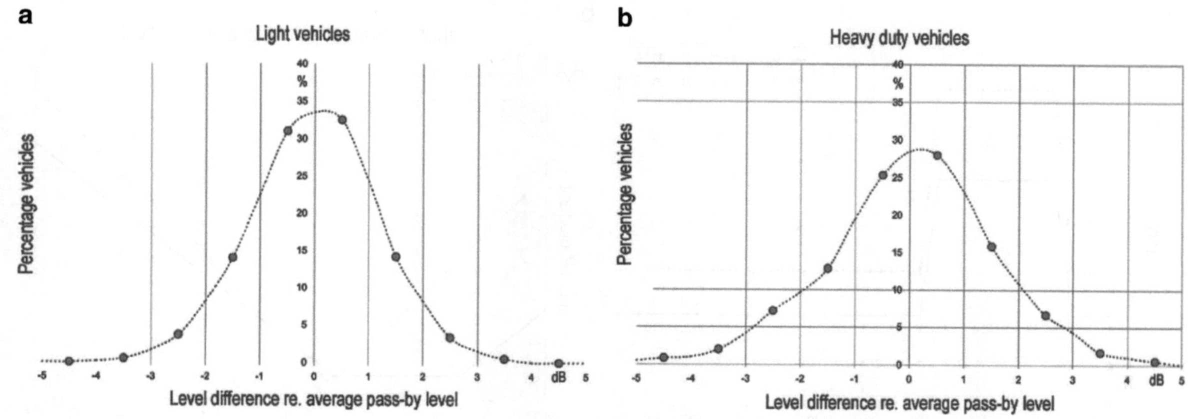

Fig. 15.3 Percentage vehicles showing a level difference of single pass-bys in dB with respect to the average statistical pass-by level; (**a**) light vehicles and (**b**) heavy duty vehicles

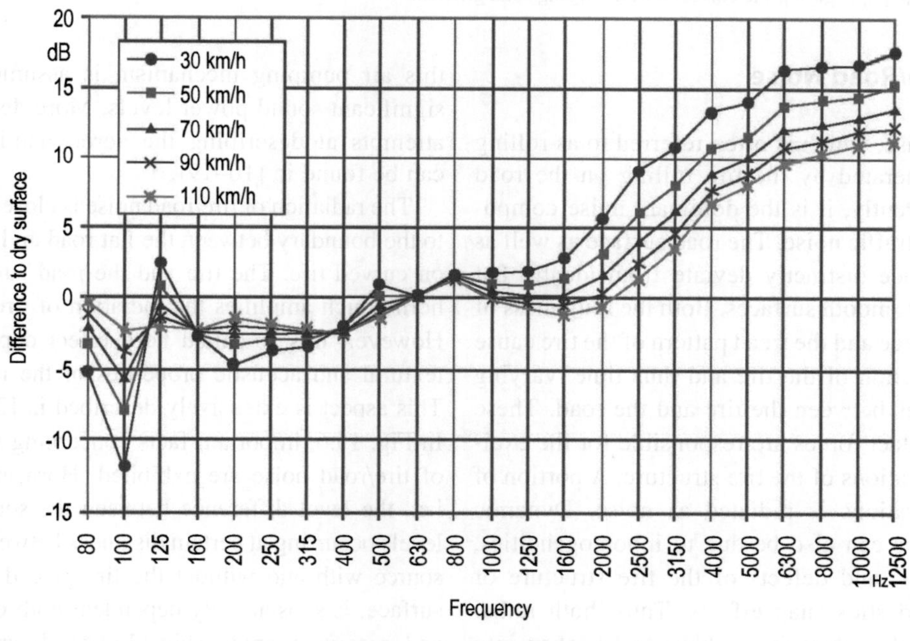

Fig. 15.4 Difference of statistical pass-by levels for passenger cars at wet driving conditions with respect to the dry condition depending on frequency and speed

respectively [9]. In recent projects, estimated corrections ΔL_{WP} were developed to be applied to the sound power values for the reference condition given in Fig. 15.5 [7]. Concerning the correction for the accelerated driving condition, it is given for moderate maximum acceleration values $a_{\max}$ of 2 m/s^2 for vehicle category 1, 1 m/s^2 for categories 2 and 3, and 4 m/s^2 for category 4. The correction is formulated as follows in (15.1). In Fig. 15.5a, the factor C_P is plotted against frequency.

$$\Delta L_{\mathrm{WP,acc}} = \left\{ \begin{array}{ll} C_P \cdot a & \text{for } a \geq -1\,\mathrm{m/s^2} \\ C_P \cdot a & \text{for } a < -1\,\mathrm{m/s^2} \end{array} \right\}, |a| \leq a_{\max}. \qquad (15.1)$$

The correction for the road gradient is due to the change in engine load. Figure 15.5b represents estimated values for the sound power level difference $\Delta L_{\mathrm{WP,gradient}}$, depending on the road gradient for different vehicle categories.

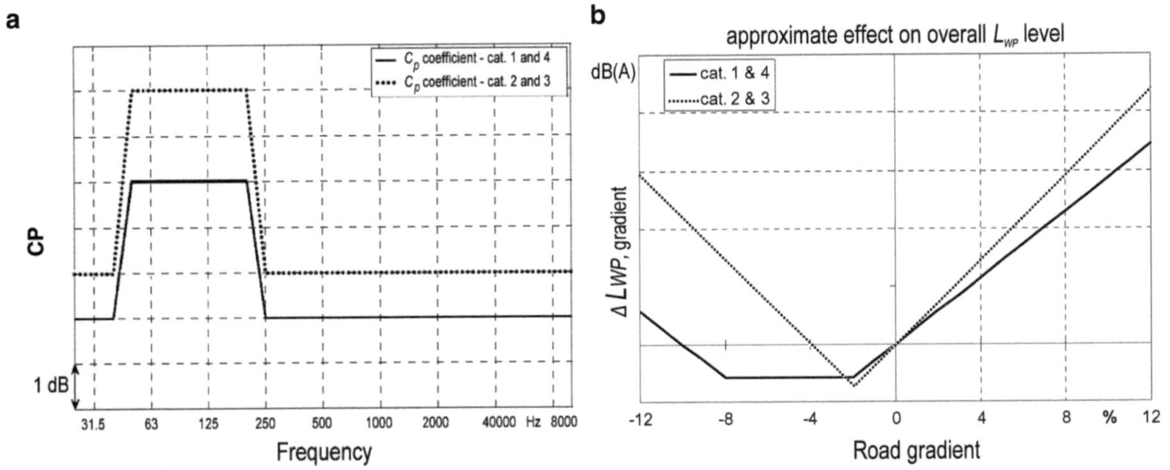

Fig. 15.5 (**a**) Coefficient C_P applied to the correction for accelerating vehicles depending on frequency; (**b**) sound power level difference $\Delta L_{\mathrm{WP,gradient}}$ giving the correction for varying road gradients

15.2.3 Tire/Road Noise

Tire/road noise, which is often referred to as rolling noise, is generated by the tire rolling on the road surface. Currently, it is the dominant noise component of road traffic noise. The road surface as well as the tire surface distinctly deviate from ideally flat and perfectly smooth surfaces. Both the roughness of the road surface and the tread pattern of the tire cause a local deflection of the tire and thus time, varying contact forces between the tire and the road. These dynamic contact forces are responsible for the excitation of vibrations of the tire structure. A portion of the tire vibrations is radiated as noise. Dynamic contact forces can also be due to inhomogeneities, imperfections, and defects of the tire structure or stick-slip and stick-snap effects. Thus, both radial and tangential contact forces have to be taken into account. As a rule of thumb, one can summarize that tire vibrations mainly influence the tire/road noise below 1 kHz [10–15]. Aerodynamic sources within the tire/road contact provide a second and quite important noise component. However, source description as well as sound radiation of these sources is currently not well known or explained in different ways unlike noise generation, which is caused by mechanical processes. Currently, there is a general agreement that, during rolling, a portion of the air in the vicinity of the tire/road contact is pumped out at the leading edge and sucked in at the trailing edge. Accelerated air flow which is due to this air pumping mechanism is assumed to cause significant sound power levels. More details on first attempts at describing the aerodynamic processes can be found in [16–19].

The radiation of tire/road noise is closely connected to the boundary between the flat road and the pressed-on curved tire. The tire and the road surface form a horn which amplifies the radiation of tire/road noise. However, this so-called horn effect depends on the textural and acoustic properties of the road surface. This aspect is extensively described in [20] and [21]. In Fig. 15.6, important facts concerning the radiation of tire/road noise are exhibited. Horn amplification, i.e., the level difference between the sound pressure levels occurring at certain distances between the sound source with and without the tire placed on the road surface, is substantially dependent both on frequency and receiving angle (Fig. 15.6b). Despite the fact that the horn is open at its lateral endings, the horn amplification is quite effective, yielding values of up to 20 dB. Conversely, it is also quite sensitive as regards the properties of the road surface. The roughness of the road surface, i.e., the road surface texture, makes the tire lift off the reflecting plane, thus reducing the horn amplification significantly, as can be seen in Fig. 15.6c. On rough surfaces with bigger distances between the tire and the road, the horn amplification is not as effective as on smooth ones. A target-oriented measure to reduce the horn amplification is the application of sound-absorbing road surfaces. Figure 15.6d shows that the horn amplification can be reduced

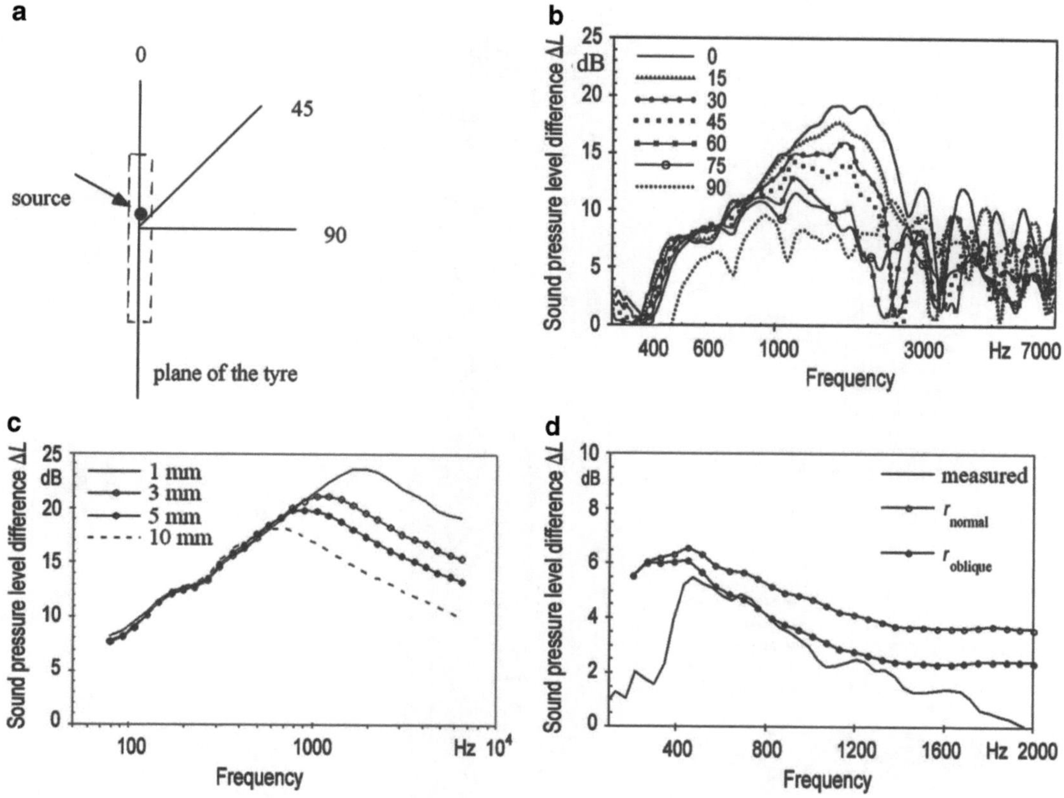

Fig. 15.6 (**a**) Co-ordinate system for measurement points at different angles outside the tire plane; (**b**) horn amplification for different receiving angles at 1.2 m over ground and 7.5 m distance from the center of the tire/road contact; (**c**) average horn amplification for different distances between tire and road; and (**d**) horn amplification due to the horn effect for a tire placed on an sound absorbing surface. All figures taken from [21]

significantly to 7 dB, assuming a sound absorption coefficient of 1 in the respective frequency band.

Figure 15.7 shows the effect of road surface roughness on tire/road noise. The figure summarizes the texture magnitude spectra of the roughness of typical impervious rigid road surfaces tending to more convex texture shapes (a) or more concave texture shapes (b). On the right, the corresponding spectra of the coast-by noise for an average passenger car tire at 80 km/h can be seen. The acoustical level difference around 1 kHz is in both cases not more than 5 dB, whereas the maximum roughness differs by 15 dB on a logarithmic scale for the surface dressings and by 9 dB for the hot rolled surfaces. This means that only a small part of the pavement roughness is acoustically effective. In other words, the tire only perceives a portion of the roughness when rolling on the surface. This signifies that there is no linear relation between road surface roughness and generated tire/road noise [22]. The shape of the roughness of the road surface is determined by the method of laying and the kind of surface treatment of the bituminous or cement-based surface course. Surface dressings applied on mastic asphalt, exposed aggregate cement concrete, and some types of road surface maintenance measures which are intended to improve the skid resistance of a surface course clearly lead to convex shapes of the texture. Single mineral grains form raised roughness elements which are separated by pits of finer material and binder, thus forming small hills and dales. In contrast, hot rolled road surfaces like stone mastic asphalt, asphalt concrete, or porous asphalt tend to more concave texture shapes. This shape is characterized by a plateau-like surface which is interspersed with small dents and ducts. According to the findings in [23], the shape can be quantified by means of the shape factor *g* which takes values between 0 and 1. Clearly, convex texture shapes take values of up to 60%; definitely, concave texture shapes take values of clearly more than 60%. Both the grain size distribution

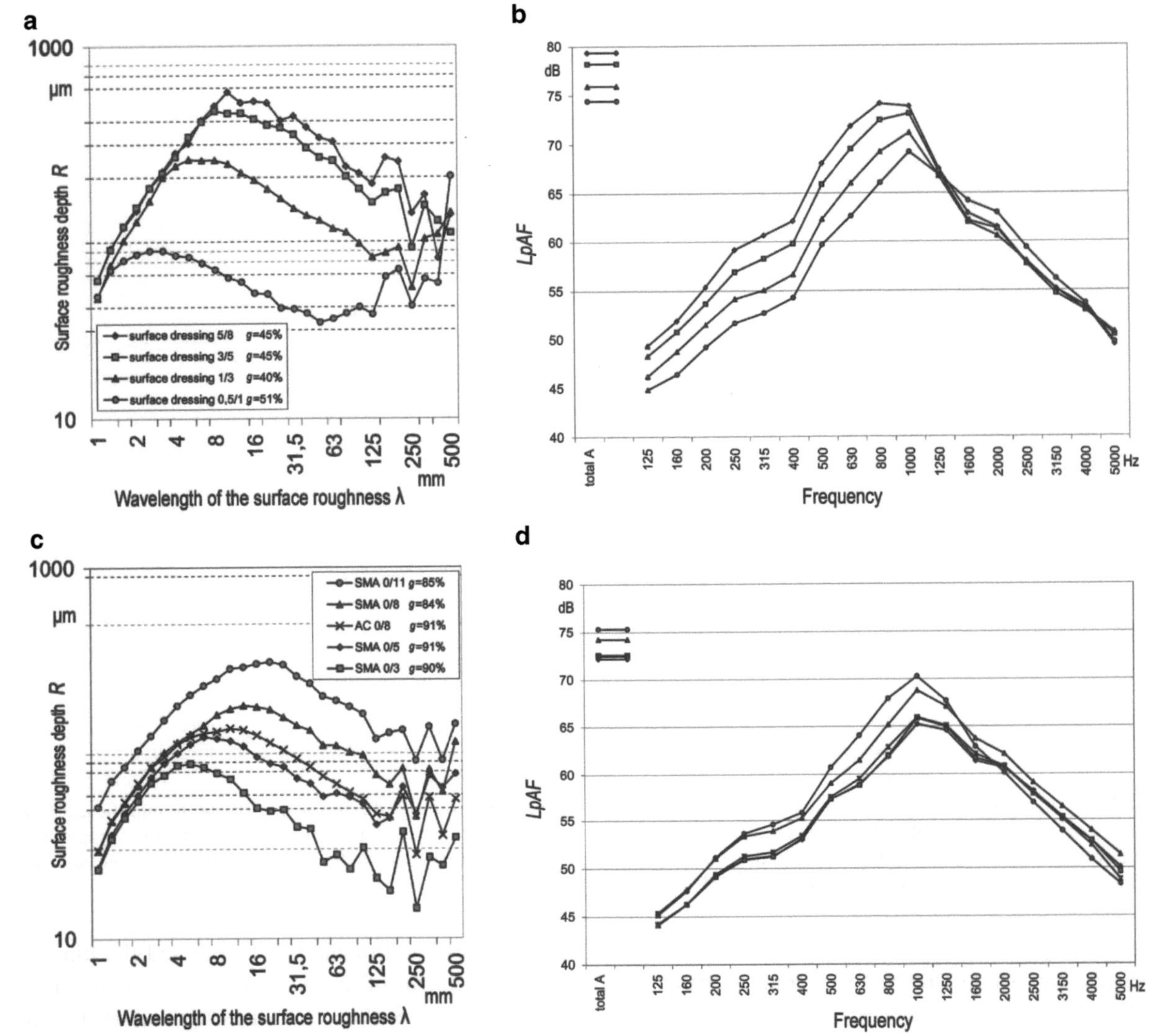

Fig. 15.7 Typical texture spectra (*left*) and coast-by level spectra (*right*) in third octave bands for dense road surfaces; upper row surface dressings with different stone size distributions; lower row hot rolled asphalt 0/8, *SMA*: stone mastic asphalt and *AC*: asphalt concrete. Shape factors *g* as indicated in the legends. *R*: rms value of the surface roughness depth in the respective wavelength band, and λ: wavelength of the surface roughness

of the mineral aggregate and the surface treatment directly affect the texture of a road surface and the generation of tire/road noise.

Acoustical and mechanical impedance of a road surface are two additional parameters which affect tire/road noise considerably. Road surfaces which show a void content of more than 8% by volume yield an absorption coefficient that is clearly greater than 0.1. Depending on frequency, the absorption coefficient can rise up to 1 when the void content is greater than 20% by volume. Figure 15.8 shows the absorption coefficient depending on frequency for three different sound absorbing road surfaces. Sound absorption of a road surface affects both horn amplification and sound propagation. This means that sound absorption causes an add-on reduction of tire/road noise independent of the texture; this holds not only for sound propagation but also for the noise emission of a single tire. In Fig. 15.9, the third-octave band spectrum of the mean coast-by levels L_{pAF} for 12 car tires measured at a speed of 80 km/h on surfaces with different textures, void content, and sound absorption

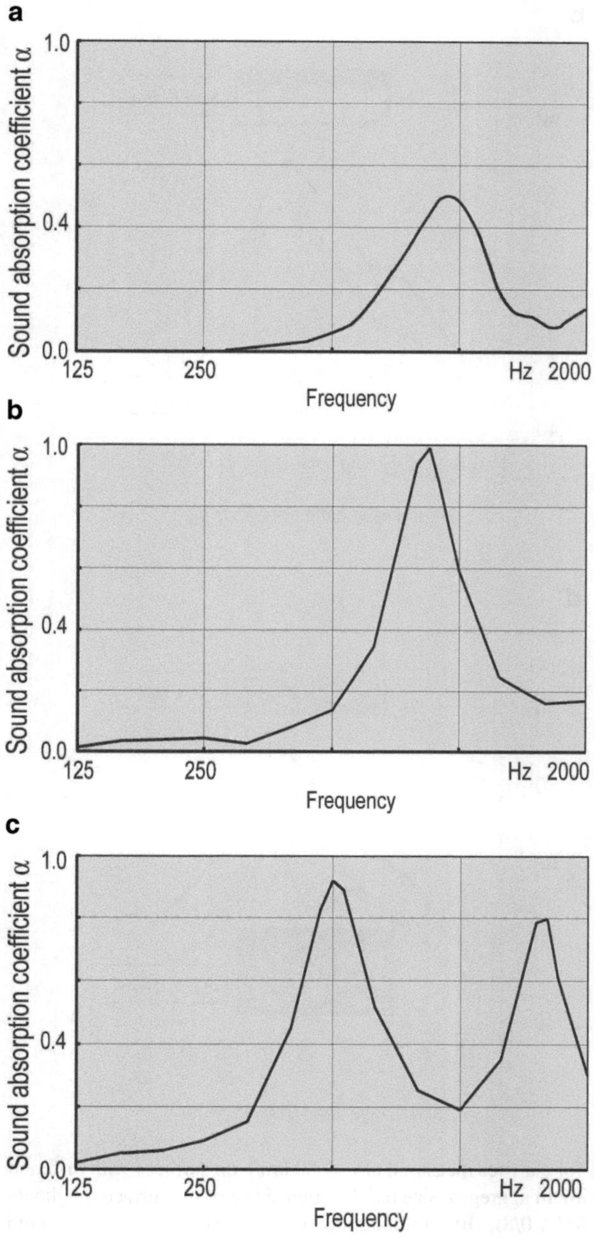

Fig. 15.8 Sound absorption coefficient α depending on frequency of (**a**) 2.5 cm thin noise reducing stone mastic asphalt with 12 Vol% void content, (**b**) 4 cm thick single layer open porous asphalt, and (**c**) 7 cm thick double layer open porous asphalt

coefficients are shown. Comparing a surface dressing with maximum aggregate size of 8 mm (SD 0/8), the spectra for (a) a dense surface with the same maximum aggregate size but different shape of the surface roughness (convex: surface dressing SD 5/8, concave: stone mastic asphalt (hot rolled) SMA 0/8), (b) a thin (2.5 cm thickness) hot rolled porous layer (TL 0/8), (c) a single layer porous asphalt (PA 0/8), and (d) two-layer porous asphalt (2PA 0/16-0/8). Since maximum aggregate size is always 8 mm, the influence of different textures and absorption characteristics can be seen quite clearly. The inserted graphics typify the texture and void structures of the road surfaces.

Tire/road noise is not merely a matter of road surface characteristics but also of the tire properties. The vibroacoustic characteristics of tire construction and tread compound and, mainly, the three-dimensional tread form, i.e., the tire texture, are considered to be important for tire/road noise emission. In Fig. 15.10, taken from [22], the coast-by levels of a representative set of 12 car tires (year 1997) of different brands, operation purposes (summer and winter), and widths as well as three different truck tires for different purposes (trailer, traction, and mud + snow) are plotted for different speeds versus various representative road surfaces. The first group of road surfaces on the left comprises four pavements made from a porous compound with different thickness and void content. The fifth pertains to a surface according to ISO 10844 [23] which is stipulated for car and tire type noise approval testing. The following three groups of surfaces comprise pavements, each made from the same type (hot rolled asphalt, surface dressing, and mastic asphalt) but with varying maximum aggregate size and thus showing different textures, varying from fine (1 or 3 mm maximum aggregate size) to coarse (8 mm maximum aggregate size). The level differences of the tire/road noise caused by different tires are significant. Concerning the car tires, the level range due to the tire is up to 8 dB. However, the dynamic of the road surface is even bigger. A level range of up to 13 dB reveals the noise reduction potential of the road surface.

Except for the porous road surfaces, the ranking of the car tires is irrespective of the speed. A "noisy"/"quiet" tire at low speed remains a "noisy"/"quiet" tire at high speed. The level range due to the car tire decreases slightly with increasing speed, which is due to the fact that air flow-related tire/road noise is less influenced by tire properties, shows a higher speed exponent and therefore becomes more dominant at higher speeds. Porous sound-absorbing road surfaces yield the lowest level differences between different tires, particularly at higher speeds. Truck tires of different types also behave in a varying manner.

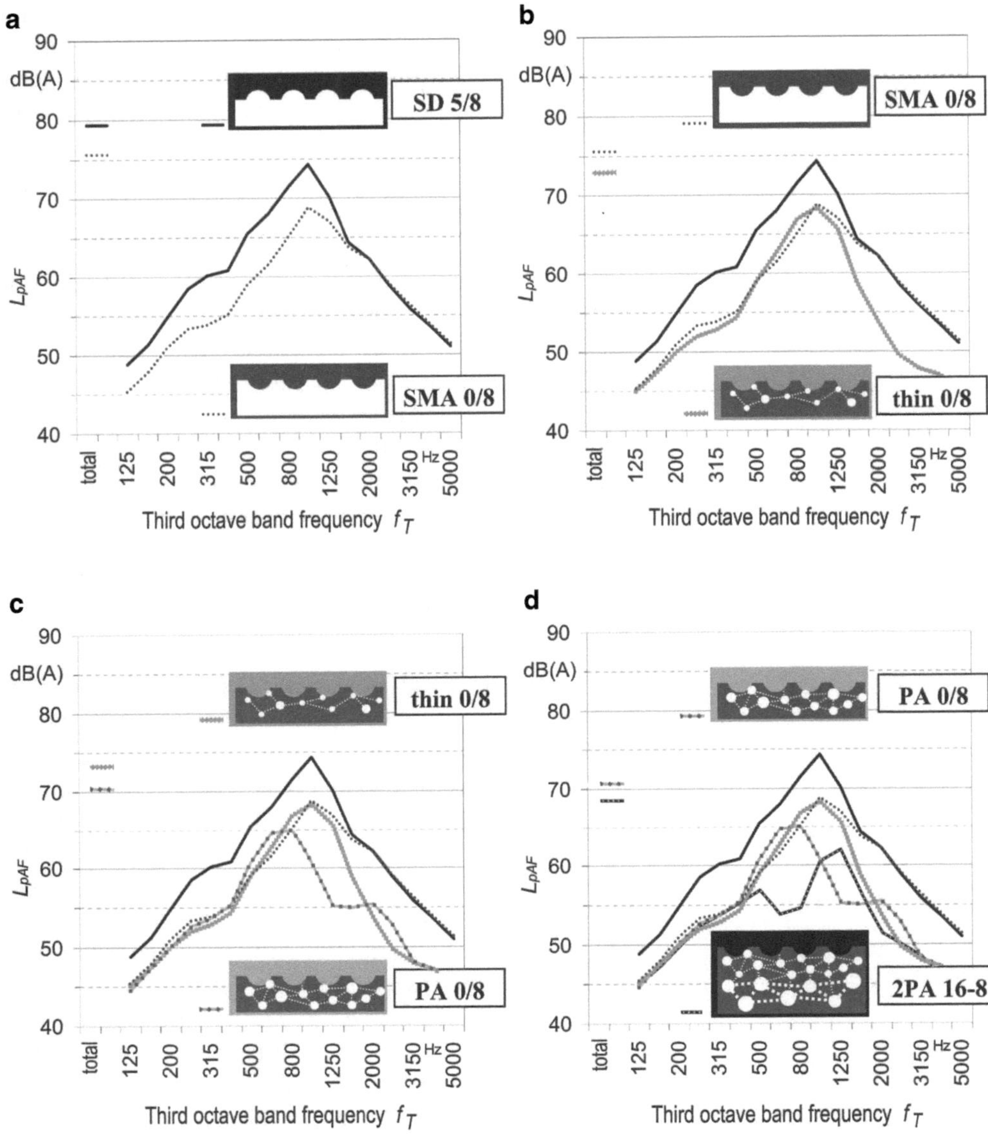

Fig. 15.9 Third-octave band spectra of mean coast-by levels L_{pAF} for car tires measured at $v = 80$ km/h on surfaces with different texture and sound absorption. (**a**) dense surfaces with the same maximum aggregate size but different shape of the surface roughness (convex: surface dressing SD 5/8, concave: stone mastic asphalt SMA 0/8), (**b**) thin hot rolled porous layer with 18 Vol% void content (TL 0/8), (**c**) single-layered porous asphalt with 26 Vol% void content and 4 cm layer thickness (PA 0/8), (**d**) double-layered porous asphalt with 27 Vol% void content and an overall layer thickness of 7 cm (2PA 0/16-0/8)

In general, tires being designed for the use on the steering axle of the tractor or on trailers reach the lowest noise levels. Truck tires with a pronounced block tread profile such as tires for the driving axle or mud + snow tires are much more noisy, independently of the road surface, except for the sound absorbing pavements. Therefore, porous road surfaces represent the only noise control measure referring to road construction, which is able to reduce significantly the tire/road noise of such truck tires.

15.2.4 Aerodynamic Noise

Aerodynamic noise is due to turbulent airflow and pressure fluctuation around exterior components of

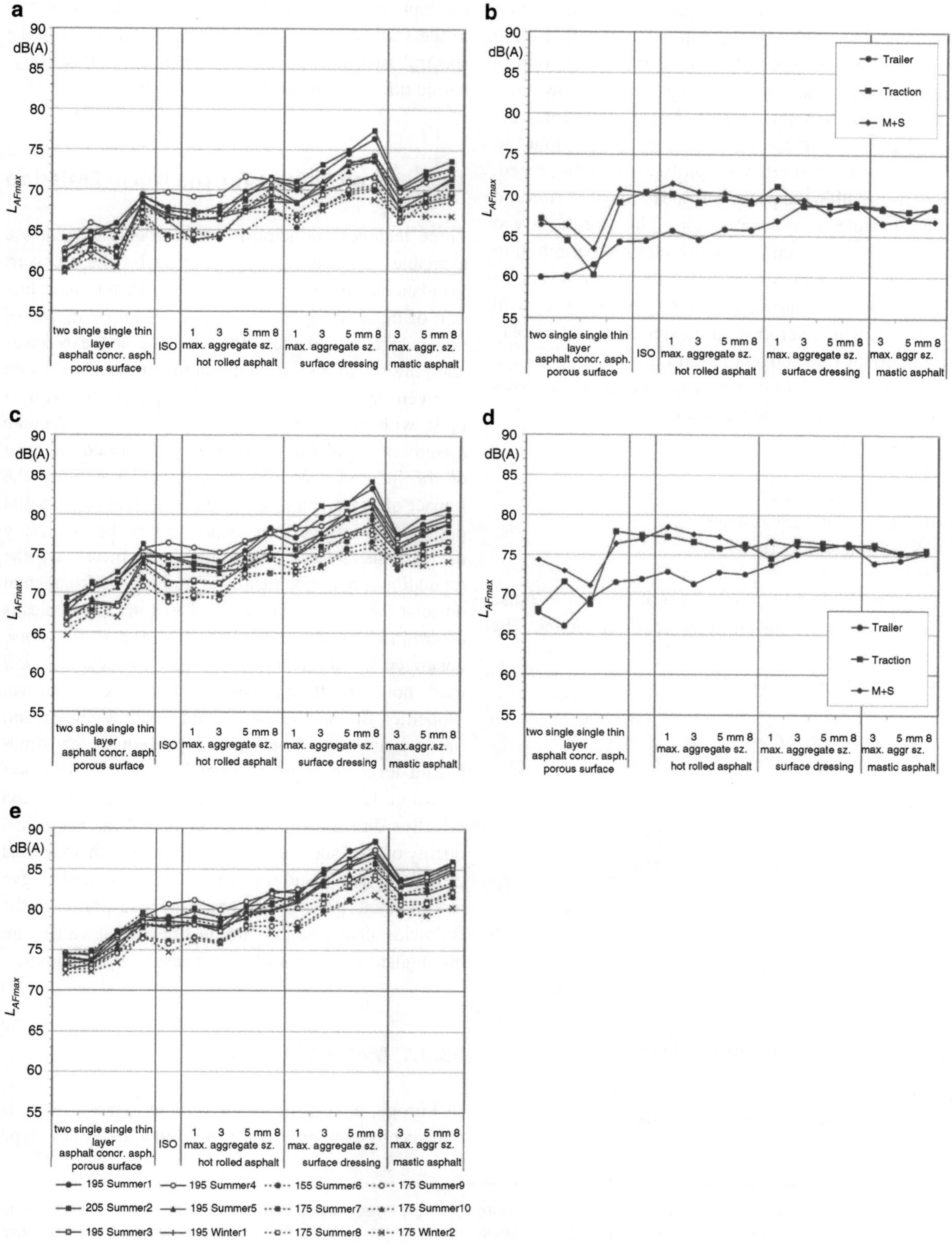

Fig. 15.10 Coast-by levels of 12 car tires and three truck tires for different speeds and road surfaces. **a**, **c**, and **e** car tires at 50, 80, and 120 kmh; **b** and **d** truck tires at 50 and 80 kmh. The road surfaces from the left hand to the right: two-layer porous asphalt (7 cm

the vehicle body and attached parts; it is an important component of a vehicle's exterior noise at high driving speed v. Various types of aerodynamic noise sources are involved in the generation of exterior airflow noise [24, 25]. Due to the large velocity exponents, the sound power of these noise sources (proportional to about v^5 up to v^6) increases rapidly with driving speed. However, within the permitted speed range of cars and heavy vehicles aerodynamic noise plays a role only at the highest speed limits that vehicles are permitted to drive at and hardly surpasses the tire/road noise (except for the unlimited driving speed for cars in Germany, where aerodynamic exterior noise can play an important role above 130 kmh).

The main vehicle components and attaching parts leading to aerodynamic noise are:

- Wheel houses
- Front vehicle
- A-pillar with the drip rail
- Exterior rear view mirrors
- Wind-shield wipers
- Rod antenna

Figure 15.11 shows the spectra of the airflow noise of a car measured in an acoustic wind tunnel. The

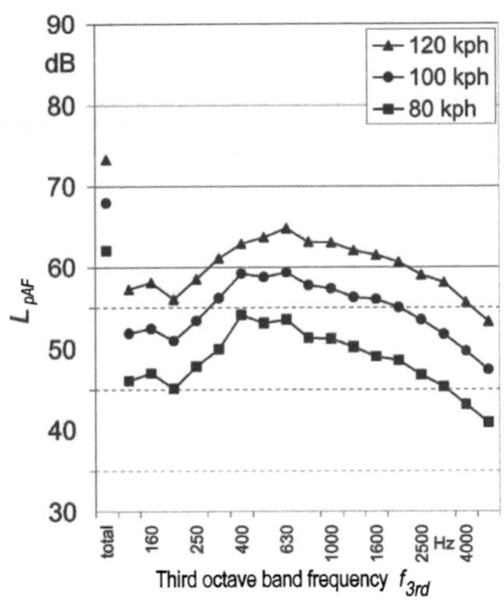

Fig. 15.11 Measured A-weighted sound pressure levels of the aerodynamic noise of a car in a laminar (acoustic) wind tunnel, distance: 7.5 m

microphone was positioned outside the airflow lateral to the vehicle body. They reflect representatively the spectral distribution and the magnitude of aerodynamic noise of cars at different speeds.

15.3 Limit Values for the Noise Emission

Three technical principles can be identified that are available to minimize noise impact on the receiver: avoidance or minimization of noise generation, reduction of noise along the propagation path by means of noise screens, and reduction of noise at the receiver applying sound insulation measures. In order to ensure a given level of environmental protection against noise without causing economic damage, an adequate system of regulations has been established in most of the industrialized countries. In order to limit the impact of noise sources as well as to reduce potential public economic efforts which might be necessary to take secondary noise protection measures, the regulations are aimed mainly at limiting environmental noise at the source. The noise of motorized vehicles is determined by the propulsion noise and the noise generated by the dynamic tire/road contact. The tire/road noise itself depends on both the acoustical properties of the tire and of the road. Vehicles and tires are subject to type approval procedures that stipulate in legal descriptions strict limit values for noise emission. However, there is a lack of harmonized and effective directives for the assurance of the acoustical quality of road surfaces, so far. This means that the road traffic system suffers from a systematic gap incorporated in the noise control of road traffic. In the following chapters, the current noise limit values are highlighted with regard to the measuring procedures.

15.3.1 Motor Vehicles

In Europe, control of motor vehicle exterior noise is realized by type-approval procedures. Each new type

Fig. 15.10 (Continued) overall thickness, 27 Vol% void content); single-layer porous asphalt (4 cm thickness; 26 Vol% void content); single-layer cement concrete asphalt (8 cm thickness; void content $\geq$ 22 Vol%); thin porous layer (2.5 cm thickness, 18 Vol% void content); ISO surface; hot rolled asphalt mixes with 1, 3, 5, and 8 mm maximum aggregate size; surface dressings on asphalt with 1, 3, 5, and 8 mm maximum aggregate size; and mastic asphalt with surface treatment using 3, 5, and 8 mm maximum stone size of vehicle which is to be introduced into the market must fulfill the requirements published in legal provisions. Acoustical requirements introduced in the EU in 1970 are part of these regulations. Since 1970, the limit value for the noise emission of passenger cars has been lowered by 8 dB(A), by 11 dB(A) for heavy trucks (more than 150 kW engine power), and by 2–6 dB(A) for motor cycles, the latter depending on the engine power.

The noise limits as well as the measuring procedures which are related to the driving noise and the stationary noise are recorded in the following EU directives.

This directive has been amended several times.

- 70/157/EEC Driving noise and stationary noise of passenger cars and trucks as well as the noise of compressed air that is released over a control valve [26].
- 77/212/EEC Replacement of the vehicle categories and introduction of reduced limit values for the noise level [27].
- 81/334/EEC Amendment of the method of measuring the noise of moving and stationary vehicles in order to bring them more into line with real operating conditions [28].
- 84/372/EEC Amendment of the method of measuring the noise emitted by high-performance vehicles and vehicles with automatic transmission equipment with a manual override, in order to bring them more into line with actual operating conditions [29].
- 84/424/EEC Replacement of the limit noise level values in force by lower ones for every category of vehicle [30].
- 92/97/EEC Comprehensive amendments by reducing the limit noise level values for each category of vehicle and by improving the test method for high-powered vehicles; because this type of vehicle had been found as being increasingly designed so as to produce a higher ratio between engine power and vehicle mass and the curve representing the torque as a function of engine speed had been modified to produce greater motive force at low engine speed, these new designs consequently gave rise to greater use of gear ratios in urban traffic and had a major influence on the noise emitted by the mechanical parts as compared with road noise [31].
- 2007/34/EEC Introduction of a new test cycle that brings the driving conditions for carrying out the noise test closer into line with real-life driving operations. The new test cycle is described in UN/ECE Regulation No 51, 02 series of amendments [32].
- 97/24/EEC Directive on certain components and characteristics of two- or three-wheel motor vehicles [33].

No modifications have been applied so far to this directive with respect to the permissible noise levels and the measuring procedure.

Table 15.2 contains a complete list of the current (2009) noise limit values.

For vehicles with a maximum permissible mass of over two tons designed for off-road use, the limit values are increased by 1 dB(A) if their engine power is less than 150 kW and 2 dB(A) if their engine power is 150 kW or more.

In Fig. 15.12, the history of the permissible noise levels of motor vehicles since 1970 is shown. Time stamps and slopes of the curves mark the year of adoption, the transitional period, and the year the limit values came into force, according to the respective EU directive (Table 15.3). The noise-limiting values described in the previous paragraphs are linked to specified measuring procedures. For passenger cars as well as for trucks with a gross vehicle weight of not more than 3.5 *t*, the vehicle is moved at a constant speed of 50 kmh. As soon as the vehicle reaches line AA′, it is accelerated at full throttle. This driving condition is retained as long as the vehicle has entirely passed line BB′. Cars with a manually operated gearbox and four forward gears at most are tested in the second gear. If such a vehicle is configured with more than four forward gears, it is successively tested in the second and the third gear. Vehicles with automatic transmission equipped with a manual override are tested with the selector in the normal driving position ("D").

In principle, heavy duty vehicles with more than 3.5 *t* gross weight are tested the same way. The velocity of the vehicle at line AA′ depends on the gear under test and the nominal rotation speed of the engine. The gear has to be selected in such a way that the nominal rotation speed can be attained within the measuring track between the lines AA′ and BB′ without activating the rotation speed limiter. The measuring method with acceleration at fully opened throttle between the lines AA′ and BB′ described for motor vehicles is also true for two-wheelers and three-wheel mopeds or tricycles.

Table 15.2 Vehicle noise limit values cars and trucks

Vehicle category	Noise level limit
Passenger cars with not more than nine seats including the driver's seat	
Direct injection diesel engine	75 dB(A)
All other engines	74 dB(A)
For vehicles in this category, equipped with a manually operated gear box having more than four forward gears andwith an engine developing a maximum power exceeding 140 kW/t and whose maximum power/maximum mass ratio exceeds 75 kW/t, the limit values are increased by 1 dB(A) if the speed at which the rear of the vehicle passes the line BB' in third gear is greater than 61 kph.	
Passenger cars with more than nine seats including the driver's seat and a maximum permissible mass of more than 3.5 *t*	
Engine power < 150 kW	78 dB(A)
Engine power ≥150 kW	80 dB(A)
Passenger cars with more than nine seats including the driver's seat and trucks	
Maximum permissible mass ≤2 *t*	
Direct injection diesel engine	77 dB(A)
All other engines	76 dB(A)
Maximum permissible mass > 2 *t* and ≤ 3.5 *t*	
Direct injection diesel engine	78 dB(A)
All other engines	77 dB(A)
Trucks with a maximum permissible mass of more than 3.5 *t*	
Engine power < 75 kW	77 dB(A)
Engine power ≥ 75 kW and < 150 kW	78 dB(A)
Engine power ≥ 150 kW	80 dB(A)
Compressed air noise	72 dB(A)

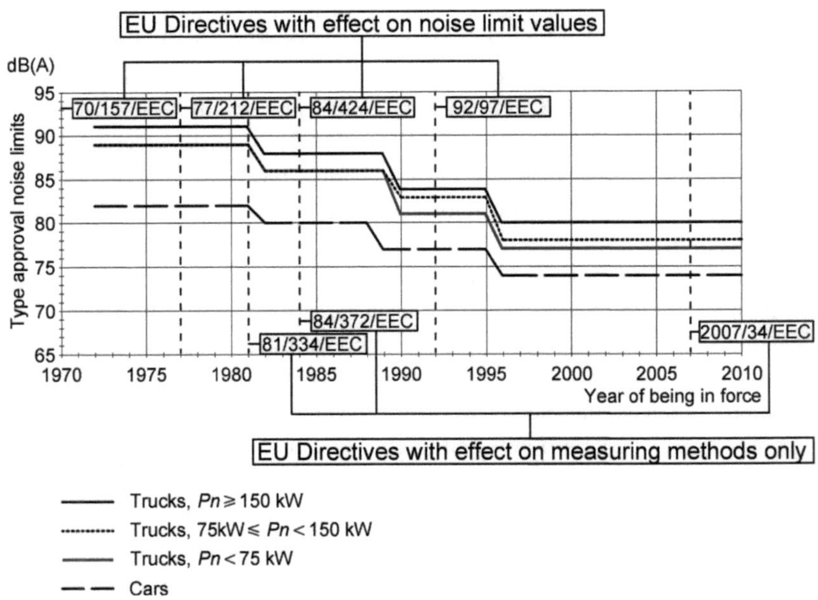

Fig. 15.12 History of the permissible noise levels of motor vehicles

However, initial speed and gear selection depend on the type of the gearbox (manual or automatic) as well as the possible gear ratios, the cylinder capacity, and the maximum road speed of the vehicle under test. The method of measurement for two-wheel mopeds is described in annex II of the EU directive 97/24/EEC [33]. Motorcycles are dealt with in annex III and three-wheel mopeds and tricycles in annex IV.

The specific vehicle noise level is defined as the overall maximum pass-by noise level measured at two microphone positions, one on each side of the vehicle, at 7.5 m distance to the center line, and a height of 1.2 m above the surface of the driving lane. Figure 15.13 shows the configuration of the test track. The noise of the stationary vehicle is also discussed in the EU directives. It is measured at nearfield positions of the microphone with respect to the outlet of the exhaust silencer. However, there are no permissible noise level values the measured noise levels are compared with. The maximum measured sound pressure level is registered in the type approval certificate, thus facilitating subsequent noise tests on vehicles in use. The configuration of the measurement is shown in Fig. 15.14.

Table 15.3 Vehicle noise limit values two- and three-wheelers

Vehicle category	Noise level limit
Two-wheel mopeds	
$v_{max} \leq 25$ kmh	66 dB(A)
$v_{max} > 25$ kmh	71 dB(A)
Three-wheel mopeds	76 dB(A)
Motorcycles	
$cc \leq 80$ cm^3	75 dB(A)
$80 < cc \leq 175$ cm^3	77 dB(A)
$cc > 175$ cm^3	80 dB(A)
Tricycles	80 dB(A)

For the type approval test of the stationary car or truck, engine speed must be stabilized at 75% of the speed at which the engine develops rated maximum power. For mopeds, motorcycles, and three-wheelers, the conditions of the engine operation are different. When constant engine speed is reached, the throttle must be rapidly returned to the idling position. The maximum noise level during the constant engine speed and the entire deceleration period is taken as the test result. The noise that is caused by the activated air pressure valve of the braking system of heavy vehicles and the all-around noise is measured according to the configuration shown in Fig. 15.15, where microphone positions 2 and 6 are used for the braking noise. The highest noise level value measured at these two positions is taken as the test result. The levels measured at all of the eight positions are used to determine the all-around noise of the stationary vehicle. This test is not mandatory for type approval but is required for the labeling of "low noise vehicles" which, for instance, get special concessions in alpine transit transport.

15.3.2 Tires

Since 2001, tires are subject to separate permissible noise levels. Limiting values as well as the measuring procedure are governed by the EU directive

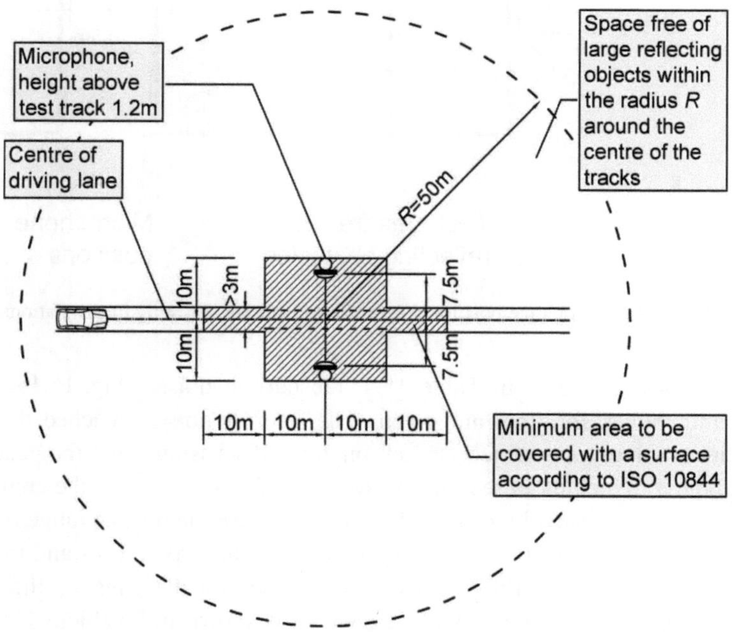

Fig. 15.13 Configuration of the test track for the type approval testing of moving vehicles

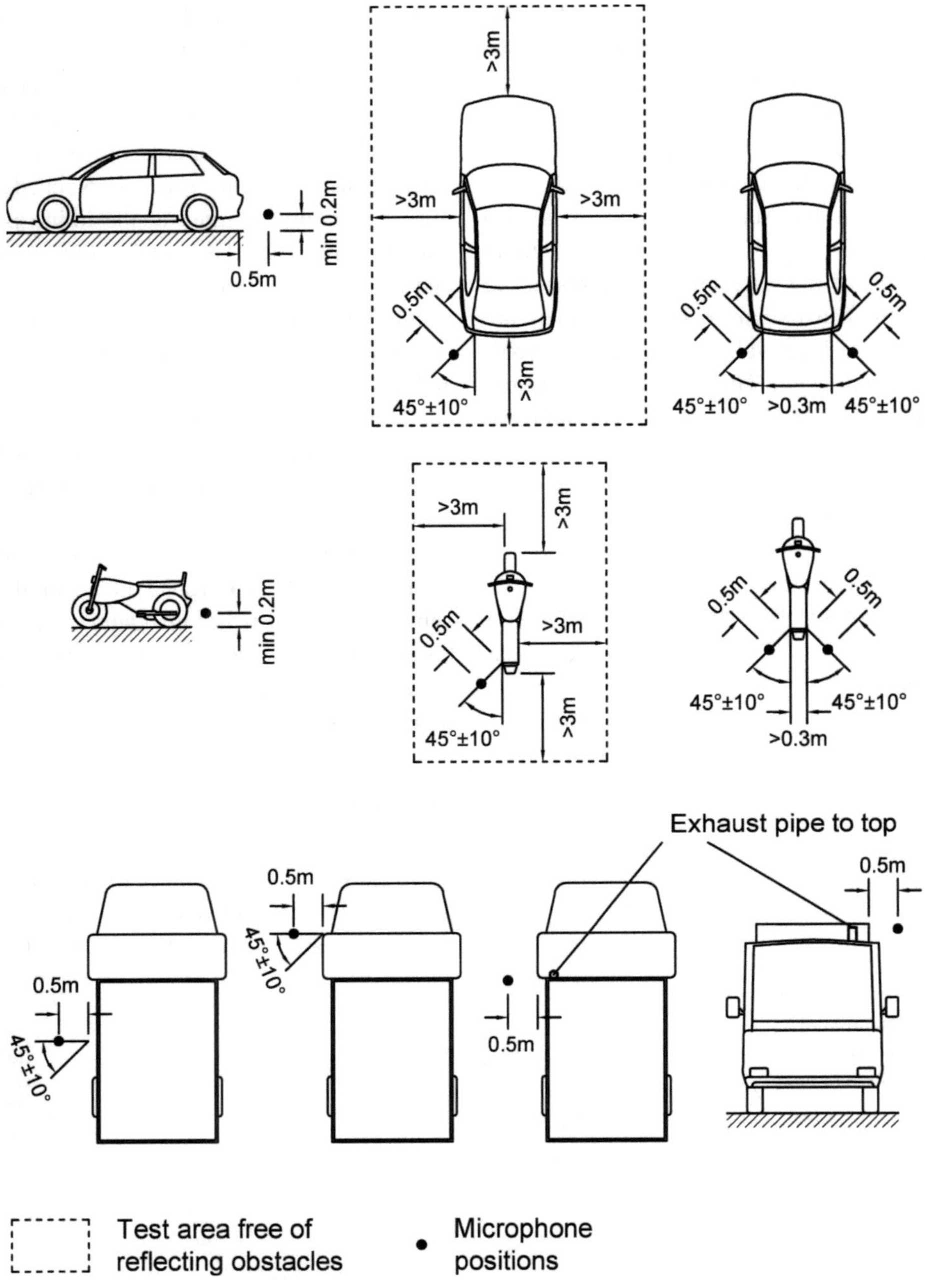

Fig. 15.14 Microphone positioning for the type approval testing of the stationary vehicle

2001/43/EEC [34]. In Table 15.4, the current noise limit values for tires are given. The interim noise limit values are subject to an ongoing discussion about a tightening of the limit values since 2004.

Tires are tested by means of a coast-by noise measurement. The test track configuration for the type approval testing of tires agrees with that used for the type approval testing of vehicles, which is shown in Fig. 15.13. When the front end of the test vehicle has reached the line AA′, the vehicle's driver must have put the gear selector in neutral position and switched off the engine. The test vehicle speeds shall be within the range from 70 to 90 kmh for class C1 and class C2 tires and from 60 to 80 kmh for class C3 tires. As test result, the maximum noise level measured as the vehicle is coasting between lines AA′ and BB′ is

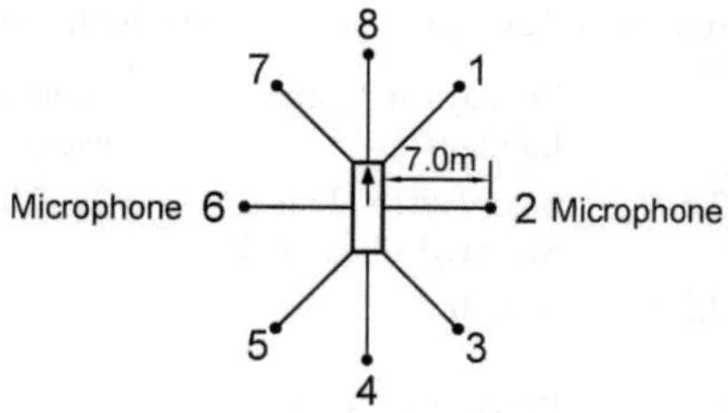

Fig. 15.15 Microphone positioning for the measurement of the compressed air and all around noise

taken. The average value at the speed of 80 kmh for class C1 and class C2 tires and the speed of 70 kmh for class C3 tires, which is derived from a regression analysis of all single measuring results, is compared with the limit value.

15.3.3 Road Surfaces

The type approval test for vehicles and tires are performed on a test track paving, whose civil engineering and acoustic properties agree with ISO 10844. In most cases however, normal road surfaces show a different acoustical behavior, leading to different tire/road noise levels. Different acoustical behavior is related to different civil engineering properties of the road pavement. These differences may be either due to the type of road pavement or due to different quality of the road pavement of the same type. Up to now, neither coordinated international legal regulations nor voluntary quality assurance measures concerning the noise-related properties of road surfaces have been established. This represents a systematic gap within the permissible noise level scheme for the road traffic noise sources consisting of the vehicle, the tire, and the road surface. Within the framework of the EU project SILVIA [35], a first attempt was made to outline a type approval test procedure for road pavings.

15.3.4 Additional Regulations for Road Traffic

In contrast to the vehicle, the road traffic itself can be seen as noise source. Primary noise control measures aim at limiting the speed or limiting the traffic volume or the traffic mix, respectively. Both are the responsibility of the local road authorities. Limiting the traffic volume and traffic mix by imposing a ban on driving during night time became an effective noise abatement measure in two European countries, Austria and Switzerland. A ban on driving on Sundays and public holidays is not primarily intended to reduce traffic noise. Most of these regulations allow for restrictions of heavy vehicle traffic. Table 15.5 gives an overview of driving regulations concerning ban on driving in Europe. Another legal measure to reduce the noise impact caused by road traffic is the limitation of the driving speed. Initially, speed limits were intended to safeguard road users from severe accidental damages, yet today they also serve as noise controls. Typically, the sound power of the tire/road noise of passenger

Table 15.4 Tire noise limit values

Tire class	Nominal section width w (mm)	Current noise limit value in dB(A)	Interim noise limit values in dB(A)
Passenger car tires			
C1a	$w \leq 145$	72	70–71
C1b	$145 < w \leq 165$	73	71–72
C1c	$165 < w \leq 185$	74	72–73
C1d	$185 < w \leq 215$	75	74
C1e	$w > 215$	76	75
Commercial vehicle tires with load capacity index in single formation $\leq$ 121 and speed category symbol $\geq$ "N"			
C2 normal		75	
C2 winter (M + S)		77	
C2 special		78	
Commercial vehicle tires with load capacity index in single formation $\leq$ 121 and speed category symbol $\leq$ "M" or commercial vehicle tires with load capacity index in single formation $\geq$ 122			
C3 normal		76	
C3 winter (M + S)		78	
C3 special		79	

Table 15.5 European countries with general ban on driving regulations (those which have none are not shown in the table)

Country	Road categories	Gross vehicle weight	Sundays and public holidays	Night time every day
Austria	Motorways	>7.5 *t* and trucks with trailer >3.5 *t*	Saturday 15–24 h Sunday/Holiday 0–22 h	22–5 h
Czech Republic	Motorways and state roads	>7.5 *t* and trucks with trailer >3.5 *t*	0–22 h	
Croatia	State roads without motorways	>7.5 *t* and all trucks with more than 14 m total length	Friday 16–23:00 h Saturday 5–14 h Sunday/Holiday 12–23 h	
France	All	>7.5 *t*	22–22 h	
Germany	All	>7.5 *t* and all trucks with trailer	0–22 h	
Hungary	All	>7.5 *t*	8–22 h	
Italy	All	>7.5 *t*	June–September 7–24 h Remaining period 8–22 h	
Liechtenstein	All	>3.5 *t* except for Campmobiles	0–24 h	
Luxemburg	Trunk roads		Previous day: 21:30–24 h Sunday/Holiday 0–21:45 h	
Poland	All	>12 *t*	Previous day: 18–22 h Sunday/Holiday 7–22 h	
Romania	Motorways and important state roads	>7.5 *t*	Differentiated regulations	
Slovakia	Motorways and state roads	>7.5 *t*	0–22 h	
Slovenia	Motorways and trunk roads	>7.5 *t* and all trucks with more than 14 m total length	Sunday/Holiday 8–22 h	
Switzerland	All	> 3.5 *t* except for Campmobiles	0–24 h	22–5 h
Turkey	Main roads	All trucks	7–10 h 17–22 h	

cars increases by the power of 3.2. This means that a reduction of the driving speed by 20%, e.g., from 100 to 80 kmh, results in a noise level reduction by 3 dB.

15.3.5 Comparison with Real Traffic Situations

Results of a comprehensive investigation on the noise of vehicles in real road situations have been published in [36]. Figure 15.16 shows the pass-by noise of vehicles in free rolling condition on road surfaces made of asphalt concrete with 11 mm maximum aggregate size (AC 0/11), depending on the speed. The diagrams show runs of functions which are derived from a great deal of measurements of the pass-by level, recorded at 7.5 m distance and 1.2 m height above the road surface. The functions are set up for different vehicle categories. Because this measurement campaign was performed in the past in a similar manner it is possible to compare the noise production of vehicles depending on the year of registration. As can be seen in Fig. 15.16, the noise production of vehicles did not really differ over the years as regards the reduction of vehicle noise in real road situations. Contrary to the development of the noise limiting values for the type approval of vehicles, almost no reduction of the vehicle noise can be noticed in real traffic situations. There is absolutely no difference between the groups of passenger cars registered at different years. Whereas only a moderate relief of some 2 dB(A) could be gained for light duty vehicles, a clear level difference of 4–5 dB(A) could be achieved for heavy duty vehicles. However, it is restricted to the speed range below 60 kmh. This situation is due to the fact that the acoustic type approval testing procedure for vehicles causes an overemphasizing of propulsion noise. The tire/road noise, which is most important for the noise of the

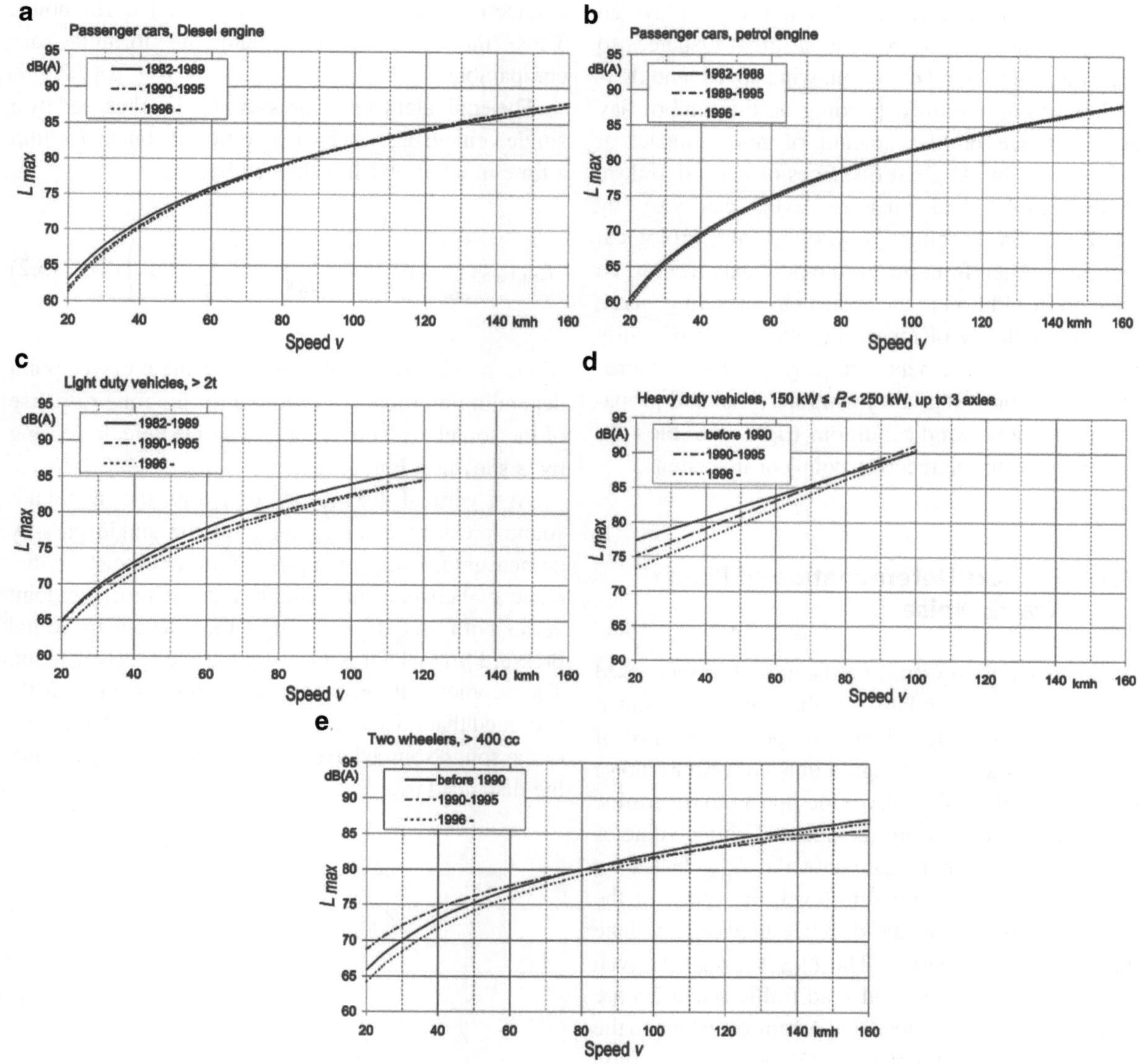

Fig. 15.16 Pass-by noise of various vehicles depending on speed and on the year of registration

freely rolling vehicle, is irrelevant in the type approval testing. The coming into force of the EU directive for the acoustic type approval testing of tires in 2001 is expected to improve or resolve this unfavorable situation. However, a sustainable formulation of the noise limiting values is needed as well.

15.4 Measurement of Road Traffic Noise

Measurements of road traffic noise are not intended to record the noise of single vehicles or even particular noise sources of vehicles but to assess the over-all noise of the traffic on a road. This purpose leads directly to some restrictions and imponderable effects on the results of such measurements. In order to be able to compare the results from different measurements, it is important to know the conditions on which the measurements are based; the following data are needed in any case:

- The number of vehicles
- The driven speed
- The type and condition of the road surface
- Meteorological data
- The influences on sound propagation and the distance to the road

Especially, meteorological conditions play an important role in measurements at great distances to traffic facilities. Wind direction, wind speed, and thermal layering in the air depending on the time of day have influence on the amount of noise impact at a receiver point. At great distances of several 100 m, the measured sound pressure levels can vary by 20–30 dB on account of varying meteorological conditions. Therefore, meteorological data need to be collected in addition to the acoustical ones. In general, the reproducibility of measuring data is significantly improved if measurements are carried out at meteorological conditions that are favorable for sound propagation, like downwind conditions (light wind blowing from the road to the receiver point) or inversion.

15.4.1 Indirect Determination of Road Traffic Noise

Reference and limit values for the noise impact caused by road traffic are related to the equivalent sound pressure level L_{Aeq} for given time periods in most of the guidelines and legal regulations concerning noise control of road traffic noise. One option to determine L_{Aeq} values is to measure the noise of single vehicles at the roadside and to calculate the L_{Aeq} based on the maximum sound pressure level, the speed of the different vehicles, and the distance between the lane and the receiver position. The L_{Aeq} values for given time periods, locations, and road traffic scenarios are derived then from the measured parameters. Due to the fact that the L_{Aeq} is not measured directly, this method is called indirect determination of road traffic noise. This method is also described in Sect. 18.2 in connection with the determination of air traffic noise. Thus, the basic approach and the formulas are comparable.

The equivalent sound pressure L_{eq} level caused by a single vehicle passing by at a vertical distance d within a time interval of 1 h is defined by:

$$L_{\mathrm{eq,1veh,1\,h}} = 10\ \lg\left(\frac{1}{3,600\,\mathrm{s}}\int_{-\infty}^{+\infty}\frac{p(t)^2}{{p_0}^2}\mathrm{d}t\right)\mathrm{dB}, \quad (15.2)$$

where $p(t)$ is the sound pressure at the receiver point depending on time. An example for the time response of the sound pressure level of a single vehicle passing by is shown in Fig. 15.17.

At a vertical distance d_m of 7.5 m, the maximum sound pressure level $L_{\mathrm{max,7.5m}}$ of the single pass-by is measured. During the pass-by of a vehicle the distance r between the vehicle and the receiver point varies with time thus causing a time dependent sound pressure $p(t)$ which is measured at the receiver point. The geometrical definitions are shown in Fig. 15.18. Provided that no additional damping and no directivity of the sound source have to be considered, $p(t)$ follows the distance law:

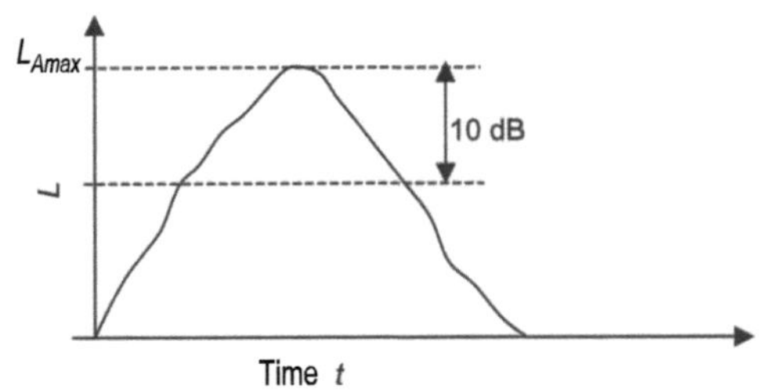

Fig. 15.17 Time response of the sound pressure level of a single car passing by

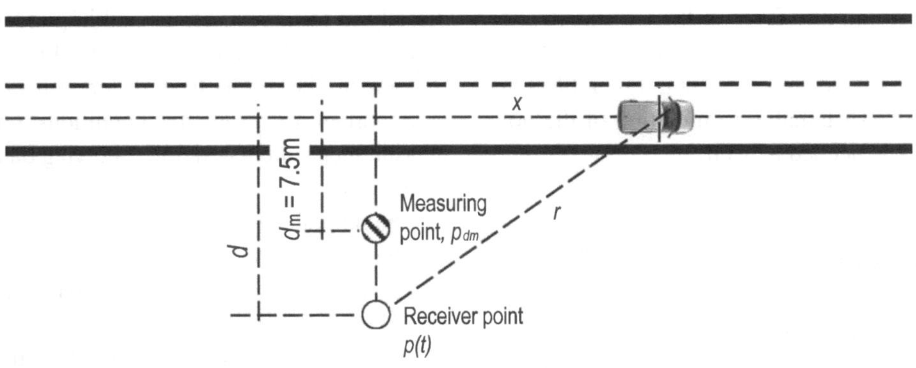

Fig. 15.18 Co-ordinates for the indirect determination of traffic noise

$$\frac{p(t)}{p_{dm}} = \frac{d_m}{r(t)}, \quad (15.3)$$

with

$$x = x(t) = vt, \quad (15.4)$$

and

$$r(t) = \sqrt{v^2t^2 + d^2},\ d = \text{const.} \quad (15.5)$$

$p(t)$ gets

$$p(t) = p_{dm}\frac{d_m}{\sqrt{v^2t^2 + d^2}}. \quad (15.6)$$

Consequently, the equivalent sound pressure level L_{eq} can be derived from the maximum sound pressure level $L_{\text{max,7.5m}}$ for single pass-bys as follows:

$$v = \frac{\mathrm{d}x}{\mathrm{d}t}, \mathrm{d}t = \frac{\mathrm{d}x}{v}, \ v = \text{const.} \quad (15.7)$$

$$L_{\text{eq,1veh,1h}} = 10\ \lg\left(\frac{1}{3{,}600\,\text{s}}\frac{p_{dm}^2}{p_0^2}\frac{d_m^2}{v}\int_{-\infty}^{+\infty}\frac{1}{x^2+d^2}\mathrm{d}x\right)\text{dB}, \quad (15.8)$$

$$\int_{-\infty}^{+\infty}\frac{1}{x^2+d^2}\mathrm{d}x = \frac{1}{d}\text{arctg}\frac{x}{d}\Big|_{-\infty}^{+\infty} = \frac{\pi}{d}, \quad (15.9)$$

$$L_{\text{eq,1 veh,1 h}} = 10\ \lg\left(\frac{1}{3{,}600\,\text{s}}10^{\left(\frac{L_{\text{max,7.5 m}}}{10\,\text{dB}}\right)}\frac{{d_m}^2}{v}\frac{\pi}{d}\right)\text{dB}. \quad (15.10)$$

With $d_m = 7.5$ m and v in kmh it gets

$$L_{\text{eq,1 veh,1 h}} = L_{\text{max,7.5 m}} - 7.5\,\text{dB} - 10\ \lg\left(\frac{v}{1\,\text{km/h}}\right)\text{dB} - 10\ \lg\left(\frac{d}{1\,\text{m}}\right)\text{dB}. \quad (15.11)$$

To be more precise, the damping caused by sound absorption in the air and ground attenuation which in turn depend on the distance between source and receiver, must also be taken into account. Moreover, these additional dampings depend on frequency as well resulting in a more complex relation between the $L_{\text{max,7.5m}}$ and the $L_{\text{eq,1veh,1h}}$ than that given in (15.11). In terms of A-weighted sound pressure levels, the spectral emphasis of road traffic noise is close to 1 kHz on free noise propagation conditions. Based on this assumption, the German guidelines for noise control at roads [37] provide the following equations for the noise attenuation D_s, which is due to distance and air absorption and the ground attenuation D_{BM}. The relations are true for long and straight roads.

$$D_s = 15.8\,\text{dB} - 10\ \lg(d/1\,\text{m})\text{dB} - 0.0142d^{0.9}, \quad (15.12)$$

$$D_{\text{BM}} = -4.8\exp\left(-\frac{h_m}{d}\left(8.5 + \frac{100}{d}\right)^{1.3}\right), \quad (15.13)$$

where h_m is the mean height between the road and the receiver point.

To take $d = 25$ m and $h_m = 2.25$ m, which results from a source height of 0.5 m and a receiver height of 4 m, as an example and substituting the distance term -10 lg(d/1 m) dB in (15.11) for the sum of (15.12) and (15.13), the $L_{\text{eq,1veh,1h}}$ is

$$L_{\text{eq,1 veh,1 h}} = L_{\text{max,7.5 m}} - 10\ \lg\left(\frac{v}{1\,\text{km/h}}\right)\text{dB} - 21.5\,\text{dB}. \quad (15.14)$$

For the rating of the noise impact caused by road traffic, the long-term equivalent sound pressure level is an important quantity. This means that measurements have to be repeated at different times of the day and on different days. The German standard DIN 45642 describes an appropriate procedure [38]. The traffic situation has to be observed very carefully. If there is a traffic jam within the local road network during the measurements, the lack of vehicles on the road under investigation could distort the measuring result.

15.4.2 Direct Determination of Road Traffic Noise

Another option to determine the noise impact of road traffic on a particular receiver point is to directly measure the equivalent sound pressure level L_{Aeq} which is caused by the road traffic. This alternative is

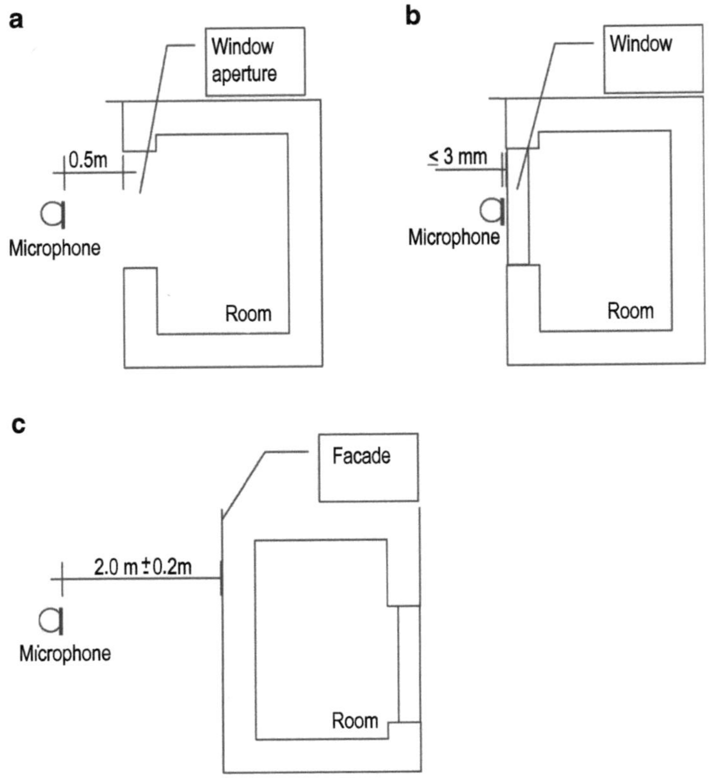

Fig. 15.19 Measuring procedures of road traffic noise

restricted to situations where the road under investigation is the only or dominating sound source affecting the receiver point. However, such a measurement is typically highlighting, but not representing the noise situation. This means that either the road traffic, meteorological conditions, and the condition of the road which were found during the measurements have to be described very carefully or the measurements must be carried out either over a long period or repeatedly to cover different meteorological and traffic conditions. Conversely, measuring the equivalent sound pressure level directly is the simplest way to characterize the noise situation.

Measuring the traffic noise impact in front of buildings or rooms needs to take care of reflections caused by walls or windows next to the microphone, which affect the measuring results. In order to avoid these influences, the microphone is often placed at a distance of 0.5 m in front of the open window (Fig. 15.19a). In this case, the measured L_{Aeq} value is comparable to that measured under free-field conditions. In ISO 140–5 [39], two measuring procedures are described which allow to quantitate the sound pressure level of road traffic noise with the window closed or in front of the façade. In this case, a free-field correction has to be applied to the measuring result of the L_{Aeq}. Using the method whereby the microphone is mounted directly on a sound reflecting surface like the glass pane of a window (Fig. 15.19b), 6 dB shall be subtracted from the measuring result. When the microphone is mounted at a distance from the sound reflecting surface, the free-field correction is -3 dB. In order to avoid spectral distortion, the dimensions of the sound reflecting surface must be homogeneous and large compared to the distance between the microphone and the surface. In general, glass panes of windows are too small to meet this requirement.

15.5 Calculation of Road Traffic Noise

The sound pressure level of road traffic noise fluctuates strongly with time. The strength of the noise impact can thus only be described by statistical quantities like exceedance levels or mean levels over time. Most of the noise control regulations distinguish between the time intervals "Day," which is 6 a.m. until

10 p.m. and "Night" from 10 p.m. until 6 a.m. The so called rating levels, i.e., equivalent sound pressure levels for these time intervals corrected for local characteristics of the road traffic situation, are calculated based on nationally stipulated calculation procedures for road traffic and are compared with legal noise limit values or guidance levels.

Various simplified as well as advanced calculation schemes are in use in different countries or have been developed in international projects. Although one may assume that each scheme is validated, one observes big differences between calculated results for identical situations due to the different levels of detail, approximations, and assumptions. Differences up to 6 dB have been reported. The assumptions related to the road surface, for instance, are treated unequally. In general, tire/road noise depending on speed is different, according to the type of road surface. The acoustical behavior of road surfaces is different, according to the vehicle type. Acoustically, road surfaces are subject to some deterioration, depending on age. Not every calculation scheme takes all of these aspects into account. It would go beyond the scope of this book to describe the calculation schemes and their differences in detail. However, an overview and some aspects should be provided at this point. The main difference between the various procedures is the description of the noise emission of a road. It is based on either the equivalent sound pressure level L_{eq} of the traffic at a defined distance and height relative to the road resp. a single lane or on the total or linear sound power level L_W resp. L_W' of the traffic on a given road section. In each case, the following parameters are necessary to specify the noise emission:

- The number of vehicles within a defined time interval, separated by different vehicle types
- The driven speed, characteristic for the respective road section, separated by different vehicle types
- The type and condition of the road surface
- The gradient of the road
- The traffic condition (free flowing or accelerated)

For each category of vehicle, the calculation schemes contain fundamental values for the sound emission of single vehicles. The road surfaces are also categorized with different levels of detail. The calculation of the noise level at a receiver point, depending on atmospheric conditions, sound reflecting, absorbing, and screening objects like buildings or noise barriers as provided in the calculation schemes, is carried out based on well-known and well-described approaches (Table 15.6). However, acoustic road surface characteristics and topological properties of the road space and its adjacent areas are the crucial factors for noise emission, radiation, and propagation, even in the far field, and not covered in each case. Particularly in connection with the tire/road noise that is emitted next to the road surface and propagates at grazing angles of sound incidence, these aspects play an important role. Comprehensive work has been done to develop a calculation scheme and a database for the propagation of road traffic noise, taking various road situations into account. In [40], ten typical road situations were defined in order to analyze the following effects on sound propagation:

- The effect of a homogeneous, flat ground for short and large distances
- The effect of an acoustic impedance discontinuity on flat ground for large and short distances
- The effect of an embankment (positive slope) with and without an acoustic impedance discontinuity for short and large distances
- The effect of a terrain depression (negative slope) with and without an acoustic impedance discontinuity for short and large distances
- The effect of a barrier with and without an acoustic impedance discontinuity for medium distances
- The effect of a positive vertical sound speed gradient for large distances

15.6 Regulations and Measures for the Protection Against Road Traffic Noise

In general, guidelines and legal regulations for protection against road traffic noise are intended to save road building projects and urban planning not only from adverse environmental and health effects but also to try to guard people from being annoyed. The scheme of noise limit values and guidance levels complies with the possible zoning of an area. Different land occupancy results in different noise limit values. The definition of limit or guidance values for the protection against traffic noise is to some degree a matter of administrative discretion and, therefore, can yield different levels in different countries, which could either be strictly required by law or merely be recommended. For residential areas, the European Commission

Table 15.6 Important calculation schemes for road traffic noise and their differences

Emission model	RLS-90 [37]	RVS [42]	NMPB08 [43]	SonRoad [44]	Nord2000 [45]	IMAGINE [7]
Traffic model						
Number of vehicle categories	2	5	2	2	3	4
Tire/motor noise separate	No	No	Yes	Yes	Yes	Yes
Propulsion noise						
Speed dependent	–	–	Yes	Yes	Yes	Yes
Dependent on traffic type	No	No	Yes	No	No	Yes
Dependent on gradient	Yes	Yes	Yes	Yes	Yes	Yes
Different for vehicle type	No	Yes	Yes	No	Yes	Yes
Up/downhill different	No	Yes	Yes	Yes	Yes	Yes
Arriving/departing traffic	No	No	Yes	No	Yes	No
Rolling noise						
Surface types	9	3	3	13	8	n
Different for vehicle type	No	Yes	Yes	No	Yes	Yes
Speed dependent	Yes	Yes	Yes	No	Yes	Yes
Dependence on age	No	No	Yes	No	Yes	Yes
Emission values						
Quantity (distance in m)	$L_{eq,25m}$	$L_{eq,1m}$	$L_W{}'$	L_W, L_{eq}	L_W	L_W, $L_W{}'$
Space angle	2π	2π	2π	4π	4π	4π
Sources						
Number	1	1	1	1	2	2
Height(s)	0.5 m	0.5 m	0.05 m	0.45 m	0.01...0.75 m	0.01...0.75 m
Directivity	No	No	No	No	Yes	Yes
Frequencies						
Frequency spectrum	No	Yes	Yes	Yes	Yes	Yes
Band width re. 1 octave	–	1/1	1/3	1/3	1/3	1/3
Number of bands	1	7	18	24	27	27
Frequency range (Hz)	–	63...4,000	100...5,000	50...10,000	25...10,000	25...10,000
Different for rolling/motor	No	No	No	No	Yes	Yes
Per vehicle category	No	No	No	No	Yes	Yes
Depending on speed	No	No	No	No	Yes	Yes
Per surface type	No	No	No	No	Yes	Yes
Other aspects						
Structural radiation of bridges	No	No	No	No	No	(Yes)
Correction for tunnel openings	No	No	No	No	Yes	(Yes)
Number of lanes	2	2 (1)	Each	(Each)	Each	Each

published target values as reference for the future research on noise and noise policy in Europe [41]. Based on findings from the World Health Organization and national expert groups, the targets are given in terms of day–evening–night equivalent sound pressure levels L_{den} and equivalent sound pressure levels L_{night} for the nighttime at the receiver point. The night-time noise indicator L_{night} is an A-weighted long-term equivalent sound pressure level as determined over all the night periods of a year where the night is from 11 p.m. until 7 a.m. by default. The L_{den} is defined as follows in (15.15):

$$L_{den} = 10\ \lg\left(\frac{1}{24\,h}\left(12\,h\ 10^{\left(\frac{L_{day}}{10\,dB}\right)} + 4\,h\ 10^{\left(\frac{L_{evening}+5\,dB}{10\,dB}\right)} + 8\,h\ 10^{\left(\frac{L_{night}+10\,dB}{10\,dB}\right)}\right)\right) dB, \tag{15.15}$$

with L_{day} and $L_{evening}$ as A-weighted long-term equivalent sound pressure levels determined over all the day resp. evening periods of a year where the day is from

Table 15.7 Target values as reference for the future research and noise policy in Europe

Target class	L_{den}	L_{night}
Minimum target	65 dB(A)	55 dB(A)
Medium target	55 dB(A)	45 dB(A)
Optimum target	50 dB(A)	40 dB(A)

7 a.m. until 7 p.m and the evening is from 7 p.m. until 11 p.m.

The targets are summarized in Table 15.7, which are intended to be achieved in the future by target-oriented research on noise in the long term. For further details on regulations and measures for the protection of noise and their systematics, including road traffic noise, see Chaps. 5 and 20.

References

1. Commission of the European Communities (1996) Green Paper on Future Noise Policy, COM(96) 540 final, 4.11.1996
2. den Boer LC, Schroten A (2007) Traffic noise reduction in Europe, CE Delft, March 2007 http://www.transportenvironment.org/Publications/prep_hand_out/lid:495
3. van den Berg M (2006) Targets for Noise Immission. CALM Workshop "Road Maps for Future Research in Environmental Noise", Brussels, 16 March 2006. http://www.calm-network.com
4. Verkehr in Zahlen (2005/2006) Bundesministerium für Verkehr, Bau und Stadtentwicklung
5. de Graaff DF (2001) A speed and acceleration limit in the noise type approval of vehicles will enable silent cars to reveal their silence. Proc. inter noise 2001, Den Haag, The Netherlands, paper no. 590
6. Barsikow B, Hellmig M (2003) Schallquellenlokalisierung bei Vorbeifahrten von Kraftfahrzeugen mittels eines zweidimensionalen Mikrofon-Arrays. Fortschritte der Akustik, Proc. DAGA 2003, Aachen, 282–283
7. Peeters B, van Blokland GJ (2007) Improved Methods for the Assessment of the Generic Impact of Noise in the Environment. The Noise Emission Model for European Road Traffic. European 6th research framework project IMAGINE, Deliverable no. 11, January 2007, http://www.imagine-project.org
8. Ejsmont J et al (2002) Influence of road surface wetness on tyre/road noise, Harmonoise report HAR11TR-020901-TUG02.doc
9. Ejsmont J, Ronowski G (2006) Propulsion and acceleration noise coefficients for HARMONOISE Model, IMAGINE report IMA05TR-060301-TUG03, Gdansk (PL)
10. Kropp W et al (2000) On the sound radiation from tyres. Acustica – acta acustica 86(5):760
11. Kropp W et al (2003) The Generation of the Tyre/Road Noise — Mechanisms and Models. Conference proceedings of the 10th International Congress on Sound and Vibration, July 7–10 2003, Stockholm, Sweden
12. Beckenbauer T (2000) Akustische Eigenschaften von Fahrbahnoberflächen. Deutscher Straßen- und Verkehrskongress 2000
13. Wullens F, Kropp W (2004) A three-dimensional contact model for tyre/road interaction in rolling conditions. Acustica – acta acustica 90:702–711
14. Andersson P, Kropp W (2008) Time domain contact model for tyre/road interaction including nonlinear contact stiffness due to small-scale roughness. J Sound Vibr 318 (1–2):296–312
15. Sabiniarz P, Kropp W (2008) A model to evaluate the importance of tangential contact forces for tyre/road noise generation. In: Proceedings Acoustics 08 conference, Paris, France, June 29-July 4
16. Hayden, RE (1971) Roadside noise from the interaction of a rolling tyre with the road surface. In: Proceedings of the Purdue Noise Control Conference, Purdue University, West Lafayette, USA, 62–67
17. Ronneberger D (1984) Experimentelle und theoretische Untersuchungen spezieller Mechanismen der Rollgeräusche. Mitteilungen des Instituts für Straßen-, Eisenbahn- und Felsbau an der ETH Zürich 57:55
18. Ronnebeger D et al (1984) Rolling noise of tyres – fluctuation of pressure and flow velocity in the vicinity of the contact area. Publ. Wiss. Film, Sektion Techn. Wiss./Naturwiss., Ser. 8, Nr. 25/C 1503
19. Hamet JF, Deffayet C and Pallas MA (1990) Air-pumping phenomena in road cavities. In: Proceedings International Tire/Road Noise Conference, August 8–10 1990, Gothenburg, Sweden, 19–29
20. Graf R, Kuo C, Dowling A, Graham W (2002) On the horn effect of a tyre/road interface, part 1: Experiment and computation. J Sound Vibr 256(3):417–431
21. Bécot FX, Barrelet S, Kropp W (2000) On the sound radiation from tyres. Acta Acustica United Acustica 86(5):769
22. Beckenbauer T et al (2002) Einfluss der Fahrbahntextur auf das Reifen-Fahrbahn-Geräusch. Forschungsberichte aus dem Forschungsprogramm des Bundeministeriums für Verkehr, Bau und Wohnungswesen und der Forschungsgesellschaft für Straßen- und Verkehrswesen e.V., Heft 847
23. ISO 10844 (1994) Acoustics – Specification of test tracks for the purpose of measuring noise emitted by road vehicles
24. Hucho WH, (ed) (1998) Aerodynamics of road vehicles. Society of Automotive Engineers Inc; 4 ed
25. McCallen R, Browand F, Ross J (eds) (2004) The Aerodynamics of Heavy Vehicles: Trucks, Buses, and Trains. Lecture Notes in Applied and Computational Mechanics, vol 19, Springer, Heidelberg
26. Council Directive 70/157/EEC (6 February 1970) On the approximation of the laws of the Member States relating to the permissible sound level and the exhaust system of motor vehicles
27. Council Directive 77/212/EEC (8 March 1977) amending Directive 70/157/EEC, Relating to the permissible sound level and the exhaust system of motor vehicles
28. Commission Directive 81/334/EEC (13 April 1981) adapting to technical progress Council Directive 70/157/EEC on the approximation of the laws of the Member States

relating to the permissible sound level and the exhaust system of motor vehicles
29. Commission Directive 84/372/EEC (3 July 1984) adapting to technical progress Council Directive 70/157/EEC on the approximation of the laws of the Member States relating to the permissible sound level and the exhaust system of motor vehicles
30. Council Directive 84/424/EEC (3 September 1984) amending Directive 70/157/EEC on the approximation of the laws of the Member States relating to the permissible sound level and the exhaust system of motor vehicles
31. Council Directive 92/97/EEC (10 November 1992) amending Directive 70/157/EEC on the approximation of the laws of the Member States relating to the permissible sound level and the exhaust system of motor vehicles
32. Commission Directive 2007/34/EC (14 June 2007) amending, for the purposes of its adaptation to technical progress, Council Directive 70/157/EEC concerning the permissible sound level and the exhaust system of motor vehicles
33. Directive 97/24/EC of the European Parliament and of the Council (17 June 1997) on certain components and characteristics of two or three-wheel motor vehicles
34. Directive 2001/43/EC of the European Parliament and of the Council (27 June 2001) amending Council Directive 92/23/EEC relating to tyres for motor vehicles and their trailers and to their fitting. Official Journal L 211, 04/08/2001, 25–46
35. Phil Morgan (Editor) (2006) Guidance manual for the implementation of low noise road surfaces. FEHRL Report 2006/2, Brussels
36. Steven H (2005) Investigations on noise emission of motor vehicles in road traffic. Final report, Research project 20054135, German Environmental Agency
37. RLS-90 (1990) Richtlinie für den Lärmschutz an Straßen. Allgemeines Rundschreiben Straßenbau Nr. 8/1990 des Bundesministers für Verkehr
38. DIN 45642 (2004) Messung von Verkehrsgeräuschen
39. ISO 140–5 (1998) Acoustics – Measurement of sound insulation in buildings and of building elements – Part 5: Field measurements of airborne sound insulation of façade elements and façades
40. http://deufrako.bast.de
41. European Commission (2007) Research for a quieter Europe in 2020. An update strategy paper of the CALM II Network, September 2007
42. Richtlinien und Vorschriften für das Straßenwesen (RVS) 04.02.11 Lärmschutz, (2006) Forschungsgesellschaft Straße Schiene Verkehr. Wien, Austria
43. Besnard F, LeDuc E (eds) (2009) Road noise prediction – Noise propagation computation method including meteorological effects (NMPB 2008), Service d'études sur les transports, les routes et leurs aménagements (SÉTRA). Bagneux, France
44. Heutschi K (2004) Berechnungsmodell für Straßenlärm. Schriftenreihe Umwelt Nr. 366. Bundesamt für Umwelt, Wald und Landschaft, Bern, Switzerland
45. Jonasson HG, Storeheier S (2001) Nord 2000. New nordic prediction method for road traffic noise, SP Swedish National Testing and Research Institute, Report No. 2001:10, Boras, Sweden

Noise and Vibration from Railroad Traffic 16

Rüdiger G. Wettschureck, Günther Hauck, Rolf J. Diehl†,
and Ludger Willenbrink

16.1 Introduction

The overdue beginning of new construction and development of train tracks since the 1960s coincided in the 1970s with an increasing sensitivity of the population towards noise, especially road traffic noise. This sensitivity has constantly increased since then and has significantly influenced the legislation concerning traffic noise. Today, in Germany all construction plans for traffic routes must also include an ensured prediction of the effect of noise on the residents.

In the mid-1950s, this task was facilitated for railway noise by the foundation of a Noise Measuring Group in the Research Institute in Munich of the former "*Deutsche Bundesbahn*" (German Federal Railways, today: Deutsche Bahn AG, DB Systemtechnik). That group which was directed and formed over two decades by Stüber [1] tested all types of rail traffic noise, so that the required knowledge regarding many open questions could quickly be developed. Thus, the German Federal Ministry for Traffic let develop the Calculation Regulations for Sound Emissions of Track Operation, in the scope of which the "Guidelines for the Calculation for Sound Tests in the Planning of Shunting or Freight Yards" [2] and the "Guidelines for Calculating Sound Emissions of Train Tracks" – "Schall 03" were developed [3].

The research on noise impact had already led to the insight that various types of traffic noise (e.g. road, rail, aircraft) with the same noise level created different degrees of annoyance [4]. For the calculation of the noise immission from train tracks, a "bonus" of 5 dB (A) in favour of the train tracks [5] was applicated. For calculations according to the "Guidelines for the Calculation of Sound Emission in the Planning of Shunting or Freight Yards" [2] extra charges were added for tonal or pulse noises. The results of such calculations are the rating levels *(Beurteilungspegel)* according to [6]. They are compared with respect to each problem with limit, reference or orientation values, for example, according to [5].

Also with respect to structure-borne sound and vibrations, assessment regulations are increasingly being required. There is a significant need for research in this field, in order to make predictions regarding this matter more reliable. In addition, in this area, as already stated in the second edition of this book [260], it has been confirmed that there is in Germany no legal provision concerning this topic until now.

The following topics regarding rail traffic noise as well as structure-borne sound and vibrations due to rail traffic have been divided into four sections:

16.2 Airborne noise from railway lines in compliance with train construction and operation regulations according to [8].

16.3 Structure-borne noise and vibrations from railway lines according to [8].

[Submitted 11 September 2006]
The authors thank all the colleagues of Deutsche Bahn AG, Systemtechnik München, as well as of various acoustic engineering companies and institutes, who have given active support in procuring data and in drawing up the manuscript.

R.G. Wettschureck (✉)
Gstädtstrasse 36, 82439 Großweil, Germany
e-mail: post@wettschureck-acoustics.eu

G. Müller and M. Möser (eds.), *Handbook of Engineering Acoustics*,
DOI 10.1007/978-3-540-69460-1_16,

16.4 Airborne and structure-borne noise and vibrations caused by short-distance trains according to tramway, construction and work regulations [9].

16.5 Prediction models for airborne noise, structure-borne noise and vibration from railway lines.

16.2 Airborne Noise from Railway Lines

The relevant systems according to the train-construction and work regulations [8] are mostly trains owned by the Federal Republic of Germany (FRG) and trains owned by the Länder of the FRG (*Bundesländer*). Included are the rapid transit trains (*S-Bahn*) and harbour trains, too. Railway companies of other countries are also included at suitable places in this chapter.

16.2.1 Definitions

Below some basic definitions are given which are required for the characterization of the noise situation in the environment of railway lines or facilities. Relevant literature or the corresponding chapters of this book are used as references for the general premises of acoustics.

16.2.1.1 Equivalent Continuous Sound Level

The equivalent continuous sound level L_{eq} in dB (A) according to [10] is generally used for the description of the noise with time-dependent alternating sound emission levels. L_{eq} is determined by the sound level and duration of the sound event, which is averaged over time intervals.

The numeric values are mostly:

(a) The equivalent continuous sound level for the time of the passage of a train, or a locomotive (train length divided by speed), also called the average passage level, or
(b) The equivalent continuous sound level for a single hourly event $L_{eq,1h}$, e.g. a train passage including approach and departure, a coupler or buffer contact during shunting, etc., referred to one whole hour.

According to most recent standards for the acoustic characterization of the train, the transit exposure level (*TEL*) is preferred [11].

16.2.1.2 Emission Level

The sound emission of a railway line (line emission source) is defined by the emission level $L_{eq,E}$ in dB(A). It is the equivalent continuous sound level for the time interval in question at a horizontal distance of 25 m from the axis of the track under consideration and a height of 3.5 m above top of rail, assuming free undisturbed sound propagation.

In case of point sound sources, such as buffer contacts or screeching of rail brakes during shunting operations, the emission level describes the equivalent continuous sound level in dB(A), produced by the source at a distance of 25 m from its centre, considering undirected sound emission.

16.2.1.3 Basic Value (Grundwert)

The basic value *(Grundwert)* is the emission level $L_{eq,E}$ in dB(A) of a train with 100 m length, a speed of 100 km/h, 100% of the vehicles equipped with disk brakes (including locomotives), a ballasted track with wooden sleepers and a rail running surface in average condition, related to the period of 1 h [3]. Its numeric value under these conditions is 51 dB(A).

16.2.1.4 Rating Level

The rating level L_r is used to identify the sound immission affecting an area or a point of an area.

It is determined at the basis of the vehicle- and track-type particularities of the emission, the propagation attenuation on the respective paths of sound propagation, and – where necessary – the correction values concerning noise impacts and differences in the annoyance of impact compared with other types of traffic noise (see Sects. 16.2.4 and 5.4.2.1).

16.2.2 Noise Emissions

The noise emissions of trains are basically determined by:

- The rolling noise – (speed range 50 km/h < v < 350 km/h)
- The machine noise – (speed range v < 60 km/h)
- The aerodynamic noise – (speed range v > 350 km/h, according to Schall 03 [3], BImschV-16 [5])

The rolling noise is largely influenced by:

(a) The running speed
(b) The train length

(c) The type of braking (block brakes, disk brakes), which affects the roughness of the wheel surface
(d) Special features on the vehicle (such as sound-reducing wheel-mounted disk brakes)
(e) The condition of the running surface (roughness of rail and wheel running surfaces)
(f) Special track features (ballasted superstructures with wooden or concrete sleepers, ballastless track,[1] with or without sound absorbing treatment)
(g) Special features on the line (e.g. bridges and railroad crossings)

For example, assuming the same speed and the same condition of the running surface of the rail, due to their even wheel surfaces, trains with disk brakes and electronic rolling protection (anti-slide devices) cause lower noise emissions than trains with cast iron block brakes.

The major sources for the radiation of the rolling noise are wheels and rails. The wheel basically emits in the frequency range above 1,000 Hz, and the rails mostly below 1,000 Hz. It has to be considered that also the type of the wheel has an effect on the sound radiation. The sound emission of railway lines is characterised by the sound emission level $L_{eq,E}$.

For a more detailed description or as a basis for the explanation of mechanisms caused by noise and radiation mechanisms, spectral analyses of the train passage noise, preferably third-octave band analyses [12] are required. For the determination of tonal noise components, narrow frequency band analyses are also necessary.

The sound pressure level L_{measure}, which was measured at the speed v_{measure} of a certain train, can be converted with Eq. (16.1) in the respective level L_v for other train velocities v:

$$L_v = L_{\text{measure}} + k \cdot \lg\left(\frac{v}{v_{\text{measure}}}\right) \text{ dB}. \qquad (16.1)$$

Here, $k = 20$ for the calculation of equivalent continuous sound level referred to time intervals, and $k = 30$ for calculating average passing train levels. The values for k are based with a certain scatter on many measurements with various types of tracks and vehicles at a very large range of speed between approximately 30 and 300 km/h.

The further treatment of the sound emission emanating from the operation of trains is divided into the following sections:

16.2.2.1 Vehicles (conventional rolling stock and magnetically levitated trains)

16.2.2.2 Track system (ballasted track, ballastless track, bridges, etc.)

16.2.2.3 Railway facilities covering large areas like shunting yards or passenger stations

16.2.2.1 Vehicles

The vehicles are divided into

(a) *Traction units*, such as electric locomotives, diesel locomotives, power cars of train sets like the German Intercity Express ICE 1 and ICE 2, multiple-unit train sets like the ICE 3 and rapid transit train sets *(S-Bahn)* Class (ET420, ET423 up to ET426), as well as diesel and electric railcars
(b) *Passenger coaches* (with block or disk brakes)
(c) *Freight wagons* (different types, with block brakes or, rarely, disk brakes)

Separately is dealt with

(d) *Magnetically levitated trains – Maglev* (Transrapid 07 – TR 07, see, e.g. [13, 14]).

In Fig. 16.1, ranges of the average train passing levels are indicated. They were measured at a distance of 25 m to the middle of the track, and a height of 3.5 m above top of rail (based on [15]) for rolling stock of Deutsche Bahn AG (German Rail). The type of brakes and the speed of the trains are indicated.

Figure 16.2 shows the average value of the passing level of the rolling stock for a selection of high-speed trains, measured in the same manner as described in connection with Fig. 16.1 with various countries of origin [15, 16]. The dependency upon the train speed is indicated.

Figure 16.3 compares the 1/3 octave-band spectra for the high-speed trains ICE 1, TGV Lyon,[2] TR 07 and X 2000 [15].

Traction Unit Details

Compared with the older electric locomotives with cast iron block brakes, the traction units of the high-

[1] In German: Feste Fahrbahn = ballastless track (slab track), different types of constructions; see, for example, [24, 95].

[2] TGV = Train à Grande Vitesse (high-speed train of French National Railway SNCF)

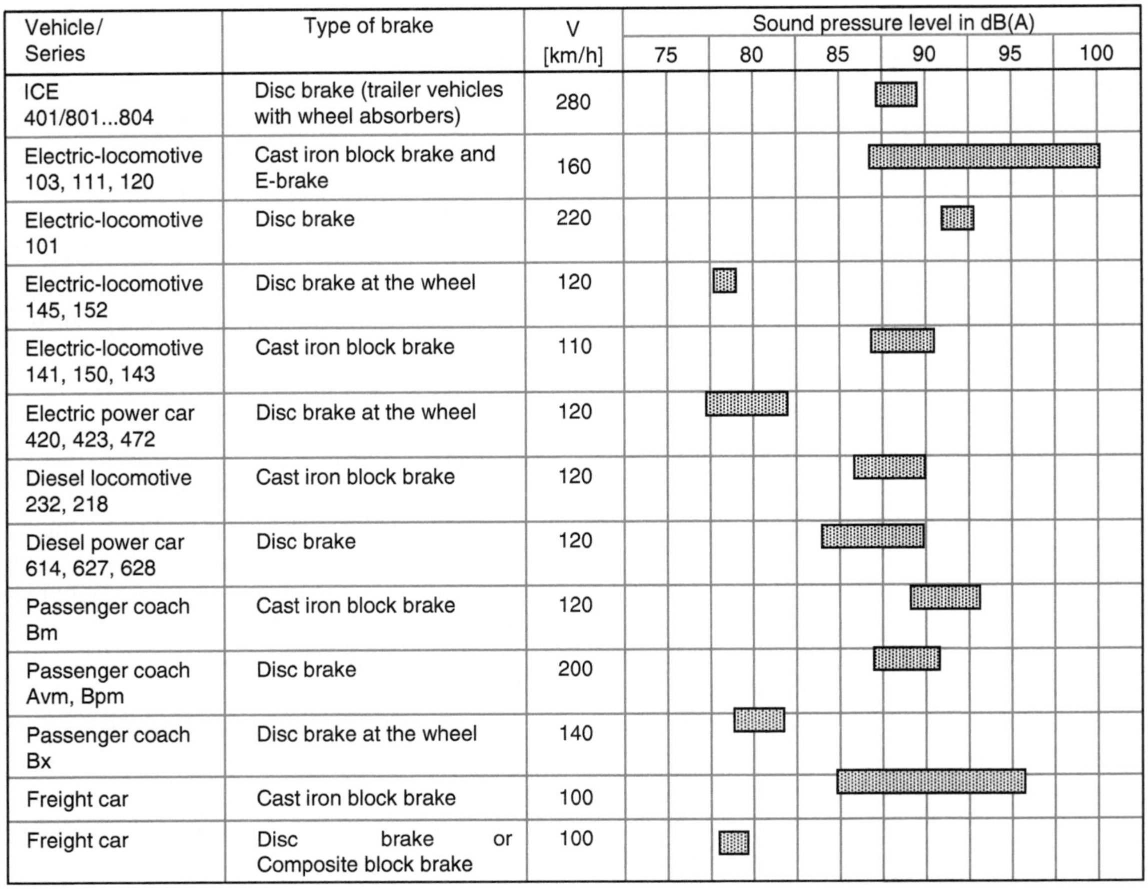

Fig. 16.1 Average transit noise level ranges of rail vehicles from German Rail (DB AG), measured at a distance 25 m to the side of open tracks (3.5 m above top of rail) while running at the respective typical train speed on rail running surfaces with a depth of short-pitch corrugation of < 20 μm (s. Fig. 16.10)

speed trains ICE 1 and ICE 2 are quiet, because they are fitted with disk brakes. However, they are louder than the passenger coaches of the ICE, since those have additional noise absorbers on the wheels. In addition, the traction units emit mechanical noise and, at a speed of ≥ 250 km/h, they emit relevant aerodynamic noise (see Fig. 16.4). Depending on the aerodynamic characteristics, the aerodynamic noise at a speed of more than approximately 300 km/h contains more sound energy than the rolling noise.

A particular acoustic problem is created when the pantographs pass by noise control barriers at such high speed, because, as a considerable aerodynamic sound source, they reduce the acoustic effect of noise control barriers calculated for rolling noise radiated from the wheel/rail area. With the help of simulation calculations and subsequent validations, designs were developed in a very fast and silent wind tunnel that promise an improvement of approximately 12 dB(A) compared with conventional pantographs. Such pantographs must be actively driven to control the pressure forces between pantograph and catenary.

Wheels with wheel noise absorbers cause approximately 4 dB(A) lower emission levels than wheels without wheel noise absorbers (measured 4–6 dB(A) on Deutsche Bahn AG passenger coach). Thus, in case of trains with a permissible speed of $v > 100$ km/h and with wheel noise absorbers according to [3], a corrector of 4 dB is subtracted from the basic value (*Grundwert*).

Comparing the ICE 1 with the TGV-Atlantique (Fig. 16.2), it has to be considered that both trains are actually equipped with disk brakes. However, on the wheels of the ICE 1 noise absorbers are mounted, whereas the TGV-Atlantique at that time had cast iron block brakes, which – depending on the use of these

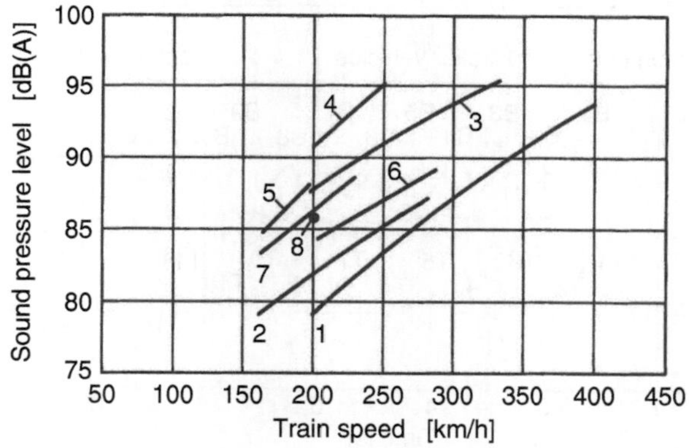

No.	Type of train	Measurement height above track or above top of rail, respectively	Country of origin
1	Maglev Transrapid 07	- 4.7 m	Germany
2a	Talgo-Pendular	3.5 m	Spain
3	TGV-Atlantique	3.5 m	France
4	TGV Lyon	3.5 m	France
5a	IC/EC	3.5 m	Germany
6a	ICE	3.5 m	Germany
7	X2000	3.5 m	Sweden
8b	Shinkansen	?	Japan

[a]running on ballasted track of German Railway (DB AG)
[b]single measurement value

Fig. 16.2 Transit noise level of different high-speed trains at a distance of 25 m to the side of the track in dependence of the train speed

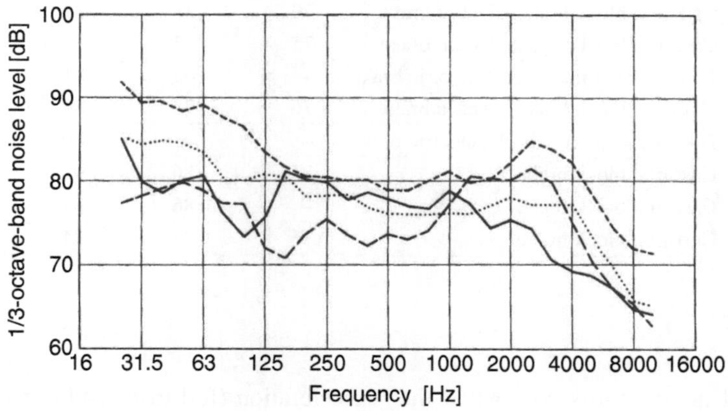

Symbol	Type of train	Measurement height in metre: Above track or top of rail, respectively	Measurement height in metre: Above ground	Train speed km/h	Overall sound pressure level dB(lin)	Overall sound pressure level dB(A)
———	Maglev Transrapid 07	-4.7	3.5	305	92	86.5
---------	TGV-A[a]	3.5	4.0	305	99	92.5
....	ICE[b]	3.5	5.0	280	94	87.5
– – – –	X2000[b]	3.5	5.0	220	92	89.5

[a]At that time with cast-iron block brakes, travel on ballasted track with biblock sleepers of French National Railway (SNCF);
[b]Travel on ballasted track with monoblock sleepers of German Railway (DB AG)

Fig. 16.3 Airborne noise at a distance of 25 m to the side of an open track during the pass-by of different European high speed trains

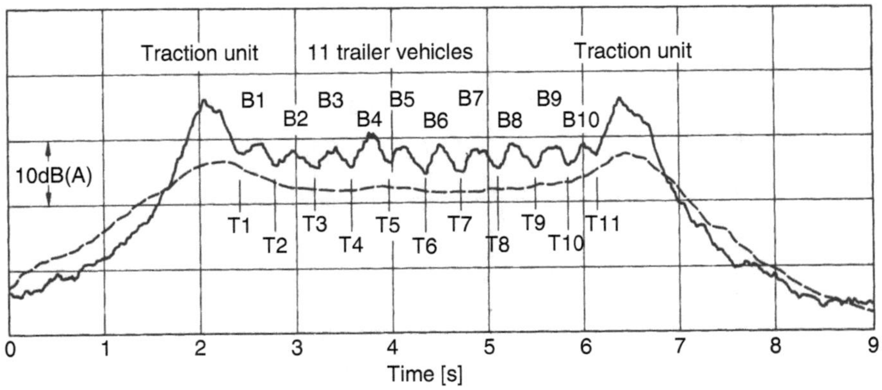

Fig. 16.4 Time history of the A-weighted noise level, measured at different distances from the track during the pass-by of the high-speed train *ICE 1* at a speed of 250 km/h. ——— 7.5 m to the side, 1.2 m above top of rail; - - - - - - 25 m to the side, 3.5 m above top of rail; traction units without wheel absorbers, but including the aerodynamic noise of the pantograph; B1–B10: "level hills" = passing of pairs of bogies (wheels with wheel absorbers); T1–T11: "level valleys" = passing of centre parts of coaches

Table 16.1 Sound radiation of different locomotives, measured at a distance of 25 m to the side (mean of up to three different specimens)

Type of locomotive class	Brake system	Noise level in dB(A)			Passing speed (km/h)
		Standing	Starting	Passing	
101[a]	Disc brake, electric brake	77	–	91	220
103[a]	Cast iron block brake, electric brake	76	79	84	150
111[a]	Cast iron block brake, electric brake	77	80	86	140
120[a]	Cast iron block brake, electric brake	79	83	89	200
143[a]	Cast iron block brake, electric brake	75	75	84	120
145[a]	Disc brake at the wheels, electric brake	–	–	81	140
151[a]	Cast iron block brake, electric brake	76	78	89	120
152[a]	Disc brake at the wheels, electric brake	–	–	80	140
218[b]	Cast iron block brake	84	86	91	140
232[b]	Cast iron block brake	–	86	88	120
290[b]	Cast iron block brake	75	83	84	80

[a]Electric locomotives
[b]Diesel locomotives

additional brakes – had more or less corrugated the running surface of the wheels. Nowadays those TGV have in addition to the disc brakes composite block brakes (see later on at freight wagons), which avoid corrugations so that their noise level is similar to that of the ICE.

Figure 16.1 shows the range of passing level of diesel locomotives and electric locomotives with various types of braking systems at their maximum speeds. For some types of locomotives, Table 16.1 indicates furthermore the sound pressure level at standstill with the maximum power of the auxiliary equipment (cooling fans, etc.), the noise emission under acceleration (full load) and at pass-by. All values are measured at 25 m distance and 3.5 m above the rail surface.

Some 1/3 octave-band spectra of those noises are shown in Figs. 16.5–16.7 [251].

As the emergency use of cast iron brake blocks at higher speeds creates corrugated wheels, the electric locomotives with stronger electric brakes but still additionally equipped with cast iron brake blocks, e.g. the locomotives Class 103, 111, 120 and 151, can temporarily emit up to 10 dB(A) higher noise levels. However, by means of the positive slip with acceleration or with the negative slip during electric

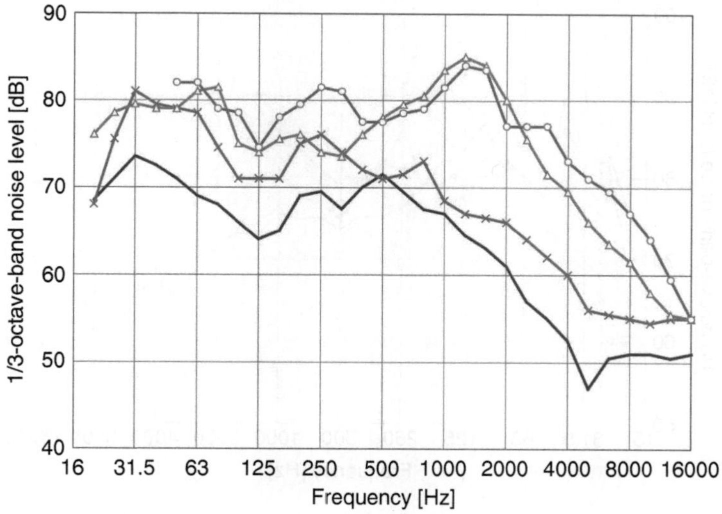

Fig. 16.5 Airborne noise at a distance of 25 m to the side (3.5 m above top of rail) of high speed locomotives, in case of free field sound propagation (average value of 3 locomotives at a time): ——— Class 103, standing noise 82 dB(lin), 76 dB(A); ×——× Class 103, starting noise 88 dB(lin), 79 dB(A); Δ——Δ Class 103, pass-by noise (v = 150 km/h) 93 dB(lin), 91 dB(A); o——o Class 101, pass-by noise (v = 220 km/h) 93 dB(lin), 91 dB(A)

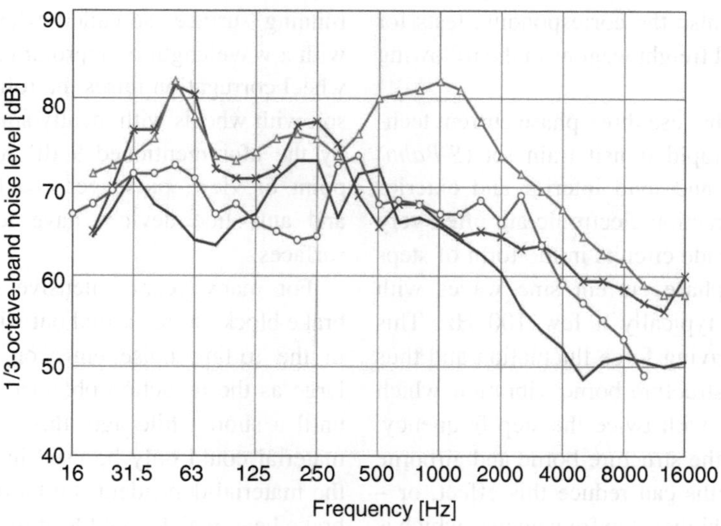

Fig. 16.6 Airborne noise at a distance of 25 m to the side (3.5 m above top of rail) of electric locomotives for freight trains, in case of free field sound propagation (average value from 2 to 3 locomotives): ——— Class 151, standing noise 82 dB(lin), 76 dB(A); ×——× Class 151, starting noise 88 dB(lin), 78 dB(A); Δ——Δ Class 151, pass-by noise (v = 120 km/h), Cast iron block brake 91 dB(lin), 89 dB(A); o——o Class 152, pass-by noise (v = 120 km/h, radiator fan full load), disc brake at the wheels 83 dB(lin), 78 dB(A)

braking after some 100 km running the corrugations are ground off again [17].

In a test a very complicated, acoustically optimised housing of the entire area of the bogie at an electric locomotive of the 103 series only succeeded in a slight reduction in the rolling noise by 2 dB(A) [18, 19]. However, the investigated housing would be unsuitable for a regular train operation. Thus, the effect would obviously be smaller with a housing suitable for typical operations. Such a housing can only

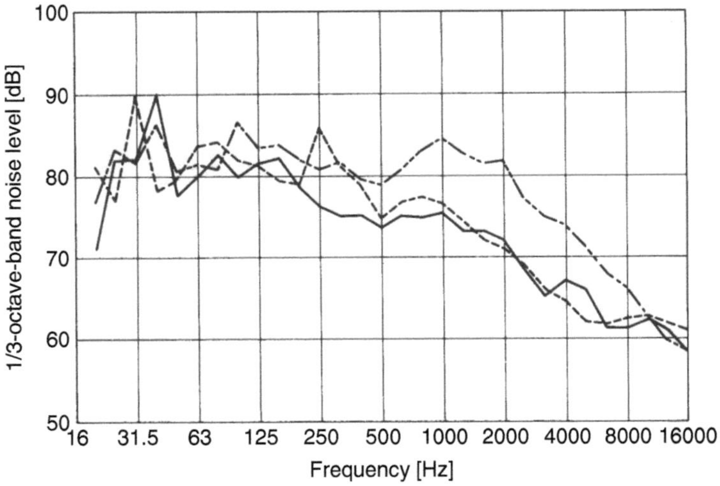

Fig. 16.7 Airborne noise at a distance of 25 m to the side (3.5 m above top of rail) of the Diesel locomotive, Class 218, in case of free field sound propagation (average value from 3 locomotives): ——— standing noise 94 dB(lin), 84 dB(A); - - - - - starting noise 95 dB(lin), 86 dB(A); —-—-— pass-by noise 96 dB(lin), 92 dB(A), (v = 140 km/h)

develop an obvious effect together with low noise control barriers positioned close to the edges of the tracks clearance (see also the corresponding tests for passenger coaches and freight wagons in the following sections).

With power units that use three-phase current technology, such as the rapid transit train set (*S-Bahn*) Class ET 423, tonal and loud interior and exterior noise caused by the traction electronic are often very disturbing. Here, cascade circuits in the form of steps reproduce the three-phase current sine waves with a step frequency of typically a few 100 Hz. This stepping leads to a driving force fluctuation and thus to the excitation of structure-borne vibration which causes airborne noise with twice the step frequency. Careful insulation of the structure-borne and airborne noise transmission paths can reduce this effect, or – more effectively – by higher step frequencies, which is possible today with modern electronics; with higher frequencies (two octaves typical) on one hand side the steps and thus the force fluctuations are much smaller and on the other hand side the insulation is simpler.

Passenger Coach Details

Compared with older passenger coaches with brake blocks, the rolling noise of today's passenger coaches with disk brakes at the same speed and on smooth track surface is approximately 9 dB(A) lower. The reason for this was found in 1975 [17]: due to the impact of the cast iron brake blocks on the wheel's running surface, so-called wheel corrugation arises with a wavelength of approximately 2 cm–6 cm. This wheel corrugation raises the rolling noise in comparison with wheels with ideally smooth running surfaces by the aforementioned 9 dB(A). From the acoustic point of view, passenger coaches with disk brakes and anti-slide devices have ideally smooth wheel surfaces.

For many years, intensive tests with composite brake blocks were carried out which led to a reduction in the rolling noise emission, which is almost as large as the reduction obtained with disk brakes. Up until a short while ago, this composite brake block material could only be used in limited cases, due to the material-dependent limitation of the discharge of brake heat, which could lead to a thermal overloading of the wheels. In order to cope with this, however, suitable wheel profiles and wheel manufacturing procedures have been developed.

Some European coaches have, in addition to the disk brakes, only rarely used cast iron block brakes (the so-called cleaning block). In this case, the wheel running surface is corrugated similar to coaches with conventional block brakes. Thus, such vehicles with disk brakes often emit rolling noise similar to coaches with cast iron block brakes.

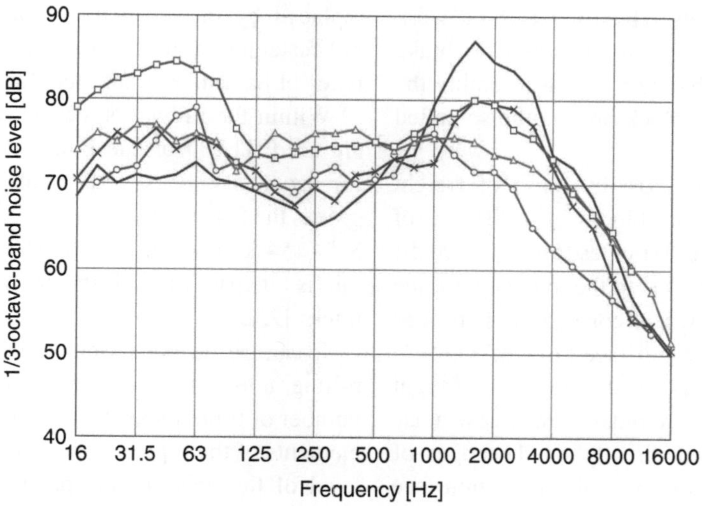

Symbol	Type of train	Number of trains	$\overline{v}$ km/h	Overall noise level dB(lin)	Overall noise level dB(A)
□———□	ICE1 and ICE2 (traction units without, trailer vehicles with wheel absorbers)	6	250	93.5	88
———	Passenger trains with cast iron block brakes	10	140	92	92.5
x———x	IC with disc brakes, (inclusive locomotive with cast iron block brakes)	10	160	88.5	87.5
Δ———Δ	Freight trains with cast iron block brakes	28	100	89.5	85
o———o	Rapid transit electric trains Class ET 420 and ET472 with disc brakes at the wheels	12	120	86	82.5

Fig. 16.8 Airborne noise at a distance of 25 m to the side of an open track (3.5 m above top of rail) during the pass-by of different types of trains at their typical mean train speed

Figure 16.8 contains the spectra of passing noises of passenger coaches with block brakes or with disk brakes [20].

The significant increase in the level of the frequency range 800–2,500 Hz can be traced back to the wheel corrugation. When the speed is halved, the position of this level increase moves one octave below.

Tests with wheel covers, which cover the wheel disks on the inside and the outside, which only have contact with the wheel at the wheel hub and are designed such that just the outer ring of the wheel can be seen, showed a level decrease from 2 to 4 dB (A), depending on the design (with or without absorption material in the space) [18]. This result was achieved on vehicles with block brakes, as well as on vehicles with disk brakes.

This level decrease by more than 3 dB(A) indicates that the wheel disk emits more than 50% of the A-weighted sound energy.

Complicated, sound-proofing housings of the bogie of passenger coaches (acoustically optimal, but unsuitable for continuous operation) [19] produced a level decrease of only approximately 2 dB(A).

Freight Wagon Details

The numerous types of freight wagons differ in their sound emission due to various wheel constructions, running gears, bogies and braking systems. Due to very rough operating conditions for freight wagons in international traffic, they and, in particular, their axle holders or bogies must be particularly robust and simply constructed. The technically demanding disk brakes have not been considered for such vehicles up until now. They are just used in special high-speed freight wagons with speeds up to 160 km/h. In the future, composite brake blocks will increasingly be used for freight wagons as well, so that the average sound level decrease by 9 dB(A) described for the above-mentioned passenger coaches can be expected. Due to the different friction behaviour of the so-called “K-blocks”, they require, however, extensive mechanical conversion work with the retrofitting and the insertion of adapted wheels. There is still hard work being done on the development of synthetic brake blocks with a friction behaviour, which resemble the behaviour of the cast iron block brakes, the so-called "LL-brake blocks".

The sound level of the pass-by noise of freight wagons with cast iron brake blocks at a distance of 25 m (3.5 m above top of rail) ranges from 84 to 90 dB (A) at a speed of 80 km/h. Due to the lower speed, the level increase caused by wheel corrugation is moved from higher to mid-frequencies (see Fig. 16.8).

Tests with sound covers on the bogies of freight wagons produced only in combination with low noise control barriers positioned directly on the edges of the track clearance a significant noise reduction (see above: traction unit details).

Magnetically Levitated Trains: Maglev

For the magnetically levitated trains that have been constantly improved over 25 years [13, 14], today on the test course in *Emsland* driven at speeds above 400 km/h, the aerodynamic noise is considerable. Figures 16.2 and 16.3 show levels and spectra of the passing noise of the German magnetically levitated train, "Transrapid 07".

For further noise data of the magnetically levitated train, see, e.g. [21, 22], for most recent information on Maglev trains in general, see [23].

16.2.2.2 Track System

Permanent Way

In European railway systems, the so-called ballast track is still mostly used: in this system, the rails are mounted on wooden or concrete rail sleepers with relatively stiff rail pads (stiffness ≥500 MN/m, normally approximately 6 mm thick) in between. Thus, the rails are pretty rigidly connected to the sleepers. The track grid is vibrated (filled) into the ballast bed so that under the sleepers remains a ballast height of 30 cm.

Since 1996 in Germany, particularly on tracks with speeds ≥200 km, an increasing amount of superstructures without ballast, ballastless tracks or slab tracks, in Germany the so-called *"Feste Fahrbahn"* have also been built. In this case, the rails are attached by means of elastic rail fastenings on or in a monolithic block consisting of cement or asphalt-compounded concrete. The elasticity of the ballast is replaced by the elasticity of the base plates in the rail fastening system (for the construction of various types of ballastless track, see, for example, [24]).

Within the railway systems, various types of rails are used. The Deutsche Bahn AG, for example, uses for secondary lines with small axle load and low speeds the S 49 type (49 kg/m), for major lines the type S 54 (54 kg/m) and UIC 60 (60 kg/m). The UIC 60 rail is, in particular, installed in newly constructed tracks [25].

It can be deduced from the results of the lateral rolling noise measurements carried out at a large number of train tracks, that the influence of individual elements of the type of superstructure on the A-sound level of the noise of the passing train can approximately be described as follows:

Type of rail	Small
Running surface of rail	Very large (corrugation growth!)
Rail fastening on ballasted track	Rather small[3]
Rail fastening on ballastless track *(Feste Fahrbahn)*	Considerably
Type of sleeper	Rather small[4]
Quality of ballast bed	Small

As far as the requirements for the superstructure technology are concerned, the travel safety and the travelling comfort were typically the basis for the layout rather than acoustical aspects.

However, the superstructure in connection with the vehicle and the substructure represents a very complex dynamically behaving system, whose structural characteristics (such as masses and stiffnesses) should be coordinated with each other according to acoustical aspects.

From the newer, not yet finished theoretical tests, it can be concluded, that the sound emissions of lines with ballasted or with ballastless track can be positively influenced by an acoustically optimised layout of the individual components and their mutual tuning (compare the qualitative assessment above).

Trains on ballastless tracks with conventional construction cause higher sound emissions compared with trains on ballast tracks. This is due to the fact that the

[3] If the required elasticity is largely supplied from the ballast bed, as usual.

[4] Dependent on the structure and mass of the sleeper.

vibration of the rail increases (caused by the softer rail fixing point and the lower masses connected to the rails) and that the absorptive behaviour of the ballast bed surface is missing. Absorbing constructions of the track surface can mostly compensate this phenomenon. Sound-absorbing ballastless tracks on the high-speed line Hannover–Berlin have no higher sound emission values in the meaning of [5] than concrete rail sleeper tracks in the ballast bed [26]. The requirements of the absorptive layer as well as the geometric configuration of the construction components are described in [27] and in the standards catalogue of the Deutsche Bahn AG for building ballastless tracks [252]. Consequently, the acoustic absorption coefficient α determined in a reverberant room in the frequency range between 400 Hz and 5 kHz must be at least 0.8. In the 600 Hz range, deviations of 0.2 to lower values are considered acoustically not critical.

Although the noise emissions with the current construction of the ballastless tracks with no sound-absorbing track surface are higher compared with the ballasted track, the installation of this track structure on bridges leads to a significant reduction in the track vibrations. This is due to the decoupling of the rail and sleeper (concrete supporting plate) by means of the elastic rail fastening systems (see Fig. 16.9). As a result, it clearly reduces the radiated noise from bridges in the typical frequency range of approximately 80–630 Hz. This is shown in the comprehensive tests for the construction of a ballastless track of the modified type "Rheda" on a reinforced concrete hollow box girder bridge [28].

Uneven wheel running surfaces and uneven (long-wave or corrugated) rail running surfaces have a considerable influence on the emission of noise. In the course of several years (in some cases even quicker), corrugation can develop on the rail surfaces. The wavelengths of the corrugation are between approximately 2 and 10 cm, whereby always corrugations of different wavelengths are combined. Within a few meters, the amplitudes of corrugation can vary considerably. The noise emission caused by the corrugation is primarily in the frequency range between 500 and 3,000 Hz (depending on the wavelengths and the train speed).

Figure 16.10 shows the increase in the noise emission with increasing corrugation depth. The figure is separated for passenger trains with cast iron block brakes (wheel corrugation) and for passenger trains with disk brakes (smooth running surfaces). As expected, the corrugated wheel surfaces react clearly less to the additional rail corrugation than the wheels with smooth running surfaces.

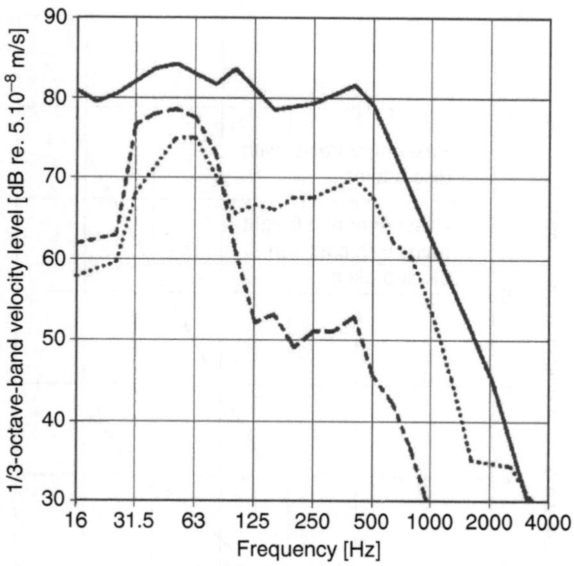

Fig. 16.9 Structure-borne noise, measured at the deck slab of a reinforced concrete hollow-box girder bridge during the passage of a test freight train on different kinds of track at a speed of 80 km/h; ——— ballasted track W54B70 (before modification); ········ slab track "Rheda modified", nonelastically supported (first stage); ------ slab track, type "Rheda modified", full surface elastically supported on Sylomer® L12 (second stage)

Figures 16.11 and 16.12 show how the spectra are changed: Both with and also without wheel corrugation a wide "hump" in the spectrum between 300 and 3,000 Hz is caused by the rail corrugation at the train speed driven. In case of wheel corrugations, however, this "hump" contributes little to the total noise emission, as already the short wavelength wheel corrugation produces a strong level increase between 800 and 4,000 Hz (Fig. 16.12). This increase is considerably less developed with smooth wheel running surfaces (Fig. 16.11). In both figures, the abscissa show in addition to the frequency the corrugation wavelength.

For the elimination of rail corrugations by a special rail grinding procedure, which is used within the framework of "acoustic track maintenance", see Section *Specially Monitored Track ("Besonders überwachtes Gleis – BüG")*.

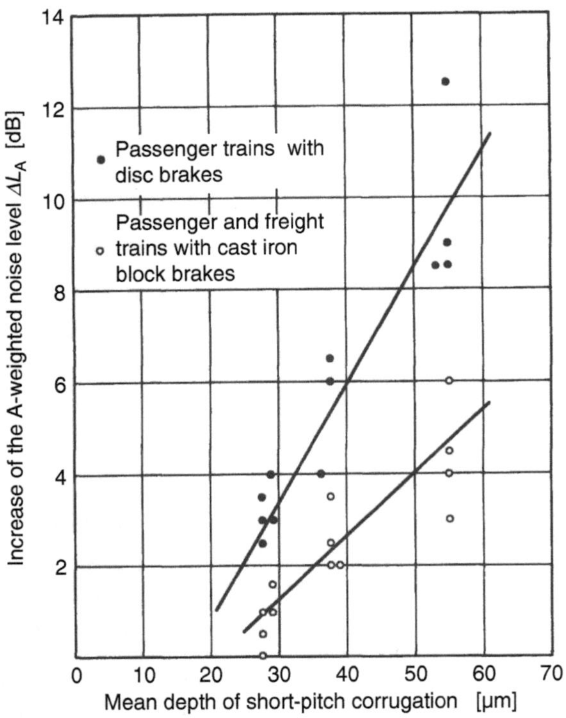

Fig. 16.10 Increase of the A-weighted noise level caused by trains passing on a track with short-pitch corrugation versus a track without corrugation as a function of the depth of corrugation (results from measurements in a distance of 25 m to the side of an open track during the passage of different types of trains running at different train speed)

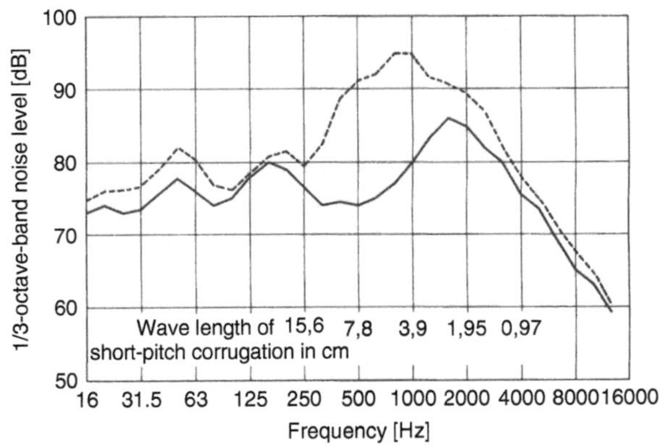

Fig. 16.11 Airborne noise at a distance of 7.5 m to the side of an open track (1.5 m above top of rail) without and with short-pitch corrugation (depth of short pitch corrugation ≤ 50 μm), during the passage of passenger trains with disc brakes, with relation between wavelength of short-pitch corrugation and frequency at a speed of 140 km/h. ——— without short-pitch corrugation: 93 dB(lin), 92.5 dB(A); ------- with short-pitch corrugation: 102 dB(lin), 101.0 dB(A)

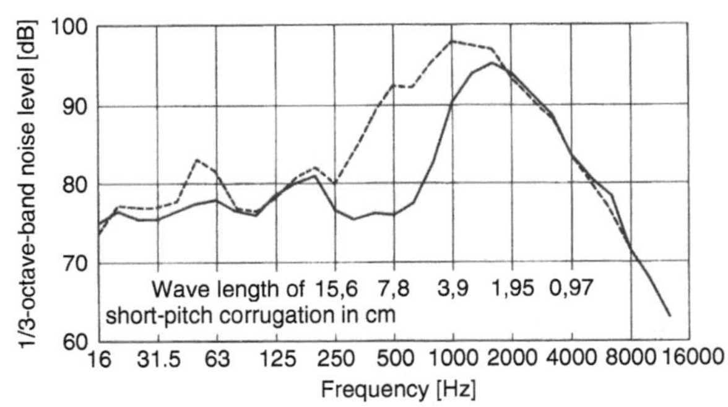

Fig. 16.12 Airborne noise at a distance of 7.5 m to the side of an open track (1.5 m above top of rail) without and with short-pitch corrugation (amplitude up to 50 μm), during the passage of passenger trains with cast iron block brakes, running at a speed of 140 km/h (also showing the relation between wavelength of short-pitch corrugation and frequency) ——— without corrugation: 101 dB(lin), 102 dB(A); ------- with corrugation: 105 dB(lin), 105 dB(A)

When driving around narrow curves, with radii of approximately 300 m and less under certain weather conditions, the so-called curve squealing occurs, which leads to a strong level increase in the range of high frequencies (Fig. 16.13). Countermeasures to this are reported in Sect. 16.2.2.3. For details about the occurrence of the curve squealing (catchword: "stick-slip effect"), see [29–31].

The already mentioned difference regarding sound emission on ballastless and on ballasted tracks can be taken out of Fig. 16.14 (results from [15]).

It can be seen that even with an absorbing surface of the ballastless track still a selective level increase in comparison with the ballast track is observed, which in the present case leads to an increase in the interior noise level within the passenger coach by 3 dB(A).

This selective level increase has been proven by means of measurements of the structure-borne sound on the rail during train runs [32]. The increase is closely correlated with the corresponding increase in the vibration levels of the rails of ballastless tracks compared with ballasted tracks (see Fig. 16.15). Of course, in the scope of the comparative measurements, the other conditions influencing the excitation of the wheel/track system (above all the roughness of the wheel and rail running surfaces) were hold unchanged within technical possibilities.

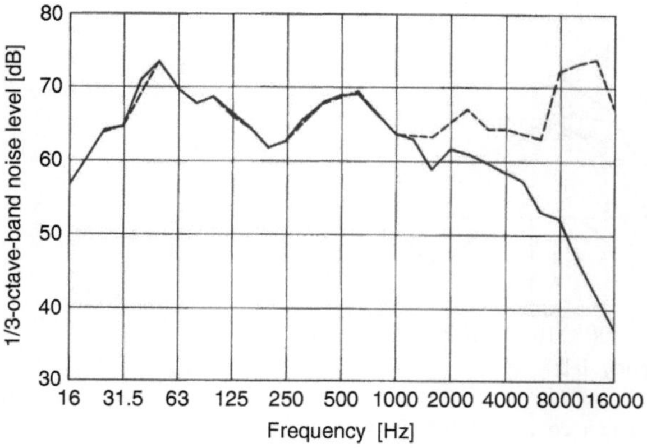

Fig. 16.13 Airborne noise at a distance of 8 m to the side (1.2 m above top of rail) of a curved open track (radius $r = 300$ m), during the passage of rapid transit trains, Class ET 420, at a speed of 60 km/h. ——— without curve squeal: 80.5 dB(lin), 74.5 dB(A); ------- with curve squeal: 83.0 dB (lin), 79.0 dB(A)

To understand this difference, the cross section of a typical rail fastening for the ballastless track is shown in Fig. 16.16.

This system differs in two important aspects from the corresponding system at the ballasted track:

(a) The damping of the elastic base plate pad of the ballastless track (3 in Fig. 16.16) is very low ($\eta \approx 0.1$–0.2) compared with the ballast with a loss factor going from $\eta \approx 1$–2.

(b) The mass of the ground plates (2) with 6–10 kg is very small compared with the sleepers with a mass of approximately 200–300 kg. Furthermore, in the range of the resonance frequencies, the sleeper damped by the ballast serves as a structure-borne sound absorber for the rail.

More precise tests of the ballastless track showed that in addition to the level increase compared with

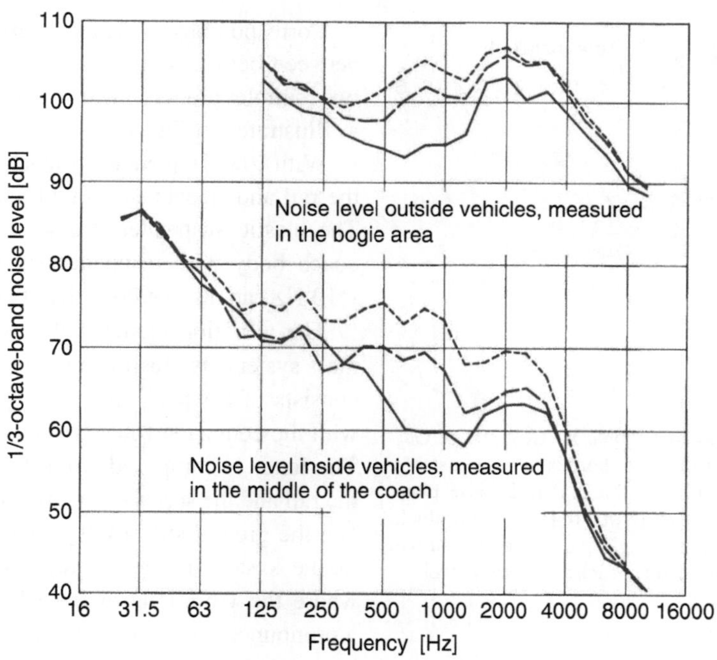

Symbol	Type of superstructure	Overall sound pressure level		
		Outside dB(A)	Inside dB(lin)	dB(A)
———	Ballasted track, type W60B70*)	111	91	73
---------	Slab track without sound absorbing layer	115	92	81
– – – –	Slab track with sound absorbing layer	114	91	76

*)Rail UIC60, w-shaped clip fastening, concrete sleeper, type B70

Fig. 16.14 Airborne noise below (bogie area) and inside an open saloon coach, Class Apmz 122 (centre of the coach), running at a speed of 200 km/h on different types of permanent way in a tunnel

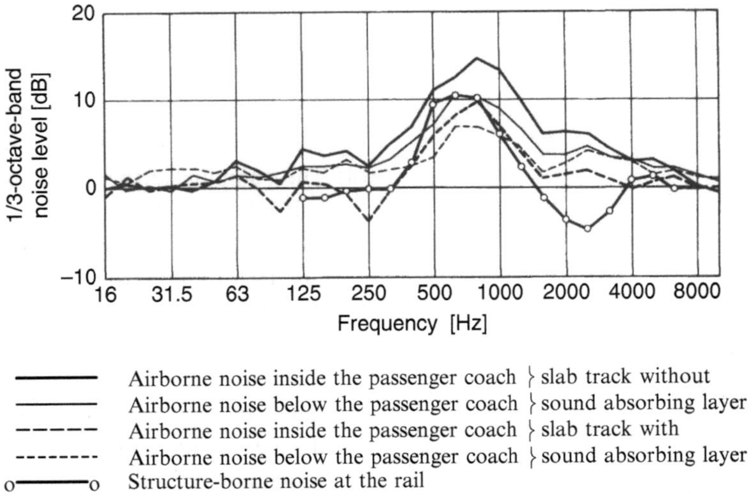

Fig. 16.15 Average 1/3-octave-band level difference between trains travelling on slab track and on ballasted track. Measurements in a tunnel at the same train speed, each in the area of 160–280 km/h

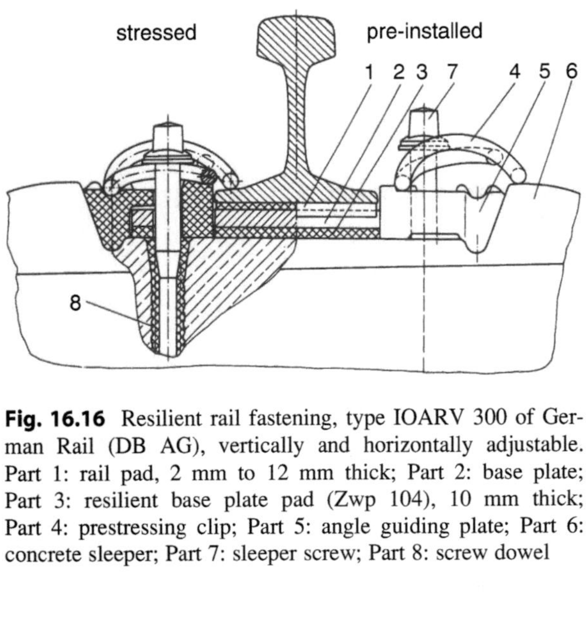

Fig. 16.16 Resilient rail fastening, type IOARV 300 of German Rail (DB AG), vertically and horizontally adjustable. Part 1: rail pad, 2 mm to 12 mm thick; Part 2: base plate; Part 3: resilient base plate pad (Zwp 104), 10 mm thick; Part 4: prestressing clip; Part 5: angle guiding plate; Part 6: concrete sleeper; Part 7: sleeper screw; Part 8: screw dowel

ballasted track already illustrated in Fig. 16.15 at the ballastless track an additional increase in the frequency range around approximately 40–100 Hz occurs, for which the varying stiffness of the track support, i.e., especially the elastic base plates in the rail fastening, have a considerable influence (see Fig. 16.17).

This can be mathematically proven with the help of the wheel/track impedance model (RIM) [33] (see Sect. 16.5, Fig. 16.87).

Corresponding to the relatively large frequency gap between both level increases according to Fig. 16.17, two simple models each for one frequency range, as illustrated in Fig. 16.18, can be established.

With low frequencies, the effect of the masses of the rail and sleeper or ground plate can be neglected. The elastic suspended masses of the bogie and the coach body with resonances in the frequency range <10 Hz can also be considered as uncoupled.

The vibration model is thus simplified to a single mass system, as illustrated in Fig. 16.18, left side. This consists of a wheel set mass m_R connected in series with the contact stiffness s_K and the rail stiffness s_{Sch}. The latter is composed from the bending stiffness of the rail and the series connection of the pad stiffness s_1 and the ground stiffness s_2. The resonance frequency of the system is the so-called wheel/rail resonance, where the vibrations of the wheel and the rail have a pronounced maximum, which is strongly influenced by the damping conditions.

Regarding level and frequency on ballastless track, the level increase in the range of the wheel/rail resonance is significantly influenced by the damping and the stiffness of the elastic base plate pads of the rail fastening.

The right hand part of Fig. 16.18 describes the system at higher frequencies. The masses of the rail and the sleepers or ground plates can no longer be ignored. The wheel impedance at this frequency is

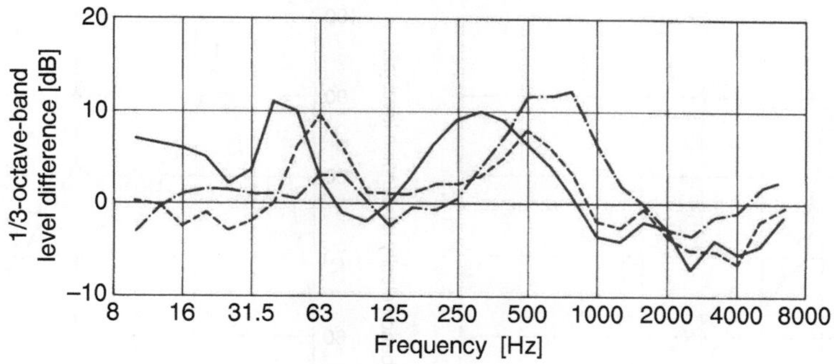

Fig. 16.17 Mean difference of 1/3-octave-band velocity levels at the rail $\Delta L = L_{v1} - L_{v2}$, determined from three constructions of slab track in a tunnel featuring different stiffnesses of the base plate pad (L_{v1}) and the adjacent ballasted track outside the tunnel (L_{v2}). Mean value from different types of train and train speeds in the range of 60 km/h to 280 km/h (single differences determined at the same train speed each time). Stiffness of the base plate pads: ——— approx. 20 kN/mm; - - - - - - - approx. 70 kN/mm; —·—·— approx. 140 kN/mm

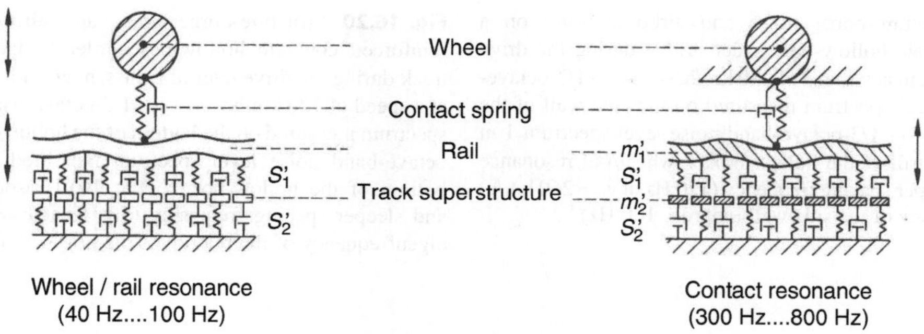

Fig. 16.18 Simplified models for interpretation of the frequency ranges of selective level increase of the rail vibration level of slab tracks, derived from the wheel/track impedance model "RIM" (see Fig. 16.87)

so large compared with the rail impedance, that the wheel appears to be a rigid termination for the contact spring.

The system corresponds now to a double mass vibrating system with a maximum vibration of the rail at two resonance frequencies.

These resonance frequencies are also called "contact resonances" [34, 35]. The relative position of both resonances of the ballastless track compared with the ballasted track is an essential reason for the resulting level increases.

At the contact resonance (between approximately 300 and 800 Hz), in addition to the difference in the bedding stiffness of ballasted track compared with ballastless track (stiffness of the base plate pads), the difference in the masses connected to the rail (see above) in connection with the stiffness of the rail pads plays an important role.

Based on the described knowledge, by means of laboratory tests as well as of a test construction with a scale 1:1 and of numerical simulations, a ballastless track was developed, which could avoid a higher noise emission compared with the ballast track [36]. The concept of this "acoustically innovative ballastless track (*AIFF*)" consists of a track-grid with damped sleepers on elastic sleeper pads on a concrete plate. The sleepers act as absorbers for structure-borne noise. By means of the results of the simulation calculations as well as lab tests and the tests on a prototype, the acoustic efficiency and the compliance to the operational test was proven [36]. The test in operating track has yet to be done.

Bridges

If a train is crossing a bridge, in addition to the sound emission of the train, the radiation due to the vibration of the bridge has to be taken into account. The excitation of the bridge depends strongly on its construction and thus can be influenced.

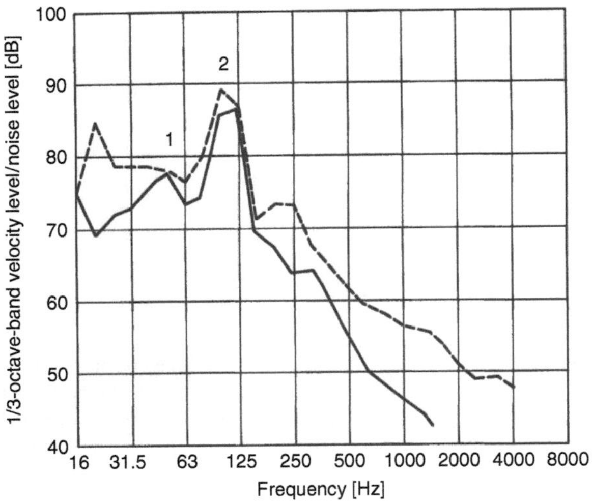

Fig. 16.19 Structure-borne noise and airborne noise on a reinforced concrete hollow-box girder bridge during the drive over of an ICE train at a speed of 280 km/h. ——— 1/3-octave-band velocity level spectrum measured on the side wall of the hollow-box; ------- 1/3-octave-band noise level spectrum 1 m beside the side wall of the hollow-box; *1* wheel/rail resonance frequency; *2* sleeper spacing frequency (120 Hz at $v = 260$ km/h) and eigenfrequency of the side wall (approx. 100 Hz)

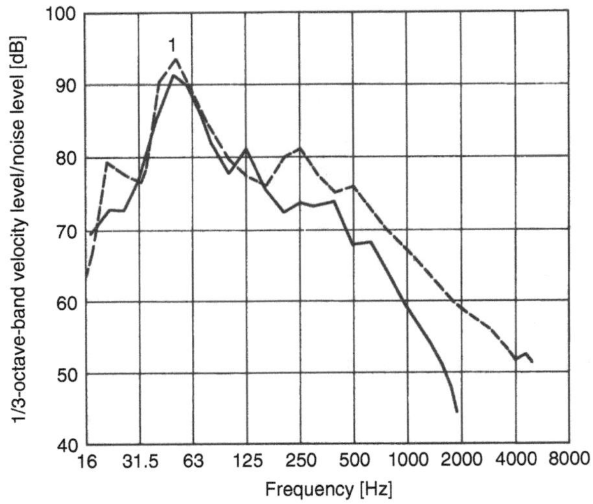

Fig. 16.20 Structure-borne noise and airborne noise at a reinforced concrete hollow-box girder bridge with ballasted track during the drive over of a passenger train with disc brakes at a speed of 120 km/h. ——— 1/3-octave-band velocity level spectrum measured on the bottom of the hollow-box; ------- 1/3-octave-band noise level spectrum measured 2 m below the bottom of the hollow-box; *1* wheel/rail resonance frequency and sleeper spacing frequency (56 Hz at $v = 120$ km/h) and eigenfrequency of the bottom of the hollow-box

The speed-independent excitation frequencies caused by *the train* result from the eigenvibrations of the two systems bogie/coach and bogie/superstructure. Particularly critical in this case is the frequency range between 40 and 100 Hz.

The excitation frequencies caused by *the track* are primarily determined by the speed-dependent *"sleeper-spacing frequency"* f_s.[5] Referring to the standard sleeper distance (approximately 60 cm), for the speed range of 50–300 km/h, the sleeper response frequency is between 23 and 140 Hz. Conformity with the above-mentioned speed-independent frequency components leads to increased vibration of bridges. Thus, the speeds leading to a *sleeper-spacing frequency* in the range between 40 and 100 Hz are particularly critical with regard to the vibration excitation of the bridge at low frequencies (Figs. 16.19 and 16.20). The excitation is particularly strong if the exciting frequencies correspond with the natural frequencies of the bridge components (e.g. track, side wall).

[5] $f_s[\text{Hz}] = \frac{\text{train speed (km/h)}}{3.6 \cdot \text{spacing between sleepers(m)}}$,

In the airborne sound spectrum of trains passing bridge constructions, the sound emission of bridge components in comparison with the open line leads to a shifting of the main sound energy towards lower frequencies (see Fig. 16.21 in comparison with Fig. 16.11, for example).

Figure 16.21 shows the spectra of noises that were measured at a distance of 25 m from three types of bridges during the passing of passenger trains with disk brakes at similar velocities. It is indicated that the steel–concrete block girder bridges are excited in the range around 60 Hz (*sleeper-spacing frequency* at the indicated speed) to stronger sound emission.

In order to decrease the excitation of vibration and with it the sound emission measures on the tracks (avoidance of unsteadiness of the superstructure at the transition open line – bridges; assurance of a faultless condition of the rail running surfaces within the bridge area) and measures at the supporting construction (increase in the masses, e.g. by means of the installation of ballast; frequency detuning by changing the mass or stiffness, increase in the stiffness of the

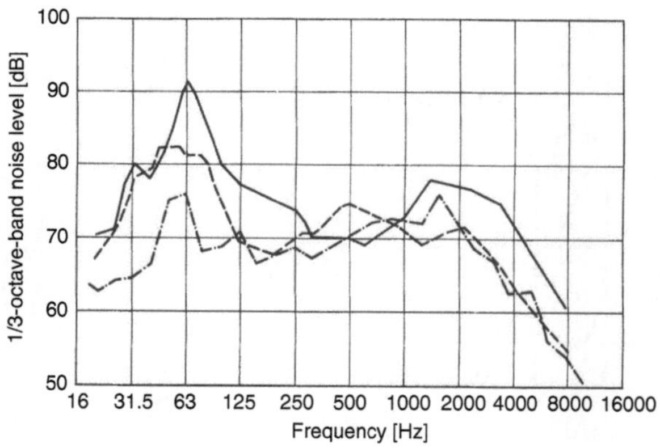

Fig. 16.21 Airborne noise at a distance of 25 m to the side of three bridges of different types of construction with ballasted track during the drive over of passenger trains with disc brakes at a speed of approx. 130 km/h: ——— steel hollow-box girder bridge, measurement height 1.5 m above top of rail: 97 dB(lin), 87 dB(A); - - - - - - - steel lattice-girder bridge, measurement height 3.5 m above top of rail: 89 dB(lin), 80 dB(A); —·—·— reinforced concrete hollow-box girder bridge, measurement height 3.5 m above top of rail: 85 dB (lin), 82 dB(A)

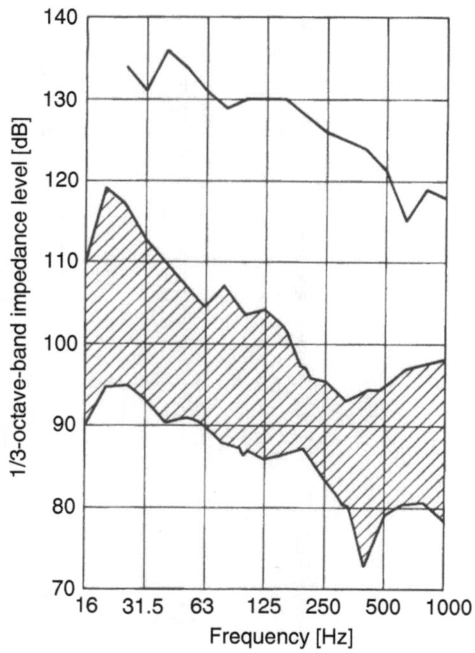

Fig. 16.22 Average amount of the mechanical input impedance of the running surface of various types of railway bridges, determined from single measurement values of local varying point impedances. ▨ Scatter from ten steel bridges of different types of construction; ——— reinforced concrete bridge in composite construction (double-T-steel framework with a 40 cm thick reinforced concrete deck slab)

area around the deck slab) have to be taken into account.

The results of the investigations on a slab track installed on a reinforced concrete hollow box girder bridge [28] give valuable indications for the reduction in the vibrations of bridge structures.

By observing some of the basic rules – already in the construction phase – it is possible to influence the sound emission from steel bridges [37]. Thus, e.g., it must absolutely be avoided that the first resonance frequencies of bridge components that contribute significantly to the total noise emission are located in the range of the excitation frequencies of the vehicle/superstructure between 40 and 100 Hz. If possible, they should be outside the frequency range between 20 and 140 Hz (*sleeper-spacing frequency*). In addition, by means of a favourable distribution of the stiffness, the input impedance of the concrete track slab has to be set as high as possible. The dynamic behaviour of the deck plates is critical for the propagation of the vibration energy to the other parts of the bridge construction. This means that, basically, high mass and damping as well as high rigidity or small deformation must be intended, too.

Figure 16.22 shows the average mechanical input impedance of the slab for various steel bridges. The top curve shows in comparison the input impedance of a 40-cm-thick concrete deck slab of a reinforced concrete bridge in composite construction [32].

Figure 16.22 shows the typical difference between bridges with steel or concrete deck slabs with respect to the vibration characteristics of the slab. Figure 16.22 also shows that the conditions for the efficiency of measures for decoupling the superstructure from the bridges, e.g. by the installation of ballast mats, usually are much more favourable for bridges with concrete track slabs. This is due to the high input impedance of their slab track compared with the steel bridges. Due to the lower sound emission of these bridges even without any additional measure, the noise reduction achieved with a ballast mat on the entire passing noise is nonetheless smaller than at steel bridges.

Today, there are no longer steel bridges without ballast beds built; the so-called directly driven (rails directly mounted on the superstructure) bridges are the loudest bridges. The sound level emitted by these is 15 dB(A) above that of the open line. A possible

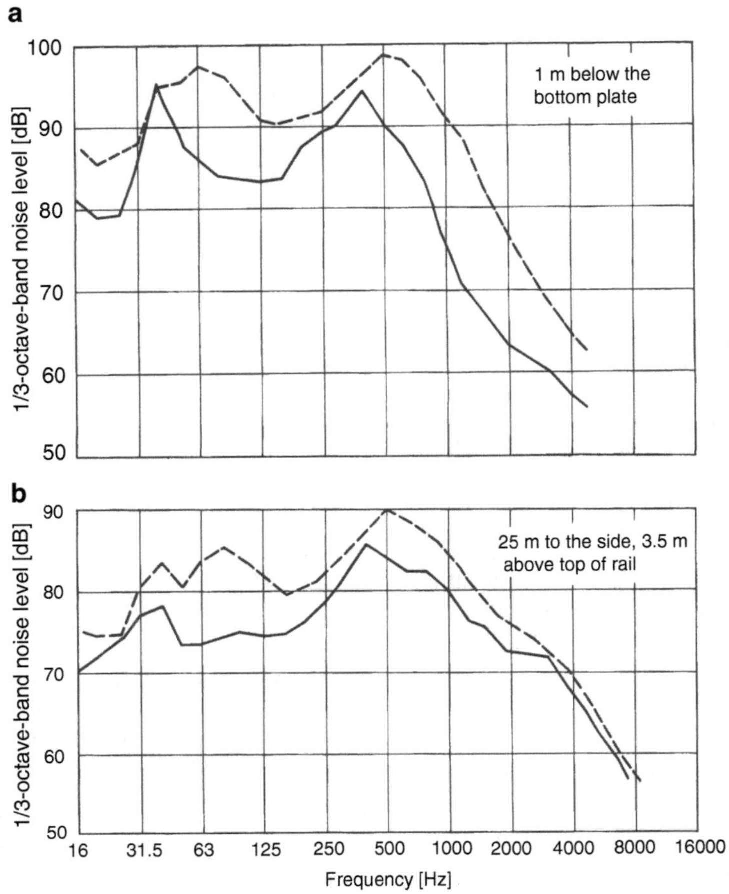

Overall noise level		1 m below the bottom plate		25 m to the side of the bridge	
		Before installation	After installation	Before installation	After installation
Unweighted	dB(lin)	108	101	88	84
A-weighted	dB(A)	102	94	84	79

Fig. 16.23 Airborne noise 1 m below the bottom plate (**a**) and 25 m to the side (3.5 m above top of rail) (**b**) of a directly driven steel hollow-box girder bridge during the drive over of a freight train at a speed of 85 km/h: ------- before installation of elastic rail fastenings; ——— after installation of elastic rail fastenings

measure to reduce the noise emission from a directly driven bridge after construction is the installation of an elastic rail fastening system (see Fig. 16.23), the installation of a sandwich coating or the installation of the track in a ballast bed (see Fig. 16.24).

A detailed illustration of the basically possible noise reduction measures to railway bridges, in particular to steel bridges with details about the measurement results concerning the investigated objects can be found in [38, 39].

New knowledge about the development and the use of elastic rail fastening systems as a means of reducing the noise emission from steel bridges without ballast are given in [40]. The successful use of elastic rail fastening systems that were installed on the reconstructed Berlin City Line for the reduction in the noise emission of an auxiliary steel railway bridge is described by [41, 42]. In these articles, the measurement results of the dynamic stiffness of the rail fastening system of the type Ioarg 314 (see, e.g., [40]) are presented. Those measurements had been carried out on a test rig according to the elaborated special acceptance test procedures for rail fastening systems [43] closely related with [44]. The dynamic

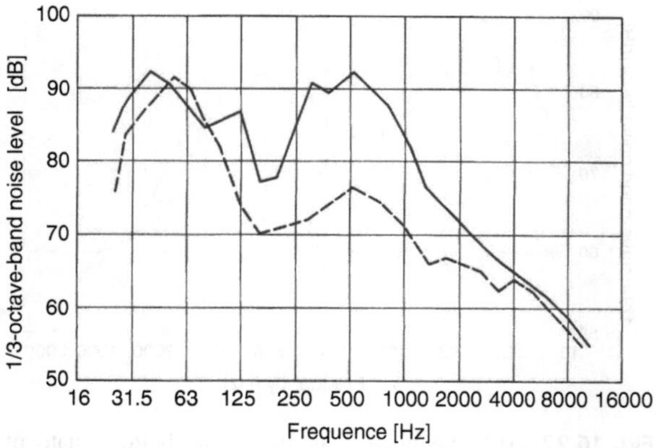

Fig. 16.24 Airborne noise at a distance of 25 m to the side (3.5 m above top of rail) of a steel plate girder bridge during the drive over of a locomotive Class 141 at a speed of 80 km/h, before and after installation of a ballast bed. ——— without ballast bed: 101 dB(lin), 94.5 dB(A); ------- with ballast bed: 96 dB(lin), 81.5 dB(A)

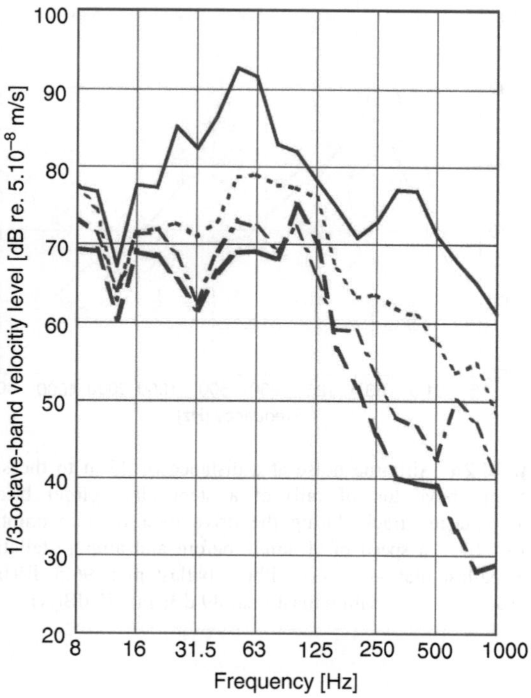

Fig. 16.25 1/3-octave-band velocity level spectra on the deck plate of three reinforced concrete bridges with different types of superstructure during the drive over of five ICE1 trains at a speed of 217 km/h $\leq v \leq$ 240 km/h. ——— bridge 1: ballasted track (BT) with rail pad Zw 900; ·········· bridge 2: BT + Zw 900 + sleeper pad "stiff" (70 MN/m); —·—·— bridge 2: BT + Zw 900 + sleeper pad "soft" (30 MN/m); ------- bridge 3: BT + Zw 900 + ballast mat (bedding modulus $c = 0.10$ N/mm^3 according to [DB-TL [45])

stiffness evaluated according to this process is to be considered as a physical parameter for the effectiveness of a rail fastening system with respect to the vibration and noise reduction. It can be used as the most important physical parameter in suitable simulation models for the characterization of the elastic properties of a rail fastening system (see Sect. 16.5).

In connection with a noise reduction programme with respect to existing steel railway bridges, the effect of elastic sleeper bearings (so-called soled sleepers with sleeper pads) was tested with respect to the reduction in the noise emission of bridges. Results of numerical simulation showed that soled sleepers with an acoustic optimised suspension layer have similar effects as ballast mats.

On soled sleepers, the elastic layer is attached tightly to the underside of the sleeper. Contrary to elastic rail fastening systems, the elastic material is thus located not under the rail but under the sleeper. Thus, less rail vibration and depending on the stiffness of the rail fastening system, only insignificant stronger vibrations of the sleepers can be expected. Therefore, no significant increase in the rolling noise has to be expected. The direct contact surface with the ballast is located at a load distribution layer. This layer provides mechanical protection for the elastic layer and distributes the forces occurring onto a larger surface of the elastic layer.

First general measurements on three bridges with very similarly constructed concrete deck slabs, but with different superstructures have confirmed the results of the model calculations. The results of these measurements are shown in Fig. 16.25.

Thereafter, one can conclude:

- Soled sleepers and ballast mats can be considered as equivalent decoupling measures[6]
- The acoustic effect of the soled sleeper clearly exceeds the one of elastic rail fastening systems.

[6] The prerequisite is especially that the requirements for the reduction of structure-borne space desired, are not so high and, as in the present case of the steel bridge to be redevelopment, the vehicle impedance is comparably low (see above Fig. 16.22), so that acoustically optimised ballast mats could not reach their full effect (see also below, Fig. 16.79: ballast mats on cement-bonded gravel sublayer of an open surface line).

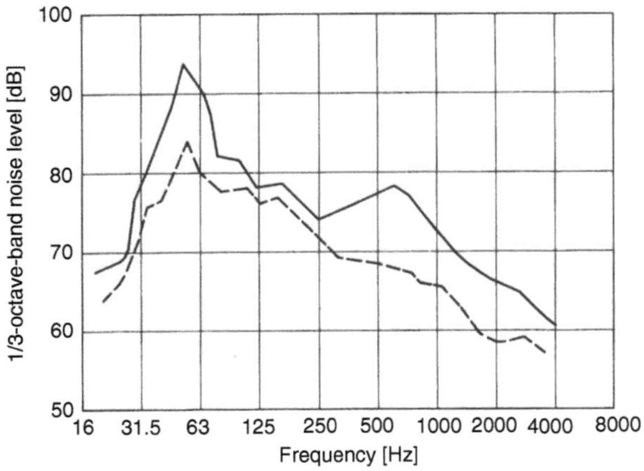

Fig. 16.26 Airborne noise at a distance of 25 m to the side (3.5 m above top of rail) of a steel plate girder bridge with ballasted track during the drive over of a locomotive Class 110 at a speed of 90 km/h, before and after installation of a ballast mat. ——— without ballast mat: 96.5 dB(lin), 83 dB(A); ------- with ballast mat: 89 dB(lin), 75 dB(A)

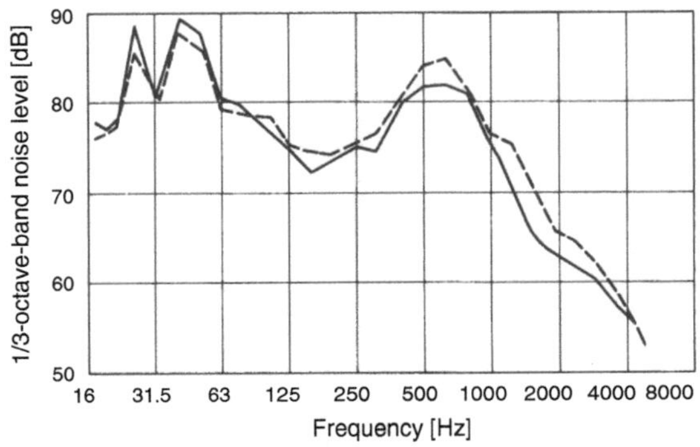

Fig. 16.27 Airborne noise 1 m below the bottom plate of a steel hollow-box girder bridge with ballasted track during the drive over of a passenger train with disc brakes and cast iron block brakes, respectively at a speed of 90 km/h. ——— passenger train with disc brakes: 95 dB(lin), 86 dB (A); ------- passenger train with cast iron block brakes: 95 dB (lin), 88 dB(A)

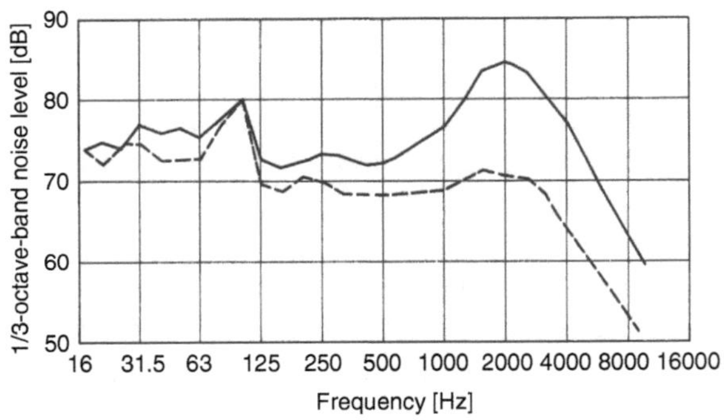

Fig. 16.28 Airborne noise at a distance of 25 m to the side (3.5 m above top of rail) of a reinforced hollow-box girder bridge with ballasted track during the drive over of a passenger train with disc brakes at a speed of 200 km/h, without and with noise control barrier (NCB), with a height of 2 m related to the top of rail. ——— bridge without NCB: 92.5 dB(lin), 90 dB (A); ------- bridge with NCB: 86 dB(lin), 81 dB(A)

The exchange of the stiff rail pads (e.g. standard pads Zw 668a) by soft pads (as in this case, the pads Zw 900) is not a suitable measure for an appreciable reduction in the noise emission of bridges.

Although on the newer bridges the tracks are positioned in ballast beds [level reduction >10 dB(A)], there are also steel bridges with ballast beds with a significant level increase in the low-frequency ranges (see Fig. 16.24).

By equipping the bridge with an acoustically appropriate ballast mat, the sound emission of such bridges can be considerably reduced (Fig. 16.26). As the vibration introduction through the ballast bed in the sidewalls is very small, the side mat is usually omitted [46].

The reduction in the structure-borne noise induced into the bridge construction with a ballast mat, i.e., the insertion loss of the ballast mat, can be calculated by means of a model of the wheel/rail/ballast bed/ballast mat/deck slab system [32]. By this way, the ballast mat can be optimised for each bridge construction by adapting its static and especially dynamic characteristics.

For bridges with ballasted track, the excitation and the noise emission of the bridge structure due to high excitation frequencies (corrugated wheels on trains with cast iron brakes) are only slightly affected. The difference, which can be observed directly under the bridge where the rolling noise is strongly shielded, is only about 2 dB(A) (Fig. 16.27).

By means of noise control barriers on bridges (especially on steel bridges), the entire noise emitted is less reduced than in open line, because the barriers reduce the rolling noise of the train itself as usual, but do not shield the noise of the bridge components. The low frequencies become even more apparent (Fig. 16.28).

Due to their higher input impedance of the deck slab compared with steel bridges, reinforced concrete bridges in composite and/or massive construction (always with ballast bed) are in most cases acoustically equivalent with steel bridges, which in addition to the ballast bed are equipped with an optimised ballast mat. This holds particularly concerning the A-weighted noise emission. Although the installation of a ballast mat can also be necessary with these types of bridges in particular cases, e.g., if residential areas are located in the vicinity of very high bridges so that the direct rolling noise of the trains is strongly shielded, and thus the low-frequency bridge noise dominates. This effect (however, in extreme measuring position) is shown in Fig. 16.29.

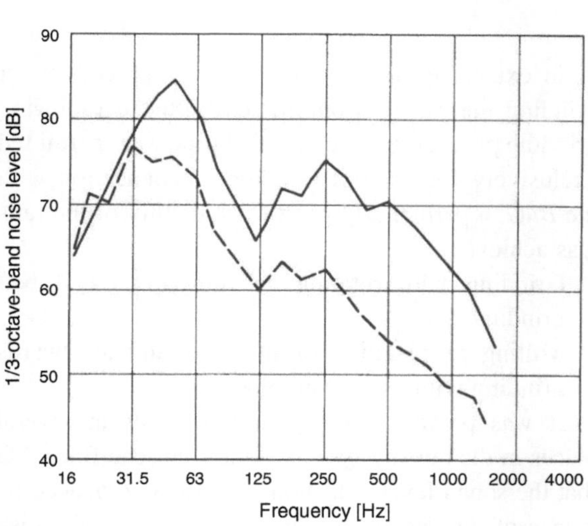

Fig. 16.29 Airborne noise 1 m below the deck plate of a reinforced concrete bridge in composite construction with continuous ballasted track, without and with ballast mat, during the drive over of a Class ET 420 multiple unit train at a speed of 100 km/h. ——— without ballast mat: 90 dB(lin), 75 dB(A); ------- with ballast mat: 82 dB(lin), 62 dB(A)

In a lateral measuring position above rail level, the effect below 80 Hz still exists; for higher frequencies however, it is fully covered up by the rolling noise directly emitted by the train (Fig. 16.30). The effect would be partially revealed, if a noise control barrier on the bridge would reduce the direct noise emission (rolling noise) of the train.

In this case, the acoustic effect of the ballast mat (its insertion loss) could be calculated numerically in good conformity with the measurement results [47].

Railway Crossings

The poor condition of the rail running surface, caused by soiling (e.g. grit), which road vehicles carry onto the tracks, as well as reflections of the sound waves on the road surface cause an increase in the rolling noise from 6 to 11 dB(A) close to railway crossings. At lateral distances >100 m, they no longer affect the immission level. The difference in the spectrum in comparison with the open line is shown in Fig. 16.31 (results from [15]).

Tunnels

On high-speed lines, the so-called tunnel sonic boom can occur. It is caused by the shock wave that runs with the speed of sound in front of the train through the tunnel to the other end. One part of this shock wave is emitted as a thumping bang from the mouth of the tunnel, and the other reflected into the tunnel again.

The shock wave is generated by the entry of the train into the tunnel; it depends on the velocity and the head form of the train, the configuration of the tunnel mouth and the relation of tunnel cross section to train cross section, the sound absorbing qualities of the

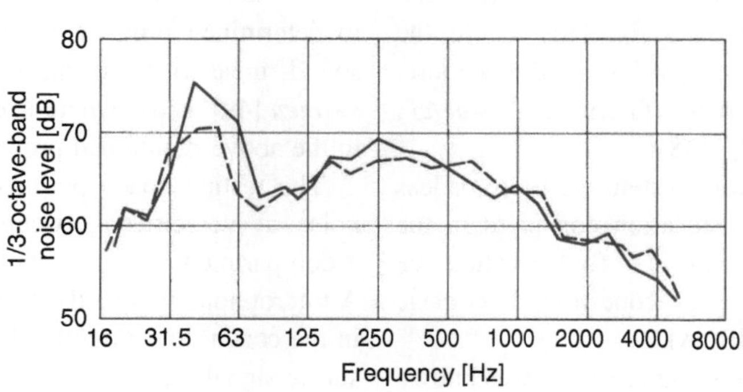

Fig. 16.30 Airborne noise at a distance of 25 m to the side (3.5 m above top of rail) of a reinforced concrete bridge in composite construction with continuous ballasted track, without and with ballast mat, during the drive over of a Class ET 420 multiple unit train at a speed of 100 km/h. ——— without ballast mat: 81 dB(lin), 73.5 dB(A); ------- with ballast mat: 79 dB(lin), 73 dB(A)

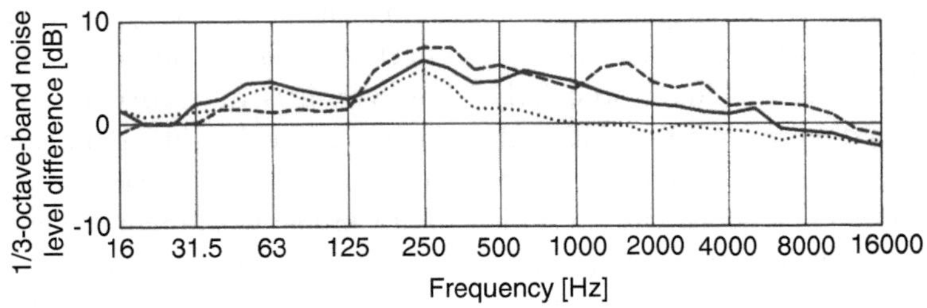

Fig. 16.31 Noise level differences between level crossings and open track, measured at a distance of 25 m from the centre of the track, 3.5 m above top of rail (average value of approx. 10 pass-by's of each train-class, respectively and then mean of up to seven level crossings): ——— passenger trains with disc brakes (v = 100–160 kmh), ·········· passenger trains with cast iron block brakes (v = 100–160 km/h); ------- rapid transit trains, Class ET 420, with wheel–disc brakes (v = 60–120 km/)

tunnel and its length. The shock wave is influenced by the time gap between the first plunging of the train nose into the tunnel until the full train cross section is positioned in the constant cross section of the tunnel (head form of the train and shape of the tunnel mouth).

The sonic boom just occurs because the slope of the shock wave is steepened up during running through the tunnel. This steepening-up occurs due to the pressure-dependent sound velocity, which ensures that the parts of the shock wave in the high-pressure range propagate more quickly than the parts in the lower range of pressure. Hence, the sound energy of the steepening-up shock wave is passing into higher frequencies, i.e. the audible range. A certain minimum length of the tunnel as well as low absorption in the tunnel are prerequisites for the development of a critical steepening-up.

Specially Monitored Track

The level of excitation documented above by Figs. 16.10–16.12 caused by wheel corrugation can be removed by grinding the rails. This led to the development of the subsequently described procedure, "Specially Monitored Track (German: *Besonders überwachtes Gleis – BüG*)" [48].

Since parts of the railway system are more or less corrugated, the German legislation considers in the determination of the "*Grundwert*" (basic value, see Sect. 16.2.1), an average track condition. This basic value was fixed with 51 dB (A).

By means of many measuring results, it was known that with ground, faultless rail surfaces, noise emission levels are emitted which are by 3 dB(A), by some types of trains up to 7 dB(A) below the values emitted with an average track condition.

In extensive and lengthy tests, the procedure for grinding the rail was optimised. With the following grinding procedures, which at that time were permitted exclusively for the rail grinding in connection with the *BüG*, a particularly good track quality of the rails was achieved:

1. Grinding with rotating grinding discs and belt grinding
2. Milling or planning of the rails and afterwards grinding with oscillating blocks

It was proven in comprehensive tests at several places in the railway system of the Deutsche Bahn AG that the sound level reduction of 6 dB(A) produced by means of the "acoustic grinding" for trains with disk brakes and 3 dB(A) for trains with cast iron block brakes lasted at least 2 years and frequently even longer [15].

In sound calculations related with new constructions, upgrading and reconstructions of tracks the application of the *BüG* permits a reduction of 3 dB. However, this requires that the acoustic characteristics of the track are monitored regularly according to determined provisions with a specially equipped sound measuring coach, the so-called *Schallmesswagen* [49], and – where needed – ground according to the above-mentioned procedures.

The testing coach permits to measure the rolling noise at velocities between 80 and 200 km/h in a compartment equipped with an absorbing surface. A microphone is directly positioned over the opening in the coach floor directly above the bogie. The measuring signals are led via filters and electronic devices (amongst others, for assurance of the constant compensation of the velocity influence on the measuring results) to a processor and there prepared for further application or illustration.

Table 16.2 Emissions level $L_{eq,25,1}$ of shunting yard noises in dB(A), as stipulated in [2]

No.	Source of noise	Level	Special characteristics
1	Shunting run	54	$v = 65$ km/h, $l = 100$ m
2	Acceleration/deceleration	56	–
3	Push-up	45	$l = 170$ m
4	Buffing impacts	45	$v = 1$ m/s
5	Buffing impacts	58	$v = 4$ m/s
6	Drag shoe impacts	54	$v = 2$ m/s
7	Drag shoe impacts	61	$v = 4$ m/s
8	Curve squeal[a]	60	Radius $R \leq 300$ m
9	Retarder brake	56	Delay section[b]
10	Retarder brake	46	Continuous braking section[b]
11	Jaw-type rail brake, dual	51	With segmentation noise-optimised[b]
12	Jaw-type rail brake, dual	54	With segmentation[b]
13	Jaw-type rail brake, dual	71	Without segmentation[b]
14	Jaw-type rail brake, single	58	With segmentation[a, b]
15	Jaw-type rail brake, single	51	With segmentation noise-optimised[b]
16	Rubber fulling brake, single	49	–

[a]Anti-squeal development not yet completed
[b]Single passage

In extensive comparison tests, the direct connection between the acoustic quantity obtained in the measuring coach and the emission level aside the track line was proven.

16.2.2.3 Large Area Railway Facilities

Shunting Yards

Shunting yards (*Rbf = Rangierbahnhof)* are train formation facilities covering large areas with noise sources different from those of regular rail traffic.

They basically consist of a reception area to receive the incoming trains, an area to sort and collect wagons (direction group) and a departure area to collect and finish the new trains.

Especially in the reception area and in the direction group, sound-emitting procedures take place, such as buffering wagons, passing through track brakes, curve squealing or shunting with drag shoes between wheel and rail [50].

Below, the emission levels of the most important *Rbf* noise sources are listed. The A-weighted emission levels $L_{eq,25,1}$, i.e. the equivalent continuous sound levels for 1 h, measured at a distance of 25 m from the noise source, with one occurrence per hour are given.

Each result represented here is based on a wagon group consisting of two wagons according to the average situation in such facilities.

Noise sources according to no. 5 and no. 7 in Table 16.2 [2] today do not occur in new, automated installations. The speed at which the wagons bump each other, which definitively determines the emission level of these events, is 1 m/s for new installations and 4 m/s for old installations. The dependency of the noise level of "bumping" on the wagon velocity is shown in Fig. 16.32.

Rail brakes are an important component of an automated *Rbf* equipment. They show – dependent on the type of construction – an intensive noise emission in the frequency range above 3 kHz ("brake squealing") with a sound level up to 120 dB(A) at a distance of 7.5 m. This holds particularly for rail brakes without noise-optimised segmented wear and tear jaws. Today, however, in new installations, the rail brakes shown in Table 16.2 under nos. 11, 12, 14 and 15 are predominantly installed [257]. These brakes, which are equipped with the above-mentioned wear and tear jaws (Fig. 16.33), just seldom lead to sound emission in the higher frequency range. Since the development of this optimised sound brake is not yet finished, still further level reductions may be possible.

Today, one of the relatively annoying sound sources of an *Rbf* is curve squealing, which occurs, when the wagons run on curved tracks (radius ≤300 m) under some not yet clarified conditions. Various measures to reduce or avoid these noise

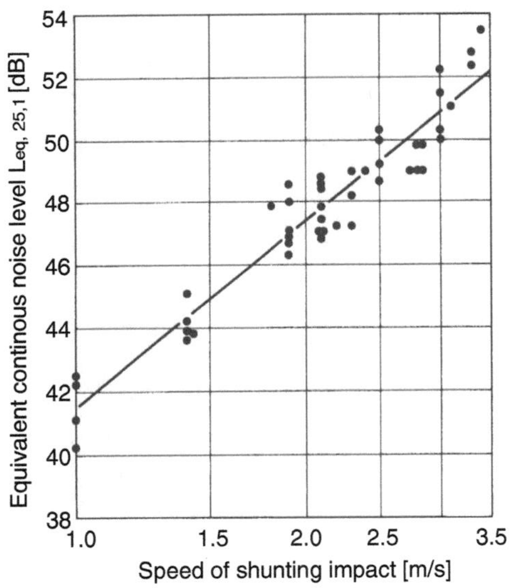

Fig. 16.32 Connection between the A-weighted energy-equivalent noise level of shunting impacts measured in a distance of 25 m referred to one shunting-impact-event per hour $L_{eq,\ 25,1}$, and the speed of the shunting impact

emissions have been tested, with the result that with all the countermeasures available a reduction in the frequency of occurrence of the squealing as well as in the sound level during squealing could be obtained. A complete suppression of this high-frequency noise emission (compare Fig. 16.13), however, could not be realised.

A measure against curve squealing is the lead alloy rail with a Pb content of approximately 0.05%. Because of the low durability, this variant is not suitable for the *Rbf*. Tests with other alloys, which do not show these limitations, could be of interest.

Another possibility is the use of head-hardened rails respectively a lubrication of the running surface with a special lubrication procedure. The latest tests took place in the year 2000 in the *Rbf* of Munich and led to a positive effect.

The sound level of the curve squealing could be reduced by approximately 4 dB(A) and, in addition, the frequency of occurrence of curve squealing was reduced.

Intermodal Stations

Intermodal stations *(Ubf = Umschlagbahnhof)* refer to facilities with horizontal and vertical loading of freight, without changing the containment of transportation (large container, semi-trailers, trucks and articulated vehicles).

The significant sound sources of an *Ubf* include, in addition to shunting movement with the already discussed noise emissions occurring in shunting yards (see resp. Section on *Shunting Yards*), the container crane, the mobile handling equipment and the devices for horizontal loading of long trucks in connection with the *"Rollende Landstraße"* (loaded trucks transported on low-floor freight trains).

Table 16.3 contains the emission levels of the most important *Ubf* noise sources (A-weighted emission level $L_{eq,25,1}$).

In addition to the rolling noise of the crane, the major sound source when unloading large containers with container cranes is the noise of the rolling crane trolley (Fig. 16.34). The noise events of the crane are related with lifting, turning, driving and setting down. Depending on the type of crane, the noise source "crane trolley" is mounted up to 17 m over the surface level.

Curve squealing is an annoying noise source also in the *Ubf* facilities. It occurs when rolling on curved tracks with a radius $R \leq 300$ m (see the previous section).

Further Railway Facilities (Passenger Stations, etc.)

In passenger stations, the following noise sources occur additional to rolling noises:

(a) Braking noises of trains with cast iron block brakes
(b) Starting noises and idling noises (fan, diesel motors idling) of traction units
(c) Rail joint (among others insulated rail joints, crossing gaps in points)
(d) Movement of baggage carts
(e) Door slamming
(f) Loudspeaker announcements

These sources influence the total emission of the facility. The stopping trains have low speed and as a result low rolling noise.

In [51], it was proven that calculations assuming that all trains pass the train station area with unreduced speed, neglecting other sound sources, result in correct equivalent continuous sound levels with a tendency to too large values. This simplifies the calculation of the level in the train station area quite significantly.

Fig. 16.33 Shunting yard track brake (segmented wear-down beams at only one rail) for destination tracks. (**a**) Overall view; (**b**) detailed view showing a squeezed-in freight-car wheel

Other train station facilities, such as freight train stations, and depots, are often not critical regarding noise sources in the environment, since large shielding halls frequently surround them or because loud work is only done inside the halls.

16.2.3 Noise Immissions

16.2.3.1 Noise Immissions Emitted by Large Area Railway Facilities

German legislation demands noise immission calculations in the scope of the planning of new or modified train facilities covering large areas. The applicable calculation procedure is described in Part 2 of [5]. The guideline included therein [2] contains all the details required for the calculation of the noise immission as well as an introduction to the processing of the individual numerical calculation steps (see also [52]).

Noise with high frequencies, such as it occurs in shunting yards and intermodal stations (e.g. curve squealing, drag block squealing, brake block squealing), shows a higher sound level decrease with increasing distance from the source than noise with low frequencies.

The shielding calculation for higher frequency noise sources shows a very good conformity with the reality, if in the equation mentioned in the following Sect. 16.2.3.2 for the calculation of the shielding effect the calculation is done with 120 z instead of 60 z.

Special cases of multiple diffractions as well as the influence of reflections are taken into account according to [2]. With respect to the calculation of

Table 16.3 Emissions level $L_{eq,25,1}$ from an intermodal station in dB(A)

No.	Source of noise	Level	Special characteristics
	Movement of container crane	52	For one load cycle
	Container crane, trolley moving	47	For one load cycle Height of source = 15 m
	Mobile loading equipment		
	Side loader, Type 1 Side loader, Type 2	60 55	400 s load carry duration 120 s load carry duration
	Horizontal loading – Rolling road		
	Head ramp for loading Stowing/removing the truck	50	Two times per train
	Driving trucks onto low-level wagon	43	Once per truck
	Fastening the truck	53	Once per truck
	Driving semi-trailer onto low-level wagon	57	Once per semi-trailer
	Starting the truck	57	Once per train as line noise source
	Driving trucks off the cars	52	Once per truck

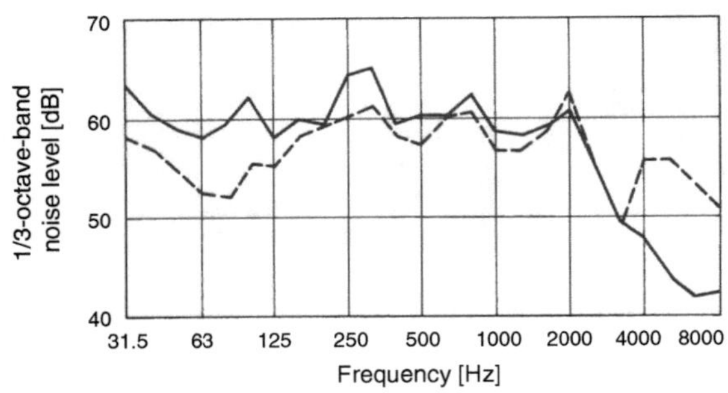

Fig. 16.34 Noise level in a distance of 25 m from a container crane at different working states: ------- whole crane moving along; ——— noise of trolley while lifting/lowering

the level reduction within built-on areas, only the shielding of the buildings lying closest to the train facility is taken into account. Other possibly existing attenuations in built-on areas are ignored.

Regarding the impact of noise sources containing tonal components and/or pulses at the place of immission, corresponding level additions can already be taken into account during the data acquisition depending on the loudness and the frequency of the sound. The procedure offers the possibility to add up to a maximum of 8 dB(A) to the immission level (see Sect. 16.2.4.2).

Since the main noise sources occurring have another time structure than the rail traffic of the open line, the "rail bonus" (see Sect. 16.2.4.1) is not applied in the planning of shunting and freight train stations. However, trains that pass by shunting yards or intermodal stations or that run through such facilities without stops should be provided with a "rail bonus" instead.

16.2.3.2 Noise Immissions Emitted by Railway Lines

In Appendix 2 of the 16th BImSchV [5], for noise immission calculations near tracks in Germany, a calculation procedure [3] is stipulated. The *"Schall 03"* contains all data required for this purpose, so that there is no need to go here into details.

Free Field Propagation

The level reduction on the propagation path can strongly fluctuate. The reasons for this are, e.g., varying types of ground vegetation and meteorological conditions in the propagation path [53].

The noise emission from trains is preferably directed laterally [53, 249].

The directivity index can be described approximately by the following Eq. (16.2):

$$D_I = 10 \cdot \lg\left(0.22 + 1.27 \cdot \sin^2\delta\right), \qquad (16.2)$$

δ is the angle between the connection line between emission and immission point and the track line.

Figure 16.35 shows the typical passing level as a function of time, measured at various distances according to [17]. The increasing steepness of the leading edges (with decreasing lateral distances) shows the influence of the directivity index, and,

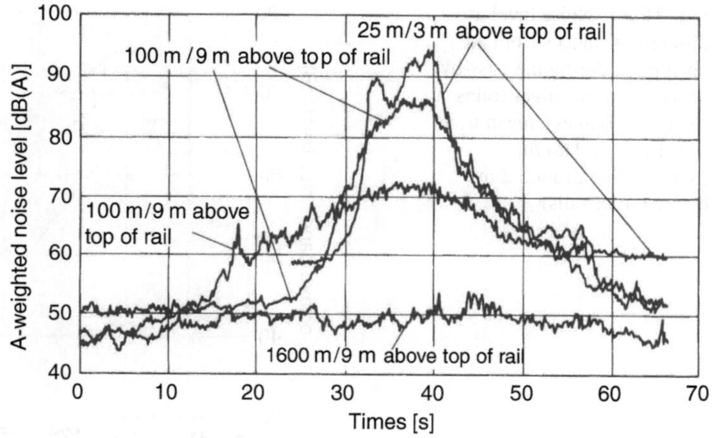

Fig. 16.35 Time histories of A-weighted noise levels at different distances from the track during the pass-by of a passenger train consisting of an electric loco Class 103 and 12 coaches (mix of coaches with disc brakes or cast iron block brakes), at a speed of 140 km/h

furthermore, that with increasing distance longer track sections influence the immission.

Figure 16.36 shows the spectrum of the pass by noise of InterCity trains at various distances from the track according to [20].

Active Noise Control Measures

The most important elements of active[7] noise control on tracks or generally in train facilities are noise control barriers and noise control embankments.

In [55] "Noise control by shielding in the free field" is treated in detail.

Valuable indications to general principles can also be found in Maekawa [56], and to shielding along train facilities in particular in Kurze [57]. When determining the shielding effect of noise control barriers and embankments, it must be taken into account that the barrier attenuation depends on the frequency. This is particularly important where higher frequency sound sources are to be expected.

Noise Control Barriers

Noise control barriers are manufactured from a variety of materials. The Deutsche Bahn AG has barriers made out of concrete, synthetic materials, glass, aluminium, bricks, wood and a mixture of these materials. They must comply with the requirements according to [58]. Here are the most important points:

(a) The noise control barriers must be able to withstand certain wind load levels
(b) The noise control barriers made of metal must be grounded electrically for safety reasons
(c) For safety reasons (for rail workers, because of aerodynamic pressure) and track construction reasons, according to [58], the following minimum clearances between the noise control barriers and the adjoining track (centreline of track) must be observed

$160 < v < 300$ km/h	3.80 m
$v \leq 160$ km/h	3.30 m
Pure S-Bahn tracks	3.20 m

(a) Regarding noise control barriers with a noise level reduction of up to 15 dB(A), the following minimum values of the sound insulation of the barriers must be observed. Considering all noise control barriers, the following minimum values of the sound absorption coefficient – along the side of the barrier which faces the sound source have to be provided:

Frequency (Hz)	125	250	500	1,000	2,000	4,000
Sound reduction index R (dB)	12	18	24	30	35	35
Sound absorption coefficient α_s	0.3	0.5	0.8	0.9	0.9	0.8

In case a larger noise reduction is required, correspondingly higher insulation values must be guaranteed. In the case of concrete barriers with concrete supporting

[7] The term "active" was introduced for railway traffic for nearby sound source secondary measures.

Fig. 16.36 Noise level at different distances from an open track during the pass-by of Intercity passenger trains (with disc brakes), mean train speed $v = 160$ km/h. (Surrounding ground 2 m below top of rails)

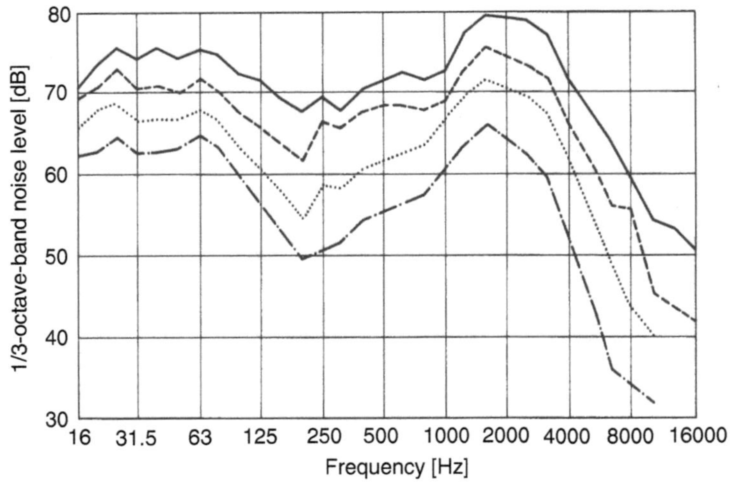

Symbol	Measurement distance from the centre of track	Height of microphone above top of rail	Overall noise level	
			dB(lin)	dB(A)
———	25 m	0m	88	87
---------	50 m	1.6 m	84	82
....	100 m	1.6 m	80	78
–·–·–	200 m	1.6 m	75	72

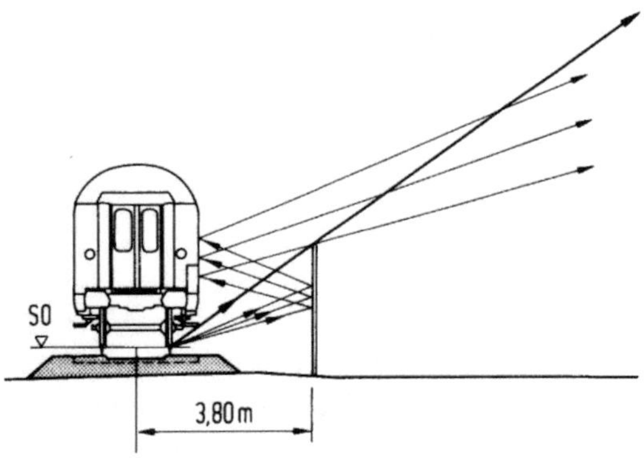

Fig. 16.37 Illustration of the influence of multiple reflections on the screening effect of a non-absorbing noise control barrier

layers of at least 8 cm thickness, the sound insulation in every case is sufficient.

The procedures for testing the above-mentioned requirements are specified in [58].

As can be seen from Fig. 16.37, reflecting barriers along the tracks (due to the small distance between a coach side and the noise control barrier) are unfavourable, since by the occurrence of multiple reflections between the noise control barrier and the train, the effect of the noise control barrier is significantly reduced. A reduction in the effect of the barrier at around 3 dB(A) in normal situations (but even up to 7 dB with a very large shield value z, see Fig. 16.40) was proven.

It was proven by measurements that sound level increases on the opposite side of the noise control barrier, caused by reflections on the noise control barrier, are not to be worried about, since these reflections are, to a large extent, shielded by the passing train itself. Passenger trains completely shield these reflections, freight trains do it partly.

Figure 16.38 shows the average spectrum of the rolling noise from freight trains with and without noise control barrier [15].

Figure 16.39 shows the level recording of the passing of a freight train for an immission site at a distance of 25 m, with and without a noise control barrier with a height of 2 m related to the surface of the rail located at a distance of 4.5 m between the barrier and the centreline of the track.

The level reduction ΔL by means of a noise control barrier depends on the shielding value z and the overhanging length. The determination of the z value is illustrated in Fig. 16.40.

The following Eq. (16.3) gives the shielding effect in a simplified way:

$$\Delta L = 10 \cdot \lg(3 + 60 \cdot z). \tag{16.3}$$

Due to the directivity of rail traffic noises, railroads require inferior additional lengths of noise control barriers at a given situation than roads [57].

For the exact calculation procedure for the shielding effect of noise control barriers, see, e.g. Schall 03 [3]. Also, for small negative z values (the barrier is not high enough to reach the connecting line between emission and immission site), there is still a measurable barrier effect (e.g. approximately 3 dB where $z = 0$).

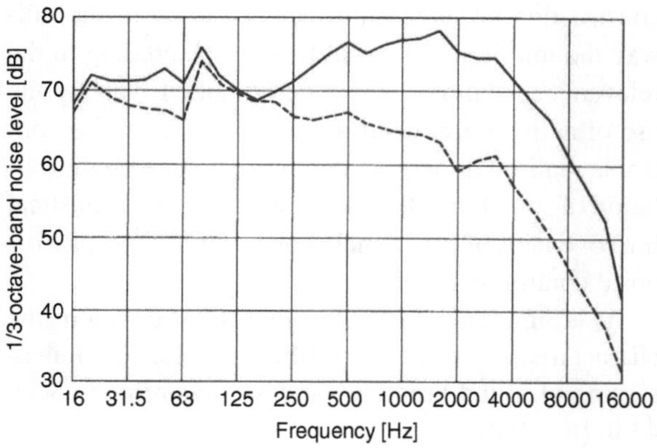

Fig. 16.38 Airborne noise at a distance of 25 m to the side of an open track (3.5 m above top of rail) with and without a noise control barrier (NCB), during pass-by of freight trains with speeds between 85 km/h and 100 km/h. ——— without NCB: 88 dB(lin), 86 dB(A); - - - - - - - with NCB (upper edge 2.0 m above top of rails): 81 dB(lin), 74 dB(A)

Using a noise control barrier with a profile showing an inward bend in its upper part, the diffraction edge is brought closer to the track even though the barrier is built at the same distance, and thus the level reduction is increased. Figure 16.41 shows the minimum values to be observed.

Tests with "interference barriers" or with interference absorbers, which are placed on top of normal noise control barriers, showed no significant additional effect until now.

"*Low noise control barriers*" have a height of only 0.75 m above the top of rail. As a result, they can be set up substantially closer to the track without infringing on the side clearance, if possible.

The only possibility to improve the effect of noise control barriers, whose side facing the sound source is coated with absorbing material, lies – presuming unchanged height of the noise control barriers – in the acoustic optimisation of the diffraction edge.

Therefore, some manufacturers of noise control barriers have developed various shapes of diffraction edges, which were also tested by the Deutsche Bahn AG, although never with any significant success.

In recent years, the influences on the diffraction field have been thoroughly tested through impedance loading of the diffraction edge [59, 60]. Based on this research, the *Deutsche Bahn AG, DB Systemtechnik München* has developed a practice-oriented top shield optimised to the typical spectra of passing trains. An effect of up to 3 dB(A) is to be expected.

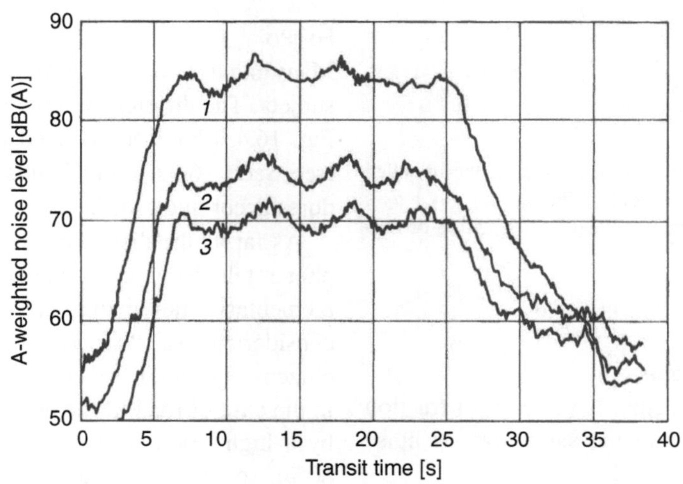

Fig. 16.39 Time history of A-weighted noise level at a distance of 25 m to the side of an open track with and without a noise control barrier (NCB) during the pass-by of a freight train at a speed of 85 km/h (height of NCB: upper edge 2 m above top of rails (TOR)). *1* without NCB, height of microphone 3.5 m above TOR; *2* with NCB, height of microphone 3.5 m above TOR; *3* with NCB, height of microphone 0,9 m below TOR

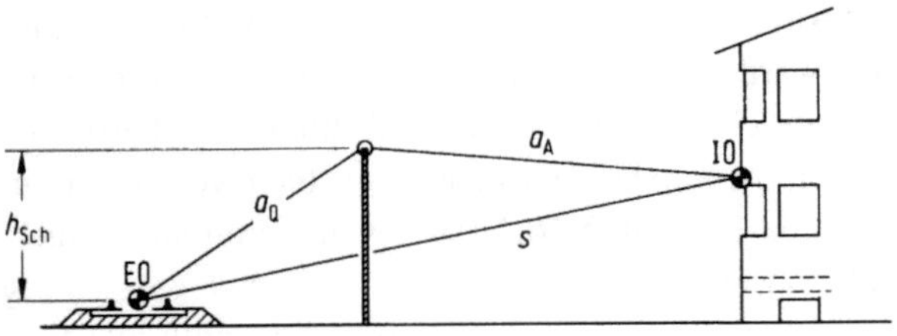

Fig. 16.40 Geometrical distances for calculating the so-called screening value z of a noise protecting barrier. $z = a_Q + a_A - s[m]$; *EO* emission point (centre of track at top-of-rail-level); *IO* immission point

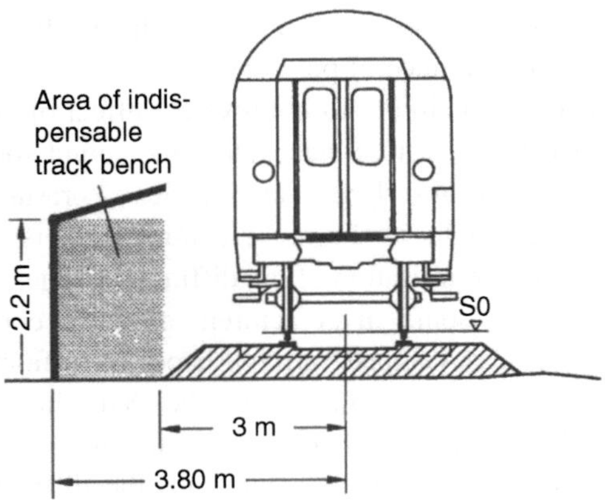

Fig. 16.41 Sketch to illustrate the allowed geometry of a track-side-angled noise protecting barrier for high-speed-lines ($v \geq 160$ km/h) of German Rail (DB AG)

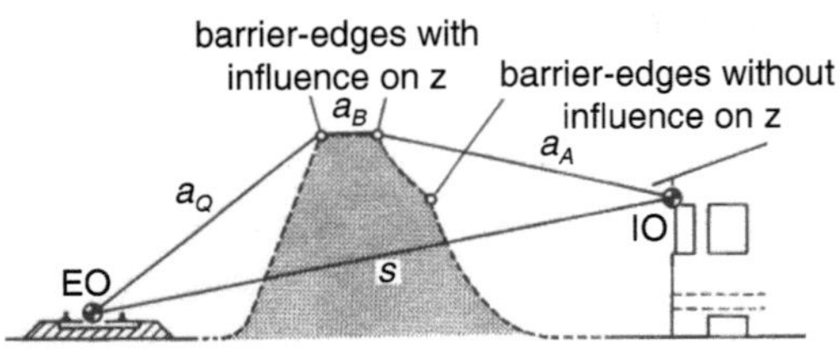

Fig. 16.42 Geometrical distances for calculating the so-called screening value z of a noise control barrier. $z = a_Q + a_B + a_A - s[m]$; *EO* emission point (centre of track at top-of-rail-level); *IO* immission point

Noise Control Embankments

Noise control embankments have two diffraction edges and are calculated in the same way as noise control barriers (see Fig. 16.42).

Vegetation on the embankment, which should be as dense as possible, increases the noise reduction by absorption.

If trees tower over the diffraction edge, reflections on leaves and branches may reduce the shielding effect.

The shielding effect of cuttings must also be taken into account according to [3].

When a train passes by a noise control barrier, the train itself shields the reflections to the opposite side considerably.

Passive Noise Control Measures

If technical reasons prohibit the construction of noise control barriers or embankments, or if in case of a new construction or substantial reconstruction of the railway the immission threshold values (according to the relevant legislation) cannot be complied with by the use of active noise control measures, then noise control is realised directly at the immission site by installation of sound-proofing windows, or by the acoustical improvement of other enclosing components (passive noise control measures).

It is important to take into consideration the characteristics of railway traffic noise for the dimensioning of sound-proofing windows according to Sect. 43 in [61, 250].

These characteristics are:

1. Less annoyance of rail traffic noise compared to road traffic noise ("rail bonus")
2. Better noise control effect of windows against rail traffic noise compared to road traffic noise at the same sound insulation index, due to the different noise spectra

For Point 1

Many studies [62–65] have been conducted on this subject. The difference in the level of annoyance (see Fig. 16.43) has been defined as a reduction of 5 dB (see Sect. 16.2.4.1 rail bonus) in calculation procedures according to [5].

As far as the disturbance of communication (telephone, television, etc.) is concerned a discussed supplementary "rail malus" has been rejected [63] by considering the "window position habits" of the citizens consulted (see also [66]). It was shown that in the case of road traffic noise most people disturbed by a high external level close the window (approx. 80%), so that then – due to the now lower internal level – the communication disturbances are significantly lower. In the case of rail traffic noise, by contrast, even with increasing external level, the windows remain predominantly open (only some 20% of those disturbed closed the window in this case), so that due to the higher internal noise when a train is passing, the communication disturbances are larger (see Fig. 16.44). However, the disturbance here is not so important that the windows are closed more often. This difference in the window position habits is also reconfirmed in the new studies, in particular, in Liepert et al. [67] (see Sect. 16.2.4.1).

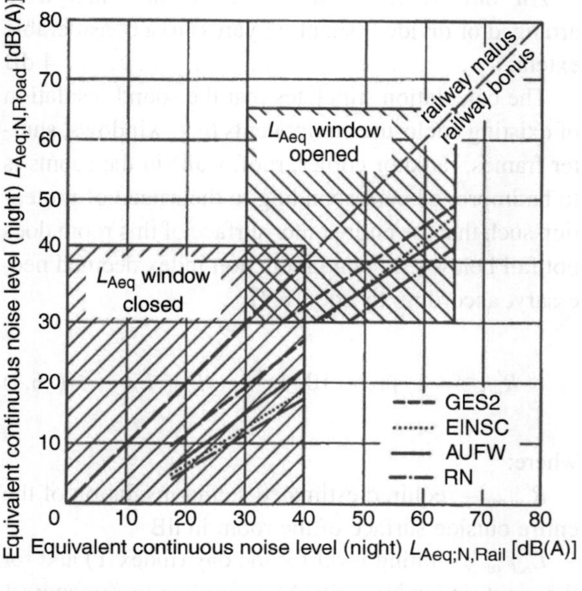

Fig. 16.43 Graphs of the connection between the equivalent continuous noise level at night of rail traffic and road traffic leading to the same annoyance in bedrooms, illustrated for selected variables of reactions. GES2: Annoyance by noise at night in general; EINSC: Annoyance by noise when going to sleep; AUFW: to be woken up by noise; RN Disturbances at night in general (mean of GES2, EINSC and AUFW)

For Point 2

Since the sound insulation of a sound-protected window depends on frequency, the spectrum of the external noise has an influence on the noise control effect.

Figure 16.45 shows an average spectrum measured outside of an apartment for rail and road traffic, respectively (upper curves). Subtraction of the sound reduction index (dotted curve) obtained with sound-proofing windows of various types of construction yields the respective spectra inside the apartment with their corresponding internal noise level.

One can see that the resulting difference in the mean level L_i (inside) of 6 dB, in this example with the same mean level L_a outside for both road an rail traffic favours the rail traffic noise.

This superior sound-proofing effect for rail traffic sound is taken into account in the calculation of the required sound insulation by means of a corrector (the so-called *E-Summand*) according to [68] [see the Eq. (16.4) below].

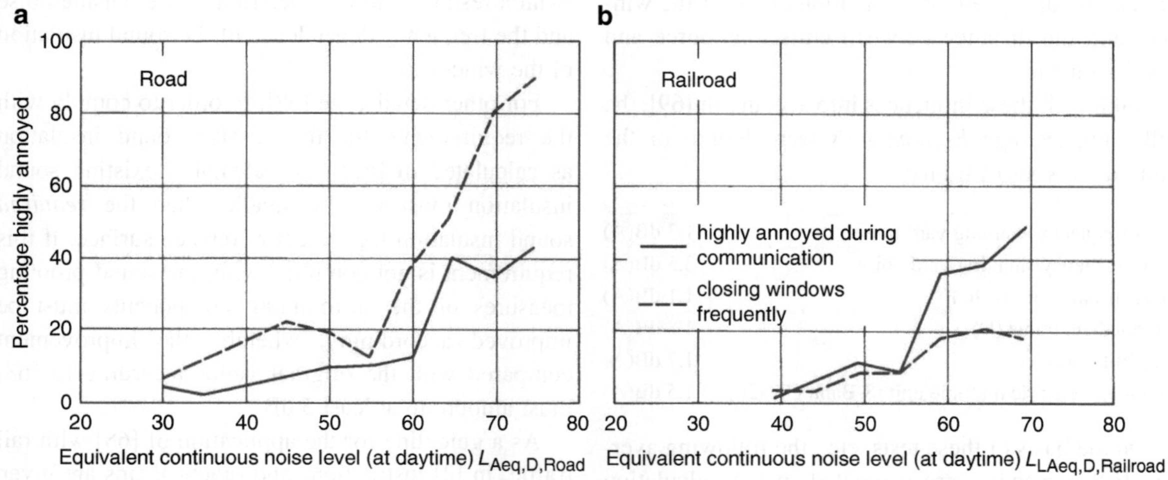

Fig. 16.44 Graphs of connections between the annoyances during communication or window closing activity, respectively, and the equivalent continuous noise level of either road traffic noise (**a**, *left*) or rail traffic noise (**b**, *right*)

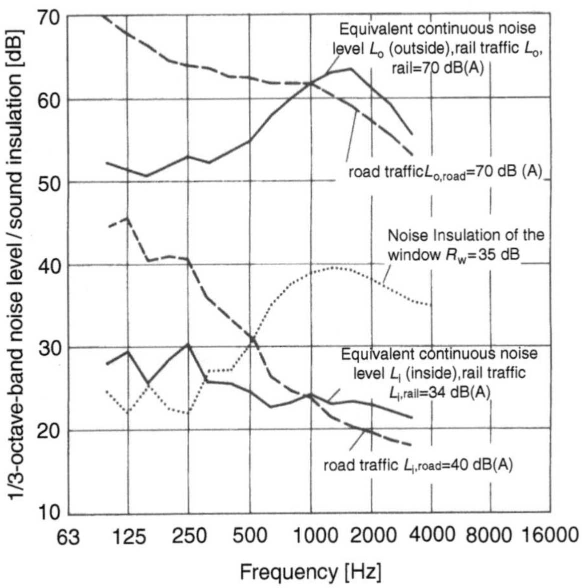

Fig. 16.45 1/3-octave-band noise level spectra of rail traffic and of road traffic noise outside of living rooms (measured same noise levels $L_A = 70$ dB(A)) and within living rooms (calculated by means of the specified sound insulation of a given window), shown for the typical frequency range of architectural acoustics

In a study [69], the *E-Summand* for rail traffic noise was determined dependent on the sound insulation of the window and the distance of the sound source. It was obvious that the E-Summand is influenced by a considerable number of parameters, such as the type of vehicle, the speed, the insulation curve of the window and the distance between emission source and immission site.

Taking all these influences into account in [69], the following average *E-Summands* were found for the individual types of trains:

High-frequency shunting yard noise	−1.3 dB(A)
Low-frequency shunting yard noise	3.5 dB(A)
Long distance trains (IC)	−1.1 dB(A)
Intermediate trains (D)	−1.8 dB(A)
Freight trains (G)	1.7 dB(A)
Commuter electric multiple unit (S-Bahn ET 420)	1.5 dB(A)

On the basis of these tests, etc., the following average E-Summands were estimated in the calculation procedures according to [68] described below:

Railway lines in general: 0 dB

Railway lines, in which in the rating period of time more than 60% of the wagons are freight trains with cast iron block brakes, as well as traffic routes of the magnetically levitated (Maglev) trains: ...... 2 dB

All railway lines on which freight trains were arranged or divided (shunting yards) to a considerable extent: .... 4 dB

The calculation stipulates that the sound insulation of existing enclosure components (e.g. windows, shutter frames, radiator corner, roof, wall) in the rooms is to be improved corresponding to their need of protection such that the entire outer surface of this room does not fall below the sound reduction index deemed necessary, according to Eq. (16.4):

$$R'_{w,\mathrm{res}} = L_{r,T/N} + 10 \cdot \lg \frac{S_g}{A} - D + E, \qquad (16.4)$$

where:

$R'_{w,\mathrm{res}}$ – required estimated sound insulation of the entire outside surface of the room in dB

$L_{r,T \text{ or } N}$ – rating level for the day (index T) and for the night (index N) in dB(A), according to Appendix 1 and 2 of [5]

S_g – entire outside surface in m^2 seen from the inside of the room (sum of all partial surfaces)

A – equivalent absorption surface of the room in m^2

D – correction according to Table 1 in [68] in dB (to take into account the use of the room),

E – correction according to Table 2 in [68] in dB (which results from the spectrum of the outside noise and the frequency dependence of the sound insulation of the windows).

For other details, see [68]. In order to comply with the requirements for the passive sound insulation as calculated in [68], the calculated existing sound insulation must not be smaller than the *required* sound insulation for the entire outside surface. If this requirement is not complied with, the sound-proofing measures of the surrounding components must be improved accordingly, whereby the improvement compared with the original value according to [68] must amount to at least 5 dB.

As a guideline for the application of [68] with rail traffic, in [2] instructions and practical tips are given for the execution of measures for passive sound insulation against rail traffic noise in the area of the Deutsche Bahn AG.

16.2.3.3 Noise Inside Rail Vehicles

A part of the airborne noise originated during rolling hits the floor and the lower areas of the sidewalls. When passing through a tunnel, it hits also the windows, walls and the roof, and excites these components to structure-borne vibrations, which in turn radiate airborne noise towards the inside. Another part of energy migrates as structure-borne noise from the wheel via the bogie to the upper vehicle components and is radiated likewise as airborne noise into the passenger compartments, etc.

The following descriptions are divided into two sections: travelling noise in passenger coaches and in the driver's cabins of the traction units.

Passengers Area

Over the last few decades, significant improvements have been obtained in the reduction of the internal sound level. Especially due to the introduction of disk brakes for passenger coaches, the internal levels (as well as the external levels) have been drastically reduced.

Internationally, the railway companies have set the following reference values for passenger trains with speeds up to 160 km/h:

In first class	65 dB(A)
In second class	68 dB(A)

Today, compliance with these values is being aspired as well for vehicles of the fast passenger trains (>200 km/h). Table 16.4 gives the internal level in the ICE 1 when travelling on an open line with ballasted track in dB(A).

In big-saloon coaches without small compartments, these values should not fall significantly below these figures, because otherwise conversations of passengers even sitting at a distance would no longer be masked by the – now too low – train noise and could become an annoyance.

Table 16.4 Noise levels inside of ICE 1 in dB(A)

Measurement location/passenger section	Speed (km/h)	
	200	280
Over the bogie	66	70
At mid-car between the bogies	62	66

When travelling through a tunnel, in the ICE these values are increased by only approximately 4 dB(A), due to the good sound insulation of the sidewalls, the windows and the roof, which have been designed especially for tunnel passages, too.

In older passenger coaches, the internal level increases significantly more while passing through tunnels, due to an insufficient insulation of the roof area. Figure 16.46 shows typical spectra as per [15].

Figure 16.47 shows the sound pressure level at different positions on and in a passenger coach while passing through a tunnel and on the open track at a speed of 250 km/h.

According to this figure, the sound pressure level in the bogie area under the coach at a speed of 250 km/h amounts to 120 dB when rolling on an open track as well as when passing through a tunnel. When rolling on the open track, the sound level decreases in direction to the centre of the roof over the bogie to 96 dB, and towards the centre of the roof in the middle of the coach, even to 90 dB, whereas in tunnel there are 115 dB(A) up there. The big differences between internal and external sound levels according to Fig. 16.47 require highly elaborate floor and wall constructions. In new passenger coaches also elaborate window and roof constructions are employed due to the large portion of tunnels on the new railway lines (see above).

Driver's Cabin

According to Code 651 of the International Railroad Association UIC [70], in the driver's cabin the equivalent continuous sound level L_{eq} should not exceed 78 dB(A), and 75 dB(A) should be the target value. In tunnels, these threshold values are 5 dB(A) higher. These values are related to a measurement duration of 30 min, with closed doors and windows, at a speed of up to 300 km/h on well-maintained rails.

These values – particularly with diesel locomotives with levels up to 120 dB(A) in the engine room – can only be maintained with very extensive sound insulations between the engine room and the driver's cabin. At train speeds ≥200 km/h, the compliance with these values requires an aerodynamically correct external configuration of the driver's cabin area.

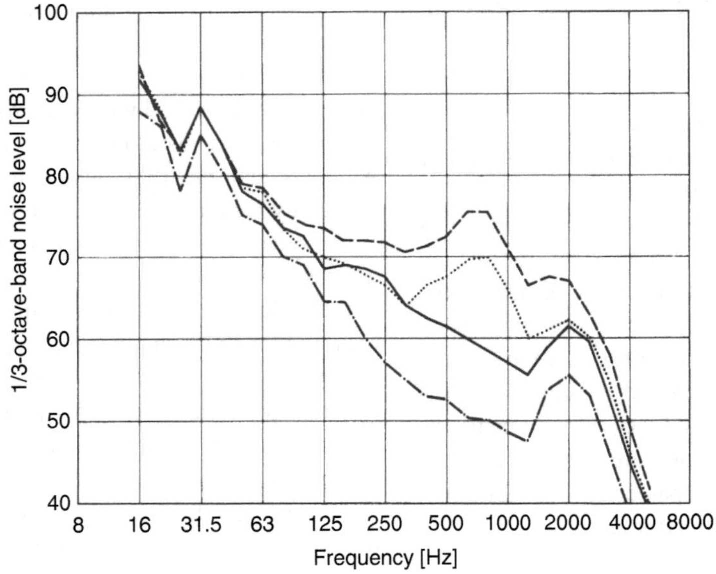

Symbol	Section of track	Type of superstructure	Overall noise level	
			dB(lin)	dB(A)
–·–·–·–	Open track	Ballasted track, type W60B70	92	64
———	Tunnel	Ballasted track, type W60B70	96	71
---------	Tunnel	Slab track, no absorption surface layer	97	81
....	Tunnel	Slab track, with absorption surface layer	96	75

Fig. 16.46 Airborne noise inside a coach (saloon in the middle of the wagon) when running on different track constructions on open track and in tunnel at a speed of 200 km/h

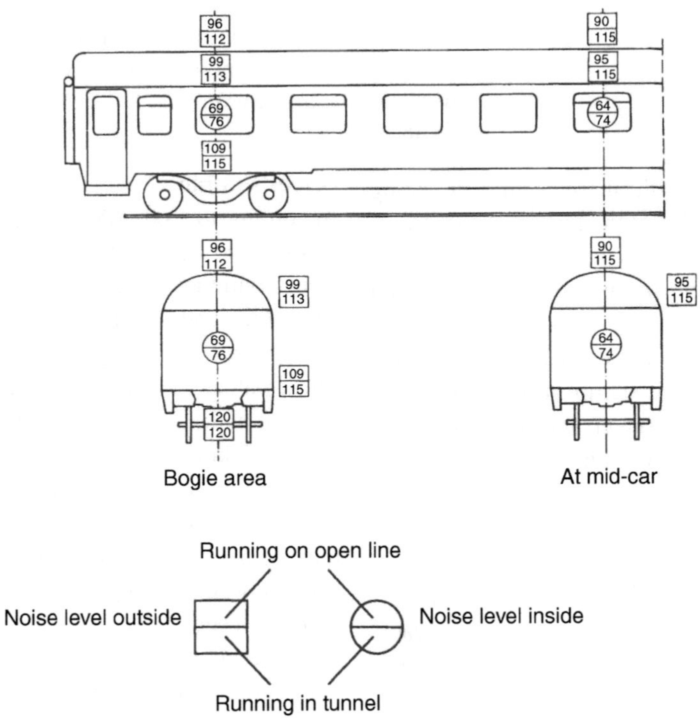

Fig. 16.47 Distribution of A-weighted noise levels at the outside sheets of a coach, Class Avmz 207, and inside that coach when running on ballasted track on an open line and in a tunnel at a speed of 250 km/h

Table 16.5 A-weighted noise level L_A in the driver's cab of various locomotives and trainsets on open ballasted track

Class	Speed (km/h)	L_A dB(A)	Notes
101[a]	220	81	–
103[a]	160	81	–
111[a]	120	74	–
120[a]	160	82	–
143[a]	120	74	–
151[a]	100	78	–
218[b]	130	80	–
232[b]	120	78	–
401 (ICE 1)[c]	250	79	–
423 (suburban rail)[c]	140	68	–
605 (ICE-VT)[d]	200	71	73 dB(A) in tunnel
644[d]	120	72	–

[a]Electric locomotive
[b]Diesel locomotive
[c]Electric trainset
[d]Diesel trainset

Individual sounds, such as warning signals, may occur in the driver's cabin with a sound level of $L_A > 80$ dB(A).

In the driver's cabins of the traction units of the Deutsche Bahn AG, the following rating levels L_{ra} are to be used for calculations:

Diesel locomotives	75–80 dB(A)
Diesel multiple-unit sets	65–75 dB(A)
Electric locomotives	70–75 dB(A)

Table 16.5 shows internal levels of the driver's cabin during full load operation and at maximum speed during open track running as per [15].

The sound pressure levels have been measured at the ear of the locomotive driver. In the driver's cabin of the steam locomotives used in former times temporary sound levels occurred briefly up to 110 dB(A) and in few cases up to 120 dB(A) (for instance when whistling), which could lead to rating levels L_{ra} in these vehicles of approximately 90 dB(A).

16.2.4 Effects and Assessment of Railway Noise

Undoubtedly, different noises with the same mean level widely vary in their degree of annoyance. A decisive influence on the annoyance caused by a noise is its sound character, the tonality, the duration (continuous noise or noise with long pauses), or opinions of the affected persons on the origin of the noise.

16.2.4.1 Railway Bonus

When the German Federal Ministry for Traffic in 1973 started to specify sound emission threshold values in the form of A-weighted equivalent continuous sound levels L_{eq} for rail traffic noise (SVL) and road traffic noise (StVL) the comparative assessment of the annoyance caused by both types of traffic noises had to be dealt with in detail, since the specification of the same numeric values as threshold values for different kinds of traffic noise would have meant to permit different annoyance levels for the noise caused by the different traffic systems.

In addition to the studies being conducted or having been terminated in Germany and abroad at that time which, with the exception of [64], only paid marginal attention to this matter, a study was conducted upon commission by the German Federal Ministry for Traffic [4].

The annoyance difference between railway traffic noise and road traffic noise for the disturbances at night was determined with some 10 dB(A) in favour of the railway traffic noise by interviewing over 1,600 inhabitants living in railway traffic noise and road traffic noise areas concerning numerous disturbance parameters (see also [71]). For the procedures and results of such studies, see [72].

This proven rail bonus with +10 dB(A) for the night time see also [73] was adopted by the legislature with only 5 dB(A) according to [5].

It is presumed that for the following reasons annoyance of railway traffic noise is significantly lower than for road traffic noise:

(a) The noise pauses between the individual sound events (trains passing)
(b) The regular occurrence of noises (timetable)
(c) Few different sound characters and sound levels perceived by a certain resident
(d) The loudness of the trains is not affected by an individual
(e) The spectrum of railway traffic noise being different in comparison with road traffic noise.

Investigations were conducted concerning the last point [65]. Here in order to investigate the differences between the C- and A-weighting of different kinds of traffic noise numerous measurements were carried out on many immission points with a broad variety of traffic situations for road, rail and air traffic. The noise impact times analysed were 23 h for rail traffic (without the pauses) and 41 h for road traffic. The study showed that road traffic noise has a considerably higher part of its sound energy in low frequencies than railway traffic noise, so that already on the basis of the A-weighting, a rail bonus of +5 dB(A) is justifiable.

Since in recent times the rail bonus has increasingly been criticised and challenged in particular within the framework of project approval procedures and the associated public relations regarding new constructions and upgrades of railway lines, the Deutsche Bahn AG initiated a comprehensive research programme on the annoying effect of rail traffic noise with the participation of the German Federal Ministry of Traffic, the Federal Environmental Agency, the Federal Railway Agency, the Austrian Federal Railway, etc., from 1996 until 2001. In the studies, it had to be verified whether the rail bonus determined in [4] retains its validity even under the current, changed traffic conditions.

In several field studies, based on interviews among involved residents, the annoyance difference between rail and road traffic for the nighttimes (the wake-up study) [74], the special features of high-speed traffic [75], and the varying noise effect of freight trains and passenger trains were examined [261]. Thus, measurements and interviews to the wake-up behaviour were conducted with the test persons in road traffic noise (StVL) and railway traffic noise areas (SVL) with the same equivalent continuous sound level L_{eq} in order to reliably detect noise-induced differences concerning the wake-up reactions.

Results from interviews of 1,600 test persons within the wake-up study led to the average annoyance differences (railway traffic noise/road traffic noise) of +13.6 dB(A) for "the sleep disturbance" question and by +8 dB(A) for those questioned about "disturbance all night long". The result of the communications disturbances resulted, also in consideration of the different window positioning habits, in annoyance differences of 3 dB(A) up to –8 dB(A) [67], whereby the "overall disturbance during the day" reached a value of +3.4 dB(A).

Concerning the communications disturbances in the inside (telephone, TV, etc.), the so-called "rail malus" [63] was further analysed by means of the window positioning habits of the persons questioned [67]. This investigation was carried out after having discovered that with increasing external noise, a larger portion of those interviewed at the street closed the window (> 50%), while at railway lines, only some 10% of the affected persons kept the windows closed. Due to the resulting difference in inner room levels along railway lines and roads, the "rail malus" shown in [63] was once again subject to scrutiny.

The result of these current investigations of noise effect shows that with predominantly closed windows along railway lines and roads the annoyance difference for communications disturbances must not necessarily be in the "malus" range.

Within the study on high-speed traffic, it was also determined on the basis of interviews completed with 315 residents that the noise disturbances provoked by the fast ICE trains in tendency are not superior to those from the conventional rail traffic [75].

In general, these studies confirm the results of earlier studies and show that also under the changed traffic conditions it is still justified to calculate with a rail bonus of +5 dB(A).

In other European countries, the rail bonus is set likewise, but partially at different values: Austria +5 dB(A), France +3 dB(A), Netherlands +7 dB (A), Switzerland, between +5 dB(A) (with a high line load) and +15 dB(A) (with a very low line load).

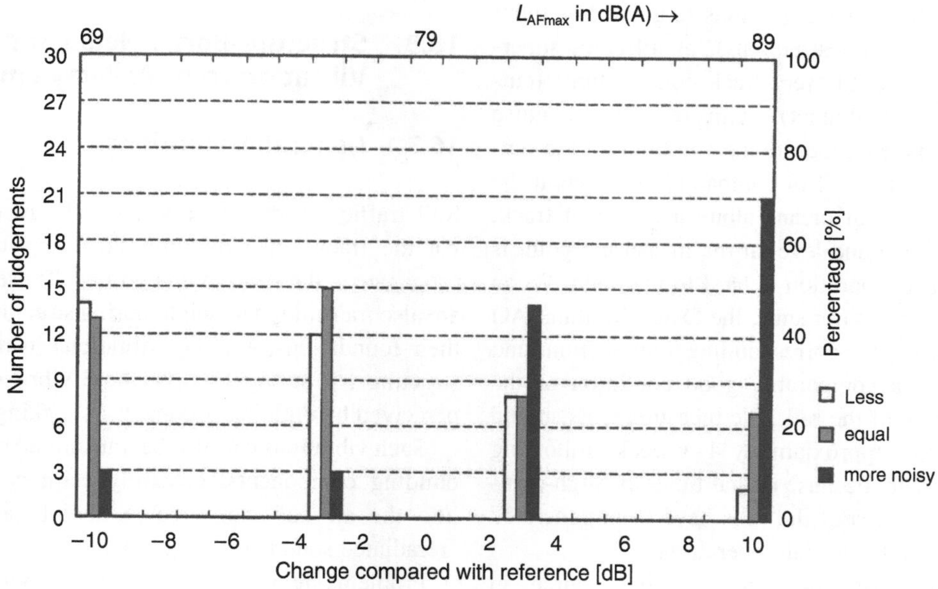

Fig. 16.48 Perception of noise level differences in case of train passages

In the meantime, the legislature has adopted in the Noise Protection Ordinance for magnetic levitation trains in 1997 a rail bonus of +5 dB(A) up to a speed of 300 km/h [76].

16.2.4.2 Level Additions (Pulse/Tonal Addition)

When rolling on narrow, curved tracks, squeaks in the frequency range between 2 and 10 kHz may appear due to longitudinal and transverse sliding of the wheels on the rail.

If at the immission site sound sources containing tonal and/or pulse components are relevant, certain level additions must be added in the course of the emission level calculation according to [2, 3].

16.2.4.3 Perception of Level Differences Regarding Passing Trains

According to the general perception, sound reductions and increases of 3 dB(A) are just about perceived as a change in the noise emission. In the tests conducted by the Technical University of Munich, test persons in the Sound Laboratory in several test series were exposed to a given rail traffic noise of a passing train; that noise was changed with pause periods of 3 min and of 6 min electrically by ±3 dB and ±10 dB. The test persons, distracted by other activities, were asked to specify whether they perceived a change in the sound level or not [77, 78]. By means of the distraction, it was intended to simulate a normal situation of living in an apartment close to a track.

The test results show that an increase or decrease in the sound level by 3 dB(A) is not perceived by the majority of the affected persons (see Fig. 16.48).

Even a noise reduction of 10 dB was not recognised as a reduction in half of the judgements given! This result is of great importance in particular with the planning of noise control measures (e.g., noise control barriers), since an increase in the height of noise control barriers by 1 or 2 m, as it is frequently demanded, has an additional acoustic effect in the range of 3 dB(A) to 5 dB(A), which – as the above-mentioned results show – is hardly perceived acoustically, but can be a strong additional optical disturbance.

16.2.5 Noise Measurements on Railway Vehicles

Measurements of noise inside and outside of railway vehicles are described in detail in standards [11, 79]. An exactly defined condition of the running surface of the rail and of the superstructure in the respective measuring section are the most significant points for reliable measuring results. Measurements of the noise emission of passing trains for the comparison of the emission at different rail sections (e.g., with/without lateral residential constructions), at different superstructure variants, different track constructions (cutting, level, embankment), with or without noise control barriers, etc., require the condition of the running surface of the rail in comparable sections to be reliably equal – that means along a length of track, from which the sound level at the measuring point is influenced. This condition is hard to maintain. When planning such measurements, the Deutsche Bahn AG grinds the rails of the corresponding track sections and thus produces a comparably good condition of the running surface of the rail. The measurements should take place after approxiamtely 4–8 weeks following the grinding date, because at that time, the high-quality running surfaces of the rails have been generated and normalised due to train operation.

Efforts to abate the noise at the source in those years resulted at the Deutsche Bahn AG in the development of a new measuring procedure to register the acoustic quality of the running surfaces of the rail near housing areas with a specially equipped measuring vehicle [49] (see Sect. 16.2.2.2, superstructure).

To locate and investigate the sound source, sound measuring procedures beyond the standards [11, 79] are required frequently. In case the sound source is stationary, as e.g., on the test bench or in wind channels, the sound-emitting areas can be located accurately to a few centimetres with the procedure of the "spatial conversion of sound fields" [80] using grid-shaped microphone arrays of typically 50–100 microphones. For the location of individual sound sources of passing trains, such as pantographs, that may emit aeroacoustic noise at speeds $\geq$250 km/h (see Sect. 16.2.2.1), microphone arrays can be used, which, for example, also typically consist of 50–100 microphones mounted on a coil of 4 m diameter in quasistochastic array or also in a cross-shaped array. With these arrays, each point – resolution according to each frequency range typically a few decimetres – of the passing train is scanned for a sufficient amount of time to allow a 1/3 octave-band sound level analysis and its contribution to the 1/3 octave-band level is determined [81]. In this manner, indentations for door handles, and steps, or cables crosswise to the air flow have already been identified as important aeroacoustic sound sources at speeds around 300 km/h.

16.3 Structure-Borne Noise and Vibration from Railway Lines

16.3.1 General, Definitions

Rail traffic causes vibrations, which are transmitted via the track superstructure into the subgrade and propagate in the surrounding ground. These vibrations are also transmitted to neighbouring structures through their foundations, exciting vibrations in the affected structures. If strong enough, these vibrations can be perceived by building occupants as shaking (tremors).

Such vibrations can also be transmitted by vibrating building components, generally ceilings and walls, into the air and can become audible as so-called "reradiated sound" (see Fig. 16.49).

Problems related to emissions of vibration and reradiated sound have become increasingly acute, as new railway lines are built and existing lines upgraded, especially in densely populated residential areas where there is often little distance between the railway lines and nearby residential structures.

The term "structure-borne noise" is frequently used as a synonym for the expression vibration, even though strictly speaking this term is, as indicated by the reference to "noise" reserved for use with regard to vibrations in solid objects within the range of audible frequencies ($f > 16$ Hz), see, for example [82].[8]

The most important quantity for the description of vibrations/structure-borne noise is the level of vibration velocity/structure-borne noise velocity, referred to as the velocity level L_v:

$$L_v = 20 \cdot \lg \frac{v}{v_0}\ \text{dB}. \qquad (16.5)$$

In which:

v – the effective value of the vibration velocity in m/s,

v_o – 5×10^{-8} m/s (reference velocity).

[8] According to Cremer et al. [82]), "structure-borne noise" denotes the field of physics, "... related to the generation, transmission and radiation of (usually very small) periodic oscillations and forces in solid bodies." In this respect, the word "noise" indicates that the main focus is on higher frequencies, roughly in the range of 16 Hz to 16,000 Hz. Oscillations and waves in lower frequencies are generally covered by the field of mechanical vibration or seismic waves.

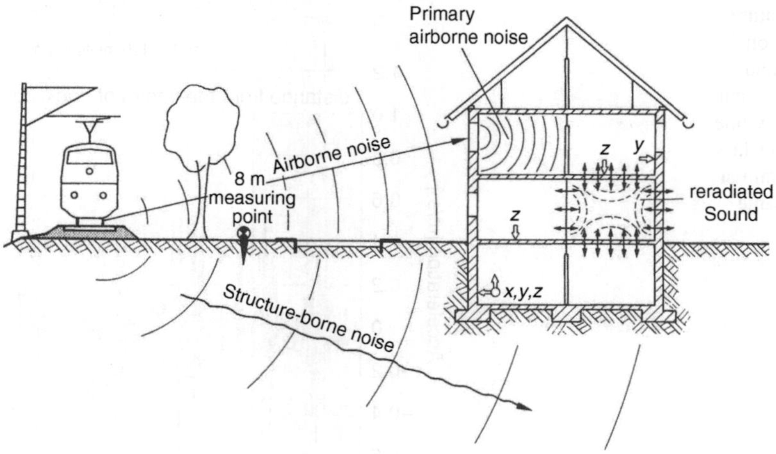

Fig. 16.49 Sketch to illustrate structure-borne and airborne noise immissions in the vicinity of an open railway line, with typical measurement points for determining emissions (8-m point) and immissions: x,y,z = direction of vibration, x parallel to the track axis (horizontal), x perpendicular to the track axis (horizontal), z perpendicular to the surface (vertical)

The velocity level L_v is used to describe oscillating structures, such as the ground, the foundations and ceilings of buildings, in terms of structure-borne noise, and to describe the propagation of structure-borne noise as well as in the measurement and numerical calculation of the efficiency of abatement measures (e.g. insertion loss, difference in velocity level).

In general, the unit measured is vibration acceleration, which is recorded using piezo-electric acceleration pick-ups the sensitivity of which has been set to the specific application being measured.[9]

In respect of periodic processes, which can practically always be assumed here, vibration acceleration a and vibration velocity v are interrelated when using Eq. (16.6) as follows (see [83] or Sect. 1.2 of this handbook):

$$|a| = \left|\frac{dv}{dt}\right| = \left|\frac{d(\hat{v} \cdot e^{j\omega \cdot t})}{dt}\right| = |j\omega \cdot v| = \omega \cdot |v|. \quad (16.6)$$

This means that the differentiation by time and integration over time passes over in the frequency range to multiplication or division by the radian frequency ω.

For the representation of the acceleration level

$$L_a = 20 \cdot \lg \frac{a}{a_0} \mathrm{dB}, \quad (16.7)$$

the reference value $a_o = 10^{-6}$ m/s is now generally used. In the past, other reference values were more common, such as $a_o = 10^{-2}$ and 10^{-4} and 9.81 m/s or $a_o = \pi \cdot 10^{-4}$ m/s.

The last of these was commonly employed in the field of structure-borne noise, as the acceleration spectrum and the velocity spectrum of an oscillation at a frequency of 1,000 Hz have the same value when it is used.

Measurements to determine the structure-borne noise situation in the vicinity of railway lines are to be performed in accordance with [84, 85, 119].

According to [84], the measurement point for determining emissions from surface lines is preferably the so-called "8-metre mark" (see Fig. 16.49), i.e. a surface measurement site 8 meters from the track axis on a metal stake (also referred to as a "spike" in [85]), the entire length of which (ca. 500 mm) has been completely driven into the ground and which generally has an L-shaped or X-shaped cross section.

[9] In particular, electro-dynamic measuring devices, so-called geophones are used to measure very-low-frequency vibrations. These devices are capable of directly measuring the vibration velocity.

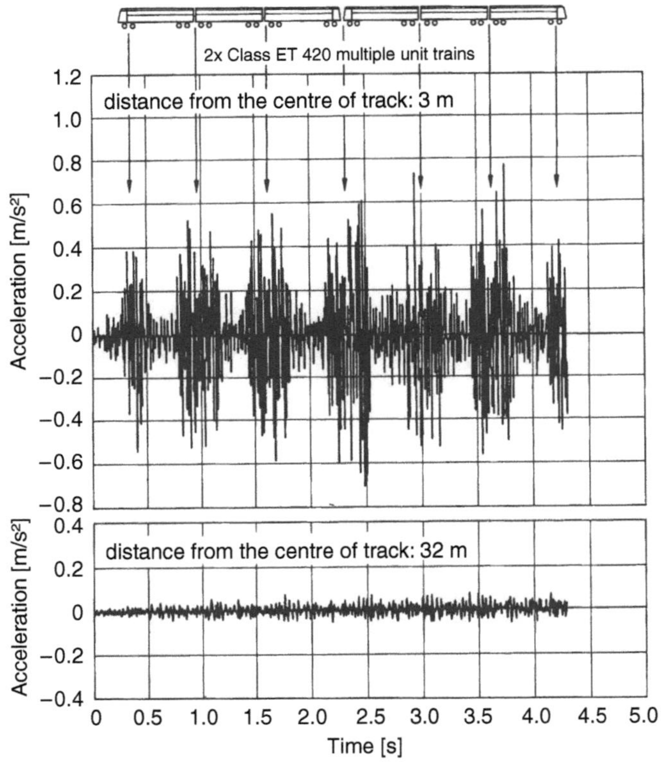

Fig. 16.50 Time signature of peak value of vibration acceleration in the ground, measured on a stake near and far from an open railway line during the passage of a Class ET 420 multiple unit train at a speed of 120 km/h, same scale

In respect of subsurface railway lines, the points for measuring emissions are usually located on the tunnel walls (for practical purposes, i.e. accessibility), generally at a height of 1.6–2 m above the tunnel floor, whereby it must be taken into consideration that under certain conditions differences in the materials (overburden) behind the tunnel wall can have a significant impact on the level of structure-borne noise measured at the tunnel wall [84].

Measurement points for determining immissions in neighbouring buildings, e.g. at the foundations, in load-bearing structural components in upper stories or in ceilings (see Fig. 16.49) should be selected in accordance with the specific requirements as per [85–87].

Figures 16.50–16.52 illustrate typical results of structure-borne noise measurements in the vicinity of a railway line, whereby in this specific case the measurement point for determining the emissions was 3 m from the track axis. Accordingly, in Fig. 16.50 one can clearly discern the passage of the bogies (wheel sets) and mid-sections of the cars in the periodically high and low maximum vibration acceleration values.

16.3.2 Generation of Structure-Borne Noise and Vibrations

Structure-borne noise is generated at the wheel/rail (wheel–rail) contact point and propagates from there into the vehicle and the subgrade.

Excitation at the wheel–rail contact point is primarily caused by roughness of the rails and the wheel tread. These deviations from ideal conditions (smooth rail, round wheels) result in corresponding contact forces, which excite vibrations in the entire wheel–rail system (cf. wheel–rail model in Fig. 16.18).

Deviations in shape have two different causes:

(a) On the one hand, they stem from geometric deviations in the wheels and rails, in the form of corrugations or roughness of the running surfaces. This type of excitement is referred to as *displacement or speed excitation.*
(b) On the other hand, they stem from local and thus time-related fluctuations in stiffness of the track superstructure system during the passage of trains, which leads to localised changes in the static and dynamic subsidence of the rolling surfaces of the rails. This form of excitement is referred to as *parametric excitation of vibrations.*

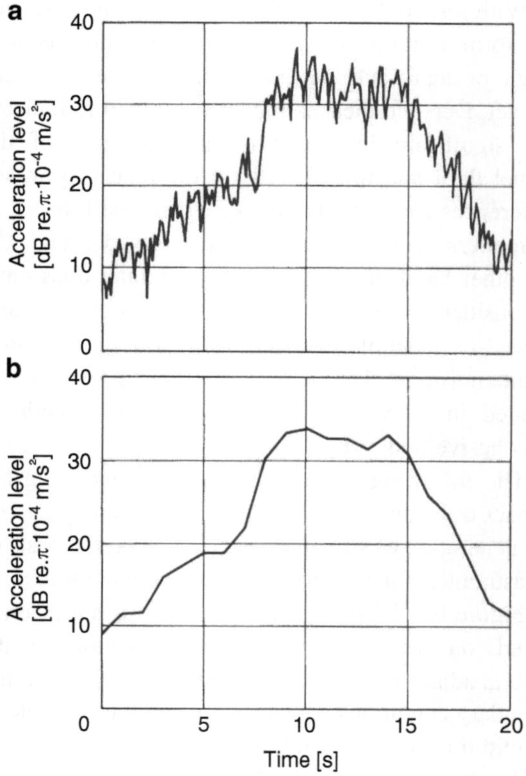

Fig. 16.51 Time signature of acceleration level 20 lg a/a_o in dB, measured on the ceiling of the third floor of a residential building located roughly 35 m to the side of an open railway line, during passage of a Class ET 420 multiple unit train on ballasted track at a speed of 120 km/h. Legend, *a* effective value of the vibration acceleration in m/s^2, $a_o = \pi \cdot 10^{-4}$ m/s^2 reference value for the vibration acceleration, time constant for the effective value: *a* $\tau = 0.125$ s ("FAST"), *b* $\tau = 1.0$ s ("SLOW")

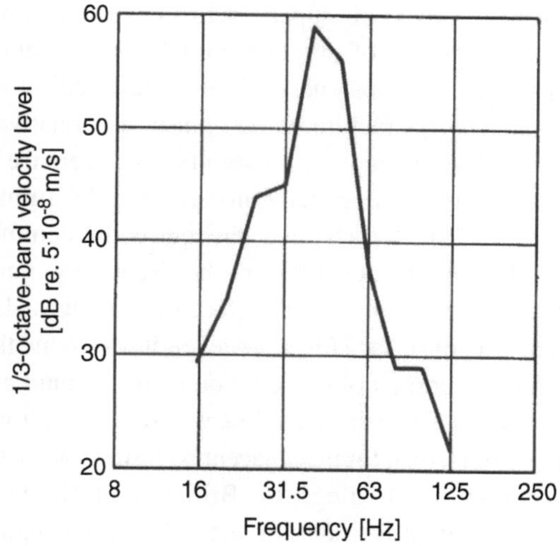

Fig. 16.52 1/3 octave-band spectrum of structure-borne noise and vibration velocity 20 lg v/vo in dB, measured on the ceiling of the third floor of a residential building located roughly 35 m to the side of an open railway line, during passage of a Class ET 420 multiple unit train on ballasted track at a speed of 120 km/h. *v* effective value of vibration velocity in m/s (time averaging "SLOW"), $v_o = 5 \cdot 10^{-8}$ m/s (reference value for vibration velocity)

Fluctuations in stiffness occur periodically according to the distance between the sleepers (secondary flexing) and stochastically according to changes in the bedding stiffness (ballast on subgrade).

The Hertzian contact stiffness of the wheel–rail interface is subject to fluctuations as a system parameter. This is influenced by the wheel–rail radius of curvature in the contact area (changes according to sinusoidal run) and by the static/dynamic contact forces, which are primarily subject to strong fluctuations in terms of the natural frequencies of the system (natural frequencies of the vehicle in the range of just few Hz, and wheel–rail natural frequencies in the range of 50–100 Hz).

Another form of excitement is found in the vibration excitation of unbalanced wheels. This falls under the category of *inertial force excitation.*

Numerous theoretical and experimental studies on the generation of structure-borne noise in wheel/rail systems have been carried out over the last few decades, and consequently our understanding of the fundamental aspects and key mechanisms of generation can broadly be viewed as well underpinned (see, for example, [88–90], [217–218], [259]).

In practice, the vehicle and superstructure as well as the subgrade and line design have proven to be key variables with regard to the generation of structure-borne noise.

In respect of vehicles, the main factors with an impact on the generation of structure-borne noise are speed, unsprung wheel mass, car weight, distance between bogies and axles and defects on the wheels (out-of-round wheels, flat spots on the running surfaces).

The main parameters for the track superstructure are the masses (rails, sleepers, ballast, slabs, etc.) and stiffness (ballast, elastic rail pads) of the superstructure components, which are involved in vibration as well as roughness on the rail surface (short-wave corrugations with the length > 8 cm, switches, insulated rail joints) and the distances between sleepers and seats.

"Heavy" types of superstructures, for example, exhibit lower levels of structure-borne noise excitation. The impact of the thickness of the ballast bed on the generation of structure-borne noise is relatively restricted [94]. Moreover, no significant difference in structure-borne noise generation has been determined between ballasted track and the currently common types of "slab track" [95] in the frequency range <80 Hz in the vicinity of subsurface railway lines [15].

In respect of surface lines, however, it was found that the structure-borne noise excitation of the ground at a distance of 20–70 m from the line was somewhat lower with slab track than for the adjacent ballasted track [96].

According to findings by British Rail Research (BRR), the improvement in the vibration situation resulting from slab track can be ascribed to less rail roughness and to generally better track geometry as compared to ballasted track [97]. By contrast, the results of studies carried out within the project RENVIB II (Reduction of Ground borne Noise and Vibration from Railways) by the European Railway Research Institute (ERRI, previously ORE) indicate that concrete slab on "soft" ground can lead to better adaptation, i.e. to enhanced transmission of vibrations into the subgrade and thus possibly to a considerably higher level of structure-borne noise at a farther distance away due to reflections from the boundaries between layers.

The maintenance condition of the superstructure, in particular the quality of the ballast bed and the subgrade, has a significant impact on the excitation of structure-borne noise to the extent that good maintenance conditions lead to lower emissions.

Even with the same vehicle and superstructure components, excitation of structure-borne noise can vary greatly depending on whether the rail line is located on the surface in flat terrain, on a viaduct, in a cut, whether the track is straight or curved or if it is in a tunnel. Rocky subgrade is more resistant to excitation than soft subgrade, whereby it should be noted that the dominant spectral component in soft subgrade is usually found at lower frequencies than in rocky subgrade.

The main excitation frequencies arising from the interaction of the vehicle, superstructure and subgrade are generally found in the frequency range of 40–80 Hz, both for ballasted track and slab track. Due to the very stiff bedding, the main excitation frequencies of newly constructed ballasted track in Germany range from 80 to 100 Hz.

With respect to underground line sections, the tunnel form (circular, oval, rectangular cross section, single or dual-track) was not found to have a relevant impact. Detailed measurements also failed to indicate any significant impact from the thickness of the tunnel floor and tunnel walls in the relatively minor differences found in the measurements (at 0.6–1.2 m in modern tunnels on newly constructed lines). On the other hand, the bedding of the tunnel does have a considerable impact on the surrounding ground. Thus, for example, greater excitation of structure-borne noise in the vicinity was found for tunnels bedded in loose rock than for tunnels in bedrock or cohesive rock [98].

The following section focuses on illustrating the impact of some of the aforementioned parameters on the generation of structure-borne noise on the basis of measurements performed on tracks in operation.

Figure 16.53[10] shows the impact of various passage speeds on the structure-borne noise spectrum in the ground adjacent to a surface railway line with all other boundary conditions unchanged, using the example of a rapid transit train (S-Bahn ET 420).

Analogously, Fig. 16.54 illustrates the same for a standard gauge line with the passage of an ICE 1 train [15].

Even though one should avoid making a direct comparison between these two figures, as the measurements were performed on different line sections (ergo subgrade conditions) with different superstructures[11] as necessary for suburban rail traffic and standard gauge traffic, based on the results it is still possible to elucidate the impact of the most important speed-dependent parameters leading to periodic oscillations, such as sleeper and axle spacing and wheel size (in the case of out-of-round wheels).

[10] Scheme for identifying different superstructure types, for example: W54 number B58 (short form: W54 B58), i.e. (a) (b) (number) (c). (a) fastening type: W for angled guide plate with clip, K for rib plate with clamp or clip; (b) rail type: 54 or 60 for rail S54 or UIC 60; (number) number of sleepers per 1000 meters of track, usually 1667 (this number is often left out); (c) sleeper type: B58/B70 for concrete sleepers, H for timber sleepers; for basic information on superstructure and permanent way, see Fiedler [246] for example.

[11] See the footnote 10 to Fig. 16.53.

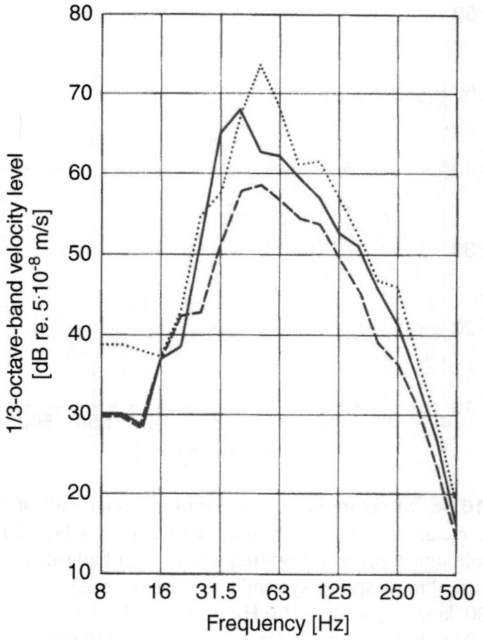

Fig. 16.53 Structure-borne noise in the ground (on a stake) at 8 m to the side of the track axis of an open railway line during passage of a Class ET 420 multiple unit train on ballasted track (type W54 B58) at various speeds - - - - - - - 40 km/h, ——— 80 km/h, ·········· 120 km/h

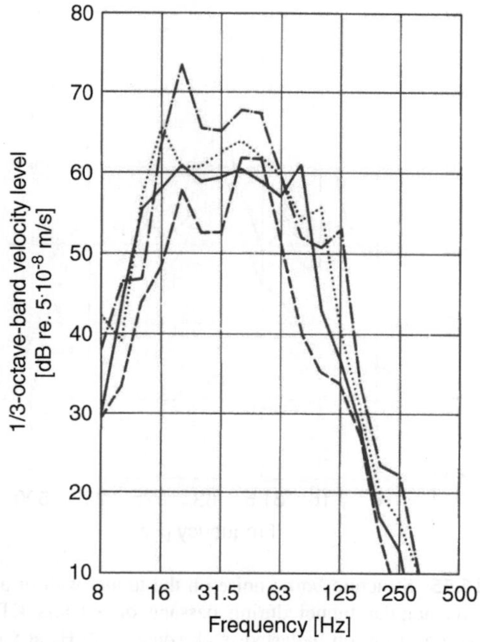

Fig. 16.54 Structure-borne noise in the ground (on a stake) at 8 m to the side of the track axis of an open railway line during passage of an ICE 1 train on ballasted track (type W60 B70) at various speeds - - - - - - 100 km/h, ——— 160 km/h, ·········· 200 km/h, —·—·— 250 km/h

According to Eq. (16.8)

$$f_s = \frac{v}{3,6 \cdot s}, \qquad (16.8)$$

in which:

f_s is the frequency in Hz,

v is the speed in km/h,

s is the sleeper spacing, axle spacing and wheel perimeter in m,

the frequency position of the maximum values for structure-borne noise excitation related to the parameters mentioned can be calculated.

If one takes the usual value of 0.6 m for the sleeper spacing and 2.5–3 m for axle spacing and wheel perimeter, it is found, as illustrated in Figs. 16.53 and 16.54, that in the frequency range in question (approxiamtely 16–125 Hz) only the "sleeper-spacing frequency"[12] plays a role for speeds under 100 km/h, while for the speed range 100 km/h < v < 200 km/h both the "sleeper-spacing frequency" and the "axle-spacing (wheel rpm) frequency" are important, whereas in the speed range over 200 km/h only the latter of these is decisive.

When speed-dependent excitation occurs in a frequency range, which includes the resonance of the vehicle/superstructure system, in particular, for example, the "wheel/rail natural frequency" described in more detail in Section 16.2.2.2, there is always a considerably higher level of excitation of structure-borne noise. This is reflected in the results of the measurements shown in Figs. 16.55 and 16.56, which were carried out in two sections of the same tunnel structure (from [15]), of which one section was fitted with a traditional ballasted superstructure (in this case, K 60 H with wooden sleepers) and the other with a low-frequency tuned, ballastless slab track system

[12] Furthermore, it should be taken into account that the fifth Harmonic of the wheel rotation frequency at a wheel radius of roughly 3 m is very close to the sleeper-spacing frequency; for example, this the case with the German ICE trains.

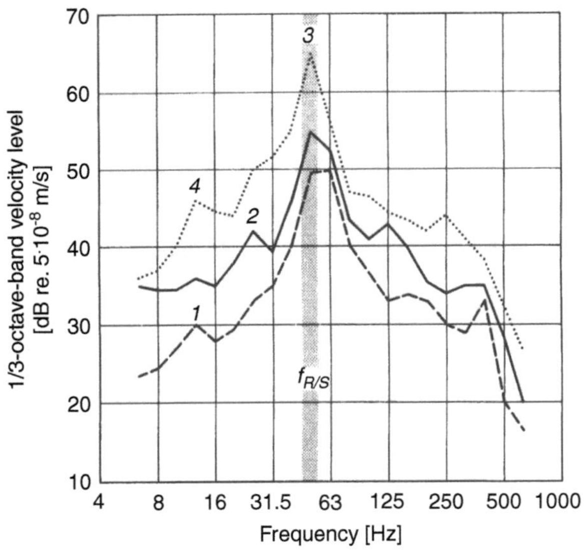

Fig. 16.55 Structure-borne noise on the tunnel wall of a dual track, rectangular tunnel during passage of a Class ET 420 multiple unit train on ballasted track (type K 60 H) at various speeds v: ------- $v = 30$ km/h: $f_{s,30\ \mathrm{Hz}} \approx 14$ Hz 1, ——— $v = 60$ km/h: $f_{s,60\ \mathrm{Hz}} \approx 28$ Hz 2, ········· $v = 120$ km/h: $f_{s,120\ \mathrm{Hz}} \approx 56$ Hz, $f_{R/S} \approx 55$ Hz 3. $f_{a,120\ \mathrm{Hz}} \approx 13$ Hz 4. Where f_s is the sleeper spacing frequency (with a sleeper spacing of $s = 0.6$ m), $f_{R/S}$ wheel–rail resonant frequency, f_a axle spacing frequency (with axle spacing of $a = 2.5$ m)

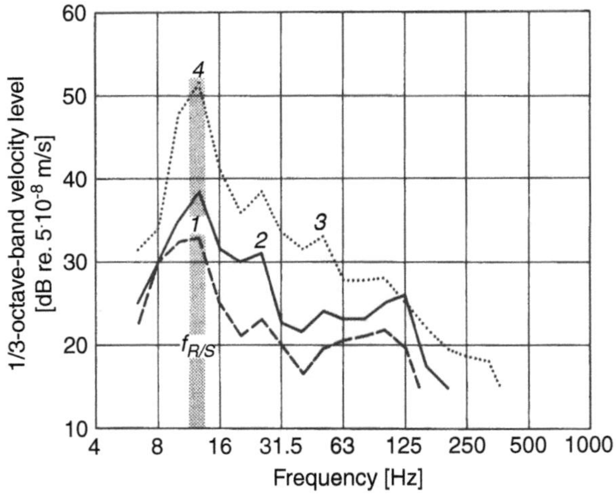

Fig. 16.56 Structure-borne noise on the tunnel wall of a dual track, rectangular tunnel during passage of a Class ET 420 multiple unit train on a low-frequency tuned ballastless superstructure ("mass-spring system") at various speeds v: ------- $v = 30$ km/h: $f_{s,30\ \mathrm{Hz}} \approx 14$ Hz 1, $f_{R/S} \approx 12$ Hz 1, ——— $v = 60$ km/h: $f_{s,60\ \mathrm{Hz}} \approx 28$ Hz 2, ········· $v = 120$ km/h: $f_{s,120\ \mathrm{Hz}} \approx 56$ Hz 3, $f_{R/S} \approx 12$ Hz 4, $f_{a,120\ \mathrm{Hz}} \approx 13$ Hz 4. Where f_s seat spacing frequency (with spacing $s = 0.6$ m),), $f_{R/S}$ wheel–rail resonance frequency, f_a axle spacing frequency (with axle spacing of $a = 2.5$ m)

(also known as a "mass-spring system"[13] see [99, 100] for example).

It is clear that for the ballasted track the sleeper-spacing frequency $f_{s,120}$ and the wheel/rail resonance $f_{R/S}$ lies within the 1/3 octave-band with the centre frequency of 50 Hz, for example, at a speed of 120 km/h, as a result of which the velocity level at this particular frequency is particularly high (Fig. 16.55: (3)).

For the "mass-spring" system, on the other hand, the wheel/rail resonant frequency $f_{R/S}$ is roughly 12 Hz, i.e. two octaves lower than in the case of the ballasted track and at the same speed of 120 km/h, it coincides with the axle-spacing frequency of $f_{a,120\ \mathrm{Hz}}$ [Fig. 16.56: (4)]. The sleeper-spacing frequency $f_{s,30}$ and the wheel/rail resonance frequency $f_{R/S}$ only fall within the same 1/3 octave-band with the centre frequency of 12.5 Hz [Fig. 16.56: (1)] at a running speed of 30 km/h, which is lower by a factor of four.

In addition to the speed-related effects described above, Figs. 16.55 and 16.56 clearly illustrate the impact of the type of superstructure construction on the generation of structure-borne noise. It can be seen that the frequency position of the maximum structure-borne noise generation in the range of the wheel–rail resonant frequency $f_{R/S}$ is, on the one hand, dependent on the running speed, but is also, on the other hand, strongly dependent on the type of superstructure. In this respect, the crucial parameters are the dynamically effective masses (unsprung mass of the wheel sets plus parts of the superstructure) and the stiffness of the superstructure.[14] The latter of these differs by

[13] This expression is not particular apt for drawing the distinction to the other superstructure types, as fundamentally speaking every normal kind of track superstructure, including ballasted track, is a form of "mass-spring system". Nevertheless, this term is also used here, as it has become rooted in the relevant literature [247] and is now widely used.

[14] This is often referred to as "bedding stiffness". In this respect, it should be noted that this actually only refers to the stiffness of the ballast bed including the subgrade or concrete foundation (see Sect. 16.2, Fig. 16.18), while the spring stiffness of the superstructure also includes the bending stiffness of the rail and

a factor of more than 10 in the two types of superstructure compared here.

Moreover, Figs. 16.55 and 16.56 can be seen as clear indications that excitation of natural frequencies within the tunnel structure cannot be the cause for the maximum structure-borne noise levels at the tunnel walls, as the two types of superstructure were located in two tunnel sections with identical cross-section dimensions.

Figure 16.57 presents an example of the influence of the rail running surface on the generation of structure-borne noise [15].

This figure shows the spectral increase in velocity level at the tunnel wall of a heavily used urban rail line resulting from so-called corrugations on the running surface of the rail.

Short-wave corrugations generally occur in tight curves with radii of less than 500 m from “skipping of the wheel set due to the lack of radial adjustment”.[15] These irregularities on the rail head are approximately 8–25 cm in length and generally occur on the interior rail of the curve; in the case of deeper corrugations (of roughly 0.5 mm) they can also be transmitted to the exterior rail.

Rail corrugation is removed by grinding the running surface of the rail using so-called “rail grinding trains” (see Sect. 16.2.2.2, specially monitored track – *BüG*). As one can see in Fig. 16.57, 15 months after grinding the rail head again exhibits corrugation with an even greater impact on structure-borne noise at the tunnel wall. In the case in question, the rail corrugation led to an increase of approximately 10 dB in the frequency range of up to roughly 125 Hz which is relevant for vibration, as compared with smooth running surfaces, while the velocity level at higher frequencies of approximately 125–315 Hz, which can be critical in terms of reradiated sound, increased by up to roughly 20 dB.

Examples of the influence of track design on the structure-borne noises emissions of railway lines can be found in Figs. 16.58 and 16.59. These present typical spectra for structure-borne noise measured in the ground for the passages of various types of trains[16] with their characteristic running speeds. In the one case, the train passages were measured on an open track section located on the newly constructed *Würzburg-Fulda* line, while in the other case the measurements are for passages in a tunnel with low overburden in a track section on the new line Mannheim-Stuttgart; both sections are ballasted track (type W60 B70) [15].

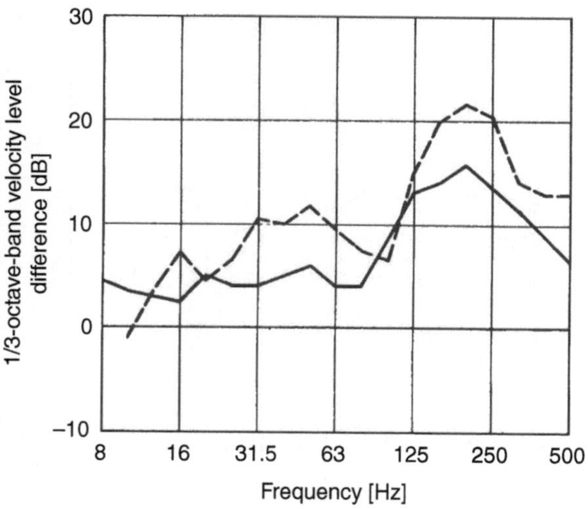

Fig. 16.57 Impact of rail corrugation with a wavelength of approx. 8–10 cm on the interior rail of a curved track section (radius $r = 420$ m) on the velocity level at the tunnel wall during passage of Class ET 420 multiple unit trains at a speed of 60 km/h. ——— Difference prior to rail grinding, ------ Difference 15 months following/immediately following rail grinding

Figure 16.58 can be considered as a supplement to Figs. 16.53 and 16.54 in respect of the typical emissions spectra for open lines.

Aside from the obvious differences between the spectra for open track and underground track, which can in part be ascribed to the significant differences in the superstructure–subgrade interface on the one hand and the superstructure–tunnel floor interface on the other (for more on this, see [101]), the characteristic structure-borne noise spectra typical for train passages in tunnels depicted in Fig. 16.59 should be noted.

the stiffness of the baseplate pads between the rail and the sleepers.

[15] Quotation from: “Advanced Training in Track Superstructure and Running Surface - Irregularities on Rails”. Information sheet of the former Central Office of the Federal Railways (BZA), Munich, Department 86.

[16] InterRegio train (qualified express train) and InterCity train are deemed as identical in terms of technical aspects related to structure-borne noise in respect of the coach materials.

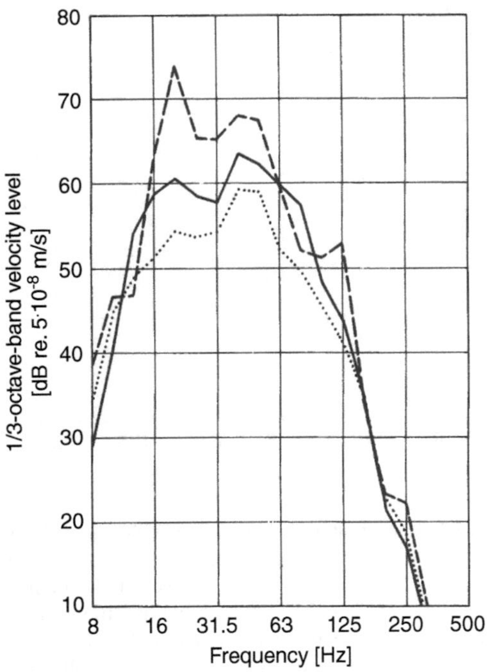

Fig. 16.58 Structure-borne noise in the ground (on a stake) at 8 m to the side of the track axis of an open railway line upon passage of various train types on ballasted track (type W60 B70) at various typical speeds. ·········· Freight train, $v = 100$ km/h, ——— InterRegio, $v = 160$ km/h, ------- ICE, $v = 250$ km/h

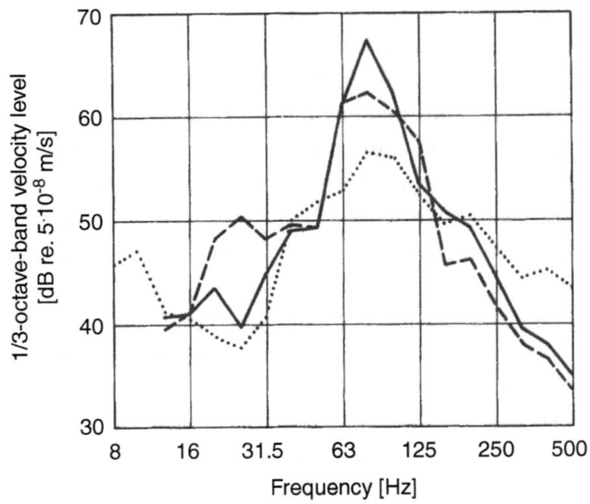

Fig. 16.59 Structure-borne noise in the ground (on a stake) over a dual-track rectangular tunnel with approximately 2 m of overburden, 16 m to the side of the connection line tunnel wall/surface during passage of various train types on ballasted track (W60 B70) at typical speeds for the respective train type, ·········· Freight train, $v = 100$ km/h, ——— InterRegio, $v = 200$ km/h, ------- ICE, $v = 250$ km/h

These spectra generally feature a well-defined maximum excitation, which is independent of speed in the area of the wheel–rail resonant frequency $f_{R/S}$ defined above.

If one compares Fig. 16.59 with Fig. 16.55, which presents the corresponding structure-borne noise spectra for tunnel passage of an ET 420 self-propelling suburban rail train (albeit measured at the tunnel wall, which is of lesser significance in respect of the fundamental aspect being discussed here), it can be seen that the area of the maximum level is at roughly 50 Hz, i.e. at a frequency at least two 1/3 octave-bands lower in comparison with Fig. 16.59.

The reasons for this can primarily be found in the different unsprung masses of the wheel sets of the types of trains compared, which are relatively high for the ET 420 and all axles are under traction, *and* the differences in the stiffness of the superstructures, which is relatively high on newly constructed lines of the Deutsche Bahn AG, as was mentioned above. Nevertheless, the differences between the two tunnels should not be seen as the main factor behind the observed effect (see above).

16.3.3 Propagation of Structure-Borne Noise in the Ground

Structure-borne noise transmitted through the ground decreases as the distance from the source of the noise increases. This decrease occurs according to certain natural regularities and depends on the type of waves involved in the propagation of the noise (free space waves, surface waves or combinations of such, see [102–105]. No clear natural laws could be determined in the immediate vicinity of railway lines (<8 m for surface lines and <15–20 m for underground lines). In addition to a geometric decrease, structure-borne noise also declines due to material damping. This phenomenon depends strongly on the soil type, layering and the groundwater level. Bedrock, for example, exhibits low level of damping, while marshy ground has a strong damping effect.

Other parameters that can have a significant impact on the radiation of structure-borne noise include frost, utility pipelines that transect the railway lines, buttress walls, concrete roadways and the like, which form a direct link between the source of emissions and the location of immissions. In underground line sections, material injections between the tunnel structures and buildings can lead to noise bridges for structure-borne noise.

For information on fundamental aspects of soil dynamics, such as wave propagation theory (one-dimensional, elastic half-spaces, etc.), dynamic soil characteristics (shear strength, density, Poisson number, etc.) and field and laboratory experiments to determine such, readers are referred to [104, 106, 107].

The results of the most comprehensive study on the radiation of structure-borne noise from railway lines can be found in ARGE [108].

A summary of the various results with notes on their relevance in practical applications can be found in Hölzl [109] and Hölzl et al. [110].

With regard to practical calculation rules, simple linear regressions of Eq. (16.9)

$$L_v = L_0 + k \cdot 20\lg(s/20), \qquad (16.9)$$

were calculated using the data from ARGE [108] in Müller-BBM [32]; for a summary thereof, see [111]. In this formula, L_0 represents the 1/3 octave-band velocity level at $s = 20$ m from the mid-point of the track and k represents the frequency-dependent change in the level over distance.

According to ARGE [108], the distance of 20 m is of particular interest for practical purposes, as identical vibration strength can be determined at that point for different kinds of soil, for example, when a strong excitation source is combined with a strong damping of propagation or when a weak emission source is combined with a weak damping of propagation, such as bedrock for instance.

As similar vibration phenomena were observed with only minor variations of k in geologically quite different kinds of ground types [e.g. low cohesion soils to soft solid bedrock, cohesive and non-cohesive soils and solid bedrock (*Buntsandstein*)] and moreover as observations which showed that the level reduction depends on the train type were limited to particular exceptions which could be explained, all measurement areas and train types were evaluated together. The result from Kurze [111] is shown in Fig. 16.60, in relation to the distance of 8 m to the track axis commonly used to characterise emissions from railway lines nowadays, including the 1/3 octave-band centre frequencies as a parameter (see also [110]).

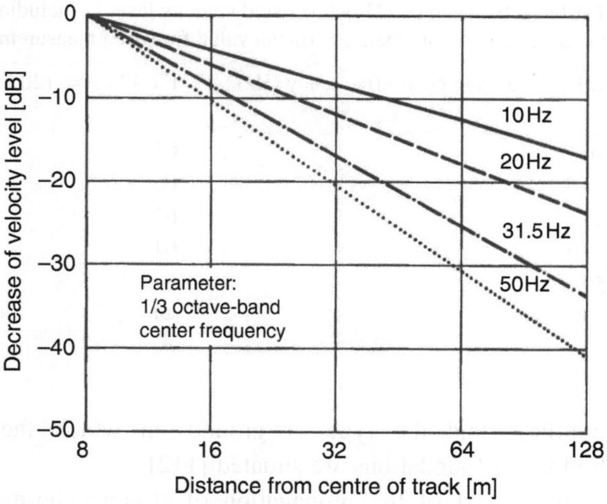

Fig. 16.60 Decrease in velocity level in the ground to the side of railway lines, at a distance of 8 m (mean value for passage of urban light rail trains, freight trains and passenger trains, at the typical speeds for the respective train types, and at several measurement sites)

Another result of a comprehensive regression analysis using the data from ARGE [108] was the increased 1/3 octave-band velocity level shown in Table 16.6, which is two times greater than the residual standard deviation 2σ and was found at all measurement sites at a distance of 20 m from the track [32].

Together with the reference values for velocity level decrease as per Fig. 16.60, these values can be employed as a basis for cautious, approximate vibration prognoses for unknown ground conditions (see also Sect. 16.3.10).

For residential structures which are close to the track axis (10–20 m), one can expect that the level of structure-borne noise, in particular at lower frequencies, will not decrease or will only decrease to small degree, if, for example, the area has exposed bedrock outcroppings, a high level of groundwater (foundations of the structure submersed in ground water) or if there are layers of a higher density at shallow depths.

16.3.4 Initiation and Propagation of Structure-Borne Noise in Buildings

Structure-borne noise initially declines at the transition point between the ground and the building foundations. In general, this depends on the type of foundation, the mass of the foundation and the

Table 16.6 Average 1/3-octave-band velocity level L_0 including doubled residual standard deviation of 2σ at a distance 20 m from the track axis for an open line (mean value for seven measurement areas)

1/3-octave-band centre frequency (Hz)	ET 420 $v = 120$ km/h	$L_0 + 2\sigma$ in dB	
		IC trains $v = 140$ km/h	Freight trains $v = 80$ km/h
10	63	67	66
12.5	66	70	68
20	62	65	62
31.5	64	65	66
50	67	65	64
100	49	49	48

building, and the type of ground on which the building's foundations are situated [112].

In respect of the propagation of structure-borne noise within the building, the levels of noise generally increase as a result excitation of the natural frequencies of the building's structural components, in particular, ceilings. The increase in the level caused by ceiling components depends strongly on their design. The key parameters are the mass, span, flexural strength, attenuation characteristics and restraint conditions of the ceiling assembly.

Floor assemblies such as floating floors can cause problems as these constructions are oscillatory systems and the natural frequency of such assemblies often lies in the main excitation frequency range of structure-borne noise originating from rail traffic.

The height of a building is generally of lesser importance in terms of the propagation of structure-borne noise [112].

Despite these and other factors, and the ensuing uncertainty they cause, in Müller-BBM [32] an attempt was made to describe the transmission of structure-borne noise from the ground into the building foundations and to ceilings in the form of average velocity levels of the peak value using standard deviation, based on recalculation and standardisation of the results of some 20 measurement protocols performed by different institutions. The results are presented in Fig. 16.61, broken down for the three directions of vibration.

Recalculation to the same distance of 20 m from the track axis was, when necessary, carried out in accordance with the distance-related reduction in levels presented in Fig. 16.60. The values for the three vibration directions at the measurement location are grouped closely together. As expected, the highest value was found at the measurement location "ceiling" in the z-direction.

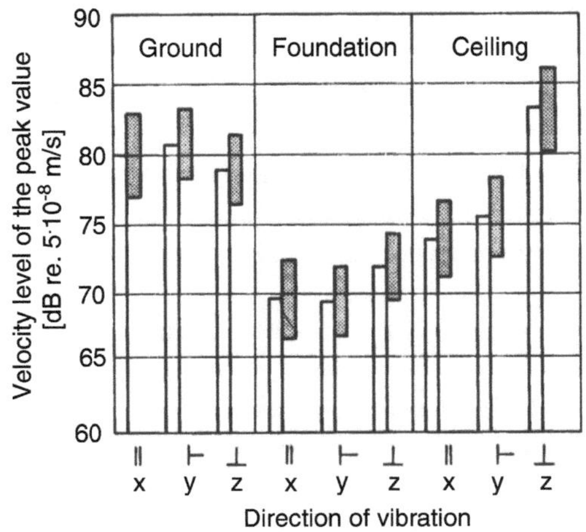

Fig. 16.61 Structure-borne noise in the next to buildings, at building foundations and on ceilings adjacent to a railway line with mixed traffic. Mean velocity level of the peak value $20 \cdot \lg \hat{v}/v_0$ with standard deviation, based on min. 25 and max. 130 measurements, at a distance of 20 m from the track axis, reference velocity $v_o = 5 \times 10^{-8}$ m/s. Vibration directions: x parallel to the track axis (horizontal), y perpendicular to the track axis (horizontal), z perpendicular to the surface (vertical)

The difference compared with the average levels in the ground (z-direction) was approximately 4 dB. However, if the levels of the three vibration directions are added together in terms of energy, the difference between the measurement locations "ground" and "ceiling" is almost 0 dB (see also Hölzl et al. [110]). This means that for rough estimates it tends to be correct to assume that the sum level from measurements of the three vibration directions in the ground, i.e. next to the site where a structure is to be built, will be close to the velocity level that can be expected in the z-direction on the ceilings.

Results of studies focusing on the resonant behaviour of concrete ceilings and wood beam ceilings are presented in Table 16.7 (evaluation from [69]).

Table 16.7 Statistics on the dimensions and resonant frequencies of ceilings in single unit/multiple unit dwellings at various distances from open railway lines

Statistical quantity → Measurement quantity ↓	Minimum	Mean value	Maximum	Standard deviation
(a) Concrete ceilings (270 ceilings in 135 buildings)				
Distance to track in m	5.0	21.4	48.0	10.3
Ceiling surface area in m^2	6.0	20.7	64.0	3.8
Ceiling length in m	3.0	5.2	10.7	1.3
Ceiling width in m	2.0	3.9	8.0	0.8
Aspect ratio (—)	1.0	1.3	2.7	0.3
Resonant frequency in Hz Median 1/3 octave band frequency	10.0	31.5	80.0[a]	1.6 1/3 octave bands
(b) Wood beam ceilings (172 ceilings in 86 buildings)				
Distance to track in m	4.0	16.5	82.0	12.2
Ceiling surface area in m^2	5.8	18.5	63.0	7.3
Ceiling length in m	2.4	4.9	9.0	1.2
Ceiling width in m	2.1	3.7	7.0	0.6
Aspect ratio (—)	1.0	1.3	2.4	0.3
Resonant frequency in Hz Median 1/3 octave band frequency	6.3	20.0	80.0[a]	2.3 1/3 octave bands

[a]It is highly probably that this is not due to the resonances of the ceilings but rather to resonance of floating floors or other floor assemblies

One can see that, with roughly the same geometric dimensions, the natural frequency of wood beam ceilings are, as expected, lower on the average than that of concrete ceilings, but that the variation range for these two types of ceilings has approximately the same band width.

Finally, Fig. 16.62 presents results of new studies on the transmission of structure-borne noise from the ground to foundations and on ceiling level increase functions, based on measurements conducted in 135 buildings with concrete ceilings [69].

Based on Part (b) of the diagram, it can be seen that on the average the difference between velocity levels at the ceilings is some 20 dB compared with the levels measured at the foundations in the resonant frequency range, i.e. at $f/f_0 = 1$, representing an average increase in the velocity level by a factor of 10 at the resonant frequencies. The frequency range of the increase function, which is important for the sum velocity level on ceilings, is roughly three to four 1/3-octave-bands wide.

Looking at the frequency response of the level difference between the building foundation and the soil in Part (a) of Fig. 16.62 one can see that this exhibits a relative minimum (cf. Table 16.7, Sect. a) in the frequency range in which the natural frequency of the ceilings studied lies, namely 31.5 Hz. The explanation for this may be that the ceilings function as a sort of "resonance absorber", drawing energy out of the structure (foundations) in this frequency range.

In principle, the results depicted in Fig. 16.62 can be applied to wood beam ceilings as well. It should, however, be taken into account that far greater scatter in transmission functions, in particular for differing floor heights, was observed in [69].

16.3.5 Reradiated Sound in Buildings

As was noted in Sect. 16.3.1, vibrations in structures are also radiated by the building's structure (mainly the walls and ceilings) and may be perceptible as low-frequency airborne noise (so-called reradiated sound, see Fig. 16.49).

The level of reradiated sound is determined both by the velocity of the structure-borne noise in the structure's components and by the radiation and absorption characteristics of the rooms.

Even structural vibrations which are far below the perception threshold [86, 113] can, depending on the frequency composition of the vibrations caused, generate reradiated sound emissions which can be perceived by building occupants.

Reradiated sound emissions caused by underground railway lines, such as urban rail and subway

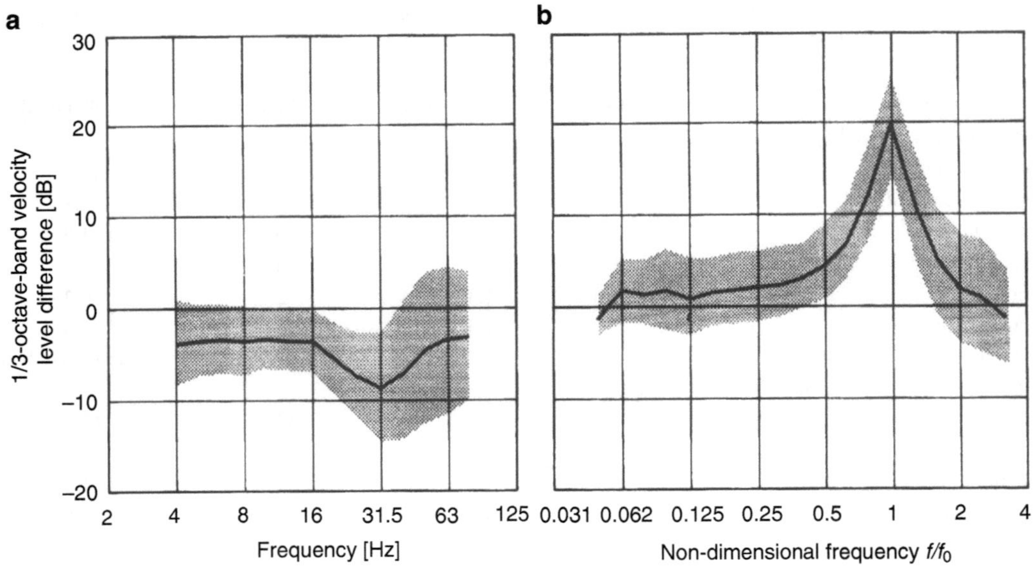

Fig. 16.62 Transmission of structure-borne noise from the ground into the foundation and from the foundation into the ceilings. Mean value and standard deviation of measurements in 135 two-story and 3-story single-family homes and multi-unit dwellings with concrete ceilings. (**a**) 1/3 octave band level difference foundation-ground, (**b**) 1/3 octave band level difference ceiling foundation, Abscissa of the individual differences standardised to the 1/3 octave band centre frequency, in which the respective ceiling resonant frequency f_o is found

lines, can cause disturbance even at very low levels, as there is no interaction with direct airborne noise from the rail traffic (masking effects, etc.).

As an example of this, Fig. 16.63 illustrates the results of measurements in a tunnel on a urban rail line in Hamburg and in the cellar of a neighbouring house, based on which one can see the transmission path of the structure-borne noise from the rail into the tunnel wall and on to the cellar wall and finally manifested as reradiated sound in the cellar itself; the A-noise level in this case was approximately 35 dB(A).

The fact that the spectral peak of the reradiated sound in Fig. 16.63 does not correspond to the structure-borne noise at the cellar wall is not surprising if one takes into account that reradiated sound is mainly determined by radiation from ceilings, which both in terms of levels and frequency positions of the maximum oscillations (natural frequencies of the ceilings) may be considerably different from those of the walls (in this case, no measurement results were available for a location on the ceiling).

Determination of reradiated sound (by measurement or calculation) is still subject to considerable uncertainties. Focused studies should bring further developments in this field in the near future, as efforts are made to formulate uniform measurement and calculation methods, and readers are hence referred to the relevant literature on the procedures currently in use [114–118].

In addition to reradiated sound, structure-borne noise can also result in secondary effects such as audible rattling of loose door fillings, glass panes, loose hardware, and glasses in cupboards. At times, these effects can occur at vibration levels, which are not perceptible to building occupants. Due to the unusual nature of these effects, they are often experienced as being annoying. Nevertheless, generally these kinds of effects can be prevented or resolved with simple means (fixing loose fillings or hardware, moving glasses farther apart, etc.).

16.3.6 Assessment of Structure-Borne Noise, Vibrations and Reradiated Sound

16.3.6.1 Assessment of Structure-Borne Noise and Vibrations

So-called KB values are used to determine the impact of vibrations on humans. As per [119], the KB signal is the signal KB(t) that is obtained by frequency evaluation and standardising of the unevaluated structure-borne noise signal $v(t)$.

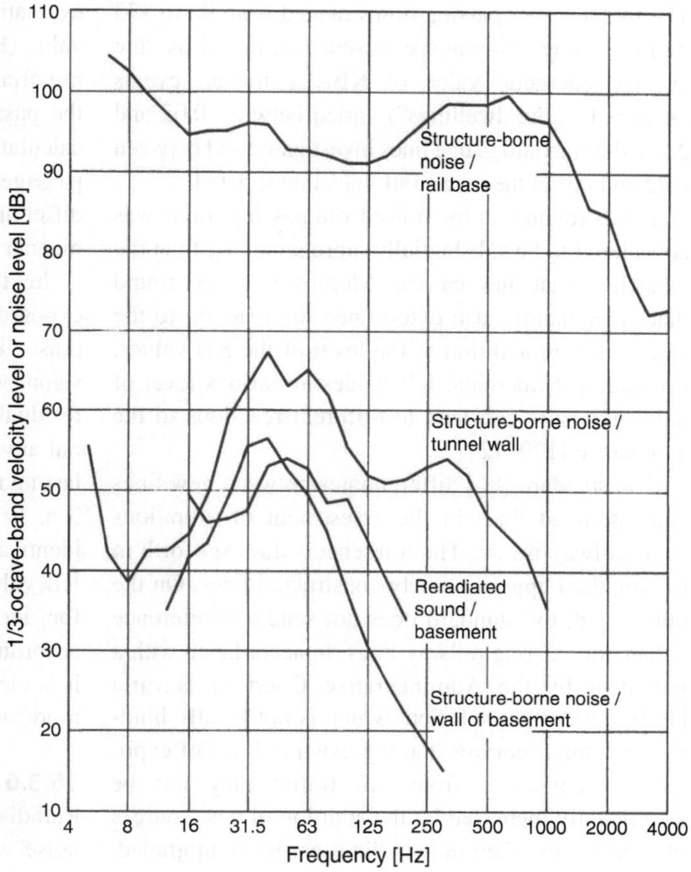

Fig. 16.63 Structure-borne noise in a tunnel as well as structure-borne noise and reradiated noise in the basement of a building situated approximately 10 m adjacent to the tunnel with 3.5 m overburden during passage of Class ET 471 multiple unit trains of the Hamburg rapid transit system at a speed of 60 km/h

The frequency response of the KB evaluation approximates that of a single pole highpass with a critical frequency of 5.6 Hz.

The KB value KB_F is the momentary value of the floating effective value weighted with the time constant "Fast" ($\tau = 125$ ms) of the KB signal KB(t). The highest KB_F value occurring during the observation period is referred to as the $KB_{F\max}$ value. A maximum interval process is proposed for the evaluation of vibrations in DIN 4150-2 [86]. In applying this process, the observation period is divided into intervals of 30 s with the corresponding maximum $KB_{F\max}$ values. Then, taking into account the periods of immission and assessment, the evaluation oscillation strength KB_{FTr} is formed from the maximum interval values and compared to reference value or evaluation criteria.

There are currently no legal regulations stipulating that certain limits be complied with in terms of the impact of vibrations from rail traffic. As a result, generally, the standard DIN 4150-2 [86] is consulted in the rating of vibration immissions. The reference values contained therein for vibrations are based on results from a study on the impact of vibrations from rail traffic on residents in Germany, which was primarily aimed at answering the following questions:

- What impact do vibrations from rail traffic have on residents living in the vicinity of railways?
- What is the relationship between physical data on vibrations and the perceived extent of annoyance?
- To what extent does noise from rail traffic influence the reactions to vibration immissions?
- What differences are there between the vibration effects from standard gauge lines and urban rail lines?

The study was carried out using vibration measurements (as per DIN 4150-2) taken in 284 dwellings in the vicinity of standard gauge lines and 102 dwellings located near urban rail lines, noise measurements and interviews based on a standardised questionnaire on sociological variables.

All of the dwellings involved in the study were situated between 5 and 60 m from the railway lines.

The frequency of passing trains ranged from 92 to 373 trains per day. Vibration exposure (expressed as "the average energetic value of $KB_{F\max}$ for all events measured in the dwellings") varied between 0.02 and 2.7 in the standard gauge lines investigated and between 0.02 and 0.9 in the urban rail lines investigated.

It was found that the noise from passing trains was considered to be substantially more annoying than the vibrations that they caused. Moreover, it was found that other factors also determined the reaction to the annoyance, in addition to the level of the KB values. For example, identical KB values at various level of noise exposure can lead to different reactions to the annoyance [120, 121].

The standard [86] differentiates between new lines and upgraded lines in the assessment of vibrations from railway traffic. The reference values set forth in the standard apply to newly constructed lines. On the other hand, the standard does not state any reference values for existing railway lines. In accordance with a judgment by the Administrative Court of Bavaria [122], the current solution, which is not legally binding in nature, requires that the existing level of exposure to emissions from rail traffic may not be significantly increased by the addition of new sources of immissions when railway lines are being upgraded. Consequently, it is important to determine what level of increase in vibration immissions is clearly perceptible in order to be able to evaluate the situation with regard to vibrations in the future.

This question was investigated in a laboratory experiment [123]. This experiment involved erecting a test room, in which it was possible to compare samples of railway-specific vibration signals of various intensities under realistic conditions. Twenty seated persons were exposed to combination of a total of four levels of vibration intensity ($KB_{F\max}$ from 0.2 to 1.6) and three levels of interior noise [30, 45, and 55 dB(A)]. Detection of the predetermined differences in stimuli was measured using the psychophysical investigation method SDT (Signal Detection Theory). The experiment found that two stimuli cannot be differentiated with certainty if the maximum effective value ($KB_{F\max}$) varies by up to 25%. In this respect, no particular combination of the stimuli was found to have a clear impact.

A more recent experiment [124] was carried out to research the applicability of the maximum interval process proposed in DIN 4150-2 [86]. With this evaluation method only the highest measured KB value ($KB_{F\max}$) in the time-related vibration signal is important. The transient peak values occurring during the passage of a train thereby finally determine the calculated level of vibration immissions. Train passages with identical $KB_{F\max}$ values but with quite different energy content are evaluated in the same manner with this method.

In this experiment, 22 test participants were exposed to pairs of vibration signals (structure-borne noise, in the *z*-direction) consisting of a reference signal with a well-defined peak value (6 dB over the residual time duration) and a comparison signal (without any well-defined peaks in the time duration) each lasting roughly 10 s with similar spectra at intervals of 3 s. It was found that the stimuli are perceived as identical if they exhibit the same energy-equivalent KB values. By contrast, the impact was quite varied for identical maximum effective values $KB_{F\max}$ according to DIN 4150-2. According to these results, it is clear that the energy-equivalent KB value is far more suitable for characterising the intensity.

16.3.6.2 Assessment of Reradiated Sound

Reradiated sound is a relatively low-frequency traffic noise which is radiated from all surfaces of a room as a result of vibration excitation of buildings by rail traffic and which has no specific identifiable directionality. The provisions of BImSchV-16 [5] (German Regulation on Traffic Noise), however, cannot be applied in this respect, and consequently there is currently no legal regulation specifying maximum limits in terms of acceptable exposure to reradiated sound.

Guideline values for acceptable indoor levels of reradiated sound can be derived from BImSchV-24 [68], (German Regulation on Noise Protection Measures for Traffic Routes) and VDI 2719 "Noise Insulation of Windows and their Accessories" [125].

The reference values set forth in VDI 2719 [125] apply for noise, which penetrates rooms directly and can thus only be used to a limited degree. Based on the regulations on primary airborne noise, one could derive a value of 3 dB (A) as a considerable increase in the level of reradiated sound for a major construction project on an existing railway line, which may in turn call for noise abatement measures to be taken.

Information on the evaluation of reradiated sound from railway traffic can also be found in DIN 45680

"Measurement and Evaluation of Low-Frequency Noise Immissions in the neighbourhood" [126] and most recently in Said et al. [127], Krüger [128].

16.3.7 Vibration Mitigation Measures in the Area of Structure-Borne Noise Generation

16.3.7.1 Mitigation Measures on Underground Railway Lines

In general, it can be stated that one important mitigation measure is regular maintenance of the track superstructure, including rail grinding in the early stages of corrugation, replacement of worn rails and maintenance or renewal of the ballast bed.

For railways lines located in tunnels there is a well-founded, trusted range of measures, which is thoroughly underpinned by calculations due to the clear set of boundary parameters in tunnels [100, 129–134].

In the field of suburban railways and in some applications on main railway lines, the installation of ballast mats and so-called mass-spring systems has proven to be an effective method of mitigating structure-borne noise.

Ballast mats are resilient mats usually made of polyurethane or rubber-based materials, which are used to isolate the entire surface of the tunnel floor (or bridge structure or subgrade) from the ballast. Figure 16.64 illustrates the basic design of a ballasted superstructure with ballast mats, based on the example of a tunnel with a circular cross section.

Figure 16.65 presents typical spectra of structure-borne noise measured at the wall of a suburban rail tunnel in Munich, Germany, before and after installation of ballast mats [133].

With a dynamic stiffness of approximately 0.04 N/mm in the critical frequency and load range, the ballast mats have a static stiffness (also referred to as the "bedding modulus") of 0.02 N/mm, in accordance with the train speeds and axle loads commonly found in suburban rail operations [45].

According to Wettschureck [132] and Wettschureck et al. [134], the insertion loss of a ballast mat, which is only effective with its spring stiffness s_M, can be calculated using the simplified dynamic model presented in Fig. 16.66, according to Eq. (16.10):

$$\Delta L_e = 20 \cdot \lg \left| 1 + \frac{j\omega / s_M}{1/Z_i + 1/Z_T} \right| \text{dB}. \qquad (16.10)$$

in which, in addition to s_M:

Z_i – the source impedance effective from the upper side of the mat to the emission source of the structure-borne noise

Z_T – the final impedance effective on the bottom of the mat (at the tunnel floor)

ω – the radian frequency

j – the imaginary unit

For normal tunnel floors, Z_T is large compared with Z_i and can thus be ignored in the equation for ΔL_e [101, 135].

The spring stiffness of the ballast mat derives from Eq. (16.11)

$$s_M = s''_M \cdot S_w \cdot (1 + jd_M), \qquad (16.11)$$

in which s''_M is the dynamic stiffness, d_M is the loss factor of the ballast mat and S_w is the effective surface,

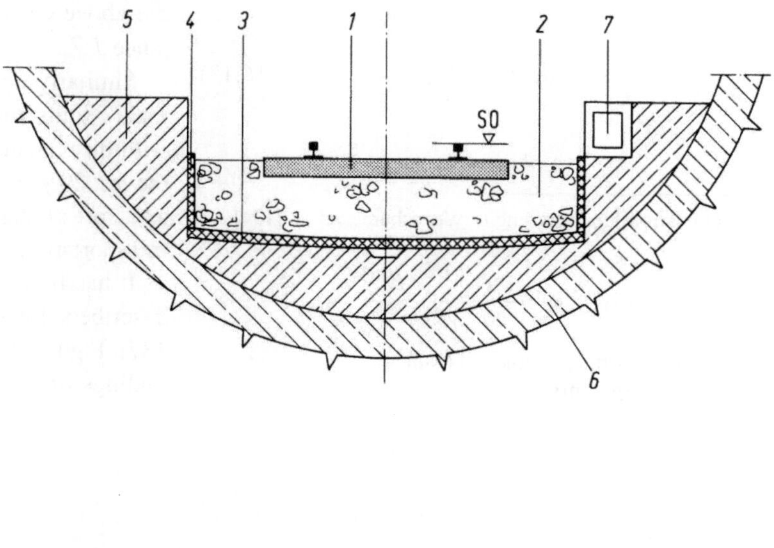

Fig. 16.64 Outline sketch of ballasted track construction with ballast mats based on the example of a single track light rail tunnel with a circular cross section. *1* track frame, *2* ballast bed, *3* ballast mat (BM), *4* side mats, *5* top concrete (escape route on side), *6* tunnel walls, 7 cable routing duct

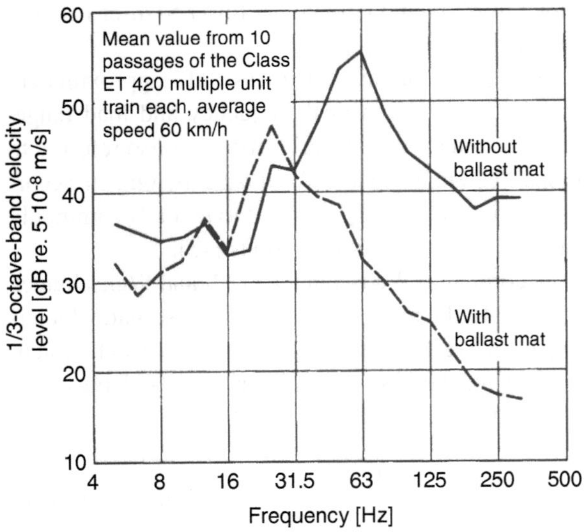

Fig. 16.65 Structure-borne noise on the tunnel wall of a single track light rail tunnel with a circular cross section during passage of Class ET 240 multiple unit trains on ballasted track (K 54 H) without and with ballast mats, Sylomer® B 851, with a static bedding modulus of $c = 0.02$ N/mm^3 and a dynamic stiffness of $s'' \approx 0.04$ N/mm^3 (characteristic values in the relevant load and frequency range)

which can be calculated from the effective load zone under the sleeper [134].

Equation (16.12) can be used as an approximation for the source impedance $1/Z_i$:

$$\frac{1}{Z_i} = \frac{j\omega}{s_S}\left[1 - \left(\frac{\omega_0}{\omega}\right)^2\right], \qquad (16.12)$$

in which s_s is the stiffness of the ballast and ω_0 is the natural radian frequency, which is primarily determined for usual track frame (for example, rails of the type S 49, S 54 or UIC 60) and a normal ballast bed in Eq. (16.13)[17]

$$\omega_0 \approx \sqrt{\frac{s_S}{m}}, \qquad (16.13)$$

by the ballast stiffness s_s and the unsprung wheel set mass m.

Using these equations, the insertion loss of a ballast mat is determined as follows:

$$\Delta L_e = 20 \cdot \lg\left|1 + \frac{s_S/s_M}{1 - \left(\frac{\omega_0}{\omega}\right)^2}\right| \text{dB}, \qquad (16.14)$$

In the upper section (a) of Fig. 16.67, the measured insertion loss for the ballast mats installed in the suburban rail tunnel in Munich is shown, whereas the lower section (b) illustrates the calculated insertion loss for this ballast mat, according to the above equation using two input impedances for the tunnel floor (final impedance Z_T).

At lower frequencies, ΔL_e is negative at first, reaching a minimum at

$$f_1 = f_0\sqrt{\frac{s_M}{s_S}} = 65\sqrt{\frac{55 \times 10^6}{5 \times 10^8}} \approx 22 \text{ Hz}, \qquad (16.15)$$

and turns positive at $\sqrt{2}f_1 = 32$ Hz, peaks at $f_0 = 65$ Hz and then declines to a constant value $20 \cdot \lg(1 + s_S/s_M) = 20$ dB.

The results of measurements and calculations no longer correspond at very low frequencies, due to the fact that some of the key conditions of the simplified calculation model are no longer fulfilled.

Figure 16.67 makes it clear that the way in which the final impedance Z_T (i.e. the input impedance of the tunnel floor) is modelled (in this case it is modelled as a highly stiff elastic half-space and as an infinite slab) has practically no influence on the results of the calculation, as long as this is large enough so that the final admittance $1/Z_T$ approaches 0 and can be ignored in the above equation compared with the source admittance $1/Z_i$.

Similarly good correspondence between measurement results and calculated results has been obtained with this calculation model in other applications over the last 20 years, leading this formula to be considered as a sort of standard by some European railways and transportation companies in tendering procedures.

It has been used successfully in the recent past as described, for example, in Wettschureck et al [136, 137]. Figures 16.68 and 16.69 present some of the key findings from these publications.

[17] Exact relationship according to Wettschureck et al. [134]:

$$\omega_0 = 1.7 \cdot \frac{(s_S/l)^{3/8} \cdot B^{1/8}}{M^{1/2}},$$

with B bending stiffness of the rail (N·m^2)
l reference length (m)

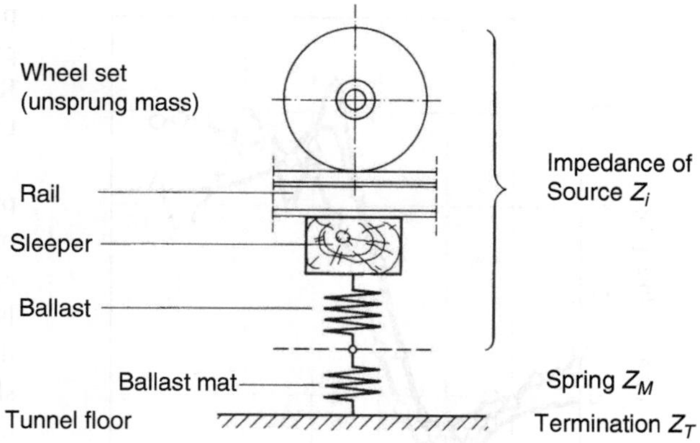

Fig. 16.66 Single degree of freedom (SDOF)model for calculation of the insertion loss of ballast mats in tunnels

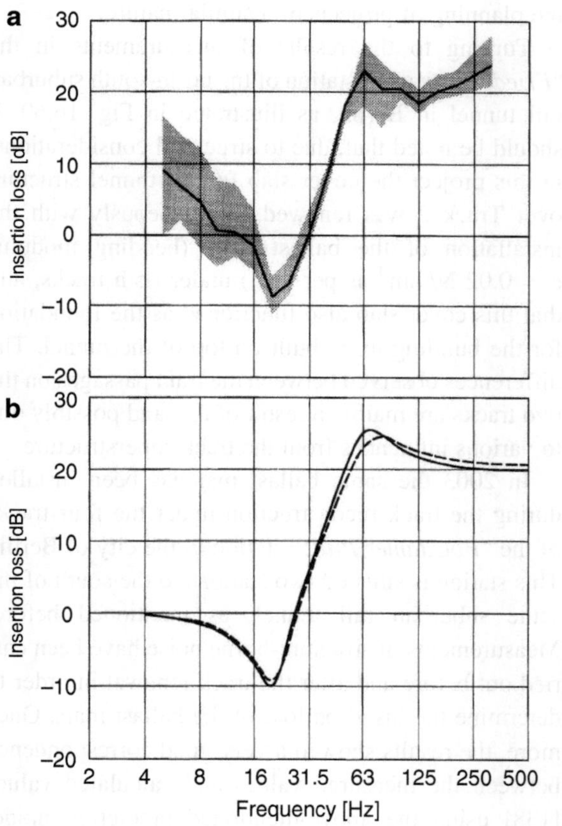

Fig. 16.67 Measured and calculated insertion loss of a ballast mat (BM) (**a**) measurement results for the ballast mat, type Sylomer® B 851, installed in the rapid transit railway tunnel in Munich, Germany. Mean value and scatter of measurements of passages of Class ET 420 multiple unit trains on six measurement points on the tunnel walls before and after installation of the ballast mats (**b**) calculated result for the same ballast mat with a stiffness of $s_M = 5.5 \times 10^7$ N/m·(1 + *j*0.2) installed under a ballast layer with a stiffness of $s_S = 5 \times 10^8$ N/m·(1 + *j*0.5) and an unsprung wheel set mass of $m = 3{,}000$ kg. Parameter: terminating impedance Z_T, ——— elastic semi-space with $s_T = 5 \times 10^9$ N/m, ------ 0.8 m thick concrete slab with $Z_T = 10^7$ Ns/m

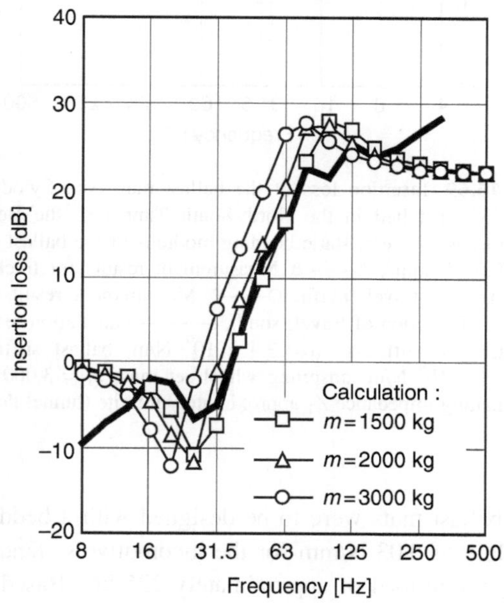

Fig. 16.68 Insertion loss of the ballast mats installed in the tunnel on the Cologne–Chorweiler line, Sylodyn® DN 335, with a static bedding modulus of $c = 0.03$ N/mm³. ——— measurement results: average value for train passages on track 1 and track 2. Calculated results for: ballast mat stiffness $s_M = 4.8 \times 10^7$ N/m, ballast stiffness $s_S = 5 \times 10^{-8}$ N/m, terminating impedance Z_T approximately infinite (tunnel floor), Parameter: unsprung wheel set mass *m*

The results presented in Fig. 16.68 are well suited for further discussion of a particular feature, which had to be taken into account in the Cologne–Chorweiler tunnel project.

The trains operated on the Rhine/Ruhr suburban rail system are so-called reversible "push-pull" trains with an electric locomotive (Class 143) and several passenger coaches. In accordance with the requirements related to suitability for use as per [45],

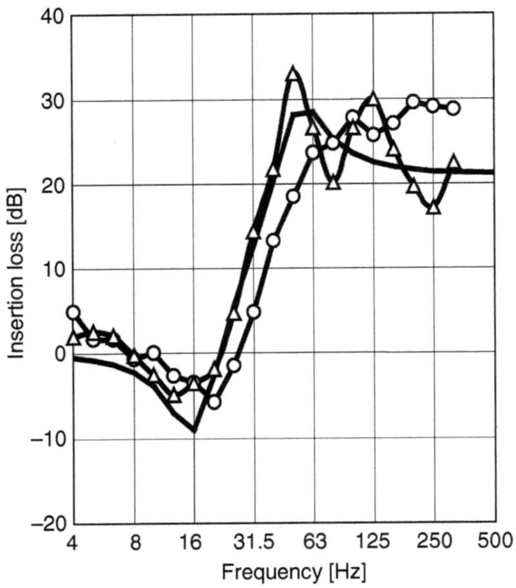

Fig. 16.69 Insertion loss of the ballast mat, type Sylodyn® CN 235, installed in the North-South Tunnel of the Berlin rapid transit system; static bedding modulus of the ballast mat $c = 0.02$ N/mm^3, Δ——Δ Measurement results for track 2: Direction of travel: north, ○——○ Measurement results for track 1: Direction of travel: south, ——— calculation result: Ballast mat stiffness $s_M = 3.4 \times 10^7$ N/m, ballast stiffness $s_S = 3 \times 10^8$ N/m, unsprung wheel set mass $m = 3{,}000$ kg, terminating impedance Z_T approximately infinite (tunnel floor)

the ballast mats were to be designed with a bedding module of 0.03 N/mm for the locomotive's standard gauge axle load of approximately 225 kN. Based on Fig. 16.68, however, it can be clearly seen that the calculated insertion loss corresponds quite well with the measured insertion loss, when the low unsprung wheel mass of 1,500 kg is used in the calculation. Experience has shown that this value is common for passenger coaches, whereas for locomotives and self-propelling trains (e.g. Class ET 420) an axle load of approximately 3,000 kg per axle can be expected. Consequently, the measurement results should be interpreted in such a manner that the effectiveness of the ballast mats – in respect of the impact of the dynamic, unsprung wheel set mass on the level of insertion loss [132] – is determined by the lighter passenger coaches and not by the heavy locomotive. The dominant role of the passenger coaches in terms of the insertion loss in this case is primarily due to the fact that when the measurements were evaluated, an average level was calculated for the entire train passage, which is determined roughly 87% by the passing coaches (cf. above a reversible train on the Rhine/Ruhr suburban rail system consisting of an electric locomotive and five passenger coaches).

If several coupled locomotives or an ET 420 self-propelling train with a higher unsprung wheel set mass of approximately 3,000 kg was used for the tests, experience with comparisons of measured and calculated insertion loss as per [132, 134] would indicate that the sharp increase in the measured curves would shift to lower frequencies and would thus correspond to the curve calculated for 3,000 kg. This note should help in the interpretation of these results with regard to the planning of projects of a similar nature.

Turning to the results of measurements in the "*Friedrichstrasse*" station of the north-south suburban rail tunnel in Berlin, as illustrated in Fig. 16.69, it should be noted that, due to structural considerations, in this project the cover slab for the tunnel structure over Track 2 was renewed simultaneously with the installation of the ballast mats (bedding modulus $c = 0.02$ N/mm^3 as per [45]) under both tracks, and that this cover slab also functioned as the foundation for the building to be built on top of the tunnel. The differences observed between the train passages on the two tracks are mainly a result of this and possibly due to various influences from the track superstructure.

In 2003 the same ballast mat has been installed during the track reconstruction under the four tracks of the "*Potsdamer Platz*" station in the city of Berlin. This station is situated two stations to the south of the same suburban rail tunnel, as mentioned before. Measurements of structure-borne noise have been carried out before and after the track renewal in order to determine the insertion loss of the ballast mats. Once more, the results showed a very good correspondence between the measured values and calculated values [138] using the above-mentioned prediction model according to Wettschureck [132].

With regard to the long-term effectiveness of ballast mats, measurements of structure-borne noise caused by the passage of urban rail trains were carried out in a suburban rail tunnel in Munich near the philharmonic concert hall "*Am Gasteig*" some 18 years after installation of Sylomer® B 851 ballast mats at the "historical" measurement locations [139]. No deterioration in the effectiveness of the ballast mats was found even after this long period of use with an extremely high operational load of 760×10^6 load

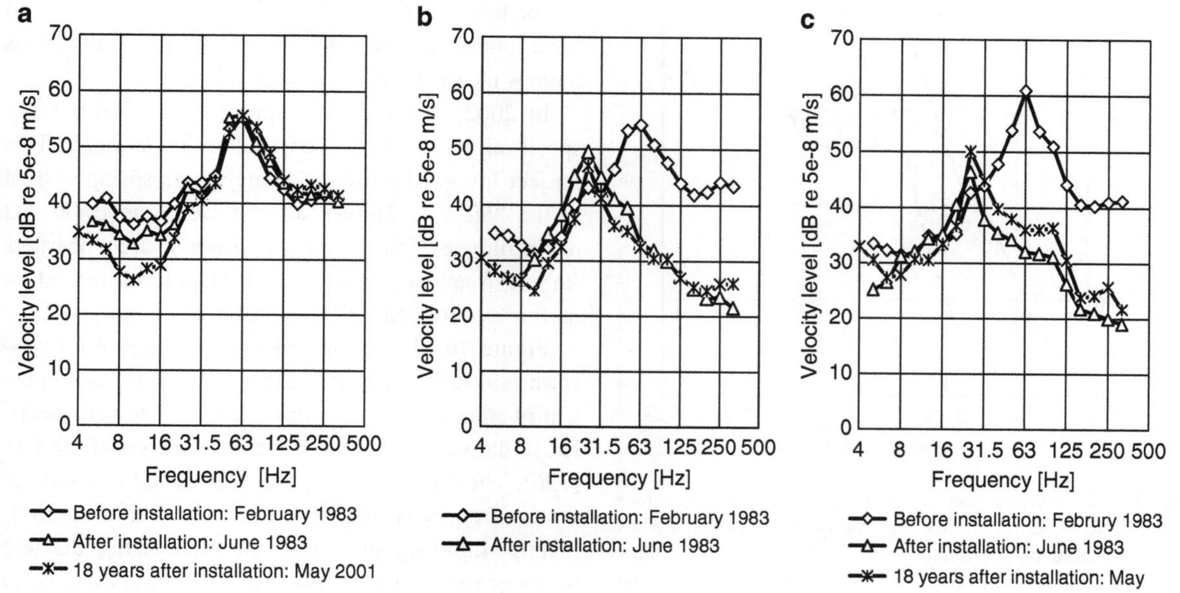

Fig. 16.70 1/3-octave-band spectra of vibration velocity levels measured at the tunnel wall of the rapid transit railway tunnel near the Philharmonic Hall "Am Gasteig" in Munich during passages of trains, Class ET 420, at a speed of 60 km/h, before and after installation of the ballast mat, type Sylomer® B 851. (**a**) Measurement point *outside* the ballast mat area in the south tube of the tunnel (**b**) Measurement point *within* the ballast mat area in the north tube of the tunnel (**c**) Measurement point *within* the ballast mat area in the south tube of the tunnel

tons (see Fig. 16.70). Based on the results of studies performed on samples of the ballast mat on a testing stand, the conclusion is reached in Wettschureck et al. [139] that the differences observable in Fig. 16.70 as compared with the measurement results immediately following installation cannot be attributed to changes in the elastic properties of the ballast mat (e.g. stiffening, ageing).

Figure 16.71 shows the principle of a "mass-spring system" constructed as a concrete trough superstructure [99].

In this case a concrete slab (pre-fabricated components or in situ slab), on which the track frame is either situated on ballast or cast into lean-mixed concrete,[18] is resiliently mounted onto the tunnel floor using elastomer bearings (discrete bearings or strip bearings) [100, 130].

Using track superstructures of this design, it is possible to achieve tuning frequencies of 10 Hz and lower, depending on the requirements and the construction design parameters.

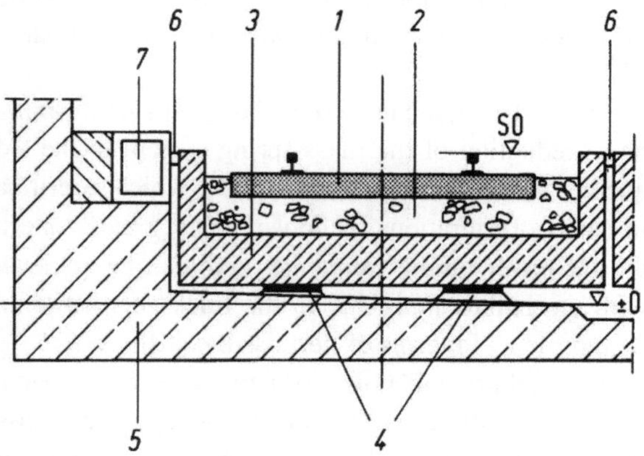

Fig. 16.71 Outline sketch of track superstructure designed as a "mass-spring system" with trough track design, based on the example of a dual-track light rail tunnel with rectangular cross-section. 1 track frame, 2 ballast or filler concrete, 3 concrete trough, 4 resilient bearing with bearing mount, 5 tunnel floor, 6 elastic seam seal, 7 cable routing duct

Figure 16.72 presents the insertion loss of the "mass-spring system" installed in the Frankfurt airport tunnel on ballastless track, the tuning frequency of which is slightly lower than 12 Hz [15].

For the purposes of comparison, the insertion loss of the ballast mats installed in the suburban rail tunnel in Munich is also depicted, the static spring stiffness of

[18] For ballastless "mass-spring systems" the construction height is generally considerably lower as compared to the outline sketch in Fig. 16.71.

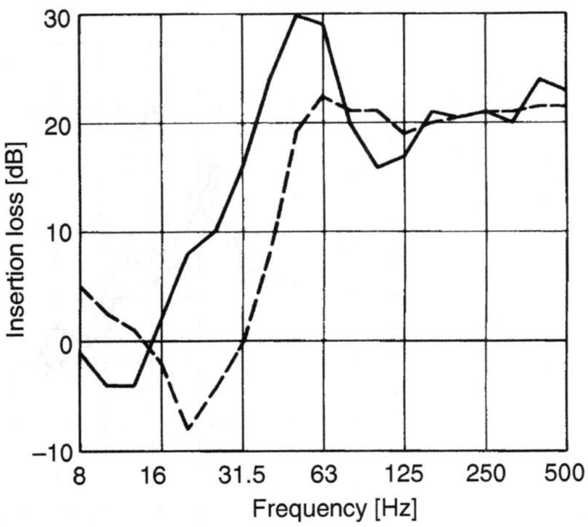

Fig. 16.72 Insertion loss of a ballast mat and a track superstructure designed as a "mass-spring system" without ballast. each measured on the tunnel wall during passage of a Class ET 420 multiple unit train - - - - - - - Ballast mat: rapid transit railway tunnel in Munich, Germany, ——— Mass-spring system: Airport tunnel Frankfurt/Main, Germany

which (as noted above) is the lowest value allowed for suburban rail applications ($c = 0.02$ N/mm as per [45]), resulting in a tuning frequency of approximately 22 Hz.

In Fig. 16.72, it can be seen that the structure-borne noise reduction of the mass-spring system in the frequency range of roughly 20–63 Hz, which is critical in terms of vibrations, is considerably greater due to the far lower tuning frequency than can be achieved with the ballast mats due to the track superstructure requirements in terms of permissible stiffness.

This "additional" noise reduction can be ascribed to the substantially lower tuning frequency of the mass-spring system. Nevertheless, as Fig. 16.72 shows, as the insertion loss of both superstructures is roughly equal in higher frequency ranges (meaning that the ratio of ballast stiffness to ballast mat stiffness must be equivalent to the ratio of ballast stiffness to the stiffness of the elastomer bearings used in the mass-spring system), the lower tuning frequency must be attributed exclusively to the far higher mass per metre of the mass-spring system. A result similar to that depicted in Fig. 16.72 can also be found in [129] for similar studies in a subway tunnel.

More recent results on the effectiveness of generally ballastless mass-spring system with discrete or full-surface bearings designed for use on standard gauge lines can be found, for example, in Wettschureck et al. [136] for a tunnel in Germany and in [140–143] for various tunnels in Austria and Switzerland.

In 2002, the first mass-spring system on a high-speed railway line was constructed in Germany. This project involved a low-frequency mass-spring system in the "*Siegaue Tunnel*" on the Deutsche Bahn AG new Cologne–Rhine/Main line using discrete bearings (tuning frequency $f_0 \approx 10$ Hz). Detailed information on this system can be found in Enoekl et al. [144].

Figure 16.73 presents the effectiveness of a mass-spring system installed in a tunnel in Cologne–Chorweiler and of two ballast mats, which were installed in the same tunnel in different sections of the line [136]. The mass-spring system was tuned to a natural system frequency of 11 Hz, with a weight of approximately 4,000 kg per metre of track using discrete bearings made of PUR elastomer (bearings spaced at intervals of 1.5 m along the track axis). The two ballast mats have a static bedding modulus of $c = 0.03$ N/mm in accordance with the requirements for axle load and train speed as per [45]. The mats, however, also have varying dynamic stiffness, in accordance with the various noise reduction requirements in the sections where they were installed.

Figure 16.73 clearly shows that the level of structure-borne noise reduction achieved from ballast mats (1) to the "dynamically softer" ballast mat (2) to the relatively expensive low-frequency mass-spring system exhibits a clearly discernable increase, in accordance with the project planning requirements.

Rubi et al. [145] provides a report on two different designs of a full-surface bearing mass-spring system[19] also intended for use on a standard gauge line which was installed in the Zurich suburban rail system. A detailed discussion on this subject can also be found in the closing report of the RENVIB project [142].

Figure 16.74 depicts cross sections of these designs, whereas Fig. 16.75 shows the levels of measured insertion loss.

Systems of this type, which generally have a tuning frequency of roughly 20 Hz, are designed for applications in which the requirements in terms of the

[19] Referred to as "Light Mass Spring Systems" (LMSS) for trams and urban railways. The corresponding terms in German and French are "Leichtes Masse-Feder-System" and "Dalle flottante", respectively.

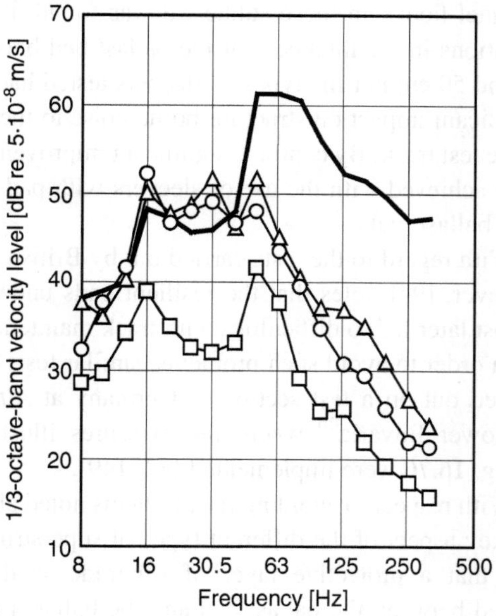

Fig. 16.73 1/3-octave-band spectra of vibration velocity levels, measured on the tunnel wall of a tunnel on the Cologne–Chorweiler line during passage of the Rhine/Ruhr light rail trains, before and after installation of a ballastless mass-spring system and two ballast mats (BM) 1 and 2 with identical static bedding modulus $c = 0.03$ N/mm, but with different dynamic stiffness. - - - - - - - ballasted track *prior to BM installation*: mean value of 24 measurement points on both tracks, *Following BM installation*: mean value of six measurement points per reconstructed section. Δ——Δ Ballasted track with BM 1, type Sylodyn® DN 325, o——o Ballasted track with BM 2, type Sylodyn® DN 335 (dynamically softer), □——□ Ballastless mass-spring system with discrete bearings, type Sylodyn® N 70690/50, tuning frequency $f_0 = 11$ Hz

amount of insertion loss to be achieved are not particular high compared with low-frequency systems with tuning frequencies of $f_0 \leq 10$ Hz.

Up to now the lowest tuning frequencies of approximately 7.5 Hz have been achieved using discrete bearings made of cellular PUR elastomers (see also Zammer Tunnel, Austria Federal Railways in [141, 142]). It should be noted, however, that in the case above this was achieved with a hefty slab weight of 10,500 kg/m of track.

As a part of the new high-speed rail link between Seoul – Pusan in South Korea, a tuning frequency of 5 Hz is projected for the area around the Chonan station, where the concrete frame construction is to be used to house shops and offices. The mass-spring system planned for the station uses steel springs and buffer elements. Based on their design (raised track),

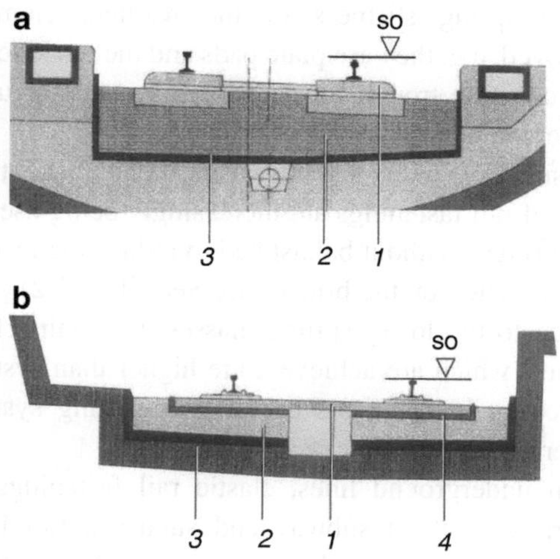

Fig. 16.74 Cross-sections of two designs of "light mass-spring systems" (LMSS) with full surface bearing in use on the Zurich rapid transit system **a** vicinity "First Church", **b** vicinity "Shopville". *1* Bi-block sleeper (a) or plastic sleeper (b), *2* reinforced concrete slab, *3* elastic bottom and side mats made of cellular PUR materials, *4* elastic sleeper plate

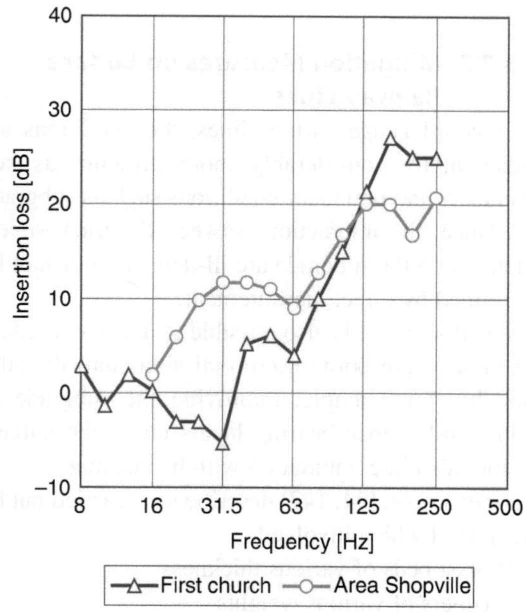

Fig. 16.75 Measured insertion loss of two types of "light mass-spring systems" in a light rail tunnel in Zurich

one can essentially view these systems as heavy bridges mounted on springs [97, 146].

Highly elastic rail fastenings can also be used to achieve a notable reduction in levels of structure-borne noise. This, however, depends on a great degree to the spring stiffness of the resilient elements employed, e.g. the base plate pads and the track design (e.g. concrete trough for slab track systems or using sleepers in a ballasted superstructure).

Aside from a few test lines using "slab track", these kinds of rail fastenings are increasingly being used on steel bridges without ballast bed to reduce the level of noise radiated by the bridge (see Sect. 16.2.2.2).

Due to the lower sprung masses, the tuning frequencies, which are achieved, are higher than systems employing ballast mats or for mass-spring systems (generally in the range $\geq$30 Hz).

On underground lines, elastic rail fastenings are generally used on subway and suburban rail lines. In such applications, it has proven possible to achieve a substantial improvement in the frequency range above 50 Hz using special rail fastenings on sleepers in ballasted track, as compared to normal ballasted track [129], i.e. this approach is suitable for resolving problems stemming from reradiated sound, but not for resolving vibration problems.

16.3.7.2 Mitigation Measures on Surface Railway Lines

On standard gauge surface lines, the conditions are fundamentally considerably more difficult, as key parameters and boundary conditions such as subgrade impedance, the interaction between the track superstructure and the subgrade are ill-defined or cannot be determined by direct measurement.

Nevertheless, it is also possible to achieve a reduction in structure-borne noise emissions into the subgrade by, for example, improving the subgrade or adding further load-bearing layers under the ballast, possibly also in combination with ballast mats.

For example, [94, 147] describe tests carried out by British Rail which involved

(a) Ballast beds of various thickness
(b) Sleepers of various weights
(c) Resilient rail pads under the sleepers (sleeper pads)
(d) Resilient mats (now referred to as ballast mats) between the ballast bed and a 50 mm thick layer of sand on the subgrade

In the 1980s, the Swiss Railways (SBB) also carried out similar tests with elastic mats on a gravel subgrade or concrete slab [148] but in this particular case involving tunnel without a bottom (without a tunnel floor). In these studies, it was found neither variations in the thickness of the ballast bed between 25 and 50 cm nor the types of sleepers tested had any significant impact on structure-borne noise to the side of the test track. By contrast, significant improvements were achieved with the use of sleepers with pads and with ballast mats.

With regard to the tests carried out by British Rail, however, [94] notes that the resilient pads under the ballast later led to difficulties with track maintenance.

In order to avoid such problems, similar tests were carried out on a test section in Germany at *Altheim* in Lower Bavaria[20] where the measures illustrated in Fig. 16.76 were implemented [32, 149].

With respect to avoiding the problems noted above, the key aspect of the different types of superstructure was that a protective layer of subgrade sand was placed between the ballast bed and the ballast mat to prevent damage to the mats during maintenance of the superstructure (tamping, cleaning of the ballast, etc.).

These tests were carried out on a line section in which a so-called subgrade protection layer was built in due to the very poor track geometry when the superstructure was renewed [replacement of the track frame and ballast, see Fig. 16.76, Part (a)].

In addition, a cement-bonded gravel support layer was built with and without ballast mats under a 100-m stretch of track in the load-bearing zone to reduce the structure-borne noise transmitted into the subgrade [see Part (b) and (c) of Fig. 16.76]. The ballast mats, made of PUR elastomer, featured a dynamic stiffness of approximately 0.09 N/mm in the frequency range under investigation, and complied with the speed and load (axle load) limitations set forth in [45], featuring a "bedding modulus" of 0.06 N/mm^3.

Measurements of structure-borne noise performed prior to and 1 year after installation at identical measurement locations 8 m to the side of the track for the passage of a test train (S-Bahn, Class ET 420) initially provided the results shown in Fig. 16.77 for superstructure renewal with the installation of a subgrade protection layer.

[20] The test track was located on a section at an abandoned station, i.e. even after removal of the platforms the ballast bed still did not have "open" shoulders, but was rather "confined". As a result, the superstructure cross section set forth in Fig. 16.76 is different from the outline sketch presented in [260], Fig. 16.67, in accordance with the field conditions.

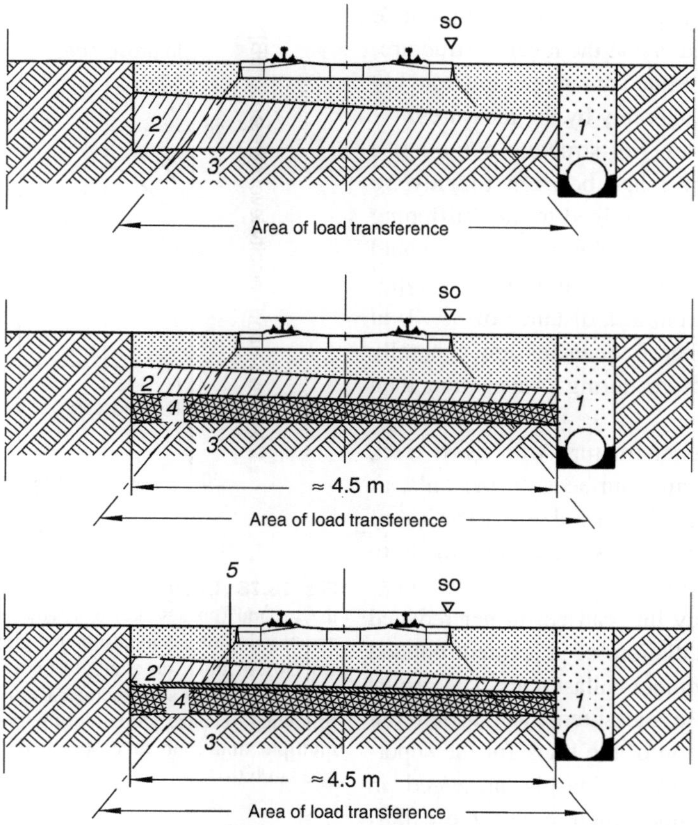

Fig. 16.76 Outline sketch on the fundamental design of various measures for the mitigation of structure-borne noise adjacent to surface railways with deep drainage (1). **a** ballasted superstructure with protective subgrade layer (2) on compacted subgrade (3), **b** ballasted superstructure with substructure (2) and cement-bonded gravel support layer (4) on compacted subgrade (3), **c** ballasted superstructure with substructure (2), ballast mat, type Sylomer® D 220 (5) and cement-bonded gravel support layer (4) on compacted subgrade (3)

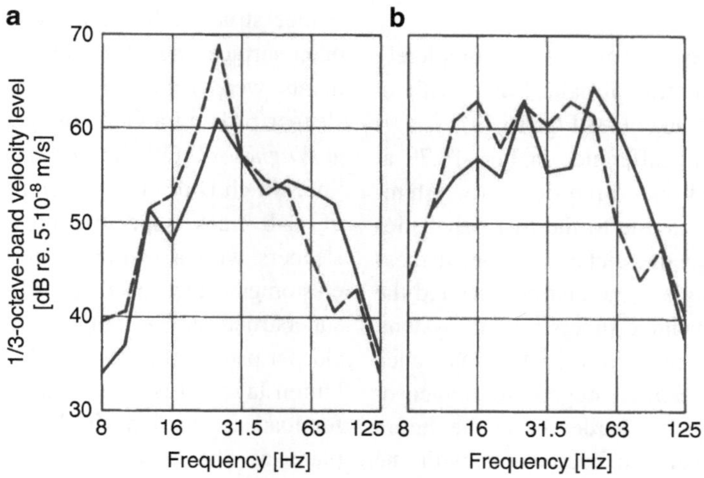

Fig. 16.77 Structure-borne noise in the ground (on a stake) at 8 m to the side of the track axis during passage of a Class ET 420 multiple unit train prior to and 1 year following superstructure renewal and construction of a substructure. ------- prior to reconstruction, ballasted superstructure K49 B58, ——— following reconstruction, ballasted superstructure W54 B70, train speed v: **a** $v = 60$ km/h, **b** $v = 120$ km/h

In terms of practical application, one positive note that should be mentioned is that the level of structure-borne noise in the low-frequency range below roughly 50 Hz was reduced considerably by the measures.

The increase in the level of structure-borne noise in the higher frequency range between roughly 50 and 100 Hz, which can be ascribed to the "stiffening of the subgrade" caused by the measures, would not have a particularly negative impact on a hypothetical immission location at a distance of 30–40 m from the track, as higher frequencies are generally damped considerably more during propagation than low frequencies are.

If one assumes a superstructure with good track geometry and good running surfaces on the rails, as was the case following track renewal as per Part (a) of Fig. 16.76, then it is also interesting to investigate to what extent the emissions of structure-borne noise emitted from the railway line can be further reduced by the additional installation of a cement-bonded gravel layer with and without ballast mats. The results of structure-borne noise measurements carried out 1 year after reconstruction of the track for the superstructure without the ballast mat is presented in Fig. 16.78 and for the superstructure with the ballast mats in Fig. 16.79.

As is evident, the levels of structure-borne noise at very low frequencies have declined even further, whereas for the test line without ballast mats the levels in the frequency range which is critical for the perception of reradiated sound increased by up to 8 dB to the side of the track.

This increase, however, was almost completely neutralised by the installation of ballast mats with an effect in this frequency range (see Fig. 16.79).

The drop in the level difference in Fig. 16.79 at 40 Hz must be accepted due to physical and system-related conditions. This is due to the fact that it lies within the resonant frequency dictated by the stiffness of the superstructure (ballast mat and ballast) and the dynamic mass of the vehicle/superstructure system. In principle, however, this drop in the resonance could be shifted to a limited degree to higher or lower frequencies, by using harder or in the future possibly even softer mats, in accordance with the specific requirements.

More recent, unpublished studies have shown that the thickness of the cement-bonded support layer or a concrete slab has a significant influence on the effectiveness of such measures. For practical purposes, the cement-bonded support layer should be at least 0.6 m thick, while a concrete support layer (without reinforcement) should be at least 0.4 m thick.

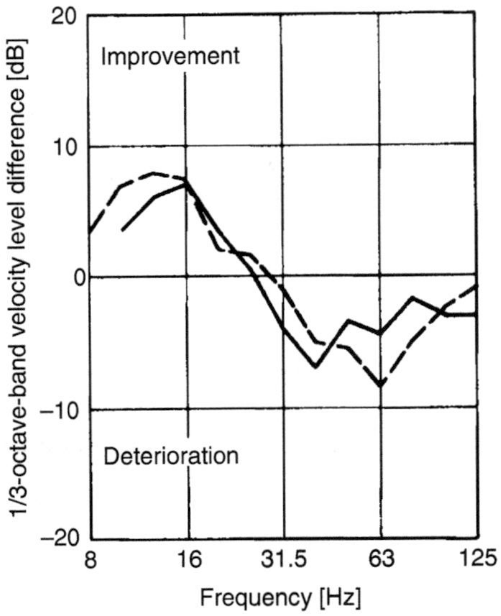

Fig. 16.78 Difference in the 1/3 octave band velocity level in the ground (on a stake) 8 m adjacent to track axis following construction of a cement-bonded gravel sublayer (lean concrete B5, 30 cm thick on average) under a ballasted superstructure (W54 B70) with substructure. Results of measurements 1 year following reconstruction during passage of a Class ET 420 multiple unit train at a speed of 60 km/h (- - - - - - -) and 120 km/h (———)

In the meantime, results of other measures to reduce structure-borne noise have been obtained for open surface lines. For example, significant improvements were achieved through the use of resilient sleeper pads on a German Federal Railways test line at *Waghäusel* [69]. On one special section of this test line, which is primarily designed to test various types of slab track superstructures [150], concrete B70 sleepers with a resilient sleeper pad made of PUR elastomer were installed in a W60 B70 ballasted superstructure. The static stiffness of the 9-mm-thick sleeper pads (6 mm resilient layer + 3 mm load distribution layer) was 0.08 N/mm^3. This was determined for load application via the so-called normal ballast plate as the secant modulus in the load range 0.02–0.16 N/mm as per [255].

Typical results from measurements of structure-borne noise at a measurement location 8 m adjacent to a test track are shown in Fig. 16.80.

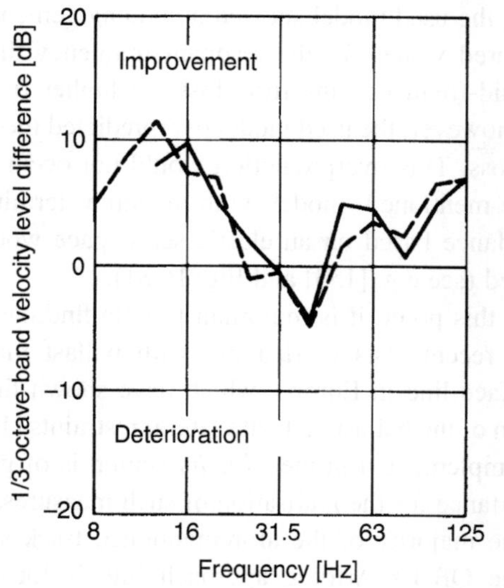

Fig. 16.79 Difference in the 1/3-octave-band velocity level, measured in the ground (on a stake) 8 m adjacent to track axis following construction of a cement-bonded gravel sublayer (lean concrete B5, 30 cm thick on average) *and* ballast mat, type Sylomer® D 220 (dynamic stiffness $s'' \approx 0.09$ N/mm³), under a ballasted superstructure (W54 B70) with substructure. Results of measurements 1 year following reconstruction during passage of a Class ET 420 multiple unit train at a speed of 60 km/h (-------) and 120 km/h (———)

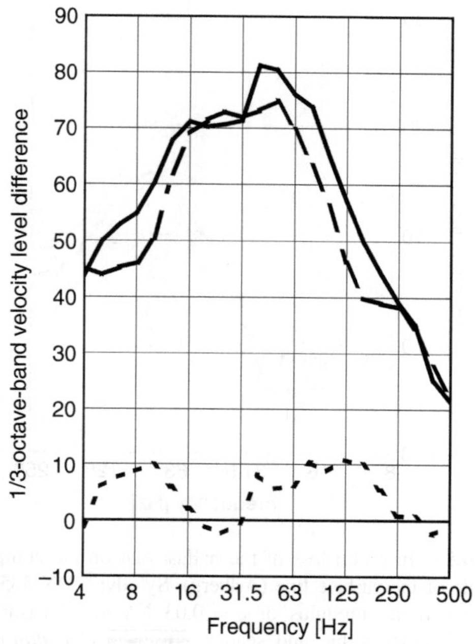

Fig. 16.80 Structure-borne noise in the ground (on a stake) at 8 m to the side of the test track with elastic sleeper pads during passage of an ICE 1 train at a speed of $v = 160$ km/h. ——— ballasted superstructure W60 B70 (1), — — — — ballasted superstructure W60 B70 with elastic sleeper pads (2), ------- velocity level difference (1)–(2): improvement due to the elastic sleeper pads

It can be seen that the structure-borne noise was reduced by up to 10 dB by the "padded sleepers" in certain frequency ranges. As is generally true for measures involving the track superstructure, the reduction of structure-borne noise that can be achieved with this measure depends on the permissible elasticity, in the case under discussion the stiffness of the sleeper pads (i.e. on the permissible level of rail subsidence which is defined by the total permissible subsidence of the rails under the wheels) *and* by technical superstructure parameters such as the thickness of the ballast bed, existence and compaction level of the subgrade protection layer.

Müller-Boruttau et al. [151] also discuss the installation of padded sleepers on a high-speed line operated by the Deutsche Bahn AG and the relevant aspects of superstructure design and construction. Kopp [152] presents the positive results of tests to reduce the formation of rail corrugation in relation to a test line operated by the Austrian Federal Railways (ÖBB), which was constructed to study the impact of padded sleepers on the formation of short waves on the rail surface.

There are also more recent results available in relation to the installation of ballast mats in surface lines as well as the results on the effectiveness of a particular project by the ÖBB. This particular application involved the installation of ballast mats on compacted subgrade in a 450-m-long line section to protect a nearby residential area, which was performed as a part of adding a second track on the Arlberg approach [153, 154]. The compaction level of the subgrade, described numerically by the E_{v2} modulus as per [254], which is commonly used in the construction of permanent way, was $E_{v2} = 180$ MN/m in this case.

Figure 16.81 illustrates the measured efficiency of the mitigation measure.

Figure 16.81 also shows the calculated level of insertion loss for the installation project as per [134]. In this case, the input impedance of the subgrade was used in the above Eq. (16.10) for calculation of the insertion loss for the terminating impedance $1/Z_T$, as this was indeed significant in comparison with

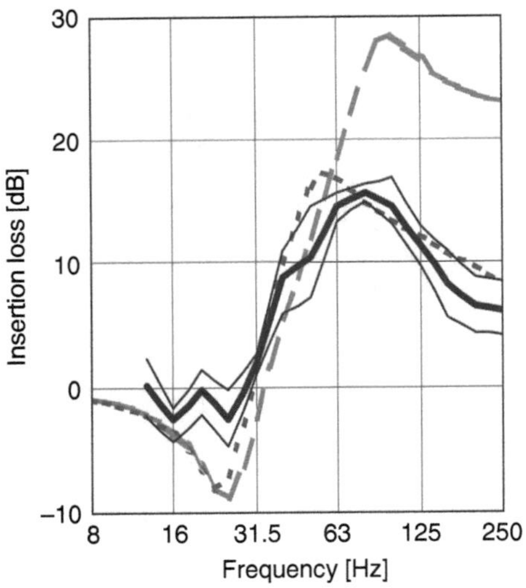

Fig. 16.81 Insertion loss of the ballast mat on the compacted subgrade of the surface line (Arlberg), Sylodyn® DN 335, with a static bedding modulus of $c = 0.03$ N/mm³. Measurement results: Average value and scatter. ▬▬ *Calculation* for: ballast mat stiffness $s_M = 3.8 \times 10^7$ N/m, ballast stiffness $s_S = 4.7 \times 10^8$ N/m, unsprung wheel set mass $m = 1{,}600$ kg, ------ terminating impedance Z_T finite (compaction of subgrade as per [254]: $E_{v2} = 180$ MN/m²), — — — — terminating impedance Z_T approximately infinite (e.g. tunnel floor)

installation conditions in a tunnel setting. This was modelled as an elastic semi-space, as per [101, 107] for example, whereby the (dynamic) subgrade stiffness was estimated based on the E_{v2} modulus noted above.

For the purposes of comparison, Fig. 16.81 also shows the insertion loss that would be expected for installation of the same mat in a tunnel with a very high input impedance Z_T for the tunnel floor [i.e. a minimal final admittance $1/Z_T$, which can be ignored compared to the source admittance $1/Z_i$ in the above formula Eq. (16.10) for calculating insertion loss]. From this diagram, one can see that the impact of the finitely stiff, self "cushioning" subgrade is noticeable in terms of a decline in the insertion loss as frequency increases.

More recent experience with respect to the installation and effectiveness of ballast mats on surface lines in the United States of America (USA) is presented in [155]. The authors used a modified prediction procedure based on the model of Wettschureck and Kurze [134], but with a finite termination impedance based on a flat plate model. The described prediction procedure was tested against measurements on at-grade installations on light rail transit and commuter railway installations in Baltimore and Boston, USA. In both cases, the used model showed a good agreement with measured values for the resonant frequency dip and the mid-frequency insertion loss. At higher frequencies, however, the used model overpredicted the insertion loss. This overprediction would not occur if the above-mentioned model with a finite termination impedance based on an elastic semi-space would be applied (see e.g. [153] and Fig. 16.81).

At this point, it is important to note findings from more recent tests carried out with ballast mats on a surface line in Europe, which have shown that the design of the ballast bed with lateral restraints [149] as was implemented in the *Altheim* station is of general importance for the realisation of such measures.

The majority of the aforementioned track section on the ÖBB's Arlberg approach line is located at a station where the ballast bed is bordered by the platforms along the track, similar to the solution employed in the *Altheim* station, Germany. At one of the end of the track section, which was outside of the area with lateral restraints from the platforms, the observation was made that the free ballast shoulder had a tendency to drift away laterally.

In relation to the installation of ballast mats on surface lines (on embankments), the Swiss Federal Railways (SBB) also found that the ballast tended to "flow away" to the sides, unless measures were taken to shore it up. In order to prevent this phenomenon, i.e. to stabilise the ballast bed, the SBB installed vertical "concrete slats" held by mini-posts and "glued the edges of the ballast bed" [97, 156]. After 3 years of operation, these measures to laterally support the ballast have proven effective.

Another design option for ballast mats on surface lines is to install a low concrete trough on a compacted subgrade, into which the mats are placed and which laterally holds and stabilises the ballast bed [157]. The bottom plate of the concrete trough also provide a considerable level of final impedance, so that the full effect of the dynamic characteristics of the ballast mats is obtained, similar to the situation when ballast mats are installed on a tunnel floor.

A detailed presentation of the measures tested up until 1997 for reducing the emission of structure-borne noise from standard-gauge surface railway lines was presented within the framework of the project RENVIB II, mentioned in Sect. 16.3.2 [97]. By way of comparison, Fig. 16.82, which is taken from the aforementioned project, shows levels of measured

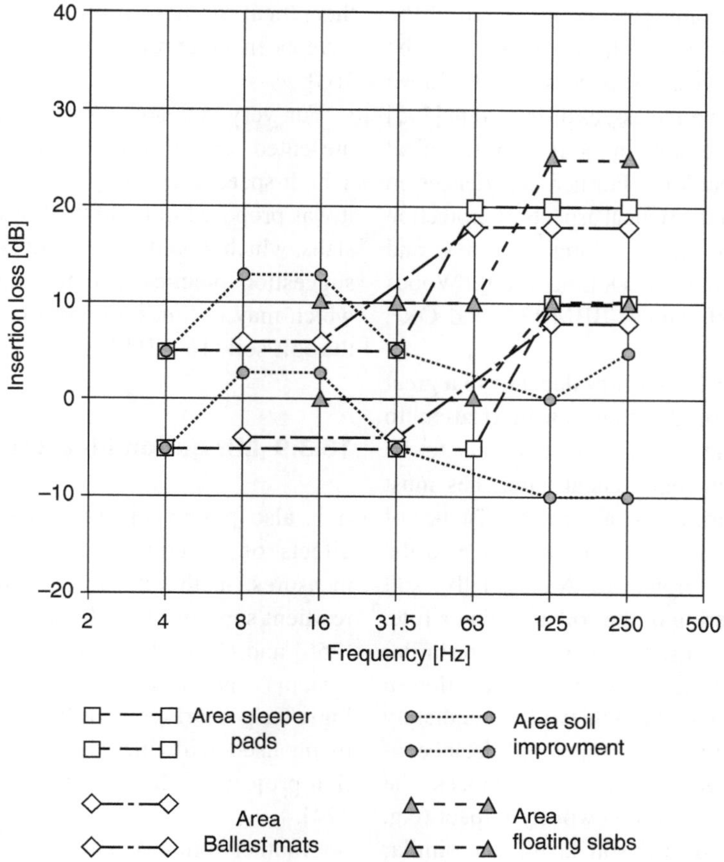

Fig. 16.82 Areas of measured insertion loss for measures to mitigate structure-borne noise on open standard gauge railway lines as per [RENVIB [97]]

insertion loss for various measures for the reduction of structure-borne noise and vibrations implemented on surface railway lines.

16.3.8 Mitigation Measures in the Area of Structure-Borne Noise Propagation in the Ground

There are essentially two physically different principles for achieving additional reduction in structure-borne noise, above and beyond the distance-dependent reduction, in the path of propagation to the sides of railway tracks: On the one hand, reduction in the structure-borne noise by absorbing vibration energy using suitable absorbers, and on the other hand, reflection or isolation of the structure-borne noise at discontinuities (impedance jumps) in the path of propagation, in the form of layers with a higher specific mass or with a lower dynamic stiffness as compared with the surrounding soil.

Tests with vibration absorbers (concrete blocks weighing approximately 1,200 kg per metre track in various arrangements), studied with a scale model and at a scale of 1:1 on a track section, as reported by [158], resulted in no significant success and are only mentioned here for the sake of completeness.

In terms of practical applications, measures based on the principle of isolation are more promising. A number of studies have been carried out in this field. These basically can be divided into the following categories:

- Heavy shielding walls, made of concrete for example
- Vertical trenches filled with elastic (or gas-filled) mats or open ditches

Shielding walls composed of rows of bore holes represent a sort of transitional solution [159]. Under favourable conditions, namely when the depth of the bore holes is more than twice the wavelength of the undisturbed surface waves, it was possible to achieve significant shielding; nevertheless, as discussed in [105] such measures are not feasible in the light of technical and economic considerations. Further experiences in relation to theoretical and practical aspects of protection against structure-borne noise using trenches and ditches are also presented in Dolling [160], Woods [161], Massarsch [162], Müller-BBM [32] and ORE D 151 [158].

The efficiency of such measures depends to a great degree on the depth of the trenches in relation to the wavelength of the oscillations that are to be shielded (depending on requirements, trenches must be 10–15 m deep in some cases) and on the distance of the trenches from the tracks and the distance to the object which is to be protected. Additionally, soil conditions and the layering of the soil also play a role.

Most of the experiments, however, come to similar conclusions in that at times a significant reduction in structure-borne noise can be achieved immediately behind the shielding measures (up to a distance of roughly 8 m), whereas at greater distances the measures seem to have little noteworthy impact (e.g. due to reflection from deeper soil layers and similar phenomenon, cf. [15, 32, 162]). In terms of their practical application, it can be stated that trenching requires very careful design, taking into account the respective geological conditions, track parameters and other boundary conditions pertaining to the objects involved.

Accordingly, no specific details and concrete results from the experiments that have been performed are included here.

Based on the current state of knowledge, one should however note that following consideration of the technical and economic aspects, use of trenches is most effective in terms of shielding if they are situated as close as possible to the object to be isolated and are designed with dimensions that take adequate account of the prevailing conditions.

In the relevant literature on this subject, one also finds reference to proposals to install concrete blocks under the track or under the foundations of the structures affected. The efficiency of these so-called "wave impedance blocks" (WIBs), however, is presented in a far more optimistic fashion in the theoretical studies, compared with the results that have been determined in practical applications [97, 163].

For very soft soil conditions, which for example presented difficulties in the new construction of a high-speed line along the west coast of Sweden, it was proposed to build the track on concrete bearing slabs, which would then be supported by pilings. This suggestion practically represents a form of bridge, which may at times not even come into contact with the soft soil [164–168].

16.3.9 Mitigation Measures at Buildings

It is also possible to protect structures against the effects of structure-borne noise with construction measures. In this respect, the main focus is on the resilient support of the buildings (cf. e.g. [169–172], [256] and Chap. 22 in this handbook). Such bearing systems are typically designed as discrete bearings, strip bearings or full surface bearings and are being used with increasing frequency in construction projects in the vicinity of railway lines [7, 173, 174].

Another option is to reduce the amount of structure-borne noise that is transmitted into the building by selecting forms and sizes of foundations and components that help prevent the transmission of structure-borne noise. In this respect, special attention should be paid to ceiling and floor structure components. For example, it is important to pay ensure that – if possible – the natural frequency of ceilings and floating floors does not lie within the range of the maximum of the spectrum of structure-borne noise excitation from the rail traffic [112].

Of course, it is possible to install resilient support systems in existing buildings, if they are suitable. Nevertheless, the related costs for this type of solution are generally quite considerable.

16.3.10 Prediction of Structure-Borne Noise Immissions

In contrast to the prognosis of immissions of airborne noise [2, 3], there is still no generally accepted method available for forecasting immissions of structure-borne noise and vibrations. One of the main reasons for this is that no generally applicable assumptions could be determined for the underlying fundamental parameters, which are necessary to be able to generate such forecasts (e.g. conditions of propagation in soil, transmission of structure-borne noise in buildings, etc.). Indeed, in order to achieve an adequate degree of certainty, such parameters generally have to be determined by carrying out measurements conducted at the immission site.

One commonplace method frequently used by the engineering offices active in this field of work to forecast vibrations stemming from rail traffic is an empirical calculation procedure. This procedure combines the existing boundary conditions, which ideally are determined for the specific application in question based on measurements, with parameters that have been found in a number of fundamental studies.

In the past, a number of attempts have been made to formulate a mathematical description of certain aspects of vibration prognoses [33]. Nevertheless, the reliability of such mathematical models is extremely dependent on knowledge of the parameters used in the calculations.

In the following, certain parameters are described, the knowledge or determination of which has a crucial impact on the accuracy of the vibration forecasts both in terms of the application of empirical methods or mathematical models.

16.3.10.1 Prediction in the Area of Structure-Borne Noise Emission

In the field of *emissions of structure-borne noise and vibrations*, the parameters described in Sect. 16.3.2 are of key importance. More recent investigations [175], however, have shown that the structure of the track superstructure (substructure, anti-frost layer and earthen foundation) and the connection of such to the surrounding subgrade are decisive in terms of the magnitude and distribution of vibrations emissions. For example, according to [175] it is not merely sufficient to determine the magnitude of emissions depending on train type and speed. In addition to this so-called "emissions spectrum" (with regard to the 8 m measurement point), in terms of the applicability of the results to other sites for which prognoses are to be prepared, a factor or a correction factor should be determined which describes the local peculiarities of the site such as the substructure, the anti-frost layer, the earthen foundation and its linkage to the surrounding subgrade (e.g. thickness, degree of compaction).

The most important factors for describing the soil conditions are damping and the shear modulus G, as well as the speed of propagation c_R of the Rayleigh waves. A pilot study [69] investigated the influence of the ground's shear modulus on the emission level. This study found that the amplitude of the structure-borne noise acceleration of the ground (z-direction at the 8 m point) was roughly inversely proportional to the shear modulus G or the square of the Rayleigh wave speed c_R^2 in the ground at the measurement site. Accordingly, it is possible to derive the correction factor K_{Ground} according to Eq. (16.16), which is independent of the frequency, as an approximation to take into account the stiffness of the ground:

$$K_{\text{Ground}} = 20 \cdot \log \frac{G_{\text{Measurement site}}}{G_{\text{Prediction site}}} = 40 \cdot \log \frac{c_R(\text{Measurement site})}{c_R(\text{Prediction site})}. \tag{16.16}$$

The propagation speed of the Rayleigh waves at the measurement and the prognosis site can be determined without great technical difficulty by conducting appropriate measurements.

On the other hand, determining the impact of soil layering is very difficult, as it is only possible to determine the necessary relevant parameters with relatively complex technical measures (boring, etc.). Soil layering leads to resonance-like amplifications in the amplitude of the vibrations, which depend on the mass (oscillating mass of the soil layer) and the stiffness of the soil (shear modulus). According to Rücker and Said [176], a soil layer of thickness H[21] and a shear wave speed of v_s exhibits a clear boundary frequency f_g that can be calculated based on Eq. (16.17):

$$f_g = \frac{v_s}{2 \cdot H} \tag{16.17}$$

Below this boundary frequency, there is no propagation of vibrations. Above the frequency f_g the

[21] H = layer thickness at the thinnest point in the layer, also known as critical layer thickness, cf. Melke [179].

amplitudes in the layer rapidly approach the amplitudes of a homogenous elastic semi-space (for more on this subject, see [176, 177]).

In the future, one can expect a considerable increase in the accuracy of prognoses in the field of *emissions of structure-borne noise and vibrations* by using the aforementioned formulas for estimates (cf. Sect. 16.5.4).

16.3.10.2 Prediction in the Area of Structure-Borne Noise Propagation

In 1979–1980, detailed studies on the *propagation of structure-borne noise and vibrations* in soil were carried out [108], a summary of the results of which as per Hölzl and Fischer [110], Kurze [111] is presented in Fig. 16.60. Results of later studies entitled "On the Excitation and Propagation of Vibrations from Rail Traffic" [178] include formulas on laws governing the distance between amplitudes, which can be very helpful in estimating the decrease in vibrations from rail traffic over distances. Nevertheless, in order to ensure adequate certainty in prognoses, in almost all practical cases it is necessary to perform measurements at the actual sites.

The reasons for this include:

- Hidden factors disturbing the propagation path, e.g. soil layering, bedrock, marshy spots, and such
- Non-quantifiable transmission conditions due to buttress walls, remains of foundations, utility pipelines, noise bridges, etc.

Quite frequently it is only possible to determine the transmission and propagation conditions with a suitable degree of accuracy after full or partial completion of the structures. For example, the connection conditions for tunnels to the surrounding soil and the propagation conditions of vibrations from tunnels can only be determined reliably after the tunnel has been completed. In spite of this, however, these parameters must be known as accurately as possible to be able to establish the dimensions of the protective measures properly.

In such cases, it is often only possible to determine the transmission and propagation conditions with the aid of artificial excitation. In this respect, unbalance vibration machines, hydraulic vibration machines, vibratory rollers, shaker plates and impulse exciters have proven useful (e.g. [179–181]). Of course, for all kinds of artificial vibration, there is a problem with the transferability of the excitation magnitude to the scale commonly found in rail operations.

16.3.10.3 Prediction in the Area of Structure-Borne Noise Immission

The greatest uncertainty in forecasting structure-borne noise and vibrations is found within the *area of immissions*. The reason for this is that almost all buildings react differently to vibrations caused by rail traffic.

This is the case because of many factors. The construction type and mass of the foundations, the connection of the foundations to the ground, the thickness of the walls, the thickness and design of the ceilings, the span width of the ceilings, and the number of stories all play an important role, not to mention other less central factors such as the furniture in a particular room, which has an effect on the excitation of vibrations in ceilings. Another critical aspect of immissions is that it is generally necessary to measure the transmission conditions, necessitating artificial excitation of the building if excitation by rail traffic is not yet possible.

In one study, 21 residential buildings of various construction types were investigated in terms of the structure-specific transmissions factors from rail traffic-induced vibrations and from artificially induced vibrations (plate vibrators, tampers) and were compared with each other [180]. Figure 16.83 illustrates the mean transmission factors found in this study for excitation from rail traffic as compared to artificial excitation with the standard deviation as a function of the 1/3 octave-band frequency.

In the frequency range above the 31.5 Hz 1/3 octave-band, one can see the spectral differences in that the transmission factors found with artificial excitation are greater than those determined for rail traffic-induced excitation.[22] This means that the prognosis formulated on this basis was on the safe side.

Steinhauser [182, 183], and Unterberger and Steinhauser [184] also discuss practical experience

[22] In interpreting Fig. 17.83, at frequencies under roughly 16 Hz it must be taken into consideration that with artificial excitation the applied vibration energy is not always sufficient to excite the structure to an adequate degree, resulting in limitations on the reliability of measurement results in this frequency range.

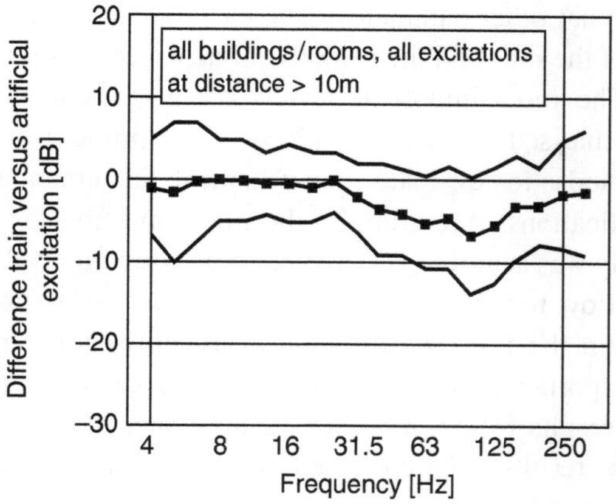

Fig. 16.83 Building specific transmission functions with excitation from rail traffic in relation to artificial excitation. Average value ± standard deviation from measurements in 21 dwellings of various construction types (Said et al. [180])

with forecasting structure-borne noise and vibrations from railway lines.

Further helpful information for formulating prognoses can also be found in the planning basis and planning data in Melke [179], Krüger et al. [146], as well as in [103] and in VDI Guideline 2716 "Airborne and structure-borne noise from rail traffic in public transport applications" [185], depending on the specific goals being pursued.

16.4 Airborne Noise and Structure-Borne Noise, Vibration from Suburban Railway Lines

16.4.1 General

In this respect "suburban railway applications" is understood as rail traffic, which is covered, by the "Tramway Construction and Operational Regulations" [9]. This essentially covers trams, suburban railways and subways.

In general, the aspects discussed in this chapter up to now are valid with regard to the inception, propagation and evaluation of airborne and structure-borne noise caused by rail traffic. This also follows from the procedure for forecasting noise conditions in the vicinity of transport routes as set forth in [186]. According to Sect. 6.1.2 of this standard, the same procedure is to be employed for trams and subways as for standard gauge railways with a correction factor of ΔL_{Fi} to take into account the type of train. For trams, for example, this amounts to +3 dB(A) (cf. Table 5 in [186]). This value corresponds to the added value D_{Fz} for trams and urban railways as per [3].

While the latest studies by the Federal Ministry of the Environment on "Noise Emissions from Trams" [187] (which will be discussed in the following) have confirmed this value, it was also found that differentiation between trams and suburban railways on the one hand and low-floor trams on the other hand was necessary (for more details on this subject, see Sect. 16.4.2.2).

A comprehensive discussion of the key transport, economics and environmental aspects of public transport can be found in "Suburban railways in Germany" [188]. In this work, however, the term "suburban railways" covers all means of public transportation, i.e. subways, trams and urban railways *and* regional railways and light rail systems. It also contains a chapter on the main focus of this section, namely "Reduction of Noise and Vibrations". Another new publication with the title "Reduction of noise and vibrations in railway applications" [146] summarises numerous examples of "practical applications involving measures to reduce noise and vibrations".

The work of Melke [179], which was already mentioned in Sect. 16.3.10 with regard to forecasting structure-borne noise and vibrations and measures for the reduction of such, also deserves to be noted here in light of the wide range of fundamental information and practical results that are presented and the comprehensive scope of the literature cited.

Furthermore, the following works are also valuable sources of information as some of them provide overviews while others discuss specific subfields of the reduction in noise and vibrations from rail traffic: (see [91, 189–198, 258]).

16.4.2 Specific Characteristics of Suburban Railway Lines in Comparison with Normal Railway Lines

Despite the in-depth discussion of the topic "Airborne and Structure-Borne Noise in Suburban Rail Applications" in the literature cited above and the fundamental similarity between railways noted at the beginning of this section [8] and suburban railways as per [9], it seems worthwhile to focus more close at this point on some of the special aspects stemming from the specific technical conditions of such railways, and consequently the technical options available when it comes to measures designed to reduce structure-borne noise.

With regard to trams, this particularly pertains to technical conditions stemming from the route of the lines in densely built-up urban areas and the mixture of road and rail traffic.

16.4.2.1 Curve Squeal

In urban areas, it is frequently necessary to design tramlines with considerably smaller curve radii than is the case for standard gauge railways. This often leads to the annoying phenomenon of "curve squeal" which frequently poses a serious problem (cf. Sect. 16.2.2.3, Shunting Yards). Curve squeal is caused by transverse slippage of the fixed wheels and the associated high-frequency stick-slip movements that result. To a great extent, this phenomenon can be avoided by using individual wheel suspension, single axle bogies and by using steerable wheel sets [92]. Secondary measures to combat curve squeal include measures to damp structure-borne noise at the wheels, spraying the rails [93, 199] and damping measures on the rail by way of special rail dampeners [200].

In Schall 03 [3], the correction value D_{Ra} is included in the calculations of the emission level $L_{\mathrm{eq},E}$ to take into account curve squeal (see Table 6 in Schall 03 [3]), when this occurs. Accordingly, for track curves with a radius of $R < 300$ m a correction value of 8 dB is applied in these cases. For curves with radii of 300 m $< R < 500$ m the correction value is 3 dB, whereas for larger radii it is $D_{\mathrm{Ra}} = 0$ dB.

Both the correction values D_{Ra} [3] which are *not* differentiated according to the train type and the special correction value for trams proposed in 1990 [201] exhibit relatively large jumps in the numeric values for curve radii which are larger or smaller than certain values. This "problem" was the basis point for the study [202], promoted by the Federal Minister for Transport, which investigated the following questions:

- What is the functional dependence between the correction value D_{Ra} and the curve radius R?
- How can the occurrence of curve squeal be measured in an objective manner?
- What other parameters (such as climate, condition of the rails and the vehicle) influence curve squeal?

The most important results of this research are summarised in Krüger [203] with proposals for methods to evaluate curve squeal in urban rail applications. Accordingly, the main objective of the study was achieved, i.e. to determine correction values to allow for differentiated assessment of curve squeal within the framework of noise prognoses for public transportation railways. Further statistical testing of the results would, however, be necessary in order for these results to be compiled into a guideline, such as that found in [3].

The results of other research on curve squeal and approaches to the abatement of such can be found in Krüger [204], Lenz [205], for example.

16.4.2.2 Specific Track and Vehicle Characteristics

An extremely wide range of noise emissions must be expected in the case of trams in particular, due to a great degree to soiling of the running surfaces of the rails with dust, grit and the like (with a corresponding effect on the running surfaces of the wheels) and to the wide variety of different types of track design (in cobblestones, smooth asphalt, slab track, grassed track, etc.).

Moreover, due to the rough running surfaces the higher noise emissions described in Schall 03 [3] for rail vehicles with cast iron block brakes generally apply for trams, even though such vehicles are hardly ever outfitted with this type of brakes anymore. In contrast to standard gauge lines, trams also generally ride on the wheel flanges in the cross frogs of switches with grooved rails, which also generates higher emissions of structure-borne and airborne noise.

Emission levels for trains in the vicinity of passenger stations can be calculated in a simplified manner as for open line sections according to Sect. 8.1 of Schall 03 [3], whereby the train speed in the area of the passenger station for the various train types can be found in Table 2 of Schall 03 [3]. Factors reducing the level of emissions (such as the platform edges) and other factors contributing to the overall level of emissions, such as secondary sources of noise (PA announcements, luggage trolleys, etc.) no longer need to be taken into account if this method is employed (cf. Sect. 16.2.2.3, Other Facilities – Passenger Stations, etc.). The reference value for train speed for trams is

Table 16.8 Proposal for correction values for taking into account the vehicle type D_{Fz} and the track type D_{Fb} for tramways as per [187] and a comparison of the corresponding correction values for tramways as per [3]

Factor	Type of vehicle/type of track	Correction value as per [187] in dB(A)	Correction value as per [3] in dB(A)
D_{Fz}	Tramway, urban railway	+3	+3
	Low-floor tram	+1	+3
D_{Fb}	Wooden sleepers	+2	0
	Concrete sleepers	+2	+2
	Slab track	+7	+5
	Track with recessed turf	+4	−2
	Track with raised turf	0	−2

Correction value for regular maintenance of the wheel-rail system −3 dB(A)

60 km/h, according to Schall 03 [3]. In practice, this value has proven to be too high with regards to the calculation of noise emissions at tram stops. As a result, a lower speed of 30 km/h is assumed in justified cases, in order to take into account the special characteristics of tram stops, when calculations are prepared nowadays for consultative purposes.

As was mentioned at the beginning of this section, a more detailed review of the findings of the basic research performed by the German Federal Ministry of the Environment "Noise Emissions of Trams" [187] is given in the following. The purpose of this study was to investigate to what extent the findings based on earlier measurements in Schall 03 [3] corresponded to the emissions recorded for trams under current conditions. Very detailed measurements were conducted at 29 public transportation companies in Germany (of the total of 56 which operate trams), with the participation of 17 different institutions and engineering offices. Based on the measurement data a recommendation for revised correction values D_{Fz} and D_{Fb} was formulated for vehicles and track systems as per Schall 03 [3] *for tram applications*. This involved drawing a distinction between types of vehicles, in accordance with the current state of technical knowledge, separating them into the categories trams/suburban railways and low-floor trams and drawing in line in terms of track type, distinguishing between grassed track with recessed or raised turf. This recommendation, including a comparison with the currently recommended correction values set forth in Schall 03 [3] is shown in Table 16.8 [3, 187].

In summary, one can see from the table that, in cases when the wheel–rail system is not regularly maintained, noise emissions from traditional tram systems are sometimes substantially underestimated by the calculation method set forth in Schall 03 [3], with the exception of concrete sleeper track, which means that a correction of the values involved would appear to be reasonable when the guideline is revised in the future.

16.4.3 Special Vibration Mitigation Measures for Tramways in the Area of City Centres

Fundamentally speaking, the measures for the reduction of structure-borne noise discussed in Sect. 16.3 can be applied to tramways, suburban railways and subways. Nevertheless, in the application of such measures it is important to take into account the different boundary conditions as compared to standard gauge operations, primarily in respect of lower axle loads and train speeds. This means that in most cases lower values can be used for the stiffness of the resilient components in the track superstructure when it comes to the usual predefined limit for permissible track subsidence.

For example, on track with ballast mats a static bedding modulus of $c = 0.01$ N/mm^3 is now commonly used for these types of trains, based on the guidelines for ballast mats issued by the German Federal Railways [45]. Due to the especially low axle loads, for trams even lower values are permitted (cf. $c = 0.02$ N/mm^3 for self-propelled rapid transit trains; $c = 0.03$ N/mm^3 for rapid transit trains with locomotives ("push-pull" trains) as per [45]).

Due to the mixed nature of traffic with both rail and road vehicles in urban centres, special requirements for construction designs to minimise structure-borne noise are necessary, as was mentioned earlier. In this respect, essentially two different construction designs have been developed which are used when reduction of structure-borne noise is a prime concern (see [146, 198, 206–210]).

These special measures for inner-city trams with mixed traffic are:

- Continuous (highly) elastic rail support
- Slab track on resilient bearings, i.e. so-called "Light Mass-Spring System" or LMSS.[23]

16.4.3.1 Continuous Highly Elastic Rail Fastening

The track superstructure design "continuous highly elastic rail fastening" is distinguished from standard designs for grooved track for trams, which also feature continuous elastic fastening, but which are not suited as measures for enhanced reduction of structure-borne noise (see e.g. [188, 198, 210]).

One of the key characteristics of continuous highly elastic rail fastening systems as opposed to classic (grooved) track support systems with discrete bearings is that the rail foot is continuously mounted on a resilient elastomer strip and/or continuously embedded in resilient material along the entire length of the rail. One of the main advantages of this method of rail mounting in terms of the acoustic properties is the avoidance of the so-called "seat spacing frequency" (referred to as the sleeper-spacing frequency in Sects. 16.2 and 16.3), which arises with discrete bearings and is dominant in the spectrum, occurring in proportion to the train speed and in inverse proportion to the spacing of the rail seats.

Due to the low amount of sprung mass (rail), in order to achieve the lowest possible tuning frequency which is a prerequisite for sufficient damping of structure-borne noise, such systems require very soft supports which results in quite considerable track subsidence typically up to 6 mm and in extreme cases up to 10 mm (this is, however, associated with a very long bending line for the rail thus limiting the danger of overstressing the rail and causing rail breakage). This leads to large relative movements, and hence with regard to ensuring long-term serviceability this results in very stringent requirements in respect of the construction of the elastic seam between the rail and the road surface *and* at transition points to standard track

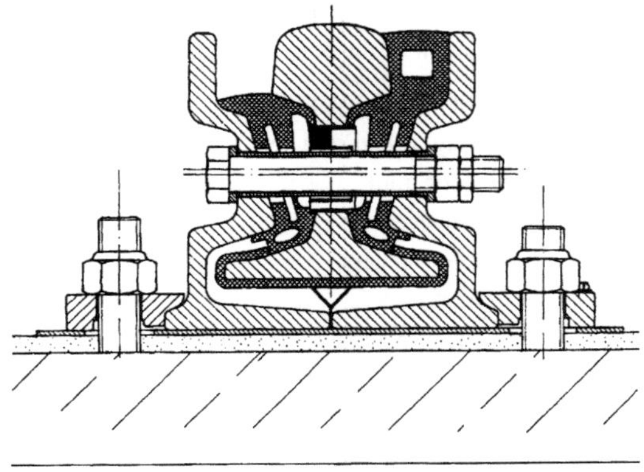

Fig. 16.84 Outline sketch of continuous highly elastic rail fastening with a grooved rail (from Lenz [210])

superstructure. Simple designs employ guard rail tie bars embedded in elastic materials to maintain gauge and prevent excessive rail deflection, which could lead to excessive gauge widening. More complex designs, as illustrated in Fig. 16.84, feature both rails without a gauge tie bar, embedded in so-called self-centering elastomer packets (see [188, 206, 210, 211]).

Up to now, however, very little in the way of detailed results of systematic measurements is available, and accordingly information on the efficiency of this type of track superstructure is not yet adequately underpinned (as mentioned in [210] as well). As a result, it is not yet possible to state with a reasonable degree of certainty the level of insertion loss, as is necessary for the purposes of formulating prognoses and as is standard procedure for ballast mats, for example (s. Sect. 16.3.7). Nevertheless, readers are also referred to the measurement results obtained with such track superstructures and the interpretation of such (see [206, 210, 211]).

16.4.3.2 Light Mass-Spring System

From an acoustics and construction perspective, the most effective method of reducing structure-borne noise for tramways in inner-city areas is the Light Mass-Spring System (LMSS). The underlying principle behind such a system, drawing on the example of the tram system in Munich, is presented in Fig. 16.85 (cf. Appendix to [255]).

In an LMSS the entire mass of the slab track including the superstructure is mounted on a

[23] As regards the adequacy of the commonly used term "Light Mass Spring System", please refer Footnote 13 in Sect. 16.3.2 for an explanatory note on the expression "Mass-Spring System".

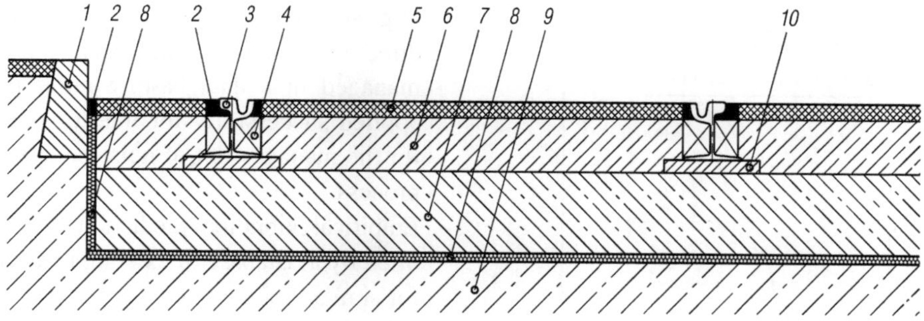

Fig. 16.85 Full surface resiliently mounted slab track for a tramway, so-called light mass-spring system (LMSS). *1* curb, *2* elastic seam seal, *3* grooved rail, *4* filler element, *5* asphalt (or cobblestones or concrete), *6* filler concrete, *7* slab track made of (reinforced) concrete, *8* elastic bottom and side mats (e.g. cellular PUR materials, Sylomer® LR 2501), *9* bonded sublayer, *10* levelling layer (elastic rail mount, continuous highly elastic rail fastening)

full-surface elastomer mat, which usually also functions as sacrifice formwork when the slab is poured. The acoustic efficiency of this solution is determined by the tuning frequency of the system, which is a function of the dynamic stiffness of the elastomer mats and the sprung mass of the system, both in terms of the area per metre of track. This means that the mounting of the grooved track can be designed with a relatively high level of stiffness. Accordingly this provides the advantage of a low level of track subsidence (usually <1 mm), resulting in a low amount of relative movement between the rail and the roadway. The risk of defective construction and *acoustic bridges* between the slabs and the subgrade and with the adjacent roadway, which reduce the efficiency of the system, is comparatively low with this type of track design. Of course, this presupposes that the "elastic liner" consisting of the bottom and side mats is continuous and free of any gaps and that the seams at the top of the side mats is properly designed and constructed in a manner which ensures its elasticity over the long term.

Another advantage of this type of system is that the track slabs (see Part 7 in Fig. 16.85) can be fitted with any of the various types of tracks used at the different public transport companies (e.g. various forms of track frame and sleepers as well as discrete rail seats). Light Mass Spring Systems have been installed in France and in Switzerland (Grenoble, Nantes, Strasbourg, Paris, Geneva, etc.) for many years now [212, 213], with the various types of bi-block sleeper track frames that are commonly used in those countries (essentially similar to the construction design depicted in Part a of Fig. 16.74).

Measurement results on the efficiency of LMSS's in tram applications, e.g. which could be used for forecasting in planned projects, are only available for the systems installed in France [212, 213]; for measurement results of LMSS's with light rail systems, see [142, 145] and Fig. 16.75).

Recently, the data illustrated in Fig. 16.86 have become available, which present the results of measurements before and after the installation of a LMSS in the Munich tram system [214]. Based on these results, the insertion loss for a LMSS shown in Fig. 16.86 was calculated, referring to a grooved track superstructure prior to reconstruction.

One can see the expected typical signature with a tuning frequency of roughly 20 Hz and a reduction of structure-borne noise of approximately 9 dB at 63 Hz and up to 20 dB in the frequency range which is relevant for the perception of secondary airborne noise (reradiated sound) in buildings.

16.5 Prediction Models for Airborne Noise, Structure-Borne Noise and Vibration from Railway Lines

16.5.1 Overview

As in other fields of acoustics, methods for prediction and model calculations have also become more important regarding railways and tracked transport systems. The fields of application range from legal procedures which are carried out in the context of a plan regulation process for new railway lines, the engineering grade predictions made in the context of vehicle and track design or scientific model calculations of partly even academic character.

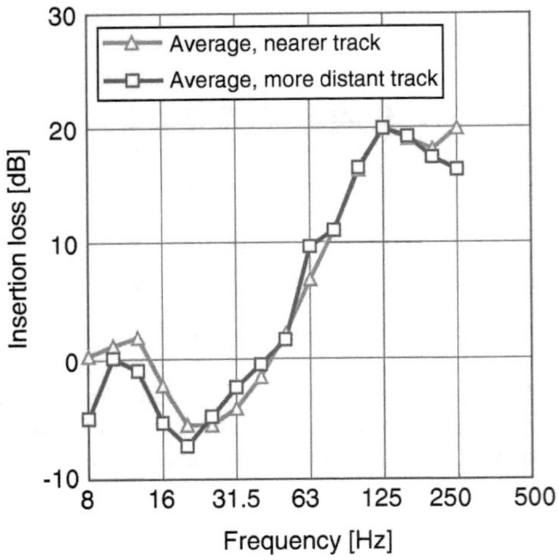

Fig. 16.86 Insertion loss of a light mass spring system (LMSS) installed in the tram system of Munich, Maximilian Street. Results of measurements before/after installation of the LMSS at identical measurement point on the side of the roadway, depending on the local site characteristics at five measurement locations each 4–8.5 m in front of the buildings

16.5.1.1 Empirical Methods

- The *Schall 03* [3] is an empirical method for the prediction of airborne noise, which is regulated by law in Germany according to [5] for plan regulation processes for railway lines (see Sect. 16.2.3.2). Design types of vehicles and tracks as well as operating conditions are considered by means of level reductions and level additions which are verified by measurements. To take account of technical progress and changes, the procedure is checked for its validity and will be complemented if necessary. Procedures similar to the German procedure (*Schall 03*) are also common in other countries and also for road traffic (see also e.g. [186] and Chap. 15, road traffic noise). The European Commission is requesting harmonisation of the methods. In countries where no adequate method is available the Dutch method is recommended until a common European method is established. To take account of recent technical developments in vehicle technology presently the German *Schall 03* is undergoing a redesign, based on the Dutch approach and several new aspects [215, 216].
- In the research field, approaches have been established according to which the pass-by levels of whole trains can be calculated on the basis of measured or calculated characteristics (e.g. sound power level, spectral distribution and directivity) of individual noise sources which are managed in a data base. Similar procedures are applied successfully for interior noise of vehicles for the management of noise sources in the design process.
- Empirical vibration prediction methods are supported by a database of existing measurement results for similar boundary conditions (i.e. geology, track, vehicles, operation). Predictions are made on the basis of the database as well as considering possibly available actual measurement results and experience regarding the buildings. Due to the more complex boundary conditions, these predictions are considerably less precise than for airborne noise.

16.5.1.2 Numerical and Analytical Methods

- In recent years, different noise prediction models with roughness excitation have been developed and have been applied successfully for studies of rolling noise regarding vehicle and track optimization (e.g. Remington [217, 218], Springboard, TWINS [219], RIM [220, 253]). The software has been developed especially for these models.
- For the examination of special questions regarding the aerodynamic excitation, which is at higher velocities responsible for important contributions to the pass-by noise of a train, the generally available CFD methods have been used successfully (CFD = Computational Fluid Dynamics).
- Furthermore, specific analytic approaches have been developed to deal with problems of mainly academic concern regarding ground-borne vibration, structure-borne noise and airborne noise. Concerning this subject, a large number of publications can be found in the acoustics journals, e.g. the *Journal of Sound and Vibration Research* (JSVR), specifically the editions of JSVR published after the *International Workshops on Railway Noise* (IWRN), the *Journal of the Acoustical Society of America* (JASA) or the *Acta Acustica united with Acustica* (the Journal of the European Acoustics Association (EAA)).

The numerical and analytical methods are mainly applied in the field of research and development, since, as in other fields, they easily allow calculating a large number of parameter constellations by means of simulations. However, apart from the possibly very precise description of the physical phenomena, the specified methods are very complex to use so that they do not seem suitable for general planning use and, regarding this field, empiric models are favoured.

16.5.2 Rolling Noise

16.5.2.1 General

Rolling noise dominates in the speed range of 60 km/h to approximately 300 km/h, aerodynamic noise is relevant starting at higher speeds, whereas machinery noise is of interest regarding interior noise, still stand and lower velocities, the precise transition speeds being dependant on the vehicle and track design (see also Sect. 16.2.2). In the following sections, the general features of the common rolling noise models are described as well as the problem of the determination of model parameters and the application of the models.

16.5.2.2 Rolling Noise Models

The basic principle of rolling noise models was first published by Remington [217, 218]. The basic models take account of the interaction between one wheel and the rail.

The vehicle as well as the track is described as impedances. For excitation, a roughness band is drawn through the contact zone between wheel and rail of the static system (see Fig. 16.87).

The wheel/rail contact geometry is considered by means of a contact filter to reduce the excitation by roughness wavelengths in the size of or shorter than the contact area. In most cases, the rolling mechanism itself is not considered to be important. The effective exciting roughness is derived from the energetic sum of the wheel and rail contribution.

The impedance model of the track is implemented using a serial connection of complex stiffnesses, including loss factors, the rail as a continuously supported beam, the rail pad and the ballast as complex stiffness, the sleeper as a mass or a complex impedance of a beam. The basic principle easily allows the inclusion of additional elements as e.g. sleeper soffit pads or ballast mats.

The vehicle is only considered related to one wheel and in the most simple case is described by the mass of the wheel, with a modal extension for the inclusion of wheel eigenfrequencies, as well as the proportionate mass of the bogie and the car body, also considering the primary and secondary suspensions including the dampers.

The calculation process is shown in Fig. 16.88: determination of the separate impedances of wheel, rail and contact, determination of the velocities of the individual elements by combination of the impedances and excitation with the contact-filtered combined roughness of wheel and rail, radiation of the velocities from the surfaces including the radiation efficiency and determination of the sound pressure levels at the microphone positions, or respectively the emitted sound power of the elements.

Figure 16.89 shows an example of calculated impedances for the elements of ballasted track. Using

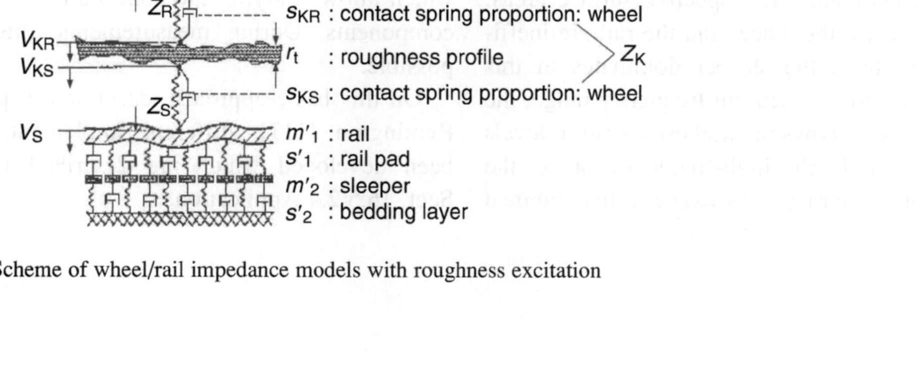

Fig. 16.87 Scheme of wheel/rail impedance models with roughness excitation

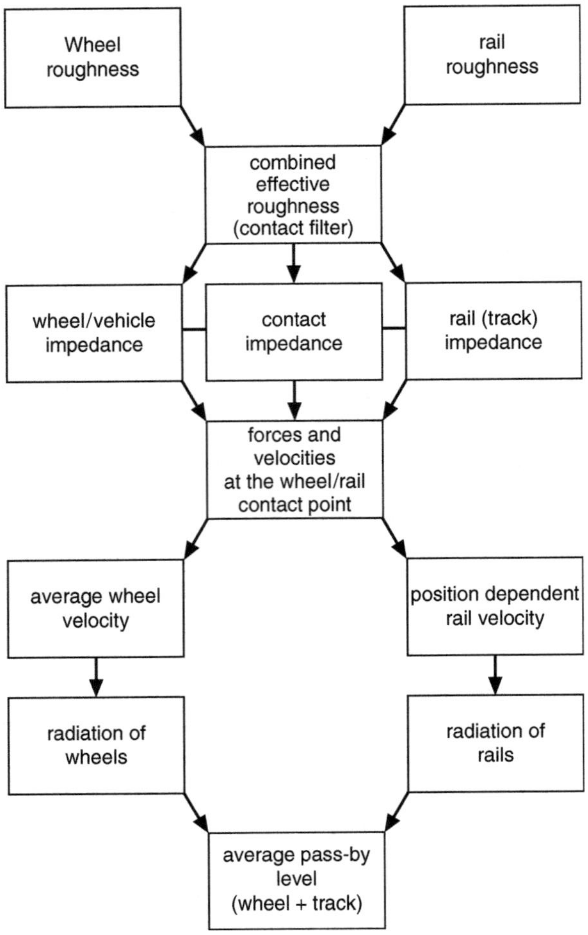

Fig. 16.88 Scheme of the calculation procedure of the roughness models

this figure, the basic behaviour of the model or respectively the wheel/rail system can be explained: in the low-frequency range, the wheel is the element with the lowest stiffness, in the medium frequency range the rail and for higher frequencies the contact. Due to the excitation the highest vibration levels will be excited on the softest element. The contribution to the airborne noise is then determined by means of the radiation efficiency and the respective surface areas: for low frequencies, the wheel and the rail are inefficient radiators, thus, the sleeper dominates in this frequency range. In the medium frequency range, the rail dominates as it shows the highest vibration levels and radiates well. In the high-frequency range, the wheel dominates due to its easy excitability combined with the surface and radiation efficiency, although a large part of the kinetic energy remains in the contact spring.

Figure 16.90 shows an example of the contributions to the radiated sound for the impedances shown in Fig. 16.89.

The above description shows one of the big advantages of working with simulation models, which allow studying the contributions of the single components. During measurements, this is hardly possible.

On this basic approach, described in principle by Remington [217], different implementations have been developed which are described in detail in Sect. 16.5.2.4 Application.

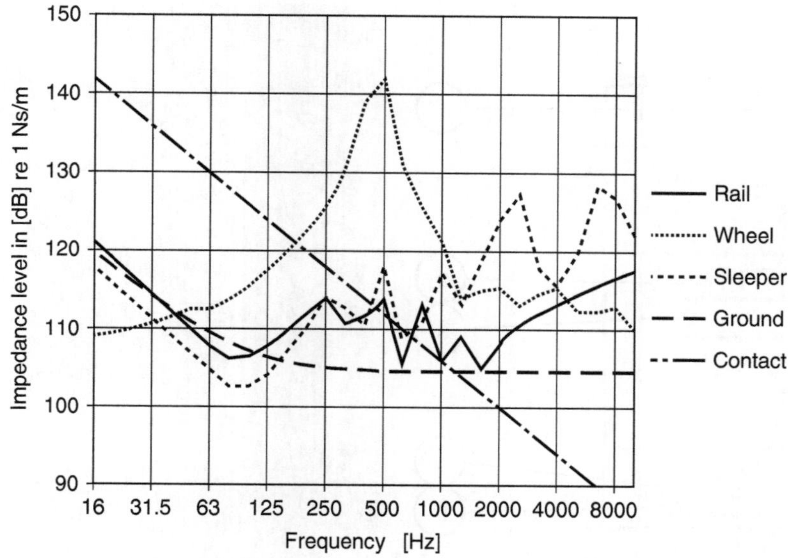

Fig. 16.89 Example for calculated impedances of the single components of the wheel/rail impedance model, here: ballasted track system

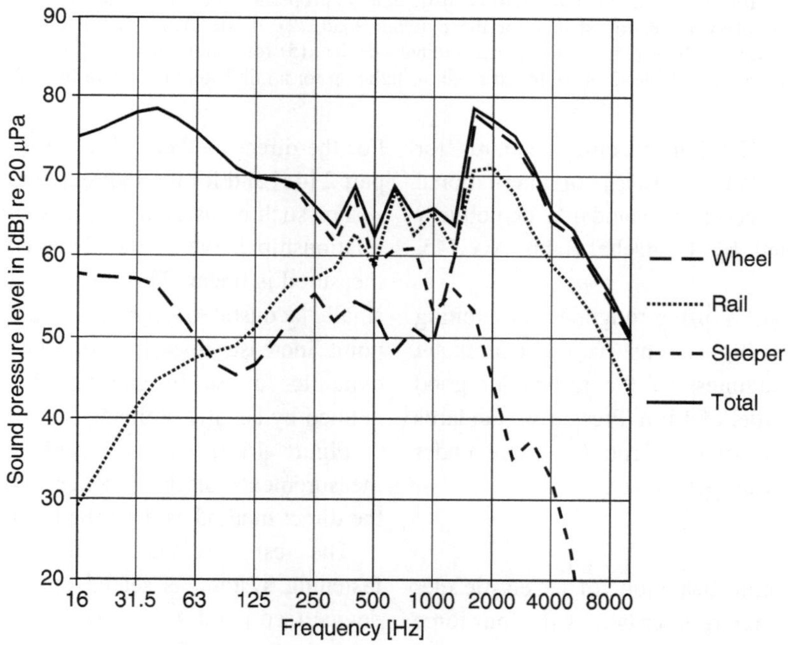

Fig. 16.90 Example for the calculated contributions of the components to the pass-by levels of a passenger train at a running speed of 200 km/h

16.5.2.3 Determination of Model Parameters

General

A prerequisite for successful predictions and model calculations is the provision of a suitable database for the model parameters. The geometric dimensions and the masses of the elements are easy to determine. The determination of the parameters for elastic elements, i.e. their complex dynamic stiffness, is clearly more difficult. For this purpose, laboratory tests for the measurement of the stiffness of elastic elements on the test rig and field measurements have proved to be useful to derive the corresponding parameters. Test rig measurements to determine the dynamic stiffness are described in general acoustic

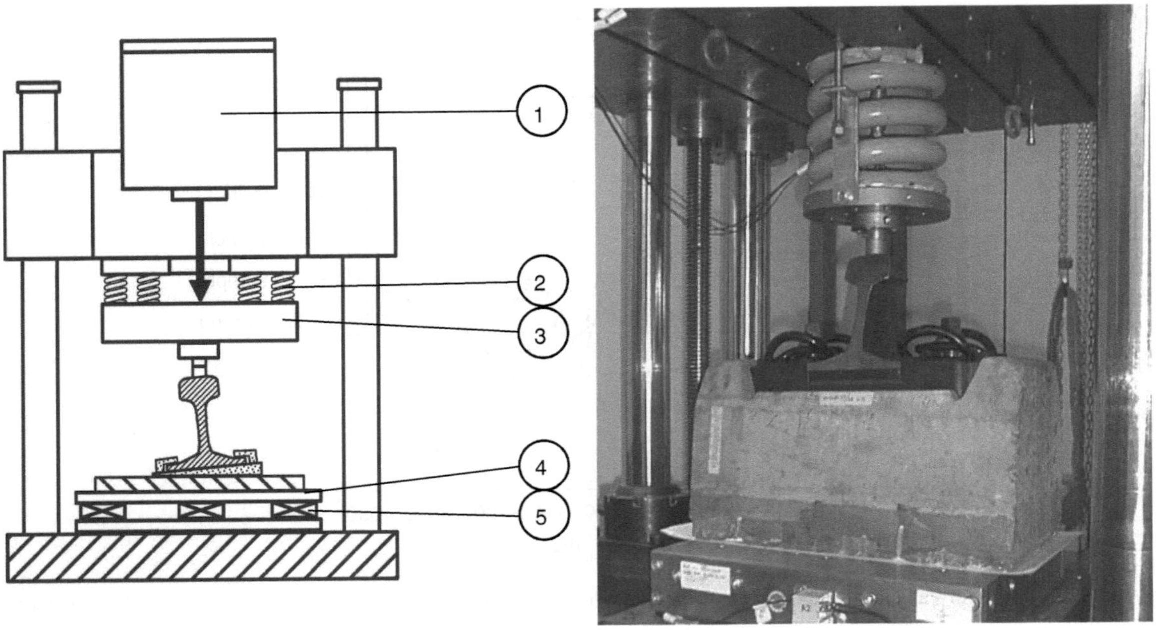

Fig. 16.91 *Left*: Test rig for the measurement of vibro-acoustic transfer properties of elastic elements. Direct method according to [ISO 10846–2 [44]]: (1) electro-dynamic shaker for the dynamic load, (2) elastic elements for the decoupling of the static preload, (3) preloading unit, (4) base plate of force measurement device, (5) force transducers. *Right*: Test rig set up for direct measurements according to [ISO 10846–2 [44]], featuring an installation for a rail fastening system for slab track

standards [44], [224, 225] in specific standards for the German railway industry [255], or international [43] and also in the company standards or delivery specifications defined by Deutsche Bahn AG [45, 222, 223].

For the determination of the roughness of exciting rail running surfaces, highly sensitive procedures are necessary as the roughness of the rail under good conditions is in the order of 1 μm. Presently, standards for the measurements of track roughness are under development on the European level.

Laboratory Tests

In general, the dynamic behaviour of an elastic element under normal force is described by the four force and deflection quantities on the top and bottom side of the element in direction of the force. Instead of deflections, also velocities or accelerations can be considered. If the stiffness character dominates the behaviour of the elements to be tested, the transfer stiffness as relation between the force at the output to the deflection at the input is sufficient as characteristic parameter.

Procedures to measure this transfer stiffness are described in general in part 1 of ISO 10846 [224]. For the direct method, they are described in detail in part 2 [44] and for the indirect method in Part 3 [225].

For such measurements, it is fundamental that the relationship between the dynamic parameters to be measured is linear. This means that the amplitudes of small "acoustic movements" for a certain working point increase linearly with the amplitudes of the dynamic forces. In general, the working point is defined by a static preload.

Figure 16.91 shows a scheme of a test rig for measurements of the transfer stiffness according to the direct method as described in ISO 10846-2 [44].

The test specimen, here represented by a rail fastening system, is seated on a force measurement device (see point 4 + 5) which is fixed rigidly to the mounting table of a heavy-duty 4-column test rig. Above this assembly, there is a traversable girder for application of the static preload. On the lower side a preloading unit (see point 3) is attached elastically (see point 2). On the upper side of the girder an electro-dynamic shaker (see point 1) is installed via elastic mounts in order to minimise reacting forces transmitted into the test rig when working. By moving the girder downwards the static preload is applied to the test specimen by means of a heavy-duty push rod.

The oscillating (dynamic) force, generated with the electro-dynamic shaker is superimposed to the static preload by exciting the preloading unit (see point 1 + 3).

The evaluation of the dynamic transfer stiffness k_{21} of resilient elements according to ISO 10864-2 [44] (direct method) is based on Eq. (16.18)

$$F_2 = k_{21}\frac{v_1}{j\omega} - k_{22}\frac{a_2}{\omega^2}, \qquad (16.18)$$

where the force F_2 is determined from a measured force F'_2 underneath a force measuring platform of mass m. The input velocity v_1 results from the measured input acceleration a_1 after integration, the output acceleration a_2 should be as small as possible but needs to be controlled, and the stiffness component k_{22} can frequently be approximated at low frequencies and for homogeneous elements by k_{21}. Since the assumption of a blocked output in ISO 10846-2 ($a_2 \approx 0$) is too restrictive for practical applications, the evaluation of k_{21} from laboratory tests is generally done with two correction terms, one accounting for the mass m of the force measuring platform, particularly at higher frequencies, and the other for the relative displacement at both ends under the assumption $k_{22} \approx k_{21}$, which holds best at low frequencies:

$$k_{21} = j\omega\frac{F'_2 + a_2\frac{Z}{j\omega}}{v_1\left(1 - \frac{k_{22}}{k_{21}}\frac{a_2}{j\omega v_1}\right)} \approx j\omega\,\frac{F'_2}{v_1}\,\frac{\left(1 + \frac{m\,a_2}{F'_2}\right)}{\left(1 - \frac{a_2}{j\omega v_1}\right)}. \qquad (16.19)$$

The numerator in the right side fraction can be called the mass correction term and the denominator the relative displacement correction term. In general the output a_2 should be as small as possible by appropriate design of the test rig.

Inaccuracies from the evaluation of the approximation generally arise when

1. The test specimen, including the load distribution device, is inhomogeneous so that the assumption $k_{22} \approx k_{21}$ is poor
2. The force F'_2 is relatively small due to a soft test element as compared to the static pre-loading springs

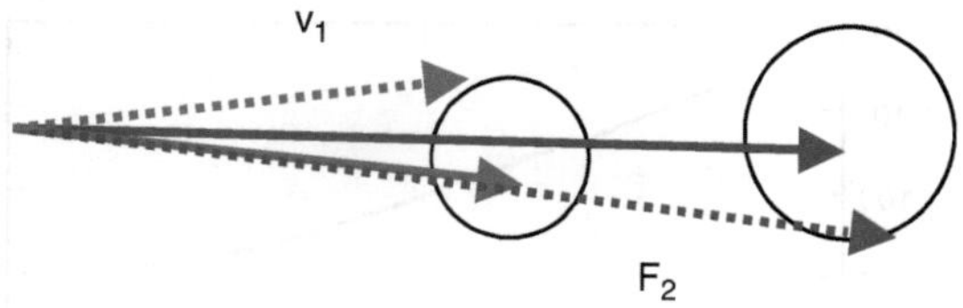

Fig. 16.92 Force and velocity vectors in the complex plane

3. The difference in velocities at both ends $v_1 - v_2$ is relatively small due to a rigid test element on the finite foundation impedance and/or due to flanking transmission via the pre-loading test rig
4. The impedance Z of the force-measuring platform is insufficiently modelled by a mass impedance $j\omega\, m$

The first and the third point are most important for the high-frequency limit of measurements on rail fastening systems.

Numerator and denominator of Eq. (16.19) can be understood as two vectors with some uncertainty about the exact lengths and phases as schematically shown in Fig. 16.92. Small uncertainties have a minor effect on the ratio of the magnitudes. Thus, the absolute value of the transfer stiffness is still well determined. A small phase difference, however, may substantially change, particularly in sign.

For large uncertainties, the circles may include the origin. In this case, the magnitude of the ratio and the phase assume completely unreasonable values.

The standard procedure for determining the loss factor η of resilient elements involves the imaginary and real parts of the transfer stiffness:

$$\eta = \tan\varphi = \frac{\text{Im}\,(k_{21})}{\text{Re}\,(k_{21})}. \qquad (16.20)$$

The above rules apply to the tests in general and shall now be demonstrated for a rail fastening system. Compared to the base plate pad the thin rail pad is relatively stiff (factor ≥ 20). The base plate pad is the primary subject of the investigations. Force and acceleration measurements have been carried out according to the direct method of ISO 10846-2 [44]. An example for measured transfer functions is depicted in Fig. 16.93.

Figure 16.94 shows the results of the evaluation of the basic data and the application of the mass and relative displacement correction terms separately and in conjunction. The displacement correction is

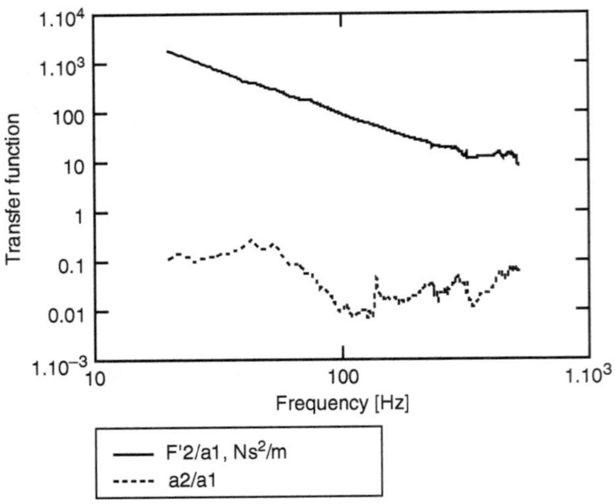

Fig. 16.93 Transfer functions from rig tests for direct procedure: basic transfer impedance F'_2/a_1 and correction ratio a_2/a_1

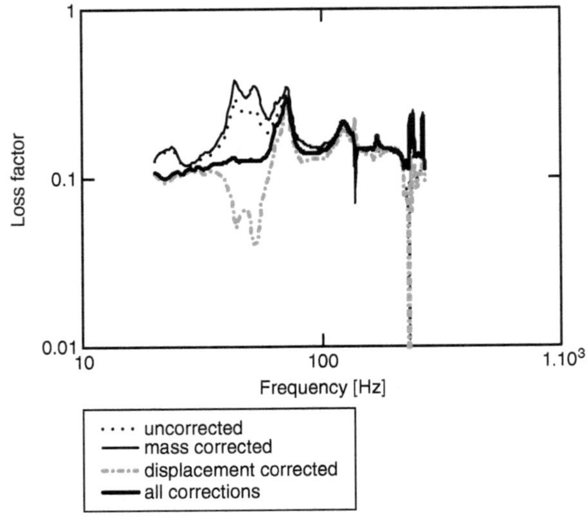

Fig. 16.95 Loss factor, uncorrected and with corrections applied separately and in combination

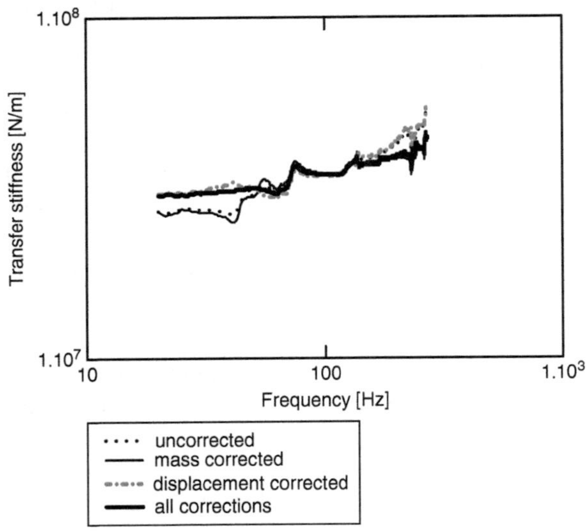

Fig. 16.94 Transfer stiffness, uncorrected and with corrections applied separately and in combination

effective in the low-frequency range and the mass correction for higher frequencies. For further smoothing the stiffness is converted from narrow band to 1/3-octave bands and presented as final results for further use.

Figure 16.95 shows the loss factor calculated according to Eq. (16.20). It can be seen that the inconsistencies are quite larger than for the magnitude shown in Fig. 16.94. A theoretical study showed, however, that for the noise radiation from rails the loss factor of the elastic elements is not crucial, as the behaviour is dominated by pass and stop bands [226].

According to ISO 10846, a measuring inaccuracy of approx. ± 1.5 dB must be expected (nearly 20%). For such stiffness measurements, it is extremely important to choose the boundary conditions correctly, like e.g. static preload and temperature, as they have an important influence on the results [227, 228].

Field Measurements

Track Characterisation

The advance of the research has now made it well known that the dynamic behaviour of the tracks, more precisely the rate of decay of vibration along the rail, has an influence on the pass-by noise of trains. Researchers have used the decay rates of vibration along the rail as an intermediate, measurable parameter by which to test and improve the accuracy of prediction models. The experimental characterisation is determined by measurements of point and transfer impedances in the track.

Nowadays the track decay rate has been advanced to a definition parameter for test tracks on the European level in the frame of the definition of the measurement sites for the TSI (Technical Specifications for the Interoperability of Trans-European trains) [229, 230]. The measurement and limit values for the track decay rates have been defined for the test site to allow type testing of the interoperable trains.

Roughness of the Running Surfaces

As the roughness of the running surfaces is the predominant source of noise of tracked transport systems in the major speed range applicable today, it is therefore crucial to be capable of reliable measurements of the roughness of rails and wheels. In the past the incentive was the understanding of corrugation growth and the understanding of rolling noise generation. Initiated by Deutsche Bahn AG the rail roughness measurement device RM1200E was available in the early-1990s and has since then gained a good reputation in the scientific community [231, 232]. In addition to the necessities of the noise research the need for the advanced study of corrugation growth had been an incentive for German Federal Railways to support the development of the measurement device RM1200E, which was based on the structure of a mechanical corrugation measurement recorder.

The RM1200E, which is widely used in the railroad acoustics community, allows the measurement of the roughness of a length of 1.2 m of rail surface on one line at a time with a discretization of 0.5 mm. The measurement principle is a straight edge with a relative displacement transducer. The basic evaluation algorithm comprises a correction using the calibration data derived from a measurement on a calibration stone, a Fourier transform including windowing and a summation to 1/3-octave-bands. The resulting 1/3-octave-band spectrum includes values for wavelengths up to 10 cm due to the requirement that a minimum of three lines in the FFT spectrum should be available for summing up the band value. Theoretically a value for the wavelength of 120 cm is available in the FFT spectrum, however, the statistical uncertainty for such a value is rather high compared to the values based on averages as for the shorter wavelengths.

Measurement experience showed that frequently raw data did contain spikes or pits caused e.g. by surface irregularities too small for a wheel to "see" on the rail or by dust buildup on the point of the sensor. These spikes or pits result in a broad banded spectrum hiding the desired results. To improve the situation the measurement data has to be corrected to remove these effects. The concept was first published in van Lier [233]. Figure 16.96 shows example spectra for two cases with and without pits and spikes correction.

Meanwhile several "pits and spikes removal" algorithms are in use in the scientific community developed by different users of the instrument (Deutsche Bahn AG, Müller-BBM, SNCF, AEA-TR). The underlying concepts vary greatly, from the practical approach of defining limits for certain geometric and derivative properties, including user intervention and manipulation, to the automatic model calculation based on an imaginary wheel checking the surface geometry and thus defining a filter for the suppression of the pits and spikes.

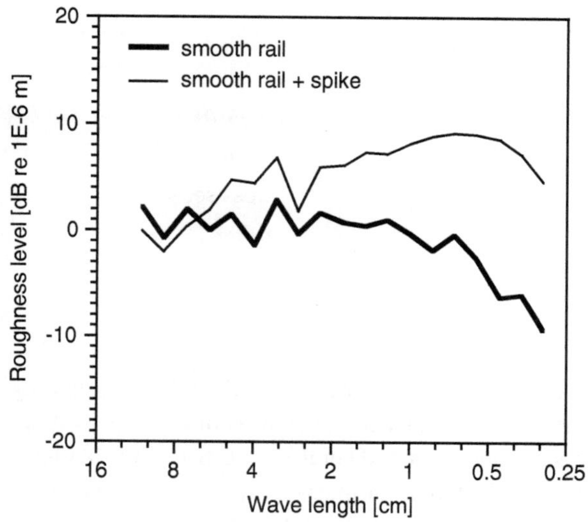

Fig. 16.96 Influence of a single spike on the 1/3-octave-band roughness spectrum

Based on the experience with the RM1200E a wheel roughness device RMR1435 was built, which uses the axle bearing as a reference point for turning the wheel round (see, e.g. [234]). Due to this fact the precision of the results is lower than for the rail instrument, as the roughness measured includes the unroundness of the bearing and axle construction. The data evaluation algorithm is similar to that for rail roughness; in addition to the 1/3-octave-band spectra the so-called wheel harmonics are extracted from the FFT spectra, attributing a value to the unroundness (first FFT line) and the following *n*-th order polygonisation of the wheels.

British Rail Research and its successors have been using an acceleration measurement based trolley system for rail roughness measurements. The advantage of the system is the ability to measure unlimited length sections, the drawback is the need for double integration from acceleration to displacement to arrive at the desired quantity needed to calculate the 1/3-octave-band roughness spectra.

With growth of experience about roughness excitation the necessity to establish values for the

Fig. 16.97 Position of traces for roughness measurements according to [EN ISO 3095 [11]]

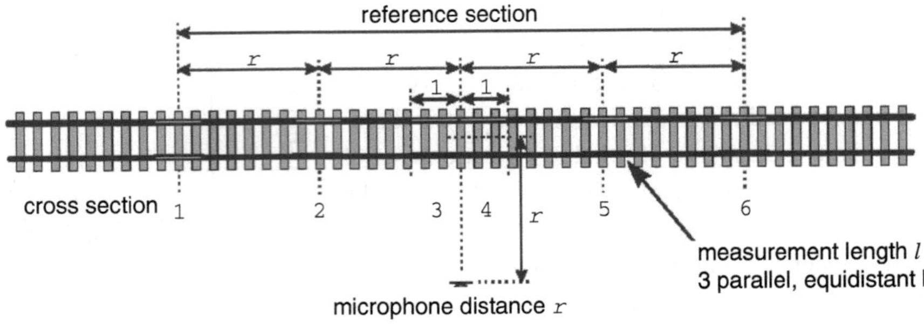

quantitative characterization of rail and wheel running surfaces, especially in the context of e.g. type testing, was acknowledged. Therefore the two ISO standards [11, 79] included the measurement of rail roughness and a limit spectrum not to be exceeded for type testing (see Fig. 16.98). The measurement method described in the standards is based on the available 1.2 m straight edge devices and asks for the measurement of one to three traces depending on the width of the running band at six positions defined by the distance of the microphone from the track as shown in Fig. 16.97.

The idea behind the limit spectrum is to insure that the contribution of the track to the noise creation on interior and pass-by noise is small compared to the vehicle contribution. This approach insures that the vehicle manufacturers have the noise creation under their control and the responsibility in case of exceeding of limit values described in technical specifications for the order of the vehicle is clearly defined. Furthermore it facilitates the reduction of legal limits, as the trackside noise has to be treated separately with the networks.

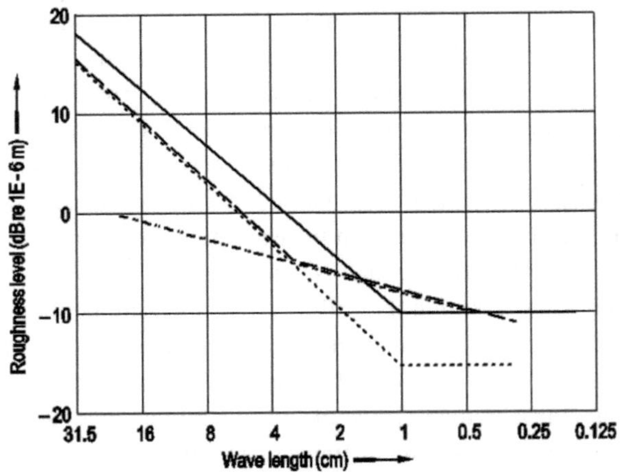

Fig. 16.98 1/3-octave-band roughness limit spectra according to [11], [79] and the European Technical Specifications for Interoperability of the Trans-European High-Speed system, "TSI-HS" as per [EC [229]] and the Trans-European Conventional Rail system, "TSI-CR" as per [EC [230]]; ——— EN ISO 3095, —·—·— TSI according to TSI-HS; – – – – – alternative proposal ATSI according to TSI-CR; OOOOOOO latest proposal TSI + (compromise between TSI and ATSI)

Within the scope of the development of the European high-speed network the *T*echnical *S*pecifications regarding *I*nteroperability of the Trans-European *H*igh-*S*peed system, "TSI-HS" [229], and the Trans-European *C*onventional *R*ail system, "TSI-CR" [230] were defined. Concerning the method for roughness measurements both, TSI-HS and TSI-CR refer to [11]. In the TSI-HS the curve labelled "TSI" in Fig. 16.98 is asked for, whereas in the TSI-CR the curve labelled "ATSI" was proposed. The definitions for the TSI-HS and TSI-CR limit curves are still under discussion and the latest proposal is a compromise between TSI and ATSI called TSI + as shown in Fig. 16.98.

The definition of the limit curves poses demands on the measurement technology concerning the precision necessary and the wavelength range to be measured. The background noise of the instrument should be 3 dB or more below the limit curve to be compared with. A comparison with Fig. 16.99 shows that this is achievable with available instrumentation [226].

A new generation of rail roughness measurement devices, such as the type *mbbmRM1200* [226], are equipped with digitally working displacement transducers. This kind of measurement devices do no longer need a change in the measuring range, and they are featuring a background noise level of approx. −25 dB re. 1 μm.

The wavelength range necessary depends on the frequency range to be considered and the speed of the vehicles. Table 16.9 shows the evaluation of the

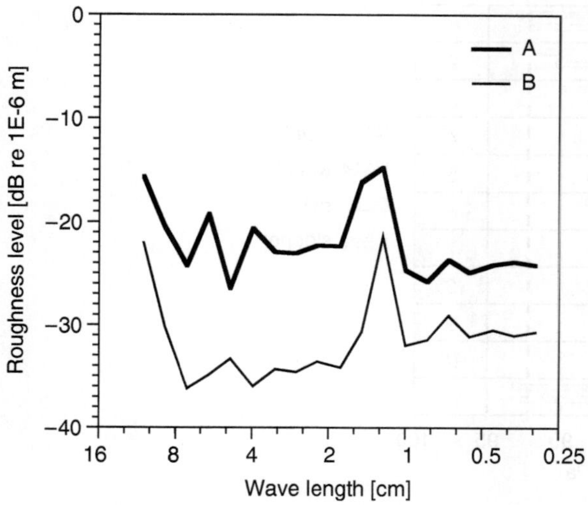

Fig. 16.99 1/3-octave-band background noise level of the system versus band centre wavelength. Parameter: measurement range of the analogue displacement transducer of the rail roughness measurement device RM1200E as per [231, 232]; graph A: ±2.0 mm, graph B: ±0.1 mm

Table 16.9 Relation between frequency f for selected wavelengths λ and train speeds v

Wavelength λ (cm)	f (Hz) at different train speeds v (km/h)			
	80 km/h	160 km/h	240 km/h	320 km/h
200	11	22	33	44
100	22	44	67	89
50	44	89	133	178
25	89	178	267	356
12.5	178	356	533	711
6.3	353	705	1,058	1,411
3.14	708	1,415	2,123	2,831
1.6	1,389	2,778	4,167	5,556

formula $f = v/\lambda$ for relevant speeds v, wavelengths λ and corresponding frequencies f.

Such measurements of the running surface of the rail are indispensable for the understanding of many questions of railway acoustics apart from simulation calculations like e.g. comparative tests on superstructure elements, acceptance tests on rail vehicles etc., as the question of an exact knowledge of the excitation by the running surface of the rail in most cases has not been answered which makes it difficult to solve the problem.

Figure 16.100 shows measuring results for rough and plain running surfaces of wheel and rail.

In [11] methods for the measurement of pass-by levels of trains are described regarding in particular the roughness of the running surfaces of wheels and rails.

16.5.2.4 Application

The roughness model described above has been implemented in various ways. Without claiming to be complete, two implementations should be pointed out: The TWINS package [219] has been developed by order of UIC by ERRI. In contrast to the method which has been previously described in general, some further selected features are listed in the following: Consideration of FE calculations for the modelling of the wheel, model of the rail to consider cross-sectional deformations and discrete bases, extensions so that the contact can be represented better as well as the determination of the effective roughness from measurement results of wheel and rail.

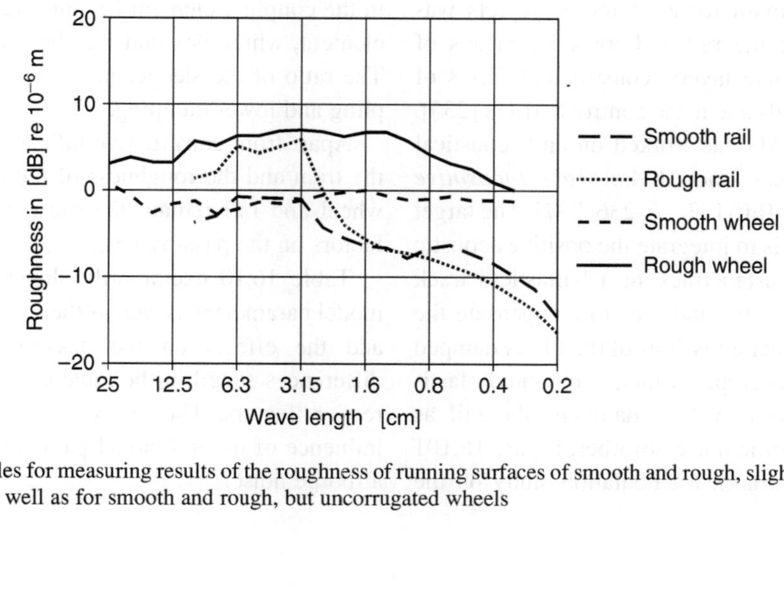

Fig. 16.100 Examples for measuring results of the roughness of running surfaces of smooth and rough, slightly corrugated running surfaces of the rail as well as for smooth and rough, but uncorrugated wheels

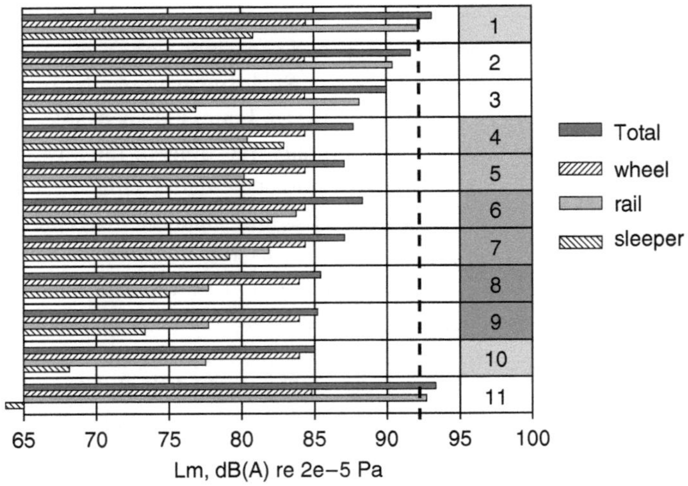

Fig. 16.101 Comparison of the sum and the ratios of the components of the A-weighted pass-by levels of a passenger train when driving on different tracks with a velocity of 200 km/h: *1* Ballastless track; *2* Ballastless track with an increased mass of the base plate; *3* Ballastless track with base plate pad (Zwp) of high damping; *4* Ballastless track with booted Bi-block sleepers; *5* Ballastless track with booted Bi-block sleepers on base plate pads with increased damping; *6* Ballastless track with booted monoblock sleeper; *7* Ballastless track with monoblock sleeper on a base plate pad with increased damping; *8* Ballastless track with monoblock sleeper with inner damping on soffit pads; *9* Ballastless track with monoblock sleeper with inner damping on base plate pad with increased damping; *10* Ballasted track on stiff subgrade; *11* Ballasted track with very soft rail pad (Zw)

The wheel/rail impedance model RIM [220] has been developed for Deutsche Bahn AG: This is marked by its extensions to determine vibration excitation and propagation (see also Sect. 16.3).

These models have mainly been developed in the context of research work dealing with the acoustical optimisation of the rail/wheel system. Regarding the work, which has been coordinated by ERRI, especially the projects "silent freight" and "silent track" should be pointed out which are promoted by the European Commission. The main focus of these projects was the optimisation of the rail and track by means of optimised intermediate layers, construction forms of wheels, bogie shrouds and noise control barriers [235].

Deutsche Bahn AG has worked on an "acoustical innovative ballastless track ("*Akustisch Innovative Feste Fahrbahn*"- AIFF) [33, 36, 236, 237]. The target of this construction is to integrate the positive acoustic features of the ballasted track in a ballastless track ("*Feste Fahrbahn*" (FF)) and thus to compensate the disadvantage of higher emissions of the FF: A damped sleeper has been developed which – mounted elastically on sleeper soffit pads – dampens the rail as a heavy structure-borne noise absorber. Figure 16.101 summarises the results of a calculation study on the basis of the A-weighted overall levels of passing trains, which have been calculated for a velocity of 200 km/h with RIM. The Figure also shows the principal connection between parameters of the superstructure elements and the emitted noise: The contribution of the overall value of the wheel is not much influenced by the superstructure although quite some differences in the frequency range may occur. The ratio of the rail becomes smaller when the coupling to the sleeper becomes stiffer, when the damping in the coupling element becomes larger and when the element, which lies under it, becomes more damped. The ratio of the sleeper increases with a higher coupling and lower damping.

Apart from constructive influences, the velocity of the train and the roughness of the running surface of wheel and rail remain the most important influence factors on the pass-by level.

Table 16.10 exemplarily shows a compilation of model parameters as well as their usual range of values and the effects on the pass-by level. The level differences stated in the table cannot be used as correction factors. They only serve for explaining the influence of the individual parameters on the emitted airborne noise.

Table 16.10 Impact of the track superstructure and vehicle parameters on noise radiation of rail vehicles calculated as per TWINS (from [METARAIL [238], [11], [79]), cf. [239]

Model parameter	Value for minimum noise level	Value for maximum noise level	A-weighted noise level difference between parameters for minimum and maximum values
Rail profile	UIC54	UIC60	0.7 dB
Static stiffness of the rail pad	5,000 MN/m	100 MN/m	5.9 dB
Loss factor of the rail pad	0.5	0.1	2.6 dB
Sleeper type	Bi-block sleeper	Timber sleeper	3.1 dB
Sleeper spacing	0.4 m	0.8 m	1.2 dB
Ballast stiffness	100 MN/m	30 MN/m	0.2 dB
Ballast loss factor	2.0	0.5	0.2 dB
Lateral offset of the rolling line on the wheel	0 m	0.01 m	0.2 dB
Lateral offset of the rolling line on the rail	0 m	0.01 m	1.3 dB
Roughness of wheel running surface	Smooth	Rough	8.5 dB
Roughness of rail running surface for rails without any corrugation	Smooth	Rough	0.7–3.9 dB
Speed of train	80 km/h	160 km/h	9.4 dB
Wheel load	12,500 kg	5,000 kg	1.1 dB
Air temperature	10 °C	30°C	0.2 dB

16.5.3 Interior Noise and Aggregate Noise

16.5.3.1 Noise Management

For the development of vehicles, values according to the specification and limiting values given by law, if existing, must be observed.

Prediction of the outside noise and the interior noise facilitate the design process and make it possible to determine the requirements regarding the sound power of aggregates and the necessary sound insulation measures for the suppliers. Furthermore, it is possible to determine sensitivities for an optimised noise reduction at exactly those sources, which dominate the situation [220, 240].

16.5.3.2 Calculation Models

As a basis for such an investigation, the sources for the development of airborne and structure-borne noise have to be considered as well as their characteristics and the features of the transmission paths. As descriptive values for airborne noise sources, the corresponding sound power levels are necessary, for structure-borne noise sources the corresponding force levels are required. The vehicle is described by its geometry, the position of its sources and the acoustic parameters of its construction (sound insulation of the surfaces limiting the interior, reverberation time in the interior, structure-borne noise insulation of the structural elements etc.). On the basis of the location of the sources and the structural features in longitudinal direction of the vehicle, a segmentation is made, for which the input of airborne noise through outside walls, floor and roof and the borders of the segments as well as the input of structure-borne noise into the segments and the following transmission within the segments is considered.

Figure 16.102 schematically shows a simplified cross section of the vehicle as well as a segment for the model with a listing of the source positions and the relevant structural elements, which have to be considered.

In Tables 16.11 and 16.12 the evaluation of results of such simulation calculations is shown exemplarily for a certain multiple unit. From this, it can be seen which sources generally have to be considered and which are important in the special case of this vehicle, which sources are important for the respective driving situation in this case and which transfer elements are weak points. Such analyses of sensitivity are required for an economical optimisation of the noise situation in and around the vehicle.

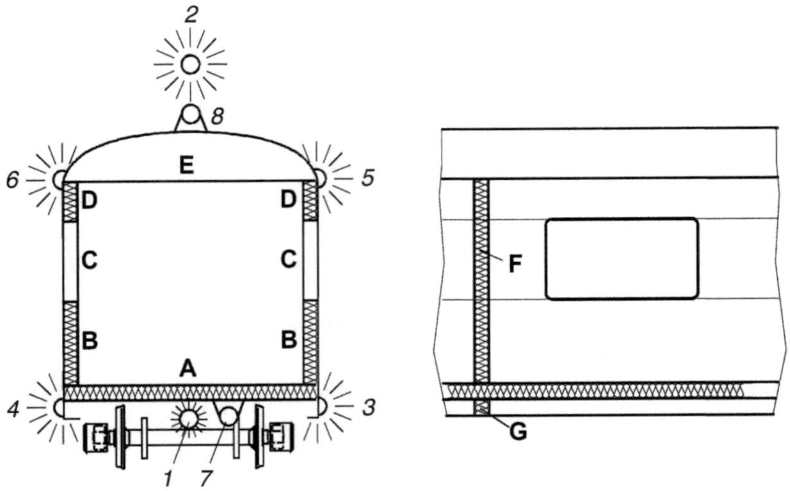

Fig. 16.102 Schematic model for noise management: Vehicle structure as well as structure-borne and airborne noise sources: *1* Airborne noise source at bottom surface of the vehicle floor; *2* Airborne noise source at the top (e.g. pantograph); *3,4* Airborne noise sources at the bottom, lateral (e.g. fan); *5,6* Airborne noise sources at the top, lateral (e.g. air conditioning system); *7* Structure-borne noise source at the bottom surface of the vehicle floor; *8* Structure-borne noise source on the roof, *A* floor, *B* Lateral wall below the window, *D* Lateral wall above the window, *C* Window, *E* Roof, *F* Interior wall and *G* Bulkhead at the bottom surface of the vehicle floor

Table 16.11 Significance of the impact of the transmission element parameters on the interior noise, based on the example of a certain trainset (the evaluation of components presented here is *only* valid for this vehicle)

Transmission element	Open track	Tunnel
Sound insulation of the floor	++++++	+++
Sound insulation of exterior doors	+	+++
Sound insulation of exterior walls	+	++
Sound insulation of the roof		+++
Reverberation time in the driver's cab	+++++	+++++
Reverberation time in the passenger compartment	++++++	++++++
Terminating impedance of the floor	+	+
Sound insulation of the corridor connection	+++	+++
Damping between floor and wall	++++++	++++++

Table 16.12 Significance of the sources for the noise based on the example of a certain trainset (the evaluation of components presented here is *only* valid for this vehicle)

Source of noise	Interior noise	Exterior noise
Rolling noise (rail component)	+++++	+++
Rolling noise (wheel component)	++++	++
Compressed air supply	+	+
Pantograph	+	++
Drive motors	++	++++
Air conditioning	+	
Structure-borne noise from the bogies	+++++	

16.5.3.3 Determination of Model Parameters

Regarding these calculations, the quality of the model parameters is decisive for the quality of the prediction. Suitable databases of measuring results should be used as a basis. The sound power of the components, the sound insulation of the building elements, the input impedances and the loss factors of the structural elements have to be determined for these calculations. For the determination of the parameters of the structure-borne noise coupling elements (dynamic stiffness) the same information can be applied as has already been stated for rolling noise models (see Sect. 16.5.2.3).

In [79] methods for the measurement of interior noise are described. Here, the roughness of the running surfaces is especially important as a significant boundary condition.

16.5.4 Structure-Borne Noise and Vibrations

Regarding vibrations and structure-borne noise, the prediction calculations are much more complex than for airborne noise. A cause for this is e.g. the inhomogeneity of the soil and in comparison with air the fact that several different wave types can appear. The inhomogeneity of the soil can both be caused by the bedding of different types of soil as well as by inhomogeneities in a normally rather uniform type of rock. However, such inhomogeneities only have to be considered, if their dimensions are in the range of a wavelength or above as individual smaller embeddings (as e.g. erratic blocks) are not "noticed" by waves, which are propagating within the soil, since they virtually integrate the conditions of the soil of a whole wavelength. Thus, individual small inhomogeneities do normally not influence the propagation of vibrations.

Another problem in the detection and prediction of vibrations is the coupling of buildings to the soil as well as of structures in buildings. Thus, it is easily possible that by a connection of unfavourable circumstances the vibrations excited in the ground by a train are intensified, e.g. by resonances of room ceilings or floating floors.

Owing to its significance and complexity, the subject of vibrations in the ground was also recognised as significant by UIC and was examined in the context of projects of ERRI [97] and ORE [147]. Generally approved simulation models regarding the subject of vibrations are not known at present, in contrast to e.g. models for the development of rolling noise for railways.

16.5.4.1 Empirical Models

At present, in most cases empiric models are used for the practical work regarding the planning [32, 241]. These models are based on measuring results for the excitement as well as for the transfer into the ground and the coupling to buildings as well as on partly calculated transfer functions in buildings (see Sect. 16.3.10). Therefore, for a new project it always has to be assessed if a prediction can be made with existing data of a similar project or if measurements are necessary. For new buildings at existing tracks, this is normally not a problem as measurements can be carried out on the building ground or in comparable buildings. For new tracks a higher inaccuracy must be expected, as in this case both the excitation as well as the propagation has to be determined from similar cases. As target values for the planning, KB values and A-weighted secondary airborne noise levels are calculated depending on the eigenfrequencies of the ceilings.

16.5.4.2 Physical Models

As the enquiries made in the context of the project RENVIB II [97] showed, there are many approaches for the calculation of the coupling of tracks to the ground [97, 236, 237, 242, 243], of the propagation of vibrations in the ground, of the coupling of buildings to the ground as well as for the propagation in buildings. The complexity of the models varies a lot and thus also the accuracy of the calculations.

However, the main problem regarding exact calculations is not the physical models or their use, but the parameters applied in these models. Regarding the determination of the dynamic parameters of the track, the same is valid as has already been stated for rolling noise models.

Additionally, the dynamic parameters for the ground and its bedding must be determined. Experience shows that one can rather rely on the classification of soil and look-up parameters in the respective tables. The static characteristic parameters which have been determined for the building ground are in most cases useless for the prediction calculation if the connection with dynamic soil parameters is not known [236, 237, 244, 245]. Owing to the inaccuracy of the parameters describing the soil, it is easier to make a prediction calculation for the insertion loss of a system for a certain geological situation than to determine absolute values e.g. of the velocity level.

For studies and for the planning of measures to reduce structure-borne noise/vibrations depending on vehicle, track and soil type, this is absolutely sufficient and target oriented (see Sect. 16.3.7). Also, this saves the user from having to specify exactly an excitation for the model, which is normally difficult to define.

References

1. Stüber C (1975) Geräusche von Schienenfahrzeugen, Kapitel 15. In: Heckl M, Müller H A (Hrsg) Taschenbuch der Technischen Akustik, 1. Auflage, Springer, Berlin
2. Akustik 04 (1990) Richtlinie für schalltechnische Untersuchungen bei der Planung von Rangier- und Umschlagbahnhöfen. Information Akustik 04 der Deutschen Bundesbahn, Ausgabe 1990
3. Schall 03 (1990) Richtlinie zur Berechnung der Schallimmissionen von Schienenwegen – Schall 03; Information Akustik 03 der Deutschen Bundesbahn, Ausgabe 1990
4. IF-Studie (1983) Interdisziplinäre Feldstudie II über die Besonderheiten des Schienenverkehrslärms gegenüber dem Straßenverkehrslärm. Forschungs-Nr. 70081/80 des Bundesministers für Verkehr, München/Bonn 1983. Planungsbüro Obermeyer (Hrsg), München
5. BImSchV-16 (1990) Sechzehnte Verordnung zur Durchführung des Bundes-Immissionsschutzgesetzes (Verkehrslärmschutzverordnung – 16. BImSchV) – 1990
6. DIN 45645–1 (1977) Einheitliche Ermittlung des Beurteilungspegels für Geräusch-Immissionen, Teil 1, April 1977
7. Wettschureck RG (1994) Vibration and Structure-Borne Sound Isolation by means of Cellular Polyurethane (PUR) Elastomers. Proc. SVIB-Vibrationsdagen 1994, Edition Swedish Vibration Society, Stockholm, 1994, pp. 21–52
8. EBO (1967) Eisenbahn- Bau- und Betriebsordnung vom 08.05.1967 (Bundesgesetzblatt 1967, Teil II, S. 1563)
9. BOStrab (1987) Verordnung über den Bau und Betrieb der Straßenbahnen (Straßenbahn- Bau- und Betriebsordnung – BOStrab) vom 11.12.1987 (Bundesgesetzblatt 1987, Teil I, S. 2648)
10. DIN 45641 (1990) Mittelung von Schallpegeln, Juni 1990
11. EN ISO 3095 (2005) Railway applications – Acoustics – Measurement of noise emitted by railbound vehicles, August 2005
12. DIN 45652 (1964) Terzfilter für elektroakustische Messungen, Ausgabe Januar 1964
13. Maglev (1989) Die neue Dimension des Reisens, Hrsg.: MVP Versuchs- und Planungsgesellschaft für Magnetbahnsysteme, Transrapid International, Gesellschaft für Magnetbahnsysteme, Darmstadt, Hestra, 1989
14. Maglev (1992) Die Fahrzeug- und Fahrwerktechnik Transrapid. Forschungsinformation Bahntechnik, ETR 41 (1992): 275–278
15. DB (1979–2001) Berichte (nicht veröffentlicht) der Deutschen Bundesbahn, Versuchsanstalt (VersA) München im Auftrag des BZA München, Dezernat 103/103a, bzw. der Deutschen Bahn AG, Forschungs- und Technologiezentrum München (jetzt DB Systemtechnik, München)
16. NN (1972) Noise control on Shinkansen. Railway Gazette International, July 1973: 249–251
17. Willenbrink L (1979) Neuere Erkenntnisse zur Schallabstrahlung von Schienenfahrzeugen. ETR 28:355–362
18. BMFT (1976) Abschlußbericht zum Forschungsvorhaben TV 7420 "Ermittlung und Erprobung von passiven Maßnahmen zur Verminderung von Schallemissionen bei hohen Geschwindigkeiten". Hrsg.: BMFT, März 1976
19. BMFT (1980) Technischer Schlussbericht zum Forschungsvorhaben TV 7630 "Passive Schallschutzmaßnahmen für das Rad/Schiene-System bei hohen Geschwindigkeiten". Hrsg.: BMFT, 1980
20. Hölzl G, Hafner P (1980) Schienenverkehrsgeräusche und ihre Minderung durch Schallschutzwände. Z Lärmbekämpfung 27:92–99
21. Barsikow B (1989) Schallabstrahlung spurgebundener Hochgeschwindigkeitsfahrzeuge bis 500 km/h. Fortschritte der Akustik, DAGA '89, Duisburg, 1989: pp 607–610
22. Barsikow B, Müller H (1992) Entwurf einer Richtlinie zur Berechnung der Schall-Immission an Strecken der Magnetschnellbahn Transrapid in Anlehnung an die Richtlinie "Schall 03" der Deutschen Bundesbahn. Fortschritte der Akustik, DAGA 92, Berlin, 1992: 333–336
23. Maglev (2006) Proc. MAGLEV'2006 – The 19th International Conference on Magnetically Levitated Systems and Linear Drives, incorporating the "6. Dresdner Fachtagung Transrapid", Dresden, Germany, 2006, http://www.maglev2006.de/
24. Darr E, Fiebig W (1999) Feste Fahrbahn – Konstruktion, Bauarten, Systemvergleich Feste Fahrbahn – Schotteroberbau". Tetzlaff, Hamburg
25. Kaess G, Schultheiss H (1984) Der Oberbau der Neubaustrecken der Deutschen Bundesbahn. Eisenbahningenieur 35:421–428
26. DB Netz AG (1998–2001) Messberichte zu akustischen Untersuchungen an Schall absorbierenden Festen Fahrbahnen unter fachlicher Leitung des Forschungs- und Technologiezentrums der Deutschen Bahn AG, im Auftrag der DB Netz AG (1998 bis 2001, nicht veröffentlicht)
27. DB AG (1997) Antrag auf Anerkennung des Nachweises der schalltechnischen Wirkung einer Schall absorbierenden Gestaltung der Oberfläche der Festen Fahrbahn, Deutsche Bahn AG, 22.08.1997
28. Wettschureck RG, Altreuther B, Daiminger W, Nowack R (1996) Körperschallmindernde Maßnahmen beim Einbau einer Festen Fahrbahn auf einer Stahlbeton-Hohlkastenbrücke. ETR 45(H6):371–379
29. Fingberg U (1990) Ein Modell für das Kurvenquietschen von Schienenfahrzeugen. Fortschr. VDI 140, Reihe 11, 1990: pp 1–96

30. ORE C137 (1975) Geräuschbelästigung durch Bremsen und Befahren enger Gleisbogen. Grundlagen und erste Versuchsergebnisse. Bericht Nr. 2, Utrecht, 1975
31. Pratt TK, Williams R (1981) Non-linear analysis of stick-slip motion. J Sound Vib 74:531–542
32. Müller-BBM (1979–2001) Müller-BBM Berichte im Auftrag des früheren Bundesbahn-Zentralamtes München, Dezernat 103/103a und anderer Dienststellen der Deutschen Bundesbahn, bzw. des Forschungs- und Technologiezentrums München der Deutschen Bahn AG (nicht veröffentlicht)
33. Diehl RJ, Görlich R, Hölzl G (1997) Acoustic Optimization of Railroad Track Using Computer Aided Methods. In Proc.*WCRR97 – World Congress Railroad Research*, Florence 1997, Vol. E: 421–427
34. Grassie SL (1983) Comments on "Surface irregularities and variable mechanical properties as a cause of rail corrugation" von Kalker JJ in: Rail Corrugation (Symposium, Berlin, Juni 1983), ILR-Bericht Nr. 56, S 107–110
35. Grassie SL, Gregory RW, Harrison D, Johnson KL (1982) The dynamic response of railway track to high frequency vertical excitation. J Mech Eng Sci 24:77–90
36. Nowack R, Hölzl G, Diehl JR, Bachmann H, Mohr W (1999) Die Akustisch Innovative Feste Fahrbahn. ETR 48 (H9):571–582
37. Bridge Study (1987) Untersuchungen zur Verringerung der Schallabstrahlung von stählernen Eisenbahnbrücken durch kontruktive Maßnahmen. Abschlußbericht zum Projekt 104 der Studiengesellschaft für Anwendungstechnik von Eisen und Stahl e.V., Düsseldorf, December 1987
38. Hölzl G, Nowack R (1996) Experience of German Railways on noise emission of railway bridges. Proc. Workshop on Noise Emission of Steel Railway Bridges, Rotterdam 1996. NS Technisch Onderzoek (Hrsg), Utrecht
39. Wettschureck RG (1996) Measures for reduction of the noise emission of railway bridges. Proc. Workshop on Noise Emission of Steel Railway Bridges, Rotterdam 1996. NS Technisch Onderzoek (Hrsg), Utrecht, 1996
40. Nowack R (1998) Elastische Schienenbefestigungssysteme als schallmindernde Maßnahme bei Stahlbrücken ohne Schotterbett. ETR 47(H4):215–222
41. Wettschureck RG, Heim M (1998) Reduction of the Noise Emission of a Steel Railway Bridge by means of Resilient Rail Fastenings with Dynamically Soft Baseplate Pads, Proc. Euro-Noise ´98, Munich, 1998, vol I, pp 289–294
42. Wettschureck RG, Diehl RJ (2000) The dynamic stiffness as an indicator of the effectiveness of a resilient rail fastening system applied as a noise mitigation measure: laboratory tests and field application. Rail Engineering International, Edition 2000, No. 4: 7–10
43. EN 13481–6 (2002) Railway applications – Track Performance requirements for fastening systems – Part 6: Special fastening systems for attenuation of vibration, 2002
44. ISO 10846–2 (1997) Acoustics and vibration – Laboratory measurements of vibro-acoustic transfer properties of resilient elements Part 2: Dynamic stiffness of elastic supports for translatory motion – Direct method, 1997
45. DB-TL (1988) Technische Lieferbedingungen "Unterschottermatten" der DB AG – DB-TL 918 071, Ausgabe 1988
46. Akustik 22 (1990) Verringerung der Schallabstrahlung von Eisenbahnbrücken durch zusätzliche Maßnahmen. Information Akustik 22 der Deutschen Bundesbahn, Ausgabe Januar 1990
47. Wettschureck RG (1987) Unterschottermatten auf einer Eisenbahnbrücke in Stahlbeton-Verbundbauweise. Fortschritte der Akustik, DAGA '87, Aachen, 1987: 217–220
48. Jäger K, Onnich H (2000) Fortschritte und Besonderheiten bei der Reduzierung des Schienenverkehrslärms. Z f Lärmbekämpfung 47:206–210
49. Hauck G, Onnich H, Prögler H (1997) Entwicklung eines Messwagens zur Erfassung der Fahrge-räuschanhebungen durch Schienenriffeln. ETR 46:153–159
50. Jäger K, Hauck G (1986) Neue Erkenntnisse und Berechnungsverfahren bei der Schallpegelermittlung im Umfeld großflächiger Rangieranlagen. AET – Archiv für Eisenbahntechnik 4:16–23
51. Bahnhofstudie 2 (1986) Studie über die Schallemission von Bahnhöfen im Vergleich mit der freien Strecke (Bahnhofstudie 2). Forschungsvorhaben im Auftrag des Bundesministers für Verkehr und des Bundesbahn-Zentralamtes München; Müller-BBM GmbH, Planegg 1986
52. Jäger K (1991) Schalltechnische Untersuchungen bei der Planung von Rangier- und Umschlagbahnhöfen; Berechnung nach Akustik 04 in der Neufassung von 1990. Z Lärmbekämpfung 38:144–150
53. VDI 2714 (1988) Schallausbreitung im Freien, Januar 1988
54. Kurze U J (1987) Long range barrier attenuation of railroad noise. Proc. Inter-Noise '87, Beijing, China, 1987: 379–382
55. VDI 2720–1 (1991) Schallschutz durch Abschirmung im Freien, Blatt 1 (E), Februar 1991
56. Maekawa Z (1968) Noise reduction by screens. Appl Acoust 1:157–173
57. Kurze UJ (1980) Abschirmung an Bahnanlagen. Acustica 45:304–315
58. DB AG (2000) Richtlinie 800.2001 der Deutschen Bahn AG, DB Netz: Netzinfrastruktur Technik Entwerfen; Lärmschutzanlagen an Eisenbahnstrecken, Januar 2000
59. Möser M (1995) Die Wirkung von zylindrischen Aufsätzen an Schallschirmen. Acustica 81:565–586
60. Möser M, Volz R (1999) Improvement of sound barriers using headpieces with finite acoustic impedance. J Acoust Soc Am 106(6):3049–3060
61. BImSchG (1994) Gesetz zum Schutz vor schädlichen Umwelteinwirkungen durch Luftverunreinigungen, Geräusche, Erschütterungen und ähnliche Vorgänge (Bundes-Immissionsschutzgesetz – BImSchG vom 15. März 1974, geändert am 14.5.1990, am 22.4.1993 und am 30.6.1994
62. OPB (1985) Die unterschiedliche Lästigkeit von Schienen- und Straßenverkehrslärm innerhalb und außerhalb von Wohngebäuden – "Fensterstudie", Planungsbüro Obermeyer (jetzt OPB – Obermeyer Planen + Beraten), München, April 1985
63. OPB (1986) Kommunikationsstörungen durch Schienenverkehrslärm – "Ergänzungsstudie" (Ergänzende

Auswertungen zur "Fensterstudie"). Planungsbüro Obermeyer (jetzt OPB – Obermeyer Planen + Beraten), München, Januar 1986
64. Heimerl G, Holzmann E (1978) Ermittlung der Belästigung durch Verkehrslärm in Abhängigkeit von Verkehrsmittel und Verkehrsdichte in einem Ballungsgebiet. Forschungsarbeiten des Verkehrswissenschaftlichen Instituts der Universität Stuttgart, Bericht 13, Stuttgart
65. Klosterkötter W, Gono F (1978) Bericht über Untersuchungen von Schienenverkehrs-, Flug- und Straßenverkehrslärm im Hinblick auf Differenzen ihrer A- und C-Schallpegel, Essen
66. Möhler U (1987) Zum Einfluss der Fensterstellung auf die Lästigkeitswirkung von Verkehrslärm. Fortschritte der Akustik – DAGA '87, Aachen, 1987: 761–764
67. Liepert M, Möhler U, Schreckenberg D, Schuemer R (2000) Lästigkeitsunterschied von Straßen- und Schienenverkehrslärm im Innenraum. SGS, München
68. BImSchV-24 (1997) Vierundzwanzigste Verordnung zur Durchführung des Bundes-Immissionsschutzgesetzes (Verkehrswege-Schallschutzmaßnahmenverordnung – 24. BImSchV) vom 4.2.1997, Bundesgesetzblatt Jahrgang 1997 Teil I Nr.8
69. OPB (1979–2001) Berichte von Obermeyer Planen + Beraten, München [früher Planungsbüro Obermeyer (PBO)], im Auftrag des früheren Bundesbahn-Zentralamtes München, Dezernat 103/103a bzw. der Deutschen Bahn AG, Forschungs-und Technologiezentrum München (nicht veröffentlicht)
70. UIC (1994) Internationaler Eisenbahnverband, UIC-Kodex 651 VE "Gestaltung der Füherräume von Lokomotiven, Triebwagen, Triebwagenzügen und Steuerwagen" (Ausgabe 01. 01. 1994)
71. Möhler U, Schuemer R, Knall V, Schuemer-Kohrs A (1986) Vergleich der Lästigkeit von Schienen- und Straßenverkehrslärm. Z Lärmbekämpfung 33:132–142
72. Hauck G (1991) Lästigkeitsunterschied zwischen Geräuschen des Straßen-verkehrs und des Schienenverkehrs. Z Lärmbekämpfung 38:162–166
73. VDI 3722-1 (1988) "Wirkungen von Verkehrsgeräuschen" Blatt 1, August 1988
74. Griefahn B, Möhler U, Schuemer R (Hrsg) (1999) Vergleichende Untersuchung über die Lärmwirkung bei Straßen- und Schienenverkehr. Studiengemeinschaft Schienenverkehr, Hrsg. FTZ der DB AG, München
75. Zeichart K, Kilcher H, Hermann W, Hils T, Gawlik M (1999) Untersuchung zur Lästigkeit von Hochgeschwindigkeitszügen am Beispiel der Neu- und Ausbaustrecke Hannover-Göttingen. Studiengemeinschaft Schienenverkehr, Hrsg. FTZ München der DB AG, 1999
76. Maglev (1997) Magnetschwebebahn -Lärmschutzverordnung, Bundesgesetzblatt, Jahrgang 1997, Teil 1-Nr. 76, vom 25. September 1997, Bonn, pp 2338–2344
77. Jäger K, Schöpf F, Gottschling G, Fastl H, Möhler, U (1996) Wahrnehmung von Pegeldifferenzen bei Vorbeifahrten von Güterzügen. Forschungsbericht der TU München im Auftrag der DB AG (ZBT 512), München
78. Jäger K, Schöpf F, Gottschling G, Fastl H, Möhler, U (1997) Wahrnehmung von Pegeldifferenzen bei Vorbeifahrten von Güterzügen. Fortschritte der Akustik – DAGA ́97, Kiel, 1997: 228–229
79. EN ISO 3381 (2005) Railway applications – Acoustics – Measurement of noise inside railbound vehicles, August 2005
80. Hald J (1989) STSF – a unique technique for scan-based Near-Field Acoustic Holography without restrictions on coherence, Brüel & Kjaer Technical Review (1989), No.1
81. Nordborg A (2000) Optimum Array Microphone Configuration, Proc. Inter-Noise 2000, Nizza: 2474–2478
82. Cremer L, Heckl M (1995) Körperschall, Physikalische Grundlagen und Technische Anwendungen. 2., völlig neubearbeitete Auflage. Springer, Berlin
83. Meyer E, Guicking D (1974) Schwingungslehre. Vieweg, Braunschweig
84. DIN 45672–1 (1991) Schwingungsmessungen in der Umgebung von Schienenverkehrswegen, Teil 1: "Meßverfahren", September 1991
85. DIN 45669–2 (1995) "Messung von Schwingungsimmissionen, Teil 2: Meßverfahren", Juni 1995
86. DIN 4150–2 (1999) Erschütterungen im Bauwesen, Teil 2: "Einwirkung auf Menschen in Gebäuden", Juni 1999
87. DIN 4150–3 (1999) Erschütterungen im Bauwesen, Teil 3: "Einwirkungen auf bauliche Anlagen", Februar 1999
88. Heckl M (1990) Körperschallentstehung bei Schienenfahrzeugen. Fortschritte der Akustik, DAGA '90, Wien, 1990: 135–140
89. Heckl M, Feldmann J, Fischer HM, Munjal M (1980) Grundlegende Untersuchungen zur Körperschallentstehung beim Rad/Schienesystem, Teil 1 und Teil 2. BMFT-Vorhaben TV-7722/9, TU Berlin
90. Thompson DJ (1991) Theoretical modelling of wheel-rail noise generation. Proc. Instn. Mech. Engrs, Part F (J Rail Rapid Transit 205 (1991): 137–149
91. Möhler U, Prestele G, Giesler H-J, Hendlmeier W (1998) Schallemissionen von Schienennahverkehrsbahnen. Z. Lärmbekämpfung 45, 1998 Nr. 6: 209–215
92. Bugarcic H, Thevis P, Breznovsky M, Lierke K (1986) Primärunterdrückung der Bogenlaufgeräusche durch alternative Radsatzstell- und steuermechanismen. Bericht des Instituts für Fahrzeugtechnik, Fachgebiet "Spurgebundene Fahrzeuge" TU Berlin, Dezember 1986
93. Raquet E (1986) Untersuchungen zur Schallminderung durch absorbergedämpfte Räder. Krupp Stahl AG, Essen, 1986
94. ORE D151 (1986) Schwingungsschutzmaßnahmen auf der freien Strecke: Auswirkung zusätzlicher elastischer Bettung des Gleises. Bericht RP 10, Utrecht, April 1986
95. Oberweiler G (1989) Die Feste Fahrbahn. Entwicklung und Beurteilung aus der Sicht des Anwenders. ETR 38:119–124
96. Prange B, Vrettos Ch, Huber G, Tamborek A (1987) Erschütterungsabstrahlung der Festen Fahrbahn im Vergleich zum Schotteroberbau. 7. Technischer Bericht (Meilensteinbericht B7c) zum Forschungsvorhaben des BMFT TV8227 Teil B, Karlsruhe, Dezember 1987
97. RENVIB II (1997–2) Phase 1 – "State of the art review", Task 5, Mitigation Measures for Surface Railways. Müller-BBM Report No. 34 441/3, für ERRI – European Railway Research Institute, Utrecht, Oktober 1997
98. Krüger F (1988) Parametereinfluß auf die Schwingungsemissionen an der Tunnelsohle. Verkehr und Technik 1988, H 9

99. DB (1983) Schutz gegen Körperschall und Erschütterungen bei unterirdisch geführten S-Bahnen. Information Körperschall/Erschütterungen 01 der Deutschen Bundesbahn, Ausgabe 1983
100. Eisenmann J, Deischl F (1986) Körperschalldämmung bei unterirdischen Bahnanlagen – Ausführungsbeispiele. Eisenbahningenieur 37:101–110
101. Kurze UJ, Wettschureck RG (1985) Erschütterungen in der Umgebung von flach liegenden Eisenbahntunneln im Vergleich mit freien Strecken. Acustica 58:170–176
102. Auersch L (1983) Ausbreitung von Erschütterungen durch den Boden. Forschungsbericht Nr. 92 der BAM, Berlin, 1983
103. DIN 4150–1 (2001) "Erschütterungen im Bauwesen, Grundsätze, Teil 1: Vorermittlung und Messung von Schwingungsgrößen", Juni 2001
104. Haupt W, Hrsg (1986) Bodendynamik. Grundlagen und Anwendung. Vieweg, Braunschweig, 1986
105. Prange B, Huber G, Triantafyllidis Th (1982–2) Dynamisches Rückwirkungsmodell des Gleisoberbaus: Feldmessung und analytisches Modell. Abschlußbericht zum Forschungsvorhaben TV 78150B des BMFT, 1982
106. Eibl J, Henseleit O, Schlüter, F-H (1988) Baudynamik. In: Betonkalender 1988. Ernst & Sohn, Berlin
107. Studer J, Ziegler A (1986) Bodendynamik. Grundlagen, Kennziffern, Probleme. Springer, Berlin
108. ARGE (1980/81) "Arbeitsgemeinschaft Schwingungsausbreitung" – Müller-BBM, IGI Niedermeyer, LGA Bayern – Schwingungsausbreitung an Schienenverkehrswegen. Hauptstudie 1, Juli 1980 und Hauptstudie 2, Juli 1981, im Auftrag des Bundesbahn-Zentralamtes München
109. Hölzl G (1982) Körperschall- bzw. Erschütterungsausbreitung an Schienenverkehrswegen. Ergebnisse des Forschungsvorhabens und praktische Anwendungen bei der DB. ETR 31:881–887
110. Hölzl G, Fischer G (1985) Körperschall- bzw. Erschütterungsausbreitung an oberirdischen Schienenverkehrswegen. ETR 34:469–477
111. Kurze UJ (1982) Erschütterungen von Eisenbahnen. Forschritte der Akustik, FASE/DAGA '82, Göttingen, 1982: 329–332
112. Auersch L (1984) Durch Bodenerschütterungen angeregte Gebäudeschwingungen – Ergebnisse von Modellrechnungen. Forschungsbericht Nr. 108 der BAM, Berlin, 1984
113. VDI 2057–3 (1987) "Einwirkung mechanischer Schwingungen auf den Menschen, Blatt 3: Beurteilung", Mai 1987
114. Grütz H-P, Said A (1992) Zur Ermittlung des sekundären Luftschalls aus oberirdischem Schienenverkehr. Fortschritte der Akustik, DAGA '92, Berlin, 1992: 353–356
115. Krüger F (1990) Verfahren zur Abschätzung des zu erwartenden sekundären Schalldruckpegels im Rohbaustadium von Tunnelstrecken. Fortschritte der Akustik, DAGA '90, Wien, 1990: 465–468
116. NN (1980) Proc. Conference "Low Frequency Noise and Hearing", Aalborg, Denmark, May 1980
117. Volberg G (1980) Tieffrequenter Luftschall in Gebäuden Fortschritte der Akustik, DAGA '80, München, 1980: 305–308
118. Wietlake KH (1983) Beurteilung und Minderung tieffrequenter Geräusche. LIS-Berichte, H. 38, Essen
119. DIN 45669–1 (1995) "Messung von Schwingungsimmissionen, Teil1: Schwingungsmesser; Anforderungen, Prüfungen", Juni 1995
120. Zeichart K, Sinz A, Schuemer R, Schuemer-Kohrs A (1993) Erschütterungswirkungen aus dem Schienenverkehr. Bericht über ein interdisziplinäres Forschungsvorhaben im Auftragedes Umweltbundesamtes, Berlin und des Bundesbahn-Zentralamtes, München – Kurzfassung -. Obermeyer Planen + Beraten, München, Februar 1993
121. Zeichart K, Sinz A, Schuemer-Kohrs A, Schuemer R (1994) Erschütterungen durch Eisenbahnverkehr und ihre Wirkungen auf Anwohner. Teil 1: Zum Zusammenwirken von Erschütterungs- und Geräuschbelastung. Z. Lärmbekämpfung 41: 43–51. Teil 2: Überlegungen zu Immissionsrichtwerten für Erschütterungen durch Schienenverkehr. Z. Lärmbekämpfung 41 (1994): 104–111
122. BayerVGH (1995) Urteil des Bayerischen Verwaltungs-Gerichtshofes vom April 1995, Az. 20 A 93 400 80
123. Said A, Fleischer D, Fastl H, Grütz H.-P, Hölzl G (2000) Laborversuche zur Ermittlung von Unterschiedsschwellen bei der Wahrnehmung von Erschütterungen aus dem Schienenverkehr. Fortschritte der Akustik – DAGA 2000, Oldenburg, 2000: 496–497
124. Said A, Fleischer D, Fastl H, Kilscher H, Grütz H.-P (2001) Einfluss des Spitzenwertes (KB_{Fmax}) bei der Wahrnehmung von eisenbahnspezifischen Erschütterungen. Erscheint in: Fortschritte der Akustik – DAGA 2001, Hamburg, 2001
125. VDI 2719 (1987) "Schalldämmung von Fenstern und deren Zusatzeinrichtungen" August 1987
126. DIN 45680 (1997) "Messung und Beurteilung tieffrequenter Geräuschimmissionen in der Nachbarschaft", März 1997
127. Said A, Grütz H-P, Garburg R (2006) Determination of Groundborne Noise in Buildings due to at-grade Railway Traffic (in German). Z Lärmbekämpfung 53(1):12–18
128. Krüger F (2006) Groundborne noise – Prediction and Assessment (in German): Proc. Congress Baudynamics, VDI-Reports No. 1941, Kassel, 2006: pp 85–102
129. BMFT (1982) Untersuchung verschiedener Oberbauformen in einem U-Bahntunnel im Hinblick auf Schall- und Erschütterungsemissionen. Bericht Nr. 8 der Berichtsreihe "Lärmminderung Schienennahverkehr" des BMFT, STUVA e.V., Köln, März 1982
130. Eisenmann J (1985) Oberbauforschung – Oberbautechnik, Stand und Weiterentwicklung. ETR 34:715–722
131. Krüger F (1985) Minderung der Schwingungsabstrahlung von U-Bahntunneln durch hochelastische Gleisisolationssysteme unter verschiedenen Tunnelrandbedingungen. Bericht Nr. 17 der Berichtsreihe "Lärmminderung Schienennahverkehr" des BMFT, STUVA e.V. Köln und IBU Essen, Februar 1985
132. Wettschureck RG (1985) Ballast mats in tunnels – Analytical model and measurements. Proc. Inter-Noise 1985, München, 1985: 721–724
133. Wettschureck RG, Doberauer D (1985a) Unterschottermatten im Münchner S-Bahntunnel. Fortschritte der Akustik, DAGA '85, Stuttgart 1985: 211–214
134. Wettschureck RG, Kurze UJ (1985b) Einfügungsdämmaß von Unterschottermatten. Acustica 58:177–182
135. Heckl M (1981) Körperschallübertragung bei homogenen Platten beliebiger Dicke. Acoustica 49:183–191

136. Wettschureck RG, Breuer F, Tecklenburg M, Widmann H (1999) Installation of highly effective vibration mitigation measures in a railway tunnel in Cologne, Germany. Rail Engineering International, Edition 1999 No. 4:12–16
137. Wettschureck RG, Daiminger W (2001) Installation of high-performance ballast mats in an urban railway tunnel in the city of Berlin. Proc. 4th European Conference on Noise Control (on CD-ROM), Euro-Noise´01, Patras, Greece, 2001
138. Achilles S, Wettschureck RG (2004) Track isolation in a light rail tunnel in downtown Berlin. Proc. French/German Joint Meeting CFA/DAGA´04 (on CD-ROM), Strassbourg, 2004
139. Wettschureck RG, Heim M, Tecklenburg M (2002) Long-term properties of Sylomer® ballast mats installed in the rapid transit railway tunnel near the Philharmonic Hall of Munich, Germany. Rail Engineering International, Edition 2002 No. 4, pp 6–11
140. Pichler D, Mechtler R, Plank R (1997) Entwicklung eines neuartigen Masse-Feder-Systems zur Vibrationsminderung bei Eisenbahntunnels. Bauingenieur 72:515–521
141. Pichler D, Zindler R (1999) Development of artificial elastomers and application to vibration attenuating measures for modern railway superstructures. In: Constitutive Models for Rubber, Dormann&Muhr (Editors), Balkema, Rotterdam, 1999: 257–266
142. RENVIB II (1997–1) Phase 1 – "State of the art review", Task 4, Reduction measures for tunnel lines, Bericht von VCE-Vienna Consulting Engineers, Wien und Rutishauser Ingenieurbüro, Zürich, für ERRI – European Railway Research Institute, Utrecht, 1997
143. Wenzel H, Pichler D Rutishauser G (1998) Reduktion von Lärm und Vibrationen durch Masse-Feder-Systeme für Hochleistungseisenbahnen. Schweizer Ingenieur- und Architektenverein, Dokumentation 0145 (1998): 123–131
144. Enoekl V, Lenz U (2003) Erstes Masse-Feder-System auf einer HGV-Strecke. ETR 52(H. 9):527–538
145. Rubi H-P, Hejda G, Rutishauser G, Kleiner P (1991) S-Bahn-Technik – Gleisoberbau und Körperschallschutzmaßnahmen. Schweizer Ingenieur und Architekt 109, 29:701–706
146. Krüger F et al (2001) Schall- und Erschütterungsschutz im Schienenverkehr: Grundlagen der Schall- und Schwingungstechnik – praxisorientierte Anwendung von Schall- und Erschütterungsschutzmaßnahmen. Krüger F (Hrsg) und 5 Mitautoren, Renningen-Malmsheim: Expert Verlag, 2001
147. ORE D151/151.1 (1989) Vibrations transmitted through the ground, Final report on a study of ground vibrations due to railways. ORE Report No. 12, Utrecht, April 1989
148. Zach A, Rutishauser G (1989) Maßnahmen gegen Körperschall und Erschütterungen. Erfahrungen bei Projekten der SBB. Technical Document DT 217 zu Frage D151 des ORE, Utrecht, 1989
149. Fischer G, Wettschureck RG, Hölzl G, Temple Ph (1988) Reduction of railway vibration propagation by means of rigid layers and flexible ballast mats. Technical Document DT 212 zu Frage D 151 des ORE, Utrecht
150. Darr E, Schaaf B (1996) Betriebserprobung Feste Fahrbahn zwischen Mannheim und Larlsruhe. ETR 45(H. 12): 772–784
151. Müller-Boruttau FH, Kleinert U (2001–1) Betonschwellen mit elastischer Sohle. Erfahrungen und Erkenntnisse mit einem neuen Bauteil. ETR 50 (2001), H3: 90–98
152. Kopp E (2001) Erfahrungen mit harten und weichen Zwischenlagen und Schwellenbesohlungen unterschiedlicher Steifigkeit in Ihren Auswirkungen auf die Schlupfwellenbildung. In: Tagungsband "Erfahrungsaustausch zum Einsatz von elastischen Komponenten im Eisenbahnoberbau", Brand bei Bludenz/Vorarlberg. Getzner Werkstoffe (Hrsg), Bürs/Bludenz, Mai
153. Wettschureck RG (1997) Measures to reduce Structure-Borne Noise Emission induced by Above-Ground, Open Railway Lines. Rail Engineering International, Edition 1997, No. 1, pp 12–16
154. Wettschureck RG, Heim M, Mühlbachler S (1997) Reduction of structure-borne noise emission from above-ground railway lines by means of ballast mats – Analytical model and measurements. Proc. Inter-Noise ´97, Budapest, 1997: pp 577–580
155. Hanson CE, Singleton Jr. HL (2004) Performance of Ballast Mats on Passenger Railroads: Measurement vs. Projections. Proc. IWRN8 – 8th International Workshop on Railway Noise (on CD-ROM, SESSION 6: GBV – Mitigation), Buxton, Derbyshire, UK, 2004
156. Müller R (2001) Die Erfahrungen der SBB mit Unterschottermatten zur sekundären Luftschall- und Erschütterungsminderung. In: Tagungsband "Erfahrungsaustausch zum Einsatz von elastischen Komponenten im Eisenbahnoberbau", Brand bei Bludenz/Vorarlberg. Getzner Werkstoffe (Hrsg), Bürs/Bludenz, Mai
157. Müller-Boruttau FH, Rosenthal V, Breitsamter N (2001–2) So trägt das Schotterbett Lasten ab – Messungen am Oberbau Systeme Grötz BSO/MK. ETR 50 (2001), H11:658–667
158. ORE D 151 (1988) Schwingungsabsorber an der freien Strecke. Bericht Nr. 11, Utrecht, September 1988
159. Prange B, Huber G (1982–1) Abschrimung von Untergrunderschütterungen durch Bohrlochreihen. Abschlussbericht zum F + E-Vorhaben B I 5-800180-48 des Bundesministeriums für Raumordnung, Bauwesen und Städtebau, Karlsruhe, 1982
160. Dolling HJ (1970) Die Abschirmung von Erschütterungen durch Bodenschlitze. Die Bautechnik, 47 (1970): H 5, pp 151–158 und H 6, pp 193–204
161. Woods RD (1967) The Screening of Elastic Surface Waves by Trenches. Dissertation University of Michigan, 1967
162. Massarsch KR (1986) I solation of Vibrations in Soil. Report Nr. 3/86 der FIT (Franki International Technology), Lüttich
163. Forchap E, Siemer T, Schmid G, Jessberge H (1994) Experiments to investigate the reduction of soil wave amplitudes using a built in block. In: Earthquake Resistant Construction and Design, Savidis Editor, Rotterdam
164. Kaynia A M (2001) Measurement and prediction of ground vibration from railway traffic. Proc. 15. International Conference on Soil Mechanics and Geotechnical Engineering, Istanbul, August 2001
165. Kaynia AM, Madshus C, Zackrisson P (2000) Ground vibration from high-speed trains: prediction and countermeasure. J Geotech Geoenviron Eng 126(6):531–537
166. Madshus C (2001) Modelling, monitoring and controlling the behaviour of embenkments under high speed train

loads. In: Geotechics for road, rail tracks and earth structures. Proc. 15. International Conference on Soil Mechanics and Geotechnical Engineering, Istanbul, August 2001
167. Madshus C, Kaynia AM (2001) High-speed trains on soft ground: track-embankment-soil response and vibration generation. Chapter 11 in: Krylov VV (Hrsg) Noise and vibrations from high speed trains. Telford, London: 314–346
168. Madshus C, Kaynia AM (2000) High-speed railway lines on soft ground: Dynamic behaviour at critical train speed. J Sound Vib 231(3):689–701
169. Wietlake KH (1985) Körperschallisolierte Gründung eines Wohnhauses oberhalb einer U-Bahn-Trasse. Bauingenieur 60:235–238
170. ORE D151 (1982) Schwingungen, die durch den Boden übertragen werden. Bericht Nr. 2: Bewertung der zur Zeit angewandten Erschütterungsschutzmaßnahmen. Bericht Nr. 2, Utrecht, April 1982
171. Westerberg G (1990) Körper- und Luftschallisolierung in zwei Stufen von Mehrfamilienhäusern mit "Eisenbahnverkehr im Keller". Fortschritte der Akustik, DAGA '90, Wien, 1990: 469–472
172. NN (1992) Comparison of Vibration and structure-borne noise control efficiency for elastic systems. Noise and Vibration Worldwide, Edition 1992: 21–24
173. Lenz U (1996) Körperschallisolierende Gebäudeabfederung. Bautechnik 73(H10):701–710
174. Zindler R (2000) Der Einsatz von zelligen Elastomer-Werkstoffen für die Körperschalldämmung im Hoch- und Tiefbau. Veröffentlichungen der FH Stuttgart – Hochschule für Technik, Bd. 51 – Bauphysikertreffen 2000: 25–45
175. Peters J, Prestele G (1999) Prediction of railway-induced vibrations by means of tranfer functions. In: Collected papers Joint Meeting (CD-ROM) – 137th ASA meeting, 2nd convention EAA, Forum Acustcum, DAGA 99 – Berlin99: 2PNSA_12
176. Rücker W, Said S (1994) Erschütterungsübertragung zwischen U-Bahn-Tunneln und dicht benachbarten Gebäuden. Forschungsbericht Nr. 199 der BAM, Berlin
177. Auersch L (1981) Wellenausbreitung durch eine Bodenschicht. DIE BAUTECHNIK 7:229–236
178. Auersch L (1988) Zur Entstehung und Ausbreitung von Schienenverkehrserschütterungen: Theoretische Untersuchungen und Messungen am Hochgeschwindigkeitszug Intercity Experimental. Forschungsbericht Nr 155 der BAM, Berlin, 1988
179. Melke J (1995) Erschütterungen und Körperschall des landgebundenen Verkehrs. Prognose und Schutzmaßnahmen. Materialien Nr. 22 des Landesumwelt-amtes Nordrhein-Westfalen (Hrsg), Essen
180. Said A, Fischer G, Hölzl G, Fleischer D (1997) Vergleich zwischen Bahn- und Fremdanregung bei der Ermittlung der gebäudespezifischen Übertragungsfunktionen. Fortschritte der Akustik, DAGA ´97, Kiel, 1997: 268–270
181. Steinhauser P (1994) VibroScan – A special seismic method for environmental vibration protection projects. Proc. 56th EAEG Meeting, Wien, 1994: 1052–1053
182. Steinhauser P (1996–1 Zur Treffsicherheit von Erschütterungs- und Körperschall-Immissionsprognosen beim österreichischen Bahntunnelbau. Österreichische Ingenieur- und Architekten-Zeitschrift (ÖIAZ) 141 (1996), H2: 46–50
183. Steinhauser P (1996–2) Zur Vorhersage und Beurteilung von Erschütterungs- und Körperschallimmisionen des Schienenverkehrs. Österreichische Ingenieur- und Architekten-Zeitschrift (ÖIAZ) 141 (1996), H1: 7–12
184. Unterberger W, Steinhauser P (1997) Bekämpfung von Erschütterungen und sekundärem Luftschall zufolge Schienenverkehr. Z Felsbau 15(2):88–96
185. VDI 2716 (2001) Luft- und Körperschall bei Schienenbahnen des städtischen Nahverkehrs, März 2001
186. DIN 18005–1 (2002) Schallschutz im Städtebau, Teil 1: Berechnungsverfahren, Juli 2002
187. Giesler H-J (2000) Geräuschemissionen von Straßenbahnen. Deutschlandweite messtechnische Erhebung. DER NAHVERKEHR 4:10–14
188. VDV (2000) Stadtbahnen in Deutschland: innovativ – flexibel – attraktiv = Light rail in Germany. VDV, Verband Deutscher Verkehrsunternehmen, VDV-Förderkreis e.V. (Hrsg.), Düsseldorf: Alba-Fachverlag, 2000
189. Heckl M (1988) Maßnahmen zur Lärmminderung am Schienenverkehr. Internationales Symposium, Forschung und neue Technologien im Verkehr Bd. 3 (Öffentlicher Nahverkehr), S. 225–244. Köln, TÜV Rheinland, 1988
190. Fritz P, Eilmes H (1994) Planung der immissionsgerechten Gestaltung von Gleisoberbauten für Stadtbahnstrecken. Verkehr und Technik 47(H. 4):129–142
191. Imelmann Chr (1994) Luft- und Körperschallprobleme beim Schienennahverkehr, Teil 1 und Teil 2. Verkehr und Technik 47 (1994), H. 1: 3–9 und H. 2: 43–48
192. Kasten P, Krüger F (1994) Geräuschsituation bei neuen Schienenfahrzeugen des Stadtverkehrs. U-Bahnen, Stadtbahnen, Straßenbahnen. Eine Bestandsaufnahme aus den alten Bundesländern, Teil I und Teil II. Verkehr und Technik 47 (1994), H. 3: 83 – 90 und H. 4: 123 – 128
193. Krüger F (1996) Minderung von Straßenbahngeräuschen. Wirkungen von Schallminderungsmaßnahmen an Tatra-Straßenbahnen – Meßergebnisse und Empfehlungen. Der Nahverkehr 14 (1996), Nr.6: 41–46
194. Krüger F (1997) Schallminderung bei Schienenfahrzeugen für den Regionalverkehr. Verkehr und Technik 50(H. 7): 327–330
195. Krüger F (2000) Leiser Schienennahverkehr – Ergebnisse 16-jähriger Forschung zur Minderung von Schall und Erschütterungen. DER NAHVERKEHR 5:36–44
196. Lenz U (1995) Luftschallimmissionen bei Stadtbahnanlagen des ÖPNV. Beurteilung und Prognose entsprechend den Festlegungen der Verkehrslärmschutzverordnung. Verkehr und Technik 48 (1995), Nr.3: 75–84
197. Lenz U (2001) mmissionsgerechte Oberbauplanung für den Stadtbahntunnel Bensberg. Verkehr und Technik (2001), H9: 369–374
198. Lenz U (2000) Immissionsgerechte Planung des Umbaus der Stadtbahnanlage in Frechen, Einsatz unterschiedlicher schwingungsisolierender Oberbauformen entsprechend der jeweiligen örtlichen Situation, Teil 1 und 2. Verkehr und Technik (2000), H. 2: 63–68 und H. 3: 93–96
199. Studt P (1986) Entwicklung von Reagenzien zur Reibungsbeeinflußung und Minderung des Verkehrslärms von

Schienenfahrzeugen. BAM-Bericht Nr. 6 in der Reihe Schienennahverkehr. Berlin, BAM 1986
200. Krüger F (1995–2) Kurvenquietschen und seine Minderung durch Schienendämpfungselemente. Grundlagen, Anordnungsoptimierung und vorbereitende Untersuchungen für einen Feldversuch. Verkehr und Technik 48 (1995), Nr.9: 364–367
201. Hendlmeier W (1990) Messung und Prognose von Schienenverkehrslärm unter Berücksichtigung des Kurvenquietschens. Z Lärmbekämpfung 37:166–169
202. STUVA (1994) Kurvenquietschen im Nahverkehr – Schall 03 – Ermittlung von Korrekturwerten zur Berücksichtigung des pegelerhöhenden Kurvenquietschens in der Schall 03 beim Durchfahren enger Gleisbögen im Schienennahverkehr. FE-Vorhaben des Bundesministers für Verkehr, Forschungsbericht FE-Nr. 70 413/93, STUVA e. V., Köln
203. Krüger F (1995–1) Das Kurvenquietschen im Schienennahverkehr – Ermittlung von Korrekturgrößen zur Berücksichtigung in der Schall 03. DER NAHVERKEHR 7-8/1995: 62–65
204. Krüger F (1998) Statistische Erfassung von Kurvenquietschen bei Nahverkehrsbahnen. Z. Lärmbekämpfung 44 (1998), Nr. 6: 216–219
205. Lenz U (1993) Kurvenquietschgeräusche, Messung und Prognose – Minderungsmaßnahmen. Verkehr und Technik 43(H1):9–13
206. Rieger Th, Lenz U (1995) Immissionmindernde Oberbauvarianten für Stadtbahnanlagen des ÖPNV – Versuchsstrecke Berliner Strasse/Gliesmaroder Strasse der Braunschweiger Verkehers-AG. Verkehr und Technik 48(H5):163–170
207. Reinauer R, Döbeli E (1998) Erschütterungs- und Körperschallminderung bei Trambahnen. Schweizer Ingenieur und Architekt 116 (1998), Nr. 25: 4–8
208. Krüger F (1999–1) Erprobung einer einfachen kontinuierlich elastischen Schienenlagerung auf einer Tunnelstrecke. Verkehr und Technik 7/1999: 306–312
209. Krüger F (1999–2) Grundlagen zur Entwicklung einer kontinuierlichen elastischen Lagerung für den Schienennahverkehr, Teil 1 und Teil 2. Verkehr und Technik 52 (1999), H. 4: 142–148 und H. 5: 200–203
210. Lenz U (1999) Schwingungsisolierende Rillenschienenlagerung bei Stadtbahnanlagen. Verkehr und Technik 52 (H6):266–271
211. Rieger Th, Lenz U (1998) Immissionsmindernder Oberbau am Magnitorwall in Braunschweig. Verkehr und Technik 51(H11):453–458
212. Chanel G (1995) Des performances qui évoluent au cours du temps. CSTB MAGAZINE 84:24–25
213. CSTB (1991) Tramway 2eme Ligne: Mesure Acoustiques. Etude No. 2.90.162 des Centre Scientifique du Batiment (CSTB), Centre de Recherche de Grenoble, Etude faite à la demande de G.M.S., Januar 1991
214. Müller-BBM (2001) Erschütterungstechnische Untersuchungen zur Wirksamkeit von Leichten Masse-Feder-Systemen im Bereich der Straßenbahn München. Müller-BBM Bericht Nr. 42 599/2 vom 26.01.2001, im Auftrag von Getzner Werkstoffe GmbH
215. Diehl RJ, Onnich J, Kurze UJ (2003) New concepts for the description of railway noise in Germany. Proc. Euro-Noise 2003 (on CD-ROM), Neapel, 2003, paper ID: 122
216. Kurze UJ, Diehl RJ, Onnich J (2006) Anpassung der Schall 03–2006 an Anforderungen europäischer Regelwerke zum Schienenverkehrslärm, Fortschritte der Akustik – DAGA 2006, Braunschweig, 2006: 257–258
217. Remington PJ (1987) Wheel/rail rolling noise, I: Theoretical analysis. J Acoust Soc Am 81:1805–1823
218. Remington PJ (1987) Wheel/rail rolling noise, II: Validation of the theory. J Acoust Soc Am 81:1824–1832
219. Thompson DJ (1993) Wheel–/–rail noise generation, parts I–V. J Sound Vib 161(3):387–482
220. Diehl RJ, Müller GH (1998) An engineering model for the prediction of interior and exterior noise of railway vehicles. Proc. Euro-Noise ′98, Munich, 1998: pp. 879–882
221. Thompson DJ (2000) A review of the modelling of wheel/rail noise generation. J Sound Vib 231(3):519–536
222. BN 918145 (2003) Technische Lieferbedingungen; Spannbetonschwellen mit elastischer Sohle: Schwellensohlen, Deutsche Bahn AG, Edition January 2003
223. BN 918235 (2004) Technische Lieferbedingungen; Elastische Zwischenlagen und Zwischenplatten, Edition January 2004
224. ISO 10846–1 (1997) Acoustics and vibration – Laboratory measurements of vibro-acoustic transfer properties of resilient elements Part 1: Principles and guidelines
225. ISO 10846–3 (2002) Acoustics and vibration – Laboratory measurements of vibro-acoustic transfer properties of resilient elements Part 3: Dynamic stiffness of elastic supports for translatory motion – Indirect method
226. Diehl RJ, Kurze UJ, Hofmann P (2004) Laboratory Testing of Elastic Layers for Railway Application. Proc. ICSV11 – 11th International Congress on Sound and Vibration, St. Petersburg, Russia, 2004, pp. 3071–3078
227. Diehl RJ, Hofmann P (2001) Bestimmung von Parametern elastischer Oberbauelemente In: Tagungsband "Erfahrungsaustausch zum Einsatz von elastischen Komponenten im Eisenbahnoberbau", Brand bei Bludenz/Vorarlberg. Getzner Werkstoffe (Hrsg), Bürs/Bludenz, Mai 2001
228. Thompson DJ, Verheij JW (1997) The dynamic behaviour of rail fasteners at high frequencies. Appl Acoust 52(1): 1–17
229. EC (2002) 2002/735/EC Commission Decision of 30 May 2002 concerning the technical specification for interoperability relating to the rolling stock subsystem of the Trans-European high-speed rail system referred to in Article 6(1) of Directive 96/48/EC
230. EC (2006) 2006/66/EC Commission Decision of 23 December 2005 concerning the technical specification for interoperability relating to the subsystem "rolling stock – noise" of the Trans-European conventional rail system
231. Hölzl G, Redmann M, Holm P (1990) Entwicklung eines hochempfindlichen Schienenoberflächen-messgerätes als Beitrag zu weiteren möglichen Lärmminderngsmaßnahmen im Schienenverkehr. ETR 39(11):685–689
232. Holm P (1999) Roughness measuring devices. In: Roughness Workshop, NSTO and Müller-BBM (ed.), Utrecht

233. Van Lier AA (1997) The measurement, analysis and presentation of wheel and rail roughness, NSTO report No. 9571011, August 1997
234. Van Beek A, Verheijen E (2003) Harmonised Accurate and Reliable Methods for the EU Directive on the Assessment and Management of Environmental Noise – Definition of track influence: roughness in rolling noise. Final report on HARMONOISE project, AEA Technology Rail BV (Editor), Utrecht, July 2003
235. Jones R, Beier M, Jones CJC, Maderböck M, Diehl, RJ, Middleton C, Verheij J (2000) Shields and Barriers, Proc. Inter-Noise 2000, Nizza, 667 – 672
236. Diehl R J, Hölzl G, Beier M, Waubke H (2000) Prediction of railway induced ground vibration. Proc. Inter-Noise 2000, Nizza 3721–3726
237. Diehl RJ, Nowack R, Hölzl G (2000) Solutions for acoustical problems with nonballasted track. J Sound Vib 231(3):899–906
238. METARAIL (1999) Final Report: Methodologies and Actions for Rail Noise and Vibration Control: For European Commision DG VII
239. Thompson DJ, Jones CJC (1999) The effects of rail support stiffness on railway rolling noise. In: Collected papers Joint Meeting (CD-ROM) – 137th ASA meeting, 2nd convention EAA, Forum Acusticum, DAGA 99 – "Berlin99": 1PNSC_1
240. Meyer G, Broschart T (1998) Körperschallverhalten und akustische Prognose moderner Hochgeschwindigkeitszüge. ZEV + DET Glasers Annalen + DET Glasers Annalen, 122 (1998), 9/10: 587–601
241. DB AG (1996) Körperschall- und Erschütterungssschutz – Leitfaden für den Planer, München, Deutsche Bahn AG – FTZ 81, Ausgabe 1996
242. Auersch L (2001) Zur praxisorientierten Berechnung der Steifigkeit und Dämpfung von Fundamenten und Fahrwegen. Proc. D-A-CH-Tagung 2001, Innsbruck. Institut für Baustatik, Festigkeitslehre und Tragwerkslehre (Hrsg) der Leopold Franzens Universität Innsbruck: 225–232
243. Müller-Boruttau FH, Ebersbach D, Breitsamter N (1998) Dynamische Fahrbahnmodelle für HGV-Strecken und Folgerungen für Komponenten. ETR 47(H. 11):696–702
244. Müller GH, Diehl RJ, Dörle M (1998) Assessment of the insertion loss of mass spring systems for railway lines and methods for the prediction of noise and vibration quantities, Proc. Euro-Noise '98, Munich vol I:97–102
245. Temple BP, Block JR (1998) Practical experience of a model for groundborne noise and vibration from railways. Proc. Euro-Noise ´98, Munich, 1998: 323–328
246. Fiedler J (1991) Grundlagen der Bahntechnik. Eisenbahnen, S-, U- und Straßenbahnen. 3., neubearb. und erw. Auflage. Werner, Düsseldorf, 1991
247. Lenz U (1997) Masse-Feder-System – alt, bewährt und noch modern. Herrn Dipl.-Ing. D. Uderstädt zum 70. Geburtstag gewidmet. Z. Lärmbekämpfung 44, Nr 4:113–114
248. Diehl RJ, Holm P (2004) Roughness measurements – Have the necessities changed? Proc. IWRN8 – 8th International Workshop on Railway Noise (on CD-ROM, Session 5: Roughness), Buxton, Derbyshire, UK, 2004
249. Kurze UJ, Donner U, Schreiber L (1982) Vergleich der Schallausbreitung von Schiene und Straße. Z Lärmbekämpfung 29:71–73
250. Akustik 23 (1990) Schalldämmung von Fenstern bei Schienenverkehrslärm, Information Akustik 23 der Deutschen Bundesbahn, Ausgabe 1990
251. BMV (1976) Schallemission von Schienenfahrzeugen. Abschlußbericht zum Forschungsvorhaben F91 des Bundesministers für Verkehr, September 1976
252. DB AG (2001) Anforderungskatalog zum Bau der Festen Fahrbahn, Deutsche Bahn AG (Hrsg), 4. Auflage, Entwurf 2001
253. Diehl R J, Hölzl G (1998) Prediction of wheel/rail noise and vibration – validation of RIM. Proc. Euro-Noise ´98, Munich, 1998: pp 271–276
254. DIN 18134 (1992) "Baugrund" Versuche und Versuchsgeräte – Plattendruckversuch, Januar 1992
255. DIN 45673–1 (2000) „Mechanische Schwingungen. Elastische Elemente des Oberbaus von Schienenfahrwegen, Teil 1: Ermittlung statischer und dynamischer Kennwerte im Labor, Mai 2000
256. Haupt W, Köhler W (1990) Gebäudeisolierung gegen U-Bahn-Erschütterungen. Bautechnik 67:159–166
257. Jäger K, Möhler U (1991) Der Lärmschutz für den Rangierbahnhof München Nord. Rangiertechnik und Gleisanschlußtechnik (RT + GT) 51: 61–66
258. Lenz U, Waßmann R (2000) Ermittlung des Lärmminderungspotentials von Straßenbahnen durch Optimierung das Fahrweges, Ableitung von Bauempfehlungen. Fortschritte der Akustik – DAGA 2000, Oldenburg, 2000:452–453
259. Remington PJ, Rudd MJ, Vér IL, Ventres CS, Myles MM, Galaitsis AG, Bender KE (1976) Wheel/Rail Noise, Part I to Part V. J Sound Vibr 46:359–451
260. Wettschureck RG, Hauck G (1994) Geräusche und Erschütterungen aus dem Schienenverkehr. Kapitel 16 in: Heckl M, Müller HA (Hrsg) Taschenbuch der Technischen Akustik, 2. Auflage. Springer, Berlin, 1994
261. Zeichart K, Sinz A, Schweiger M, Kilcher H, Hermann W (2001) Untersuchung zur Lästigkeit von Güter- und Reisezügen. Studiengemeinschaft Schienenverkehr, Hrsg. FTZ München der DB AG, 2001

Aircraft Noise

17

Ulf Michel, Werner Dobrzynski, Wolf Splettstoesser, Jan Delfs, Ullrich Isermann, and Frank Obermeier

Aircraft industry is exposed to increasing public pressure aiming at a continuing reduction of aircraft noise levels. This is necessary to both compensate for the detrimental effect on noise of the expected increase in air traffic and improve the quality of living in residential areas around airports. Therefore, advances in source noise reduction are required in the first place, which is the subject of Sect. 17.1 in this chapter. While engine noise dominates during take-off and the climb-out flight phase (Sects. 17.1.1 and 17.1.2), during approach and landing in addition, airframe noise contributes to the total aircraft noise signature (Sect. 17.1.4). Noise nuisance from helicopters is mainly due to their low-flight-level and -speed operations and the typical rotor impulsive noise characteristic, which is described in Sect. 17.1.3. The prediction of noise contours around airports for defined aircraft operation scenarios is a prerequisite for land use planning in the vicinity of airports. Related aspects are presented in Sect. 17.2, while Sect. 17.3 is dedicated to a discussion of the controversial issue of human noise perception. Finally, the characteristics and effects of the sonic boom originating from supersonic aircraft operation are summarized in Sect. 17.4.

U. Michel (✉)
Deutsches Zentrum für Luft- und Raumfahrt DLR, Müller-Breslau-Straße 8, 10623 Berlin, Germany
e-mail: ulf.michel@dlr.de

17.1 Noise Emission

17.1.1 Noise of Jet Engines

17.1.1.1 Overview

The majority of large fixed-wing aircraft are powered by jet engines. The first engines used were turbojet engines (Fig. 17.1a) in which the whole mass flow passes through its compressor, participates in the combustion process, and expands in the turbine, whose power is exclusively used to drive the compressor. After leaving the last turbine stage, the remaining large specific energy of the flow is used to accelerate the flow to a very high jet speed, which causes the very high noise emission of this engine variant.

All present day jet engines (including those of combat aircraft) are turbofan engines. The mass flow is divided into two separate flow streams after the low-pressure compressor, the so-called fan. The primary stream passes the core engine like in a turbojet. An additional low-pressure turbine is used to drive the fan. The remaining mass flow, the secondary stream passes around the engine core and is either discharged through a separate annular nozzle (short cowl nozzle, staggered nozzle) (Fig. 17.1c) or is mixed with the primary stream before being discharged through a common nozzle (long-cowl nozzle) (Fig. 17.1d). With a turbofan engine, a given thrust is achieved with a larger mass flow rate and a smaller jet speed, yielding better propulsive efficiency and lower noise. In engines for combat aircraft, the combined mass flow may be reheated for a short period in an afterburner before being discharged through the nozzle. This heating increases the flow speed and subsequently the thrust of the engine when needed.

G. Müller and M. Möser (eds.), *Handbook of Engineering Acoustics*,
DOI 10.1007/978-3-540-69460-1_17, © Springer-Verlag Berlin Heidelberg 2013

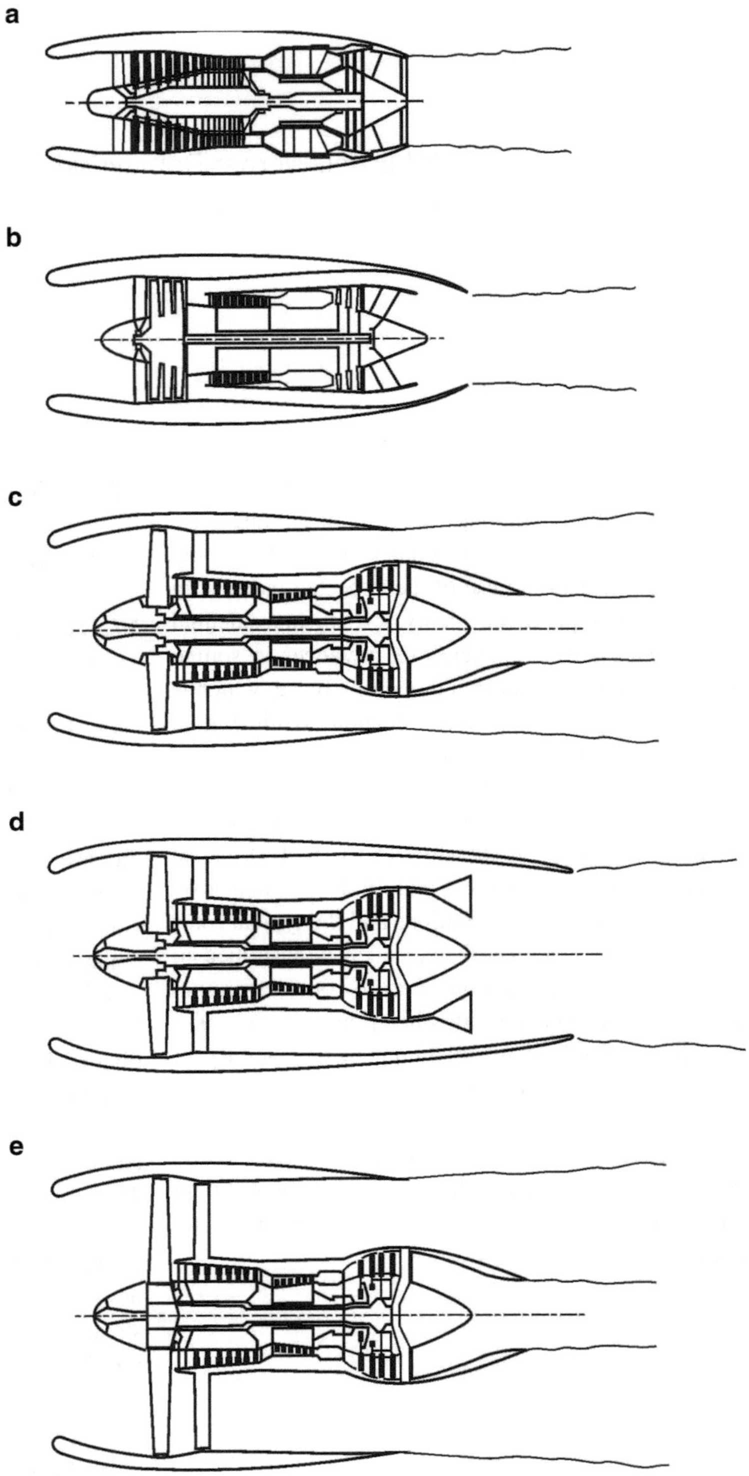

Fig. 17.1 Longitudinal sections of jet engines: (**a**) turbojet engines, (**b–e**) turbofan engines, (**b**) with low bypass ratio, (**c**) with high bypass ratio and staggered nozzles, (**d**) with high bypass ratio and common nozzle, (**e**) with ultra-high bypass ratio and gearbox in the hub of the fan

The ratio between the mass flow rates of the secondary and primary streams is called bypass ratio μ. This ratio was still small $\mu < 2$ in turbofan engines of the first generation (since about 1960) (Fig. 17.1b). The low-pressure compressor required several stages to achieve the required fan pressure ratio. Current turbofan

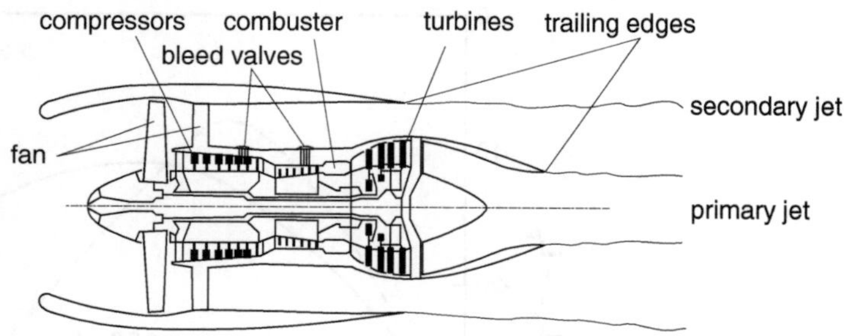

Fig. 17.2 Sound sources of a modern turbofan engine

engines of the second generation (since about 1970) feature bypass ratios of $\mu = 4 - 7$ (Fig. 17.1c, d), the newest engines for long range aircraft reach values $\mu = 9 - 10$. The low-pressure compressor consists of only a single stage without inlet guide vanes. The transition to turbofans of the third generation with bypass ratios $\mu > 10$ is imminent (Fig. 17.1e). They feature very large low-speed fans which may require a gearbox to allow the low-pressure turbine to run at higher speeds (geared turbofan).

The sound sources of modern turbofans are indicated in Fig. 17.2. During take-off and climb-out, when the engines operate close to their full power, the noise is dominated by the jet and the fan. During approach, when the fan rotor operates at about 50–60% of the maximum speed, the other noise sources turbine, combustion chamber, and at times also the compressor play a role. The trailing edges of the nozzles are also noise sources (see Sect. 17.1.4). Other engine noise sources during approach are the bleed valves, which have to be opened at approach power for a stable operation to reduce the core mass flow in later stages of the compressor.

Numerous authors have dealt with the noise emission of aircraft engines. The book of Smith [258] may be recommended as introduction. It contains a multitude of references. A compilation of contributions of many experts was published by Hubbard [127, 128].

17.1.1.2 Jet Mixing Noise

Jet mixing noise is generated by the turbulent mixing process of the free jet with the ambient air. This noise source is also described in Sect. 20.1.4. The flow field of a subsonic jet with constant flow speed in the nozzle exit plane is described in Fig. 20.20. The turbulence has small scales in the region close to the nozzle, where the higher frequencies are emitted. The sizes of the turbulent structures increase with increasing distance from the nozzle resulting in lower frequencies. The maximum source strength is located close to the end of the potential core (see Fig. 20.20) at about five to six jet diameters. With U the jet speed and D the nozzle diameter, the peak frequency emitted to the side and into the forward arc (opposite to the jet direction) is about 0.2 U/D to 0.3 U/D in the narrow-band spectra, or about 0.5 U/D to 1.0 U/D (see Fig. 17.3) in the one-third octave-band spectra. The peak frequency of current large engines is located only slightly above 100 Hz in one-third octave spectra. The directivity is heart shaped (Fig. 20.22), where the maximum sound pressure level is measured under an angle of 135° to 150° relative to the engine inlet. The minimum on the jet axis for an angle of 180° has two causes, the first one being refraction of the sound waves during their propagation through the hot and fast mean-flow field [7], since the refraction is reduced, when the wavelengths are long in comparison to the jet diameter, the sound emission close to the jet axis is dominated by low frequencies. The second cause is radial interference in the source region [192, 193].

The relative one-third octave-band spectrum of a stationary jet is almost constant over a large angular range [287, 288]. Lower frequencies are observed in the direction of the jet (Fig. 17.3). The frequencies of a flying aircraft observed on the ground are Doppler shifted according to the flight speed.

A causal relation between the emitted sound field and the turbulent flow field can be established with the acoustic analogy [167, 168]. It is based on an inhomogeneous wave equation that is derived from the equations for the conservation of mass and momentum. The terms on the right-hand side of this equation are considered to be the acoustic noise sources. The power-spectral density in the far field can be expressed in terms of a double integral [195]. The acoustic analogy was studied in numerous variants (e.g., [51, 80, 170, 193, 213, 219, 226, 233]. See review articles for details [81, 171, 234].

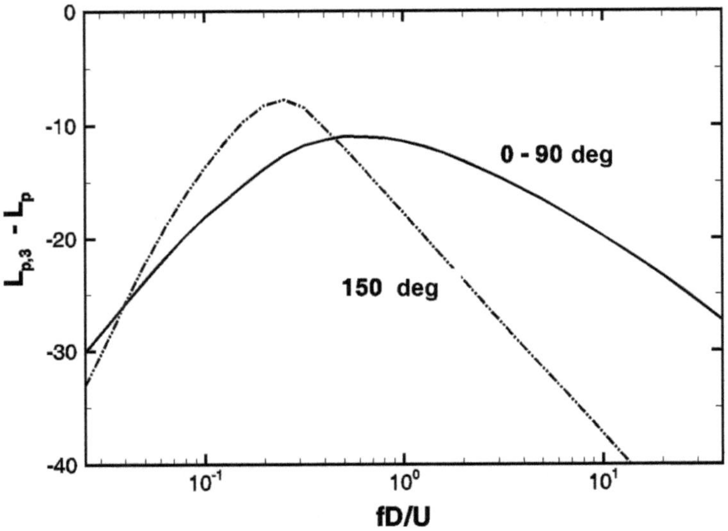

Fig. 17.3 Typical one-third octave-band spectrum of a free jet (after [205]) for the angle range 0° to 90° and for 150° relative to the engine inlet. The differences between the one-third octave levels and the corresponding overall sound pressure level are plotted as a function of Strouhal number *fD*/*U*. The peak frequency becomes substantially smaller for angles close to the jet axis (→180°). In addition the spectrum decays rapidly with rising frequency as shown for 150°. The ratio of the total temperatures of jet and ambience is 2

The first publications based on the acoustic analogy yielded quadrupole source terms (terms with a second spatial derivative), and the important finding that the sound power of a free jet is proportional to the eighth power of the jet speed and that the sound emission is larger in the rear arc of the jet. (see Fig. 20.22). When large density gradients are present in the flow, an additional dipole source term (one spatial derivative) may become important. The sound emission of this term is proportional to the sixth power of the jet speed [194, 197], a result that is found experimentally in directions normal to the jet axis in model jets [287, 288] as well as in engine jets. In the limit of very high speeds, the sound power of a jet can only rise with the third power of the jet speed due to energy conservation.

When the jet's Mach number exceeds unity, there are additional noise source mechanisms, broadband shock noise, screech, and crackling. These are absent during take-off of modern civil aircraft, but are substantial with combat aircraft with their very high exhaust Mach numbers. Broadband shock noise is important for passenger aircraft during cruise, when it may dominate the noise inside the rear cabin. Cause of broadband jet noise is the cell structure of the mean-flow field of supersonic jets [117, 196, 208, 209, 275, 276, 283, 288], which appears, when the pressure at the nozzle exit differs from the pressure in the ambient air. As a consequence of these cells, the mean and the fluctuating flow quantities as well as the sound sources change as function of the distance from the nozzle [196]. Characteristic for this noise is a hump in the frequency spectrum, whose frequency depends on the length of the cells and the flow speed and changes with emission angle as a consequence of interference effects [117, 208, 209, 288]. The frequency reaches a minimum in the direction of flight [196, 276]. For subsonic flights with high Mach numbers the frequency can become very small and equal to resonance frequencies of structures in the nozzle region. This can lead to rapid material failure due to acoustic fatigue. Screech occurs, when broadband shock noise emitted toward the nozzle induces the generation of instability waves with the same frequency, which triggers the whole jet column to oscillate with this frequency [208, 277]. The screech frequency of a stationary jet is constant for all emission angles and is Doppler shifted in flight. The amplitudes in the cell structure of a supersonic jet can be reduced or even be completely suppressed by convergent–divergent (Laval) nozzles. Variable nozzles would be required to adapt to different nozzle pressure ratios. Crackling is an especially unpleasant component of mixing noise of jets with high Mach numbers. It is visible in the microphone signals in form of sudden rapidly rising pressure signals [85].

An especially strong sound emission can appear, when a free jet hits an obstacle [72, 203]. In such a case an acoustic feedback process is possible even for subsonic jets [155, 202].

An approximate Eq. (20.32) was already given for the sound power P of a jet in Sect. 20.1.4. This equation has to be modified for a jet with nonconstant density as follows [205, 206, 307]:

$$P = 6.67 \times 10^{-5} \rho_0 {a_0}^3 S \left(\frac{U_s}{a_0}\right)^8 \left(\frac{\rho_s}{\rho_0}\right)^\omega F. \quad (17.1)$$

Here, ρ_s and ρ_0 are the densities of the jet and the ambient air, respectively; a_0 is the speed of sound in the ambience; U_s is the jet speed at the nozzle exit; and S is the nozzle exit area. The exponent ω and the function F consider the deviation of the sound power from the U^8 relation. Some values are given in Table 17.1. The directivities can be determined with the aid of [205, 206].

Equation (17.1) predicts the noise of engine jets slightly too low. Cause might be the amplification of jet mixing noise by noise from inside the engine [16]. Coaxial jets of bypass engines exhibit three source regions. The shear layer close to the nozzle lip of the secondary jet determines the high frequencies and the fully developed jet far downstream the low frequencies. The intermediate frequency range is dominated by the mixing between the two jets [87, 88]. Further instruments for the jet noise prediction are presented in [278, 286, 307].

Table 17.1 Density exponent ω and power factor F for the computation of the sound power of a free jet [204]

$\log_{10} U_s/a_0$	ω	$\log_{10} F$
−0.45	−1.0	−0.13
−0.40	−0.9	−0.13
−0.35	−0.76	−0.13
−0.30	−0.58	−0.13
−0.25	−0.41	−0.13
−0.20	−0.22	−0.13
−0.15	0	−0.12
−0.10	0.22	−0.10
−0.05	0.5	0
0	0.77	0.1
0.05	1.07	0.21
0.10	1.39	0.32
0.15	1.74	0.41
0.20	1.95	0.43
0.25	2.0	0.41
0.30	2.0	0.31
0.35	2.0	0.14
0.40	2.0	0.14

Flight speed reduces the emission of jet mixing noise into the rear arc substantially yet hardly into the forward arc. With jet engines, sometimes even higher levels than in the static case are observed in the forward arc [65, 267], albeit this could never be observed in carefully controlled model tests in wind tunnels [39]. The influence of flight speed can be considered with empirical methods [65]. The relatively strong sound emission into the forward arc can be explained with the acoustic analogy [194]. Broadband shock noise [1, 196, 276], the noise of the nozzle lip, and core noise from inside the engine is also amplified during flight into the forward arc. Important for jet noise of aeroengines are also installation effects. An engine mounted below a wing emits noticeably more noise than when it is installed on the rear fuselage [295, 297]. This can be explained with reflections of the sound waves on the underside of the wing or by an interaction of the jet turbulence with the trailing edge of the extended flaps.

According to Eq. (17.1), the most effective way to reduce jet mixing noise is the reduction of jet speed. The relevant value is the specific thrust (thrust divided by mass flow rate). While the engines of the supersonic transport aircraft Concorde achieved jet speeds of about 700 m/s during take-off with afterburners (without afterburner about 600 m/s), the specific thrusts of first generation bypass engines were about 480 m/s and current engines achieve about 300 m/s [258]. In order to achieve the same thrust with a smaller specific thrust, the mass flow rate must be increased, which in turn requires an increase of the nozzle exit area. Thus, the sound power of a free jet with constant thrust is proportional to the sixth power for quadrupole sources (instead of eighth power for a jet with constant exit area). Smaller jet speeds and larger nozzle diameters yield smaller frequencies of the emitted sound, which is beneficial for the weighted sound pressure levels, like the internationally standardized perceived noise level (PNL) [136] or the A-level (see Sect. 2.3.2).

The reduction of jet speed accompanied by an increase of the nozzle exit diameter is only a choice for new engine designs. However, there are also possibilities to reduce the emitted sound power of existing engines, some of them are based on intensifying the mixing process between the jet and the ambient air. This can be achieved by subdividing the nozzle exit area into many separate nozzles or by corrugated nozzles [90, 98]. Such nozzles have been retrofitted on turbofan or turbojet engines (Fig. 17.1a, b). They reduce primarily the broadband shock noise and crackling of supersonic jets [258]. The thrust losses of these solutions are so large that they are no longer used. A recently developed noise reducing variant is the serrated nozzle (chevron nozzle) (Fig. 17.4). Two counter-rotating longitudinal vortices are generated at each serration leading to an improved mixing with a thrust loss of less than 1%. Reductions of the noise level of up to 3 dB (EPNL) are reported [242].

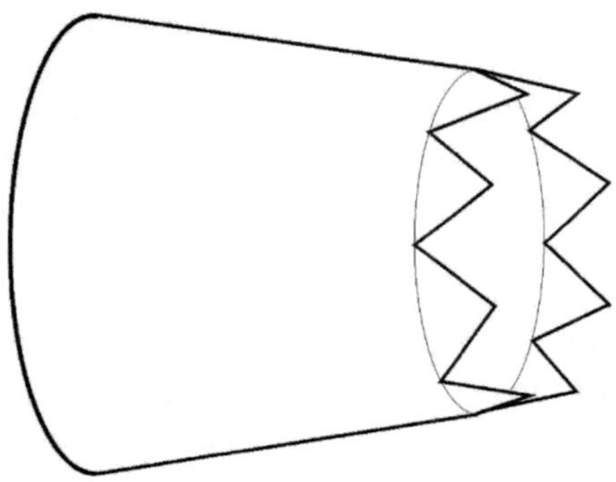

Fig. 17.4 Serrated nozzle for jet noise reduction

With bypass engines, there are solutions with separate (short cowl or staggered) nozzle (Fig. 17.1c) or common (long cowl) nozzle (Figs. 17.1d and 17.5). Engines with long-cowl nozzle are noticeably quieter. One reason is that the two streams are mixed inside the nozzle, which also yields a small thrust gain. This process can be enhanced by an internal forced mixer (Fig. 17.5).

17.1.1.3 Sound Emission by Fan, Compressor, Turbine, and Combustor

Next to the free jet, the fan is the most important sound source during take-off, whereas the turbines and compressors of the core engine may play a role during the landing approach. An overview over the turbomachinery noise can be found in [100, 258]. The combustor plays a role during the landing approach at low frequencies up to 800 Hz [179, 256, 258].

The fans and compressors of aeroengines differ from the fans described in Sect. 20.3 primarily by the much larger blade-tip Mach numbers and stage pressure ratios. Typical circumferential blade-tip Mach numbers of modern aeroengine fans are 1.4. The sound emission of a fan, compressor, or turbine consists of broadband noise and tones. Tones normally appear at the blade passing frequencies of the various rotors and their harmonics. As soon as the relative flow velocity at the rotor tip exceeds the speed of sound, tones are emitted also at multiples of the shaft rotational frequency composing the annoying buzz-saw noise [188, 189, 258].

Fig. 17.5 Engine BR710 of Rolls-Royce Deutschland with internal forced mixer (visible on the *right end* in the cut-out part) for the mixing of primary and secondary streams

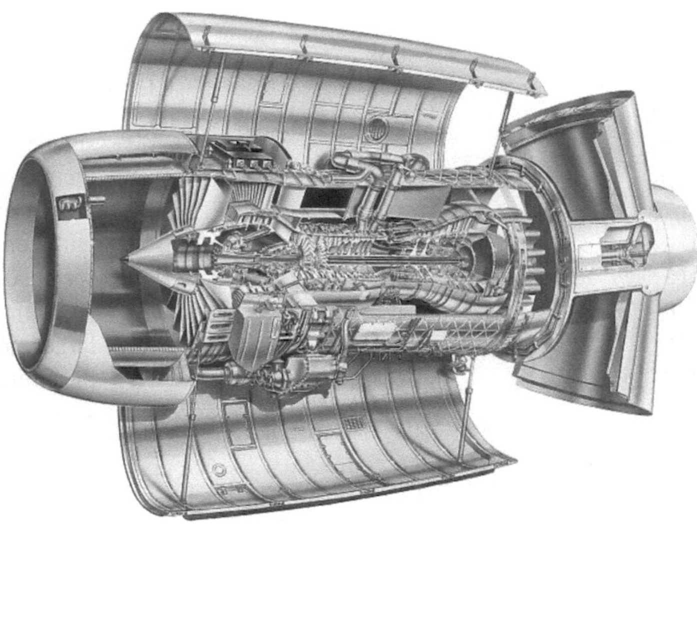

For all internal noise sources, the propagation capability of the generated sound waves in the attached duct plays an important role. For example, the sound field in a circular symmetric duct can be expanded into axisymmetric and helical duct modes, and only a part of these is propagable [71, 200, 290]. For a hard-walled duct (radius R), the fluctuating pressure as function of position (x, r, θ) in cylindrical coordinates and time t can be expanded into

$$p(x,r,\theta,t) = \sum_{m=-\infty}^{\infty}\sum_{n=0}^{\infty} P_{mn}^{\pm} J_m\left(\sigma_{mn}\frac{r}{R}\right)\exp[\mathrm{i}(\omega t - m\theta - k_{mn}^{\pm}x)],$$

$$k_{mn}^{\pm} = \frac{\omega/a}{1-M^2}\left[-M \pm \sqrt{1-(1-M^2)\left(\frac{\sigma_{mn}}{\omega R/a}\right)^2}\right]. \quad (17.2)$$

Here $P_{mn}^{\pm}$ are free constants, J_m is the Bessel function of first kind with order m, σ_{mn} is the nth zero of the first derivative $J'_m(\sigma_{mn}) = 0$ of the first derivative; the first ten zeroes are $\sigma_{00}=0; \sigma_{10}=1.841; \sigma_{20}=3.054; \sigma_{01}=3.832; \sigma_{30}=4.201; \sigma_{40}=5.318; \sigma_{11}=5.331; \sigma_{50}=6.416; \sigma_{21}=6.706; \sigma_{02}=7.016$; $\omega=2\pi f$ is the circular frequency, M the Mach number in the duct, which is the averaged flow speed normalized with the sound speed a in the duct with $M>0$, when the flow is directed toward the positive x-axis. The parameter $k_{mn}^{\pm}$ is the axial wave number (real or complex), which depends on M, σ_{mn}, and ω.

For $m \neq 0$, the waves propagate as spinning modes. For a fixed axial position x in the duct, the pressure can be described by a wave into the azimuthal direction θ

$$p \propto \exp[\mathrm{i}(\omega t - m\theta)] \quad (17.3)$$

and for a fixed azimuthal position, θ, we have waves in x-direction,

$$p \propto \exp[\mathrm{i}(\omega t - k_{mn}^{\pm}x)]. \quad (17.4)$$

Waves can only propagate in the x-direction if $k_{mn}^{\pm}$ is real. This requires that the argument in the root of above equation for $k_{mn}^{\pm}$ is positive or that the so-called cut-off ratio

$$\beta_{mn} = \frac{\omega R/a}{\sqrt{(1-M^2)}\sigma mn} > 1. \quad (17.5)$$

The waves propagate in the positive direction for positive $k_{mn}^{\pm}$ and in the negative direction for negative values. For $\beta_{mn}<1$ (for low frequencies ω), $k_{mn}^{\pm}$ is complex and the amplitudes of the waves decay exponentially, the waves are called "cut-off." Such waves are not connected with an acoustic power [70].

With the above-mentioned zeroes σ_{mn} of the derivative of the Bessel function, one can determine the first ten cut-on frequencies $f = \omega/2\pi$, above which the corresponding radial modes (m, n) in a duct with hard wall and radius R are propagable.

In the limiting case $\beta_{mn} = 1$, we obtain at the wall (radius R) for the phase velocity in circumferential direction normalized by the sound speed a

$$\frac{U_{\mathrm{pu}}}{a} = \frac{R}{a}\frac{\partial\theta}{\partial t} = \frac{\sqrt{(1-M^2)}\sigma_{mn}}{m}, \quad m \neq 0. \quad (17.6)$$

For mode $(m, n) = (1, 0)$ $\sigma_{mn}/m = 1.841$. For $n = 0$ and large m σ_{mn}/m goes asymptotically toward 1. Waves with smaller circumferential phase velocity than defined in this equation are "cut-off."

For a fan we have to distinguish between rotor-alone noise and rotor–stator interaction noise.

Examples for rotor-alone noise are the waves generated by a rotor with constant blade forces in the rotating coordinate system. These waves rotate with the same speed as the rotor and the azimuthal mode m is equal to the rotor blade count and multiples thereof. It can be concluded from Eq. (17.6) that a ducted propeller does not emit sound into the far field if the helical blade-tip Mach number is subsonic. This conclusion for rotor-alone noise contrasts to an open propeller (see Sect. 17.1.2). The modal composition of the rotor–stator interaction tones is determined by the blade counts of rotor and stator [290]. The following azimuthal modes m can be generated,

$$m = hB_{\mathrm{r}} + sB_{\mathrm{s}}, \quad (17.7)$$

where B_{r} is the rotor blade count and B_{s} is the stator blade count. $h = 1$ describes the modes for the blade passing frequency, $h = 2, 3\ldots$ their higher harmonics and s is an arbitrary positive or negative integer. The azimuthal angular speed of the generated interaction modes is given by

$$\frac{\partial\theta}{\partial t} = \frac{hB_{\mathrm{r}}\Omega}{hB_{\mathrm{r}} + sB_{\mathrm{s}}}, \quad (17.8)$$

where Ω is the rotation frequency of the rotor shaft and $B_r\Omega$ is the blade passing frequency of the rotor. For each of the generated interaction modes m the cut-off ratio (Eq. 17.5) determines, if the frequency $hB_r\Omega$ is propagable. The interaction theory was extended to counter-rotating rotors [122] and to multistage compressors and turbines [68].

Important for a low-noise design is that there are no propagable modes for the blade passing frequency $B_r\Omega$ of the rotor. This is achieved, when the stator blade count B_s is slightly larger than twice the rotor blade count B_r. The exactly required blade count of the stator also depends on the flow Mach number, the hub-to-tip ratio (ratio of inner to outer radius of the annular duct), the wall impedance, and the propagation direction in the duct. The propagation direction of a helical interaction mode according to Eq. (17.8) determines the ability to propagate upstream through the fan rotor. Waves with same propagation direction as the rotor propagate much better through the rotor than those with counter-rotating direction [100].

Broadband noise plays a large role with current engines and will likely become even more important with further success in the reduction of tonal sound emission. The biggest contributors to broadband noise are the trailing edges of stationary and moving blade cascades [4, 25, 28, 97] and the leading edge noise by turbulent inflow [113, 180, 254]. The flow through the gap at the blade tip is a further source of broadband noise [152]. Broadband noise is also generated by the interaction of the rotor blade tips and the turbulent boundary layer at the casing wall. The noise of the fan stator is dominated by the turbulent wake of the rotor and has lower frequencies than the rotor noise [198]. A narrow peak in the frequency spectrum with a peak frequency below the blade passing frequency can be emitted by slowly rotating highly loaded rotors [152] due to rotating instabilities. These are generated by the flow through the gaps at the blade tips and are precursors to rotating stall of the rotor.

For a minimal noise emission of a fan, a disturbance free inflow is required [2, 154, 258]. This is the reason that none of the modern civil aeroengines has inlet guide vanes. For the reduction of rotor–stator interaction noise, the distance between rotor and stator must be selected as large as possible because the blade wakes become smaller with increasing that distance. Such a design is not possible for multistage compressors or turbines for weight and size reasons. Therefore, it is especially desirable to design a turbomachine such that sound waves below 5 kHz are not propagable. Higher frequencies are highly attenuated in flight by the atmosphere. For a further reduction of the rotor–stator interaction noise the stator blades of new engines are swept in the axial direction and some are also leaned in the circumferential direction [301].

A very effective measure to reduce the emitted noise of turbomachines is the lining of the flow ducts with sound absorbing surfaces [199, 258]. Such linings consist generally of honeycomb-like structures covered with a porous cover in form of perforated metal plates or wire meshes. The lining is tuned to the most important frequencies in the spectrum. The damping bandwidth of liners with perforated plates is quite narrow. Liners with wire meshes are more broadband. The bandwidth can be increased by liners with two honeycomb layers. Liners with three layers have already been tested [151]. They cover an even larger frequency range.

The prediction of fan, compressor, and turbine noise is mostly based on empirical methods [100, 258]. However, numerical simulations are becoming more and more important.

17.1.1.4 Noise Certification of Jet Aircraft

The manufacturer of each aircraft type has to demonstrate to the authorities that the internationally agreed noise limits are satisfied. The methods and limits are defined by the International Civil Aviation Organization [136]. Flight tests have to be carried out in a prescribed way for three defined measuring positions as shown in Fig. 17.6. The sideline (lateral) measuring point is located on a line with a distance of 450 m parallel to the runway. The maximum noise level has to be determined on this line. The noise level is dominated by the engines. The maximum noise level is generally measured, when the aircraft reaches an altitude of about 300 m. In the flyover measuring position 6,500 m after start of roll the engine power is already reduced. The measured noise levels depend very much on the achieved flyover altitude, which is a function of the length of the ground roll and the climb performance of the aircraft. Since the performances of aircraft with two, three, and four engines differ, the noise limits depend on the number of engines. The approach measuring point is located 2,000 m before the landing threshold where the aircraft has an altitude

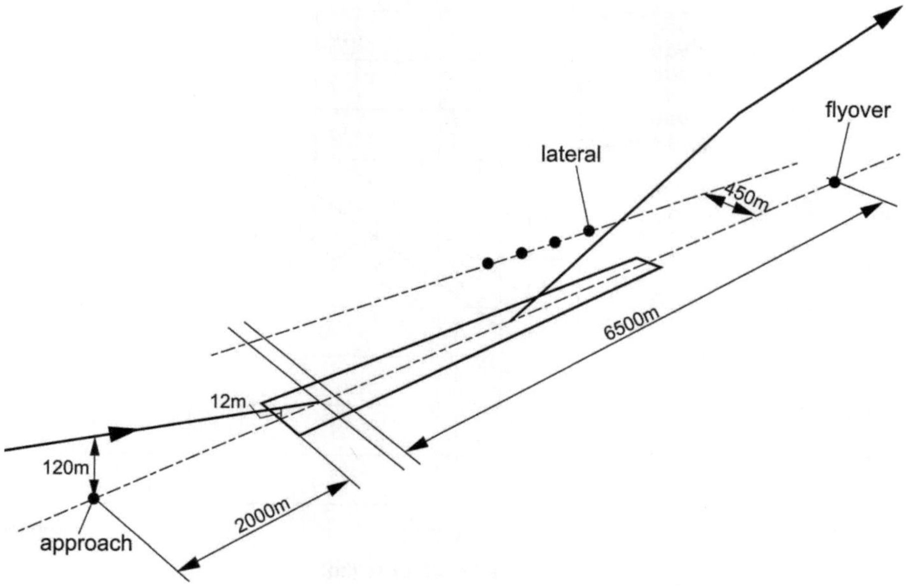

Fig. 17.6 The three measuring points for the certification of subsonic fixed wing aircraft

of 12 m. With a glide angle of three degrees, the flyover altitude in the approach measuring point is 120 m. Because of this low altitude, the measured approach noise levels are high. However, these high levels are limited to a narrow band below the flight path. The area affected with high sound pressure levels is much larger during take-off.

The measured sound pressure signals are subject to an evaluation procedure to determine the "effective perceived noise level" (EPNL) [136]. The limiting EPNL values are a function of the maximum take-off mass of the aircraft. It has become common to add the three certification values to a cumulative noise level, which is shown in Fig. 17.7 for some aircraft. This figure contains the new limits that are valid since 2006. The EPNL values are determined from the time-dependent one-third octave levels of whole flyovers. For each aircraft type, there are generally several certification levels, which depend on maximum take-off mass and on the engine choice. With the example of the Airbus family, A319/A320/A321, one can recognize that the noise level of an aircraft rises much more rapidly with increasing take-off mass than the associated noise limit. For the A321 and the Boeing 777 it becomes apparent, how much the noise limits may depend on the engine choice.

In Fig. 17.8 are shown examples of one-third octave spectra during take-off and during approach. The departure noise spectrum is dominated by jet mixing noise with a peak frequency of about 150 Hz. The spectrum decays rapidly with increasing frequency, supported by the atmospheric attenuation of the higher frequencies. In contrast, the approach noise is very broadband. The level maxima and minima in the range 60–500 Hz are a consequence of frequency and angle-dependent ground reflections, which are a consequence of the prescribed microphone height of 1.2 m above ground.

17.1.2 Propeller Driven Aircraft

The majority of General Aviation (GA) aircraft (leisure, sporting and small business aircraft) are driven by propellers and powered by piston- or turboshaft engines. Compared to turbojet engines, propellers exhibit a higher propulsion efficiency, since the propeller thrust is achieved by a high mass flow but a comparatively low through flow velocity.

17.1.2.1 Propeller Noise for Steady Operational Conditions

Propeller noise is composed from a broadband noise floor and rotational tone noise components. Broadband noise originates from the interaction of the turbulent flow with the solid propeller blade surfaces and the

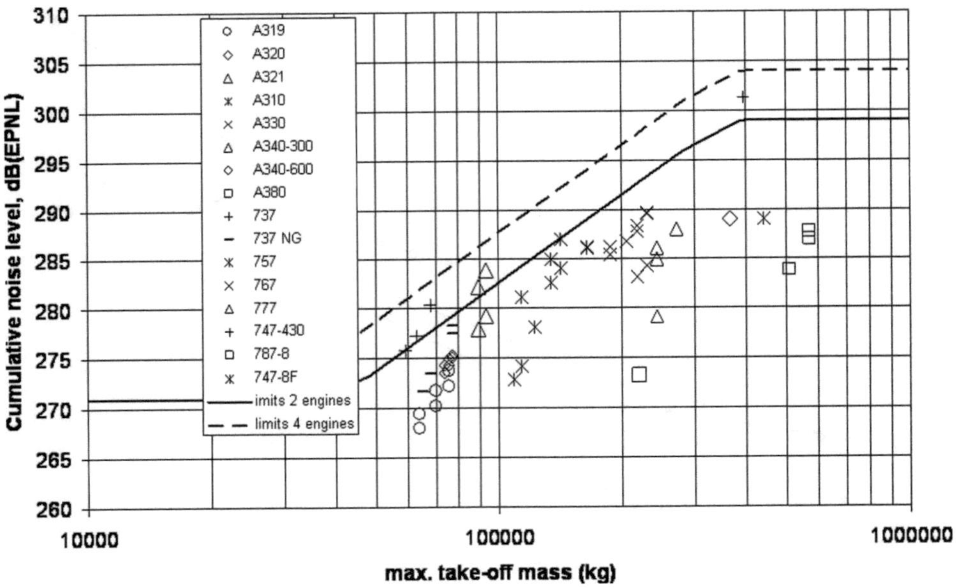

Fig. 17.7 Cumulative noise level (sum of three certification levels) for jet aircraft. The values for some aircraft types (sources: Luftfahrt-Bundesamt and European Aviation Safety Agency) are compared with the limits of 2006 for aircraft with two and four engines

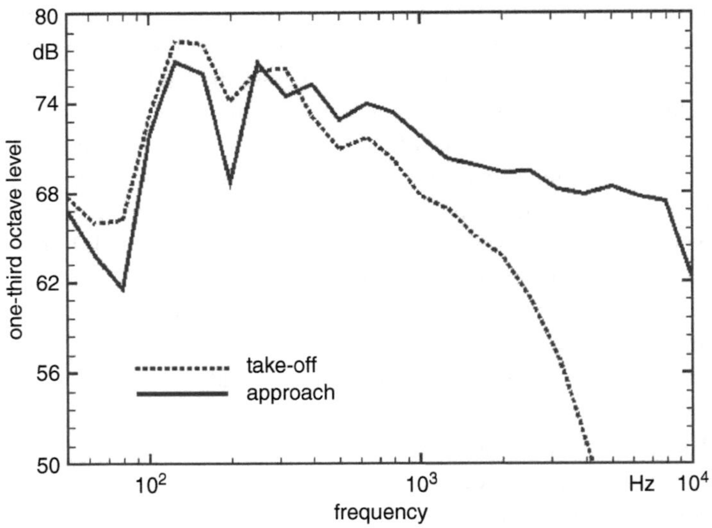

Fig. 17.8 Typical one-third octave spectra of a Boeing 717 with BR715 engines during take-off and during landing approach. The blade passing frequencies above 1.5 kHz hardly influence the one-third octave levels of this modern engine

blades' trailing edges. Accordingly, the corresponding noise radiation characteristics can be modeled by a Dipole source with its axis orientated perpendicular to the blade chord. The broadband noise directivity, thus, shows maximum levels close to the propeller axis (depending on the actual blade pitch angle) and the noise intensity increases corresponding to the sixth power of local flow velocity. Therefore, broadband propeller noise is of minor importance with respect to the total aircraft noise signature for a horizontal aircraft flyover.

The propeller rotational noise comprises tonal components at the fundamental frequency (or blade passing frequency) and its harmonics. The fundamental frequency is determined by the product of the propeller rotational frequency and the number of blades. For blade-tip Mach numbers below 0.8, highest tone levels are observed to occur at the fundamental frequency and

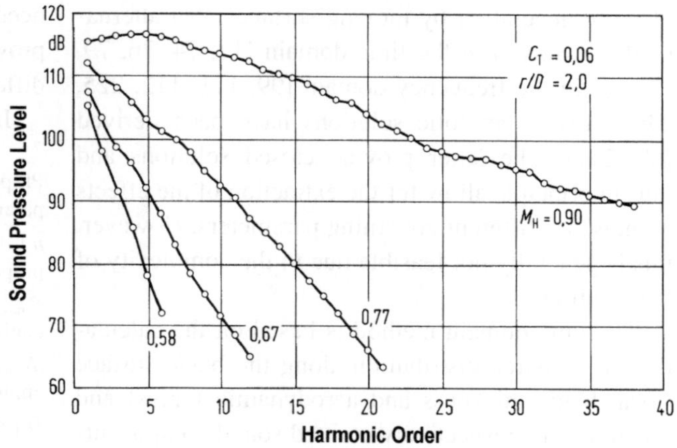

Fig. 17.9 Increase in propeller rotational harmonic noise levels with helical blade-tip Mach number M_H for constant blade loading C_T (thrust coefficient), radiation distance r, propeller diameter D

the levels of successive harmonics decrease almost linearly with increasing harmonic order. In contrast, the envelop of rotational tone noise levels exhibits modulations in the frequency domain and additional subharmonics come up in case of asymmetries in azimuthal blade spacing, blade geometry, or power distribution among the different blades. The latter can be a result of slight differences in blade pitch angle settings.

For propellers operating at subsonic tip speeds, the generation of rotational noise is due to the following two mechanisms:

- Periodic (with 1/revolution) flow displacement by the moving blades due to their finite thickness
- Periodic (with 1/revolution) variation of the aerodynamic blade force (lift and drag) vector component in the direction toward a fixed observer position

The effect of periodic flow displacement (thickness noise) can be described as an acoustic monopole, while the reaction of aerodynamic forces on the fluid (loading noise) corresponds to an acoustic dipole source (see also Sect. 17.1.3.2). The effective local flow velocity (vector sum of rotational and axial velocity components) at a selected radial blade station represents the most important parameter for these two sources of rotational noise. Therefore, rotational noise levels are essentially governed by both the blade thickness and the magnitude of aerodynamic forces close to the blade tip (typically data at an 80% radial station are taken as representative integral values for the complete blade), and the helical blade-tip Mach number constitutes the most important operational parameter with respect to propeller noise. With increasing values of blade-tip Mach number, the corresponding sound pressure time signatures exhibit steeper gradients (Doppler amplification) which in turn results in a rapid increase of higher rotational harmonic noise levels (Fig. 17.9).

For a typical GA propeller geometry (with about 7% blade thickness re blade chord at a 75% radial station), loading noise is the dominating source mechanism up to helical blade-tip Mach numbers of about 0.6–0.7, while thickness noise typically dominates for higher Mach numbers. The polar radiation directivity of loading noise exhibits level maxima both in forward and (most pronounced) in rear arc directions, while the thickness noise radiation directivity peaks close to the propeller rotational plane. The latter, seemingly strange, source characteristic for a monopole source is due to a kinematic effect, namely the relative motion of the source (blade element) with respect to the observer. Since this relative motion is zero in the direction of the propeller axis, no rotational noise is radiated along the propeller axis in case of straight and uniform inflow conditions. A more detailed description of these effects can be found in [272].

Most of the currently available propeller noise prediction schemes only tackle the rotational noise component, while no comprehensive and validated scheme for propeller broadband noise prediction has been published yet. However, some semiempirical approaches have been made to predict the broadband noise originating from selected source mechanisms [2–4, 25, 95].

The prediction of propeller rotational noise is based on the acoustic analogy as developed by Lighthill and Ffowcs-Williams [82, 167]. The so-called Ffowcs-Williams/Hawkings (FW–H) equation [84] (see Eq. (17.22) in Sect. 17.1.3.2); i.e., an integral equation of noise generation by moving surfaces, can alternatively be solved in the time domain [17, 74, 76, 79, 273] or in the frequency domain [99, 111, 112, 225, 249]. Also asymptotic solutions have been derived [69, 217]. The latter provide closed solutions and thus in principle allow for the extraction of the effects on noise of different governing parameters. However, this is generally not feasible due to the complexity of the solutions.

The time domain method is based on the calculation of a source distribution along the blade surface (local blade thickness and aerodynamic forces) and integrates the respectively emitted sound components from all source elements which would arrive at the same time at a selected observer location. As a result, the total pressure time signature is obtained and the corresponding frequency spectrum can be calculated via a Fourier transformation. A detailed description of the time domain noise prediction method for subsonic propellers is published in [308]. The prediction method is divided into several segments, starting with the discretization of the blade surface and the computation of the steady aerodynamic blade load distribution, which is a prerequisite for the calculation of the loading noise component.

To obtain initial propeller noise estimates also, semi-empirical prediction schemes have been developed, such as the method of Hamilton Standard [110], the scheme published in SAE/AIR 1407 [260], or the procedure developed by [54]. The latter is based on a data set as obtained from propeller noise calculations utilizing the time domain method as described above, however, related to a “representative” constant blade geometry (planform, twist and specifically a 7% blade thickness at the 75% radial blade station), a fixed radial blade loading distribution and typical blade lift and drag coefficients for a GA propeller. Respective calculations were performed for an extended matrix of propeller design (diameter and number of blades) and operational parameters. With reference to general physical principles of propeller noise generation and radiation, the respectively calculated overall A-weighted noise levels were empirically approximated by simple equations which contain only easily accessible design and operational data. Thus, a tool is available which allows for a quick estimate of propeller noise and provides additional information on the effects on noise of different propeller parameters. Since, in practice, the overall A-weighted noise level is the most frequently needed information this simple prediction scheme is provided below. The physical background behind the different equations is described in [54].

Input parameters are:

Propeller design parameters:	Operational parameters:	Environmental parameters:
n [–] Number of propellers	N [1/min] Propeller rpm	T [K] Ambient air temperature
$P_{\max}$ [kW] max. engine power	H [m] Flyover height	p [Pa] Ambient air pressure
$N_{\max}$ [1/min] max. engine rpm	V [m/s] Flight speed	
D [m] Propeller diameter		
B [–] Number of propeller blades		

In the first step, the following basic operational and aerodynamic parameters are calculated from this input data.

Power consumption of the propeller for the actual operational conditions:

$$P = [N/N_{\max}] \cdot P_{\max} \tag{17.9}$$

Ambient air speed of sound:

$$a_0 = \sqrt{\kappa \cdot R \cdot T} \tag{17.10}$$

with the adiabatic exponent $\kappa = 1.4$ and the specific gas constant $R = 287.1\, m^2/s^2\, K$.

Helical blade-tip Mach number:

$$M_{\mathrm{H}} = \frac{\sqrt{U^2 + V^2}}{a_0} \quad \text{with} \quad U = (\pi \cdot D \cdot N)/60 \tag{17.11}$$

Ambient air density:

$$\rho_0 = 1.252 + 1.250 \times 10^{-5} \cdot (p) - 0.0045 \cdot (T) \tag{17.12}$$

Aerodynamic blade loading (power coefficient) per blade:

$$c_{P,B} = \frac{31,006 \cdot P}{\rho_0 \cdot U^3 \cdot D^2 \cdot B}. \tag{17.13}$$

In the following, the effects on rotational propeller noise of different design and operational parameters are determined in terms of overall A-weighted noise level differences relative to respective reference conditions. Some of the equations contain exponents which depend on Mach number and thus indicate the respective balance between thickness and loading noise contributions:

Effect of blade loading:

$$\Delta L_1 = 10 \cdot \lg \left[\frac{c_{P,B}}{c_{P,B,\mathrm{ref}}}\right]^{E_1} \quad \text{with} \quad c_{P,B,\mathrm{ref}} = 0.01 \quad \text{and} \quad E_1 = 2.36 - 1.25 \cdot M_\mathrm{H} \tag{17.14}$$

Effect of number of blades:

$$\Delta L_2 = 10 \cdot \lg \left[\frac{B}{B_\mathrm{ref}}\right]^{E_2} \quad \text{with} \quad B_\mathrm{ref} = 2 \quad \text{and} \quad \text{for } \begin{array}{l} B = 2 : E_2 = 1.00 \\ B = 3 : E_2 = 0.98 \\ B = 4 : E_2 = 0.86 + 0.118 \cdot M_\mathrm{H} \\ B = 5 : E_2 = 0.36 + 0.706 \cdot M_\mathrm{H} \\ B = 6 : E_2 = -0.26 + 1.441 \cdot M_\mathrm{H} \end{array} \tag{17.15}$$

Effect of propeller rotational speed:

$$\Delta L_3 = 10 \cdot \lg \left[\frac{N}{N_\mathrm{ref}}\right]^{E_3} \quad \text{with} \quad N_\mathrm{ref} = \frac{1{,}000 \times 1}{\mathrm{min}} \quad \text{and} \quad E_3 = 3.39 - 1.75 \cdot M_\mathrm{H} \tag{17.16}$$

Effect of flyover height:

$$\Delta L_4 = 10 \cdot \lg \left[\frac{D}{H}\right]^2 \tag{17.17}$$

Effect of helical blade-tip Mach number:

$$\Delta L_5 = 10 \cdot \lg \left[\frac{M_\mathrm{H}^{13.4}}{(1 - M_\mathrm{H})^{1.5}}\right] \quad \text{for} \quad M_\mathrm{H} \leq 0.85. \tag{17.18}$$

Correction for climb-out conditions:

$$\Delta L_6 = 3\ \mathrm{dB}\ (\Delta L_6 = 0\ \mathrm{dB\ for\ level\ flight}) \tag{17.19}$$

Effect of number of propellers:

$$\Delta L_7 = 10 \cdot \lg(n). \tag{17.20}$$

The maximum overall A-weighted noise level is determined by the summation of an empirical constant and the above calculated level differences according to:

$$L_\mathrm{A} = 108.6 + \sum_{i=1}^{7} \Delta L_i. \tag{17.21}$$

The obtained noise level corresponds to a measurement with a ground-based microphone arrangement. The accuracy of this noise estimate is in the order of ±2 dB. The scheme is valid for the following values of input parameters:

Maximum engine power:	40–640 kW
Maximum propeller rpm:	1,500–4,000 × 1/min
Propeller blade number:	2–6
Propeller diameter:	1.0–3.0 m

It must be emphasized, however, that due to physical reasons only parameter combinations are allowed which would result in subsonic helical blade-tip Mach numbers [according to Eq. (17.11)] in the range $0.45 \leq M_\mathrm{H} \leq 0.85$. Since this noise estimation method does not account for excess noise due to unsteady propeller operational conditions and/or to additional noise contributions from the engine the calculated noise levels represent a lower noise limit.

From Eqs. (17.14) through (17.20), it is obvious that a decrease in blade-tip Mach number represents the most effective means for propeller noise reduction. This can be achieved through a reduction in blade diameter for constant engine rotational speed. In order to maintain the propeller thrust the number of blades can be increased [55]. The design of a potentially low noise propeller should also avoid high blade loads to occur in the blade-tip area. For propellers operating at blade-tip Mach numbers in excess of 0.7, the blade thickness of the outer blade section should be as small as possible and rounded blade-tip shapes should be realized. Moreover unsymmetrical blade spacing was found to provide some reduction of overall A-weighted propeller noise levels for certain radiation directions. This effect is due to interferences between the sound signatures originating from individual blades which cause a redistribution of sound energy both in the overall noise spectrum and the radiation directivity [53].

For commuter aircraft, operating at flight Mach numbers up to 0.8, advanced propellers have been developed (propfans with 8–12 blades) which feature supersonic blade-tip speeds. In order to reduce both aerodynamic losses and additional noise, a bow-type blade shape (blade sweep) is applied. By such means the phase lag of sound signatures originating from individual radial blade segments, which would arrive simultaneously at the observer position, is staggered in such a way that otherwise generated steep pressure gradients are diminished, thus limiting an increase of higher harmonic noise levels.

17.1.2.2 Propeller Noise for Disturbed Inflow Conditions

Nonuniform inflow conditions generate additional unsteady blade forces and thus excess propeller noise. As a result the azimuthal propeller noise is no longer unidirectional and the dependence of noise levels on blade-tip Mach number differs from the ideal uniform inflow case. In practice, the latter is not likely to occur since already a slight oblique attitude of the propeller plane with respect to the inflow direction causes cyclic changes of the blades' pitch angle and thus a periodic variation of blade forces. In addition oblique propeller plane attitude causes cyclic changes of the helical blade Mach number. Oblique inflow conditions to the propeller typically occur during climb out (increased aircraft angle-of-attack) but can also occur in twin engine aircraft when the propellers are attached to the wing and therefore are exposed to the wing's lifting circulation flow (installation effect). Numerous experimental and theoretical studies have been performed to quantify the effects on noise of oblique propeller plane attitude [52, 58, 92, 243].

Propeller operation with the aircraft at rest causes unsteady recirculation flow conditions to occur at the blade tips, which in turn result in significantly increased propeller noise levels compared to those in flight conditions for the same blade loading. Therefore, the assessment of noise radiation from different types of propellers should never be based on static test results.

The effects of both nonuniform and unsteady inflow conditions on noise as they typically occur for pusher propeller configurations are of major importance [18, 19, 149, 274]. The propeller may either operate in the turbulent wake flow of a pylon and/or the wing. Even worse, pusher propellers installed on ultralight aircraft often operate in the wake flow of the pilot and his seat. The latter situation causes complex unsteady propeller inflow conditions which cannot easily be modeled theoretically. The ensuing effects on noise, therefore, can best be determined from full-scale wind tunnel experiments (Fig. 17.10) [43].

In contrast, propeller excess noise generation due to a defined inflow wake deficit originating from an upstream located strut can be calculated. The characteristics of corresponding unsteady blade forces depend both on the magnitude and the velocity

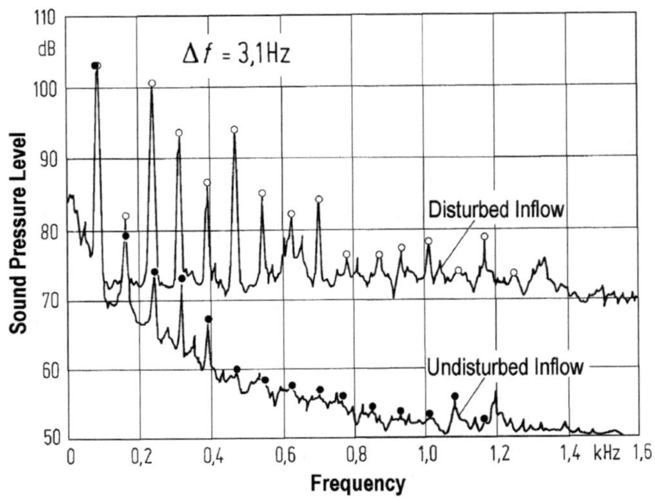

Fig. 17.10 Effect on propeller rotational and broadband noise of disturbed inflow conditions for a helical blade-tip Mach number of $M_H = 0.5$ (resolution $\Delta f = 3.1$ Hz)

gradients of the wake deficit. The latter is most important since it determines the higher harmonic excess noise levels. The presence of correspondingly generated unsteady blade forces result in the following changes in propeller rotational noise characteristics compared to those for uniform and undisturbed inflow conditions:

- Increase in higher harmonic noise levels, while the level at the fundamental frequency remains almost unaffected (see Fig. 17.10).
- Increased noise radiation both for forward and rear arc radiation directions (Fig. 17.11).
- The azimuthal noise directivity exhibits a noise maximum in the direction of the rotational blade motion when it passes through the wake (Fig. 17.12).
- Reduced noise level increases with helical blade-tip Mach number.

In contrast to the rotational noise source mechanism of moving blades in a steady and uniform flow sound generation through unsteady blade forces is not inherently related to the blades' motion. This causes the latter stated effect of a reduced noise level gradient vs. Mach number in case of unsteady inflow conditions.

In total, therefore, unsteady/nonuniform inflow conditions result in significant excess noise levels at low helical blade-tip Mach numbers, while at high Mach numbers this detrimental effect diminishes continuously.

In summary, in order to reduce inflow disturbances – and thus excess noise generation – the distance between an upstream located obstacle (e.g., a strut) and the propeller rotational plane must be increased. Care should be taken to avoid wake flows with inherent steep velocity gradients. If an upstream obstacle only extends over half the propeller diameter (e.g., a pylon) then the direction of the blades' rotational motion must be selected in such a way that the blade does not cut through the wake when it moves towards noise sensitive directions (i.e., toward the ground). Also upstream located struts or the like may be bow shaped or installed in an off-axis position in order to

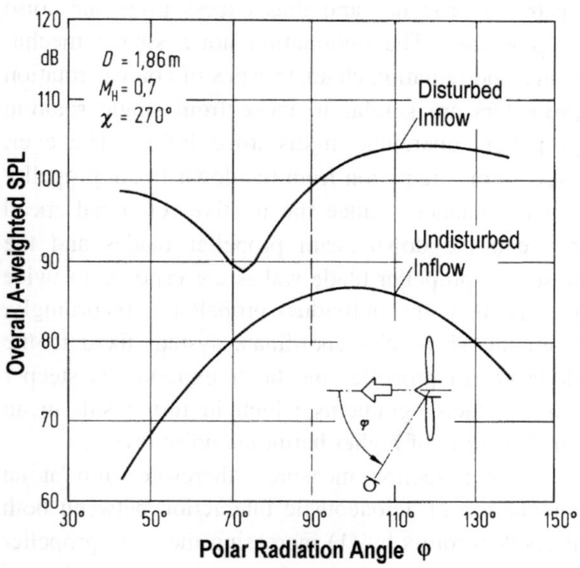

Fig. 17.11 Typical polar directivities of propeller noise for undisturbed and disturbed inflow conditions, respectively (for azimuthal radiation angle χ see Fig. 17.12)

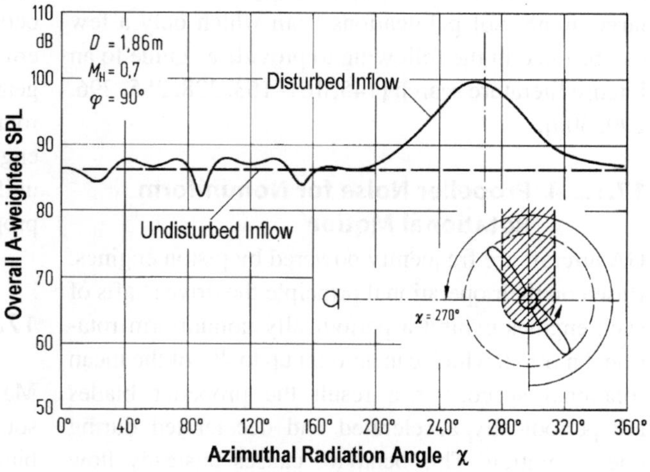

Fig. 17.12 Un-symmetry in azimuthal directivity for an inflow disturbance (wake deficit) extending into the propeller plane on one side only. A sound pressure level maximum occurs in the direction of the propeller blades' motion when it cuts through the wake (for polar radiation angle φ see Fig. 17.11)

achieve a continuous passage of the blade through the respective wake.

17.1.2.3 Noise from Counter Rotation Coaxial Propellers

Coaxial counter rotation propellers were developed due to their greater propulsion efficiency compared to single rotation propellers: The swirl energy in the upstream propeller wake is no longer lost but rather utilized by the downstream propeller. However, in such arrangements the downstream propeller is exposed to the turbulent blades' wake flows from the upstream propeller and thus excess propeller noise is generated. The dominating noise source mechanisms and radiation characteristics of counter rotation propellers are similar to those from single rotation propellers operating in disturbed inflow. However, excess noise radiation from the downstream propeller is still enhanced, since the relative rotational speed between the downstream propeller blades and the upstream propeller blade wakes corresponds to twice the speed of the individual propellers. Accordingly, from the view of a coordinate system fixed to the downstream propeller the latter experiences steeper wake velocity gradients which in turn result in an amplification of higher harmonic noise levels.

Noise reduction measures, therefore, aim at an attenuation of aeroacoustic interaction between both propellers/rotors by (1) increasing the axial propeller separation or (2) reducing the aerodynamic loading of the downstream propeller. Both means diminish the amplitudes of unsteady blade forces. Numerous (mostly experimental) investigations in the noise characteristics of counter rotation propellers or propfans have been performed. Accordingly, there exists a large number of publications from which only a few can be cited in the following to provide as guide to an intense literature search [145, 161–163, 238, 285, 296, 299, 300].

17.1.2.4 Propeller Noise for Nonuniform Rotational Motion

GA aircraft are frequently powered by piston engines. Based on their operational principle the drive shafts of such engines exhibit a periodically nonuniform rotational motion, which can amount up to 2% of the mean rotational speed. As a result the propeller blades are periodically accelerated and decelerated during one revolution. This behavior causes unsteady flow conditions at the very blade and, thus, unsteady blade forces. Since this unsteadiness is periodic with the drive shaft revolution, excess rotational harmonic noise is generated [56, 303].

In case that there is no interference between frequencies of the propeller noise (fundamental and harmonics, "steady" rotational noise) and a frequency of excess engine rotational nonuniformity, then additional harmonics ("unsteady" excess noise) rise up in the rotational propeller noise spectrum as a result of the periodic propeller kinematics. Corresponding frequencies are determined from the sum of blade passing frequencies and the respective frequency of dominating engine nonuniformity (Fig. 17.13). Vice versa, in case of coincidence (and this typically occurs for four- or six-cylinder engines in combination with even numbers of propeller blades), the spectrum of rotational harmonic noise levels is modulated and does no longer show a monotonous level decrease with increasing harmonic order. Engine drive shaft nonuniform rotation has the following effects on the propeller rotational noise directivity (compared to that for uniform rotation):

- Increased noise radiation both for forward and rear arc radiation directions (similar as for disturbed inflow conditions according to Fig. 17.11).
- Azimuthal noise directivity with distinct level maxima and minima.

Also similar as for disturbed inflow conditions a reduced noise level increase with helical blade-tip Mach number is obtained when compared to that for uniform rotation conditions.

In order to reduce excess noise due to nonuniform propeller rotational speed torsional damping devices may be installed at the engines' crank shaft or inserted between the drive shaft flange and the propeller. Modern aircraft propellers are increasingly powered by geared piston engines with reduced capacity but high rotational speed (compared to standard high capacity engines). Fortunately, the typical rotational speed nonuniformity of such drive systems does not cause major propeller noise problems.

17.1.3 Helicopters

Main and tail rotors represent the dominant noise sources of helicopter external noise. Engine (gas turbine or piston engine) and gearbox noise are of less

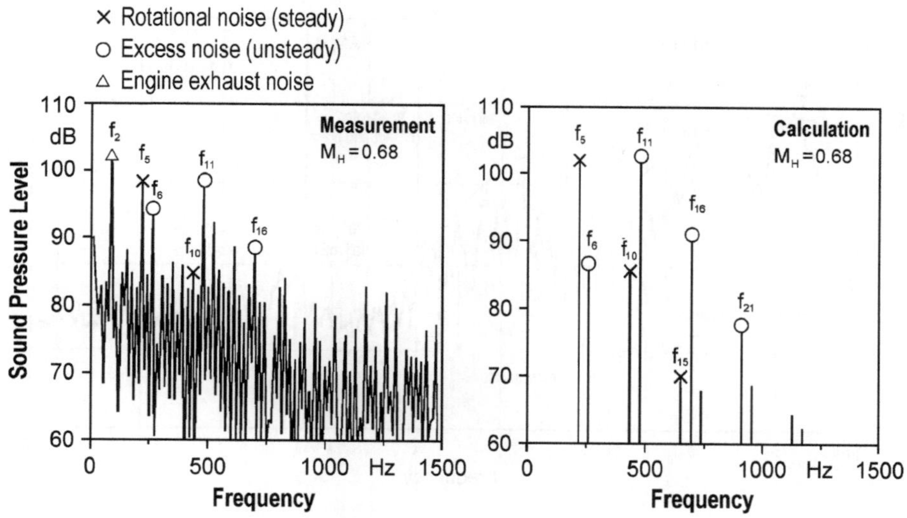

Fig. 17.13 Example for the effect of non-uniform propeller rotational speed (dominating at the sixth engine rational frequency) on the rotational noise spectrum of a 5-blade propeller driven by a 4-cylinder piston engine. The frequency numbering indicates multiples of the engine rotational frequency. *Left*: Measured data; *right*: calculation (based on FW–H equation) for non-uniformity at the sixth engine order

importance. And if necessary, the noise of the drive unit could be easily treated (e.g., by encapsulation) than that of the rotors which naturally are operating in the free atmosphere. An example for the sound pressure spectrum of a helicopter in the far field is shown in Fig. 17.14.

At first the aeroacoustic mechanisms of rotor noise generation and radiation as well as potential noise reduction measures will be presented. The chapter then concludes with a brief description of the procedures for rotor noise prediction.

17.1.3.1 Rotor Noise Generation and Reduction Measures

Within the last 20 years, the understanding of the aeroacoustic mechanisms of sound generation and radiation of helicopters has significantly been improved. Apart from theoretical investigations with modern numerical procedures, flight tests and model rotor experiments in aeroacoustic (anechoic) wind tunnels (like the Deutsch-Niederländische Windkanal "DNW") play an essential role [22, 245, 247].

A classification of the very complex rotor noise is given in Table 17.2. As shown in this representation, the rotor rotational noise comprises thickness noise (due to volume displacement) and aerodynamic loading noise (due to rotating blade forces, i.e., lift and drag forces). The thickness noise is important for the lower frequency range, i.e., the blade passage frequency and the first few harmonics which amplitudes rapidly decrease when impulsive noise events are absent. At low blade-tip Mach numbers and small aerodynamic blade loadings the discrete frequency spectrum can be masked by stochastic broadband noise components [31, 73, 93, 174].

Virtually of more importance for the noise signature of helicopter rotors, however, are the impulsive noise mechanisms on the rotor blade surface, which generate the typical rotorcraft blade slap. These are firstly the high-speed (HS) impulsive noise and secondly the blade–vortex interaction (BVI) impulsive noise. As shown in the flight envelope of Fig. 17.15, representing climb/descent rate versus flight velocity, the first one occurs at high flight speed independently of climb or descent angle, the latter one predominately at certain descent flight regimes.

In absence of impulsive noise events stochastic broadband noise can represent the primary source of rotor noise. It is caused by different mechanisms: turbulent on-flow, blade–wake interaction (BWI), and blade self-noise generated by interaction of the blade airfoil with its boundary layer and its near wake. Only of this, five different mechanisms are known which are based on turbulent boundary layer separation at the blade tailing edge, laminar boundary layer separation and vortex formation at the trailing edge,

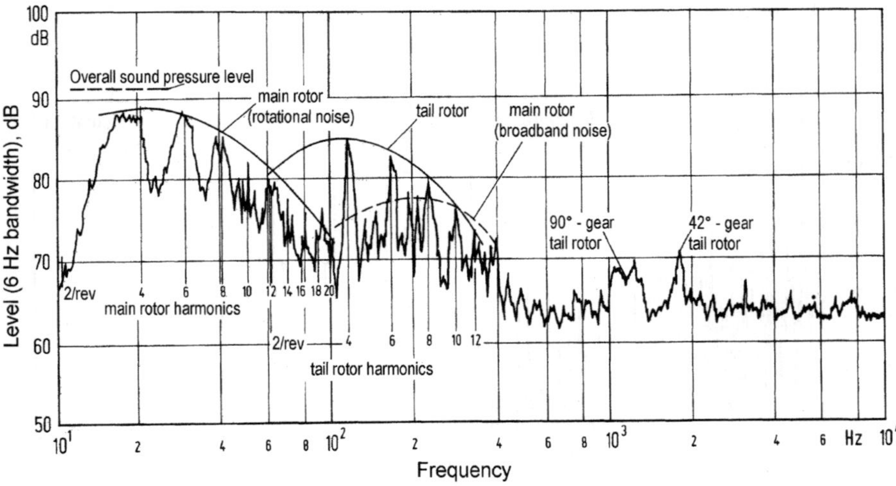

Fig. 17.14 Sound level spectrum of the helicopter UH-1A. Rotor thrust 27,000 N, blade-tip speed 220 m/s, measurement distance 60 m

Table 17.2 Rotor noise classification

Physical mechanism	Acoustic source type	Spectral character	Rotor-specific noise type
Volume displacement	Thickness noise	Discrete frequency	Rotational noise
Non-linear effects (at high blade-tip Mach numbers)	Shock delocalization	Discrete frequency	HS impulsive noise
Rotating blade forces	Deterministic loading noise	Discrete frequency	Rotational noise
Blade–vortex interaction	Deterministic loading noise	Discrete frequency	BVI impulsive noise
Turbulent on-flow, flow separation, blade–wake interference	Nondeterministic loading noise	Stochastic	Broadband noise

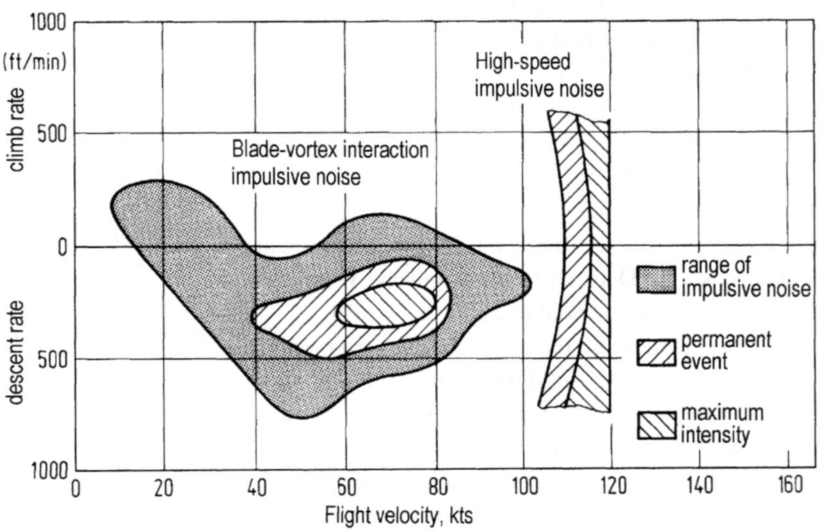

Fig. 17.15 Occurrence of rotor impulsive noise at moderate speed descent and at high speed flight

flow separation (stall), vortex shedding at blunt trailing edges, and vortex formation at the blade tip. Character and importance of the broadband noise are described in detail in [30, 32, 33, 293].

Main Rotor High-Speed Impulsive Noise

At fast forward flight the rotor blade on the advancing side ($\psi \approx 90°$) is subjected to the highest local on-flow velocity during each rotor revolution. The combination of flight velocity and blade-tip speed can result in blade on-flow velocities near the speed of sound. Since a rotor blade is of finite thickness and furthermore of convex curvature on the blade upper surface in chordwise direction, the flow velocity near the upper surface increases. Here the flow Mach number M can reach values exceeding 1, even when the on-flow Mach number M_{AT} of the advancing blade tip is still less than 1. That means the flow may be supersonic for a very short period during which the advancing blade passes through the azimuthal angle range of $\psi \approx 60°$–$90°$. In the flow region near the blade surface, an aerodynamic shock occurs each time when the flow rapidly turns from supersonic to subsonic flow. The periodic occurrence of the shock on the blade and its propagation across the blade tip (shock delocalization) leads to an impulsive acoustic wave formation. The product of number of blades and rotations per second yields the fundamental frequency (blade-passage frequency, bpf) of the periodic impulsive noise signature (typically 10–30 Hz).

This so-called high-speed (HS) impulsive noise is characterized by negative spikes in the sound pressure time signature (Fig. 17.16) which shows a saw-tooth-like shape with a shock-like pressure rise for advancing blade-tip Mach numbers near 0.9 and above. The frequency spectrum indicates a large number of harmonics of the blade passage frequency. The advancing blade-tip Mach number M_{AT} is the dominant parameter for high-speed impulsive noise. HS impulsive noise is strongly focused and directed forward with maximum intensity in the rotor plane [21, 142, 226, 246, 247]. All measures suitable for decreasing the supersonic flow region on the rotor blades (e.g., sweep, thickness reduction) contribute to a reduction of HS impulsive noise.

Main Rotor Blade–Vortex Interaction Impulsive Noise

On a lifting blade moving through air, a strong vortex is formed and shed at the blade tip. This is caused by air flow around the blade tip from the pressure (lower) side to the suction (upper) side and by roll-up of the free vortex layer shed along the trailing edge of the blade. While for airplanes these tip vortices form straight trailers, the traces of the blade-tip vortices for rotors are nearly epicyclical. Under certain conditions this fact can cause a blade to collide or interact with the shed tip vortex of a preceding blade. Thereby, for each BVI intensive local velocity and blade surface pressure changes are induced which can be observed in the acoustic waveform as significant positive impulses (Fig. 17.16) or negative ones depending on the interaction geometry and the observer position. The related frequency spectrum indicates a significant increase in the midfrequency range of sound pressure (about 6th to 40th bpf harmonic). The physical reason for the generation of pressure impulses during the aerodynamic interaction of a single vortex with a fixed blade airfoil, as explained in [191], is thought to be the break-down of the stagnation point near the airfoil leading edge as well as a shock delocalization on the pressure side of the airfoil induced by the closepassage of the vortex. Numerical simulations employing Euler methods could illustrate these aeroacoustic phenomena, e.g., [12].

Such BVI are observed at low to moderate speed descent flight (at velocities of typically 25–50 m/s and descent angles of about 5°–8°), representing a flight condition which is unavoidable during landing approach flights. In this flight regime, numerous vortex filaments penetrate the rotor plane resulting in multiple interactions. Accordingly, this noise phenomenon is called BVI impulsive noise.

The intensity of BVI impulsive noise is dependent on parameters like the ratio of flight to descent velocity, i.e., the flight-path angle and the rotor plane inflow both defining the rotor tip-path-plane angle α_{TPP}, the advance ratio μ (ratio of flight to blade-tip speed) and increases with both the rotor thrust coefficient C_T, and the advancing blade-tip Mach number M_{AT} [20, 94, 120, 121, 129, 181, 183, 262, 263]. Since these interactions are extremely short in time, they only have marginal effects on the rotor thrust behavior. Intensive BVI impulsive noise may occur at maneuver flight as well. It should be noted here that the BVI phenomenon is crucial for the excitation of rotor vibrations and for the vibration level of the total helicopter fuselage.

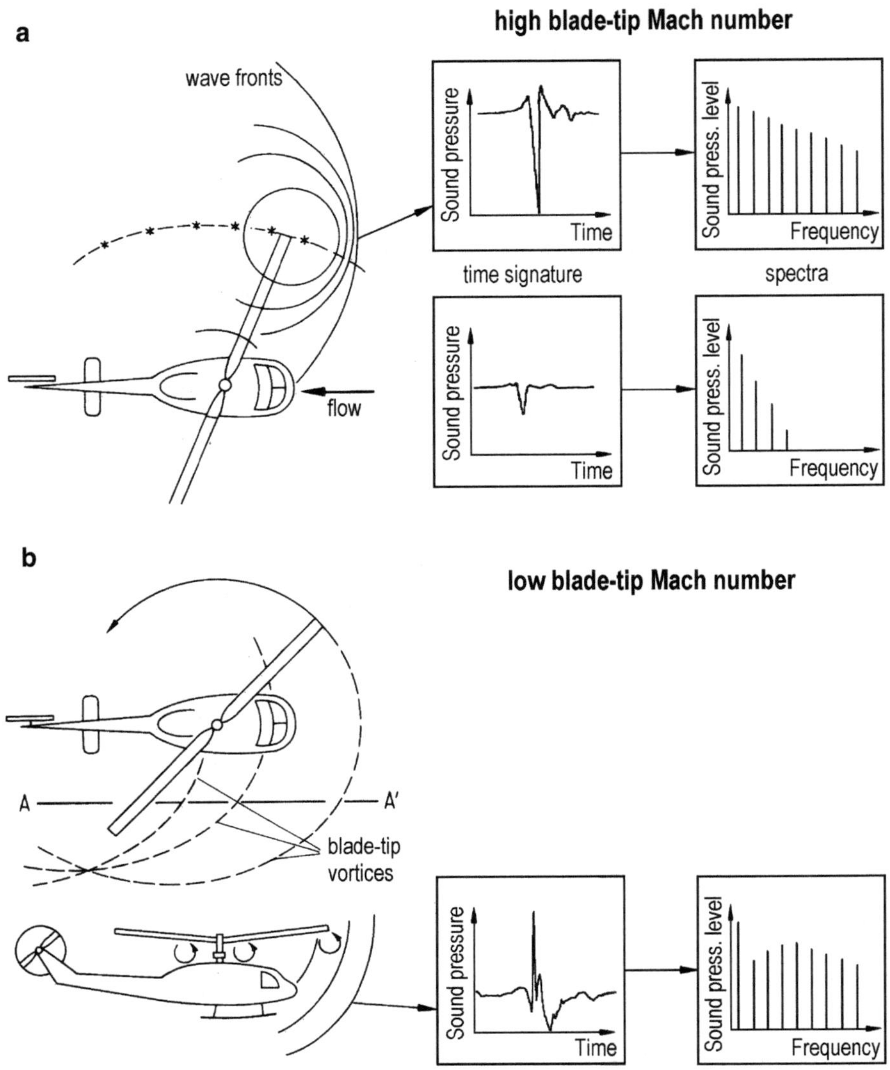

Fig. 17.16 (**a**) Generation mechanism of high-speed impulsive noise, (**b**) Generation mechanism of blade–vortex interaction impulsive noise with typical sound pressure signatures and frequency spectra

Intensive BVI predominantly occurs in the first rotor quadrant (I), i.e., on the advancing side ($\psi \approx 45^\circ \pm 20^\circ$), where the vortex segments are essentially parallel to the blade leading edge. This is distinctly visible in the blade surface pressure fluctuations shown in Fig. 17.17a. The corresponding intensive BVI noise radiation can be observed underneath the advancing side (Fig. 17.17b). Similar interactions occur in the fourth quadrant (IV) of the rotor plane on the retreating side [265].

Active Noise Reduction Measures. From a physical point of view, there are several potentialities to reduce BVI impulsive noise. These may consist of reducing the tip vortex strength of that specific local vortex segment which downstream interacts with a rotor blade, minimizing the interaction intensity by reduction of the local aerodynamic blade loading and increasing the blade-vortex miss-distance during the interaction. All these measures can be realized by so-called active blade control, whereby the blade angle-of-attack during a rotor revolution will be locally changed at the "correct moment." This momentary (eventually, during a revolution multiple) variation of the blade angle of attack can be realized in different ways depending on the arrangement of the active control actuators with respect to the rotor swashplate.

At higher harmonic control (HHC), the actuators are placed underneath the swashplate in the fuselage-fixed system, and the blade pitch variation is performed by the stationary part of the swashplate,

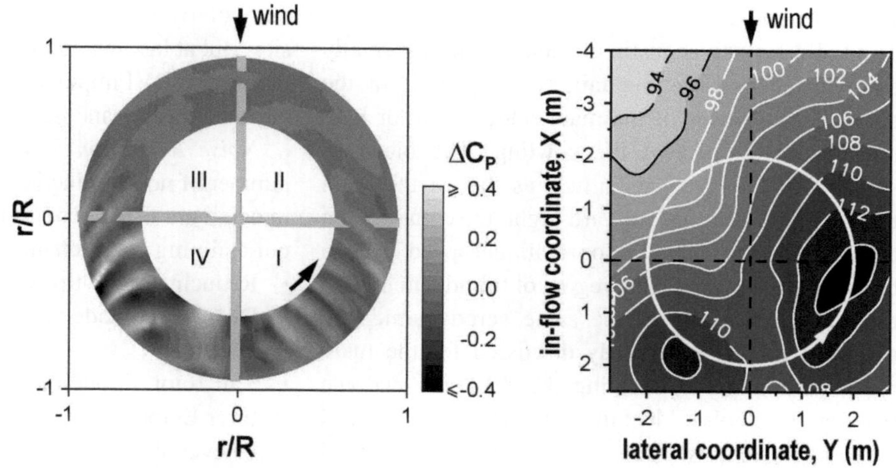

Fig. 17.17 BVI characteristics. (**a**) Leading edge blade surface differential pressure fluctuations due to BVI and (**b**) band-pass filtered (6th to 40th bpf) sound pressure level field (dB) underneath the rotor with intensity maxima underneath the advancing and the retreating side

so that all blades simultaneously experience the identical pitch angle variation. Concerning simultaneous noise and vibration reduction this is not always beneficial. Furthermore, the control frequency in the rotational system is limited to $(B - 1)\Omega$, $B\Omega$, and $(B + 1)\Omega$ with number of blades B and rotor rotational frequency Ω, i.e., for a 4-bladed rotor to 3Ω, 4Ω,and 5Ω. Reductions of BVI impulsive noise up to 50% (6 dB) have been verified in wind tunnel and flight tests, but mostly at the cost of increased vibration levels [27, 264].

More beneficial in this respect is the individual blade control (IBC), where the actuators are placed above the swashplate in the rotating system. It allows for both harmonic (typically 2Ω to 10Ω) and nonharmonic blade pitch variation. With the IBC technique applied in wind-tunnel and flight tests, noise and vibration reductions have been demonstrated at a similar magnitude as for HHC. In particular, it is possible to reduce simultaneously both BVI noise and vibrations [143, 266].

A survey on the status of the active rotor control technology providing numerous references is given in [305]. This also summarizes the state of the research work on novel active blade control techniques (i.e., without blade root control) like via aerodynamically effective blade trailing edge flaps or by controlled change of the blade geometry (e.g., blade twist) by means of "smart" materials.

Passive Noise Reduction Measures. Investigations on BVI noise reduction by modification of the blade-tip geometry yield only limited results at the order of 1–2 dB [26]. Significant reductions of BVI noise can only be obtained by an aerodynamic/aeroacoustic optimization of the rotor in total including the blade geometry. By means of extensive parametric studies and optimization procedures and subsequent wind-tunnel tests, noise reductions of 4–7 dB(A) at simultaneously improved rotor performance could be obtained, compared to a reference rotor of the 1990 decade. This was accomplished by means of selection of modern high-performance blade airfoils, shaping of the blade planform with the objective to decrease the intensity of the blade-tip vortex by shifting the maximum blade lift location toward the rotor centre, and a special design of the blade quarter-chord line (e.g., forward/backward sweep) to avoid or reduce the parallel BVIs [228].

Noise Reduction by Optimized Approach Flight Procedures. Flight tests employing numerous ground microphones have shown that controlled multi-segment landing approach flight procedures with variable flight speed and descent velocity offer a significant noise reduction potential. By suitable selection of deceleration rate and descent rate (flight-path angle), it is possible to avoid those flight conditions most critical for BVI noise. Compared to a 6° landing approach flight with constant velocity (flight condition for noise certification), noise reductions up to 8 dB have been demonstrated [144]. But for the implementation of such low-noise multisegment approach flights, it will be necessary to provide a flight guidance system in order to reduce the pilot's workload.

Tail Rotor Noise

The tail rotor aeroacoustic phenomena are especially complicated because the tail rotor operates in the aerodynamic wakes of the main rotor, the rotor hub, and eventually part of the cowling. The blade-tip speed is of the same magnitude as that of the main rotor. Therefore at fast forward flight, the combination of flight velocity and tail rotor rotational speed leads to transonic on-flow at the tip region of the advancing tail rotor blade resulting in the same aerodynamic and acoustic phenomena already discussed for the main rotor, particularly concerning the thickness and the HS impulsive noise. Most likely tail rotor self-induced BVI impulsive noise is also of importance, but a final proof still needs to be provided. By the impact of the different aerodynamic wakes on the tail rotor blades, additional interaction noise (typically broadband noise) is generated; but the effects of the highly disturbed on-flow have not yet been fully understood. By theory increased tail rotor noise is predicted, but experimental results partly have shown less noise radiation.

Due to the high rotational speed, the acoustic signature of the tail rotor indicates some similarity with that of a fast spinning airplane propeller. Corresponding to its installation on the helicopter, the tail rotor radiates the aerodynamic loading noise (inclusively possible BVI noise) in flight direction forward/down on both sides of the tail rotor plane, while its thickness noise radiation (inclusively possible HS impulsive noise) occurs in-plane with a maximum in flight direction [89, 166, 182]. In climb condition for an observer on ground, the tail rotor noise signature dominates the total noise radiation. At this flight condition, there exists no main rotor impulsive noise and the tail rotor thrust demand is particularly high, increasing the loading noise intensity. But at landing approach flight for which normally main rotor BVI impulsive noise prevails, the tail rotor noise is masked and negligible.

Noise Reduction Measures. For the tail rotor, a number of noise reduction potentialities not explained in detail are listed here (without explanation and without claiming completeness):

- Reducing blade-tip speed
- Decreasing blade air load by increasing number of blades
- Tail rotor shrouding (e.g., Fenestron of manufacturer Eurocopter)
- Uneven blade spacing
- Use of a no-tail-rotor configuration, for which the required thrust is generated by a lateral air jet and taking advantage of the Coanda effect on the tail boom (e.g., NOTAR of the MD500)

17.1.3.2 Rotor Noise Computation

For the computation of rotor noise different methods have been developed and termed

(a) Acoustic Analogy
(b) Kirchhoff Formulations
(c) Computational Aeroacoustics (CAA)

Of these the Acoustic Analogy is the mostly used method. The possible procedures of this method are listed in Table 17.3.

(a) *Acoustic Analogy method.* The theoretical basis of the rotor noise calculation consists of an integral equation describing the sound radiation of moving sources, known as Ffowcs-Williams and Hawkings (FWH) equation [84]. This represents a result of a special evolution of Lighthill's Acoustic Analogy [82, 168]. For solving the FWH

Table 17.3 Schematic of the acoustic analogy method for rotor noise calculation (after F. Farassat)

Given data: rotor trim values, blade-geometry and -motion			
Aerodynamic theory →	Aeroacoustic source terms →	Acoustic theory →	Result
e.g., • Navier-Stokes equations • Euler equations • Full potential equ. • Lifting surface theory • Thin airfoil theory • Lifting line theory	• Calculated temporal blade pressure and velocity distributions or unsteady blade forces Alternatively: • Measured aerodynamic data	e.g., • "Ffowcs Williams and Hawkings" equation • Curle's equation	• Sound pressure time-history • Frequency spectrum

equation, information on the blade surface pressures and on aerodynamic near-field data are required, which needs to be provided either by wind tunnel measurements or by aerodynamic calculations.

The FWH equation is given by

$$4\pi p'(\vec{x},t) = \frac{\partial}{\partial t}\int_S \left[\frac{\rho_0 V_n}{r|1-M_r|}\right]_{\text{ret.}} dS(\vec{y}) + \frac{\partial}{\partial x_i}\int_S \left[\frac{P_{ij}n_j}{r|1-M_r|}\right]_{\text{ret.}} dS(\vec{y}) + \frac{\partial^2}{\partial x_i \partial x_j}\int_V \left[\frac{T_{ij}}{r|1-M_r|}\right]_{\text{ret.}} dV \times (\vec{y}) \quad (17.22)$$

Herein is the sound pressure p' at time t at the observer-point $\vec{x}$ a function of three aerodynamic source terms, which emit sound from source location $\vec{y}$ at the retarded time ($t_{\text{ret.}} = t - |\vec{x} - \vec{y}|/a_0$) with speed of sound a_0 in air at rest; r is the source–observer distance at the retarded time and M_r the component of the blade-tip velocity in observer direction normalized with the ambient speed of sound; V_n is the normal velocity component at the source point, P_{ij} the stress tensor, and T_{ij} the "Lighthill tensor." The motion of the sources is taken into account by the Doppler amplification factor $1/(|1-M_r|)$. The first of the three source terms on the right-hand side of the FWH Eq. (17.22) – the surface integral (monopole) – represents the so-called thickness noise or volume displacement noise, which originates from the sudden displacement of the air caused by the fast rotating blade. The second term (dipole), a surface integral as well, represents the aerodynamic loading noise which in forward flight is generated by the unsteady aerodynamic forces (lift and drag) on the rotor blade. The third term (quadrupole), a volume integral, comprises the nonlinear composites of the distorted flow field near the rotor blade, expressed by the Lighthill tensor

$$T_{ij} = \rho u_i u_j + P_{ij} - a_0^2 \rho \delta_{ij}. \quad (17.23)$$

Equation (17.22) simplifies, when the sound radiation is considered for an observer located in the acoustic farfield:

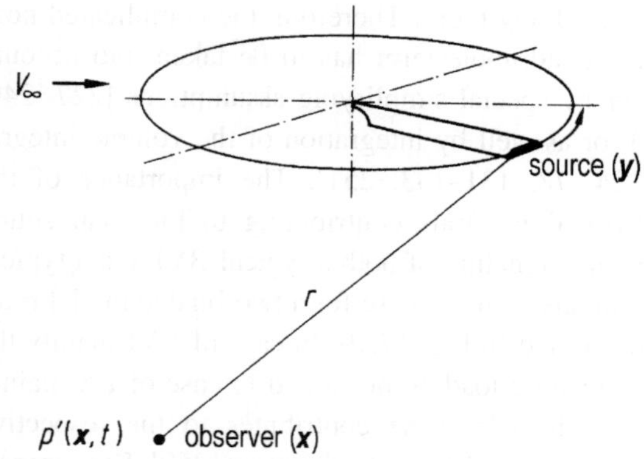

Fig. 17.18 Geometry concerning the noise generation of a rotor blade (see text)

$$4\pi p'(\vec{x},t) = \frac{\partial}{\partial t}\int_S \left[\frac{\rho_0 V_n}{r|1-M_r|}\right]_{\text{ret.}} dS(\vec{y}) + \frac{1}{a_0}\frac{\partial}{\partial t}\int_S \left[\frac{P_r}{r|1-M_r|}\right]_{\text{ret.}} dS(\vec{y}) + \frac{1}{a_0^2}\frac{\partial^2}{\partial t^2}\int_V \left[\frac{T_{rr}}{r|1-M_r|}\right]_{\text{ret.}} dV(\vec{y}). \quad (17.24)$$

Herein P_{ij} has been replaced by P_r (the force component of the blade surface pressure in observer direction, Fig. 17.18) and $T_{ij}$by T_{rr} (the component of the Lighthill tensor in observer direction).

For very accurate calculations, especially in the acoustic nearfield, additional integral terms must be considered. An exact analysis of the FWH equation is provided, e.g., in [77]. The equation is mainly solved in the "time domain," but also used in the "frequency domain" applying Fourier analysis techniques. Initially published [174] rotor noise calculations deal only with the loading noise component. Further evolutions and applications based on the linear source terms (thickness and loading noise) are given in [73, 75, 173, 293]. An intelligible introduction into aeroacoustics and the Lighthill acoustic analogy as well as a derivation of the FWH equation and its solving approaches for wind turbine noise (corresponding to a rotor in hover) is provided in [293].

For the computation of HS impulsive noise, observed at high advancing blade-tip Mach numbers (about ≥ 0.85), the nonlinear effects cannot be neglected any more. Therefore, the complicated non-linear quadrupole term has to be taken into account either by special simplifying assumptions [227, 246, 250] or as well by integration of the volume integral [9, 24, 78, 131–133, 251]. The importance of the individual integrals contributing to the total sound pressure signature of both a typical BVI and a typical HS impulsive noise case for a two-bladed model rotor is illustrated in Fig. 17.19. In case of BVI mainly the aerodynamic loading noise and in case of HS mainly the quadrupole noise contributes to the respective impulsive sound pressure signature [250]. For comparison also the corresponding experimental results are also given.

A fundamental preposition for the numerical solution of the FWH equation is the knowledge about the highly unsteady input parameters like the distribution of the absolute blade surface pressure and that of the velocity field around the blade. Due to the lack of aerodynamic computer codes suitable for unsteady, real flight conditions with BVI (in the 1990s), in some cases experimental blade surface pressure data have been used [150, 173, 201, 250].

Initially, the aerodynamic input parameters have been calculated for simpler conditions like rotors in hover or forward flight without lift. Different computational fluid dynamics (CFD) methods have been applied comprising potential equations assuming small perturbations, full potential (Euler) equations, and more recently Navier–Stokes equations [8, 10, 12, 157, 229, 244], but also singularity methods. Initial approaches to calculate cases including blade lift with the corresponding problem of BVI are reported in [10, 12, 244]. Thereby it appeared that for a reliable BVI noise prediction, an accurate calculation of the unsteady blade surface pressure (or airload) distribution with high temporal resolution is a necessary preposition (azimuth resolution $<0.5°$). A comparison of different aerodynamic methods, applied to the generic case of an exact parallel BVI is published in [37] (Caradonna et al. 2000) and provides in comparison with experimental results the present state of the numerical simulation. However,

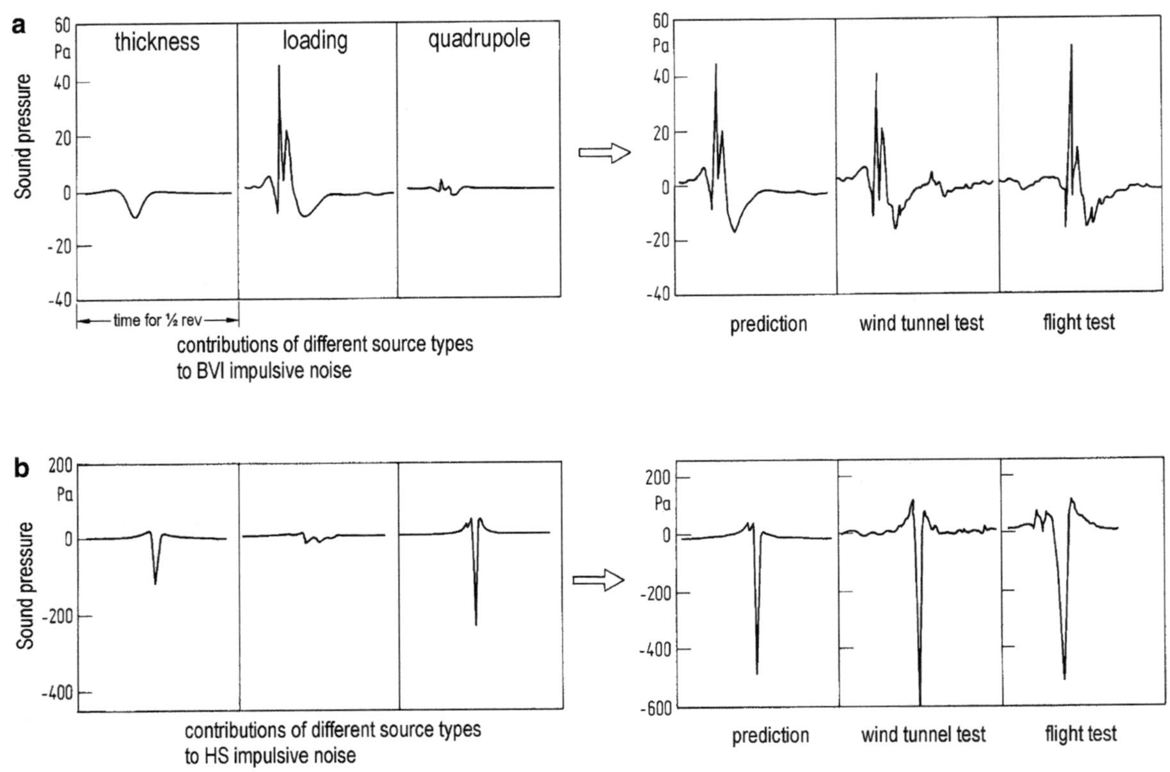

Fig. 17.19 Contribution of the three aeroacoustic source types thickness noise (monopole), loading noise (dipole), and HS noise (quadrupole) to the total sound pressure waveform and comparison with wind tunnel measurement results (AH-1/OLS rotor) and full-scale flight test results. (**a**) BVI impulsive noise ($M_{AT} = 0.79$, $\mu = 0.194$, $C_T = 0.0054$); (**b**) HS impulsive noise ($M_{AT} = 0.90$, $\mu = 0.33$, $C_T = 0.0054$)

if the blade-tip vortex (position, diameter, and strength) is not given, like for a real rotor, then a precise rotor free wake code with accurate modeling of the vortex roll-up process is required in order to precisely calculate the vortex position, BVI geometry, particularly the blade–vortex miss-distance during the interaction, but also the vortex strength and the vortex core diameter. Furthermore, the potential of multiple vortex formation on a rotor blade has to be included in the modeling [15, 33, 147, 294]. With the refined aerodynamic and aeroacoustic simulation codes, it will then be possible to evaluate the effectiveness of measures for BVI noise reduction by active rotor control (HHC, IBC, ABC) or by blade geometry optimization [27, 228, 266].

(b) *Kirchhoff Formulations.* The Kirchhoff method is based on the idea to split the flow field into two parts. The inner nonlinear area, closely surrounding the noise source(s), is computed by employing a CFD field procedure, e.g., an Euler method. The external linear sound propagation area then be treated by linear methods. The exterior envelope of the nonlinear area, called Kirchhoff surface, can be defined as a fixed surface as well as a surface rotating with the rotor. Starting point and basic parameter for the calculation of the far field sound pressure is the time-dependent pressure distribution effective on the Kirchhoff surface [177, 231]. More recent investigations [78], however indicated that the solutions, e.g., for HS impulsive noise, are significantly dependent on the choice of the Kirchhoff surface. Particularly, the penetration of the surface by shear layers or blade-tip vortices should be avoided. Therefore, some care must be taken when this method will be applied.

(c) *Computational Aeroacoustics (CAA).* Most of the CFD methods mentioned above can be applied not only for the near-field calculation of the transonic flow at the rotor blade and of the unsteady pressure distribution on the blade, but theoretically also for computation of the far field of these parameters. In this way, aerodynamics and acoustics will simultaneously be determined with the same formalism [8, 10, 12, 158]. Due to required high temporal and spatial resolution within the computation area, however, the far-field calculations of sound pressure today are largely restricted because of the limited availability of adequate computer time and storage capacity. On the present status of the CAA methods is reported in Sect. 17.1.4.2.

17.1.4 Airframe Noise of Commercial Aircraft

Aerodynamic noise – or flow noise – is generated by the interaction of turbulent flow pressures with solid boundaries. In case of an infinitely extended surface the correspondingly generated aerodynamic noise is called "boundary layer noise," and represents a lower noise limit. Turbulent flow which is shed off the edge of a bounded surface is called "trailing edge noise." In contrast to "boundary layer noise" the edge noise mechanisms is much more efficient, i.e., a relatively larger portion of the unsteady hydrodynamic pressures is converted into travelling acoustic pressure waves. The same is true for aerodynamic noise which is generated through turbulent flows impinging on solid bodies. Accordingly, aerodynamic noise as generated through the turbulent flow around an aircraft ("airframe noise") in its landing configuration, i.e., with deployed landing gears and flaps; it is up to about 10 dB higher compared to that for the aircraft in its cruise configuration.

Analytical solutions for the prediction of aerodynamic noise are available only for very simple geometries and flow fields, such as for flat plate trailing edge noise [123] and vortex shedding noise from flow around cylinders. In contrast, aircraft landing gears and high-lift devices represent extremely complex structures and related aerodynamic noise generation cannot yet be predicted by analytical means. A comprehensive overview of the major airframe noise sources and the overall physical background is provided in [41]. In short terms, airframe noise prediction methods for technical applications can only be developed on the basis of dedicated experiments. Until 1977, an empirical model had been developed [86, 207] for the prediction of airframe noise originating from commercial aircraft. This model is still in use, although it provides only rough noise estimates and does not allow differentiation between the contributions from the various individual flow noise sources on an aircraft. Such prediction models are, therefore, useless for the development of low-noise aircraft components.

17.1.4.1 Experimental Analysis of Flow Noise Sources

In the recent past, both flyover noise measurements [221, 222] and aeroacoustic wind tunnel experiments [60, 109, 118, 218, 255, 270, 298] were performed taking advantage of modern source localization techniques. These tests aimed at a detailed analysis and understanding of airframe noise source mechanisms. As one first important result it turned out that today's production type aircraft landing gears and high-lift devices feature a variety of open cavities and holes in structural elements which are exposed to the surrounding flow. As a consequence, tones may be generated due to cavity resonances which, however, could easily be avoided through an adequate design of such airframe components. In contrast broadband aerodynamic noise originating from the flow around the complex structures of landing gears and high-lift devices is the subject of interest and will be discussed in the following.

Aircraft Landing Gears

Landing gear structures are composed from a large number of struts, joints, axles, and wheels/tires. Major sources of broadband aerodynamic noise are (1) flow separation off the different individual structural components (bluff bodies) and (2) the interaction of turbulent wake flows from upstream components with downstream located structural elements.

Initial experiments on idealized scale model gears showed that the results from such tests do not reproduce high frequency noise contributions originating from flow around the small-scale details of real landing gear structures. Therefore noise tests were performed on full-scale landing gears in the German-Dutch Wind Tunnel (Fig. 17.20) [59, 61].

These experiments revealed that today's landing gears represent a 3D cluster of broadband noise sources. The effect of flow velocity on the emitted noise level spectra can be presented on a nondimensional basis, i.e., normalized sound pressure levels vs. Strouhal number (Fig. 17.21). Since no characteristic source dimension can be defined for a complex gear structure, the Strouhal number is defined with the tire diameter as a reference dimension. Sound intensities

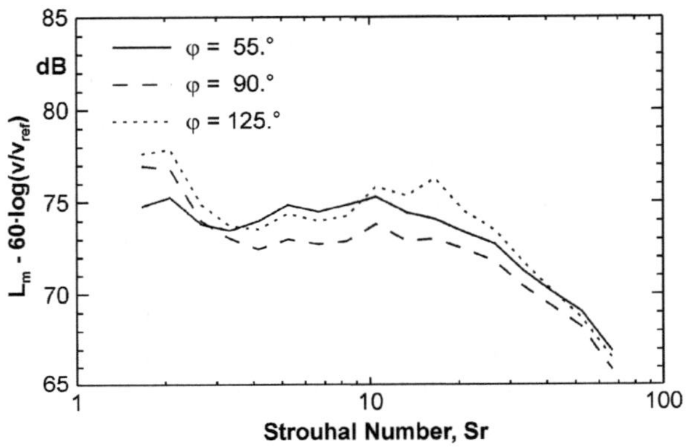

Fig. 17.21 Normalized 1/3-oct. band flow noise spectra for a typical landing gear configuration and for different radiation directions, respectively (V_{ref} arbitrary reference speed, *Sr* based on arbitrary constant reference length)

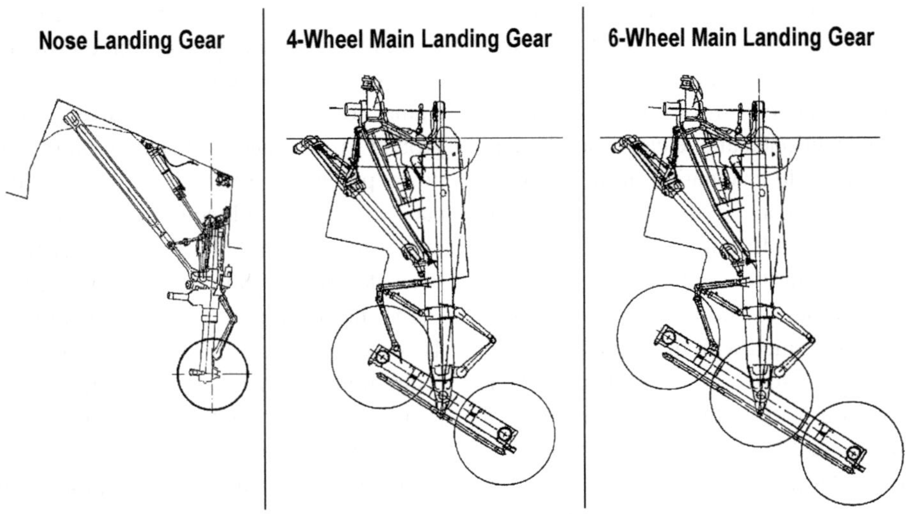

Fig. 17.20 Schematic representation of different landing gear configurations

increase according to the sixth power of flow velocity. The noise source directivity is almost omnidirectional except for a slight noise increase for both forward (60°) and rear (140°) arc radiation directions. Due to the effect of convective amplification for real flight conditions, the forward arc radiation will, therefore, dominate the maximum flyover noise level. For under the wing installed gears, noise levels are affected by the aircraft angle-of-attack through corresponding changes of the relevant local velocity at the gears' position. An increase in angle-of-attack results in an increase in lift. As a result, local flow velocities under the wing decrease and thus landing gear noise. This installation effect is most pronounced in close proximity to the wing.

From these findings, it is concluded that broadband landing gear noise can be described as a superposition of the noise contributions from a variety of individual compact dipole sources (wavelength > strut diameter) with different orientation in space. The noise database as obtained from wind tunnel testing of different full-scale landing gear configurations was used to develop a semiempirical noise prediction tool [259] designed to account for individual noise contributions from the major landing gear components and thus is expected to be capable to predict the effects on noise of changes in gear configuration. For today's landing gears a noise reduction potential of up to 3–4 dB can be utilized through the protection of complex gear structures from the flow by means of "streamlined" fairings.

High-Lift Systems

"Handley page" (slotted) slats and flap side edges represent the major noise sources of today's high-lift systems of commercial aircraft [45, 62, 269]. While flap side edge noise is most important at high frequencies, slat noise dominates the low to mid frequency spectral levels. However, the respective noise source mechanisms are not yet fully understood.

When the slats are deployed, a strong vortex is generated in the rear slat cove for typical aircraft angles-of-attack during approach flight conditions (Fig. 17.22) [240]. This vortex is separated from the high-speed slot flow through a free shear layer. The latter contains and propagates flow instabilities (turbulent eddies) toward the slat trailing edge where this turbulent flow is shed off the edge.

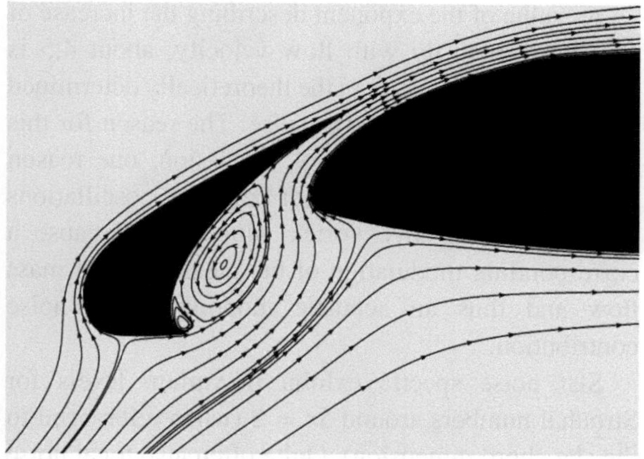

Fig. 17.22 Typical flow field in the slat/wing leading edge area for a high lift configuration [241]

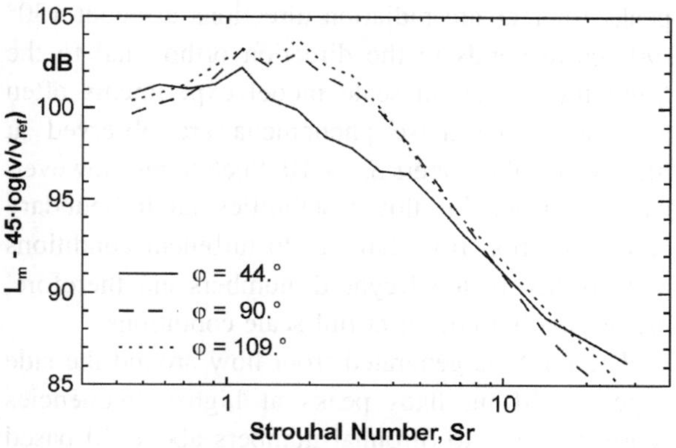

Fig. 17.23 Normalized 1/3-oct. band slat noise spectra (Handley page type slat) for different radiation directions, respectively (V_{ref} arbitrary reference speed, Sr based on arbitrary constant reference length)

The resulting trailing edge noise generation is considered the major slat noise source mechanism. Trailing edge noise intensity is directly related to the local flow turbulence kinetic energy and increases correspondingly to the 5th power of the local flow velocity component tangential to the wall surface and orthogonal to the edge [123]. The effect of flow velocity V on the broadband slat trailing edge noise spectrum can be accounted for by a Strouhal number $Sr = fl/V$ (Fig. 17.23) with slat chord l as characteristic source dimension. This definition assumes that the dimensions of typical turbulence eddies scale with the size of the slat cove vortex and thus with the slat geometry for the limited range of aircraft angles-of-attack in the order of 5° for approach conditions [57, 62, 105, 218].

As value of the exponent describing the increase of slat noise intensity with flow velocity, about 4.5 is adopted which is less than the theoretically determined value of 5 for trailing edge noise. The reason for this deviation is still a matter of discussion; one reason might be the superposition of periodical oscillations of the rear slat cove vortex which would cause a corresponding modulation of the mean slat-slot mass flow and thus an acoustic monopole type noise contribution.

Slat noise spectra exhibit maximum levels for Strouhal numbers around $Sr = 2$ (when referenced to the slat chord dimension). Quite often also tonal noise components are observed in this Strouhal number range which may be related to the above-mentioned vortex oscillations [60, 237]. Broadband slat noise peaks for rear arc radiation directions at about 130° (90° corresponds to the direction orthogonal to the flight trajectory). In scale model experiments often additional tone noise phenomena are observed at high Strouhal numbers $Sr > 10$. Such tones, however, can be attributed to flow instabilities due to boundary layer transition from laminar to turbulent conditions at, respectively, low Reynolds numbers and, therefore, are not likely to occur at full-scale conditions.

The noise as generated from flow around the side edges of lifting flaps peaks at higher frequencies (corresponding to Strouhal numbers above 40 based on flap chord) than that noise originating from the slat. Flap side edge noise intensity is higher compared to slat noise when based on a unit source area. While side edge noise sources are concentrated at the very flap edges only, slat noise sources are distributed over the entire wing span (line source). Therefore, the overall aircraft high-lift noise signature is governed typically by slat noise.

Flap side edge noise source mechanisms are not yet fully understood [114, 115, 124, 253]. A strong edge vortex develops at the flap side edges. This vortex is driven by the flap surface static pressure distribution, which inherently is related to the lift force acting on the flap. For typical flap deflection angles of an aircraft in landing configuration this vortex separates close to a 70% flap chord position. The virtual surface which envelops this vortex represents a free shear layer with embedded flow instabilities which thus rotate around the vortex core. The development of such edge vortex flows has been investigated both experimentally and within dedicated CFD studies [230, 271]. Two potential edge noise mechanisms can be considered: (1) interaction of the vortical flow with the solid flap surface and (2) the acceleration of vorticity in the very vortex flow. The latter mechanism of edge noise generation has in fact been established within initial numerical CAA simulations (*CAA = Computational Aeroacoustics*) [63] (see Sect. 17.1.4.2). Noise measurements show flap side edge noise intensities to increase with the sixth power of flow speed. The noise radiation directivity exhibits a complex 3D pattern depending on sound frequency [29, 107].

Flap side edge noise can be reduced by means of edge fences or absorbing edge liners [60, 106, 108, 239, 268]. However, the detailed physical effects of such devices on the edge noise generation mechanism are still not completely known.

17.1.4.2 Numerical Simulation of Airframe Noise Sources

During the mid nineties of the last century, the development of so-called CAA methods started. These methods serve to simulate numerically the aeroacoustic *source processes* and the *sound propagation through media in arbitrary motion.* The computation of sound is not based on the solution of an acoustic wave equation but on the solution of the *balance equations of continuum mechanics.* A view on the noise generation mechanism of airframe noise shows, which physical phenomena any *source simulation method* needs to describe. The source process for *airframe noise* consists in the conversion of hydrodynamic (i.e., vorticity related) fluctuating pressure in acoustic (i.e., propagable) pressure fluctuations at abrupt changes in geometries as for instance trailing, leading and side edges, steps, slots, etc. In a CAA computation of the aeroacoustic source both the dynamics of the vorticity fluctuations and the related sound generation represent parts of the simulation result. In comparison the source term of an inhomogeneous wave equation (e.g., Lighthill's stress tensor as volume source, surface pressure fluctuations as surface source) is considered known in advance (e.g., by modeling) along with all the vortical processes at the noise generating geometric component. By definition the perturbation of the medium as described by an acoustic wave equation does not contain any vortical (and entropic) degrees of freedom; the source term embodies all of these. Only for comparatively simple flows, the vortex dynamics is either known beforehand or could be modeled in a simple way. In case of feedback processes as for instance self-excited oscillation of cavities in tangential flow, the wave equation approach poses principle problems because here the vortex dynamics and hence the source term itself depends on the acoustic signals, i.e., the (unknown) solution.

In order to make the vortical degrees of freedom computable with a CAA code while fully taking into account *flow effects on the sound propagation* the equations of mass Eq. (17.25), momentum Eq. (17.26), and energy Eq. (17.27),

$$\frac{\partial \rho}{\partial t} + \vec{v} \cdot \nabla \rho + \rho \nabla \cdot \vec{v} = 0 \qquad (17.25)$$

$$\rho \frac{\partial \vec{v}}{\partial t} + \rho \vec{v} \cdot \nabla \vec{v} + \nabla p = \nabla \cdot \vec{\vec{\tau}} \qquad (17.26)$$

$$\frac{\partial p}{\partial t} + \vec{v} \cdot \nabla p + \kappa p \nabla \cdot \vec{v} = (\kappa - 1)(\vec{\vec{\tau}} : \nabla \vec{v} - \nabla \vec{q}) \qquad (17.27)$$

are solved numerically to determine the field variables densityρ, velocity $\vec{v}$ and pressure$p$ in a similar way as in the methods for numerical aerodynamics (CFD). In these equations κdenotes the isentropic exponent, $\vec{\vec{\tau}}$ is the viscous stress tensor, $\vec{q}$ represents the heat flux, a dot denotes a simple contraction, and a column denotes a double contraction (scalar product and double scalar product).

In CAA methods, the domain which is occupied by the (flowing) acoustic medium is covered by a computation grid, whose intersection points or nodes define the locations at which the field variables are represented (*discretization of continuum*). Various such *discretization schemes* have been developed [6, 38, 46, 165, 282, 284, 302], of which the finite difference *DRP scheme* (DRP = Dispersion Relation Preserving) of Tam and Webb [284] or Tam and Shen [282], respectively, has become quite popular. The highly accurate formulation of *numerical boundary conditions* at geometries and artificial free field boundaries of the computation domain are of particular significance for CAA simulations. Artificial reflexions of sound signals must be avoided [125, 279, 280]. The first three "NASA CAA-Workshops on Benchmark Problems" [42, 116, 281] show the accuracy and efficiency of various CAA schemes.

In the course of a CAA simulation, the distribution of field variables is advanced from an initial state at time $t = t_0$ by successive time incrementsΔt with a highly accurate *time integration scheme* (e.g., fourth order Runge–Kutta scheme RK4 or modifications with minimum dispersion error [126]). At the end of the simulation the variables are available at discrete points in time $t_n = t_0 + n\Delta t$. In spite of all the conceptual similarity to the established CFD methods, the schemes implemented in CAA methods are different in two respects, which are of essential importance for aeroacoustics: a) high *spectral fidelity* (global approximation) and b) high *consistency order* (local approximation). A high spectral fidelity ensures for a given resolution of a wave (number of points per wavelength, PPW), the best numerical representation of its dynamics. A high consistency order (typically $\geq$4 in contrast to 2 for CFD) is necessary to adequately represent strong gradients of the flow field (boundary layers, flow separation zones ...) employing as low a resolution as possible.

Even with such methods for realistic technical problems (usually associated with high flow Reynolds numbers) a complete solution of Eqs. (17.25)–(17.27) while resolving temporally and spatially, all turbulence scales does not seem feasible for the foreseeable future. The requirement on computer storage and computation time even on today's high performance computers would be orders of magnitudes too high. For this reason, CAA methods are often used for the simulation of perturbations $\mathbf{g}'(t, \vec{x}) = [\rho', \vec{v}', p']$about the time averaged flow field $\mathbf{g}^0(\vec{x}) = [\rho^0, \vec{v}^0, p^0]$ only. The steady mean flow $\mathbf{g}^0(\vec{x})$ is usually taken from a CFD simulation and contains all time averaged effects of viscosity and (if a suitable turbulence model is used) the turbulence eddy viscosity. The flow variables $\mathbf{g}(t, \vec{x}) = \mathbf{g}^0(\vec{x}) + \mathbf{g}'(t, \vec{x})$ are represented as mean flow plus perturbation and are substituted in Eqs. (17.25)–(17.27) resulting in a nonlinear equation system for the perturbations.

For many technical applications as for instance airframe noise, the presence of acoustic boundary layers is of minor importance such that the viscous terms may be neglected in the *perturbation equations* (inviscid perturbation dynamics). If only linear terms in the perturbation variables are retained, the perturbation equations are called *linearized Euler equations* (LEE). In the LEE, the kinematic effects of the mean flow on the sound propagation are accounted for as well as the dynamics of linear (i.e., weak) vorticity perturbations (e.g., unstable shear layer waves) and feedback phenomena of the perturbations onto themselves with and without the influence of surface boundaries. Based on the LEE the mechanisms of the noise generation at the side edge of a *Flowler flap* (although reduced to a generic case) were simulated for the first time in 1999 [63].

Since due to the multiscale nature of turbulence, the direct numerical prediction of technical airframe noise is currently not feasible, it is reasonable for CAA simulations to use a *model for the excitation of perturbations*. In contrast to aeroacoustic source terms of wave equations which have to be modeled in the complete source domain accounting for the detailed vorticity dynamics here only the *excitation* of the vorticity perturbations is modeled. Their very dynamics being the origin of the generation of airframe noise is then part of the numerical simulation. Several such models for the excitation of perturbations were developed [11, 257]. Depending on the purpose of the CAA simulation, often it is sufficient to introduce vorticity perturbations into the mean flow $\boldsymbol{g}^0$ only upstream of the source domain. Given the mean flow, the vorticity perturbation is carried into the source domain and generates sound by interacting with flow gradients and changes in the geometry [63, 101, 257, 302].

In contrast to the technically relevant airframe noise source at abrupt changes of the geometry, the sound generation in quasihomogeneous subsonic turbulent flows (e.g., free shear layers, free jet flows, turbulence in attached weakly curved boundary layers) cannot be simulated on the basis of the linearized equations, because the turbulence self-noise is governed by nonlinear dynamics.

As an example for the numerical simulation of airframe noise solving the LEE, Fig. 17.24 shows the 3D sound generation at airfoils (spanwise direction$z$, streamwise direction x) upon interaction with a 3D test vortex which was seeded upstream of the airfoil's leading edge [101] representing a perturbed freestream. The acoustic response $p'(t)$ to the vortex is recorded at virtual microphone positions on a circle around the airfoil's nose in the plane$z = 0$. Then $p'(t)$ is Fourier transformed and compared for various airfoil thicknesses. The simulation confirms the property of thick airfoils to convert freestream vorticity perturbations less efficiently into sound as thin airfoils. This result quantifies how much less noise thick airfoils produce compared to thin airfoils subject to the same turbulent freestream (see also [102]). By means of the nonlinear perturbation equations it is possible to verify that even strong, say some 20% perturbations of the freestream velocity in the abovementioned airfoil study changes the acoustic result

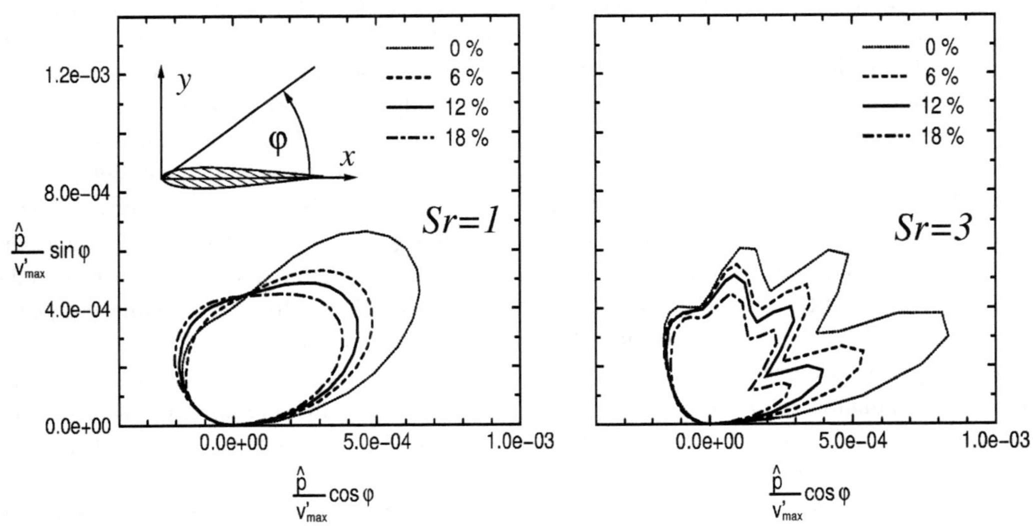

Fig. 17.24 Acoustics response of airfoils of various thicknesses for frontal interaction with 3D test vortex in Mach-0.5 flow. Initial perturbation velocity distribution $\vec{v}'/a_\infty = \varepsilon \cdot (x, -y, 0)\exp[-(x^2 + y^2 + z^2)/c^2 \cdot \ln 2]$ far upstream of airfoil, $\varepsilon \ll 1$, $c = l/10$, l-airfoil chord length. Sound pressure$\hat{p}(Sr)/|\vec{v}'_{\max}|$referenced to freefield impedance as function of Strouhal number $Sr = f \cdot l/U_\infty$with$f$-frequency evaluated for band width $\Delta Sr = 0.2$

only slightly. In the context of an analytical investigation of the generation of *trailing edge noise* of a plate Ffowcs-Williams and Hall [83] emphasized the dominant role of the linear components of the source term. These findings are confirmed in numerical computations, which are as well based on linear source terms [216]. As one of the potential uses of CAA for the prediction of aerodynamically generated sound the exemplary mentioned "vortex test" may serve as a design tool for geometry components with minimum airframe noise.

17.2 Noise Impact

17.2.1 Single Noise Events

If an aircraft passes an observer located at a point *P*, the sound level *L* recorded at *P* is a function of time *t*, as shown schematically in Fig. 17.25. The level-time-history *L*(*t*) can be characterized by two parameters: the maximum sound level L_{max} and the noise durationτ. For aircraft noise, the level *L* is usually expressed as an A-weighted sound pressure level L_A, measured with time weighting *SLOW*, or as a *PNL* (which is denoted as *PNLT* in case that it includes a tone correction). For the noise duration τ , there are two common definitions:

- The so-called "10 dB-downtime," t_{10}, where *L*(*t*) is at most 10 dB below the maximum level L_{max} (see Fig. 17.25)
- The "effective duration," t_{eff}, which is defined by the equation

$$t_{eff} \cdot 10^{L_{max}/10} = \int_{-\infty}^{\infty} 10^{L(t)/10} dt. \qquad (17.28)$$

In practice, a finite integration interval is chosen, usually the interval t_{10}. According to ISO 3891 [140] and DIN 45643 [47], the approximate relationship $t_{eff} = t_{10}/2$ can be used.

Both parameters, L_{max} and τ, are often combined to a "single event noise level" L_E:

$$L_E = L_{max} + k \cdot \lg(\tau/t_{ref}). \qquad (17.29)$$

Examples for the most common single event noise levels can be found in Table 17.4.

17.2.2 Dependence of the Parameters Describing a Single Noise Event on Distance to Observer and Aircraft Speed

An approximate dependence of the parameters L_{max} and τ on the perpendicular distance *d* to the observer (usually called "slant distance" or "distance of closest approach," see Fig. 17.26) and on the aircraft speed *V* can be derived using the following simplifying approach:

$$10^{L(t)/10} = \frac{A}{r^2} e^{-2\alpha r} \cdot \sin^2\theta. \qquad (17.30)$$

The constant *A* depends – for a particular aircraft type – on the actual engine power setting. The distance

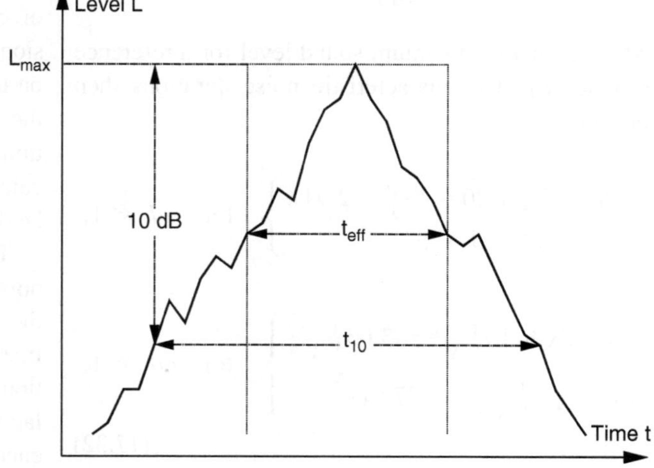

Fig. 17.25 Schematic noise level-time history for an aircraft fly-by

Table 17.4 Examples of single event noise levels L_E with the corresponding parameter values according to Eq. (17.29). The Single Event Exposure Level L_{AX} is also designated as Sound Exposure Level L_{AE} [202]

Single event noise level L_E	k	L_{max}	τ	t_{ref}
Effective Perceived Noise Level *EPNL*[134, 140]	10	*PNLT*	t_{eff}	10 s
Single Event Exposure Level L_{AX}[47, 140]	10	L_{AS}	t_{eff}	1 s
Single Event Level L_{AZ} [47]	13.3	L_{AS}	t_{10}	20 s

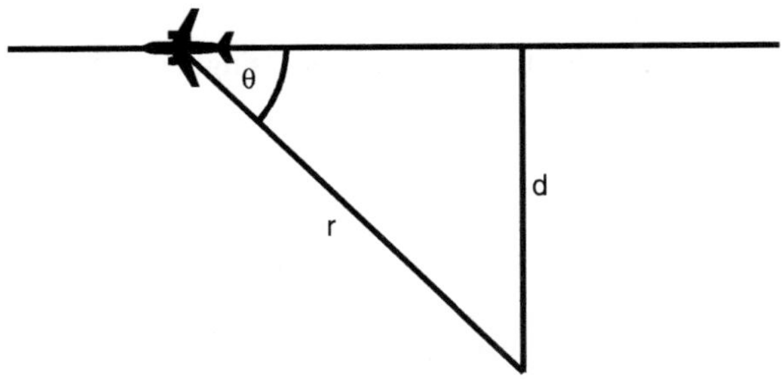

Fig. 17.26 Level fly-by of an aircraft *F* passing the observer location *P* in a distance *d*

r between aircraft and observer location as well as the radiation angle θ (see Fig. 17.26) have to be set for the retarded time $t' = t - r(t')/a_0$ (where a_0 is the speed of sound). The absorption constant α has a value of about 5×10^{-4} m^{-1} for frequencies close to 1,000 Hz. This frequency range is dominant for the estimation of L_A and *PNL*. The term $\sin^2\theta$ describes a simplified directional characteristic. Following this approach yields the maximum sound level as [184]:

$$L_{max} = L_1 - 20\,\lg\left(\frac{d}{d_1}\right) - 8.69\alpha(d - d_1), \quad (17.31)$$

where L_1 is the maximum sound level for a reference distance d_1. The characteristic noise durations then become

$$\left.\begin{array}{l} t_{10} = 2\sqrt{\sqrt{20}-1}\cdot\frac{d}{V} = 2.94\frac{d}{V} \\ t_{eff} = \frac{\pi}{2}\frac{d}{V} = 1.57\frac{d}{V} \end{array}\right\} \quad \text{for} \quad \alpha d \ll 1,$$

$$\left.\begin{array}{l} t_{10} = 2\sqrt{\ln 10}\frac{1}{V}\sqrt{\frac{d}{\alpha}} = 3.03\frac{1}{V}\sqrt{\frac{d}{\alpha}} \\ t_{eff} = \sqrt{\pi}\frac{1}{V}\sqrt{\frac{d}{\alpha}} = 1.77\frac{1}{V}\sqrt{\frac{d}{\alpha}} \end{array}\right\} \quad \text{for} \quad \alpha d \ll 1. \quad (17.32)$$

The overall relationship t_{10} *(d, V)* can be expressed in good approximation by the equation

$$t_{10} = \frac{2.94}{\sqrt{1 + 0.939\alpha d}}\cdot\frac{d}{V}. \quad (17.33)$$

It may be noticed that the effective duration t_{eff} is estimated for the whole integration interval $(-\infty, \infty)$. Integrating only over the 10 dB-downtime t_{10} yields (in case of moderate distances) $t_{10} = 2.04 \cdot t_{eff}$. This confirms the approximate relationship $t_{eff} = t_{10}/2$ given in the previous section.

17.2.3 Aircraft Noise Calculation Procedures

The noise around airports can be estimated by measurement as well as by calculation. However in case of forecasted air traffic scenarios (e.g., for planned airports) the calculation remains as the only available tool. So different national and international aircraft noise calculation procedures have been developed within the last decades. Examples are the German "Anleitung zur Berechnung von Lärmschutzbereichen nach dem Gesetz zum Schutz gegen Fluglärm – AzB" [36] or the integrated noise model (INM) [91, 104], which was developed by the US Federal Aviation Administration (FAA).

Aircraft noise prediction procedures are a combination of a calculation algorithm (sometimes called the "noise engine" and a corresponding database, consisting of acoustic as well as of operational datasets. These datasets can be defined for particular aircraft types or for aircraft categories (usually a result of grouping aircraft of same size and/or noise emission). The specific structure of the database depends on the calculation algorithm and – in some cases – on the specific field of application for which the prediction procedure is designed for (e.g., forecasting or calculation of past scenarios, where the air traffic can be described much more exactly).

To perform an aircraft noise study, any aircraft noise calculation procedure needs a description of the airport and its air traffic as the basic input. The first outlines the airport geometry (i.e., runways location, shape of approach, and departure tracks), the latter lists the aircraft operating within a defined reference time period to the particular flight tracks.

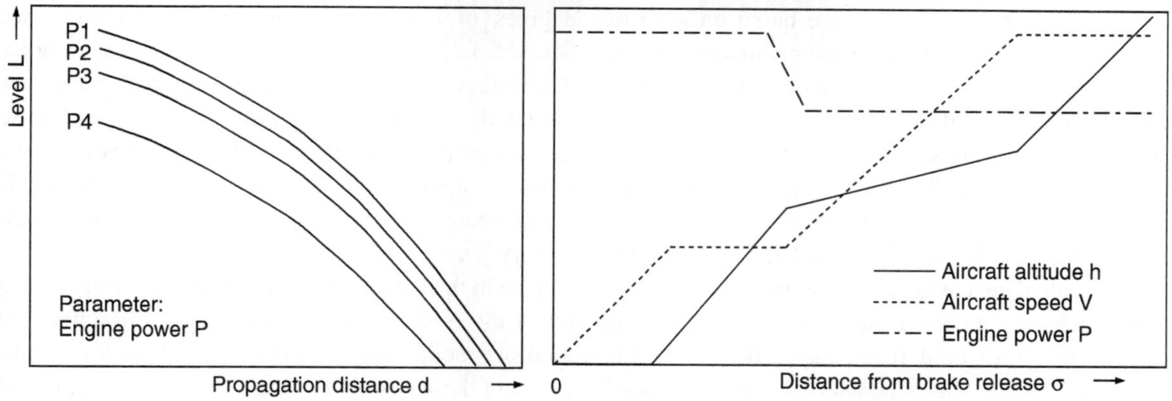

Fig. 17.27 Databases for aircraft noise calculation procedures: Level versus distance tables for different engine powers (NPD-tables) on the *left*, aircraft performance data (for a departure) on the *right*

For most of the "conventional" aircraft noise procedures (i.e., for those used in broad practice), an acoustical data set consists of tables where maximum as well as single event noise level data are stored as functions of the slant distance d for different engine power settings P (so-called "noise-power-distance or NPD data"). A flight operational dataset is a more or less exact description of an aircraft's real flight path by a series of straight segments. At the endpoint of each segment, flight altitude and flight speed as well as the engine power are defined (see Fig. 17.27), hence describing the vertical shape of the flight trajectory. The projection of this trajectory onto the horizontal plane then is matched a flight track, which is usually described by a series of straight segments and circular arcs. The effect of a lateral flight track spreading can be modeled defining a swathe of subtracks building a flight corridor around the idealized backbone track.

Based on this information, the geometry between aircraft and observer location is known and the maximum and single event noise levels can be estimated by a sound propagation calculation. All conventional calculation procedures account at least for the following propagation effects:

(a) Geometric spreading of spherical waves (6 dB per doubling of distance)
(b) Atmospheric attenuation (under standardized atmospheric conditions)
(c) Excess attenuation $E(d)$ of over-ground sound propagation

The directional characteristic of the emitted sound is usually taken into account implicitly in the NPD tables. For single event noise levels, these tables are normalized to an idealized fly-by performed on a horizontal flight path of infinite length with a defined reference speed. Effects of different aircraft speed as well as of flight-path segments of finite length are taken into account by suitable correction terms.

The German AzB procedure is differing from this scheme in so far as maximum level $L_{\max}$ and noise duration t_{10} are calculated separately:

$$L_{\max} = L(d) - c(\beta) \cdot E(d) + Z(\sigma) \tag{17.34}$$

$$t_{10} = \frac{a \cdot d}{V(\sigma) + \frac{d}{b}}. \tag{17.35}$$

The maximum level, $L_{\max}$, is calculated from a reference sound spectrum. For each aircraft category in the AzB, such a spectrum is defined for departure as well as for approach. $L(d)$ includes the effects of geometric spreading and atmospheric absorption. The term $c(\beta)$ equals 0 for an elevation angle β (between aircraft and observer) above 15°. With decreasing $\sin\beta$ $c(\beta)$ increases to 1 for β=0°, hence describing in full overground excess attenuation $E(d)$. Changes in engine power setting are modeled by an additive correction $Z(\sigma)$, σ being the coordinate along the flight-path projection into the horizontal plane (see Fig. 17.27). The coefficients a and b [s] in Eq. (17.35) are depending on the particular aircraft category (see Sect. 17.2.2).

Since many years all these conventional calculation procedures have been proven in practice. However, it must be remembered that they are based on a set of simplifying assumptions concerning modeling flight path, directional characteristic of the source, and propagation effects. A different approach is the use of simulation procedures, which are based on a discretization of the flight path and a more exact description of the sound emission characteristics of the aircraft. The use of simulation procedures allows to calculate a level-time-history at the observer location which – in the ideal case – is as close as possible to a measured time shape. Based on this level-time-history, the characteristic noise related parameters can be derived (see Fig. 17.25).

Although already existing [137, 223] simulation procedures are not implemented in broad practice. This is mainly related to the fact that these procedures require very detailed input data which in practice are not available in the necessary amount. Current "best practice models" are based on the segmentation technique described in [67]. Further information on aircraft noise calculation can be found in [138].

Currently, international (and especially European) activities are dealing with the harmonization of aircraft noise calculation procedures and that of the corresponding databases.

17.3 Assessment of Aircraft Noise

One intention of actual noise related research activities is the harmonized assessment and management of different kinds of transportional and industrial noise (ISO [141], DIN [50], EU [40, 66]). This is a special topic of interest in cases where noise combined from different sources has to be assessed. It must be noticed that the reliability of such universal noise descriptors is limited: noises of equal loudness, but from different sources, usually result in different degrees of annoyance (e.g., "railway bonus," see Sects. 5.4.2.1 and 16.2.4.1). Nevertheless such noise descriptors provide a practical approach, necessary especially as a basis for land-use planning and noise legislation (DIN 18005 [48], EU [66]). However, the problem of a common assessment of noise from different sources is currently not solved yet in a satisfactory way.

Due to the very special character of aircraft noise, a lot of different specific aircraft noise descriptors are used in practice since several decades; field of application of these descriptors are situations where aircraft noise is the dominating source, or cases where legal reasons or aspects of land-use planning require a separate assessment of aircraft noise.

The world-wide most frequently used noise descriptors can be expressed as a weighted equivalent continuous sound level $L_{\mathrm{eq,w}}$:

$$L_{\mathrm{eq,w}} = k \cdot \lg\left[\frac{t_{\mathrm{ref}}}{T}\sum_{i=1}^{N} g_i \cdot 10^{L_{\mathrm{E},i}/k}\right] + C. \qquad (17.36)$$

The summation is performed over all noise events occurring at the observer location during the reference time period T. $L_{\mathrm{E},i}$ is the single event noise level of the ith noise, t_{ref} is the reference noise duration used for the definition of L_{E} (see Sect. 17.2.1), and g_i is a weighting factor depending on the time of day. The "trade-off factor" k usually equals to 10 ("energetic" summation). C is a normalizing constant. Examples of common derivates of weighted equivalent continuous sound levels are discussed below (see also Table 17.5):

(a) The equivalent continuous sound level $L_{\mathrm{eq(4)}}$ according to the German Air Traffic Noise Act [35], sometimes denoted as "Störindex," has to be calculated for two cases A and B with different weighting factors g_i (see Table 17.5). The $L_{\mathrm{eq(4)}}$ is

Table 17.5 Example of weighted equivalent continuous noise levels $L_{\mathrm{eq,w}}$ and the corresponding parameter values according to Eq. (17.36)

				g_i for time of day period from			
$L_{\mathrm{eq,w}}$	$L_{\mathrm{E},i}$	k	C	6–7 h	7–19 h	19–22 h	22–6 h
$L_{\mathrm{eq(4)}}-A$	L_{AZ}	13.3	0	1.5	1.5	1.5	0
$L_{\mathrm{eq(4)}}-B$				1	1	1	5
L_{DN}	L_{AX}	10	0	10	1	1	10
L_{DEN}	L_{AX}	10	0	10	1	3.162	10
NEF	$EPNL$	10	−48.63	16.67	1	1	16.67

the maximum of both values. Reference time period is the 6 busiest months of a year.

(b) The "Day–Night Average Sound Level" L_{DN} [135] is probably the word-wide mostly used aircraft noise descriptor. In most cases the reference time period is 1 year.

(c) The "Day–Evening–Night Sound Level" L_{DEN} [135] is a modification of the L_{DN}, accounting for the additional noise sensitivity during the three evening hours from 19–22 h by a level offset of 5 dB. This is equivalent to a weighting factor g_i of 3.162. In practice the L_{DEN} is normalized to an average 24 h day – the corresponding reference time period is not explicitly defined.

(d) The "noise exposure forecast" (*NEF*) [135] is based upon the tone-corrected PNL. This descriptor is used in Australia and partly also in the USA. The reference time period is usually 1 year.

During the process of harmonization of the European noise policy, the European Commission has prescribed a modified Day–Evening–Night Sound Level L_{DEN} as the harmonized descriptor for all kinds of transportation noise [66]. This L_{DEN} uses in the same weighting factors as the L_{DEN} defined by the ICAO, but is based on a division of the whole day into 16 h for daytime, 4 h for the evening, and 8 h for the night period. The exact definition of these three time slices is free for each member state.

The choice of the trade-off factor $k = 13.3$ for the German Air Traffic Noise Act was based on the assumption, that halving the noise duration or the number of noise events is – with respect to annoyance – equivalent to a decrease of the maximum sound level by $q = 4$ dB (q usually denotes the halving parameter). This special choice was derived from a laboratory study performed by K.D. Kryter und K.S. Pearsons [160] and from a field study performed by A.C. McKennell [190]. However, the energetic averaging ($k = 10$) has been established as the standard procedure, as well as the use of the A-weighting as the standard frequency weighting for aircraft noise. Relationships for the approximate conversion between different aircraft noise descriptors can be found in [138, 185].

It is often criticized that (weighted) equivalent continuous noise levels are not describing a noise situation in an adequate way, e.g., the frequency of the noise events *N*/*T* tends to be small. However, results of noise effect studies – which have to be the basis for a suitably modified noise descriptor – are not easy to obtain. Current approaches (e.g., DIN 4109 [49]) where an averaged maximum sound level is proposed) are developed more or less ad hoc. It has to be noted that any modified assessment procedure should at least fulfill the condition that the corresponding modified noise descriptor will change steadily to the value of the unmodified descriptor with an increasing number of noise events.

A serious problem is the assessment of nocturnal aircraft noise: wake-up reactions are a fundamental aspect of the disturbances produced by aircraft noise during the night. Such reactions are usually correlated with maximum sound levels. Hence it is questionable whether equivalent noise levels may serve as suitable descriptors for the assessment of nocturnal aircraft noise. So in practice sometimes so-called NAT criteria (NAT = number above threshold) are used. Such a criterion is defined by the number of excesses of a defined maximum level within a reference time period. A NAT criterion commonly used in Germany was introduced by Jansen [146]. A different approach which is also recently developed in Germany [13] is based on a dose–response relationship between the maximum sound level and the probability for a wake-up reaction induced by aircraft noise.

17.4 Sonic Boom

17.4.1 Definition and Description

Pressure waves caused by flying aircraft propagate in all directions approximately at the local speed of sound. When the aircraft itself flies faster than the speed of sound, the sonic waves are confined within a nearly conical region that is dragged behind the aircraft. In a linear approach, this cone is called Mach cone with a half angle $\sin\alpha = M^{-1}$, the Mach number M is the flight speed of the aircraft measured in units of the local speed of sound. The supersonic flow around the aircraft is caused by the thickness (displacement) and the weight (lift) of the plane, it consists mainly of systems of compression waves which originate from the leading edges and trailing edges of fuselage and wings. The compression waves are separated by expansion waves. As the later compression waves from the leading edges are somewhat faster than the first one and the former compression waves from the trailing edges are somewhat slower than the last one due to nonlinear effects, the compression waves merge with increasing distance from the aircraft, Fig. 17.28, and result in a sudden increase of pressure at the front as well as at the end of the system of pressure waves [224]. These pressure jumps are called shock waves. Since the shock waves are generated during the entire supersonic flight they usually propagate down to the ground along rays, normal to the Mach cone. Depending on the character of the ground, they are reflected more or less diffusively. The total system of the incoming and the reflected waves is perceived as sonic boom (sonic bang). A schematic profile of the pressure signature of a sonic boom caused by an aircraft flying at constant speed and constant altitude is displayed in Figs. 17.28 and 17.29.

The profile starts with a sudden pressure increase within the rise time τ defined as the interval between the onset of the jump and its peak value $\Delta p = \Delta p_{\max}$. Then the pressure decreases more or less steadily to negative values ahead of the tail wave behind which the pressure jumps back to the ambient pressure. The time interval between the onsets of both jumps is called "length" of the sonic boom. Small variations of the signature shown are mainly attributed to atmospheric inhomogeneities (wind turbulence, temperature fluctuations). These variations can result in so-called spiky or rounded signatures. The ideal pressure signature is called N-wave because of its shape similar to the capital letter N.

For future generations of supersonic aircrafts there are theoretical designs that do appear to generate modified pressure signatures at the ground where either the peak pressure is reduced compared to the one of an ordinary N-wave or that part of the pressure jump with

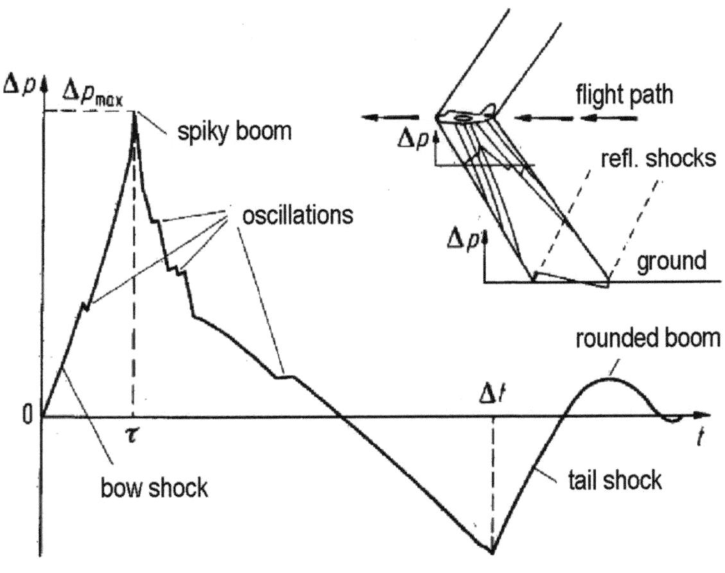

Fig. 17.28 Pressure profile of a sonic boom at the ground surface as a function of time

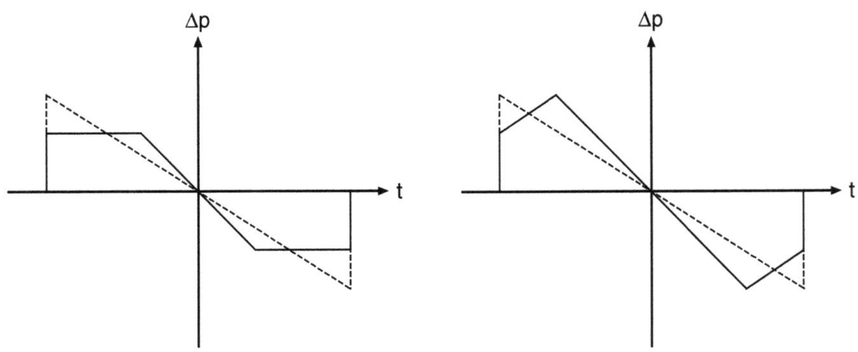

Fig. 17.29 Ideal minimized pressure profiles at the ground for projected aircraft (*continuos line*) und for present ones (*dashed line*)

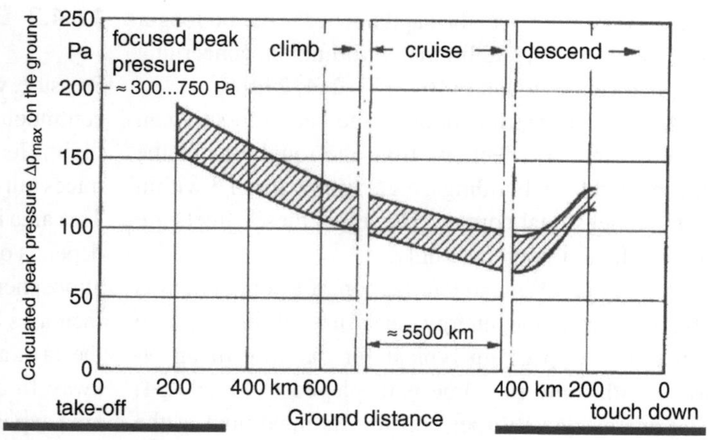

Fig. 17.30 Calculated peak pressure Δp_{max} on the ground along the projection of the flight path of aircraft like Concorde (*lower curve*) and previously projected aircraft like Boeing 2707-300 with a take-off mass of 341 t (*upper curve*)

the sudden pressure increase is reduced. Unfortunately, the aerodynamic drag of such projected aircrafts would be larger than the one of aircrafts like the Concorde [44] – at least accotrding to the present state of knowledge.

The pressure signature of a boom generated by a single aircraft at constant flight velocity and altitude still can vary rapidly along the flight path, in fact the recorded shape can change from a spiky form to a rounded one along a distance of less than 100 m. With it also the rise time and the peak pressure vary considerably [164, 178]. Typical values for τ can vary between 1 ms and 30 ms and for Δt between 100 ms and 400 ms. Due to the separation between the bow shock and the tail shock in time, the sonic boom usually is experienced outdoor as a double boom. Nevertheless, reflection of the boom at the ground and at buildings and scattering by inhomogeneities in the atmosphere (clouds, wind turbulence, temperature variations) can entangle the shock waves so that they are experienced on the ground as a mainly deep rumble even at constant flight conditions. Indoor, the sonic boom often is heard as a single deep bang.

17.4.2 Explanation of Characterizing Quantities of a Sonic Boom

The peak pressure is the most important parameter characterizing a sonic boom – especially with respect to outdoor auditory experiments on human behavioral response to sonic booms. Effects of booms on buildings and structural materials correlate probably better with the characteristic pressure $\Delta p_c = 4I/\Delta t$; here*I*being the maximum of the indefinite pressure pulse integral $\int_0^t p dt$. In the case of an ideal N-wave with zero rise time, Δp_{max} equals Δp_c.

When an aircraft flies at constant supersonic speed and constant altitude, Δp_{max} depends on the weight (lift), the shape (drag), and the Mach number of the plane. Typical values for aircraft like the Concorde are summarized in Fig. 17.30.

The dependence becomes more complex when the aircraft accelerates, maneuvers, or passes into a nose dive. In all these cases, Δp_{max} increases by focusing. However, in the case of civil supersonic transport, only the acceleration from subsonic to supersonic velocities is relevant, where peak pressures double or quadruple the normal ones along a distance of few hundred meters and roughly 200 km after take-off.

The length between the bow and the back shock increases mainly with the length of the aircraft, in addition slowly with flight altitude [224].

Measured rise times (see above) are often larger than those predicted by theories of plane shock waves even when viscosity and heat conduction are taken into account. Of crucial importance seems to be the increasing air density along the downward propagation of the boom [169] and the wind turbulence in the lower atmosphere [164, 172, 178]. Additionally, the rise time is also affected by focusing of the boom due to atmospheric gradients of the temperature and the wind, due to flight maneuvers, and by relaxation effects (N_2 and O_2) which depend very much on the vapor content of the air [220].

Furthermore, it is important to mention that the experimentally recorded signature as well as the theoretically predicted one of a focused sonic boom is shaped very much like the capital letter U but no longer like the letter N. Additional information concerning focusing can be found in [64, 103, 214, 292].

Additional amplifications of the peak pressure can be attributed to reflections from the roughness of the ground and from buildings, e.g., by a factor of 3 within three-dimensional corners. However, these effects are of very local importance only.

A more precise characterization of a sonic boom is obtained from the energy spectrum of its pressure signature. A spectrum typical for the case of an N-wave with finite rise time is displayed in Fig. 17.31. The energy density (per Hz) has its maximum at the rather low frequency of $0.55\Delta t^{-1}$. That can be important to excite vibrations of buildings and of the ground. The high frequency contributions are mainly attributed to the rise time (bent of the envelope of the spectrum). These frequencies are of crucial importance to the loudness of the sonic boom and its startling effects.

17.4.3 Boom Carpet

Pressure waves causing the sonic boom are generated continuously during the entire time of the supersonic flight. Hence the double cone of the front and tail shocks traces out a path on the ground where the boom is heard. This area is called "boom carpet." The size of this carpet depends on the size of the aircraft, its flight path, and on atmospheric conditions. Since the speed of sound increases with increasing temperature near the ground, the rays along which shock waves propagate are bent away from the ground and, hence, reduce the ground area exposed to the sonic boom, Fig. 17.32. This lateral extension can be 100 km wide, Fig. 17.33.

The width of the carpet grows with the velocity and the altitude of the aircraft. An additional result of refraction effects occurs. When flying in the stratosphere with $M \leq 1.15$, the boom is refracted by the atmosphere and does not reach the ground.

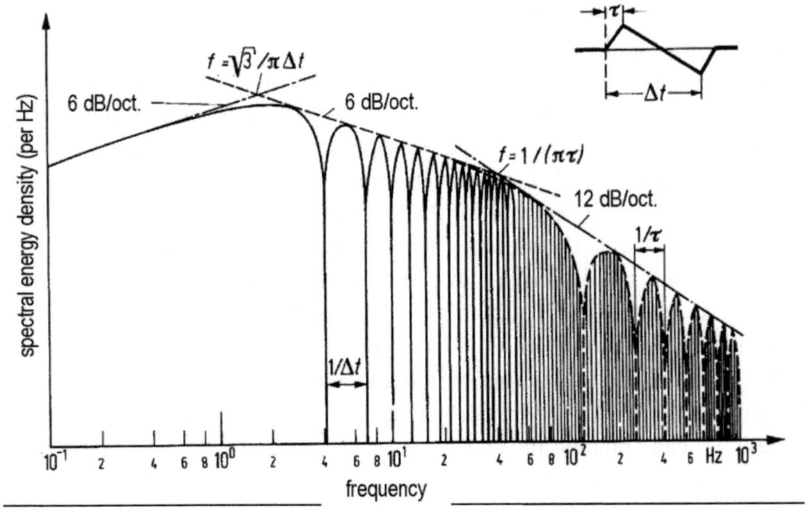

Fig. 17.31 Example of the spectral energy density of a sonic boom [148]. $\Delta t = 350$ ms, $\tau = 8$ ms

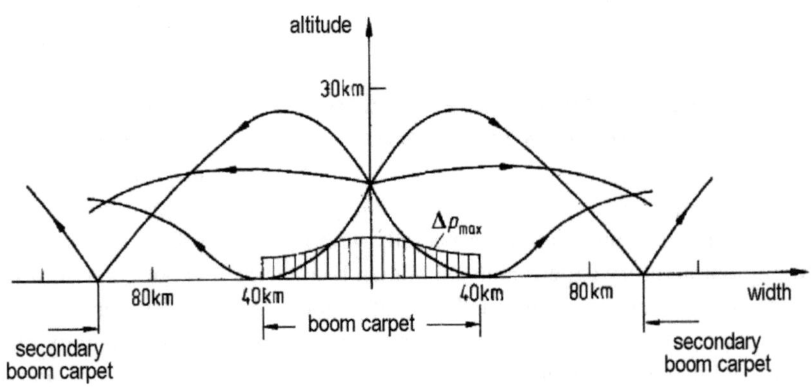

Fig. 17.32 The paths of rays along which the sonic boom propagates bent by refraction in the atmosphere (geometric acoustics)

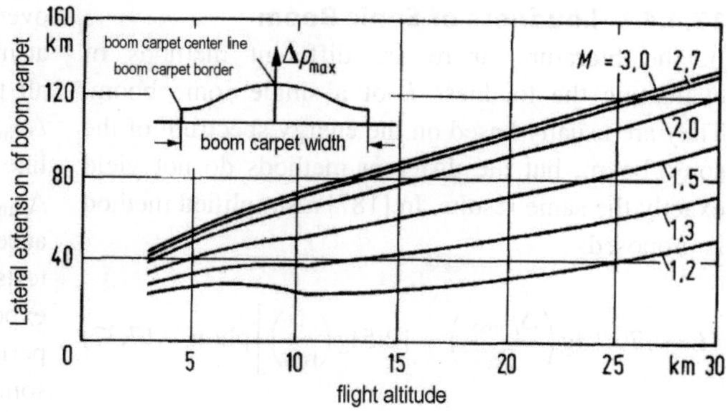

Fig. 17.33 Width of boom carpet [153] for standard atmosphere [291] at sea level, wind effects neglected

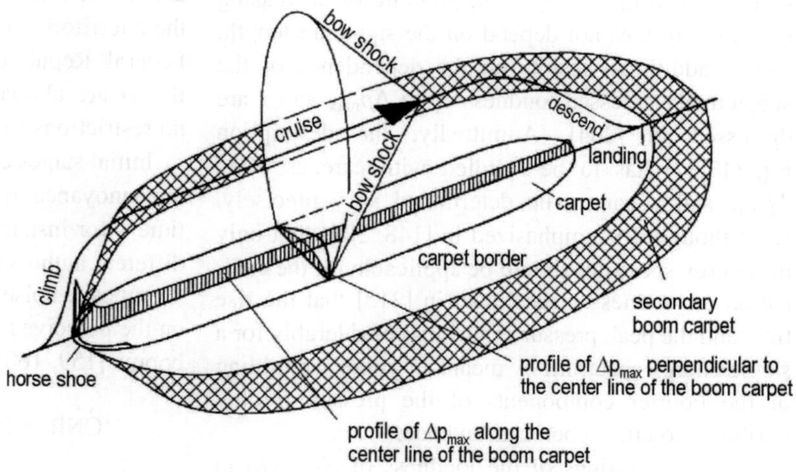

Fig. 17.34 Boom carpet (qualitatively) [235]

A comprehensive survey of a boom carpet is displayed in Fig. 17.34. Here especially the area of focusing is pointed out. It occurs when the carpet reaches the ground first; it is horseshoe shaped; it extends along the centreline of the carpet is 30–100 m; and it reduces to zero at the borders of the carpet. Here the lateral extension of the boom carpet is approximately 25–35 km. The peak pressure values at the centre line can be read qualitatively from Fig. 17.34, at the border of the carpet these values are essentially smaller. New experiments indicate that these values may even be lower.

As the precise location of the horseshoe-shaped area cannot be predicted in the case of real flights – for civil supersonic transport it is uncertain by ± 6 km – the risk of high peak pressures exists for a rather large area.

Finally, even originally upwards propagating shock waves can be refracted downward to form a secondary boom carpet at a larger distance from the main boom carpet. Fortunately, here the peak pressure is much smaller than the peak pressure along the original carpet. Hence, any adverse human responses in this area are not expected.

The physical properties of a sonic boom should be measured and described in accordance with the norm ISO 2249 [139].

17.4.4 Effects of Sonic Booms on Human Beings

Based on many different studies and experiences, especially with stronger booms caused by explosions and artillery barrages, it seems to be very unlikely that sonic booms generated by civil supersonic transport will cause any hearing loss. However, strong startling response and psychological and physiological sequelae (impact on breathing, heart function, or muscle tension) are expected.

17.4.4.1 Loudness of Sonic Boom

In the literature, there are different methods of evaluating the loudness L of a single sonic boom. They are usually based on the energy spectrum of the sonic boom, but the different methods do not yield exactly the same results. In [187] a simplified method is proposed:

$$L = \left[79 + \lg\left(\frac{\Delta p_{\max}}{\text{Pa}}\right) - 12.5 \lg\left(\frac{\tau}{\text{ms}}\right)\right] \text{phon.} \quad (17.37)$$

From this estimate, one sees that the loudness will increase with increasing peak pressure or decreasing rise time. It does not depend on the signature length. Newer additional results on the dependence of the subjectively assessed loudness upon Δp_{max} and τ are discussed in [204]. Admittedly, the description Eq. (17.37) has to be handled with care as often Δp_{max} and τ cannot be determined very precisely. Even though it is emphasized in [148, 306] that only the energy spectrum should be applicable for the evaluation of loudness, it is shown in [215] that the rise time and the peak pressure can vary considerably for a single energy spectrum by means of phase scrambling of the Fourier components of the pressure profile attributed to atmospheric turbulence.

Subjective ratings of the loudness of N-waves in laboratory tests indicate a reasonably good agreement with calculated loudness at least when the peak pressure varies between 40 Pa and 115 Pa. Nevertheless, one has to realize – generally speaking – that it is still an open question which method (loudness, PNL, or A-rated sound level) correlates best with the annoyance due to sonic booms. In the US modified C-rated sound pressure levels are used for explosions and sonic booms [248].

17.4.4.2 Annoyance of Sonic Booms

If regular civil supersonic transport (SST) were to occur, people below flight paths will be exposed to several sonic booms per day. The annoyance by series of booms has been investigated experimentally in comparison with the annoyance caused by subsonic flights as well as absolutely. In the probably most significant investigation [210] test persons were subjected to outdoor sonic booms of different peak pressures, Δp_{max}. They had to decide whether they would accept 10–15 booms per day of a given peak overpressure during day time and the evening (but not at night). Booms of $\Delta p_{max} = 36$ Pawere accepted by all test persons, about 60% still accepted booms of $\Delta p_{max} = 100$ Pa (some kind of average of former SST like Concorde), but no one accepted booms of $\Delta p_{max} = 172$ Pa. Similar results with respect to the acceptability of sonic booms were obtained during tests in Oklahoma City in 1964 when people were exposed at daytime to eight booms per day over a period of 6 months. Here the peak pressures were somewhat lower, i.e., 75–100 Pa [119]. These and additional tests made governments to take legal actions in order to ban civil supersonic transport on their territories [176, 252]. Over the territory of the Federal Republic of Germany, military supersonic flights are allowed with some restrictions, there are no restrictions for flights above 15,000 m [175].

Initial stages concerning a theoretical estimation of the annoyance by series of sonic booms within a given time – for instance during a day – were presented by different authors [186, 304]. In the US, a rating called "composite noise rating" (CNR) is used which is based on the perceived noise level (PNdB) assigned to sonic booms [159, 160]:

$$\text{CNR} = \text{PNdB} - 12 + 10 \lg N, \quad (17.38)$$

N is the number of flights at daytime.

The issue of sleep disturbance or startling response is discussed in [236, 241, 289]. An epidemiological study on long-term effects by sonic booms caused by military aircrafts is described in [5].

17.4.5 Effects of Sonic Booms on Animals

Sonic boom cause mainly startling response on animals – especially on flocks of animals – e.g., sudden shying or hasty run-away with corresponding consequences. Newer investigations [232] indicate that also hearing losses may occur in the cases of very intense exposure to sonic booms ($\Delta p_{max} = 200$ Pa, $\Delta t = 200$ ms, $\tau = 10$ ms). Further results can be found in the survey [235] and the bibliographies [96, 211]. Possibly adverse response to noise by very sound sensitive sea animals, e.g., whales swimming close to the ocean surface could neither be proved nor be excluded completely.

It is known that sonic booms can penetrate through the ocean surface when the Mach angle of the Mach cone is smaller than $13.1°(M > 4.4)$. In the ocean, the peak pressure of a sonic boom decays exponentially with increasing depth [261].

17.4.6 Effects of Sonic Booms on Buildings and on the Ground

Effects of sonic booms on buildings depend on the properties of the building and its building materials beside the actual pressure signature of a boom. The number of not well-known parameters makes it difficult to predict the damage caused by a particular supersonic flight on a single building. Hence, only some kind of averaged total damages along a flight path can realistically be estimated. With reference to [156, 235], the following can be stated: Effects on load-bearing parts of a building are negligibly small in the case of flights at constant altitude and constant speed. There are small risks of breakage glass when their strength is below the norm or when they were mounted unprofessionally, i.e., mounted with preload. The sonic boom can only trigger other failures – the fall off of jointing from pantiles or the cracking of plaster – when these parts are already endangered by strong winds. There is no fear of damage to sound plaster and stucco. Measured accelerations of vibrations in buildings and structural components indicate that slamming doors causes accelerations of the same order of magnitude. However, when sonic booms occur frequently and have an effect on buildings extending over years then the lifetime of structural components can possibly be reduced. With respect to further results it is referred to the literature survey [212]. Effects on other aircrafts – including on full size gliders – are not expected neither in the air nor on the ground as they are designed to withstand much larger dynamical loads than those caused by sonic booms. The same is valid for ships.

Vibration velocities of grounds of different properties that are exposed to sonic booms with Δp_{max} between 24 and 240 Pa [235] add up to 5×10^{-5} m/s to 5×10^{-4} m/s (corresponding velocities caused by foot steps of a person weighing 90 kg are 10^{-4} m/s). These induced velocities are confined to a very thin layer of the ground surface and they are by two orders of magnitude smaller than the legal limit for blastings fixed by the US Bureau of Mines. Hence, considerable effects are hardly to be expected [14, 23]. Triggering of avalanches could not be observed in experiments with 18 sonic booms and peak pressures up to 500 Pa [235]. Nevertheless, it cannot completely be excluded that the onset of sliding of unstable masses of mud or snow can be triggered by sonic booms.

17.4.7 Bibliographies

The bibliography [130] contains 519 papers on measurements, prediction, propagation, and simulation of sonic booms, in addition, on their effects on human beings, animals, buildings, and ground. The bibliographies [96, 211, 212] concentrate on the effects of sonic booms on life and buildings.

References

1. Ahuja KK, Tanna HK, Tester BJ (1979) Effects of simulated forward flight on jet noise, shock noise and internal noise. AIAA Paper 79-0615
2. Amiet RK (1975) Acoustic radiation from an airfoil in a turbulent stream. J Sound Vib 41:407–420
3. Amiet RK (1976) Noise due to turbulent flow past a trailing edge. J Sound Vib 47:387–393
4. Amiet RK (1978) Effect of incident surface pressure field on noise due to turbulent flow past a trailing edge. J Sound Vib 57:305–306
5. Anton-Guirgis H, Culver BD, Wang S, Taylor TH (1986) Exploratory study of the potential effects of exposure to sonic boom on human health, Volume 2: Epidemiological study. Wyle Labs, El Segundo, CA. Report AAMRL-TR-86-020-Vol-2
6. Atkins HL (1999) A high-order method using unstructured grids for the aeroacoustic analysis of realistic aircraft configurations. AIAA Paper No 99-1945
7. Atvars J, Schubert LK, Grande E, Ribner HS (1965/1966) Refraction of sound by jet flow or jet temperature. University of Toronto, Institute for Aerospace Studies, TN 109 (1965), NASA CR-494 (1966)
8. Baeder JD (1990) Euler solution to non-linear acoustics of non-lifting hovering rotor blades. Paper No II, 3.3, 16th European Rotorcraft Forum, Glasgow
9. Baeder JD (1994) The role and status of Euler solvers in impulsive rotor noise computations. AGARD Symposium on Aerodynamics and Aeroacoustics of Rotorcraft, Berlin
10. Baeder JD, McCroskey WJ, Srinivasan GR (1986) Acoustic propagation using computational fluid dynamics. In: Proceedings 42nd annual forum of the American Helicopter Society, vol 1. Washington, DC, pp 551–562

11. Bailly C, Juvé D (1999) A stochastic approach to compute subsonic noise using linearized Euler's equations. AIAA-Paper No 99-1872
12. Ballmann, Kocaaydin (1990) Some aerodynamic mechanisms of impulsive noise during blade/vortex interaction. Paper No II 10, 16th European Rotorcraft Forum, Glasgow
13. Basner M, Buess H, Elmenhorst D, Gerlich A, Luks N, Maaß H, Mawet L, Müller EW, Müller U, Plath G, Quehl J, Samel A, Schulze M, Vejvoda M, Wenzel J (2004) Effects of nocturnal aircraft noise – Volume 1: Executive summery. DLR-Forschungsbericht, 2004-07/E
14. Battis JC (1983) Seismo – acoustic effects of sonic booms on archaeological sites, Valentine military operations area. Air Force Geophysics Lab, Hanscom AFB, MA, USA, Report AFGL-TR-83-0304
15. Beaumier P, Spiegel P (1995) Validation of the ONERA aeroacoustic prediction methods for blade-vortex interaction using HART test results. 51st annual forum of the American Helicopter Society, Fort Worth, TX
16. Bechert D, Pfizenmaier E (1975) On the amplification of jet noise by a pure tone excitation. J Sound Vib 43:581–587
17. de Bernardis E, Tarica D (1992) Surface and volume quadrupoles in the prediction of propeller noise. DGLR/AIAA Paper No 92-02-0651992
18. Block PJW, Gentry Jr GL (1986) Directivity and trends of noise generated by a propeller in a wake. NASA TP-2609
19. Borchers IU, Scholten R, Gehlhar B (1986) Experimental results of the noise radiation of propellers in non-uniform flows. AIAA-86-1928, Seattle/Washington
20. Boxwell DA, Schmitz FH (1982) Full-scale measurements of blade/vortex interaction noise. J Am Helicopter Soc 27; also (1980) Preprint 8061, Proceedings 36th annual forum of the American Helicopter Society
21. Boxwell DA, Schmitz FH, Splettstoesser WR, Schultz K-J (1983) Model helicopter rotor high speed impulsive noise – measured acoustics and blade pressures. NASA TM-85850 and USAAVRADCOM Technical Report-83-A-14
22. Boxwell DA, Schmitz FH, Splettstoesser WR, Schultz K-J, Lewy S, Caplot M (1986) A comparison of the acoustic and aerodynamic measurements of a model rotor tested in two anechoic wind tunnels. Paper No 38, 12th European Rotorcraft Forum, Garmisch-Partenkirchen
23. Bradley J, Stephens RWB (1973) Seismic vibrations induced by Concorde sonic booms. Acustica 28:191–192
24. Brentner KS (1997) An efficient and robust method for predicting helicopter rotor high-speed impulsive noise. J Sound Vib 203(1):87–100
25. Brooks TF (1981) Trailing edge noise prediction using Amiet's method. J Sound Vib 77:437–439
26. Brooks TF (1993) Studies of blade-vortex interaction noise reduction by rotor blade modification. In: Proceedings Noise-Con 93, Noise Control in Aeroacoustics, Williamsburg, VA
27. Brooks TF, Booth ER (1993) The effects of higher harmonic control on blade-vortex interaction noise and vibration. J Am Helicopter Soc 35(3)
28. Brooks TF, Hodgson TH (1981) Trailing edge noise prediction using measured surface pressures. J Sound Vib 78:69–117
29. Brooks T, Humphreys Jr W (2000) Flap edge aeroacoustic measurements. AIAA/CEAS-2000-1975
30. Brooks TF, Schlinker RH (1983) Progress in rotor broadband noise research. Vertica 7:287–307
31. Brooks TF, Jolly RJ, Marcolini MA (1988) Determination of noise source contributions using scaled model rotor acoustic data. NASA TP-2825
32. Brooks TF, Marcolini MA, Pope DS (1989) Main rotor broadband noise study in the DNW. J Am Helicopter Soc 34(2):3–12
33. Brooks TF, Pope DS, Marcolini MA (1989) Airfoil self noise prediction. NASA RP-1218
34. Brooks TF, Boyd DD, Burley CL, Jolly JR (1996) Aeroacoustic codes for rotor harmonic and BVI noise – CAMRAD.Mod1/HIRES. 2nd AIAA/CEAS Aeroacoustics Conference, State College, PA
35. Bund (1971) Gesetz zum Schutz gegen Fluglärm. Bundesgesetzblatt, Jahrgang 1971, Teil I, Nr 28, 282–287, Bonn
36. Der Bundesminister des Innern (1975) Bekanntmachung vom 27.2.75, Durchführung des Gesetzes zum Schutz gegen Fluglärm, hier: Bekanntmachung der Datenerfassungssysteme für die Ermittlung von Lärmschutzbereichen an zivilen (DES) und militärischen Flugplätzen (DES-MIL) sowie einer Anleitung zur Berechnung (AzB). Gemäß Ministerialblatt 26, Ausgabe A, Nr 8, 126-227, Bonn, 10.3.1975, Ergänzung der Anleitung zur Berechnung von Lärmschutzbereichen an zivilen und militärischen Flugplätzen – AzB vom 27. Februar 1975, U II 4-560 120/43, Bonn, 20.2.1984
37. Caradonna F et al (2000) Methods for the prediction of blade-vortex-interaction noise. J Am Helicopter Soc 45 (4):303–317
38. Chang SC (1995) The method of space-time conservation element and solution element – a new approach for solving the Navier-Stokes and Euler-equations. J Comput Phys 119:295
39. Cocking BJ, Bryce WD (1975) Subsonic jet noise in flight based on some recent wind-tunnel tests. AIAA Paper 75-462
40. Commission of European Communities, Directorate General XI. Working Group on Noise Indicators (1999) Position Paper on EU noise indicators
41. Crighton D (1991) Airframe noise. In: Hubbard HH (Hrsg) Aeroaoustics of flight vehicles: theory and practice: Volume 1: noise sources. NASA RP-1258, 391–447
42. Dahl MD (Hrsg) (2000) Third computational aeroacoustics (CAA) workshop on benchmark problems. NASA CP 2000-209790
43. Dahlen H, Dobrzynski W, Heller H (1988) Aeroakustische Untersuchungen zum Lärm von Ultraleichtflugzeugen. DFVLR – FB 88-03
44. Darden CM, Powell CA, Hayes WD, George AR, Pierce AD (1989) Status of sonic boom methodology and understanding. NASA CP-3027
45. Davy R et al (1998) Airframe noise characteristics of a 1/11 scale airbus model. AIAA/CEAS-98-2335
46. Delfs J (2001) An overlapped grid technique for high resolution CAA schemes for complex geometries. AIAA-Paper No 2001-2199
47. DIN 45643 (1984) Messung und Beurteilung von Flugzeuggeräuschen
48. DIN 18005 (1987) Schallschutz im Städtebau, Teil 1

49. DIN 4109 (1989) Schallschutz im Hochbau, Anforderungen und Nachweise
50. DIN 45645 (1996) Ermittlung von Beurteilungspegeln aus Messungen, Teil 1: Geräuschimmissionen in der Nachbarschaft
51. Doak PE (1998) Fluctuating total enthalpy as the basic generalized acoustic field. Theor Comput Fluid Dyn 10:115–133
52. Dobrzynski W (1986) The effect on radiated noise of nonzero propeller rotational plane attitude. AIAA Paper 86-1926, Seattle/Washington
53. Dobrzynski W (1993) Propeller noise reduction by means of unsymmetrical blade-spacing. J Sound Vib 163 (1):123–136
54. Dobrzynski W (1994) Ermittlung von Emissionskennwerten für Schallimmissionsrechnungen an Landeplätzen. DLR-IB 129-94/17
55. Dobrzynski W, Gehlhar B (1993) Untersuchungen zur Propellerlärmminderung durch kleineren Durchmesser bei höherer Blattzahl. DLR-FB 93-48
56. Dobrzynski W, Gehlhar B (1997) The noise from piston engine driven propellers on general aviation airplanes. AIAA/CEAS Paper No 97-1708
57. Dobrzynski W, Pott-Pollenske M (2001) Slat noise source studies for farfield noise prediction. AIAA/CEAS Paper No 2001-2158
58. Dobrzynski W, Heller H, Powers J, Densmore J (1986) DFVLR/FAA propeller noise tests in the German-Dutch wind tunnel DNW. DFVLR-IB 129-86/3 or FAA Report No AEE 86-3
59. Dobrzynski W et al (1997) Full scale noise testing on airbus landing gears in the German Dutch Wind Tunnel. AIAA/CEAS-97-1597
60. Dobrzynski W et al (1998) Airframe noise studies on wings with deployed high-lift devices. AIAA/CEAS-98-2337
61. Dobrzynski W, Chow LC, Guion P, Shiells D (2000) A European study on landing gear airframe noise sources. AIAA/CEAS Paper No 2000-1971, Lahaina/Hawaii
62. Dobrzynski W, Gehlhar B, Buchholz H (2000) Model- and full scale high-lift wing wind tunnel experiments dedicated to airframe noise reduction. 7th international congress on sound and vibration, Garmisch-Partenkirchen, 4–7 July
63. Dong ThZ (1999) Direct numerical simulations of flap side edge noise. AIAA-Paper No 99-1803
64. Downing JM et al (1988) Controlled focused sonic booms from maneuvering aircraft. J Acoust Soc Am 104:112–121
65. Drevet P, Duponchel JP, Jacques JR (1977) The effect of flight on jet noise as observed on the Bertin Aérotrain. J Sound Vib 54:173–201
66. The European Parliament and the Council of the European Union (2002) Directive 2002/49/EC of the European Parliament and of the Council of 25 June 2002 relating to the assessment and management of environmental noise. Official Journal of the European Communities, L 189/12, June 2002
67. European Civil Aviation Conference (2005) Methodology for Computing Noise Contours around Civil Airports. Vol 1: Applications Guide, Volume 2: Technical Guide. ECAC.CEAC Doc.29, 3rd Edn, Dec 2005
68. Enghardt L, Tapken U, Neise W, Kennepohl F, Heinig K (2001) Turbine blade/vane interaction noise: acoustic mode analysis using in-duct sensor rakes. AIAA-Paper 2001-2153
69. Envia E (1992) An asymptotic theory of supersonic propeller noise. DGLR/AIAA Paper No 92-02-064
70. Eversman W (1971) Energy flow criteria for acoustic propagation in ducts with flow. J Acoust Soc Am 49: 1717–1721
71. Eversman W (1991) Theoretical models for duct acoustic propagation and radiation. In: Hubbard H (Hrsg) Aeroacoustics of flight vehicles: theory and practice: Volume 2: noise control. NASA Reference Publication 1258, vol 2, NASA, 101–163
72. Evertz E et al (1976) Noise generation by interaction between subsonic jets and blown flaps. DLR-FB 76-20
73. Farassat F (1975) Theory of noise generation from moving bodies with an application to helicopter rotors. NASA TR R-451
74. Farassat F (1981) Linear acoustic formulas for calculation of rotating blade noise. AIAA J 19:1122–1130
75. Farassat F (1982) Rotor noise prediction technology – theoretical approach. NASA CP-2234
76. Farassat F (1983) The prediction of the noise of supersonic propellers in time domain – new theoretical results. AIAA Paper 83-0743, Atlanta
77. Farassat F, Brentner KS (1987) The uses and abuses of the acoustic analogy in helicopter rotor noise prediction. Paper, AHS National Specialists' Meeting on Aerodynamics and Aeroacoustics, Arlington, TX
78. Farassat F, Brentner KS (1998) Supersonic quadrupole noise theory for high-speed helicopter noise. J Sound Vib 218(3):481–500
79. Farassat F, Succi GP (1980) A review of propeller discrete frequency noise prediction technology with emphasis on two current methods for time domain calculations. J Sound Vib 71:399–419
80. Ffowcs-Williams JE (1963) Noise from turbulence convected at high speed. Trans R Soc A225:469–503
81. Ffowcs-Williams JE (1969) Hydrodynamic noise. Annu Rev Fluid Mech 1:197–222
82. Ffowcs-Williams JE (1984) Acoustic analogy. IMA J Appl Math 31:113–124
83. Ffowcs-Williams JE, Hall LH (1970) Aerodynamic sound generation by turbulent flow in the vicinity of a scattering half plane. J Fluid Mech 40:657–670
84. Ffowcs-Williams JE, Hawkings DL (1969) Sound generation by turbulence and surfaces in arbitrary motion. Philos Trans R Soc Lond 264A:321–342
85. Ffowcs-Williams J E, Simpson J, Virchis VJ (1975) Crackle – an annoying component of jet noise. J Fluid Mech 71
86. Fink MR (1977) Airframe noise prediction method. FAA RD-77-29
87. Fisher MJ, Preston GA, Bryce WD (1998) A modelling of the noise from simple coaxial jets, part I: with unheated primary flow. J Sound Vib 209:385–403
88. Fisher MJ, Preston GA, Bryce WD (1998) A modelling of the noise from simple coaxial jets, part II: with heated primary flow. J Sound Vib 209:405–417
89. Fitzgerald J, Kohlhepp F (1988) Research investigation of helicopter main rotor/tail rotor interaction noise. NASA CR-4143

90. Fitz Simmons RD et al (1980) Flight and wind tunnel test results of a mechanical jet noise suppressor nozzle. AIAA-80-0165
91. Fleming GG, Olmstead JR, D'Aprile JR, Gerbi PJ, Gulding JM, Plante JA (1997) Integrated noise model (INM). Version 5.1 Technical Manual, Department of Transportation, Federal Aviation Administration, Report No FAA-AEE-97-04
92. Frota J, Lempereur P, Roger M (1998) Computation of the noise of a subsonic propeller at an angle of attack. AIAA/CEAS Paper No 98-2282
93. George AR (1977) Helicopter noise state of the art. AIAA-Paper 77-1337, Atlanta, GA
94. George AR, Chang SB (1983) Noise due to blade/vortex interactions. Paper No A-83-39, Proceedings of the 39th annual forum of the American Helicopter Society
95. George AR, Chou S-T (1984) Broadband rotor noise analysis. NASA CR-3797
96. Gladwin DN, Manci KM, Villella R (1988) Effects of aircraft noise and sonic booms on domestic animals and wildlife: Bibliographic abstracts. Air Force Engineering and Services Center, Tyndall AFB, FL, Report AFESC-TR-88-14
97. Glegg SAL, Jochault C (1997) Broadband self noise from a ducted fan. 3rd AIAA/CEAS Aeroacoustics Conference, Paper No 97-1612, Atlanta, GA
98. Gliebe PR, Brausch JF, Majjigi RK, Lee R (1991) Jet noise suppression. In: Hubbard H (Hrsg) Aeroacoustics of flight vehicles: theory and practice: Volume 2: noise control. NASA Reference Publication 1258, vol 2, NASA, pp 207–269
99. Gounet H, Lewy S (1988) Prediction of propfan noise by a frequency-domain scheme. J Aircr 25:428–435
100. Groeneweg JF, Sofrin TG, Rice EJ, Gliebe PR (1991) Turbomachinery noise. In: Hubbard H (Hrsg) Aeroacoustics of flight vehicles: theory and practice: Volume 1: noise sources. NASA Reference Publication 1258, vol 1, NASA, pp 151–209
101. Grogger HA, Lummer M, Lauke Th (2001) Simulating the interaction of a three-dimensional vortex with airfoils using CAA. AIAA-Paper No AIAA-2001-2137
102. Guidati G. Wagner S (1999) The influence of airfoil shape on gust-airfoil interaction noise in compressible flows. AIAA/CEAS Paper 99-1843
103. Guiraud JP (1969) Focalisation dans les ondes courtes non linéaires; application au bruit balistique de focalisation. AGARD Conference Proceedings No 42, Aircraft Engine Noise and Sonic Boom, Paper 12
104. Gulding JM, Olmstead JR, Fleming GG (1999) Integrated noise model (INM). Version 6.0 User's Guide. Department of Transportation, Federal Aviation Administration, Report No FAA-AEE-99-03
105. Guo YP (1997) A model for slat noise generation. AIAA/CEAS- 97-1647
106. Guo YP (1999) Modeling of noise reduction by flap side edge fences. NASA CR CRAD-9402-TR-5767
107. Guo Y (1999) Prediction of flap side edge noise. AIAA/CEAS-99-1804
108. Guo Y (2000) Modeling of noise reduction by flap side edge fences. AIAA/CEAS-2000-2065
109. Guo YP, Hardy BA, Bent PH, Yamamoto K, Joshi MC (1998) Noise characteristics of DC-10 aircraft high lift system. NASA CR CRAD-9310-TR-4893
110. Hamilton-Standard Inc (1971) Generalized propeller noise estimating procedure – Revision D. Windsor Locks, CT, USA
111. Hanson DB (1980) Influence of propeller design parameters on farfield harmonic noise in forward flight. AIAA J 18:1313–1319, see also AIAA Paper 79-0609 1979
112. Hanson DB (1983) Compressible heliccoidal surface theory for propeller aerodynamics and noise. AIAA J 21:881–889
113. Hanson DB (1999) Influence of lean and sweep on noise of cascades with turbulent inflow. AIAA-Paper 99-1863
114. Hardin JC (1980) Noise radiation from the side edge of flaps. AIAA J 18(5):549–552
115. Hardin JC, Martin JE (1996) Flap side-edge noise: acoustic analysis of Sen's model. AIAA-96-1674
116. Hardin JC, Ristorcelli JR, Tam CKW (Hrsg) (1995) ICASE/LaRC Workshop on benchmark problems in computational aeroacoustics (CAA). NASA CP 3300
117. Harper-Bourne M, Fisher MJ (1973) The noise from shock waves in supersonic jets, noise mechanisms, AGARD-CP-131. In: Proceedings of an AGARD Conference
118. Hayes JA, Horne WC, Soderman PT, Brent PH (1997) Airframe noise characteristics of a 4.7% scale DC-10 model. AIAA/CEAS-97-1594
119. Hilton DA, Huckel V, Steiner R, Maglieri DJ (1964) Sonic boom exposures during FAA community-response studies over a 6-month period in the Oklahoma city area. NASA TN D-2539
120. Hoad DR (1980) Helicopter model scale results of blade/vortex interaction impulsive noise as affected by tip modification. Paper No 80-62, 36th Annual Forum, American Helicopter Society
121. Hoad DR (1987) Helicopter blade/vortex interaction locations-scale-model acoustics and free-wake analysis results. NASA TP-2658
122. Holste F, Neise W (1997) Noise source identification in a propfan model by means of acoustical near field measurements. J Sound Vib 203(4):641–665
123. Howe MS (1975) Contributions to the theory of aerodynamic sound, with application to excess jet noise and the theory of the flute. J Fluid Mech 71(Pt 4):625–673
124. Howe MS (1980) Aerodynamic sound generated by a slotted trailing edge. Proc R Soc Lond A373:235–252
125. Hu FQ (1996) On absorbing boundary conditions for linearized Euler equations by a perfectly matched layer. J Comput Phys 129:201–219
126. Hu FQ, Hussaini MY, Manthey J (1995) Low-dissipation and – dispersion Runge-Kutta schemes for computational aeroacoustics. J Comput Phys 124(1):177–191
127. Hubbard H (Hrsg) (1991) Aeroacoustics of flight vehicles: theory and practice: Volume 1: noise sources. NASA Reference Publication 1258, Vol 1, NASA
128. Hubbard H (Hrsg) (1991) Aeroacoustics of flight vehicles: theory and practice: Volume 2: noise control. NASA Reference Publication 1258, vol 2, NASA
129. Hubbard JE, Leighton JE (1983) A comparison of model helicopter rotor primary and secondary blade/vortex

interaction blade slap. AIAA 8th Aeroacoustics Conference, Paper AIAA-83-0723
130. Hubbard HH, Maglieri DJ, Stephens DG (1986) Sonic-boom research: selected bibliography with annotation. NASA TM-87685
131. Ianniello S (1999) An algorithm to integrate the FW-H equation on a supersonic rotating domain. AIAA J 37 (9):1040–1047
132. Ianniello S (1999) Quadrupole noise predictions through the Ffowcs Williams-Hawkings equation. AIAA J 37 (9):1048–1054
133. Ianniello S (2001) Acoustic analysis of high tip-speed rotating blades. Aerosp Sci Technol 5:179–192
134. International Civil Aviation Organization (ICAO) (1981) Environmental protection. Annex 16 to the Convention on International Civil Aviation, vol 1, Aircraft noise, 1st edn
135. International Civil Aviation Organization (ICAO) (1988) Recommended method for computing noise contours around airports. ICAO Circular 205-AN/1/25
136. International Civil Aviation Organization (ICAO) (1988) International standards and recommended practices. Environmental protection. ICAO, ANNEX 16 to the Convention on International Civil Aviation, vol I, 2nd edn
137. Isermann U (1988) Berechnung der Fluglärmimmission in der Umgebung von Verkehrsflughäfen mit Hilfe eines Simulationsverfahrens. MPI für Strömungsforschung, Bericht 7/1988, Göttingen
138. Isermann U, Schmid R (1999) Bewertung und Berechnung von Fluglärm. Im Auftrag des Bundesministeriums für Verkehr, FE-Bericht Nr L-2/96-50144/96, DLR Institut für Strömungsmechanik, Göttingen
139. International Standard ISO 2249-1973 (E) (1973) Acoustics description and measurement of physical properties of sonic booms
140. International Standard ISO 3891-1978 (E) (1978) Acoustics – procedure for describing aircraft noise heard on the ground
141. International Standard ISO (1996) Acoustics – description and measurement of environmental noise – (1982) Part 1: Basic quantities and procedures. (1987) Part 2: Acquisition of data pertinent to land use. (1987) Part 3: Application to noise limits
142. Isom MP (1980) Acoustic shock waves generated by a transonic helicopter blade. Paper 63, 36th Annual National Forum of the American Helicopter Society
143. Jacklin SA, Blaas A, Teves D, Kube R (1995) Reduction of helicopter BVI noise, vibration and power consumption through individual blade control. In: Proceedings of the 51st annual forum of the American Helicopter Society
144. Jacobs EW, Prillwitz RD, Chen RT, Hindson WS, Santa Maria OL (1997) The development and flight test demonstration of noise abatement approach procedures for the Sikorsky S-76. AHS Technical Specialists Meeting for Rotorcraft Acoustics and Aerodynamics, Williamsburg, VA
145. Janardan BA, Gliebe PR (1989) Acoustic characteristics of counterrotating fans from model scale tests. AIAA Paper 89-1142, San Antonio, TX
146. Jansen G, Linnemeier A, Nitsche M (1995) Methodenkritische Überlegungen und Empfehlungen zur Bewertung von Nachtfluglärm. Zeitschrift für Lärmbekämpfung 42:91–106
147. Johnson W (1995) A general free wake geometry calculation for wings and rotors. 51st annual forum of the American Helicopter Society, Forth Worth, TX
148. Johnson DR, Robinson DW (1967) The subjective evaluation of sonic bangs. Acustica 18:241–258
149. Jonkouski GJ, Home WC, Soderman PT (1983) The acoustic response of a propeller subjected to gusts incident from various inflow angles. AIAA-83-0692, Atlanta/Georgia
150. Joshi MC, Lin SR, Boxwell DA (1987) Prediction of blade/vortex interaction noise. In: Proceedings of the 43rd annual forum, American Helicopter Society, pp 405–420
151. Julliard J, Antoine H, Riou G (2001) Development of a three degree of freedom liner. AIAA Paper 2001-2203
152. Kameier F, Neise W (1997) Rotating blade flow instability as a source of noise in axial turbomachines. J Sound Vib 203:833–853
153. Kane EJ, Palmer TY (1964) Meterological aspects of the sonic boom. FAA SPDS Report RD 64-180
154. Kellner A (1980) Experimentelle und theoretische Untersuchungen über den Einfluß inhomogener Geschwindigkeitsverteilung in der Zuströmung auf die Lärmerhöhung von Mantelschrauben. Dissertation RWTH Aachen
155. Klöppel V (1976) Schallabstrahlung durch akustische Rückkopplung bei rechtwinklig umgelenkten Luftstrahlen. Dissertation RWTH Aachen
156. Koch HW, Weber G (1970) Flugzeugknalle und ihre Wirkung auf Gebäude. Die Bautechnik 7:238–244
157. Kroll N (1986) Comparison of the flow field of propellers and hovering rotors using Euler-equations. Paper 28, 12th European Rotorcraft Forum, Garmisch-Partenkirchen
158. Kroll N, Lohmann D, Schöne J (1987) Numerical methods for propeller aerodynamics and acoustics at DFVLR. 69th AGARD-Symposium on Gasturbine Components, Paris
159. Kryter KD (1969) Sonic boom transport. Science 163: 357–367
160. Kryter KD, Pearsons KS (1963) Some effects of spectral content and duration on perceived noise level. J Acoust Soc Am 35:866–883
161. Laur MN, Squires RL, Nagel RT (1992) Forward rotor vortex location effects on counter rotating propeller noise. DGLR/AIAA Paper No 92-02-153
162. Laurence JH, Woodward RP (1989) Unsteady blade pressure measurements on a model counterrotation propeller. AIAA Paper 89-1144, San Antonio, TX
163. Lavrich PL, Simonich JC, McCormick DC (1992) An assessment of wake structure behind forward swept and aft swept propfans at high loading. DGLR/AIAA Paper No 92-02-154
164. Lee RA, Downing JM (1996) Comparison of measured and predicted lateral distribution of sonic boom overpressures from United States Air Force sonic boom database. J Acoust Soc Am 99:768–776
165. Lele SK (1992) Compact finite difference schemes with spectral-like resolution. J Comput Phys 103:16–42
166. Leverton JW (1980) Reduction of helicopter noise by use of a quiet tail rotor. Paper No 24, 6th European Rotorcraft Forum
167. Lighthill MJ (1952) On sound generated aerodynamically. Part I: general theory. Proc R Soc Lond A 211:564–587

168. Lighthill MJ (1954) On sound generated aerodynamically. Part II: turbulence as a source of sound. Proc R Soc Lond A 222:1–32
169. Lilley GM (1969) The generation and propagation of shock waves leading to the sonic boom. Report in 5 parts on the sonic boom, prepared for the OECD Conference on Sonic Boom Research, Part 1
170. Lilley GM (1974) On the noise from jets. In: Noise mechanisms. AGARD-CP 131, 13.1–13.12
171. Lilley GM (1991) Jet noise classical theory and experiments. In: Hubbard H (Hrsg) Aeroacoustics of flight vehicles: theory and practice: Volume 1: noise sources. NASA Reference Publication 1258, vol 1, NASA, pp 211–289
172. Lipkens B, Blackstock DT (1998) Model experiments to study sonic boom propagation through turbulence. Part I: general results. J Acoust Soc Am 103:148–158
173. Lowson MV (1965) The sound field for singularities in motion. Proc R Soc A 286:559–572
174. Lowson MV, Ollerhead JB (1969) A theoretical study of helicopter. J Sound Vib :187–222
175. Luftfahrthandbuch Deutschland (1984) Überschallflüge militärischer Strahlflugzeuge. RAC-3-3-1
176. Luftverkehrs-Ordnung (1986) § 11a, § 11b
177. Lyrintzis AS (1994) Review, the use of Kirchhoff's method in computational aeroacoustics. J Fluid Eng-T ASME 116:665–676
178. Maglieri DJ (1967) Sonic boom flight research: some effects of airplane operations and the atmosphere on sonic boom signatures. NASA SP-147, pp 25–48
179. Mahan JR, Karchmer A (1991) Combustion and core noise. In: Hubbard H (Hrsg) Aeroacoustics of flight vehicles: theory and practice;, Volume 1: noise sources. NASA Reference Publication 1258, vol 1, NASA, pp 483–517
180. Mani R (1971) Noise due to interaction of inlet turbulence with isolated stators and rotors. J Sound Vib 17: 251–260
181. Martin RM, Splettstoesser WR, Elliott JW, Schultz K-J (1998) Advancing side directivity and retreating side interactions of model rotor blade/vortex interaction noise. NASA TP 2784, AVSCOM TR 87-B3
182. Martin RM, Burley CL, Elliott JW (1989) Acoustic test of a model rotor and tail rotor. Results for the isolated rotors and combined configuration. NASA TM-101550
183. Martin RM, Marcolini MA, Splettstoesser WR, Schultz KJ (1990) Wake geometry effects on rotor blade/vortex interaction noise directivity. NASA TP-3015
184. Matschat K, Müller E-A (1979) Effektivpegel und Geräuschdauer bei Flugzeugvorbeiflügen. Festschrift zum 100-jährigen Bestehen der Versuchs- und Forschungsanstalt Wien, Hrsg von der Stadtbaudirektion Wien, 145–147
185. Matschat K, Müller E-A (1981) Vergleich nationaler und internationaler Fluglärmbewertungsverfahren. Aufstellung von Näherungsbeziehungen zwischen den Bewertungsmaßen. Umweltforschungsplan des Bundesministers des Innern, Forschungsbericht 81-10501307, UBA-FB 82-025, Hrsg v. Umweltbundesamt, Berlin
186. Matschat K, Müller E-A, Obermeier F (1970) On the assessment of the annoyance of a series of sonic boom exposures. Acustica 23:49–50
187. May DN (1971) The loudness of sonic booms heard outdoors as simple functions of overpressure and rise time. J Sound Vib 18:31–43
188. McAlpine A, Fisher MJ (2000) On the prediction of 'buzz-saw' noise generated by an aero-engine. AIAA Paper 2000-2095
189. McAlpine A, Fisher MJ (2001) On the prediction of 'buzz-saw' noise generated in aero-engine inlet ducts. J Sound Vib 248:123–149
190. McKennell AC (1963) Aircraft noise annoyance around London (Heathrow) airport. Central Office of Information, 337, London
191. Meier GEA, Lenth H-M, Löhr KF (1988) Sound generation flow interaction of vortices with an airfoil and a flat plate in transonic flow. Fluid Dyn Res 3:344–348
192. Michalke A (1972) An expansion scheme for the noise from circular jets. Z Flugwiss 20:229–237
193. Michalke A (1977) On the effect of spatial source coherence on the radiation of jet noise. J Sound Vib 55:377–394
194. Michalke A, Michel U (1979) Prediction of jet noise in flight from static tests. J Sound Vib 67:341–367
195. Michel U, Michalke A (1981) Prediction of flyover jet noise spectra, AIAA Paper 81-2025, AIAA 7th Aeroacoustics Conference
196. Michel U (1995) Broadband shock noise: theory vis-a-vis experimental results, CEAS/AIAA Aeroacoustics Conference, CEAS/AIAA Paper 95-071, in: DGLR-Bericht 95-01, 545–554
197. Morfey CL (1973) Amplification of aerodynamic noise by convected flow inhomogeneities. J Sound Vib 31:391–397
198. Morin BL (1999) Broadband fan noise prediction system for gas turbine engines. AIAA-Paper 99-1889
199. Motsigner RE, Kraft RE (1991) Design and performance of duct acoustic treatment. In: Hubbard H (Hrsg) Aeroacoustics of flight vehicles: theory and practice: Volume 2: noise control. NASA Reference Publication 1258, vol 2, NASA, pp 165–205
200. Munjal ML (1987) Acoustics of ducts and mufflers. Wiley, New York, NY
201. Nakamura Y (1981) Prediction of blade/vortex interaction noise from measured blade pressure. Paper 32, 7th European Rotorcraft and Powered Lift Aircraft Forum, Garmisch-Partenkirchen
202. Neuwerth G (1972) (Deutscher Titel unbekannt) Deutsche Luft- und Raumfahrt, DLR-FB 72-72. Englische Übersetzung: (1974) Acoustic feedback of a subsonic and supersonic free jet which impinges on an obstacle. Royal Aircraft Establishment, Libraray Translation 1739
203. Neuwerth G (1982) Flowfield and noise of jet impingement on flaps and ground surface. AGARD-CP-308, 13.1-13.7
204. Niedzwieki A, Ribner HS (1978) Subjective loudness of N-wave sonic booms. J Acoust Soc Am 64:1617–1621
205. NN (1985) Gas turbine jet exhaust noise prediction. SAE ARP 876 C
206. NN (1985) Gas turbine coaxial exhaust flow noise prediction. SAE AIR 1905
207. NN (1990) Airframe noise prediction. ESDUpac A9023
208. Norum TD, Seiner JM (1982) Broadband shock noise from supersonic jets. AIAA J 20:68–73

209. Norum TD, Seiner JM (1984) Measurement of mean static pressure and far-field acoustics of shock-containing jets. NASA TM 84521
210. National Sonic Boom Evaluation Office (1967) Sonic boom experiments at Edwards Air Force Base. Interim Report NSBEO-1-67
211. NTIS (National Technical Information Service Data Base) (1988) Aircraft sonic boom. "Biological effects". Jan 1970 – Mar 1988, Springfield, VA
212. NTIS (National Technical Information Service Data Base) (1988) Aircraft sonic boom. "Effects on buildings". Jan 1970 – Mar 1988, Springfield, VA
213. Obermeier F (1979) On a new representation of aeroacoustic source distribution. I. General theory. Acustica 42:56–61
214. Obermeier F (1989) Ausbreitung schwacher Stoßwellen – Stoßfokussierung und Stoßreflexion. Z Flugwiss Weltraumforsch 13:219–232
215. Obermeier F, Zimmermann G (1971) Das Streuverhalten eines Überschallknalles beim Durchgang durch eine turbulente Schicht. In: Proceedings of the 7th international congress on acoustics, Budapest, pp 457–460
216. Ostertag JSD, Guidati S, Guidati G, Wagner S, Wilde A, Kalitzin N (2000) Prediction and measurement of airframe noise on a generic body. AIAA Paper No AIAA-2000-2063
217. Parry AB, Crighton DG (1989) Asymptotic theory of propeller noise. Part I: subsonic single-rotation propeller. AIAA J 27:1184–1990
218. Pérennès S et al (1998) Aerodynamic noise of a two-dimensional wing with high-lift devices. AIAA/CEAS-98-2338
219. Phillips OM (1960) On the generation of sound by supersonic shear flows. J Fluid Mech 9:1–28
220. Pierce AD, Kang J (1990) Molecular relaxation effects on sonic boom waveforms. Frontiers in nonlinear acoustics. In: Proceedings of the 12th ISNA, Elsevier, Amsterdam
221. Piet JF et al (1997) Airframe noise source localization using a microphone array. 3rd AIAA/CEAS-97-47
222. Piet J et al (1999) Localization of acoustic source from a landing aircraft with a microphone array. AIAA/CEAS-99-1811
223. Pietrzko S, Hofmann RF (1988) Prediction of A-weighted aircraft noise based on measured directivity patterns. Appl Acoust 23:29–44
224. Plotkin KJ (1989) Review of sonic boom theory. AIAA Paper No 89-11 05
225. Polacsec C, Spiegel P, Boyle F, Eaton J (2000) Noise computation of high-speed propeller-driven aircraft. AIAA/CEAS Paper No 2000-2086
226. Powell A (1964) Theory of vortex sound. J Acoust Soc Am 36:177–195
227. Prieur J (1987) Calculations of transonic rotor noise using a frequency domain formulation. 43rd AHS-Forum Proceedings, St. Louis, MI, pp 469–479
228. Prieur J, Splettstoesser WR (1999) ERATO – an ONERA-DLR cooperative programme on aeroacoustic rotor optimisation. In: Proceedings of the 25th European Rotorcraft Forum, Rome, Italy
229. Purcell T (1989) A prediction of high-speed rotor noise. AIAA, 12th Aeroacoustics Conference, San Antonio, TX
230. Radezrsky RH, Singer BA, Khorrami MR (1998) Detailed measurements of a flap side-edge flow field. AIAA/CEAS-98-0700
231. Rahier G, Prieur J (1997) An efficient Kirchhoff integration method for rotor noise prediction starting indifferently from subsonically or supersonically rotating meshes. American Helicopter Society 53rd annual forum, Virginia Beach, VA
232. Reinis S, Weiss DS, Featherstone JW, Tsaros C (1987) Long-term effects of simulated sonic booms on hearing in rhesus monkeys. J Sound Vib 113:355–363
233. Ribner HS (1959) New theory of jet-noise generation, directionality, and spectra. J Acoust Soc Am 31:245–246
234. Ribner HS (1981) Perspectives in jet noise (Dryden lectureship in research). AIAA Paper 81-0428
235. Ribner HS, Balazard J, Müller E-A, Obermeier F, Lundberg Bo KO, Warren CHE, Foster CR, Ingerslev F (1970) Report on the sonic boom phenomenon, the ranges of sonic boom values likely to be produced by planned SSTs and the effects of sonic boom on humans, property, animals and terrain. Sonic Boom Panel 2nd Meeting, Montreal, ICAO Doc 8894, SBP/II
236. Rice CG (1972) Sonic boom exposure effects II.2: sleep effects. J Sound Vib 20:511–517
237. Roger M, Pérennès S (2000) Low-frequency noise sources in two-dimensional high-lift devices. AIAA/CEAS-2000-1972
238. Rose GE, Jeracki RJ (1989) Effect of reduced aft diameter and increased blade number on high-speed counter-rotation propeller performance. NASA TM-102077
239. Ross JC, Storms BL, Kumagai H (1995) Aircraft flyover noise reduction using lower-surface flap-tip fences. NASA CDTM-21006
240. Rudnik R, Ronzheimer A, Schenk M, Rossow C-C (1996) Berechnung von 2- und 3-dimensionalen Hochauftriebskonfigurationen durch Lösung der Navier-Stokes Gleichungen. DGLR-Jahrestagung, Dresden
241. Rylander R, Dancer A (1978) Startle reactions to simulated sonic booms: influence of habituation, boom level and background noise. J Sound Vib 61:235–243
242. Saiyed HN, Mikkelsen KL, Bridges JE (2000) Acoustics and thrust of separate-flow exhaust nozzles with mixing devices for high-bypass-ratio engines. NASA/TM-2000-209948
243. Sarin SL, Donelly RP (1992) Angle of incidence effects on the far-field noise of an isolated propeller. DGLR/AIAA Paper No 92-02-050
244. Schaffar M, Haertig J, Gnemmi P (1990) Effect of non-rectangular blade tips on BVI noise for a two-bladed rotor. 16th European Rotorcraft Forum, Glasgow
245. Schmitz FH, Boxwell DA (1976) In-flight farfield measurement of helicopter impulsive noise. J Am Helicopter Soc 21(4)
246. Schmitz FH, Yu YH (1981) Transonic rotor noise – theoretical and experimental comparisons. Vertica 5:55–74
247. Schmitz FH, Yu YH (1983) Helicopter impulsive noise: theoretical and experimental status. NASA TM-84390
248. Schome PD, Sias JW (1998) On spectral weightings to assess human response indoors to blast noise and sonic booms. Noise Control Eng J 46:57–71

249. Schulten JBHM (1987) A spectral method for the computation of propeller acoustics. AIAA Paper 87-2674, Palo-Alto, CA
250. Schultz KJ, Splettstoesser W (1987) Measured and predicted impulsive noise directivity characteristics. Paper 1.2, 13th European Rotorcraft Forum, Arles
251. Schultz KJ, Lohmann D, Lieser JA, Pahlke K (1994) Aeroacoustic calculations of helicopter rotors at DLR. AGARD Symposium on Aerodynamics and Aeroacoustics of Rotorcraft, Paper No 29, Berlin
252. Schwenk W (1976) Das Verbot von zivilen Flügen mit Überschallgeschwindigkeit für die Bundesrepublik Deutschland. Kampf dem Lärm 23:57–61
253. Sen R (1996) Local dynamics and acoustics in a simple 2D model of airfoil lateral-edge noise. AIAA-96-1673
254. Sharland IJ (1964) Sources of noise in axial flow fans. J Sound Vib 1:302–322
255. Sijtsma P et al (1999) Source location by phased array measurements in closed wind tunnel test sections. AIAA/CEAS-99-1814
256. Siller H, Arnold F, Michel U (2001) Investigation of aero-engine core-noise using a phased microphone array. AIAA Paper 2001-2269
257. Singer BA, Brentner KS, Lockard DP, Lilley GM (1999) Simulation of acoustic scattering from a trailing edge. AIAA-Paper No 99-0231
258. Smith MJT (1989) Aircraft noise. Cambridge University Press, Cambridge
259. Smith M et al (1998) Prediction method for aerodynamic noise from aircraft landing gear. AIAA/CEAS-98-2228
260. Society of Automotive Engineers Inc (1977) Prediction procedure for near-field and far-field propeller noise. SAE-AIR 1407
261. Sparrow VW (1995) The effect of supersonic aircraft speed on the penetration of sonic boom into the ocean. J Acoust Soc Am 97:159–162
262. Splettstoesser WR, Schultz K-J, Boxwell DA, Schmitz FH (1984) Helicopter model rotor-blade/vortex interaction impulsive noise: Scalability and parametric variations. Paper No 18, 10th European Rotorcraft Forum, The Hague, The Netherlands, also NASA TM-86007
263. Splettstoesser W, Schultz K-J, Martin R (1987) Rotor blade/vortex interaction impulsive noise source identification and correlation with rotor wake predictions. AIAA-87-2744, AIAA 11th Aeroacoustics Conference, Palo Alto, CA
264. Splettstoesser WR, Schultz KJ, Kube R, Brooks TF, Booth ER, Niesl G, Streby O (1994) A higher harmonic control test in the DNW to reduce impulsive BVI noise. J Am Helicopter Soc 39(4)
265. Splettstoesser WR, Kube R, Wagner W, Seelhorst U, Boutier A, Micheli F, Mercker E, Pengel K (1997) Key results from a higher harmonic control aeroacoustic rotor test (HART). J Am Helicopter Soc 42(1):58–78
266. Splettstoesser WR, Schultz KJ, van der Wall B, Buchholz H, Gembler W, Niesl G (1998) The effect of individual blade pitch control on BVI noise – comparison of flight test and simulation results. In: Proceedings AC07, 24th European Rotorcraft Forum, Marseilles, France
267. Stevens RCK, Bryce WD, Szewczyk VM (1983) Model and fullscale studies of the exhaustnoise from a bypass engine in flight. AIAA Paper 83-0751
268. Storms BL, Takahashi TT, Horne WC, Ross JC, Dougherty RP, Underbrink JR (1996) Flap-tip treatments for the reduction of lift-generated noise. NASA CDTM-21006
269. Storms BL, Ross JC, Horne WC, Hayes JA, Dougherty RP, Underbrink JR, Scharpf DF, Moriarty PJ (1998) An aeroacoustic study of an unswept wing with a three-dimensional high lift system. NASA/TM-1998-112222
270. Storms B et al (1999) Aeroacoustic measurements of slat noise on a three-dimensional high-lift system. AIAA/CEAS-99-1957
271. Streett CL (1998) Numerical simulation of fluctuations leading to noise in a flap-edge flowfield. AIAA-98-0628
272. Stuff R (1982) Propellerlärm bei Unterschallbattspitzen-Machzahlen, Umfangskraft und Axialkraft. DFVLR-Mitteilung 82-17
273. Succi GP (1979) Design of quiet efficient propellers. SAE Paper 790584
274. Takallu MA, Block PJW (1987) Prediction of added noise due to the effect of unsteady flow on pusher propellers. AIAA 25th Aerospace Sciences Meeting, AIAA-87-0255, Reno/Nevada
275. Tam CKW (1987) Stochastic model theory of broadband shock associated noise from supersonic jets. J Sound Vib 116:285–302
276. Tam CKW (1991) Broadband shock-associated noise from supersonic jets in flight. J Sound Vib 151:55–71
277. Tam CKW (1995) Supersonic jet noise. Annu Rev Fluid Mech 27:131–147
278. Tam CKW, Auriault L (1999) Jet mixing noise from fine scale turbulence. AIAA J 37:145–153
279. Tam CKW, Dong Z (1994) Wall boundary conditions for high-order finite-difference schemes in computational aeroacoustics. Theor Comput Fluid Dyn 6:303–322
280. Tam CKW, Dong Z (1995) Radiation and outflow boundary conditions for direct computation of acoustic and flow disturbances in a nonuniform mean flow. AIAA Paper No 95-007
281. Tam CKW, Hardin JC (Hrsg) (1997) Second computational aeroacoustics (CAA) workshop on benchmark problems. NASA CP 3352
282. Tam CKW, Shen H (1993) Direct computation of nonlinear acoustic pulses using high-order finite difference schemes. AIAA Paper No 93-4325
283. Tam CKW, Tanna HK (1982) Shock associated noise of supersonic jets from convergent-divergent nozzles. J Sound Vib 81:337–358
284. Tam CKW, Webb JC (1993) Dispersion relation preserving finite difference schemes for computational acoustics. J Comput Phys 107:262–281
285. Tam CKW, Salikuddin M, Hanson DB (1988) Acoustic interference of counter-rotation propellers. J Sound Vib 124:357–366
286. Tam CKW, Golebiowski M, Seiner JM (1996) On the two components of turbulent mixing noise from supersonic jets. AIAA Paper 96-1716
287. Tanna HK (1977) An experimental study of jet noise. Part 1: jet mixing noise. J Sound Vib 51:429–444
288. Tanna HK (1977) An experimental study of jet noise. Part 2: shock associated noise. J Sound Vib 51:429–444
289. Thackray RI (1972) Sonic boom exposure effects II.3: starle responses. J Sound Vib 20:519–526

290. Tyler JM, Sofrin TG (1962) Axial flow compressor noise studies. SAE Trans 70:309–332
291. U.S. standard atmosphere (1962) Prepared under sponsorship of NASA, USAF. US Weather Bureau, Washington, DC
292. Vallée J (1969) Etude expérimentale des focalisations de bangs soniques engendrés par le vol supersonique en accélération rectiligne ou en virage d'um avion Mirage IV à l'altitude de 11000 m. Opération Jéricho-Virage, Rapport d'études No 277, Centre d'essais en vol, annexe d'Istres
293. Wagner S, Bareiss R, Guidati G (1996) Wind turbine noise. Springer, Berlin
294. van der Wall B, Roth M (1997) Free-wake analysis on massively-parallel computers and validation with HART test data. 53rd annual forum, Virginia Beach, VA
295. Wang ME (1980) Wing effect on jet noise propagation. AIAA-Paper 80-1047
296. Watanabe T, Kawachi K (1987) Noise prediction of counter rotation propeller. AIAA Paper 87-2658, Palo Alto, CA
297. Way DJ, Turner BA (1980) Model tests demonstrating under-wing installation effects on engine exhaust noise. AIAA-Paper 80-1048
298. Wood T et al (1999) Aeroacoustic predictions of a wing-flap configuration in three dimensions. AIAA/CEAS-99-1893
299. Woodward RP, Hughes CE (1990) Aeroacoustic effects of reduced aft tip speed at constant thrust for a model counterrotation turboprop at takeoff conditions. AIAA Paper 90-3933, Tallahassee
300. Woodward RP, Loeffler IJ, Dittmar JH (1989) Measured far-field flight noise of a counterrotation turboprop at cruise conditions. NASA TM-101383
301. Woodward RP, Gazzaniga JA, Bartos LJ, Hughes CE (2002) Acoustic benefits of stator sweep and lean for a high tip speed fan. AIAA-Paper 2002-1034, 40th Aerospace Science Meeting, Reno, NV
302. Yin J, Delfs J (2001) Sound generation from gust-airfoil interaction using CAA-chimera method. AIAA-Paper No 2001-2136
303. Yin JP, Ahmed SR, Dobrzynski W (1999) New acoustic and aerodynamic phenomena due to non-uniform rotation of propelleres. J Sound Vib 225(1):171–187
304. Young RW (1989) Day-night average sound level (DNL) and sound exposure level (SEL) as efficient descriptors for noise compatibility planning. Internoise 89 Proceedings, pp 1289–1292
305. Yu YH, Gmelin B, Splettstoesser W, Philippe JJ, Prieur J, Brooks TF (1997) Reduction of helicopter blade-vortex interaction noise by active rotor control technology. Prog Aerosp Sci 33:647–687
306. Zepler EE, Harel JRP (1965) The loudness of sonic booms and other impulsive sounds. J Sound Vib 2:249–256
307. Zorumski WE (1982) Aircraft noise prediction program. Part I: theoretical manual. NASA TM-83199
308. Zorumski WE, Weir DS (1986) Aircraft noise prediction program. Theoretical manual, part 3: propeller aerodynamics and noise. NASA TM-83199

Construction Noise

18

Achim Böhm, Olaf-Tobias Strachotta, and Volker Irmer

18.1 Introduction

Construction is, as one of the more meaningful tasks of mankind, indeed of a positive nature. With the growth of the world population (approximately 2 billion in 1950 to already 6 billion in 2000), construction activities will increase as well and so will the number of persons being affected by noise and vibration.

Construction sites of all kinds are, in contrast to stationary installations, time limited and thus not subject to licensing. In the list of complaints by European authorities however, complaints about noise and vibration range from third to sixth place. In areas of high population, the violations of exposure limits are particularly high. Noise induced hearing loss still is the number one occupational disease among construction workers. Approximately 25% of all pensions paid to occupational diseases are paid in the construction industry.

Therefore, reducing the emission of construction noise remains important in the future. All limitations of construction equipment noise emission and, especially, the timely limitation of construction activities are troublesome, expensive and, at this point of time, sometimes not even realizable for construction companies. Therefore, it is necessary to strive for low-noise construction equipment and for low-noise techniques.

It is considered desirable of legislation to adjust the noise emission values to the state-of-the-art technology.

In view of the globalization and the cross-border international market for construction equipment, a harmonization of regulations and laws is becoming increasingly more important.

Practical experience shows that the sound power level, depending on the construction equipment and on the number of workers, is an important criterion for the expected noise level in the vicinity of a construction site. Decisive factors, however, are the handling and the technical skills of the workers. The individual factor of the technical skills at construction sites is not to be underestimated.

That is why construction site workers should not only have the technical skills but should also know that there is a significant difference between construction sites in rural areas and those in areas of high population. Information on the relationship of "distance between noise source and noise impact site" and "operating duration of noise equipment" should be imparted to the workmen. It would be unfortunate if the efforts to lower the noise level of construction equipment and techniques, according to the state of the art, were nullified by inefficient or unskilled handling [1, 2].

Already in the early 1980, a standardized noise labeling of construction equipment was introduced by European legislation in the member states. The so-called "Outdoor Noise Directive 2000/14/EG" of July 3rd 2000 [3, 4] approximates the laws of the member states relating to the noise emission in the environment by equipment for use outdoors, especially valid for construction equipment.

A. Böhm
Fliederstrasse 23, 85757 Karlsfeld

O.-T. Strachotta
TÜV NORD AG, Am TÜV 1, 30519 Hannover, Germany

V. Irmer (✉)
Prinz-Handjery-Str. 29b, 14167 Berlin, Germany
e-mail: volkerirmer@yahoo.de

G. Müller and M. Möser (eds.), *Handbook of Engineering Acoustics*,
DOI 10.1007/978-3-540-69460-1_18, © Springer-Verlag Berlin Heidelberg 2013

There are still neither European nor international regulations concerning vibration emissions (For further information on construction sites, see Chap. 22.)

18.2 Noise Exposure

Noise exposure caused by construction activities occur at workplaces as well as in the vicinity of construction sites. Both areas are covered by different regulations and measuring procedures as well as on different levels (European and national). European directives for the workplace have been transformed and supplemented into national law. Since the European member states refer to the subsidiary principle, there are no equivalent European regulations in the environmental sector; the member states have enacted national legislations, if necessary.

18.2.1 Noise Exposure at the Workplace

In the entire German industrial sector, about 1,000 new cases of occupational diseases caused by noise are admitted every year. According to the Federal Occupational Safety & Health and Medicine Agency of Germany [5], approximately 200 of them are caused by construction site noise. In order to ensure health, safety, and work ability of the workers, noise at the workplace has to be kept as low as possible.

18.2.1.1 Legislative Regulations

In Directive 86/188 EEC of 12 May 1986 relating to the protection of employees from the risks of noise at the workplace [6], the European Council lays down requirements protecting the worker from extreme noise exposure in order to ensure a situation in which risk to hearing ability can be avoided to a large extent. These requirements have been transposed into German law by the "Arbeitsstättenverordnung" [7] and the "UVV Lärm" [8]. The Standard DIN 45645 Part 2 [9], the VDI-Guideline 2,058 parts 2 and 3 [10, 11], and the ISO-Standard 1999 [12] contain additional regulations that supplement the regulations of the "Arbeitsstättenverordnung" and the "UVV Lärm."

Some standard examples for measured noise levels at workplaces are listed in Table 18.1. The values listed in the left column were obtained during the years 1970–1980, while the values in the right column are results of measurements of construction equipment that was awarded the "Blue Angel" (Table 18.1).

18.2.1.2 Noise Reduction Measures

The noise exposure at the workplace on the building and construction sites has been reduced considerably during the past years. This is partly due to a series of noise emission limits that have been fixed. These limits have reduced the general noise level on construction sites. This is particularly perceptible for equipment of small power, with which substantial reduction was reached by noise reduction measures used at the source. In addition, operator's cabins have been greatly improved acoustically and in the sense of technical vibrations, to provide the workers with more comfort – not only at the handling level, but also and most of all in terms of noise exposure. Nowadays, average sound pressure levels of 70 dB (A) at the workplace are no longer a rarity. Such low values, however, can only be achieved on equipment of a higher performance, because only this provides enough space for the use of optimal sound protection techniques.

18.2.2 Noise Exposure in the Neighborhood of Construction Sites

In the neighborhood of construction sites, noise exposure generally only occurs for a limited time period (weeks or months). Depending on the distance between the construction site and the neighborhood, noise exposures can be very high. It is, however, impossible to prohibit the operation of construction sites because of high emission or exposure, as it is possible with the setup and the operation of installations subject to licensing. Therefore, the legislator can only pass regulations for the operation of construction sites and the noise exposure caused thereby and keep the exposure as low as possible.

18.2.2.1 Target Values for the Exposure in the Neighborhood, Germany

According to the German Federal Immission Control Act (BImSchG) [13], construction sites are regarded as installations not subject to licensing. They are therefore covered by § 22 BImSchG. They have to be set up and operated as follows:

Table 18.1 Rating levels at the equipment's workstation

Type		Rating level in dB(A) 1970–1980 without special noise reduction measures	1990 to 2002 with special noise reduction measures[a]
Excavator	16–145 kW	86	70
Wheeled loader	13–155 kW	91	72
Vibrating roller	32–82 kW	89	–
Vibratory plate	1.5–6.6 kW	89	–
Self-erecting crane	3–10 kW	70	69[a]
Tower crane	10–35 kW	75	65
Mobile crane	120–315 kW	85	71
Dumper	25–200 kW	90	81
Dozer	70–150 kW	85	75
Grader	50–150 kW	88	78
Hammer, handheld	18–33 kg	97	89
Truck concrete mixer	7–9 m^3	–	74

[a]Measured in the closed driver's cabin, ventilation running

- Harmful effects on the environment, which are, according to the state-of-the-art technology, considered avoidable, have to be prevented.
- Harmful effects on the environment, which are, according to the state-of-the-art technology, considered not avoidable, have to be reduced to a minimum.

These, rather general, requirements are put in concrete terms in the General Administrative Regulation (AVwV) for the protection against construction noise [14]. The responsible authority of each state of the Federal Republic of Germany has to follow this General Administrative Regulation during the evaluation of construction noise and the fight against it. The AVwV lays down the target value for the exposure of construction sites for the protection of the neighborhood on the one hand, and gives advice to the authorities on what measure to take for preventing harmful effects upon the environment, in case of limit violations, etc., on the other.

The exposure value limits of the AVwV correspond in most terms to those of the "TA Lärm" [15]. In contrast to the "TA Lärm," however, some particularities have to be commented on:

- The average level is determined as "Taktmaximalpegel" (maximum level in a 5 s time period).
- For the determination of the rated energy equivalent level from these "Taktmaximalpegel," assessment periods of 13 h at daytime (from 7 a.m. to 8 p.m.) and 11 h at night-time (from 8 p.m. to 7 a.m.) have been fixed.
- For the determination of the rated level, an adjustment of up to 5 dB (A) can be taken into account, if distinctly audible tones occur in the noise.
- No daytime maximum level criteria exist.
- Operation hours of less than 11 or 13 h are taken into account for the assessment by means of a simplified regulation.
- The target values should be kept, if possible. The responsible authority should only intervene, if the rated level determined exceeds the target value by more than 5 dB (A). This is to be seen as a reaction to the limited time of exposure to the noise of the construction site. In this case, measures to reduce the noise are to be issued.

All the following measures should be implemented at the source or close to it:

- Measures for the set-up of construction sites
- Measures for the construction equipment
- Use of low-noise construction equipment
- Use of low-noise construction techniques
- Timely limitation of operating hours for high-noise level construction equipment
- Instructions for the workers on how to minimize work performance-related noise (hammer blows, throwing of boards, and loud shouting)

With regard to noise emission, more advanced equipment of the same design and of comparable power should be used, in order to assess whether noise of construction equipment is, according to the state-of-the-art technology, avoidable. This essentially corresponds to the definition of the state-of-the-art technology of the BImSchG.

The state-of-the-art technology is – among other things – described by the criteria that construction equipment has to meet in order to be awarded the "Blue Angel" [16].

At inner-city construction sites, it is often impossible to avoid the exceeding of the target value of the AVwV, even if the equipment according to the state-of-the-art technology is used. Restrictions of the operating time and the information regarding the neighborhood might be helpful in order to minimize conflicts between parties.

18.2.2.2 Exposure Values in the Neighborhood, Other Countries

The spectrum of different approaches elsewhere, on the basis of selected countries, is pointed out in the following:

Austria

The abatement of construction noise is a matter for each individual Austrian Federal State having found various solutions. In the states of Kaernten, Salzburg, and Vienna, a general regulation stipulates that construction noise has to be avoided (however, it does not lay down any limits or guidelines). In the states of Tyrolia and Upper Austria, limits for the individual regional categories have been set. There are no corresponding regulations in other federal states of Austria.

Switzerland

A Construction Noise Regulation contains instructions for the responsible authorities [17]. According to the regulation, in the case of heavy and long-term construction noise exposure, measures have to be taken that depend on

- The distance between construction site and the closest noise-exposed areas
- The time of the day during which the operation is undertaken
- The noise sensitivity of the areas involved
- The "noisy" construction phase, more precisely the duration of the noise-intense construction work

The measures are divided into three levels, A, B, and C, as shown in Table 18.2.

The "Construction Noise Regulation" is way too complex to be presented here in detail. Depending on the situation:

- In distances of more than 300 m (328 yards, 984 feet) to the neighborhood, construction equipment of any kind is allowed during the daytime (no measure to be taken)
- In distances of less than 300 m (328 yards, 984 feet) to the noise-sensitive neighborhood, noisy construction operation (piling and saws) can only be undertaken with the state-of-the-art construction equipment (e.g., equipment that has been awarded the "Blue Angel"), during the daytime, even if these measures hinder the construction process (C-level measurement).

Sweden

The exposure values (average level) applicable to the neighborhood of construction sites are compiled in Table 18.3.

If the construction period is shorter than a month (6 months), the daytime and evening values may be exceeded by 10 dB (A) (5 dB (A)).

Denmark

Legislative regulations explicitly concerning construction noise do not exist in Denmark. In cases of complaint within the neighborhood, however, measures can be taken. The legislative regulations for noise-emitting industrial and commercial facilities can, with special regard to the individual construction noise characteristics, also be applied. They lay down indicators for intolerable noise pollution.

Hungary

Regulation 4/1984, Attachment 4 of the Ministry of Health lays down the permissible average level of construction noise in residential areas. See Table 18.4.

The Netherlands

There are no legislative regulations concerning construction noise in residential neighborhoods in the Netherlands. Local authorities might have enacted corresponding regulations.

In 1981, the government published an enactment including the statement that no work should be done in the evening and at night-time and that the average

Table 18.2 Requirements concerning different levels of measures in Switzerland

Level	Measures taken may hinder or otherwise construction work and transport on the construction site	Equipment and vehicles have to meet
A	Not hinder	The normal standard
B	Hinder	The established state of noise reduction technique[a]
C	Hinder considerably	The best available state of noise reduction technique[b]

[a]Corresponds to existing European directives
[b]Corresponds to the German Blue Angle Award

Table 18.3 Exposure levels in the neighborhood of construction sites in Sweden

Area	Exposure levels in dB(A)		
	07:00 a.m. to 6:00 p.m.	06:00 p.m. to 10:00 p.m. (Saturday and Sunday)	10:00 p.m. to 07:00 a.m.
Residential areas, hospitals, and recreational areas	60	50	45
Office areas, other areas without loud activities	70	65	–
Industrial areas	75	70	70

Table 18.4 Permissible exposure levels in the neighborhood of construction sites in Hungary

Area	Permissible average level in dB(A) for construction work					
	Less than 1 month		1 month to 1 year		More than 1 year	
	06:00 a.m. to 10:00 p.m.	10:00 p.m. to 06:00 a.m.	06:00 a.m. to 10:00 p.m.	10:00 p.m. to 06:00 a.m.	06:00 a.m. to 10:00 p.m.	10:00 p.m. to 06:00 a.m.
Recreational area, near hospitals, and natural reserve	60	45	55	40	50	35
Residential area not densely built up	65	50	60	45	55	40
Residential area densely built up	70	55	65	50	60	45
Industrial area, densely built up areas	70	55	70	55	65	50

noise level of 60 dB (A) should not be exceeded between 7 a.m. and 7 p.m.

Great Britain

Throughout the country, no standardized exposure values exist for the neighborhood of construction sites. It is regarded more efficient to discuss and solve construction-site problems on a local or communal basis. Therefore, it is left to the local authorities to decide upon noise exposure values of construction sites in residential areas on a case-by-case basis. The builder can obtain a declaration of consent for the planned construction measure from the authorities. More explicit details about the procedures and about the basics of construction noise are included in the British Standard BS 5228 [18].

Hong Kong

During the night-time (7 p.m. to 7 a.m.) and throughout the day on holidays, construction operations require permission in order to be carried out. Permission is granted by the responsible authorities (Environmental Protection Resort), who set forth the appropriate exposure values and check adherence to it. There are no restrictions, however, concerning construction noise during all other times (i.e., weekday's daytime) in Hong Kong.

18.3 Noise Emission of Equipment for Use Outdoors and at Construction Sites

During the years of 1979 and 1995, eight European directives on noise emission of various types of construction equipment have been enacted. They included the obligation to have noise emission measured by institutions appointed by the member states in order to adhere to the emission values and to label equipment with the guaranteed sound power level L_{WA} in dB/pW.

In 1996, the European Community announced in its green paper "Future Noise Policy" that it was about to prepare a Directive on the regulation of noise of a large number of equipment for use outdoors that would replace the existing EC Directive [19].

18.3.1 EC Directive on the Limitation of Noise Emissions

On July 3rd, 2000, the Directive 2000/14/EC of the European Parliament and the Council of 8 May 2000 on the approximation of laws of the member states relating to noise emission in the environment by equipment for use outdoors (in the following "Outdoor Noise Directive") was published in the Official Journal of the European Communities and, thus, came into force at the European level [3, 4].

The EC "Outdoor Noise Directive" aims at:

- Guaranteeing the smooth functioning of the European market
- Harmonizing the common laws regarding noise emission of equipment for outdoor use
- Protection of human health and well-being

Measures to reach those aims in the European Community are:

- Harmonization of the noise emission by limits
- Harmonization of the conformity assessment procedures
- Harmonization of noise labeling of equipment
- Collection and publication of data on noise emission of equipment with an impact on the environment, put on the market

The Outdoor Noise Directive contains provisions for the noise emission in the environment. The regulation on noise emission at the workplace included in the "EC Machinery Directive 98/37/EC" [20] remains unaffected. The transposition into German law by means of the "Third Ordinance to the Law concerning the Safety of Machinery" (Engine Noise Information Ordinance – 3. GSGV) has still to be followed [21]. This means that the principle established in the EC Machinery Directive is still valid, which requires equipment to be designed and built in a way that reduces the risks of noise emission to a minimum. Furthermore, data on the equipment's noise emission at the workplace have to be included in the instructions accompanying each product. It is important that for measurements pursuant to the EC Machinery Directive, for the same type of equipment, the same type of measurement procedure is applied.

Thus, the measurement procedures laid down in the Outdoor Noise Directive have an impact on both, the measurement procedures of the EC Machinery Directive as well as the determination of the sound pressure level L_{pA} in dB/20μPa at the operator's workplace. A continuation of the obligatory labeling of the sound pressure level at the workplace is possible on a voluntary basis and has to be agreed upon by the notified body. The industry's desire is understandable since the lowest sound pressure level possible with which the equipment is labeled is a reference to the quality of the equipment. Thus, it presents a distinct sales argument for construction equipment.

18.3.2 Content of the EC Directive 2000/14/EC

The provisions of the Outdoor Noise Directive apply, if the equipment is placed onto the European market or put into service in Europe for the first time. Placing on the European market in this case means that the equipment is made available, commercially and noncommercially, for the use within the European Community. The directive applies to the first use by the end-user. If a product is first placed on the market and put into operation on the European common market, it has to meet the criteria of the new directive. The directive is valid for both, the products manufactured in the EC or manufactured elsewhere (e.g., USA and Japan). It does, however, not apply to equipment that has already been placed on the market (e.g., were sold) or put into operation earlier.

Generally, the regulations are valid for all types of equipment manufactured and put into operation outside the EC, if it is first imported to the EC or Common Market.

18.3.2.1 Equipment Subject to Limitation and Labeling

The Outdoor Noise Directive [3] applies to 63 types of equipment (37 types of construction equipment), most of them being designated for outdoor use. The types of equipment are listed and defined in the Directive.

Tables 18.5 and 18.6 discuss the equipment with and without the limits. All products have to be labeled with both, the EC conformity marking and the manufacturer's indication of the guaranteed noise emission level of all equipment produced. Twenty-two of these types of equipment (18 types of construction equipment) also have to meet the permissible sound power level L_{WA} in dB/pW. The directive fixes limits on two levels, the first having been put into force on 3 January 2001 and the second with a limit of about 3 dB(A) less, coming into force on 3 January 2006.

Table 18.5 Equipment covered by noise emission limits [3]

Type of equipment	Net installed power (P in kW) Electric power (P_{el} in kW) Mass of appliance (m in kg) Cutting width (L in cm)	Permissible sound power level (L_{WA}/1 pW in dB(A))	
		Stage I as from 3 January 2002	Stage II as from 3 January 2006
Compaction machines (vibrating rollers, vibratory plates, and vibratory rammers)	$P \leq 8$	108	105
	$8 < P \leq 70$	109	106
	$P > 70$	$89 + 11 \lg P$	$86 + 11 \lg P$
Wheeled dozers, wheeled loaders, wheeled excavator-loaders, dumpers, graders, loader-type landfill compactors, combustion-engine driven counterbalanced	$P \leq 55$	104	101
Lift trucks, mobile cranes, compaction machines (non-vibrating rollers) paver finishers, and hydraulic power packs	$P > 55$	$85 + 11 \lg P$	$82 + 11 \lg P$
Tracked dozers, tracked loaders, and tracked excavator-loaders	$P \leq 55$	106	103
	$P > 55$	$87 + 11 \lg P$	$84 + 11 \lg P$
Excavators, builder's hoists for the transport of goods, construction winches, and motor hoes	$P \leq 15$	96	93
	$P > 15$	$83 + 11 \lg P$	$80 + 11 \lg P$
Hand-held concrete breakers and picks	$m \leq 15$	107	105
	$15 < m < 30$	$94 + 11 \lg m$	$92 + 11 \lg m$
	$m \geq 30$	$96 + 11 \lg m$	$94 + 11 \lg m$
Tower cranes	–	98 + lg P	$96 + \lg P$
Welding and power generators	$P_{el} \leq 2$ $2 < P_{el} \leq 10$ $P_{el} > 10$	$97 + \lg P_{el}$ $98 + \lg P_{el}$ $97 + \lg P_{el}$	$95 + \lg P_{el}$ $96 + \lg P_{el}$ $95 + \lg P_{el}$
Compressors	$P \leq 15$	99	97
	$P > 15$	$97 + 2 \lg P$	$95 + 2 \lg P$
Lawn mowers, lawn trimmers, and lawn-edge trimmers	$L \leq 50$	96	94
	$50 < L \leq 70$	100	98
	$70 < L \leq 120$	100	98
	$L > 120$	105	103

Table 18.6 Equipment that has to be labeled with the guaranteed sound power level [3]

Arial access platforms with combustion engine
Brush cutters
Builder's hoists for the transport of goods (electric motor)
Building site band-saw machines
Building site circular-saw benches
Chain saws, portable
Combined high pressure flushers and suction vehicles
Compaction machines (explosive rammers only)
Concrete and mortar mixers
Construction winches (electric motor)
Conveying and spraying machines for concrete and mortar
Conveyor belts
Cooling equipment on vehicles
Drill rigs
Equipment for loading and unloading silos or tanks on trucks
Glass recycling containers
Grass trimmers/grass-edge trimmers
Hedge trimmers
High pressure flushers
High pressure water jet machines
Hydraulic hammers
Joint cutters
Leaf blowers
Leaf collectors
Lift trucks, combustion-engine driven
Mobile waste containers
Paver finishers (equipped with a high compaction screed)
Piling equipment
Pipe layers
Piste caterpillars
Power generators (more than 400 kW)
Power sweepers
Refuse collection vehicles
Road milling machines
Scarifiers
Shredders/chippers
Snow removing machines with rotating tools
Suction vehicles
Trenchers
Truck mixers
Water pump units (not for use under water)

18.3.2.2 Noise Emission Measurement Methods

Since the noise emission value is dependent on the measurement method applied, a special measurement method is laid down for every type of equipment including general rules concerning the measurement method (basic noise emission standard with the number and place of microphone positions, measuring distance, averaging method, measurement surface, etc.) on the one hand and operating conditions during the measuring process on the other.

The basic noise emission standard of the Outdoor Noise Directive is the DIN EN ISO 3744 [22]. In order to achieve the required accuracy, at least the following criteria have to be taken into consideration:

- Grade of accuracy: 2
- Measuring environment: outdoors or indoors
- Criteria for the suitability of the measuring environment: $K_2 \leq 2$ dB
- Distance to background noise: $>$ 6 dB (if possible $>$ 15 dB)
- Number of microphone positions: ≥ 9
- Sound level meter: class 1 (IEC 6551/804)
- Accuracy of measuring L_{WA} in dB/1 pW: $\sigma_R \leq 1.5$ dB

For the determination and application of measurement accuracy and for the determination of the required statistical data in accordance with [23], see Chap. 5.

The positioning of the microphones and the measuring paths pursuant to DIN EN ISO 3744 [22] or [9] for different equipment, which are given in the Outdoor Noise Directive, are referred to in Fig. 18.1. Figure 18.2 shows the arrangement of the six microphones according to ISO 6395 [24, 25]. The arrangement has been proven over the years and can especially be applied to noise emission measuring of construction equipment – measurements taken during the operation cycle with and without a drive-through path.

The choice of the noise measurement must be approved by the Notified Body. This method should also be recommended to manufacturers, whose products do not yet have a noise emission limit but have to be labeled with their guaranteed sound power level. Incorrect noise emission measurements can lead to the interpretation that the admission to the free movement of goods is no longer available. As has been explained in Sect. 18.3.2.1, continuation of the obligatory labeling with the sound pressure levels at the workplace on a voluntary basis is highly recommendable. Generally, the criteria that have to be met for the determination of sound pressure levels at the workplace as well as for the determination of sound power levels and guaranteed values for the EC marking are the same.

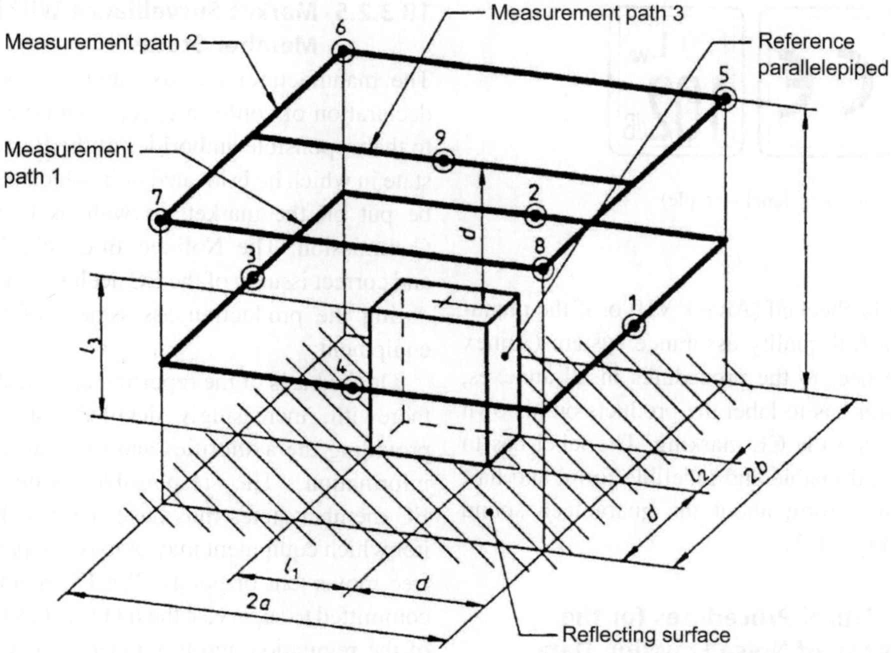

Fig. 18.1 Model of a measuring surface and of the microphone positions according to ISO 3744 (1995)

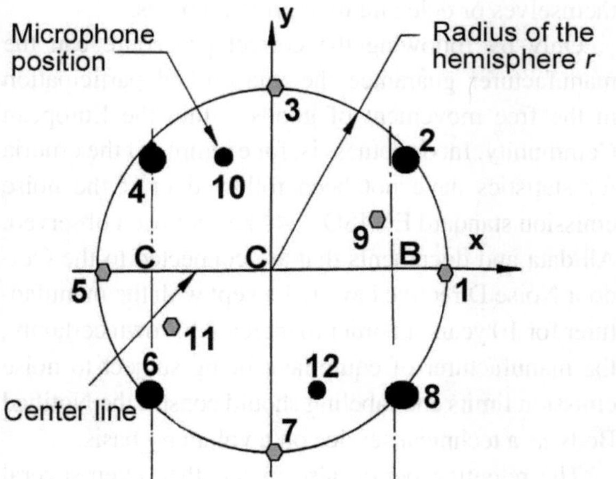

Fig. 18.2 Microphone array on the hemisphere with microphone positions

The procedure is to be included into the mandatory EC conformity assessment and agreed upon by the notified body. For the determination of the sound pressure level at the workplace of construction equipment, ISO 9396 [23] has to be applied.

18.3.2.3 EC Conformity Assessment Procedures

The manufacturer must subject all types of equipment to an EC Conformity Assessment Procedure. All products that are not subject to noise limits will be subject to an "internal production control." The manufacturer can measure the noise emission of his equipment himself or he can delegate the measurement to others (Annex V of the Outdoor Noise Directive [3]).

For all products that are subject to noise limits, the manufacturer has to have the measurement checked for plausibility by a Notified Body. The Notified Body should do this by means of the technical documents, but it can also conduct its own assessments on a regular basis, or, in an ad hoc manner, e.g., in case of doubt. Furthermore, the Notified Body supervises the keeping of the value during the production. Assessment by the Notified Body is not necessary, if every

Fig. 18.3 *CE* Conformity label (sample)

single product is checked (Annex VII) or if the manufacturer runs a full quality assurance system (Annex VIII). With respect to the procedures in all annexes, the manufacturer has to label the products on his own responsibility with the CE-marking. The label has to be fixed in a well visible and indelible form, and has, furthermore, to inform about the guaranteed sound power level. (Fig. 18.3).

18.3.2.4 Statistical Procedures for the Control of Noise Emission Data

Detailed information on the procedure will be given in Chap. 5. The manufacturer has – on the basis of noise emission measurements – to estimate, which value in the series, with regard to the production uncertainty and the inaccuracy of the measurement, can be guaranteed.

In case of a validation of the values by the Notified Body or by any other official institution, operating on behalf of the member states, the values must be verifiable. This means that the manufacturer has to label the equipment in such a manner that he can pass a verification procedure of his values in case of an inspection. Therefore, it became necessary for the member states to lay down statistical testing procedures. Furthermore, the manufacturer has to use internationally recognized and harmonized statistical procedures. The statistical methods to be applied are, following a general expert's and legislative reasoning, the following standards: EN ISO 4871 [26] or ISO 7574 (corresponds to EN 2754-1-4 [27]).

According to Article 16 of the Outdoor Noise Directive and Annex II, the manufacturer is committed to add an EC declaration of conformity to each product. In doing so, he assures that the product meets all legal criteria. With respect to the noise emission of the equipment, the declaration has to state the sound pressure level and the sound power level that has been measured on a representative equipment.

18.3.2.5 Market Surveillance Within the EC Member States

The manufacturer has to submit a copy of the EC declaration of conformity for each type of equipment to the responsible authorities of the European member state in which he is located or in which the product will be put on the market, as well as to the European Commission. The Notified Body checks the proper and correct issuing of the EC declaration of conformity during the production assessment of the respective equipment.

On the basis of the reporting and the data submitted therewith, increasingly detailed lists will develop, providing the authorities and the market with further information. The responsible authorities of the EC member states thus have the possibility to monitor which equipment may or may not participate in the free movement of goods. The EC member states are committed to supervise the manufacturer's compliance of the regulation and to cooperate with the other EC member states. This also includes the inspection of the noise emission limits and the guaranteed sound power level, a task that the member states either perform themselves or delegate to approved bodies.

Only by following the correct procedure can the manufacturer guarantee the undisturbed participation in the free movement of goods within the European Community. Incorrectness is, for example, if the criteria for statistics have not been followed or if the noise emission standard EN ISO 3744 has not been observed. All data and documents that are connected to the Outdoor Noise Directive have to be kept with the manufacturer for 10 years. In order to increase his own certainty, the manufacturer of equipment being subject to noise emission limits and labeling should consult the Notified Body as a technical service on a voluntary basis.

The retention period also means that, even several years later, the member states are able to remove the equipment from the market for the sake of improvement. If it is impossible for the manufacturer or his representative in the EC to deposit the data, a fine to absorb the profits that have been achieved without justification can be counted on. This may lead to unforeseeable economical consequences for the manufacturer.

The EC member states are not allowed to prohibit, limit, or hinder the placing on the market or putting into service of equipment, which complies with the regulations of the Outdoor Noise Directive, which are labeled with the CE marking and the guaranteed sound power level, and that is included in the release list of the free-movement-of-goods products.

18.3.3 Consequences for Manufacturers and Consumers

According to the Outdoor Noise Directive, manufacturers of equipment predominantly designed for outdoor use in the future have to check:

- Whether the type of equipment falls, according to the lists in Articles 12 and 13, under the EC Directive
- Whether the type of equipment has to be labeled only or is subject to noise limits, too and have to determine:
- The noise emission of the equipment; for types with limits look for a Notified Body and, depending on the procedure desired, (Annexes VI, VII, and/or VIII) call the Notified Body
- Statistically, which sound power level L_{WA} in dB/pW can be guaranteed for the equipment of a certain series Manufacturers have to:
- Label each piece of equipment with the CE marking and the marking giving the guaranteed sound power level L_{WA} in dB/1 pW, according to Annex IV
- Include an EC declaration of conformity with the data laid down in Annex II
- Send a copy of the EC declaration of conformity, pursuant to Article 16, for each type of equipment to the responsible national authority and to the European Commission, and
- Be prepared, by means of statistical production control (e.g., following EN ISO 4871 [26]), for the inspection of products by member states or by the Notified Body.

The Outdoor Noise Directive improves the information on the noise emission of equipment to the consumer and, thus, ensures for more transparency on the market.

18.3.4 Present and Future User Advantages for Low-Noise Equipment

Noise emission limits for equipment for use outdoors may only be laid down by European legislation. Member states are, however, allowed to limit the use of this equipment in areas considered noise sensitive. The member states can lay down that, in the neighborhood of hospitals and resorts or in other residential areas, only equipment quieter than is required by the Outdoor Noise Directive is to be used. This procedure is an openly declared policy of the European industrial nations of a high population density. In the following, for two member states advantages will be displayed through examples.

18.3.4.1 Germany

Equipment awarded the German environmental label "Blue Angel," RAL UZ-53 [14] provably represent the state-of-the-art noise reduction techniques. In order to lay down a standardized basis, the following procedure has been established:

- In an evaluation process, the criteria for the granting of the "Blue Angel Award" are being drafted (the Federal Environmental Agency, UBA, as well as experts in science, the economy, of trade associations, and official authorities will be called upon)
- A jury of independent public figures decides upon the criteria for the awarding of the "Blue Angel"
- Manufacturers have to render proof that for each type of equipment, for which they desire the Blue Angel Award, the criteria are being met. In a contract with the RAL, the German Institute for Quality Assurance and Labeling Association, they are legally bound to use the Blue Angel Award only for those products that meet the strict criteria of the jury
- Manufacturers are allowed to use the "Blue Angel" label for products that meet the criteria for the awarding of the Blue Angel.

Fig. 18.4 German environmental label RAL UZ-53 for low-noise equipment "Blue Angel"

Table 18.7 Requirements for Blue Angel award RAL-UZ 53

Type of equipment	Power P in kW	Permissible sound power level (L_{WA} in dB/1 pW)	Lowest sound power level (L_{WA} in dB/1 pW)	Highest sound power level (L_{WAmax} in dB/1 pW)
Tracked machines (no excavators)	All	$L_{WA} = 80 + 11 \lg P$	99	101
Dozers, loaders, excavator- loaders graders, wheeled loaders and excavator-loaders, landfill compactors, dumpers, and mobile cranes	All	$L_{WA} = 79 + 11 \lg P$	97	101
Compaction machines (vibratory rollers)	All	$L_{WA} = 82 + 11 \lg P$	97	101
Excavators	All	$L_{WA} = 78 + 11 \lg P$	91	101

Table 18.8 Requirements for Blue Angel award RAL-UZ 53

Type of equipment	Power related range	Sound power level L_{WA} in dB/1 pW
Compressors	Net capacity (Q in m^3/min) $Q \leq 5$ $5 < Q \leq 10$ $10 < Q \leq 30$ $Q > 30$	88 89 91 93
Power generators	All	91
Welding generators	All	91
Combined power and welding generators	All	91
Paver finishers	$Q < 300$ t/h	90 100
	$Q \geq 300$ t/h	94 104
Truck mixers	Net capacity < 8 m^3 Net capacity ≥ 8 m^3	98 100
Tower cranes –Lifting device	$P \leq 15$ kW $15 < P \leq 30$ kW $P > 30$ kW	86 88 90
–Power device	All	91
–Unit of lifting and power device	All	91
Concrete pumps	≤50 kW >50 kW	99 101

Figure 18.4 shows the German environmental label RAL UZ-53 for low-noise equipment that is awarded to manufacturers and their products that represent the state-of-the-art sound protection techniques in Europe.

Tables 18.7 and 18.8 show the criteria for the presently regulated types of equipment that may be awarded the "Blue Angel."

That the use of ecofriendly equipment does not only stand for non-economic advantages but also for financial advantages to the user will be demonstrated by means of the positive example of the Netherlands.

18.3.4.2 The Netherlands

Up to 1990, the so-called WIR Environment Protection Award existed. In 1991, VAMIL, a tax deduction possibility for environmentally friendly investment was established and finally, since 1999, the MIA, an investment deduction for environmental friendly investments was founded.

A so-called environment list, which is being updated annually, contains products subject to tax allowances. The following environmental areas are covered by the list:

1. Water
2. Air
3. Soil
4. Refuse
5. Noise
6. Energy

Under Sect. 18.5, Noise, the following equipment is included in the Environmental List during 2002:

- Generators

- Compressors
- Loaders
- Excavators
- Pumps
- Shredders
- Forklift trucks
- Mobile cranes
- Hydraulic power packs

The procedure of getting a tax advantage is so easy (informing the tax office is all) that the amount of environmental investments has increased ten times between 1991 and 2001 and amounts to € 1.1 billion (VAMIL) and € 0.7 billion (MIA).

Spurred on by government aid for ecofriendly investments, the sound power levels of (for example) forklift trucks could be lowered from 105 dB(A) to 102 dB(A) (median level), as Kruithof and Werring stated [28]. It would be desirable, if the Netherlands' environmental politics with its tax advantages for users would be imitated by other EC Member States. The tax incentives as well as the requirements are listed in VAMIL [29].

18.3.5 Sound Power Level of Construction Equipment: Relative Spectra

Table 18.9 lists the sound power level L_{WA} in dB/pW, which can be used for the prediction of noise emission and of noise exposure due to construction equipment. This table is based upon the results of several research projects concerning the state-of-the-art sound reduction techniques of construction equipment (particularly promoted by the UBA) as well as on the data of noise emission measurements in connection with the EC type approval procedure, which were valid before the Outdoor Noise Directive was put into force.

18.4 Noise Limits and Noise Reduction Measures

18.4.1 Complaints About Insufficient Noise Limits and Noise Reduction Measures

The most common complaints reported to noise protection agencies are:

- Exceeding of the operating hours that have been agreed upon by authorities, especially the too early start of noise-intense construction work (e.g., 6 a.m. instead of 7 a.m.)
- Use of hand- and engine-operated hammers during reconstruction of occupied buildings
- Use of non-silenced construction site drain pumps during night-time
- Driving by of construction site heavy vehicles at neighboring houses
- Operation of demolition equipment breaking up building rubble (according to the new Waste Recycling Act, waste should no longer be taken to the disposal site, but should be broken up on the spot)

In case of legitimate claims and infringements, sanctions up to the closing down of the construction site by interim injunctions are possible. Furthermore, fines may be imposed and in especially severe cases, legal proceedings because of physical injury might be the consequence.

It is important to plan and set-up construction sites under sound technical aspects in order to exclude the risk of contravention and to avoid unnecessary trouble with the neighbors.

18.4.2 Noise Planning, Setup, and Clearing of Construction Sites

Especially prior to the set-up of large-size construction sites, but also for smaller ones in agglomerations, competent authorities will often demand proof of sufficient noise protection in the form of a prognosis [1, 30].

The acoustical advisor is confronted with the question as to which noise levels (noise emission limits, sound power level data, state-of-the-art noise reduction techniques, company data, or in-house data) he should use for the calculations. Since the coming into force of the Outdoor Noise Directive, there is an extended number of construction equipment, for which, since 3 January 2002, the permissible sound power levels are given. Furthermore, there is an extended list of the equipment requiring noise labeling.

However, neither the use of equipment meeting the lowest emission limits nor the use of low-noise equipment that has been awarded the "Blue Angel" (environmental mark) guarantee the keeping of the area and operating times-related target values for the exposure. For noise planning and the setting up of a construction site, additional measures have to be considered:

Table 18.9 Sound power levels L_{WA} and relative spectra of equipment and working processes from in situ measurements on building sites

Nr	Equipment	L_{WA}	Relative octave spectra in dB frequency in Hz						
			63	125	250	500	1,000	2,000	4,000
1	Compaction machines (vibratory rollers)	100–108	−30	−19	−15	−7	−6	−5	−10
2	Compaction machines (vibratory plates and vibratory rammers)	100–115	−32	−24	−15	−7	−7	−6	−10
3	Dozers	105–115	−20	−15	−12	−9	−5	−6	−12
4	Wheel loaders	95–115	−25	−16	−12	−7	−4	−7	−12
5	Paver finishers	100	−21	−7	−15	−7	−6	−7	−12
6	Excavators	110	−20	−15	−10	−7	−5	−7	−10
7	Concrete breakers, hand-held	110	−37	−25	−19	−14	−10	−9	−5
8	Tower cranes	95–103	−23	−15	−8	−6	−4	−10	−13
9	Power and welding generators	95–105	−26	−19	−13	−8	−5	−8	−10
10	Compressors	92–102	−18	−8	−11	−8	−7	−8	−11
11	Circular saws (electric motor)	108–112	−40	−35	−27	−18	−9	−5	−4
12	Chain saws (electric motor)	105	−40	−16	−18	−6	−6	−7	−9
13	Concrete and mortar mixers	96–108	−18	−14	−8	−11	−9	−8	−11
14	Drill rigs	110–115	−30	−22	−14	−9	−4	−6	−10
15	Percussion drills (drilling into rock)	110	−40	−32	−24	−12	−6	−3	−7
16	Trenchers (cutting asphalt)	115	−40	−23	−17	−13	−8	−9	−5
17	Piling equipment (impact hammer)	123–134	−31	−26	−18	−11	−5	−4	−8
18	Piling equipment (vibratory rammer)	125	−22	−17	−12	−7	−7	−8	−11
19	Truck mixers	100	−17	−19	−13	−6	−4	−7	−13
20	Concrete pumps	105	−19	−18	−13	7	−4	−7	−12
21	Concrete shaker	100–108	−46	−36	−15	−10	−12	−7	−7
22	Heavy trucks	103–112	−20	−16	−9	−5	−5	−9	−18
23	Truck, noise from	115	−37	−32	−22	−21	−6	−4	−6
24	Breaker for building rubble (mobile)	108–115	−23	−19	−11	−8	−4	−6	−12
25	Cutting of concrete by water jet (noise from air compressor and water jet)	110–114	−30	−13	−15	−9	−6	−4	−11
26	Building a framework on building sites	114–118	−23	−19	−13	−10	−7	−4	−7

- Limitation of operating hours
- Use of additional low-noise techniques
- Measures such as screens, encapsulations, and silencers
- In-time recognition of conflicts and planning of appropriate measures
- In-time information of those who might be affected by the noise
- Quick response in dealing with complaints from those affected by noise

Construction noise-reduction plans should not only entail the calculation of expected noise exposure for the respective construction period and appropriate measures for noise reduction, but also a cost/benefit analysis for additional noise reduction measures in order not to cause economically unbearable costs.

The detailed VDI regulation 3758 on acoustical planning and setting up of construction sites, which existed as a draft, was unfortunately not published in time.

In coordination with the Federal Environmental Agency (UBA), a "Standards Performance Book of Construction Engineering" (LB 898) for the construction noise and vibration sector has been published in April 1996 and since then has been regularly updated [31]. It contains detailed advice and standard performance descriptions with reference to the reduction of noise and vibration.

This LB 898 version has been prepared for electronic data processing. Suggestions for improvement, supplementation, and extension are being received by the UBA on a regular basis. For the prognosis of noise from large-sized construction sites, the following aspects generally have to be observed:

- Determination of permissible sound pressure levels in the area of impact of the construction site in cooperation with the competent authority. If necessary, definition of possible short-term time periods during which the permissible noise levels may be exceeded (pursuant to the AVwV Construction Noise)

- Calculation of the expected noise emission of equipment and work process
- Calculation and assessment of the expected noise exposure at fixed points in the neighborhood
- If necessary, establishing of the average daily operating hours at daytimes (7 a.m. to 8 p.m.) and nighttimes (8 p.m. to 7 a.m.), and determination of the appropriate time adjustments in dB(A)
- Calculation of the expected sound exposure due to construction heavy vehicles on public roads

If the noise-affected neighbors have agreed, it might even be better not to limit the noise-intense daily operating hours, if, as a result, the overall duration of the construction can be reduced.

Switzerland has a different method for noise planning and set-up of construction sites: the construction noise guideline [17], which came into force on 3 February 2000. According to the guideline, concrete stipulations for noise reduction measures, instead of limits, have to be established. A list of measures shall serve as a basis for the planning and the performance. These measures will be suited to the respective noise emission of the construction site, the noise-sensitivity of the neighborhood, and the duration of the construction work [32].

An additional help for the noise planning of construction sites is the VDI guideline 3765 [33] "Characteristic noise emissions of typical work phases at construction sites," whose draft was submitted in December 2001.

The noise indices mentioned in the guideline do not only reflect noise of the machines used, but also the influence of the material to be handled and the noise emission due to the behavior of the workers. A rough calculation of the sound pressure level in the vicinity of the construction site requires at least the following:

- The noise emission values of the three to five most noise-relevant construction machines, tools, and devices and
- Their distance to the exposed places.

The factor "operating hours" has to be constantly checked and improved by inspections.

18.4.3 Noise Reduction of Equipment Used Outdoors

During the past years, a large number of studies concerning construction equipment have been conducted. Because of the precise measurements of single components (occasionally with the assistance of sound intensity measurements), the main noise sources of most of the construction equipment are well known.

The UBA financed a series of research projects for the determination of the state of the art of noise reduction. In academies and universities, as well as by supervising institutions and acoustical advisors, specific investigations have been carried out. Some manufacturers of construction equipment carried out acoustical research projects concerning the reduction of their equipment's noise emission and the gain with low-noise equipment in competition [34].

If the legislative body will put even more pressure on the manufacturers, additional reduction of noise annoyance by equipment for outdoor use and construction sites can be expected.

With the increasing mechanization of construction activities, however, more construction equipment, which has not yet been investigated will be used. Thus, the noise reduction of equipment and construction sites remains an ongoing topic.

Examples for noise reduction measures for construction equipment are given in the following.

18.4.3.1 Wheel Loaders, Backhoe Loaders, and Excavators

Figure 18.5 shows the main noise sources of a wheel loader.

Examples of noise reduction measures are:

1. Encapsulation of the main sound source (e.g., engine, pumps, and transmission)
2. Use of low-noise aggregates
3. Use of optimized exhaust silencer
4. Use of silencers in cooling fans, oil coolers, and suction filters
5. Isolation and encapsulation of all hydraulic and driving components
6. Buffers for hydraulic cylinders
7. Use of a mechanically separated, vibration-isolated driver's cabin for a considerable noise reduction at the operator's position

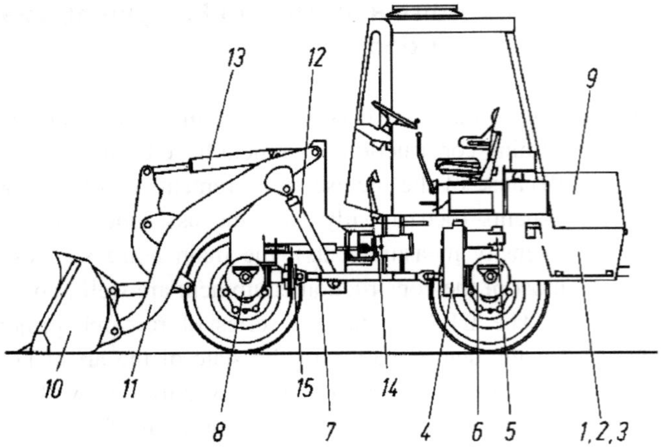

Most important	Less important
1 Diesel engine	10 Shovel
2 Driving pump	11 Lifting device
3 Hydraulic pump	12 Lifting cylinder
4 Transfer gear	13 Tilting cylinder
5 Driving engine	14 Bending cylinder
6 Rear axle	15 Steering cylinder
7 Articulated shaft	
8 Front axle	
9 Oil cooler	

Fig. 18.5 The most important noise sources of a loader

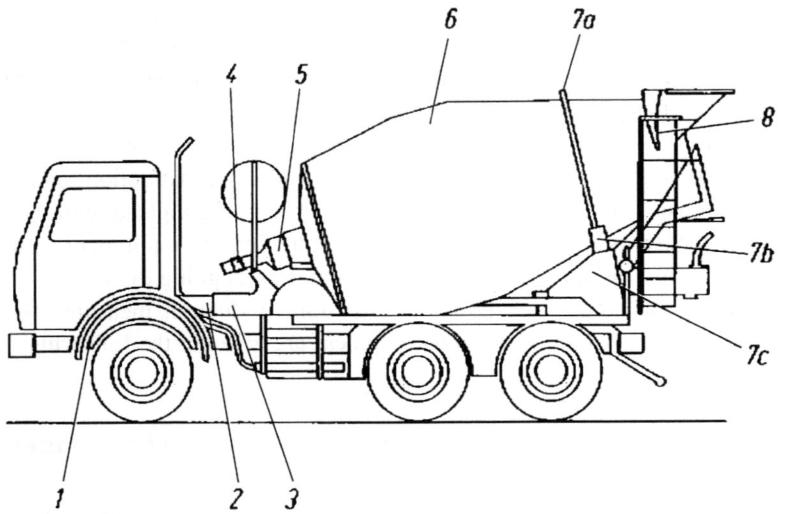

1 Driving engine
2 Articulated shaft
3 Hydraulic pump
4 Hydraulic motor
5 Gear box
6 Mixing drum
7 Drum bearing
8 Drum opening
9 Oil cooler (not shown)

Fig. 18.6 The most important noise sources of a truck mixer

18.4.3.2 Concrete Mixers

Figure 18.6 shows the main noise sources of concrete mixer trucks.

Examples for noise reduction measures are:

1. Use of a low-noise chassis
2. Use of noise-minimized load transmission by means of optimized construction and isolation against airborne and structure-borne noise
3. Reduction of the sound radiation of the mixing drum by means of optimized bearings

18.4.3.3 Piling Equipment (Impact Hammers and Hydraulic Pilings)

With sound power levels of 125 ± 10 dB(A), piling equipment belong to the loudest sound sources at construction sites.

1. A noise reduction of approximately 5 dB(A) can be achieved for example, by insertion of Teflon™ or plastic disks on top of the pile
2. Noise reduction of 5–8 dB(A) can be achieved by encapsulation

3. Values of 5 up to 15 dB(A) less can be reached by using different techniques such as:
   - Vibrating of piles
   - Drilling instead of ramming or vibrating
   - Supporting the excavation by means of drilling poles
   - Supporting the excavation by means of slit walls
   - Pressing of sheet piles

18.4.3.4 Circular Saws

Examples of noise reduction measures are:

1. Encapsulation by means of sound protection hoods
2. Shielding by means of portable absorbing walls
3. Use of saw blades with a low tooth height and diamond technique
4. Use of saw blades with a damping layer (sandwich blades)

18.4.3.5 Other Measures of Noise Reduction

- Use of electric engines instead of internal combustion engines
- If internal combustion engines are used, they should be low-noise ones (e.g., those with water cooling instead of air cooling)
- If construction site pumps or underground water pumps have to be operated at night-times, they should be low-noise type and should be installed as far away as possible from the sensitive neighborhood (even though longer cables and hoses might be needed)
- Demolition work at noise-sensitive construction sites should not be done with pneumatic equipment (concrete breakers or picks), but with diamond-type cutting techniques or with hydraulic concrete crushers
- Setting up of self-supporting screens; their use makes a reduction of 3 up to 6 dB(A) possible.

Additional valuable suggestions for all types of general noise reduction measures on construction equipment are given by Pickartz and Hecker [35].

The Federal Environmental Agency, Berlin, Germany, has an extensive collection of research reports on low-noise construction equipment at its disposal. The reports are based on research projects that were financed by the Minister of Environment, Nature Conservancy, and Nuclear Safety in the name of the Federal Government.

Noise emission data of low-noise equipment can be found in [36].

References

1. Neuhofer R, Wippermann R (2001) Schalltechnische Untersuchungen zur Ermittlung der Geräuschimmissionen während des Aufbaus des Umweltbundesamtes Dessau ausgehend vom Baustelleneinrichtungsplan für die 1. bis 8. Bauphase. Umweltbundesamt, Berlin
2. Neuhofer R (2002) Schalltechnische Untersuchungen, theoretische Überlegungen und Ansätze für die Ablage von Emissionsdaten für Baustellen. Umweltbundesamt, Berlin
3. Directive 2000/14/EC of the European Parliament and the Council of 08. May 2000 on the approximation of the laws of the Member States relating to the noise in the environment by equipment for use outdoors, OJ No L 162 of 03.07.2000, p.1
4. Corrigendum to Directive 1999/103/EC of the European Parliament and of the Council of 24 January 2000 amending Council Directive 80/181/EEC on the approximation of the laws of the Member States relating to units of measurement, OJ L 311 of 12 December 2000, p 50
5. Sicherheit und Gesundheit bei der Arbeit 1999. Unfallverhütungsbericht Arbeit, Bundesministerium für Arbeit und Sozialordnung, Ausgabe 53105, Bonn
6. Directive 86/188/EEC of the Council of 12. May 1986 on the protection of workers from the risks related to exposure to noise at work, OJ No L 137 of 24. 05. 1986, p. 28
7. Verordnung über Arbeitsstätten Arbeitsstättenverordnung – ArbstättV vom 20. März 1975, BGBl. 1 Seite 729, zuletzt geändert durch Artikel 4 der Verordnung zur Umsetzung von EG-Einzelrichtlinien zur EG-Rahmenrichtlinie Arbeitsschutz vom 04. December 1996, BGBl. 1, Seite 1841
8. Unfallverhütungsvorschrift: Lärm BGV B 3 vom 01. Januar 1997, Carl Heymanns-Verlag, Ausgabe Köln 1990
9. DIN 45 645 (1997) Teil 2: Einheitliche Ermittlung des Beurteilungspegelsfür Geräuschimmissionen – Geräuschimmissionen am Arbeitsplatz
10. VDI 2058 (1988) Blatt 2: Beurteilung von Lärm hinsichtlich Gehörgefährdung
11. VDI 2058 (1999) Blatt 3: Beurteilung von Lärm am Arbeitsplatz unter Berücksichtigung unterschiedlicher Tätigkeiten
12. ISO 1999 (1990) Acoustics: Determination of occupational noise exposure and estimation of noise-induced hearing impairment
13. Gesetz zum Schutz gegen schädliche Umwelteinwirkungen durch Luftverunreinigungen, Geräusche, Erschütterungen und ähnliche Vorgänge, Bundes-Immissionsschutzgesetz – BImSchG in der Fassung vom 14. Mai 1990, BGBl. 1 Seite 880 zuletzt geändert am 27. Juli 2001, BGBl. 1, p. 1950
14. Allgemeine Verwaltungsvorschrift zum Schutz gegen Baulärm – Geräuschimmissionen (1970) (Beilage zum BAnz. Nr. 160 vom 01. September 1970)

15. Sechste Allgemeine Verwaltungsvorschrift zum Bundes-Immissionsschutzgesetz (1998) Technische Anleitung zum Schutz gegen Lärm – TALärm, GMBl. 1998, Nr. 26, 503
16. Umweltzeichen, Produktanforderungen, Zeichenanwender und Produkte (1999) RAL, Deutsche Institut für Gütesicherung und Kennzeichnung e. V.
17. Baulärmrichtlinie (1999) Richtlinie über bauliche und betriebliche Maßnahmen zur Begrenzung des Baulärms. Bundesamt für Umwelt, Wald und Landschaft
18. BS 5228–1,2,3,4 und 5: Noise and vibration control on construction and open sites, Editions 1992-03-01, 1999-05-15, 1997-11-15
19. Green Paper of the European Commission Future Noise Policy, Brussels, November 1996, COM(96) 540 final
20. Directive 98/37/EC of the European Parliaments and the Council of 22. June 1998 on the approximation of the laws of the Member States relating to machinery, OJ No L207/of 07.12.1998, p 1
21. Dritte Verordnung zum Gerätesicherheitsgesetz (Maschinenlärminformations-Verordnung - 3.GSGV) vom 18.01.1991 (BGBl. 1, Seite 146), letzte Änderung durch Artikel 2 Absatz 3 der Verordnung vom 12.05.1993 BGBl, Seite 704
22. ISO 3744 (1995) Acoustics: Determination of sound power levels of noise sources using sound pressure – Engineering method in an essentially free field over a reflecting plane
23. ISO 6396 (1992) Acoustics: Measurement of exterior noise emitted by earthmoving machinery – Dynamic test conditions
24. ISO 6395 (1988) Acoustics: Measurement of exterior noise emitted by earthmoving machinery at the operators position – simulated work cycle test conditions
25. ISO 6395 AMD1 (1996) Measurement of exterior noise emitted by earthmoving machinery – simulated work cycle test conditions
26. EN ISO 4871 (1997) Acoustics: Declaration and verification of noise emission values of machinery and equipment
27. EN 27574 parts 1 – 4 (1989) Acoustics: Statistical methods for determining and verifying stated noise emission values of machinery and equipment
28. Kruithof PJ, Werring FJ (2002) Steuerliche Vergünstigungen von Umweltinvestitionen. Ministerie van Volkshuisvesting, Ruimtelijke Ordening en Milieubeheer, Den Haag, Nederland, Ausgabe, 02
29. VAMIL-afschrijving, Milieu-investeringen, Ministerie van Volkshuisvesting, Ruimtelijke Ordening en Milieubeheer, Den Haag, Nederland, Milienlist 2001
30. Bartl M, Gilg J (1997) Kraftwerk Niederaußem, Block K. Ermittlung der Geräuschimmission während der Baufeldvorbereitung sowie der Errichtung des neugeplanten Kraftwerkblockes.Müller-BBM GmbH (nicht veröffentlichter Bericht Nr. 32532/13)
31. Standardleistungsbuch für das Bauwesen, Leistungsbereich: Schutz gegen Baulärm und Erschütterungen, Herausgeber Umweltbundesamt Berlin Ausgabe April 1996
32. Meloni t, Fischer F (2000) Wenn Grenzwerte nicht greifen: Eine Lösung für Baulärm. Fortschritte der Akustik DAGA 2000
33. VDI 3765 (2001) Kennzeichnende Geräuschemissionen typischer Arbeitsabläufe auf Baustellen, Ausgabe 2001–12
34. Goldemund K (2001) Messtechnisch ermittelte Schallleistungspegel einer Bohranlage. Und: Vorschläge zur Reduzierung der Geräuschemission der Bauer Maschinen GmbH. Müller-BBM GmbH (nicht veröffentlichter Bericht Nr. 51 600/1)
35. Prickartz R, Hecker R (1991) Möglichkeiten der Lärmminderung an Baumaschinen I und II, VDI Berichte 900, Tagungsband Schalltechnik '91
36. VDI 3765 E: Kennzeichnende Geräuschemission typischer Arbeitsabläufe auf Baustellen. Ausgabe 2001–12

Urban Noise Protection

19

Michael Jäcker-Cüppers

19.1 Introduction

Noise belongs to the severest environmental impairments in towns, with road traffic being the most annoying noise source. The reduction of these impairments and the precaution against new noise impacts is an important task of the communities. However, many of the potential abatement measures are not in the responsibility of the communities. In most European countries, noise emission regulations for road and rail vehicles and outdoor machinery are nowadays enforced by the European Union. Noise reception limits are generally enforced by national laws. Therefore, efficient noise abatement in towns has to be coordinated with the regional, national and supranational, i.e. European noise policy. The most important fields of action for the urban noise abatement are the roads, railways and airports with heavy traffic. For the avoidance of health risks due to noise here short-term reductions are needed, which can generally be achieved only by a combination of measures for which different stakeholders are responsible. This underlines the importance of integrated and coordinated noise abatement concepts.

19.2 Impairments Due to Noise in Towns

Recent surveys on the environmental awareness in Germany in 2004 show that in towns road traffic noise is judged to be most disturbing, even more than air pollution from road vehicles. Only 40% of the interviewees indicate not to be disturbed or annoyed at all, at least 10% feel themselves highly disturbed or annoyed. Noise from neighbours is the second important source, followed by noise from air traffic, rail traffic and industry [1] (Fig. 19.1).

Also, the noise impacts on the German population show the dominant role of road traffic. Figures 19.2 and 19.3 show the noise impacts in the 'old' German Länder from road and rail traffic in the years 1999 resp. 1997 (outdoor equivalent sound pressure levels in dB(A)) [2].

The results were gained with a simulation model of the German Federal Environmental Agency (UBA), which determines the total exposures in Germany through a projection of the exposures in representative communities. The simulation model – developed in the early eighties – is restricted to noise from to road and rail traffic, industry and construction sites in the former Federal Republic due to methodological reasons.

An analysis of the data shows that high exposures above 65 dB(A) are found along inner-urban major roads (69% of the highly exposed population), a quarter of the highly exposed live along minor roads and 6% along federal motorways [3]. Many communities got the noise levels in their area calculated in the framework of noise action plans. The Environmental Map of Berlin, for example, shows – on the basis of traffic volumes of 1998 – road traffic exposures up to 80 dB(A) at daytime and 75 dB(A) at night-time [4]. Along railways, night-time exposures above 75 dB(A) can be found. The comparison of these figures with the noise abatement targets in Sect. 3.1 demonstrates the high amount of necessary noise reductions.

M. Jäcker-Cüppers (✉)
Ludwigkirchstr. 1, 10719 Berlin, Germany
e-mail: jaecker.cueppers@t-online.de

G. Müller and M. Möser (eds.), *Handbook of Engineering Acoustics*,
DOI 10.1007/978-3-540-69460-1_19, © Springer-Verlag Berlin Heidelberg 2013

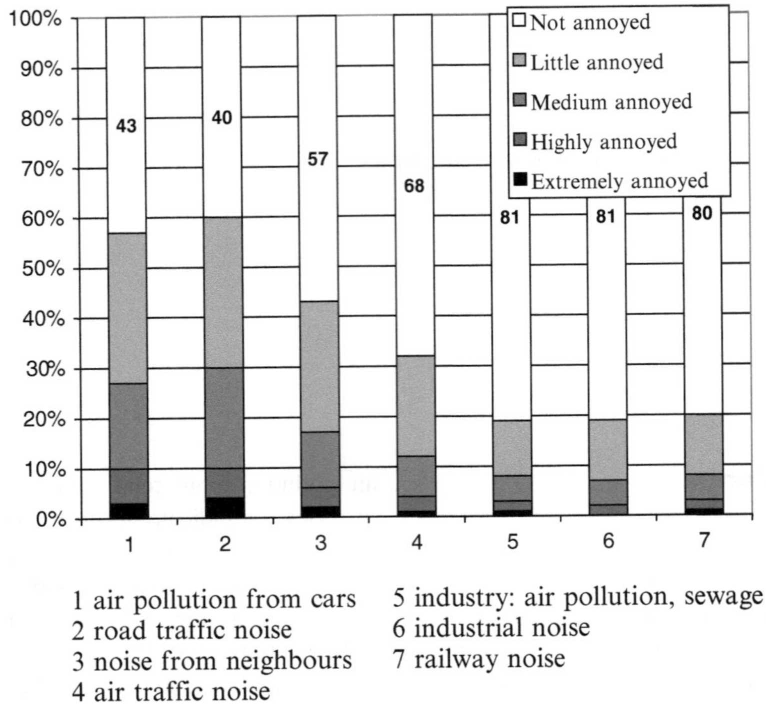

Fig. 19.1 Impairments in towns in Germany 2004

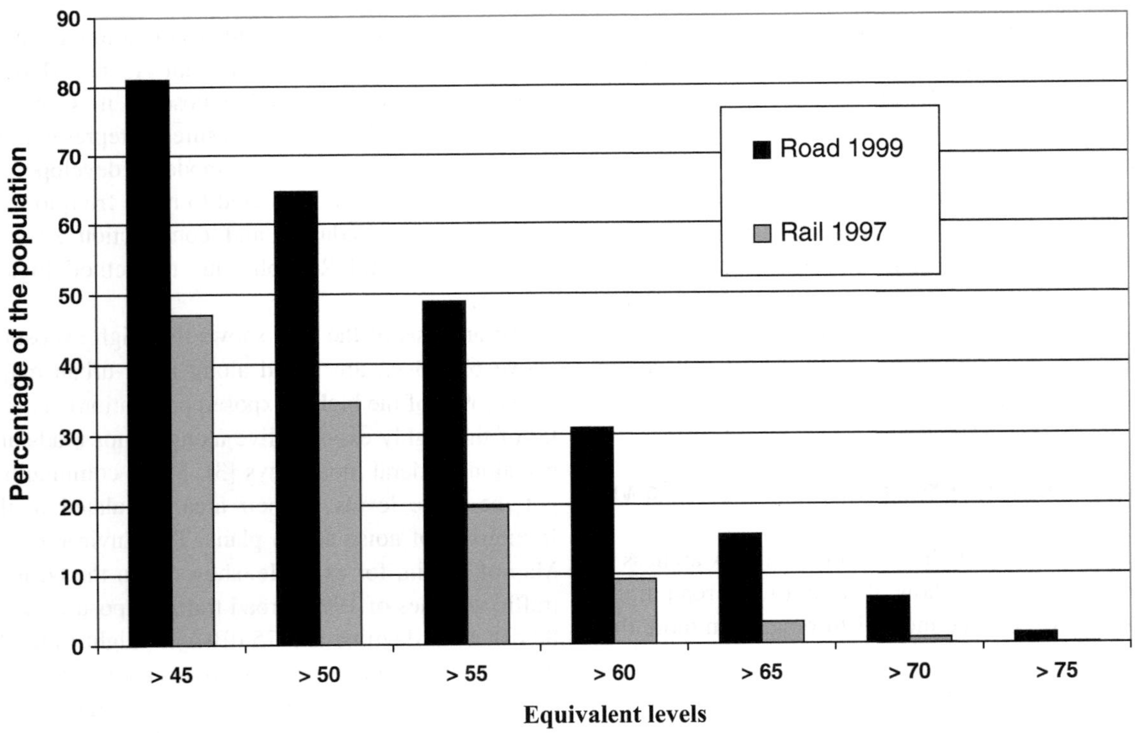

Fig. 19.2 Cumulative daytime noise impacts from road and rail traffic in Germany

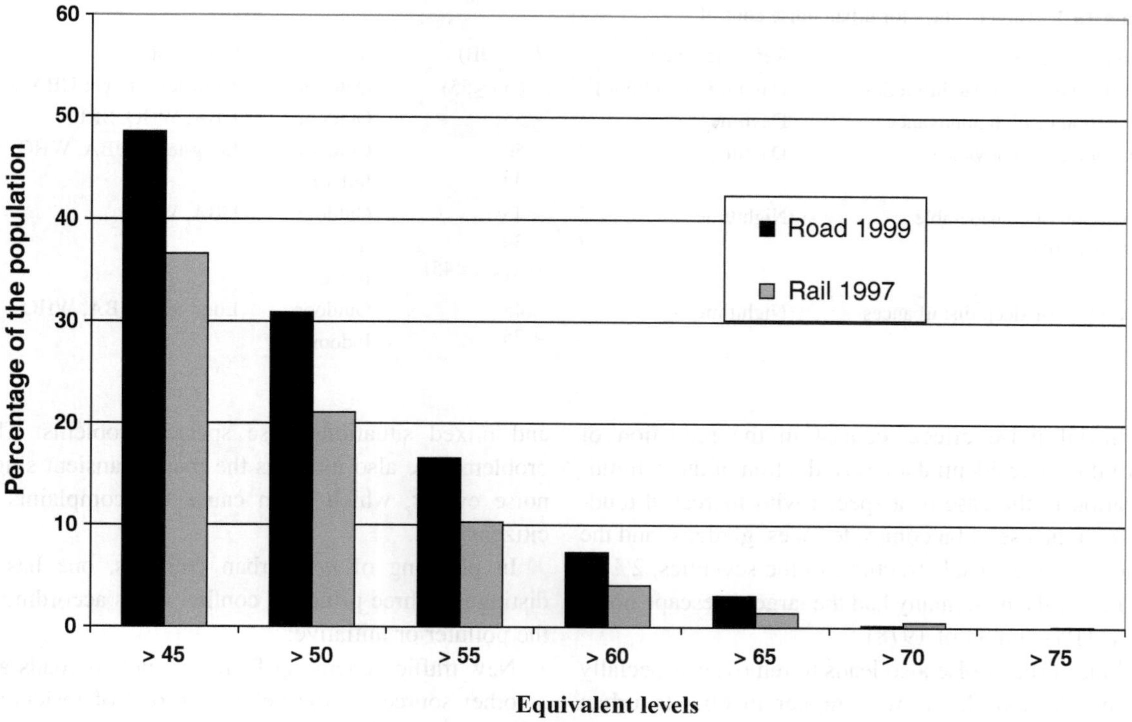

Fig. 19.3 Cumulative nighttime noise impacts from road and rail traffic in Germany

19.3 Principles in Urban Noise Protection

19.3.1 Noise Effects and Targets

The impacts of noise exposure are determined by research on noise effects. In general, the outdoor equivalent sound pressure level is chosen as indicator. Additional indicators such as maximum levels of short time events, i.e. pass-by of noisy vehicles and air planes are particularly relevant for sleep disturbances.

In the field of urban noise protection, one has to distinguish among health, psychic and social noise effects.

Health[1] effects consist above all in an increased risk of cardiac infarction due to traffic noise. Epidemiological studies[2] on the correlation between noise and cardiac infarction show a consistent tendency of increased risks at exposures above 65–70 dB(A) (outdoor equivalent sound pressure levels L_{eq} at daytime). Daytime road traffic noise exposures above $L_{eq} = 65$ dB(A) may lead to an increased risk of cardiac infarction by about 20%.

Sleep disturbances may also constitute negative health effects, they consist in the change of the sleep stages with and without awakening, the impediment of getting to sleep, the shortening of the deep sleep time and in vegetative reactions (emission of hormones, etc.). With equivalent levels below 25 dB(A) in sleeping rooms, no significant disturbances can be expected, with equivalent levels not above 30 dB(A) and maximum levels not above 45 dB(A) sleep disturbances due to noise can be avoided. The corresponding outdoor levels are about 15 dB(A) higher if the sleeping with partly opened windows should be possible.

Psychic effects consist above all in the impairment of recreation, relaxation and communication as well as in reduced performance. These effects can be avoided if the levels are below 50 dB(A) outdoors and 35 dB (A) within buildings. Outdoor levels above 55 dB(A) will lead to considerable annoyance.

[1] Following the human health definition of the WHO (state of complete physical, mental and social well-being) psychic and social noise effects are relevant to health.

[2] See z. B. H. Ising et al. Risikoerhöhung für Herzinfarkt durch chronischen Lärmstreß Zeitschrift für Lärmbekämpfung 44 (1997) 1–7 ("Risk increase for cardiac infarction by chronic noise stress").

Table 19.1 Target values for urban noise control

Protection target	Reference time	L_{Am} (dB)	Receiver	Comment
Health (risk of cardiac infarction)	Day time (nighttime)	≤65 (≤55)	Outdoor	Short-term target UBA, SRU
Prevention of high annoyance	Daytime	<55	Outdoor	UBA, WHO, SRU
Prevention of annoyance	Daytime	<50	Outdoor	Long-term, UBA, WHO
		<35	Indoor	
Prevention of considerable sleep disturbances	Nighttime	<45	Outdoor	UBA, WHO
		<30	Indoor	
		(L_{AFmax}<45)	Indoor	
Prevention of sleep disturbances	Nighttime	<45	Outdoor	Long-term UBA, WHO
		<25	Indoor	

Social noise effects consist in the reduction of activities in residential areas (reduction of the communication in the case of a speech with increased loudness, in the use of balconies, terraces, gardens) and the change in the social structure. In the seventies, 2% of all removals in Germany had the target to escape noisy areas (1%-sample of 1978).

Inner-urban noise also leads to removals especially of mobile households with higher income to urban outskirts. Rents and real estate prices are reduced by noise (in 1981 in Cologne land prices dropped by 1.5% per dB(A) with exposures between 50 and 70 dB(A)). This leads to a concentration of low income households in noisy areas (see [5]).

On the basis of the above mentioned research results on noise effects, the following targets in urban noise abatement are proposed from different institutions [2, 6, 7] Table 19.1.

Priority has the compliance with the targets for outdoor levels, as noise protection should include the outdoor residential areas (balconies, terraces, gardens); besides, the reduction of the outdoor levels improves the quality of the use of inner-urban streets, parks and squares.

19.3.2 Principal Conflicts in Urban Noise Protection

Comparing the existing exposures and the targets, it is obvious that the most difficult problem with the highest priority is the noise reduction along traffic lines with high impacts, that is the remediation of an *existing* situation, for example in urban redevelopments (case 'noise remediation'). Also for other sources (industry, leisure facilities, etc.) grown and mixed situations pose special problems. This problem type also includes the many transient single noise events, which often cause the complaints of citizens.

In planning of *new* urban facilities, one has to distinguish three principal conflict cases according to the polluter or initiative:

- New traffic generating facilities such as roads and other sources in the neighbourhood of residential areas (Case 'Noise precaution')
- New residential facilities near to existing sources (Case 'Approaching residential areas')
- Joint planning of residential facilities and sources (streets, etc.)

Methodology, responsibilities, legal framework and measures differ for these four cases. Generally, the targets of urban noise protection are not sufficiently observed in the evaluation of new traffic generating structures. Urban noise protection must therefore become an integral part of urban land use and traffic planning.

19.3.3 Methodology in Urban Noise Protection

The task of urban noise protection consists of the following major steps:

- Definition of the area of conflicts and investigation
- Assessment of the existing and future noise levels for the area of investigation
- Evaluation of the exposures on the basis of limit or target values
- Analysis of the driving forces
- Development of an action plan
- And its implementation

The solution of noise conflicts should be based on the participation of all stakeholders, a broad participation of the persons affected must be guaranteed at all steps. Besides, for efficient and citizen-oriented noise action plans, sufficient and qualified urban staff must be available.

19.3.3.1 Definition of the Area of Conflicts and Investigation

The impairments due to noise often seem to be a local problem, which might be solved by local measures. The German Guideline for industrial noise (TA Lärm) [8] for example defines the affected area of an industrial installation as the one where the noise exposure due to the installation is 10 dB(A) or less below the guideline values. But a deeper analysis of noise exposures often shows far reaching driving forces (local traffic may have relevant regional or even international contributions). Some measures such as the shift of disturbing traffic from a road will increase the exposure elsewhere. Therefore, a sufficiently large area of investigation should be chosen, especially in the case of traffic noise, at least the community area should be included.

19.3.3.2 Assessment of the Existing and Future Noise Levels for the Area of Investigation

Generally, the noise levels will be determined for each source separately with its proper assessment scheme. These schemes define the exposure indicators, the reference time, emission data and the propagation model. In most cases, the exposures are calculated, because only then the assessment of future exposures is possible. But some schemes contain procedures for the measurement of exposures as well (such as the German TA Lärm [8]). In the German standard DIN 45 682 'Schallimmissionspläne' ('Sound immission maps') of September 2002, the source related schemes are listed and supplementary procedures for the visualisation of the exposures are defined.

The most important assessment schemes for urban noise abatement in Germany are:

- DIN 18005 Part 1 Urban Noise Control; calculation scheme [9]
- Instructions for road traffic noise protection (Richtlinien für den Lärmschutz an Straßen; edition of 1990 (RLS-90) [10])
- Instructions for the calculation of railway noise exposures (Schall 03 – edition of 1990 [11]).

RLS-90 and Schall 03 have a special legal character, as they are an integral part of regulations for noise reception limits for newly built or substantially altered traffic lines (roads and railways, see Chap. 4).

In the road traffic guideline, simplified average traffic situations are used, which are classified according to the upper speed limit. For the urban noise abatement sometimes a more detailed assessment of the traffic situation is needed (for example, 'noise reduction measures for a major road in the city centre with two lanes, few congestions and with bus traffic'). This can be calculated with the computer program CITAIR.[3]

For new and substantially altered roads and railways in Germany, the design of noise protection measures has to be based on the predicted traffic volumes. For roads this is the traffic volume in 10–20 years,[4] for railways this is the full capacity of a line.[5]

In June 2002, the *European Union* (EU) enforced a Directive of the Council and the Parliament with which the assessment, evaluation and management of environmental noise shall be harmonised in the EU [13]. In the medium and long term, this will considerably influence the national calculation schemes, as they differ in indicators, emissions and in the propagation model from the future harmonised schemes. The Directive prescribes that the exposure of the European population has to be assessed in so-called 'strategic noise maps' ('map designed for the global assessment of noise exposure in a given area due to different noise sources or for overall predictions for such an area') for agglomerations, major roads, major railways and major airports. The time schedule has the following steps:

[3] CITAIR: Computergestütztes Instrument zur Prognose the Auswirkung verkehrlicher Maßnahmen zur Immissionsreduzierung (Computer-based instrument for the prediction of the effects of traffic management for the reduction of pollutions [air and noise]) can be bought from the German Federal Environmental Agency.

[4] See the official explanations for §3 of the 16. BImSchV [12] in the Bundesratsdrucksache 661/89, S. 37.

[5] See the "Application guidelines for the 16. BImSchV" version 6 September 1993 of Deutscher Reichsbahn and Bundesbahn.

- *Step 1*: By 30 June 2007 strategic maps have to be made for:
  - Agglomerations with more than 250,000 inhabitants
  - Major roads with more than 6,000,000 vehicle passages per year
  - Major railways with more than 60,000 train passages pro Jahr per year
  - Major airports with more than 50,000 movements per year
- *Step 2*: By 30 June 2012 strategic maps have to be made for
  - Agglomerations with more than 100,000 inhabitants
  - Major roads with more than 3,000,000 vehicle passages per year
  - Major railways with more than 30,000 train passages per year

  (see Chap. 4)

19.3.3.3 Evaluation of the Exposures with Respect to Limit and Target Values

Exposures have to be compared with limit, target or guideline values. In general, these values depend on the particular land use, as defined in corresponding plans (in Germany defined in the so-called Land Use Ordinance – Baunutzungsverordnung [14]). They are visualised in so-called sensitivity maps, exposure above the limits can be shown in conflict maps [15].

The prioritisation of the conflict cases may be accomplished by a weighting of the number of the exposed persons with their exposure or with the difference between their exposure and the target levels.

19.3.3.4 Analysis of the Driving Forces

An efficient noise reduction must be based on the analysis of the driving forces. This shall be exemplified for road traffic. The noise levels in a road, for example, depend on the following influential factors:

- Categories and numbers of the emitting vehicles
- Emissions from the road traffic
- Transmission of the emissions to the receiver

This shall be explained for the vehicle categories and their emissions.

Categories of the Emitting Vehicles

It is well known that noisy vehicles such as heavy trucks even in small portions determine the noise exposure. According to the emission assumptions in the German road traffic calculation scheme (RLS-90 [10]), a truck above 2.8 *t* generates a noise exposure (upper speed limit of 50 km/h), which is 13.6 dB(A) above that of a car, thus 23 cars are as noisy as one truck. Equivalent levels are, therefore, determined by trucks if their percentage exceeds 4.2%. The reduction of the number of trucks and the reduction of their emissions is, therefore, of special importance.[6]

The emissions from *road traffic* are mainly determined by the two sources propulsion and road/tyre interaction. For a car (which runs on petrol and with a motor power of 55 kW), Fig. 19.4 shows the sound pressure levels and the principal parameters, which determine the influence of these two sources[7]:

- Road/tyre noise increases with the vehicle speed
- Propulsion noise increases with the engine speed and the engine load. Engine speed and vehicle speed are linked with each other via the gear transmission ratio

Figure 19.4 shows that the emissions at full load (for example, due to acceleration) can be up to 7 dB (A) above operations without load (constant speed with low driving resistance). Therefore, these two operational conditions have to be distinguished.

A certain constant vehicle speed can be reached in different gears. The differences in emissions in total vehicle noise are very high, at 30 km/h, for example, the emissions in the second gear are about 11 dB(A) higher than in the fourth gear. Therefore, the manner of driving is of high importance for the noise reduction.

Figure 19.4 furthermore shows the typical relationship between road/tyre noise and propulsion noise, which of course may be different for other technical designs. Road/tyre noise is already dominant for constant speed between 30 and 50 km/h in the third gear.

---

[6] According to the assumptions of the RLS-90 ([10], Table 19.3) only on municipal roads the average truck percentages are lower (3%).

[7] See Kemper, G, Steven H Geräuschemissionen von Personenwagen bei Tempo 30, Zeitschrift für Lärmbekämpfung 31 (1984), 36–44 (Noise emissions from cars in 30 km/h zones).

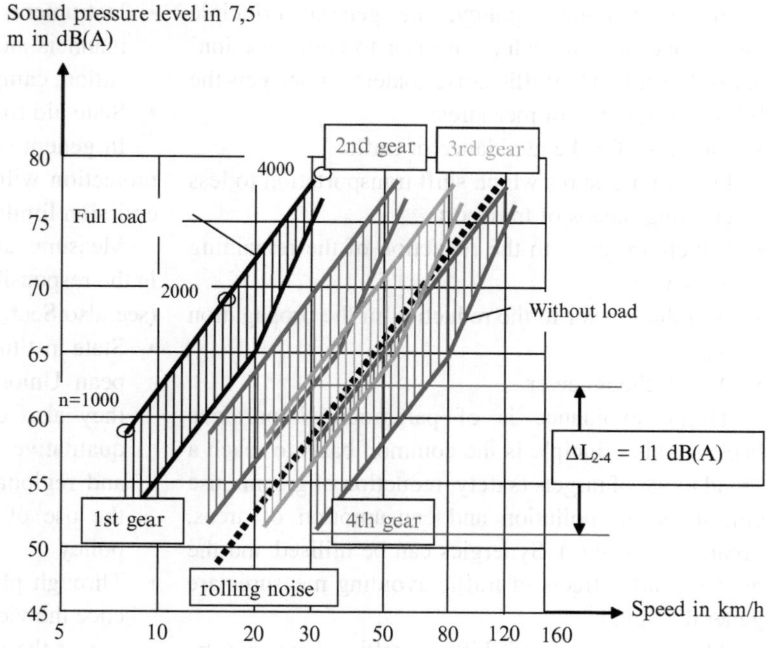

Fig. 19.4 Propulsion and road/tyre noise of a car as function of speed and gear

The road/tyre noise is generated by the interaction of tyre and road surfaces; thus for the reduction of the emissions, the following actions and stakeholders are involved:

- The vehicle manufacturers determine the propulsion noise through the technical design of the vehicles
- Legislation limits the maximum allowable emissions of the involved sources
- Manufacturers of tyres and road surfaces determine the rolling noise through the technical design of their products
- The drivers will considerably influence the emissions by gear selection, vehicle speed and engine load
- Car owners (among them also the communities and public transportation) could purchase low noise vehicles and tyres
- The local road construction and traffic management authorities can influence the driving manner through road design and upper speed limits

19.3.3.5 Development of noise action plan

On the basis of conflict maps and the analysis of the influential factors concepts for measures are to be developed to reduce the exposures (see Chap. 5).

The legal framework and the responsibilities have to be observed.

Integrated concepts on communal level are, for example, the noise action plans according to §47a of the German Federal Immission Control Act (see Chap. 4). As many measures are in the responsibility of national or European authorities, the communal noise action plans have to be supplemented and supported with higher-ranking concepts such as further steps for the reduction of allowable emissions from the involved sources.

On the European level such noise action plans are now obligatory via the Directive on Environmental Noise [13]. The action plans have to be presented about 1 year after the strategic noise maps (see Sect. 3.3.2), this is by 18th July 2008 resp. 2013.

19.3.4 Principles, Instruments, Measures and Stakeholders in Urban Noise Protection

The paramount problem of high exposure situations can generally be solved only through a combination of *measures* (these are concrete implementations of noise reduction actions such as the construction of a noise barriers) and *instruments* (they induce such measures as in the case reception limits). It is useful to structure the potential measures and instruments according to priorities, degree of obligation and responsibilities.

In environmental policy, the general principle 'avoidance prior to reduction prior to compensation' is valid. Applied to traffic noise abatement one gets the following ranking of measures:

- Measures for the avoidance of traffic
- Prior to measures which shift transportation to less emitting means of transport
- Which are prior to the reduction of the remaining emissions
- And these prior to the reduction in the propagation path
- And at the receiver

Traffic avoidance is of paramount importance because this principle is the common basis to reach a broad range of targets (safety, reduction of greenhouse emissions, air pollution and consumption of areas, urban quality, etc.). Synergies can be utilised and the negative side effects of traffic avoiding measures are generally small.

Also for the case of shifting traffic, there may be synergies (for example, with respect to energy saving and noise). But the principle to shift traffic requests the identification of vehicles, which are – with respect to transport performance – quieter than others.

Emissions reduction as measure at the source corresponds to the polluter-pays-principle. It may cause conflicts between different targets (as, for example, increase of energy consumption due to increased weight caused by encapsulations of vehicle sources).

Finally, measures in the propagation path are often not compatible with urban quality such as inner-urban noise barriers; measures at the receiver such as sound insulation windows do not protect outdoor living areas.

These principles – here explicated for traffic noise abatement – can be transposed to other sources in an analogous way.

With respect to the *degree of obligation*, the following instruments might be distinguished:

- Setting of noise emission limits and receptions through acts, ordinances and administrative instructions
- Obligation for noise reduction in the framework of legally prescribed balancing of concerns as in the urban land use planning
- According to the polluter-pays-principle attribution of the costs of noise exposures to the polluter
- Privileges and preference for low noise products and processes as in public procurement or on the part of the consumers
- Instruments which aim at the modifications of manners and behaviours such as information, education, campaigns
- State aid for noise abatement

In general, the highest and most concrete level of protection will be achieved with the setting of noise emission limits and noise reception limits.

Measures and instruments for noise abatement are in the responsibility of different actors or stakeholders (see also Sect. 3.3.4):

- State institutions set limits for the emissions (European Union) or the reception (nations, regions); they also define the general framework for the qualitative demands on noise reduction. National and regional governments influence the costs for the use of noisy sources through tax and fiscal policy
- Through planning regions and communities influence the vicinity of noise sources and sensible land use or the traffic generation. They are responsible for local action plans
- Different local authorities supervise the extent of noise levels as regulatory bodies (industrial regulatory bodies, plan approval bodies)
- The road traffic authorities are responsible for the noise reduction via traffic management
- The citizens will generate or avoid noise through their behaviour and consumption
- The manufacturers define the noise emissions of their products
- Police and courts will be addressed in many conflicts due to noise impairments.

In many conflicts, this variety of responsibilities aggravates a fast solution for the affected persons. Urban noise control can improve this situation through the installation of a centralised body of contact ('noise phone').

19.4 Legal Framework in Urban Noise Protection

For urban noise abatement all acts, ordinances and administrative instructions are relevant, which regulate or influence traffic generation or avoidance, the shift of traffic, the control of the emissions and receptions. The latter are of paramount importance as *direct* regulations for the emission and pollution

Table 19.2 Noise reception limits and guideline values for the noise control along traffic lines in Germany

	Noise precaution (new lines)[a]		Noise remediation (existing traffic lines)[b,c]	
	Daytime	Nighttime	Daytime	Nighttime
Hospitals, etc.	57	47	70	60
Residential areas	59	49	70	60
Mixed areas	64	54	72	62
Industrial areas	69	59	75	65

[a]New or substantially altered according to the Traffic Noise Ordinance
[b]Federal roads and railways (in accordance with the availability of financial means)
[c]Targets of the German Federal Environmental Agency (UBA): 65/55 dB(A) daytime/nighttime for residential areas

control. Among the traffic related legal framework, the example of the German Regional Land Use Planning Act (ROG) [16] will be cited which gives clear rules for the avoidance of traffic and the shift to less-emitting transport means.

§2 ROG states:

1. The principles of regional land use planning have to be applied in the spirit of the governing concept of sustainable spatial development according to §1 Abs. 2.
2. Principles of the regional land use planning are:

5. Condensed areas have to be safeguarded as focal points for living, production and services. The development of land use shall be oriented at an integrated transport system and the safeguarding of free space. The attractivity of public transport has to be expanded by the development of public transport associations and the creation of efficient interfaces. Green areas shall be safeguarded as elements of the free space and shall be merged. Environmental damages have to be reduced.

12. A good accessibility of all space elements among themselves through passenger and goods traffic has to be assured. Above all in spaces and corridors which are heavily burdened with traffic, the prerequisites for the shift of traffic to more environmentally friendly means of transport such as rail or waterways shall be improved. The development of land use has to be organised through allocation and mixture of different land use types in such a way that the traffic impacts are reduced and additional traffic is avoided.

Among the noise reception limits and noise emission limits, the German *Federal Immission Control Act (BImSchG)* of 1974 [17] is the most important one. It regulates – together with corresponding ordinances or administrative instructions

- The noise control for the establishment and operation of installations (§4ff) with setting of reception guideline values in the general administrative instruction 'Technische Anleitung zum Schutz gegen Lärm' (Technical instruction for noise control) (TA Lärm) in the version of 1998 [8]
- Noise limits for vehicles (§38) in combination with the Straßenverkehrs-Zulassungs-Ordnung (StVZO) (Road Traffic Licencing Ordinance), which transposes the respective European Directives into national law (limits and measurement procedures) since 1970
- The noise control for the new construction and the substantial alteration (due to construction!) of traffic lines (noise precaution) (§§41–43) with the setting of limits and calculation schemes in the Verkehrslärmschutzverordnung (VlärmSchV) (Traffic Noise Ordinance) of 1990 [12]. The VlärmSchV defines the admissible *outdoor* levels[8] in Table 19.2, the exceedance of which has to be avoided (§41 (1)), as long as the costs of the protective measures commensurate with the desired objective for protection (§41 (2)); if not, only so-called passive protective measures shall be applied, i.e., measures which guarantee a certain indoor level and which will be designed according to the Verkehrswege-Schallschutzmaßnahmenverordnung of 1997 [18] (Noise Protection Ordinance for Traffic Lines). The limits for railways are 5 dB(A) higher than those for roads due to the fact that railway noise is assumed to be less annoying.
- The VlärmSchV also defines the substantial alteration of a traffic line: this implies a significant change in the construction, which increases the levels by 3 dB(A)[9] and more or to levels above

[8] Rating levels, i.e. with a correction of the equivalent levels representing noise nuisance effects such as the railway "bonus" (railway noise considered to be less annoying than road noise).

[9] Due to the rounding rules this corresponds to 2.1 dB(A) and more.

70 dB(A) at daytime/60 dB(A) at night-time. Furthermore, the VlärmSchV lays down the corresponding calculation schemes for road and railways.

- The noise action plans of the communities (§47a of 1990): local authorities have to determine the noise exposures in areas in which harmful environmental effects are caused by noise or are expected to be caused thereby and, if necessary, to draw up noise action plans. Even if §47a does not contain any deadlines, it is the most important instrument for the noise remediation of existing harmful situations. Representatives of the regional pollution control in Germany have passed a model administrative instruction [19] defining the noise exposures, which represent harmful effects on the environment. For roads and railway lines, these thresholds are identical with the limits in the VlärmSchV (see Table 19.2).
- And with §50 for the regional planning stating that areas for specific types of use shall be zoned in such a manner that harmful effects on the environment are kept to a minimum.

It is planned to integrate the transposition of the EU-Directive on environmental noise into national law – which was due by 18th July 2004 – within the Federal Immission Control Act.

For the *urban land use planning* the German Construction Code (Baugesetzbuch BauGB, in the version of 1997) is the most important law [20].

The urban land use planning shall "guarantee a sustainable urban development and a socially equitable land use, which corresponds to the public well-being; it shall contribute to safeguard a humane environment and to protect and develop the natural basis of life" (§1 BauGB). The general demands for healthy living and working conditions and the environmental concerns have to be observed.

'In the development of urban land use plans the public and private concerns have to be fairly balanced against and among each others'. (§1 (6) BauGB)

The planning instruments for these targets are among others the type and degree of the land use and the definition of areas for special installations and provisions in order to protect against harmful environmental impacts in the sense of the Federal Immission Control Act (§9, contents of the land use plan).

In the Construction Code itself, no limit values are given to avoid harmful environmental impacts due to noise. In the supplementary sheet 1 of the German standard DIN 18005 'Urban Noise Control' ('Schallschutz im Städtebau') of May 1987 [21], the guideline values in Table 19.3 are recommended for urban planning. The guideline values are no limits, they have to be balanced with other concerns. 'The exposures should be not higher or even below the guideline values in order to fulfil the expectations for an adequate noise protection, which are linked to the specific characteristics of the concerned areas for land use'.

The guideline values are of paramount importance for the planning of new residential areas. They are not valid for the authorization of single buildings according to §§30 and 34 BauGB. For the especially problematic case of a new building along an existing road with high exposures the standard does not give any guideline value. But in this case at least 'the demands for healthy living and working conditions must be fulfilled' (§34(1) BauGB).

DIN 18005 has not been introduced in all German regions, for example, not in Hamburg, here 'such low' targets in agglomerations are thought to be unrealisable [22]. But anyhow two principles should be observed:

- Exposures which cause negative health effects have to be excluded (see Sect. 3.1: rating levels of 65/55 dB(A) at daytime/night-time)

Table 19.3 Noise guideline values for urban planning

Land-use	Guideline values in dB(A) daytime	night-time[a]
1. Pure residential, weekend cottage and holiday home areas	50	40/35
2. General residential areas, camping sites	55	45/40
3. Special residential areas	60	45/40
4. Village and mixed areas	60	50/45
5. Centre zones and industrial areas	65	55/50
6. Special sensitive areas according to their sensitiveness	45–65	35–65
7. Cemeteries, allotments, parks	55	55

[a]Lower values for industrial and leisure noise

- Low exposures for living- and especially sleeping rooms have to be safeguarded by appropriate orientation within the buildings, averting them from the sources. In [22], the target value of 49 dB(A) at nighttime is proposed (compare Traffic Noise Ordinance).

The German Road Traffic Ordinance Straßenverkehrs-Ordnung (StVO) in §45 authorises the urban traffic authorities to restrict traffic in order to protect the population against noise and air pollution.[10] The following measures can be applied:

- Speed limits as in so-called Tempo-30-zones (upper speed limit 30 km/h) or in traffic calming zones (upper speed limit about 7 km/h)
- Parking management
- Installation of pedestrian zones
- Traffic bans for certain times and vehicle categories.

The application of §45 STVO for noise control has been put in concrete terms in 1981 by instructions by the German ministry of transport [23]. According to that, its application might be especially considered if in residential areas daytime resp. nighttime levels exceed 70/60 dB(A). But jurisdiction[11] clarified that such measures are not only justified if certain high levels are exceeded but already then if exposures not customary in a place can be reduced by reasonable restrictions (such as the ban of through traffic in roads, which are dedicated to local accession).

19.4.1 The Directive of the European Union on Environmental Noise [13]

This Directive for the first time introduces deadlines for the assessment of strategic noise maps (see Sect. 3.3.2) and the design of noise action plans (see Sect. 3.3.5) it is, therefore, more stringent than the above-mentioned regulations in §47a BImSchG.

19.4.2 Noise emission limits in the European UNION

As is generally known the EU is responsible for the harmonisation of product regulations – including noise limits. In several Directives limits for:

- Cars, trucks and busses [24] since 1970, motor cycles[25] since 1978, with the last changes of the emission limits in 1992 [26] (development of the limits see [2])
- Tyres of road vehicles [27] since 2001
- Outdoor machinery [28], last version of 2000

have been enforced.

Noise limits for rail vehicles are in force for high speed trains since 1 December 2002 and in preparation for conventional interoperable rail vehicles.[12]

The effectiveness of the Tyre Directive will be rather limited according to the evaluation of many organisations (such as the German Federal Environmental Agency): the limits are already today observed by very many tyre types, furthermore, for all tyres on the market, they are enforced only by 1 October 2009.

19.4.3 Evaluation of the Legal Instruments for (Traffic) Noise

The protection against noise in Germany, especially traffics noise is *not comprehensively regulated*:

- The German Traffic Noise Ordinance is only valid for newly built or substantially altered roads and railways
- Substantial alterations require substantial constructive changes. Increase of levels due to increased traffic volumes or speeds are disregarded
- The limits in the Traffic Noise Ordinance related to outdoor living areas are not compulsory if the costs of the active measures do not commensurate with the objectives for protection
- §47a BImSchG does not envisage deadlines and noise limits

[10] §45 STVO (1) The traffic authorities are allowed to restrict or forbid the use of certain roads or parts of them for traffic safety or orderliness reasons and they are allowed to reroute the traffic. They have the same right for the protection of the population against noise and air pollution.

[11] On §45 StVO decision of the Bundesverwaltungsgericht (Federal Court of the Administrative Law) of 4 June 1986 (7C76.84) (NJV 1986, 2655ff.) Guiding principle : "1. StVO § 45 Abs. 1 S 2 Nr 3 provides protection against road traffic noise not only if a certain level is exceeded; it is sufficient that the exposures are above the levels which have to be accepted as customary taking the concerns of traffic into account."

[12] These will be defined within the Technical Specifications for Interoperability.

- §45 StVO is subject to be balanced with other concerns (such as ensuring mobility)
- Also in land use planning this balancing of concerns has to carried out; regulations for single projects are missing
- In the German regulations the sources are treated separately. Appropriate protective measures for multiple exposed persons or those who do not have any quiet façade cannot be demanded
- The noise emission limits of the EU do not always correspond to the state of art in noise reduction.

Especially at the existing inner-urban roads and railways with high exposures concrete protective regulations are missing (the problem of the 'noise remediation'). For this case, the gravest problems are tackled by the noise remediation programmes for federal roads (since 1978) and railways (since 1999). The limits applied in these programmes (see Table 19.2) are above the levels, which are necessary for the protection against negative health effects. In addition these programmes are financed by voluntary federal contributions depending on the available budgets. This situation has been criticised for a while now,[13] corresponding political initiatives have not been successful yet.

19.5 Measures

In the following, measures for noise reduction will be presented. They will be structured according to the ranking described in Sect. 3.4. Often certain sets of measures comprise several ranking related elements. The concept of 'area wide traffic calming' (see Sect. 5.3.2), for example, not only aims at low noise driving manners but also seeks to promote the use of transport means without emissions (walking, cycling). Also, the noise action plans according to $47a BImSchG or the Environmental Noise Directive should be based on the coordinated integration of many single measures.

19.5.1 Traffic Avoidance

Traffics avoidance more exactly means the avoidance of motorised traffic (mt). This means in a broader sense the reduction of the mileage (m) of the motorised traffic according to the following tautology (compare [29] p. 434):

$$m = \frac{m}{tp}\frac{tp}{a}\frac{a}{b}b,$$

(a) Reduction of the mileage intensity $\frac{m}{tp}$
(b) Reduction of the transport intensity $\frac{tp}{a}$
(c) Reduction of the activity intensity $\frac{a}{b}$
mit tp – transport performance (in passenger-kilometres or tons-kilometres)
a – activities
b – benefit

(a) Means increase of the occupancy or the workload[14] of the vehicles.
(b) Means the reduction of the transport performance (tp) per activity, for example for the ways to and from work (traffic avoidance in the *narrower* sense through low traffic land use structures).
(c) Means the reduction of activities linked with a certain benefit (for example, benefit recreation: one long vacation instead of many short trips).

Analogous relationships can be used for freight traffic, whereby activities have to be substituted by commodity units. Reductions of the commodity intensity according to (c) would then, for example,

[13] See, for example, the German Council for Environmental Questions in its special report "Environment and Health" of 31 August 1999 (extract from the short version [18]:). In contrast to installations which are subject to the Federal Immission Control Act (§§ 17, 25 BImSchG) the legal regulations do not provide for the remediation of existing traffic installations. This legislative state where noise control nearly completely relies on fiscal considerations is not acceptable in the long term even under budgetary restrictions. The attitude of refusal of the fiscal policy does not only deviate from the individual preferences of many citizens. The obligation to protect the health of the citizens according to the German Constitution (Article 2, Paragraph 2) rather demands appropriate actions against noise exposure from existing installations, at least if they are on the verge to endanger health what is to be expected for long-lasting considerable annoyances. As above all low income households are exposed to unacceptable noise a reduction of the noise exposure is also a requirement of the welfare state.

[14] Thus, cars used for the journeys to and from work are occupied by 1.1 persons/vehicle on the average, in public transport the occupancy rate is only 20%.

be the use of long-lasting commodities instead of ex and hopp products or renting instead of buying.

Traffic avoidance according to (c) means the renunciation of a life style with high use of commodities and a high travel intensity; it is a more general cultural task but it can be supported on the urban level through the promotion of living without a car (for example, by developing car free residential areas[15]).

But the growth of the transport performance of the total motorised traffic in Germany[16] from about 88 billion passenger-kilometres (Pkm) in 1950 to 905,7, billion Pkm in 2002 shows how rather ineffective strategies of traffic avoidance have been up to now. Indeed, the reasons for growing transport demands such as prosperity, individualization of lifestyles, globalisation, economic concentration are unbroken.

In [30], a lot of traffic avoidance strategies have been described, here only some more important examples will be presented.

The effects of measures to avoid traffic are generally not easy to be quantified and of relative low efficiency. A quite ambitious reduction of the mileage by 50% corresponds – as is generally known – to an average reduction of exposures by only 3 dB(A).

An example for the reduction of the commodity intensity according to (c) (renting instead of buying/owning) is car-sharing:

Car-sharing supports a way of life without one's own car. One shared car corresponds to 5–8 private cars[17] with a corresponding reduction of materials and transportation in the car production. Persons who share cars reduce their car mileage from 7,000 to 4,000 km/year on average (whereas private cars have an average mileage of 12,700 km/year). Car-sharing companies with the German Environmental Label oblige themselves to utilise low noise vehicles (the type approval noise emissions of the vehicles shall be $\leq$71 dB(A) i.e. 3 dB(A) below the limit, in addition low noise tyres have to be used according to the German Environmental Label).

Low traffic land use according to (b) or the '*short distance city*' will be promoted by urban land use planning with the instruments 'mixture of use',[18] 'decentralisation'[19] and' condensation'[20]. In [31] it is recommended to safeguard urban quarters with high historic urban quality, to build dwellings preferably in down town areas and to preserve the greenbelts around the cities. A good example for a 'compact' city is Delft (92,000 inhabitants) in the Netherlands; here 90% of the residential area are within a radius of 2.2 km around the city centre, whereas for comparable cities in Germany the figures are 3.0 (Tübingen) to 5.0 km (Trier). Land use structures can only slowly be changed; therefore, the conservation or extension of structures with mixed land use should be promoted with additional instruments such as a housing policy, which eases moves nearer to the workplace, the decentralisation of urban services (see the example of the city of Heidelberg), the conservation of service facilities near to the dwellings such as the village store through corresponding trade and commerce policy, etc.

The *short distance city* ideally allows to make the necessary journeys by foot or bicycle. After all 50% of the journeys today made by car are shorter than 5 km, thus there is a high potential for a mobility without emissions.

Generally, noise reception limits and guideline values depend on the type of land use (see Tables 19.2 and 19.3), areas with a high mixture of uses are less protected than, for example, pure residential areas. This dilemma between pollution control and traffic avoiding mixtures of land use is only partly solvable, one example in Germany is the definition of 'special residential areas' which allow a high degree of mixture but at the same time have ambitious noise guideline values for the night.

An important instrument for traffic avoidance is the rise in prices of the *motorised traffic* internalising its 'true' or external costs, which comprise the cost of

[15] Compare http://www.carfree.com/.

[16] ifo Institut, Verkehrskonjunktur 2000; BMV: Verkehr in Zahlen 1999; (Traffic in Figures) with 1950 0.5 journeys and 1997 3 journeys per day with motorised vehicles, 1950 about 9,076 km mileage per person, 1997 15,330 km

[17] All figures from Germany.

[18] via definition of the type of land use according to the Baunutzungsverordnung BauNVO (Land Use Ordinance).

[19] via allocation of the land use zones in the Land Use Plan.

[20] via definition of the degree of the utilisation of the land area (BauNVO §17),here again the allowed value of the Floor Space Ratio corresponds to the reception limits.

Table 19.4 Modal Split in Münster (% of the journeys)

	Walking	Cycling	Public Transport	Car
Münster 1972		10		
Münster 1990	27	34	5	34
Germany 1992	27	9	10	53
Netherlands 1991/1993		27		
Groningen 1990		39		

noise exposure.[21] This instrument will influence the mileage if the cost elements reflect this parameter.

This instrument is implemented in different forms:

- *Petroleum tax* (taking mileage into account) in Germany supplemented by ecological elements
- *Motor vehicle tax* (independent from mileage but which may take the specific emissions of a vehicle type into account)
- Road pricing as an urban instrument, which aims at the limitation of traffic in the case of insufficient road capacities and excessive environmental impacts (applied in Singapore, Norway, London [congestion charge][22]), in Germany in the framework of pilot projects (Stuttgart) but with relatively small effects
- *Lorry toll* (taking mileage into account) as applied in Switzerland and planned in Germany by 1st January 2005 substituting the mileage independent vignettes. In Switzerland, the toll money is used for the construction of railway tunnels and the railway noise remediation programme
- Parking management, a measure with potentially high efficiency.

On the other hand, in countries such as Germany people get a tax reduction depending on the length of their way to work. This subvention promotes or at least supports commuting.

The pedestrian and bicycle friendly city promotes this mobility without emissions via

- The increase of safety (especially through the reduction of the speed of the motorised traffic)
- Granting privileges at traffic lights (such as reducing waiting time)
- Increase of the travel speed and the comfort (through direct and shortest ways, sufficient, safe and independent areas such as bike lanes, no pedestrian and biker underpasses)
- Safe parking lots
- Public relations
- Availability of service bikes
- And the personal example in administration and policy.

The example of the city of *Münster in* Westfalia (270,000 inhabitants) shows the potential of transfer to biking (see Table 19.4). The portion of ways by bike was more than tripled within 18 years.

With traffic restrictions, traffic volumes can locally be reduced. But this contributes only then to traffic avoidance if the traffic is not merely shifted to other roads.

The traffic ban of the historic Old Town of *Lübeck* (216,000 inhabitants) from 10 to 18 Uhr (with exemptions for its inhabitants, taxis and public transport) reduced the levels between 4 and 6 dB(A) (Table 19.5).

19.5.2 Measures for the Shift to Low Emission Sources

This measure is implemented in the following forms:

- Shift of the motorised private traffic (cars, etc.) to busses and rail vehicles of public transport
- Shift of long-distance travels to rail traffic (freight, passengers)

[21] For the external costs of the noise exposure, there are some estimates:

- UBA estimates costs in 1993 to be 10.4 billion .DM per year for the road and 4.5 billion DM for the rail (Huckestein, B, Verron H "Externe Effekte des Verkehrs in Deutschland" in: Fortbildungszentrum Gesundheits- and Umweltschutz "Mobilität um jeden Preis?", 382. Seminar, 611.1995.
- *INFRAS/IWW* estimates the costs in 1995 to be 5.69 billion € for the cars, 2.61 billion € for the trucks, 0.404 billion € rail passenger and 0.325 billion € rail freight transport („External Costs of Transport", March 2000).

[22] See the official website on the congestion charge, https://www.cclondon.com/whatis.shtml# "Congestion charging is a way of ensuring that those using valuable and congested road space make a financial contribution. It encourages the use of other modes of transport and is also intended to ensure that, for those who have to use the roads, journey times are quicker and more reliable. The London scheme requires drivers to pay £5 per day if they wish to continue driving in central London during the scheme's hours of operation."

Table 19.5 Modal split in Zürich (% of the journeys)

	Walking	Cycling	Public transport	Car
Zürich 1989	25	4	42	29
Stuttgart 1989	29	5	23	35 + 8[a]
Essen 1996	27	5	15	42 + 11

[a]First number: drivers, second number : passengers

- Shift to low emission vehicles within one vehicle category (for example, via incentives)
  The shift can be promoted by
- Land use planning
  The lines and stations/nodes of public transport as nuclei of urban development
- Freight logistics
  (Freight stations near to consumers, cargo transport centres with connection to the rail net)
- Economic instruments
  Reduction of the charges for public transport with respect to the costs of cars and trucks (so-called environmental tickets, rail cards, job tickets, etc.)
- Promotion of combined transport forms
  Walk/Bike/Park and Ride, combined road/rail transport, car-sharing + public transport, transport of bikes in trains, trams and busses
- Privileging public transport
  (Areas, bus lanes, priority at traffic lights)
- Improved offers in public transport
  (Frequency, density of stops and stations, service, punctuality. . .)
- Public relations, information.

The example of the city of *Zürich* (385,000 inhabitants) shows the potential for the use of public transport if the latter is promoted and privileged, expressed in the modal split (% of the journeys). By the way, among European major cities Zürich has the lowest portion of private car travels.

Measures for the shift of traffic aimed at noise reduction must be based on the knowledge of quieter alternatives. In [32], the environmental impacts of the different transport means have been compared in relation to the same *transport performance:* with respect to noise buses are right now the quietest means of transport (2–4 dB(A) lower exposures than from cars; whereas trams operating in streets generate 2 dB(A) higher exposures than cars, but trams are supposed to be less annoying than cars).

19.5.3 Reduction of the Emissions

Noise emissions are reduced mainly by technical measures at the sources and by low noise operation. For road traffic noise, this means technical measures at the vehicles and the road surfaces as well as low noise driving manners.

19.5.3.1 Technical Measures

Technical measures for the noise reduction generally are not applied automatically (e.g. the demand for vehicles, which create less noise into the environment is limited); therefore, they must be induced by the following instruments:

- Statutory law (highest degree of obligations)
  e.g., noise limits for vehicles and their components
  e.g., reception regulations for new roads
  (incentives for application of low noise road surfaces) (see Chap. 4)
  e.g., bans for noisy vehicles
- Subventions and duties (voluntary)
  e.g., financial incentives for the purchase of low noise vehicles
  (vehicle taxes, compare the introduction of the catalytic converter in Germany)
- Incentives for low noise vehicles (voluntary)
  e.g., exemptions from traffic bans with respect to times and areas
- Information on low noise products
  e.g., via an Environmental Label ('Blue Angel' in Germany) for such products
- *Preference* for low noise machinery and vehicles within the environmentally friendly procurement of local authorities
- *Construction* of low noise road surfaces.

Traffic bans for noisy vehicles as well as the last four instruments are also in the responsibility of the communities and shall therefore be explained.

Incentives for Low Noise Vehicles [33]

Incentives for low noise Vehicles are a measure, which has been applied in communities. The most common implementation is the exemption of low noise vehicles from bans (in Germany according to § 45 STVO (see Chap. 4)). In Germany this measure was first introduced in Bad Reichenhall. Here the bans for trucks – already existing since 1954 – were abolished

Fig. 19.5 Traffic sign in Heidelberg for the Truck Noise Control Zone (second step)

or eased in October 1981 for 'low noise trucks'. In November 1984, the 'low noise vehicle' was defined in the Annex XXI of the German Road Traffic Licencing Ordinance (StVZO) thus establishing the legal basis for such incentives.

In December 1989, Austria introduced a night time lorry ban (22:00 to 5:00) with the exemption of low noise version (with a definition slightly different from the German one) on the transit expressways crossing the Alps, which above all caused a considerably increased offer of low noise truck types on the market. In the framework of a pilot project of the German Environmental Protection Agency (UBA) the city of Heidelberg introduced lorry bans with the exemption for low noise lorries in several quarters of the town in two steps in 1991 and 1994 ('Truck noise control zones' see Fig. 19.5). Since 1999, this instrument was applied in one street with high exposures in the city of Berlin.

The effectiveness of the measure depends on the percentage of trucks of the total traffic volume and on the year of introduction. The noise emission limits of new truck types since 1996 differ only slightly from the corresponding limits of the 'low noise truck' as defined in 1984. At least this measure substantially contributed to the fact that the today's 'normal truck' corresponds to the former low noise truck.

The German Länder Berlin and Baden-Württemberg subsidised the purchase of low noise trucks.

Information on Low Noise Products

The purchase/procurement and the operation of low noise products imply the knowledge of this segment of the market. The most important identification of low noise products in Germany is the Environmental Label. The example of the label for vehicles tyres shall demonstrate the importance of this instrument. The Environmental Label RAL – UZ 89 for low noise and fuel saving road vehicle tyres of 1997 demands a noise emission level of 72 dB(A), which is worldwide the most ambitious limit for tyres and which is up to 4 dB(A) below the limits in the Directive 2001/43/EG [27] (see Chap. 4) (if taking into account the different rounding rules even up to 6 dB(A)). Furthermore, due to the prescribed low rolling resistance one can reduce fuel consumption by 5% with labelled tyres.

Labelled low noise products should be preferred in public procurement. Information on these products should be part of the public awareness campaigns accompanying the implementation of noise action plans.

Preference for Low Noise Equipment and Vehicles in the Environmentally Friendly Procurement of Local Authorities

In the framework of urban noise reduction concepts, it is important that the local authorities themselves set a good example. They can do this in the procurement of equipment and vehicles for the urban institutions. In Germany, the manual 'Umweltfreundliche Beschaffung' ('Environmentally friendly procurement') makes some recommendations [34]. The manual proposes among others that the noise emission from cars should be 5 dB(A) below the EU-limits (p. 177). Low noise lorry for urban services (with the Environmental Label RAL – UZ 59 a and b) and tyres (RAL – UZ 89) should be preferred.

Construction of Low Noise Road Surfaces

For roads, squares and railway lines, which are in the communal responsibility the local authorities can directly influence the emissions (rolling noise from road and rail-bound vehicles).The following rules should be observed in the design of road surfaces:

- The application of paving stones can result in a considerable increase of levels (up to 10 dB(A)). For vehicle speeds above 20 km/h low noise paving stones should be applied. There are concrete paving stones on the market, which are not noisier than asphalt surfaces. They are relatively large, without chamfers and as smooth as possible but with a certain micro roughness and a diagonal lay-out. Above 50 km/h, no paving stones should be used.
- For inner-urban roads with speeds above 70 km/h fine-grained asphalt road surfaces with open voids (DA = 'drainage asphalt', porous asphalt) should be installed, e.g. DA 0/8 (i.e. with a maximum grain size of 8 mm). These are currently the most silent surfaces; the open voids help to reduce the high frequency aerodynamic parts of the rolling noise, the so-called air-pumping. With respect to mastic asphalt, the reductions are about 4 dB(A). Further reductions (up to additional 4 dB(A)) through optimised drainage asphalts seem to be possible.[23] Drainage asphalts lose their noise reducing properties in the course of time due to the clogging of the voids. Then they have to be renewed.
- Unfortunately, drainage asphalts on urban roads with lower speeds have not been effective, here the clogging is relatively fast. Currently two layer drainage asphalts are investigated but the prove of permanent reductions is still missing. Therefore, fine-grained stone mastic asphalts are currently the best and most stable solution (i.e. SMA 0/5, maximum grain size 5 mm).

19.5.3.2 Low Noise Driving Manners

Figure 19.4 shows the principal potential of a low noise driving pattern. This is characterised by

- A driving pattern with low engine speed
- At reduced vehicle speed
- Changing into higher gears as soon as possible, changing into lower gears as late as possible.
- And accelerating and braking as little as possible.

The reduction potential of a driving pattern with low engine speed is about 6 dB(A) with respect to an average driving behaviour for pass-by levels during acceleration.

A reduction of the upper speed limit from 50 to 30 km/h will reduce pass-by levels by 5–8 dB(A), depending on the chosen gears. The reduction of the equivalent levels is 2 dB(A) lower.

In practice, the effectiveness of speed restrictions is lower, it depends on the degree of observance. According to the German calculation scheme RLS-90, a reduction of the allowable speed from 50 to 30 km/h will reduce the equivalent level by only 2 dB(A). The introduction of 'speed 30' (Tempo 30) should, therefore, be accompanied by speed damping road lay-out and additional traffic management regulations

[23] s. S. Ullrich. Noise reduction potential of motorway pavements. International Workshop Further Noise Reduction for Motorised Road Vehicles, Umweltbundesamt 17./18.Sept. 2001.

Table 19.6 Noise reduction through speed reduction

	Noise reductions for cars – comparison with original situation (50 km/h) in dB(A)	
	Noise emissions (pass-by levels)	Noise exposure (equivalent levels)
'Traffic calming zones' (maximum speed 7 km/h)	Up to 6	Up to 4
30 km/h zone (Tempo-30-Zone)	Up to 5	Up to 3
Speed reduction at major roads (compliance with 50 km/h)	Up to 5	–

such as equal right-of-way for crossing roads. The combination of all the regulatory and constructive measures has been called 'traffic calming',[24] this concept has intensively been tested in the German model project 'Area wide traffic calming' [35]. Table 19.6 shows the resulting possible noise reductions through speed restrictions.

In [36], road lay-out elements, which are suitable for speed reductions are presented and evaluated.

19.5.4 Measures in the Propagation path

These measures types consist of four essential elements:

- Increase of the distances between source and receiver (including the spatial shift of traffic)
- Shielding of the source (or the receiver) through constructive measures (encapsulation, barriers, etc.)
- Increase of the absorption in the propagation space
- Prevention of reflections, especially for the façades averted from the source.

19.5.4.1 Noise Reduction Through Increase of Distances

As it is well known the sound pressure levels with increased distance from the source are reduced due to the increase of the enveloping surface of the sound wave ('geometric' sound reduction). For point sources, the reduction is 6 dB(A), for line sources 3 dB(A) per doubling of the distance . In addition, the sound pressure levels are reduced by air absorption. Above all this measure is applied in the allocation of sources and sensitive forms of land use according to §50 BImSchG and in urban land use planning. Hereby sources should be grouped together as much as possible (major roads and industrial areas, etc.); adjacent forms of land use should be allocated according to their degree of sensitivity.

In inner-urban areas, this measure is due to the limited and expensive space-generally only implementable in the form of shifting traffic flows. Often the concentration of traffic volumes on roads with high exposures will be justified because a considerable level reduction in the minor roads corresponds to an only slight level increase along the major road.[25] But such an approach complicates the noise reduction at roads with high exposures, which has priority for health reasons. Concentration of traffic volumes should, therefore, be implemented only on traffic lines without sensitive dwellings or in conjunction with noise reducing measures.

Also, the construction of ring roads for the relief of cross-town links belongs to this type of measure. Here, one has to compare the reductions at the cross-town link with the new exposures along the ring road. There are no binding procedures for that, recommended is the calculation of the annoyed persons for the two planning alternatives.[26]

19.5.5 Shielding

Industrial sources can be shielded through encapsulations and enclosures, traffic lines through buildings, noise barriers, walls and tunnels or through the

[24] Also see http://www.trafficcalming.org/.

[25] If from a minor road with a traffic volume of 6,000 vehicles/day 50% are shifted to a major road with 60,000 vehicles per day, the reduction at the minor road is 3 dB(A), the increase at the major road 0.2 dB(A) instead.

[26] The number of highly annoyed persons can be calculated according to the VDI-Guideline 3722 "Wirkungen von Verkehrsgeräuschen" ("Effects of traffic noise") if the noise exposure of the affected persons is known.

geometry of the cross-section (road in a cut, etc.). The effects of shielding along traffic lines can be calculated with the schemes mentioned above (see Sect. 3.3.2). The use of barriers is limited in inner-urban areas.[27]

Especially through complete enclosures such as tunnels high reductions can be achieved but generally they are also very expensive. A special form of a tunnel is its combination with dwelling above it such as the dwellings built around and above an inner-urban motorway in Berlin (Schlangenbader Straße).

19.5.5.1 Increase of the Sound Absorption in the Propagation Space

The use of absorbing coatings or elements in the propagation space contributes to the noise reduction. Most important application is the coating of noise barriers with highly absorbing materials in order to avoid reflections. Tramways can be built in the form of lawn tracks. But often the effects of absorbing materials or elements are overestimated as in the case of plants.

19.5.5.2 Prevention of Reflections

In the case of scattered buildings, one has to prevent that façades within the acoustical shadow get reflections from other buildings as this may lead to a considerable increase of levels.

19.6 Measures at the Receiver

These measures comply

- Noise control in building sand rooms in the true sense through improvement of their sound insulation
- Orientation of rooms within dwelling according to their sensitivity
- And the shielding effect of buildings or parts of them.

The sound insulation of rooms also yields protection against sources in the building itself, such as noise from neighbours or installations within buildings.

The limits of the sound control in buildings has already been mentioned: the outdoor living areas remain unprotected. Sound insulation windows as most important application of the noise control in buildings are, however, very effective compared to their costs. Therefore, they are applied in Germany for the noise control of new roads according to the Traffic Noise Ordinance [12] when active measures are not justifiable as well as in the noise remediation programmes, here on an equal basis with active measures.

The dimensioning of the sound insulation of buildings along newly built or substantially altered traffic lines follows the regulation in [18], information on the minimum requirements for sound insulation in buildings against sources within and outside the buildings are provided by the German standard DIN 4109 'Sound insulation in building construction'. This standard is supplemented by the VDI-guideline 4100 'Sound insulation of buildings' which defines improved levels of sound protection (see also [37]).

Noise control through orientation of the rooms [38] within buildings aims at the use of rooms with a quiet façade – i.e., which are in the acoustal shadow – for sensitive activities such as sleeping. In this sense, the construction of groups of dwellings in such a way that they enclose quiet courtyards is a very efficient means to guarantee reception levels, which allow a nearly sleep disturbance even along roads with high exposures (reductions between 30 and 35 dB(A)).

[27] Here, a shielding through –for example – industrial buildings is preferable.

References

1. Bundesministerium für Umwelt, Naturschutz und Reaktorsicherheit (BMU) (2000) Umweltbundesamt (UBA) Umweltbewusstsein in Deutschland 2000 Berlin 2000
2. Umweltbundesamt (2000) Jahresbericht 1999 Berlin
3. Wende H, Ortscheid J, Kötz WD, Jäcker-Cüppers M, Penn-Bressel G (1998) Schritte zur Reduzierung gesundheitlicher Beeinträchtigungen durch Straßenverkehr in: Bundesministerium für Umwelt, Naturschutz und Reaktorsicherheit Gesundheitsrisiken durch Lärm Tagungsband zum Symposium Bonn
4. Senatsverwaltung für Stadtentwicklung Berlin (2002) Digitaler Umweltatlas Berlin http://www.stadtentwicklung.berlin.de/umwelt/umweltatlas/
5. Umweltbundesamt (1989) Lärmbekämpfung '88 Erich-Schmidt-Verlag, Berlin
6. Der Rat von Sachverständigen für Umweltfragen (1999) Pressemitteilung vom 31. August 1999 zum Sondergutachten 1999 "Umwelt and Gesundheit"
7. WHO (1999) Executive Summary of the Guidelines for Community Noise, Genf
8. Sechste Allgemeine Verwaltungsvorschrift zum Bundes-Immissionsschutzgesetz Technische Anleitung zum Schutz

gegen Lärm – TA Lärm vom 26.August 1998 Gemeinsames Ministerialblatt 49. Jahrgang, Nr. 26, 503–515
9. DIN 18005 Teil 1 (1987a) Schallschutz im Städtebau; Berechnungsverfahren
10. Bundesminister für Verkehr (1990) Richtlinien für den Lärmschutz an Straßen; Ausgabe 1990 (RLS-90)
11. Deutsche Bundesbahn – Bundesbahn-Zentralamt (1990) Information Akustik 03 Richtlinie zur Berechnung the Schallimmissionen von Schienenwegen – Schall 03 – Ausgabe 1990
12. Sechzehnte Verordnung zur Durchführung des Bundes-Immissionsschutzgesetzes (Traffic Noise Ordinance – 16. BImSchV) vom 12. Juni 1990 BGBl. I S. 1036
13. Directive of the European Parliament and of the Council of 25 June 2002 relating to the assessment and management of environmental noise Official Journal of the European Communities L 189/12ff 18 July 2002
14. Verordnung über die bauliche Nutzung der Grundstücke (1990) (Baunutzungsverordnung BauNVO) BGBl. I S. 132
15. DIN 45682 (1997) Schallimmissionspläne Entwurf Juni 1997
16. Raumordnungsgesetz (ROG) (1997) BGBl. I S. 208
17. Bundes-Immissionsschutzgesetz (BImSchG) (1974) in der Fassung der Bekanntmachung vom 14. Mai 1990 BGBl. I S. 880
18. Vierundzwanzigste Verordnung zur Durchführung des Bundes-Immissionsschutzgesetzes (Verkehrswege-Schallschutzmaßnahmenverordnung- 24. BImSchV) vom 4. Februar 1997 BGBl. I S. 172
19. Unterausschuss Lärmbekämpfung des Länderausschusses f'r Immissionsschutz (1992) Musterverwaltungsvorschrift zur Durchführung des $47a BImSchG Aufstellung von Lärmaktionsplänen
20. Baugesetzbuch (BauGB) (1997) BGBl. I S. 2141
21. Beiblatt 1 zur DIN 18005 Teil 1 (1987) Schallschutz im Städtebau; Berechnungsverfahren Schalltechnische Orientierungswerte für die städtebauliche Planung
22. Bönnighausen G (1995) Die Berücksichtigung des Lärms in der Bauleitplanung – Aufstellungsgrundsätze für Bebauungspläne in: Lärmkontor GmbH Informations- and Beratungssystem noise (INFOSYS) 1997 (CD-Rom)
23. Bundesminister für Verkehr (1981) Vorläufige Richtlinien für straßenverkehrsrechtliche Maßnahmen zum Schutz der Bevölkerung vor Lärm (Lärmschutz-Richtlinien-StV) vom 6.11.1981 Verkehrsblatt 1981: 428ff
24. Richtlinie 70/157/EWG des Rates vom 6. Februar 1970 zur Angleichung der Rechtsvorschriften der Mitgliedstaaten über den zulässigen Geräuschpegel und die Auspuffvorrichtung von Fahrzeugen
25. Richtlinie 87/56/EWG des Rates vom 18.Dezember 1986 zur Änderung der Richtlinie 78/1015/EWG zur Angleichung der Rechtsvorschriften der Mitgliedsstaaten über den zulässigen Geräuschpegel und der Auspuffanlagen von Krafträdern Amtsblatt L 24 vom 27 Januar 1987
26. Richtlinie 92/97/EWG des Rates vom 10. November 1992 zur Änderung der Richtlinie 70/157/EWG zur Angleichung der Rechtsvorschriften der Mitgliedstaaten über den zulässigen Geräuschpegel und die Auspuffvorrichtung von Kraftfahrzeugen Amtsblatt L 371 vom 19.Dezember 1992, S0001–0031
27. Richtlinie 2001/43/EG des Europäischen Parlaments und des Rates vom 27. Juni 2001 zur Änderung der Richtlinie 92/23/EWG des Rates über Reifen von Kraftfahrzeugen and Kraftfahrzeuganhängern und über ihre Montage Amtsblatt L 211 vom 8.April 2001
28. Richtlinie 2000/14/EG des Europäischen Parlaments und des Rates vom 8. Mai 2000 zur Angleichung der Rechtsvorschriften der Mitgliedstaaten über umweltbelastende Geräuschemissionen von zur Verwendung im Freien vorgesehenen Geräten und Maschinen Amtsblatt L 162 vom 3.Juli 2000
29. Der Rat von Sachverständigen für Umweltfragen (1994) Umweltgutachten 1994 Für eine dauerhaft-umweltgerechte Entwicklung Februar 1994
30. Prehn M, Schwedt B, Steger U (1977) Verkehrsvermeidung – aber wie? Verlag Paul Haupt Bern Stuttgart Wien 1997
31. Apel D, Lehmbrock M, Pharoa T, Thiemann-Linden J (1997) Kompakt, mobil, urban Stadtentwicklungskonzepte zur Verkehrsvermeidung im internationalen Vergleich DIFU Beiträge zur Stadtforschung 24 Deutsches Institut für Urbanistik
32. Verkehrsclub Deutschland (2001) Bus, Bahn and Pkw im Umweltvergleich Bonn
33. Jäcker M, Rogall H (1993) Benutzervorteile für lärmarme Lastkraftwagen – das Heidelberger Modell Zeitschrift für Lärmbekämpfung 40:161–168
34. Umweltbundesamt (Hrsg.) (1999) Handbuch Umweltfreundliche Beschaffung 4. Auflage Verlag Franz Vahlen, München
35. Bundesministerium für Raumordnung, Bauwesen und Städtebau (Hrsg.) (1992) Bundesministerium für Verkehr (Hrsg.) Bundesministerium für Umwelt, Naturschutz und Reaktorsicherheit (Hrsg.) Forschungsvorhaben Flächenhafte Verkehrsberuhigung Folgerungen für die Praxis 2. Auflage, Bonn, September 1992
36. Richard J, Steven H (2000) Planungsempfehlungen für eine umweltentlastende Verkehrsberuhigung Minderung von Lärm- and Schadstoffemissionen an Wohn- and Verkehrsstraßen UBA-Texte 52/2000 Berlin 2000
37. Kötz WD (2000) Vorbeugender Schallschutz im Wohnungsbau Bundesbaublatt 12/2000: 42–45
38. Sonntag H (1984) Verkehrslärmschutz durch Gebäudeplanung in: P. Klippel u.a, Straßenverkehrslärm-Immissionsermittlung und Planung von Schallschutz expert verlag Grafenau

Flow Noise

20

K.R. Fritz, C.-C. Hantschk, S. Heim, H. Nürnberger, E. Schorer, B. Stüber, and D. Vortmeyer

In this chapter, sound sources are discussed where sound is generated by aerodynamic or hydrodynamic flow or where flow is at least an essential influencing quantity. To start with, sound generation originating from flow should be explained considering some typical examples.

20.1 Sound Generation by Flow

20.1.1 Source Terms

When solving the Lighthill equation [1], it results that a sound field generated by subsonic flow can be created by an arrangement distribution of elementary emitters with monopole, dipole and quadrupole character [see first, second and third integral in equation (17.22)].

Regarding *monopole sources*, a variable volume flow is responsible for sound generation, e.g. a pulsating exhaust flow or a collapsing cavity.

Concerning *dipole sources*, on spatial average no volume is supplied at any time. However, there are alternating forces, e.g. on the surface of rigid objects hit by the flow. These forces are caused by vortex shedding or by irregularities in the incident flow.

If all alternating forces are compensated, like in case of free vortex pairs or free turbulences, the sound sources are of *quadrupole character*.

E. Schorer (✉)
Müller-BBM GmbH, Robert-Koch-Strasse 11, 82152 Planegg, Germany
e-mail: Edwin.Schorer@MuellerBBM.der

In the following, the sound powers P of different sound sources are compared with each other, namely breathing sphere and breathing cylinder as well as sources consisting of point forces and line forces. It is assumed that the length l of the breathing cylinder and the line forces are large compared with the sound wave length, $\lambda_0 < l$, i.e. $k_0\, l \gg 1$ with

$q(t) = q_0\, e^{i\omega t}$ – volume flow in m^3/s
$F(t) = F_0\, e^{i\omega t}$ – force in N
ω – angular frequency in s^{-1}
t – time in s
c – sound velocity in the flow field in m/s
ρ – density of the flow medium in kg/m^3
$k_0 = \omega/c$ – wave number
l – length of the cylinder in m
Δr – distance between the point and line forces in m
a – radius of the breathing sphere or breathing cylinder

20.1.1.1 Monopole

Breathing sphere with $k_0\, a \ll 1$ (three-dimensional source):

$$P = \frac{\rho c}{8\pi} q_0^2 k_0^2. \qquad (20.1a)$$

Breathing sphere with $k_0\, a \gg 1$ (one-dimensional source):

$$P = \frac{\rho c}{8\pi} q_0^2 k_0^2 \frac{1}{k_0^2 a^2}. \qquad (20.1b)$$

Breathing cylinder with $k_0\, l \gg 1$ and $k_0\, a \ll 1$ (two-dimensional source):

G. Müller and M. Möser (eds.), *Handbook of Engineering Acoustics*,
DOI 10.1007/978-3-540-69460-1_20, © Springer-Verlag Berlin Heidelberg 2013

$$P = \frac{\rho c}{8\pi} q_0^2 k_0^2 \frac{\pi}{k_0\, l}. \tag{20.2a}$$

Breathing cylinder with $k_0\, l \gg 1$ and $k_0\, a \gg 1$ (one-dimensional source):

$$P = \frac{\rho c}{8\pi} q_0^2 k_0^2 \frac{\pi}{k_0\, l} \frac{2}{\pi k_0 a}. \tag{20.2b}$$

20.1.1.2 Dipole

Point force (three-dimensional source):

$$P = \frac{1}{12\pi} \frac{F_0^2}{\rho c} k_0^2. \tag{20.3}$$

Line force with $k_0\, l \gg 1$ (two-dimensional source):

$$P = \frac{1}{12\,\pi} \frac{F_0^2}{\rho c} k_0^2 \frac{3\pi}{2 k_0 l}. \tag{20.4}$$

20.1.1.3 Lateral Quadrupole

Parallel point forces with $k_0\ \ \Delta r \ll 1$ (three-dimensional source):

$$P = \frac{1}{60\,\pi} \frac{F_0^2}{\rho c} (k_0 \Delta r)^2 k_0^2. \tag{20.5}$$

Parallel line forces with $k_0\ \Delta r \ll 1$ and with k_0 $l \gg 1$ (two-dimensional source):

$$P = \frac{1}{60\,\pi} \frac{F_0^2}{\rho c} (k_0 \Delta r)^2 k_0^2 \frac{15\pi}{8 k_0 l}. \tag{20.6}$$

It results from an analysis of the dimensions that for the sound power P of the three different elementary sources, the following velocity dependencies can be expected if the typical angular frequency ω (or k_0) characterising the noise and the velocity u (e.g. typical flow velocity) are proportional, if the sources are "compact" and, in addition, the following proportionalities are valid for volume flow and force: $q_0 \sim u$ *and* $F_0 \sim u^2$.

20.1.1.4 Two-Dimensional Flow

$$P \sim \rho c^3 \cdot M^{2m+1}. \tag{20.7}$$

20.1.1.5 Three-Dimensional Flow

$$P \sim \rho c^3 \cdot M^{2m+2}. \tag{20.8}$$

$m = 1$ – *volume flow (monopole)*
$m = 2$ – *alternating force (dipole)*
$m = 3$ – *free vortices (quadrupole)*

The Mach number gives the velocity in units of the sound velocity,

$$M = \frac{u}{c}. \tag{20.9}$$

A breathing sphere and a breathing cylinder with $k_0\, a \gg 1$ [Eqs. (20.1b) and (20.2b)] are one-dimensional sound sources with $P \sim \rho c^3 \cdot M^2$.

For Mach numbers $M \ll 1$, the relations (20.1a)–(20.8) show:

- The emitted sound power is determined mainly by the flow velocity. Therefore, the essential sound sources in a flow field can normally be found in the areas of the highest flow velocities.
- The dependency of the emitted sound power from the flow velocity increases with an increasing order m of the sound sources. This is the reason why the direct sound emission of quadrupole sources is practically of no importance in water – where Mach numbers usually are very low.
- Although alternating velocities in the near field of a sound source usually increase with the order m of the source, the sound power emitted in the far field is reduced.

Disturbing objects can considerably influence the sound field. In which way the sound field is influenced depends on whether the disturbing object is located in the near field or in the far field of the sound sources. Disturbing objects in the far field led to a dispersion of the emitted sound power without changing its quantity. By contrast, disturbing objects in the near field of a sound source can lead to an increase or to a reduction in sound emission. Such a behaviour can be observed not only for resonating objects located near to a sound source, but also for disturbing objects which are rigid and immobile. A distinctly increased sound generation by disturbing objects is restricted to sources with a strong near field, i.e. especially to aerodynamic and hydrodynamic sound sources. This disturbance causes an increased transformation of the flow energy in the near field into sound energy.

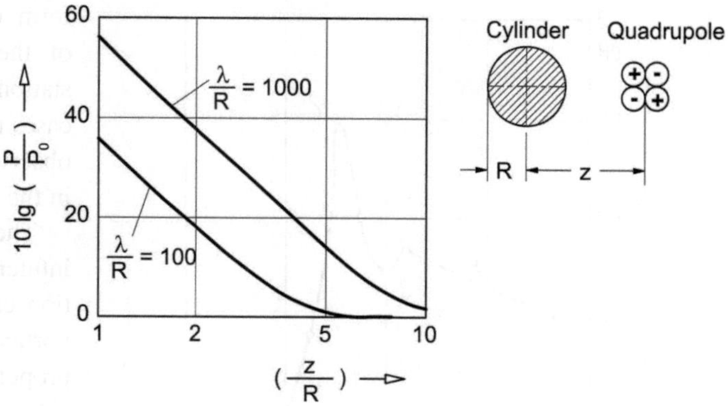

Fig. 20.1 Influence of a circular cylinder near a two-dimensional, lateral quadrupole on the emitted sound power [2]. P sound power with cylinder, P_0 sound power without cylinder, z distance between the axis of the cylinder and the punctiform quadrupole source, R radius of the cylinder, λ sound wavelength

The effect of a disturbing object is shown in Fig. 20.1. A rigid circular cylinder is put near a lateral two-dimensional quadrupole. In this way, the emitted sound power can increase considerably, whereas the sound emission of a monopole does not change.

With regard to the facts stated above, the following rules should be observed when noise reduction is aimed for:

- Reduce flow velocities
- Put areas with antiphase movements as close to each other as possible, i.e. increase the order m of the sound source
- For a sound source with high order m (e.g. quadrupoles), avoid disturbing objects closest to the source (Fig. 20.1)

20.1.2 Cavitation

As soon as in the area of a flowing fluid a certain critical pressure, which is approximately the vapour pressure p_D of the fluid, is reached or if the pressure falls below this critical pressure, cavities filled with gas (vapour) will develop if there are seeds.[1] This effect, which is typical for fluids, is called *cavitation*, more exactly flow cavitation. If the ambient pressure rises above the critical pressure, these cavitation bubbles will suddenly collapse (Fig. 20.2). Locally, this leads to very high-pressure peaks (for strong cavitation more than 10^5 bar [5], a reason for material damage). Apart from that, this results in a considerable noise generation. Usually, cavitation causes a characteristic crackling broadband noise with only a small content of low frequencies. Extensive bibliographical references on this subject can be found in [6–9].

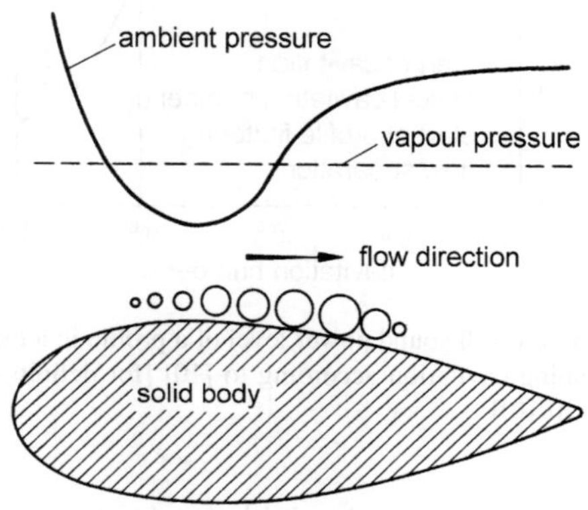

Fig. 20.2 Development and collapse of cavitation bubbles at a profile in the flow [3, 4]

20.1.2.1 Cavitation Initiation

Figure 20.3 shows that in case of a profile exposed to a fluid flow sound emission can increase by more than 40 dB when cavitation starts [10] (similar observations can also be made for other flow processes [11]). The cavitation number σ used in Fig. 20.3 is defined by:

$$\sigma = \frac{p_s - p_D}{0.5\,\rho_0\,u^2}. \qquad (20.10)$$

p_s – static pressure of the undisturbed flow in Pa
p_D – vapour pressure of the fluid in Pa
ρ_0 – density of the undisturbed fluid in kg/m^3
u – typical flow velocity (e.g. velocity of approach of the profile) in m/s

[1] In water, cavitation seeds are above all air bubbles, which are usually there.

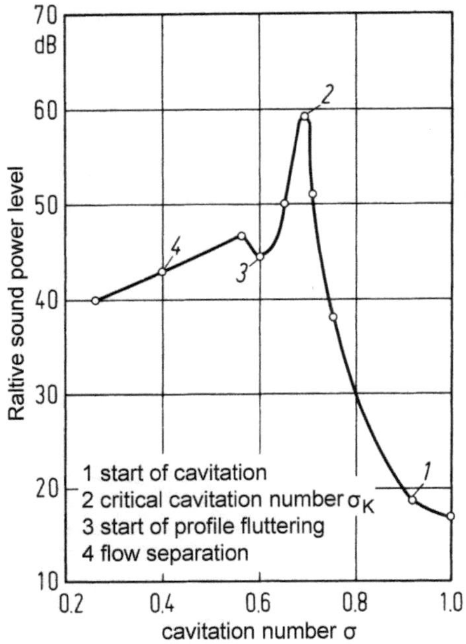

Fig. 20.3 Overall sound power level of a profile in a fluid flow for beginning cavitation according to [10] [for definition of σ, see Eq. (20.10)]

The point of cavitation initiation for a certain flow process, e.g. in water, depends not only on the cavitation number, but also very much on the air content (seed content) of the water. For a high air content, cavitation starts already at low flow velocities, more precisely at higher cavitation numbers. In this case, cavitation noise increases only gradually with a decreasing cavitation number. For a low air content, cavitation starts a little bit later. However, it will then lead to a sudden increase in noise.

20.1.2.2 Types of Cavitation

In a stationary non-separated flow, two types of cavitation can be distinguished, i.e. *bubble cavitation* (travelling bubbles [12], transient cavities [13]) and *sheet cavitation* (fixed cavitation [12], steady state cavities [13]). In [6], these two types of cavitation are defined as follows:

Bubble cavitation: The cavitation bubbles move continuously with the flow along the wall while expanding, imploding and decaying. The life of an individual bubble can be observed cinematographically [14].

Sheet cavitation: A multitude of small bubbles – the life of an individual bubble cannot be observed – creates two-phase areas attached to the wall in the form of sheets, heaps or streaks. The outer shape of these two-phase areas seems to be essentially stationary to the naked eye. However, in isolated cases it could be proven by means of cinematographic observations that there are strong temporary changes in the shape of the cavitation areas [15–17].

The appearance of cavitation can be considerably influenced by flow separation and turbulence. Cavitation can also be observed in the centre of separated vortices (e.g. in the tip vortex and root vortex of a ship propeller).

20.1.2.3 Theory

For beginning cavitation or a small seed content, there are mainly individual cavitation bubbles which hardly influence each other (see Fig. 20.4, conditions 1 and 2).

The temporal behaviour of the volume of individual bubbles can be approximately determined by means of the Rayleigh–Plesset differential equation [4, 18, 19]. If the time dependency $V(t)$ of the bubble volume is known, the monopole term of sound emission, which is essential for Mach numbers $M \ll 1$, can be determined [3, 20]. For the sound pressure p in the distance r from the bubble cloud, the following equation is valid:

$$p(r,t) = \frac{\rho_0}{4\pi r} \frac{\partial^2}{\partial t^2} V(t - r/c_0). \qquad (20.11)$$

ρ_0 – density of the undisturbed fluid in kg/m^3
t – time in s
c_0 – sound velocity in the undisturbed fluid in m/s

For a fully developed cavitation, usually proper bubble clouds with many small bubbles can be found (see Fig. 20.4, conditions 4 and 5).

As shown in Fig. 20.5, initially the sound power emitted from a cavitation area increases with the number of cavitation bubbles (individual bubbles) and decreases when a maximum is exceeded (bubble cloud). It can be concluded from this behaviour that the sound emission of an individual bubble is strongly influenced by the neighbouring bubbles.

For the theoretical treatment of the fully developed cavitation, it would hardly be practical and often not even possible to observe the life of individual bubbles. Therefore, some authors [21–25] regard the cavitating fluid as continuum. Thus, Eq. (1.28) for a one-phase

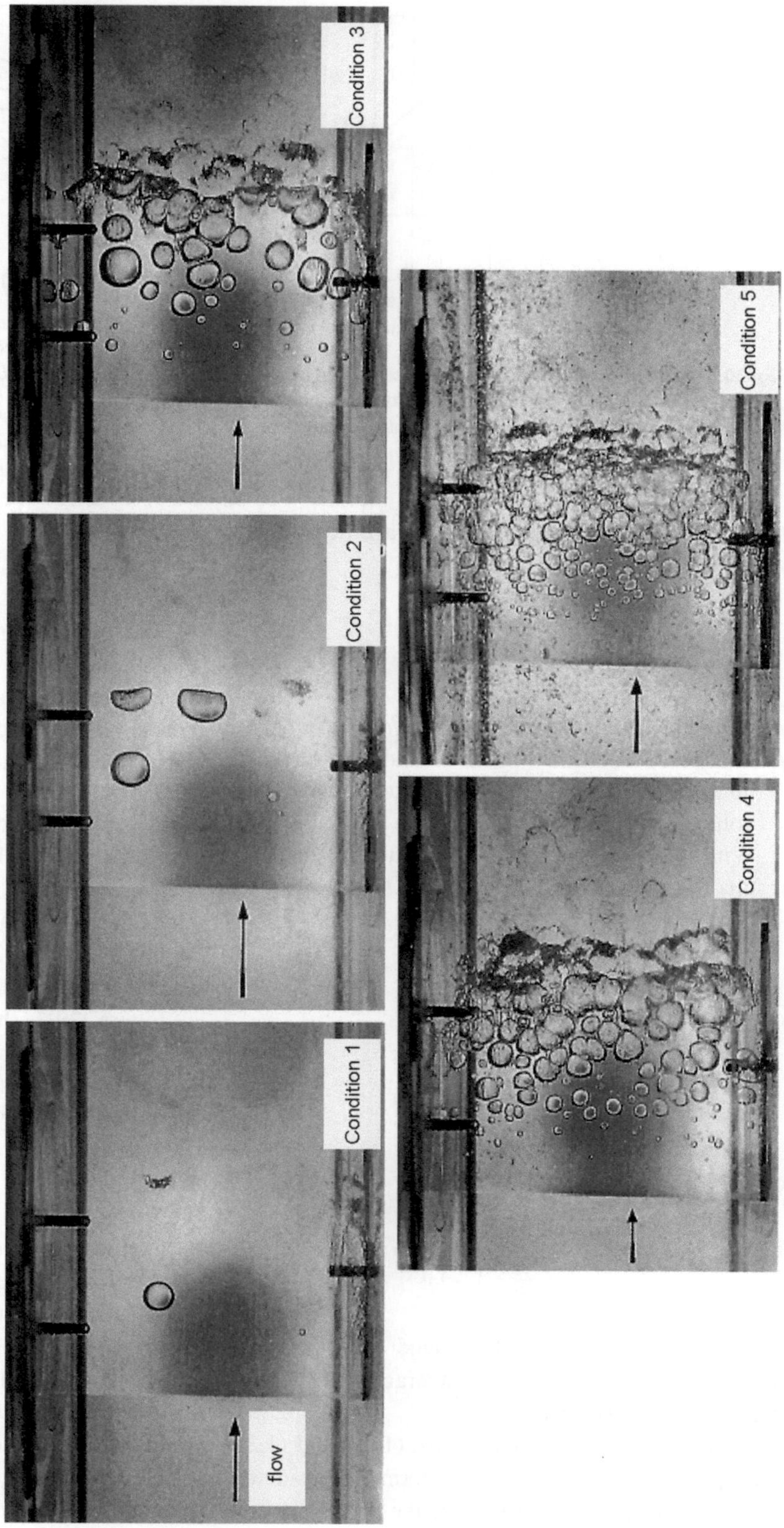

Fig. 20.4 Development of a cavitation cloud at a profile in a fluid flow for five different seed contents of the fluid and constant cavitation number according to [10] (see also Fig. 20.5)

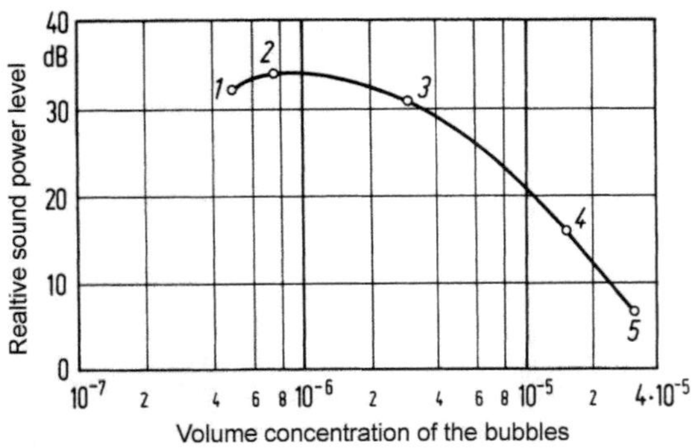

Fig. 20.5 Overall sound power level of the cavitation noise of a profile in a fluid flow depending on the volume concentration of the bubbles according to [10]. The bubble cloud structures of the conditions 1 to 5 are shown in Fig. 20.4

fluid continues to be valid if an average density ρ is introduced [22, 23]:

$$\rho = \rho_0(1 - \beta) - \beta\rho_B, \tag{20.12}$$

ρ_B is the density of the bubbles in kg/m^3.

β is a place and time function and describes the volume occupied by the cavitation bubbles per volume unit ($0 < \beta < 1$).

With $\rho_B \ll \rho_0$ and $\beta \ll 1$, the following equation results from Eq. (20.12)

$$\rho = \rho_0(1 - \beta)$$

or

$$\frac{d\rho}{dp} = \frac{d\rho_0}{dp} - \rho_0\frac{d\beta}{dp}$$

or also

$$\frac{1}{c^2} = \frac{1}{c_0^2} - \rho_0\frac{d\beta}{dp}. \tag{20.13}$$

In this equation, c_0 is the sound velocity in the undisturbed fluid and c the average local sound velocity in the two-phase flow.

If Eq. (20.13) is inserted in Eqs. (1.28) and (1.29) and only the essential monopole term is considered, the following equation results for the sound pressure p in the distance r of a cavitation area with the volume V:

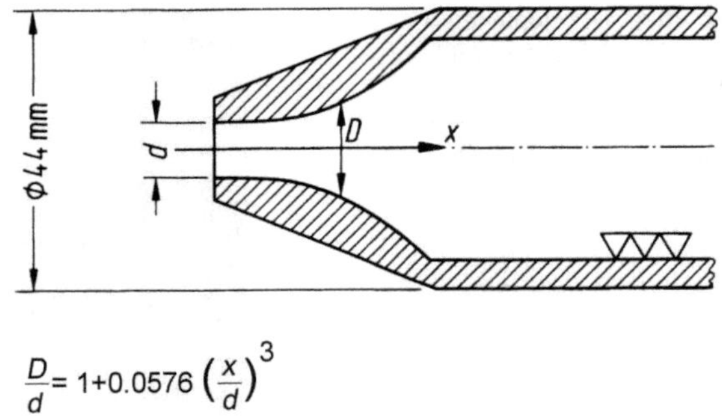

Fig. 20.6 Investigated nozzle type without pressure minimum according to [27]. $D/d = 1 + 0.0576\,(x/d)^3$

$$\begin{aligned} p(r,t) &= \frac{1}{4\pi}\int_V \frac{1}{r}\frac{\partial}{\partial t}\left[\left(\frac{1}{c_0^2} - \frac{1}{c^2}\right)\frac{\partial p}{\partial t}\right]dV \\ &= \frac{\rho_0}{4\pi}\int_V \frac{1}{r}\left[\frac{\partial^2\beta}{\partial t^2}\right]dV. \end{aligned} \tag{20.14}$$

The functions in the square brackets [...] have to be taken at the time $t - r/c_0$.

This relation shows that the noise of a cavitation area is caused by an expansion and collapse of more or less independent steam bubbles (monopole sources).

20.1.2.4 Cavitating Nozzles

At cavitating nozzles, extensive sound measurements were carried out [11, 26]. The form of the nozzles was chosen in a way that no pressure minimum could occur at the nozzle (Fig. 20.6) and thus cavitation within the nozzle was impossible (free cavitation).

Figure 20.7 shows the water sound power P standardised with $\rho_0\, d^2\, u^4/c_0$ as function of σ_k/σ with:

ρ_0 – density of water kg/m^3

c_0 – sound velocity in the undisturbed, not cavitating water in m/s

u – flow velocity at the narrowest position in the nozzle in m/s

u_k – critical flow velocity, when the respective nozzle starts to cavitate, in m/s

d – diameter of the nozzle (6.3 mm $\leq d \leq$ 38 mm) in m/s

σ – cavitation number in Eq. (20.10)

σ_k – critical cavitation number, when the respective nozzle starts to cavitate

For the investigated nozzles, the critical cavitation number σ_k is between 0.3 and 0.7.

For the water sound power P of the investigated nozzles, the following approximation is valid:

Fig. 20.7 Standardised total water sound power level of water nozzles according to Fig. 20.6

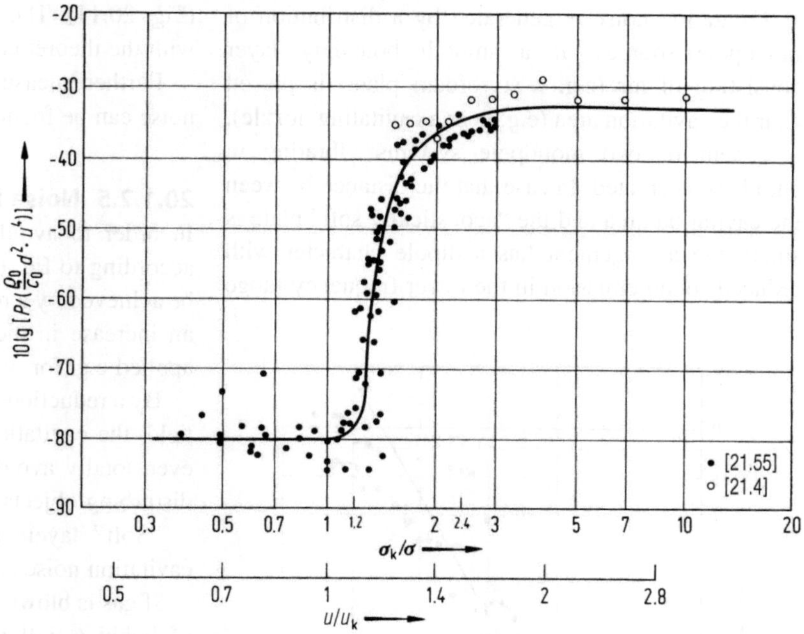

No cavitation

$$P = 10^{-8}\frac{\rho_0}{c_0}d^2u^4 \quad \text{for} \quad \sigma > 0.8\,\sigma_k. \tag{20.15}$$

Fully developed cavitation

$$P = 5 \cdot 10^{-4}\frac{\rho_0}{c_0}d^2u^4 \quad \text{for} \quad \sigma < 0.4\,\sigma_k. \tag{20.16}$$

The acoustic efficiency ξ is the ratio of emitted sound power P and flow power P_F:

$$\xi = \frac{P}{P_F} \tag{20.17}$$

with

$$P_F = \frac{\rho_0}{2}u^3\frac{\pi}{4}d^2. \tag{20.18}$$

It follows from Eqs. (20.15) and (20.16):
No cavitation

$$\xi = 2.5 \cdot 10^{-8}\frac{u}{c_0} \quad \text{for} \quad \sigma > 0.8\,\sigma_k. \tag{20.19}$$

Fully developed cavitation

$$\xi = 1.3 \cdot 10^{-3}\frac{u}{c_0} \quad \text{for} \quad \sigma < 0.4\,\sigma_k. \tag{20.20}$$

For cavitation, the acoustic efficiency of a water nozzle (according to Fig. 20.6) increases by a factor of >30,000 (45 dB).

The sound spectra measured during cavitation are broadband and have a distinct maximum [11, 28], approximately at the frequency:

$$f_m = 0.03 \cdot \frac{u}{d}10^{5.5\times\sigma} \quad \text{for} \quad \sigma < \sigma_k. \tag{20.21}$$

In line with theory [3], with an increasing flow velocity u of the jet the frequency of the maximum level f_m shifts towards lower frequencies (Fig. 20.8). All 1/3 octave sound power spectra measured during cavitation ($\sigma < 0.8\,\sigma_k$), which are related to the total water sound power, are within the hatched area in Fig. 20.9.

For simple orifice-type injectors, the cavitation area already develops within the nozzle, and cavitation already starts at cavitation numbers of about 3.0, i.e. at about 2.5 times lower jet velocities than for the nozzles according to Fig. 20.6.

If, for example, a cylinder is put in front of the nozzle, as shown in Fig. 20.10, the following observations can be made:

- Cavitation already starts at higher cavitation numbers (lower jet velocities)
- The sound pressure levels increase sharply
- The maximum in the sound pressure spectra shifts to lower frequencies

Cavitation noise is generated by a distribution of monopole sources. If a smooth boundary layer consisting of air (e.g. a styrofoam plate) is placed near the cavitation area (e.g. near a cavitating nozzle), a system of two monopole systems vibrating in antiphase is created. In case that the distance between the cavitation area and the "acoustically soft" plate is small, the arrangement has a dipole character with reduced sound emission in the lower frequency range (Fig. 20.11). The measured level reduction is in line with the theoretically expected reduction.

Further measuring results concerning cavitation noise can be found e.g. in [10, 29–33].

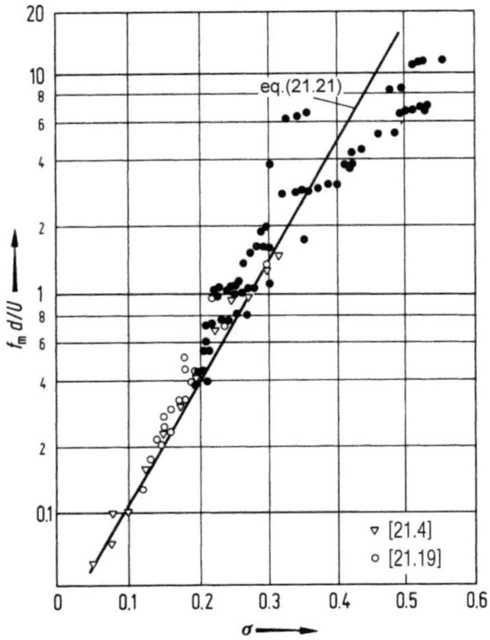

Fig. 20.8 Frequency of the maximum sound emission f_m for cavitating nozzles. d diameter of the nozzle, u flow velocity at the narrowest position of the nozzle

20.1.2.5 Noise Reduction

In order to avoid cavitation, the cavitation number according to Eq. (20.10) must be increased. This can be achieved by a reduction in the flow velocity u or by an increase in the static pressure p_s; this method is applied e.g. for water pipe fittings.

By a reduction of turbulences and vortices in a flow field, the cavitation noise can be reduced or perhaps even totally avoided. If possible, there should be no disturbing objects in the cavitation area (Fig. 20.10).

"Soft" layers near a cavitation area can reduce cavitation noise considerably (Fig. 20.11).

If gas is blown into the cavitation area, the collapse of bubbles will be cushioned which often leads to a considerable noise reduction.

20.1.3 Circular Cylinders Exposed to Flow

If a rigid object is approached by flow of a gaseous or liquid medium with subsonic velocity, it can be observed from a certain Reynolds number onwards, which is characteristic for the respective process [Eq. (20.22)], that vortices separate from the surface of the object (Fig. 20.12). This vortex shedding leads to local and temporal pressure changes at the surface of the rigid object and thus also sound is generated.

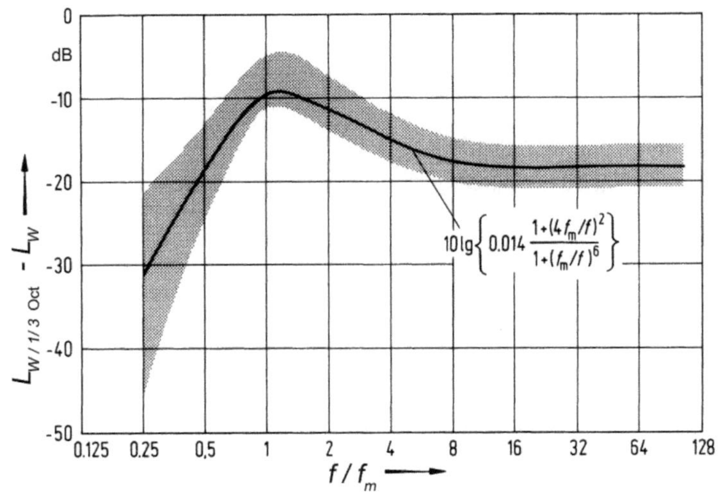

Fig. 20.9 Maximum spread of all 1/3 octave sound power spectra at cavitating nozzles according to Fig. 20.6 for $\sigma < 0.8\,\sigma_k$. L_W total sound power level, $L_{W/1/3oct}$ 1/3 octave sound power level, f_m see Eq. (20.21)

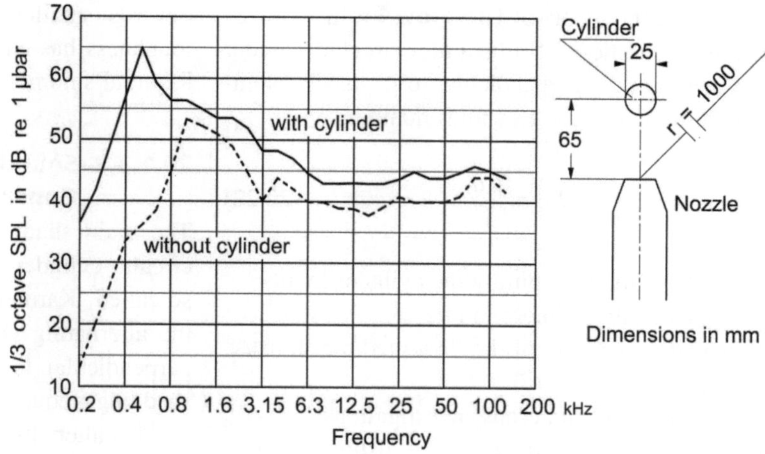

Fig. 20.10 Influence of a cylinder on the underwater noise level of a cavitating water nozzle according to Fig. 20.6. Diameter of the nozzle $d = 14.1$ mm, cavitation number (related to the jet velocity) $\sigma = 0.2$, diameter of the cylinder 25 mm, measuring distance $r = 1$ m

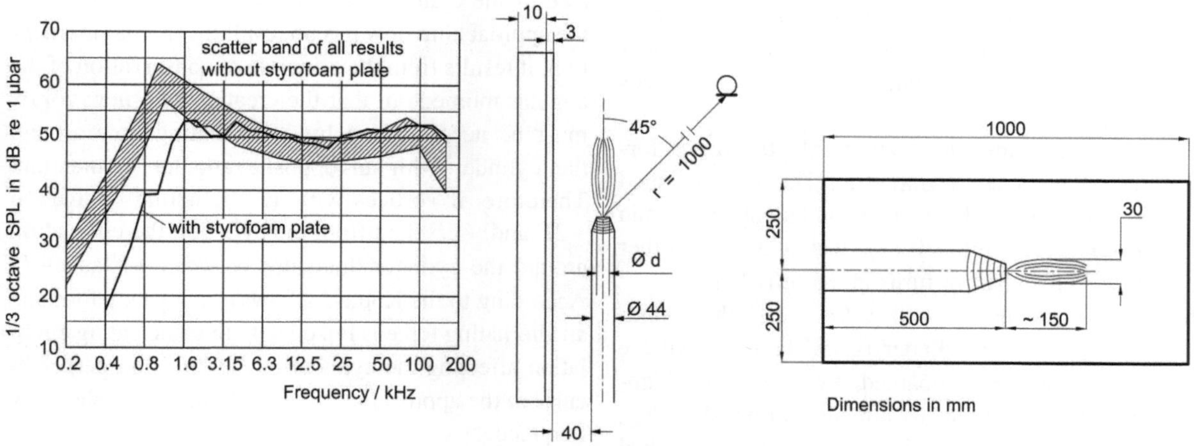

Fig. 20.11 Influence of a styrofoam plate on the underwater noise level of a cavitating water nozzle according to Fig. 20.6. Diameter of the nozzle $d = 14.1$ mm, cavitation number $\sigma = 0.2$, measuring distance $r = 1$ m

Fig. 20.12 Kármán vortex street [28]

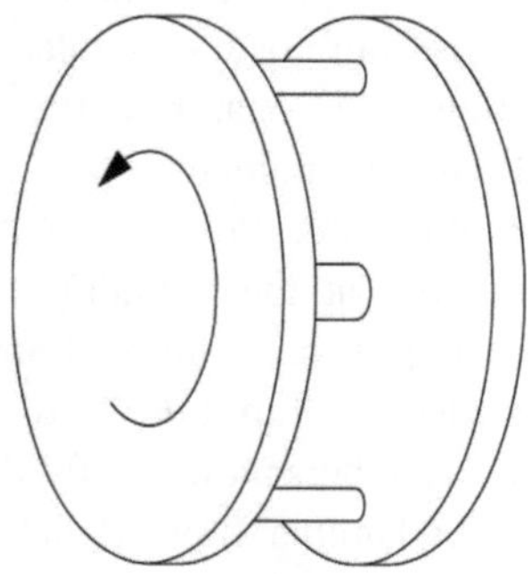

Fig. 20.13 Incident flow to circular cylinders in a rotor

The vortex shedding was investigated experimentally and theoretically in detail on the basis of a rigid circular cylinder approached by flow perpendicular to its axis and the connected sound field (see also Blake [7], Chap. 4). The experimental investigations were carried out either in wind tunnels or at rotors (Fig. 20.13), e.g. consisting of two discs with circular cylinders fixed in between [34, 35].

20.1.3.1 Description of the Flow Field

The form of the flow field behind a circular cylinder approached by flow perpendicular to its axis is essentially determined by the Reynolds number.

$$Re = \frac{u\,d}{\nu}. \tag{20.22}$$

u – velocity of the flow hitting the cylinder in m/s
d – diameter of the cylinder in m
ν – kinematic viscosity of the flow medium in m^2/s (Table 20.1)

On the basis of numerous investigations [36–40], following types of flow fields can be distinguished:

(a) In the symmetric area ($4 < Re < 40$), there is no vortex shedding. Only one stable pair of vortices is created, which does not separate from the cylinder and, thus, does not create either alternating forces or sound.
(b) In the stable area $40 < Re < 200$, the vortex formation becomes asymmetric. The vortices shed regularly, move downstream and create a Kármán vortex street (Fig. 20.12), which in the further course falls apart or turns into a turbulence.
(c) In the unstable area $400 < Re < 10^5$, the boundary layer of the cylinder is still laminar, but the vortices are accompanied by low- and high-frequency disturbances already during shedding, i.e. the regular shedding of the free individual vortices is superimposed by turbulent changes in velocity leading to a rapid disintegration of the vortices (turbulence) in the further course. The transition of the more or less regular vortex street to full turbulence is nearer to the cylinder the larger the Reynold's number.
(d) In the overcritical area $Re > 3 \times 10^5$, the flow is already turbulent at the cylinder. Nevertheless, several authors [40–42] could still observe a periodic vortex shedding (up to $Re = 8 \times 10^6$).

The Reynold's numbers indicated for the individual areas are valid for laminar flow. If the approaching flow is already slightly turbulent, the four areas shift towards smaller Reynold's numbers [36]. Surface roughness has an effect similar to an increase in the Reynold's number [43].

Table 20.1 Kinematic viscosity ν for air and water

Kinematic viscosity in 10^{-6} m^2/s				
Temperature in °C	0	20	40	60
Air	13.20	15.10	16.90	18.90
Water	1.79	1.01	0.66	0.48

20.1.3.2 Sound Generation (Karman Tone Generation)

The main share of the sound power radiated by a circular cylinder approached by flow results from the so-called "Karman tone" which is caused when during the alternating vortex shedding an alternating force perpendicular to the flow direction with the vortex shedding frequency is created.

This alternating force can be illustrated as follows (Fig. 20.14). The vortices shed alternately from each side of the cylinder carry an angular momentum. As the approaching flow has no resulting angular momentum, it results from the principle of conservation of the angular momentum that the creation of a new vortex must be accompanied by a circulating flow around the cylinder with an opposite angular momentum. Therefore, if vortices with an alternating circulation -2Γ and $+2\Gamma$ shed from the cylinder, the circulation around the cylinder fluctuates between $+\Gamma$ and $-\Gamma$. According to the Kutta–Joukowsky buoyancy formula, an alternating force is linked with this fluctuating circulation affecting the cylinder in the direction perpendicular to the approaching flow with the vortex shedding frequency.

The vortex shedding frequency and thus the frequency of the Karman tone is described by the Strouhal relation [44]:

$$f_{\text{Karman}} = St\frac{u}{d}. \tag{20.23}$$

For $2 \cdot 10^2 < Re < 2 \cdot 10^5$, the Strouhal number St is relatively constant for a laminar flow (Fig. 20.15) and is approximately 0.2.

The "Karman tone" is not a pure tone, but a narrow-band noise with a bandwidth of at least 1/3 octave. The bandwidth of the Karman tone is increasing

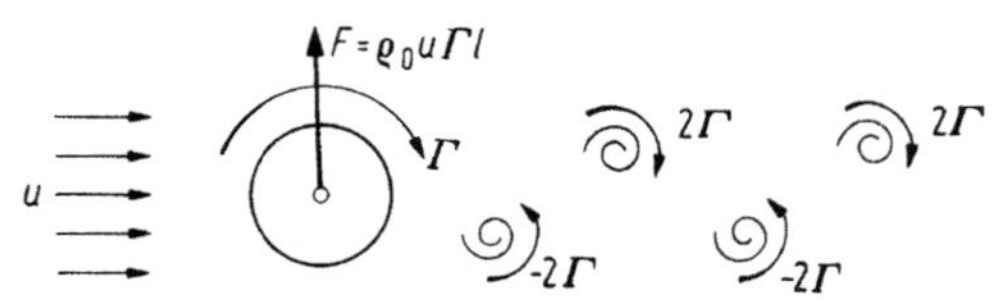

Fig. 20.14 Alternating force affecting a cylinder in a flow in perpendicular direction to the flow

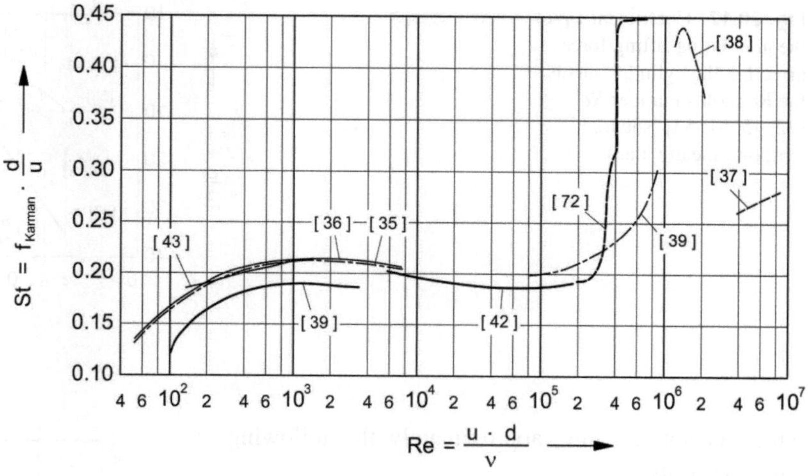

Fig. 20.15 Dependence of the Strouhal number St on the Reynolds number *Re* [38–42, 45–47]

with the Reynolds number and the turbulence of the approaching flow. In addition, outside of the rotational axis, Doppler broadening also becomes apparent in rotors (Fig. 20.13) [34, 35].

Apart from periodic vortex shedding, alternating forces (i.e. sound emission with dipole character) at obstacles in a flow also occur when the flow is not completely laminar and, thus, leads to small changes in the flow resistance (force parallel to flow direction) and the buoyancy (force perpendicular to the flow direction). Finally, a sound excitation mechanism with quadrupole character must be mentioned which can be explained by the free turbulence in the slip stream behind the solid (see Sect. 20.1.4).

20.1.3.3 Calculation of Sound Emission

As stated above, the vortex shedding at a cylinder perpendicular to the flow direction leads to an alternating force F on the cylinder in the same direction. The coherence length of vortex shedding strongly depends on the construction of the investigated arrangement at the ends of the cylinder [34, 48]. On the basis of the investigations in [34], for the arrangement shown in Fig. 20.16 for $1 < l/d < 10, Re < 10^5$ and $M < 0.6$ ($M = u/c$ Mach number), the vortex shedding might be coherent practically over the entire length of the cylinder (two-dimensional vortex shedding) and thus the amplitude and phase of the alternating force F should also be nearly constant over the whole length l of the cylinder.

In this case, for the sound pressure p at a large normal distance r from the cylinder axis, the following equation results:

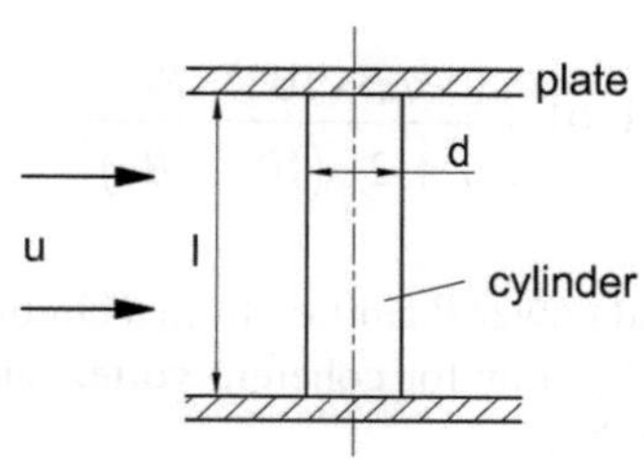

Fig. 20.16 Incident flow on a rigid cylinder clamped on both sides

$$p(r,\theta,t) = \frac{1}{4\pi r}\sin(\theta)\frac{\partial}{\partial r}F(t - r/c). \qquad (20.24)$$

where θ is the angle between the direction of the approaching flow and the line connecting the cylinder and the reference point. The alternating lifting force F is often replaced by the coefficient C_{L}:

$$F(t) = C_{\mathrm{L}}(t)\, l\, d \frac{1}{2}\rho u^2. \qquad (20.25)$$

The squared temporal mean $\overline{C_{\mathrm{L}}^2}$ was measured in flow channels (Fig. 20.17) by several authors [48–55].

The tolerance range according to [34] indirectly resulted from sound emission. As it can be seen here, there is a good correspondence with the directly measured values of C_{L}. The only exception is the results of Gerrard [51] for small Reynolds numbers. This difference can possibly be explained by the fact that Gerrard carried out his measurements for a very small turbulence of the incident flow [36, 55–57]. According to Fig. 20.17, for a turbulence of the flow

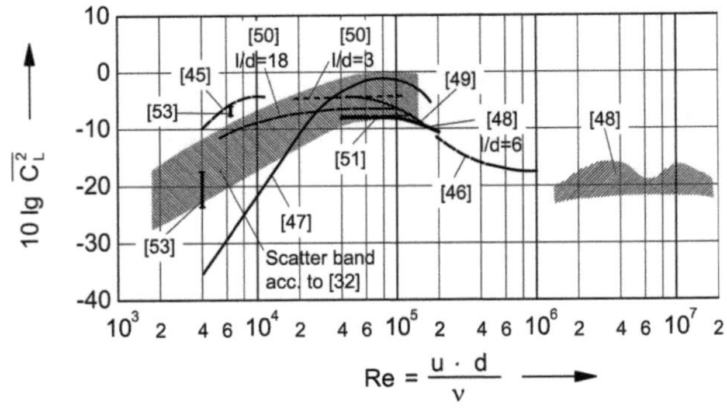

Fig. 20.17 Coefficient C_L of the alternating lifting force affecting the cylinder versus the Reynolds number Re [32, 45–51, 53] (squared temporal mean value)

which is not too low, approximately the following equation results:

$$\overline{C_L^2} \approx 0.01 + \frac{7.5 \cdot 10^{-6} \cdot Re}{1 + 2 \cdot \left(10^{-5} \cdot Re\right)^3}. \tag{20.26}$$

The sound power P emitted from a circular cylinder approached by flow for coherent vortex shedding can be described by:

$$P = \frac{\pi}{16} \rho \, St \, \overline{C_L^2} \, l \, d \, M^2 \, u^3 \, G(k_0 \, l). \tag{20.27}$$

For the function $G(k_0\, l)$, the following simplification is valid:

$$G(k_0\, l) \approx \begin{cases} \frac{2}{3\pi} k_0\, l, & k_0\, l \leq \frac{3\pi}{2} \\ 1, & k_0\, l > \frac{3\pi}{2} \end{cases}. \tag{20.28}$$

where $k_0 = 2\pi f_{\text{Karman}}/c$ is the wave number with f_{Karman} according to Eq. (20.23).

The converted flow power P_F for a circular cylinder in the direction perpendicular to the flow is:

$$P_F = C_D \frac{1}{2} \rho \, u^3 \, l \, d. \tag{20.29}$$

$C_D \approx 1.1$ resistance coefficient of a cylinder approached by flow (see e.g. [7], Chap. 4).

For the acoustic efficiency ξ, it follows from Eqs. (20.17) and (20.26) to (20.29):

$$10^{-3} M^2 < \xi < 2 \cdot 10^{-2} M^2 \quad \text{for} \quad 4d/l < M. \tag{20.30}$$

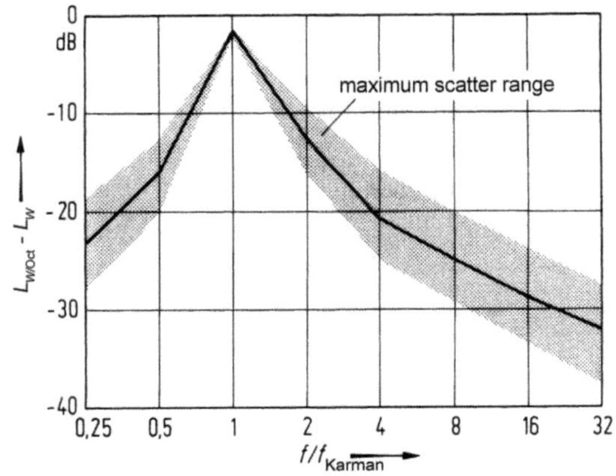

Fig. 20.18 Standardised mean octave sound power spectrum of a circular cylinder in an air flow according to [34]. $L_{W/\text{Oct}}$ octave sound power level, L_W total sound power level

Applying the noise spectra stated in [34], the standardised octave sound power spectra shown in Fig. 20.18 approximately result for the circular cylinder approached by an air flow for coherent vortex shedding.

If the cylinder is not approached perpendicularly to its axis but at an angle ϕ, in Eqs. (20.23), (20.27) and (20.28) u must be approximately replaced by $u \cos \phi$.

If the vortex shedding is incoherent, i.e. for a very large ratio l/d or for a changed fixation of the cylinder [34, 48] or for flow direction angles other than perpendicular to the cylinder axis, in the frequency range of the Karman tone and thus also in total, a smaller sound emission results than might be expected according to Eqs. (20.27) and (20.28).

For a known coherence length l_c of the vortex shedding, for $l_c < l$ the emitted sound power can be estimated with Eqs. (20.27) and (20.28), if in the iteration Eq. (20.28) l is replaced by l_c. For the

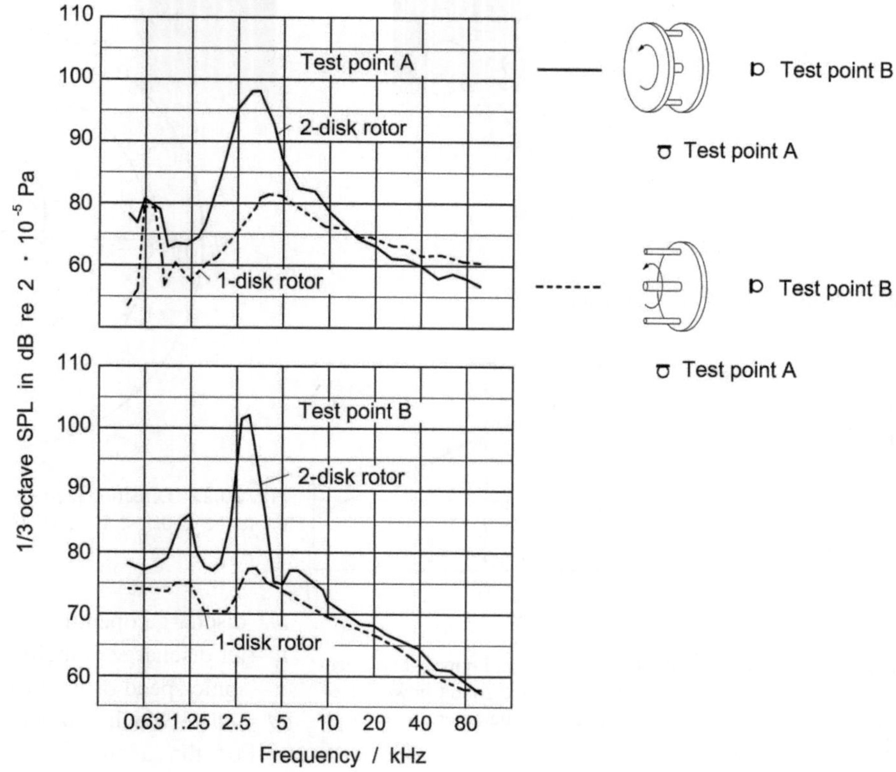

Fig. 20.19 Measured 1/3 octave sound pressure level for rotors with two discs or one disc. Diameter of the rotor disc: 200 mm, diameter of the cylinders: 5 mm, length of the cylinders: 28.5 mm, track speed of the cylinders: 105 m/s, measuring distance from rotor axis: 0.5 m

arrangement in Fig. 20.15, the following equation is valid:

$$l_c = \frac{l}{\sqrt{1 + 0.01 \cdot (l/d)^2}}. \qquad (20.31)$$

For $4 < l/d < 10$ and for a circular cylinder fixed on one side, about 13 dB lower sound levels were measured compared with the cylinder fixed on both sides [34]. This is shown in Fig. 20.19 for the rotor according to Fig. 20.13.

An increased sound emission can occur in case that the frequency of the Karman tone corresponds to the eigenfrequency of the cylinder [55] or to the eigenfrequency of the pipeline where the cylinder is located. For example, very high sound levels can occur in heat exchangers if there is a correspondence between the vortex shedding frequency of the tube bundles and the eigenfrequency of the adjacent pipe or duct (see [58] with many recommendations for further reading).

20.1.3.4 Noise Reduction

Noise emission of circular cylinders subject to flow can be reduced e.g. by the following measures:

- Reduction in the velocity of the incident flow [see Eq. (20.27)]
- Spoiling of coherent vortex shedding, e.g. by a corresponding arrangement at the ends of the cylinder [34, 48] (Fig. 20.19) or by an angular flow direction to the cylinder at an highest possible angle
- Selection of a streamlined form of the object in the flow (vortex shedding only starts at a higher Reynolds number *Re*, small C_L)

20.1.4 Turbulent Open Jet

Apart from the ideal case of a dancing vortex pair [59, 60], the details of sound generation by free vortices are subject to intensive investigations. The focus of attention is on the turbulent open jet (Fig. 20.20) as this is an important sound source regarding jet engines [61–66].

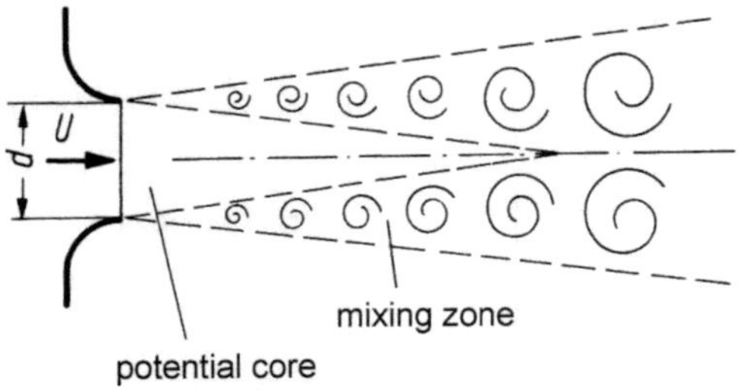

Fig. 20.20 Geometry of the open jet

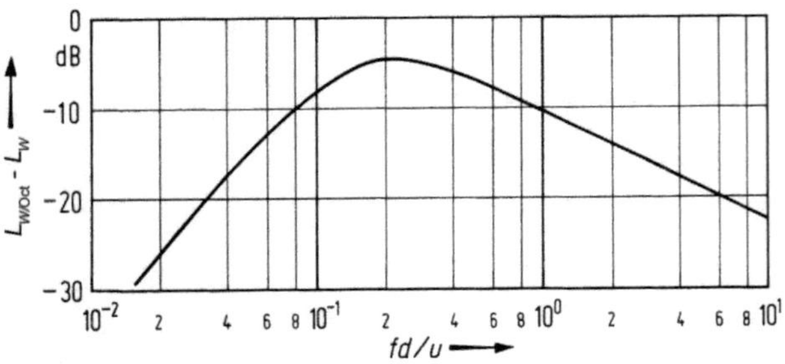

Fig. 20.21 Standardised mean octave sound power level spectrum of a turbulent open jet. $L_{W/Oct}$ octave sound power level, L_W total sound power level, d nozzle diameter in m, u jet discharge velocity in m/s

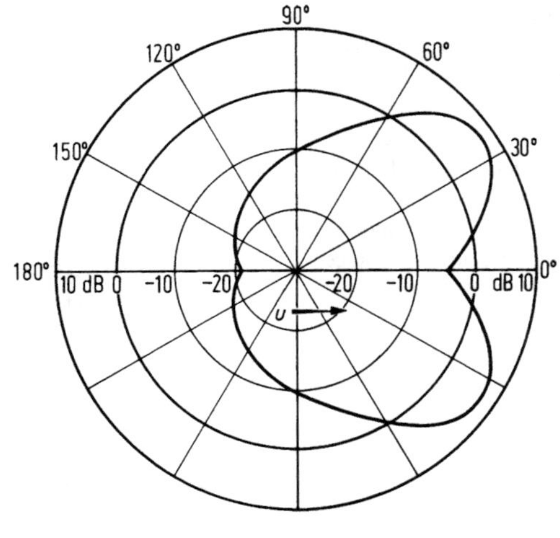

Fig. 20.22 Directional characteristic of a turbulent open jet in the range of $0.7 < M < 1.6$ ($M = u/c$ Mach number)

While for the sound generation mechanisms discussed so far the position of the sound source coincides with a distinct limiting surface (fluid – cavity for cavitation, solid – flowing medium for the Karman tone), for the turbulent open jet the position of sound generation is much harder to locate.

Concerning the turbulent open jet, the sound is generated in a so-called mixing zone, i.e. in a distance equal to several nozzle diameters outside of the discharge opening (Fig. 20.20). If there are no disturbing objects in the potential core and in the mixing zone, the generated noise is a very broadband noise (Fig. 20.21) with a distinct directional characteristic (Fig. 20.22).

In the range of $0.7 < M < 1.6$, the sound power P of a turbulent open jet increases with the eighth power of the flow speed [67]; in the range $M > 2$ it increases with the flow speed to the third power [68]. For the sound power, the following approximation is valid:

$$P = \begin{cases} 5 \cdot 10^{-5} \rho S u^3 M^5, & M \leq 1.82 \\ 10^{-3} \rho S u^3, & M > 1.82 \end{cases}. \quad (20.32)$$

ρ – density of the flowing medium at the jet opening in kg/m^3
S – discharge opening cross-sectional area in m^2
u – jet discharge velocity in m/s
c – sonic speed of the flowing medium in m/s
$M = u/c$ – Mach number, see Eq. (20.9)

For the acoustic efficiency ξ, it follows from Eqs. (20.17) and (20.32):

$$\xi = \frac{P}{0.5 \rho u^3 S} = \begin{cases} 10^{-4} M^5, & M \leq 1.82 \\ 2 \cdot 10^{-3}, & M > 1.82 \end{cases}. \quad (20.33)$$

In the range $M < 0.7$, open jet noise (three-dimensional quadrupole sources) is very often masked by other noise sources, especially by turbulent emanation. The sound power of the noise generated in this way increases with the sixth power of the discharge velocity. In such a case, the location of the sound generation is not the mixing zone but the exhaust opening (three-dimensional dipole sources). For diameters d of the discharge opening which are not too large ($d < 200/u$, d in m, u in m/s) the spectrum is shown quite well in Fig. 20.21.

If there is a disturbing object in the potential core, mostly distinct individual tones develop which lead to an increase in the sound power by 10–20 dB[2] [70].

[2] Very loud discrete sounds can also be observed when a jet blows into the opening of a cavity. Connected with this is a rise of temperature in the cavity. Sometimes, dangerously high values can be reached [69].

For supersonic jets, individual tones (screech tones) can also appear without disturbing objects [71, 72]. In both cases, feedback mechanisms are involved, i.e. phenomena similar to the generation of edge tones in flue pipes.

If an open jet hits a large even plate, the radiated sound power is slightly increased compared with an undisturbed jet. If the plate is at a distance of 5 nozzle diameters from the discharge opening, the increase is about 5 dB and for 10 diameters it is only about 2 dB. At smaller distances of the plate, individual tones may be generated.

The different methods of noise reduction for jet engines (application of a large bypass ratio, installation of "corrugated nozzles") will be briefly discussed in Sects. 14.1.1.2 and 14.1.1.3.

20.1.5 Turbulent Boundary Layer

For turbulent flow along a wall, high alternating pressures occur in the boundary layer. The thickness of the boundary layer can be estimated by:

$$\delta \approx 0.37 \cdot l \cdot Re^{-0.2}. \qquad (20.34)$$

where l is the length of run from the beginning of the wall and Re is the Reynold's number calculated with the length l and the free flow velocity u [see Eq. (20.22)]. The alternating pressure can be measured with a small pressure receiver installed in the wall [73–80]. From this, standardised spectra result with a course as shown in Fig. 20.23. The mean square of the alternating pressure over all frequencies is approximately

$$\overline{p^2} \approx \left(6 \cdot 10^{-3} \cdot \frac{1}{2} \rho U^2\right)^2 \quad \text{or} \quad \overline{p^2} \approx (2\,\tau)^2 \qquad (20.35)$$

with τ being the local wall shear stress.

The course of the spectrum at low frequencies is somewhat controversial as it is very difficult to separate the pressure fluctuations from other disturbing influences in this area. For high frequencies, the measured alternating pressure strongly depends on the

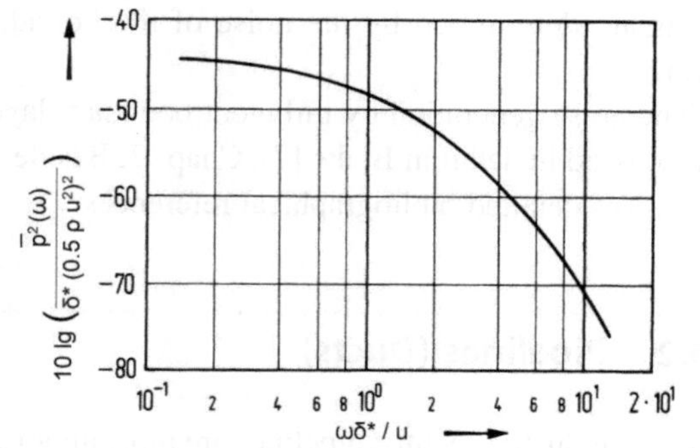

Fig. 20.23 Standardised spectrum of the alternating pressure in a turbulent boundary layer, measured with very small pressure receivers. $\overline{p^2}(\omega)$ mean square of the alternating pressure at angular frequency ω and a bandwidth of $2\pi \cdot 1$ Hz, $\delta^* \approx \delta/8$ size of displacement

dimensions of the applied pressure receiver. Pressure fluctuations in the boundary layer can be assumed to be created because small "turbulence bales" (eddies) with dimensions comparable with the thickness of the boundary layer "roll" along the wall (and are interrupted from time to time by sudden "bursts") [78, 79]. The rolling speed is approximately 60–80% of the free flow velocity. In moving, the turbulence bales fall apart and new ones are permanently created. For this reason, the correlation coefficient between two pressure receivers separated in flow direction rapidly decreases and already reaches zero at a distance of more than 20 δ.

As areas of different pressure are very close to each other in a boundary layer – the correlation length is always small compared with the sound wavelength – the direct sound emission of a turbulent boundary layer (e.g. in front of a completely rigid wall) is usually negligible. Nevertheless, turbulent boundary layers can indirectly lead to a considerable sound emission, that is, if they border on a very light wall which can be slightly set into motion by the alternating pressures. For infinitely large homogenous walls, the resulting movements would have a similar three-dimensional structure as the wall pressure fluctuations [81] and thus do not emit any sound. However at all discontinuities (reinforcements, edges etc.) free waves occur [82–84] emitting much more sound. This can, for example, lead to the fact that inside an airplane (at least in the front part where the engine noise level is low), the sound levels are determined by the noise of the boundary layer.

The noise generation by turbulent boundary layers is discussed in detail in Blake [7], Chap. 7. Besides, it comprises detailed bibliographical references.

20.2 Pipelines (Ducts)

In many industrial plants, pipelines are most important sound sources. The noise generated inside of a pipeline by flow is normally not important for its external sound emission. Pipeline noise is generated, e.g. by fans, compressors, control valves and pumps connected to the pipeline. The generated gas or liquid sound (fluid sound) spreads in the flow medium in the pipeline, excites structure-borne sound vibrations of the pipe walls and is then partly emitted to the outside in the form of airborne sound. Apart from that, the pipe wall can also be excited directly by structure-borne sound transmission from a sound source and by the aerodynamic or hydrodynamic near field behind the sound source. However, these types of excitation are normally of minor importance in practise.

This chapter deals with sound emission of pipelines excited by gas or liquid sound (fluid sound). In the following, the term "pipeline" is used for both circular pipelines and rectangular ducts.

20.2.1 Sound Emission into the Pipeline

Below the following limiting frequencies, only even sound waves can spread in a long, straight pipeline or duct (modes of vibration of zeroth order):

$$\text{Circular pipe} \quad f_G = 0.58\,\frac{c}{d_i},$$
$$\text{Rectangular duct} \quad f_G = 0.5\,\frac{c}{b}. \qquad (20.36)$$

c – sound velocity of the flow medium inside of the pipeline in m/s
d_i – interior diameter of the circular pipeline in m
b – largest side of the rectangular duct in m

Above these limiting frequencies, sound transmission inside of the pipe additionally occurs by numerous other types of waves. Above all in the low-frequency range $f < f_G$, a sound source in the free field and in a pipeline behaves very differently.

Figure 20.24 shows the difference between the sound power P in a circular pipeline and the sound power P_0 in the free field for a monopole source. In this figure, the ratio P/P_0 is shown depending on $k \cdot d_i$ ($k = 2\pi f/c$ wave number, f frequency, d_i interior diameter of the pipe). The parameter is the distance r_0 between the point sound source and the pipe axis.

A monopole in a pipe in the area $k_0 \cdot d_i < 2$ emits a sound power which is distinctly higher than in the free field. The following approximation is valid:

$$P/P_0 \approx 1 + (0.77 f_G/f)^2. \qquad (20.37)$$

This approximation is also valid for dipole and quadrupole sources when averaged over all radiation directions of these sound sources.

20.2.2 Inner Sound Power Level and Sound Pressure Level

The relation of the inner sound power level L_{Wi} (re 10^{-12} W) to the sound pressure level L_i

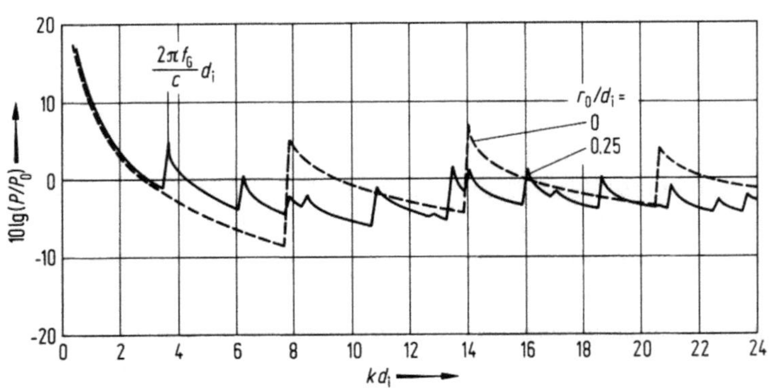

Fig. 20.24 Ratio of the sound power P in a circular pipe and the sound power P_0 in the free field for a monopole and averaged over a bandwidth of $\Delta(k\,d_i) = 0.2$. $k = 2\pi f/c$ wave number, d_i interior diameter of the pipe, r_0 distance between the monopole and the pipe axis

(re $2 \cdot 10^{-5}$ Pa) inside of a pipe is described by the following equation:

$$L_{\mathrm{Wi}} = L_{\mathrm{i}} + 10\ \lg\left(\frac{\rho_0\, c_0}{\rho\, c}\right) + 10\ \lg(S) - K_{\mathrm{d}}. \quad (20.38)$$

$\rho_0\, c_0$ – characteristic impedance of air ($\approx$410 Ns/m^3)
ρ – density of the flow medium in kg/m^3
c – sound velocity in the flow medium in m/s
S – inner cross-sectional area of the pipeline in m^2

In the characteristic K_{d}, it is considered that the different sound waves above the limiting frequency f_G do not all pass normal the cross-sectional area of the pipe. If the sound pressure level L_{i} is measured at the pipe wall, the following approximation is valid [85]:

$$K_{\mathrm{d}} = \begin{cases} 0 & \text{for } f \leq 0.77 \cdot f_G \\ 3 + 10 \lg\left[1 + 0.1 \cdot (f/f_G)^{1.3}\right] & \text{for } f > 0.77 \cdot f_G \end{cases}. \quad (20.39)$$

20.2.3 Sound Power of Pipelines Filled with Gas

At the position $l = 0$ of a very long pipeline filled with gas, an inner sound power $P(0)$ is fed into the pipeline (e.g. by a fan). This inner sound power is more or less continuously decreasing along the pipeline. For pipelines filled with gas, this happens mainly owing to dissipative processes in the flowing medium as well as due to the sound absorption at the pipe wall. A small part of the sound power is reduced when sound is transmitted to the pipe wall and then to the open air by airborne sound emission, to other building elements by structure-borne sound transmission and because of the damping in the pipe material. For the sound power passing through the position l of a pipeline filled with gas, the following equation is valid:

$$P(l) = P(0)\exp\left(-(\alpha + \tau)\frac{U}{S}l\right). \quad (20.40)$$

α – sound absorption coefficient of the flow medium and the pipe wall
τ – sound transmission coefficient of the pipe
U – circumference of the pipeline in m
S – inner cross-sectional area of the pipeline in m^2

Apart from the absorbed sound power inside the pipe, for the sound power emitted per unit of length to the outside of the pipeline at the position l, the following equation is valid:

$$\frac{\mathrm{d}P_{\mathrm{a}}(l)}{\mathrm{d}l} = P(l)\tau\frac{U}{S}. \quad (20.41)$$

For a sufficiently long pipeline filled with gas, it follows for the total airborne sound power emitted to the outside:

$$P_{\mathrm{a}} = \frac{\tau}{\alpha + \tau}P(0). \quad (20.42)$$

The sound transmission coefficient τ and the sound absorption coefficient α are discussed in detail in Sects. 20.2.5 and 20.2.7.

20.2.4 Sound Power of Pipelines Filled with Fluid

For a fluid-filled pipeline, there is a strong coupling between the interior of the pipe and the pipe wall. By means of the Statistical Energy Analysis (SEA), the airborne sound power P_{a} emitted to the outside of the pipeline can be estimated from the fluid sound power P_{e}:

$$P_{\mathrm{a}} = \frac{\eta_{\mathrm{wa}}}{\eta + \eta_{\mathrm{wa}}}P_{\mathrm{e}}. \quad (20.43)$$

η_{wa} – coupling loss factor between the pipe wall (w) and the outside (a),
η – loss factor of the pipe wall by dissipative processes in the pipe wall as well as by structure-borne sound transmission via the pipe brackets to the supporting structure and to building elements connected to the pipe wall with regard to structure-borne sound.

In this equation, it is assumed that the pipeline is sufficiently long and has several bends where a transformation of fluid sound into structure-borne sound is easy. The factor η includes the loss by sound emission from the pipe wall neither to the interior of the pipe nor to the outside of the pipe. The loss factor η and the coupling loss factor η_{wa} are discussed in Sects. 20.2.6 and 20.2.8.

20.2.5 Sound Level Reduction in Pipelines Filled with Gas

With Eqs. (19) and (20) in [86], the sound level reduction ΔL_L per unit of length owing to dissipative processes in pipelines filled with gas (with smooth inner wall) can be calculated, if losses owing to flow effects are neglected. From this relation, for all gases the following simplified numerical value equation results [87]:

$$\Delta L_L \approx 0.13 \frac{1}{\sqrt{S/1\,\text{m}^2}} \sqrt{\frac{f/1\,\text{Hz}}{p_s/1\,\text{Pa}}} \sqrt[4]{\frac{T}{293\,\text{K}}}\,\text{dB/m}. \quad (20.44)$$

S – inner cross-sectional area of the pipe in m²
T – absolute temperature in K
f – frequency in Hz
p_s – pressure in the pipe in Pa

In Fig. 20.25 measured sound level decreases in circular pipelines (at environmental conditions) are compared with the values calculated according to Eq. (20.44).

The level reductions per unit of length stated in Eq. (20.44) are valid for straight pipelines without cross-sectional jumps, bends or junctions. In order to estimate these influencing variables, the statements made in [86] can be applied. The level reduction by a pipe bent is certainly smaller than 1 dB, as long as the sound wave length is larger than the interior diameter of the pipe (compare pipelines 4 and 5 in Fig. 20.25).

The turbulent flow in a pipeline causes an additional sound level decrease [88]. This can be considered approximately by multiplying the right side of Eq. (20.44) by

$$(1 + 11 \times u/c). \quad (20.45)$$

u – average flow velocity in the pipeline in m/s
c – sound velocity in the flow medium in m/s

Between the sound absorption coefficient α and the sound level decrease ΔL_L per unit of length (in dB/m) in a pipeline filled with gas, there is a following connection:

$$\alpha = 0.065 \cdot \sqrt{S} \cdot \Delta L_L. \quad (20.46)$$

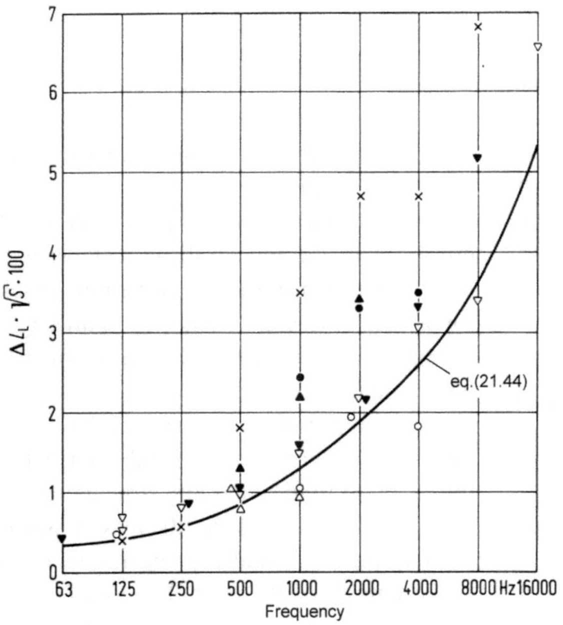

symbol		reference	diameter, mm	measured distance, m
1	△	[85]	50	
2	▲	[83]	75	
3	×	[86]	220	
4	▽		50	42 1)
5	▼		50	42 2)
6	○		200	91
7	●		300	85

1) pipe without elbows
2) pipe as No.4 but with 10 90°-elbows

Fig. 20.25 Sound level reduction in circular pipelines without flow for air at 1 bar and 20°C [88–90]

20.2.6 Sound Level Decrease in Pipelines for Structure-Borne Sound Excitation

In a petrochemical plant under construction, five non-insulated circular steel pipelines filled with air [91] were excited at one end with a tapping machine and the decrease in the octave acceleration levels along the pipelines were determined. For the octaves with mid frequencies of 125–4,000 Hz, a frequency-independent average structure-borne sound level decrease of

$$\Delta L_K \approx 0.24\,\text{dB/m} \tag{20.47}$$

was determined.

On the basis of these results (octaves 125–4,000 Hz) and additional measurements in a test stand (octaves 31.5–8,000 Hz), the total loss factor η of a sufficiently long pipeline made of steel or aluminium

- By dissipative processes in the pipe wall
- By structure-borne sound transmission to building elements connected with the pipe wall with regard to structure-borne sound

can be calculated approximately by:

$$\eta \approx \frac{0.02}{1 + 0.002\,(f/1\,\text{Hz})}. \tag{20.48}$$

This approximation is valid for pipelines without special damping measures.

Decisive for the damping of a pipeline in the lower frequency range is the structure-borne sound transmission from the pipe wall via the pipe brackets to the supporting structure.

20.2.7 Sound Insulation of Pipelines Filled with Gas

According to a suggestion in [92], in order to define the transmission loss R_R for pipelines filled with gas the relation between the sound power P_i passing through the pipe cross section, referred to the inner cross-sectional area S, and the sound power emitted to the outside P_a, referred to the emitting surface $U \cdot l$ (U circumference and l length of the pipeline), can be applied:

$$R_R = 10\,\lg\left(\frac{1}{\tau}\right) = 10\,\lg\left(\frac{P_i}{P_a}\frac{U\,l}{S}\right). \tag{20.49}$$

τ – sound transmission coefficient.

In this equation, the pipeline is assumed to be so short that no sound level decrease occurs in the pipeline, and the interior of the pipe ends is assumed to have a sound absorbing termination.

20.2.7.1 Circular Pipelines

Basic experimental and theoretical investigations regarding the sound insulation of cylinder shells were carried out by Lothar Cremer [93] and Manfred Heckl [94, 95]. Figure 20.26 shows the typical frequency course of the transmission loss for circular pipelines filled with gas. An important quantity characterising the sound insulation behaviour of a pipeline is the so-called ring expansion frequency:

$$f_R = \frac{c_L}{\pi\,d_i}. \tag{20.50}$$

c_L – longitudinal wave velocity in the pipe wall material in m/s

d_i – inner diameter of the pipe in m

For example in [86], Table 20.3, statements on the longitudinal wave velocity of different materials can be found.

Close to the ring expansion frequency, the transmission loss is small. In the frequency range above the ring expansion frequency, the pipeline behaves approximately like an even plate. For frequencies smaller

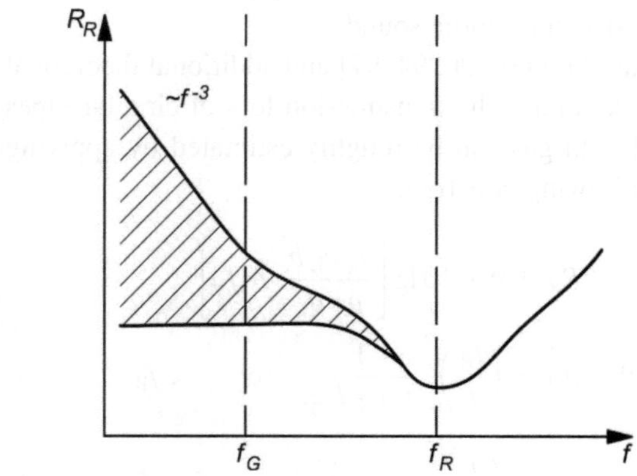

Fig. 20.26 Transmission loss of circular pipes filled with gas

than the ring expansion frequency, the sound insulation of a circular pipe filled with gas strongly depends on the type of the sound field inside of a pipeline (more exactly: on the alternating pressure field).

Below the limiting frequency f_G [see Eq. (20.36)] in a straight and long pipe, only even waves can spread, which theoretically leads to a high transmission loss. However, this high transmission loss is normally not reached in practise as the pipe wall is generally *not* excited by vibration modes of zeroth order (even waves), but by higher modes for which the sound insulation is considerably smaller. The modes of higher order occur near the sound source (near field) and near unbalances in the pipeline (i.e. pipe bends, pipe junctions, pipe contractions, etc.). Therefore, in these areas mainly the pipe wall is excited. If it is possible to prevent a structure-borne sound transmission from these excitation points to the straight parts of the pipeline, it would actually be possible to achieve approximately the theoretically expected high transmission losses in the frequency range below the first limiting frequency f_G – as shown in [96]. Thus, especially in the low-frequency range $f < f_G$ pipe bends lead to a strong reduction in sound insulation of circular pipelines.

In the whole frequency range, the sound insulation increases with an increasing damping of the pipe wall. The structure-borne sound power transmitted to the pipe wall from the interior of the pipe decreases by:

- Dissipative processes in the pipe wall
- Airborne sound radiation to the outside
- Structure-borne sound transmission via the pipe brackets to the supporting structure
- Structure-borne sound transmission to the building elements connected with the pipe wall with regard to structure-borne sound

Based on [86, 90, 94–97] and additional theoretical considerations, the transmission loss of circular pipes filled with gas can be roughly estimated by applying the following equation:

$$R_R = 9 + 10\lg\left[\frac{\rho_w c_L h}{\rho c d_i} \cdot A(f)\right]$$

$$\text{with } A(f) = \left(\frac{f_R}{f}\right)^2 \frac{1}{1 + z\frac{d_i}{l}\frac{f_R}{f}} \quad \text{for } f < f_R$$

$$A(f) = \left(\frac{f}{f_R}\right)^2 \quad \text{for } f \geq f_R \quad (20.51)$$

ρ_w – density of the pipe material in kg/m^3
h – wall thickness of the pipe in m
c – sound velocity of the flow medium in m/s
ρ – density of the flow medium in kg/m^3
f – frequency in Hz
z – number of pipe bends
l – pipe length in m

This approximation is valid on the condition that:

- The sound excitation of the interior of the pipe is broadband.
- The excitation of the pipeline by the acoustic near field of the sound source can be neglected.
- The pipeline is not mounted resiliently.
- The flow velocity is smaller than one-third of the sound velocity.

Equation (20.51) is an approximation, which was deliberately set to the lower limit of existing measuring results with regard to a desirable safety.

In the low-frequency range, transmission losses of circular pipelines are distinctly larger than those of rectangular ducts of the same cross-sectional area and wall thickness.

20.2.7.2 Rectangular Ducts

Based on [98] and complementary experimental and theoretical investigations, the transmission loss of a thin-walled, elastically mounted steel duct (without reinforcement and without a special damping) can be estimated by applying the following equation:

$$R_R = 22 + 10\lg\left[\frac{\rho_w h f}{\rho c}\right] \quad \text{for } f \leq f_g/2. \quad (20.52)$$

ρ_w – density of the duct wall material in kg/m^3
h – wall thickness of the duct in m
c – sound velocity in the flow medium in m/s
ρ – density of the flow medium in kg/m^3
f – frequency in Hz

The critical frequency of coincidence f_g in an even plate of the same thickness as the duct wall (sound radiation to the outside) can be described by:

$$f_g = \frac{c_0^2}{1.8\, c_L h}. \quad (20.53)$$

c_0 – sound velocity of the outside air in m/s
c_L – longitudinal wave velocity in the duct wall material in m/s

If structure-borne sound is noticeably transmitted from the duct wall to other building elements (duct is not elastically mounted), the loss factor of the duct wall is increased and thus also its transmission loss according to Eq. (20.52), especially in the low-frequency range.

20.2.8 Radiation Factor

By means of the radiation factor σ, a connection can be made between structure-borne sound velocity and the sound power of a pipeline radiated to the outside. There is the following relation between the radiation factor σ of a pipeline and the radiation loss factor η_{wa} between the pipe wall and the outside air [Eq. (6.16) in [99]]:

$$\eta_{wa} = \frac{\rho_0\, c_0\, \sigma}{\rho_w\, h\, \omega}. \qquad (20.54)$$

ρ_0 – density of the medium at the outside in kg/m^3
c_0 – sound velocity of the medium at the outside in m/s
ρ_w – density of the pipe wall material in kg/m^3
h – wall thickness of the pipe in m
ω – angular frequency in Hz

20.2.8.1 Circular Pipelines

According to a suggestion of Manfred Heckl, the radiation factor σ of a circular pipeline for airborne sound radiation to the outside can be estimated as follows [92, 94, 97, 100, 101], if the critical frequency of coincidence f_g Eq. (20.53) and the second ring frequency f_2 of the pipeline

$$f_2 = 0.49\, \frac{c_L\, h}{d_a^2}. \qquad (20.55)$$

as well as the functions

$$G(f) = \left[1 + \frac{2}{\pi}\left(\frac{c_0}{\pi f\, d_a}\right)^3\right]^{-1} \quad X(f) = \frac{\lg(f/f_g)}{\lg(f_2/f_g)} \qquad (20.56)$$

are introduced (see also Fig. 20.27):

$$\sigma = \begin{cases} G(f) & \text{for} \le f_2 \\ G(f_2)^{X(f)} & \text{for} \quad f_2 < f < f_g\,. \\ 1 & \text{for} \quad f \ge f_g \end{cases} \qquad (20.57)$$

c_L – longitudinal wave velocity in the pipe wall material in m/s
d_a – outside diameter of the pipeline in m

For pipes with $h/d_a > c_0/c_L$ is valid: $f_2 > f_g$. Here, it is recommended to apply the relation stated for $f < f_2$ to the whole frequency range.

If $f_2 < f_g$, three frequency ranges must be distinguished (Fig. 20.27):

- Below the frequency f_2 Eq. (20.55), only bending waves are excited on a pipeline [94]. In this frequency range, a long pipeline behaves like a cylinder radiator of first order. The radiation factor increases with the frequency to the power of three.
- Above the frequency f_2, the radiation factor increases much more slowly with the frequency, i.e. – depending on the respective pipeline – to the power of 0.5 up to 1.2.
- Above the frequency f_g Eq. (20.53), the radiation factor of the pipeline is approximately 1.

For $f_2 \ge f_g$, the pipeline behaves like a cylinder radiator of first order in the whole frequency range.

The radiation indices of a circular pipeline for structure-borne sound excitation do not differ substantially from those for airborne sound excitation.

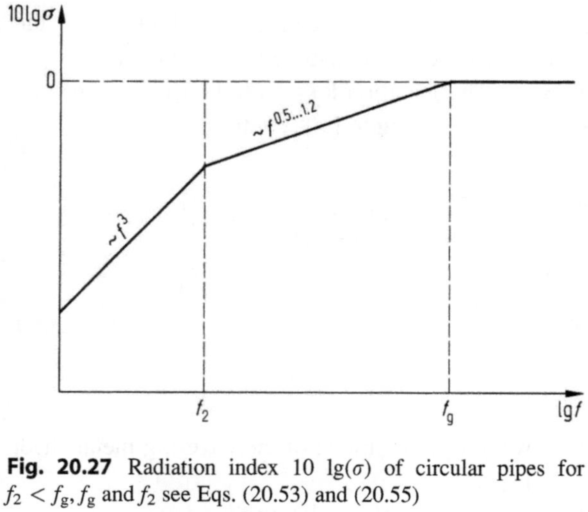

Fig. 20.27 Radiation index 10 lg(σ) of circular pipes for $f_2 < f_g$, f_g and f_2 see Eqs. (20.53) and (20.55)

20.2.8.2 Rectangular Ducts

Based on the investigations [98, 101], the radiation factor σ of a thin-walled elastically mounted steel duct (without reinforcements and without special damping) for airborne sound excitation from the interior of the duct and sound radiation to the outside can be estimated as follows:

$$\sigma = \begin{cases} f/f_g & \text{for } f \le f_g/4 \\ 1/4 & \text{for } f_g/4 < f < 0.8 f_g \,. \\ 1 & \text{for } f \ge f_g \end{cases} \tag{20.58}$$

f_g – critical frequency of coincidence [see Eq. (20.53)] of an even plate of the same thickness as the duct wall in m.

For *structure-borne sound excitation* of a rectangular duct, the radiation index is smaller than for airborne sound excitation. Besides, it very strongly depends on the type of excitation.

20.2.9 Sound Insulating Lagging of Circular Pipelines

One of the most important measures to reduce sound radiation of circular pipelines is sound insulating laggings. They usually consist of some kind of insulating fibre material covered by a sound-deadened sheet metal coating (thickness approximately 0.5–1 mm). Extensive investigations regarding the sound-insulating effect of pipe claddings have already been carried out using the measuring method described in [102]. The effectiveness of a sound-insulating cladding does not only depend on the buildup of the insulation system, but also depend on the diameter of the pipe. The insertion loss D_e to be expected can be estimated according to [102] with:

$$D_e = \frac{40}{1 + 0.12/d_a} \lg\left(\frac{f}{2.2 f_0}\right) \quad \text{for } f > f_0$$

with

$$f_0 = \frac{60}{\sqrt{m'' h_F}} \,. \tag{20.59}$$

m'' – weight per unit area of the covering metal sheet in kg/m^2

h_F – thickness of the insulating fibre layer in m

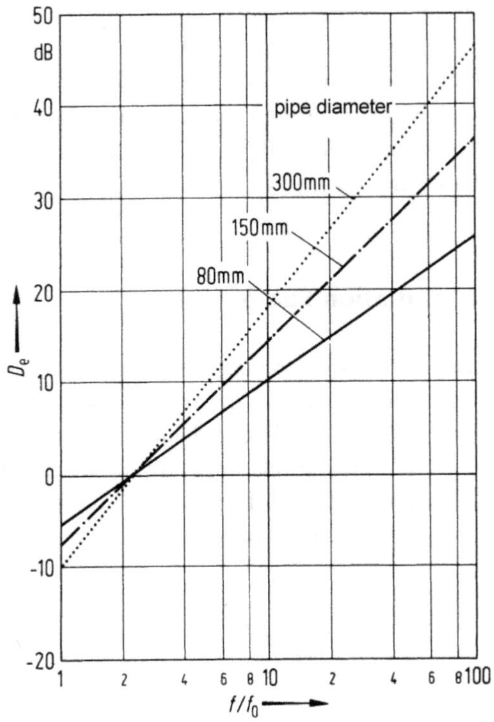

Fig. 20.28 Approximation for the insertion loss of sound insulating pipe laggings

d_a – outside diameter of the pipe in m

f – frequency in Hz

This equation is valid for laggings without rigid spacers.

The standard deviation of the measured values to the values determined by the approximation is 4 dB. Figure 20.28 shows the course of the damping curves for pipe laggings of 100 mm thickness on pipelines with nominal widths of 80, 150 and 300 mm.

20.3 Fans

The most common fan types are centrifugal fans and axial fans (Fig. 20.29). Extensive literature can be found in [103].

20.3.1 Characteristics

Fans are usually characterised by their $\Delta p - \dot{V}$ diagram. In this diagram, Δp (Pa) is the total pressure, i.e. the pressure difference between the discharge and

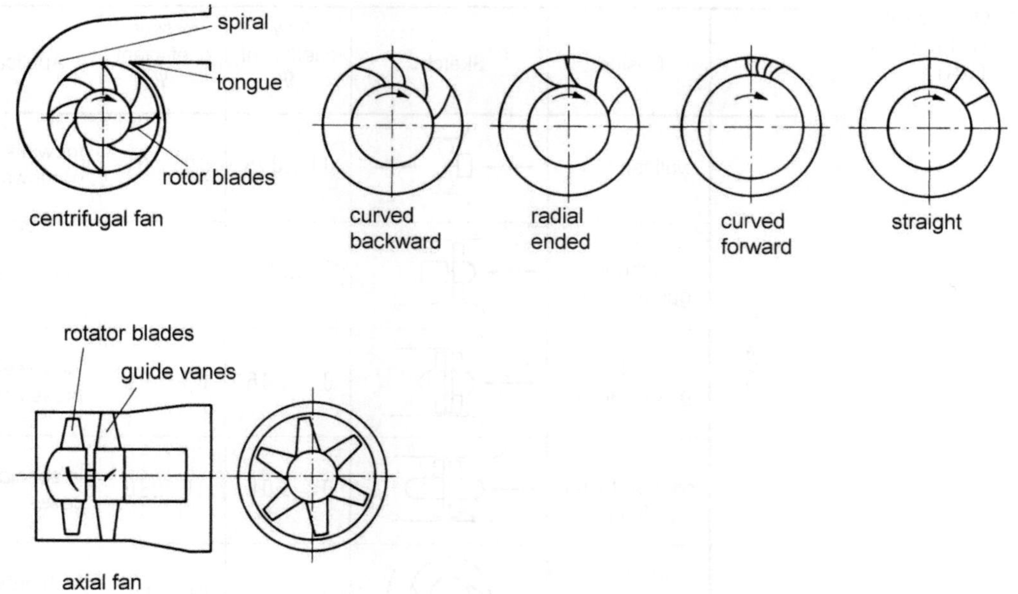

Fig. 20.29 The most important types of fans

suction side of the fan. $\dot{V}(m^3/s)$ is the volume of the flow medium transported per unit of time. In this diagram, the functions $\Delta p(\dot{V})$ (fan characteristics) are shown with the revolutions per minute of the fan as parameter.

The delivery performance $P_L(W)$ of a fan can be calculated using the following equation:

$$P_L = \Delta p\, \dot{V}. \quad (20.60)$$

Its efficiency η is defined as quotient of the output P_L to the consumed driving power P_e:

$$\eta = \frac{P_L}{P_e}. \quad (20.61)$$

The output pressure generated by the fan is necessary to overcome pressure losses caused by internal parts, cross-sectional jumps, deflections, etc., in the pipe system in front of and behind the fan. More detailed results on the pressure losses to be expected in ducts with a flow medium can be found, e.g., in [104, 105]. In most cases, the output pressure is proportional to the square of the conveyed amount of air:

$$\Delta p = W\, \dot{V}^2. \quad (20.62)$$

where W is the total resistance of the ductwork connected to the fan. Equation (20.62) describes the characteristic curve of the duct system, i.e. the plant.

In order to further characterise the different types of fans, the following characteristic numbers are normally introduced using the diameter D and the circumference speed U of the impeller:

Flow coefficient

$$\varphi = \frac{\dot{V}}{\frac{\pi}{4} D^2 U}. \quad (20.63)$$

Pressure coefficient

$$\psi = \frac{\Delta p}{0.5\, \rho\, U^2}. \quad (20.64)$$

ρ – density of the flow medium in kg/m^3.

Figure 20.30 gives an overview of different types of fans according to [104] with indication of φ and ψ. Figure 20.31 shows typical characteristics of the main types of fans (see [104, 106]).

20.3.2 Sound Generation

Noise spectra of fans consist of a broadband noise component and superimposed tones. Normally, sound sources with dipole character are responsible for the broadband noise. These sound sources can be

Fig. 20.30 Overview of different types of fans according to [104]

	Design	Sketch	Flow coefficient φ	Pressure coefficient ψ	Application
axial fans	wall fan	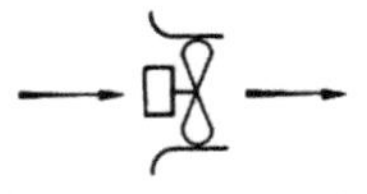	0.1…0.25	0.05…0.1	for walls and windows
axial fans	without guide vanes	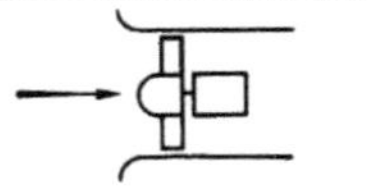	0.15…0.30	0.1…0.3	low pressures
axial fans	with guide vanes	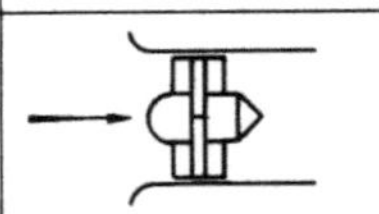	0.3…0.6	0.3…0.6	increased pressures
axial fans	contrarotating fans	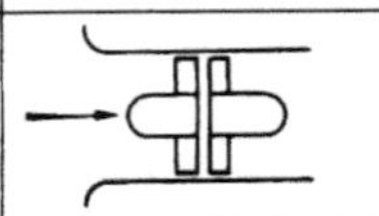	0.2…0.8	1.0…3.0	in special cases
centrifugal fans	backward curved blades	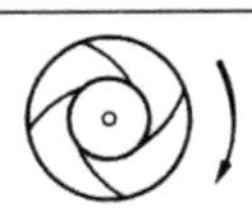	0.2…0.4	0.6…1.0	high pressures and efficiencies
centrifugal fans	straight blades	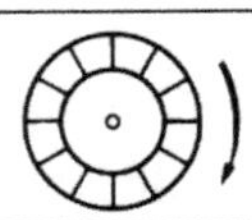	0.3…0.6	1.0…2.0	in special cases
centrifugal fans	forward curved blades (drum rotor)	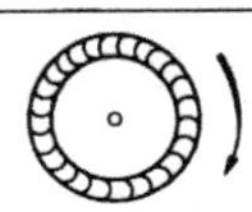	0.4…1.0	2.0…3.0	low pressures and efficiencies
cross flow fans		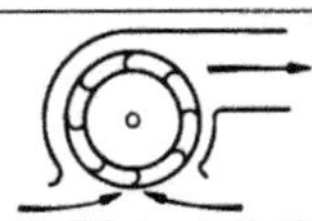	1.0…2.0	2.5…4.0	high pressures low space requirement

explained by vortex shedding and turbulent flow impinging solid structural components. Single tones in the spectrum are usually caused by periodically alternating forces, which result from an interaction of moving and stationary solid structural components of fans and the connected periodically varying flow these components are exposed to.

In Table 20.2, the most important sound generation mechanisms of fans are listed. Moreover, the influence of the most important quantities on the emitted sound is outlined (see also Sects. 17.1.2 and 17.1.3). The following parameters are used:

u – relative velocity between the blades and the flowing gas
c – sound velocity in the flow medium
$M = u/c$ – Mach number
b – impeller width
Tu – turbulence factor of the flow approaching the blades
Θ – projection of the blade perpendicular to the flow direction
S_L – total surface of the blades
z – number of blades
P – emitted sound power
R – impeller radius
n – rotational speed of the fan in s^{-1}

20.3.3 Approximate Calculation of Sound Emission

Based on numerous measurements, different empirical relations regarding the sound power level L_W of fans have been determined [117–125].

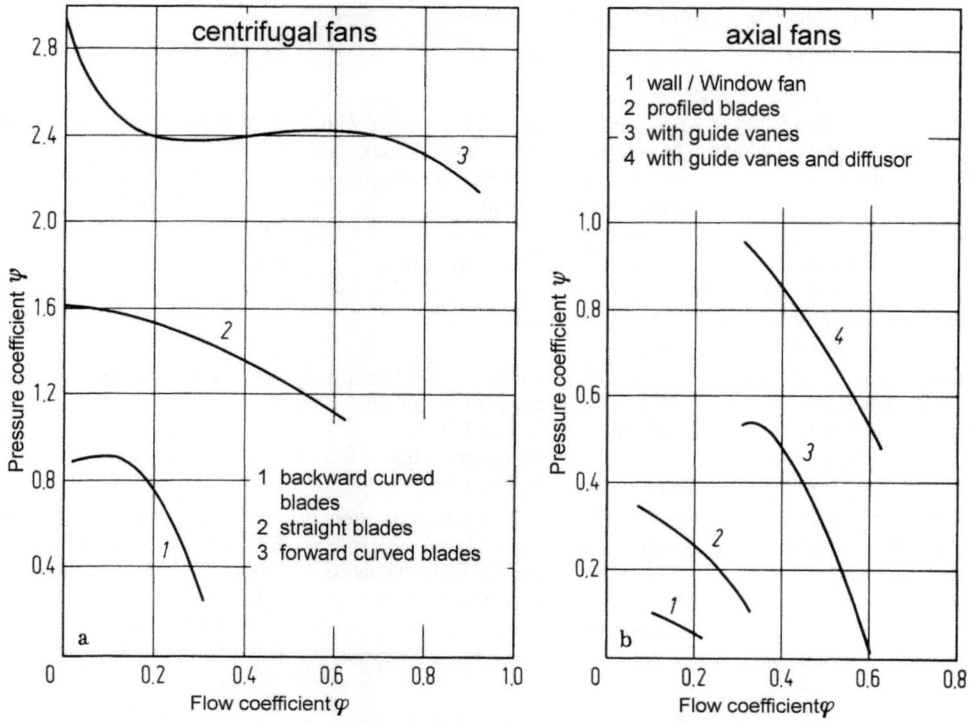

Fig. 20.31 Typical dimensionless characteristics of centrifugal and axial fans used in ventilation systems according to [104]

The octave sound power level $L_{W/Oct}$ (re 10^{-12} W) of the broadband noise transmitted into the connected discharge duct can be estimated with the following equation:

$$L_{W/Oct}(f_{oct}) = 79 + 10\lg\left\{\frac{P_e(1-\eta)\cdot(U/c)^{1.5}}{1+(St/St_O)^m}\right\}\ \text{dB}. \quad (20.65)$$

with

	St_O	m
Centrifugal fan	2.5	1.5
Drum rotor	5	1.5
Axial fan with outlet guide vanes (single stage)	17	1.7

P_e – power consumed by the fan in W
f_{oct} – octave centre frequency in Hz
η – fan efficiency
U – tip speed of the impeller in m/s
c – sound velocity in the discharge pipe in m/s
$St = f_m\, D/U$
D – diameter of the rotor blades in m

If a centrifugal fan does not intake from a long straight pipeline but from a suction box or a 90° elbow in front of the suction flange, the low-frequency turbulence of the incoming flow increases. This is associated with an increase in the octave sound power level by approximately 3 dB. For centrifugal fans, single tones in the spectrum, which determine the overall level, can be caused by an interaction between rotor blades and the volute tongue; for axial fans with the outlet guide vanes by an interaction between rotor blades and guide vanes. The frequencies of the single tones can be described by:

$$f_\nu = (\nu+1)zn \text{ with } \nu = 0, 1, 2, \ldots \quad (20.66)$$

z – number of rotor blades
n – rotational speed in s^{-1}

In order to consider the approximate contribution of these single tones to the emitted sound power, for the octaves containing single tones ($f_m > 0.7\, f_0$), the following allowances ΔL are made for the octave sound power levels $L_{W/Oct}$ calculated according to Eq. (20.65) (empirical connection):

Table 20.2 The most important sound generation mechanisms of fans

Mechanisms of sound generation	Emitted sound power	Characteristics of the spectrum
Centrifugal fan		
Sound generation by alternating forces at the blade surfaces due to vortex shedding at the trailing edge of the blade	$P \sim \rho u^3 M^{(2\ldots3)} z\, b^{(1\ldots2)}$, P increases with Tu and the coherence rate of vortex shedding. The coherence rate depends on b and the type of lateral blade termination (see Fig. 20.19)	Broadband spectrum with a broad frequency maximum (Doppler effect) [34, 35] at $f_{max} \approx 0.18\, u/\Theta$ [35, 107], above this maximum decrease by about 4 dB/octave (see Fig. 20.18)
Sound generation by alternating force on the blade surfaces caused by the turbulent wake behind the blade leading edge, inlet guide vanes, struts, intake nozzles, etc.	$P \sim \rho u^3 M^3 S_L$ and $P \sim \mathrm{Tu}^2$ [108]	Broadband spectrum
Sound generation by alternating forces on the surface of the volute tongue due to the very turbulent incident flow caused by vortex shedding	$P \sim \rho u^3 M^3$, P increases when the tip clearance (distance between tongue and impeller) is reduced [109], see Fig. 20.32; the tongue radius is of minor importance [109]	Broadband spectrum
Sound generation by alternating on the tongue and the blades caused by interaction between them	$P \sim u^3 M^3$ [109], P increases very strongly when the tip clearance is reduced [109–111], see Fig. 20.32; an influence of the tongue radius on P is existing but small [109]	Spectrum consists of single tones f_υ ($v = 0, 1, 2, \ldots$); for z symmetrically arranged blades, the following equation is valid: $f_v = (v+1)zn$, with increasing tip clearance, the content of harmonics is reduced [109], see Fig. 20.32
Axial fan		
Sound generation by alternating forces at the rotor blades, caused by vortex shedding at the trailing edge	$P \sim \rho u^3 M^{(2\ldots3)} z\, R$, P increases with Tu and depends on the blade angle	Broadband spectrum with broad frequency maximum at $f_{max} \approx 0.18\, u/\Theta$ [35, 107], above this maximum decrease by about 4 dB/octave (see Fig. 20.18)
Sound generation by alternating forces on the rotor blades due to the turbulent wake behind inlet guide vanes, struts, intake nozzles, etc.	$P \sim \rho\, u^3 M^3 S_L$ and $P \sim \mathrm{Tu}^2$ [108, 112]	Broadband spectrum
Sound generation by alternating forces on downstream guide vanes caused by the turbulent wake of the rotor blades	$P \sim \rho u^3 M^3$, P increases when the distance between rotor and stator is reduced	Broadband spectrum
Sound generation by alternating forces on guide vanes and rotor blades caused by interaction between them [108, 113–115]	$P \sim \rho u^3 M^{(2\ldots3)}$, P increases very strongly when the distance between rotor and stator is reduced [108, 112], see Fig. 20.33; P decreases with an increasing z [116] and depends on the angle between stator and rotor [108]	The spectrum consists of single tones $f_v \sim n$ which depend on the number of guide vanes and blades and on their arrangement; with an increasing distance between stator and rotor, the content of harmonics is reduced

Centrifugal fan

$$\Delta L = 10\lg\left\{1 + 0.4\left(\frac{f_0}{f_m}\right)^{2.5}\frac{(U/c)^{1.5}}{\Delta r/D}\right\}. \quad (20.67)$$

Axial fan with guide vane

$$\Delta L = 10\lg\left\{1 + 4\Big/\left(1 + \left(\frac{f_m}{8f_0}\right)^2\right)\right\}. \quad (20.68)$$

Δr – minimum distance between rotor blades and tongue in m

$f_0 = z\, n$ in Hz [Eq. (20.66)]

In order to estimate the octave sound power levels of the noise emitted into the suction duct, the levels calculated according to Eqs. (20.65)–(20.68) are reduced by 2 dB.

The approximate calculation of the airborne sound power emitted to the outside from the suction and discharge piping is carried out considering the equations described in Sect. 20.2.

If the suction or discharge pipe has an opening to the outside, the airborne sound power radiated from this opening can be calculated applying a correction for the sound reflection at the orifice according to [124, 125].

In *air coolers and forced-draught cooling towers*, single-stage axial fans are used. The noise spectra are usually broadband without prominent pure tones [122]. Tones can occur:

– If solid obstacles are close to the rotating fan blades.
– If from the gear box or structure-borne sound from directly driving motors is transmitted via the drive shaft to the hub and blades of the fan and is then radiated from these elements as airborne sound.

According to [122], the total octave sound power levels (re 10^{-12} W) radiated by the suction and discharge side of such an axial fan can be calculated approximately by applying the following equation:

$$L_{\mathrm{W/Oct}}(f_{\mathrm{oct}}) = 87 + 10\,\lg\left\{\frac{P_e (U/c)^3}{1 + (St/17)^{1.5}}\right\}\ \mathrm{dB}. \tag{20.69}$$

With this equation, the total A-weighted sound power level of axial fans for air coolers and forced-draught cooling towers can be calculated with a precision of +2/−4 dB(A). For axial fans with very broad, airfoiled and pressure-balanced blades sound levels result which are by up to 5 dB lower than those calculated with Eq. (20.69).

Equation (20.69) can also be used for an approximate calculation of other single-stage axial fans without guide vanes emitting into the free field (e.g. wall fans).

20.3.4 Noise Control

Constructive measures at fans are only useful if sound emission is reduced without or only with a small decrease in the output. Furthermore, the following items must be observed (see also [103, 125–129]):

Centrifugal fan

– The relative velocity u between blades and flow medium should be as small as possible, but with the same output (φ and ψ should be as large as possible). This can be achieved by:
  An increase in the number of blades, thus reduction in u for the same U
  An extension of fan dimensions
  Favourable design of the inlet section with regard to flow in order
  To reduce pressure losses (diffuser, be careful with guide vanes!)
– The distance between tongue and impeller (tip clearance) should always be as large as permissible with regard to efficiency (see Fig. 20.32)
– Obstacles and disturbances in the intake section should be avoided [130]
– A fixation of the rotor blades on both sides should preferably be avoided, see Fig. 20.19

Axial fan

– The relative velocity u between rotor blades and flow medium should be as small as possible for the

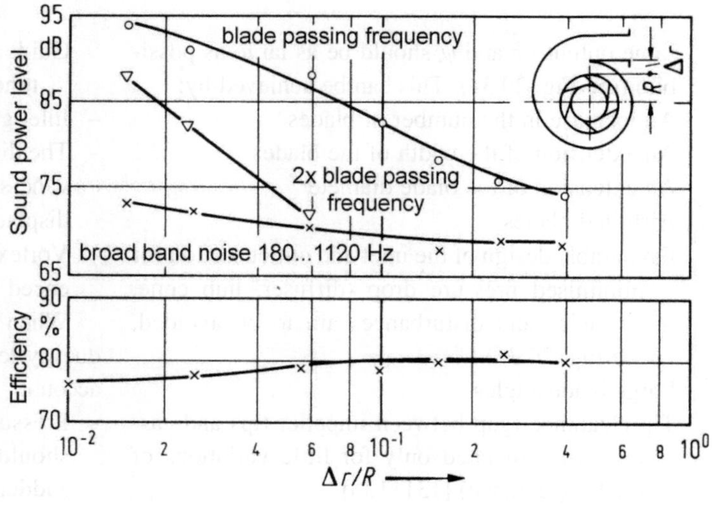

Fig. 20.32 Influence of the tip clearance Δr on the most important contributions to the noise and efficiency of a centrifugal fan according to [109] for an optimal flow coefficient φ_{opt}

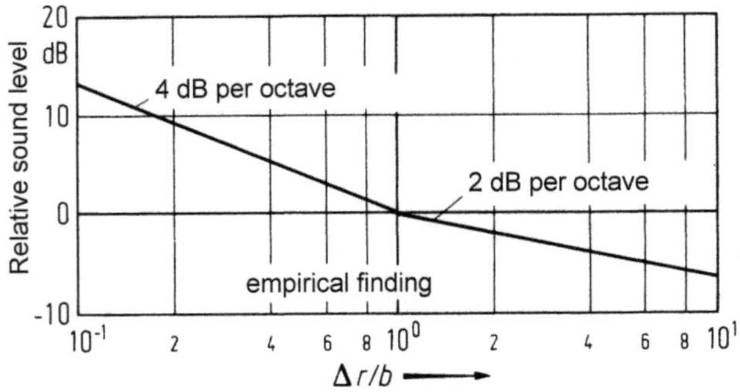

Fig. 20.33 Influence of the distance between rotor and stator on the first fundamental tone which is caused by an interaction between blades and guide vanes of an axial fan, according to [112]. Δr distance between rotor and stator, b length of the guide vanes or rotor blades, measured in low direation

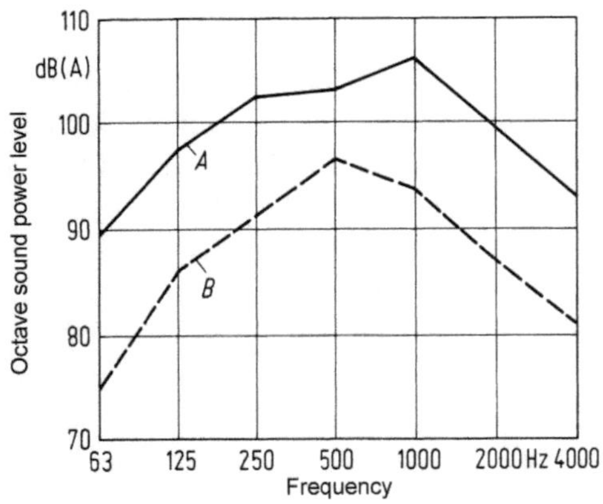

	axial fan	
	A	B
volume flow, m³/s	335	400
pressure, Pa	162	157
fan diameter, mm	7100	7100
number of blades	5	6
tip speed, m/s	65	38
blade width	small	large
sound power level, dB(A)	110	99.5

Fig. 20.34 Octave sound power level spectra of two axial fans for cooling towers for similar operating conditions

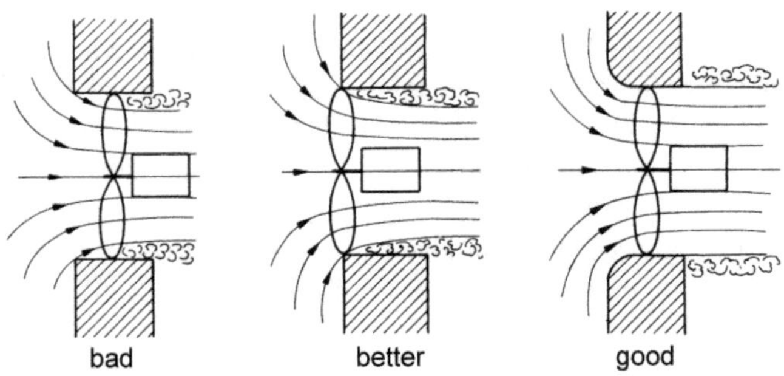

Fig. 20.35 Good and bad installation of an axial fan, from [106]

same output (φ and ψ should be as large as possible) (see Fig. 20.34). This can be achieved by:

An increase in the number of blades

An extension of the width of the blades

An extension of the blade diameter

Airfoiled blades

Favourable design of the inlet and outlet section for minimised pressure drop (diffuser, hub cone, obstacles and disturbances are to be avoided, see Fig. 20.35)

Large blade angles

Tip clearance (gap between impeller tips and casing, recommended only for little variations of the tip clearance) [131–133]

Guide vanes downstream of the impeller (attention, tones!)

- Inlet guide vanes should be avoided
- The distance between blade wheel and outlet guide vane should be at least 20 to 30 times the size of displacement (see Fig. 20.23)
- Vortex shedding at the rotor blades can be influenced by modifications of their trailing edge

When selecting a fan and designing the connected duct system, the following factors have to be taken into account:

- Pressure drop in the connected duct system should be as small as possible, unnecessary sudden cross-section changes or deflections are

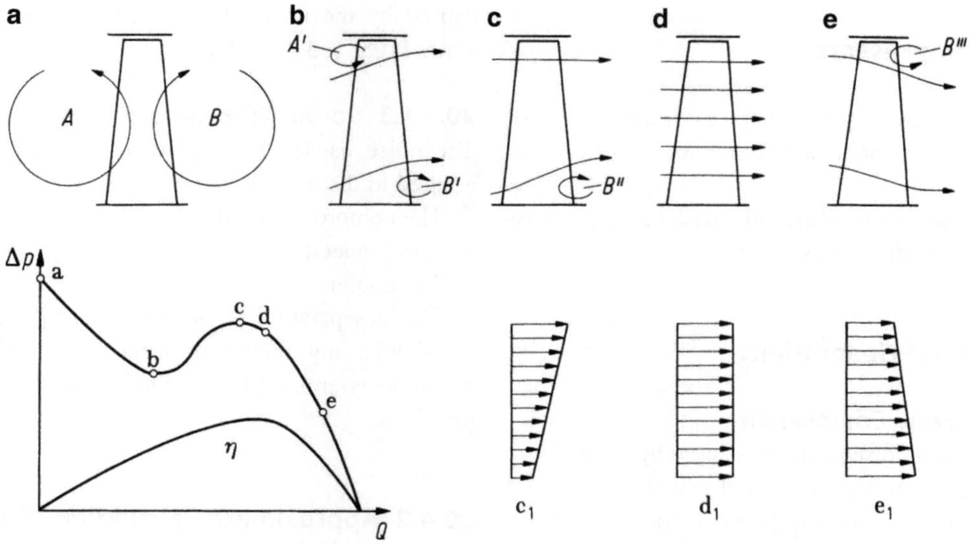

Fig. 20.36 Diagram showing the different flow conditions of an axial fan for different throttling conditions, from [106]

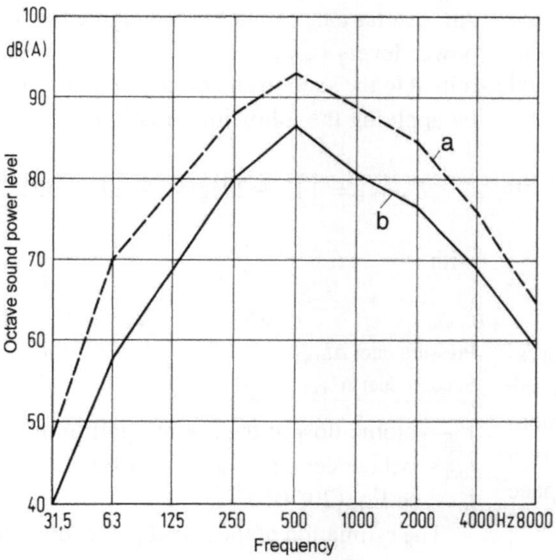

graph	*a*	*b*
speed, rpm	1470	1200
shaft power, kW	1670	1130
vane control adjustment, degrees	-55	0
volume flow, m³/s	70	70
total pressure, mbar	134	135
SPL $L_{A,1m}$, dB(A)	95.5	88.5

Fig. 20.37 Octave sound pressure level at 1 m distance of a centrifugal fan. Adjustment of the volume flow by vane control (**a**) or speed regulation (**b**)

to be avoided, convection should be used, when possible.

- For a given flow and pressure (an exact knowledge of the "plant resistance" is required) the fan should work in the range of maximum efficiency (point d in Fig. 20.36). The fan must not be operated at an ascending branch or at a saddle point of the characteristic curve (area between points b and c in Fig. 20.36) [103], as this would likely provoke low-frequency oscillations resulting in a high noise level and "pumping" of the fan.
- If possible, the volume flow should be adjusted by speed control, not by vane control or throttle valves [130, 134] (Fig. 20.37).
- An additional sound level decrease in the duct system at low frequencies can be achieved using thin-walled rectangular ducts.

The most efficient measure to reduce fan noise is to install silencers (see Chap. 11).

20.4 Compressors

In this chapter, two typical compressors are discussed:
- *Screw compressors* are used to achieve high pressures.
- *Axial turbo compressors* are used for large mass flows in a medium pressure range.

20.4.1 Sound Generation

20.4.1.1 Screw Compressor

The compressor noise is mainly caused by the periodic conveyance of gas and, to a small extent, by flow and expansion processes during pressure compensation at the discharge flange. The spectra of the tonal noise components transmitted to the connected pressure line can be calculated with good accuracy if the volume flow take and owing to the construction data of the compressor is known. Essentially, noise spectra consist of harmonic single tones with the fundamental frequency:

$$f_0 = z_H\, n_H. \qquad (20.70)$$

z_H – teeth number of the main rotor
n_H – rotational speed of the main rotor in s^{-1}

If at the end of the compression process, the pressure in the tooth space volume is smaller than the pressure in the pressure line, after opening the tooth space volume to the pressure duct there will be a backflow from the pressure duct to the tooth space volume. This process leads to particular high sound levels.

20.4.1.2 Axial Turbo Compressor

The noise generated in the compressor by flow consists of a broadband noise and superimposed tones. The broadband noise is caused by:
- Alternating forces at the rotor blades caused by vortex shedding at the trailing edge and the turbulent wake of the inlet guide vanes struts and inlet nozzles
- Alternating forces on the downstream guide vanes due to the turbulent wake of the rotor blades

The single tones are mainly caused by periodic alternating forces at the guide vanes and rotor blades, caused by the interaction of rotor blades and guide vanes [108, 113–115, 135].

20.4.1.3 Sound Transmission Paths

The noise caused by both types of compressors is emitted to the outside via:
- The compressor casing
- The connected pipelines
- The cooler
- The oil separator (for oil-flooded screw compressors)

Without any noise control measures, most part of the noise is emitted to the outside via the connected pipelines.

20.4.2 Approximate Calculation of Sound Emission

20.4.2.1 Screw Compressor

An estimation of the non-weighted octave sound power levels $L_{W/Oct}$ (re 10^{-12} W) of the noise transmitted to the suction and discharge piping can be made by applying the following equation:

$$L_{W/Oct} = 147 + 20\lg(\dot{V}) + \Delta L_{Oct}\,\text{dB}. \qquad (20.71)$$

with

f_{oct}/f_0	0.5	1	2	4	8	16
Pressure duct ΔL_{Oct}	−20	−4	−4,5	−10	−18	−21
Suction duct ΔL_{Oct}	−31	−16	−17	−18	−19.5	−21

$\dot{V}$ – volume flow at the suction port in m^3/s
f_{oct} – octave centre frequency in Hz
f_0 – see Eq. (20.70)

The estimation of the sound power emitted from the suction and pressure duct to the outside is made by means of the procedures discussed in Sect. 20.2.

20.4.2.2 Axial Turbo Compressor

A rough estimate of the A-weighted sound power levels L_{WA} of the compressor noise emitted via the different sound transmission paths (without noise control measures) can be made with the following equation (see also Fig. 20.38):

$$L_{WA/Oct} = 112 + 20\,\lg(P_N/1\,\text{MW})\,\text{dB(A)}. \qquad (20.72)$$

P_N – nominal driving power in MW.

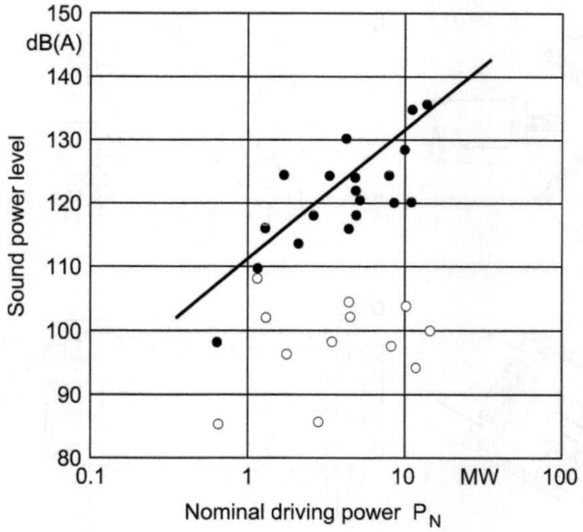

Fig. 20.38 A-weighted sound power level of axial turbo compressors and connected pipelines. *Filled circle* Total sound emission of the compressor casing and the connected pipelines, *open circle* Sound emission from the compressor casing only

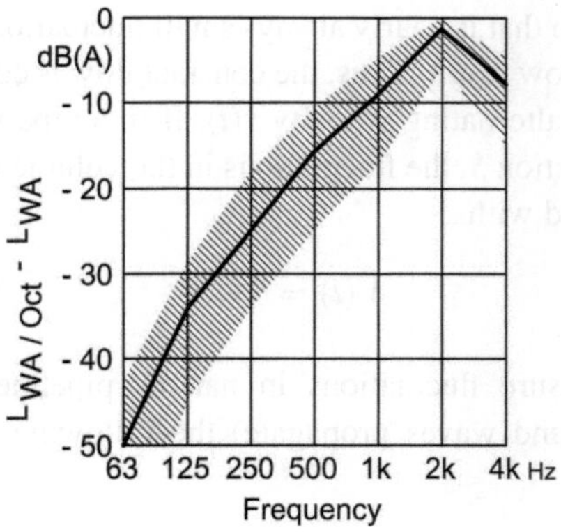

Fig. 20.39 Average standardised A-weighted octave sound power level of axial turbo compressors including the connected pipelines without noise control measures. $L_{WA/Oct}$ A-weighted octave sound power level, L_{WA} total A-weighted sound power level

The A-weighted sound power levels of the compressor casing are normally by 5–35 dB(A) lower than those in the connected pipelines (Fig. 20.38). In order to estimate the A-weighted octave spectrum $L_{WA/Oct}$, Fig. 20.39 can be used.

20.4.3 Noise Control

Both for the screw compressors and for the turbo compressors, silencers are used to reduce the sound transmission to the connected pipelines, which are effective against gas-borne sound in the pipelines as well as against structure-borne sound in the pipe walls.

However, the most common noise control measures at pipelines are sound-insulating claddings (see Sect. 20.2.9). In order to reduce airborne sound emission of compressors, they are installed in buildings or provided with sound-insulating enclosures.

20.5 Pumps

In the following, at first the most important types of pumps are described. More detailed information on basic principles of pumps can be found in [136–140].

Hydrostatic pumps are used to create very high pressures. Characteristic for these pumps is the mechanical closure between suction and pressure side and that, depending on the tightness of this closure, the volume flow depends only very little on the counter-pressure. The most important hydrostatic pumps are vane pumps, gear pumps, axial piston pumps and screw pumps (basic sketches see Fig. 20.40).

For *hydrodynamic pumps* (centrifugal pump, Fig. 20.41), there is no mechanical separation between suction and pressure side, and the pressure difference between the two ports must be maintained by hydrodynamic forces. The blades of the rotating blade wheel transfer speed energy to the fluid coming from axial direction, which is converted to a large part into pressure energy. The conveyance is continuous.

20.5.1 Sound Generation

The main cause for the noise emission of a pump is alternating pressures in the fluid created during pumping. The structure-borne sound levels caused by the movement of mechanic transmission elements (bearings, gear, drive, etc.) are normally essentially lower.

20.5.1.1 Hydrostatic Pumps

For hydrostatic pumps, there are three main mechanisms of noise generation:

- Cavitation and separation of gas during the suction process
- Periodic fluctuations of the output and changes in flow velocity connected with it

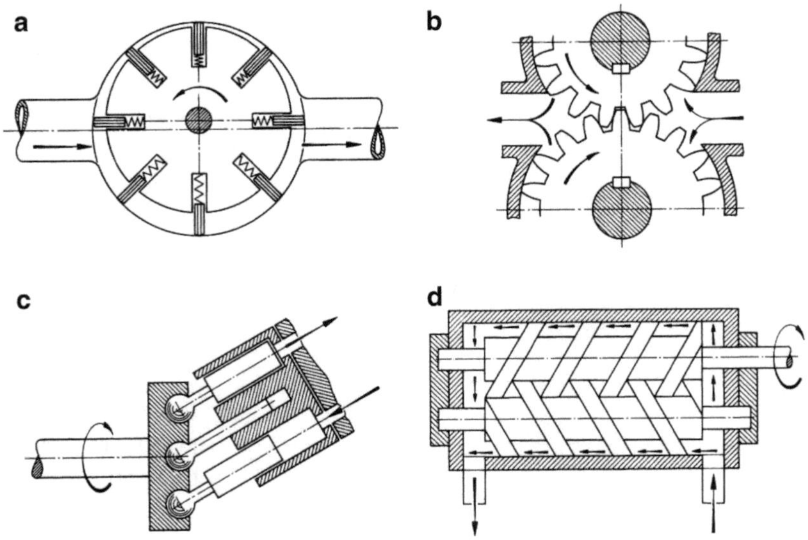

Fig. 20.40 Basic sketches of the most important types of hydrostatic pumps (**a**) vane pump, (**b**) gear pump, (**c**) axial piston pump, (**d**) screw pump or helicoidal gear pump

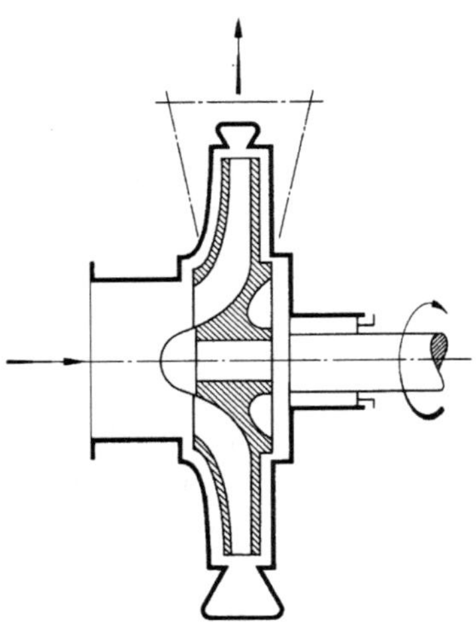

Fig. 20.41 Sketch of a centrifugal pump, from [136]

– Impulsive pressure compensation when fluid volumes of different pressures meet. This mechanism does not occur with the screw pump

Separation of Gas and Cavitation During the Suction Process

With decreasing suction pressure, the gas dissolved in the fluid separates in the form of bubbles. A gas/fluid mixture is created. As a result, the emitted airborne sound increases (Fig. 20.42). Much more problematic than the noise of gas separating from the fluid is cavitation noise (see Sect. 20.1.2). During normal operation, the mechanisms stated in the following are important.

Periodic Fluctuations of the Discharge Flow

With most pumps separate small volumes in closed cavities (piston stroke, tooth pockets, etc.) are transported so that is nearly always small fluctuations in the output flow result. Thus, the constant flow is combined with an alternating velocity v(t). If in a pipe with the cross section S, the fluctuations in the volume flow are described with

$$\dot{V}(t) = S\,\mathrm{v}(t) \tag{20.73}$$

for pressure fluctuations in narrow pipelines (only even sound waves propagate) the following relation results

$$p(t) = \rho\, c\, \mathrm{v}(t) \tag{20.74}$$

or

$$p(t) = \rho\, c\, \frac{\dot{V}_0}{S} \varepsilon. \tag{20.75}$$

ρ – density of the fluid in kg/m^3
c – sound velocity in the fluid in m/s
$\dot{V}_0$ – average volume flow in m^3/s
S – inner cross-sectional area of the pipe in m^2
ε – the relative temporal change of the volume flow

$$\varepsilon = \frac{\dot{V}(t)}{\dot{V}_0}. \tag{20.76}$$

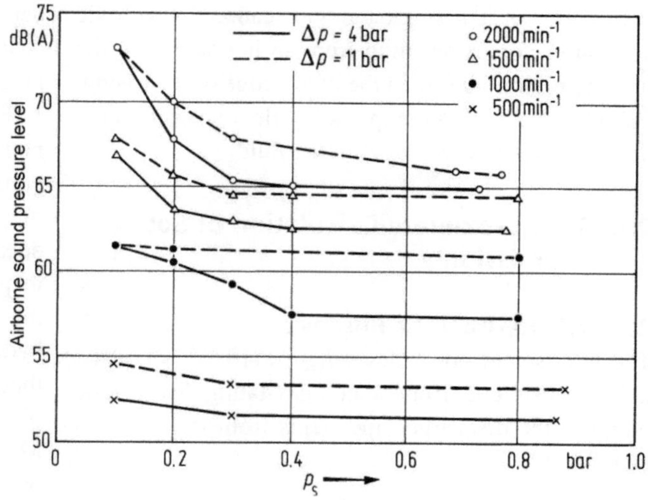

Fig. 20.42 Influence of the suction pressure on the airborne sound level emitted by a gear pump according to [141]

ε depends on the geometry of the pump. For gear pumps, ε is approximately 0.02, for axial piston and vane pumps it is approximately 0.2–0.005.

Impulsive Pressure Compensation Processes

This mechanism will take effect if during operation of the pump a small volume of fluid V_o is taken up at the suction side (displaced volume), e.g. between the blades of a vane pump or between the teeth of a gear pump and transported to the pressure side. There, it suddenly connects with the liquid column, which is under high pressure, and the fluid in the volume V_0 is compressed. The resulting volume reduction is

$$\Delta V = V_0 \frac{\Delta p}{\rho\, c^2} \tag{20.77}$$

and Δp is the difference of the static pressure in the volume V_0 immediately before and after connection with fluid on the pressure side. If the fluid is not compressed during the pumping process, Δp then accounts for the pressure difference between discharge and suction side (discharge pressure). For compression of the displaced volume V_o by an amount ΔV, some fluid must flow back from the pressure side. The associated velocity of the fluid on the pressure side is

$$\mathrm{v}(t) = \frac{1}{\Delta t}\frac{\Delta V}{S}. \tag{20.78}$$

Δt – duration of the compression in s.

From Eq. (20.74) follows the partial tone amplitudes of the alternating pressure $p(t)$:

$$p_\nu = \frac{\rho\, c}{S}\frac{2}{T}\int_0^T \frac{\Delta V}{\Delta t}\cos(2\,\pi\,\nu\, t/T)\,\mathrm{d}t \text{ with } \nu = 0, 1, 2, \ldots. \tag{20.79}$$

T – $1/(z\ n)$ cycle duration in s
z – number of discharges
n – rotational speed of the pump in 1/s

When considering that the displaced volume is compressed very fast ($\Delta t \ll T$), for the first partial tones with $\dot{V}_0 = V_0/T$ and Eq. (20.77) the following equation results:

$$p_\nu = \rho\, c\frac{2}{T}\frac{\Delta V}{S} = 2\,\dot{V}_0\frac{\Delta p}{c\, S} \quad \text{with } \nu = 0, 1, 2, 3, 4. \tag{20.80}$$

This relation is valid e.g. for the first five partial tones, which essentially determine the total level.

20.5.1.2 Hydrodynamic Pumps (Centrifugal Pumps)

Similar to a centrifugal fan, the fluid sound spectrum of centrifugal pumps consists of broadband noise and superimposed tones. The broadband noise results from vortices and turbulences caused at fix construction structures in the pump, from the asymmetrical flow towards the blades and possibly also from cavitation caused at the rotor blades during the suction- or outlet-process. The single tones are caused by periodic alternating forces resulting from an interaction of the moving rotor blades with the outlet edge of the casing. For a non-cavitating pump, the single tones determine the total sound power level in the fluid.

20.5.2 Approximate Calculation of Sound Emission

20.5.2.1 Hydrostatic Pumps

The total sound power level L_W (re 10^{-12} W) of the noise transmitted from a non-cavitating hydrostatic pump to the discharge pipe stems from Eqs. (20.75) and (20.80):

$$L_W = 117 + 10\lg\left\{\rho\, c \frac{\dot{V}_0^2}{S}\left[\varepsilon^2 + 40\left(\frac{\Delta p}{\rho\, c^2}\right)^2\right]\right\} \text{dB.} \tag{20.81}$$

The sound power level at the suction side is by 10–30 dB lower than at the pressure side.

Figure 20.43 shows the total sound pressure level L (re $2 \cdot 10^{-5}$ Pa) resulting from Eq. (20.81)

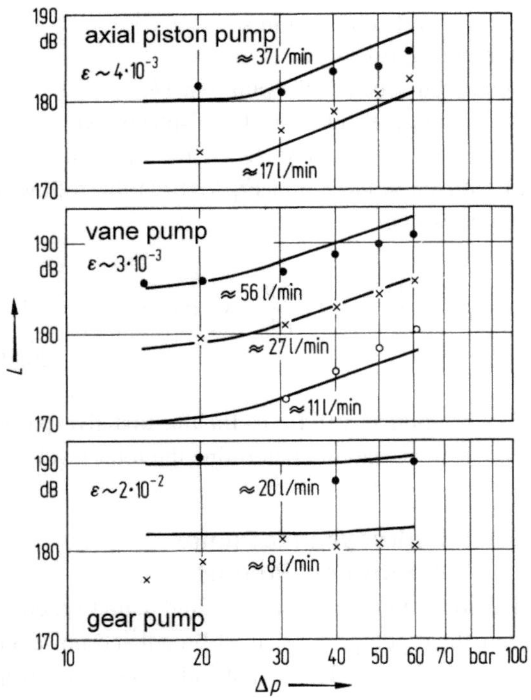

Fig. 20.43 Dependence of the total sound power level L in oil from the discharge pressure and the volume flow (internal diameter of the pipe system 24 mm). *Continuous line* calculated with Eq. (20.81); *Symbols*: measuring results

depending on the discharge pressure Δp and the discharge rate $\dot{V}_0$ in comparison with measuring results [142].[3]

Equation (20.81) cannot be applied for the *screw pump*, as the leakage loss is much larger than for the pumps discussed before and, therefore, an impulsive pressure compensation between two fluid volumes does not occur. There will rather be a gradual rise in pressure on the way to the pressure side. Hence, the screw pump is halfway between hydrostatic and hydrodynamic pumps. On the other hand, it is possible that the fluctuations of the volume flow are comparably large and grow with increasing discharge pressure. Measurements have shown that the total sound pressure level in the fluid exclusively depends on the discharge pressure. Figure 20.44 shows the measured dependency of the total sound pressure level from the discharge pressure in the fluid of a screw pump.

Figure 20.45 shows schematic octave spectra in the conveyed liquid on the pressure side of some pumps.

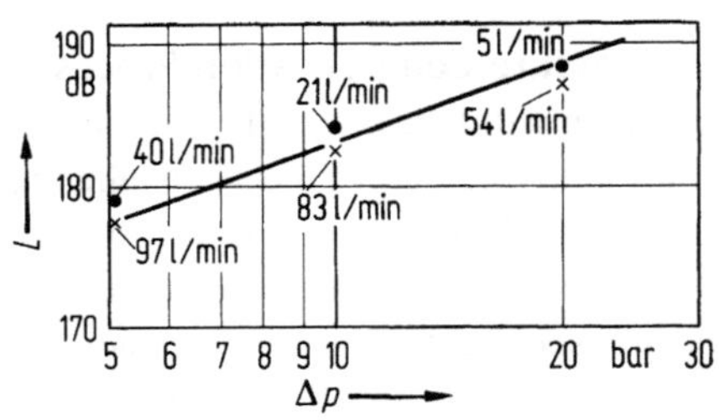

Fig. 20.44 Dependency of the total sound pressure level in oil from the discharge pressure and volume flow of a screw pump at the pressure side (internal diameter of the pipe system 24 mm)

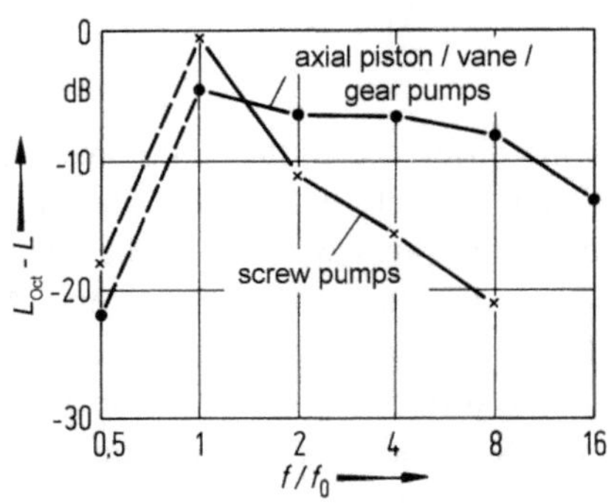

Fig. 20.45 Schematic normalised octave sound levels in the pressure pipe of hydrostatic pumps

[3] It was considered here that owing to reflections at the load resistor (valve) and the pump the level will increase by approximately 6 dB on the measuring length.

The frequency of the fundamental tone, which shows in all spectra, can be calculated by the following equation:

$$f_0 = z \cdot n. \qquad (20.82)$$

z – number of output volumes per revolution
n – rotational speed of the pump in s^{-1}

The noise of the vane pump, axial piston pump and gear pump has many harmonics. Therefore, for narrow-band analyses a pure tone spectrum results where lines show up to the 50th order.

If cavitation is the main noise source, the broad-band spectrum shows fewer individual lines.

The sound power emitted from the pressure pipe to the outside can be estimated considering the relations described in Sect. 20.2 using the octave sound power spectrum in the fluid. For an approximate calculation of the airborne sound emitted from the pump casing, resort to [143].

20.5.2.2 Hydrodynamic Pumps (Centrifugal Pumps)

In industrial plants, one finds a large number of centrifugal pumps. The noise generated (by a centrifugal pump) is mainly emitted to the outside via the connecting pipelines and not via the casing. As shown in Fig. 20.46, the total A-weighted sound power level of a centrifugal pump including the pipelines can be estimated from the rated power of the motor P_N in kW [144]:

$$L_{WA} = 68 + 16\ \lg(P_N/1\,\text{kW})\ {}^{+6}_{-9}\ \text{dB(A)}. \qquad (20.83)$$

Ninety percent of all measuring results are in the indicated tolerance range (Fig. 20.46). According to Schmitt, Klein [145], the A-weighted sound power level emitted by standard chemistry pumps (series CPK) via the casing can be calculated using the following equation:

$$L_{WA} = 60 + 12\ \lg(P_N/1\,\text{kW})\ \text{dB(A)}. \qquad (20.84)$$

This equation is valid for rated powers between 1 and 100 kW [146, 147]. Further information on the airborne noise emitted by the pump casing is given in [148].

In order to estimate the A-weighted octave sound power level $L_{WA/Oct}$, Fig. 20.47 can be used.

20.5.3 Noise Control

20.5.3.1 Primary Measures

As stated in Sect. 20.5.1, in order to reduce fluid sound of *hydrostatic pumps* the important thing is to keep discharge fluctuations and pressure differences when coupling the pumped fluid volumes at the pressure side

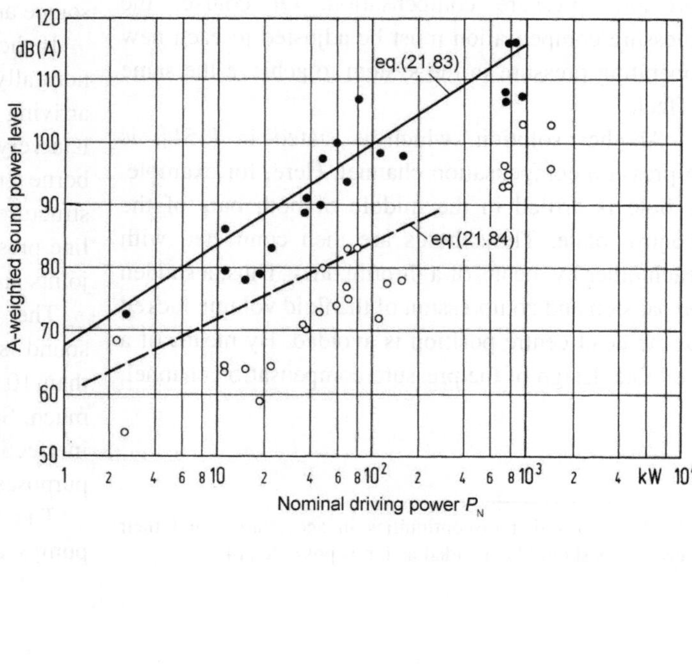

Fig. 20.46 A-weighted sound power level of centrifugal pumps and the connected pipelines [144]. *Filled circle* sound emission of the casing and the connected pipelines, *open circle* sound emission of the pump casing only

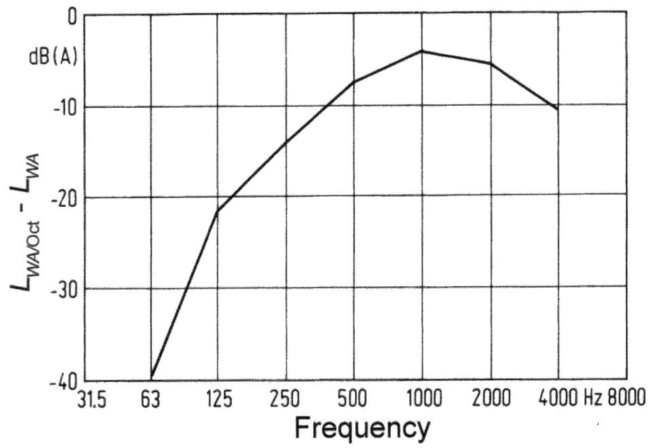

Fig. 20.47 Average standardised A-weighted octave sound power level spectrum of centrifugal pumps including the connected pipelines [144]

as small as possible and to make all processes as "soft" as possible.[4]

For *gear pumps*, the pressure differences between the pumped fluid volume and the fluid volume to which it is coupled can be essentially reduced by installing a pilot groove in the saddle apron as well as a groove for pressure relief [141, 150]. Thus, a reduction in the emitted airborne sound level by approximately 5 dB can be achieved.

For *axial piston pumps*, the pressure difference between the pumped volume and the fluid on the pressure side can be reduced by twisting the control plate. Thus, the pressure in the pumped volume can be adjusted so that the pressure in the cylinder volume in the moment of coupling is equal to the pressure of the system. As a result, pressure peaks are largely reduced and the sound pressure level decreases by approximately 5 dB compared with an operation without previous pressure compensation. Of course, the pressure compensation must be adjusted to each new operating pressure in the system to achieve the same effect.

Another solution, which is stated in [151], is a pressure compensation channel. Here, for example, a hole is drilled in the middle of both bars of the control plate. These holes are then connected with each other by means of a throttle duct. Thus, a sudden expansion and compression of the fluid volume locked at the dead-centre position is avoided. By means of a suitable design of the pressure compensation channel, a reduction in the emitted airborne sound by more than 10 dB can be achieved.

The tonal noise of a *centrifugal pump with guide vanes* can, above all, be reduced by a suitable selection of the number of rotor and guide vanes, as well as by an increase in the distance between rotor and guide vanes [152].

At all costs, a pump operation free of cavitation must be aimed for.

20.5.3.2 Measures at the Pipe System

A very effective measure is to attach a flexible layer in the immediate vicinity of the position, where the alternating pressure is generated. This layer compensates pressure surges, i.e. spreads them over a longer period (at least some milliseconds). Such pulsation dampers are more effective the closer they are located to the sound source and the "softer" they are.

If the sound can only be damped in the pipeline, normally also soft layers are used which reflect the arriving fluid sound, i.e. prevent its transfer. As there is a very close coupling of fluid sound and structure-borne sound in the pipe wall, the propagation of structure-borne sound must also be prevented. If the line pressure is less than 5 bar, soft rubber expansion joints are suitable.

The construction of silencers for liquid-borne sound is more difficult, if the static pressure is more than 10 bar and thus the soft layers are strained too much. So-called reinforced discharge hoses are rather ineffective, as they are already too hard for acoustic purposes.

The most important noise control measures for pumps are sound-insulating enclosures or claddings

[4] "Soft" means that discontinuities in acceleration and their derivations should be avoided as far as possible [149].

of the pump casing and sound-insulating lagging of the connected pipes (see Sect. 20.2.9).

20.6 Electric Motors

Three-phase current low-voltage motors (usually 400 V) are used for rated powers of approximately 1–400 kW, and three-phase current high-voltage motors (prevalently 6,000 V) for rated powers from approximately 160 kW upwards.

In recent decades, motor manufacturers have successfully reduced the noise emission of their electric motors. The most important share of noise emitted by air-cooled motors is caused by the fans. The noise caused by magnetostriction and by the bearings is usually negligible. Only for speed-controlled motors powered by a frequency converter, magnetostrictive noise can be important.

For low-voltage motors with quiet fans and for high-voltage motors with integrated noise control, the following A-weighted sound power levels L_{WA} (re 10^{-12} W) can be expected at sinusoidal power supplies:

Low-voltage motor

$$L_{WA} = 60 + 10\ \lg(P_N/1\,\text{kW}) \quad \text{dB(A)}. \qquad (20.85)$$

High-voltage motor

$$L_{WA} = 85 + 5\ \lg(P_N/1\,\text{kW}) \quad \text{dB(A)}. \qquad (20.86)$$

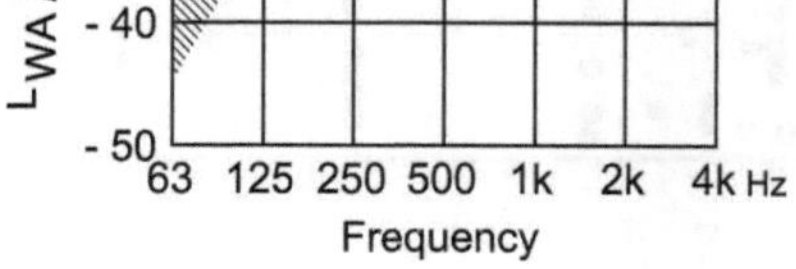

Fig. 20.48 Average standardised A-weighted octave sound power level spectrum of air and water-cooled electric motors. $L_{WA/Oct}$ A-weighted octave sound power level, L_{WA} total A-weighted sound power level

P_N – rated power of the motor in kW (1–3,000 kW).

For water-cooled high-voltage motors, the A-weighted sound power levels fall below the values according to Eq. (20.86) by approximately 5 dB.

Figure 20.48 can be used to estimate the A-weighted octave spectrum $L_{WA/Oct}$.

Further noise reduction can be achieved by a sound-insulating enclosure of the motors.

20.7 Wind Turbine Generator Systems

The first German highly advanced wind turbine generator system was developed by Hütter from 1955 to 1957 and was tested in the Swabian Jura Mountains until 1968. The energy crisis in the early 1970s led to a strong public support of the development of wind turbine generator systems (WTGS). As a result, the large wind turbine generator system Growian was developed with a rated power of 3 MW.

20.7.1 Design and Operation

Regarding about 30 existing different turbine systems, only the high-speed axial-flow rotor with three rotor blades has gained final acceptance. The rotor blades are mounted on the upwind side so that the wind hits the blades first and afterwards the tower or mast. The electric generator is housed within an engine car at the level of the rotor hub and is either directly (gearless WTGS) or through the intermediary of a gear transmission coupled to the shaft of the wind rotor.

WTGS are switched on at a hub wind speed of about 3–4 m/s, reach their rated power at about 12–14 m/s and are regulated to their rated power at higher wind speeds up to 25 m/s to protect them from overload. Two control mechanisms are common, either "*pitch*" *control* or "*stall*" *control*. For the pitch control, the rotor blades' angle of incidence is adjusted dynamically, so that the point of attack is smaller at higher wind speeds. In the stall control, the form of the rotor blades leads to a flow separation at the back side of the rotor blades when the rated power is reached and thus to an increase in the resistance at the rotor blades.

The electric power P_e that can be generated by a WTGS can be calculated with the following equation [153]:

$$P_e = \eta c_p S \frac{\rho}{2} v^3. \tag{20.87}$$

P_e – electric power that can be generated, in W
η – electric efficiency ($\eta \approx 0.75$)
c_p – coefficient of performance ($c_p \approx 0.45$)
S – area swept over by the rotor in m^2
ρ – density of air, ≈ 1.2 kg/m^3
v – wind speed at upwind side in front of the rotor in m/s.

The data of four wind turbine generator systems is shown in Table 20.3.

20.7.2 Noise Generation

The noise emitted by the WTGS consists of aerodynamically and mechanically generated noise. The aerodynamic noise, which in most cases governs the sound level, is caused by the rotor blades by vortex shedding at the blade tips and at the blade's trailing edge. Mechanical noise is generated by the gear unit (for gear-equipped WTGS only), by the generator and by auxiliary equipment like fans, yaw drives and power converters. Above all, the gear noise can cause an emission of loud single tones from the machine car and from the steel tower.

20.7.3 Approximate Calculation of Sound Emission

For a rated power of 0.5–3 MW, sound power levels of 103 dB are typical during operation of WTGS, as shown in Table 20.3. For plants with a rated power below 0.5 MW, the sound power level can be less than 100 dB(A). Some empirical values are given in Fig. 20.49.

The noise emission of a WTGS depends on the wind speed. The noise emission is lowest at the onset-wind speed and rises with increasing wind speed. Once the rated power is reached, the noise emission remains nearly constant for WTGS with pitch control, whereas it still increases for WTGS with stall control. The increase in the A-weighted sound power level with wind speed up to the rated power is about 1–2.5 dB(A) for every 1 m/s of wind speed increase [155, 156].

The A-weighted sound power level for WTGS without gear box and with pitch control can be estimated according to the following equation:

Table 20.3 Technical characteristics of wind turbine generator systems

Rated power (MW)	Rotor diameter (m)	Hub height (m)	Revolutions per minute at rated power (1/min)	Sound power level at rated power (dB(A))
4.5	113	124	12	103
1.5	66	70	23	103
0.6	46	60	24	95
0.25	30	40	40	95

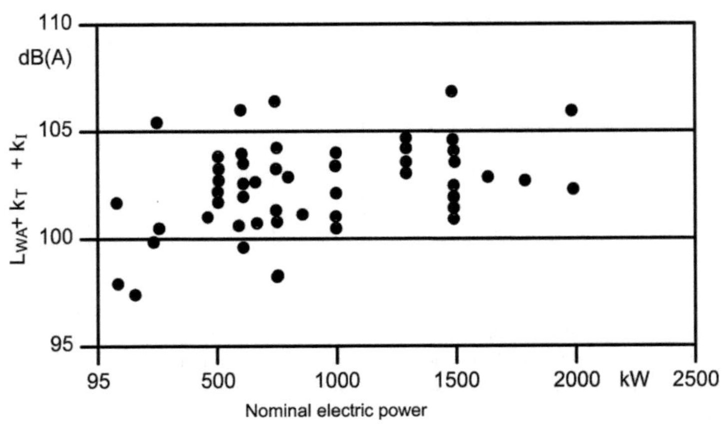

Fig. 20.49 A-weighted sound power levels of wind turbine generator systems including penalties k_T and k_I for tonal and impulsive noise, according to [154]

$$L_{WA} = -9 + 50\lg\left(\frac{U}{1\ \text{m/s}}\right) + 10\lg\left(\frac{D}{1\ \text{m}}\right)\ \text{dB(A)} \tag{20.88}$$

U – circumferential speed in m/s
D – rotor diameter in m

20.7.4 Noise Reduction

Important measures to reduce the noise emission of WTGS are:

– Reduction in the rotational speed. Experience shows that a noise emission reduction of 4 dB(A) results in a halving of the generated electric power. This method is applied if, for example, noise reduction is required during nighttime.
– Application of special planetary gears with small gear errors and high contact ratio.
– Vibration isolation of rotor and gear box.
– Vibration isolation of the machine car and the mount base of the generator by means of resilient elements.
– Reduction in the noise emission of the tower by reinforced concrete construction.
– Application of silencers in the cooling air circulation of the machine car.

20.7.5 Measurement and Evaluation of Noise

The noise emission of the WTGS is determined according to the European standard [157]. Comments to this standard and additional details are stated in [158]. This information helps to reduce the measurement uncertainty and to increase the reproducibility of noise test results.

According to the standard [157], noise emission measurements (and the determination of the actual electric power) should be carried out at wind speeds between 6 and 10 m/s (measured at a height of 10 m) or, at most, at a standardised wind speed for 95% nominal power of the WTGS. The background noise from vegetation or buildings is determined separately and is considered in the determination of the sound power level by a correction factor of up to 1.3 dB(A).

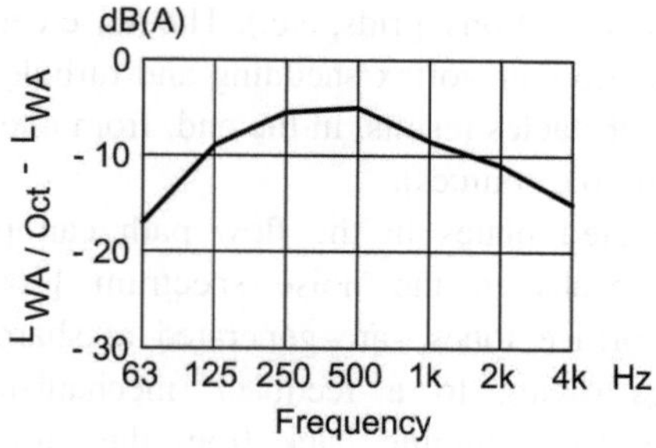

Fig. 20.50 Typical standardised A-weighted octave sound power spectrum of WTGS. $L_{WA/Oct}$ A-weighted octave sound power level, L_{WA} A-weighted sound power level

The essential acoustic characteristics to be determined are:

– The A-weighted sound power level and its 1/3 octave band spectrum (Fig. 20.50)
– A penalty for tonality if the noise emitted by the WTGS contains single tones

Optionally, the following WTGS acoustic data may be documented:

– Directivity of the sound emission
– Impulsiveness, if the noise emitted by the WTGS is impulsive
– Infrasound and low-frequency components, if existent
– Amplitude modulation of the emitted noise, if existent

Infrasound measurements, which were carried out in the past, show that the sound at frequencies below 20 Hz generated by the mainly used upwind turbines is far below the human perception threshold and is therefore not important [159].

The fact that WTGS noise is sometimes regarded as annoying can be explained by the pitch variations caused by gusty wind, which are perceived as very disturbing.

20.8 Turbulent Flow

20.8.1 Sound Generation

If air or another gas flows in a pipeline or flows out of the pipeline, vortices result at all technically interesting flow velocities and generate sound. Intensive sound is also generated by obstacles in the flow (orifices, deflections, grids, etc.). The noise caused by turbulent flow or vortex shedding and turbulent flow towards obstacles results, in the end, from alternating forces (dipole sources).

Perforated plates in the flow path can produce distinct peaks in the noise spectrum [160–162]. These "orifice tones" are generated at sharp-edged openings owing to a feedback mechanism: The pressure wave running back from the outlet edge periodically controls the ring vortex shedding at the inlet edge. The same tone generation was investigated by Heller [163] at sharp-edged openings for high subsonic flow velocities.

The frequency of the fundamental tone f_L is approximately:

$$f_L = \frac{1}{\frac{h}{\alpha u_L} + \frac{h}{c - \beta u_L}} \quad \text{with } 0.8 < \alpha < 1 \text{ and } 0.9 < \beta < 1. \tag{20.89}$$

h – thickness of the plate (=length of the openings) in m

u_L – average flow velocity in the opening in m/s

c – sonic speed in the undisturbed gaseous medium in m/s

20.8.2 Approximate Calculation of Sound Emission

The total sound power level L_W (re 10^{-12} W) of the noise emitted by a single-stage throttle in a gas flow can be estimated using the following equation [87]:

$$L_W = 93 + 10\,\lg\left(q \times c^2 \frac{1}{10^{-12}\mathrm{W}}\right) - 10\,\lg\left[1 + 6\left(\frac{p_2}{p_1 - p_2}\right)^{2.5}\right] \mathrm{dB}. \tag{20.90}$$

q – throughput in kg/s

c – sound velocity of the flow medium in m/s

p_1 – pressure in front of the obstacle in Pa

p_2 – pressure behind the obstacle in Pa

Figure 20.51 shows the sound power levels of single-stage throttles (obstacles, valve types, nozzles and ring nozzles in a pipeline as well as ventilation grids, perforated plates, bar grates, flow rectifiers and wire meshes) depending on $(p_1 - p_2)/p_2$. For comparison purposes, Eq. (20.90) is also drawn in.

As shown in [160], an increase in the turbulence of the approaching flow can amplify the emitted sound power significantly (by more than 10 dB).

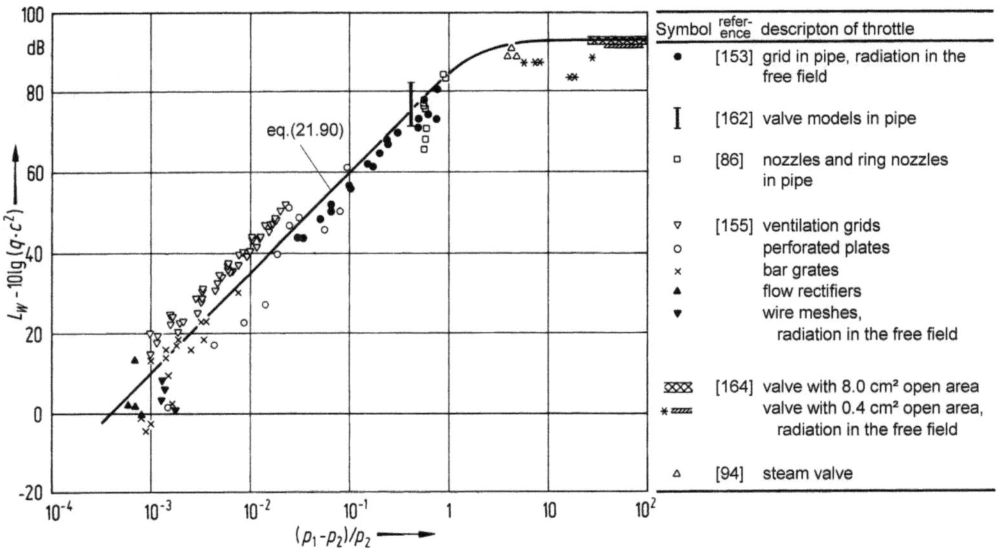

Fig. 20.51 Total sound power level of a single-stage throttle. L_W total sound power level in dB, q throughput in kg/s, c sound velocity in the flow medium behind the obstacle in m/s, p_1 pressure in front of the obstacle in Pa, p_2 pressure behind the obstacle in Pa

References regarding the pressure difference $p_1 - p_2$ can be taken e.g. from [104], and information on noise spectra can be found e.g. in [160, 164, 165].

According to [160], for ventilation grids and emission to the free field, the noise spectrum can be calculated e.g. with the average standardised octave sound pressure level shown in Fig. 20.52.

If the noise generated at a single-stage throttle is emitted to the free field via an exhaust opening, the noise spectrum is about that sketched in Fig. 20.21. For sound emission into a pipeline, the change in the noise spectrum owing to the influence of the pipe according to Sect. 20.2.1 must be considered.

20.8.3 Noise Control

As noise rapidly increases with pressure losses and thus with the rise of flow velocity, it is recommended to use low velocities and to avoid forms that are unfavourable with regard to flow (sudden changes in the cross section, sharp edges, etc.). For perforated plates in the flow sharp edges should be avoided to avoid orifice tones. Obviously, attention must also be paid to self-excited vibrations and cavity resonances.

If perforated plates are located in a pipeline with a relatively low flow velocity (Mach number <0.2), tones may be generated. These tones are caused by vortex shedding at the perforated plates and synchronised by transverse resonances in the room behind the perforated plate. A generation of such tones can be avoided by means of sound-absorbing linings of the duct walls behind the perforated plates.

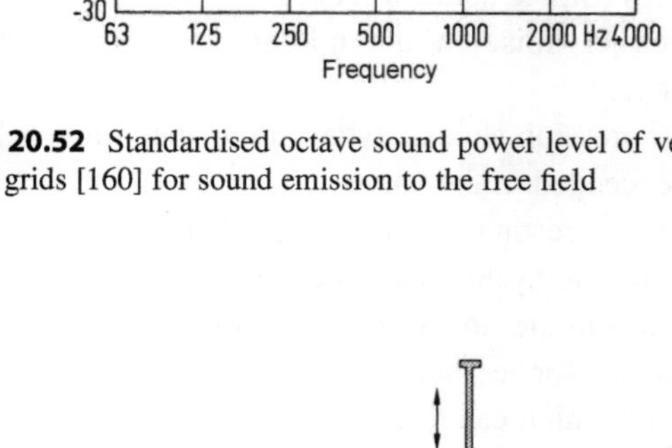

Fig. 20.52 Standardised octave sound power level of ventilation grids [160] for sound emission to the free field

20.9 Valves

For flow and pressure control of flowing gaseous and fluid media, fittings (valves, control elements) are used to vary the flow resistance by varying the cross section of the opening. Some types of fittings are shown in Fig. 20.53.

20.9.1 Sound Generation

Figure 20.54 shows typical pressure and velocity distributions in a fitting. In the vicinity of the smallest cross section of the fitting, the highest flow velocities occur. Therefore, this area is the location where most sound is generated.

20.9.1.1 Liquids

If the pressure p_1 in front of the fitting is constant and the pressure p_2 after the fitting is reduced so that the pressure p_3 drops to a value near the steam pressure p_D, cavitation occurs, together with a strongly growing noise. If the pressure p_2 is further reduced, the pressure p_3 remains constant, i.e. about at the steam pressure p_D. This happens, until all the liquid is vaporised.

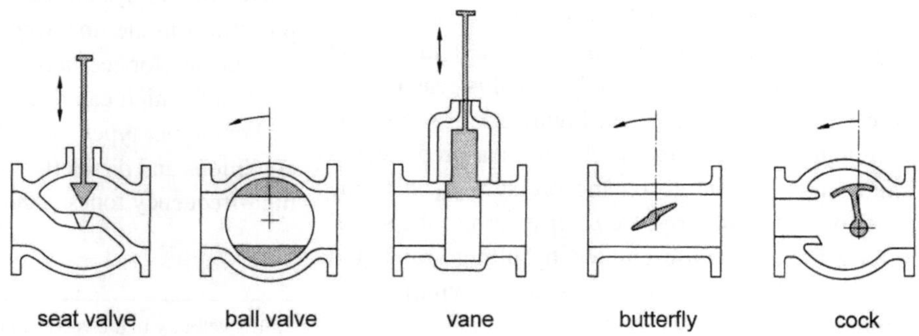

Fig. 20.53 Types of fittings [166]

Fig. 20.54 Sound generation in fittings

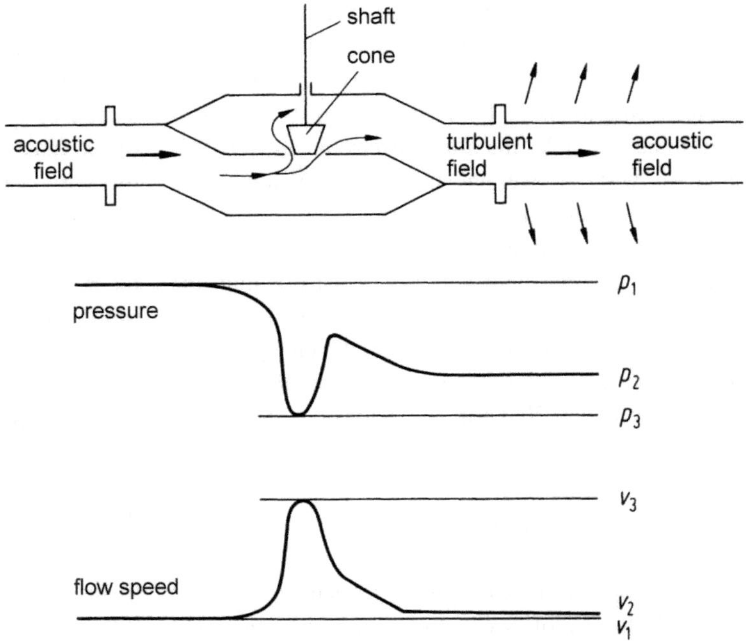

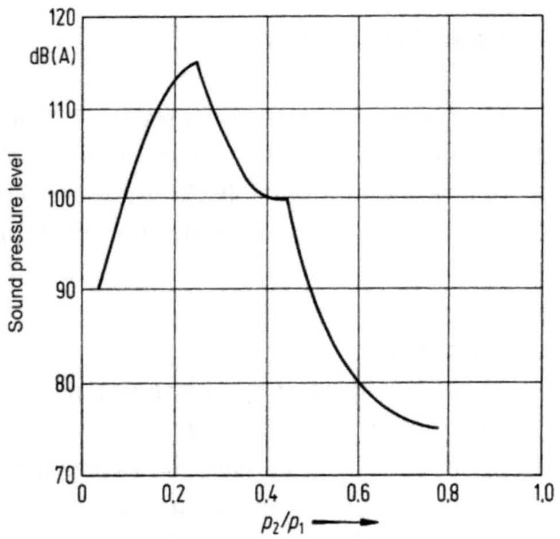

Fig. 20.55 Sound pressure level of a liquid valve according to [167]

As soon as p_2 has also reached the value of the steam pressure, cavitation bubbles cannot collapse any longer, as the medium behind the valve is already in a gaseous state ("evaporation"). For this reason, less noise will be generated then. Figure 20.55 shows the A-weighted sound pressure level measured near a liquid valve depending on the pressure p_2 after the valve for a constant pressure p_1 upstream of the valve.

As a rule, the noise emitted by fittings for liquid can be neglected as long as neither cavitation nor evaporation occurs.

20.9.1.2 Gas

If the ratio p_1/p_2 of a single-stage relief exceeds a value of approximately 2, the flow velocity in the narrowest part reaches sonic speed. Downstream of this narrowest spot, also supersonic speed and, thus, shock waves can occur. The flow in the valve is very turbulent.

For fittings in a gas flow, noise mainly occurs owing to the flow field in the area of the narrowest cross section (the so-called veha contracta), caused by:

- Vortex shedding and turbulent flow approaching solid structural components, so that alternating forces (dipole sources) occur
- Open-jet noise in the mixing zone (quadrupole sources)
- Compression shocks, when sonic speed is reached or exceeded in the valve

Besides, resonance vibrations of the valve cone – valve spindle system can occur in a valve and, thus, pulsations in the flow (monopole sources). This effect can occur for certain valve settings and lead to a "rattle" which can destroy the valve.

For some types of valves, for certain operating conditions and distinctly subcritical[5] flow, occasionally high-frequency tones occur which can be explained by

[5] Flow velocity in the valve smaller than half the sound velocity.

a control of vortex shedding owing to eigenfrequencies of the valve cavity. In general, small changes in the valve cone are sufficient to avoid this effect.

According to [87, 168–170], the pipeline connected to a fitting is mainly excited by the acoustic alternating pressure field is generated by the fitting and not by the turbulent flow field behind the fitting, as long as relief only takes place within the fitting.

20.9.2 Approximate Calculation of the Sound Emission

In this chapter, the estimation of the sound power level of fittings with single-stage relief is discussed. The noise caused by a fitting and transmitted to the outside is mainly emitted via the pipe downstream of the valve. For this reason, the estimation of the sound emission is restricted to the pipe after the valve. At first, the sound power level spectrum is determined, which is emitted by the fitting under free-field conditions (i.e. the diameter of the pipe after the valve is very large). The statements made in Sect. 20.2, can be used to calculated the sound power level in the pipeline after the valve, from which the sound power level spectrum of the noise emitted to the outside can be calculated.

Further calculation methods of the sound emission from fittings are described in [171–173]; they require specific information on the valve in contrast to those following next.

20.9.2.1 Liquid

The total sound power level (re 10^{-12} W) of the noise transmitted to the pipeline downstream of the valve is estimated considering the results stated in Sect. 20.1.2.4.

20.9.2.2 Without Cavitation

$$L_W = 41 + 10\,\lg\left(\frac{q}{c_0}u^3\frac{1}{10^{-12}\mathrm{W}}\right)\ \mathrm{dB}\quad \text{for}\quad \sigma > \sigma_k. \tag{20.91}$$

20.9.2.3 Fully Developed Cavitation

$$L_W = 88 + 10\lg\left(\frac{q}{c_0}u^3\frac{1}{10^{-12}\mathrm{W}}\right)\ \mathrm{dB}\ \text{for}\ 0 < \sigma < 0.5\sigma_k. \tag{20.92}$$

20.9.2.4 Evaporation

$$L_W \approx 88 + 10\,\lg\left(\frac{q}{c_0}u^3\left[\frac{p_2}{p_D}\right]^3\frac{1}{10^{-12}\mathrm{W}}\right)\ \mathrm{dB}\ \ \text{for}\ \ \sigma < 0 \tag{20.93}$$

with

$$\sigma = \frac{p_2 - p_D}{0.5\,\rho\,u^2}. \tag{20.94}$$

u – flow velocity at the narrowest cross section of the fitting in m/s
c_0 – sound velocity in the undisturbed fluid in m/s
q – flow volume in kg/s
σ – cavitation number
σ_k – critical cavitation number when the respective valve starts to cavitate
p_1 – pressure upstream of the fitting in Pa
p_2 – pressure downstream of the fitting in Pa
p_D – steam pressure of the fluid in Pa
ρ – density of the fluid in kg/m^3

In the range $\sigma_k > \sigma > 0.5\,\sigma_k$, the sound power strongly increases with a decreasing σ (Fig. 20.7), such that definite calculation of the sound power level cannot be made, whereas the sound power level for evaporation can be roughly estimated by Eq. (20.93).

An approximate calculation of the flow velocity u at the smallest passage in the fitting can be made using the following equation:

$$u = \sqrt{1.4\cdot\frac{2}{\rho}(p_1 - p_2)}. \tag{20.95}$$

The sound power spectrum can be estimated with Fig. 20.9 in connection with Eqs. (20.21) and (20.37), if

$$d = \sqrt{\frac{4q}{\pi \rho u}}. \quad (20.96)$$

20.9.2.5 Gas

The total sound power level L_W of the noise transmitted to a pipeline downstream of a fitting in a gas flow can be estimated for a standard adjustable valve with single-stage relief by application of Eq. (20.90) for sound emission into the free field (very large pipe diameter). Figure 20.56 shows the measured total sound power levels L_W (re 10^{-12} W) – standardised with $q\,c^2$ – versus $(p_1 - p_2)/p_2$ for different standard valves. The solid curve represents Eq. (20.90) for comparison.

According to [87], the 1/3 octave spectrum in the pipeline downstream of the fitting in a gas flow can be calculated approximately with the following equation:

$$L_{W/\mathrm{Terz}} = L_W + \Delta L_1 - \Delta L_2 - 5\,\mathrm{dB} \quad (20.97)$$

with [see Eq.(20.37)]

$$\Delta L_1 = 10\lg\left(1 + (0.77 f_G/f)^2\right)\mathrm{dB} \quad (20.98)$$

with f_G as in eq. (8.36).
and

$$\Delta L_2 = \begin{cases} 10\lg\left(1 + (f_0/f)^{2.5}\right)\mathrm{dB} & \text{for} f < f_0 \\ 10\lg(1 + f/f_0)\mathrm{dB} & \text{for} f \geq f_0 \end{cases}$$

$$\text{with } f_0 = 0.4\frac{u}{\sqrt{S_3}}. \quad (20.99)$$

In the narrowest cross section of the fitting, the flow velocity u, the density ρ and the surface S_3 can be estimated as follows:

$$u/c = \begin{cases} \sqrt{(p_1 - p_2)/p_2} & \text{for } p_1 < 2p_2 \\ 1 & \text{for } p_1 \geq 2p_2 \end{cases}$$

$$\rho_3/\rho_1 = \begin{cases} [1 - 0.47(p_1 - p_2)/p_2]^{1/\kappa} & \text{for } p_1 < 2p_2 \\ [2/(\kappa + 1)]^{1/(\kappa - 1)} & \text{for } p_1 \geq 2p_2 \end{cases}$$

$$S_3 = \frac{q}{\rho_3 u}. \quad (20.100)$$

ρ_1 – density of the gas in the pipeline in front of the fitting in kg/m^3
κ – isentropic coefficient of the gas
c – sound velocity of the pipe downstream of the fitting in m/s

20.9.3 Noise Control

For *fittings for liquids*, the main task is to avoid cavitation. This can be achieved by:

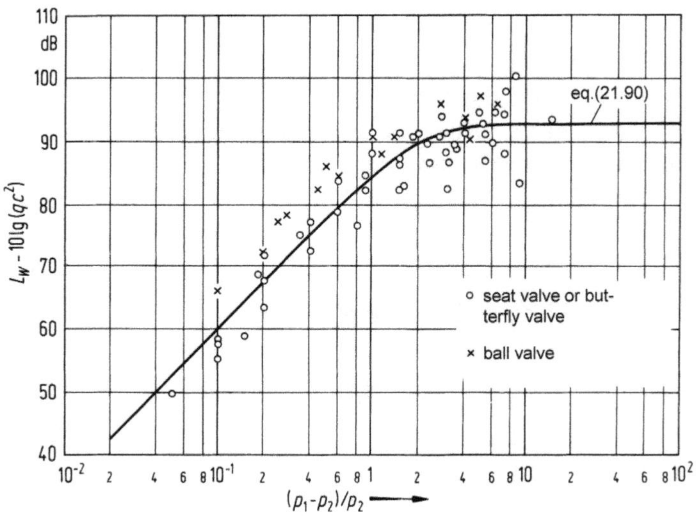

Fig. 20.56 Total sound power level in a pipeline behind a single-stage standard valve in a gas flow [87] (very large pipe diameter). *Open circle* single-seat adjustable valve with parabolic cone or butterfly valve, *crosses* ball valve

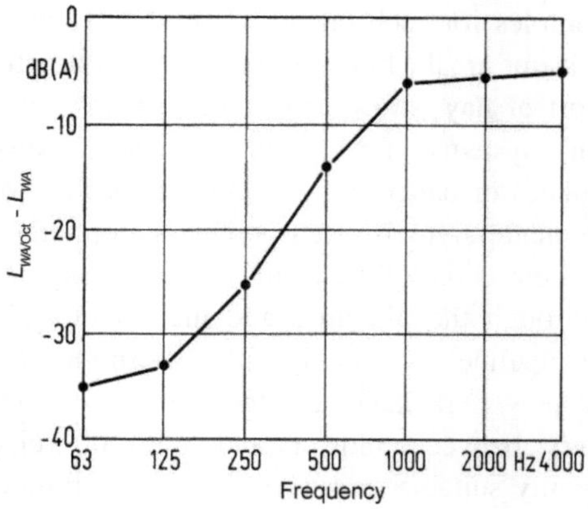

Fig. 20.57 Average standardised A-weighted octave sound power spectrum of the water noise of a cooling tower [179]. $L_{WA/Oct}$ A-weighted octave sound power level, L_{WA} total A-weighted sound power level

- Selection of a large counter-pressure, e.g. by installing the fitting at the lowest possible location.
- Using a fitting with a small pressure recovery (multistage fitting). Sometimes it is sufficient to change the flow direction or to install two valves in series.
- Installation of a flow resistance downstream of the fitting, e.g. perforated plates.

Regarding *fittings for gas*, the main task is to keep the maximum flow velocity in the fitting as low as possible and to shift the noise spectrum to very high frequencies. This can be achieved by:

- Relief in several steps
- Installation of flow resistances, e.g. perforated plates
- Partition of the free cross section of the opening into several small openings

The level reduction achieved by these measures can be up to 25 dB.

Further noise control measures for fittings are:

- For fittings in a gas flow, installation of silencers[6] between the fitting and the connected pipelines. For noise in the pipelines, level reductions by 20 dB(A) or more are possible.
- Sound insulating laggings for the pipelines and the valve casing. A level reduction of up to 35 dB(A) can be achieved.

[6] Reduction of gas sound in the pipelines as well as of structure-borne sound in the pipe walls.

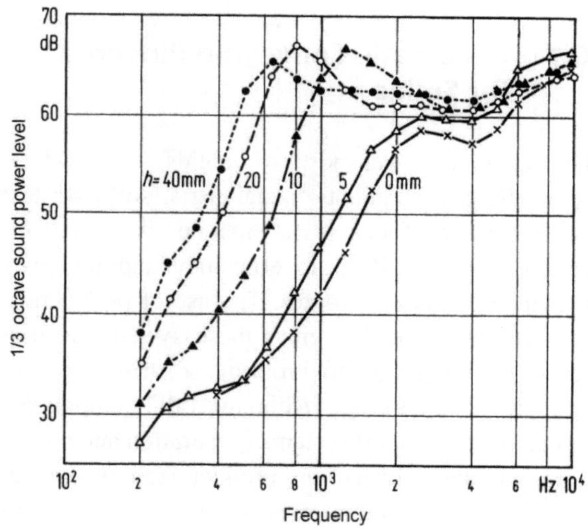

Fig. 20.58 One-third octave sound power spectra of impact noise emitted by water drops hitting the surface of water with different depths h

20.10 Water Noise in Cooling Towers

For large cooling towers (in power stations, refineries and petrochemical plants), noise is generated by water falling freely from several metres of height [174–179]. The emitted sound power is assumed to be proportional to the water throughput and the impact velocity of the drops.

An estimation of the A-weighted sound power level of the water noise of cooling towers, which has proved to be effective in practise, can be made with the following relation together with Fig. 20.57:

$$L_{WA} = 33 + 10\ \lg\left(\frac{q}{1\,\mathrm{m^3/s}}\right) \pm 2\ \mathrm{dB(A)}. \quad (20.101)$$

q – water throughput in m^3/s.

According to [177], the resulting noise spectra strongly depend on the depth h of the cold water basin, as shown in Fig. 20.58. From this, the following possibilities for water noise reduction in cooling towers exist:

- Constructive measures at the basin so that the water level is as shallow as possible.
- Impact reduction, e.g. by means of close-meshed grids or webbings floating on the water surface.

With such measures, the sound power level of water noise can be reduced by 10 dB(A).

20.11 Pneumatic Conveying Pipelines for Solids

The pneumatic transport of solids – powders, chemicals, grains, pellets, metal parts, solid wastes, coal, rocks, and foods – via pipelines is a common method owing to its high efficiency and the low-maintenance transport routes. The use of pipe bends, switches and branching gives these systems a large degree of flexibility for routing, distribution and automated networking. Although all components of pneumatic transport systems (generation and conditioning of the delivery air, product feeding device, transport pipeline and product discharge device) may be important from the acoustic point of view, often the pipelines are problematic sound sources because of their large extension and their exposed position.

Apart from the type of the conveying system, the type of the fluid flow engine which supplies the delivery air is also decisive for the sound emission of a pneumatic transport system. Therefore, from the acoustic point of view, there are two basically different types of pneumatic transport systems which differ from each other with regard to the pressure level of the delivery air.

20.11.1 Low-Pressure Conveying Systems

As a rule, low-pressure transport systems have conveyor fans which are flown through by the delivery air and the bulk material to be transported. Conveyor fans are radial fans with a specially designed impeller. The total pressure difference is normally <0.1 bar. The transport lines are normally thin-walled sheet metal pipes or ducts with a length of up to 100 m. Product feeding is normally direct aspiration of the material to be transported by the suction port of the fan. The delivery air speed is normally above 20 m/s. During the transport, the low-concentrated material is spread evenly over the pipeline's diameter. This state of transport is called dilute- or lean-phase pneumatic conveying [180, 181]. The product discharge is usually made by ejection of the delivery air and the transported material within a storage container or a storeroom. The delivery air can then be separated by means of a textile filter or a cyclone separator.

Examples for such pneumatic low-pressure transport systems are the blowers used in agriculture for the transport of hay, straw or chopped straw as well as suction systems for wood shavings, sawdust, particulates or paper cuttings. As the specific weight of the material to be transported is normally very small, there is only little contact noise during its passage through the pipelines and the sound emission of the pipelines is actually only determined by the sound power radiated by the blower. Therefore, silencers between blower and transport pipeline are usually suitable to reduce the sound emission of pneumatic low-pressure transport pipelines or their discharge ports, if the type of the silencer allows the unhindered passage of the material to be transported. For low-pressure lean-phase pneumatic conveying systems, the relation of the mass flow rate of the material to be transported $\dot{M}_P$ and of the conveying air $\dot{M}_L$, which is also called load factor μ is normally <1. Therefore, the sound absorption of the air loaded with solids is in usual applications only marginally higher than that of pure air and the sound level decrease along the pipeline actually corresponds to that in regular ventilation ducts. Thus, there are normally no specific acoustic phenomena for pneumatic low-pressure transport systems and possible sound reduction measures are the same as for standard blowers applications.

20.11.2 High-Pressure Conveying Systems

For pneumatic high-pressure transport systems, the required delivery air is generated by a rotary blower (roots compressor or rotating screw compressor) or is taken from a compressed air net. The pressure is between about 0.5 and 5 bar. The transport pipelines consist of steel, stainless steel or aluminium. The pipelines can have a length of several 100 m up to about 2 km. Owing to the high pressure of the delivery air, the material to be transported must be fed into the transport pipe by a sealed feeding device. For grainy bulk goods, e.g., rotary-vane feeders are used, and for powdered material screw feeders are used. The product discharge at the end of the transport pipeline is made either directly or by a cyclone separator. The product is filled into a storage or transport container or is discharged directly at the place of consumption.

Concerning the sound emission of pneumatic transport pipelines, the transport condition, i.e. the type of flow in the pipeline, is decisive. In the literature as well as in practise, a variety of terms can be found to describe the transport condition, for instance dilute- or dense-phase, plug, streak, cyclic or slow motion conveying. These terms are not used in a standardised way by manufacturers and experts of pneumatic transport plants and can therefore lead to confusion. For a better understanding of the sound generation in transport pipelines, the two possible extremes of the transport state should be discussed, *lean-phase pneumatic conveying* and *dense-phase pneumatic conveying* [182].

At a high air speed and with a small material fraction in the delivery air, the gaseous and solid phases are transported through the pipe in a nearly completely mixed state. The velocity of the solid particles is approximately 50–80% of the air speed. This state of flow clearly belongs to the domain of lean-phase conveying and is acoustically characterised by a large number of collisions of the solid particles with the pipe wall which causes a considerable structure-borne sound excitation – and thus sound emission – of the transport pipeline (Fig. 20.59). Therefore, the velocity and mass of the particles as well as the thickness and material of the pipe wall are essential parameters of the sound emission.

For low air speeds and a high solid loading of the delivery air, especially for coarse-grained bulk goods, an almost completely separated gas/solid flow results. In this case, the bulk goods are transported in the form of streaks or dunes at the bottom of the pipe with the air sweeping across it. The velocity of the solid particles is much lower than the air speed, which means that this is pure dense-phase conveying (Fig. 20.59). Owing to the slow movement of the solid accumulations and the accompanying rubbing and rolling, the transport pipe is much less excited to vibrations than in lean-phase conveying.

For technically used transport systems, there is no clear parting line between the above mentioned extremes of lean-phase and dense-phase conveying. The sound emission of the transport pipes is determined by the percentage of lean-phase flow in the mass flow of the bulk goods.

Depending on the concentration of the solid particles and the type of the transported goods, an increased sound reflection and sound absorption can occur in the transport pipeline. Thus, as a rule the noise transmitted to the transport pipe by the delivery air blower or the feeding device is reduced so much at some metres of pipe length that the sound emission of the transport pipe is practically only determined by the collision excitation of the pipe wall by the solid particles. It is obvious that a sound reduction in these transport pipelines cannot be achieved with silencers, but only with a sound-insulating cladding, whereas a silencer at the exit of the delivery air blower may be useful owing to the often considerable length of air piping between blower and solid feeding point, if the sound emission of the air conduit is essential and a sound insulation of the pipe is more expensive.

20.11.2.1 Sound Generation at Bends

Owing to the high momentum change in the material to be transported caused by obstacles and changes of direction, elbows and bends of transport pipes are object to intense wear and attrition, especially in lean-phase conveying systems. Therefore, in pipelines for highly abrasive goods, special welded constructions with a reinforced outward curve, often called "box

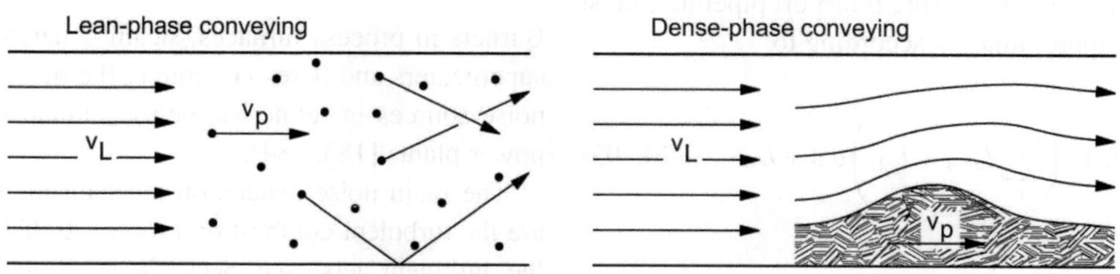

Fig. 20.59 Conditions of the gas–solid flow in the pipeline of a lean-phase or a dense-phase pneumatic conveying system

Lean-phase conveying	Dense-phase conveying
(left) completely mixed	(right) completely separated
Gas-solid flow	Gas-solid flow

elbows" due to their rectangular shape, are commonly used instead of normal pipe elbows.

As in this case the collision excitation of the pipe wall is much more vigorous, the sound emission of bends is much higher than that of a straight transport pipe. Owing to the structure-borne sound excitation of the bends, the adjacent parts of the transport pipes also have a higher sound emission than those pipe sections which are more distant from the bends. Compared with the straight pipe, an increased sound emission can be assumed in a section about double the length of the bend.

As compromise between the smallest possible bending radius and a minimal wear of material, the relation of bending radius R and pipe diameter d_a of transport pipe bends is normally dimensioned to $R/d_a = 6$. In this case, the "acoustical" length $l_{ak,B}$ (in m) of a bend, i.e. the length of a straight transport pipe of equal sound emission, can be estimated according to:

$$l_{ak,B} = k_b d_{\mathrm{a}} \tag{20.102}$$

with

$k_b = 80$ for bends in lean-phase conveying pipelines
$= 160$ for bends in dense-phase conveying pipelines
$= 320$ for box elbows in lean-phase conveying pipelines

d_{a} – outside diameter of the transport pipeline in m

These recommended values are meant for 90° bends. For directional changes by 45°, the above listed values for k_b can be halved.

20.11.2.2 Approximate Calculation of the Sound Emission

The A-weighted sound power level L_{WA} (re 10^{-12} W) of a pneumatic high-pressure transport pipeline can be calculated approximately according to:

$$L_{\mathrm{WA}} = 10\ \lg\left(\sum_n l_{ak,B} + l_{\mathrm{g,L}}\right)\mathrm{dB} + L_{\mathrm{W''A}}. \tag{20.103}$$

$\Sigma l_{ak,B}$ – sum of the "acoustic lengths" of n bends, elbows and switches in the pipeline according to Eq. (20.102)

$l_{\mathrm{g,L}}$ – geometric length of the straight sections of the transport pipeline

$L_{\mathrm{W''A}}$ – the length-specific A-weighted sound power level per metre of the transport pipeline, which can be estimated for granular material with a Young's modulus of 200–2,000 N/mm^2 and metal transport pipelines as follows:

$$\begin{aligned} L_{\mathrm{W''A}} = {} & 28 + 13\lg\left(\frac{d_{\mathrm{a}}}{1\,\mathrm{m}}\right) - 30\lg\left(\frac{h}{1\,\mathrm{mm}}\right) \\ & - 10\lg\left(\frac{\rho_{\mathrm{w}}}{1\,\mathrm{kg/m^3}}\right) + 3.5\lg\left(\frac{\dot{M}_{\mathrm{p}}}{1\,\mathrm{t/h}}\right) \\ & + 7\lg\left(\frac{m_{\mathrm{p}}}{1\,\mathrm{mg}}\right) + 10\lg\left(\frac{\mathrm{v}_L}{1\,\mathrm{m/s}}\right)\mathrm{dB}. \end{aligned} \tag{20.104}$$

d_{a} – outside diameter of the transport pipeline in m
h – wall thickness of the transport pipeline in mm
ρ_{w} – density of the pipe wall in kg/m^3
v_L – air speed in the transport pipeline in m/s
$\dot{M}_{\mathrm{P}}$ – product mass flow in the transport pipeline in t/h
m_{P} – particle mass of the material to be transported in mg

For dense-phase conveying, the significant product velocity v_{P} has to be inserted in Eq. (20.104) instead of the air speed v_L. For conventional applications $v_p \approx 1$ m/s.

Figure 20.60 shows the standardised octave sound power spectrum of pneumatic transport pipelines of lean-phase or dense-phase pneumatic conveying systems.

20.12 Industrial Burners

Burners in process furnaces, steam generators, steam superheaters and flares are among the most important noise sources in refineries, petrochemical plants and power plants [183, 184].

The main noise generation mechanisms in burners are the turbulent combustion process itself [185–190], the turbulent jets (see Sect. 20.8) of gaseous fuel injected through the burner nozzles and, if applicable, the jets of injected steam or air to atomise liquid fuel or to optimise the combustion process. Combustion noise itself is mainly dependent on the fired duty (heat release rate) and the degree of turbulence in the flame.

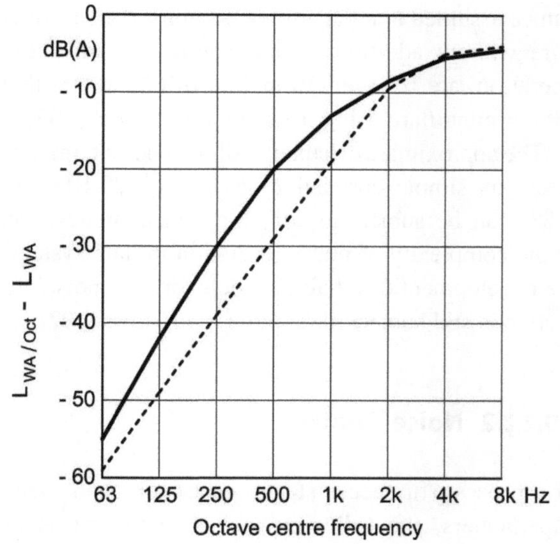

Fig. 20.60 Standardised A-weighted octave sound power spectrum of high-pressure pneumatic transport pipelines. $L_{WA/Oct}$ A-weighted octave sound power level, L_{WA} A-weighted sound power level. *Continuous line* dense-phase conveying, *dotted line* lean-phase conveying

From an acoustics point of view, the burners in process furnaces, steam generators, steam superheaters and ground flares are firing into an enclosure. In that case, a typical burner noise spectrum is characterised by broadband noise to which a number of peaks with higher noise levels are superimposed, the latter ones being caused by resonances in the combustion chamber and, sometimes, in the connected ductwork of the supply systems for gaseous fuel and/or combustion air.

In situations, where an undesired feedback between the resonances of the combustion chamber and the combustion process leads to self-excitation of pulsations in the system, tonal noise components with very high intensity can be generated (see Fig. 20.63 and Sect. 20.13).

20.12.1 Approximate Calculation of the Noise Emissions

The noise generated by *burners in process furnaces, steam generators and steam superheaters* is mainly radiated into the air supply system and into the combustion chamber. The noise radiated into the air supply system is either emitted directly into the open – in case of self-aspiring natural draft burners – or transported upstream in the air supply duct system in case of forced draft burners. Noise in the air supply system is transmitted through the duct walls and from there radiated into the open. Likewise, noise in the combustion chamber is transmitted through the combustion chamber walls and casing, through the walls of the flue-gas duct system and via the flue-gas stack into the open.

Burner noise emissions are mainly influenced by

- The fired duty (heat release rate) of the burner
- Composition, temperature and pressure of the injected fuel gas
- Flow rate and exit velocity of injected steam or air for fuel oil atomisation or combustion enhancement
- Flow rate and flow velocity of the combustion air

Noise originating from sources in the gas or air supply system – from control valves or fans, for example – can be amplified by the combustion process and contribute to the overall noise emissions of the burner.

Generally applicable calculation models that allow to predict reliably the noise emissions of burners, taking into account the relevant process parameters, are still subject of current research.

For process furnaces, steam generators and steam superheaters provided with high-quality noise control, it is in most cases possible to comply with an A-weighted sound power level L_{WA} (re 10^{-12} W) of

$$L_{WA} = 83 + 10\,\lg\left(\frac{N_0}{1\,\text{MW}}\right)\text{dB(A)}. \qquad (20.105)$$

N_0 – nominal heat release rate of a burner in MW; $0.1\ \text{MW} < N_0 < 5\ \text{MW}$

for the noise generated in the burners and radiated into the open. By adding $25\ \lg(N/N_0)$ to the right-hand side of Eq. (20.105), the A-weighted sound power level at the actually fired heat release rate N can be estimated.

So-called *ground flares* typically consist of a number of burners firing into a kind of stack open at the top with a large diameter. In typical ground flares, noise from the burners is radiated to the open either from the top of the flare enclosure – the flare stack – or through the air intake openings at the bottom of the flare – the burner "windows". For locations at ground level and at short to medium distances from the flare, the noise received from the burner windows is usually up to 6 dB higher than the noise coming from the stack top if the flare is not equipped with any secondary noise control. To reduce the noise emissions from the burner windows, state-of-the-art ground flares are usually equipped with noise barriers around the base. Often, for this purpose, the so-called wind fence that prevents wind from blowing into the burner windows is improved with respect to its efficiency as a noise barrier by adding a sound-absorbing lining on its inner side and by increasing its sound transmission loss.

For a standard steam-assisted ground flare with good secondary noise control, in normal operation and firing a waste gas with little tendency to smoke, a rough estimate of the total A-weighted sound power level can be obtained as

$$L_{\mathrm{WA}} = 100 + 15\lg\left(\frac{q}{1\,\mathrm{t/h}}\right)\begin{matrix}+3\\-5\end{matrix}\,\mathrm{dB(A)}. \qquad (20.106)$$

q – waste gas mass flow rate in t/h (3–100 t/h).

A typical corresponding octave-band spectrum is shown in Fig. 20.61 [191].

Ground flares are normally built for waste gas flow rates of up to 100 t/h.

So-called *elevated flares* are generally characterised by a stack with the discharge point at the top end where the waste gases burn in a single large flame in the open. The range of possible designs, operating conditions, smoke suppression systems and fired waste gases is even wider than for ground flares. Accordingly, the noise emitted by elevated flares will strongly depend on these and a number of additional parameters, so that no simple approximate equation for their calculation is available. As a rule of thumb, it

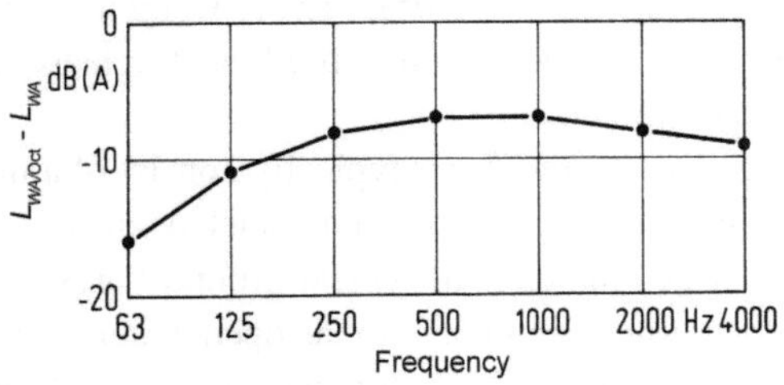

Fig. 20.61 Example for the A-weighted octave-band spectrum of the sound power level of the noise emitted by a typical steam-assisted ground flare with good secondary noise control, in normal operation and for a waste-gas composition with little tendency to smoke [144, 191]. $L_{\mathrm{WA/Oct}}$ A-weighted octave-band sound power level, L_{WA} total A-weighted sound power level

can be assumed that the noise emissions of an elevated flare without additional noise control, in smokeless operation, are between 10 and 20 dB(A) higher than for a ground flare firing the same rate of fuel [183].

The approximate calculation of flare noise emissions based on simple correlations as in Eq. (20.106) and [183] can be subject to very high uncertainties. Due to the complexity of noise generation in flare systems, the development of reliable models for flare noise prediction is still subject to research in progress [192].

20.12.2 Noise Control

For process furnaces, steam generators and steam superheaters, the following noise control measures are usually installed to reduce the emission of burner noise into the open:

- For natural draft burners: Installation of silencers (mufflers) at the air intakes of the burners.
- For forced draft burners: Installation of combustion air ducts with circular cross section instead of rectangular cross section (higher sound transmission loss, see Sect. 20.2.7); application of an exterior acoustic insulation to the combustion air ducts (see Sect. 20.2.9) and/or installation of a silencer between the burner and the combustion air duct.
- Application of an exterior acoustic insulation to the burner casing (wind box).
- Optimisation of the furnace walls from an acoustics point of view: higher sound transmission loss by increased mass per unit area or acoustic double-shell design [193]; increased sound absorption inside the furnace combustion chamber, for example, by a lining with ceramic fibre.
- Installation of a silencer between furnace and flue-gas duct in the stack.

For flare systems in general, the following noise control principles may be considered [191]:

- For flares with steam injection for smoke suppression: limitation of the steam flow rate and the steam exit velocity to the minimum necessary to maintain smokeless combustion.
- Minimisation of the exit velocity of the waste gases.
- If noise from inside the flare gas system plays a role: installation of a silencer in the gas piping.

For ground flares, the following additional noise control measures are possible:

- Increased sound absorption in the flare stack, for example by an absorptive gravel bed.
- Installation of sound barriers or acoustically optimised wind fences at the burner windows.

20.13 Self-Excited Combustion Oscillations

In combustion systems of all kinds and sizes – from domestic heating devices to process furnaces, jet engines and stationary gas turbines – so-called self-excited combustion oscillations are observed. They are characterised by loud humming or whistling noise that can also show beats phenomena and strong vibrations of the affected unit. The pressure fluctuations associated with self-excited combustion oscillations show very high amplitudes at discrete frequencies, which is the main difference from normal broadband combustion noise. Amplitudes reach far higher values than in case of pure resonance effects, i.e. when the natural frequencies of the combustion chamber are excited by combustion noise. Combustion oscillations are accompanied by an increased heat transfer to the system walls; they often have a negative effect on flame stability and system efficiency and can lead to increased pollutant emissions. In addition to the unacceptable noise emissions, the increased mechanical and thermal loads brought about by the oscillations lead to premature wear and can, in extreme cases, result in the complete destruction of the burner and/or connected system components.

In contrast to the noise emissions originating exclusively from the turbulent flame, self-excited combustion oscillations are characterised by a feedback between the oscillations and the combustion process (the flame). Oscillations are maintained by a periodic energy supply from the combustion reaction, with that periodic energy supply being caused and timed by the oscillations themselves. Thus, the oscillations can excite and intensify themselves.

Although described for the first time already more than 200 years ago [194], it was only much later that significant problems caused by self-excited combustion oscillations were observed in technical applications [195, 196]. Since then, their importance has constantly grown. Also in these days, self-excited combustion oscillations ever more often cause problems in industrial combustion systems which, among other reasons, is mainly owed to the continually increasing power densities of modern burners and the intensified development towards lean premixed combustion – both effects increasing the susceptibility for combustion oscillations.

Avoiding or eliminating self-excited combustion oscillations is made difficult by the often extreme sensitivity of the effect to minimal variations of individual system parameters, such as system duty, temperature, pressure, flow velocity, vorticity, geometry, and fuel mixing. Often are minor variations of these parameters critical for self-excitation.

20.13.1 Self-Excitation Mechanisms

From case to case, the mechanism leading to self-excitation of combustion oscillations can differ significantly in its details. However, in most technically relevant cases, it is based, in principle, on a feedback interaction between the sound field in the system, the flow through the system and the energy supplied by the exothermal chemical reaction.

An example for such an interaction shows Fig. 20.62 as simplified block diagram. Pressure fluctuations p'

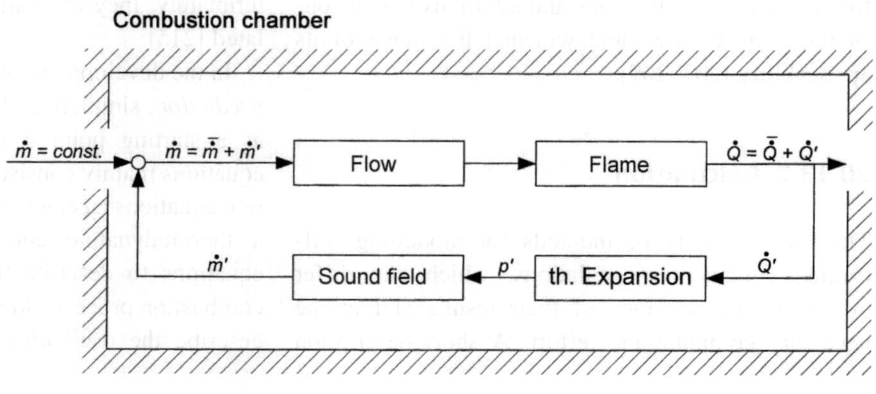

Fig. 20.62 Self-excited combustion oscillations: simplified block diagram of a possible feedback mechanism

excite a characteristic sound field in the combustion chamber and the connected air and gas ducts. These pressure fluctuations and the fluid movement associated with them can induce a fluctuating mass flow rate $\dot{m}'$ of the air and/or gas flow entering the combustion chamber. Carried by the mean flow to the reaction zone, these mass flow rate fluctuations result in an unsteady supply of the flame with combustible mixture, which in turn leads to fluctuations of the chemical conversion rate and thus of the heat released by the combustion process $\dot{Q}'$. Via the volumetric expansion of the gases, the oscillating heat release rate $\dot{Q}'$ acts as an acoustic monopole source and generates new pressure fluctuations p'. Under suitable circumstances, these pressure fluctuations further excite the sound field in the combustion chamber and a positive feedback loop is established: the oscillation amplitudes excite themselves.

In the above example, flow rate fluctuations of the fuel/air mass cause the unsteady heat release rate in the flame. Alternatively, other mechanisms can lead to an unsteady heat release rate as well, for example, fluctuation in the fuel/air mixture composition [197–199], unsteady fuel conditioning (e.g. [200]), periodic vortex shedding (e.g. [201, 202]) and fluctuations of the heat transfer rate from the flow to solid surfaces [203, 204].

A periodic heat release rate $\dot{Q}'$ will only excite pressure fluctuations p' in a fluid flow if the heat is added to the flow at the right moment, i.e. with the proper timing. A more specific wording for this fact was found already in 1777 by Lord Rayleigh with the criterion later named after him: according to Rayleigh's criterion, pressure oscillations are excited by periodic heat addition, if the phase shift between pressure and heat oscillations is no more than $\pm 90°$ [205, 206]. Originally, Rayleigh applied his criterion to harmonic oscillations only, but it has been the basis for numerous modifications and additions which considerably generalised and widened the range of its applicability [207–209].

20.13.2 Calculation

There is a variety of methods for modelling self-excited combustion oscillations, which can differ greatly in the accuracy of their results and in the necessary computational effort. A short description of the most important approaches and an extensive overview of the existing literature (until 2001) can be found, for example, in [210].

Many calculation approaches use forms of the acoustic wave equation to describe the sound field excited by the oscillations. This usually works well. However, modelling the unsteady behaviour of the flame into the wave equation – which is necessary to properly describe the self-excitation mechanism – is a critical problem.

To account for the flame behaviour, a large group of calculation models uses so-called time lags, which consider the time it takes until fluctuations of the pressure in the burner lead to corresponding fluctuations of the heat released by the flame. Incorporated into the acoustic equations, an equation system is obtained for which solutions with amplitudes growing over time can be found (e.g. [195, 211, 212]). The total magnitude of the fully excited system amplitude (limit cycle) can usually not be calculated with this approach.

The flame behaviour can be accounted for in more detail with so-called flame transfer functions. Simplified, flame transfer functions reproduce the frequency-dependent amplification and phase shift that a sound wave is subjected to when it moves through the flame, i.e. the "acoustic transmission behaviour" of the flame. The flame transfer function in turn can be incorporated into the acoustic field equations (e.g. [213, 214]).

The main problem of the described models – and of most models for the calculation of self-excited combustion oscillations in general – is that they do not describe the oscillating system as a whole with all effects involved in the self-excitation mechanism. Accordingly, they require input parameters – as time lags or flame transfer functions – that are not known in advance and that are difficult to determine since, ultimately, they are part of the solution to be calculated [215].

In the development of methods that allow for a real *prediction*, simulations have been performed that use as a starting point a very fundamental system of equations mainly consisting of the unsteady conservation equations (species, mass, momentum and energy), a thermodynamic equation of state and suitable equations to describe the chemical reaction in the combustion process. Solutions of this equation system describe the oscillating system more or less in its completeness so that interactions between combustion, flow and sound field and the associated self-excitation mechanism are directly obtained as a result.

In spite of the considerable progress that has been made in the numerical calculation of such equation systems [210, 216–221], the described procedures cannot yet practicably be used for the calculation and prediction of self-excited combustion oscillations in industrial systems. Even today the enormous computational efforts required for such simulations are usually not economically justifiable.

20.13.3 Counter Measures

Self-excited combustion oscillations in industrial firing systems can still not be reliably predicted to date. They usually show up unexpectedly and must then be eliminated or at least mitigated afterwards. The most important principles applied for this purpose are briefly outlined in the following. Because of the large number of publications on this topic, given references must remain a selection of examples only.

Passive measures dampen or affect the oscillations without requiring an external energy supply. Examples are ordinary silencers in gas, air and/or flue-gas ducts of the system or acoustic insulation and sound absorbing linings, in the first place. In the example shown in Fig. 20.63, the experienced problems with strong oscillations in a process furnace could be eliminated by the installation of a sound-absorbing lining of the furnace floor.

Other important passive measures are the Helmholtz resonator [222] and the quarter-wave

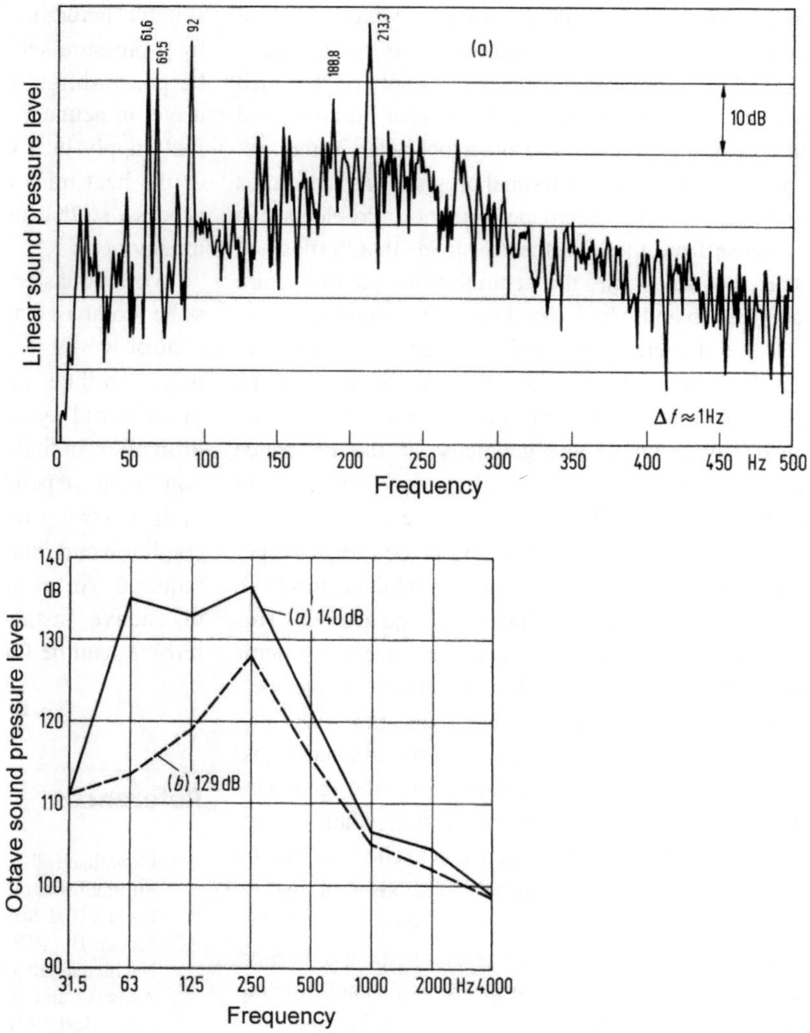

Fig. 20.63 Sound pressure levels in a process furnace (**a**) standard non-absorbing furnace floor, large pressure amplitudes by self-excited combustion oscillations, (**b**) furnace floor with sound-absorbing lining

resonator (λ/4-resonator). Since the effect of these two special damping elements is based on excitation of their fundamental natural frequency, they are only effective within a very narrow frequency band and become very large if they are designed to operate at low frequencies. Therefore, it is often difficult to integrate these resonators into existing combustion systems.

To interfere with the sound propagation in the system and to restrict the number of eigenmodes at lower frequencies, baffles can be installed [223]. Usually designed as fin plates and orifice plates, these elements subdivide the volume in the system and lead to fewer low natural frequencies. By increasing the pressure drop across the fuel gas injection ports – e.g. by orifice plates – the susceptibility of the gas supply system to fluctuations of the pressure in the combustion chamber can be reduced. An additional effect is brought about by all kinds of flow restrictions because of the dissipative losses at these obstacles which increase sound absorption. In contrast to the measures mentioned in the beginning, which can usually be designed in a way such as to create only little or no additional flow resistance, baffles and other obstacles in the system inevitably create additional pressure drops and thus limitations to the system performance. Problems can also arise from the high thermal loads that baffles and other obstacles in the combustion chamber are subjected to and which can destroy these components.

Since the eigenfrequencies of the gas volume in the combustion system play an important role for the self-excitation mechanism, the system can be "detuned" by changing the geometry. If the resonance frequencies can be shifted – e.g. by modifying the length of the air supply or flue-gas system – such as to disturb the interaction between fluctuations in the energy supply and the pressure pulsations of the sound field (Rayleigh criterion), self-excited oscillations can be avoided. This approach is often very successful in smaller systems with only few eigenmodes at lower frequencies. Disadvantages of this method are the risk that oscillations are only shifted to another frequency and the effort required for variations in the geometry of systems already in operation that often cannot be justified. In cases in which the excited oscillation modes are highly symmetrical, methods that disturb this symmetry have been very successful [224]. Examples are a slightly irregular arrangement of burners in boilers and incinerators or an irregular load distribution over the side-wall burners of process furnaces.

Another possible starting point for passive counter measures is the combustion process itself. For example, a combustion system can be influenced by equivalence ratio and fired duty, the flame length ["soft" flame, see also Eqs. (20.1a) and (20.1b)] or the burner design (axial- or swirl burner, staged combustion etc.), as well as the path to be taken by the combustible mixture towards the reaction zone – e.g. by changing the location of fuel injection [220]. Since modifications of the combustion process – as far as they are possible at all – often have undesired side effects (e.g. lower efficiency, increased pollutant immissions etc.), other approaches try to disturb the interaction between flame, flow and sound field via the flow field mainly [225, 226].

Active counter measures are based on the principle of feedback control and compensate the self-excited oscillations by automatically counteracting with a suitable actuator. For example, the signal measured by a pressure sensor in the combustion chamber can be phase-shifted and amplified suitably and used to drive an actuator. This actuator modulates the air or fuel supply in a way as to level out the fluctuations of the heat release rate caused by the self-excitation process, so that no self-excited oscillations can build up anymore.

While it has been shown many times in laboratory-scale combustion systems that oscillation feedback control works well (for an overview see [227–229]), it has, to date, only occasionally been implemented in industrial systems. The main reasons for this are difficulties in finding suitable actuators, a strong, often non-linear, dependency of the oscillations – and thus of the control parameters – on the respective operating condition and finally the considerable technical efforts required. An example for successful implementation of "active instability control" with stationary gas turbines can be found in [224, 230].

References

1. Oswatitsch K (1959) Physikalische Grundlagen der Strömungslehre. In: Flügge S (ed) Handbuch der Physik. Band VIII/1 Strömungsmechanik I. Berlin Springer
2. Stüber B (1970) Schallabstrahlung von Quellen in der Nachbarschaft von starren Körpern. Gemeinschaftstagung Akustik und Schwingungstechnik Berlin, VDI-Verlag Düsseldorf 1971, 231

3. Fitzpatrick HM, Stasberg M (1956) Hydrodynamik sources of sound, 1st Symposium on Naval Hydrodynamics, Washington, pp 241–282
4. Plesset MS (1949) The dynamics of cavitation bubbles. J Appl Mech 16:277–282
5. Radek U (1972) Kavitationserzeugte Druckimpulse und Materialzerstörung. Acustica 26:270
6. Baiter HJ (1971) Anmerkungen zur Literatur über die Entstehung von Strömungskavitation, 2. Teil: Bericht der Forschungsgruppe Hydroakustik der Fraunhofer-Gesellschaft zur Förderung der angewandten Forschung e.V. München
7. Blake WK (1986) Mechanics of flow-induced sound and vibration. Academic, London
8. Lauterborn W, Heinrich G (1971) Literaturverzeichnis: Kavitation und Blasenbildung in Flüssigkeiten. 3. Physikalisches Institut der Universität Göttingen
9. Baiter HJ (1971) Geräusche der Strömungskavitation. Bericht 9/71 der Forschungsgruppe Hydroakustik der Fraunhofer-Gesellschaft zur Förderung der angewandten Forschung e.V. München
10. Erdmann H et al (1969) Untersuchungen über die mit der Geräuscherzeugung durch den Propeller, insbesondere mit der Kavitation verbundenen akustischen Probleme. Arbeitsabschnitt II und III, Battelle-Insitut e.V. Frankfurt/Main
11. Jorgensen DW (1961) Noise from cavitating submerged water jets. J Acoust Soc Am 33:1334–1338
12. Knapp RT (1952) Cavitation mechanics and its relation to the design of hydraulic equipment. Proc Inst Mech Eng A 166:150
13. Eisenberg P (1953) Kavitation. Schiffstechnik 1, Heft 3, 111, Heft 4, 155, (1954) Heft 5, 201
14. Knapp RT, Hollander A (1948) Laboratory investigations of the mechanism of cavitation. Trans ASME 70:419–435
15. Hunsaker JC (1935) Cavitation research. Mech Eng 57:211–216
16. Knapp RT (1955) Recent investigations of the mechanics of cavitation and cavitation damage. Trans ASME 77:1045–1054
17. Knapp RT (1956) Further studies of the mechanics and damage potential of fixed type cavities. In: Cavitation in hydrodynamics. H.M. Stationery Office, London
18. Gallant H (1962) Untersuchungen über Kavitationsblasen. Österr. Ing.-Z. 5:74
19. Lord Rayleigh JW (1971) On the pressure developed in a liquid during the collapse of a spherical cavity. Philos Mag 34:94–98
20. Fitzpatrick HM (1958) Cavitation noise. In: 2nd Symposium on naval hydrodynamics, Washington, pp 201–205
21. Lyamshev LM (1970) On the theory of hydrodynamic cavitation noise. Sov Phys Acoust 15:494
22. Akulichev VA (1969) Experimental investigation of an elementary cavitation zone. Sov Phys Acoust 14:284
23. Crighton DG, Ffowcs Williams JE (1969) Sound generation by turbulent two-phase flow. J Fluid Mech 36:585–603
24. YuYa B (1968) Propagation of sound waves in a liquid during cavitation. Sov Phys Acoust 14:151
25. Boguslavskii YuYa, Ioffe AI, Naugol'nykh KA (1970) Sound radiation by a cavitation zone. Sov Phys Acoust 16:17
26. Frank P (1976) Untersuchungen an kavitierenden Düsen im Freifeld. DAGA Düsseldorf VDI, 441
27. Rouse H, Hassan MM (1949) Mech Eng 71:213
28. Esipov JB, Naugol'nykh KA (1976) Cavitation noise in submerged jets. Sov Phys Acoust 21(4):404
29. Mellen RH (1954) Ultrasonic spectrum of cavitation noise in water. J Acoust Soc Am 26:356
30. Mellen RH (1956) An experimental study of the collapse of a spherical cavity in water. J Acoust Soc Am 28:447–454
31. Lesunovskii VP, Khokha YuV (1969) Characteristics of the noise spectrum of hydrodynamic cavitation on rotating bars in water. Sov Phys Acoust 14:474
32. Branns D (1970) Effekte der Strömungskavitation an rotierenden Förderorganen. Dissertation an der TH Aachen
33. Aleksandrov IA (1962) Physical nature of the "rotation noise" of ship propellers in the presence of cavitation. Sov Phys Acoust 8:23–28
34. Stüber B (1967) Untersuchung aerodynamisch erzeugter Schallfelder mit Hilfe der Modellmethode. Dissertation TH München
35. Holle W (1938) Frequenz- und Schallstärkemessungen an Hiebtönen. Akust Z 3:321–331
36. Bloor MS (1964) The transition to turbulence in the wake of a circular cylinder. J Fluid Mech 19:290–304
37. Hama FR (1957) Three-dimensional vortex pattern behind a circular cylinder. J Aero Sci 24:156–158
38. Kovasznay LSG (1949) Hot-wire investigation of the wake behind cylinders at low Reynolds numbers. Proc R Soc A 198:174–190
39. Roshko A (1954) On the development of turbulent wakes from vortex streets. NACA. Techn. Report 1191
40. Roshko A (1961) Experiments on the flow past a circular cylinder at very high Reynolds number. J Fluid Mech 10:345–356
41. Delany NK (1953) Sorensen, N.E.: Low-speed drag of cylinders of various shapes. NACA Techn Note 3038
42. Relf EF, Simmons LFG (1925) The frequency of the eddies generated by the motion of circular cylinders through a fluid. Philos Mag 49:509
43. Surry D (1969) The effect of high intensity turbulence on the aerodynamics of a rigid circular cylinder at subcritical Reynolds number. University of Toronto UTIA Report No. 142, Toronto
44. Strouhal V (1878) Über eine besondere Art der Tonerregung. Annalen der Physik und Chemie 5:216–251
45. Etkin B, Korbacher GK, Keefe RT (1956) Acoustic radiation from a stationary cylinder in a fluid stream. University of Toronto UTIA Report 39, Toronto; (1957) J Acoust Soc Am 29:30–36
46. Lehnert R (1937) Acoustic measurements of vortex streets behind cylinders and flat plates. Phys Z 38:476–498
47. Bearman PW (1967) On vortex street wakes. J Fluid Mech 28:625–641
48. Keefe RT (1961) Investigation of the fluctuating forces acting on a stationary circular cylinder in a subsonic stream and of the associated sound field. University of Toronto UTIA Report No. 76, Toronto; (1962) J Acoust Soc Am 34:1711–1714
49. Bishop RED, Hassan AY (1964) The lift and drag forces on a circular cylinder in a flowing fluid. Proc R Soc A 277:32

50. Fung YC (1960) Fluctuating lift and drag acting on a cylinder in a flow at supercritical Reynolds numbers. J Aero Sci 27:801–814
51. Gerrard JH (1961) An experimental investigation of the oscillation lift and drag of a circular cylinder shedding turbulent vortices. J Fluid Mech 11:244–256
52. Jones GW (1968) Unsteady lift forces generated by vortex shedding about large stationary, and oscillating cylinder at high Reynolds number. ASME Symp Unsteady Flow Pap 68 FE-36
53. Hamphreys JS (1960) On a circular cylinder in a steady wind at transition Reynolds numbers. J Fluid Mech 9:603–612
54. McGregor DM (1957) An experimental investigation of the oscillating pressures on a circular cylinder in a fluid. University of Toronto UTIA Report Note 14, , Toronto
55. Leehey P, Hanson CE (1971) Aeolian tones associated with resonant vibration. J Sound Vib 13:465–483
56. Gerrard JH (1965) A disturbance-sensitive Reynolds number range of the flow past a circular cylinder. J Fluid Mech 22:187–196
57. Bloor MS, Gerrard JH (1966) Measurements on turbulent vortices in a cylinder wake. Proc R Soc A 294:19
58. Chen SS (1987) Flow-induced vibration of circular cylindrical structures. Hemisphere, London
59. Obermeier F (1967) Berechnung aerodynamischer Schallfelder mittels der Methode der "Matched Asymptotic Expansions". Acustica 18:238–240
60. Stüber B (1970) Schallabstrahlung und Körperschallanregung durch Wirbel. Acustica 23:82–92
61. Ffowcs Williams JE (1969) Hydrodynamic noise. Annu Rev Fluid Mech 1:197–222
62. Ribner HS (1964) The generation of sound by turbulent jets. Adv Appl Mech 48:105–182
63. Crow SC, Champagne FH (1971) Orderly structure in jet turbulence. J Fluid Mech 8:547–591
64. Mollo-Christensen F (1970) Jet noise and shear flow instabilities seen from an experimenters viewpoint. J Appl Mech 89:1–7
65. Michalke A (1972) An expansion scheme for the noise from circular jets. Z Flugwiss 20:229–237
66. Fuchs HV (1971) Eigenschaften der Druckschwankungen im subsonischen Freistrahl. In: Proceedings of 7th international congress on acoustics, Budapest, vol 4, p 449
67. Lighthill MJ (1952) On sound generated aerodynamically. Proc R Soc A 211:564–587; (1954) Proc R Soc A 222:1–32
68. Ffowcs Williams JE (1963) The noise from turbulence convected at high speeds. Phil Trans R Soc Lond A 225:469–503
69. Sprenger H (1954) Über thermische Effekte in Resonanzrohren. Mittelungen aus dem Institut für Aerodynamik der ETH Zürich 21, 18
70. Böhnke W (1970) Schallerzeugung durch einen gestörten Freistrahl. Gemeinschaftstagung Akustik und Schwingungstechnik, Berlin 1970. VDI, Düsseldorf, pp 237–240
71. Powell A (1953) On the mechanism of choked jet noise. Proc Phys Soc B 66:1039–1056
72. Powell A (1954) The reduction of choked jet noise. Proc Phys Soc B 67:313–327
73. Willmarth WW, Wooldridge CW (1962) Measurements of the correlation between the fluctuating velocities and the fluctuating wall pressure in a thick boundary layer. J Fluid Mech 14:187–210; (1965) 22:81–94
74. Schloemer HH (1967) Effects of pressure gradients on turbulent-boundary-layer wall pressure fluctuations. J Acoust Soc Am 42:93–113
75. Blake WK (1970) Turbulent boundary layer wall pressure fluctuations on smooth and rough walls. J Fluid Mech 44:637–660
76. Corcos GM (1964) The structure of the turbulent pressure field in boundary layer flows. J Fluid Mech 18:353–378
77. Wills JAB (1970) Measurements of the wave number/phase velocity spectrum of wall pressure beneath a turbulent boundary layer. J Fluid Mech 45:65–90
78. Kim HT, Kline SJ, Reynolds WC (1971) The production of turbulence near a smooth wall in a turbulent boundary layer. J Fluid Mech 50:133–160
79. Emmerling R (1973) Die momentane Struktur des Wanddrucks einer turbulenten Grenzschichtströmung. Mitt. Max-Planck-Institut f. Strömungsforschung Nr. 56, Göttingen
80. Kistler AL, Chen WS (1962) The fluctuating pressure field in a supersonic turbulent boundary layer. Jet Propulsion Lab. Techn. Report No. 32–277
81. Ffowcs Williams JE (1965) Sound radiation from turbulent boundary layers formed on compliant surfaces. J Fluid Mech 22:347–358
82. Crighton DC (1970) Radiation from turbulence near a composite flexible boundary. Proc R Soc A 314:153–173
83. Heckl M (1969) Körperschallanregung von elastischen Strukturen durch benachbarte Schallquellen. Acustica 21:149–161
84. Ffowcs Williams JE (1966) The influence of simple supports on the radiation from turbulent flow near a plane compliant surface. J Fluid Mech 26:641–649
85. Sebald A (1988) Schallleistungsmessung mit kreisförmigen Strömungskanälen mit wandbündig eingebauten Aufnehmern. Diplomarbeit Fachhochschule München, Fachbereich 03: Maschinenbau
86. VDI-Richtlinie 3733 (1996) Geräusche bei Rohrleitungen (Noise at pipes)
87. Stüber B, Fritz KR, Lang F (1986) Gasdurchströmte Stellventile, Näherungsweise Berechnung der Schallabstrahlung der angeschlossenen Rohrleitungen. Müller-BBM Bericht 11461 mit Anhang, gefördert durch das Ministerium für Umwelt, Raumordnung und Landwirtschaft des Landes Nordrhein-Westfalen
88. Ahrens C, Ronneberger D (1971) Luftschalldämmung in turbulent durchströmten schallharten Rohren bei verschiedenen Wandrauhigkeiten. Acustica 25:150–157
89. Michael PL, Hogan D (1967) Comparison of experimental and theoretical sound-attenuation values in a buried network of long pipes. J Acoust Soc Am 41:593–596
90. Sinambari GR (1981) Ausströmgeräusche von Düsen und Ringdüsen in angeschlossenen Rohrleitungen: ihre Entstehung, Fortpflanzung und Abstrahlung. Dissertation Universität Kaiserslautern
91. Stüber B (1980) Verlustfaktoren von gasgefüllten Rohrleitungen. VDE, DAGA, Berlin, pp 439–442
92. Fritz KR, Stüber B (1980) Schalldämmung und Abstrahlgrad von gasgefüllten Stahlrohren. VDE, DAGA, Berlin, pp 357–360

93. Cremer L (1955) Theorie der Luftschalldämmung zylindrischer Schalen. Acustica 5:245–256
94. Heckl M (1957) Schallabstrahlung und Schalldämmung von Zylinderstrahlen. Dissertation TU Berlin
95. Heckl M (1958) Experimentelle Untersuchungen zur Schalldämmung von Zylindern. Acustica 8:259–265
96. Kuhn GF, Morfey CL (1976) Transmission of low-frequency internal sound through pipe walls. J Sound Vibr 47:147
97. Stüber B, Fritz KR, Lang F (1978) Schalldämmende Rohrleitungsummantelungen hoher Pegelsenkungen. Müller-BBM Bericht Nr. 5500, erarbeitet im Auftrag des Ministers für Arbeit, Gesundheit und Soziale des Landes Nordrhein-Westfalen, Gesch.-Zeich. III B 2–8824.3 RE
98. Kozlik W (1983) Experimentelle Bestimmung der Schalldämmung und Schallabstrahlung von gasgefüllten Rechteckkanälen. Diplomarbeit aus der technischen Akustik an der Fachhochschule München
99. Cremer L, Heckl M (1995) Körperschall. Springer, Berlin
100. Heckl M (1959) Schallabstrahlung von punktförmig angeregten Hohlzylindern. Acustica 9:86–92
101. Stüber B, Lang F (1983) Abstrahlmaße verschiedener Bauteile in verfahrenstechnischen Anlagen. DGMK-Projekt 312
102. Michelsen R, Fritz KR, v. Sazenhofen C (1980) Wirksamkeit schalldämmender Ummantelungen von Rohren. VDE, DAGA, Berlin, p 301
103. Muheim JA, Rathé EJ (1968) Geräuschverhalten von Ventilatoren, kritische Übersicht über bisherige Erfahrungen und Erkenntnisse. TH Zürich
104. Recknagel-Sprenger (1977) Taschenbuch für Heizung. Lüftung und Klimatechnik. Oldenbourg, München
105. Eck B (1961) Technische Strömungslehre. Springer, Berlin
106. Eck B (1962) Ventilatoren. Springer, Berlin
107. Wolf H (1958) Akustische Wirkung von Propellern, Stahltriebwerken und Freistrahlern. Maschinenbautechnik 7(11):573–580
108. Sharland IJ (1964) Sources of noise in axial flow fans. J Sound Vib 1:302–322
109. Leidel W (1969) Einfluss von Zungenabstand und Zungenradius auf Kennlinie und Geräusch eines Radialventilators. Deutsche Versuchsanstalt für Luft- und Raumfahrt, Forschungsbericht 69–76
110. Embleton TW (1963) Experimental study of noise reduction in centrifugal blowers. J Acoust Soc Am 35:700–705
111. Simpson HC, Macaskill R, Clark TA (1966–1967) Generation of hydraulic noise in centrifugal pumps. Proc. Inst. Mech. Eng. 181, Part 3A
112. Lowson MV (1968) Reduction of compressor noise radiation. J Acoust Soc Am 43:37–50
113. Kemp NH, Sears WR (1953) Aerodynamic interference between moving blade rows. J Aero Sci 20:585–598
114. Kemp NH, Sears WR (1955) The unsteady forces due to viscous wakes in turbomaschines. J Aero Sci 22:478–483
115. Lowson MV (1970) Theoretical analysis of compressor noise. J Acoust Soc Am 47:371–385
116. NleS F (1966) An investigation of axial flow fan noise. J Sound Vib 3:147–165
117. Allen CH (1957) Noise from air conditioning fans. Noise control 3:28–34; and Noise Reduction (ed. L.L. Beranek) oder Handbook of Noise Control, Chap. 25 (ed. H.C. Harris), McGraw-Hill, New York
118. Bommes L (1961) Geräuschentwicklung bei Ventilatoren kleiner und mittlerer Umfangsgeschwindigkeit. Lärmbekämpfung 5:69–75
119. Beranek LL, Kampermann GW, Allen CH (1955) Noise of centrifugal fans. J Acoust Soc Am 27:217–219
120. Zeller W, Stange H (1957) Vorausbestimmung der Lautstärke von Axialventilatoren. Heiz Lüft Haustechn 8:322
121. Wickström B (1964) Beitrag zur zweckmäßigen Bestimmung und Darstellung des Ventilatorgeräusches als Grundlage für akustische Berechnungen von Lüftungsanlagen. Diss. TU Berlin
122. Stüber B, Ludewig H (1980) Schallabstrahlung von Axialventilatoren für Luftkühler und Kühltürme. Z Lärmbekämpfung 27:104–108
123. Fritz KR, Ludewig H (1985) Schallabstrahlung von Radialventilatoren hoher Förderleistung. Z Lärmbekämpfung 32:73–78
124. VDI-Richtlinie 3731, Blatt 2 (1990) Emissionskennwerte technischer Schallquellen. Ventilatoren
125. VDI-Richtlinie 2081, Blatt 1 (2001) Geräuscherzeugung und Lärmminderung in Raumlufttechnischen Anlagen
126. Yudin EJ (1958) Untersuchungen des Lärms von Lüfteranlagen und die Methode zu seiner Bekämpfung. ZAGI Bericht Nr. 713, Moskau
127. Nemec J (1967) Noise of axial fans and compressors: study of its radiation and reduction. J Sound Vib 6:230–236
128. Duncan PE, Dawson B (1974) Reduction of interaction tones from axial flow fans by suitable design of rotor configuration. J Sound Vib 33:143–154
129. Zeller W (1967) Lärmabwehr bei Lüftungsanlagen. Forschungsbericht des Landes Nordrhein-Westfalen Nr. 1117, Köln: Westdeutscher Verlag
130. Barsikow B, Neise W (1978) Der Einfluss ungleichförmiger Zuströmung auf das Geräusch von Radialventilatoren. VDE, DAGA, Berlin, pp 411–414
131. Longhouse RE (1978) Control of tip vortex noise of axial flow fans by rotating shrouds. J Sound Vib 58:201–214
132. Fukano T, Kodama J, Takamatsu J (1986) The effects of tip clearance on the noise of low pressure and mixed-flow fans. J Sound Vib 105:291–308
133. Marcinowski H (1958) Der Einfluss des Laufradspaltes bei leitradlosen frei ausblasenden Axialventilatoren. Voith Forschung und, Konstruktion, 3
134. Wieland H (1986) Vergleich verschiedener Systeme zum Verändern der Förderleitung bei Radialventilatoren. VDI-Berichte 594:267–281
135. Lowson MV, Potter RS (1966) Potential noise reduction methods for axial flow compressors. Wyle Lab. Res. Staff Rept. WR 66–9
136. Chaimowitch EM (1959) Ölhydraulik, 3rd edn. VEB Verlag Technik, Berlin
137. Schulz H (1977) Die Pumpen, 13th edn. Springer, Berlin
138. Leuschner G (1967) Kleines Pumpenhandbuch für Chemie und Technik. Verlag Chemie, Weinheim
139. Sulzer Pumpen (1990) Kreiselpumpen-Handbuch, 3. Aufl.: Vulkan-Verlag
140. Autorenkollektiv (1987) Technisches Handbuch Pumpen, 7. Aufl. Berlin: Verlag Technik

141. Hübsch HG (1969) Untersuchung des Geräuschverhaltens und konstruktive Möglichkeiten zur Geräuschminderung an nicht druckkompensierten Zahnradpumpen. Dissertation Univ. Stuttgart
142. Heckl M, Mühle C (1970) Geräuscherzeugung durch Hydraulikanlagen. Müller-BBM GmbH, Report No 2177
143. VDI-Richtlinie 3743, Blatt 2 (1989) Emissionskennwerte technischer Schallquellen, Pumpen, Verdrängerpumpen
144. Stüber B, Lang F (1979) Stand der Technik bei der Lärmminderung in der Petrochemie. Umweltbundesamt Berlin
145. Schmitt A, Klein V (1979) Schallemission von Chemie-Normpumpen, Chemie-Anlagen und Verfahren 6
146. Saxena SV, Wonsak G, Nagel W (1978) Geräuschemission von Kreiselpumpen, Forschungsbericht Nr. 184 der Bundesanstalt für Arbeitsschutz und Unfallforschung. Dortmund, Wirtschaftsverlag NW, Bremerhaven
147. Lehmann W, Melhorn P (1978) Untersuchung der Geräuschemission von Kreiselpumpen in Abhängigkeit der konstruktiven Ausführungen und Betriebsbedingungen. Pumpentagung Karlsruhe
148. VDI-Richtlinie 3743, Blatt 1 (2003) Emissionskennwerte technischer Schallquellen, Pumpen, Kreiselpumpen
149. Föller D (1972) Untersuchung der Anregung von Körperschall in Maschinen und der Möglichkeiten für eine primäre Lärmbekämpfung. Dissertation TH Darmstadt
150. Hagen K (1958) Volumenverhältnisse, Wirkungsgrade und Druckschwankungen in Zahnradpumpen. Dissertation TH Stuttgart
151. Kahrs M (1968) Die Verbesserung des Umsteuervorganges schlitzgesteuerter Hydro-Axialkolbenmaschinen mit Hilfe eines Druckausgleichskanals. Ölhydraulik und Pneumatik 12:9–15
152. Domm U, Dernedde R Über eine Auswahlregel für die Lauf- und Leitschaufelzahl von Kreiselpumpen. KSB Technische Berichte Nr. 9
153. Guglhör M (1986) Problematik heutiger Windkraftanlagen. Diplomarbeit FH München
154. Information des Staatlichen Umweltamtes Bielefeld (March 2002)
155. Landesumweltamt Nordrhein-Westfalen (2002) Sachinformation zu Geräuschemissionen und –immissionen von Windenergieanlagen
156. Piorr D (1991) Schallemissionen und –immissionen von Windkraftanlagen. Fortschritte der Akustik – DAGA
157. DIN EN 61400-11 (2009) Wind turbines – Part 11: Acoustic noise measurement techniques
158. Technische Richtlinien für Windenergieanlagen, Teil 1 (2000) Bestimmung der Emissionswerte. Revision 13, Fördergesellschaft für Windenergie e.V., Kiel
159. Ising H, Markert B, Shenoda F, Schwarze C (1982) Infraschallwirkungen auf den Menschen. VDI-Verlag
160. Hubert M (1970) Untersuchungen über die Geräusche durchströmter Gitter. Dissertation TU Berlin
161. Sondhauss CFJ (1854) Über die Ausströmung der Luft entstehende Töne. Pogg Ann Phys Chem 91:126–214
162. v. Gierke H (1944) Über Schneidetöne an kreisrunden Gasstrahlern und ebenen Lamellen. Dissertation Karlsruhe
163. Heller H (1965) Tonbildung bei der Durchströmung scharfkantiger Düsen mit hoher Unterschallgeschwindigkeit. Dissertation TU Berlin
164. Gordon C (1968) Spoiler-generated flow noise. J Acoust Soc Am 43:1041–1048; (1969) 45:214–223
165. Heller H, Widnall SE (1970) Sound radiation from rigid flow spoilers correlated with fluctuating forces. J Acoust Soc Am 47:934–936
166. VDI 3738 (1994) Emissionskennwerte technischer Schallquellen, Armaturen
167. Allen EE (1972) Valves can be quiet. Hydrocarbon Process 51:137–141
168. Izmit A, McDaniel, OH, Reethof G (1977) The nature of noise sources in control valves. In: Proceedings of Inter-noise 77, Zurich, Switzerland, B 183
169. Bach M (1983) Strömungsbild, Wanddruck, Körperschallanregung und Luftschallabstrahlung bei einem Ventilmodell mit verschiedenen Drosselkörpern. Dissertation TU Berlin
170. Reethof G (1977) Control valve and regulator noise generation, propagation and reduction. Noise Control Eng 9(2):74–85
171. VDMA 24422 (1989) Armaturen; Richtlinie für die Geräuschberechnung; Regel- und Absperrarmaturen
172. DIN EN 60534-8-3 (2001) Industrial-process control valves – Part 8–3: Noise considerations; Control valve aerodynamic noise prediction method (IEC 60534-8-3:2000)
173. DIN EN 60534-8-4 (2005) Industrial-process control valves – Part 8–4: Noise considerations – Prediction of noise generated by hydrodynamic flow (IEC 60534-8-4:2005 + Corrigendum 1:2006)
174. Böhm A, Hubert M (1970) Geräuschminderung an Kühltürmen. Gemeinschaftstagung Akustik und Schwingungstechnik Berlin, Düsseldorf: VDI, pp 331–334
175. Bublitz D (1970) Die Geräuschemission großer Rückkühlanlagen. Gemeinschaftstagung Akustik und Schwingungstechnik Berlin, Düsseldorf: VDI, pp 325–329
176. Böhm A, Bublitz D, Hubert M (1971) Geräuschprobleme bei Rückkühlanlagen. Mitteilungen der VGB 51:235
177. Hubert M (1972) Geräusche fallender Wassertropfen. Gemeinschaftstagung Akustik und Schwingungstechnik Stuttgart, Berlin: VDE, pp 410–413
178. Riedel E (1976) Geräusche aufprallender Wassertropfen. Dissertation TU-Berlin
179. Reinicke W (1975) Über die Geräuschabstrahlung von Naturzug-Kühlern. TÜ, Nr. 7/8
180. Wirth KE (1986) Pneumatische Förderung – Grundlagen, in: Preprints Technik der Gas/Feststoff-Strömung, VDI-Gesellschaft Verfahrenstechnik und Chemieingenieurswesen, 157
181. Brauer H (1971) Grundlagen der Einphasen- und Mehrphasenströmungen, in: Grundlagen der Chemischen Technik. Verlag Sauerländer, Aarau und Frankfurt am Main
182. Krambrock W Dichtstromförderung, in: Preprints Technik der Gas/Feststoff-Strömung, VDI-Gesellschaft Verfahrenstechnik
183. VDI 3732 (1999) Emissionskennwerte technischer Schallquellen, Fackeln
184. VDI 3730 (1988) Emissionskennwerte technischer Schallquellen, Prozessöfen (Röhrenöfen)
185. Briffa FEJ, Clark CJ, Williams GT (1973) Combustion noise. J Inst Fuel 46:207–216

186. Beer JM, Gupta AK, Syred N (1973) The reduction of noise emission from swirl combustors by staged combustion. Sheffield University Report
187. Putnam AA (1976) Combustion roar of seven industrial gas burners. J Inst Fuel 49:135
188. Bertrand C, Michelfelder S (1976) Experimental investigation of noise generated by large turbulent diffusion flames. (results obtained during the Ap-5-trials) IFRF Doc. nr. K 20/a/87
189. Pauls D, Günther R (1975) The noise level of turbulent diffusion flames. 2. Symposium (Eur.) on combustion, Orleans, 426 ff
190. Pauls D (1977) Geräuschentstehung turbulenter Diffusionsflammen. Dissertation TH Karlsruhe
191. Pauls D, Stüber B, Horns H, Pröpster A (1982) Messung, Berechnung und Verminderung der Schallabstrahlung von Bodenfackeln. Forschungsbericht 82-105-03-407/01 des Umweltbundesamtes
192. Hantschk C-C, Schorer E (2004) Flares – noise prediction and thermo-acoustic efficiency. AFRC-JFRC 2004 Joint International Combustion Symposium
193. Stüber B, Lang F (1983) Schalldämmmaße von Prozessofenwänden. DGMK-Projekt 313
194. Higgins B (1802) On the sound produced by a current of hydrogen gas passing through a tube. J Nat Philos Chem Arts 1:129–131
195. Crocco L, Cheng S-I (1956) AGARDograph. Bd. 8: Theory of combustion instability in liquid propellant rocket motors. Butterworth's Scientific, London
196. Putnam AA (1971) Combustion driven oscillations in industry. American Elsevier, New York
197. Richards GA, Janus MC (1998) Characterization of oscillations during premix gas turbine combustion. Transactions of the ASME. J Eng Gas Turbines Power 120:294–302
198. Lieuwen T, Zinn BT (1998) The role of equivalence ratio oscillations in driving combustion instabilities in low NO_x gas turbines. In: 27th International symposium on combustion, The Combustion Institute, pp 1809–1816
199. Sattelmayer T (2000) Influence of the combustor aerodynamics on combustion instabilities from equivalence ratio fluctuations. The American Society of Mechanical Engineers (2000-GT-0082). ASME-paper
200. Dressler JL (1991) Atomization of liquid cylinders, cones, and sheets by acoustically-driven, amplitude-dependent instabilities. In: International conference on liquid atomization and spray systems. Institute for Liquid Atomization and Spray Systems, Gaithersburg, MD, USA, pp 397–405
201. Schadow KC et al (1989) Large-scale coherent structures as drivers of combustion instability. Combust Sci Tech 64:167–186
202. Büchner H, Külsheimer C (1997) Untersuchungen zum frequenzabhängigen Mischungs- und Reaktionsverhalten pulsierender, vorgemischter Drallflammen. Gas Wärme International 46:122–129
203. Schimmer H, Vortmeyer D (1977) Acoustical oscillation in a combustion system with a flat flame. Combust Flame 28:17–24
204. Schimmer H (1974) Selbsterregte akustische Schwingungen in Brennräumen mit flacher Flamme, Technische Universität München, Dissertation
205. Baron Rayleigh JWS (1878) The explanation of certain acoustical phenomena. Nature 18:319–321
206. Baron Rayleigh JWS (1945) Theory of sound, vol 2, 2nd edn. Dover, New York
207. Putnam AA, Dennis WR (1954) Burner oscillations of the gauze-tone type. J Acoust Soc Am 26:716–725
208. Lawn CJ (1982) Criteria for acoustic pressure oscillations to be driven by a diffusion flame. In: 19th International symposium on combustion, The Combustion Institute, pp 237–244
209. Lang W (1986) Dynamik und Stabilität selbsterregter Verbrennungsschwingungen beim Auftreten mehrerer Frequenzen. Ein erweitertes Stabilitätskriterium, Technische Universität München, Dissertation
210. Hantschk C-C (2000) Numerische Simulation selbsterregter thermoakustischer Schwingungen/ VDI-Verlag. Düsseldorf, (441)-Fortschritt-Berichte, Reihe 6
211. Tsujimoto Y, Machii N (1986) Numerical analysis of a pulse combustion burner. In: 21st International symposium on combustion, The Combustion Institute, 539–546
212. Lieuwen T, Torres H, Johnson C, Zinn BT (1999) A mechanism of combustion instability in lean premixed gas turbine combustors. The American Society of Mechanical Engineers (99-GT-3), Washington. ASME-paper
213. Krüger U, Hüren J, Hoffmann S, Krebs W, Bohn D (1999) Prediction of thermoacoustic instabilities with focus on the dynamic flame behaviour for the 3A-series gas turbine of Siemens KWU. The American Society of Mechanical Engineers (99-GT-111), Washington. ASME-paper
214. Hobson DE, Fackrell JE, Hewitt G (1999) Combustion instabilities in industrial gas turbines – measurements on operating plant and thermoacoustic modelling. The American Society of Mechanical Engineers (99-GT-110), Washington, ASME-paper
215. Dowling AP (1995) The calculation of thermoacoustic oscillations. J Sound Vib 180:557–581
216. Hantschk C-C, Vortmeyer D (2002) Numerical simulation of self-excited combustion oscillations in a non-premixed burner. Combust Sci Tech 174:189–204
217. Hantschk C-C, Vortmeyer D (1999) Numerical simulation of self-excited thermoacoustic instabilities in a Rijke tube. J Sound Vib 227:511–522
218. Menon S, Jou W-H (1991) Large-eddy simulations of combustion instability in an axisymmetric ramjet combustor. Combust Sci Tech 75:53–72
219. Smith CE, Leonard AD (1997) CFD modeling of combustion instability in premixed axisymmetric combustors. The American Society of Mechanical Engineers (97-GT-305), Washington. ASME-paper
220. Steele RC, Cowell LH, Cannon SM, Smith CE (1999) Passive control of combustion instability in lean premixed combustors. The American Society of Mechanical Engineers (99-GT-52), Washington. ASME-paper
221. Murota T, Ohtsuka M (1999) Large-eddy simulation of combustion oscillation in premixed combustor. The American Society of Mechanical Engineers (99-GT-274), Washington. ASME-paper
222. Gysling DL, Copeland GS, McCormick DC, Proscia WM (1998) Combustion system damping augmentation with Helmholtz resonators. The American Society of Mechanical Engineers (98-GT-268), Washington. ASME-paper

223. Culick FEC (1988) Combustion instabilities in liquid-fuelled propulsion systems: an overview. In: AGARD Conference Proceedings No. 450, Combustion instabilities in liquid-fuelled propulsion systems, Advisory Group for Aerospace Research and Development, 1-1 – 1-73
224. Seume JR et al (1998) Application of active combustion instability control to a heavy duty gas turbine. Trans ASME J Eng Gas Turbines Power 120:721–726
225. Straub DL, Richards GA (1999) Effect of axial swirl vane location on combustion dynamics. The American Society of Mechanical Engineers (99-GT-109), Washington. ASME-paper
226. DVGW Bonn, Büchner H, Leuckel W (1995, 1995, 1997) Verfahren und Vorrichtung zur Unterdrückung von Flammen-/Druckschwingungen bei einer Feuerung. Deutsche und Europäische Patentanmeldungen. Offenlegungsschriften DE 195 26 369 A 1, DE 195 42 681 A 1 und 97100753.9
227. Hermann J (1997) Anregungsmechanismen und aktive Dämpfung (AIC) selbsterregter Verbrennungsschwingungen in Flüssigkraftstoffsystemen. VDI-Verlag, Düsseldorf (364). Fortschritt-Berichte, Reihe 6
228. Candel SM (1992) Combustion instabilities coupled by pressure waves and their active control. In: 24th International symposium on combustion, The Combustion Institute, pp 1277–1296
229. McManus KR, Poinsot T, Candel SM (1993) A review of active control of combustion instabilities. Prog Energy Combust Sci 19:1–29
230. Hermann J et al (1997) Aktive Instabilitätskontrolle an einer 170 MW Gasturbine. VDI Berichte 1313:337–344. 18. Deutsch-Niederländischer Flammentag

Ultrasound

21

H. Kuttruff

21.1 Introduction

Sound with frequencies above the range of the human hearing is called ultrasound. Although the upper frequency limit of hearing differs from one person to another and also changes in the course of life, its average is usually assumed as 20 kHz. Accordingly, there is general agreement that the lower limit of ultrasound frequencies is about 20 kHz.

Basically, the propagation of ultrasound is governed by the same laws as that of sound of all other frequencies (see Chap. 1). In particular, this is true for the radiation of sound from vibrating surfaces, for the reflection of a sound wave at a boundary, and for its refraction when it enters another medium. Likewise, the diffraction or scattering of ultrasonic waves by obstacles follows the general laws. However, the latter phenomena are less prominent in the ultrasonic range. This holds especially for sound waves with frequencies above 1 MHz. On the other hand, sound attenuation or absorption becomes more significant than in the audio range where it is often neglected. The attenuation of ultrasound will be briefly discussed in Sect. 21.2 which contains also some remarks about the reflection and refraction of ultrasound, including the behaviour of transverse waves (shear waves) in solids (see Sect. 1.7).

Regarding the various practical uses of ultrasound, one has to distinguish between the applications of low-amplitude ultrasound and those for which high sound intensities or large particle velocities are required. In the former case (Sect. 21.5), the ultrasound is used as a carrier of information, for instance, to obtain insight into an opaque body. Most important in this field are the nondestructive testing of materials and medical diagnosis with ultrasound. In high power applications, as treated in Sect. 21.6 the ultrasound is utilised to effect certain changes of the object treated with ultrasound. Of particular importance is ultrasonic cleaning and joining. Certain special fields such as underwater sound or the techniques of surface wave (SAW) devices will not be discussed here.

Since this chapter is to present a brief overview on the field of ultrasonics, only few original papers will be cited. Instead, reference will be given mainly to comprehensive presentations of the whole field and to some special topics.

21.2 Propagation and Radiation

21.2.1 Attenuation

In general, the absorption of sound energy and hence the attenuation of sound waves increase strongly with frequency; therefore, they play a more prominent rôle in the ultrasonic range than in the audio range. As a rule, it is higher in gases than in liquids, and here it is higher as in solids. For this reason, ultrasound propagation in air plays a minor role only; hence, the propagation of ultrasound in solids and liquids lies in the foreground of interest.

Ultrasound attenuation is caused by manifold quite different physical processes which cannot be described here in detail [1, 2]. It may just be mentioned

H. Kuttruff
Institute of Technical Acoustics, RWTH Aachen University, Templergraben 55, 52056 Aachen, Germany
e-mail: hku@akustik.rwth-aachen.de

G. Müller and M. Möser (eds.), *Handbook of Engineering Acoustics*,
DOI 10.1007/978-3-540-69460-1_21, © Springer-Verlag Berlin Heidelberg 2013

that the relevant loss processes in gases are viscosity, heat conduction and thermal relaxation. In liquids the same processes may occur, but often additional dissipative effects become more significant. They are linked to the internal structure, for instance, to the molecular constitution of the medium. Accordingly, the observation of the ultrasonic attenuation and its frequency dependence is a valuable tool for the physical examination of materials. In solids the number of effects responsible for the sound absorption is even larger. A fact of great practical importance is the polycrystalline structure of many metallic materials: the crystallites which the material is composed of scatter the sound waves; the attenuation caused by this effect rises steeply with the frequency and hence makes these materials practically opaque for ultrasound beyond a certain frequency. Table 21.1 lists the ultrasonic attenuation in various materials.

21.2.2 Reflection and Refraction

A plane sound wave impinging on the plane boundary between two different media of infinite extent will be partially reflected at this boundary while another portion of the wave will be refracted, i.e., it will enter the adjacent medium [1]. Figure 21.1 depicts the most general case of two solids rigidly connected to each

Table 21.1 Sound attenuation in some media

Material	Frequency (MHz)	Attenuation (dB/cm)	Source	Remarks
Air (20°C, 1 at)	1	1.7	[3]	At 70% relative humidity
Water (20°C)	1	0.0022	[2]	
Ethyl alcohol (20°C)	1	0.0047	[2]	
Steel	2	0.3	[4]	
Aluminium	2.5	0.02	[4]	
Polyethylene (20°C)	1.46	9.7	[4]	
Plexiglass (20°C)	1.46	2.3	[4]	
Soft tissue	1	0.5–2	[5]	Frequency dependence about linear
Bone	1	4–10	[5]	

other. We assume first that the incident wave is longitudinal arriving at an angle ϑ_{1L}.

In general, both the reflected and the refracted waves contain a longitudinal and a transverse component. The latter is produced by "wave-type conversion" at the boundary; the direction of particle oscillation lies in the drawing plane. The reflection and refraction angles involved are related to each other by Snell's law

$$\frac{\sin\vartheta_{1L}}{c_{1L}} = \frac{\sin\vartheta'_{1L}}{c_{1L}} = \frac{\sin\vartheta'_{1T}}{c_{1T}} = \frac{\sin\vartheta_{2L}}{c_{2L}} = \frac{\sin\vartheta_{2T}}{c_{2T}}. \tag{21.1}$$

The dashed quantities refer to the reflected components; c_{iL} und c_{iT} ($i = 1$ or 2) denote respectively the longitudinal and transverse wave velocities in both media (Table 21.2). An analogous relation holds, if the incident wave is a transversal wave with particles oscillating in the plane of incidence; in this case, the first fraction in Eq. (21.1) has to be replaced with $\sin\vartheta_{1T}/c_{1T}$. If, on the contrary, the particle motion is perpendicular to the plane of incidence, no wave-type conversion occurs. The same holds for normal incidence of the primary wave, be it longitudinal or transverse.

It may happen that one or the other of the partial equations above cannot be fulfilled because the value of the sine function is always within the range ± 1. Then the corresponding secondary wave vanishes. If, for instance, $c_{2L} > c_{1L}$, no refracted longitudinal wave will occur if the angle of incidence ϑ_{1L} exceeds $\arcsin(c_{1L}/c_{2L})$. In this case, the refracted wave is purely transverse. If additionally $c_{2T} > c_{1L}$, then for $\vartheta_{1L} > \arcsin(c_{1L}/c_{2T})$

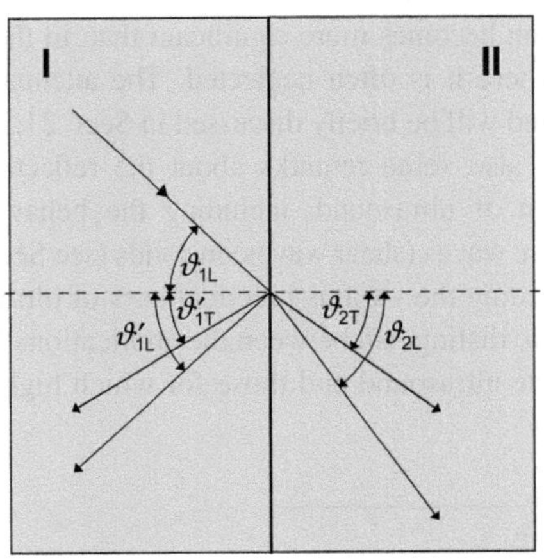

Fig. 21.1 Reflection and refraction at the boundary between two rigidly fixed solids

Table 21.2 Sound velocity of some materials

Material	Density (kg/m^3)	Sound velocity (longitudinal, m/s)	Sound velocity (transverse, m/s)
Oxygen	1,429	316	–
Hydrogen	0.090	1,284	–
Nitrogen	1,251	334	–
Carbondioxide	1,977	259	–
Air	1,293	331	–
Water	998	1,483	–
Mercury	13,500	1,451	–
Benzene	878	1,324	–
Carbon tetrachloride	1,594	938	–
Aluminium (rolled)	2,700	6,420	3,040
Iron	7,900	5,950	3,240
Brass (70% Cu, 30% Zn)	8,600	4,700	2,110
Steel (stainless)	7,900	5,790	3,100
Glass (flint)	3,880	3,980	2,380
Quartz, fused	2,200	5,968	3,764
Plexiglass	1,180	2,680	1,100
Polyethylene	900	1,950	540

The data for gases are for a temperature of 0°C, those for liquids are for 20°

there will be no refracted wave whatsoever (total reflection).

If one of both media is fluid, i.e., a liquid or a gas, then no reflected or refracted transversal wave can exist in it. In the case of two fluid media, the reflection factor R and the transmission factor T are given by the equations

$$R = \frac{Z_2 \cos \vartheta_{1L} - Z_1 \cos \vartheta_{2L}}{Z_2 \cos \vartheta_{1L} + Z_1 \cos \vartheta_{2L}}, \quad (21.2)$$

$$T = 1 + R = \frac{2Z_2 \cos \vartheta_{1L}}{Z_2 \cos \vartheta_{1L} + Z_1 \cos \vartheta_{2L}}. \quad (21.3)$$

Both quantities are defined as the sound pressure of the secondary wave divided by that of the incident wave. Z_1 and Z_2 are the characteristic impedances, $Z_i = \rho_i c_{iL}$, of both materials.

21.2.3 Radiated Sound Field

In this section the discussion is restricted to the radiation into a fluid medium from a plane oscillating disk,

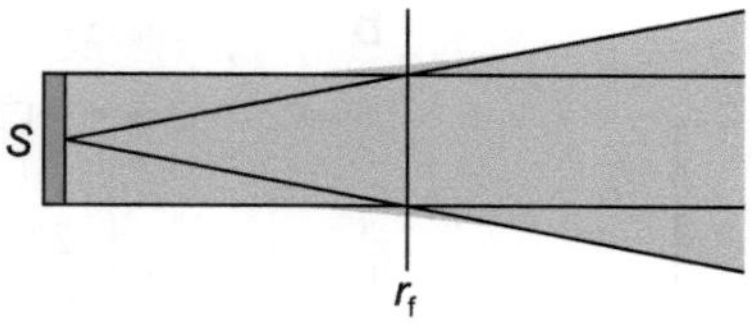

Fig. 21.2 Sketch of the sound field produced by a plane oscillating disk of radius large compared to the wavelength λ (r_f = far-field distance after Eq. (21.4))

whose lateral dimensions are large compared with the wavelength λ. In this case, one has to distinguish between the near field and the far field of the radiator; the limit between both is roughly given by

$$r_f \approx \frac{S}{\lambda}, \quad (21.4)$$

where S is the area of the disk. Figure 21.2 is to give an idea of the spatial extension of the sound field. For $r < r_f$ sound field is contained within a tube whose cross section equals roughly the oscillating disk; the sound intensity in this range, however, is by no means constant but exhibits rather strong spatial fluctuations. Beyond the far-field distance r_f, the tube expands into a conic spreading focussed on the centre of the disk, the aperture of the cone being the smaller the shorter the wavelength. A more accurate calculation of the sound field shows that in the near field the region filled with sound has a constriction (see Fig. 21.9), which is sometimes denoted erraneously as a "focus."

21.3 Generation of Ultrasound

Nowadays, technical ultrasound is almost exclusively generated by electrical means, i.e., by electroacoustic transducers. Among those, the piezoelectric transducer principle (Sect. 14.4.1.1) is proved to be particularly useful and versatile.

In the simplest case, a piezoelectric sound generator consists of a layer or a disk of some piezoelectric material, suitably oriented between two metal electrodes forming an electric capacitor (see Fig. 21.3a). If an electric voltage is applied to it, a certain strain will be produced in the material due to the piezoelectric effect, i.e., the thickness of the disk will undergo a relative change s:

$$s = d \times E, \quad (21.5)$$

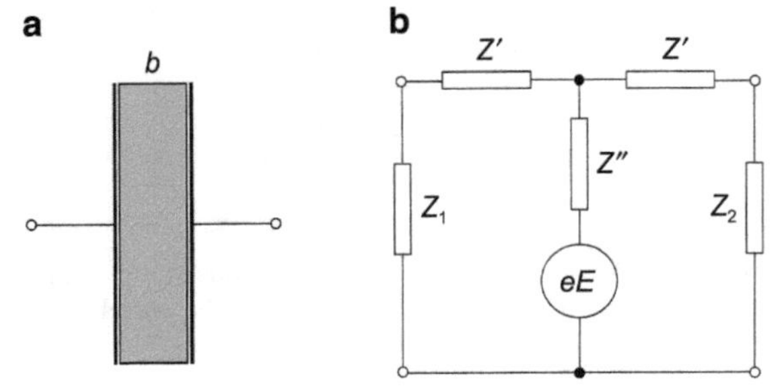

Fig. 21.3 (**a**) Schematics of piezoelectric thickness transducer. (**b**) Equivalent electrical circuit (after Mason): $Z' = jZ_0 \quad \tan(\pi b/\lambda_L), Z' = Z_0/j\sin(\pi b/\lambda_L)$, Z_0 = characteristic impedance of the transducer material, λ_L = wavelength, both for longitudinal waves in the piezoelectric material

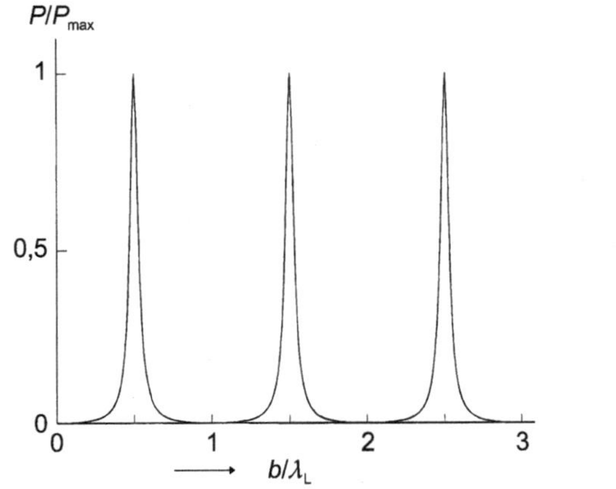

Fig. 21.4 Power radiated from a symmetrically loaded, piezoelectric thickness transducer as a function of the ratio b/λ_L, $Z_0/Z_1 = 10$

provided the surfaces of the disk are free of external forces (E = electrical field strength which is assumed as homogeneous). The factor d depends on the type of the material and is called the piezoelectric modulus. Alternatively, the piezoeffect can be described by the elastic stress caused by the electrical field inside the disk, which now is assumed as clamped in such a way that its thickness is kept constant:

$$\sigma = e \times E. \tag{21.6}$$

The piezoelectric constant e is related to the piezoelectric modulus d by the elastic constants of the material.

Apart from this longitudinal piezoeffect considered so far, there is also a transverse effect, i.e., along with the change of thickness, the lateral dimensions of the disk will also be altered. Furthermore, under certain conditions, a shear deformation or shear stress of the disk material may be generated. In the following, however, we shall restrict ourselves to the longitudinal effect, as it is the most important one in view of the practical applications. Accordingly, we call the device sketched in Fig. 21.3a a thickness transducer.

If the thickness b of the disk is a not small compared with the acoustical wavelength in the piezoelectric, the disk must be considered an acoustical waveguide. Consequently, any change of the elastic state due to piezoelectric interaction will be propagated in the form of a longitudinal wave travelling parallel to the electrical field. It will be reflected repeatedly between the disk surfaces, giving rise to a standing wave if the electrical signal is stationary. The dynamical properties of such an ultrasound transducer can be discussed using, for instance, Mason's equivalent electrical circuit [6] as shown in Fig. 21.3b. Here, all mechanical quantities are treated as electrical ones. The piezoelectric force per unit area, for instance, corresponds to an electrical voltage eE. The adjacent media are represented by their characteristic impedances Z_1 und Z_2. The forces acting onto their boundaries are easily calculated from the equivalent circuit, as well as the acoustical power radiated towards both sides. In Fig. 21.4 the power radiated from a symmetrically loaded thickness transducer is presented as a function of b/λ_L. The most obvious feature of this diagram is the regular sequence of maxima, which are the more pronounced the greater the mismatch between the disk material and the surrounding medium. They occur at the frequencies

$$f_n = (2n + 1) \cdot \frac{c_L}{2b} \; (n = 1, 2, ...), \tag{21.7}$$

and can be interpreted as thickness resonances of the disk. If the transducer is used for ultrasonic power applications, these resonances are quite welcome, but they are annoying if one is interested in the generation of acoustic broadband signals, for instance, of short ultrasonic impulses. The bandwidth of a transducer can be increased to a remarkable extent, however, by applying some damping material to one of its surfaces which combines high internal losses with a relatively good acoustical match to the transducer material ($Z_2 = Z_0$).

The transducer described so far is the prototype of a piezoelectric ultrasound transmitter. According to the practical requirements, it can be modified in many ways. For producing high acoustic power output at relatively low frequencies, very often compound transducers are used (see Fig. 21.5a). It consists of at least two piezoelectric disks in the middle and two end pieces of metal. The latter may be formed in a suitable way, for instance, so as to improve the sound radiation into the region where the ultrasound is needed (see for instance Fig. 21.11). For many applications, the transducer is attached to a so-called velocity transformer consisting of a rod with varying cross section in which the transducer excites a standing extensional wave. The simplest version of such a transformer is a "stepped horn" as presented in Fig. 21.5b. In this case, the velocities at both its ends are inversely proportional to their cross-sectional areas. Both halves have the length of a quarter wavelength. Besides this, velocity transformers with continuously varying cross section, for instance, of conical or exponential shape, are in use; their transformation ratios differ from the above-mentioned rule [7, 8].

Although quartz is the classical piezomaterial, it is no longer used for sound generation to any large extent. Practical piezoelectric transducers consist mostly of ceramic materials such as barium titanate, lead circonate titanate (PZT) or lead metaniobate. They can be given almost any shape, for instance, spherical or cylindrical. However, prior to use, they must be electrically polarised. The same holds for foils of piezoelectric high polymers such as polyvinylidenfluoride (PVDF) which are sometimes used too for the construction of ultrasound transducers. Table 21.3 lists the properties of some piezoelectric materials.

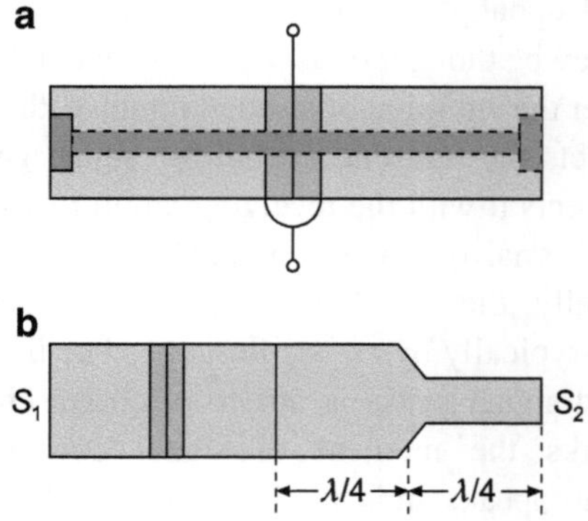

Fig. 21.5 (**a**) Piezoelectric compound transducer; both piezoelectric disks have opposite polarisation. (**b**) Piezoelectric compound transducer with attached velocity transformer (step transformer)

Besides the piezoelectric transducer, magnetostrictive transducers can also be used for the generation of intense ultrasound in the frequency range of up to *ca.* 50 kHz. They are based upon the change of dimensions which a ferromagnetic body undergoes when it is magnetised. However, in contrast to the piezoelectric effect, magnetostriction is basically a nonlinear effect; therefore, particular measures are needed to obtain linear transduction. Nowadays, the magnetostrictive transducer principle has been widely superseded by the piezoelectric one.

21.4 Measurement and Detection

Like most electroacoustical transducers, the piezoelectric transducer is reversible, i.e., it can be employed not only for the production of ultrasound but also for detection and reception of ultrasonic signals. This fact is the basis for the construction of ultrasound microphones, mostly named hydrophones. Furthermore, one and the same transducer can be employed to generate ultrasound fields and to receive the echoes caused by obstacles therein.

As before we consider a transducer consisting of a plate of piezoelectric material (see Fig. 21.3a). The relation corresponding to Eq. (21.6) is $D = -es$, where s is the strain, i.e., the relative change of the plate's thickness enforced from outside, and D is the dielectric displacement observed at short-circuited electrodes. From a practical point of view, however, the open circuit voltage of the plate is more interesting than the displacement. It is given by

Table 21.3 Properties of some piezoelectric materials

Material	Sound velocity c_L(m/s)	Density (kg/m^3)	Relative dielectric constant	Piezo constant e(As/m^2)
Quartz (x-cut)	5,700	2,650	4.6	0.17
Lead zirconate titanate (PZT-5A)	4,350	7,750	1,700	15.8
Lead metaniobate	3,300	6,000	225	3.2
Polyvinylidene fluoride (PVDF)	2,200	1,780	10	0.14

$$U_e = -\frac{e}{\varepsilon_S} \cdot \delta b. \qquad (21.8)$$

Here δb denotes the thickness change of the piezoelectric disk and ε_s is its dielectric constant at constant thickness (clamped disk).

Figure 21.6 shows two forms of piezoelectric hydrophones. At the left hand, a small hollow cylinder of a piezoelectric material serves as the active element; inside, it is damped by some lossy material. Other common forms use one or several small piezoelectric disks (Fig. 21.6b). If the hydrophone is to act as a pressure receiver, it must be small compared with the acoustical wavelength in the surrounding medium. A particularly wide frequency bandwidth is achieved with the so-called needle hydrophone [9]; it consists of a metal needle whose tip is covered by a thin layer of polyvinylidenfluoride (PVDF).

For the detection of ultrasound, mechanical, thermal or optical effects can also be employed. A limited, parallel sound beam in an otherwise resting fluid exerts a constant pressure on an obstacle, the so-called radiation pressure. If the obstacle consists of a sound absorbing plate arranged perpendicularly to the axis of the sound beam, the radiation pressure equals numerically the energy density in the sound wave; with a reflecting plate, it is twice that value. From the radiation force acting on the target the energy density and the intensity can be determined. Instruments used for this purpose are known as radiation balances [10]. Thermal ultrasound sensors determine the sound intensity from the temperature rise of a small sound absorbing body placed in the sound field [12]. Optical methods are based on the fact that harmonic (progressive or standing) ultrasound waves propagating in a transparent material act as optical diffraction gratings due to the density changes produced by the sound wave. By comparing the brightness of the various diffraction orders, the sound intensity can be determined absolutely, i.e., without any additional calibration [13]. Finally, an extended sound field can be visualised by using Toepler's Schlieren method. Again, it is the density gradients associated with the sound waves upon which this method is based.

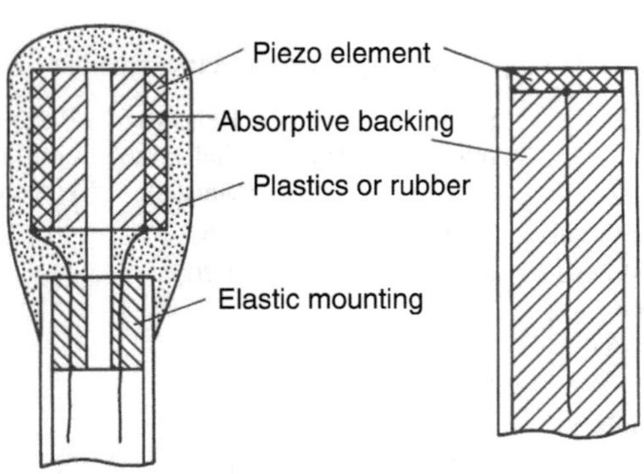

Fig. 21.6 Piezoelectric hydrophones, after [7] (courtesy S. Hirzel Verlag)

21.5 Small Signal Applications

21.5.1 Impulse Echo Method

In the applications to be described in this section, the ultrasonic wave is used as a carrier of information. Most of these applications are diagnostic, particularly in nondestructive testing of materials and technical components [4, 14] and in medical sonography [5, 15]. In both fields, the task is the inspection of an optically opaque body and its inhomogenities.

Apart from special purposes, the standard procedure for ultrasound diagnostics is the impulse echo method: An ultrasound impulse is emitted in the form of a well-collimated beam into the object to be examined. Any inhomogenities of sufficient size, for instance, voids, inclusions, or boundaries, will cause secondary waves due to scattering or reflection; the part travelling back towards the sending transducer will be detected for diagnosis, either by this transducer itself or by a second transducer nearby.

The principle of the impulse echo method is depicted in Fig. 21.7a. The transducer T is fed by an electrical signal generator IG via a separating circuit S which may be thought of as a kind of switch. Immediately after the emission of a sound impulse, this circuit disconnects the transducer from the signal generator and connects it with the receiving circuitry where the received signal is properly amplified and rectified in AR. Finally, the signal is visualised by means of a monitor, typically by an oscilloscope. The first of the displayed signal peaks is due to electrical cross-talk and marks the moment of signal emission; the subsequent peaks indicate echos received from the inspected area. The distance of a particular signal peak from the initial peak is proportional to the transit time of the corresponding ultrasound signal from which the location of the indicated inhomogenity can

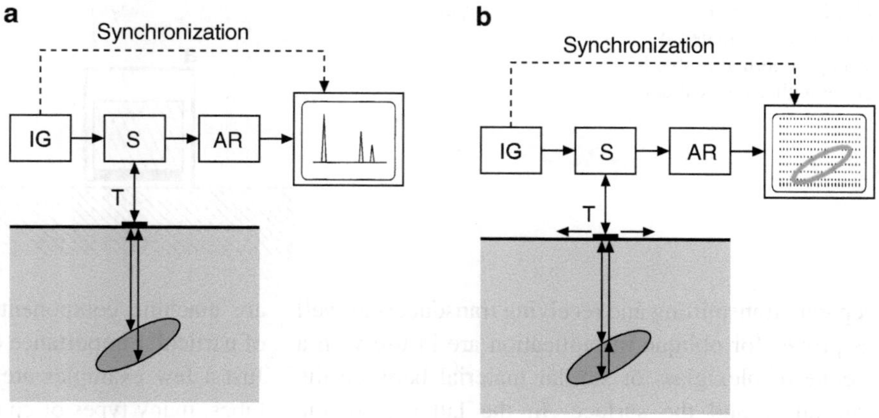

Fig. 21.7 Impulse echo method. **a** A-scan; **b** B-scan. (*IG* impulse generator, *S* separating circuit, *AR* amplification and rectification)

be determined provided the velocity of sound in the medium is known.

Often an interior boundary reflects only a part of the primary sound energy; therefore, the remaining sound wave can penetrate it and produce echos from boundaries or obstacles behind it. If the test object is a plate, a tube, a container, etc., its rear wall will produce in general a strong echo. Hence, the impulse echo method is also an excellent tool for measuring the thickness of walls.

The method outlined so far is the so-called A-scan. The more illustrative B-scan is mainly applied in medical diagnosis; it produces a two-dimensional mapping of the inhomogeneities embedded in a given object area (see Fig. 21.7b): Whenever the transducer emits a sound impulse, the electron beam on the screen starts a downward trip with constant speed over the screen of the monitor; the lateral position of this trace is controlled by a voltage corresponding to the lateral position of the transducer. The received echo signals are used to control the energy of the electron beam and hence the actual brightness of the spot on the screen; if no signal is received, the spot remains dark. By scanning the surface of the test object in this manner, a more or less faithful map of the included obstacles is obtained.

The centre frequency of the test impulse determines the extension of the smallest, just detectable object. For instance, if the object were a rigid sphere or circular disk with $ka \ll 1$ (a = radius), the intensity of the backscattered secondary wave would increase with the fourth power of the frequency. For other objects, the law of increase is different; nevertheless, it can be stated as a general rule that the detectability of an object of given shape and size is the better the higher the test frequency. However, with increasing frequency the attenuation of ultrasound waves also grows, which can only partially be compensated for by electronic depth compensation. In any case, a suitable compromise with respect to the operating frequency must be met. In flaw detection of materials, the testing frequencies are mostly in the range of 1–10 MHz. A similar statement holds for the medical sonography, although for the examination of small organs such as the eye or the skin, ultrasound of a much higher frequency is employed.

The lateral resolution of the impulse echo method depends on the width of the sound beam emitted by the transducer (see Fig. 21.2). In certain situations, it can be improved by using convergent sound beams. They are produced with suitably curved piezoelements or by combining the transducer with a collecting lens. The axial resolution depends on the duration of the ultrasound impulse, i.e., again on its centre frequency.

21.5.2 Nondestructive Testing of Materials

Figure 21.8a depicts a typical probe, i.e., an ultrasonic transducer for material testing. On its front side, the piezoelectric disk is provided with a protective cover. If properly designed, it can improve the impedance match between the transducer and the material under test. On the rear side of the piezoelement, a block of damping material is usually attached to it in order to increase the bandwidth (see Chap. 3). For detecting flaws near the surface, the probe may be combined with an acoustical delay line, for instance, of plexiglass (see Fig. 21.8b). Furthermore, probes with

Fig. 21.8 Transducers for nondestructive material testing. **a** normal probe, **b** probe with acoustical delay

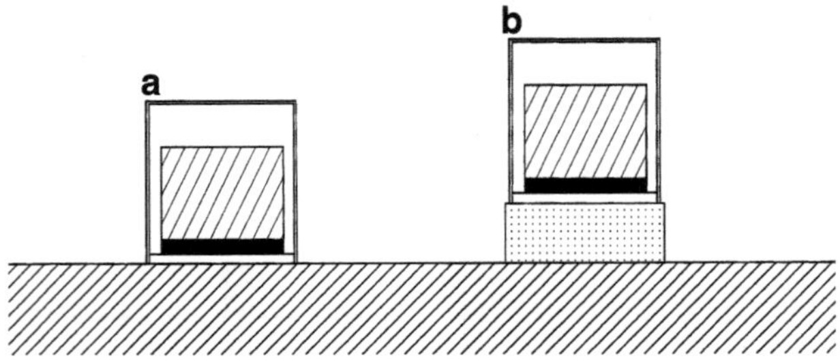

separate transmitting and receiving transducers as well as probes for oblique insonification are in use with a wedge of plexiglass or similar material between the transducer and the surface. In the latter case, the refraction and wave-type conversion at the surface has to be taken into account (see Sect. 21.2.2); if the angle of incidence is sufficiently large, a purely transversal wave is emitted into the test volume.

Usually the probe is directly set onto the surface of the material under test with a liquid film in between, for instance, oil or water. Alternatively, both the test piece and the transducer may be immersed in a water tank. In this case, the water serves as a delay line and, at the same time, as a coupling of the wave to the test piece. Also, coupling via a water beam is occasionally employed. Scanning the surface of an extending test piece is achieved either manually or automatically.

Nondestructive flaw detection with ultrasound can be applied to nearly all materials, however, with different success. An important factor is the interior structure of a material: many materials are not homogeneous but have a polycrystalline structure, or they contain small inclusions (for instance carbon in cast iron). Diffraction by such inhomogeneities attenuate the sound signals on the one hand and contributes to a kind of background noise consisting of many tiny echoes on the other. Fortunately, most sorts of steel are well suited for ultrasonic testing. About the same holds for light metals such as aluminium and magnesium and their alloys. Examples of materials which are difficult to test with ultrasound are cast iron as well as copper and its alloys such as brass or bronze. Because of the coarse structure, ultrasonic testing of concrete and artificial stones is possible only at very low frequencies.

Ultrasound is used for the inspection of raw materials and semifinished products as well as for detecting flaws in finished workpieces, the latter both right after manufacture and in the course of maintenance periods. In the foreground of interest are machine components and assemblies which are of particular importance or which are heavily stressed. Just a few examples are sheet metal, bars, axles and tubes, many types of containers, welds, rails and railway wheels, etc. – an enumeration that could easily be continued.

21.5.3 Medical Diagnosis

The most important medical application of ultrasound is the sonography, which permits the inspection of organs in the human body and their state of health. It employs almost exclusively the impulse echo method. Sonography is particularly well-suited for the investigation of soft tissue structures as the characteristic impedances show only slight variations and thus permit large penetration depths. In contrast, air-filled organs such as the lungs cannot be examined with ultrasound since their boundaries are almost perfect sound reflectors. Hence, regions covered by the lungs are shielded inaccessibly to sonography from outside. About the same holds for bones.

It is a particular advantage of sonography that it does not use ionising radiation and hence is free of the health risks accompanied with that kind of radiation. Moreover, the average sound intensity can be kept low enough to prevent excessive mechanical stress or intolerable temperature increase, damaging the tissues.

In sonography methods are preferred which yield a visual display of the examined region, mostly in the form of a B-scan. For systematically scanning a certain region, transducer arrays consisting of 60–200 small piezoelements arranged equidistantly side by side are widely used (see Fig. 21.9). The width of the single transducer element corresponds to few wavelengths. Such a transducer array is slightly pressed onto the skin to which some coupling gel has been applied before. The active sound radiator (and

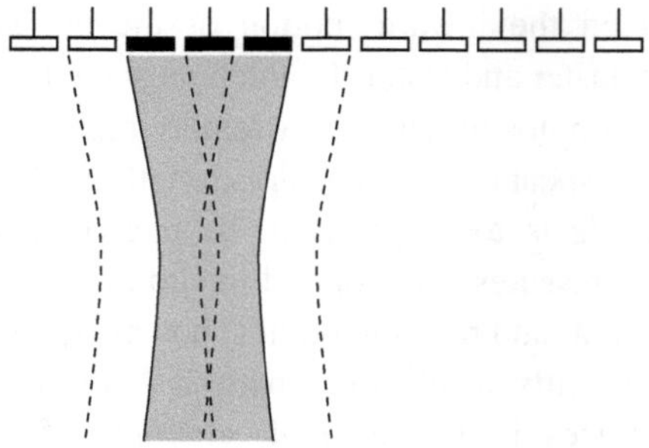

Fig. 21.9 Linear transducer array for B-scan. (For shifting the active zone elements are switched on and off)

receiver) consists of a group of elements which are connected in parallel. After each transmitting-receiving cycle, the active region of the transducer is shifted laterally by one unit. This is effected by electronically switching on and off elements at both sides of the active group. Another commonly used device is the sector scanner which contains a rotating head carrying several transducer elements. They scan a sector of the region under test or can even produce a panorama display whose resolution depends, of course, on the distance. In any case, the number of frames per second is high enough to visualise time-dependent processes such as the motion of heart valves.

B-scans of biological tissue produce usually as an appearance of irregular patches, the so-called "speckles." These are not images of the tissue structure itself but are caused by interference of numerous weak echo components. Nevertheless, the structure of the speckles yields information on the kind of tissue and its pathological changes; moreover, it marks the boundaries between different tissues or organs and facilitates the recognition of their position and size.

Nowadays, sonography is employed in nearly all branches of medicine like in internal medicine, gynaecology and obstetrics, cardiology, ophthalmology, urology, etc.

The speed V_0 of moving structures, for instance of cardiac valves, and especially of blood corpuscles in blood vessels can be measured by means of Doppler sonography. It is determined from the frequency shift

$$\delta f = \pm 2\frac{V_0}{c}f, \tag{21.9}$$

which a sound signal undergoes when it is reflected or back-scattered from a moving object (Doppler effect); here, f denotes the frequency of the primary signal; the upper sign holds for an object moving towards the transmitter. If combined with the regular B-scan, the Doppler sonography produces a two-dimensional overview of the flow velocity of the blood within a certain region, for instance, in the heart; the different velocities are represented by applying a suitable colour code.

21.5.4 Further Applications

To conclude this section, some further applications of low-intensity ultrasound may be mentioned.

As already remarked in Sect. 21.4, an ultrasound wave in a transparent material may be conceived as a diffraction grating and therefore be used for the fast deflection of a light beam which penetrates the sound wave perpendicularly Moreover, the light frequency undergoes a Doppler shift when it is diffracted by a sound wave since the latter moves relative to the observer. These effects have gained some importance in optical communication.

The possibility to transfer information via ultrasound is exploited for the remote control of certain devices such as television sets, garage doors, miniaturised hearing aids inserted into the auditory channel, etc. (However, in these cases ultrasound has to compete with other carriers, for instance, with infrared.) Furthermore, ultrasound is used to determine the distance from a reflecting body by measuring the transit time from the transducer to the reflector and back to the transducer. Examples of such applications are the measurement of the level of some liquid in a large tank, or of the distance between automobiles, for instance, when entering a parking space, or as a collision warning in a tailback. In these applications, the frequency of the ultrasound is in the range of 20–100 kHz.

Ultrasound of much higher frequency, say 1 GHz or higher, is used in ultrasound microscopes [16]. Figure 21.10 depicts the principle of such a microscope. For generating the sound impulse and for detecting its echo, a piezoelectric thin film transducer is used, typically of zinc oxide. Basically, it emits a plane wave into a solid substrate with particularly low attenuation such as sapphire. This wave enters the coupling liquid underneath the substrate through a concave boundary which acts as an acoustic lens and focuses the sound beam. The impedance match

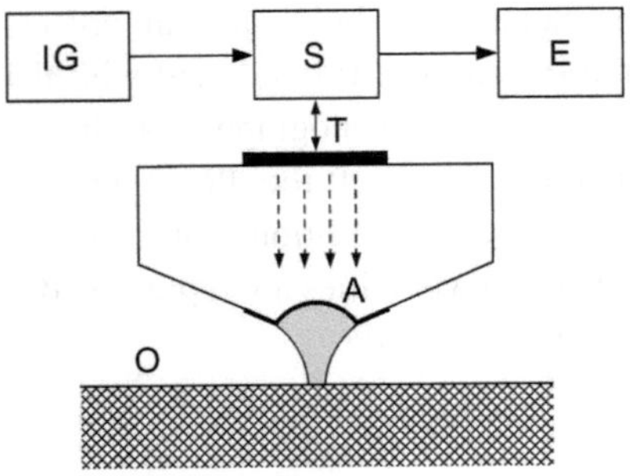

Fig. 21.10 Ultrasound reflection microscope (*G* signal generator, *S* separating circuit, *E* receiver and monitor, *A* matching layer, *O* object)

between the solid substrate and the liquid can be greatly improved by an anti-reflection coating A on the lens, a technique commonly applied to optical lenses. The focal spot of the sound beam is located on or slightly below the surface of the object O, its diameter is of the order of 1 acoustical wavelength; at 1 GHz and with water as a coupling liquid, it is about 1 μm. The separation of the reflected or backscattered signal from the transmitted one is achieved either by means of a circulator or with a fast electronic switch S. A two-dimensional display of the object is obtained by scanning its surface. This device is called a reflection microscope. A transmission microscope consists of two separate transducer units in confocal arrangement, one for sending the sound signal and one for receiving it. Both of them are situated opposite to each other with a thin object in between.

21.6 Effects and Applications of High-Intensity Ultrasound

21.6.1 Cavitation

One of the most remarkable effects of strong ultrasound waves in liquids is cavitation and the phenomena associated with it [17, 18] (see also Sect. 20.1.2). Ultrasonic cavitation means the formation of voids or cavities during the phase of negative sound pressure. In contrast to more or less stable gas bubbles, these voids contain only small amounts of gas. Under the influence of the sound field, they either perform strongly nonlinear pulsations or implode as soon the negative pressure vanishes by which they have been created initially. Such a collapse starts with very low speed, then the inward motion of the bubble wall becomes faster and faster. In the final state, the velocity of the implosion will be extremely high.

Strictly speaking, the tensile strength of physically pure liquids is too high as to be overcome by the negative pressures encountered in common ultrasonic fields. In real and realistic liquids, however, numerous microscopically small solid particles are suspended with crevices in which small amounts of gas are stabilised and which are believed to act as nuclei for the onset of cavitation. As a consequence, the cavitation threshold, i.e., the minimum sound pressure amplitude needed to produce cavitation, is dramatically reduced. At frequencies below 30 kHz, its order of magnitude is 1 bar (corresponding to an intensity of about 0.3 W/cm^2 in water); at higher frequencies, it grows monotonically with the frequency.

In an imploding cavitation void, the remaining gas is highly compressed, at the same time its temperature is raised dramatically, and as a consequence, a short light impulse is emitted during the very last state of implosion. This is the reason for the faint light sometimes observed in strong ultrasound fields in certain liquids. However, it seems that an exact explanation of this light emission, which is known as sonoluminescence, is still lacking. The same holds for the fact that certain chemical reactions are initiated or accelerated in cavitation fields which is the basis of the so-called sonochemistry. Another important fact is that each strong bubble collapse is associated with the emission of a short and strong pressure impulse into the surrounding liquid. Hence, cavitation effects a strong temporal and spatial concentration of energy. This plays an important role in several applications.

21.6.2 Ultrasonic Cleaning

To clean a specimen or an object with ultrasound, it is immersed into a vessel or tank filled with a cleaning liquid that is exposed to an intense ultrasound field. The sound frequency is usually between 20 and 50 kHz. The cleaning process is brought about by cavitation produced on the contaminated surface which provides the nuclei needed for cavitation inception. On the one hand, strong and short pressure impulses emerging from the imploding cavities erode the surface layer and loosen insoluble dirt particles. On the other hand, strong local flows occur in the direct vicinity of pulsating or collapsing cavitation bubbles which do not implode synchronously. These currents remove the dirt particles from the surface and provide for a quick exchange of cleaning liquid.

Ultrasonic cleaning is carried out in cleaning tanks of quite different sizes which are made of stainless steel or of plastics. The electrical power needed lies in the order of 10 watts/litre. Today, the sound field is mostly generated by piezoelectric compound transducers (see Sect. 21.3), which insonify the liquid from the bottom or from one wall (see Fig. 21.11).

Since the walls of the tank and the surface of the liquid reflect the sound waves, a standing wave field is established in the tank. Therefore, the cleaning effect is not uniformly efficient at all locations. Furthermore, the standing waves are accompanied with more or less pronounced resonance effects, particularly in small tanks. By superposition of several different frequencies, more homogeneous sound fields can be produced.

The appropriate choice of the cleaning liquid depends mainly on the kind of contamination. It may be aqueous (alkaline or acid) or organic. Examples of organic solvents are alcohols, or chlorine or fluorine hydrocarbons such as trichloroethylene or freon. Since many of these liquids are toxic or cause environmental hazards, their application is subject to rigorous regulations. Often ultrasonic cleaning is carried out at elevated temperatures in order to reduce the surface tension of the liquid and hence to facilitate the onset of cavitation and, at the same time, to increase the solubility of the contaminations.

Ultrasonic cleaning proves to be particularly useful whenever highest cleanness must be achieved, or when the objects to be cleaned are mechanically delicate or very small, or if they have irregularly shaped surfaces inaccessible to brushes, etc. Examples are the products of fine mechanics and precision engineering, medical instruments, optical lenses, jewellery of all kind, television screens, electronic circuits and many more.

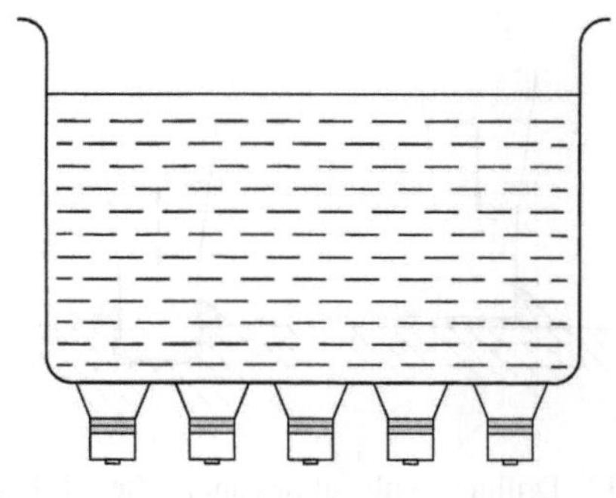

Fig. 21.11 Ultrasonic cleaning tank

21.6.3 Ultrasonic Joining

Another application of high-intensity ultrasound firmly established in industrial production is joining, mainly welding and bonding of materials and parts. In this case, the sound is applied to the work not by an intermediate wave medium but by direct action of a vibrating tool. Welding is achieved by thermal plastification or liquification due to sound energy dissipated in the work material. Therefore, this method is well-suited for the treatment of thermoplastics but not of duroplastics.

For welding, the two parts are pressed together between the "anvil" and the actual welding tool, the "sonotrode." The latter serves at the same time for introducing the vibrational energy which is generated by an efficient ultrasonic vibrator, typically a compound transducer, and fed to the sonotrode via a velocity transformer (see Fig. 21.5) in such a way that the sonotrode vibrates perpendicularly to the surface of the work piece. The frequency is about 20 kHz. For "near-field welding" as shown in Fig. 21.12a, the sonotrode whose geometry determines the shape of the joint is pressed directly onto one of the partners next to the joint. In contrast, for "far-field welding" (see Fig. 21.12b), the ultrasound is led to the mating surfaces over a certain distance within one of the partners to be joined. In any case, the welding process is initiated by some kind of compressional friction at isolated contact points. Here a small amount of the

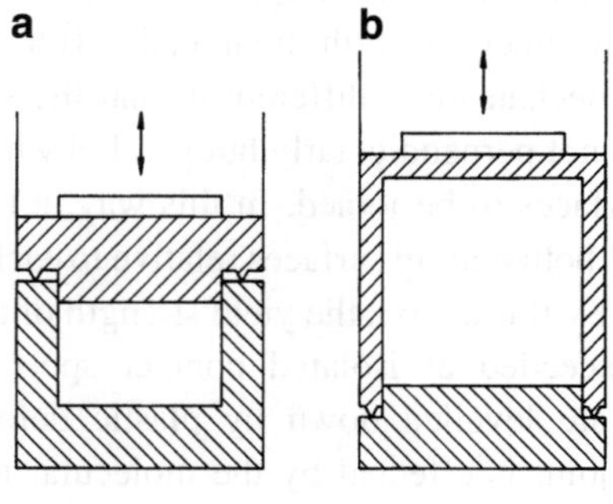

Fig. 21.12 Ultrasonic welding. **a** near field welding; **b** far field welding, after [7] (courtesy S. Hirzel Verlag)

material is plastified at first since the contact points must transfer the total alternating forces and hence the total energy flux. Since in general sound absorption in plastics increases with rising temperature, those parts of the material which have already been plastified will be heated faster and faster until extended areas will become liquid and finally both components are joined to each other. The entire welding process takes only a fraction of a second. In order to obtain more regular welds, one of the partners is often manufactured with a small elevation ("energy concentrator") in the form of a cone or a ridge as shown in Fig. 21.12. In any case, the heat is produced exactly where it is needed which is a particular advantage of ultrasonic welding.

Not all thermoplastics are equally well-suited for welding with ultrasound. Polystyrene and its copolymers can easily been welded; the same holds for polycarbonates and polymethylmetacrylates. Other materials such as polyolefines are suited only for near-field welding since their internal losses are too high for far-field operation.

Nowadays, ultrasonic welding of plastics is employed in nearly all branches of the plastics-processing industry, particularly for the series production of a great variety of parts. Just a few examples are electrical plugs, switches and housings, components of automotive industry like instrument boards, fuel tanks, housings of rear lights, etc. Likewise, ultrasonic welding plays an important role in the packaging industry.

Riveting with ultrasound is based on a similar principle as ultrasonic welding. Here the contact area of the sonotrode must correspond to the shape of the rivet head. The sonotrode acts immediately onto the latter and plastifies it. Furthermore, the closing of housings by ultrasonic edging and the insertion of metal parts into thermoplastics should be mentioned.

Ultrasound can also be applied for joining metals with each other or with nonmetals. But here the welding mechanism is different in that the sonotrode oscillates not perpendicularly but parallel with respect to the surfaces to be joined. In this way, a tangential motion of both mating surfaces relative to each other is effected. By this action, the yield strength of the material is exceeded at isolated contact spots and the surfaces are levelled down by plastic deformation; the final joint is effected by the molecular attraction forces. Thus, welding is not or not predominantly a thermal process as in welding of plastics. Best suited for ultrasonic welding is copper as well as aluminium and its alloys, either with themselves or with other metals. Joints of metals with semiconductor materials, with glass or ceramic materials are also possible.

21.6.4 Drilling and Cutting

Another useful application of ultrasound is machining hard or brittle materials such as glass, ceramics, hard metal, and gems. Although we speak here mainly of drilling or cutting, the process in reality is some kind of grinding.

For drilling holes into a work piece, a tool shaped according to the desired hole – similarly as with ultrasonic welding of plastics – is set in vigorous vibrations perpendicular to the surface of the work piece. These vibrations are generated by a power transducer in combination with a velocity transformer, usually a conical horn. Between the drilling tool and the work piece, an aqueous suspension of an abrasive is applied as shown in Fig. 21.13 (silicon carbide, boron carbide, or diamond powder). When the vibrating tool approaches the surface, an alternating, horizontal flow in the abrasive slurry underneath the tool is created. Additionally, strong cavitation is produced in the liquid. Both effects set the grains of the abrasive in fast motion which erodes the material under the tool. By continuously lowering the tool, a dip in the surface is gradually produced. This motion must be, however, sufficiently slow to avoid direct contact between tool and workpiece.

Of course, ultrasonic drilling is not restricted to making circular holes; holes of any cross section can be made in this way. To drill large holes, hollow tools are customary as they erode less material. This has the additional advantage that fresh abrasive can be supplied continuously to the cutting region through the

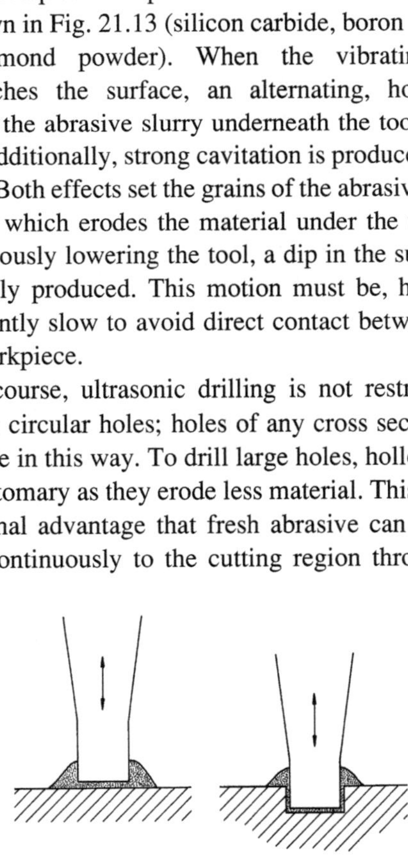

Fig. 21.13 Drilling with ultrasound, after [7] (courtesy S. Hirzel Verlag)

tool. Furthermore, with hollow tools small disks can be cut out of plates. Thin slices are cut from bar of stock by using a thin steel blade as a cutting tool.

The tool is usually soldered to the velocity transformer; it does not need to be particularly hard; so it can be made of soft steel or brass. As in ultrasonic welding it is important that the transforming piece together with the drilling or cutting tool is exactly tuned to the frequency at which the transducer operates.

21.6.5 Atomisation of Liquids

With ultrasound fine dispersions or suspensions can be produced of solid materials in liquids as well as mixtures of liquids which are mutually insoluble (emulsions). The latter are very stable because of the small diameter of the droplets. The equipment used for such applications is similar to that employed for ultrasonic drilling, i.e., it consists of a powerful transducer with an attached velocity transducer as shown in Fig. 21.5b, its, narrow end dips into the liquid to be treated. It appears that in these applications cavitation plays a decisive role too.

Another useful application is the atomisation of liquids, i.e., the production of aerosols. In this case, the ultrasound is generated in the liquid and is directed from below towards the liquid surface (see Fig. 21.14). At a certain vibrational amplitude, a capillary wave is formed on the surface. With increasing amplitude, this wave will become increasingly distorted, and finally tiny droplets forming a mist will be propelled from the crests of the wave. It is clear that the size of the droplets is in a certain relation to the wavelength of the capillary wave and hence to the frequency. For water, the following relation holds:

$$D \approx 0.2 \times f^{-2/3}, \tag{21.10}$$

where D denotes the diameter of the droplets in mm and f is the frequency in kHz.

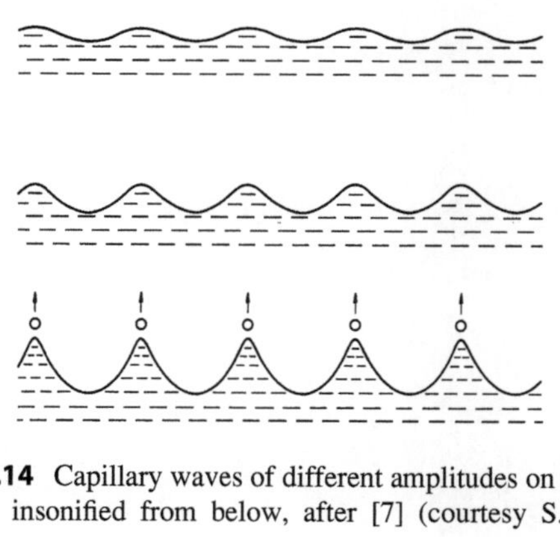

Fig. 21.14 Capillary waves of different amplitudes on a liquid surface, insonified from below, after [7] (courtesy S. Hirzel Verlag)

For the production of aerosols, the liquid can be supplied continuously through a hole to the narrow end of a vertical velocity transformer in which extensional waves are excited. A still better effect is achieved if the transducer is connected with a conical end plate, the lowest bending resonance of which is excited by the vibration. The frequency is about 20 kHz until 50 kHz at maximum; accordingly, the size of the droplets is in the range of about 10–20 μm. Finer aerosols are obtained with a concavely shaped piezo-transducer which is arranged below the liquid surface and which concentrates the emitted sound near the surface. These devices are operated with frequencies in the order of about 1 MHz.

Ultrasonic atomisers have the advantage of producing about equally sized droplets; hence, the produced aerosols are very stable. They are mainly used as air humidifiers and for the preparation of medicines to be inhaled by the patient. The atomisation of fuels is in discussion since long but it seems that this application has not yet left the experimental state.

21.6.6 Medical Therapy

In medical therapy, ultrasound is employed since long to improve blood circulation, for the relief of pain and cramp and also for retarding inflammatory processes. One relevant effect in these applications is absorption of sound energy in the tissue and the local heating accompanied with it. Furthermore, the high acceleration in ultrasonic fields seems to increase the permeability of cell membranes and to improve the perfusion.

The sound projectors used for ultrasonic therapy are designed in a way similar to those employed for nondestructive testing of material (see Fig. 21.8a) with the difference that the transducer is operated at its resonance; hence, the damping back material is omitted. The operating frequency is typically a few

100 kHz to about 1 MHz; sound of still higher frequency is less applicable because of its reduced penetration depth. The active area of the sound transmitter is a few square centimetres; the transmitted acoustical power is in the range of few watts/cm^2.

References

1. Mason WP (1958) Physical acoustics and the properties of solids. Van Nostrand, New York
2. Schaaffs W (1963) Molekularakustik. Springer, Berlin
3. Bass HE, Sutherland LC, Zuckerwar AJ, Blackstock DT, Hester DM (1995) Atmospheric absorption of sound: Further developments. J Acoust Soc Am 97:680
4. Deutsch V, Platte M, Vogt M (1997) Ultraschallprüfung – Grundlagen und industrielle Anwendungen. Springer, Berlin
5. Hill CR (ed) (1986) Physical principles of medical ultrasonics. Ellis Horwood, Chichester
6. Edmonds PD (ed) (1981) Methods of experimental physics, Vol 19: Ultrasonics. Academic, New York
7. Kuttruff H (1988) Physik und Technik des Ultraschalls. S. Hirzel Verlag, Stuttgart
8. Millner R (1987) Ultraschalltechnik – Grundlagen und Anwendungen. Physik Verlag, Weinheim
9. Platte M (1985) A polyvinylidene fluoride needle hydrophone for ultrasonic applications. Ultrasonics 23:113
10. Beissner K (1985) Ultraschall-Leistungsmessung mit Hilfe der Schallstrahlungskraft. Acustica 58:17
11. Kuttruff H (1991) Ultrasonics – Fundamentals and applications. Elsevier Applied Science, London
12. Rinker M, Fay B (1994) Wärmeproduktion in Festkörpern durch Ultraschallabsorption als Grundlage für ortsauflösende Ultraschalleistungssensoren. Acustica 80:300
13. Kwiek P, Reibold R (1992) Light diffraction by ultrasonic waves for normal and Bragg incidence. Acustica 77:193
14. Krautkrämer J, Krautkrämer H (1986) Werkstoffprüfung mit Ultraschall. Springer, Berlin
15. Wells PNT (1975) Physical principles of ultrasonic diagnosis. Academic, London
16. Lemons RA, Quate CF (1979) Acoustic microscopy. In: Mason WP, Thurston RN (eds) Physical acoustics – Principles and methods, Vol XIV. Academic, New York
17. Leighton TG (1994) The acoustic bubble. Academic, London
18. Young FR (1989) Cavitation. McGraw-Hill, London
19. Platte M (1984) Ultraschallwandler aus dem piezoelektrischen Hochpolymer Polyvinylidenfluorid. Diss. Technische Hochschule Aachen

Vibrations

22

Johannes Guggenberger and Gerhard Müller

22.1 General Aspects, Definitions

Low frequent vibrations may cause from disturbing up to damaging effects. There is no precise distinction between structure-borne sound and vibrations. However – depending on the frequency range – measurements and predictions require different techniques. In a wide frequency range, the generation, transmission and propagation of vibrations can be investigated similar to structure-borne sound (see Chap. 9).

Vibrations and structure-borne sound related to rail traffic are discussed in Chap. 16. Excitations caused by wind and earthquakes are treated in [1–7] and in the standards [8–11], bell towers and the excitation forces of bells in DIN 4178 [12].

An overview over structural dynamic issues in civil structures is given in [2, 13–18].

J. Guggenberger (✉)
Müller-BBM GmbH, Robert-Koch-Strasse 11, 82152 Planegg, Germany
e-mail: Johannes.Guggenberger@MuellerBBM.de

G. Müller
Lehrstuhl für Baumechanik, Technische Universität München, Arcisstrasse 21, 80333 München, Germany
e-mail: gerhard.mueller@bv.tum.de

22.2 Reference Values for the Assessment of Vibrations

22.2.1 Human Exposure to Mechanical Vibrations

22.2.1.1 Preliminary Remarks

The strain affecting an individual is subdivided on the following criteria: "perception", "comfort", "performance" and "health". The human perception of vibrations depends on individual constraints (age, sex, state of health, shape), on the activity (professional life, school, leisure) and on the environment (vehicles, residences, offices, schools, hospitals). Whole body vibrations are often stochastic or contain different frequencies or shock signals. Therefore thresholds describing a limit above which a decrease of the human performance or of the felt comfort has to be expected cannot generally be given.

The following summary is restricted to exposure in buildings which are treated in the ISO 2631-2 [19, 20], the German DIN 4150-2 [21] and VDI 2057-3 [22], the British BS 6472 [23, 24] and the Austrian "Ö-Norms" S9010 [25] and S9012 [26]. A survey of European standards for the assessment of human exposure to mechanical vibrations is given in [27–29].

Reference values for the comfort in train coaches are given in ISO 10056 [30], ISO 2631–4 [31] and UIC 513 [32], on ships DIN ISO 6954 [33]. Rules for measuring, evaluating and assessing hand-arm vibrations are treated in DIN EN ISO 5349 [34] and VDI 2057-2 [35]. Reference [36] describes a simplified method for assessing the risk for employers exposed to vibrations.

G. Müller and M. Möser (eds.), *Handbook of Engineering Acoustics*,
DOI 10.1007/978-3-540-69460-1_22, © Springer-Verlag Berlin Heidelberg 2013

22.2.1.2 Perception of Vibrations

In a frequency range from 1 to approximately 4 Hz, the perception of vibrations is – depending on the posture –roughly proportional to the amplitudes of the acceleration. Between 10 and approximately 80 Hz, the perception is rather proportional to the amplitudes of velocity. Above 80 Hz, the sensitivity of persons decreases due to the interior insulation of the human body.

Below approximately 0.6 Hz, amplitudes of acceleration of around ≥ 1 m/s^2 may cause sickness within 30 min.

Secondary effects related to vibrations often are more disturbing than the vibrations themselves (e.g., rattling of glasses, bouncing of hanging objects). However, it is not possible to deduce the intensity of the vibration from those effects, as they even might occur at levels far below the limit of perception.

In German standards, the description of the impact of vibrations on humans is still based traditionally on the rated intensity of perception KB. In order to assess the frequency dependency, the value KB is derived from a frequency weighted velocity signal according to DIN 45669 [37, 38]. The attenuation curve is defined as

$$|H_{\mathrm{KB}}(f)| = \frac{1}{\sqrt{1 + (5.6\,\mathrm{Hz}/f)^2}}. \tag{22.1}$$

In more recent standards like ISO 2631 [19] and VDI 2057-1 [39] the vibration intensity is expressed by a weighted acceleration $a_w(t)$. Regarding the direction-dependent human perception, different weighting curves for vertical and horizontal vibrations are introduced. For example, for upright standing or sitting persons the weighting curve W_k refers to vertical and W_d to horizontal body posture, while for undefined body posture on building floors, a weighting curve W_m is introduced.

The magnitude of the weighting curves exhibit a high pass H_h, a low pass H_l, an acceleration–velocity transition H_t and an upward step H_s

$$|H(f)| = |H_h(f)| \cdot |H_l(f)| \cdot |H_t(f)| \cdot |H_s(f)|, \tag{22.2}$$

The individual filters are summarized in Table 22.1.

The different cutoff frequencies f_i and parameters Q_i are:

Table 22.1 Bandwidth filters and frequency weighting for the weighting curves W_k, W_d, W_m [19, 35]

$|H_h(f)| = \sqrt{\frac{f^4}{f^4+f_1^4}}$ $|H_l(f)| = \sqrt{\frac{f_2^4}{f^4+f_2^4}}$ $|H_t(f)| = \sqrt{\frac{{f_3}^2}{f^2+f_3^2}}$ for $\boldsymbol{W_m}$
$|H_t(f)| = \sqrt{\frac{f^2+f_3^2}{f_3^2}} \times \sqrt{\frac{f_4^4 \times Q_4^2}{f^4 \times Q_4^2 + f^2 \times f_4^2(1-2Q_4^2) + f_4^4 \times Q_4^2}} \cdots$ for $\boldsymbol{W_k}, \boldsymbol{W_d}$
$|H_s(f)| = \frac{Q_6}{Q_5} \times \sqrt{\frac{f^4 \times Q_5^2 + f^2 \times f_5^2(1-2Q_5^2) + f_5^4 \times Q_5^2}{f^4 \times Q_5^2 + f^2 \times f_5^2(1-2Q_5^2) + f_6^4 \times Q_6^2}}$

W_k: $f_1 = 0.4$ Hz; $f_2 = 100$ Hz; $f_3 = f_4 = 12.5$ Hz; $f_5 = 2.37$ Hz; $f_6 = 3.35$ Hz; $Q_4 = 0.63$; $Q_5 = Q_6 = 0.91$

W_d: $f_1 = 0.4$ Hz; $f_2 = 100$ Hz; $f_3 = f_4 = 2.0$ Hz; $Q_4 = 0.63$; $H_s = 1$

W_m: $f_1 = 0.7493$ Hz; $f_2 = 100$ Hz; $f_3 = 5.684$ Hz; $H_s = 1$

In Table 22.2, the relationship between the limit levels of vibration and the subjective perceptions are given. Humans can clearly distinguish an increase of the weighted intensity of vibration if it rises by a factor of 1.6.

The weighting curves are shown in Fig. 22.1, where also the curves of equivalent level of vibration are compared with the threshold of perception (KB = 0.1, $a_w = 0.01$ m/s). More recent studies [28] show that the threshold of perception may be significantly lower in the frequency range above 10 Hz than indicated by the weighting curves according to ISO 2631 [19, 20, 31].

22.2.1.3 Comfort: Assessment Criteria for Residential and Equivalent Use

The assessment according to the German standard DIN 4150/2 [21] is based on two limits:

- $\mathrm{KB}_{F\mathrm{max}}$, maximum weighted intensity of vibration
- $\mathrm{KB}_{F\mathrm{Tr}}$, rated intensity of vibration

The weighted level of vibration $\mathrm{KB}_F(t)$ is defined as a sliding RMS value of the frequency weighted velocity signal, band-filtered between 1 and 80 Hz (time constant $\tau = 0.125$ s, "FAST").

$$\mathrm{KB}_F(t) = \sqrt{\frac{1}{\tau}\int_{\xi=0}^{t} e^{-\frac{t-\xi}{\tau}} \mathrm{KB}^2(\xi)\,d\xi}. \tag{22.3}$$

From this, the energy-equivalent $\mathrm{KB}_{\mathrm{eq}}$ is obtained as time average of $\mathrm{KB}_\tau(t)$ Eq. (22.4):

Table 22.2 Relationship between level of vibration and subjective perception, according to [39, 40]

Description of the perception	Weighted intensity of vibration (KB)	Weighted acceleration a_w (m/s^2) (RMS)
Not perceptible	<0.1	<0.01
Threshold of perception	0.1–0.4	0.015–0.02
Easily perceptible	0.4–1.6	0.02–0.08
Strongly perceptible	1.6–6.3	0.08–0.315

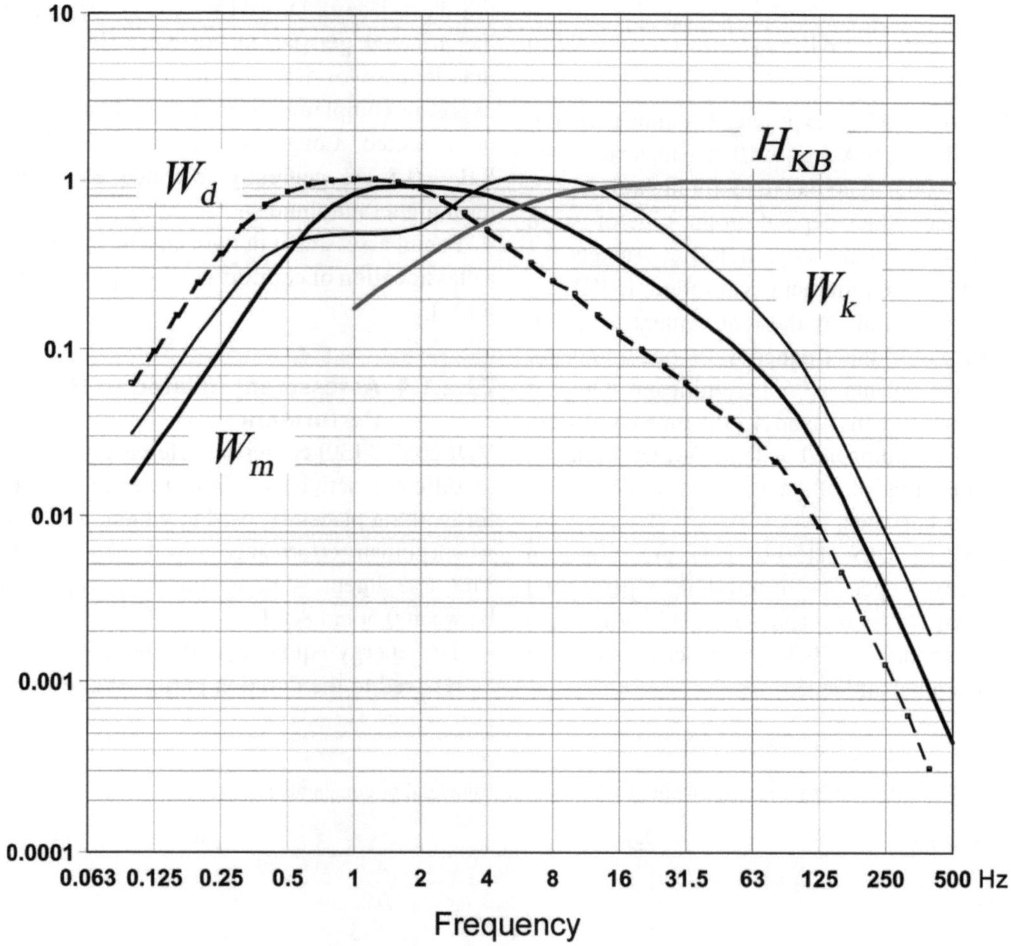

Fig. 22.1 Weighting curves H_{KB} [37], W_d and W_k [19] and W_m [22]

$$\mathrm{KB_{eq}} = \sqrt{\frac{1}{\mathrm{T}_e}\int_0^{\mathrm{T_e}} \mathrm{KB}_F(t)^2 \mathrm{d}t} \qquad (22.4)$$

The maximum weighted intensity of vibration $\mathrm{KB}_{F\max}$ describes the maximum of the weighted vibration $KB_F(t)$ according to Eq. (22.3) which occurs once or several times during the considered observation time.

The assessment level of vibration KB_{FTr} considers the frequency of occurrence and the duration of the events. The duration period of exposure is divided in cycles of $T = 30$ s. Within each cycle, i KB_{FTi} is determined from the corresponding maximum of $|KB_F(t)|$. The time equivalent intensity of vibration KB_{FTm} results from the energy-equivalent average of the individual events:

$$KB_{FTm} = \sqrt{\frac{1}{N}\sum_{i=1}^{N} KB_{FTi}^2}. \quad (22.5)$$

If $KB_{FTi} < 0.1$, it will be neglected for the respective cycle. The assessment level of vibration KB_{FTr} results in

$$KB_{FTr} = KB_{FTm}\sqrt{\frac{T_e}{T_r}}, \quad (22.6)$$

(T_e – duration period of exposure; T_r – time period of assessment, 16 h from 6:00 a.m. to 10:00 p.m. for day time and 8 h from 10:00 p.m. to 6:00 a.m. for night time).

The reference values depend on the type of usage zone and the time of the day. If $KB_{F\max}$ meets the lower limit A_u the requirement is fulfilled, if $KB_{F\max}$ exceeds the upper limit A_o the requirement is exceeded. In both cases, the frequency of occurrence or duration of the events is not significant. Only if $KB_{F\max}$ ranges between A_o and A_u, the time equivalent KB_{FTr} has to be compared with a further limit A_r, considering the duration of the exposure or the events´ frequency of occurrence.

The limits of DIN 4150–2 [21] are given in Table 22.3. For most persons the threshold of perception ranges between $KB = 0.1$ and $KB = 0.2$. Vibrations with values of about $KB = 0.3$ typically cause annoying sensations in a quiet ambience.

The reference values depend on the source, e.g., rail vehicles, blasting, construction sites etc. The upper limit for aboveground rail vehicles is specified independently from the considered usage zone at $A_o = 0.6$.

For short period vibrations at construction sites, $A_o = 5$ is considered to be still acceptable. The limits for long period exposures are set lower to $A_u =$ 0.8–1.6 and depend on the length of the event (78 days–1 day). DIN 4150–2 [21] suggests notifying the affected persons in advance. Experience shows that in case of KB = 1 with a longer duration of exposure complaints from the neighbourhood have to be expected. Long-term vibrations in the range of KB = 0.5 are generally tolerated even with short-term higher amplitudes.

For an assessment in train coaches (see Table 22.4), a classification of comfort values is given in ISO 2631-4 [31].

22.2.1.4 Assessment Related to Human Performance

VDI 2057-3 [39] comprises reference values according to different activities. These references can be used for working places exposed to vibration as limits under which human performance at work places is not affected. The assessment is based on three values evaluated between 0.5 and 80 Hz:

- The energy-equivalent weighted acceleration a_{we} referred to the duration period of exposure T

Table 22.3 Reference values for the assessment of vibrations in residential or comparable use [21]

Location of exposure	Day			Night		
	A_u	A_o	A_r	A_u	A_o	A_r
Industrial area	0.4	6	0.2	0.3	0.6	0.15
Commercial area	0.3	6	0.15	0.2	0.4	0.10
Mixed area, villages	0.2	5	0.1	0.15	0.3	0.07
Exclusive residential area	0.15	3	0.07	0.1	0.2	0.05
Very sensitive locations, special area	0.1	3	0.05	0.1	0.15	0.05

Table 22.4 Relationship between frequency weighted acceleration and subjective comfort [31]

Level of comfort	Effective value of the frequency weighted acceleration a_w (m/s^2)
Comfortable	<0.315
Slightly uncomfortable	0.315–0.63
Quite uncomfortable	0.5–1
Uncomfortable	0.8–1.6
Very uncomfortable	1.25–2.5
Extremely uncomfortable	>2

$$a_w = \sqrt{\frac{1}{T}\int_0^T a_w^2(t)\mathrm{d}t}, \qquad (22.7)$$

$$a_{we} = k_1 a_w, \qquad (22.8)$$

- The coefficient k_1 depends on the axis direction of the exposure ($k_1 = 1$ for vertical, $k_1 = 1.4$ for horizontal exposure). If the exposure duration T_e is made up of i exposure segments T_i, a_{we} is calculated from:

$$a_{we} = k_1\sqrt{\frac{1}{T_e}\sum_i a_{wi}^2 T_i}. \qquad (22.9)$$

- The assessment acceleration a_w Eq. (22.8) is calculated from the energy-equivalent mean value:

$$a_{w(8)} = a_{we}\sqrt{\frac{T_e}{8\,\mathrm{h}}}, \qquad (22.10)$$

- The maximum $\max\{a_{wF}(t)\}$ of the sliding RMS value of the frequency-weighted acceleration $a_{wF}(t)$ (FAST) (time constant $\tau = 0.125$ s, "FAST") during the duration of an exposure:

$$a_{w\tau}(t) = \sqrt{\frac{1}{\tau}\int_{\xi=0}^{t} a_w^2(\xi)e^{\frac{\xi-t}{\tau}}d\xi}. \qquad (22.11)$$

The assessment is based on the indicator values in Table 22.5 for each exposure direction x, y, z.

It is common practice to also assess vibration exposures in buildings according to DIN 4150-2 [21] since office use can be considered as comparable to residential use for day time. Here single peaks during a duration period are better represented by the maximum in each 30-s cycle of which the assessment intensity of vibration KB_{Ftr} is composed. The corresponding reference values are proposed in Table 22.6.

The range depends on various factors, e.g., on the location of the origin within the rooms of the own (higher values) or third-party (lower values) office areas.

In [41], a catalog of representative noise and vibration data on working places is assorted. The effect of a simultaneous impact of noise and vibration leading to possible hearing loss is discussed in [42].

22.2.1.5 Assessment Related to Health Damages

In order to assess vibrations which could possibly lead to health damages according to ISO 2631-1 [19] the weighted equivalent acceleration a_{we} is determined Eq. (22.7). If the vibration exposure occurs in periods of variable magnitudes and durations, the energy equivalent vibration magnitude, corresponding to the total duration of exposure, is evaluated according to:

$$a_{we} = \left(\frac{\sum a_{wi}^n T_i}{\sum T_i}\right)^n, \qquad (22.12)$$

whereby a_{we} is assessed for both exponents $n = 2$ and 4.

Horizontal accelerations are multiplied with a factor $k = 1.4$. The largest value – in horizontal resp. vertical direction – is assessed. If all components are comparably high, the vector sum may be used.

Table 22.6 Reference values for office use

A_u	A_o	A_r
0.35–0.4	6.0	0.15–0.2

Table 22.5 Indicator values for whole-body vibrations on working places depending on various kinds of demand, according to [31]

Location	a_{we} (m/s^2)	$a_{w(8)}$ (m/s^2)	$\max\{a_{wF}(t)\}$ (m/s^2)
Recreation rooms, rest rooms, sanitary rooms	0.01		0.03
Workplaces with high demands to fine motor skills (e.g. research laboratory	0.015		0.015
Workplaces predominantly with exercitation (e.g. switch rooms, office rooms)		0.015	0.045
Workspaces with enhanced attention (e.g. shops)		0.04	0.12
Workspaces with simple or predominantly with mechanical activities		0.08	
Other workspaces		0.15	

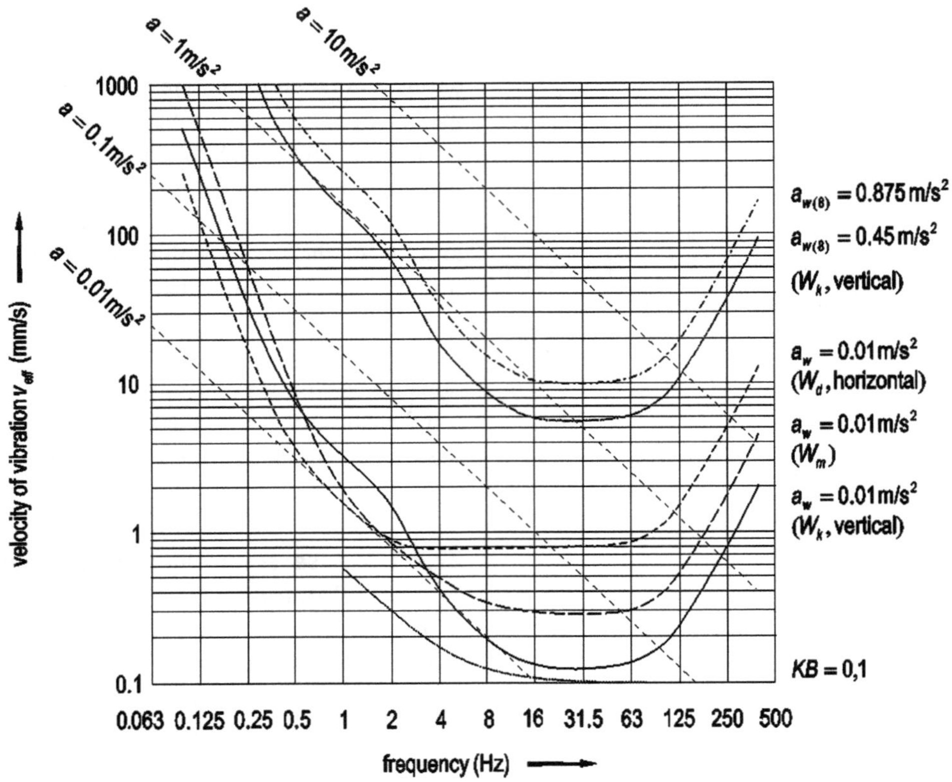

Fig. 22.2 Curves of equivalent vibration level [37], and acceleration level [19, 22]

Referring to a daily duration of the impact, the assessment is split into two steps: Vibrations above $a_{w(4)} = 0.6$ m/s^2 ($T = 4$ h) and $a_{w(8)} = 0.45$ m/s^2 ($T = 8$ h) are expected to cause a significant threat of the exposed sitting persons; vibrations above the value $a_{w(4)} = 1.14$ m/s^2 ($T = 4$ h) and $a_{w(8)} = 0.875$ m/s^2 are hazardous to sitting persons. Figure 22.2 shows curves of weighted accelerations for an assessment time of $T = 8$ h. The threshold values of an assessment between 4 and 8 h are defined by

$$a_{w(T)} = a_{w(T_0)} \left(\frac{T_0}{T}\right)^{\frac{1}{n}}. \qquad (22.13)$$

For exposures from 4 to 8 h, the threshold for hazardous vibrations is defined by the minimum of two curves defined by Eq. (22.13) and the coefficients:

Curve 1: $n = 2$; $a_{w(8)} = 0.5$ m/s^2, $T_0 = 8$ h

Curve 2: $n = 4$; $a_{w(8)} = 0.45$ m/s^2, $T_0 = 8$ h

A significant threat exists if the maximum of two curves with the coefficients:

Curve 1: $n = 2$; $a_{w(8)} = 0.875$ m/s^2 $T_0 = 8$ h

Curve 2: $n = 4$; $a_{w(8)} = 0.8$ m/s^2 $T_0 = 8$ h is exceeded. For short-time duration ($T < 10$ min) the absolute maximum weighted acceleration may not be higher than $a_w = 6$ m/s^2.

ISO 2631 part 5 [43] contains procedures to evaluate multiple shock signals.

22.2.1.6 Reference Values for the Assessment of Human Induced Vibrations

The reference values, given in Sect. 22.2.1.3, can also be considered suitable for an assessment of human-induced low-frequency vibrations in apartments or similarly used interior rooms.

In regular buildings, noticeable vibrations are generally not expected. Therefore even small amplitudes in such rooms can cause uncomfortable feelings and – in extreme cases – panic reactions. In some cases, even the structural safety for the dynamic loads has to be proved. In general, however, the threshold, above which persons may react with discomfort or unease, is generally significantly lower than the amplitudes that may affect the buildings structural safety.

According to [2], tolerable values for office buildings are established: In regular offices, a maximum acceleration of 0.05 m/s^2 should not be exceeded. Quiet rooms or sensitive working areas should not be charged with amplitudes higher than 0.02 m/s^2 for repetitive events with 20–30 cycles.

The tolerable values for sport stadiums and halls, in which moderate events are organized is referenced to 0.5 m/s^2. However, in euphoric atmospheres, e.g., during a rock concert, no complaints have been filed even with accelerations up to 3 m/s^2. Uninvolved sitting spectators should not be exposed to accelerations greater than 0.2 m/s^2. For fixed seating, the reference values given are between 0.1 and 0.5 m/s^2 depending of the nature of the presentations. Further information can be found in [44–46].

22.2.2 Impact of Vibrations on Buildings

The impact of vibrations on buildings is addressed in the German DIN 4150–3 [47]. The Swiss code SN640312a [48], the British Standard BSI 7385 [49], the Austrian standard Ö-Norm S9020 [50], ISO 4866 [51] and ISO 10137 [46] are further standards for the assessment of vibrations as far as the serviceability is concerned.

The standards provide empirical values below which damages causing a reduction of the serviceability are not expected. These values refer not only to the safety of the construction, but also to slight damages ("cosmetic damages") like cracks or fissures in the plaster, fissures in masonry structures, not permissible settlings, an enlargement of already existing fissures, a drop of tiles and exceeding deformations. Events exceeding the reference values do not necessarily lead to structural damage. In this case, it is recommended to further investigate the structural safety of the building. In the Ö-Norm S9020 [50], the risk of damages is related explicitly to the level of exceedance of the reference values (e.g., fissures in the plaster at a max. exceedance of 70% and cracks in masonry at a max. exceedance of 120%).

The codes of practice distinguish generally between short- and long-term excitation. The Swiss code [48] discerns between frequently occurring, permanent (much more than 100,000 events) and sporadic (less than 1,000 events) impacts. An impact is defined as short term, if neither fatigue nor resonance effects play a role. In other words, the reference value for stationary vibrations of civil constructions has to be taken into account as soon as resonant excitations – even for a short-time period – cannot be excluded. On the other hand, repetitive impulsive excitations provided that the impulse reaction can decay between two events and if the number of load changes is sufficiently small. Explicitly, e.g., short-term excitations caused by pile drivers, vibrators or shakers must be considered as a continuous excitation, if resonant effects on slabs cannot be excluded.

Furthermore, the codes take the importance of the building and the relevance of slight damages into account. In case of protected historical buildings, the reference values are lower than for residential buildings. The least restricted reference values are attributed to industrial buildings. The British Standard BSI 7385 [49] increases the reference values by a factor 2 or 4, depending on the permissible – not safety relevant – damages.

The assessment is based on different values as e.g.:

- The maximum value of the three components of the velocity of vibrations at the foundations (DIN 4150–3 [47])
- The total velocity of vibrations at the foundations (Ö-Norm S9020 [50], SN640312a [48])
- The velocity of vibration at a chosen structural element (plate) (DIN 4150–3 [47], SN640312a [48])

The reference values also differ with respect to the location of the measurement points within a building. Its dynamic response caused by an excitation can be evaluated by measurements at various points, ranging from the foundations (all standards) to the horizontal vibration in the highest floor (Germany).

The reference values for different buildings and permanent excitations range between $v_{max}(t) = 2.5$ mm/s (sensitive buildings, Germany) and 25 mm/s (industrial buildings, Great Britain). For short-term excitations, there is a frequency-dependent range, e.g., for the most sensitive category between 3 and 10 mm/s (Germany), 3–15 mm/s (Switzerland), 5 mm/s (Austria) and 15–50 mm/s (Great Britain). The German reference values for the levels of vibration are given exemplarily for different types of buildings in DIN 4150–3 [47], Table 22.7.

Table 22.7 Reference values for the assessment of vibrations in buildings [47]

Type of building	Level of vibration max($v(t)$) (mm/s)				
	Short time vibration				Continuous vibration
	Foundation			Top slab, horizontal direction	Top slab, horizontal direction
	1–10 Hz	10–50 Hz	50–100 Hz	All frequencies	All frequencies
Commercial-, industrial-buildings	20	20–40	40–50	40	10
Residential buildings	5	5–15	15–20	15	5
Buildings with high requirements on conservation	3	3–8	8–10	8	2.5

Table 22.8 Reference values for the assessment of short time vibrations for pipes in the soil [47]

Material	Assessment value for the level of the vibration on the pipe (mm/s)
Steal welded	100
Concrete, pre-stressed concrete, metal with or without flanges, vitrified clay	80
Masonry, synthetic material	50

In addition to this table, a maximum value of 10 mm/s for continuous vibrations on slabs in vertical direction is referenced. The standard contains furthermore material-dependent values for the admissible amplitudes of the vibration velocity on pipes in the soil (Table 22.8).

22.2.2.1 Damaging in Case of the Excitation of Vibrations Via the Soil

When the source of vibration is located outside of the construction (external excitation), the structural response depends significantly on the soil properties [52, 53]. The vibrations are transmitted by waves, influencing the vibration behaviour of buildings and industrial plants. The soil even may change its properties due to vibrations (liquefaction, settling).

Damages on buildings thus may result from the direct impact of vibrations or from vibration-induced changes of the soil properties. Settling of neighbouring foundations due to soil compression can occur immediately or after a certain time delay and thus can cause damages.

Furthermore, substructure and excavation work like e.g., the installation of sheeting in the soil can create vibrations in the adjacent buildings. Following a primary impact of the vibrations also secondary effects may arise, related to other deficiencies of a structure or its foundations [54]. The general condition of the building is decisive for those damages. Monitoring measurements can just exclude damages that are primarily related to the impact of vibrations presuming a structure without major deficiencies.

Sand with low relative density in ground water and silt are sensitive to vibrations. On the other hand, the vibration amplitudes which may lead to permanent settling typically are high and occur only in the near field of a source. It is therefore recommended not to exceed a minimum distance from the vibration source to the next foundation. This distance depends on the soil situation. If the vibrations amplitudes exceed 1/3 g, a redistribution of the grain structure cannot be excluded in sensitive soil [55].

A soil investigation report has to verify beforehand that the soil is appropriate for the planned construction work. Before starting the work, a preservation of evidence along the adjacent neighbouring buildings is recommended together with a detailed protocol of the works.

22.2.3 Impact of Vibrations on Sensitive Equipment and Processes

22.2.3.1 Classification of Vibration Sensitive Equipment

Maximum admissible amplitudes at the installation site of highly sensitive equipment can be orders of magnitude lower than the threshold of human perception. The specifications typically depend on the required precision of the device and the necessary resolution.

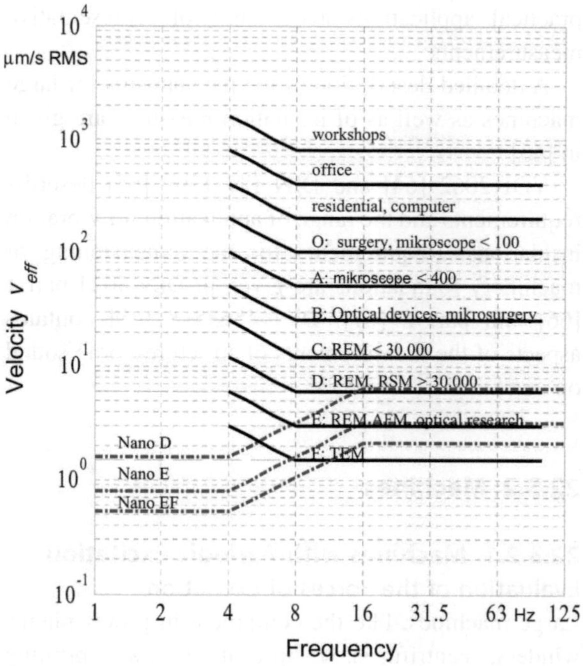

Fig. 22.3 References for the admissible vibrations for highly sensitive devices according to [56, 57] with extensions [58, 59]

Typical disturbing interferences caused by vibrations are observed at measurement devices with high resolution, microscopy, manufacturing devices of chips (beamer, stepper, photolithography etc.), mass spectrometers as well as at laser optical applications. For this kind of devices, investigations led to a classification of the sensitivity by means of Vibration Criteria (VC)-curves (Fig. 22.3, which were established together with the overview given in Table 22.9 [56, 57]. The curves that incline to the low frequencies (<8 Hz) are valid for devices without interior insulation. With interior passive insulation, the curves are extended horizontally down to 1 Hz reflecting a higher sensitivity at resonance. Reference [58] provides an overview of the vibration criteria for critical areas in healthcare facilities.

Advances in Nano-technology demand more severe restrictions on the influences of vibration. The BBN-criteria extended to "Nano-classes" are shown in Table 22.9 and Fig. 22.3 [59, 60]. The lower admissible amplitudes in the frequency range below 20 Hz are required to permit a low-tuned insulation measure.

Table 22.9 Reference values (third octave band, RMS) for the permissible vibrations for highly sensitive devices, partly according to [56–60]

Class	Level of vibration (µm/s)	Type of devices
	200	Residences, computers
0	100	Operating rooms, surgery, bench microscopes at up to 100× magnification, laboratory robots
VC A	50	Bench microscopes at up to 400× magnification, optical and other precision balances, coordinate measuring machines, metrology laboratories, optical comparators, microelectronics manufacturing equipment, inspection, probe tests, and other manufacturing support equipment
VC B	25	Micro-surgery, eye-surgery, neuro-surgery, bench microscopes at magnification greater than 400×, optical equipment on isolation tables, microelectronics manufacturing equipment, aligners, steppers and other critical equipment for photolithography with line widths of 3 µm or more
VC C	12	Electron microscopes at up to 30,000× magnification, microtomes, magnetic resonance imagers, microelectronics manufacturing equipment, aligners, steppers and other critical equipment for photolithography with line widths of 1 µm
VC D	6	Electron microscopes at greater than 30,000×magnification, mass spectrometers, cell implant equipment, microelectronics manufacturing equipment, aligners, steppers and other critical equipment for photolithography with line widths of 1/2 µm includes electron-beam systems
VC E	3	Microelectronic manufacturing equipment, not isolated laser and optical research systems, aligners, steppers and other critical equipment for photolithography with line widths of 1/4 µm includes electron-beam systems
VC F	1	Tunnel electron microscopes, atomic force microscopes
VC D	1.6/6.4	REM for nano-technical applications , resolution: 1 nm
VC E	0.8/3.2	Nano-technical applications with resolutions of 0.2–0.5 nm
VC EF	0.53/2.1	Nano-technical applications of REM and TEM < 0.1 nm (sub- Ångström domain)

In typical urban surroundings, it is very difficult to guarantee the most severe conditions. Structural and insulation measures can often not sufficiently reduce these deficiencies.

This overview serves as initial indicator for assessing a location planned for sensitive equipment. The detailed planning, however, has to be based on the specification of the manufacturer, measurements at comparable devices as well on the experience of the users. The requirements do not only depend on the precision or resolution, but also on the structural dynamic behaviour of the device, its operating mode and the typical task it is used for.

In comparison to precision devices, data storage systems are less sensitive. However, in the case of extraordinarily high vibrations, e.g., due to construction work or blasts in the vicinity, monitoring – with a communication feedback to the construction site – or specifically designed measures might become necessary.

The impact of vibrations and the necessity of insulation have to be assessed in relation to the risk of damage in case the reference values are exceeded (e.g., just a short interference of a process which can be repeated in an uncomplicated way or devices where an interference can cause severe damages e.g., loss of important data in storage systems).

Specifications for the measurement at the location of devices are referred to in ISO 10811 [61].

The time of observation as well as the effort for a reliable statistical evaluation of the data obtained during a site survey must be adapted to the importance and category of the respective vibration sensitive operation.

22.2.3.2 Binding Process of Concrete

Extensive investigations [62] show that the binding process of fresh, young concrete is not deteriorated up to a velocity of vibrations with a peak value around 20 mm/s; in this case, the targeted compressive strength is not affected.

22.3 Sources and Insulation of Vibrations

22.3.1 General Remarks

DIN 4150–1 [40] describes relevant sources of vibration and their mechanisms. Furthermore, the transmission of vibrations is illustrated by means of practical applications and results of representative measurements.

A detailed description of the mechanisms for large machines as well as of insulation measures are given in [63].

VDI 2062 [64] and DIN EN 1299 [65] describe requirements and the range of application for vibration insulations. Requirements for the dimensioning of machinery foundations are given in DIN 4024 part 1 [66] and part 2 [67]. DIN 45635–8 [68] contains aspects of the measurements of structure-borne sound on machines.

22.3.2 Machines

22.3.2.1 Machines with Periodic Excitation

Evaluation of the Forces of Excitation

Large machines, like the equipment in power plants, winders, centrifuges, reciprocating saws, printing machines, turbines, power sets, conveying troughs, turning lathes, ventilators, pumps and compressors, generally spoken machines with rotating masses, and also transformers cause periodic excitations. The excitation is generally caused by forces of non-balanced masses or pistons. The dynamic forces increase in a quadratic relation with the rotational frequency. They may further cause elastic deformations (deformations of the shaft) and thus a further increase of the unbalanced forces.

After crossing the critical frequencies shaft rotations are recentered by gyroscopic forces and the dynamic excitation decreases in relation to the rotational speed.

It is possible that the effect of a harmonic excitation is not sensed in the immediate vicinity of the source, but causes significant vibrations at a larger distance, e.g., due to resonant coupling with adjacent structures, e.g., a floor slab. During start-up and run-down of machines, frequencies of excitation can cross natural frequencies of the surrounding structure and the tuning frequencies of the mounting. If the sweeping is too slow, resonant effects may lead to considerable amplitudes.

ISO 1940 [69, 70] gives hints for assessing induced dynamic forces and the balance quality for various types of machines. For specific types of machines, individual standards exist (e.g., fans). The balance quality has a range from drives of e.g., high precision gyroscopes (Q 0.4) up to inherently unbalanced large slow marine diesel engines (G 4000). The amplitude of the dynamic vertical and horizontal force F_{max} is given by:

$$F_{max} = m\,e_{per}\,\Omega^2 = m\,G\,\Omega = U_{per} \cdot \Omega^2, \qquad (22.14)$$

(m – rotating mass; e – eccentricity = distance of the centre of gravity to the axis of rotation; Ω – angular frequency of the rotational speed [rad/s]; $G = e\ \Omega$ – balance quality grade [mm/s]; $U_{per} = m\ e_{per}$ – residual unbalance [gram mm]).

To account for operation-induced wear, it is recommended to assume the next higher balance quality than indicated in the specifications for new devices.

ISO 10816 [71] contains general criteria for acceptance tests and for monitoring of vibrations at machines by the help of measurements on non-rotational parts. Appropriate positions of measurement points in order to describe the impact of oscillating forces in a significant way are indicated (e.g., bearing, housing). The machinery type and the quality of installation is reflected by zones of vibration thresholds:

- Zone A: The vibration of newly commissioned machines would normally fall within this zone.
- Zone B: Machines with vibration within this zone are normally considered acceptable for unrestricted long-term operation.
- Zone C: Machines with vibration within this zone are normally considered unsatisfactory for long-term continuous operation. Generally, the machine may be operated for a limited period in this condition until a suitable opportunity arises for remedial action.
- Zone D: Vibration values within this zone are normally considered to be of sufficient severity to cause damage to the machine.

Those criteria mainly should help to guarantee a long term, reliable and safe use of the machine. Even if the limiting values are fulfilled they do not imply that the vibrations affecting in the surroundings are not necessarily within the permissible range.

The individual parts of the standard ISO 10816 refer to

- Part 2: land-based steam turbines and generators in excess of 50 MW with normal operating speeds of 1,500–1,800 min^{-1}, 3,000 min^{-1} and 3,600 min^{-1}.
- Part 3: industrial machines with nominal power above 15 kW and nominal speeds between 120 and 15,000 min^{-1} when measured in situ.
  - Group 1: Large machines with rated power above 300 kW; electrical machines with shaft height $H \geq 315$ nm.
  - Group 2: Medium-size machines with a rated power above 15 kW up to and including 300 kW; electrical machines with shaft height 160 nm $\leq$ H $\leq$ 315 nm.
  - Group 3: Pumps with multivane impeller and with separate driver (centrifugal, mixed flow or axial flow) with rated power above 15 kW.
  - Group 4: Pumps with multivane impeller and with integrated driver (centrifugal, mixed flow and axial flow) with rated power above 15 kW.
- Part 4: gas turbine driven sets excluding aircraft derivatives
- Part 5: machine sets in hydraulic power generating and pumping plants
  - Group 1: Horizontal machine sets with pedestal or end-shield bearings mounted on a rigid foundation, usually with operational speeds of above 300 min^{-1}.
  - Group 2: Horizontal machine sets with bearing housings which are only braced against the casing of the hydraulic machine, usually with operational speeds of less than 300 min^{-1}.
  - Group 3: Vertical machine sets with bearing housings which are all braced against the foundation, usually with operational speeds of between 60 and 1,800 min^{-1}.
  - Group 4: Vertical machine sets with lower bearing housings braced against the foundation and upper bearing housings braced against the generator stator only, usually with operational speeds of between 60 and 1,000 min^{-1}.
- Part 6: reciprocating machines with power ratings above 100 kW. The limits are given for seven classes of machines. However, the standard does not assign the corresponding machines by objective characteristics. It is noted that a certain machine may be assigned to more than one class. The final classification is left to the experience of experts. The limits are subdivided in the frequency ranges as follows (exemplarily given for zone C): $f <$ 10 Hz: 71 μm–1,125 μm < zone C < 113 μm–1,784 μm; 10 Hz $< f <$ 250 Hz: 4.46 mm/s –70.7 mm/s < zone C < 7.07 mm/s–112 mm/s; $f >$ 250 H: 7.01 m/s^2–111 m/s^2 < zone C < 11.1

m/s^2–176 m/s^2. The ranges refer to class 1–class 7. Zone A/B and zone D are correspondingly defined by the limits.

The limits for the different zones according to ISO 10816 parts 1–5 are summarized in Table 22.10.

Rotating electrical machines and power generators with combustion engines are described in DIN EN 60034–14 [72] respectively in DIN ISO 8528–9 [73].

In case of reciprocating engines also, higher harmonics emit significant excitations. They can be estimated from the machinery parameters. Reciprocating saws have a fundamental frequency around 4–8 Hz and show a series of harmonics.

Two simultaneously operated machines with approximately equal rotational frequencies can cause beating with twice the amplitude compared to a single machine. For a large number of N machines the maximum velocity of vibration v_N on the soil outside of the industrial plant can be approximately calculated according to Eq. (22.15) [40]:

$$v_N = v_B N^{0.5}\left(e^{-0.1N-1.9} + 0.103\right)\left(\frac{N_B}{100}\right)^{-0.284}, \tag{22.15}$$

(v_B – measured or calculated velocity of vibration for the simultaneous use of N_B machines with $N_B > 3$).

Measures for Insulation

In case the impact on neighbouring areas caused by vibrations is exceedingly high, first of all it should be investigated whether the dynamic forces can be reduced (e.g., by balancing the machine, by changing the rotational frequency, by eliminating resonance-phenomena at the machine, by aligning axial couplings, etc.). A reduction of the dynamic forces also increases the lifetime of the machine.

In a further step it has to be clarified if resonances resulting from the mountings' tuning frequency can be avoided. In principle, a machine can be mounted "high" or "low tuned". Often the installation of a machine can be modelled by means of a system with one single degree of freedom. Then the tuning frequency is expressed by

$$2\pi \cdot f_1' = \sqrt{\frac{k}{m}\left(1-\zeta^2\right)}, \tag{22.16}$$

(f_1' – resonant frequency of the damped system, k – spring stiffness, m – mass, ζ – damping ratio)

Mounting with a high eigenfrequency "high tuning" ("rigid mounting [66]"). The fundamental frequency of the operational rotational speed is considerably lower than the tuning frequency of the mounting. The related dynamic forces excite the structure without or with little amplification. The velocities of vibration at the machine's mountings are

Table 22.10 Vibrations induced by several machines, max($v(t)$)

ISO 10816 Part	Zone	A/B	B/C	C/D	A/B	B/C	C/D
	Group	Velocity amplitude, (mm/s)			Displacement amplitude (μm)		
2	1,500 and 1,800 min^{-1}	2.8	5.3	8.5	–	–	–
2	3,000 and 3,600 min^{-1}	3.8	7.5	11.8	–	–	–
3	Group 1, rigid support	2.3	4.5	7.1	29	57	90
3	Group 1, flexible support	3.5	7.1	11	45	90	140
3	Group 2, rigid support	1.4	2.8	4.5	22	45	71
3	Group 2, flexible support	2.3	4.5	7.1	37	71	113
3	Group 3, rigid support	2.3	4.5	7.1	18	36	56
3	Group 3, flexible support	3.5	7.1	11	28	56	90
3	Group 4, rigid support	1.4	2.8	4.5	11	22	36
3	Group 4, flexible support	2.3	4.5	7.1	18	36	56
4	3,000–20,000 min^{-1}	4.5	9.3	14.7	–	–	–
5	Group 1	1.6	2.5	4	30	50	80
5	Group 2	2.5	4	6.4	–	–	–
5	Group 3	1.6	2.5	4	30	50	80
5	Group 4, top bearing	2.5	4	6.4	65	100	160
5	Group 4, main bearings	1.6	2.5	4	30	50	80

proportional to the input impedance at the mounting position. A sufficient distance to the resonances in relation to the impedance at the location of installation according to Eq. (22.17) has to be guaranteed [2].

$$f_l \geq f_B \cdot n_h \cdot 2 \cdot s_f, \tag{22.17}$$

(f_l – tuning frequency, f_B – frequency of operation, n_h – order number of the relevant dynamic forces, $s_f = 1.1–1.2$ – factor for the evaluation of uncertainties in the impedance at the mounting position such as uncertainties in the equivalent soil spring or in the parameters of the construction).

Especially, in the case of table-like foundations, the structural natural frequencies have to be considered. In the case of block foundations, however, equivalent soil springs according to Sect. 22.4.2 can be assumed. In order to bridge softer soil layers the soil has to be reinforced by pile foundations onto a deeper stiff layer.

Mounting with a low eigenfrequency "low tuning" [66]. In this case, the tuning frequency is sufficiently lower than the operational frequency. In order to obtain an insulating effect, the tuning frequency is typically chosen at $f_l < 1/4\, f_B$ of the operational frequency. For an insulating measure, generally the excitation in the lowest frequency is decisive. If the impedance at the location of installation is much larger than the spring impedance, the insertion loss of the transmitted dynamic force is evaluated according to the spectral transmissibility function. It gives the relationship between the sum of the spring and damping force to the exciting force Eq. (22.18):

$$T(f_B) = \frac{\sqrt{1 + \left(2\zeta\eta_f\right)^2}}{\sqrt{\left(1 - \eta_f^{\,2}\right)^2 + \left(2\zeta\eta_f\right)^2}}, \tag{22.18}$$

($\eta_f = f_B/f_l$, $f_l = \sqrt{k/m}$ – tuning frequency, f_B – frequency of excitation, ζ – damping ratio).

Figure 22.4 shows the resulting level difference of the forces for various damping ratios. For small damping ratios, the insulation efficiency increases approximately with 12 dB/octave. With increasing damping, the insulation efficiency decreases. The adjustment of the damping ratio is thus a matter of optimization. Damping has to be low enough to provide good insulation efficiency, but high enough to avoid exceeding vibrations at resonance. Resonances can usually not be avoided during a start-up or run-down when the frequencies of excitation caused by the rotational speed cross the natural frequencies of the

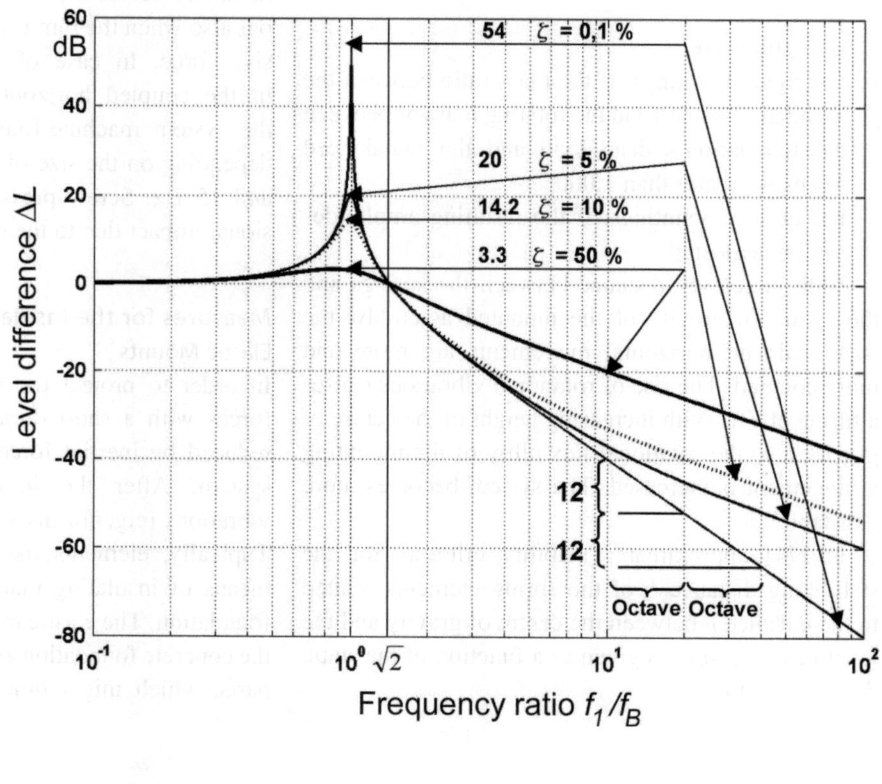

Fig. 22.4 Level difference for a force excited one-single-degree of freedom-system with respect on the frequency relationship

mounting. The latter effect can also be influenced by the speed of the start up or run down of a machine. If the crossing time is smaller than the number of periods, a system needs to reach its maximum resonant amplitude, the damping ratio may be adjusted correspondingly.

Above the tuning frequency, the vibration measured at the mountings is approximately proportional to the sum of the vibrating mass of the machine and its elastically mounted foundation. The mass of the mounted foundation has to be designed in such a way that the permissible vibrations of the machine – above the elastic support – are not exceeded. Assuming a certain tuning frequency and a stiff foundation below the soft spring the mass of the elastically mounted foundation has no impact on the transmissibility [see Eq. (22.18)].

DIN 4024 [66] and [67] distinguishes rigid and elastic mountings with the following requirements to avoid resonances

$$0.8 \cdot f_{B1} \geqslant f_1 \geqslant 1.25 \cdot f_{B1}, \tag{22.19}$$

additionally for elastic mountings also for all higher orders ($n > 1$) the range to avoid is given by:

$$0.9 \cdot f_{Bn} \geqslant f_1 \geqslant 1.1 \cdot f_{Bn} n > 1. \tag{22.20}$$

Exceptions are:

- For rigid mountings: If the mass ratio between the foundation's mass and the rotating mass or between the foundation's dead load and the unbalanced forces is greater than 1,000
- For elastic mountings: if the resulting amplitudes can be neglected

With increasing distance between the springs and the centre of gravity of the mounted assembly, the rotational and horizontal movements are more and more coupled. The rise of rotational vibrations causes a tilting effect. With increasing height of the centre of gravity also the rotational flexibility of the mounting arrangement is increased. The system becomes more "unstable".

In [74] approximate "stability criteria" for the stabilizing distance b of the spring elements related to the distance h between the centre of gravity and the level of the springs is given as a function of the static displacement δ_{stat}.

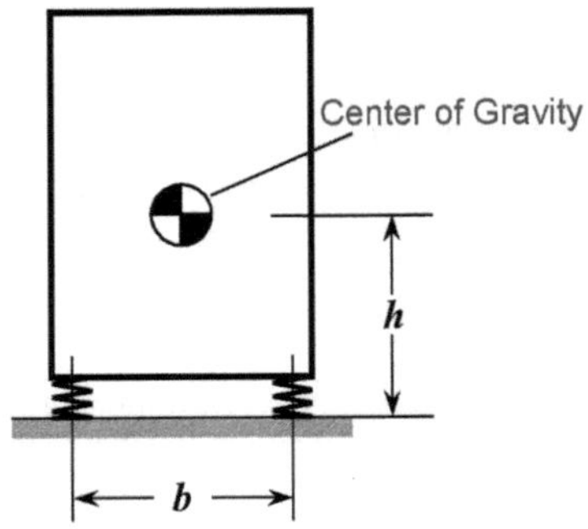

Fig. 22.5 "Stability criteria" for the stabilizing distance b of the spring elements

$$b > \sqrt{40\,\mathrm{h}\delta_{\text{stat}}}. \tag{22.21}$$

In this case, a sufficiently "stable" arrangement can be assumed (Fig. 22.5), in which mainly the vertical vibration has to be considered.

22.3.2.2 Large Machines with Impulsive Loads

Large machines, e.g., forging hammers, drop weights, presses, and large shears with impulsive loads can create significant vibrations in their vicinity. Presses work with much higher masses and forces than hammers, however, with lower velocities. Forging hammers create not only vibrations at the impact but also when the ram is accelerated by a large impulsive force. In case of rigid mountings, vibrations in the coupled horizontal/tilting eigenfrequency of the system machine-foundation-soil occur, which – depending on the size of the press – range between 5 and 15 Hz. Screw presses cause additionally a torsional impact due to the deceleration of the masses.

Measures for the Insulation

Elastic Mounts

In order to protect the surroundings, high dynamic forces with a short duration of the impulse can be reduced by inertial forces of an elastically mounted system. After the impulse, the system's eigenvibrations (eigenforms with eigenfrequencies) decay. Typically, elements like the anvil are mounted by means of insulating mats on an elastically mounted foundation. The elastic mounts serve as a protection of the concrete foundation and as a filter for high frequent parts, which might otherwise damage the concrete.

The main part of the impact energy is assimilated by the elastically mounted foundation.

If the tuning frequency is kept unchanged the remaining transmitted "rest forces" are largely independent from the mass of the foundation. Thus – as in case of periodic excitation – the foundation's mass results from the admissible amplitudes for the respective operational process. Modern devices are often mounted without additional attenuating masses. Smaller machines can be mounted directly on elastomer material, larger and midsize machines are typically mounted on steel springs with a tuning frequency $f < 4$ Hz.

The reaction to an impulsive load is mathematically derived from the Duhamel or convolution integral for a single degree of freedom system:

$$w(t) = \int_0^t F(\tau)\frac{1}{m\omega_D}e^{-\delta(t-\tau)}\sin\omega_D(t-\tau)\,d\tau, \quad (22.22)$$

($\omega_D > \omega\sqrt{1-\zeta^2}$ – circular natural frequency of the damped system ($\omega = 2\pi f$); ζ – damping ratio; m – total mass of the foundation, machine etc.)

The impulse I generated by the mass of the hammer m_0 can be calculated implying the conservation of energy:

$$I = m_0 v_0 = m_0\sqrt{2h\frac{m_0 g + F_A}{m_0}}, \quad (22.23)$$

(v_0 – hammer velocity shortly before impact, h – the initial height of the hammer)

In this formula, the driving force F_A of pneumatically or hydraulically driven hammers is considered. $F(\tau)$ in Eq. (22.22) can be deduced out of the duration of the impulse Δt and the impulse form (half sinus, rectangular shape etc.).

The maximum force F_R which is transmitted to the foundation is shown in Fig. 22.6. It is related to the product of the duration time of the impulse and the frequency and the maximum impulse force F_I respectively.

In most cases the period of the vibration in the eigenfrequency is much larger than the duration of the impulse. As can be deduced from Fig. 22.6 (left) the response can be directly calculated from the impulse I modeled as a Dirac-function. Then the initial velocity v is given by:

$$v = v_0\frac{m_0}{m}(1+\varepsilon). \quad (22.24)$$

The value ε describes the elastic contribution of the impact. Whereas for a completely plastic impact ε equals to zero, for typical impingements $\varepsilon = 0.8$ may be assumed.

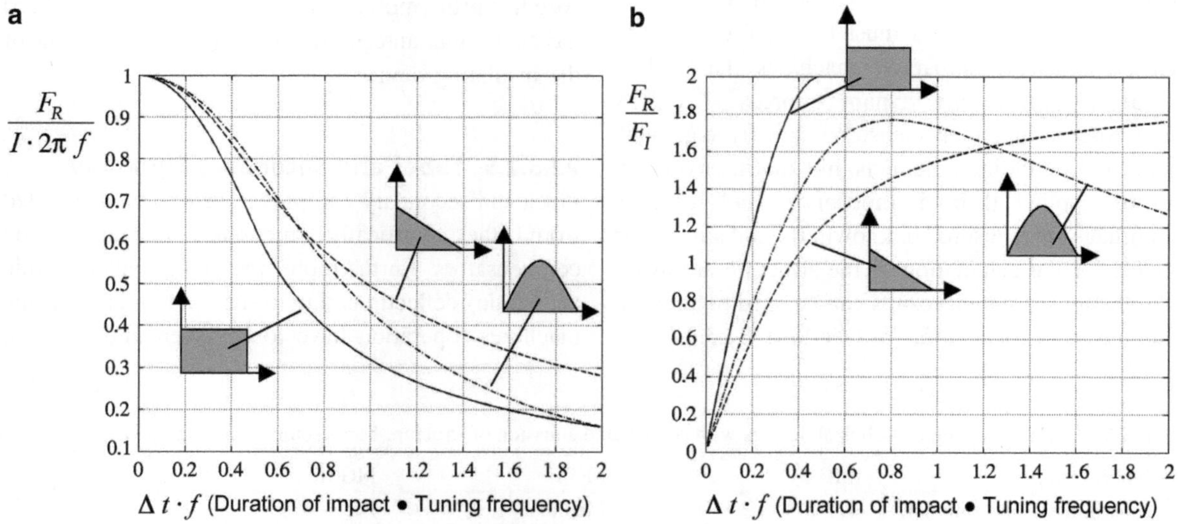

Fig. 22.6 Transmitted force related to the duration time of the impulse multiplied with the circular tuning frequency (*left*), transmitted force related to the maximum impulsive load F_I (*right*), depending on the product of tuning frequency and impact time

If the duration time of the impulse is unknown, the Dirac-impulse is linked with a conservative assumption. If the damping is much smaller than 1, the transmitted force can be calculated by:

$$F_R = v\, m\, \omega = \sqrt{2h\frac{m_0 g + F_A}{m_0}} \cdot m_0 \omega (1 + \varepsilon). \quad (22.25)$$

For damping ratios, $\zeta < 0.5$ the result is conservative. The transmitted force is proportional to the eigenfrequency. In case of counter blow hammers and airhammers, the relationships Eqs. (22.24) and (22.25) cannot be directly applied.

Damping

Generally viscous damper elements are applied. The transmitted force can be minimized to approximately 80% compared to the undamped system by an optimum damping ratio of 25%. The design of the eigenfrequency and the damping should guarantee a sufficient decay between the individual strokes.

22.3.2.3 Large Machines with Stochastic Excitations

Large machines, as for example mills, crushers, crane runways, etc. create stochastic excitations. This excitation is characterized by the power spectral density of the excitation force. However, in general, the power spectral density cannot directly be deduced from usually given machinery data. In order to obtain an adequate dataset for the dimensioning and for possible insulating measures, data must be acquired e.g., by measurements at comparable machines. The task is to identify the acting dynamic forces. Typically those forces cannot be measured directly e.g., with force sensors. Indirect methods, like matrix inversion, can be applied if both impedances and response amplitudes are measured or known. If valid simulation models for the calculation of the structure are available, the measurement results can be used to approximate the input forces [75, 76] (see Sect. 22.3.2.6).

The tuning frequencies of an insulation measure are typically chosen low enough to avoid disturbing frequency components in the response. However, an excitation in the vicinity of the resonance frequencies of the mounting cannot generally be avoided. Therefore, the damping should be sufficiently high, optimally around damping ratios of $\zeta = 0.5$.

22.3.2.4 Heating, Ventilation, Air Conditioning and Refrigeration

An overview over various technical devices in buildings [heating, ventilation, air conditioning and refrigeration (HVACR)] is given in [74]. Here appropriate insulating measures are listed considering location and machinery type. For office or residential use, the compliance with the requirements on structure-borne sound is mostly a guarantee for the conformance with the limits of vibration amplitudes.

MG I to MG III (Table 22.11) describe the type of the machinery, EL1 to EL7 (Fig. 22.7) the required type of insulation – including flexible tubes and cables.

In Fig. 22.8, the appropriate types of mounting are indicated with respect to the distance to the next room to be protected [74].

The tuning frequency is typically designed between 10 and 20 Hz aiming at a reduction in the frequency range of the human perception. However, the tuning frequencies must not coincide with the frequencies of excitation given by the rotational speed. Furthermore, here the stability (Fig. 22.5) has to be guaranteed by an adequate positioning of the insulating devices.

22.3.2.5 Tubes and Secondary Equipment

Pipes and secondary equipment have to be decoupled from the elastically mounted machinery by compensators. Furthermore, the additional (dynamic and static) deflections due to the loads related to the machines' operation have to be taken into account.

Table 22.11 Classification of technical devices with comparable emission of structure-borne sound according to [74]

Machinery group	MG I	MG II	MG III
Structure borne sound emission	Little	Medium	High
Machine and devices	Ventilators, RLT – devices	Pumps, rotator compressors	Elevators piston compressors, emergency diesels

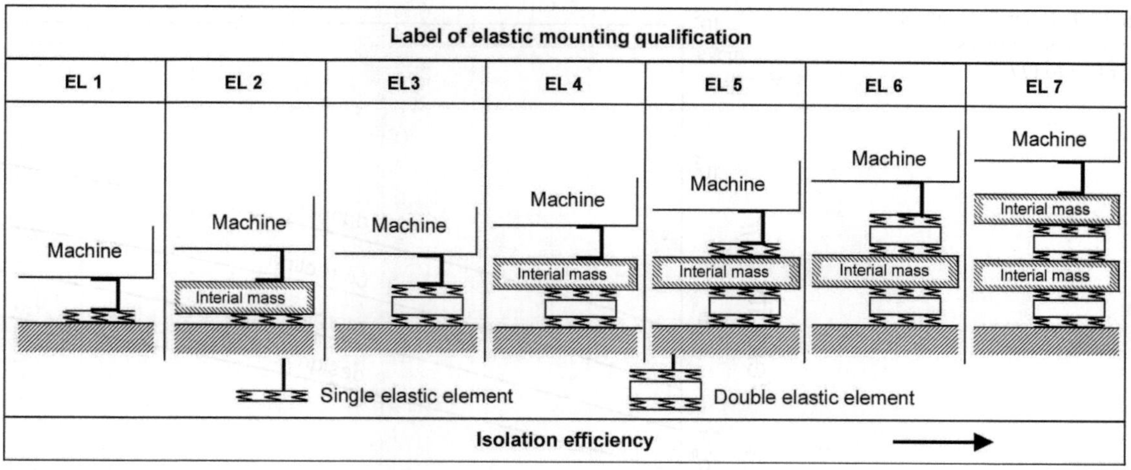

Fig. 22.7 Classification of elastic insulation [74]

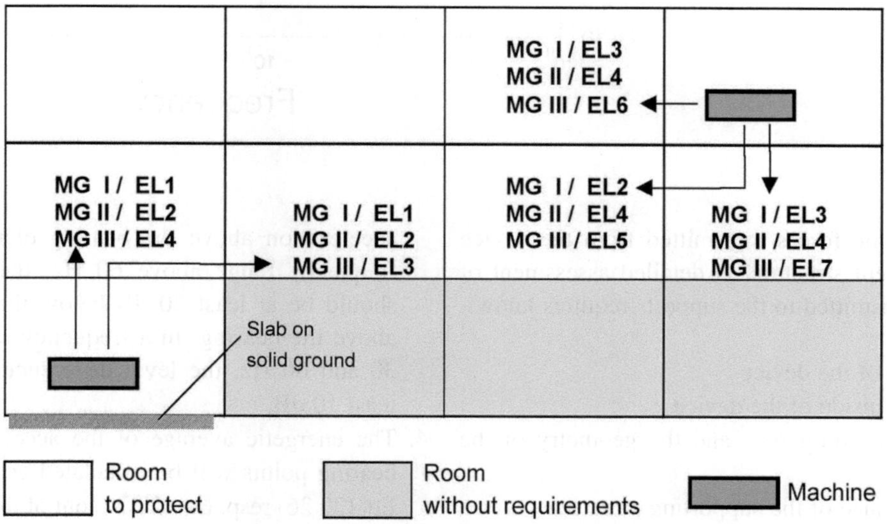

Fig. 22.8 Schematic design of measures for the insulation of structures borne sound and machines [74]

When using tube compensators, the length of the tube has to be chosen sufficiently long, so that a resonant excitation of the tube due to pulsation is avoided.

Especially high pressure ducts (as used, e.g., with hydraulic devices) lead to high structure borne sound and vibration. The mounting points of the pipes should be chosen at positions with high-input impedances (e.g., not in the centre of a slab).

In [77], experience-based indicative values for the maximum permissible vibrations of steel tubes in industrial plants are given. They are shown in Fig. 22.9.

22.3.2.6 Identification of Dynamic Forces by Vibration Tests

In case dynamic forces are not known, measurements can be carried out to describe them indirectly. In the limit case of an infinitely rigid mounting, the dynamic forces occurring inside the device are transmitted without attenuation into the adjacent structure. In the limit case of an infinitely soft support, the dynamic forces will be in equilibrium with the forces of inertia due to the machine's vibration. For the assessment of dynamic forces, we distinguish between the forces occurring inside the device

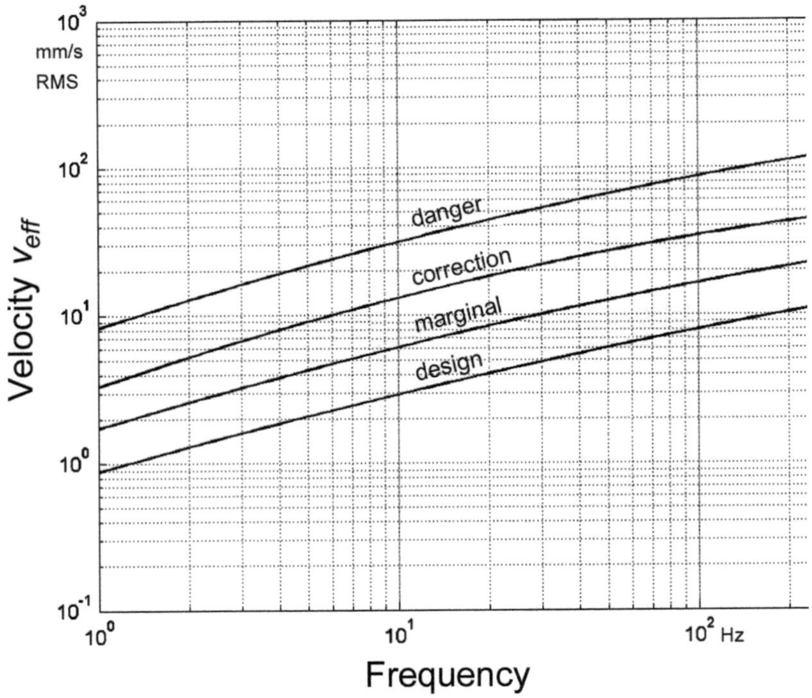

Fig. 22.9 Reference values for vibrations of tubes, according to [77, 78]

and the dynamic forces transmitted from the device into the adjacent structure. A detailed assessment of the forces transmitted to the supports requires knowledge of

1. The inertia of the device
2. The forces inside of the device
3. The dynamic properties and the geometry of the bearing
4. The impedance of the supporting structure

For the approximate assessment of the forces inside of the device, three general procedures are proposed.

Test of the device on a flexible bearing positioned on rigid support

1. In this procedure, the device is mounted on elastic bearings which are located on a rigid support (e.g., plate rigidly mounted on the adjacent soil). The tuning frequency of the elastic support should be below 7 Hz and 30% of the principal frequency of the excitation.
2. The device is put in function with an appropriate simulation of its load.
3. The acceleration is measured above (device) and below (rigid support) the elastic bearing. It can be thus controlled whether the acceleration below the bearing is significantly lower than the acceleration above the bearing elements. In the frequency range above 60 Hz, the acceleration should be at least 20 dB below the acceleration above the bearing. In a frequency range between 30 and 60 Hz, the level difference should be at least 10 dB.
4. The energetic average of the accelerations for n bearing points will be calculated according to the Eq. (22.26) resp. Eq. (22.27) out of the acceleration in third octaves or in narrow band:

$$\bar{L}_a = 10 \log \left(\sum_{i=1}^{n} \left[10^{\frac{L_{ai}}{10}} \right] \right) - 10 \log n, \quad (22.26)$$

$\bar{L}_a$ – level of the energetically averaged accelerations, L_{ai} – third octave level of the acceleration at the bearing i, n – number of bearings, L_a – third octave level of the acceleration $L_a = 20 \log\left[\frac{a}{a_0}\right]$, a – effective value of the acceleration in the third octave, a_0 – reference value of the acceleration).

In case the excitation is predominated in discrete frequencies, the assessment can be reduced on the effective values of the acceleration in the respective frequencies:

$$\bar{a} = \frac{1}{n}\sqrt{\sum_{i=1}^{n} a_i{}^2}, \tag{22.27}$$

$\bar{a}$ – energetically averaged acceleration for a discrete frequency of excitation, a_i – effective value for the acceleration at the ith bearing, n – number of bearings

5. An estimate for the level of the induced forces can be obtained from the inertia by means of Eq. (22.28) resp. Eq. (22.29) in third octaves or in narrow band:

$$L_F = 20\log\left[\frac{F}{F_0}\right] = \bar{L}_a + 20\log\left[\frac{m}{m_0}\right] + 20\log\left[\frac{m_0 \cdot a_0}{F_0}\right] \tag{22.28}$$

6. In parallel to the investigations described above the transmitted forces can be calculated at the basis of the bearing's stiffness k by Eq. (22.30)

(F – effective value of the induced force in the third octave, F_0 – reference value of the force, $F_0 = 1$ N, m – total mass above the elastic mounting, m_0 – reference value of the mass, a_0 – reference value of the acceleration [see Eq. (22.26)]. Accordingly, in case of an excitation with predominating single frequencies, the assessment can be confined to the frequencies of excitation.

$$F = m \cdot \bar{a}, \tag{22.29}$$

$$L_F = \bar{L}_a + 20\log\left[\frac{k}{(2\pi f)^2 \cdot m_0}\right] + 20\log\left[\frac{m_0 \cdot a_0}{F_0}\right], \tag{22.30}$$

and in the case of predominating excitation at discrete frequencies by Eq. (22.31)

$$F = k \cdot \frac{\bar{a}}{(2\pi f)^2}, \tag{22.31}$$

(f – frequency of excitation, k – total dynamic stiffness of the elastic bearing, which can differ significantly from the static stiffness e.g., deduced from the static displacement).

The estimates according to Eq. (22.30) or (22.31) define a lower bound. They are typically below the results calculated according to Eq. (22.28) or (22.29), as generally – especially in case of higher frequencies – not the total mass of the device is dynamically active. Then the results obtained by Eq. (22.28) or (22.29) may be significantly above the forces to be expected.

The quality of the experimental assessment can be enhanced by measurements of the impedances. Also here, the device should be mounted on elastic bearings with a tuning frequency of less than 7 Hz resp. 30% of the frequency of excitation. The impedance $Z_{\mathrm{device}i}$ will be measured on the decoupled device. Furthermore, the impedance at the location Z_{location}, where the device shall be positioned, will be measured. The forces induced into the structure can be approximated by the relation between the impedances and the measured velocity at the elastically mounted device by:

$$F \approx \sqrt{\sum_{i=1}^{n}\left(\frac{Z_{\mathrm{location}} \cdot Z_{\mathrm{device}i}}{Z_{\mathrm{location}} + Z_{\mathrm{device}i}} \cdot v_{\mathrm{device}i}\right)^2}, \tag{22.32}$$

(Z_{location} – impedance at the location where the device shall be positioned, $Z_{\mathrm{device}\ i}$ – measured impedance at the bearing i of the device, $v_{\mathrm{device}\ i}$ – measured velocity at the bearing i of the device, n – number of bearings)

Measurement at a device mounted on a typical impedance.

In case the system cannot be mounted on an elastic layer, the force can be deduced from the velocity measured at the bearing. For the measurement, the device should be mounted on a structure with input impedances similar to the construction. The velocity of vibration will be measured at the connection points to the structure. Furthermore, the input impedance of the supporting structure will be measured without the device.

- In case the dimension of the device is smaller than approximately half of the bending wave length, the induced force can be approximated by:

$$F \approx Z_{\text{Struktur}_i} \cdot v_{\text{Struktur}_i}, \qquad (22.33)$$

- In case the dimension of the device is larger than approximately half of the bending wave length, i.e., there is a considerable phase shift between the various points of the structure, the total force can be approximated to:

$$F \approx \sqrt{\sum_{i=1}^{n} \left(Z_{\text{Struktur}_i} \cdot v_{\text{Struktur}_i}\right)^2}, \qquad (22.34)$$

(Z_{Struktur_i} – measured structure's input impedance on the support point i, v_{Struktur_i}– measured velocity of vibration on the support point i, n – number of support points).

22.3.3 Excitation of Vibrations at Construction Work

Significant excitation of vibration can occur at construction work, e.g., during joggling, driving and pulling of sheet piles, girders, piles etc. as well as during soil compaction, demolition works, blasting and due to the traffic at construction sites.

In case of sensitive areas in the vicinity, a monitoring system with alarm for the operator of the vibrator can be installed. As soon as a certain threshold is exceeded, the operator obtains a signal and consequently can change the operation conditions of the device. Monitoring is recommended for the preservation of evidence and for limiting the vibrations in the area of sensitive equipment.

The propagation shown in Fig. 22.10 can approximately be described by the law of diffusion in homogeneous soil:

$$v(r) = v(r_0)\left(\frac{r_0}{r}\right)^n e^{-\alpha_R(r-r_0)}, \qquad (22.35)$$

($v(r)$, $v(r_0)$ – amplitude of velocity at the distance r and at reference distance (r_0), $\alpha_R = 2\pi\zeta/\lambda$ – damping parameter depending on the material damping ratio ζ and the characteristic wavelength of the soil $\lambda = c/f$; (c – wave speed, f – frequency), here typically the wave-length of the Rayleigh-wave is chosen (see Sect. 1.8.2).

According to DIN 4150–1 [40] the damping ratio must be set to $\zeta = 0.01$ unless proved data exist.

The exponent n depends on the type of source and of the type of wave:

surface waves: point-source – $n = 0.5$, line-source – $n = 0$

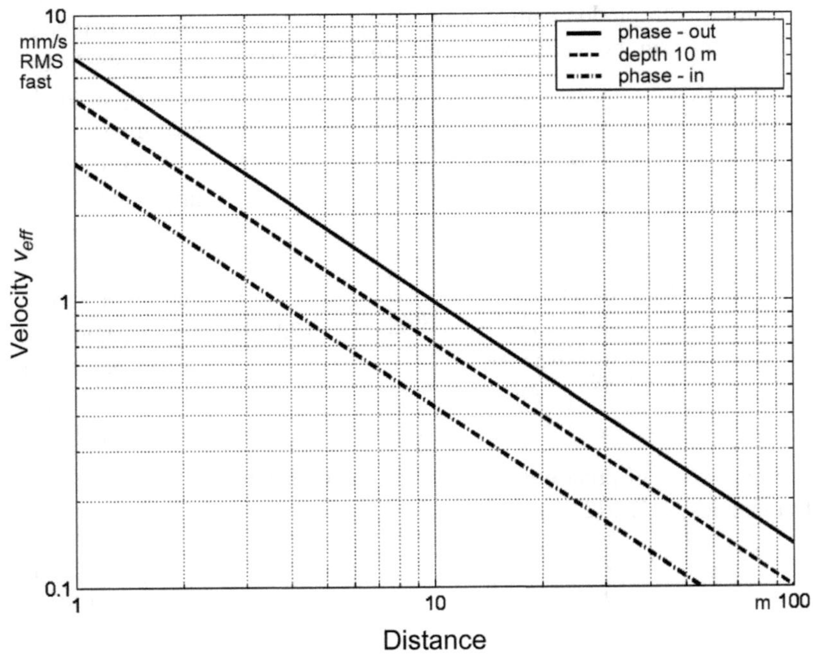

Fig. 22.10 Diagram for the prediction of the radiation of vibrations, double sheet pile l =15 m, frequency $f = 35$–40 Hz, unweight mass of the device 30–50 kg m, gravel soil, according to [54]

spatial waves: point-source – $n = 1$, line-source – $n = 0.5$

Bachmann [2] recommends to set $n_I = n + 0.5$ for impact loads.

The statistic evaluation of numerous field data [79, 80] led to simplified practical formulae applicable to various devices for pile ramming, vibration rams, pile drivers, vibrating plates and rollers. For rams, usually the power is given in the data sheet of the machine.

$$v(r) = K_E \frac{\sqrt{E}}{r^n}, \qquad (22.36)$$

with K_E in $\frac{\text{mm}}{\text{s}} \cdot \frac{\text{m}^n}{\sqrt{\text{kNm}}}$.

The energy E is derived from the input power W per cycle ($E = W/f$).

For compacting machines, the total mass (weight G) of the machine serves as a good parameter reflecting the input energy:

$$v(r) = K_G \frac{\sqrt{G}}{r^n}, \qquad (22.37)$$

with K_G in $\frac{\text{mm}}{\text{s}} \cdot \frac{\text{m}^n}{\sqrt{t}}$.

The resulting velocities refer to the soil surface and also to the foundations of buildings at the distance r (given in meter). The proposed values of K_E and K_G are assembled in Table 22.12. Here the foundation-related parameters K were evaluated with an exceeding probability of exceedance of 50% respectively 2.25%.

The radiation of vibration caused by the following machines was evaluated for the statistic survey (the vibration amplitudes refer to a foundation at 10 m distance):

Vibration rollers: $G = 1.7$ t–32 t, max(v(t, 95%)) = 0.7 mm/s–4 mm/s, Vibration rams: $E = 3$ kNm–7.9 kNm, max(v(t, 95%)) = 2 mm/s–5 mm/s, Diesel pile hammers: $E = 31.3$ kNm–67.1 kNm, max(v(t, 95%)) = 2 mm/s–3 mm/s

A survey of machines of these types is given in [82]. The largest ram which is described has got a maximum ram energy of $E = 3{,}000$ kN/m per strike.

The most frequent source of relevant vibrations is the driving of sheet pile walls using a vibrator pile driver. Phenomena, effects and measures are described in detail in [54]. The vibrations transmitted into the soil depend on the soil characteristics, the depths and dimensions of the pilings as well as on the vibrator. All types of vibrators permit to tune the frequency of excitation. Modern devices also allow adjusting the unbalanced mass during the operation (type HFV or HF-VAR). Typically, the vibrations are smaller and shorter when pulling the sheet piles than when inserting them. However, if the sheet piles have been in place for a long period of time, or if they have been welded or partly cast in concrete, an even higher vibration can occur during pulling.

The frequencies of the vibrators typically range between 28 and 40 Hz, thus in the resonant area of slabs of apartment buildings, but above the resonant frequencies of slabs with very large span. In order to check alternative measures, monitoring tests should be performed before hand. Higher requirements in the nearer neighbourhood (e.g., hospitals, sensitive technical devices) can be met by building the construction pits with methods causing smaller excitation forces, like bored pile walls, diaphragm walls, freezing techniques, injections or by pressing in sheet pile walls.

Figure 22.10 [54] shows exemplarily the diagram for predicting the propagation of vibrations in the vicinity of a vibrator with unbalanced masses of 30–50 kg m, vibration frequencies $f = 35$–40 Hz for a double sheet pile with a length of $l = 15$ m. The data have been recorded on gravel soil. The curves represent Eq. (22.38) with the exponent $n = 0.85$:

Table 22.12 Reference values for the constant K [80] ($n = 1$, peak velocity)

Device	Soil surface	Foundation 50% probability	Foundation 2.25% probability
Vibration ram	$K_E = 23.7$	$K_E = 7.9$	$K_E = 18.52$
Pile drivers: diesel pile hammers	$K_E = 15.8$–31.6 (soft or unconsolidated soil – stiff or compact soil)	$K_E = 2.45$	$K_E = 3.82$
Pile drivers: tups	$K_E = 31.6$–94.6	($K_E = 11.07$) [81]	
Vibration plate vibration roller	$K_E = 100$	$K_G = 4.31$	$K_G = 10.87$

$$\begin{aligned} \text{phase out} : v_{F,\text{eff}}(r) &= 7\,\text{mm/s}\,1/r^n m^n, \\ \text{depth 10 m} : v_{F,\text{eff}}(r) &= 5\,\text{mm/s}\,1/r^n m^n, \\ \text{phase in} : v_{F,\text{eff}}(r) &= 3\,\text{mm/s}\,1/r^n m^n. \end{aligned} \quad (22.38)$$

All the mentioned empiric data may only serve for a first approximate assessment. Experience shows that the vibrations depend to a greater extent on the soil conditions rather than on the power or mass of the machines.

If the vibrations exceed the limit values, in a first step the unbalanced mass can be reduced; in addition the frequency of excitation and the heading rate can be changed. Modern systems allow to adjust the unbalanced masses after reaching and before leaving the operational speed when starting or stopping the device. The transition of resonances is avoided. There is furthermore the possibility to prepare the soil by predrilling or rinsing. It is important to insert the sheet piles directly after predrilling. Thus larger areas should not be predrilled as the vibration in the adjacent areas causes a soil compaction and therefore the conditions are possibly worsened. Alternatively, it is possible to predrill with casing. Here the soil material difficult to treat is replaced by an appropriate material.

22.3.4 Blasts

The guideline KDT 046/86 [83] contains hints how damages on constructions by blasts can be avoided. Permissible vibration amplitudes at foundations and the relationship between distance and explosive charge are given. Vibrations can be assessed according to DIN 4150–3 [47] in combination with Table 22.13 [83, 84] (see Sect. 22.2.2). Blast-induced vibrations are hardly relevant at distances larger than 1,500 m from quarries or 400 m from construction sites [40].

22.3.5 Road Traffic

22.3.5.1 Mechanisms of Excitation

Truck-induced vibrations in adjacent buildings depend on a number of parameters that are typically difficult to obtain. Besides the vehicle's velocity, especially the surface quality of the road is decisive. Surface irregularities such as damages, steps, duct covers, pavements, potholes etc. can increase the vibrations significantly. They create transient impulsive and in periodic loads. The latter are due to the vehicle's vibration in its natural frequencies and the vibration of the suspended masses which are decoupled by the tires.

The measurement data published in the literature vary significantly. Extensive investigations [85, 86] show velocities of vibration on slabs of adjacent buildings between 0.1 and 1 mm/s. In case of very good road conditions, the amplitudes may fall below the lower value; in cases of very bad conditions, values above 1 mm/s are possible.

Reference values about the surface roughness of roads are given in [85]. In Germany the spectral surface roughness is defined according to Eq. (22.39) as a function of the wave number. The wave number spectrum ranges between $\Omega = 0.06\,\text{m}^{-1}$–$\Omega = 30\,\text{m}^{-1}$ corresponding to a wave length of $\lambda = 2\pi/\Omega = 0.2$–$1.0$ m.

$$\Phi_h(\Omega) = \Phi_h(\Omega_0)\left(\frac{\Omega}{\Omega_0}\right)^{-w}, \quad (22.39)$$

Table 22.13 Velocities ($\max(v(t))$) of vibration for the assessment of explosion induced vibrations on building, according to [83, 84]

Type of building	Measurement position	Small damages	Medium damages	Heavy damages
		For velocity of vibration above mm/s		
A: framework	Foundation $f < 30$ Hz ($f = 100$ Hz)	5 (22)	10 (44)	25 (110)
	upper floor horizontal	10	20	50
B: wall panels	Foundation $f < 30$ Hz ($f = 100$ Hz)	10 (44)	25 (110)	50
	upper floor horizontal	20	50	100
C: skeleton	Foundation $f < 30$ Hz ($f = 100$ Hz)	30 (135)	50 (205)	
	upper floor horizontal	60	100	

Small damages Cracks in the plaster, cracks in not supporting walls, resp. elements, *medium damages*: Cracks in supporting structures without danger for the safety, *heavy damages*: Endangering the structure's safety
The permissible velocities of vibration in the area of the foundations can be interpolated between $f = 30$ Hz and $f = 100$ Hz in a double logarithmic scale

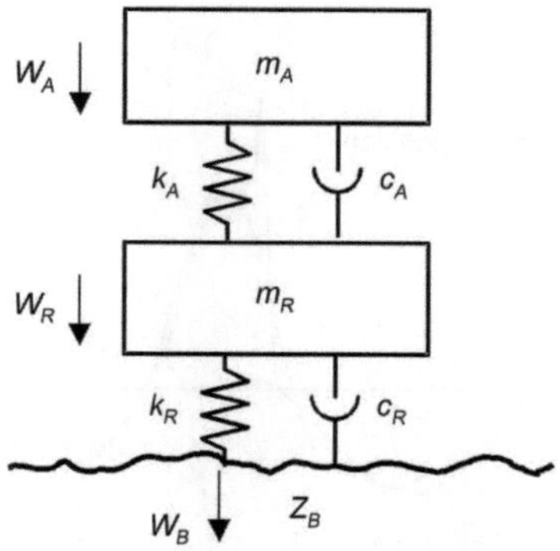

Mass of the vehicle per wheel: $m_A = 3500...9000\,\text{kg}$

Spring stiffness, damping ratio and tuning frequency of the upper structure:

Air cushion springs: $s_A = 0.15\,\text{MN/m}$, $\zeta_A = 30...40\%$, $f_A = 1...2\,\text{Hz}$

Steel springs/multiple leaf springs: $s_A = 0.2 - 1.0\,\text{MN/m}$, $\zeta_A = 8...15\%$, $f_A = 1.5...4\,\text{Hz}$

With multiple leaf springs at slight excitation the contact friction may not be exceeded. The spring stiffness than can be give by $k_A = 6\,\text{MN/m}$.

Spring stiffness, damping ratio of the tyres and the tuning frequency of the axis mass:

Stiffness of the tyre: $s_R = 1...1.5\,\text{MN/m}$, $\zeta_R = 5...10\%$, $f_R = 8...15\,\text{Hz}$

Mass of the tyre: $m_R = 200...600\,\text{kg}$

Distance of the axis: $e = 4.8...7.8\,\text{m}$ multi axes $e = 1.15...3.2\,\text{m}$

Fig. 22.11 Two-degree of freedom model for the assessment of the dynamic load induced by the tyre of a truck with given reference parameters

The dimension of the wave number is $\Omega_0 = 1$ (m^{-1}). The parameter w "waviness" ranges between $w = 1.8$–2.4 (highways with a concrete slab – ancillary roads). Depending on the surface quality (good to bad), the average value of the surface irregularity varies between $\Phi_h(\Omega_0) = 0.8$–$9.8\ \text{cm}^3$.

At a given vehicle's speed V, the frequency of excitation f is related to the wave number Ω by $f = V\Omega/2\pi = V/\lambda$.

The power spectral density $S_{FF}(\omega)$ of the dynamic wheel load (which is transmitted between wheel and surface) can be calculated from the spectral surface roughness $\Phi_h(\Omega)$ and the square of the complex dynamic stiffness function of the vehicle $H_{fy}(\omega)$ according to

$$S_{FF}(\omega) = \frac{1}{V}\left|H_{fy}(\omega)\right|^2 \Phi_h(\Omega)\,. \qquad (22.40)$$

The passing by of heavy vehicles causes an additional "quasi static" deformation of the surface. Periodically changing stiffness under the surface, e.g., due to ribbed slabs, can lead to resonant excitation at certain vehicle speeds (e.g., fork lift trucks moving on slabs).

For an approximate prediction for the dynamic loads transmitted through the tires, a system with two degrees of freedom (Fig. 22.11) can be applied. Ref. [87] describes a validated model for the prediction of the vehicle-induced vibration in the vicinity of roads.

22.3.5.2 Measures

In the vicinity of sensitive equipment, surface irregularities in form of steps etc. should be avoided. Furthermore, vibrations can be reduced by means of a speed limit. In addition – using the same principle as for elastically mounted track systems (see Chap. 16) – an elastic mounting of the road on a heavy mass system can be considered.

22.3.6 Human Induced Vibration

22.3.6.1 Mechanism of Excitation

Bachmann [2] carried out intensive investigations on human-induced vibrations. The following explanations refer among others basically to his work.

Walking, dancing, sport activities like jumping, running, spinning etc. create periodic excitations.

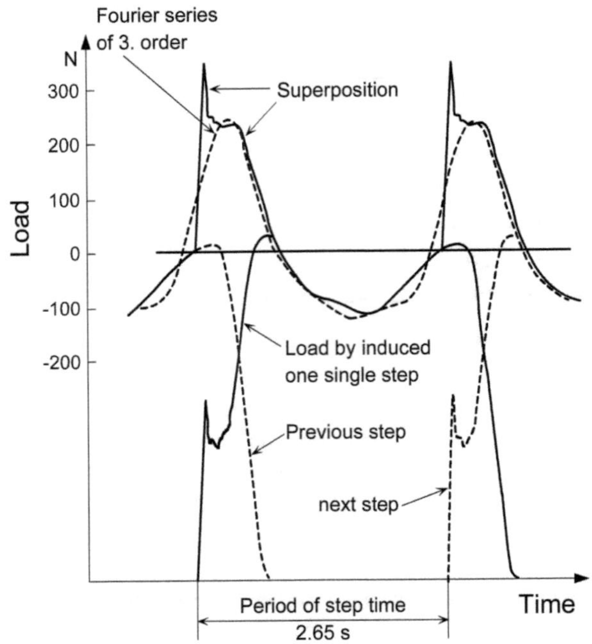

Fig. 22.12 Example for the superposition of the oscillating forces due to walking an approximation by Fourier series of 3 order

At resonance with the natural frequencies of a bridge or a slab these excitations might cause significant amplitudes. The fundamental frequency of walking ranges with a probability of 95% from $f = 1.65$ Hz to $f = 2.35$ Hz around the mean value of about $f = 2$ Hz. Running fast can cause fundamental frequencies up to 3.5 Hz. Furthermore, rhythmic clapping and trampling at resonance can cause significant amplitudes. Besides vertically induced dynamic forces on various structures e.g., bridges, tribunes, balconies in theatres, etc. also horizontal loads must be considered. As the excitation is not purely sinusoidal, higher harmonics of the dynamic forces are relevant as well (see Fig. 22.12). The forces can be approximated by three components of a Fourier-series. Above the third harmonic (above 8–10 Hz) the contributions are typically small; only very light weighted constructions are significantly excited. Figure 22.13 shows the corresponding spectral contributions.

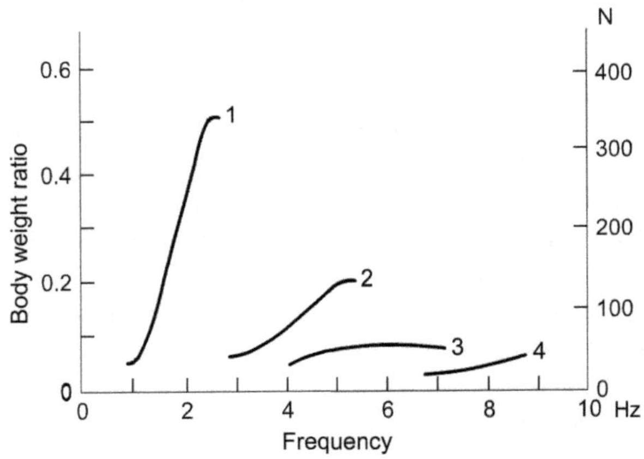

Fig. 22.13 Spectral components for men induced vibrations depending on the walking frequency [2]

22.3.6.2 Reference Values for Structural Eigenfrequencies

With respect to human-induced vibration, only slender constructions such as pedestrian bridges [88–90], diving platforms, large span slabs in theatres or gymnasiums with a first eigenfrequency below $f = 10$ Hz have to be investigated. A first criterion for the evaluation of the sensitivity for vibrations is the first eigenfrequency of the structural component (Table 22.14).

When the excitation is related to walking only (little frequented pedestrian bridge), it suffices mostly to tune the eigenfrequency above the range of the walking frequency (first harmonic $f > 3$ Hz).

The first natural frequency of the respective structural element should be above the second harmonic of

Table 22.14 Reference values for critical natural frequencies depending on the use, according to [2]

Pedestrian bridges	$f_1 < 4.5$ Hz
Floors in offices, hotels etc.	$f_1 < 5$ Hz ($f_1 > 7.5$ Hz)[a]
Floors for sport	$f_1 < 7.5$ Hz–9 Hz[b]
Dancing floors, concert halls for pop-/rock concerts	$f_1 < 6.5$ Hz–8 Hz[b]
Concert halls with fixed chairs for classical music and moderate pop music	$f_1 < 3.4$ Hz
Concert halls with fixed chairs, pop music	$f_1 < 6.5$ Hz
Concert halls with fixed chairs, horizontal	$f_1 < 2.5$ Hz

[a]Just in case of lightweighted constructions (wooden slab, steel frame) and higher requirements
[b]Depending on the damping ratio $\zeta = 0.0025$ (concrete) – 0.012 (steel)

the walking frequency for slabs with a large span and/or lightweight structures resp. in case of small damping. For slabs in gymnasiums or dancing halls, on which a rhythmical movement has to be expected, Eurocode 3 [90] defines a minimum eigenfrequency of $f = 5$ Hz. In order to avoid a resonant excitation due to the third harmonic of the fundamental frequency, in [2, 91] even higher natural frequencies depending on the type of construction (see Table 22.14) are recommended. This recommendation has been adapted in the Swiss norm SIA 260 [92].

Above 10 Hz the structural resonances are still excited by strikes of heels. The response decays completely between two steps. The structural response depends on the dynamically relevant mass: Lightweight constructions can thus show disturbing vibrations even if the natural frequency is far above the fundamental frequency of the excitation.

For lightweight timber structures, the first natural frequency should be higher than 8 Hz (Eurocode 5 [93]). Otherwise a detailed investigation is required. In addition, criteria pertaining to an impact load and a static point load must be fulfilled.

The maximum deflection is limited to $w = 1.5$ mm under a point load of 1 kN. The impact criteria refer to the response to a unitary impact of 1 Ns where all modes up to a frequency of 40 Hz must be considered. The resulting velocity is then limited to:

$$v < b^{(f_1 \xi - 1)} \text{ m/s}. \tag{22.41}$$

If the first natural frequency f_1 is lower then 8 Hz, the response to a heel impact function with $I = 55$ Ns within an impact time of $t_I = 0.05$ s must not exceed the maximum velocity v given in Eq. (22.42).

$$v < 6 \cdot b^{(f_1 \xi - 1)} \text{ m/s}, \tag{22.42}$$

f_1 – first natural frequency of the floor structure; $\xi = 0.01$ – damping coefficient (the given value may be replaced by a realistic value, e.g., obtained by a measurement); $b = 100$ – comfort parameter.

Furthermore, the maximum acceleration is limited to $a_{vert} < 0.1$ m/s^2 for one walking person with a body weight of 700 N in resonance with the slab. If the slab is sufficiently small, a reduction factor (>0.4) may be considered to account for the limited amount of steps when crossing a slab. The verifications are discussed in detail in [94, 95].

22.3.6.3 Resonant Vibration Response

The dynamic load induced by one person (average weight of a person $G = 800$ N) is given by Eq. (22.43)

$$F_p(t) = G + \sum_i G\alpha_i \sin(2\pi i f_p t - \Phi_i), \tag{22.43}$$

(G – weight of the person, α_i – Fourier-coefficient of the i-th harmonic, f_p – fundamental frequency of the activity, Φ_i – phase shift with respect to the excitation in the fundamental frequency).

In the model, the dynamic force is distributed over the area of excitation according to the number of persons per unit area. The mass of the persons has to be added to the dynamic forces. For the assessment of the critical frequency of excitation, the uncertainties and the variation of the structural natural frequencies have to be considered. If a precise assessment is not possible, the worst case according to Table 22.15 [2] has to be considered.

Table 22.15 Frequency range and the Fourier coefficients as a function of the type of movement

Type of movement	Frequency (Hz)	a_1	a_2	a_3	Density of persons/m^2
Walking	1.6–2.4	0.4–0.5	0.1[a]		
Running	2.0–3.0	1.6	0.7[b]	0.2[b]	
Jumping	2.0–3.0	1.7–1.9	1.1–1.6	0.5–1.1	0.25–0.5
Dancing	2.0–3.0	0.5	0.15	0.1	4 (–6)
Applause(without chairs, standing)	1.6–2.4	0.17–0.38	0.1–0.12	0.02–0.04	4 (–6)
Applause(with chairs)	1.6–2.4	0.024–0.17	0.01–0.05	0.01–0.04	2–3

[a] $\Phi_2 = \frac{\pi}{2}$
[b] $\Phi_2 = \Phi_3 = \pi(1 - f_p \cdot t_{\text{contact with the floor}})$.

Table 22.16 Integral values for the generalized forces and masses (hinged support on four sides)

Type of construction	γ_j	m_j^*
One-sided spanned slabs (e.g., timber beam slabs)	$2/\pi$	$0.5m''A$
Two sided-spanned slabs	$4/\pi^2$	$0.25\, m''A$

The highest amplitudes occur if one of the first harmonics ($i = 1, 2, 3$) of the lowest frequency of walking, running or jumping coincides with one of the structure's natural frequency ($i f_p = f_n$). The maximum acceleration a_j in the resonance j can be calculated by Eq. (22.44):

$$a_j = \frac{F_j^*}{m_j^*} \cdot \frac{1}{2\zeta}. \tag{22.44}$$

The generalized modal forces F_j^*caused by n_p persons and the generalized masses m_j^* of the eigenform can be calculated according to Eqs. (22.45) and (22.46):

$$F_j^* = G\alpha_i n_p \int_A \Psi_j dA = G\alpha_i n_p \gamma_j, \tag{22.45}$$

and

$$m_j^* = \int_A m'' \Psi_j^2\, dA. \tag{22.46}$$

For the determination of the total distributed mass m'', the mass of the dynamically relevant life loads has to be added to the mass of the slabs. In the surface integral Eq. (22.46), the total mass is weighted by the geometric mode shape Ψ_j . In Table 22.16, the integrals are evaluated for slabs with hinged supports on four sides and a constant mass distribution m''.

The uncertainty of the damping ratio ζ is relevant for the achievable quality of the prediction of the amplitudes of vibration. In [2], average values of the damping are listed, depending on the type of construction.

In most cases, it is sufficient to consider just the first natural frequencies. In [96], an approximate method to determine the first natural frequencies is described for slab-systems with main and secondary girders in composite constructions.

For pedestrian bridge simple rules are given, e.g., in the British standards [97, 98]. An overview is given in [99].

22.3.6.4 Excitation by a Group of Persons

Without external rhythmical signals, synchronization of several persons walking occurs above amplitudes of vibration of 10–20 mm. However, synchronization might occur above a critical number of persons also at lower levels of vibration. This is considered in the so-called synchronization factor [44, 100]. The activities of a larger number of persons might be synchronized with signals (e.g., music). The liberty of movement and thus the excitation is reduced with more than two persons/m^2.

The excitation forces depend on the number of persons and also their coordination. To account for the scatter of the phase shift of the excitation, a reduction factor $C(N)$ is introduced to compute the total forces (Table 22.17):

Table 22.17 Damping ratios depending on the type of construction, according to [2]

Type of construction	Structure without finish	Structure with finish Minimum	Medium	Maximum
Concrete	0.007–0.04	0.014	0.025	0.035
Pre stressed concrete	0.004–0.012	0.010	0.020	0.030
Composite construction	0.002–0.003	0.008	0.016	0.025
Steel	0.001–0.002	0.006	0.012	0.020

Table 22.18 Recommendation for the coordination factor *C*(*N*) for synchronized jumping of up to more than 50 persons [44]

Activity	Fundamental frequency	Second harmonic	Third harmonic
Gymnastics	0.80	0.70	0.50
Sports event	0.70	0.50	0.40
Pop concert	0.50	0.40	0.30

$$F(t)_N = F(t) \cdot C(N) \cdot N, \qquad (22.47)$$

(N – number of acting persons; $F(t)$ – acting forces of one person ;$F(t)_N$ – total dynamic force)

For uncoordinated acting persons, the coordination factor can be assumed to

$$C(N) = \sqrt{N}/N. \qquad (22.48)$$

Up to a group of five persons *C*(*N*) is set to 1 for synchronized acting persons. The more complex the action of the persons, the higher will be the scatter of the phase shift and the less will be the synchronization. Activities like gymnastics, ballet, etc. involve high synchronization, the audience at sports events is already less synchronized and at pop concerts an even lower synchronization can be assumed. The synchronization decreases also with the number of persons and with the harmonic order of the fundamental frequency of excitation. The reference values for *C* (*N*) proposed in [44] are given in Table 22.18.

22.4 Transmission and Insulation of Vibrations

22.4.1 Excitation and Transmission of Vibrations in the Soil

Chapter 16 summarizes a large number of – mostly empirical – investigations which serve for a first estimation of the transmission of vibration through the soil.

The wave propagation in the soil can be modeled via the Lamé-equation (1.73) in which the soil is considered as a linear-elastic isotropic continuum, described by three parameters, the shear modulus G, the Poisson value μ and the density ρ. In an infinite continuum two wave types exist, the compressional wave (also longitudinal wave or P-wave) and the shear wave (also transversal-, S- or T-wave). At free surfaces or at planar boundaries of different materials and near to excitations with short wave-lengths, additional wave-forms occur. They are called surface-waves (see Sect. 1.8.1). Their amplitudes decay with increasing normal distance from the surface. In Fig. 22.14 [101], the various wave types emerging under a harmonically loaded circular foundation are shown.

The Rayleigh-wave is induced at the stress free surface and can be compared to a free vibration (homogeneous solution) of a vibrating system. The wave form can be easily excited and thus dominates often the response. The decay of the Rayleigh-wave with depth is shown in Fig. 22.15.

In deeper soil, compressional (P-wave) and shear-waves (S-wave) dominate. All dynamic processes in an infinite horizontally layered elastic isotropic continuum can be modelled by superposing these wave types.

For arbitrary boundary conditions, the dynamic equations for the half space have to be solved numerically. The Boundary Element Method (BEM) offers a numerical solution for the half space with an irregular surface topology and irregular acting forces. Less general but often more efficient solutions are offered by Integral Transformation Methods (ITM). Here the Fourier Transform is applied with respect to the horizontal plane and time. The transformed coordinates

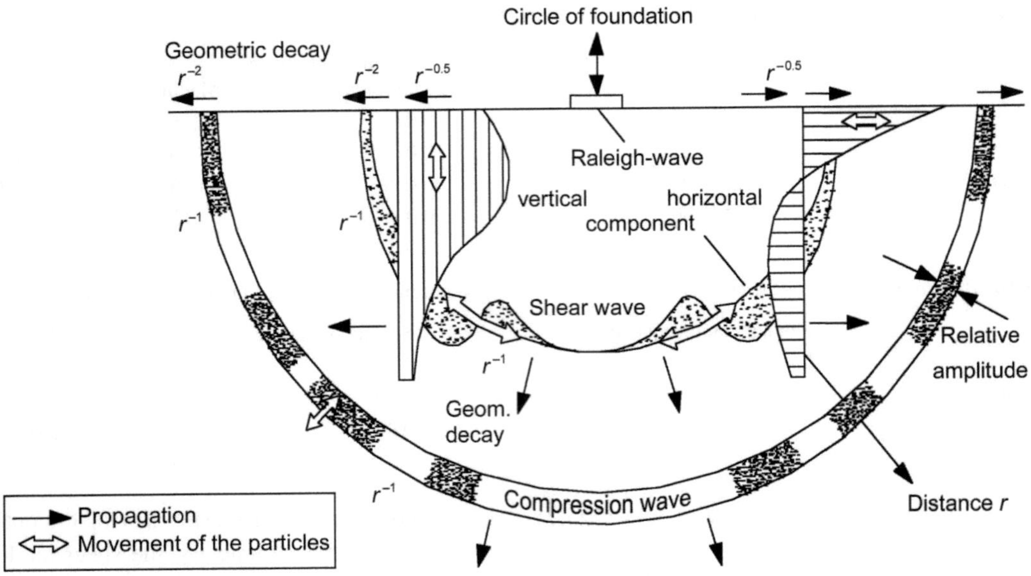

Fig. 22.14 Radiation of waves induced by a dynamically loaded circular foundation located on the elastic isotopic half space [101]

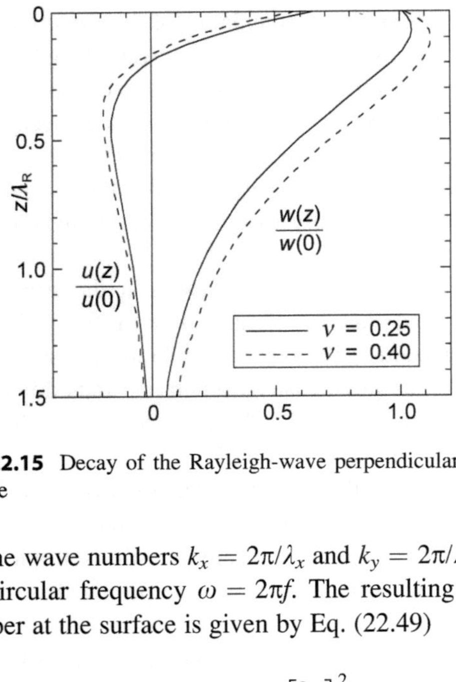

Fig. 22.15 Decay of the Rayleigh-wave perpendicular to the surface

are the wave numbers $k_x = 2\pi/\lambda_x$ and $k_y = 2\pi/\lambda_y$ and the circular frequency $\omega = 2\pi f$. The resulting wave number at the surface is given by Eq. (22.49)

$$k_0^2 = k_x^2 + k_y^2 = \left[\frac{2\pi}{\lambda_0}\right]^2. \tag{22.49}$$

Then the admittance (or impedance) functions which describe the relationship between the surface velocity and the excitation force [102–104] (or vice versa) can be calculated analytically. With the help of these transformations the original governing equations are split up into independent equations which describe the different wave types. (see Sect. 1.8).

Each combination of wave numbers k_x and k_y can be assigned to a certain angle of propagation in the horizontal plane (x, y).

This description of the wave propagation easily reveals the relevance of the various wave forms.

1. Case $\lambda_0 > \lambda_P$: the wave length λ_0 at the surface is larger than the wave length of the longitudinal wave. In this case, the excitation causes spatially propagating P- and S-waves.
2. Case $\lambda_P > \lambda_0 > \lambda_S$: the wave length λ_0 at the surface is larger than the wave length of the transversal wave, but smaller than the wave length of the longitudinal wave. In this case, the excitation causes a spatially propagating S-wave and a surface wave.
3. Case $\lambda_0 < \lambda_S$: the wave length λ_0 at the surface is smaller than the wave length of the transversal wave and the wave length of the longitudinal wave. In this case, only surface waves are induced.

The distinction of the three cases is of special interest if the attenuation of the vibration due to geometrical damping is investigated. This is discussed for a harmonic point load, respectively a line load:

1. Case: Considering the parts of the solution under a point load related to the first case, the square of the velocity of vibration decreases proportional to the square of the distance from the source. This is due to

the fact that the P- and S- waves radiate spatially and thus their energy – which is proportional to the square of the velocity of vibration – is distributed over the surface of a semi-sphere which increases with the square of the radius. The energy induced by line loads, however, is distributed over the surface of a half cylinder which increases linearly with increasing distance. Thus for line loads – due to geometrical damping – the velocity of vibration decreases in proportion to the square root of the distance.

2. Case: Here parts of the energy are transported by surface waves. As surface waves just radiate horizontally, their corresponding energy is distributed over the surface of a cylinder. In this case, geometrical damping leads to a decrease of the velocity of vibration inversely proportional to the square root of the distance. For line loads, the surface waves propagate perpendicular to the line along the surface. Thus, even with increasing distance, the area over which the energy is distributed will not increase, which means that no geometric damping of surface waves occurs.
3. Case: The energy is just radiated in surface waves. In case of a point source, the square of the velocity of vibration decreases linearly with increasing distances. In case of line sources, no geometric damping occurs.

So the damping in the second case is smaller than in the first case, but higher than in the third case. The spectrum of wave numbers of exciting forces can be calculated by a Fourier-transform. Equation (22.50) gives the relationship between the amplitudes of a distributed normal stress p at the surface (direction z) to the amplitude of the velocity of vibration v also normal to the surface. The calculation assumes a surface of the half space free of shear stress. Figure 22.16 shows the corresponding wave impedance functions Z.

In the figure, the three before-mentioned cases are indicated.

(c_T, c_L wave velocities of the transversal (S-) respectively longitudinal (P-) wave).

In order to calculate relative displacements between the free field and structures coupled to the soil, a comparison of the wave impedance Z with typical building impedances can be helpful [e.g., plates on the soil, compare Chap. 1, (1.100)].

In Fig. 22.16c, the magnitude of the impedance of a concrete plate with a thickness of 50 cm compared to the magnitude of the impedance according to Eq. (22.50) is shown exemplarily for a soil with a longitudinal wave velocity $c_{L\text{soil}} = 0.2\ c_{L\text{concrete}}$.

$$Z(k_0,\omega) = \frac{p(k_0,\omega)}{v(k_0,\omega)} = \frac{G\left[\left(\left(\frac{\omega}{c_T}\right)^2 - 2k_0^2\right)^2 + 4k_0^2\sqrt{\left(\frac{\omega}{c_T}\right)^2 - k_0^2}\sqrt{\left(\frac{\omega}{c_L}\right)^2 - k_0^2}\right]}{\omega\left(\frac{\omega}{c_T}\right)^2\sqrt{\left(\frac{\omega}{c_L}\right)^2 - k_0^2}}. \tag{22.50}$$

Also soil horizontally layered can be modeled with corresponding impedance functions. In this case, a sufficient decay of the dominating surface waves between the layers can be assumed, and hence the solution of an unlayered half-space is approximated for excitations at higher frequencies.

22.4.2 Transmission of Vibrations into Buildings

Due to soil/structure coupling, the dynamic deformation of the free field is altered by the impact of the forces occurring between structure and soil. Depending on the input impedance of the building's foundation, the vibration on the foundation will differ more or less from the vibration in the free field. The dynamic coupling of the building with the soil can be simulated with the Finite Element Method (FEM) and the Boundary Element Method (BEM) (see Chap. 3). Alternatively, equivalent stiffness and damping values are taken into account. They are given as complex values and characterize the soil impedance. The corresponding parameters can be calculated with the analytical approaches described above [103]. Equivalent stiffness and damping values are given in literature, e.g., for the ideal case of an infinitely ridged circular foundation on a homogenous half space [105]. For most buildings, the thus resulting natural frequency ranges between 8 and 15 Hz [54, 106].

Figure 22.17 shows real and imaginary part of the (dimensionless) values that follow [105] for an infinitely ridged circular foundation of radius R. The real part corresponds to the spring stiffness k_i and the imaginary part to the equivalent damping c_i, resulting

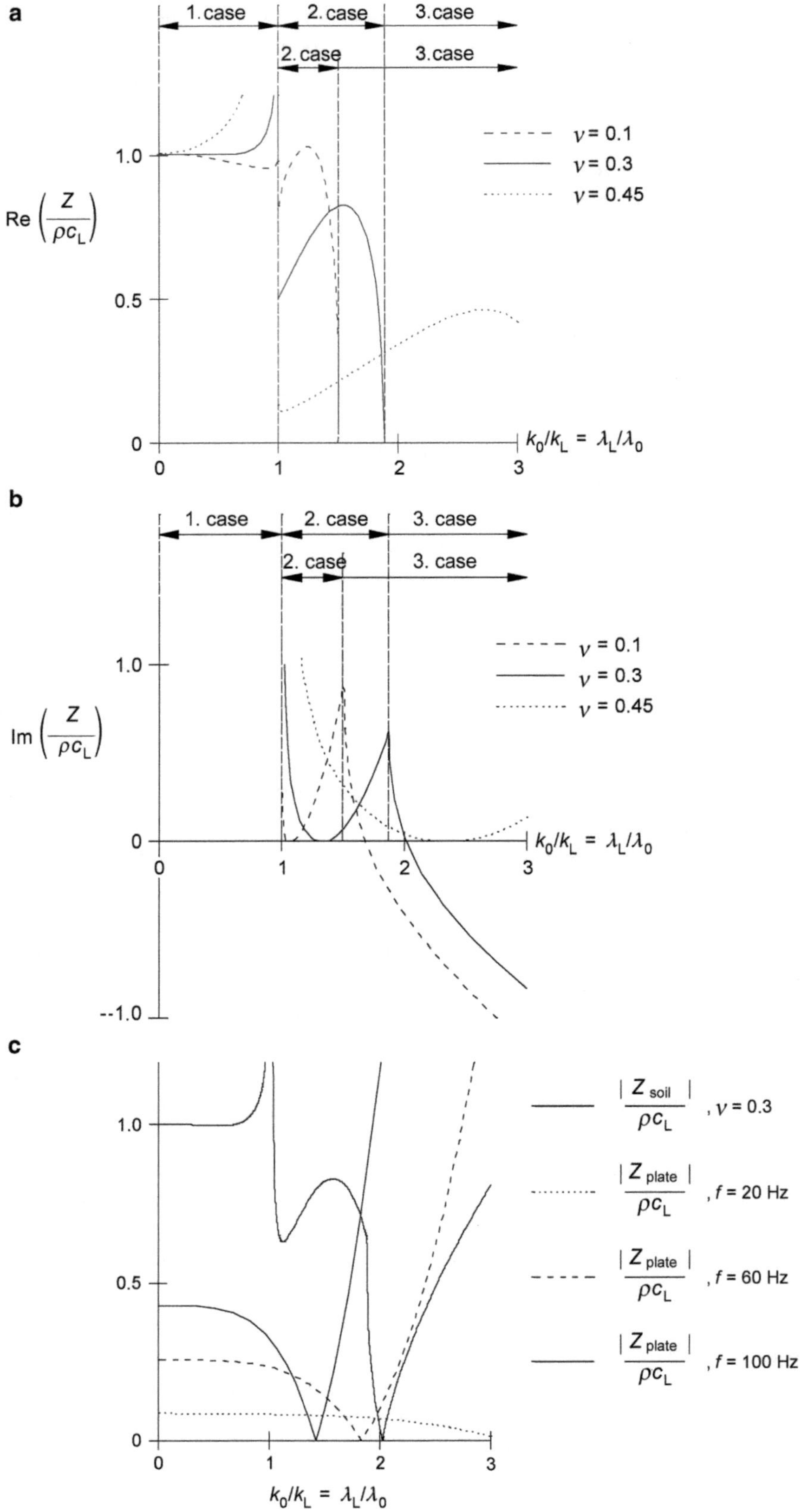

Fig. 22.16 a Real- and b imaginary part of the impedance functions, according to (22.50), c comparison of the magnitude of the impedance function of the half-space with the impedance of a plate for different Poisson numbers μ

in the complex stiffness for the different directions of vibration i:

$$S_i = K_i(k_i + ja_0c_i); a_0 = \frac{\omega R}{c_T}. \qquad (22.51)$$

Note, that in the relevant frequency range mostly the imaginary part dominates for vertical vibrations. The complex stiffness then corresponds mainly to an equivalent damper.

In [107], equivalent stiffness and damping coefficients are listed for different geometries. In [108], simple models for rectangular foundations on an elastic half space are given. The interaction of the soil with buildings on strip foundations is described in [109] and on a group of individual foundations is described in [110, 111]. The coupling of systems modelled with the Finite-Element Method with systems described by the Boundary Element Method to simulate the soil-structure interaction of tall buildings is described in [112]. In [106], an overview over various engineering models is given.

22.4.3 Transmission of Vibrations Inside of Buildings

The dynamic behaviour of buildings can be investigated by means of the Finite Element Method. In many applications, especially in stiff residential buildings, often simplified models of practice are sufficient, in which the most relevant structural elements (slabs) for vibrations are considered as vibrators that are weakly coupled to a relatively stiff main structure [109, 113]. This means that the response at the support points of the slabs is considered to be independent from the dynamic characteristics of the slabs themselves. Thus in rigid buildings, the plate vibrations can be considered superposed to the vibration of the rigid body movement of the whole building (model shown in Fig. 22.18).

Vibrations can cause substantial annoyance, if the slabs are excited at resonance. The maximum amplification of the excitation at the centre of a plate on a four-sided hinged support is given by:

$$V_{\max} \approx \frac{4^2}{\pi^2} \cdot \frac{1}{2\zeta}. \qquad (22.52)$$

The first natural frequency of the slab can be evaluated approximately according to Eq. (22.53).

$$f = \frac{\beta}{a^2}\sqrt{\frac{Ed^3}{12(1-\mu^2)m''}}, \qquad (22.53)$$

(E – Young-modulus, d – plate thickness, μ – Poisson-value, m'' – total distributed mass including the interior finishing and life load, ζ – damping ratio)

The coefficient β depends on the rotational stiffness conditions at the support (Fig. 22.19).

A survey of resonance frequencies for slabs and floating floors is given in Table 22.19 [54].

Various empirically obtained factors can be found in the literature. They describe – depending on the type of sources – the relation between the vibration v_S at the centre of a slab in the upper floors to the vibration at the foundations v_F where the excitation is induced. For railway excitation, the transmission factor can be deduced from Chap. 16. With an excitation from sheet piling, amplification factors of the vertical vibration of $v_S/v_F < 15$ for wooden floors, slabs on steel grids, and of $v_S/v_F < 10$ for concrete slabs in residential buildings are observed [114]. The amplification is less for single impacts, $v_S/v_F < 2.5$ [115]).

Horizontal amplification from the foundation to the upper floors is not significant. Depending on the soil stiffness here amplifications are given between $v_S/v_F < 0.5$ (stiff soil, $E_{s,\mathrm{stat}} = 100$ MN/m) and $v_S/v_F < 2$ (soft soil).

22.4.4 Reduction of Vibrations in Buildings

22.4.4.1 Elastic Mounting of Buildings

Functional Principle

An elastic mounting of buildings is generally the most efficient way to protect against the impact of soil vibrations. A reduction between 10 and 20 dB can be obtained, provided that the structure is well stiffened and has a minimum of relevant elastic modes in the critical frequency range. It has to be considered that for a given spring stiffness, the insertion loss of an elastic layer depends on the dynamic mass (see Chap. 9). As a weakly stiffened structure will have a dynamic mass which may be much smaller than the static mass, the structure's impedance will be much smaller than the impedance of an infinitely rigid body.

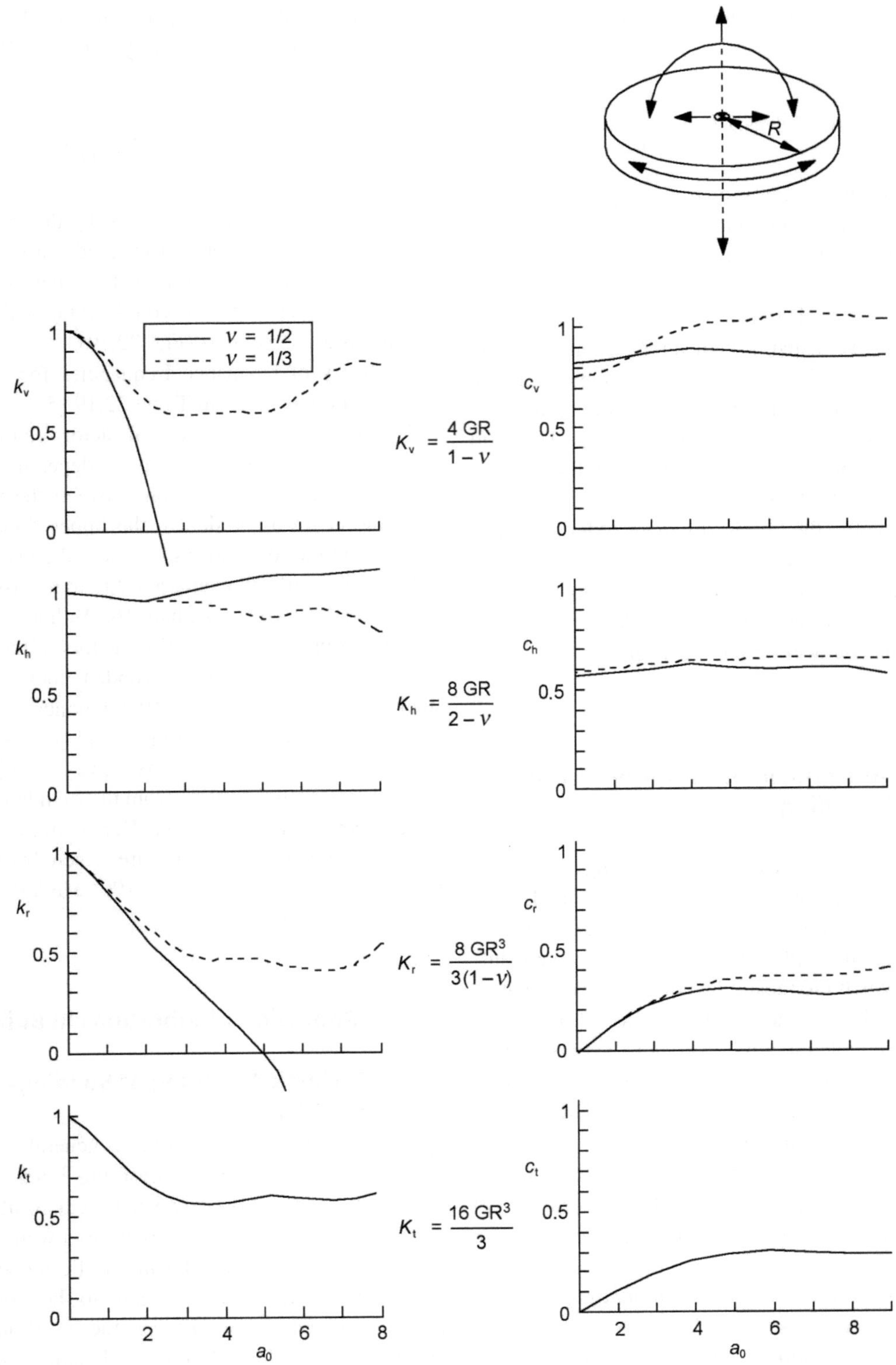

Fig. 22.17 Real- and imaginary part of the dimensionless spring stiffness and damping, according to [105]

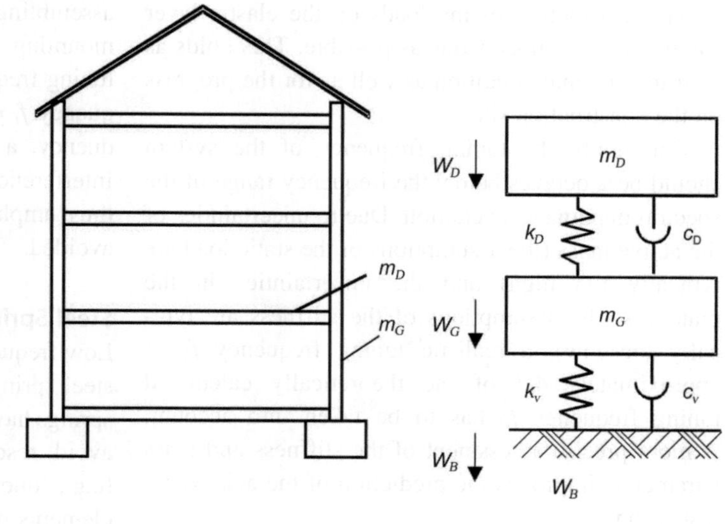

Fig. 22.18 Mechanical models for well stiffened structures

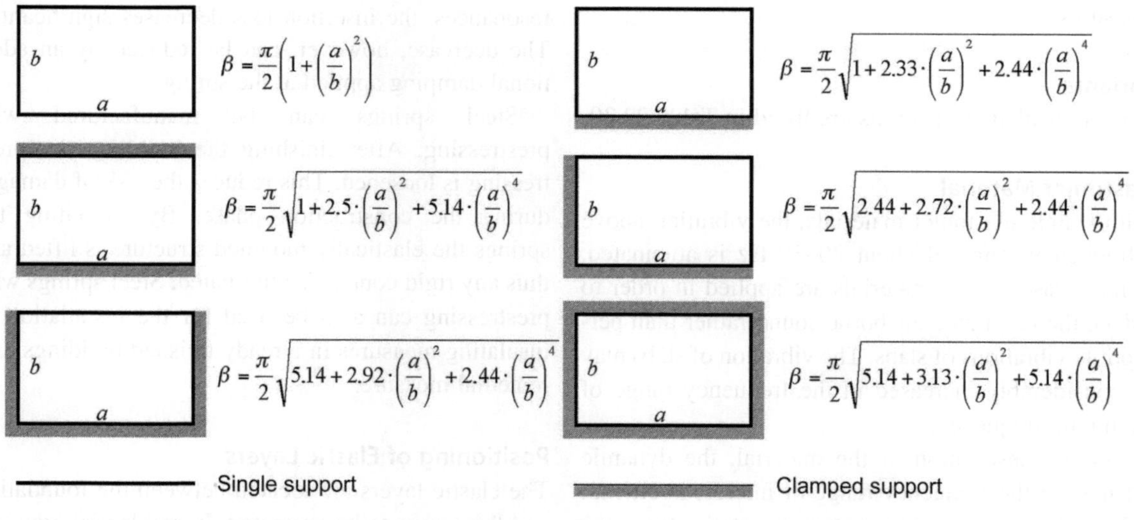

Fig. 22.19 Coefficient β for the assessment for the calculation of the first natural frequency of rectangular plates

Table 22.19 Eigenfrequencies of slabs/floating floors [54]

Construction	Eigenfrequency (Hz) Frequently	Seldom	Damping coefficient
Wooden floors	9–12	8–15	0.028
Reinforced concrete slabs residential buildings	20–25	15–35	0.035
Weight spanned reinforced concrete or compound slabs in industrial or commercial buildings	7–10	3–15	0.02
Concrete floating floor	40–50	30–70	
Asphalt floating floor	50–60	40–80	
Mastic asphalt floating floor	80–10	70–120	

The distribution of the loads on the elastic layer should be as homogeneous as possible. This holds as well for the final condition as well as for the progress on the construction site.

Commonly the tuning frequency of the system should be 2 octaves below the frequency range of the spectral maxima of excitation. Due to uncertainties of the active mass (the assumptions of the static load are typically too high) and the uncertainties in the materials (the assumptions of the stiffness are typically too low), a realistic tuning frequency f_1 of approximately 4/3 of the theoretically calculated tuning frequency f_d has to be taken into account. A more precise assessment of the stiffness and mass parameters improves the prediction of the achievable insulation.

The material must be selected to provide an approximately similar deflection of all elements. Unavoidable inhomogeneous reaction forces in the structure have to be considered by sufficient stiffening measures.

Variants

Typical insulation materials are listed in Table 22.20

Elastomer Material

With simple elastomer materials, the vibration above a frequency range of about 20–30 Hz is attenuated. In most cases, these materials are applied in order to reduce the reradiated air-borne sound rather than perceptible vibrations of slabs. The vibration of slabs may be considerably increased in the frequency range of the tuning frequency.

For the description of the material, the dynamic stiffness in the frequency range of interest is crucial. It depends on various parameters, the static stress, the frequency, and the amplitude of vibration.

Steel plates, which are vulcanized inside mounts, increase the mount's stability. Thus compared to the static stresses a softer material is possible by assembling multiple layers. In [116], a punctual mounting on elastomer elements with a calculated tuning frequency $f_{t'} < 5$ Hz and a realistic tuning frequency $f_{t'} = 6$ Hz is described. With this tuning frequency, a lower frequency range can be insulated. Interference with natural frequencies of slabs and thus amplification of susceptible vibrations can be avoided.

Steel Springs

Low frequency vibrations are typically insulated with steel springs. Compared to elastomer elements, steel springs have a smaller damping. Therefore, in order to avoid resonant amplification at lower frequencies (e.g., due to wind), additional parallel damping elements must be provided. Due to the small interior damping, the steel-springs have strong block resonances at the natural frequencies of the spring. These natural frequencies are determined by the spring's mass, length, and stiffness. Near to those resonances, the insertion loss decreases significantly. The decrease, however, can be reduced by an additional damping applied at the spring.

Steel springs can be manufactured with prestressing. After finishing the building, the prestressing is loosened. This reduces the risk of damages during the construction phase. By loosening the springs the elastically mounted structure is lifted and thus any rigid contact is eliminated. Steel springs with prestressing can also be used for the installation of insulating measures in already finished buildings as a remedial measure.

Positioning of Elastic Layers

The elastic layers are located between the foundation and the rooms to be protected. In residential housing, the basements typically do not serve as living-rooms. The elastic mountings may be positioned beneath the slab between basement and ground floor (see Fig. 22.20).

Table 22.20 Materials for isolation of buildings

Type of material	Tuning frequency (Hz)	Thickness (mm)	Static deflection w, (mm)	Damping ratio ζ
Elastomer materials on surface areas or on stripes	8–25	10–75	3–8	7–15%
Reinforced elastomer mounts (punctual mounts)	6–9	60–80	10–15	7–15%
Steel springs	3–6		10 (6 Hz)–30 (3 Hz)	

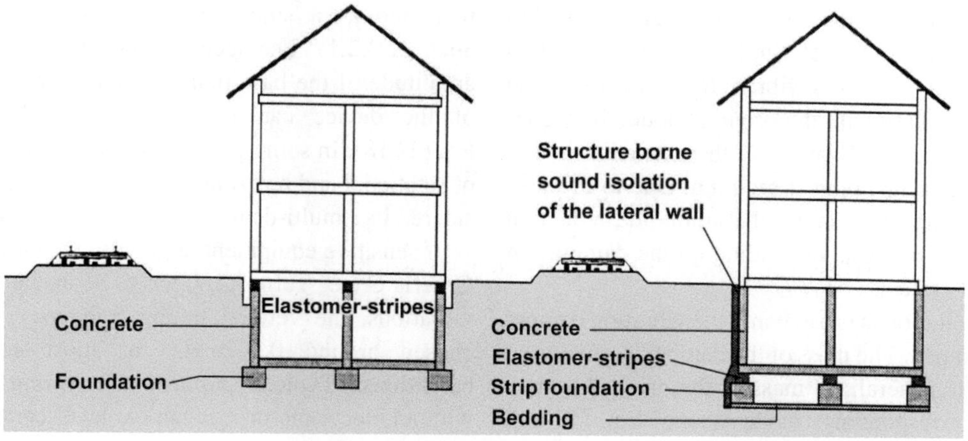

Fig. 22.20 Position of the elastic mounts for elastic mounts of buildings

Table 22.21 Qualitative assessment of measures at buildings

Measure	Assessment	Limitations
I. Extension of the distance	Depending on the source and the wave types (Sect. 22.4.1)	Reduced efficiency for line sources and in case of Rayleigh-waves.
II. Improvement of the stiffening	Just for excitations with wave length smaller than the dimensions of the building	In case of very stiff soils the impedance relationship between building and soil is small
III. Pile foundation, exchange of soil materials	If it is necessary to bridge strong vibrating soft soil layer	In homogeneous soils reduced efficiency
IV. Additional interior damping, e.g., moving joints in case of plaster board walls, sandwich structures, empty drumming, damping	Frequency-dependent effect of the measure has to be checked	Few experience
V. Tuning of the frequency for single components	For excitations, which are limited to a narrow frequency range	Uncertain parameters must be varied to account for the scatter of the tuning frequency
VI. Tuned mass damper	In case of steady state vibrations in resonance (e.g., pedestrian bridges), as remedial measure	To account for the scatter of the tuning frequency (measurement is necessary), reduced efficiency for non harmonic excitations

The joint must not be bridged by staircases, elevators, ducts, etc. In case the joint is at the level of the foundation, problems with the sealing against water might occur.

The measure requires careful site supervision, in order not to create any sound-bridges. A later repair of defects is very difficult or even impossible. Additionally, all structures, which are in contact with the surrounding soil have to be decoupled appropriately.

22.4.4.2 Secondary Measures

Table 22.21 I–IV gives an overview over the secondary measures for buildings.

The efficiency of the measures is generally difficult to predict and often quite limited (<5 dB), especially in the case of a broad-band excitation, which, e.g., exists near railway-lines.

For narrow-band excitations and weakly damped structures (e.g., excitation by pedestrians), the measures V and VI can be considered. They focus on a reduction of the resonant vibrations and need a precise knowledge of natural frequencies which have to be determined experimentally. If they are tuned properly, the reduction efficiency may reach to 50–90%.

Vibration dampers (Tuned Mass Dampers, TMD) act as a further single degree of freedom system attached to a low-damped structure. They are tuned to split up one maximum of single resonance into two lower maxima. Two new vibration modes occur, the lower one in-phase with the original mode, the higher mode in opposite 180°-phase. If the excitation is narrow-banded, no damping is required. Due to the shift of the resonance frequency, the amplitude reduction can be more than 90%, depending on the damping of the original system.

If the excitation is broad band, the vibration damper must be damped. The mass of the damper is usually set to 5% of the generalized mass of the original system and the corresponding mode shape Eq. (22.46). The optimum parameters are evaluated in order to minimize the displacement amplitudes with force excitation [117].

22.4.5 Measures for the Insulation of Sensitive Equipment

22.4.5.1 Parameters of the Insulation and Elements

The preferable solution to install sensitive equipment is to mount it on a heavy foundation mass rigidly connected to solid ground. It has to be investigated, whether the foundation block or slab must be decoupled from the surrounding structure or slab by a lateral joint. A joint decreases the influence of excitations at higher frequencies which are generally less important than excitations induced by longer low frequency waves. On a foundation separated by a joint even higher amplitudes than on the adjacent slab [59] especially in horizontal direction can occur. The advantages and disadvantages of a joint must be carefully assessed.

If a site survey reveals environmental conditions which exceed the specified vibration limits, insulation measures have to be designed. However, it is important to recognize that the device can become extremely sensitive to direct excitation, e.g., by the personnel, by internal mechanisms, vibrations, which are transmitted via ducts (cooling or electricity) and/or by low frequency pressure changes in the air. That means, the sensitive equipment shows larger amplitudes if it is directly excited; therefore it is recommendable to provide a heavy mass to stabilize the system.

Similar to the insulation of sources, the insulation at the location to be protected can also be modelled in a first step as a single degree of free system (see Sect. 22.3.2.1). The spectral transfer function of the amplitude of the base-point-excitation at the position of the device can be calculated analogously to Eq. (22.18). In some cases, also the coupled vibration of rotational and horizontal movements has to be considered in a multi-degree of freedom system model.

If sensitive equipment, e.g., with the category VC-Criteria C (see Table 22.9) has to be insulated against vibrations, the vertical tuning frequency should be chosen around 0.6–2 Hz for most surrounding conditions. Typically, pneumatic spring elements with an electronic or mechanical level control (semi-active elements) are applied.

The horizontal tuning frequency depends on the type of the elements applied. It should not be significantly higher than the vertical tuning frequency. The design of the damping has to be optimized to provide maximum insulation but also a short decay of the vibration if the device is excited in operating conditions (charging of samples, internal mechanisms). The damping ratio should be adjustable in a range between 5 and 20%.

Sufficient stability has to be guaranteed. The centre of gravity should be as low as possible compared to the level of the springs. It must be assured that the mass is not directly acceded by the personnel operating the device. A wooden grid or a metallic grid (here electromagnetic requirements must be checked) must be installed above and detached from the mounted mass to make the device accessible.

In case an internal insulation is provided, the model changes to a two degree of freedom system.

Often the tuning frequency of this mounting is within the range of the tuning frequency of the foundation. The larger the mass of the foundation compared to the mass of the device, the weaker is the coupling of the rigid body modes.

If there is not enough space for a large mass of foundation it can be replaced by actively controlled mounts. Active elements provide the advantage of insulation down to the quasi-static frequency range. For high resolution microscopes, often standardized insulating platforms are available.

Figure 22.21 shows exemplarily a section through such a foundation, which is supplied with a separated platform for the operators.

Below the foundation, the gap between the foundation and the slab below should be sufficiently large

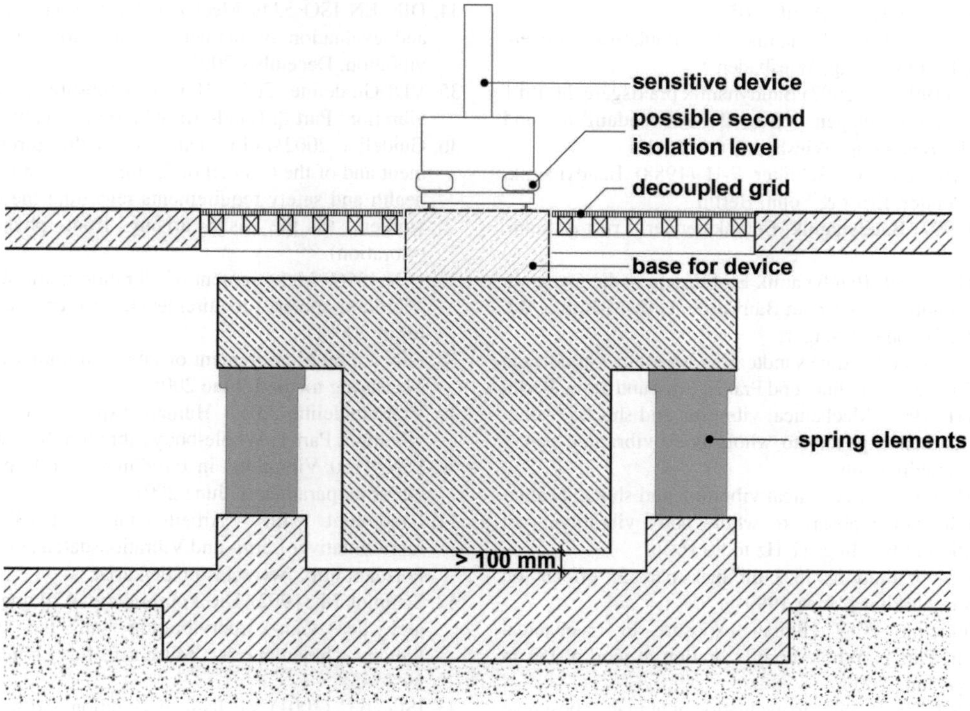

Fig. 22.21 Schematic sketch for an isolated foundation for highly sensitive equipment

(generally >10 cm). Otherwise the stiffness of the air layer in between may become significant in parallel to the springs.

Ducts and connections have to be dynamically decoupled.

22.4.5.2 Accompanying Measures for Building Acoustics

Highly sensitive devices are not just excited by vibrations, but also by pressure changes typically in a frequency range below 100 Hz. Within the range of infrasound, the devices may be much more sensitive than the human ear.

As the sound insulation of the walls, windows and doors as well as sound absorbing measures is limited in this frequency range, the rooms must be located in quiet zones. For highly sensitive equipment, the access of the sensitive areas has to be provided with door-sluices, flanked by an adequate arrangement of the adjacent rooms.

Pressure changes induced by HVRC-devices have to be minimized. In the case of high sensitivities static cooling or heating systems have to be used instead. Convection has to be considered.

References

1. Newmark NM, Rosenblueth E (1971) Fundamentals of earthquake engineering. Prentice Hall, Englewood Cliffs, NJ
2. Bachmann et al. (1995) Vibration problems in structures. Birkhäuser, Basel
3. Schueller G, Rackwitz R, Bachmann H (1997) Tragwerks-zuverlässigkeit, Einwirkungen. In: Mehlhorn (Hrsg) Der Ingenieurbau, Bd 8, Ernst&Sohn, Berlin
4. Ruscheweyh H (1982) Dynamische Windwirkung an Bauwerken. Bauverlag, Wiesbaden, Berlin
5. Bachmann A, Graubner, C-A (2002) Böenerregte Windeinwirkungen auf Hochhäuser, das Spektralverfahren und deren Umsetzung in Windlastnormen. Beton- und Stahlbetonbau 97, Heft 8, Ernst & Sohn, Berlin
6. Simiu E, Scanlan RH (1996) Wind effects on structures. Wiley, New York
7. Herzog M (2005) Baupraktische Sicherheitsnachweise gegen winderregte Schwingungen von Brücken und Türmen. Stahlbau 74, Heft 12, Ernst & Sohn, Berlin
8. DIN 1055-4, Action on structures – wind loads. March 2005
9. DIN 4149 Buildings in German earthquake areas – Design loads, analysis and structural design of buildings. April 2005
10. DIN V ENV 1998–1 Eurocode 8, Design of structures for earthquake resistance. April 2006
11. ISO/DIS 4354 Wind actions on structures. July 1997

12. DIN 4178 Belltowers. April 2005
13. Petersen C (1996) Dynamik der Baukonstruktionen. Vieweg Braunschweig, Wiesbaden
14. Flesch R (1993 und 1997) Baudynamik praxisgerecht, Bd 1 Berechnungsgrundlagen, Bd 2 Anwendungen und Beispiele. Bauverlag, Wiesbaden und Berlin
15. Eibl J, Henseleit O, Schlüter F-H (1988) Baudynamik. Betonkalender, Ernst & Sohn, Berlin
16. Eibl J (1997) Baudynamik. Betonkalender, Ernst & Sohn, Berlin
17. Natke HG (1989) Baudynamik Einführung in die Dynamik mit Anwendungen aus dem Bauwesen. LAMM-Reihe, Bd. 66, B. G. Teubner, Stuttgart
18. Kramer H (2007) Angewandte Baudynamik. Grundlagen und Beispiele für Studium und Praxis. Ernst and Sohn, Berlin
19. ISO 2631 (1997) Mechanical vibration and shock, evaluation of human exposure to whole body vibration, part 1 General requirements
20. ISO 2631 (2003) Mechanical vibration and shock, evaluation of human exposure to whole body vibration, part 2 Vibration in buildings (1 Hz to 80 Hz)
21. DIN 4150 Vibrations in buildings, part 2: effects on persons in buildings. June 1999
22. VDI-Guideline 2057 Human exposure to mechanical vibration: Part 3: Whole-body vibration at workplaces in buildings. June 2006
23. BS 6472–1 (2008) Guide to evaluation of human exposure to vibration in buildings. Vibration sources other than blasting
24. BS 6472–2 (2008) Guide to evaluation of human exposure to vibration in buildings. Blast-induced vibration
25. Ö-Norm S9010 Evaluation of human exposure to mechanical vibrations – whole body. March 1982
26. Ö-Norm S9012 "Evaluation of human exposure in buildings to immissions by railway traffic - Vibrations and secondary air-borne noise". August 1996
27. Griffin MJ, Griffin J (2001) Human response to vibrations. J Sound Vibr 240(5), 979–983
28. Morioka M, Griffin MJ (2008) Absolute thresholds for the perception of fore-and-aft, lateral and vertical vibration at the hand, the seat, and the foot. Journal of sound and vibration 314 (1–2), 357–370
29. Egger A, Rutishauser G (1998) Schutz vor Erschütterungen und sekundärem Luftschall vergleichender nationaler Normen und Richtlinien. Fortschritte der Akustik, DAGA, 234–235
30. ISO 10056 (2001) Mechanical vibration – Measurement and analysis of whole-body vibration to which passengers and crew are exposed in railway vehicles
31. ISO 2631 (2001) Mechanical vibration and shock, evaluation of human exposure to whole body vibration, part 4 Guidelines for evaluation of the effect of vibration and rotational motion on passenger and crew comfort in fixed guideway transport systems
32. UIC 513 (1994) Guidelines for evaluating passenger comfort in relation to vibration in railway vehicles
33. DIN ISO 6954 Mechanical vibration – Guidelines for the measurement, reporting and evaluation of vibration with regard to habitability on passenger and merchant ships. June 2006
34. DIN EN ISO 5349 Mechanical vibration – measurement and evaluation of human exposure to hand-transmitted vibration. December 2001
35. VDI-Guideline 2057 Human exposure to mechanical vibration: Part 2: Hand-arm vibration, September 2002
36. Guideline 2002/44/EC, Directive of the European Parliament and of the Council of 25 June 2002 on the minimum health and safety requirements regarding the exposure of workers to the risks arising from physical agents (vibration)
37. DIN 45669 Measurement of vibration immission – Part 1: Vibration meters; requirements, verification. June 1995 (draft 2008)
38. DIN 45669 Measurement of vibration immission – Part 2: Measuring method. June 2005
39. VDI-Guideline 2057 Human exposure to mechanical vibration, Part 1: Whole-body vibration. September 2002
40. DIN 4150 Vibrations in buildings, part 1: prediction of vibration parameters. June 2001
41. Landesamt für Arbeitsschutz (LAS), Katalog repräsentativer Lärm- und Vibrationsdaten am Arbeitsplatz Potsdam http://www.las-bb.de/karla/
42. Schmidt M (1992) Die kombinierte Wirkung des Lärms und der Ganzkörpervibration auf das Gehör des Landmaschinenfahrers. Zeitschrift für Lärmbekämpfung 39:43–51
43. ISO 2631 (2001) Mechanical vibration and shock, evaluation of human exposure to whole body vibration, part 5 Method for evaluation of vibration containing multiple shocks
44. Kasperski M (2001) Menschenerregte Schwingungen in Sportstadien. Bauingenieur 76:575–581
45. Kasperski M (2008) Safety and reliability of stand structures. Proc. EURODYN, Southhampton
46. ISO 10137 Bases for design of structures – Serviceability of buildings and walkways against vibrations (2007)
47. DIN 4150 Vibrations in buildings, part 3: effects on structures. February 1999
48. SN640312a (1992) Erschütterungen, Erschütterungseinwirkung auf Bauwerke
49. BS 7385 Evaluation and measurement for vibration in buildings part 1 (11–1990) and part 2 (11–1993)
50. Ö-Norm S9020 Building vibrations; blasting vibrations and comparable immissions of impulse shape 08–1986
51. ISO 4866, Mechanical vibration and shock; vibration of buildings; guidelines for the measurement of vibrations and evaluation of their effects on buildings 08–1990 (Draft 2008)
52. Haupt W (1986) Bodendynamik. Vieweg, Braunschweig
53. Studer JA, Laue J, Koller MG (2007) Bodendynamik Grundlagen, Kennziffern. Probleme und Lösungsansätze. Springer, Berlin
54. Müller-Boruttau FH (1996) Erschütterungen beim Spundwandbau: Einwirkungen auf Menschen. Bauwerke und technische Einrichtungen. Bauingenieur 71:33–39
55. Funk K (1996) Expertensystem für Lärm- und Erschütterungsprognosen beim Einbringen von Spundbohlen. Schall und Messtechnik der Universität Hannover, Eigenverlag, Mitteilungen des Curt-Risch-Institutes für Dynamik

56. Ungar EE, Sturz DH, Amick CH (1990) Vibration control design of high technology facilities. Acoustical Publications Inc, Bay Village, Ohio, USA
57. Gordon CG (1991) Generic criteria for vibration sensitive equipment. Proc. of Int. Society for Optical Engineering SPIEE Vol. 1619 San Jose CA 11/1991
58. Ungar EE (2007), Vibration Criteria for Healthcare Facility Floors, Sound and Vibration, 2007, vol 41, no 9, 26–27, Acoustical Publications, Bay Village
59. Heiland D, Beyer K (2005) Vibration criteria for nanotechnology facilities. Recent design criteria for nanotechnology production and research. VDI-Report Nr. 1920, 241–254, VDI Verlag GmbH, Düsseldorf
60. Heiland D, Beyer K (2008) Baudynamische Anforderungen an Labors und Produktionsflächen der Nanotechnik. Beton und Stahlbetonbau, Heft 7:455–463
61. ISO 10811 Mechanical vibration and shock – Vibration and shock in buildings with sensitive equipment. June 2000
62. Bonzel J, Schmidt M (1980) Einfluß von Erschütterungen auf frischen und jungen Beton. Beton 30 H.9, S.333-337 u. Beton 30 H.10, S. 372–378
63. Schwingungsisolierungen GmbH GERB, Co KG (2003) GERB Schwingungsisolierungen, 11th edn. Woeste Druck + Verlag, Essen
64. VDI-Guideline 2062 Part 1 Shock and vibration isolation conceptions and principles. January 1976 and part 2 Insulation elements. November 2007
65. DIN EN 1299/A1mechanical vibration and shock – vibration isolation of machines information for the application of source vibration (draft). April 2008
66. DIN 4024 Part 1 Machine foundations; flexible structures that support machines with rotating elements. April 1988
67. DIN 4024 Part 2 Machine foundations; rigid foundations for machinery with periodic excitation. April 1999
68. DIN 45635 Part 8 Determination of airborne noise emitted by machines by, measurement of structure borne noise; basic requirements June 1985
69. DIN ISO 1940–1 Mechanical vibration – Balance quality requirements for rotors in a constant (rigid) state – Part 1: Specification and verification of balance tolerances (ISO 1940–1:2003) April 2004
70. DIN ISO 1940–2 Mechanical vibration – Balance quality requirements of rigid rotors – Part 2: Balance errors (ISO 1940–2:1997) February 1998
71. DIN ISO 10816 Mechanical vibration – Evaluation of machine vibration by measurements on non-rotating parts – Part 1: General guidelines (ISO 10816–1:1995), parts 1–4, 6,7 1997–2007
72. DIN EN 60034–14 (VDE 0530–14:2008–03) Rotating electrical machines – Part 14 Mechanical vibration of certain machines with shaft heights 56 mm and higher – Measurement, evaluation and limits of vibration severity. March 2008
73. DIN ISO 8528–9 Reciprocating internal combustion engines driven alternating current generating sets – Part 9: Measurement and evaluation of mechanical vibrations. January 1999
74. Saalfeld M (1994) Körperschalldämmende Maßnahmen bei haustechnischen Anlagen. VDI Berichte Nr. 1121
75. Meek JW (1990) Rekursive Berechnung baudynamischer Vorgänge. Bautechnik 67, 205–210, Wilhelm Ernst und Sohn, Berlin
76. Dascotte E, Desanghere G (1991) Updating of force functions. Proceedings of the 9th IMAC Conference, Florence/Italy
77. VDI Guideline 3842 Vibrations in piping systems. June 2004
78. Daiminger W, Fritz K, Schorer E, Stüber B (1995) Noise and Vibration. In: Ullmann's Encyclopedia of industrial Chemistry, Volume B7, VCH Verlagsgesellschaft, 383–401
79. Achmus M, Kaiser J, tom Wörden F (2005) Bauwerkserschütterungen durch Tiefbauarbeiten. Mitteilungen des Instituts für Grundbau, Bodenmechanik und Energiewasserbau der Universität Hannover, Heft 61, Hannover
80. Achmus M Kaiser J Bauwerkserschütterungen infolge Tiefbauarbeiten. http://www.igbe.uni-hannover.de/arb/ac/bauwer.htm
81. Wieck J (2003) Erschütterungen im Bauwesen. In: Die vorsorgliche Beweissicherung im Bauwesen, Schadenfreies Bauen, Band 31, Fraunhofer IRB Verlag, Stuttgart
82. Buja HO (2004) Der Spezialtiefbau, Aktueller Stand der Geräte- und Verfahrenstechnik. Bautechnik, V 81 Ausgabe S1, 25–109. Wilhelm Ernst und Sohn, Berlin
83. KDT-Richtlinie 046/86 Sprengerschütterungen – Wirkung übertägiger Sprengungen auf Gebäude
84. Heinze H (1987) Sprengtechnik, Anwendungsgebiete und Verfahren, 2nd edn. Deutscher Verlag für Grundstoffindustrie, Leipzig, Stuttgart
85. Melke J (1995) Erschütterungen und Körperschall des landgebundenen Verkehrs. Prognose und Schutzmaßnahmen. Landesumweltamt Nordrhein-Westfalen LUA-Materialien Nr. 22, Essen
86. Wots GR (1988/1990) Traffic induced vibrations in buildings, TRRL-Reports 156/246
87. Lombaert G, Degrande G (2000) The validation of a numerical prediction model for free field traffic induced vibrations by in situ experiments. Proceedings of the ISMA25, International Conference on Noise and Vibration Engineering, September 13–15, 2000, Leuven
88. Kramer H (1998) Dynamische Belastung durch Fußgänger. Bauingenieur Bd 73:342–346
89. Grundmann H, Kreuzinger H, Schneider M (1993) Schwingungsuntersuchung von Fußgängerbrücken. Bauingenieur Bd 68:211–225
90. DIN EN 1993 (EC 3) Design of steel structures. July 2005
91. Ontario Highway Bridge Design Code. Ontario Ministry of Transportation, Toronto (1983)
92. SIA 260 (SN 505260) and SIA 261 (SN 505261):):2003–01 Grundlagen der Projektierung von Tragwerken (Basis of structural design). Zürich: SIA
93. Eurocode EN 1995 Design of timber structures - Part 2, Bridges, European Committee for Standardization. 2004
94. Mohr B (2001) Schwingungen von Wohnungsdecken aus Holz, Stahl und Beton; Vorschläge für eine zutreffende Bewertung. In: Proc. Ingenieurbau, Karlsruher Tage 2001 Bruderverlag Albert Bruder GmbH, Karlsruhe
95. Blaß HJ, Ehlbeck J, Kreuzinger H, Steck G (2008) Erläuterungen zu DIN 1052: 2004–08. Entwurf,

Berechnung und Bemessung von Holzbauwerken DGfH Innovations- und Service GmbH, München
96. Wyatt TA (1989) Design Guide on the Vibration of Floors. The Steel Construction Institute SCI Publication 076
97. BS 5400–2:2006-09-29 Steel, concrete and composite bridges. Specification for loads London: British Standards Institution
98. Serviceability Criteria for Deflections and Vibrations (1990) Commentary A. Supplement to the National Building Code of Canada. National Research Council Canada, Ottawa
99. Butz C (2008) Codes of Practise for Lively Footbridges: state-of-the art and required measures, Proc. of the 3rd Inten. Conference Footbridge 2008 – Footbridge for urban reneawal. Port, Portugal, 51–68
100. Schneider M (1991) Ein Beitrag zu fußgängerinduzierten Brückenschwingungen. Technische Universität München, Berichte aus dem Konstruktiven Ingenieurbau
101. Woods RD (1968) Screening of surface waves in soils. J Soil Mech Found Div ASCE 94:951–979
102. Wolf JP (1985) Dynamic soil-structure interaction. Englewood Cliffs, NJ, Prentice Hall
103. Wolf JP (1994) Foundation vibration analysis using simple physical models. Englewood Cliffs, NJ, Prentice Hall
104. Cremer L, Heckl M (1995) Körperschall, Physikalische Grundlagen und technische Anwendungen. Springer 2 Auflage
105. Gazetas G (1983) Analysis of machine foundations: state of the art. Soil Dyn Earthq Eng No 2:2–42
106. Meinhardt C (2008) Einflussgrößen für das Schwingungsverhalten von Gebäuden zur Prognose von Erschütterungsimmissionen. PHd Thesis TU Berlin Fakultät VI
107. Studer JA, Koller MG (1997) Bodendynamik, Grundlagen, Kennziffern, Probleme. Springer, Berlin
108. Knothe K, Wu Y (1999) Rechteckfundament auf elastischem Halbraum. Bauingenieur, Bd 74, Nr. 12
109. Breitsamter N (1996) Ersatzmodelle zur Bestimmung der Schwingungsantwort von Gebäuden unter Anregung durch Bodenerschütterungen. Dissertation, Berichte aus dem Konstruktiven Ingenieurbau, Technische Universität München
110. Kovacs I (1987) Zum Entwurf der Gründungen schwingungsempfindlicher Produktionsstätten, Teil 1: Dynamische Wechselwirkung zwischen Baugrund und Fundamentgruppe Bautechnik Heft 5, 145–158. Wilhelm Ernst & Sohn, Berlin
111. Kovacs I, Bertold R (1992) Teil II: Parameterstudien zu dynamisch gutmütigen eingeschossigen Gerätegründungen, Bautechnik, Heft 12. Wilhelm Ernst & Sohn, Berlin
112. Lehmann L, Antes H (2001) Numerische Simulation des Schwingungsverhaltens von Hochhäusern unter Berücksichtigung der Boden-Bauwerk-Interaktion. Bauingenieur, Bd 76, Februar 2001, 62–68
113. Grundmann H, Müller G (1994) Erschütterungseinleitung in Bauwerke und Maßnahmen zur Reduzierung von Erschütterungen und sekundären Luftschallemissionen. Bauingenieur 69:129–137
114. Funk K (1996) Expertensystem für Lärm- und Erschütterungsprognosen beim Einbringen von Spundbohlen. Schall- und Messtechnik der Universität Hannover, Mitteilungen des Curt-Riesch-Institutes für Dynamik
115. Funk K, Gerasch W (1995) Expertensystem für Lärm- und Erschütterungsprognosen beim Einbringen von Spundbohlen. Pfahl-Symposium, Mitteilungen des Instituts für Grundbau und Bodenmechanik der TU Braunschweig
116. Wilson GP(1998) Vibration Isolation Design for Benaroya Hall. Proceedings of the 16th International Congress on Acoustics and 135th Meeting of the Acoustical Society of America, Seattle, Washington / USA, 20.-26. June 1998
117. Hartog Den JP (1956) Mechanical Vibrations. Mc Graw Hil, New York

Index

G. Müller and M. Möser (eds.), *Handbook of Engineering Acoustics*,
DOI 10.1007/978-3-540-69460-1, © Springer-Verlag Berlin Heidelberg 2013

MIX
Papier aus verantwortungsvollen Quellen
Paper from responsible sources
FSC® C105338

If you have any concerns about our products,
you can contact us on
ProductSafety@springernature.com

In case Publisher is established outside the EU,
the EU authorized representative is:
Springer Nature Customer Service Center GmbH
Europaplatz 3, 69115 Heidelberg, Germany

Printed by Libri Plureos GmbH
in Hamburg, Germany